ANNUAL REVIEW OF CELL AND DEVELOPMENTAL BIOLOGY

ANNUAL REVIEW OF CELL AND DEVELOPMENTAL BIOLOGY

VOLUME 12, 1996

JAMES A. SPUDICH, *Editor*
Stanford University School of Medicine

JOHN GERHART, *Associate Editor*
University of California, Berkeley

STEVEN L. McKNIGHT, *Associate Editor*
TULARIK, South San Francisco

RANDY SCHEKMAN, *Associate Editor*
University of California, Berkeley

http://annurev.org science@annurev.org 415-493-4400

ANNUAL REVIEWS INC. 4139 EL CAMINO WAY P.O. BOX 10139 PALO ALTO, CALIFORNIA 94303-0139

ANNUAL REVIEWS INC.
Palo Alto, California, USA

International Standard Serial Number: 1081-0706
International Standard Book Number: 0-8243-3112-5

⊗ The paper used in this publication meets the minimum requirements of American National Standard for Information Sciences—Permanence of Paper for Printed Library Materials, ANSI Z39.48-1984.

TYPESET BY TECHBOOKS, FAIRFAX, VA
PRINTED AND BOUND IN THE UNITED STATES OF AMERICA

PREFACE

As I sit overlooking Eel Pond at Woods Hole, I am reminded of my personal roots in cell and developmental biology. Here in Massachussetts to give a lecture to the students in their first week of the "Physiology Course," I described the uniqueness of the experience on which they were embarking by taking this course, long-known for its everlasting effect on the lives of those it has touched. Historically, the course is called Physiology, but equally it could be called Multifaceted Molecular Approaches to Cell and Developmental Biology because it introduces students to a wide variety of the most modern approaches to problem solving in these sciences. The tools of biochemistry, physics, genetics, and molecular genetics are all used in their finest and most elegant forms. A student in the "Physiology Course," as both my wife Anna and I were when Woody Hastings, as course director, and faculty such as Shinya Inoue and Ken VanHolde worked with us from eight AM to midnight six days a week, is fortunate indeed to obtain firsthand knowledge of the importance of embracing many approaches.

A primary goal of the *Annual Review of Cell and Developmental Biology* is to express that same view by illustrating in well-chosen chapters the variety of molecular approaches, using all the tools of the sciences, that are available to cell and developmental biologists. Within these pages and those of past volumes are a variety of techniques that have been applied to studying the cell nucleus, its membrane system, and its cytoskeleton. So for the young impressionable student, read these volumes not only for the latest information on your chosen topic, but to witness the importance in your own work of what I have long called the multifaceted approach. If you understand this, you become free to choose a biological problem that fascinates you, to use any approach you need to solve it, and even to invent new approaches as you need them. This twelfth volume of the *Annual Review of Cell and Developmental Biology*, as the eleven before it, continues to serve this purpose.

JAMES A. SPUDICH
EDITOR

Annual Review of Cell and Developmental Biology
Volume 12 (1996)

CONTENTS

(continued)

OTHER REVIEWS OF INTEREST TO CELL BIOLOGISTS

From the *Annual Review of Biochemistry*, Volume 65 (1996):

Relationships Between DNA Repair and Transcription, E. C. Friedberg

DNA Excision Repair, A. Sancar

Mismatch Repair in Replication, Fidelity, Genetic Recombination, and Cancer Biology, P. Modrich, R. Lahue

DNA Repair in Eukaryotes, R. D. Wood

Mechanisms of Helicase-Catalyzed DNA Unwinding, T. M. Lohman, K. P. Bjornson

Protein Prenylation: Molecular Mechanisms and Functional Consequences, F. L. Zhang, P. J. Casey

Protein Transport Across the Eukaryotic Endoplasmic Reticulum and Bacterial Inner Membranes, T. A. Rapoport, B. Jungnickel, U. Kutay

Molecular Biology of Mammalian Amino Acid Transporters, M. S. Malandro, M. S. Kilberg

Telomere Length Regulation, C. W. Greider

The Structure and Function of Proteins Involved in Mammalian Pre-mRNA Splicing, A. Krämer

Molecular Genetics of Signal Transduction in Dictyostelium, C. A. Parent, P. N. Devreotes

Structural Basis of Lectin-Carbohydrate Recognition, W. I. Weis, K. Drickamer

Connexins, Connexons, and Intercellular Communication, D. A. Goodenough, J. A. Goliger, D. L. Paul

Rhizobium Lipo-Chitooligosaccharide Nodulation Factors: Signaling Molecules Mediating Recognition and Morphogenesis, J. Dénarié, F. Debellé, J.-C. Promé

Crosstalk Between Nuclear and Mitochondrial Genomes, R. O. Poyton, J. E. McEwen

Hematopoietic Receptor Complexes, J. A. Wells, A. M. de Vos

DNA Topoisomerases, J. C. Wang

Interrelationships of the Pathways of mRNA Decay and Translation in Eukaryotic Cells, A. Jacobson, S. W. Peltz

Recording: Dynamic Reprogramming of Translation, R. F. Gesteland, J. F. Atkins

(continued)

Biochemistry and Structural Biology of Transcription Factor IID (TFIID), S. K. Burley, R. G. Roeder

Structure and Functions of the 20S and 26S Proteasomes, O. Coux, K. Tanaka, A. L. Goldberg

From the *Annual Review of Biophysics and Biomolecular Structure*, Volume 25 (1996):

The Sugar Kinase/Heat-Shock Protein 70/Actin Superfamily: Implications of Conserved Structure for Mechanism, J. H. Hurley

Lipoxygenases: Structural Principles and Spectroscopy, B. J. Gaffney

Engineering the Gramicidin Channel, R. E. Koeppe, II, O. S. Andersen

Activating Mutations of Rhodopsin and Other G Protein-Coupled Receptors, V. R. Rao, D. D. Oprian

Visualizing Protein-Nucleic Acid Interactions on a Large Scale with the Scanning Force Microscope, C. Bustamante, C. Rivetti

Antigen-Mediated IGE *Receptor Aggregation and Signaling: A Window on Cell Surface Structure and Dynamics*, D. Holowka, B. Baird

From the *Annual Review of Immunology*, Volume 14 (1996):

Altered Peptide Ligand-Induced Partial T Cell Activation: Molecular Mechanisms and Roles in T Cell Biology, J. Sloan-Lancaster, P. M. Allen

Early T Lymphocyte Progenitors, K. Shortman, L. Wu

Immunotoxins: An Update, G. R. Thrush, L. R. Lark, B. C. Clinchy, E. S. Vitetta

Cellular Interactions in Thymocyte Development, G. Anderson, N. C. Moore, J. J. T. Owen, E. J. Jenkinson

One Step Ahead of the Game: Viral Immunomodulatory Molecules, M. K. Spriggs

Genetic Analysis of Tyrosine Kinase Function in B Cell Development, A. Satterthwaite, O. Witte

Orchestrated Information Transfer Underlying Leukocyte Endothelial Interactions, S. Shaw, K. Ebnet, E. P. Kaldjian, A. O. Anderson

Interleukin-2 Receptor Gamma Chain: Its Role in Multiple Cytokine Receptor Complexes and T Cell Development in XSCID, K. Sugamura, H. Asao, M. Kondo, N. Tanaka, N. Ishii, K. Ohbo, M. Nakamura, T. Takashita

From the *Annual Review of Microbiology*, Volume 50 (1996):

Mechanisms of Adhesion by Oral Bacteria, C. J. Whittaker, C. M. Klier, P. E. Kolenbrander

The β-Ketoadipate Pathway and the Biology of Self-Identity, C. S. Harwood, R. E. Parales

Lessons from a Cooperative Bacterial-Animal Association: The Vibrio fischeri-Euprymna scolopes Light Organ Symbiosis, E. G. Ruby

Spontaneous Mutators in Bacteria: Insights into Pathways of Mutagenesis and Repair, J. H. Miller

rRNA Transcription and Growth Rate-Dependent Regulation of Ribosome Synthesis in Escherichia coli, R. L. Gourse, T. Gaal, M. S. Bartlett, J. A. Appleman, W. Ross

Cell Biology of the Primitive Eukaryote Giardia lamblia, F. D. Gillin, D. S. Reiner, J. M. McCaffery

Census and Consensus in Bacterial Ecosystems: The LuxR-LuxI Family of Quorum-Sensing Transcriptional Regulators, C. Fuqua, S. C. Winans, E. P. Greenberg

The F_0F_1-Type ATP Synthases of Bacteria: Structure and Function of the F_0 Complex, G. Deckers-Hebestreit, K. Altendorf

Immunopathogenesis of HIV Infection, G. Pantaleo, A. S. Fauci

From the *Annual Review of Neuroscience*, Volume 19 (1996):

Human Immunodeficiency Virus and the Brain, J. D. Glass, R. T. Johnson

RNA Editing, L. Simpson, R. B. Emeson

Trinucleotide Repeats in Neurogenetic Disorders, H. L. Paulson, K. H. Fischbeck

Active Properties of Neuronal Dendrites, D. Johnston, J. C. Magee, C. M. Colbert, B. R. Christie

Neuronal Intermediate Filaments, M. K. Lee, D. W. Cleveland

Structure and Function of Cyclic Nucleotide-Gated Channels, W. N. Zagotta, S. A. Siegelbaum

Gene Transfer to Neurons Using Herpes Simplex Virus-Based Vectors, D. J. Fink, N. A. DeLuca, W. F. Goins, J. C. Glorioso

Physiology of Neurotrophins, G. R. Lewin, Y.-A. Barde

Mechanisms and Molecules that Control Growth Cone Guidance, C. S. Goodman

From the *Annual Review of Physiology*, Volume 58 (1996):

A Personal View of Muscle and Motility, H. E. Huxley

Determinants of Maximal Oxygen Transport and Utilization, P. D. Wagner

Mechanisms of Gene Expression and Cell Fate Determination in the Developing Pulmonary Epithelium, B. P. Hackett, C. D. Bingle, J. D. Gitlin

Formation of Pulmonary Alveoli and Gas-Exchange Surface Area: Quantitation and Regulation, G. D. Massaro, D. Massaro

Morphogenesis of the Lung: Control of Embryonic and Fetal Branching, S. R. Hilfer

Molecular Mechanisms of β-Adrenergic Relaxation of Airway Smooth Muscle, M. I. Kotlikoff, K. E. Kamm

Defects in G Protein-Coupled Signal Transduction in Human Disease, A. M. Spiegel

Glucokinase Mutations, Insulin Secretion, and Diabetes Mellitus, G. I. Bell, S. J. Pilkus, I. T. Weber, K. S. Polonsky

Molecular Mechanism of Growth Hormone Action, C. Carter-Su, J. Schwartz, L. S. Smit

Transgenic Approaches to Salivary Gland Research, L. C. Samuelson

Molecular Genetics of Early Liver Development, K. S. Zaret

The Trefoil Peptide Family, B. E. Sands, D. K. Podolsky

Intestine-Specific Gene Transcription, P. G. Traber, D. G. Silberg

Queer Current and Pacemaker: The Hyperpolarization-Activated Cation Current in Neurons, H.-C. Pape

Low-Threshold Calcium Currents in Central Nervous System Neurons, J. R. Huguenard

Persistent Sodium Current in Mammalian Central Neurons, W. E. Crill

Myocardial Potassium Channels: Electrophysiological and Molecular Diversity, D. M. Barry, J. M. Nerbonne

Cyclic Nucleotide-Gated Ion Channels: An Extended Family with Diverse Functions, J. T. Finn, M. E. Grunwald, K.-W. Yau

Physiology and Biochemistry of the Kidney Vacuolar H^+-ATPase, S. L. Gluck, D. M. Underhill, M. Iyori, L. S. Holliday, T. Y. Kostrominova, B. S. Lee

Thin Filament-Mediated Regulation of Cardiac Contraction, L. S. Tobacman

Inherited Diseases of the Vasculature, C. L. Shovlin, J. Scott

The Polyclonal Origin of Myocyte Lineages, T. Mikawa, D. A. Fischman

Multiple Roles of Carbonic Anhydrase in Cellular Transport and Metabolism, R. P. Henry

Downregulation of Cellular Metabolism During Environmental Stress: Mechanisms and Implications, S. C. Hand, I. Hardewig

Molecular Mechanisms of Renal Apical Na/Phosphate Cotransport, H. Murer, J. Biber

Molecular Mechanisms of NaCl Cotransport, M. R. Kaplan, D. B. Mount, E. Delpire, G. Gamba, S. C. Hebert

Molecular Motors of Eukaryotic Cells, H. L. Sweeney

The Active Site of Myosin, I. Rayment, C. Smith, R. G. Yount

The Movement of Kinesin Along Microtubules, J. Howard

The Kinetic Cycles of Myosin, Kinesin, and Dynein, D. D. Hackney

Mutational Analysis of Motor Proteins, H. L. Sweeney, E. L. F. Holzbaur

From the *Annual Review of Plant Physiology and Plant Molecular Biology*, Volume 47 (1996):

Homology-Dependent Gene Silencing in Plants, P. Meyer, H. Saedler

14-3-3 Proteins and Signal Transduction, R. J. Ferl

DNA Damage and Repair in Plants, A. B. Britt

Plant Protein Phosphatases, R. D. Smith, J. C. Walker

The Functions and Regulation of Glutathione S-Transferases in Plants, K. A. Marrs

Physiology of Ion Transport Across the Tonoplast of Higher Plants, B. J. Barkla, O. Pantoja

The Organization and Regulation of Plant Glycolysis, W. C. Plaxton

Light Control of Seedling Development, A. von Arnim, X.-W. Deng

Dioxygenases: Molecular Structure and Role in Plant Metabolism, A. C. Prescott, P. John

Phosphoenolpyruvate Carboxylase: A Ubiquitous, Highly Regulated Enzyme in Plants, R. Chollet, J. Vidal, M. H. O'Leary

Xylogenesis: Initiation, Progression, and Cell Death, J. Fukuda

Compartmentation of Proteins in the Endomembrane System of Plant Walls, T. W. Okita, J. C. Rogers

What Chimeras Can Tell Us About Plant Development, E. J. Szymkowiak, I. M. Sussex

The Molecular Basis of Dehydration Tolerance in Plants, J. Ingram, D. Bartels

Biochemistry and Molecular Biology of Wax Production in Plants, D. Post-Beittenmiller

Role and Regulation of Sucrose-Phosphate Synthase in Higher Plants, S. C. Huber, J. L. Huber

Structure and Biogenesis of Cell Walls in Grasses, N. C. Carpita

Some New Structural Aspects of Old Controversies Concerning the Cytochrome b_6f Complex of Oxygenic Photosynthesis, W. A. Cramer, G. M. Soriano, M. Ponomarev, D. Huang, H. Zhang, S. E. Martinez, J. L. Smith

Carbohydrate-Modulated Gene Expression in Plants, K. E. Koch

Chilling Sensitivity in Plants and Cyanobacteria: The Crucial Contribution of Membrane Lipids, I. Nishida, N. Murata

The Molecular Genetics of Nitrogen Assimilation into Amino Acids in Higher Plants, H.-M. Lam, K. T. Coschigano, I. C. Oliveira, R. Melo-Oliveira, G. M. Coruzzi

Membrane Transport Carriers, W. Tanner, T. Caspari

ANNUAL REVIEWS INC. is a nonprofit scientific publisher established to promote the advancement of the sciences. Beginning in 1932 with the *Annual Review of Biochemistry*, the Company has pursued as its principal function the publication of high-quality, reasonably priced *Annual Review* volumes. The volumes are organized by Editors and Editorial Committees who invite qualified authors to contribute critical articles reviewing significant developments within each major discipline. The Editor-in-Chief invites those interested in serving as future Editorial Committee members to communicate directly with him. Annual Reviews Inc. is administered by a Board of Directors, whose members serve without compensation.

ANNUAL REVIEWS OF

Anthropology
Astronomy and Astrophysics
Biochemistry
Biophysics and Biomolecular Structure
Cell and Developmental Biology
Computer Science
Earth and Planetary Sciences
Ecology and Systematics
Energy and the Environment
Entomology
Fluid Mechanics
Genetics
Immunology
Materials Science
Medicine
Microbiology
Neuroscience
Nuclear and Particle Science
Nutrition
Pharmacology and Toxicology
Physical Chemistry
Physiology
Phytopathology
Plant Physiology and Plant Molecular Biology
Psychology
Public Health
Sociology

SPECIAL PUBLICATIONS

Excitement and Fascination of Science, Vols. 1, 2, 3, and 4
Intelligence and Affectivity, by Jean Piaget

For the convenience of readers, a detachable order form/envelope is bound into the back of this volume.

Annu. Rev. Cell Dev. Biol. 1996. 12:1–26

IMPORT AND ROUTING OF NUCLEUS-ENCODED CHLOROPLAST PROTEINS

Kenneth Cline and Ralph Henry
Horticultural Sciences Department and Plant Molecular and Cellular Biology Program, University of Florida, Gainesville, Florida 32611

KEY WORDS: thylakoid membrane, protein transport, signal recognition particle (SRP), SecA, signal peptide

ABSTRACT

Most chloroplast proteins are nuclear encoded, synthesized as larger precursor proteins in the cytosol, posttranslationally imported into the organelle, and routed to one of six different compartments. Import across the outer and inner envelope membranes into the stroma is the major means for entry of proteins destined for the stroma, the thylakoid membrane, and the thylakoid lumen. Recent investigations have identified several unique protein components of the envelope translocation machinery. These include two GTP-binding proteins that appear to participate in the early events of import and probably regulate precursor recognition and advancement into the translocon. Localization of imported precursor proteins to the thylakoid membrane and thylakoid lumen is accomplished by four distinct mechanisms; two are homologous to bacterial and endoplasmic reticulum protein transport systems, one appears unique, and the last may be a spontaneous mechanism. Thus chloroplast protein targeting is a unique and surprisingly complex process. The presence of GTP-binding proteins in the envelope translocation machinery indicates a different precursor recognition process than is present in mitochondria. Mechanisms for thylakoid protein localization are in part derived from the prokaryotic endosymbiont, but are more unusual and diverse than expected.

CONTENTS

1081-0706/96/1115-0001$08.00

INTRODUCTION

Plastids are developmentally related organelles in algae and higher plants. Among the differentiated plastids are the chloroplast, etioplast, chromoplast, and amyloplast. Plastid biogenesis, similar to mitochondrial biogenesis, requires the participation of two genetic systems. The plastid genome encodes ≈ 100 plastid-localized proteins that are translated on 70S ribosomes and functionally assembled within the plastid. The remaining hundreds of plastid proteins are encoded in the cell nucleus and synthesized in the cytosol. Nucleus-encoded plastid proteins are initially synthesized as precursor proteins on free polyribosomes. Virtually all plastid precursor proteins possess transient amino-terminal transit peptides that govern their import into the organelle and occasionally direct their suborganellar routing (see Figure 1). Although the general features of the plastid protein import process have been known for over a decade, only recently have the detailed steps of translocation and the plastid machinery involved been described. Using an in vitro assay with intact chloroplasts, it has been shown that precursors bind to the chloroplast's surface, insert into a proteinaceous outer envelope translocation complex, and then proceed across outer and then inner envelope membranes in an extended conformation. While these steps of the import process are similar to those for protein import into mitochondria, the translocation machinery consists of unique polypeptide components that include at least two GTP-binding proteins. These GTP-binding proteins appear to participate in early events of import and probably regulate precursor recognition and advancement into the translocation complex.

Import into the plastid is only half of the localization process because proteins must then be routed into their proper suborganellar compartment. The routing of imported chloroplast proteins is highly complex because of the multiplicity of compartments. Whereas all plastids have in common a double-membrane envelope and an aqueous matrix space called the stroma, the chloroplast uniquely

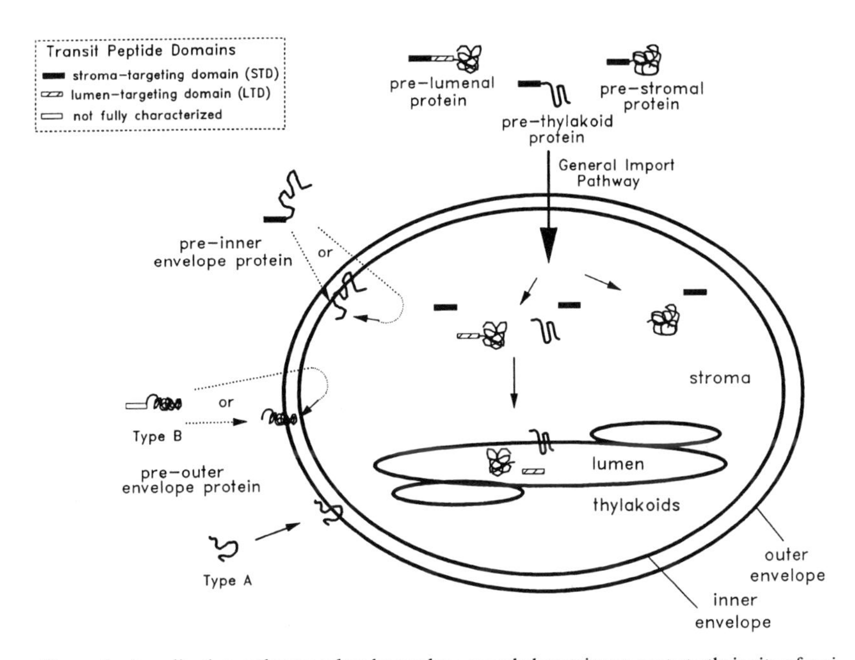

Figure 1 Localization pathways taken by nucleus-encoded proteins en route to their site of residence within the chloroplast.

contains a closed internal thylakoid membrane system that surrounds an aqueous lumen. Thus precursor proteins must be specifically and faithfully localized to one of six different subcompartments. Remarkable progress has recently been made in defining the pathways and plastid machinery involved in routing to the thylakoid membrane and thylakoid lumen. Thylakoid proteins are localized by a two-step process that involves import across the envelope into the stroma and subsequent transport/integration into the thylakoids (Figure 1). Surprisingly, four distinct mechanisms have been described for thylakoid translocation; one is homologous to the bacterial Sec system; one homologous to the signal recognition particle (SRP) system of the endoplasmic reticulum (ER); one appears unique in its sole reliance on a *trans*-thylakoid ΔpH; and the last may be a spontaneous mechanism.

There are several excellent reviews of chloroplast protein import (de Boer & Weisbeek 1991, Theg & Scott 1993, Robinson & Klösgen 1994, Gray & Row 1995, Keegstra et al 1995, Soll 1995, Schnell 1995). This review focuses most intensely on developments in the rapidly progressing areas of translocation across the envelope membranes and localization to thylakoids.

PATHWAYS FOR PLASTID ENTRY AND SUBCOMPARTMENT LOCALIZATION

The first step toward understanding precursor protein localization is defining their routes into and within the plastid. Figure 1 summarizes the current state of knowledge for the pathways taken by proteins en route to the different chloroplast subcompartments.

Import into the stroma is the major route for protein entry into plastids and has been termed the general import pathway. Proteins destined for the chloroplast interior, i.e. stromal, thylakoid, and thylakoid lumenal proteins, are carried by common machinery as evidenced by common energy requirements, the need for proteinaceous components on the chloroplast surface (see de Boer & Weisbeek 1991), and the results of competition studies with overexpressed precursor proteins and with synthetic transit peptide fragments (see Gray & Row 1995). The in vitro rates of import on this pathway under precursor saturating conditions have been estimated to be 40–50,000 precursors per chloroplast per minute (Pilon et al 1992b, Cline et al 1993), which are consistent with estimates for the in vivo rates during peak periods of chloroplast development in plants (Pfisterer et al 1982). A subset of proteins imported into the chloroplast stroma are further directed to the thylakoid membrane and lumen (see below).

Pathways for localization of integral envelope proteins have not been fully elucidated. However, preliminary studies suggest that there are several routes into the envelope. A subgroup of outer envelope proteins (OEP), the Type A proteins, appears to insert directly into the outer envelope membrane. These proteins are small to moderate in size, contain at least one extended hydrophobic segment, and lack cleavable targeting peptides (e.g. Salomon et al 1990, Li et al 1991, Fischer et al 1994). Their insertion into the outer membrane may be spontaneous because there is no apparent energy requirement nor dependence on protease-sensitive components of the chloroplast surface (but see Seedorf et al 1995).

Type B OEPs, represented by two large polypeptides of the general import apparatus (OEP75 and OEP86), possess transit peptides and are localized by a process that is dependent on nucleoside triphosphates (NTPs) and protease-sensitive outer envelope components (Hirsch et al 1994, Tranel et al 1995). OEP75 appears to use at least part of the general import apparatus because its import is competed by the precursor to the small subunit of Rubisco (preSS), a stromal protein (Tranel et al 1995). Import of OEP86 is not competed by preSS, which suggests that it employs different localization machinery (Hirsch et al 1994). Both proteins, although tightly anchored in the outer membrane, are devoid of hydrophobic segments capable of spanning the bilayer as an α helix.

Inner envelope proteins (IEPs), e.g. IEP37, the triose phosphate/phosphate translocator, and Bt1p, are synthesized with transit peptides and possess one or more hydrophobic segments capable of spanning the bilayer. Their import and localization involves translocation machinery, probably the general import apparatus, because experiments with chimeric precursors show that their transit peptides are functionally identical to stroma-targeting transit peptides (see below) (Li et al 1992, Brink et al 1995, Knight & Gray 1995). Studies by Knight & Gray (1995) argue that inner envelope targeting is contained within a hydrophobic segment of the mature protein sequence of the triose phosphate/phosphate translocator. This envelope-targeting element might function as a re-export signal to promote insertion of the imported protein from the stroma into the inner envelope membrane. This type of mechanism is referred to as conservative sorting (Hartl & Neupert 1990). Alternatively, the hydrophobic segment might function as a stop-transfer sequence and induce premature exit from the general import translocation machinery at the level of the inner envelope. If hydrophobic segments of IEPs act as stop-transfer sequences, then a specific recognition mechanism must exist because many thylakoid proteins contain hydrophobic segments yet are imported across both envelope membranes (see Knight et al 1993 for discussion). In mitochondria, both the conservative sorting pathway (Hartl & Neupert 1990) and the stop-transfer pathway (Glick et al 1992) appear to operate for proteins of the inner membrane and intermembrane space.

THE GENERAL IMPORT PATHWAY

Stroma-Targeting Domains of Transit Peptides

Access to the general import pathway is governed by stroma-targeting domains (STDs) of transit peptides. Stromal protein precursors possess transit peptides that contain only an STD, whereas thylakoid lumenal protein precursors have an additional targeting domain in their transit peptides (de Boer & Weisbeek 1991; Figure 2*a*). STDs are both necessary and sufficient for import of the passenger protein to the stroma and are usually specific for plastid protein import (but see Creissen et al 1995). Nevertheless, the manner by which this specific targeting information is encoded has eluded investigators. STDs range in size from about 30 to 120 residues and have properties that superficially resemble presequences of mitochondrial precursor proteins; i.e. they are rich in hydroxylated residues and deficient in acidic residues. STDs tend to share several compositional motifs: an amino terminal 10–15 residues devoid of Gly, Pro, and charged residues; a variable middle region rich in Ser, Thr, Lys, and Arg; and a carboxy-proximal region with a loosely conserved sequence

(Ile/Val-x-Ala/Cys*Ala) for proteolytic processing (von Heijne et al 1989, de Boer & Weisbeek 1991). However, there are no extended blocks of sequence conservation, nor any conserved secondary structural motifs. Theoretical analyses suggest that STDs adopt predominantly random coil conformation (von Heijne & Nishikawa 1991). Circular dichroism and tryptophan fluorescence quenching analysis confirm a general lack of secondary structure for the transit peptides of preferredoxin (Pilon et al 1992a) and preplastocyanin (prePC) (Endo

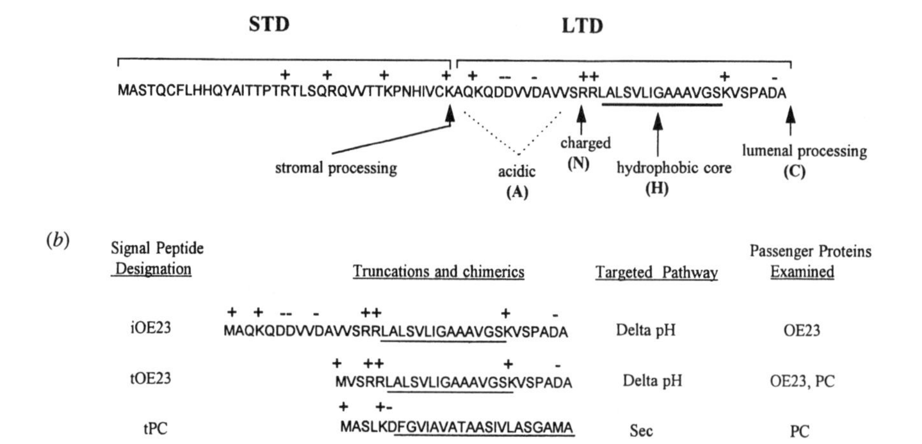

Figure 2 Domain structure and sorting elements of a transit peptide that directs protein import into the chloroplast and transport into the thylakoid lumen. (*a*) The transit peptide for the precursor to OE23 (23-kDa protein of the oxygen-evolving complex) contains a stromal targeting domain (STD) that directs import into the chloroplast stroma where the STD is cleaved as shown (Bassham et al 1991) by a processing protease. The lumen targeting domain (LTD) directs transport into the lumen where the LTD is removed by a second processing protease. The signal peptide motif (N, H, and C) is indicated, as is an acidic (A) domain. (*b*) Transport characteristics of precursors containing truncated LTDs demonstrate that the signal peptide motif of the LTD is necessary and sufficient for pathway-specific transport of OE23 and PC. A hybrid signal peptide that directs passenger proteins to both pathways shows that the N and H/C domains play a role in pathway-specific targeting. When directed by the hybrid signal peptide, transport of different passenger proteins demonstrates that delta pH mature domains were arrested in translocation across the Sec pathway (see text for discussion).

et al 1992) in aqueous solution. This general lack of structural motif differs from mitochondrial presequences, where an amphipathic helix is an essential structural element (von Heijne et al 1989). Thus interaction between STDs and plastids probably involves a subtle recognition event that could include several checkpoints to validate the targeting element.

One way that STDs may specifically interact with plastids is by inserting into the lipid bilayer. Biochemical studies of synthetic transit peptides and entire precursor proteins have shown that some transit peptides have a propensity to insert into membrane lipids (see Theg & Scott 1993). van't Hoff et al (1993) demonstrated that preferredoxin or its transit peptide inserts into monolayers of plastid envelope lipids at physiological surface pressures. Interactions were predominantly with monogalactosyldiacylglycerol, sulfoquinovosyldiacylglycerol, and phosphatidylglycerol. Interaction with the two former polar lipids is particularly notable because these lipids are present in eukaryotes only in plastid membranes. Interactions of the transit peptide and precursor with mitochondrial lipids occurred only below physiological surface pressures, indicating that the interactions with plastid lipids were specific. These studies suggest that precursors may initially bind to the lipids of the outer chloroplast envelope, but clearly more studies are required to establish definitively such an interaction during the import process. Furthermore, the generality of this model needs to be addressed, e.g. the PC (plastocyanin) STD appears not to interact with plastid lipids under comparable conditions (Endo et al 1992).

Evidence that STDs interact specifically with outer chloroplast envelope proteins includes the fact that synthetic STDs compete for a saturable component of the import apparatus (Gray & Row 1995) and the observation that STDs are in close contact with polypeptides of the translocation machinery at a very early stage in the import process (D Schnell, personal communication).

Steps of the Import Process

Defining the steps of the import process has been pivotal not only for understanding the progression of recognition and translocation but also for identification of translocon components (see Figure 3). Binding of precursors to the outer envelope membrane constitutes the first step and produces the first stable intermediate, the so-called early intermediate (Theg & Scott 1993, Schnell & Blobel 1993). Binding requires 50–100 μM of any of several NTPs and the presence of outer envelope membrane proteins (OEPs) (Olsen & Keegstra 1992, Theg & Scott 1993). Binding is saturable between 1500 and 3500 sites and represents an on-pathway step because bound precursors progress into the stroma when the ATP concentration is raised to approximately 1 mM (Theg & Scott 1993, Schnell & Blobel 1993). Such productive binding is mediated by the STD (Friedman & Keegstra 1989).

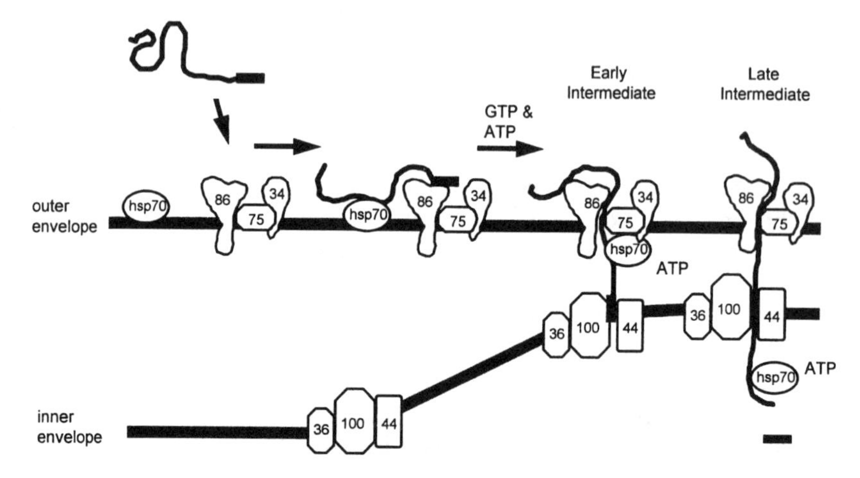

Figure 3 A model for protein import into chloroplasts. Components are identified by their molecular masses in kilodaltons (see Figure 1 for definition of symbols). See text for a discussion of the characteristics of these components and their proposed roles in the import process.

Early intermediate precursors have probably progressed deeply into the translocation machinery because they are irreversibly bound (Theg & Scott 1993) and frequently partially protected from protease treatment of chloroplasts and membranes (Freidman & Keegstra 1989, Waegemann & Soll 1991). The relatively large size of protected fragments raises the possibility that much of the early intermediate precursor has crossed the outer envelope membrane (see Figure 3).

Time-course analysis of precursor import from the bound state led to identification of a second or late intermediate (Schnell & Blobel 1993). Late intermediate proteins have translocated across the envelope sufficiently to be processed by the STD processing protease, yet are still exposed to the outside of the chloroplast as judged by susceptibility to exogenous protease. The STD processing protease is an ≈ 140-kDa metalloprotease that has recently been cloned (VanderVere et al 1995). Because this protease is released into the soluble fraction upon chloroplast lysis, it is believed to reside in the stroma. Thus the late intermediate protein probably represents a translocating precursor that spans both envelope membranes. In other words, similar to transport across mitochondrial membranes, polypeptide chain translocation proceeds amino-terminus first, in an extended conformation, and across both envelope membranes simultaneously. Consistent with this latter conclusion, Schnell & Blobel (1993) have presented electron microscopic as well as biochemical fractionation evidence that these early and late intermediates are preferentially located at sites of close contact between inner and outer envelope membrane, i.e. contact sites.

The implication is that proteins cross the two envelope membranes in regions where they are attached. A similar conclusion was originally reached regarding protein import across outer and inner mitochondrial membranes (Schwaiger et al 1987). However, conditions that enhance the arrest of intermediates would also stabilize tight precursor-mediated outer envelope/inner envelope association. Thus the possibility that proteins can cross outer and inner envelope membranes independently cannot be discounted. Indeed, Scott & Theg (1996) provide compelling evidence that precursor proteins can cross outer and inner envelope membranes sequentially. By using reduced ATP concentrations and chloroplasts that were plasmolyzed with hypertonic osmoticum, they were able to capture full-size precursor that had apparently crossed the outer, but not the inner envelope membrane. This protease-protected precursor was subsequently imported into the stroma upon addition of larger amounts of ATP. These results imply that chloroplasts, similar to mitochondria (Hwang et al 1991, Segui-Real et al 1993), have independent outer membrane and inner membrane translocation machinery. When considered together, the results of Schnell & Blobel (1993) and Scott & Theg (1996) suggest that contact sites are formed under isotonic conditions when translocating precursor proteins engage the inner envelope translocation apparatus prior to completely crossing the outer membrane.

Envelope Translocation Machinery

Several early attempts to identify components of the import machinery were inconclusive or produced misleading identifications (see Gray & Row 1995 for discussion). However, in the past several years, converging investigations of four independent research groups have led to the identification of five OEPs and three IEPs that participate in precursor recognition and translocation (Figure 3). These studies, which relied primarily on early intermediate precursor, used chemical cross-linking, immunoaffinity chromatography of proteins associated with a protein A-containing chimeric precursor, and sedimentation-purification of detergent-solubilized translocation complexes. There are several excellent reviews of these translocon components (Gray & Row 1995, Schnell 1995, Soll 1995). The location and topology of each protein, as well as the manner of identification, suggest the order in which they participate in the import process. A cytosolically exposed outer envelope Hsp70 protein (Com70), which was cross-linked to a precursor protein intermediate (Wu et al 1994), may bind the precursor as it encounters the chloroplast surface and maintain it in an unfolded conformation. OEP86 appears to interact with precursor proteins at a very early stage of the import process, i.e. prior to formation of the early intermediate.

OEP86 was cross-linked to preSS in the absence as well as the presence of NTPs (Perry & Keegstra 1994). In addition, binding and import of preSS are inhibited by antibodies to OEP86 (Hirsch et al 1994). OEP75, OEP34, OEP86, and an inter-envelope space Hsp70 protein are considered to be components of the outer membrane translocation complex because they are co-isolated as a complex with early intermediate precursor (Waegemann & Soll 1991, Schnell et al 1994). OEP86 and OEP75 are intimately associated with precursor because they are directly cross-linked to preSS (Perry & Keegstra 1994). OEP75 is notable in that it is largely buried in the outer envelope membrane but possesses no extended hydrophobic sequences. It has been speculated that, because OEP75 is predicted to contain a substantial amount of β structure, it may adopt a β barrel conformation similar to bacterial outer membrane pore proteins, i.e. OEP75 may make up part of the protein translocating channel (Schnell 1995, Tranel et al 1995). The inner-envelope space Hsp70 appears to be integrally associated with the outer envelope membrane.

OEP86, OEP75, and OEP34 appear to form a complex prior to interaction with precursor. Seedorf et al (1995) showed that a substantial amount of OEP34 and OEP75 are connected by disulfide bridges and that addition of copper chloride to further oxidize sulfhydryl groups resulted in an OEP86, OEP75, OEP34 complex (Seedorf & Soll 1995). The fact that most of these polypeptides are located in the free outer envelope membrane subfraction on sucrose gradients (Cline et al 1985) suggests that precursor binding in the presence of NTPs is required to promote this complex into contact sites. This conclusion is further supported by the fact that OEP86 cross-linked to preSS in the absence of NTPs is located in an outer envelope-enriched fraction, but when cross-linked in the presence of NTPs, it is in the contact site-enriched fraction (Perry & Keegstra 1994).

Three inner envelope proteins, IEP97, IEP36, and IEP44, are thought to participate in translocation across the inner envelope membrane because they are only associated with translocating polypeptide in the later stages of translocation (Schnell et al 1994, Wu et al 1994).

Perhaps the most intriguing aspect of the outer membrane translocation machinery concerns OEP86 and OEP34, which have been shown to bind GTP (Kessler et al 1994). Additionally, OEP34 hydrolyzes GTP (Seedorf et al 1995). GTP-binding proteins are involved in protein targeting to the ER (Walter & Johnson 1994) and in the early steps of nuclear protein import (Sweet & Gerace 1995). In these systems, binding and/or hydrolysis of GTP appears necessary for commitment to the translocation site. The case of ER transport is particularly relevant. Here, GTP hydrolysis by the SRα subunit and GTP loading of the SRP54 polypeptide subunit is necessary for transfer of the nascent chain:ribosome complex to the translocon. These roles are consistent with the known functions of other GTP-binding proteins as molecular switches that ensure directionality and accuracy of biological processes (Bourne et al 1991). Interestingly, SRP54 and SRα share significant sequence homology both within and flanking the GTP-binding domain (Walter & Johnson 1994), a characteristic that OEP86 and OEP34 also share (Kessler et al 1994, Seedorf et al 1995). It is likely that OEP86 and OEP34 act in concert, through GTP binding and hydrolysis, to recognize and then commit authentic plastid precursor proteins to the translocation complex of the outer envelope membrane, but this remains to be experimentally demonstrated. However, it does appear that GTP hydrolysis is necessary for engagement in the envelope translocation machinery because stable precursor binding to chloroplasts is inhibited by GTP-γ-S (Kessler et al 1994).

One surprise is that none of the chloroplast translocation components, with the exception of Hsp70, is homologous to the mitochondrial import apparatus (Ryan & Jensen 1995). It appears that chloroplast and mitochondrial import machinery have arisen independently and that mitochondria evolved a different means of recognizing targeting peptides. Another possibility is that in plant cells where mitochondrial targeting must be more stringent, mitochondria also employ a proof-reading system analogous to plastids.

What Drives Translocation?

NTPs appear to be the sole energy source for polypeptide translocation across the plastid envelope; transmembrane electrical or ion gradients are not involved in this process (de Boer & Weisbeek 1991, Theg & Scott 1993). ATP in the inter envelope space is required for precursor transport across the outer envelope membrane (Scott & Theg 1996); ATP in the stroma is necessary for translocation across the inner-envelope membrane (Theg & Scott 1993). This implicates *trans*-located ATPases in the translocation event.

There is now compelling evidence that protein translocation proceeds through a pore or channel through the membrane (Simon & Blobel 1991). Protein translocation channels may also exist in the plastid envelope. This is suggested by the observation that transit peptides mediate opening of aqueous channels across the chloroplast envelope, as assessed by patch-clamp studies (Bulychev et al 1994). *Trans*-located Hsp70 proteins facilitate translocation across the ER and mitochondrial membranes, presumably by a thermal ratcheting mechanism (Wickner 1994, Glick 1995), whereby successive binding and release of the polypeptide emerging from the channel serves as the thermodynamic driving force for chain movement. The presence of the inter-envelope Hsp70 in the early intermediate complex provides support for its role in transport across the outer envelope, and the observation that the stromal Hsp70 transiently binds to newly imported proteins (Tsugeki & Nishimura 1993, Madueño et al 1993) suggests that it might be involved in inner envelope translocation. Nevertheless, these proposed roles remain to be verified experimentally. In addition, the possibility that other stromal peptide-binding proteins, e.g. the Hsp60 protein, a chloroplast SecA homologue (see below), and an SRP54 homologue (see below), are involved in translocation should be explored. In the latter case, one would expect that translocation across the inner envelope membrane would depend upon GTP.

An interesting example of a stromal requirement for transport has been recently described by Reinbothe et al (1995), in which translocation of the precursor to protochlorophyllide oxidoreductase requires the presence of the substrate, protochlorophyllide, inside the plastid. One interesting possibility is that the substrate drives the translocation step by inducing folding of the enzyme on the stromal side of the inner envelope membrane.

ROUTING OF PROTEINS INTO/ACROSS THE THYLAKOIDS

Thylakoid Proteins Are Localized by Sequential Protein Translocation Events

Thylakoid precursors are imported across both envelope membranes into the stroma and are then transported into/across thylakoid membranes by additional translocation machinery (see Figures 1 and 4). This model was initially adopted based on the transient appearance of a soluble, stromal form of an imported thylakoid protein (Smeekens et al 1986) and on the reconstitution of a thylakoid translocation step (Cline 1986). A thorough analysis of stromal intermediates has now authenticated this two-step localization process. Time-course analysis combined with rapid stopping methods demonstrated that the stromal intermediates accumulate with kinetics expected for pathway intermediates (Reed et al 1990, Cline et al 1992b). The intermediates accumulate to high levels in import assays when the thylakoid transport/integration step is inhibited (Cline et al 1989, Mould & Robinson 1991, Cline et al 1993, Konishi & Watanabe 1993, Henry et al 1994, Knott & Robinson 1994). And, most importantly, accumulated intermediates can be "chased" into thylakoids in vitro (Reed et al 1990, Cline et al 1992b 1993, Konishi & Watanabe 1993, Creighton et al 1995) and in vivo (Howe & Merchant 1993). In most cases, the intermediates had been processed to remove the STD (but see Bauerle et al 1991, Cline et al 1992b, Nielsen et al 1994). Thus lumenal protein intermediates are generally intermediate in size between the full-length precursor and mature protein, whereas most membrane protein intermediates are mature size (Figures 1 and 4). This is consistent with studies showing that thylakoid targeting is governed by the lumen targeting domain (LTD) of the transit peptide (Ko & Cashmore 1989, Hageman et al 1990; Figure 2 and see below) and by elements in the mature sequence of membrane proteins (de Boer & Weisbeek 1991, Madueño et al 1994). Robinson and colleagues (Halpin et al 1989) showed that LTDs are removed in the thylakoid lumen by a processing protease that has the identical reaction specificity as bacterial and ER signal peptidases.

Integration of the membrane protein LHCP (Cline 1986) and transport of the lumenal protein OE33 (Kirwin et al 1989) into isolated thylakoids demonstrates that thylakoid localization can be uncoupled from import into chloroplasts. Subsequent studies have reconstituted thylakoid transport/integration of a variety of other thylakoid proteins (Robinson & Klösgen 1994). Pathway intermediates observed *in organello* (Cline et al 1992b), or produced in vitro (Viitanen et al 1988, Cline et al 1993, Hulford et al 1994), are efficiently transported/integrated in thylakoid transport assays, but the full-length precursors are also effective substrates (Robinson & Klösgen 1994). The fact that removal of the STD is not prerequisite to translocation is consistent with the loop model invoked for initiating transport of signal peptide precursors or signal anchor-bearing precursors (Kuhn et al 1994). These reconstituted thylakoid translocation assays provide the means for detailed analysis of the translocation process (see below).

There Are at Least Four Pathways for Protein Translocation Into or Across the Thylakoid Membrane

The discovery that different subgroups of precursors have different energy and soluble protein factor requirements for translocation suggested that more than one mechanism is involved (see Figure 4). Most striking was the fact that two lumenal proteins, OE23 and OE17, are transported in the total absence of NTPs (Cline et al 1992a). These proteins rely entirely on the *trans*-thylakoid ΔpH as an energy source and have no soluble factor requirement (see Robinson & Klösgen 1994 for review). Equally intriguing is the finding that integration of LHCP requires GTP (Hoffman & Franklin 1994) rather than ATP, as originally thought (Cline 1986). A third subgroup (OE33 and PC) requires ATP and a soluble factor and is stimulated by ΔpH (Hulford et al 1994, Yuan & Cline 1994).

The implication of the above results, i.e. that there exists group-specific translocation machinery, was demonstrated by precursor competition studies (Cline et al 1993). In a novel *in organello* competition assay with intact chloroplasts, saturating amounts of iOE23 (the stromal intermediate) selectively competed with iOE17 for thylakoid transport, whereas iOE33 selectively competed with iPC. Neither iOE23 nor iOE33 competed for LHCP integration. The *in*

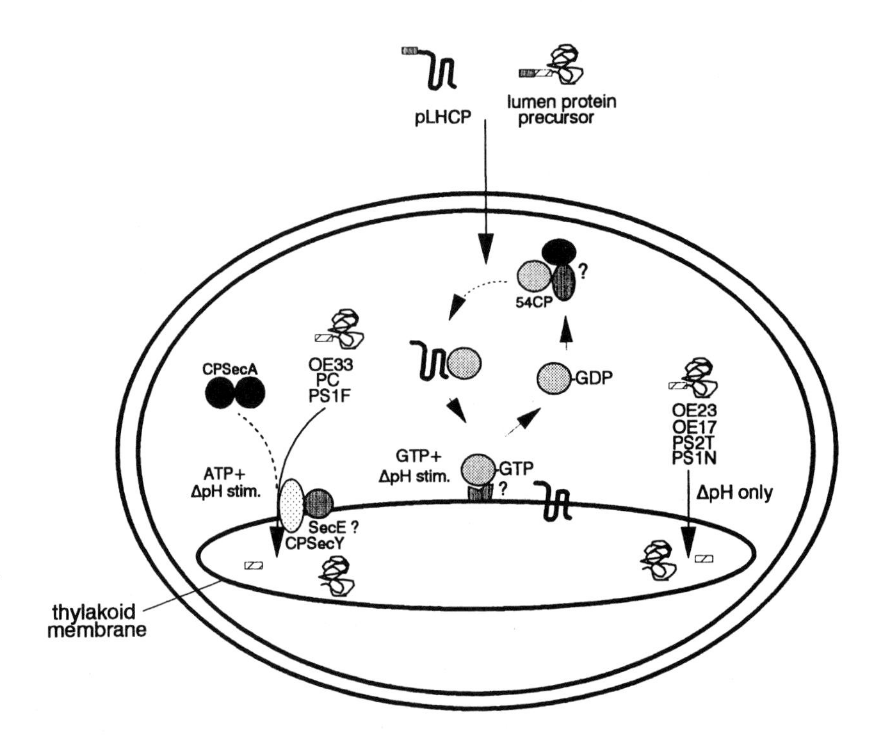

Figure 4 A working model for routing of lumen-resident and integral membrane thylakoid proteins via three precursor-specific pathways. Proteins requiring CPSecA and ATP are presumed to move across the membrane through a pore consisting of CPSecY and possibly a homologue of *E. coli* SecE. Cytochrome f, a plastid-encoded protein that also employs the CPSecA mechanism, is not shown. Integration of the light-harvesting chlorophyll-binding protein (LHCP) requires 54CP, a homologue of SRP54. 54CP is thought to pilot LHCP to the thylakoid in a targeting cycle that involves GTP hydrolysis (see text for discussion). A third pathway is unique and requires only ΔpH to power transport across the membrane. A fourth pathway that mediates the insertion of CF_0II, presumably by a spontaneous mechanism, is not shown. Evidence for pathway-specific groupings is described in the text. Biochemical studies demonstrate that PS2T (Henry et al 1994) and PS1N (Mant et al 1994) are transported by the delta pH pathway. Question marks denote that these components are hypothetical and have not been identified.

organello assay was possible because the V_{max} for import of preOE23, for example, into chloroplasts is 10 to 40 times the V_{max} for transport into thylakoids, thereby allowing concentration of intermediates within the chloroplast to ≈ 100 times the K_m for thylakoid transport. Thus by use of precursor concentrations that are subsaturating for import, competition of imported proteins for thylakoid localization occurs *in organello*. Although this type of situation is probably more physiologically relevant, it does not permit determination of the specific site of competition. Competition assays with isolated thylakoids also produce the same groupings as those obtained by energy requirements (above), and additionally reveal that iOE23 and iOE17 compete for a thylakoid membrane component, possibly a receptor. Taken together, these results lead to a model for three mechanistically different pathways for thylakoid transport/integration (Figure 4). Analysis of chimeric precursor proteins (Henry et al 1994, Robinson et al 1994) show that commitment to a particular pathway is determined by elements in the LTD, rather than by the mature passenger proteins (but see below).

Recent analysis of two maize mutants provides in vivo evidence for operation of the above three pathways (Voelker & Barkan 1995). The maize mutant *tha1* is selectively defective in thylakoid transport of OE33 and PC and of another nucleus-encoded thylakoid protein, PS1F. Interestingly, *tha1* is also impaired in thylakoid transport of a chloroplast-synthesized protein, cytochrome f. Biochemical studies reinforce the conclusion that PS1F (Karnauchov et al 1994) and cytochrome f (R Mould & J Gray, personal communication) are transported on the same pathway as OE33 and PC. The mutant *hcf106* has a complementary phenotype to *tha1*; *hcf106* plants are selectively defective in thylakoid transport of OE16 (OE17) and OE23. LHCP accumulated normally in both mutants.

Biochemical examination of the localization of CF_0II, an integral membrane protein with a bipartite transit peptide, suggests the existence of a fourth thylakoid pathway. CF_0II inserts into isolated thylakoids in the absence of ΔpH, NTPs, and soluble factors and is unaffected by saturating quantities of pre-OE23 as a competitor (Michl et al 1994). CF_0II also inserts into protease-treated thylakoids (R Klösgen, personal communication), raising the possibility that it inserts spontaneously; however, this possibility remains to be firmly established.

Translocation Components Identify Two Thylakoid Pathways as Homologues of Bacterial and ER Systems

BACTERIAL SEC-HOMOLOGOUS PATHWAY FOR TRANSPORT OF PC, OE33, PS1F, CYTOCHROME F Chloroplasts are thought to have evolved from a prokaryotic endosymbiont similar to modern-day cyanobacteria. This suggests that a conserved prokaryote-like protein transport mechanism would be operational within chloroplasts. The bacterial system for protein export across the cytoplasmic membrane employs the SecA translocation ATPase to initiate membrane insertion of precursor and a core translocon made up of the SecY/E/G polypeptides (Pugsley 1993, Wickner 1994). Translocation on this system requires ATP and a proton-motive force. Preliminary evidence for a Sec system operating in thylakoids includes the presence of *secA*- and *secY*-homologous genes in the chloroplast genomes of several algae (Robinson & Klösgen 1994), the appropriate energetics for OE33 and PC transport (Hulford et al 1994, Yuan & Cline 1994), and the fact that thylakoid translocation of OE33 and PC, but not OE23, OE17, or LHCP, are inhibited by azide (Henry et al 1994, Knott & Robinson 1994, Yuan et al 1994). Azide is a diagnostic inhibitor of *Escherichia coli* SecA (Oliver et al 1990).

A SecA-dependent thylakoid transport mechanism was demonstrated in pea chloroplasts by two approaches. In the first, antibodies prepared to a conserved region of a deduced algal SecA polypeptide were used to identify and purify a stromal 110-kDa SecA homologue, designated CPSecA (Yuan et al 1994). Purified CPSecA reconstituted transport of PC and OE33 across buffer (or urea) washed thylakoids in a concentration and azide-sensitive manner. Thus CPSecA accounts for the stromal factor requirement for PC and OE33 transport. CPSecA was unable to replace the stromal requirement for LHCP integration and neither simulated nor inhibited transport of OE23 and OE17. In a second approach, antibodies to a peptide deduced from a partial CPSecA cDNA inhibited thylakoid transport of OE33, but not OE23 (Nakai et al 1994). A similar reduction in OE33 transport occurred when stromal extract was immunodepleted of CPSecA. Together, these results indicate that one thylakoid transport pathway employs a SecA-dependent translocation mechanism (see Figure 4). Full-length cDNAs to two higher plant chloroplast SecAs have now been isolated (Berghofer et al 1995, Nohara et al 1995). The deduced protein sequence of these two nuclear-encoded proteins is strikingly homologous to that of cyanobacterial SecA. Recently, in vivo evidence for the selective operation of the SecA pathway has come from the identification of the maize *tha1* gene as a *secA* homologue (R Voelker & A Barkan, personal communication).

Laidler et al (1995) isolated a full-length cDNA to a chloroplast homologue of SecY (CPSecY) in *Arabidopsis*. Presumably, CPSecY functions in thylakoid transport in a manner similar to SecY in bacteria, forming part of the membrane translocon. Definitive evidence for CPSecY involvement in any of the thylakoid transport pathways is currently lacking. Based on the bacterial paradigm, CPSecY is expected to function with CPSecA in mediating transport of PC, OE33, PS1F, and cytochrome f (see Figure 4).

ER, SRP-HOMOLOGOUS TARGETING PATHWAY FOR THE MEMBRANE PROTEIN LHCP
Chloroplasts also contain a localization system homologous to the SRP system used for protein targeting to the ER and the bacterial cytoplasmic membrane (see Walter & Johnson 1994). Eukaryotic SRP binds nascent protein chains and, in a targeting cycle that involves GTP binding and hydrolysis, pilots the entire SRP:ribosome:nascent chain complex to the ER membrane. Franklin & Hoffman (1993) identified a chloroplast homologue (54CP) of the SRP54 protein subunit and showed that the 54CP protein resides primarily in the stroma. Two observations implicated 54CP in LHCP targeting. First, the LHCP stromal intermediate is a soluble complex of $\approx$ 120 kDa, or about four to five times the size of the LHCP polypeptide (Payan & Cline 1991). This transit complex can be formed post-translationally by a proteinaceous stromal activity that shares similar properties with the stromal activity required for LHCP integration into isolated thylakoids (Payan & Cline 1991). Both activities co-elute from gel filtration columns with a peak of 54CP (R Henry & K Cline, unpublished data). The second correlative factor is that GTP is required for LHCP membrane integration (Hoffman & Franklin 1994).

Recent experiments more directly implicate 54CP in LHCP integration (Li et al 1995). First, 54CP was shown to be present in the transit complex by immunoprecipitation and also by chemical cross-linking. More importantly, 54CP was shown to be required for LHCP integration by immunodepleting 54CP from stromal extract. The depleted extract was unable either to form transit complex or to support LHCP integration. These results demonstrate that 54CP interacts with imported LHCP in the stroma and is required for its integration.

The precise role of 54CP in this process awaits purification of the active particle. Some observations suggest that there are important differences between the 54CP system and other SRP systems. The most obvious difference is that 54CP functions post-translationally, whereas all other SRPs appear to be strictly co-translational (Walter & Johnson 1994). Another difference is that all other SRPs contain an essential RNA component; RNA has not been found associated with 54CP.

However, the chloroplast SRP homologue also shows similarities with mammalian SRP (Walter & Johnson 1994), suggesting a model for its operation (see Figure 4). Mammalian SRP must be in a guanine nucleotide-free state in order to bind signal peptides (or signal anchors of membrane proteins). Similarly, 54CP must be in a nucleotide-free state to form a transit complex because GDPox, which covalently modifies guanine nucleotide-binding sites, prevents transit complex formation and LHCP integration (R Henry & K Cline, unpublished data). Certain observations indicate that the role of 54CP extends beyond that of a molecular chaperone and argue for a role in targeting. For example, although 54CP can maintain LHCP solubility and integration competence for extended periods, it is not possible to bypass the stromal requirement for integration by urea-denaturation of LHCP or replace the stromal requirement with other chaperone proteins (Payan & Cline 1991, Yuan et al 1993). The GTP hydrolysis required for integration appears to occur at the thylakoid membrane because, although the non-hydrolyzable GTP analogue, GMP-PNP, inhibits integration (Hoffman & Franklin 1994), it does not inhibit transit complex formation (R Henry & K Cline, unpublished data). This is consistent with the mammalian SRP cycle, where GTP hydrolysis by SRP54 is necessary for release from the ER receptor and SRP recycling to the cytosol. A thylakoid receptor is implied by the fact that treatment of thylakoids with low levels of protease destroys their ability to integrate LHCP (C Dahlin & K Cline, unpublished data). Nevertheless, this targeting model for 54CP-LHCP needs to be experimentally examined.

It is clear that there are many unanswered questions regarding the chloroplast SRP-like system. The chloroplast 54CP particle is probably a multimeric complex because its native size is $\approx$200 kDa (R Henry & K Cline, unpublished data). What are the other components of the active particle and how do they function? The 54CP-binding segment of LHCP has not been identified; presumably, it is one or more of the three essential membrane-spanning (signal anchor?) domains. If so, why don't the signal peptides of the lumenal proteins (see below) also bind to 54CP? In addition, what other proteins utilize the SRP pathway? One attractive possibility is that the chloroplast SRP is dedicated to the localization of integral membrane proteins, both imported and plastid-encoded.

DOES TRANSLOCATION MACHINERY FACILITATE TRANSPORT ACROSS THE DELTA PH PATHWAY? Because of the singular use of a pH gradient to power proteins (OE23, OE17, PS2T, PS1N) across the thylakoids, this pathway is referred to as the delta pH pathway. The fact that it requires neither soluble factors nor NTPs makes this translocation pathway unique and suggests a more rudimentary mechanism than the Sec and SRP pathways. Components of the delta pH machinery have not been identified. However, several observations imply the existence of proteinaceous translocation machinery. Precursor saturation of transport at the level of the membrane argues for a specific membrane component that recognizes the LTD of delta pH precursors (Cline et al 1993). Controlled protease pretreatments of thylakoids indicates that a proteinaceous thylakoid component is necessary for transport (Creighton et al 1995). Finally, the identification of the *hcf106* mutant deficient in transport on this pathway and localization of the *hcf106* gene product to the thylakoids support a protein-mediated translocation process (Voelker & Barkan 1995; R Martienssen, personal communication).

DO PATHWAYS SHARE COMMON TRANSLOCON COMPONENTS? Although it is possible that the delta pH pathway employs a unique translocon, we think it more likely that it uses either the same translocon as the Sec pathway (a common pore) or a homologous one. Thylakoid transport via the delta pH pathway as well as via the Sec pathway employs hydrophobic targeting elements such as classical signal peptides and signal anchors (see below). The well-characterized signal peptide-based systems possess translocon components that are strikingly homologous at the sequence level; the ER Sec61α polypeptide is homologous to bacterial SecY, and Sec61γ is homologous to SecE (Dobberstein 1994). This points to a common evolutionary origin for such systems. Furthermore, where more than one pathway has been identified in the ER, the same core translocon is still utilized (Siegel 1995).

Definitive evidence for a common or homologous pore is lacking, but a recent study in *Chlamydomonas* points to the existence of common component(s) for at least two pathways. Smith & Kohorn (1994) used signal peptide mutations of cytochrome f to block its integration and isolate suppressors. One such signal peptide mutant, A15E, exhibited a dominant-negative effect on the accumulation of LHCP, the plastid-encoded D1 protein, and even a wild-type copy of cytochrome f. Although other thylakoid proteins appeared not to be affected as assessed by their steady-state levels, subtle effects on transport may have escaped detection. The identification of suppressors to A15E may shed light on which step of the translocation process is affected by this mutant. Suppressors isolated to a similar signal peptide mutation in *E. coli*, the lamB14D mutation, were alleles of *secA, secY*, and *secE* (Stader et al 1989). Because LHCP does not employ SecA, this analogy suggests that the A15E effect may be at the level of the translocon.

A Rationale for the Operation of Dual Targeting Pathways Into the Thylakoid Lumen

The underlying reason for multiple targeting pathways to the same membrane is not known for any system. However, analysis of lumen-targeting elements is beginning to shed light on this question for thylakoids. As mentioned, LTDs carry signaling information for exclusive transport in vitro via either the delta pH or the Sec pathway. They have in common the motifs of the classical signal peptides that direct targeting and transport across the ER and the bacterial cytoplasmic membrane: i.e. a charged (N) domain, a hydrophobic (H) domain, and a (C) domain for proteolytic processing (Figure 2). Extensive studies in bacteria have shown that N and H, but not C, domains are important for transport

(Pugsley 1993). Most LTDs also contain an acidic A domain amino-proximal to the signal peptide (Figure 2). Surprisingly, the A domain appears to be dispensable for pathway-specific thylakoid targeting because a truncated OE23 precursor lacking the A domain was transported exclusively on the delta pH pathway (Figure 2; R Henry et al, in preparation). This means that commitment to one of two different translocation systems relies totally on subtle differences between signal peptides.

Both the N and the H/C regions of thylakoid signal peptides play a role in pathway-specific targeting. Chaddock et al (1995) recently showed that an N domain double arginine (Arg-Arg), which is conserved in delta pH pathway LTDs, is essential for transport of delta pH precursors. Studies in our laboratory show that, although the Sec pathway can utilize Arg-Arg-containing N regions, the Sec pathway is incompatible with H/C regions derived from delta pH pathway precursors (R Henry et al, in preparation). One emerging conclusion is that exclusive sorting to the delta pH pathway is achieved with an N domain containing an Arg-Arg to gain access to that transport system and an H/C domain that is not recognized by the Sec system.

But why would delta pH pathway precursors need to avoid recognition by the Sec pathway? To address this question, a dual pathway targeting signal was constructed with an Arg-Arg-containing N domain and a Sec pathway H/C domain (Figure 2). This provided the ability to simultaneously test different passenger proteins for their ability to be transported across the two pathways. In this test, proteins normally transported by the Sec system were efficiently transported on either pathway (R Henry et al, in preparation). In contrast, proteins normally transported by the delta pH pathway were only transported by the delta pH pathway, even when purified CPSecA was added to boost Sec pathway transport. This agrees with previous studies of chimeric precursors showing that, even *in organello*, delta pH pathway proteins are very inefficiently translocated by the Sec pathway (Clausmeyer et al 1993, Henry et al 1994). In our experiments, delta pH pathway passenger proteins with a dual targeting peptide appeared to engage the Sec machinery but were arrested in the chain translocation step. One possible explanation for this is that the mature domains of delta pH proteins fold tightly (Creighton et al 1995) and cannot be unfolded by the Sec machinery. These results suggest that delta pH precursors have evolved a mechanism to avoid nonproductive interactions with Sec machinery.

The incompatibility of delta pH proteins with Sec machinery is surprising. The apparent simplicity of the delta pH mechanism, when compared with the ATPase-assisted Sec mechanism, might argue that the delta pH mechanism is a more primitive protein translocation system. Yet the above results suggest otherwise, i.e. that the delta pH mechanism can accommodate a more diverse population of proteins than the Sec mechanism. In considering the presence of the delta pH pathway, it is interesting to note that proteins carried on this pathway do not appear to be present in modern-day cyanobacteria. Thus these proteins, as well as a compatible translocation mechanism, may have been recruited into the chloroplast after the endosymbiotic event. This possibility makes the identification of the components of the delta pH machinery and their comparison with Sec machinery an even more exciting and interesting endeavor.

FUTURE DIRECTIONS

In addition to the above-mentioned avenues for further investigation, we think that studies of the import and routing problem in other developmental plastid stages is a particularly important area for further research. An intriguing problem concerns the specific manner by which thylakoid proteins may be targeted very early in chloroplast development. The progenitor of chloroplasts, the proplastid, contains very little internal membrane. The deposition of thylakoid membranes during the proplastid-chloroplast progression is accompanied by invaginations of the inner envelope membrane and the accumulation of vesicles in the stroma (Whatley et al 1982). It has long been theorized that the inner envelope directly gives rise to thylakoids. However, until recently there was little experimental evidence to support this hypothesis. Experiments in *Chlamydomonas reinhardtii* now show that at 38°C, photosynthetic complexes form rapidly following transfer to light. Extensive invagination of the envelope is observed concomitant with the acquisition of photosynthetic competence, but before any recognizable thylakoid membranes are present (Hoober et al 1991). In another study, Hugueney et al (1995) reconstituted plastid vesicle fusion in vitro. The vesicles, isolated from developing chromoplasts of *Capsicum* plants, fused in a stroma- and ATP-dependent manner. Purification of the stromal component defined a 72-kDa protein with homology to the *N*-ethylmaleimide-sensitive fusion protein (NSF) of yeast and animal cells. This protein is known to be involved in vesicle fusion. Importantly, the RNA for the plastid fusion protein also accumulated constitutively in green leaves. It is unlikely that thylakoid protein localization would differ fundamentally in the early stages of thylakoid formation, i.e. it would involve import into the stroma and subsequent translocation. But it is possible that the inner envelope membrane is the target for the final transport/integration step in early developmental stages and that this target site shifts to thylakoids as they mature. Thus it will be very important to learn how the components of the translocon are assembled in the membrane, when they are expressed, and in which plastid membrane they are located at various stages of development.

ACKNOWLEDGMENTS

We thank Vivian Fincher for critical reading of the manuscript and Alice Barkan, Rabe Klösgen, John Gray, Rob Martienssen, Danny Schnell, and Steve Theg for communicating their unpublished work. The research in the authors' laboratory is supported by National Institutes of Health grant R01 GM46951 and National Science Foundation grant MCB-9419287.

Literature Cited

Bassham DC, Bartling D, Mould RM, Dunbar B, Weisbeek P, et al. 1991. Transport of proteins into chloroplasts: delineation of envelope "transit" and thylakoid "transfer" signals within the presequences of three imported thylakoid lumen proteins. *J. Biol. Chem.* 266(35):23606–10

Bauerle C, Dorl J, Keegstra K. 1991. Kinetic analysis of the transport of thylakoid lumenal proteins in experiments using intact chloroplasts. *J. Biol. Chem.* 266(9):5884–90

Berghöfer J, Karnauchov I, Herrmann RG, Klösgen RB. 1995. Isolation and characterization of a cDNA encoding the secA protein from spinach chloroplasts. *J. Biol. Chem.* 270(31):18341–46

Bourne HR, Sanders DA, McCormick F. 1991. The GTPase superfamily: conserved structure and molecular mechanism. *Nature* 349:117–27

Brink S, Fischer K, Klösgen RB, Flügge U-I. 1995. Sorting of nuclear-encoded chloroplast membrane proteins to the envelope and the thylakoid membrane. *J. Biol. Chem.* 270(35):20808–15

Bulychev A, Pilon M, Dassen H, van't Hof R, Vredenberg W, de Kruijff B. 1994. Precursor-mediated opening of translocation pores in chloroplast envelopes. *FEBS Lett.* 356:204–6

Chaddock AM, Mant A, Karnauchov I, Brink S, Herrmann RG, et al. 1995. A new type of signal peptide: central role of a twin-arginine motif in transfer signals for the ΔpH-dependent thylakoidal protein translocase. *EMBO J.* 14(12):2715–22

Clausmeyer S, Klösgen RB, Herrmann RG. 1993. Protein import into chloroplasts: the hydrophilic lumenal proteins exhibit unexpected import and sorting specificities in spite of structurally conserved transit peptides. *J. Biol. Chem.* 268(19):13869–76

Cline K. 1986. Membrane integration of a thylakoid precursor protein reconstituted in chloroplast lysates. *J. Biol. Chem.* 261(31):14804–10

Cline K, Ettinger WF, Theg SM. 1992a. Protein-specific energy requirements for protein transport across or into thylakoid membranes. *J. Biol. Chem.* 267(4):2688–96

Cline K, Fulsom DR, Viitanen PV. 1989. An imported thylakoid protein accumulates in the stroma when insertion into thylakoids is inhibited. *J. Biol. Chem.* 264(24):14225–32

Cline K, Henry R, Li CJ, Yuan J. 1992b. Pathways and intermediates for the biogenesis of nuclear-encoded thylakoid proteins. In *Research in Photosynthesis,* ed. N Murata, 3:149–56. The Netherlands: Kluwer

Cline K, Henry R, Li C, Yuan J. 1993. Multiple pathways for protein transport into or across the thylakoid membrane. *EMBO J.* 12(11):4105–14

Cline K, Keegstra K, Staehelin LA. 1985. Freeze-fracture electron microscopic analysis of ultrarapidly frozen envelope membranes on intact chloroplasts and after purification. *Protoplasma* 125:111–23

Creighton AM, Hulford A, Mant A, Robinson D, Robinson C. 1995. A monomeric, tightly folded stromal intermediate on the ΔpH-dependent thylakoidal protein transport pathway. *J. Biol. Chem.* 270(4):1663–69

Creissen G, Reynolds H, Xue Y, Mullineaux P. 1995. Simultaneous targeting of pea glutathione reductase and of a bacterial fusion protein to chloroplasts and mitochondria in transgenic tobacco. *Plant J.* 8(2):167–75 de Boer AD, Weisbeek PJ. 1991. Chloroplast protein topogenesis: import, sorting and assembly. *Biochim. Biophys. Acta* 1071:221–53

Dobberstein B. 1994. On the beaten pathway. *Nature* 367:599–600

Endo T, Kawamura K, Naka M. 1992. The chloroplast-targeting domain of plastocyanin transit peptide can form a helical structure but does not have a high affinity for lipid bilayers. *Eur. J. Biochem.* 207:671–75

Fischer K, Weber A, Arbinger B, Brink S, Eckerskorn C, Flügge U-I. 1994. The 24 kDa outer envelope membrane protein from spinach chloroplasts: molecular cloning, in vivo expression and import pathways of a protein with unusual properties. *Plant Mol. Biol.* 25:167–77

Franklin AE, Hoffman NE. 1993. Characterization of a chloroplast homologue of the 54-kDa subunit of the signal recognition particle. *J. Biol. Chem.* 268(29):22175–80

Friedman AL, Keegstra K. 1989. Chloroplast protein import: quantitative analysis of precursor binding. *Plant Physiol.* 89:993–99

Glick BS. 1995. Can Hsp70 proteins act as force-generating motors? *Cell* 80:11–14

Glick BS, Brandt A, Cunningham K, Müller S, Hallberg RL, et al. 1992. Cytochromes c1 and b2 are sorted to the intermembrane space of yeast mitochondria by a stop-transfer mechanism. *Cell* 69:809–22

Gray JC, Row PE. 1995. Protein translocation across chloroplast envelope membranes. *Trends Cell Biol.* 5:243–47

Hageman J, Baecke C, Ebskamp M, Pilon R, Smeekens S, Weisbeek P. 1990. Protein import into and sorting inside the chloroplast are independent processes. *Plant Cell* 2:479–94

Halpin C, Elderfield PD, James HE, Zimmerman R, Dunbar B, Robinson C. 1989. The reaction specificities of the thylakoidal processing peptidase and *Escherichia coli* leader peptidase are identical. *EMBO J.* 8(12):3917–21

Hartl FU, Neupert W. 1990. Protein sorting to mitochondria: evolutionary conservations of folding and assembly. *Science* 247:930–38

Henry R, Kapazoglou A, McCaffery M, Cline K. 1994. Differences between lumen targeting domains of chloroplast transit peptides determine pathway specificity for thylakoid transport. *J. Biol. Chem.* 269(14):10189–92

Hirsch S, Muckel E, Heemeyer F, von Heijne G, Soll J. 1994. A receptor component of the chloroplast protein translocation machinery. *Science* 266:1989–92

Hoffman NE, Franklin AE. 1994. Evidence for a stromal GTP requirement for the integration of a chlorophyll a/b binding polypeptide into thylakoid membranes. *Plant Physiol.* 105:295–304

Hoober JK, Boyd CO, Paavola LG. 1991. Origin of thylakoid membranes in *Chlamydomonas reinhardtii y-1* at 38°C. *Plant Physiol.* 96:1321–28

Howe G, Merchant S. 1993. Maturation of thylakoid lumen proteins proceeds posttranslationally through an intermediate in vivo. *Proc. Natl. Acad. Sci. USA* 90:1862–66

Hugueney P, Bouvier F, Badillo A, d'Harlingue A, Kuntz M, Camara B. 1995. Identification of a plastid protein involved in vesicle fusion and/or membrane protein translocation. *Proc. Natl. Acad. Sci. USA* 92:5630–34

Hulford A, Hazell L, Mould RM, Robinson C. 1994. Two distinct mechanisms for the translocation of proteins across the thylakoid membrane, one requiring the presence of a stromal protein factor and nucleotide triphosphates. *J. Biol. Chem.* 269(5):3251- 56

Hwang ST, Wachter C, Schatz G. 1991. Protein import into the yeast mitochondrial matrix: a new translocation intermediate between the two mitochondrial membranes. *J. Biol. Chem.* 266(31):21083–89

Karnauchov I, Cai D, Schmidt I, Herrmann RG, Klösgen RB. 1994. The thylakoid translocation of subunit 3 of photosystem I, the *psaF* gene product, depends on a bipartite transit peptide and proceeds along an azide-sensitive pathway. *J. Biol. Chem.* 269(52):32871–78

Keegstra K, Bruce B, Hurley M, Li HM, Perry S. 1995. Targeting of proteins into chloroplasts. *Physiol. Plant.* 93:157–62

Kessler F, Blobel G, Patel HA, Schnell DJ. 1994. Identification of two GTP-binding proteins in the chloroplast protein import machinery. *Science* 266:1035–39

Kirwin PM, Meadows JW, Shackleton JB, Musgrove JE, Elderfield PD, et al. 1989. ATP-dependent import of a lumenal protein by isolated thylakoid vesicles. *EMBO J.* 8(8):2251–55

Knight JS, Gray JC. 1995. The N-terminal hydrophobic region of the mature phosphate translocator is sufficient for targeting to the chloroplast inner envelope membrane. *Plant Cell* 7:1421–32

Knight JS, Madueño F, Gray JC. 1993. Import and sorting of proteins by chloroplasts. *Biochem. Soc. Trans.* 21:31–36

Knott TG, Robinson C. 1994. The secA inhibitor, azide, reversibly blocks the translocation of a subset of proteins across the chloroplast thylakoid membrane. *J. Biol. Chem.* 269(11):7843–46

Ko K, Cashmore AR. 1989. Targeting of proteins to the thylakoid lumen by the bipartite transit peptide of the 33 kd oxygen-evolving protein. *EMBO J.* 8(11):3187–94

Konishi T, Watanabe A. 1993. Transport of proteins into the thylakoid lumen: stromal processing and energy requirements for the import of the precursor to the 23-kDa protein of PS II. *Plant Cell Physiol.* 34(2):315–19

Kuhn A, Kiefer D, Köhne C, Zhu HY, Tschantz WR, Dalbey RE. 1994. Evidence for a loop-like insertion mechanism of pro-Omp A into the inner membrane of *Escherichia coli. Eur. J. Biochem.* 226:891–97

Laidler V, Chaddock AM, Knott TG, Walker D, Robinson C. 1995. A secY homolog in *Arabidopsis thaliana. J. Biol. Chem.* 270(30):17664–67

Li HM, Moore T, Keegstra K. 1991. Targeting of proteins to the outer envelope membrane uses a different pathway than transport into chloroplasts. *Plant Cell* 3:709–17

Li HM, Sullivan TD, Keegstra K. 1992. Information for targeting to the chloroplastic inner envelope membrane is contained in the mature region of the maize *Bt1*-encoded protein. *J. Biol. Chem.* 267(26):18999–9004

Li X, Henry R, Yuan J, Cline K, Hoffman NE. 1995. A chloroplast homologue of the signal recognition particle subunit SRP54 is involved in the posttranslational integration of a protein into thylakoid membranes. *Proc. Natl. Acad. Sci. USA* 92:3789–93

Madueño F, Bradshaw SA, Gray JC. 1994. The thylakoid-targeting domain of the chloroplast Rieske iron-sulfur protein is located in the N-terminal hydrophobic region of the mature protein. *J. Biol. Chem.* 269(26):17458–63

Madueño F, Napier JA, Gray JC. 1993. Newly imported Rieske iron-sulfur protein associates with both cpn60 and hsp70 in the chloroplast stroma. *Plant Cell* 5:1865–76

Mant A, Nielsen VS, Knott TG, Möller BL, Robinson C. 1994. Multiple mechanisms for the targeting of photosystem I subunits F, H, K, L, and N into and across the thylakoid membrane. *J. Biol. Chem.* 269(44):27303–9

Michl D, Robinson C, Shackleton JB, Herrmann RG, Klösgen RB. 1994. Targeting of proteins to the thylakoids by bipartite presequences: CFoII is imported by a novel, third pathway. *EMBO J.* 13(6):1310–17

Mould RM, Robinson C. 1991. A proton gradient is required for the transport of two lumenal oxygen-evolving proteins across the thylakoid membrane. *J. Biol. Chem.* 266(19):12189–93

Nakai M, Goto A, Nohara T, Sugita D, Endo T. 1994. Identification of the secA protein homolog in pea chloroplasts and its possible involvement in thylakoidal protein transport. *J. Biol. Chem.* 269(50):31338–41

Nielsen VS, Mant A, Knoetzel J, Möller BL, Robinson C. 1994. Import of barley photosystem I subunit N into the thylakoid lumen is mediated by a bipartite presequence lacking an intermediate processing site. *J. Biol. Chem.* 269(5):3762–66

Nohara T, Nakai M, Goto A, Endo T. 1995. Isolation and characterization of the cDNA for pea chloroplast secA evolutionary conservation of the bacterial-type secA-dependent protein transport within chloroplasts. *FEBS Lett.* 364:305–08

Oliver DB, Cabelli RJ, Dolan KM, Jarosik GP. 1990. Azide-resistant mutants of *Escherichia coli* alter the SecA protein, an azide-sensitive component of the protein export machinery. *Proc. Natl. Acad. Sci. USA* 87:8227–31

Olsen LJ, Keegstra K. 1992. The binding of precursor proteins to chloroplasts requires nucleoside triphosphates in the intermembrane space. *J. Biol. Chem.* 267(1):433–39

Payan LA, Cline K. 1991. A stromal protein factor maintains the solubility and insertion competence of an imported thylakoid membrane protein. *J. Cell Biol.* 112(4):603–13

Perry SE, Keegstra K. 1994. Envelope membrane proteins that interact with chloroplastic precursor proteins. *Plant Cell* 6:93–105

Pfisterer J, Lachmann P, Kloppstech K. 1982. Transport of proteins into chloroplasts: binding of nuclear-coded chloroplast proteins to the chloroplast envelope. *Eur. J. Biochem.* 126:143–48

Pilon M, Rietveld AG, Weisbeek PJ, de Kruijff B. 1992a. Secondary structure and folding of a functional chloroplast precursor protein. *J. Biol. Chem.* 267(28):19907–13

Pilon M, Weisbeek PJ, de Kruijff B. 1992b. Kinetic analysis of translocation into isolated chloroplasts of the purified ferredoxin precursor. *FEBS Lett.* 302:65–68

Pugsley AP. 1993. The complete general secretory pathway in gram-negative bacteria. *Microbiol. Rev.* 57(1):50–108

Reed JE, Cline K, Stephens LC, Bacot KO, Viitanen PV. 1990. Early events in the import/assembly pathway of an integral thylakoid protein. *Eur. J. Biochem.* 194:33–42

Reinbothe S, Runge S, Reinbothe C, van Cleve B, Apel K. 1995. Substrate-dependent transport of the NADPH:protochlorophyllide oxidoreductase into isolated plastids. *Plant Cell* 7:161–72

Robinson C, Cai D, Hulford A, Brock IW, Michl D, et al. 1994. The presequence of a chimeric construct dictates which of two mechanisms are utilized for translocation across the thylakoid membrane: evidence for the existence of two distinct translocation systems. *EMBO J.* 13(2):279–85

Robinson C, Klösgen RB. 1994. Targeting of proteins into and across the thylakoid membrane: a multitude of mechanisms. *Plant Mol. Biol.* 26:15–24
Ryan KR, Jensen RE. 1995. Protein translocation across mitochondrial membranes: what a long, strange trip it is. *Cell* 83:517–19
Salomon M, Fischer K, Flügge U-I, Soll J. 1990. Sequence analysis and protein import studies of an outer chloroplast envelope polypeptide. *Proc. Natl. Acad. Sci. USA* 87:5778–82
Schnell DJ. 1995. Shedding light on the chloroplast protein import machinery. *Cell* 83:521–24
Schnell DJ, Blobel G. 1993. Identification of intermediates in the pathway of protein import into chloroplasts and their localization to envelope contact sites. *J. Cell Biol.* 120(1):103–15
Schnell DJ, Kessler F, Blobel G. 1994. Isolation of components of the chloroplast protein import machinery. *Science* 266:1007–12
Schwaiger M, Herzog V, Neupert W. 1987. Characterization of translocation contact sites involved in the import of mitochondrial proteins. *J. Cell Biol.* 105:235–46
Scott SV, Theg SM. 1996. A new intermediate on the chloroplast protein import pathway reveals distinct translocation machineries in the two envelope membranes: energetics and mechanistic implications. *J. Cell Biol.* 132:63–75
Seedorf M, Soll J. 1995. Copper chloride, an inhibitor of protein import into chloroplasts. *FEBS Lett.* 367:19–22
Seedorf M, Waegemann K, Soll J. 1995. A constituent of the chloroplast import complex represents a new type of GTP-binding protein. *Plant J.* 7(3):401–11
Segui-Real B, Kispal G, Lill R, Neupert W. 1993. Functional independence of the protein translocation machineries in mitochondrial outer and inner membranes: passage of preproteins through the intermembrane space. *EMBO J.* 12(5):2211–18
Siegel V. 1995. A second signal recognition event required for translocation into the endoplasmic reticulum. *Cell* 82:167–70
Simon SM, Blobel G. 1991. A protein-conducting channel in the endoplasmic reticulum. *Cell* 65:371–80
Smeekens S, Bauerle C, Hageman J, Keegstra K, Weisbeek P. 1986. The role of the transit peptide in the routing of precursors toward different chloroplast compartments. *Cell* 46:365–75
Smith TA, Kohorn BD. 1994. Mutations in a signal sequence for the thylakoid membrane identify multiple protein transport pathways and nuclear suppressors. *J. Cell Biol.* 126(2):365–74
Soll J. 1995. New insights into the protein import machinery of the chloroplast's outer envelope. *Bot. Acta* 108:277–82
Stader J, Gansheroff LJ, Silhavy TJ. 1989. New suppressors of signal-sequence mutations, *prlG*, are linked tightly to the secE gene of *Escherichia coli. Genes Dev.* 3:1045–52
Sweet DJ, Gerace L. 1995. Taking from the cytoplasm and giving to the pore: soluble transport factors in nuclear protein import. *Trends Cell Biol.* 5:444–47
Theg SM, Scott SV. 1993. Protein import into chloroplasts. *Trends Cell Biol.* 3:186–90
Tranel PJ, Froehlich J, Goyal A, Keegstra K. 1995. A component of the chloroplastic protein import apparatus is targeted to the outer envelope membrane via a novel pathway. *EMBO J.* 14(11):2436–46
Tsugeki R, Nishimura M. 1993. Interaction of homologues of Hsp70 and Cpn60 with ferredoxin-NADP$^+$ reductase upon its import into chloroplasts. *FEBS Lett.* 320(3):198–202
VanderVere PS, Bennett TM, Oblong JE, Lamppa GK. 1995. A chloroplast processing enzyme involved in precursor maturation shares a zinc-binding motif with a recently recognized family of metalloendopeptidases. *Proc. Natl. Acad. Sci. USA* 92:7177–81
van't Hof R, van Klompenburg W, Pilon M, Kozubek A, de Korte-Kool G, et al. 1993. The transit sequence mediates the specific interaction of the precursor of ferredoxin with chloroplast envelope membrane lipids. *J. Biol. Chem.* 268(6):4037–42
Viitanen PV, Doran ER, Dunsmuir P. 1988. What is the role of the transit peptide in thylakoid integration of the light-harvesting chlorophyll a/b protein? *J. Biol. Chem.* 263:15000–7
Voelker R, Barkan A. 1995. Two nuclear mutations disrupt distinct pathways for targeting proteins to the chloroplast thylakoid. *EMBO J.* 14(16):3905–14
von Heijne G, Nishikawa K. 1991. Chloroplast transit peptides: the perfect random coil? *FEBS Lett.* 278(1):1–3
von Heijne G, Steppuhn J, Herrmann RG. 1989. Domain structure of mitochondrial and chloroplast targeting peptides. *Eur. J. Biochem.* 180:535–45
Waegemann K, Soll J. 1991. Characterization of the protein import apparatus in isolated outer envelopes of chloroplasts. *Plant J.* 1(2):149–58
Walter P, Johnson AE. 1994. Signal sequence recognition and protein targeting to the endoplasmic reticulum membrane. *Annu. Rev. Cell Biol.* 10:87–119

Whatley JM, Hawes CR, Horne JC, Kerr JDA. 1982. The establishment of the plastid thylakoid system. *New Phytol.* 90:619–29

Wickner WT. 1994. How ATP drives proteins across membranes. *Science* 266:1197–98

Wu C, Seibert FS, Ko K. 1994. Identification of chloroplast envelope proteins in close physical proximity to a partially translocated chimeric precursor protein. *J. Biol. Chem.* 269(51):32264–71

Yuan J, Cline K. 1994. Plastocyanin and the 33-kDa subunit of the oxygen-evolving complex are transported into thylakoids with similar requirements as predicted from pathway specificity. *J. Biol. Chem.* 269(28):18463–67

Yuan J, Henry R, Cline K. 1993. Stromal factor plays an essential role in protein integration into thylakoids that cannot be replaced by unfolding or by heat shock protein Hsp70. *Proc. Natl. Acad. Sci. USA* 90:8552–56

Yuan J, Henry R, McCaffery M, Cline K. 1994. SecA homolog in protein transport within chloroplasts: evidence for endosymbiont-derived sorting. *Science* 266:796–98

Annu. Rev. Cell Dev. Biol. 1996. 12:27–54

SIGNAL-MEDIATED SORTING OF MEMBRANE PROTEINS BETWEEN THE ENDOPLASMIC RETICULUM AND THE GOLGI APPARATUS

Rohan D. Teasdale and Michael R. Jackson

R. W. Johnson Pharmaceutical Research Institute, 3535 General Atomics Court, Suite 100, San Diego, California 92121

KEY WORDS: biological transport, intracellular membranes, membrane proteins, retrograde transport, protein targeting

ABSTRACT

Each organelle of the secretory pathway is required to selectively allow transit of newly synthesized secretory and plasma membrane proteins and also to maintain a unique set of resident proteins that define its structural and functional properties. In the case of the endoplasmic reticulum (ER), residency is achieved in two ways: (*a*) prevention of residents from entering newly forming transport vesicles and (*b*) retrieval of those residents that escape. The latter mechanism is directed by discrete retrieval motifs: Soluble proteins have a H/KDEL sequence at their carboxy-terminus; membrane proteins have a dibasic motif, either di-lysine or di-arginine, located close to the terminus of their cytoplasmic domain. Recently it was found that di-lysine motifs bind the complex of cytosolic coat proteins, COP I, and that this interaction functions in the retrieval of proteins from the Golgi to the ER. Also discussed are the potential roles this interaction may have in vesicular trafficking.

CONTENTS

1081-0706/96/1115-0027$08.00

INTRODUCTION

Trafficking and sorting of proteins through the central organellar system have been a focus of interest since Palade (1975) and coworkers formulated the general outline of the secretory pathway in pancreatic exocrine cells. This pathway, common to all eukaryotic cells, defines the transport route by which secretory proteins are delivered from the endoplasmic reticulum (ER) to the plasma membrane via the Golgi apparatus. The initial targeting of proteins to the ER is accomplished using a signal sequence (Blobel & Dobberstein 1975), and export from the ER involves incorporation of proteins and membrane into transport vesicles that are destined to fuse with the next compartment along the pathway. The secretory pathway thereby consists of a series of membrane-bound organelles between which proteins and membrane are moved in a vectorial fashion by a carefully coordinated series of vesicle budding and fusion events (Rothman 1994). A major issue in cell biology today is how the distinct intracellular compartments of a cell maintain their unique composition of proteins and lipids. Presumably, the processes involved are carefully regulated and may underlie numerous disease states associated with protein mislocalization (Amara et al 1992), and it is likely that understanding these processes might help with elucidating the causes of some diseases. Over the last few years many molecules involved in the general processes of vesicle budding and fusion have been identified. In addition, molecules dedicated to specific transport steps have also been characterized. Taken together, these molecules provide the first molecular model to explain how intracellular protein transport may be accomplished (for general reviews, see Simons & Zerial 1993, Rothman 1994, Kreis et al 1995, Sudhof 1995). The specific docking and fusion of transport vesicles appears to be directed by two families of proteins—V-SNAREs on the transport vesicles and T-SNAREs on target membranes—and candidate molecules have been identified to play these roles (see review by S Pfeffer, this volume). This information, together with the molecular description of how membrane proteins and lipids are sorted at each of the transport steps, has enabled us to begin to understand how the integrity of the various compartments of the exocytic pathway are maintained and how the cargo is sorted. Of particular importance has been the identification of a retrograde pathway that retrieves both membrane and proteins from the Golgi apparatus back to the ER.

This review focuses on sorting operations between the ER and Golgi apparatus; in particular, on the characteristics of the signals directing sorting of membrane proteins to the ER, the mechanism by which these signals function, the sorting process, and the molecular machinery involved.

THE EARLY SECRETORY PATHWAY

The Endoplasmic Reticulum

Morphologically the ER is recognized as an extensive membrane bound organelle composed of a network of cisternae stretching throughout the cytoplasm. Functionally, the ER performs many housekeeping processes for the cell, and residents of this organelle can be broadly classified into five functional groups: (*a*) proteins involved in the translocation and vesicular transport machinery, (*b*) folding enzymes and chaperones that assist in generating correct tertiary structures, (*c*) proteins involved in posttranslational modifications, (*d*) enzymes involved in biosynthesis of lipids, and (*e*) proteins involved with the maintenance of an ER calcium store. Specific cell types also have ER residents dedicated to a specific task, for example, UDP-glucuronyltransferases and cytochrome P-450 are localized to the ER of hepatocytes and other cell types where they are involved with detoxification (Batt et al 1994).

The idea that the ER is a single organelle is perhaps an oversimplification. Historically, it has been divided into two structurally distinct types: rough and smooth ER (RER and SER, respectively). Electron micrographs show continuities between cisternae of RER and SER, and many proteins are common to both. However, it is also clear that protein sorting begins within the ER and that certain proteins are largely restricted to the rough or smooth ER (Sitia & Meldolesi 1992). A third subdomain of the ER, the transitional elements, consists of a network of smooth, vesicular-tubular structures in the vicinity of the Golgi apparatus, which are thought to be the sites of transport vesicle formation (Palade 1975).

It has been difficult to draw a precise distinction between the transitional elements and a so-called ER-Golgi intermediate compartment (ERGIC; Saraste

& Kuismanen 1984) or vesicular-tubular cluster (VTC; Balch et al 1994) located between the ER and the Golgi apparatus. This intermediate compartment has been characterized biochemically as the site of *N*-acetylgalactosamine (GalNAc) addition and palmitylation of proteins, and identified morphologically by localization of the proteins p53, p58, and rab2 (see references within Hauri & Schweizer 1992). It is unclear whether the ERGIC is a distinct organelle. Some experiments indicate that a round of vesicular transport is not required to reach this compartment (Krijnse-Locker et al 1994), indicating that it is contiguous with the ER. In contrast, other experiments indicate that the ERGIC may be a transient structure composed of transport vesicles fused with each other (Balch et al 1994). This topic is still hotly contested and rightly so, because at its heart lies the question of what defines an organelle. The lack of consensus on the exact nature of the first compartment after the ER has caused much confusion in the literature (Hauri & Schweizer 1992). If one defines the intermediate compartment (ERGIC) as the first acceptor compartment to which vesicles that bud from the ER fuse, then there is little to distinguish this organelle from a definition of the *cis* elements of the Golgi apparatus.

The Golgi Apparatus

The Golgi apparatus or complex was originally identified in eukaryotic cells as four to eight membrane-enclosed flattened cisternae held together as parallel stacks (see Farquhar 1985). More recently, two tubular networks of membranes at either end of the Golgi stacks have been identified. The *cis*-Golgi network (CGN; Rambourg & Clermont 1990), probably equivalent to the ERGIC, functions as an acceptor compartment of newly synthesized material from the ER and the *trans*-Golgi network (TGN; Griffiths & Simons 1986), which is the exit site of the Golgi, is responsible for sorting proteins to their next destination.

During their passage through the Golgi apparatus, proteins are subjected to a variety of posttranslational processes, for example, stepwise biosynthesis of O-linked glycans (Tooze et al 1988) and remodeling of the N-linked oligosaccharides (Kornfeld & Kornfeld 1985). Processing of carbohydrates occurs in an ordered fashion denoting the sequential exposure of proteins to the processing enzymes that are located in defined subsets of the Golgi stack. However, as our understanding of the secretory pathway moves from a morphological description to a more mechanistic and molecular view in which the dynamics of membrane fusion and vesicle traffic predominate, we are left to reconcile morphological data that provide snap shots of the steady-state conditions and biochemical analyses that reflect these dynamic processes. Indeed, the value of monitoring carbohydrate modifications as an absolute measure of a path trodden has been tempered by our lack of knowledge of just where the markers are localized!

Export from the ER: Requirement for Transport Signals?

The long-standing argument over whether trafficking and sorting of proteins in the secretory pathway rely on discrete retention or transport signals (see Rothman 1987, Klausner 1989) appears to be more complicated than first thought. Multiple mechanisms for retention of proteins in the ER have been described, and although specific signals for transport have not been identified, it is probably more appropriate to recognize that proteins need to acquire a transport-competent state before they can leave the ER. Indeed, the ER is now well recognized as a quality control station preventing proteins that have not folded correctly or have not reached their mature quaternary structure from reaching the cell surface (Rose & Doms 1988, Klausner et al 1990, Hammond & Helenius 1995). The question of whether there are specific signals/motifs that target proteins for secretion remains open and is discussed at the end of this review. However, specific signals do exist that distinguish proteins to be maintained in the ER (Munro & Pelham 1987, Nilsson et al 1989, Lotteau et al 1990), and we focus on targeting by these signals (see below).

SORTING SIGNALS THAT DIRECT LOCALIZATION OF MEMBRANE PROTEINS TO THE ER

The general criteria defining targeting motifs are that they function in a specific position of a protein, they can be disrupted through mutation, and that they confer residency in a specific organelle when engineered onto a reporter protein. For example, the carboxy-terminal tetrapeptide, KDEL, was shown to be both necessary and sufficient for the retention of a family of soluble resident ER proteins (Munro & Pelham, 1987), and transplanting the KDEL motif onto the carboxy-terminus of lysozyme (a protein that is normally secreted) results in ER residency of the chimera (Pelham 1989). Likewise, some ER resident type I, II, and III membrane proteins also contain discrete ER targeting motifs (see Table 1). The following section describes the identification and characterization of these motifs.

Properties of Functional ER Targeting Motifs

THE DI-LYSINE MOTIF The carboxy-terminal six amino acid residues (DEKKMP) of the cytoplasmic domain of the resident ER type I membrane protein E19, a protein encoded by adenovirus 3, were demonstrated to be necessary and sufficient for the retention of this protein in the ER of mammalian cells (Nilsson et al 1989). Transplantation of the E19 cytoplasmic domain sequence onto either of the T-cell surface glycoproteins CD4 or CD8 resulted in efficient targeting of the chimeric proteins to the ER. It was found that this sequence must occupy the extreme carboxy-terminal position to be functional (Nilsson et al

Table 1 Membrane proteins with potential endoplasmic reticulum localization motifs[a]

	COOH-terminal localization[b]	Subcellular localization	Protein type[c]	Accession name
Carboxy-terminal di-lysine motifs				
3-Hydroxy-3-methylglutaryl-coenzyme A reductase	LQGACT**K**K**T**A (i)	ER	III	HMDH_HUMAN
53-kDa Sarcoplasmic protein	ETPKN**R**Y**K**KH (i)	ER	I	J04480/RABBIT
C-8 Sterol isomerase (ERG2)	GKNLLQN**KK**F	ER	I	ERG2_YEAST
Calnexin	SPRNR**K**P**R**RE	ER	I	CALX_HUMAN
Ceramide UDP- galactosyltransferase	GHIKHE**KK**VK	ER ?	I	CGT_RAT
Emp47	RQEII**K**T**K**LL (ii)	Golgi	I	EM47_YEAST
ERGIC-53	QQEAAA**KK**FF (iii)	ERGIC	I	S42626/HUMAN
Glucose transporter type 7	SDQVK**K**M**K**ND	ER	III	GTR7_RAT
Glucose-6-phosphatase	VLGQPH**KK**SL	ER	III	G6PT_HUMAN
Glycerol uptake/efflux facilitator protein	SHYGNA**KK**VT	?	III	FPS1_YEAST
Glycoprotein 25L	KNFFIA**KK**LV	ER	I	G25L_CANFA
GPI: protein transamidase (GAA1)	VVVRS**K**E**K**QS (iv)	ER	III	GAA1_YEAST
High affinity Ig ε receptor α-subunit	QKTGKG**KK**KG (v)	Cell Surface	I	FCE1_RAT
HM-1 killer toxin resistance protein	TTSSME**KK**LN	ER	III	RHK1_YEAST
Longevity-assurance protein 1	ENEES**K**E**K**CE	?	III	LAG1_YEAST
Oligosaccharyl transferase, 48-kDa	LHMKE**K**E**K**SD	ER	I	OST4_CANFA
Oligosaccharyl transferase beta subunit	KKLETF**KK**TN (vi)	ER	I	OSTB_YEAST
Signal sequence receptor beta subunit	KYDTP**K**S**K**KN	ER	I	SSRB_CANFA
SN2-acylglyceride fatty acyltransferase	NEGSSV**KK**MH	ER ?	I	PLSC_YEAST
Steroid delta-isomerase	HKETL**K**S**K**TQ	ER	III	3BH1_HUMAN
Surf-4 protein	VSMDE**KKK**EW (vii)	ER	III	M62601/MOUSE
T-cell receptor beta-1 chain C region	VMAMV**K**R**K**NS	Cell Surface	I	TCB1_MOUSE
TRAM	SPRNR**K**E**K**SS	ER	III	TRAM_CANFA
UDP-glucuronosyltransferase 1A	VKKAH**K**S**K**TH (i)	ER	I	UD1A_HUMAN
UDP-glucuronosyltransferase 2B1	TANMG**KKK**KE	ER	I	UDB1_RAT
UDP-glucuronosyltransferase 2B2	VKKEK**K**M**K**NE	ER	I	UDB2_RAT
UDP-glucuronosyltransferase 2B4	VRTGK**K**G**K**RD (i)	ER	I	UDB4_HUMAN
UDP-glucuronosyltransferase 2B6	AKKQK**K**M**K**NE	ER	I	UDB6_RAT
UDP-glucuronosyltransferase 2B7	ARKAK**K**G**K**ND	ER	I	UDB7_HUMAN
UDP-glucuronosyltransferase 2B8	AKKGK**KKK**RD (i)	ER	I	UDB8_HUMAN
UDP-glucuronosyltransferase 2B10	ARKGK**K**G**K**RD	ER	I	UDBA_HUMAN
UDP-glucuronosyltransferase 2B11	VRTGK**K**G**K**RD	ER	I	UDBB_HUMAN
UDP-glucuronosyltransferase 2B12	VKKEK**K**T**K**NE	ER	I	UDBC_RAT
UDP-glucuronosyltransferase 2B13	LGAGK**KKK**RD	ER	I	UDBD_RABIT
UDP-glucuronosyltransferase 2B14	VKIGK**K**Q**K**RD	ER	I	UDBE_RABIT
Vma21	HKVDGN**KK**ED (viii)	ER	I	VM21_YEAST

(*Continued*)

Table 1 *(continued)*

Early E3 19-kDa glycoprotein				
Human adenovirus type 2	RSFIDE**KK**MP (i)	ER	I	E3GL_ADE02
Human adenovirus type 3	QSNEE**K**E**K**MP	ER	I	E3GL_ADE03
Human adenovirus type 5	RSFIEE**KK**MP	ER	I	E3GL_ADE05
Human adenovirus type 11	KNANN**K**E**K**MP	ER	I	E3GL_ADE1A
Env glycoprotein				
Human foamy virus	SWIPT**KK**KNQ	I.C. budding[d]	III	ENV_FOAMV
Simian foamy virus, type 1	SWLPG**K**P**K**KN	I.C. budding	III	ENV_SFV1
Simian foamy virus, type 3	SWLPG**KK**KRN	I.C. budding	III	ENV_SFV3L
M polyprotein precursor				
Prospect hill virus	PRRVVH**KK**SS	Golgi budding	I	VGLM_PHV
Uukuniemi virus	FTLCL**K**V**K**KS	Golgi budding	I	VGLM_UUK
Hantaan virus	LCPVR**K**H**K**KS	Golgi budding	I	VGLM_HANTB
Punta toro phlebovirus	VNSINI**KK**KN	Golgi budding	I	VGLM_PTPV
Rift valley fever virus	MWLAAT**KK**AS	I.C. budding	I	VGLM_RVFVZ
Seoul virus	LCPVR**K**H**K**KS	I.C. budding	I	VGLM_SEOU8
Peplomer glycoprotein precursor				
Berne virus	FEMNG**K**V**K**KS	I.C. budding	I	VGLP_BEV
T-cell receptor beta chain				
Feline leukemia virus	LMAKV**K**R**K**DS	Cell Surface	I	TCB_FLV
Amino-terminal di-arginine motifs				
3-Hydroxy-3-methylglutaryl-coenzyme A reductase	MDV**RR**RSEKP	ER	II	HMDH_NICSY
Dolichyl-phosphate beta-glucosyltransferase (ALG-5)	MRAL**R**FLIEN	ER	II	ALG5_YEAST
HLA Class II associated invariant chain (p33); CD74	MH**RRR**SRSCR (ix)	ER	II	HG2A_HUMAN
GPI-anchor biosynthesis protein (PIG-A)	MAN**RR**GGGQG	ER	II	A55731/MOUSE
p63	MPSA**K**Q**R**GSK (ix)	ER	II	
TRAM	MAI**RK**KSTKS (ix)	ER	III	TRAM_CANFA
UDP-N-acetyl-glucosamine-1-P transferase (ALG-7)	ML**R**LFSLALI	ER	III	GPT_YEAST
Carboxy-terminal HDEL sequence				
SEC 20	VDRIVS**HDEL** (v)	ER	II	SC20_YEAST
SED 4	VNYAGL**HDEL**	ERGIC	II	SED4_YEAST

[a]The protein databases were systematically searched and the list supplemented based on the current literature. Hypothetical proteins have not been included and only selected cross-species examples have been presented.

[b]The characterization of specific sequences as ER localization motifs is detailed in the following: (i) Jackson et al (1990); (ii) Schroder et al (1995); (iii) Itin et al (1995); (iv) see Hamburger et al (1995); (v) Letourneur et al (1995); (vi) Gaynor et al (1994); Townsley & Pelham (1995); (vii) Reeves & Fried (1995); (viii) Hill & Stevens (1994); (ix) Schutze et al (1994); (x) Sweet & Pelham (1992).

[c]All these proteins have a defined membrane orientation that will position the motif in the cell's cytoplasm.

[d]Viruses bud from a undefined intracellular membrane.

1989). Unlike the highly conserved KDEL motif (Pelham 1989) of resident ER soluble proteins, DEKKMP is not a common feature of ER resident membrane proteins. Analogous sequences do exist in a subset of ER resident proteins where a lysine, positioned three residues from the carboxy-terminus (−3), is a common feature (Jackson et al 1990, Shin et al 1991). Mutation of this lysine to a serine in these sequences destroyed the targeting motif. Additional mutagenesis demonstrated that a second lysine in the −4 position is the only other essential residue of this ER targeting motif in the E19 tail. The −4 lysine could be moved to the −5 position without disrupting the signal; however, within the context of the E19 tail sequence, arginines or histidines could not functionally replace lysines (Jackson et al 1990). Based on the requirements for two lysine residues correctly positioned relative to the carboxy-terminus, this motif has been termed the KKXX or di-lysine motif.

Although the basic consensus of this motif is present on many ER proteins, some substitutions of lysine by arginine are permitted (see Table 1). Introduction of lysine residues at various positions in a poly-serine background allowed a systematic evaluation of the di-lysine motif. The incorporation of a single lysine into an engineered poly-serine tail on CD8, normally efficiently transported to the cell surface, diminished the transport rate, with a lysine at the −3 position having the most profound effect followed by lysines in the −4 and −5 positions. Only if two lysine residues were introduced in the poly-serine tail and only when they were located at positions −3 and −4 or −3 and −5 from the carboxy-terminus was CD8 undetectable at the cell surface (Jackson et al 1990). The di-lysine motif appears to be conserved across eukaryotes, and mutagenesis studies show remarkable conservation of the functional motif between yeast and humans (Gaynor et al 1994, Townsley & Pelham 1994, Schroder et al 1995). As more sequences of ER resident membrane proteins are determined (see Table 1), the di-lysine motif has become a common feature.

THE IMPORTANCE OF RESIDUES FLANKING THE DI-LYSINE MOTIF Although the di-lysine motif is the minimum sequence requirement for targeting, the sequences flanking the lysines can influence retention efficiency. Thus lysines positioned at −3 and −4 from the carboxy-terminus, when surrounded by serine or alanine residues, function in ER targeting, whereas ER targeting is disrupted if the lysines are surrounded by glycine or proline residues. However, a functional ER targeting motif can be recovered in a poly-glycine background if two serine residues are positioned around the lysines in any combination (MR Jackson, unpublished data). Whether the sequence context is simply important for exposure of the lysines or whether the di-lysine motif needs to attain a specific three-dimensional structure, as has been suggested for endocytosis motifs (Trowbridge et al 1993), is unknown.

LOCALIZATION OF THE DI-LYSINE MOTIF RELATIVE TO THE MEMBRANE Systematic reduction of the length of the CD8/E19 cytoplasmic domain identified that a minimum of 5 amino acid residues are required between the first charged residue of the cytoplasmic tail and the −3 lysine, which indicates that the di-lysine motif needs to be exposed from the membrane lipids in order to function (RD Teasdale & MR Jackson, unpublished data). A similar distance requirement is found for the NPXY endosomal sorting signal (Collawn et al 1990). Determination of the maximum length of a cytoplasmic domain on which a di-lysine motif still functions has not been experimentally tested; however, it should be noted that the di-lysine motif of human HMG CoA reductase is located 546 amino acids from the proposed membrane-spanning region (Luskey & Stevens 1985).

STERIC MASKING Inspection of the list of proteins that possess the di-lysine motif reveals several proteins known to be localized in their mature folded state at the cell surface. For example, the mouse T-cell receptor β chain (Klausner et al 1990) or the high affinity IgE receptor α chain, which both contain di-lysine motifs, are found in protein complexes located at the cell surface (Letourneur et al 1995). The di-lysine motif on the high affinity IgE receptor α chain can function as an ER-localizing motif when attached to the interleukin-2 receptor α chain (Tac antigen), normally located on the cell surface (Letourneur et al 1995). An explanation for these findings was provided by the observation that the motif on the α chain could be sterically masked by polypeptide sequences from the γ chain of the receptor. As a result, the di-lysine motif became nonfunctional after the α chain correctly assembled with the γ chain thereby ensuring that only functional complexes exit the ER (Letourneur et al 1995). Inspection of the growing list of di-lysine-bearing proteins indicates that steric masking of this motif upon assembly of subunits is widely used to aid in the assembly of complexes in the ER prior to their export to the plasma membrane.

DI-ARGININE MOTIF FOR TYPE II MEMBRANE PROTEINS Few ER resident type II membrane proteins have been identified, and those that do exist do not have di-lysine motifs. Nevertheless, an analogous ER-targeting signal, the di-arginine or XXRR motif, has been identified as responsible for localizing the Ii p33 isoform of invariant chain to the ER. The essential elements of this motif are two arginine residues close to the amino-terminus, located at positions 2 and 3, 3 and 4, or 4 and 5, or split by a residue at positions 2 and 4, or 3 and 5 (Schutze et al 1994). Precise positioning of the arginines relative to the amino-terminus is complicated by possible processing of the initiator methionine (Kendall & Bradshaw 1992). Within the context of the Ii p33 sequence, arginines could not be substituted by lysines; however, as with the di-lysine motif, sequence context appears to be important for function and dictating if substitution of arginines by lysines is permissible. Transplanting this motif onto the human transferrin receptor or *N*-acetylglucosaminyltransferase (GlcNAc-TI) type II proteins, which are ordinarily expressed at the cell surface and in the Golgi stack, respectively, resulted in efficient ER localization of the chimeras (Nilsson et al 1994, Schutze et al 1994).

Similar ER-localization motifs, with some variation in both position and context, are also shared by other ER resident-type II membrane proteins, namely TRAM and p63 (Schutze et al 1994). Indeed, the protein translocation component, TRAM, has eight transmembrane-spanning domains, and both its amino- and carboxy-termini are located in the cytoplasm. Both cytoplasmic domains carry ER-targeting motifs (see Table 1), raising the possibility that this protein may be maintained in the ER by both the di-lysine and di-arginine motifs.

CARBOXY-TERMINAL H/KDEL MOTIF ON TYPE II PROTEINS The H/KDEL ER-targeting motif used by soluble ER resident proteins has been identified on the carboxy-terminus of two yeast type II membrane proteins, Sec20p and SED4. Removal of the HDEL sequence from Sec20p results in diminished levels of protein in the cell, presumably because of its transport to post-ER protease-containing compartments (Sweet & Pelham 1992). This mechanism of retention probably also functions for mammalian type II membrane proteins because a KDEL motif added to the carboxy-terminus of dipeptidyl peptidase IV retained this protein in the ER (Tang et al 1992).

MECHANISMS FOR MAINTAINING PROTEINS IN THE ER

Two distinct mechanisms for maintaining and concentrating proteins in the ER were proposed based on theoretical considerations: (*a*) ER proteins could be retained by active exclusion from vesicles that exit the ER, or (*b*) ER proteins could enter such transport vesicles but be subsequently retrieved from a post-ER compartment via a retrograde transport flow. In order to distinguish between these possibilities, posttranslational modifications of proteins tagged with the various ER-targeting motifs (see above) were sought that would indicate whether a molecule had left the ER.

Retrieval of H/KDEL-Bearing Proteins

Using this approach, Pelham and colleagues convincingly demonstrated that soluble KDEL-tagged (HDEL in yeast) molecules were exposed to Golgi enzymes and subsequently retrieved back to the ER. Indeed, the mechanism of retrieval of soluble H/KDEL-tagged proteins, including identification of the membrane receptor, Erd2p, which binds and retrieves these proteins, is now well established (Pelham 1990). The Erd2p protein, or KDEL receptor in mammalian cells, encodes a 26-kDa transmembrane protein with seven putative transmembrane domains, a short cytoplasmically located carboxy-terminus, and an amino-terminus oriented to the lumen. This receptor resides at steady state in the Golgi apparatus (Lewis & Pelham 1990, Griffiths et al 1994), but its ER localization may increase with high expression levels of KDEL-tagged proteins (Lewis & Pelham 1992). Thus it is proposed that upon ligand binding in the Golgi apparatus, the receptor-ligand complex undergoes a conformational change that triggers retrograde transport of the complex to the ER (Lewis & Pelham 1992). As optimal binding to KDEL occurs at acid pH, selective binding and release of the ligand may be regulated by a pH difference within these organelles (Wilson et al 1993). It is presumably via this mechanism that type II membrane proteins with H/KDEL motifs at their carboxy-terminus are maintained in the ER. The sorting signals that direct the retrieval of the occupied Erd2 receptor are not known. A single aspartic acid residue in the seventh transmembrane domain appears to be critical for retrograde movement of the receptor-ligand complex, a finding leading to the suggestion that retention of the unoccupied Erd2 receptor in the Golgi apparatus is mediated by interactions with the lipid bilayer (Townsley et al 1993).

Evidence that the Di-Lysine Motif Directs Retrieval to the ER

There is now considerable evidence that di-lysine or di-arginine targeting signals also function by retrieval of membrane proteins to the ER.

Immuno-localization of di-lysine chimeras in mammalian cells (Jackson et al 1993) showed that although the majority of the proteins were localized to the ER, peripheral vesicular structures were also detected that co-stained for the intermediate compartment antigens p58 and ERGIC-53.

Analysis of the posttranslational modifications of di-lysine-tagged marker proteins in mammalian cells and yeast also provide evidence that these proteins are exposed to enzymes located in post-ER compartments. For example, soon after synthesis, CD8/di-lysine chimeras receive GalNAc and palmitate, modifications reported to occur in post-ER location(s) (Tooze et al 1988, Bonatti et al 1989). Furthermore, after extended chase periods, both galactose and, to a lesser extent, sialic acid could be detected on CD8/di-lysine chimeras, and a small percent of the N-linked sugars on CD4/di-lysine chimeras was found to be resistant to digestion by endoglycosidase H, indicating exposure to *medial*-Golgi enzymes. Evidence that these Golgi-exposed chimeras were actually retrieved to the ER was provided by subcellular fractionation and lectin-staining experiments in which the ER of transfected cells expressing high levels of CD8/di-lysine proteins or CD8/KDEL chimeras stained strongly for the presence of GalNAc and terminal Gal(1–3)GalNAc (Jackson et al 1993).

Similar data have been provided by studies in yeast. Two groups have engineered yeast type I membrane proteins that contain a di-lysine motif. Gaynor et al (1994) produced a hybrid between invertase and Wbp1, the yeast oligosaccharide transferase α subunit. The chimeric molecule was shown to be modified by the addition of $\alpha 1,6$ mannose residues, an early Golgi modification, and yet it fractionated with ER membranes. Townsley & Pelham (1994) performed similar studies fusing a membrane domain and di-lysine motif to the reporter molecule preproalpha factor and again the presence of the di-lysine motif resulted in the ER accumulation of an $\alpha 1,6$ mannose modified protein. In addition, this chimera did not reach the late Golgi as it was not cleaved by Kex2p. The use of yeast mutants that are conditionally defective in vesicular transport definitively established that the Golgi modifications to these di-lysine chimeras did not occur in the ER and thus that the proteins must ordinarily be able to pass to the Golgi and then be selectively retrieved (Gaynor et al 1994, Townsley & Pelham 1994).

SEQUENCE CONTEXT AND OLIGOMERIZATION AFFECT THE EFFICIENCY OF DI-LYSINE MOTIFS Although di-lysine motifs transplanted from a variety of ER resident proteins were all found to be effective at accumulating proteins in the ER, a careful analysis of these chimeras showed that the rate at which the various molecules received Golgi modifications varied, as did the degree to which they co-localized to the intermediate compartment (Jackson et al 1993). The data were consistent with the fact that the chimeras that co-stained most strongly with the intermediate compartment were also the ones that received Golgi modifications most rapidly. Based on these findings, it was proposed that the sequence context of a di-lysine motif affects the efficiency of retrieval. Support for the idea of retrieval efficiency has come from the discovery that ERGIC-53 is targeted to the ERGIC by a di-lysine motif (Itin et al 1993). Mutational analysis of this motif suggests that its efficiency has been carefully titrated by the positioning of two phenylalanines (at positions -1 and -2 from the carboxy-terminus) such that it continually recycles between the ER and Golgi, with a steady-state localization in the ERGIC. If the phenylalanines are replaced with alanine, the mutated ERGIC-53 now localizes predominantly to the ER proper, whereas mutating the -3 lysine residue results in cell surface expression (Itin et al 1995). A similar situation appears to exist with Emp47p, a yeast type I membrane protein, that localizes to the Golgi apparatus at steady state but relocates to the ER in mutants that block vesicle budding, for example *sec12* (Schroder et al 1995). This finding indicates that like ERGIC-53, Emp47p is localized at steady state to the Golgi apparatus by continually recycling between the ER and Golgi.

The idea of retrieval efficiency may also extend to the number of targeting motifs per protein complex, i.e. a protein complex with two targeting motifs would be expected to be more efficiently retrieved than a complex with only one motif. Indeed, transplanting the E19 tail onto CD4, a monomer, results in a chimera that is almost exclusively co-localized with the intermediate compartment marker ERGIC-53, whereas transplanting this same sequence onto CD8, which forms homodimers, results in a chimera that is mainly ER localized, with only a minor portion co-localizing with ERGIC-53 (Jackson et al 1993). Post-ER oligomerization of molecules with weak dibasic retrieval motifs may be one means of sorting.

TARGETING OF PROTEINS TO THE ER BY OTHER MECHANISMS Many ER resident membrane proteins do not bear dibasic motifs in their cytoplasmic domains. Thus alternative mechanisms must exist that allow such molecules to be efficiently targeted. One obvious alternative would be a complex having a protein that possesses a di-lysine motif. For example, ribophorins I and II, which are ER-resident membrane proteins without di-lysine motifs, are known to associate with the oligosaccharyltransferase complex that includes the di-lysine bearing OST48/Wbp1 protein (Kelleher et al 1992).

A second alternative is one wherein these proteins use alternative ER-targeting motifs. A number of other ER-retention sequences have been identified in various proteins, for example, cytochrome P-450 (Murakami et al 1994, Szczesna-Skorupa et al 1995), the rotavirus VP7 protein (Stirzaker & Both 1989, Maass & Atkinson 1994), and the pre S1 protein of hepatitis B virus (Kuroki et al 1989). In these cases, the sequences involved are multi-component rather than short and discrete. However, a discrete ER-targeting motif has been identified close to the carboxy-terminus of the CD3-ϵ chain (Mallabiabarrena et al 1995). The properties of this motif suggest that it is functionally and structurally similar to the tyrosine-based motif required for endocytosis. Whether this motif functions by directing retrieval to the ER is not known.

A third alternative for which there is growing evidence is that molecules may be maintained in the ER by exclusion from transport vesicles, i.e. completely retained in the ER. The evidence for exclusion and possible mechanisms by which this might be achieved are discussed in the following section.

Retention in the ER by Exclusion from Transport Vesicles

Although the ER-localization of KDEL or di-lysine-tagged marker proteins is almost certainly the result of continuous recycling from post-ER compartments (Connolly et al 1994), it is less certain that the steady-state distribution of endogenous KDEL or di-lysine-bearing proteins is totally dependent on retrieval. For example, BiP, which lacks a KDEL sequence, is secreted from the ER with a half time of approximately 3 h, whereas many soluble secreted proteins have a half time of approximately 30–45 min (Munro & Pelham 1987). Similar results are found for many di-lysine-bearing ER residents, for example, removal of the di-lysine motif from the ER resident protein UDP-glucuronosyltransferase 1A results in poor, if any, transport of the truncated molecule. In addition, reporter chimeric proteins tagged with KDEL or di-lysine are modified by Golgi enzymes (see above), whereas endogenous ER proteins lack such modifications (Rosenfeld et al 1984, Kornfeld & Kornfeld 1985, Booth & Koch 1989).

These and other observations (see Brands et al 1985, Rose & Doms 1988, Hurtley & Helenius 1989) suggest that many ER proteins never leave this organelle. How such proteins are retained in the ER remains enigmatic, but various models have been proposed (Hortsch & Meyer 1985, Booth & Koch 1989, Bonifacino et al 1990, Suzuki et al 1991). Of these various models, oligomerization of proteins into complexes that are largely kinetically or physically excluded from the budding vesicle seems the most attractive. There is now strong evidence that retention of many Golgi proteins (reviewed in Gleeson et al 1994) occurs by the formation of oligomers, a process termed kin recognition (Nilsson et al 1993, Weisz et al 1993, Slusarewicz et al 1994). This was elegantly demonstrated by selectively localizing GlcNAc-TI to the ER via a di-arginine motif, which resulted in the relocalization to the ER of another *medial*-Golgi enzyme mannosidase II (Nilsson et al 1994). Oligomerization into higher ordered structures seems a likely mechanism by which components of the rough ER are segregated, because it is well recognized that they form extensive complexes that are largely resistant to salt extraction (Hortsch et al 1985). Whether kin recognition is at work here is not clear; however, weak oligomerization forces appear to maintain a subset of heterologous calcium-binding proteins in the ER (Booth & Koch 1989). In this model, it is envisaged that the lumen of the ER is a space-occupying matrix stabilized by calcium to passively exclude all proteins other than those programmed to form the matrix.

In terms of efficiency, it would make sense if the majority of the ER proteins were prevented from leaving the ER. A combination of retention and retrieval would ensure minimal leakage and would explain why marker proteins with a KDEL motif, for example, CD8/KDEL, which rely only on retrieval, are very slowly secreted (Zagouras & Rose 1989), whereas endogenous ER-soluble KDEL proteins are fully retained (Booth & Koch 1989).

COATOMER BINDING TO DI-LYSINE MOTIF: IMPLICATIONS FOR MEMBRANE TRAFFIC

Interaction of Di-Lysine Motifs with Coatomer

In vitro binding studies with cytosol and glutathione S-transferase-di-lysine fusion proteins demonstrated that coatomer binds with high affinity to these fusion proteins but only if the di-lysine motif was intact (Cosson & Letourneur

1994). Coatomer is a protein complex composed of seven subunits: α-coat protein (α-COP), β-COP, β'-COP, γ-COP, δ-COP, ϵ-COP, and ζ-COP (Kreis et al 1995, Waters et al 1991), which was first isolated from the cytosol of bovine brain. However, the finding that Golgi-derived transport vesicles are surrounded by coatomer immediately implicated this coat, now termed COP I, along with the GTPase-binding protein ARF1, in intra-Golgi transport (Rothman & Orci 1992).

Further evidence for the involvement of coatomer interaction with the di-lysine motif was provided by yeast genetic screens that used a di-lysine-tagged yeast α-factor surface receptor Ste2p (Letourneur et al 1994). This system will detect a loss in efficient retrieval of Ste2p while maintaining some degree of forward transport. Several mutants, the so-called *ret* mutants, deficient in this process were isolated. Analysis of these mutants and others showed that yeast strains in which the genes encoding α-COP (RET1), β'-COP (SEC27), or γ-COP (SEC21) were mutated resulted in defective retrieval of Ste2p/KKXX chimeras (Letourneur et al 1994). Further studies revealed that yeast mutants in which the genes encoding δ-COP (RET2) and ζ-COP (RET3) were disrupted also led to a *ret* phenotype (P Cosson, personal communication).

In vitro binding studies using the cytosol from these temperature-sensitive mutants showed that coatomer prepared from α-COP (*ret1-1*) or β'-COP (*sec27-1*) yeast mutants had lost the ability to bind to di-lysine motifs, whereas coatomer prepared from γ-COP (*sec21*) mutants still bound (Letourneur et al 1994). These data suggest that binding to di-lysine motifs is most likely via the α- or β'-COP rather than the γ-COP, a suggestion supported by the initial findings of Cosson & Letourneur (1994). These findings were later substantiated using purified partial coatomer complexes (Lowe & Kreis 1995) composed of only α-, β'- and ϵ-COPs. In contrast to these observations, Harter et al (1996) have provided cross-linking data using a di-lysine peptide, with a photoactivatable phenylalanine at position -5, that show γ-COP as the primary coatomer subunit interacting with a di-lysine motif. The same result was found whether purified coatomer or whole-cell lysates were analyzed. These data are difficult to reconcile. A loss in binding does not necessarily mean that the mutated COP subunit per se is involved in creating the di-lysine binding site. For example, a mutation in α-COP might affect the conformation of γ-COP such that it can no longer bind to the di-lysine motif. Such arguments do not, however, explain the compelling data that show partial coatomer complexes (in which γ-COP is absent) bind to di-lysine motifs.

Several of the individual COP I subunits show homology with the adaptor subunits of clathrin-coated vesicles (β-COP, Duden et al 1991; δ-COP, P Cosson, personal communication; D Faulstich et al, unpublished data; ζ-COP,

Kuge et al 1993). Recently, clathrin-associated adaptors were shown to bind specifically to the tyrosine-based sorting signals (Ohno et al 1995). These interactions are thought to be responsible for the ability of the cell to selectively include such proteins in the clathrin-coated vesicle. Likewise, the subunits of COP I with adaptor homology may regulate the types of proteins that enter COP I-coated vesicles. Although a consensus on which coatomer subunit(s) directly bind the di-lysine motif has not been reached, the evidence is overwhelming that coatomer does indeed bind this motif, and therefore COP I vesicles mediate retrieval of di-lysine tagged membrane proteins (Pelham 1994) from the Golgi to the ER.

Retrograde Transport from the Golgi to the ER

A retrograde transport pathway between the Golgi and the ER had long been assumed to exist based on theoretical grounds, as without it, the continuous flow of membrane from the ER to Golgi would presumably deplete the ER of membrane. In addition to fulfilling the requirement for recycling lipids and proteins, such a pathway would obviously be ideal to retrieve lost ER residents, both soluble (K/HDEL) and membrane-bound (KKXX), and essential to retrieve the molecular components involved in vesicle targeting (V-SNAREs). Until recently, molecular details of this pathway have been sparse, and although many molecules were identified that impact on the retrieval of HDEL-bearing proteins via the erd2 receptor, devising assays that allowed the retrograde transport route to be targeted without impacting the anterograde transport proved to be a major hurdle. The effects of the fungal metabolite brefeldin A (BFA), which reversibly blocks transport of secretory proteins by inducing a rapid redistribution of Golgi membranes, via long tubular extensions, to the ER, were initially interpreted as strong evidence for this retrograde pathway (Lippincott-Schwartz et al 1989). However, the finding that BFA treatment results in an immediate dissociation of Golgi-bound COP I coats (Donaldson et al 1992) suggests that the effects seen with BFA may simply be the result of the uncontrolled fusion of Golgi membranes with the ER in the absence of a protective COP I coat (Alcatde et al 1994) and may therefore have little physiological relevance.

The Role of COP I- and COP II-Coated Vesicles in Retrieval of Membrane Proteins

The observation that coatomer binds di-lysine motifs led to a re-evaluation of the more established role of COP I in mediating anterograde transport between Golgi stacks and from the ER to Golgi apparatus (Pelham 1994). Inhibition of COP I function may block anterograde traffic indirectly by causing depletion from the ER of essential targeting proteins (i.e. SNAREs) due to the lack of membrane retrieval. Indeed, the role of COP I-coated vesicles in ER to Golgi transport is now controversial, and it has been argued that the first transport step out of the ER may be accomplished using only COP II vesicles. The composition of the COP II coat is entirely different from that of coatomer; in yeast it is composed of four subunits, Sec31p, Sec13p, Sec23p, and Sec24p (Barlowe et al 1994), that require the small GTP-binding protein Sar1 (rather than ARF1 for COP I) and the nucleotide exchange factor Sec12p in order to form a complex. These COP II vesicles were first identified in yeast as necessary to transport cargo out of the ER (Salama et al 1993), but it is now clear that they play an equally important and presumably equivalent role in mammalian cells. Deletion of the genes encoding the components of either COP I or COP II is lethal, and temperature-sensitive mutants in components from either coat block ER to Golgi transport in both yeast (Pryer et al 1992) and mammalian cells (Guo et al 1994). Experiments designed to define which coat carries out anterograde and retrograde transport, for example, using antibodies against coat subunits to define their roles (Pepperkok et al 1993, Peter et al 1993), are proving notoriously difficult to interpret with certainty. Although a role for COP I in the retrograde pathway seems assured, COP II vesicles are now the prime candidates for accomplishing the first transport step from the ER.

Using stage-specific in vitro transport assays to synchronize movement of cargo to and from pre-Golgi intermediates and GDP- and GTP-restricted forms of ARF1 and Sar1 proteins to control coat recruitment, Aridor et al (1995) have provided evidence that COP II vesicles may be solely responsible for export from the ER, whereas COP I is required to segregate VSV-G and p58 (the rat homologue of ERGIC-53) to the anterograde and retrograde pathways, respectively. These findings complement previous in vitro experiments investigating the role of Sar1, in which antibodies to Sar1 and *trans*-dominant Sar1 mutants having a preferential affinity for GDP were found to strongly inhibit vesicle budding from the ER but did not affect transport of VSV-G between sequential Golgi compartments (Kuge et al 1994).

Immuno-electron microscopy showed that both COP I and COP II vesicles could be identified budding from an isolated yeast nuclear envelope. Analysis of the type of molecules that compose COP I and COP II vesicles showed a number of features in common. They both contain the V-SNARE molecules Sec22p, Bos1p, and Bet1p, along with a number of unidentified common polypeptides, whereas resident ER proteins, for example Kar2p (BiP), appear to be absent (Bednarek et al 1995). COP I and COP II vesicles differ, however, in the types of cargo they carry. For example, α-factor precursor, two amino acid permeases, and Gas1p, a major GPI-linked cell surface protein, are exclusively packaged by COP II vesicles. Interestingly, packaging of Gas1p into COP II vesicles was found to be dependent on its GPI-anchor (Doering & Schekman

1996). Several proteins specific to COP I vesicles have been found (Bednarek et al 1995). However, while we await their identification, there remains no hard evidence that the identified COP I vesicles budding from the ER are carrying secretory cargo. If COP II alone is responsible for transport of the majority of the cargo from the ER, we are left to speculate on the role of the COP I vesicles budding from this compartment (Bednarek et al 1995).

A candidate ER T-SNARE, Ufe1p (a homologue of Sed5p, a Golgi T-SNARE), with which COP I vesicles returning along a retrograde pathway to the ER may dock, has recently been identified in yeast (M Lewis & H Pelham, personal communication). In *ufe1-1* mutants, redistribution of the di-lysine containing Emp47p and Erd2p to the ER was blocked, and the molecules remained in the Golgi apparatus. The data are consistent with a block in retrograde but not anterograde transport caused by the absence of a functional ER T-SNARE. With use of these same assays, a *sec-21* (γ-COP) mutant was found to have a phenotype almost identical to that of *ufe1-1*, arguing that COP I vesicles may be involved only in retrograde transport.

Although models for the anterograde pathway from the ER to Golgi remain controversial, a consensus is emerging on a molecular description of retrograde transport whereby membrane proteins containing di-lysine motifs, H/KDEL-bearing proteins bound to their receptor, and recycling V-SNAREs are collected into a COP I-coated vesicle in the Golgi apparatus. These vesicles then dock and fuse with the ER in a process that involves a specific T-SNARE Ufe1p, soluble fusion components, and possibly also the Sec20p/Tip20p complex (M Lewis & H Pelham, personal communication).

WHERE DOES DI-LYSINE RETRIEVAL OCCUR? Immuno-localization studies have clearly localized di-lysine-bearing proteins to the ERGIC and, based on the posttranslational modifications of di-lysine-tagged marker proteins, some molecules appeared to have been exposed to *medial-* and *trans-*Golgi enzymes (Jackson et al 1993, Martire et al 1996). Interestingly, CD8/di-lysine chimeras, which differ only in the context of the di-lysine motif, receive these modifications with quite different kinetics. Assuming that the forward rates of transport of these different molecules are identical, then one expects, on average, that retrieval of the different chimeras occurs from different sites in the exocytic pathway. Thus it seems likely that retrieval of di-lysine-bearing proteins occurs from throughout the Golgi stack, a suggestion that is consistent with immuno-localization of coatomer (Griffiths et al 1995). It is also worth noting that the capacity to retrieve KDEL-tagged molecules appears to extend throughout the Golgi stack (Miesenbock & Rothman 1995). Whether di-lysine or KDEL-bearing molecules, which are collected into COP I-coated vesicles in the *medial-*Golgi, are delivered to an earlier Golgi stack or directly back to the ER is unknown.

MECHANISMS CONTROLLING DI-LYSINE-COATOMER BINDING If coatomer binding is dependent on di-lysine motifs, why, under physiologic conditions, is the majority of coatomer bound to Golgi membranes (Duden et al 1991, Oprins et al 1993, Griffiths et al 1995) rather than to the ER where presumably a significant proportion of proteins containing di-lysine motifs are localized? Although specialized coatomer-enriched ER structures have been reported under conditions that reduce ER to Golgi transport (Orci et al 1994), it seems likely that ordinarily the binding of coatomer to di-lysine motifs is controlled. One mechanism by which ER-located di-lysine motifs might be prevented from binding coatomer would be by the specific masking of these motifs. A candidate molecule that could achieve steric masking is tubulin. Studies using a synthetic peptide containing the di-lysine motif of the adenovirus E19 protein indicate that this peptide can bind β tubulin and promote tubulin polymerization in vitro (Dahllof et al 1991). Furthermore, over-expression of certain di-lysine and di-arginine constructs in mammalian cells leads to a complete remodeling of the ER into fine tubules that closely associate with the microtubule network (Schutze et al 1994, Schweizer et al 1994). In addition, these structures do not co-localize with COP I subunits, as assessed by immunofluorescence staining with specific anti-COP I antibodies (RD Teasdale & MR Jackson, unpublished data).

Nothing is known about the physiological conditions that allow di-lysine binding to coatomer, although in vitro studies suggest that binding occurs even in the absence of ARF1 and GTP. However, the requirement for ARF1 and GTP in the formation of anterograde COP I-coated vesicles involved in intra-Golgi transport is well established (Rothman 1994). How then does the COP I machinery distinguish between anterograde and retrograde cargo? One possible explanation is that the di-lysine proteins in the Golgi apparatus bind coatomer to form the COP I-Golgi coat independently of ARF. This could potentially trap escaped di-lysine-bearing proteins in the Golgi apparatus until they accumulate to a certain density, when a retrograde vesicle forms. Anterograde transport, however, would first require the recruitment of ARF to the membrane prior to COP I binding. Alternatively, the different types of cargo may induce recruitment of other molecules necessary for vesicular transport, for example Rab-GTPases (Pfeffer 1994). Indeed, the recruitment of rab1-GTP to ER membranes is required for the formation of a transport vesicle. This was identified by the lack of protein export from the ER when a dominant-negative rab mutant, i.e. GDP-bound, was expressed (Tisdale et al 1992, Nuoffer et al 1994).

The observation that the majority of coatomer is localized to membranes between the ER and the Golgi proper (Oprins et al 1993) suggests that there should be a significant density of di-lysine motifs in this location. The ERGIC marker proteins p53/p58 in mammalian cells and Emp47 in yeast are obvious candidates for coatomer-binding sites. Indeed, the cytoplasmic domain of these molecules might be dedicated to this function. Antibodies directed against the cytoplasmic domain of ERGIC-53/p58 block ER to Golgi transport and the subsequent recycling of p58 to the ER in vitro (EJ Tisdale & WE Balch, personal communication).

Other molecules possibly involved in providing coatomer-binding sites on Golgi membranes are proteins of the p24 family (Stamnes et al 1995). Two members of this family, p23 and p24, are recognized as major components of Golgi-derived COP I-coated vesicles. The p24 family (see Table 2) contains type I membrane proteins that show little sequence identity in their lumenal domains except for two conserved cysteines and heptad repeats of hydrophobic residues close to the membrane (i.e. coiled-coils). The cytoplasmic domains, however, are more conserved. Two features are common to the majority of members: (*a*) two basic residues at their carboxy-termini, similar to di-lysine motifs, that may function to bind coatomer, and (*b*) phenylalanine residues between the basic residues and the transmembrane. Indeed, a functional role for the cytoplasmic domain is likely based on the high degree of sequence conservation compared with the lumenal portion. Another member of this family, Emp24p, a yeast homologue of p24, is a major component of COP II vesicles, and yeast mutants that lack Emp24p show a significant delay in the transport

Table 2 Protein homologues of the 24-kDa components of coated vesicles[a]

Cytoplasmic domain	Species	Accession number
QFLFTG**R**Q**K**NYV	*S. cerevisiae*	P39704
QFFFTS**R**Q**K**NYV	*S. cerevisiae*	S54622
KRFFEV**RR**VV[b]	*C. griseus*	U26264
KNYFKT**KH**II	*S. cerevisiae*	S55107
RHFFKA**KK**LIE	*X. laevis*	X90517
RRFFKA**KK**LIE	*H. sapiens*	L40397
KSLKNFFIA**KK**LV[c]	*C. familiaris*	X53592
RHLKSFFEA**KK**LV	*H. sapiens*	X90872
DAFTF**K**SFF	*H. sapiens*	R76166
RYLKNFFV**K**Q**K**VV	*S. cerevisiae*	P38819
HLKTFFQ**KKK**LI	*A. thaliana*	T46519
RRFFEVTSLV[d]	*S. cerevisiae*	X67317

[a]The listed proteins were identified by a BLAST search of the DNA databases with the characterized members.
[b]CHOp24 (Stammes et al 1995).
[c]gp25L (Wada et al 1991).
[d]Emp24p (Schimmoller et al 1995).

of a subset of secretory proteins to the Golgi (Schimmoller et al 1995). Note that the cytoplasmic domain of Emp24 lacks the basic residues at its carboxy-terminus. These findings have led to the idea that the p24 family of proteins may sort and thereby concentrate proteins into transport vesicles (see below).

Dibasic Motifs and the Concentration of Cargo

It has been long recognized that newly synthesized secretory and plasma membrane proteins become concentrated five- to tenfold as they reach the Golgi apparatus. All attempts to identify discrete transplantable signals that direct proteins to be concentrated by direct interaction with the vesicle coat or with a specific cargo receptor have been unsuccessful. Working models on how concentration is achieved have mainly centered around the idea of bulk flow out of the ER by default with selective retrieval of ER residents and lipid (Pfeffer & Rothman 1987). A modification of this model, whereby the majority of ER residents are segregated from the bulk flow by differences in some physical property, for example oligomerization (see above), improves the overall efficiency of the process. In favor of the default movement of proteins into this pathway is the observation that when proteins normally absent from the transport pathway are artificially introduced into the lumen of the ER, they are transported to the cell surface as efficiently, if not as rapidly, as endogenous secretory proteins (Wiedmann et al 1984). Furthermore, ER resident proteins deprived of their ER retrieval/retention signals are, in many instances, slowly exported out of the ER. Because it is unlikely that such proteins contain cryptic export signals, a slow or low level default export pathway probably exists.

Although this retention/retrieval bulk flow model is attractive and may apply to some extent, strong evidence has been advanced over the past two years that cargo may in fact be selectively extracted from the ER. The most compelling data come from in vitro studies in yeast, where the secretory protein preproalpha factor is clearly found to be enriched in vesicles budding off the ER compared with the levels expected by a bulk flow mechanism (Barlowe et al 1994). Further evidence for this phenomenon has come from quantitative immuno-electron microscopy studies of VSV-G protein in infected fibroblasts and of secretory proteins in liver cells; also identified was a significant concentration of cargo at the point of budding from the ER (Mizuno & Singer 1993, Balch et al 1994). These findings have led many to re-examine the possible existence of cargo receptors that actually direct and concentrate proteins into the vesicles that leave the ER.

One clear candidate for this role is ERGIC-53. It not only recycles between the ER and Golgi, but recent data suggest it has lectin-binding properties that would allow the lumenal portion to selectively bind proteins that bear mannose residues (Arar et al 1995). As the majority of soluble ER resident proteins are not glycosylated (except GRP94), ERGIC-53 could selectively extract secretory glycoproteins from the ER by binding to their high mannose oligosaccharides. Upon arrival in the Golgi, trimming enzymes are thought to remove the mannose residues leaving ERGIC-53 to recycle back to the ER and the secretory proteins to proceed along the pathway. Although this model is attractive, there are currently no experimental data to substantiate a role for ERGIC-53 as a cargo receptor.

The other clear candidates for cargo receptors are members of the p24 family discussed above. The lumenal domains of this family of proteins are degenerate compared with their cytoplasmic tails (see Table 2), a feature that might be expected of a family of receptors that collectively are able to bind the complete spectrum of cargo molecules. Although the COP II vesicle component Emp24p, which does not have a dibasic motif, has been shown indirectly to be involved in sorting of proteins out of the ER, other family members (see Table 2) that do possess dibasic motifs may be involved in selectively concentrating molecules in COP I-mediated budding events. Although such mechanisms remain pure speculation, selective extraction processes that could sort proteins with reasonable efficiency into either anterograde or retrograde vesicles, would certainly complement the mechanisms of retention and retrieval. If such a process were to be used in a reiterative manner, as proposed in the distillation hypothesis (Rothman 1981), it seems likely that the sorting power required for an efficient secretory pathway would be achieved.

CONCLUSIONS AND PERSPECTIVE

In this review we have outlined multiple roles for dibasic ER retrieval motifs, either di-lysine or di-arginine. Figure 1 summarizes those roles that can be broadly catagorized into (1) maintaining ER resident membrane proteins in the ER, (2) assisting in the assembly of plasma membrane multi-subunit complexes by maintaining subunits in the ER until a complex is formed, (3) targeting proteins to the intermediate compartment/*cis*-Golgi, and (4) providing an anchor for coatomer to bind and thereby perform a central role in the retrograde transport and the general process of sorting in the Golgi complex.

The discovery that these dibasic motifs provide an anchor for coatomer has brought them to center stage in the general sorting operations that function in the Golgi apparatus. It now appears that the role for which these motifs were initially defined, i.e. the direct retrieval of ER residents, may be a minor one because in the majority of cases, such proteins rarely escape the ER.

With the rapid pace at which both general and specific components of sorting processes are being discovered and the elegant in vitro assays that have been developed, we can expect much progress in the next few years in defining the

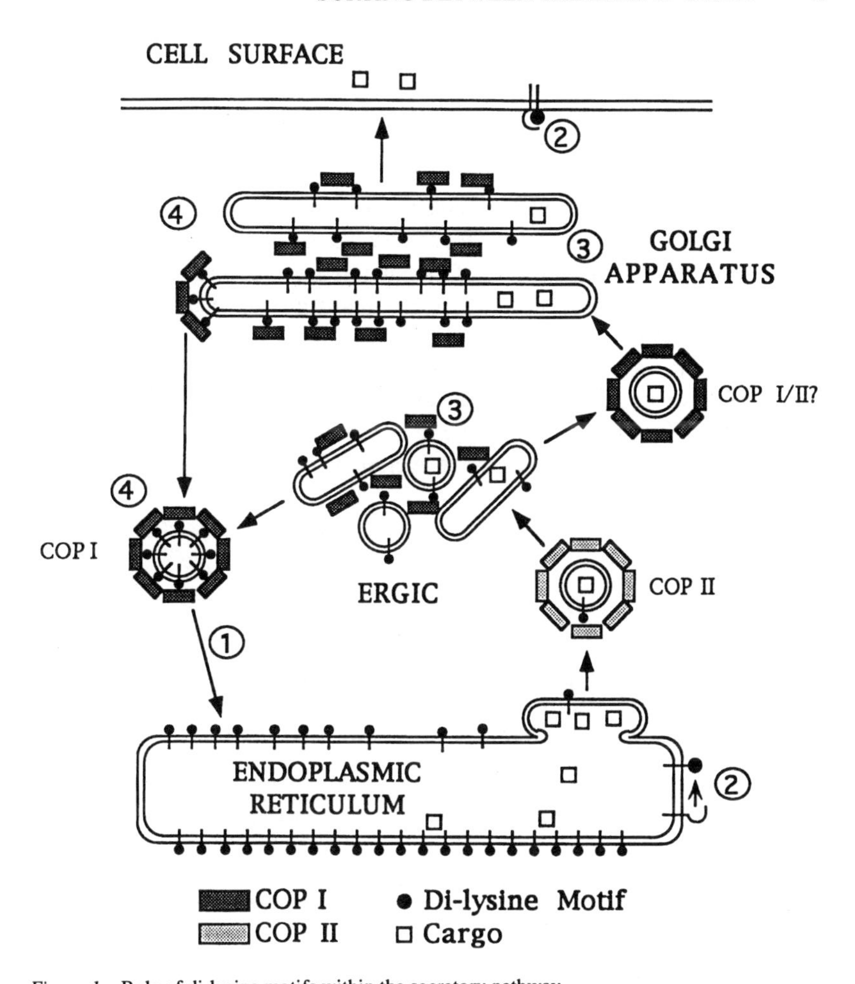

Figure 1 Role of di-lysine motifs within the secretory pathway.

sorting processes that are essential for the secretory pathway. In particular, we can expect clarification of the role of COP I; for example, is it involved in budding of vesicles from the ER and if so do these vesicles contain cargo destined for export or is COP I used only for retrograde transport? Of particular interest will be how this role fits with the newly emerging idea of cargo receptors and thereby the sorting processes in the Golgi stack. Perhaps the most fascinating process, for which no details exist, is what features of the proteins to be concentrated into the budding vesicle do these putative cargo receptors recognize.

If cargo receptors do not exist then we will be left to explain the molecular details of how cargo is apparently extracted from the ER. Presumably this will entail providing a much improved picture of how ER residents are prevented from entering the transport vesicles. Other unresolved issues in this field are (*a*) what controls the binding of coatomer to dibasic motifs, (*b*) do dibasic motifs alone provide the anchor for coatomer binding to membranes or do other sequences substitute, (*c*) what provides the anchors for the coat on COP II vesicles, and (*d*) what is nature of the intermediate compartment. Although it seems suprising that the latter remains a controversial issue, it is a reminder of the difficulties that will undoubtedly be faced in the coming years in trying to reconcile data from yeast genetics, in vitro assays, and quantitative immunoelectron microscopy. In the case of the intermediate compartment it seems likely that much of the controversy will evaporate once we have a better grip on the molecular details of the sorting operations that are accomplished in this organelle.

ACKNOWLEDGMENTS

We are grateful to MS Teasdale and EJ Tisdale for critical appraisal of the manuscript and S Sharp for assistance in searching the databases. We thank P Cosson, C Harter, J Lippincott-Schwartz, HRB Pelham, RW Schekman, EJ Tisdale, and FT Wieland for generously supplying unpublished data. Thanks also to PA Peterson and JE Rothman for helpful discussions.

Literature Cited

Alcalde J, Egea G, Sandoval IV. 1994. gp74 a membrane glycoprotein of the *cis*-Golgi network that cycles through the endoplasmic reticulum and intermediate compartment. *J. Cell Biol.* 124:649–65

Amara JF, Cheng SH, Smith AE. 1992. Intracellular protein trafficking defects in human disease. *Trends Cell Biol.* 2:145–49

Arar C, Carpentier V, Le Caer JP, Monsigny M, Legrand A, Roche AC. 1995. ERGIC-53, a membrane protein of the endoplasmic reticulum-Golgi intermediate compartment, is identical to MR60, an intracellular mannose-specific lectin of myelomonocytic cells. *J. Biol. Chem.* 270:3551–53

Aridor M, Bannykh SI, Rowe T, Balch WE. 1995. Sequential coupling between COP II and COP I vesicle coats in endoplasmic reticulum to Golgi transport. *J. Cell Biol.* 131:875–93

Balch WE, McCaffery M, Plutner H, Farquhar MG. 1994. Vesicular stomatitis virus glycoprotein is sorted and concentrated during export from the endoplasmic reticulum. *Cell* 76:841–52

Barlowe C, Orci L, Yeung T, Hosobuchi M, Hamamoto S, et al. 1994. COP II: a membrane coat formed by Sec proteins that drive vesicle budding from the endoplasmic reticulum. *Cell* 77:895–907

Batt AM, Magdalou J, Vincent-Viry M, Ouzzine M, Foumel-Gigleux S, et al. 1994. Drug metabolism enzymes related to laboratory medicine: cytochrome P-450 and UDP-glucuronsyltransferases. *Clin. Chim. Acta* 226:171–90
Bednarek SY, Ravazzola M, Hosobuchi M, Amherdt M, Perrelet A, et al. 1995. COP I- and COP II-coated vesicles bud directly from the endoplasmic reticulum in yeast. *Cell* 83:1183–96
Blobel G, Dobberstein B. 1975. Transfer of proteins across membranes. I. Presence of proteolytically processed and unprocessed nascent immunoglobulin light chains on membrane-bound ribosomes of murine myeloma. *J. Cell Biol.* 67:835–51
Bonatti S, Migliaccio G, Simons K. 1989. Palmitylation of viral membrane glycoproteins takes place after exit from the endoplasmic reticulum. *J. Biol. Chem.* 264:12590–95
Bonifacino JS, Suzuki CK, Klausner RD. 1990. A peptide sequence confers retention and rapid degradation in the endoplasmic reticulum. *Science* 247:79–82
Booth C, Koch GL. 1989. Perturbation of cellular calcium induces secretion of luminal ER proteins. *Cell* 59:729–37
Brands R, Snider MD, Hino Y, Park SS, Gelboin HV, Rothman JE. 1985. Retention of membrane proteins by the endoplasmic reticulum. *J. Cell Biol.* 101:1724–32
Collawn JF, Stangel M, Kuhn LA, Esekogwu V, Jing SQ, et al. 1990. Transferrin receptor internalization sequence YXRF implicates a tight turn as the structural recognition motif for endocytosis. *Cell* 63:1061–72
Connolly CN, Futter CE, Gibson A, Hopkins CR, Cutler DF. 1994. Transport into and out of the Golgi complex studied by transfecting cells with cDNAs encoding horseradish peroxidase. *J. Cell Biol.* 127:641–52
Cosson P, Letourneur F. 1994. Coatomer interaction with di-lysine endoplasmic reticulum retention motifs. *Science* 263:1629–31
Dahllof B, Wallin M, Kvist S. 1991. The endoplasmic reticulum retention signal of the E3/19K protein of adenovirus-2 is microtubule binding. *J. Biol. Chem.* 266:1804–8
Doering TL, Schekman R. 1996. GPI anchor attachment is required for Gaslp transport from the endoplasmic reticulum in COP II vesicles. *EMBO J.* 15:182–91
Donaldson JG, Finazzi D, Klausner RD. 1992. Brefeldin A inhibits Golgi membrane-catalysed exchange of guanine nucleotide onto ARF protein. *Nature* 360:350–52
Duden R, Griffiths G, Frank R, Argos P, Kreis TE. 1991. β-COP, a 110 kd protein associated with non-clathrin-coated vesicles and the Golgi complex, shows homology to beta-adaptin. *Cell* 64:649–65
Farquhar MG. 1985. Progress in unravelling pathways of Golgi traffic. *Annu. Rev. Cell Biol.* 1:447–88
Gaynor EC, te Heesen S, Graham TR, Aebi M, Emr SD. 1994. Signal-mediated retrieval of a membrane protein from the Golgi to the ER in yeast. *J. Cell Biol.* 127:653–65
Gleeson PA, Teasdale RD, Burke J. 1994. Targeting of proteins to the Golgi apparatus. *Glycoconjugate J.* 11:381–94
Griffiths G, Ericsson M, Krijnse-Locker J, Nilsson T, Goud B, et al. 1994. Localization of the Lys, Asp, Glu, Leu tetrapeptide receptor to the Golgi complex and the intermediate compartment in mammalian cells. *J. Cell Biol.* 127:1557–74
Griffiths G, Pepperkok R, Locker JK, Kreis TE. 1995. Immunocytochemical localization of β-COP to the ER-Golgi boundary and the TGN. *J. Cell Sci.* 108:2839–56
Griffiths G, Simons K. 1986. The *trans* Golgi network: sorting at the exit site of the Golgi complex. *Science* 234:438–43
Guo Q, Vasile E, Krieger M. 1994. Disruptions in Golgi structure and membrane traffic in a conditional lethal mammalian cell mutant are corrected by ϵ-COP. *J. Cell Biol.* 125:1213–24
Hamburger D, Egerton M, Riezman H. 1995. Yeast Gaalp is required for attachment of a completed GPI anchor onto proteins. *J. Cell Biol.* 129:629–39
Hammond C, Helenius A. 1995. Quality control in the secretory pathway. *Curr. Opin. Cell Biol.* 7:523–29
Harter C, Pavel J, Coccia F, Draken E, Wegehingel S, et al. 1996. Non-clathrin-coat protein γ, a subunit of coatomer, binds to the cytoplasmic di-lysine motifs of membrane proteins of the early secretory pathway. *Proc. Natl. Acad. Sci. USA.* In press
Hauri HP, Schweizer A. 1992. The endoplasmic reticulum-Golgi intermediate compartment. *Curr. Opin. Cell Biol.* 4:600–8
Hill KJ, Stevens TH. 1994. Vma2lp is a yeast membrane protein retained in the endoplasmic reticulum by a di-lysine motif and is required for the assembly of the vacuoler H^+-ATPase complex. *Mol. Biol. Cell.* 5:1039–50
Hortsch M, Griffiths G, Meyer DI. 1985. Restriction of docking protein to the rough endoplasmic reticulum: immunocytochemical localization in rat liver. *Eur. J. Cell Biol.* 38:271–79
Hortsch M, Meyer DI. 1985. Immunochemical analysis of rough and smooth microsomes from rat liver: segregation of docking protein in the rough membranes. *Eur. J. Biochem.*

150:559–64
Hurtley SM, Helenius A. 1989. Protein oligomerization in the endoplasmic reticulum. *Annu. Rev. Cell Biol.* 5:277–307
Itin C, Schindler R, Hauri HP. 1995. Targeting of protein ERGIC-53 to the ER/ERGIC/*cis*-Golgi recycling pathway. *J. Cell Biol.* 131:57–67
Jackson MR, Nilsson T, Peterson PA. 1990. Identification of a consensus motif for retention of transmembrane proteins in the endoplasmic reticulum. *EMBO J.* 9:3153–62
Jackson MR, Nilsson T, Peterson PA. 1993. Retrieval of transmembrane proteins to the endoplasmic reticulum. *J. Cell Biol.* 121:317–33
Kelleher DJ, Kreibich G, Gilmore R. 1992. Oligosaccharyltransferase activity is associated with a protein complex composed of ribophorins I and II and a 48 kd protein. *Cell* 69:55–65
Kendall RL, Bradshaw RA. 1992. Isolation and characterization of the methionine amino peptidase from porcine liver responsible for the co-translational processing of proteins. *J. Biol. Chem.* 267:20667–73
Klausner RD. 1989. Sorting and traffic in the central vacuoler system. *Cell* 57:703–6
Klausner RD, Lippincott-Schwartz J, Bonifacino JS. 1990. The T cell antigen receptor: insights into organelle biology. *Annu. Rev. Cell Biol.* 6:403–31
Kornfeld R, Kornfeld S. 1985. Assembly of asparagine-linked oligosaccharides. *Annu. Rev. Biochem.* 54:631–64
Kreis TE, Lowe M, Pepperkok R. 1995. COPs regulating membrane traffic. *Annu. Rev. Cell. Dev. Biol.* 11:677–706
Krijnse-Locker J, Ericsson M, Rottier PJ, Griffiths G. 1994. Characterization of the budding compartment of mouse hepatitis virus: evidence that transport from the RER to the Golgi complex requires only one vesicular transport step. *J. Cell Biol.* 124:55–70
Kuge O, Dascher C, Orci L, Rowe T, Amherdt M, et al. 1994. Sarl promotes vesicle budding from the endoplasmic reticulum but not Golgi compartments. *J. Cell Biol.* 125:51–65
Kuge O, Hara-Kuge S, Orci L, Ravazzola M, Amherdt M, et al. 1993. ζCOP, a subunit of coatomer, is required for COP-coated vesicle assembly. *J. Cell Biol.* 123:1727–34
Kuroki K, Russnak R, Ganem D. 1989. Novel N-terminal amino acid sequence required for retention of a hepatitis B virus glycoprotein in the endoplasmic reticulum. *Mol. Cell. Biol.* 9:4459–66
Letoumeur F, Gaynor EC, Hennecke S, Demolliere C, Duden R, et al. 1994. Coatomer is essential for retrieval of di-lysine-tagged proteins to the endoplasmic reticulum. *Cell* 79:1199–207
Letourneur F, Hennecke S, Demolliere C, Cosson P. 1995. Steric masking of a di-lysine endoplasmic reticulum retention motif during assembly of the human high affinity receptor for immunoglobulin E. *J. Cell Biol.* 129:971–78
Lewis MJ, Pelham HR. 1990. A human homologue of the yeast HDEL receptor. *Nature* 348:162–63
Lewis MJ, Pelham HR. 1992. Ligand-induced redistribution of a human KDEL receptor from the Golgi complex to the endoplasmic reticulum. *Cell* 68:353–64
Lippincott-Schwartz J, Yuan LC, Bonifacino JS, Klausner RD. 1989. Rapid redistribution of Golgi proteins into the ER in cells treated with brefeldin A: evidence for membrane recycling from Golgi to the ER. *Cell* 56:801–13
Lotteau V, Teyton L, Peleraux A, Nilsson T, Karlsson L, et al. 1990. Intracellular transport of class II MHC molecules directed by invariant chain. *Nature* 348:600–5
Lowe M, Kreis TE. 1995. In vitro assembly and disassembly of coatomer. *J. Biol. Chem.* 270:31364–71
Luskey KL, Stevens B. 1985. Human 3-hydroxy-3-methylglutaryl coenzyme A reductase. Conserved domains responsible for catalytic activity and sterol-regulated degradation. *J. Biol. Chem.* 260:10271–77
Maass DR, Atkinson PH. 1994. Retention by the endoplasmic reticulum of rotavirus VP7 is controlled by three adjacent amino-terminal residues. *J. Virol.* 68:366–78
Mallabiabarrena A, Jimenez MA, Rico M, Alarcon B. 1995. A tyrosine-containing motif mediates ER retention of CD3-ϵ and adopts a helix-turn structure. *EMBO J.* 14:2257–68
Martire G, Mottola G, Pascale MC, Malagolini N, Turrini I, et al. 1996. Different fate of a single reporter protein containing KDEL or KKXX targeting signals stably expressed in mammalian cells. *J. Biol. Chem.* 271:3541–47
Miesenbock G, Rothman JE. 1995. The capacity to retrieve escaped ER proteins extends to the *trans*-most cisterna of the Golgi stack. *J. Cell Biol.* 129:309–19
Mizuno M, Singer SJ. 1993. A soluble secretory protein is first concentrated in the endoplasmic reticulum before transfer to the Golgi apparatus. *Proc. Natl. Acad. Sci. USA* 90:5732–36
Munro S, Pelham HR. 1987. A C-terminal signal prevents secretion of luminal ER proteins. *Cell* 48:899–907
Murakami K, Mihara K, Omura T. 1994. The transmembrane region of microsomal cytochrome P450 identified as the endoplas-

mic reticulum retention signal. *J. Biochem.* 116:164–75
Nilsson T, Hoe MH, Slusarewicz P, Rabouille C, Watson R, et al. 1994. Kin recognition between medial Golgi enzymes in HeLa cells. *EMBO J.* 13:562–74
Nilsson T, Jackson M, Peterson PA. 1989. Short cytoplasmic sequences serve as retention signals for transmembrane proteins in the endoplasmic reticulum. *Cell* 58:707–18
Nilsson T, Slusarewicz P, Hoe MH, Warren G. 1993. Kin recognition. A model for the retention of Golgi enzymes. *FEBS Lett.* 330:1–4
Nuoffer C, Davidson HW, Matteson J, Meinkoth J, Balch WE. 1994. A GDP-bound of rabl inhibits protein export from the endoplasmic reticulum and transport between Golgi compartments. *J. Cell. Biol.* 125:225–37
Ohno H, Stewart J, Fournier MC, Bosshart H, Rhee S, et al. 1995. Interaction of tyrosine-based sorting signals with clathrin-associated proteins. *Science* 269:1872–75
Oprins A, Duden R, Kreis TE, Geuze HJ, Slot JW. 1993. β-COP localizes mainly to the *cis*-Golgi side in exocrine pancreas. *J. Cell Biol.* 121:49–59
Orci L, Perrelet A, Ravazzola M, Amherdt M, Rothman JE, Schekman R. 1994. Coatomer-rich endoplasmic reticulum. *Proc. Natl. Acad. Sci. USA* 91:11924–28
Palade GE. 1975. Intracellular aspects of the process of protein transport. *Science* 189:347–58
Pelham HR. 1989. Control of protein exit from the endoplasmic reticulum. *Annu. Rev. Cell Biol.* 5:1–23
Pelham HR. 1990. The retention signal for soluble proteins of the endoplasmic reticulum. *Trends Biol. Sci.* 15:483–86
Pelham HR. 1994. About turn for the COPS? *Cell* 79:1125–27
Pepperkok R, Scheel J, Horstmann H, Hauri HP, Griffiths G, Kreis TE. 1993. β-COP is essential for biosynthetic membrane transport from the endoplasmic reticulum to the Golgi complex in vivo. *Cell* 74:71–82
Peter F, Plutner H, Zhu H, Kreis TE, Balch WE. 1993. β-COP is essential for transport of protein from the endoplasmic reticulum to the Golgi in vitro. *J. Cell Biol.* 122:1155–67
Pfeffer SR. 1994. Rab GTPases: master regulators of membrane trafficking. *Curr. Opin. Cell. Biol.* 6:522–26
Pfeffer SR, Rothman JE. 1987. Biosynthetic protein transport and sorting by the endoplasmic reticulum and Golgi. *Annu. Rev. Biochem.* 56:829–52
Pryer NK, Wuestehube LJ, Schekman R. 1992. Vesicle-mediated protein sorting. *Annu. Rev. Biochem.* 61:471–516
Rambourg A, Clermont Y. 1990. Three-dimensional electron microscopy: structure of the Golgi apparatus. *Eur. J. Cell Biol.* 51:189–200
Reeves JE, Fried M. 1995. The surf-4 gene encodes a novel 30 kDa integral membrane protein. *Mol. Membr. Biol.* 12:201–8
Rose JK, Doms RW. 1988. Regulation of protein export from the endoplasmic reticulum. *Annu. Rev. Cell Biol.* 4:257–88
Rosenfeld MG, Marcantonio EE, Hakimi J, Ort VM, Atkinson PH, et al. 1984. Biosynthesis and processing of ribophorins in the endoplasmic reticulum. *J. Cell Biol.* 99:1076–82
Rothman JE. 1981. The Golgi apparatus: two organelles in tandem. *Science* 213:1212–19
Rothman JE. 1987. Transport of the vesicular stomatitis glycoprotein to *trans* Golgi membranes in a cell-free system. *J. Biol. Chem.* 262:12502–10
Rothman JE. 1994. Mechanisms of intracellular protein transport. *Nature* 372:55–63
Rothman JE, Orci L. 1992. Molecular dissection of the secretory pathway. *Nature* 355:409–15
Salama NR, Yeung T, Schekman RW. 1993. The Secl3p complex and reconstitution of vesicle budding from the ER with purified cytosolic proteins. *EMBO J.* 12:4073–82
Saraste J, Kuismanen E. 1984. Pre- and post-Golgi vacuoles operate in the transport of Semliki Forest virus membrane glycoproteins to the cell surface. *Cell* 38:535–49
Schimmoller F, Singer-Kruger B, Schroder S, Kruger U, Barlowe C, Riezman H. 1995. The absence of Emp24p, a component of ER-derived COP II-coated vesicles, causes a defect in transport of selected proteins to the Golgi. *EMBO J.* 14:1329–39
Schroder S, Schimmoller F, Singer-Kruger B, Riezman H. 1995. The Golgi-localization of yeast Emp47p depends on its di-lysine motif but is not affected by the retil mutation in α-COP. *J. Cell Biol.* 131:895- 912
Schutze MP, Peterson PA, Jackson MR. 1994. An N-terminal double-arginine motif maintains type II membrane proteins in the endoplasmic reticulum. *EMBO J.* 13:1696–705
Schweizer A, Rohrer J, Hauri HP, Komfeld S. 1994. Retention of p63 in an ER-Golgi intermediate compartment depends on the presence of all three of its domains and on its ability to form oligomers. *J. Cell Biol.* 126:25–39
Shin J, Dunbrack RL Jr, Lee S, Strominger JL. 1991. Signals for retention of transmembrane proteins in the endoplasmic reticulum studied with CD4 truncation mutants. *Proc. Natl. Acad. Sci. USA* 88:1918–22
Simons K, Zerial M. 1993. Rab proteins and the road maps for intracellular transport. *Neuron* 11:789–99

Sitia R, Meldolesi J. 1992. Endoplasmic reticulum: a dynamic patchwork of specialized subregions. *Mol. Biol. Cell* 3:1067–72

Slusarewicz P, Nilsson T, Hui N, Watson R, Warren G. 1994. Isolation of a matrix that binds medial Golgi enzymes. *J. Cell Biol.* 124:405–13

Stamnes MA, Craighead MW, Hoe MH, Lampen N, Geromanos S, et al. 1995. An integral membrane component of coatomer-coated transport vesicles defines a family of proteins involved in budding. *Proc. Natl. Acad. Sci. USA* 92:8011–15

Stirzaker SC, Both GW. 1989. The signal peptide of the rotavirus glycoprotein VP7 is essential for its retention in the ER as an integral membrane protein. *Cell* 56:741–47

Sudhof TC. 1995. The synaptic vesicle cycle: a cascade of protein-protein interactions. *Nature* 375:645–53

Suzuki CK, Bonifacino JS, Lin AY, Davis MM, Klausner RD. 1991. Regulating the retention of T-cell receptor alpha chain variants within the endoplasmic reticulum: Ca(2+)-dependent association with BiP. *J. Cell Biol.* 114:189–205

Sweet DJ, Pelham HR. 1992. The *Saccharomyces cerevisiae SEC20* gene encodes a membrane glycoprotein which is sorted by the HDEL retrieval system. *EMBO J.* 11:423–32

Szczesna-Skorupa E, Ahn K, Chen CD, Doray B, Kemper B. 1995. The cytoplasmic and N-terminal transmembrane domains of cytochrome P450 contain independent signals for retention in the endoplasmic reticulum. *J. Biol. Chem.* 270:24327–33

Tang BL, Wong SH, Low SH, Hong W. 1992. Retention of a type II surface membrane protein in the endoplasmic reticulum by the Lys-Asp-Glu-Leu sequence. *J. Biol. Chem.* 267:7072–76

Tisdale EJ, Bourne JR, Khosravi-Far R, Der CJ, Balch WE. 1992. GTP-binding mutants of rabl and rab2 are potent inhibitors of vesicular transport from the endoplasmic reticulum to the Golgi complex. *J. Cell Biol.* 119:749–61

Tooze SA, Tooze J, Warren G. 1988. Site of addition of *N*-acetyl-galactosamine to the El glycoprotein of mouse hepatitis virus-A59. *J. Cell Biol.* 106:1475–87

Townsley FM, Pelham HR. 1994. The KKXX signal mediates retrieval of membrane proteins from the Golgi to the ER in yeast. *Eur. J. Cell Biol.* 64:211–16

Townsley FM, Wilson DW, Pelham HR. 1993. Mutational analysis of the human KDEL receptor: distinct structural requirements for Golgi retention, ligand binding and retrograde transport. *EMBO J.* 12:2821–29

Trowbridge IS, Collawn JF, Hopkins CR. 1993. Signal-dependent membrane protein trafficking in the endocytic pathway. *Annu. Rev. Cell Biol.* 9:129–61

Wada I, Rindress D, Cameron PH, Ou WJ, Doherty JJD, et al. 1991. SSR alpha and associated calnexin are major calcium binding proteins of the endoplasmic reticulum membrane. *J. Biol. Chem.* 266:19599–610

Waters MG, Serafini T, Rothman JE. 1991. 'Coatomer': a cytosolic protein complex containing subunits of non-clathrin-coated Golgi transport vesicles. *Nature* 349:248–51

Weisz OA, Swift AM, Machamer CE. 1993. Oligomerization of a membrane protein correlates with its retention in the Golgi complex. *J. Cell Biol.* 122:1185–96

Wilson DW, Lewis MJ, Pelham HR. 1993. ph-dependent binding of KDEL to its receptor in vitro. *J. Biol. Chem.* 268:7465–68

Wiedmann M, Huth, Rapoport TA. 1984. *Xenopus* oocytes can secrete bacterial beta-lactamase. *Nature* 309:637–39

Zagouras P, Rose JK. 1989. Carboxy-terminal SEKDEL sequences retard but do not retain two secretory proteins in the endoplasmic reticulum. *J. Cell Biol.* 109:2633–40

Annu. Rev. Cell Dev. Biol. 1996. 12:55–89

AH RECEPTOR SIGNALING PATHWAYS

Jennifer V. Schmidt
Department of Molecular Pharmacology and Biological Chemistry, Northwestern University Medical School, 303 E. Chicago Avenue, Chicago, Illinois 60611

Christopher A. Bradfield
McArdle Laboratory for Cancer Research, University of Wisconsin Medical School, 1400 University Avenue, Madison, Wisconsin 53706

KEY WORDS: AHR, ARNT, bHLH-PAS, TCDD

ABSTRACT

The aryl hydrocarbon (Ah) receptor has occupied the attention of toxicologists for over two decades. Interest arose from the early observation that this soluble protein played key roles in the adaptive metabolic response to polycyclic aromatic hydrocarbons and in the toxic mechanism of halogenated dioxins and dibenzofurans. More recent investigations have provided a fairly clear picture of the primary adaptive signaling pathway, from agonist binding to the transcriptional activation of genes involved in the metabolism of xenobiotics. Structure-activity studies have provided an understanding of the pharmacology of this receptor; recombinant DNA approaches have identified the enhancer sequences through which this factor regulates gene expression; and functional analysis of cloned cDNAs has allowed the characterization of the major signaling components in this pathway. Our objective is to review the Ah receptor's role in regulation of xenobiotic metabolism and use this model as a framework for understanding the less well-characterized mechanism of dioxin toxicity. In addition, it is hoped that this information can serve as a model for future efforts to understand an emerging superfamily of related signaling pathways that control biological responses to an array of environmental stimuli.

CONTENTS

1081-0706/96/1115-0055$08.00

EARLY OBSERVATIONS

Individuals are constantly exposed to a variety of xenobiotics, among which are secondary plant metabolites, mycotoxins, venoms, pharmaceuticals, and the byproducts of industrialization. For several decades scientists have been aware that adaptive mechanisms exist to minimize toxicity from these ubiquitous environmental poisons. A metabolic response to polycyclic aromatic compounds (PAHs) was first described in the late 1950s. In these early rodent experiments, the administration of benzanthracene, benzo[a]pyrene or 3-methylcholanthrene led to the induction of a number of liver microsomal enzyme activities collectively referred to as arylhydrocarbon hydroxylase (AHH) (Conney et al 1956). This induced metabolism met the criterion of an adaptive response in that the upregulated enzymes were able to oxidize the same PAH-inducing agents upon short-term re-exposure. Similar adaptive responses were also observed for other classes of structurally unrelated xenobiotics such as phenobarbital and various pesticides. In these cases as well, initial exposures led to increased expression of microsomal and soluble enzymes with metabolic activity toward the inducing agent and the effect of reducing the pharmacological response of subsequent exposure (Snyder & Remmer 1979).

The Ah Receptor and Ahr Locus

Classical murine genetics provided initial insights into the regulation of AHH activity. First, it was observed that the inducibility of AHH activity varied significantly among inbred mouse strains, with C57 strains being highly responsive to PAHs, whereas the DBA and AKR strains were described as nonresponsive (Nebert & Gelboin 1969). Crosses and back-crosses of these strains indicated that multiple alleles at a single locus controlled inducibility of AHH. This locus initially became known as *Ah*, for aryl hydrocarbon responsiveness (Green 1973, Thomas & Hutton 1973). Although the terms responsive and nonresponsive are still widely used, their application should be limited to induction by PAHs, because halogenated aromatic hydrocarbons (HAHs) such as 2,3,7,8-tetrachlorodibenzo-*p*-dioxin (TCDD) soon were found to be orders of magnitude more potent than PAHs and were capable of eliciting AHH induction in nonresponsive strains (Poland et al 1974). These results demonstrated that nonresponsive strains were actually less responsive, requiring 10- to 100-fold higher doses of TCDD to attain the same level of enzyme induction as responsive strains.

Data from a number of laboratories led to the suggestion that a receptor existed for this large class of chemicals and that C57 mice harbored a receptor with a greater affinity for ligand than DBA mice (Poland & Glover 1975). The existence of this Ah receptor (AHR) was confirmed using radiolabeled dioxin congeners to demonstrate the existence of low-capacity, high-affinity binding sites in target tissues (Poland et al 1976). As predicted, the binding affinity for TCDD differed between mouse strains, with receptors from the responsive and nonresponsive strains displaying equilibrium dissociation constants of 6 and 37 pM, respectively (Okey et al 1989, Poland et al 1994). The segregation of these alleles, as well as structure-activity studies performed with various dioxin congeners, confirmed the existence of the AHR and its role in regulating the induction of AHH. The idea that the *Ah* locus encoded the AHR resulted in the recent renaming of this locus to *Ahr* by the Mouse Genome Nomenclature Committee (Eppig 1993).

The AHR is highly polymorphic, particularly when compared with other nuclear receptors. This polymorphism extends beyond the classical responsive and nonresponsive phenotypes described above to include significant differences in receptor primary structure. For example, marked differences in AHR molecular weight have been revealed with the use of [^{125}I]-photoaffinity ligands and antibodies (Poland et al 1986, Poland & Glover 1987). Three different *Ahr* alleles, denoted with a "b" superscript from the prototype C57BL strain, have been identified that encode high-affinity receptors in responsive strains. The allele found in C57 strains, Ahr^{b-1}, encodes a 95-kDa receptor with high affinity for agonists, whereas a 104-kDa high-affinity allele, Ahr^{b-2}, is found in most other commonly used laboratory strains such as C3H/He and BALB/c (Poland et al 1987). Several wild-mouse strains, including *Mus spretus, caroli,* and *molossinus*, harbor a third high-affinity allele, *Ahr*$^{b-3}$, encoding a 105-kDa receptor protein (Poland & Glover 1990). At present, only a single allele has been identified that encodes the low-affinity receptor in nonresponsive strains (*Ahr*^{d}) (Swanson & Bradfield 1993). This allele is denoted with a "d" superscript, from the prototype DBA strain, and encodes a receptor protein of 104 kDa (Poland & Glover 1990). The structural and functional variability of the AHR is also significant across species. Photoaffinity labeling of hepatic cytosol indicates that the AHR can vary in molecular weight by almost 30 kilodaltons, e.g. C57 mouse, 95; chicken, 101; guinea pig, 103; rabbit, 104; rat, 106; human, 106; monkey, 113; and hamster, 124 (Poland & Glover 1987). Recent cloning studies have demonstrated that this difference in molecular weight is primarily due to differences in the position of the AHR's translational termination codon, rather than differential splicing or posttranslational modification (Dolwick et al 1993a, Schmidt et al 1993, Carver et al 1994a, Poland et al 1994).

Regulation of Xenobiotic Metabolism

The receptor-mediated upregulation of xenobiotic metabolizing enzymes remains the most clearly understood aspect of AHR biology. Much of our understanding of this pathway comes from analysis of the regulatory regions controlling expression of the genes encoding the cytochrome P4501A1, 1A2, and 1B1 monooxygenases that contribute to AHH activity (Jones et al 1986, Quattrochi et al 1994; W Greenlee, personal communication). Analysis of the 5′ regulatory regions of the Cyp1A1 gene revealed the existence of a number of dioxin-responsive enhancer elements (DREs) that were required for induction and were bound by the AHR in response to ligand (Denison et al 1988a,b; Fujisawa-Sehara et al 1988). DNA footprinting of the P4501A1 promoter showed that additional regions were also necessary for transcriptional activation in this system: a TATA box, a G-box element, and two binding sites for the transcription factor NF1 (Jones & Whitlock 1990). These observations led to a model where receptor interactions with multiple upstream DREs facilitate the disruption of local chromatin structure, allowing downstream promoter elements to bind their respective factors and initiate transcription (Wu & Whitlock 1992, Okino & Whitlock 1995). Although considerable evidence indicates a role for chromatin disruption in AHR signaling, the potential for direct contacts between factors bound to upstream DREs and those bound to proximal elements through DNA looping has yet to be determined. The regulation of the Cyp1A1 gene is a useful paradigm. For example, deletion analysis of the human CYP1A2 promoter has demonstrated that two DRE regions are required for TCDD inducibility of this gene as well (Quattrochi et al 1994). Cytochrome P4501B1 is the most recently discovered TCDD-inducible P450 gene, and its promoter also appears to be regulated by a similar mechanism (Sutter et al 1994; W Greenlee, personal communication). In addition to the cytochrome P450 enzymes, other genes regulated by this mechanism are the glutathione S-transferase Ya subunit, NAD(P)H:quinone reductase 1 and 2, and class 3 aldehyde dehydrogenase (Paulson et al 1990, Favreau & Pickett 1991, Asman et al 1993, Jaiswal 1994).

AHR SIGNALING PATHWAYS

The Adaptive Response Pathway

Although these designations may be overly simplistic, it has been convenient for our laboratory to classify AHR-mediated responses as either adaptive or toxic. We define the adaptive response as being limited to the ligand-induced, DRE-driven, upregulation of genes encoding xenobiotic-metabolizing enzymes. This definition is based on the teleological argument that all genes currently known to be transcriptionally upregulated through a functional DRE encode enzymes with important roles in the metabolism of foreign chemicals. In fact, most of these enzymes appear to have metabolic activity toward PAHs. This observation supports the idea that the AHR system has evolved, at least in part, to minimize an organism's body burden of compounds with extended polycyclic aromatic structures. The reasons that such a metabolic response has arisen specifically for PAHs are still largely speculative.

The selective pressures for maintenance of this detoxification system include the possibility that polycyclic aromatic compounds with extended planarity have deleterious effects for many important cellular processes. Most notable are their capacity to act as nonspecific inhibitors of enzymatic reactions and their ability to intercalate into nuclear and mitochondrial DNA. A complication of this detoxification system is the potential for PAHs to be metabolized to electrophilic intermediates that can alkylate cellular macromolecules, leading to altered cellular function and genotoxicity. This aspect is somewhat problematic given the observation that many AHR-induced enzymes can actually contribute to the generation of these electrophiles.

The Toxic Response Pathway

Our designation of the toxic response pathway is based on the observation that many receptor-mediated effects have a negative impact on the exposed individual and are inconsistent with an adaptive response. It is also a useful classification for receptor-mediated responses that make little sense teleologically or have an unknown mechanism. Many of the agonists of the AHR are familiar environmental toxicants (Skene et al 1989). PAHs are generated from

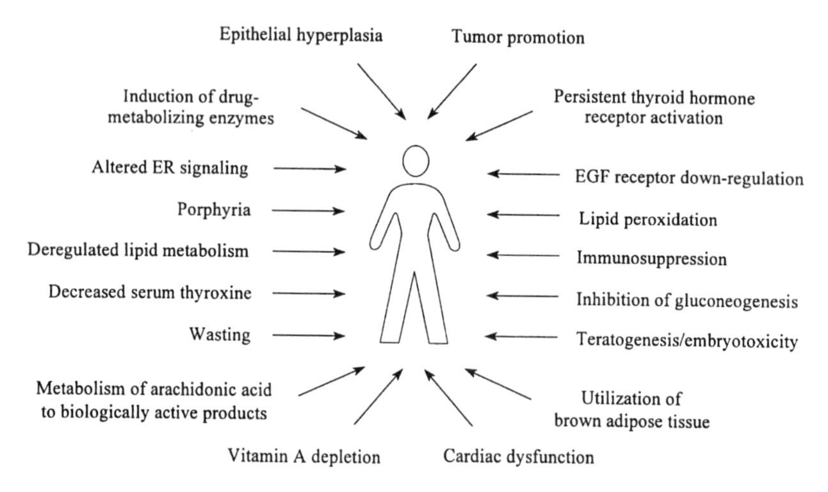

Figure 1 Biological responses to TCDD. A wide variety of cellular processes have been shown to be affected by TCDD.

the incomplete pyrolysis of various carbon sources and are typically found in diesel exhaust, cigarette smoke, and charbroiled foods. HAHs such as TCDD are trace contaminants in industrial processes that involve chlorination in the presence of phenolic substrates. Although both classes of compounds have carcinogenic potential, in recent years HAHs have received greater attention due to their environmental persistence and remarkably acute toxicity.

By the late 1970s, it became clear that in addition to its role in regulating AHH activity, the AHR was also the primary mediator of the toxicity of halogenated dioxins and dibenzofurans. Subsequently, attempts were made to identify the most sensitive endpoints of dioxin exposure and develop estimates of safe environmental levels for these compounds. The proof that a given biological endpoint is mediated by the AHR is twofold. First, the structure-activity studies should indicate that the rank-order potency of a congener to generate the endpoint of interest correlates with agonist-binding affinity for the AHR. Second, the responsiveness of the biological endpoint to agonist should segregate with the Ahr^{b} and Ahr^{d} loci in mice. A surprising number of deleterious biological responses have been shown to result from TCDD exposure, and many have met the above criteria (Figure 1) (Poland & Knutson 1982, Pohjanvirta & Tuomisto 1994 and references therein).

The fact that the same experimental proof has been used to demonstrate the role of the AHR in both adaptive and toxic responses may give the impression that the two mechanisms are the same. This has not been demonstrated and seems unlikely. The best argument that the adaptive and toxic pathways differ mechanistically comes from the observation that for adaptive responses such as the induction of AHH activity, HAHs and PAHs have parallel dose-response curves (Poland & Knutson 1982). Yet, only potent Ah receptor agonists such as the halogenated dioxins are able to elicit the classic toxic responses described below. This observation suggests the existence of distinct mechanisms for the adaptive and toxic response pathways.

Because of its potency and environmental persistence, TCDD has become the prototype agonist for the study of AHR biology. As a result, the terms TCDD toxicity and AHR biology are often used interchangeably. In most cases, this is a fair simplification, because most, if not all, of the toxicity of TCDD is the result of interactions with the AHR. TCDD toxicity observed in animal models includes tumor promotion, embryotoxicity/teratogenesis, epithelial hyperplasia and metaplasia, lymphoid involution, porphyria, and a severe wasting syndrome followed ultimately by death (Poland & Knutson 1982). In most cases, the proof described above has confirmed the role of the AHR in these events. The ultimate cause of TCDD-induced lethality is unknown, and although wasting seems to be an obvious candidate, it does not appear to be a direct cause of death. When TCDD-treated rats are fed parenterally, such that normal body weight is maintained, they die at approximately the same time as TCDD-treated controls fed ad libitum (Poland & Knutson 1982). Significant species differences are seen in the spectrum of toxicity observed and in the dose of TCDD required to elicit a particular response. For example, the LD_{50} for acute TCDD exposure varies from 1 μg/kg in the guinea pig to 20 to 40 μg/kg in the rat, 70 μg/kg in the monkey, 114 μg/kg in the mouse and rabbit, and 5000 μg/kg in the hamster (Poland & Knutson 1982).

In addition to acute toxicity, TCDD has been shown to act as a potent tumor promoter in the two-stage model of liver carcinogenesis (Pitot et al 1980). TCDD may also act as a complete carcinogen because rodents demonstrate an increased incidence of specific tumors in chronic toxicity studies, without prior exposure to an experimental initiator. Rats maintained for two years on dietary TCDD showed an increased incidence of squamous cell carcinomas of the lung, hard palate/nasal turbinates, and tongue (Kociba et al 1978, Goodman & Sauer 1992). Unlike many potent carcinogens, TCDD appears to be nongenotoxic. It does not covalently bind DNA, RNA, or protein and is not mutagenic in the Ames assay (Poland & Glover 1979, Geiger & Neal 1981). Proposed mechanisms for the carcinogenic effects of TCDD include increased cytochrome P450-mediated metabolic activation of other carcinogens or endogenous compounds such as estrogen, DNA single-strand breaks resulting from lipid peroxidation, and alterations in cell proliferation through transcriptional regulation of cytokines and growth factors (Huff et al 1994).

The Endogenous Pathway

It is possible that the AHR plays a biological role that is not yet understood. For years, many researchers have entertained the hypothesis that there is an endogenous ligand of the AHR and that toxicity is the result of inappropriate activation of this endogenous pathway. Along similar lines, the AHR may have ligand-independent or constitutive functions in biological processes distinct from the more thoroughly characterized ligand-activated pathways. As is discussed in the section on *Ahr* null mouse models, it appears that the AHR is required for normal liver development and possibly for immune system function. Such evidence could support a role for the AHR in an unknown endogenous pathway or could be indicative of the importance of the adaptive pathway in defending certain cell types from environmental toxicants.

AHR-ARNT INTERACTIONS

An Overview of the bHLH Protein Class

The basic-helix-loop-helix (bHLH) motif has been described in a wide variety of transcription factors such as the mammalian proteins Myc, Max, MyoD, and E2A, and the *Drosophila* proteins Achaete-scute and Daughterless (Murre et al 1989a, Kadesch 1993) that function as sequence-specific transcriptional regulators. This motif has been demonstrated to harbor subdomains that play roles in both DNA binding (basic region) and protein dimerization (HLH) (Murre et al 1989a,b, Davis et al 1990). A feature of many bHLH proteins is the presence of a secondary dimerization surface adjacent to the HLH domain. One well-characterized example of such a secondary dimerization domain is the leucine zipper, and bHLH proteins containing this motif are called bHLH-ZIP proteins (Kadesch 1993).

The myogenic determination protein MyoD and its relatives Myogenin, Myf-5, and MRF4 are among the most widely studied members of the bHLH proteins and illustrate many of the general features of these proteins (Olson 1990, Weintraub et al 1991). The myogenic bHLH proteins were identified based on their ability to activate muscle-specific genes and induce muscle cell differentiation in nonmyogenic cells. These factors autoregulate their own expression and cross-regulate the expression of the other family members. Studies on the regulation of skeletal muscle development provide evidence for distinct roles for each of the myogenic factors in both determination and differentiation of muscle cell phenotype (Braun et al 1992, Rudnicki et al 1992, Cheng et al 1993). All four myogenic factors form heterodimers with the E12 and E47 proteins, which are alternately spliced products of the E2A gene, to generate functional DNA-binding complexes (Murre et al 1989a, Sun & Baltimore 1991, Weintraub et al 1991). Regulation of this system is maintained under different physiologic conditions not only by the complement of dimeric partners that are expressed, but also by restricting the heterodimeric pairs that may form. Key regulators of partner availability are two dominant-negative inhibitory proteins, Id1 and Id2 (Benezra et al 1990, Sun & Baltimore 1991). These proteins have been shown to interact with E12 and E47, as well as with MyoD, forming nonfunctional complexes devoid of DNA-binding ability.

The AHR and ARNT Are bHLH-PAS Proteins

Recently a number of observations have changed our understanding of the AHR. Prior to the cloning of the AHR cDNA, the only nuclear receptors known to exist in mammals were members of the steroid/thyroxin receptor superfamily (Mangelsdorf et al 1995). This led to the assumption that the AHR would ultimately be found to contain zinc-finger DNA-binding domains and have a modular structure similar to that of steroid receptors. Although the steroid receptor homology was ultimately disproven, the presumed relationship led to the finding that the AHR was bound to a dimer of the 90-kDa heat shock protein (Hsp90) (Denis et al 1988, Perdew 1988). This association was correlated with a cytosolic location of the AHR and a receptor state that binds ligand but not DNA. In a manner similar to steroid receptors, this Hsp90 interaction was shown to be destabilized by ligand binding, a process referred to as transformation (Wilhelmsson et al 1990). Correlates of the weakened Hsp90 interaction are the appearance of the AHR in the nuclear fraction of cell homogenates and a higher affinity for DRE sequences in vitro (Pongratz et al 1992).

A finding with great impact on how we think of AHR signaling was the determination that the AHR bound to DREs as part of a heterodimeric complex (Elferink et al 1990, Reyes et al 1992, Dolwick et al 1993b). Genetics experiments in somatic cells indicated that induction of AHH activity required the product of a second locus, which encodes a protein product referred to as the Ah receptor nuclear translocator or ARNT (Hoffman et al 1991). This nomenclature is based on initial reports indicating that ARNT is required for high-affinity association of the AHR with the nuclear fraction of cells upon TCDD binding. Subsequent experiments have suggested that ARNT is not required for nuclear translocation per se, but is required to generate an AHR-ARNT complex with a greater affinity for nuclear extracts upon cell disruption. Two observations support this argument. First, immunocytochemistry was performed to visualize the location of the AHR and ARNT proteins in Hepa cells in the absence and presence of ligand (Pollenz et al 1994). The unliganded AHR is found almost exclusively in the cytoplasm of the cell and treatment with ligand causes a time-dependent movement of the AHR into the nucleus. The ARNT protein, on the other hand, is found to be exclusively nuclear without respect to ligand. Thus ligand may serve to initiate translocation of the AHR to the nucleus where dimerization of these two partners can occur. Second, in ARNT-deficient cells, the AHR can still translocate to the nucleus in vivo, a process therefore independent of ARNT.

Cloning of the AHR and ARNT allowed amino acid sequence alignments, which revealed that these two proteins are similar in primary amino acid sequence to the *Drosophila* proteins Sim and Per (Citri et al 1987, Crews et al 1988, Benezra et al 1990, Hoffman et al 1991, Sun & Baltimore 1991, Burbach et al 1992, Ema et al 1992, Huang et al 1993). The homologous domain present in all four proteins has been termed the PAS domain, for Per-ARNT-Sim. In addition to the PAS motif, the AHR, ARNT, and Sim also have adjacent bHLH domains. Sim is a bHLH-PAS protein involved in the specification of cell fate during midline cell differentiation in *Drosophila* (Nambu et al 1990, 1991). Per, known to be involved in the maintenance of circadian rhythms, is the most unusual member of this family in that it does not contain a bHLH region and may function as a dominant-negative inhibitor in a manner similar to the Id proteins of the MyoD system (Benezra et al 1990). Very recently, a number of new members of the bHLH-PAS superfamily have emerged from cloning studies. The hypoxia-inducible factor 1α (HIF-1α), a regulator of cellular responses to hypoxic stress, was purified and cloned from hepatoma cells, and two additional bHLH-PAS members are the products of the *similar* and *trachealess* genes of *Drosophila* (Wang et al 1995, Isaac & Andrew 1996, Nambu et al 1996, Wilk et al 1996). Together these cloning and sequencing studies suggest that a superfamily of PAS proteins exists in a wide variety of cell types and organisms.

The homology between bHLH-PAS superfamily members is also seen at the level of gene structure. Cloning of the structural gene of the mammalian AHR (*Ahr* gene) demonstrated the presence of 11 exons over a region of greater than 35 kb of DNA (Schmidt et al 1993). The AHR is most homologous to *Drosophila* Sim in nucleotide sequence, and analysis of the structural genes of these two proteins has revealed striking conservation of intron-exon pattern despite the evolutionary divergence of their sources (Crews et al 1988, Nambu et al 1990). Within the bHLH-PAS domain, six of eight intron-exon splice junctions are conserved. These results suggest that the *Ahr* and *sim* may have arisen from a common primordial gene. Recent unpublished work from our laboratory has shown that the genomic structure of the murine HIF-1α PAS domain is highly conserved with the *Ahr* and *sim* genes as well (G Luo & C Bradfield, unpublished data).

As mentioned above, four different *Ahr* alleles can be distinguished among different inbred and wild-mouse strains (Poland et al 1987, 1994, Poland & Glover 1990, Swanson & Bradfield 1993). The cloning of the cDNAs encoding each of these alleles has allowed molecular dissection of observed biochemical differences in ligand-binding affinity between the high- and low-affinity alleles (Ema et al 1994, Poland et al 1994). In vitro translation of the receptors encoded by the Ahr^{b-1}, Ahr^{b-2}, and Ahr^{d} cDNAs allowed measurement of saturable binding by the radioligand 2-[^{125}I]iodo-7,8-dibromodibenzo-*p*-dioxin. Equilibrium dissociation constants for the AHR^{b-1} and AHR^{b-2} receptor proteins were 6 to 10 pM, whereas that for the AHR^{d} receptor was 37 pM (Poland et al 1994). The Ahr^{d} allele is most similar in sequence to the Ahr^{b-2} allele, with only three amino acid differences. Individual mutation of each of these amino acids in the Ahr^{b-2} allele demonstrated that the reduced AHR^{d} ligand affinity was the result of an $Ala_{375} \longrightarrow Val_{375}$ polymorphism. As is discussed below, this amino acid change is localized to a region of the AHR that has been demonstrated to contain the ligand-binding domain.

Given the differences in sensitivity to TCDD toxicity across species, it has been of interest to compare the primary structures of their AHRs. Cloning of the human and rat AHR cDNAs demonstrated strong N-terminal sequence conservation with the mouse protein (Dolwick et al 1993a, Carver et al 1994a). The N-terminus of the human AHR shows 100% amino acid identity with the mouse AHR in the basic region, 97% in the HLH, and 87% in the PAS domain. The rat AHR identity with the mouse is 100% in the bHLH and 96% in the PAS domain, and with the human AHR is 98% identical in the bHLH domain and 86% in the PAS domain. The C-termini of these three proteins are more divergent, however, with 60% identity between human and mouse, 61% between human and rat, and 79% between rat and mouse. The position of the termination codons varies between these three proteins and, with similar data from other mouse alleles, appears to provide a general explanation for the high degree of molecular weight polymorphism observed (Dolwick et al 1993a, Schmidt et al 1993, Carver et al 1994a, Poland et al 1994). A comparison of the amino acid sequence of the human AHR and murine receptors shows that the $Ala_{375} \longrightarrow Val_{375}$ polymorphism responsible for the reduced ligand-binding affinity of the AHR^{d} receptor is present in the human AHR as well (Dolwick et al 1993a, Ema et al 1994, Poland et al 1994). While further investigation of the significance of this amino acid change on the human AHR is needed, the agonist-binding affinity of humans may be more like that of the nonresponsive than the responsive mouse strains. Such an observation could have a significant impact on the risk assessment of these toxicants.

The primary sequence of the ARNT protein appears to be more highly conserved between species than the AHR. Cloning of a fragment of rat ARNT corresponding to nucleotides 286-1392 of human ARNT shows 98% overall amino acid identity between the two proteins (Carver et al 1994a). A cDNA-encoding mouse ARNT has been isolated from Hepa-1 cells and is highly homologous to human ARNT, with 92% amino acid identity between the two proteins overall (Li et al 1994). Although the C-terminal halves of human and mouse ARNT display somewhat greater diversity than the N-terminal halves, this difference is considerably less pronounced than that seen in the hypervariable C-terminal regions of different species of AHR.

Tissue and Cellular Localization

Northern blot analysis of RNA from eight different human tissues showed that the human AHR mRNA was present in all tissues examined (Dolwick et al 1993a). The highest expression levels were found in placenta and lung, and the lowest levels were found in kidney, brain, and skeletal muscle. Preliminary Northern blot analysis of several human tissues showed that ARNT is expressed in the liver, placenta, and chorion (Brooks et al 1989). Ribonuclease protection assays of numerous rat tissues showed that the rat AHR was highest in lung, thymus, liver, and kidney and lowest in heart and spleen (Carver et al 1994a). In contrast to the human placenta, that of rat did not exhibit a high amount of AHR mRNA. This result may reflect a true species difference in AHR expression or may result from different gestational ages of the two tissue samples. The rat ARNT message was also ubiquitously expressed and by ribonuclease protection was found to be highest in placenta, lung, and thymus and lowest in spleen, brain, and heart (Carver et al 1994a). In general, the AHR and ARNT proteins appear to be coexpressed; however, pronounced differences in relative expression levels exist between the two dimeric partners in some tissues. The non-stochiometric distribution of these two proteins may have significance for the AHR and ARNT signaling; tissues in which one protein is present in excess over the other may indicate the existence of additional dimerization partners and signaling pathways. Additionally, low levels of ARNT could decrease the sensitivity of a particular tissue to agonist despite high AHR levels.

Developmental expression of the AHR and ARNT has been demonstrated by in situ and immunohistochemical methods in C57BL/6 mouse embryos from gestational day (gd) 10–16, although mRNA and protein were not examined in all tissues (Abbott et al 1995, Abbott & Probst 1995). The AHR and ARNT were found to be present at low levels in many embryonic tissues, with markedly higher levels in certain areas. Both the AHR and ARNT are already expressed at gd 10–11, with the highest levels in heart and neuroepithelium and neuroepithelial/neural crest-derived tissues such as visceral arches and otic and optic placodes. The AHR is also expressed in facial membranous bone and Meckel's cartilage anlagen. At gd 12–13, brain and heart levels of the AHR and ARNT have decreased, and the highest levels are now found in the liver. Areas of bone formation retain high levels of the AHR, which also appears in the epithelium of gut, lung, and kidney, and along the medial edge of the palatal shelves. ARNT is also expressed at significant levels in the tongue. At gd 14–16, both the AHR and ARNT remain strongly expressed in the liver and are also high in adrenal gland and developing bone. The AHR is also expressed in the epidermis. The authors summarize that the relative levels of the two dimeric partners are coordinate overall, with several exceptions, and display the greatest expression in tissues undergoing rapid proliferation and differentiation.

Dimerization and Domain Mapping

Initial observations that the AHR bound to DNA as a heteromeric complex were provided by UV cross-linking experiments with a synthetic DRE. A bromodeoxyuridine-substituted DRE oligonucleotide could be covalently cross-linked to a TCDD-inducible complex in rat liver and visualized by gel-shift assays. SDS-PAGE analysis of this protein complex demonstrated two unique proteins of 110 and 100 kDa; only the 100-kDa protein was able to bind the AHR photoaffinity ligand (Elferink et al 1990, Gasiewicz et al 1991). Gel-shift and immunoprecipitation experiments were used to demonstrate that both the AHR and ARNT proteins were required to form a DRE-binding complex and that the formation of this complex was greatly enhanced by the addition of ligand (Whitelaw et al 1993). Cell culture systems showed that the AHR and ARNT could transactivate a reporter gene under the control of a DRE enhancer in vivo (Matsushita et al 1993, Li et al 1994, Mason et al 1994).

Several groups have analyzed AHR deletion mutants in vitro and in vivo to demonstrate the existence of discrete functional domains (Figure 2). CNBr cleavage products photoaffinity-labeled with 2-azido-3-[^{125}I]iodo-7,8-dibromodibenzo-*p*-dioxin showed that ligand bound to a fragment containing amino acids 232–334 within the PAS domain (Burbach et al 1992). Photoaffinity labeling of AHR deletion mutants confirmed that the ligand-binding region lies between residues 166 and 425 (Dolwick et al 1993b). It was demonstrated that both the bHLH and PAS domains are required for DNA-binding and thus presumably for dimerization with ARNT (Dolwick et al 1993b). Immunoprecipitation studies have allowed the dissociation of dimerization and DNA binding and showed that the basic region of ARNT is not required for dimerization, but both helix regions, and either the N-terminal or C-terminal half of the PAS domain, are essential (Reisz-Porszasz et al 1994, Dolwick et al 1936). Another group has recently confirmed these results employing chimeras of the glucocorticoid receptor DNA-binding domain with the AHR or ARNT (Whitelaw et al 1994).

Deletion constructs and DNA-binding chimeras have been used to map the transcriptional activation domains (TADs) of the AHR, ARNT, Sim, and Per. The AHR and ARNT proteins have a single TAD in their C-terminus, comprising amino acids 521–640 in the AHR and amino acids 582–774 in ARNT (Jain et al 1994, Li et al 1994, Whitelaw et al 1994, Ma et al 1995). Both TADs contain glutamine and hydrophobic residues, whereas the AHR also has acidic amino acids, and ARNT is rich in serine, proline, and threonine residues. Despite their different compositions, in our Gal4 fusion system, the TADs of the AHR and ARNT appear to be equally potent (Jain et al 1994). Full-length AHR-ARNT activation of a DRE-driven reporter plasmid, however, may be

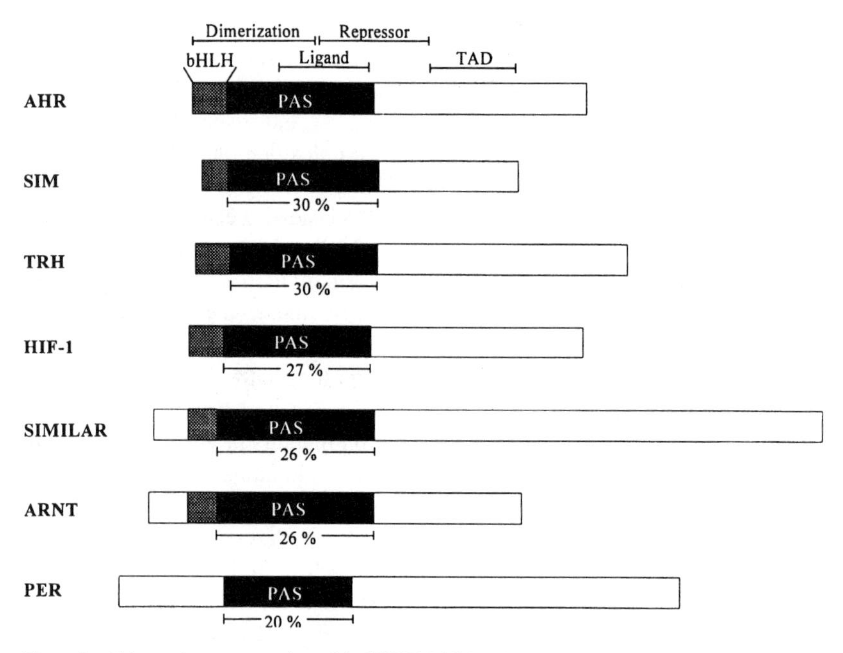

Figure 2 Schematic representation of the bHLH-PAS family proteins. The stippled areas represent the bHLH region and the black areas the PAS domain. For the AHR, the region marked Dimerization indicates bHLH and PAS sequences required for AHR-ARNT dimerization and therefore also for DNA binding. The region marked Repressor indicates the area of Hsp90 interaction, the region marked Ligand indicates the ligand-binding domain as mapped by photoaffinity labeling of deletion constructs, and the region marked TAD indicates the transactivation domain. For the other bHLH-PAS family members, percent amino acid identity to the AHR within the PAS region is indicated beneath each protein. Per does not contain a bHLH domain.

more dependent on the TAD of ARNT (Whitelaw et al 1994). In vivo, the TADs of the AHR and ARNT may synergize, because removal of the Q-rich region of ARNT did not impair AHR-ARNT dimerization but did diminish transactivation of a DRE-driven CAT reporter gene in ARNT-defective Hepa cells (Li et al 1994). Experiments have shown a strong transcriptional activation domain in the C-terminus of the Sim protein as well, whereas the Per protein was devoid of transcriptional activity (Franks & Crews 1994, Jain et al 1994, Whitelaw et al 1994).

The PAS domain of the AHR also harbors the contact region for association with Hsp90. Results from numerous laboratories have demonstrated that the AHR forms a stable complex with Hsp90 and that the region of interaction corresponds to AHR residues 340 to 422 (Denis et al 1988, Perdew 1988, Whitelaw et al 1994). Interestingly, this region colocalizes with a domain our laboratory had previously identified as imposing repression on AHR signaling (Dolwick et al 1993b). It has been hypothesized that Hsp90 functions to keep the AHR in a conformation capable of high-affinity ligand binding and represses the intrinsic DNA-binding affinity of the AHR (Pongratz et al 1992, Whitelaw et al 1994). Our laboratory and others have used a yeast expression system to study AHR-Hsp90 interactions and the role of Hsp90 in AHR signaling (Carver et al 1994b, Whitelaw et al 1995). A LexA-AHR chimera and reporter plasmid expressed in yeast were shown to be a useful model for AHR signaling in mammalian cells. This system was introduced into a yeast strain in which the level of Hsp90 could be regulated. At wild-type Hsp90 levels, normal AHR signaling was seen; however, when Hsp90 levels were decreased to 5% of wild-type, ligand-induced AHR signaling was blocked. From these results it can be concluded that Hsp90 is an essential component of the AHR-signaling pathway, and loss of Hsp90 most likely results in an improperly folded or destabilized receptor protein. More recently it has been shown that an antibody to Hsp90 can precipitate a complex of Hsp90 and the Sim protein (McGuire et al 1995). Hsp90 may thus act as a regulator of Sim activity as well.

A Model for AHR-ARNT Signaling

The work of many laboratories has combined to produce a model for AHR-ARNT signaling (Figure 3). This model holds that the unliganded AHR exists in the cytosol complexed with a dimer of Hsp90, which maintains the AHR in a ligand-binding conformation and prevents nuclear translocation and/or dimerization with ARNT. The hydrophobic AHR ligands enter the cell by diffusion and are bound by the Hsp90-associated AHR. Ligand binding causes a conformational change resulting in a receptor species with an increased affinity for DNA and a much slower rate of ligand dissociation (Bradfield et al 1988). This event is associated with nuclear translocation and an exchange of Hsp90 for

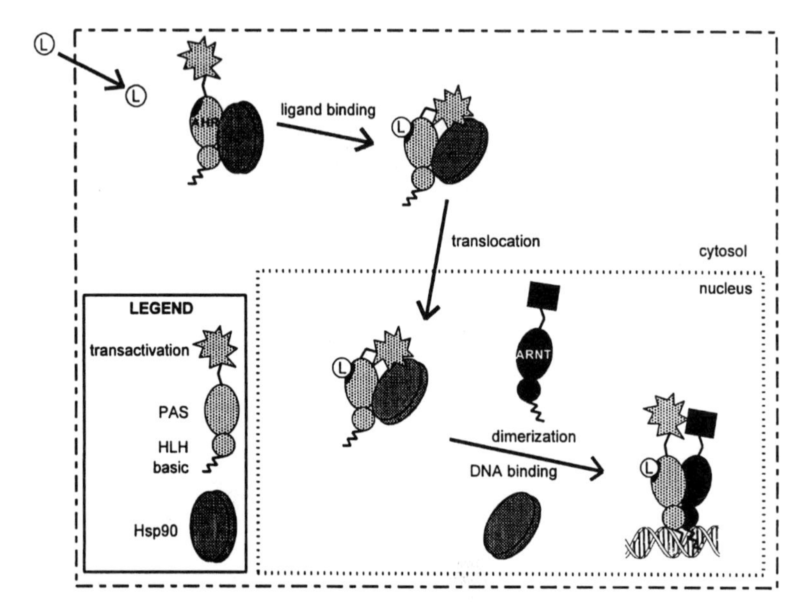

Figure 3 Model of AHR-ARNT signaling. Ligand binding by the AHR results in nuclear translocation and release of Hsp90, followed by dimerization with ARNT, and DNA binding/transactivation. The outer box represents a cell, and the inner box the cell nucleus. The AHR is indicated by the shaded partner, ARNT by the solid black partner, and Hsp90 by the shaded circle. The small circle labeled L represents the AHR ligand.

ARNT. It has been shown that in a purified system the AHR-Hsp90 complex is not dissociated by the addition of ligand (McGuire et al 1994). Addition of a cellular fraction from Hepa cells, but not ARNT-deficient Hepa mutants, can promote Hsp90 dissociation, suggesting that ARNT protein plays an active role in this process (McGuire et al 1994). Ultimately, recognition of DRE enhancer sequences by the AHR-ARNT complex results in the transactivation of target genes.

Sequence-Specific DNA Binding

Site-directed and deletion mutagenesis combined with domain swapping experiments with several bHLH proteins have defined the basic region as the DNA-binding subdomain (Davis et al 1990). This subdomain lies immediately N-terminal to the HLH domain and is generally 12 to 15 amino acids in length with two distinct clusters of basic residues (Figure 4). One of these clusters, a highly conserved ERXR sequence (where E is glutamic acid, R is arginine, and X is any amino acid), is located at a precise distance from helix 1 and is found in the basic region of all the bHLH proteins that have been shown to bind

	BASIC REGIONS	5' HALF-SITES
CAG Binding		
MyoD	ADRRKAATMRERRRL	CAG
E12	KERRVANNARERLRV	CAG
E47	RERRMANNARERVRV	CAG
AP4	RIRREIANSNERRRM	CAG
Tall	VVRRIFTNSRERWRQ	CAG
Consensus	**--RR---n-rER-RX**	
CAC Binding		
c-Myc	NVKRRTHNVLERQRR	CAC
Max	ADKRAHHNALERRRR	CAC
USF	EKRRAQHNEVERRRR	CAC
TFE3	RQKKDNHNLIERRRR	CAC
TFEB	RQKKDNHNLIERRRR	CAC
ARNT	RLARENHSEIERRRR	CAC
Consensus	**---r--Hn--ERrRR**	
TPyGC Binding		
AHR	AEGIKSNPSKRHRDR	TC/TGC
Consensus?	**----KS----R-R--**	
GTPuC Binding		
SIM	MKEKSKNAARTRRE	GTA/GC
Consensus?	**----KS--AAR-RR-**	
TAC Binding		
HIF-1	RRKEKSRDAARSRRS	TAC
SIMILAR	KRKEKSRDAARCRRS	TAC?
TRACHEALESS	LRKEKSRDAARSRRG	TAC?
Consensus	**-RKEKSRDAAR-RR-**	

Figure 4 The bHLH-PAS proteins have nonconserved basic regions. The basic regions of bHLH proteins have previously been grouped into class A with a consensus sequence ERXRX and class B with a consensus sequence ERXRR, which recognize CAG and CAC half-sites, respectively. The residue conferring site specificity is the presence of an R or a nonconserved amino acid in the fifth position. Although the ARNT protein belongs to class B group of proteins, as befits its CAC half-site, the other bHLH-PAS proteins bear little homology to other bHLH proteins in this region. We have thus grouped these proteins according to the non-E-box half-sites they recognize.

to an E-box core sequence (CANNTG) (Murre et al 1989a, Dang et al 1992). Studies have begun to examine which residues in the basic region are important for the sequence-specific DNA binding of individual bHLH proteins. MyoD binds with greatest affinity to the DNA element CAGCTG located upstream of muscle-specific genes but is unable to bind to the c-Myc binding site CACGTG. Likewise, c-Myc is unable to bind to the MyoD recognition element. X-ray crystallography studies have demonstrated that the conserved ER residues recognize the outer CA or TG nucleotides of the E-box motif (Ellenberger et al 1994). Experiments directed toward understanding the determinants for the central two E-box nucleotides have shown that the known bHLH proteins can be divided into two groups: class B proteins that have an R residue directly following the ERXR sequence (ERXRR) and recognize a CACGTG consensus element, and class A proteins that have a nonconserved residue in this position (ERXRX) and recognize a CAGCTG element (Figure 4) (Dang et al 1992).

In contrast to the classical bHLH transcription factors, which bind the symmetric E-box sequence, the AHR and ARNT bind an asymmetric recognition site. Because the core DRE consensus TNGCGTG contains one E-box half-site and one nonconsensus half-site, it was of interest to identify where each heterodimeric partner bound to the DRE. ARNT belongs to the class B group of bHLH proteins, with an ERRRR sequence in the basic region (Figure 4) (Hoffman et al 1991). Thus it could be predicted that ARNT would bind to the 3′ GTG half-site of the DRE. Because the AHR does not contain the conserved ERXR sequence, it was predicted that the AHR would contact DNA at the 5′ half of the DRE that does not conform to the E-box element. UV-cross-linking of TCDD-treated Hepa cytosolic proteins bound to variously substituted DREs allowed the use of AHR- and ARNT-specific antibodies to distinguish the ARNT protein bound to the GTG half-site and the AHR bound to the TNGC site (Bacsi et al 1995). Work from our laboratory has supported these results using an oligonucleotide selection and amplification strategy that allows the purified AHR and ARNT to select their preferred DNA-binding sequences from a pool of random oligonucleotides (Swanson et al 1995).

Analysis of the basic regions of the AHR and Sim and their asymmetric DNA recognition sequences, has resulted in their placement in a new class of bHLH proteins, designated class C (Swanson et al 1995). A comparison of basic region sequences and DNA recognition sites of previously identified bHLH-PAS proteins and several novel bHLH-PAS superfamily members has shown that assignment of these proteins to basic region homology classes is more complicated than for the other bHLH families (Figure 4). The majority of bHLH-PAS proteins, with the exception of ARNT, recognize unique sequences and show little basic region homology to the class A and B proteins. As shown in Figure 4, we have chosen to group the bHLH-PAS proteins by the DNA E-box half-sites they recognize.

Consideration of the ARNT GTG half-site suggests that an ARNT homodimer might recognize the palindromic E-box sequence CACGTG. The oligonucleotide selection and amplification strategy used by our laboratory demonstrated that the ARNT protein, in the absence of the AHR, selected a consensus-binding site conforming perfectly to this E-box sequence (Swanson et al 1995). Additionally, an ARNT homodimer recognized a CACGTG E-box in gel-shift assays and transactivated an E-box-driven reporter gene in tissue culture cells (Antonsson et al 1995, Sogawa et al 1995).

INTER- AND INTRAMOLECULAR INTERACTIONS

ARNT Dimerizes With HIF-1α In Vivo

A characteristic of many dimeric transcription factors is their capacity to participate in multiple complexes, both as homodimers and heterodimers. It appears that this may also be a feature of the bHLH-PAS superfamily of proteins, particularly the ARNT protein (Figure 5). ARNT has recently been shown to

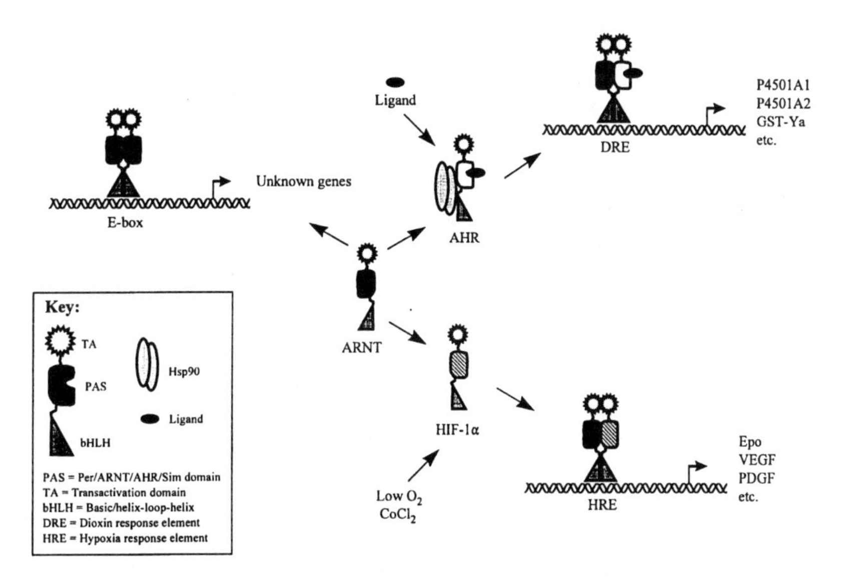

Figure 5 ARNT-mediated signaling pathways. ARNT appears to be a general dimerization partner that can homodimerize or heterodimerize with other bHLH-PAS proteins. The relative proportion of ARNT participating in these different complexes may determine the overall response of a cell to various stimuli.

interact in vivo with a recently identified bHLH-PAS protein known as HIF-1α to form the HIF-1 transcription factor complex (Wang et al 1995). HIF-1 was initially characterized as a DNA-binding activity induced in cultured human hepatocytes, or in rodent liver, under hypoxic conditions (Semenza et al 1991, Semenza & Wang 1992). This complex appears responsible for mediating the cellular response to hypoxia by transcriptional regulation of genes such as erythropoietin and other mediators of cellular oxygenation through binding to the enhancer element TACGTG (Wang & Semenza 1993, Semenza 1994). The two proteins involved in this DNA-binding complex (designated HIF-1α and HIF-1β) have now been identified as HIF-1α and ARNT (Wang & Semenza 1995), providing strong support for the hypothesis that ARNT is a general dimerization partner for multiple bHLH-PAS proteins where the identity of the partner determines the functional specificity of the complex.

Identification of the HIF-1α protein as a bHLH-PAS superfamily member and ARNT as its dimeric partner suggests that interactions might occur between the TCDD and hypoxia signaling pathways. Indeed, recent evidence from our laboratory indicates that such cross talk does occur, perhaps through the ARNT subunit or a common accessory factor (W Chan & C Bradfield, unpublished data). Gel-shift and coimmunoprecipitation experiments demonstrate that the AHR and HIF-1α compete for ARNT in vitro, with approximately equal dimerization efficiencies. Transfection of a P4501A1-driven luciferase reporter construct into the human hepatoma cell line Hep3B revealed that the HIF-1α inducer cobalt chloride can suppress TCDD-induced expression of luciferase activity. Additionally, cobalt chloride is able to down-regulate the endogenous response to TCDD in Hep3B cells because both cytochrome P4501A1 message and protein levels are suppressed.

Multiple Interactions of bHLH-PAS Proteins In Vitro

Analysis of Baculovirus-expressed ARNT has revealed that ARNT is able to homodimerize in vitro, as well as heterodimerize with Sim and Per (Sogawa et al 1995, Swanson et al 1995). Sim regulates several *Drosophila* genes through interaction with the CNS midline enhancer (CME) core sequence GTACGTG (Wharton et al 1994). This element bears some similarity to the DRE core sequence TNGCGTG, and by comparison with the DRE binding specificities of the AHR and ARNT, Sim is thought to bind the nonconsensus half-site GTAC, whereas a *Drosophila* ARNT-like protein binds the GTG half-site (Wharton et al 1994, Swanson et al 1995). This theory has been supported by the observation that a heterodimeric complex can be formed in vitro by Sim and the human ARNT protein. This complex recognizes the oligonucleotide core sequence GTGCGTG, which differs by one nucleotide from the consensus CME (Swanson et al 1995).

Although the in vivo significance of these multiple interactions is unknown, the potential physiologic importance of interactions between members of the bHLH-PAS transcription factor superfamily has been demonstrated recently by the identification of a human Sim homologue that may play a role in the cause of Down's syndrome (trisomy 21) (Chen et al 1995, Dahmane et al 1995). This human Sim protein is found in embryonic CNS and facial tissues, where it may interact with ARNT or an ARNT-like protein to regulate important developmental processes. The role of *Drosophila* Sim as a regulator of midline cell neurogenesis suggests that trisomy for the human Sim gene may be involved in the phenotype of this human disease.

The idea that bHLH-PAS proteins may participate not only in PAS-PAS interactions, but also with unrelated protein sequences, has been documented by AHR-Hsp90 interactions and more recently for the Per protein. The PAS domain of Per was shown to be involved in the formation of Per-Per homodimers and Per-Sim heterodimers (Huang et al 1993). Although these two complexes are not known to have physiological roles, their formation serves as an important model and suggests that the PAS domain is an independent dimerization surface that may cooperate with adjacent bHLH domains. This role as a secondary dimerization motif is similar to that observed for the leucine zipper found in the Myc and Max bHLH-ZIP proteins (Landschulz et al 1988, Turner & Tjian 1989). Additionally, protein interactions within the PAS domain may play a role in the regulation of pairing with other partners. In this regard, it has been shown that the PAS domain of Per can be sequestered through intramolecular interactions with regions within its own C-termini (Huang et al 1995). This idea is supported by the observation that C-terminal truncations of Per, but not full length constructs, are capable of homodimeric interactions. The PAS domain of Per has also been shown to interact in a two-hybrid system with Tim, the product of a recently identified *Drosophila* circadian rhythm gene, *timeless*. Tim is not a PAS protein and contains no other known protein dimerization motif (Gekakis et al 1995). Finally, the PAS domain may also serve as a mechanism to restrict interactions between subclasses of bHLH proteins. Although this idea is difficult to prove, interaction screens with bHLH proteins have identified members of the same subclass, suggesting restrictions exist between bHLH subclass interactions (Staudinger et al 1993).

CELLULAR MECHANISMS OF TCDD TOXICITY

TCDD Teratogenesis Results from Altered Epithelial Cell Differentiation

TCDD was shown to be a classical teratogen by the fact that it induced cleft palate in mice at doses not systemically toxic to the dam or fetus (Courtney &

Moore 1971). Mice appear to be exquisitely sensitive to induction of TCDD terata, whereas most other species display this effect only at overtly toxic doses. All species, however, display a spectrum of embryotoxicity including increased resorptions and fetal mortality and decreased fetal body weight (Couture et al 1990). The physical basis for TCDD-induced cleft palate has been studied in some detail and supports the hypothesis that TCDD acts to alter the proliferation and differentiation of epithelial tissues. Formation of the mouse secondary palate begins with the growth of the palatal shelves, which consist of mesenchyme covered by a two-cell layer of medial edge epithelium (MEE) (Fitchett & Hay 1989). Contact and fusion of the shelves occurs on embryonic day 14 to 15 and is preceded by the death and sloughing of the outer layer of MEE. Once contact occurs, the remaining basal cell MEE layer undergoes a mesenchymal transformation facilitating complete fusion (Fitchett & Hay 1989). In mice treated with TCDD, growth of the palatal shelves occurs normally, but they do not fuse, and intervening medial epithelial cells persist and adopt a stratified squamous epithelial appearance (Pratt et al 1984, Abbott & Birnbaum 1989). Induction of cleft palate has been shown to be most sensitive to TCDD given on embryonic day 10, and TCDD appears to act at this time to alter the differentiation state of the MEE, interfering with their later competency to undergo mesenchymal transformation (Pratt et al 1984, Abbott & Birnbaum 1989).

Recently, gene targeting has been used to generate a mouse line null for the TGF-β3 protein (Kaartinen et al 1995, Proetzel et al 1995). These mice display clefting of the palate resulting from a failure of the opposing palatal shelves to adhere and fuse and/or the midline epithelial seam to disappear. This mechanism of cleft palate is remarkably similar to that seen with TCDD treatment and is very different from the mechanism of other teratogens such as glucocorticoids (Pratt 1983). This similarity suggests that the AHR may be involved, directly or indirectly, in the regulation of TGF-β3 in the developing palate. TGF-β proteins have a broad range of effects on cellular proliferation and differentiation. Alternately, TGF-β3 may be positively regulated by ARNT alone or by a heterodimer of ARNT and an uncharacterized PAS protein. If such a situation exists, TCDD exposure would increase the formation of AHR-ARNT dimers, decreasing the amount of ARNT available for other interactions and resulting in decreased TGF-β3 expression. Interestingly, the expression of various isoforms of TGF-β in the palate of TCDD-treated mice has been previously investigated (Abbott & Birnbaum 1990). TCDD, when administered on embryonic day 10 to 12, decreased the expression of the TGF-β1 isoform in palatal epithelium and mesenchyme at embryonic day 14 to 16. The TGF-β3 isoform has not been examined, but future experiments may show it to be a gene whose altered expression contributes to at least one aspect of the TCDD toxic syndrome.

TCDD Immunosuppression May Result from Altered Lymphocyte Differentiation

Thymic involution and immunosuppression are the most consistently observed toxic effects of TCDD across species, occurring at doses well below those which cause systemic toxicity. Thymic atrophy has been shown to be dependent on the presence of the AHR (Poland & Glover 1980). The thymus appears to be exquisitely sensitive to TCDD because involution can be seen in mice after a single dose of 4 μg/kg TCDD, and suppression of cell-mediated immunity (CMI) is apparent using isolated lymphocytes at 4 ng/kg (Clark et al 1981). The generation of bone marrow chimeras between B6 and DBA mice demonstrates that the immune response of the grafted lymphocytes to TCDD is dependent on the genotype of the host, suggesting that the suppression of CMI in the responsive mice is dependent on a nonlymphoid tissue (Nagarkatti et al 1984). It was suggested that this tissue may be the thymic epithelium (TE), because TCDD treatment of TE cells caused altered maturation and a decreased response to mitogens such as concanavalin A or phytohemagglutinin in cocultured thymocytes (Greenlee et al 1985). The ability of several different congeners to induce this effect was shown to correlate with their affinity for the AHR (Greenlee et al 1985).

TCDD effects on humoral immunity appear to involve the suppression of B lymphocyte responses. TCDD treatment of B6 and DBA mice decreased the antibody response to sheep red blood cells and this immunosuppression segregated with the B6 genotype in B6D2F$_1$ $\times$ DBA back-crosses (Vecchi et al 1983). In mice congenic at the *Ahr* locus, TCDD and related congeners were shown to decrease antibody production in response to both T lymphocyte-dependent and T lymphocyte-independent antigens, in accordance with structure-activity relationships for AHR binding (Tucker et al 1986, Davis & Safe 1988, Kerkvliet et al 1990). Decreased antibody production resulted from a decreased number of antibody-producing cells, suggesting that TCDD inhibits B lymphocyte differentiation into plasma cells (Tucker et al 1986). Examination of the mechanism of TCDD effects on the B cell have shown that while TCDD is required early in B cell activation, it has no effect on initial proliferation but rather exerts its inhibition later at the differentiation stage (Luster et al 1988). TCDD may thus affect the B lymphocyte at a very early point in the commitment to differentiation. As is discussed below, the perturbation of cellular differentiation may play a key role in the mechanism of TCDD toxicity.

Alterations in humoral immunity appear to be one of the rare cases where some of the effects of HAHs may be independent of the AHR. For example, although subchronic treatment of B6C3F mice with TCDD suppressed the antibody response in correspondence with known structure-activity relationships, the 2,7-dichlorodibenzodioxin congener, which does not appear to bind the AHR, caused a similar immune suppression (Holsapple et al 1986b). This effect was further supported in spleen cells from *Ahr* congenic mice where in vitro exposure to 2,7-dichlorodibenzodioxin showed an equivalent effect in both strains, suggesting this compound may be acting through a mechanism independent of the AHR (Holsapple et al 1986a). Additionally, when lymphocytes from B6 and DBA mice were treated in vitro with polychlorinated dibenzofuran congeners having a range of potency varying 15,000-fold in vivo, all compounds were approximately equipotent in suppressing the antibody response in both strains of mice (Davis & Safe 1991). Although often contradictory, the weight of evidence suggests that TCDD and related HAHs may exert an inhibitory effect on the maturation or differentiation of B lymphocytes. It is still unclear, however, if the AHR is required for all or part of this effect.

MOLECULAR MECHANISMS OF TCDD TOXICITY

Possible Mechanisms for TCDD Toxicity

Despite our knowledge of AHR-regulated induction of xenobiotic metabolizing enzymes, the mechanisms by which the AHR mediates the broad spectrum of TCDD toxicity remain unknown. However, emerging knowledge of the diversity of the PAS gene superfamily along with parallels from other transcription factor families allows us to begin to speculate on possible mechanisms for these effects (Figure 6). The most evident possibility is that TCDD toxicity may be the result of persistent transcriptional activation of genes regulated by the AHR-ARNT dimeric complex. Such a mechanism would be most similar to the induction of xenobiotic-metabolizing enzymes. As yet, none of the genes known to be regulated by the AHR and ARNT proteins can explain all the diverse toxic effects seen, although additional regulated genes with distinct activities will undoubtedly be identified. Initial support for this mechanism has been provided by subtractive hybridization of a TCDD-treated human keratinocyte cell line (Sutter et al 1991). In these experiments, two RNAs were isolated, encoding interleukin-1β and plasminogen activator inhibitor-2, that were increased by TCDD treatment. TCDD toxicity might also result from hypothetical, low-affinity AHR-ARNT binding sites in the promoters of certain genes involved in toxicity. These elements may have a degenerate sequence that precludes interactions at levels of activated receptor generated by the majority of PAH-type agonists. Activation of the AHR by the potent and persistent ligand TCDD, however, could achieve a rate of receptor occupancy that allows

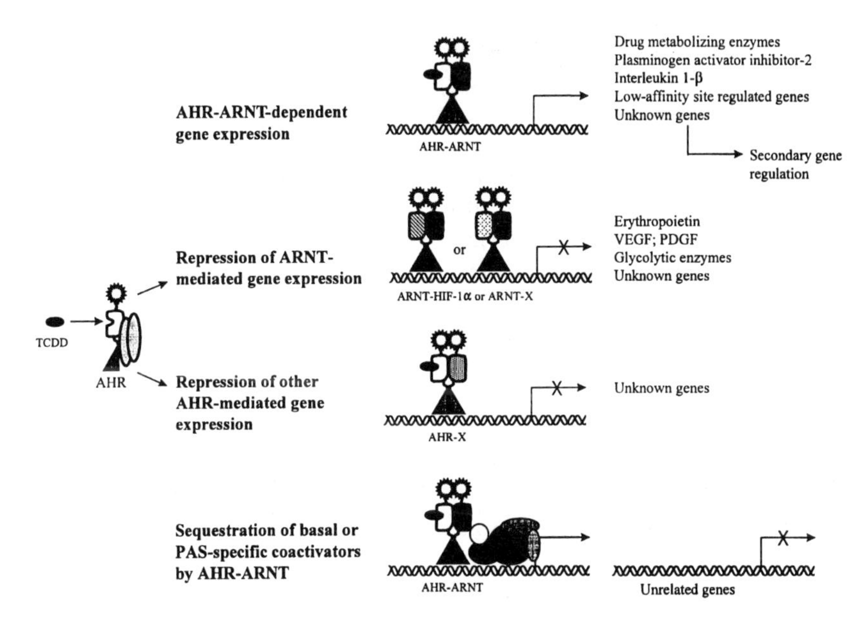

Figure 6 Possible models for the mechanism of TCDD toxicity, which probably results from alterations in gene expression induced by AHR-ARNT activity. This may be either a direct effect of the activation of AHR-ARNT-regulated genes or an indirect effect resulting from a decrease in the availability of either the AHR or ARNT to participate in different transactivation complexes.

activation from these low-affinity sites, resulting in expression of a normally restricted battery of genes. Yet another possibility is that the AHR and ARNT may recognize DRE sequences involved in transcriptional repression rather than activation, as has been shown for the glucocorticoid receptor (Sakai et al 1988). TCDD toxicity might result from altered repression of specific genes by the activated AHR-ARNT complex. Characteristics of TCDD toxicity predict that future implicated gene products, whether positively or negatively regulated by high- or low-affinity sites, will be involved in the control of cellular proliferation and differentiation.

Alternately, the mechanism of TCDD toxicity may be unrelated to direct transcriptional regulation by the AHR-ARNT complex. In fact, there is no data to date proving that ARNT is required for any of the toxic effects of dioxins. Rather, the production of this complex in response to TCDD may have repercussions for other bHLH-PAS protein pairs. Decreased concentrations of free ARNT owing to recruitment by the liganded AHR may shift the balance of this general dimeric partner away from HIF-1α or other partners, or from the formation of ARNT homodimers. In this model, TCDD toxicity could result from decreased activity of other ARNT-dependent transcription pathways. A related explanation is that the AHR may have other dimeric partners with which it regulates additional physiologic processes. Toxicity in this case could result from decreased activity of the limiting AHR with its other partners.

Emerging information about the constraints on protein dimerization has shown that these interactions need not be limited to domains of like type. For example, the zinc-finger domain of the glucocorticoid receptor has been shown to form a complex with an unrelated region of the basic/leucine-zipper (bZIP) protein Fos, preventing it from interacting with its bZIP partner Jun (Kerppola et al 1993). These considerations must be included in the types of responses that TCDD may perturb; either the AHR or ARNT may have dimeric partners outside the bHLH-PAS protein family. As mentioned above, such a phenomenon has already been shown for AHR-Hsp90 and Per-Tim interactions (Gekakis et al 1995).

Another possible mechanism for toxicity is that TCDD activation of the AHR may result in such dramatic and sustained activation of cytochrome P4501A1 and related genes that squelching of unrelated transcriptional processes occurs through saturation of coactivators. For example, overexpression of one steroid hormone receptor can diminish the activity of another, presumably through sequestration of common accessory factors or coactivator proteins (Meyer et al 1989). A similar mechanism may apply to the bHLH-PAS proteins and play a role in TCDD toxicity.

The Limited-Restricted Pleiotropic Response Model

An early model to explain TCDD toxicity proposed a limited and restricted pleiotropic response (Poland & Knutson 1982). In this model, most tissues challenged with TCDD respond with a limited response consisting primarily of the induction of xenobiotic-metabolizing enzymes (adaptive response pathway). Under certain conditions, particular tissues might be permissive for the expression of a set of genes that is normally restricted, or unexpressed. The activation of these additional genes presumably results in the characteristic toxic response to TCDD, and variations in the identity of these genes between different cell types may account for the species and tissue differences seen (toxic response pathway). This earlier model fits well with the models proposed above and recent evidence demonstrating tissue-specific differences in TCDD induction of certain genes. In human keratinocytes, for example, the plasminogen activator inhibitor-2 and TGF-α genes are induced in response to TCDD treatment; however, in TCDD-treated rat hepatocytes neither gene is induced (Choi et al 1991, Sutter et al 1991, Vanden Heuvel et al 1994). In contrast, the P4501A1 gene is strongly induced by TCDD in both cell types (Sutter et al 1991, Vanden Heuvel et al 1994). Variably expressed or inducible cell-type specific and species-specific coactivator proteins may be the molecular mechanism behind the restricted response that results in TCDD toxicity. Those species and/or cell types showing this broad spectrum response may express accessory factors that allow the activated AHR-ARNT complex to transcribe a set of genes resulting in the TCDD toxic phenotype. This type of regulation has recently been described for the POU homeodomain transcription factor Oct-1, with the identification of a B cell-specific coactivator called alternately Bob1 or OBF-1 (Gstaiger et al 1995, Strubin et al 1995). This protein interacts with Oct-1, allowing this ubiquitous transcription factor to regulate a set of B lymphocyte-restricted genes.

Ahr KNOCKOUT MICE

Despite our increased understanding of the AHR signaling pathway, fundamental questions remain concerning the endogenous function of the AHR and its role in the toxicity of TCDD. The use of gene targeting technology to inactivate murine genes in vivo (knockout mice) has been a powerful technique to elucidate protein function, confirming predicted actions in some cases while uncovering unexpected roles in others. The *Ahr* gene is an ideal candidate for targeted inactivation; *Ahr* null mice might demonstrate an unknown AHR function (endogenous pathway) and provide a valuable model system for investigation of TCDD toxicity. *Ahr* null mice have been generated independently by two groups, yielding very different phenotypes (Fernandez-Salguero et al 1995, Schmidt et al 1996). These phenotypes, as well as possible reasons for the differences between them and their implications for AHR function, are discussed below.

Our laboratory has used gene targeting to delete exon 2, which encodes the bHLH DNA-binding and dimerization domain, generating an *Ahr* null mouse line (Schmidt et al 1996). RT-PCR analysis detects the presence of a full-length alternately spliced *Ahr* transcript, lacking exon 2, produced from the targeted allele. This splicing event generates a frame-shift, and detailed Western blot, as well as functional assays, detect no AHR protein in this model system. We believe, therefore, that our $Ahr^{-/-}$ mice represent a true knockout and that the phenotype we observe results from the loss of AHR activity. Our $Ahr^{-/-}$ animals are viable and fertile; however, they exhibit a spectrum of hepatic defects suggesting that the AHR may play a previously unrecognized role in liver growth and development. $Ahr^{-/-}$ mice appear normal at birth but display slowed growth for the first few weeks of life. At 1 week of age, these animals show a dramatic yet transient liver phenotype including decreased liver weight, fatty metamorphosis, and increased residual extramedullary hematopoiesis. Detailed analysis of the decrease in liver weight has shown that this aspect of the $Ahr^{-/-}$

phenotype is present at all ages examined so far, from birth through 6 weeks. The fatty change of the liver, however, develops after birth and resolves entirely by 3 weeks of age. The residual extramedullary hematopoiesis resolves by this age as well. Older *Ahr*$^{-/-}$ mice (beyond 3 weeks of age) begin to develop mild portal hypercellularity with thickening and fibrosis, and approximately 50% of animals have enlarged spleens by 6 weeks. Although the underlying basis for this phenotype is unknown and will be the subject of much future study, we believe it may represent a hepatic developmental delay. The phenotype we have described may indicate a role for the AHR in liver growth and maturation to a functionally metabolic organ. In addition to providing additional functional roles for the AHR, these mice will serve as valuable tools to delineate receptor-mediated from nonreceptor-mediated effects of various AHR agonists.

Ahr$^{-/-}$ mice that display a quite different phenotype from that seen in our mice were first generated by Fernandez-Salguero et al (1995). This group targeted their inactivating mutation to the first exon of the *Ahr*, deleting the initiation methionine and a portion of the basic region. For convenience in contrasting the two *Ahr* null mouse lines, we have designated these mice alleles by the exon that has been deleted, i.e. the Fernandez-Salguero et al mice as *Ahr*$^{\Delta 1/\Delta 1}$ and our mice as *Ahr*$^{\Delta 2/\Delta 2}$. Demonstration of a true null allele by analysis of AHR protein or message was not included in the *Ahr*$^{\Delta 1/\Delta 1}$ report. However, these mice do not induce P4501A1 in response to TCDD. The mice display a 50% neonatal mortality rate, with inflammation of several major organ systems. Surviving *Ahr*$^{\Delta 1/\Delta 1}$ mice have decreased liver weights and portal fibrosis similar to that seen in the *Ahr*$^{\Delta 2/\Delta 2}$ mice; however, both phenotypes appear to be more severe in the *Ahr*$^{\Delta 1/\Delta 1}$ mice. Additionally, the *Ahr*$^{\Delta 1/\Delta 1}$ animals have a severely depressed immune system, with an 80% decrease in total splenic lymphoid cells at 2 weeks of age that gradually resolves over time. Despite experiments designed to specifically address these differences, we see no evidence of neonatal lethality or immune cell depletion in our *Ahr*$^{\Delta 2/\Delta 2}$ mice. The reasons underlying the phenotypic differences between the two *Ahr* null mouse lines remain unclear.

In our attempts to explain these differences, we have considered possibilities such as genetic background effects, partial inactivation of the allele, and differences in environmental factors. Because both *Ahr*$^{-/-}$ mouse lines were generated on 129 × C57BL/6 backgrounds, differences in genetic background are an unlikely cause. The possibility that the phenotypic differences result from a partial knockout in one of the mouse lines deserves consideration. We have demonstrated that the formation of any functional protein from our targeted *Ahr* allele is unlikely because we detect no AHR protein in *Ahr*$^{\Delta 2/\Delta 2}$ mice by Western blot using three domain-specific antibodies. Additionally, we have shown that our $Ahr^{\Delta2/\Delta2}$ mice are not inducible for cytochrome P4501A1 activity by TCDD. Analysis of AHR message or protein was not provided in the characterization of the $Ahr^{\Delta1/\Delta1}$ mice; however, the lack of P4501A1 induction in response to TCDD in these animals suggests that this mouse line represents a functional inactivation of the *Ahr* gene as well. A third possibility is that the differences may result from environmental factors. Two experiments argue against such differences. We have attempted to provide our $Ahr^{\Delta2/\Delta2}$ mice with an environment free of PAH compounds through the use of highly purified food and bedding material. These environmental modifications had no effect on the described pathology. Additionally, we have established $Ahr^{\Delta2/\Delta2}$ colonies at Northwestern University and the University of Chicago with no differences in $Ahr^{\Delta2/\Delta2}$ phenotype. Clarification of the differences between these two *Ahr* null mouse lines may require intercrossing and direct comparison of the $Ahr^{\Delta1/\Delta1}$ and $Ahr^{\Delta2/\Delta2}$ animals.

CONCLUSION

The AHR lies at the heart of two important toxicological problems: the regulation of xenobiotic metabolism by PAHs and the receptor-mediated toxicity of halogenated dioxins. Recent years have seen a remarkable clarification in our understanding of AHR biology. Despite these advances, our satisfaction is tempered by the realization that many of the most interesting questions remain unanswered. Why does the AHR signaling pathway exist? What are the selective pressures that have led to the conservation of this pathway in organisms from diverse environments? How does this receptor mediate toxic responses such as immune suppression, wasting, epithelial hyperplasia, cancer, terata, and death? How closely does the mechanism of xenobiotic metabolism induction mirror the pathway to toxic endpoints? An unexpected result of this research has been the realization that the AHR represents a novel mechanism for transduction of environmental signals to the nucleus. The structural similarities of the AHR with the ARNT, HIF-1α, and Per proteins suggest that this model system may yield information important to our understanding of organismal responses to environmental cues that influence biological rhythms and regulate metabolism under conditions of low oxygen.

ACKNOWLEDGMENTS

The authors thank members of the Bradfield laboratory for contributions of unpublished data and critical review of the manuscript. Support for this work was provided by the National Institutes of Health (ES06883 and CA09560) and the Pew Foundation.

Literature Cited

Abbott BD, Birnbaum LS. 1989. TCDD alters medial epithelial cell differentiation during palatogenesis. *Toxicol. Appl. Pharmacol.* 99:276–86

Abbott BD, Birnbaum LS. 1990. TCDD-induced altered expression of growth factors may have a role in producing cleft palate and enhancing the incidence of clefts after coadministration of retinoic acid and TCDD. *Toxicol. Appl. Pharmacol.* 106:418–32

Abbott BD, Birnbaum LS, Perdew GH. 1995. Developmental expression of two members of a new class of transcription factors: I. Expression of aryl hydrocarbon receptor in the C57BL/6N mouse embryo. *Dev. Dyn.* 204:133–43

Abbott BD, Probst MR. 1995. Developmental expression of two members of a new class of transcription factors: II. Expression of aryl hydrocarbon receptor nuclear translocator in the C57BL/6N mouse embryo. *Dev. Dyn.* 204:144–55

Antonsson C, Whitelaw ML, McGuire J, Gustafsson JA, Poellinger L. 1995. Distinct roles of the molecular chaperone hsp90 in modulating dioxin receptor function via the basic helix-loop-helix and PAS domains. *Mol. Cell. Biol.* 15:756–65

Asman DC, Takimoto K, Pitot HC, Dunn TJ, Lindahl R. 1993. Organization and characterization of the rat class 3 aldehyde dehydrogenase gene. *J. Biol. Chem.* 268:12530–36

Bacsi SG, Reisz-Porszasz S, Hankinson O. 1995. Orientation of the heterodimeric aryl hydrocarbon (dioxin) receptor complex on its asymmetrical DNA recognition sequence. *Mol. Pharmacol.* 47:432–38

Benezra R, Davis RL, Lockshon D, Turner DL, Weintraub H. 1990. The protein Id: a negative regulator of helix-loop-helix DNA binding proteins. *Cell* 61:49–59

Bradfield CA, Kende AS, Poland A. 1988. Kinetic and equilibrium studies of Ah receptor-ligand binding: use of [^{125}I]2-iodo-7,8-dibromodibenzo-*p*-dioxin. *Mol. Pharmacol.* 34:229–37

Braun T, Rudnicki MA, Arnold HH, Jaenisch R. 1992. Targeted inactivation of the muscle regulatory gene *Myf*-5 results in abnormal rib development and perinatal death. *Cell* 71:369–82

Brooks B, Johnson B, Heinzmann C, Monandas T, Sparkes R, et al. 1989. Localization of a gene required for the nuclear translocation of the dioxin receptor to human chromosome 1 and mouse chromosome 3 and a human RFLP with Msp1. *Am. J. Hum. Genet.* 45:A132

Burbach KM, Poland A, Bradfield CA. 1992. Cloning of the Ah receptor cDNA reveals a distinctive ligand-activated transcription factor. *Proc. Natl. Acad. Sci. USA.* 89:8185–89

Carver LA, Hogenesch JB, Bradfield CA. 1994a. Tissue specific expression of the rat Ah-receptor and ARNT mRNAs. *Nucleic Acids Res.* 22:3038–44

Carver LA, Jackiw V, Bradfield CA. 1994b. The 90-kDa heat shock protein is essential for Ah receptor signaling in a yeast expression system. *J. Biol. Chem.* 269:30109–12

Chen H, Chrast R, Rossier C, Gos A, Antonarakis S, et al. 1995. Single minded and Down's syndrome? *Nature Genet.* 10:9–10

Cheng T, Wallace MC, Merlie JP, Olson EN. 1993. Separable regulatory elements governing myogenin transcription in mouse embryogenesis. *Science* 261:215–18

Choi EJ, Toscano DG, Ryan JA, Riedel R, Toscano WA Jr. 1991. Dioxin induces transforming growth factor-α in human keratinocytes. *J. Biol. Chem.* 266:9591–97

Citri Y, Colot HV, Jacquier AC, Yu Q, Hall JC, et al. 1987. A family of unusually spliced biologically active transcripts encoded by a *Drosophila* clock gene. *Nature* 326:42–47

Clark DA, Gauldie J, Szewczuk MR, Sweeney G. 1981. Enhanced suppressor cell activity as a mechanism of immunosuppression by 2,3,7,8-tetrachlorodibenzo-*p*-dioxin. *Proc. Soc. Exp. Biol. Med.* 168:290–99

Conney AH, Miller EC, Miller JA. 1956. The metabolism of methylated aminoazo dyes. V. Evidence for induction of enzyme synthesis in the rat by 3-methylcholanthrene. *Cancer Res.* 16:450–59

Courtney DK, Moore JA. 1971. Teratology studies with 2,4,5-trichlorophenoxyacetic acid and 2,3,7,8-tetrachlorodibenzo-*p*-dioxin. *Toxicol. Appl. Pharmacol.* 20:396–403

Couture LA, Abbott BD, Birnbaum LS. 1990. A critical review of the developmental toxicity and teratogenicity of 2,3,7,8-tetrachlorodibenzo-*p*-dioxin: recent advances toward understanding the mechanism. *Teratology* 42:619–27

Crews ST, Thomas JB, Goodman CS. 1988. The Drosophila *single-minded* gene encodes a nuclear protein with sequence similarity to the *per* gene product. *Cell* 52:143–51

Dahmane N, Charron G, Lopes C, Yaspo M-L, Maunoury C, et al. 1995. Down's syndrome-critical region contains a gene homologous to *Drosophila sim* expressed during rat and human central nervous system development. *Proc. Natl. Acad. Sci. USA* 92:9191–95

Dang CV, Dolde C, Gillison ML, Kato GJ. 1992. Discrimination between related DNA sites by a single amino acid residue of Myc-related basic-helix-loop-helix proteins. *Proc. Natl. Acad. Sci. USA* 89:599–602

Davis D, Safe S. 1988. Immunosuppressive activities of polychlorinated dibenzofuran congeners: quantitative structure-activity relationships and interactive effects. *Toxicol. Appl. Pharmacol.* 94:141–49

Davis D, Safe S. 1991. Halogenated aryl hydrocarbon-induced suppression of the in vitro plaque-forming cell response to sheep red blood cells is not dependent upon the Ah receptor. *Immunopharmacology* 21:183–90

Davis RL, Cheng P-F, Lassar AB, Weintraub H. 1990. The MyoD DNA binding domain contains a recognition code for muscle-specific gene activation. *Cell* 60:733–46

Denis M, Cuthill S, Wikstrom AC, Poellinger L, Gustafsson JA. 1988. Association of the dioxin receptor with the M_r 90,000 heat shock protein: a structural kinship with the glucocorticoid receptor. *Biochem. Biophys. Res. Commun.* 155:801–7

Denison MS, Fisher JM, Whitlock JP Jr. 1988a. The DNA recognition site for the dioxin-Ah receptor complex. *J. Biol. Chem.* 263:17221–24

Denison MS, Fisher JM, Whitlock JP Jr. 1988b. Inducible, receptor-dependent protein-DNA interactions at a dioxin-responsive transcriptional enhancer. *Proc. Natl. Acad. Sci. USA* 85:2528–32

Dolwick KM, Schmidt JV, Carver LA, Swanson HI, Bradfield CA. 1993a. Cloning and expression of a human Ah receptor cDNA. *Mol. Pharmacol.* 44:911–17

Dolwick KM, Swanson HI, Bradfield CA. 1993b. In vitro analysis of Ah receptor domains involved in ligand-activated DNA recognition. *Proc. Natl. Acad. Sci. USA* 90:8566–70

Elferink CJ, Gasiewicz TA, Whitlock JP Jr. 1990. Protein-DNA interactions at a dioxin-responsive enhancer. Evidence that the transformed Ah receptor is heteromeric. *J. Biol. Chem.* 265:20708–12

Ellenberger T, Fass D, Arnaud M, Harrison S. 1994. Crystal structure of transcription factor E47: E-box recognition by a basic region helix-loop-helix dimer. *Genes Dev.* 8:970–80

Ema M, Ohe N, Suzuki M, Mimura J, Sogawa K, et al. 1994. Dioxin binding activities of polymorphic forms of mouse and human aryl hydrocarbon receptors. *J. Biol. Chem.* 269:27337-43

Ema M, Sogawa K, Watanabe N, Chujoh Y, Matsushita N, et al. 1992. cDNA cloning and structure of mouse putative Ah receptor. *Biochem. Biophys. Res. Commun.* 184:246–53

Eppig JT. 1993. Mouse nomenclature. *Mouse Genome* 91:8

Favreau LV, Pickett CB. 1991. Transcriptional regulation of the rat NAD(P)H:quinone reductase gene. Identification of regulatory elements controlling basal level expression and inducible expression by planar aromatic compounds and phenolic anitoxidants. *J. Biol. Chem.* 266:4556–61

Fernandez-Salguero P, Pineau T, Hilbert DM, McPhail T, Lee SST, et al. 1995. Immune system impairment and hepatic fibrosis in mice lacking the dioxin-binding Ah receptor. *Science* 268:722–26

Fitchett JE, Hay ED. 1989. Medial edge epithelium transforms to mesenchyme after embryonic palatal shelves fuse. *Dev. Biol.* 131:455–74

Franks RG, Crews ST. 1994. Transcriptional activation domains of the single-minded bHLH protein are required for CNS midline cell development. *Mech. Dev.* 45:269–77

Fujisawa-Sehara A, Yamane M, Fujii-Kuriyama Y. 1988. A DNA-binding factor specific for xenobiotic responsive elements of P-450c gene exists as a cryptic form in cytoplasm: its possible translocation to the nucleus. *Proc. Natl. Acad. Sci. USA* 85:5859–63

Gasiewicz TA, Elferink CJ, Henry EC. 1991. Characterization of multiple forms of the Ah receptor: recognition of a dioxin-responsive enhancer involves heteromer formation. *Biochemistry* 30:2909–16

Geiger LE, Neal RA. 1981. Mutagenicity testing of 2,3,7,8-tetrachlorodibenzo-*p*-dioxin in histidine auxotrophs of *Salmonella typhimurium*. *Toxicol. Appl. Pharmacol.* 59:125–29

Gekakis N, Saez L, Delahaye-Brown A, Myers MP, Sehgal A, et al. 1995. Isolation of *timeless* by PER protein interaction: defective interaction between *timeless* protein and long-period mutant PER[L]. *Science* 270:811–15

Goodman DG, Sauer RM. 1992. Hepatotoxicity and carcinogenicity in female Sprague-Dawley rats treated with 2,3,7,8-tetrachlorodibenzo-*p*-dioxin (TCDD): a pathology working group reevaluation. *Reg. Pharmacol. Toxicol.* 15:245–52

Green MC. 1973. Nomenclature of genetically determined biochemical variants in mice. *Biochem. Genet.* 9:369–74

Greenlee WF, Dold KM, Irons RD, Osborne R. 1985. Evidence for direct action of 2,3,7,8-tetrachlorodibenzo-*p*-dioxin (TCDD) on thymic epithelium. *Toxicol. Appl. Pharmacol.* 79:112–20

Gstaiger M, Knoepfel L, Georgiev O, Schaffner W, Hovens CM. 1995. A B-cell coactivator of octamer-binding transcription factors. *Nature* 373:360–62

Hoffman EC, Reyes H, Chu FF, Sander F, Conley LH, et al. 1991. Cloning of a factor required for activity of the Ah (dioxin) receptor. *Science* 252:954–58

Holsapple MP, Dooley RK, McNerney PJ, McCay JA. 1986a. Direct suppression of antibody responses by chlorinated dibenzodioxins in cultured spleen cells from (C57BL/6 × C3H)F1 and DBA/2 mice. *Immunopharmacology* 12:175–86

Holsapple MP, McCay JA, Barnes DW. 1986b. Immunosuppression without liver induction by subchronic exposure to 2,7-dichlorordibenzo-*p*-dioxin in adult female B6C3F1 mice. *Toxicol. Appl. Pharmacol.* 83:445–55

Huang ZJ, Curtin KD, Rosbash M. 1995. PER protein interactions and temperature compensation of a circadian clock in *Drosophila*. *Science* 267:1169–72

Huang ZJ, Edery I, Rosbash M. 1993. PAS is a dimerization domain common to *Drosophila* period and several transcription factors. *Nature* 364:259–62

Huff J, Lucier G, Tritscher A. 1994. Carcinogenicity of TCDD: experimental, mechanistic and epidemiologic evidence. *Annu. Rev. Pharmacol. Toxicol.* 34:343–72

Isaac DD, Andrew DJ. 1996. Tubulogenesis in *Drosophila*: a requirement for the *trachealess* gene product. *Genes Dev.* In press

Jain S, Dolwick KM, Schmidt JV, Bradfield CA. 1994. Potent transactivation domains of the Ah receptor and the Ah receptor nuclear translocator map to their carboxyl termini. *J. Biol. Chem.* 269:31518–24

Jaiswal AK. 1994. Human NAD(P)H:quinone oxidoreductase2: gene structure, activity, and tissue-specific expression. *J. Biol. Chem.* 269:14502–8

Jones KW, Whitlock JP. 1990. Functional analysis of the transcriptional promoter for the CYP1A1 gene. *Mol. Cell. Biol.* 10:5098–105

Jones PBC, Durrin LK, Fisher JM, Whitlock JP. 1986. Control of gene expression by 2,3,7,8-tetrachlorodibenzo-*p*-dioxin. *J. Biol. Chem.* 261:6647–50

Kaartinen V, Voncken JW, Shuler C, Warburton D, Bu D, et al. 1995. Abnormal lung development and cleft palate in mice lacking TGF-β3 indicates defects of epithelial-mesenchymal interaction. *Nature Genet.* 11:415–21

Kadesch T. 1993. Consequences of heteromeric interactions among helix-loop-helix proteins. *Cell Growth Diff.* 4:49–55

Kerkvliet NI, Steppan LB, Brauner JA, Deyo JA, Henderson MC, et al. 1990. Influence of the Ah locus on the humoral immunotoxicity of 2,3,7,8-tetrachlorodibenzo-*p*-dioxin: evidence for Ah-receptor-dependent and Ah-receptor-independent mechanisms of immunosuppression. *Toxicol. Appl. Pharmacol.* 105:26–36

Kerppola TK, Luk D, Curran T. 1993. Fos is a preferential target of glucocorticoid receptor inhibition of AP-1 activity in vitro. *Mol. Cell. Biol.* 13:3782–91

Kociba R, Keyes D, Beyer J, Wade C, et al. 1978. Results of a two-year chronic toxicity and oncogenicity study of 2,3,7,8-tetrachlorodibenzo-*p*-dioxin in rats. *Toxicol. Appl. Pharmacol.* 46:279–303

Landschulz WH, Johnson PF, McKnight SL. 1988. The leucine zipper: a hypothetical structure common to a new class of DNA binding proteins. *Science* 240:1759–64

Li H, Dong L, Whitlock JP Jr. 1994. Transcriptional activation function of the mouse Ah receptor nuclear translocator. *J. Biol. Chem.* 269:28098–105

Luster MI, Germolec DR, Clark G, Wiegand G, Rosenthal GJ. 1988. Selective effects of 2,3,7,8-tetrachlorodibenzo-*p*-dioxin and corticosteroid on in vitro lymphocyte maturation. *J. Immunol.* 140:928–35

Ma Q, Dong L, Whitlock JP Jr. 1995. Transcriptional activation by the mouse Ah receptor. Interplay between multiple stimulatory and inhibitory functions. *J. Biol. Chem.* 270:12697–703

Mangelsdorf DJ, Thummel C, Beato M, Herrlich P, Schutz G, et al. 1995. The nuclear receptor superfamily: the second decade. *Cell* 83:835–39

Mason GG, Witte AM, Whitelaw ML, Antonsson C, McGuire J, et al. 1994. Purification of the DNA binding form of dioxin receptor. Role of the Arnt cofactor in regulation of dioxin receptor function. *J. Biol. Chem.* 269:4438–49

Matsushita N, Sogawa K, Ema M, Yoshida A, Fujii-Kuriyama Y. 1993. A factor binding to the xenobiotic responsive element (XRE) of P-4501A1 gene consists of at least two helix-loop-helix proteins, Ah receptor and Arnt. *J. Biol. Chem.* 28:21002–6

McGuire J, Coumailleau P, Whitelaw ML, Gustafsson J-A, Poellinger L. 1995. The basic helix-loop-helix/PAS factor sim is associated with hsp90. *J. Biol. Chem.* 270:31353–57

McGuire J, Whitelaw ML, Pongratz I, Gustafsson JA, Poellinger L. 1994. A cellular factor stimulates ligand-dependent release of hsp90 from the basic helix-loop-helix dioxin receptor. *Mol. Cell. Biol.* 14:2438–46

Meyer M-E, Gronemeyer H, Turcotte B, Bocquel M-T, Tasset D, et al. 1989. Steroid hormone receptors compete for factors that mediate their enhancer function. *Cell* 57:433–42

Murre C, McCaw PS, Baltimore D. 1989a. A new DNA binding and dimerization motif in immunoglobin enhancer binding, daughterless, MyoD and Myc proteins. *Cell* 56:777–83

Murre C, McCaw PS, Vaessin H, Caudy M, Jan LY, et al. 1989b. Interactions between heterologous helix-loop-helix proteins generate complexes that bind specifically to a common DNA sequence. *Cell* 58:537–44

Nagarkatti PS, Sweeney GD, Gauldie J, Clark DA. 1984. Sensitivity to suppression of cytotoxic T cell generation by 2,3,7,8-tetrachlorodibenzo-*p*-dioxin (TCDD) is dependent on the genotype of the murine host. *Toxicol. Appl. Pharmacol.* 72:169–76

Nambu JR, Chen W, Hu S, Crews ST. 1996. The *Drosophila melanogaster similar* bHLH-PAS gene encodes a protein related to human hypoxia-inducible factor 1-α and *Drosophila single-minded. Gene.* In press

Nambu JR, Franks RG, Hu S, Crews ST. 1990. The *single-minded* gene of Drosophila is required for the expression of genes important for the development of CNS midline cells. *Cell* 63:63–75

Nambu JR, Lewis JO, Wharton KA Jr, Crews ST. 1991. The Drosophila *single-minded* gene encodes a helix-loop-helix protein that acts as a master regulator of CNS midline development. *Cell* 67:1157–67

Nebert DW, Gelboin HV. 1969. The in vivo and in vitro induction of aryl hydrocarbon hydroxylase in mammalian cells of different species, tissues, strains, and development and hormonal states. *Arch. Biochem. Biophys.* 134:76–89

Okey AB, Vella LM, Harper PA. 1989. Detection and characterization of a low affinity form of cytosolic Ah receptor in livers of mice nonresponsive to induction of cytochrome P1-450 by 3-methylcholanthrene. *Mol. Pharmacol.* 35:823–30

Okino ST, Whitlock JP Jr. 1995. Dioxin induces localized, graded changes in chromatin structure: implications for Cyp1A1 gene transcription. *Mol. Cell. Biol.* 15:3714–21

Olson EN. 1990. MyoD family: a paradigm for development? *Genes Dev.* 4:1454–61

Paulson KE, Darnell JE Jr, Rushmore T, Pickett CB. 1990. Analysis of the upstream elements of the xenobiotic compound-inducible and positionally regulated glutathione S-transferase Ya gene. *Mol. Cell. Biol.* 10:1841–52

Perdew GH. 1988. Association of the Ah receptor with the 90-kDa heat shock protein. *J. Biol. Chem.* 263:13802–5

Pitot HC, Goldsworthy T, Campbell HA, Poland A. 1980. Quantitative evaluation of the promotion by 2,3,7,8-tetrachlorodibenzo-*p*-dioxin of hepatocarcinogenesis from diethylnitrosamine. *Cancer Res.* 40:3616–20

Pohjanvirta R, Tuomisto J. 1994. Short-term toxicity of 2,3,7,8-tetrachlorodibenzo-*p*-dioxin in laboratory animals: effects, mechanisms, and animal models. *Pharmacol. Rev.* 46:483–549

Poland A, Glover E. 1975. Genetic expression of aryl hydrocarbon hydroxylase by 2,3,7,8-tetrachlorodibenzo-*p*-dioxin: evidence for a receptor mutation in genetically non-responsive mice. *Mol. Pharmacol.* 11:389–98

Poland A, Glover E. 1979. An estimate of the maximum in vivo covalent binding of 2,3,7,8-tetrachlorodibenzo-*p*-dioxin to rat liver protein, ribosomal RNA, and DNA. *Cancer Res.* 39:3341–44

Poland A, Glover E. 1980. 2,3,7,8-Tetrachlorodibenzo-*p*-dioxin: segregation of toxicity with the *Ah* locus. *Mol. Pharmacol.* 17:86–94

Poland A, Glover E. 1987. Variation in the molecular mass of the Ah receptor among vertebrate species and strains of rats. *Biochem. Biophys. Res. Commun.* 146:1439–49

Poland A, Glover E. 1990. Characterization and strain distribution pattern of the murine Ah receptor specified by the Ah^{d} and Ah^{b-3} alleles. *Mol. Pharmacol.* 38:306–12

Poland A, Glover E, Ebetino FH, Kende AS. 1986. Photoaffinity labeling of the Ah receptor. *J. Biol. Chem.* 261:6352–65

Poland A, Glover E, Kende AS. 1976. Stereospecific, high affinity binding of 2,3,7,8-tetrachlorodibenzo-*p*-dioxin by hepatic cytosol. *J. Biol. Chem.* 251:4936–46

Poland A, Glover E, Robinson JR, Nebert DW.

1974. Genetic expression of aryl hydrocarbon hydroxylase activity: induction of monooxygenase activities and cytochrome P1-450 formation by 2,3,7,8-tetrachlorodibenzo-*p*-dioxin in mice genetically "nonresponsive" to other aromatic hydrocarbons. *J. Biol. Chem.* 249:5599–606

Poland A, Glover E, Taylor BA. 1987. The murine *Ah* locus: a new allele and mapping to chromosome 12. *Mol. Pharmacol.* 32:471–78

Poland A, Knutson JC. 1982. 2,3,7,8-Tetrachlorodibenzo-*p*-dioxin and related halogenated aromatic hydrocarbons: examination of the mechanism of toxicity. *Annu. Rev. Pharmacol. Toxicol.* 22:517–54

Poland A, Palen D, Glover E. 1994. Analysis of the four alleles of the murine aryl hydrocarbon receptor. *Mol. Pharmacol.* 46:915–21

Pollenz RS, Sattler CA, Poland A. 1994. The aryl hydrocarbon receptor and aryl hydrocarbon receptor nuclear translocator protein show distinct subcellular localizations in Hepa 1c1c7 cells by immunofluorescence microscopy. *Mol. Pharmacol.* 45:428–38

Pongratz I, Mason GG, Poellinger L. 1992. Dual roles of the 90-kDa heat shock protein hsp90 in modulating functional activities of the dioxin receptor. *J. Biol. Chem.* 267:13728–34

Pratt RM. 1983. Mechanisms of chemically-induced cleft palate. *Trends Pharmacol. Sci.* 4:160–2

Pratt RM, Dencker L, Diewert VM. 1984. 2,3,7,8-Tetrachlorodibenzo-*p*-dioxin-induced cleft palate in the mouse: evidence for alterations in palatal shelf fusion. *Teratogen. Carcinogen. Mutagen.* 4:427–36

Proetzel G, Pawlowski SA, Wiles MV, Yin M, Boivin GP, et al. 1995. Transforming growth factor-$\beta 3$ is required for secondary palate fusion. *Nature Genet.* 11:409–14

Quattrochi LC, Vu T, Tukey RH. 1994. The human CYP1A2 gene and induction by 3-methylcholanthrene. *J. Biol. Chem.* 269:6949–54

Reisz-Porszasz S, Probst MR, Fukunaga BN, Hankinson O. 1994. Identification of functional domains of the aryl hydrocarbon receptor nuclear translocator protein (ARNT). *Mol. Cell. Biol.* 14:6075–86

Reyes H, Reisz-Porszasz S, Hankinson O. 1992. Identification of the Ah receptor nuclear translocator protein (ARNT) as a component of the DNA binding form of the Ah receptor. *Science* 256:1193–95

Rudnicki MA, Braun T, Hinuma S, Jaenisch R. 1992. Inactivation of MyoD in mice leads to up-regulation of the myogenic HLH gene Myf-5 and results in apparently normal muscle development. *Cell* 71:383–90

Sakai DD, Helms S, Carlstedt-Duke J, Gustafsson J-A, Rottman FM, et al. 1988. Hormone-mediated repression: a negative glucocorticoid response element from the bovine prolactin gene. *Genes Dev.* 2:1144–54

Schmidt JV, Carver LA, Bradfield CA. 1993. Molecular characterization of the murine *Ahr* gene. Organization, promoter analysis, and chromosomal assignment. *J. Biol. Chem.* 268:22203–9

Schmidt JV, Su GH-T, Reddy JK, Simon MC, Bradfield CA. 1996. Characterization of a murine *Ahr* null allele: animal model for the toxicity of halogenated dioxins and biphenyls. *Proc. Natl. Acad. Sci. USA.* 93:6731–36

Semenza GL. 1994. Regulation of erythropoietin production. New insights into molecular mechanisms of oxygen homeostasis. (Review). *Hematol. Oncol. Clin. N. Am.* 8:863–84

Semenza GL, Nejfelt MK, Chi SM, Antonarakis SE. 1991. Hypoxia-inducible nuclear factors bind to an enhancer element located 3′ to the human erythropoietin gene. *Proc. Natl. Acad. Sci. USA* 88:5680–84

Semenza GL, Wang GL. 1992. A nuclear factor induced by hypoxia via de novo protein synthesis binds to the human erythropoietin gene enhancer at a site required for transcriptional activation. *Mol. Cell. Biol.* 12:5447–54

Skene SA, Dewhurst IC, Greenberg M. 1989. Polychlorinated dibenzo-*p*-dioxins and polychlorinated dibenzofurans: the risks to human health. A review. *Human Toxicol.* 8:173-203

Snyder R, Remmer H. 1979. Classes of hepatic microsomal mixed function oxidase inducers. *Pharmacol. Ther.* 7:203–44

Sogawa K, Nakano R, Kobayashi A, Kikuchi Y, Ohe N, et al. 1995. Possible function of Ah receptor nuclear translocator (Arnt) homodimer in transcriptional regulation. *Proc. Natl. Acad. Sci. USA* 92:1936–40

Staudinger J, Perry M, Elledge SJ, Olson EN. 1993. Interactions among vertebrate helix-loop-helix proteins in yeast using the two-hybrid system. *J. Biol. Chem.* 268:4608–11

Strubin M, Newell JW, Matthias P. 1995. OBF-1, a novel B cell-specific coactivator that stimulates immunoglobulin promoter activity through association with octamer-binding proteins. *Cell* 80:497–506

Sun X, Baltimore D. 1991. An inhibitory domain of E12 transcription factor prevents DNA binding in E12 homodimers but not E12 heterodimers. *Cell* 64:459–70

Sutter TR, Guzman K, Dold KM, Greenlee WF. 1991. Targets for dioxin: genes for plasminogen activator inhibitor-2 and interleukin-1β. *Science* 254:415–18

Sutter TR, Tang YM, Hayes CL, Wo Y-YP, Jabs EW, et al. 1994. Complete cDNA sequence of a human dioxin-inducible mRNA identifies a new gene subfamily of cytochrome P450 that maps to chromosome 2. *J. Biol. Chem.* 269:13092–99

Swanson HI, Bradfield CA. 1993. The AH-receptor: genetics, structure and function (Review). *Pharmacogenetics* 3:213–30

Swanson HI, Chan WK, Bradfield CA. 1995. DNA binding specificities and pairing rules of the Ah-receptor, ARNT and SIM. *J. Biol. Chem.* 270:26292–302

Thomas PE, Hutton JJ. 1973. Genetics of aryl hydrocarbon hydroxylase induction in mice: additive inheritance in crosses between C3H/HeJ and DBA/2J. *Biochem. Genet.* 8:249–57

Tucker AN, Vore SJ, Luster MI. 1986. Suppression of B cell differentiation by 2,3,7,8-tetrachlorodibenzo-*p*-dioxin. *Mol. Pharmacol.* 29:372–77

Turner R, Tjian R. 1989. Leucine repeats and an adjacent DNA binding domain mediate the formation of functional cFos-cJun heterodimers. *Science* 243:1689–94

Vanden Heuvel JP, Clark GC, Kohn MC, Tritscher AM, Greenlee WF, et al. 1994. Dioxin-responsive genes: examination of dose-response relationships using quantitative reverse transcriptase-polymerase chain reaction. *Cancer Res.* 54:62–68

Vecchi A, Sironi M, Canegrati MA, Recchia M, Garattini S. 1983. Immunosuppressive effects of 2,3,7,8-tetrachlorodibenzo-*p*-dioxin in strains of mice with different susceptibility to induction of aryl hydrocarbon hydroxylase. *Toxicol. Appl. Pharmacol.* 68:434–41

Wang GL, Jiang BH, Rue EA, Semenza GL. 1995. Hypoxia-inducible factor 1 is a basic-helix-loop-helix-PAS heterodimer regulated by cellular O_2 tension. *Proc. Natl. Acad. Sci. USA* 92:5510–14

Wang GL, Semenza GL. 1993. General involvement of hypoxia-inducible factor 1 in transcriptional response to hypoxia. *Proc. Natl. Acad. Sci. USA* 90:4304–8

Wang GL, Semenza GL. 1995. Purification and characterization of hypoxia-inducible factor 1. *J. Biol. Chem.* 270:1230–37

Weintraub H, Davis R, Tapscott S, Thayer M, Krause M, et al. 1991. The myoD gene family: nodal point during specification of the muscle cell lineage. *Science* 251:761–66

Wharton K Jr, Franks RG, Kasai Y, Crews ST. 1994. Control of CNS midline transcription by asymmetric E-box-like elements: similarity to xenobiotic responsive regulation. *Development* 120:3563–69

Whitelaw M, Pongratz I, Wilhelmsson A, Gustafsson J, Poellinger L. 1993. Ligand-dependent recruitment of the Arnt coregulator determines DNA recognition by the dioxin receptor. *Mol. Cell. Biol.* 13:2504–14

Whitelaw ML, Gustafsson JA, Poellinger L. 1994. Identification of transactivation and repression functions of the dioxin receptor and its basic helix-loop-helix/PAS partner factor Arnt: inducible versus constitutive modes of regulation. *Mol. Cell. Biol.* 14:8343–55

Whitelaw ML, McGuire J, Picard D, Gustafsson JA, Poellinger L. 1995. Heat shock protein hsp90 regulates dioxin receptor function in vivo. *Proc. Natl. Acad. Sci. USA* 92:4437–41

Wilhelmsson A, Cuthill S, Denis M, Wikstrom AC, Gustafsson JA, et al. 1990. The specific DNA binding activity of the dioxin receptor is modulated by the 90 kD heat shock protein. *EMBO J.* 9:69–76

Wilk R, Weizman I, Shilo B-Z. 1996. *trachealess* encodes a bHLH-PAS protein which is an inducer of tracheal cell fates in *Drosophila*. *Genes Dev.* In press

Wu L, Whitlock JP Jr. 1992. Mechanism of dioxin action: Ah receptor-mediated increase in promoter accessibility in vivo. *Proc. Natl. Acad. Sci. USA* 89:4811

Annu. Rev. Cell Dev. Biol. 1996. 12:91–128

CYTOKINE RECEPTOR SIGNAL TRANSDUCTION AND THE CONTROL OF HEMATOPOIETIC CELL DEVELOPMENT

Stephanie S. Watowich
Department of Immunology, MD Anderson Cancer Center, Box 178, 1515 Holcombe Boulevard, Houston, Texas 77030

Hong Wu, Merav Socolovsky, Ursula Klingmuller, Stefan N. Constantinescu, and Harvey F. Lodish
Whitehead Institute for Biomedical Research, Nine Cambridge Center, Cambridge, Massachusetts 02142; Department of Biology, Massachusetts Institute of Technology, Cambridge, Massachusetts 02139

KEY WORDS: cytokine receptors, hematopoiesis, signal transduction, receptor-associated diseases

ABSTRACT

The cytokine receptor superfamily is characterized by structural motifs in the exoplasmic domain and by the absence of catalytic activity in the cytosolic segment. Activated by ligand-triggered multimerization, these receptors in turn activate a number of cytosolic signal transduction proteins, including protein tyrosine kinases and phosphatases, and affect an array of cellular functions that include proliferation and differentiation. Molecular study of these receptors is revealing the roles they play in the control of normal hematopoiesis and in the development of disease.

CONTENTS

1081-0706/96/1115-0091$08.00

INTRODUCTION

This review is intended to cover the structure and function of cytokine receptors. Because this receptor family is growing rapidly, we have chosen to limit our detailed discussions principally to cytokine receptors involved in hematopoiesis, since this was the system through which they were first identified and their essential functional roles defined. Unfortunately, for space reasons, we cannot cover several members of the family with quite interesting biological properties, such as the recently identified receptor (OB-R) for leptin, the *ob* gene product (Tartaglia et al 1995). However, we believe that the hematopoietic cytokine receptors serve as model systems for other members of the family.

Members of the cytokine receptor superfamily are defined by structural criteria (Figure 1). They are type I membrane-spanning glycoproteins, with extracellular and cytoplasmic domains of varying sizes. Their extracellular regions are comprised of one or two hematopoietin-domains, each containing four conserved cysteine residues and the sequence motif WSXWS (D'Andrea et al 1989a, Gearing et al 1989, Bazan 1990, Cosman et al 1990). The cytoplasmic regions of the cytokine receptors range from GPI anchors (CNTFα receptor subunit) to tails comprising over 200 residues (Figure 1 and references therein). Although these regions lack apparent enzymatic activity, they participate in receptor signal transduction by interacting with, and activating, cytosol-localized proteins. Cytokine receptor signal transduction is initiated by ligand-mediated receptor oligomerization. Interestingly, some members of the

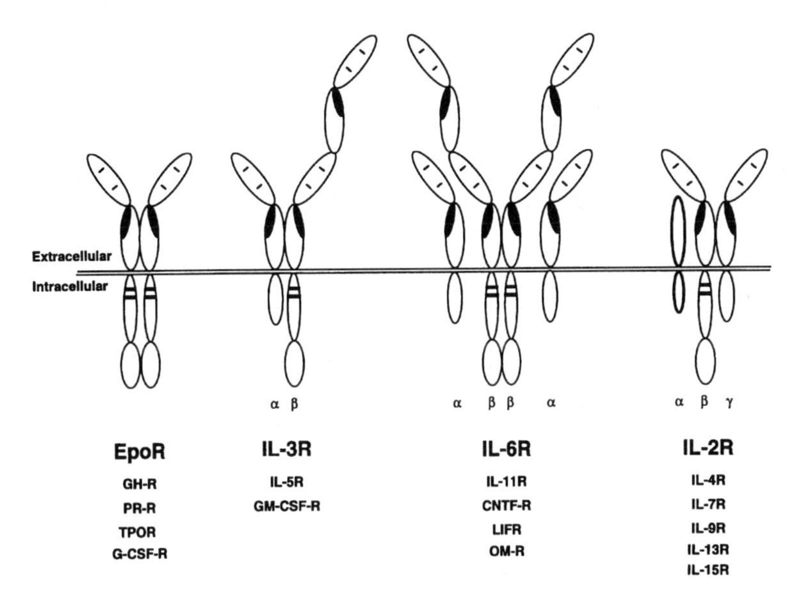

Figure 1 Schematic of the cytokine receptor superfamily. Members of the superfamily are divided into subfamilies and a representative member (indicated in bold) is drawn for each subfamily. The designation of subfamilies is based on oligomeric structure and shared components of the receptors. In the extracellular region, the hematopoietin domain is indicated by two joined, elongated ovals. Some receptors also have regions in their extracellular domains that are not related to the hematopoietin domain; these have been omitted for simplicity. Black bars in the membrane distal portion of the extracellular region represent disulfide bonds between the conserved cysteine residues, and the stippled area represents the WSXWS region; these assumptions are based on the X-ray structure of the GH-R (De Vos et al 1992). Black bars in the cytoplasmic tail of some receptor components indicate the conserved Box1/Box2 motifs (Murakami et al 1991). References for the ligand-specific subunits (α chains) are as follows: EpoR (D'Andrea et al 1989b); GH-R (Leung et al 1987); PrR (Boutin et al 1988, Davis & Linzer 1989); TpoR (Vigon et al 1992, Skoda et al 1993); G-CSF-R (Fukunaga et al 1990b); IL-3R (Kitamura et al 1991a, Hara & Miyajima 1992); IL-5R (Takaki et al 1990, Tavernier et al 1991); GM-CSF-R (Gearing et al 1989); IL-6R (Yamasaki et al 1988); IL-11R (Hilton et al 1994); CNTF-R (Davis et al 1991); LIFR (Gearing et al 1991); OM-R (Gearing et al 1992); IL-4R (Mosley et al 1989); IL-7R (Goodwin et al 1990); IL-9R (Renauld et al 1992); IL-13R (Hilton et al 1996a); IL-15R (Giri et al 1995). In the case of the IL-2R there are two ligand-specific subunits, α and β; the α subunit is not a member of the cytokine receptor superfamily (Nikaido et al 1984, Hatakeyama et al 1989b). See text in CYTOKINE RECEPTOR STRUCTURE for further references.

family are active as homodimers, whereas others are active as hetero-oligomers (Figure 1). Molecular cloning and study of the cytokine receptors have significantly increased our understanding of the biochemical events that control hematopoiesis and of diseases caused by aberrant receptor function.

BIOLOGY OF CYTOKINES AND CYTOKINE RECEPTORS

Hematopoiesis occurs continuously throughout the life of an individual. Hematopoietic stem cells, which are capable of both self-renewal and generation of multi- or single lineage-specific progenitor cells, lie at the top of this developmental hierarchy (for review, see Morrison et al 1995). Committed lineage-specific progenitor cells undergo terminal differentiation to produce at least eight major types of mature blood cells (Morrison et al 1995 and references within). Control of these diverse proliferation, differentiation, and maturation events requires cooperative actions of hematopoietic growth factors, which interact with specific receptors present on the surfaces of hematopoietic progenitor cells (Metcalf 1984, Yoshida 1994).

Purification and Cloning of Hematopoietic Growth Factors and Receptors

The identification of hematopoietic growth factors was made possible primarily by cell culture assays developed from the mid-1960s to early 1980s (Metcalf 1984). These studies revealed growth factor activities that allowed immature hematopoietic progenitor cells to survive, proliferate, and differentiate in semi-solid cultures into morphologically identifiable colonies. Because of the ability of these cells to form colonies in defined culture conditions, an operational term, colony-forming cells (CFC) or colony-forming unit (CFU), is used to describe the progenitors. For example, CFU-E, standing for colony-forming unit-erythroid, is used to describe the late erythroid progenitor cells. These culture systems also made it possible to purify and then clone the cDNAs for the hematopoietic growth factors (also known as colony-stimulating factors (CSF) or cytokines) (for reviews see Nicola 1989, Krantz 1991, Kaushansky & Karplus 1993, Kaushansky 1995). Subsequently, cDNAs for cytokine receptors were cloned, and most were found to belong to the large cytokine receptor superfamily, which now includes receptors for both hematopoietic and non-hematopoietic ligands (Bazan 1990, Cosman et al 1990, Miyajima et al 1993, Kishimoto et al 1994, Tartaglia et al 1995, Hilton et al 1996a).

In vitro culture systems also demonstrated that some growth factors support development of specific lineages, while others have profound influence on multiple blood cell lineages. For instance, erythropoietin (Epo), the primary hormone regulating the levels of circulating red blood cells, supports the growth and differentiation in vitro of erythroid cells. In contrast, interleukin-3 (IL-3) supports the growth and differentiation of multipotent progenitors, B-cell precursors, and myeloid progenitors, including those of erythroid, mast cell, and granulocyte lineages. Thus both hematopoietic cell development and hematopoietic growth factors can be arranged into a hierarchical system, with multi- or pluripotent progenitors and broadly acting growth factors at the top and lineage-specific progenitors and lineage-specific growth factors at the bottom (Metcalf 1984, Dexter & Spooncer 1987, Metcalf 1989, Nicola 1989, Morrison et al 1995). Of primary importance in understanding the molecular control of hematopoiesis are the expression patterns and functions of cytokines and their receptors.

Biological Activities of Hematopoietic Growth Factors

Cytokines and cytokine receptors support a wide array of physiological functions (Arai et al 1990, Kaushansky & Karplus 1993, Mertelsmann 1993). Despite their diverse functions in different hematopoietic lineages, one common feature is their ability to promote the survival, proliferation, or differentiation of hematopoietic progenitor cells. It appears that cytokine stimulation of cell proliferation is a process distinct from cytokine-supported cell survival (anti-apoptosis), although many of the details of these pathways remain unclear (Vaux et al 1988, Williams et al 1990). Similarly, the mechanisms by which cytokines promote or support cell differentiation are unclear, and hypotheses range from stochastic to inductive models (Metcalf 1989, Fairbairn et al 1993, Ogawa 1993, and references within). Although there are many reports supporting both models, studies on the granulocyte colony-stimulating factor receptor (G-CSF-R) provided the first evidence that a cytokine receptor can directly stimulate lineage-specific maturation events (Fukunaga et al 1993, Ziegler et al 1993). Specifically, a region in the membrane-distal portion of the G-CSF-R cytoplasmic domain is required for induction of neutrophil-specific genes or acute-phase response genes. However, this region is dispensable for cell proliferation, which is supported by the membrane-proximal region of the receptor (Fukunaga et al 1991, 1993, Ziegler et al 1993). These experiments supported an inductive role for cytokines and their receptors in hematopoietic development. Still, many questions remain. Do cytokine receptors directly signal specific differentiation events in each hematopoietic progenitor cell type? Or do activated cytokine receptors keep the cells alive so that an internal differentiation program can be expressed? A combination of molecular biological and gene targeting techniques is likely to yield answers to these questions.

Pleiotropy and Redundancy of Hematopoietic Growth Factors

All cytokines, to a greater or lesser extent, exhibit pleiotropy (multiple biological actions) and redundancy (shared biological actions). Study of cytokine receptors has revealed that redundancy may be explained by shared receptor subunits and the overlapping expression patterns of different receptors (Miyajima et al 1992, 1993, Metcalf 1993, Kishimoto et al 1994, 1995). For instance, IL-3, granulocyte macrophage colony-stimulating factor (GM-CSF), and IL-5, which share several biological activities, have hetero-oligomeric receptors with distinct ligand-binding α-subunits and a common β-subunit (Miyajima et al 1993). The β-subunit is required for signal transduction, explaining in part why these ligands have overlapping effects (Figure 1; Metcalf 1993, Miyajima et al 1993). The activation of similar signal transduction proteins by different receptors may also contribute to this phenomenon.

The pleiotropic action of a cytokine appears to be the result of expression of its specific receptor in different cell types, or its differential activity in conjunction with other cytokines. For example, GM-CSF stimulates the proliferation of a broad spectrum of progenitor cells in vitro because of the wide expression pattern of its receptor (Metcalf 1993). Similarly, interleukin-2 (IL-2) enhances B- and T-cell proliferation in vitro, in conjunction with specific antigen presented in the context of the major histocompatibility complex (Smith 1988) and can act synergistically with interferon-γ (IFN-γ) to promote the maturation of B cells (Sidman et al 1984). In addition, IL-2 directly stimulates production of IFN-γ by macrophage and NK cells (Kawase et al 1983) and augments macrophage cytotoxicity (Malkovsky et al 1987). Although many studies indicate that cytokine receptors can function in heterologous cell types (Hatakeyama et al 1989a, Yoshimura et al 1990a), the contribution of factors such as cell type-specific signal transduction proteins or other receptor/ligand-induced pathways (Wu et al 1995a) is not clear.

Animal Models: A New Era in the Study of Cytokines and Their Receptors

Because at any given time hematopoietic progenitors in different hematopoietic organs are exposed to different cytokines, and because the system of cytokines and their receptors is complex, it has been difficult to define the physiological roles of individual cytokines or receptor subunits in vivo. One approach is to study mice that are genetically deficient in specific cytokines or their receptors, either by spontaneous mutation or by targeted gene disruption (gene knock-out).

Several ligand/receptor pairs have been disrupted by targeted mutagenesis, and study of these animal models reveals their in vivo function; in some cases in vivo redundancy of cytokines and their receptors has been found (Table 1). Epo-/- and EpoR-/- mice showed similar phenotypes: severe anemia and embryonic lethality, indicating that no other ligands or receptors can replace Epo function in vivo (Wu et al 1995b). In contrast, study of mice mutant in leukemia inhibitory factor (LIF) or LIFRα; ciliary neurotrophic factor (CNTF) or CNTFRα; IL-2 or IL-2R α, β, or γ chains; and IL-7 or IL-7Rα showed that the receptor-/- mice had more severe phenotypes than ligand-/- mice (Table 1 and references therein). These results suggest that multiple ligands can activate the same receptor and generate the same signals in vivo. Indeed, besides IL-7, thymic stromal-derived lymphopoietin (TSLP) is capable of binding to the ligand-specific IL-7Rα subunit (Peschon et al 1994) and presumably subserves some of the functions of IL-7. The other factor(s) that presumably interact with the ligand-specific LIFRα or CNTFRα subunit remain to be identified.

Mice mutant in the IL-2Rγ subunit have severely reduced numbers of lymphocytes, whereas IL-2-/- mice appear to have normal in vivo immune function, although they exhibit other defects (Table 1 and references within). The IL-2Rγ subunit is also a component of the receptors for IL-4, IL-7, IL-9, and IL-15, which may explain the apparent in vivo redundancy of IL-2 and the severe effects of IL-2Rγ chain deletion (Table 1). Interestingly, deletion of the IL-2Rα or β chains results in dysregulated proliferation and activation of the lymphoid system, indicating that an essential in vivo function of the IL-2 receptor is to maintain lymphoid homeostasis (Table 1 and references within).

Both GM-CSF and IL-3 are critical for hematopoiesis in vitro. However, only a pulmonary pathological defect has been noted to date in the GM-CSF-/- mice (Dranoff et al 1994, Stanley et al 1994), indicating that GM-CSF is not essential for murine hematopoiesis in vivo. The phenotype of mice lacking IL-3, or both GM-CSF and IL-3 genes, will be of great interest, although some clues may be gleaned from the receptor-/- mice. Deletion of the β-subunit specific for the IL-3R (β_{IL-3}) had no observable effect. In contrast, mice in which β_c, the shared β-subunit for IL-3, IL-5, and GM-CSF receptors, was deleted exhibited pulmonary pathology and a decrease in the level and response of eosinophils—defects that are attributable to essential in vivo functions for GM-CSF and IL-5, and their corresponding receptors, respectively (Table 1 and references therein). These results indicate that β_{IL-3} and β_c compensate for each other in in vivo IL-3 signaling (Table 1; Nishinakamura et al 1995).

However, one must consider several points before drawing final conclusions regarding redundant cytokine and receptor functions. First, the mice used in these studies are often housed in specific pathogen-free environments, and subtle defects may not be detected. Second, some factors have both unique functions and functions that are shared with other factors, as may be the case for GM-CSF (Dranoff et al 1994, Nishinakamura et al 1995). Finally, the function of some cytokines or receptors may appear only under specific circumstances. One example is IL-6: IL-6-/- mice have essentially normal hematopoietic parameters but exhibit a slow recovery after hematopoietic ablation. This may be due to a requirement for IL-6 and/or gp130 signaling in normal stem cell

Table 1 Targeted mutations in cytokine or cytokine receptor genes

Subfamily	Ligand	Receptor	Phenotypes (homozygotes)	References
I	Epo		Embryonic lethal at E13-14; Severe anemia: complete block of fetal liver erythropoiesis and partial defect in yolk sac erythropoiesis; normal development of BFU-E and CFU-E progenitor cells.	Wu et al 1995b
		EpoR	Same as Epo knock-out.	Wu et al 1995b Lin et al 1996
	G-CSF		Viable and fertile; chronic neutropenia; deficiency in granulocyte and macrophage progenitors; impaired neutrophil mobility.	Lieschke et al 1994
		TpoR	Viable; decreased number of platelets and megakaryocytes; normal levels of other hematopoietic cell types.	Gurney et al 1994
IIa		β_{IL3}	No defects observed.	Nishinakamura et al 1995
	GM-CSF		Viable and fertile; normal basal hematopoiesis; pulmonary pathology.	Dranoff et al 1994 Stanley et al 1994
	IL-5		Viable and fertile; normal levels of B and T cells in adult mice; normal immunoglobulin levels; developmental defect in subset of B cells; decrease in basal level of eosinophils; abolished eosinophil response after infection.	Kopf et al 1996
		β_c	Viable and fertile; pulmonary pathology; lower basal numbers of eosinophils; defective eosinophilia after infection; bone marrow had no response to GM-CSF and IL-5, whereas IL-3 stimulation was normal.	Nishinakamura et al 1995
IIb	IL-6		Viable and fertile; impaired immune and inflammatory acute-phase response after tissue damage or infection; decrease in the absolute numbers of hematopoietic stem cells and early progenitors; slow recovery from hematopoietic ablation.	Kopf et al 1994, 1995 Bernad et al 1994
	LIF		Viable; decreased hematopoietic stem cells; deficient neurotransmitter switch in vitro but normal sympathetic neurons in vivo; blastocysts failed to implant in homozygous uterus.	Stewart et al 1992 Escary et al 1993 Rao et al 1993

Continues

Table 1 *Continued*

Subfamily	Ligand	Receptor	Phenotypes (homozygotes)	References
		LIF-R	Postnatal lethality; normal hematopoietic and primordial germ cell compartments but placental, skeletal, neural, and metabolic defects.	Ware et al 1995 Li M et al 1995
	CNTF		Viable and fertile; progressive atrophy and loss of motor neurons.	Masu et al 1993
		CNTF-Rα	Perinatal lethality; severe motor neuron deficits.	DeChiara et al 1995
		gp130	Embryonic lethality; myocardial and hematological disorders.	Yoshida et al 1996
III	IL-2		Viable and fertile; normal thymocyte and peripheral T-cell subset composition; normal in vivo immune responses; ulcerative colitis-like disease.	Schorle et al 1991 Kundig et al 1993 Sadlack et al 1993
		IL-2Rα[a]	Viable; normal T- and B-cell development in young mice; polyclonal expansion of lymphoid tissues and autoimmune disorders in adult mice.	Willerford et al 1995
		IL-2Rβ	Lethal by ~12 weeks; spontaneously activated T cells; dysregulated B-cell differentiation and immunoglobulin secretion; myeloproliferative disorder.	Suzuki et al 1995
		IL-2Rγ[b]	Viable and fertile; reduced absolute number of lymphocytes; lack of NK cells.	DiSanto et al 1995
	IL-4		Viable and fertile; block in Th2 cytokine responses; impaired mucosal immune responses.	Kopf et al 1993 Vajdy et al 1995
	IL-7		Viable and fertile; highly lymphopenic; block in B lymphopoiesis at the transition from pro-B to pre-B cells.	von Freeden-Jeffry et al 1995
		IL-7Rα	Viable and fertile; impaired early lymphocyte expansion; affects B-cell development at the pro-B cell stage.	Peschon et al 1994

[a]Although IL-2Rα is not a member of the cytokine receptor superfamily, it is included here because it is a component of the functional IL-2R.

[b]IL-2Rγ is also a component of the receptors for IL-4, IL-7, IL-9, and IL-15.

or pluripotent progenitor cell function (Bernad et al 1994, Yoshida et al 1996). Careful study of cytokine- or receptor-mutant mice is required to understand fully the complex functions of cytokines and their receptors. In addition, it may be necessary to generate mouse strains deficient in multiple cytokines or receptor components to understand apparent overlapping functions.

Even though most of the knock-out phenotypes are consistent with conclusions drawn from in vitro cell culture studies, there have been surprises. For instance, some primitive (yolk sac) erythropoiesis does occur in Epo-/- and EpoR-/- embryos, whereas definitive (fetal liver) erythropoiesis is completely blocked. These results suggest that another growth factor can support erythropoiesis in the yolk sac (Wu et al 1995b, Lin et al 1996). In addition, in vitro studies have indicated that IL-2 is an important T-cell growth factor. However, IL-2R α or β-deficient mice indicate that IL-2 has complex functions in controlling the immune response and maintaining lymphoid homeostasis (Suzuki et al 1995, Willerford et al 1995).

Results derived from gene deletion studies represent a new era in the study of cytokines and cytokine receptor function. Physiological roles of individual factors can be defined, and questions of functional redundancy and pleiotropy can be addressed. In combination with in vitro studies, new cytokines may be identified and cloned based on the genetic indications of the knock-out phenotypes. These animals may also be useful model systems for testing novel therapeutic modalities.

CYTOKINE RECEPTOR STRUCTURE

Extracellular Region

The extracellular domains of the cytokine receptors share a similar genomic organization, have approximately 20% amino acid identity, and are thought to form similar tertiary structures (Bazan 1990, Cosman et al 1990). The crystal structure of the growth hormone receptor (GH-R) extracellular region, which comprises one hematopoietin-like domain, demonstrates that many of the highly conserved residues maintain the three-dimensional structure of the molecule, supporting the idea that other family members adopt a similar conformation (Bazan 1990, De Vos et al 1992). GH-R has an immunoglobulin-like structure containing two subdomains, each composed of a 7-stranded β-barrel (De Vos et al 1992). The four conserved cysteine residues form two disulfide bonds in the N-terminal subdomain and appear to be required to maintain the structural integrity of this region (De Vos et al 1992). The WSXWS motif is located away from all binding interfaces. Although the function of the WSXWS motif was not clarified by the GH-R crystal structure, mutagenesis studies suggest that this region plays an essential role in folding of the receptor polypeptide and/or stabilizing its structure (Yoshimura et al 1992, Doshi & DiPersio 1994, Ronco et al 1994, Hilton et al 1995, 1996b). Analysis of the Pr-R extracellular domain structure reveals its high similarity to GH-R (Somers et al 1994). In contrast, the structure of the distantly related interferon-γ receptor (IFN-γRα), bound to IFN-γ, exhibits some interesting differences, including the orientation between the two subdomains, with implications for important distinctions in ligand-binding and receptor oligomerization (Bazan 1995, Walter et al 1995).

The GH-R structure, solved with bound GH, is also interesting because two receptor molecules are bound to a monomeric, non symmetrical ligand (De Vos et al 1992). Similar binding determinants in each receptor monomer, comprising segments from both the membrane distal and membrane proximal subdomains, interact with distinct sites (site I and site II) on GH (De Vos et al 1992). Initially Site I on GH binds to one receptor subunit on the cell surface. Subsequently Site II binds to a second receptor subunit; residues in the membrane-proximal domains of the two GH-R monomers (the dimer interface) interact with each other and stabilize the ligand-bound dimer (Cunningham et al 1991, De Vos et al 1992, Fuh et al 1992). GH mutants defective in Site I cannot bind to cell surface receptors, even though Site II is intact. In contrast, GH mutants defective in Site II are antagonists; they bind via Site I to cell surface receptors, are unable to form signaling receptor dimers, and prevent wild-type GH from binding. Recently Philo et al (1996) showed that the soluble extracellular domain of the EpoR forms a dimer that binds a single Epo molecule, similar to the situation for the GH-R/GH complex.

Cytoplasmic Region

At present there is no published information on the three-dimensional structure of the cytoplasmic region of cytokine receptors, although elucidation will be of interest because it should yield insight into their specific downstream signaling activities. Most of the information on structure/function relationships has been derived from receptor mutagenesis studies. As previously mentioned, the cytoplasmic tails of the cytokine receptors are of various lengths and their sequences are not well conserved, with the exception of Box1 and Box2 regions in the membrane-proximal portion of several receptors (see Murakami et al 1991 and references in Figure 1). This area has been defined as the minimal domain required to support proliferation of cultured transformed hematopoietic cells, and it is thought to be the binding site of essential signaling molecules such as the JAK kinases. Other common functional features are tyrosine residues present in the cytoplasmic tail of the signaling receptor subunits; many phosphorylated tyrosines provide specific docking sites for SH2 domains in cytosolic regulatory proteins and thus participate in receptor signal transduction.

Other interesting functional domains have been identified in the cytoplasmic tails of specific receptors. For example, the membrane-distal portion of the G-CSF-R is required for induction of neutrophil-specific genes and appears to be necessary for full neutrophil differentiation (Fukunaga et al 1993, Dong et al 1994). A negative regulatory domain is present in the membrane-distal portion of the EpoR; when it is deleted, the receptor is able to stimulate proliferation in one tenth the normal concentration of Epo (Yoshimura et al 1990a, D'Andrea et al 1991). This region contains the binding site for the SH2 domain of the phosphatase SH-PTP1, which is involved in termination of the proliferation signal (Klingmuller et al 1995). The expectation is that other members of the cytokine receptor family will activate multiple intracellular signaling proteins, some with positive and others with negative effects, through specific regions in their cytoplasmic tails.

Cytokine Receptor Oligomerization and Activation

The oligomeric structures of the members of the cytokine receptor family are varied and complex (Figure 1; Miyajima et al 1992, 1993, Stahl & Yancopoulos 1993, Kishimoto et al 1994). Several members of the cytokine receptor family, including the GH-R, prolactin receptor (Pr-R), G-CSF-R, EpoR and thrombopoietin receptor (TpoR), form homodimers, and a variety of studies support a critical role for receptor dimerization in signal transduction (Elberg et al 1990, Fukunaga et al 1990a, Cunningham et al 1991, Rozakis-Adcock & Kelly 1991, Fuh et al 1992, Watowich et al 1992, 1994, Hooper et al 1993, Alexander et al 1995; for review see Heldin 1995). Studies of the GH-R by biochemical, biophysical, and structural techniques demonstrated that receptor dimerization is essential for signal transduction (Cunningham et al 1991, De Vos et al 1992, Fuh et al 1992). Constitutively active (hormone-independent) forms of the EpoR and TpoR were generated by retroviral transduction (EpoR) or by mutagenesis (EpoR and TpoR) (Yoshimura et al 1990a, Watowich et al 1992, 1994, Alexander et al 1995). These receptors form ligand-independent disulfide-linked homodimers because of point mutations that generate cysteine residues at specific sites in their extracellular domains (Watowich et al 1992, 1994, Alexander et al 1995). The mutant cysteine residues were introduced into the region of EpoR or TpoR that corresponds to the GH-R dimer interface segment. Thus the disulfide-linked dimers are likely to mimic the structure of the ligand-occupied wild-type receptors, explaining their constitutive activity. At least one dimeric EpoR stimulates Epo-independent erythroid differentiation as well as proliferation of hematopoietic cells (Pharr et al 1993). Certain monoclonal antibodies to the EpoR activate the receptor, stimulating erythroid cell proliferation and differentiation (Elliot et al 1995; L Giebel, personal communication). In addition, the dominant inhibitory effects of EpoR and G-CSF-R mutants, which contain deletions of all or part of the cytosolic domain, support a critical role for ligand-induced receptor dimerization in intracellular signaling (Barber et al 1994, Dong et al 1994, Watowich et al 1994).

The majority of cytokine receptors function as heterodimers or hetero-oligomers, and subfamilies exist that share common receptor signaling subunits. The three major subfamilies are termed the IL-3, the IL-6, and the IL-2 families (Figure 1; see also Table 1 and references within for in vivo functions of specific receptor subunits). As mentioned previously, receptors for IL-3, IL-5, and GM-CSF are hetero-oligomers with ligand-specific α-subunits and a common β-subunit (β_c) (Hayashida et al 1990, Devos et al 1991, Kitamura et al 1991a,b, Takaki et al 1991, Tavernier et al 1991, Hara & Miyajima 1992). An additional β-subunit ($\beta_{IL\text{-}3}$), specific for the IL-3R, is found in mice, but not in humans (Miyajima et al 1992 and references within). The α-subunits bind ligand in the absence of the β-subunit, but oligomerization with the β-subunit is required for high-affinity ligand binding and for cell proliferation in physiological concentrations of hormone (Kitamura et al 1991b, Kitamura & Miyajima 1992). The intracellular regions of both the β and α chains appear to be required for signal transduction (Sakamaki et al 1992, Takaki et al 1993, Weiss et al 1993). At present the stoichiometry of ligand, α, and β chains in the active complex is unknown; this is important to clarify in order to understand how ligand activates the receptor complex for intracellular signaling.

The high-affinity receptors for IL-6, CNTF, IL-11, LIF, and oncostatin m (OM) also have distinct and common receptor components that oligomerize in response to ligand binding to initiate signaling (Kishimoto et al 1995). The gp130 subunit is shared among all the members of this subfamily; several members also utilize the LIFRα subunit. Although the cytoplasmic tails of some of the ligand-binding α-subunits in this subfamily are dispensable for signal transduction, the cytoplasmic regions of others are required. For example, binding of IL-6 to its specific α chain (IL-6Rα) (Yamasaki et al 1988) stimulates interaction with and homodimerization of gp130 (Taga et al 1989, Hibi et al 1990, Murakami et al 1993). Because a soluble, secreted form of IL-6Rα bound to IL-6 can stimulate signaling, gp130 homodimerization is sufficient for signal transduction (Taga et al 1989, Murakami et al 1993). The IL-6R signaling complex appears to contain two molecules each of IL-6, IL-6Rα, and gp130 (Ward et al 1994).

CNTF signals through a complex consisting of the GPI-anchored CNTFRα subunit, gp130, and LIFRα (Davis et al 1991, Ip et al 1992, Davis et al 1993a,b). Similar to IL-6Rα, a soluble form of CNTFRα can signal cell proliferation. However, in this case, a heterodimer of gp130 and LIFR is necessary (Davis et al 1993a,b). gp130 is also required, together with IL-11Rα, to generate a high-affinity, signaling-competent receptor for IL-11 (Hilton et al 1994). The cytoplasmic tail of the IL-11Rα chain contains only 39 amino acids, and we do not know whether gp130 homodimerization is involved in IL-11 signaling, or if a heterodimer of IL-11Rα and gp130 is sufficient. Interestingly, the functional, high-affinity receptor for both LIF and OM is a heterodimer of LIFRα and gp130. The difference is in the order of binding and subunit assembly: LIF binds first to LIFRα, which then recruits gp130, whereas OM binds first to gp130, then recruits LIFRα (Gearing et al 1991, 1992). Thus the shared signaling receptor components used in this subfamily appear to explain functional redundancy of the respective ligands (Kishimoto et al 1995).

The functional IL-2R has three subunits: α, β, and γ (Taniguchi & Minami 1993 and references within). The IL-2Rγ subunit is also shared with the receptors for IL-4, IL-7, IL-9, and IL-15 (Kondo et al 1993, 1994, Noguchi et al 1993a, Russell et al 1993, Giri et al 1994, Kawahara et al 1994). In addition, IL-4 and IL-13 may share receptor components (Zurawski et al 1993, 1995). The cloning of the IL-13Rα subunit supported this hypothesis: The high-affinity, functional, IL-4 receptor is a hetero-oligomer, consisting of the specific IL-4R binding chain (IL-4Rα) paired with IL-2Rγ or with IL-13Rα, whereas the functional IL-13R consists of IL-13Rα and IL-4Rα (Hilton et al 1996a). Similarly, the IL-2Rβ chain appears to be part of the IL-15 receptor (Grabstein et al 1994). This sharing of several receptor subunits is reminiscent of the LIF and OM receptors (see above).

In some cases (e.g. GH-R), it is clear that ligand binding induces receptor oligomerization; it remains to be determined if this is a general feature or if some of the receptors exist as pre-formed oligomers, which are then activated upon ligand binding (Ronco et al 1994). In addition, some receptors appear to exist as both membrane-bound and soluble forms, although the functional role of the soluble receptors is unclear (Mosley et al 1989, Goodwin et al 1990). In general, understanding the oligomeric structure of the cytokine receptors will be of great importance in understanding apparent redundant functions shared by specific cytokines, the phenotypes of genetically engineered mutant mice, and the phenotypes of several diseases.

CYTOKINE RECEPTOR FUNCTION

The majority of studies to date have focused upon the role cytokine receptor signaling plays in stimulating cell proliferation. Although we understand a great deal about which regions of the receptors are required—in most cases the membrane-proximal portion of the cytoplasmic tail is sufficient—the details of which signaling pathways are activated and how they control cell division are poorly defined. In addition, very little is understood about the molecular events involved in cytokine-stimulated cell survival or differentiation. Elucidating these pathways is critical to understanding the function of cytokine receptors.

Despite the lack of a catalytically active region in the cytoplasmic tail, tyrosine phosphorylation of the cytokine receptors and cellular substrates represents a major mode of intracellular signaling, suggesting that the receptors contact and activate intracellular tyrosine kinases. Protein tyrosine kinases in the Janus family (JAK proteins) play a major role in cytokine signaling; other protein tyrosine kinases, serine/threonine kinases and lipid kinases, also appear to be involved. Receptor mutagenesis studies have delineated regions functionally important for activation of specific signaling molecules, and in some cases, the binding sites of specific signaling proteins have been mapped. Studies that assay the effects of receptor tyrosine point mutations on cell proliferation and differentiation are required to understand fully the role of receptor tyrosine phosphorylation. In addition, it is important to clarify whether other non kinase pathways play a role in cytokine signaling.

Kinase Pathways in Cytokine Signaling: JAK Kinases

Genetic complementation experiments first demonstrated a critical role for the cytoplasmically localized JAK kinases (for review, see Ziemiecki et al 1994) in interferon signal transduction (Velazquez et al 1992, Muller et al 1993, Watling et al 1993). Several findings suggest that the JAK kinases play a pivotal role in cytokine receptor signaling (Argetsinger et al 1993, Silvennoinen et al 1993, Witthuhn et al 1993, Campbell et al 1994, Dusanter-Fourt et al 1994, Johnston et al 1994, Narazaki et al 1994, Nicolson et al 1994, Rui et al 1994, Stahl et al 1994, Witthuhn et al 1994; for review, see Ihle 1995). First, they associate with and are activated by ligand-occupied receptors. Their binding site on the receptors is in a region in the membrane-proximal portion of the cytoplasmic tail that contains the conserved Box1 and Box2 segments (Witthuhn et al 1993, DaSilva et al 1994, Narazaki et al 1994, Gurney et al 1995, Lebrun et al 1995a, Tanner et al 1995, Zhao et al 1995). As previously mentioned, the Box1/Box2 region is contained within the minimal segment of receptor essential for supporting cell proliferation. Mutations in the Box1/Box2 region, which abrogate JAK association and/or activation, also prevent ligand-mediated proliferation (Miura et al 1993). Second, kinase-deficient JAK2 or JAK3 proteins are dominant inhibitors and block signaling by Epo and GH, or IL-2, respectively (Zhuang et al 1994, Franks et al 1995, Kawahara et al 1995). Third, mice lacking JAK3 (disrupted by gene targeting) exhibit immunodeficiencies due to severe defects in lymphoid cell production (Nosaka et al 1995, Thomis et al 1995); similarly, JAK3 is mutated in some patients with immunodeficiency syndromes. Finally, in *Drosophila,* a loss-of-function mutation in the JAK homologue *hopscotch* results in lethality and under-proliferation of diploid tissue of the larvae. In contrast, a dominant, gain-of-function mutation leads to hematopoietic tumors

(Harrison et al 1995, Luo et al 1995). Although the JAK kinases seem to be required for cytokine-mediated cell proliferation, an important point to keep in mind is that interferon signaling, which also utilizes the JAKs, has largely anti-proliferative effects. Thus the overall function of the JAK proteins in cytokine signaling may be quite complex.

The central role of the JAK kinases suggests that their activation may be the initial intracellular event in cytokine receptor signal transduction. Presumably, JAK kinases are bound to the cytosolic domains of cytokine receptors in the absence of ligand. Ligand-induced receptor oligomerization would bring the associated JAKs into proximity and enable them to *trans*-phosphorylate each other. By analogy with the insulin receptor kinase (Hubbard et al 1994), phosphorylation of a specific tyrosine residue in the loop region of the JAK protein could activate the kinase. This would lead to tyrosine phosphorylation of the associated receptors as well as of downstream substrates (Figure 2).

Phosphorylated tyrosine residues on the receptors provide docking sites for SH2-containing signaling molecules (Klingmuller et al 1995). JAKs may also directly phosphorylate and activate signaling proteins (He et al 1995). This possibility is supported by experiments using receptor/JAK chimeras, which showed that some signaling pathways can become activated by JAK dimerization in the absence of receptor cytoplasmic tails (Franks et al 1995, Sakai et al 1995). It is not clear if JAKs directly tyrosine phosphorylate cytokine receptors in vivo, or if other kinases such as the protein tyrosine kinase lck, which is associated with IL-2Rβ and activated by IL-2 binding (Minami et al 1993), do so.

Kinase Pathways in Cytokine Signaling: Src and Zap70 and Relatives

Certain cytokine receptors, including the IL-2Rβ, IL-3Rβ_c and gp130 polypeptides, associate with and activate src family-related kinases (Hatakeyama et al 1991, Horak et al 1991, Torigo et al 1992, Kobayashi et al 1993, Ernst et al 1994, Matsuda et al 1995a,b, Rao & Mufson 1995), although their role in cytokine signaling is poorly defined. The best studied example is the IL-2R, which interacts with $p56^{lck}$ through the acidic ('A') region of the cytoplasmic tail of the receptor β-subunit. The 'A' region is not required for proliferation of cultured cells, although residues in the serine-rich ('S') region of IL-2Rβ, which is required for proliferation, are also necessary for $p56^{lck}$ activation (Hatakeyama et al 1991, Minami et al 1993). This suggests that another protein, possibly a JAK kinase, may be involved in $p56^{lck}$ activation. The IL-2Rβ 'A' region and $p56^{lck}$ are required for activation of ras and induction of fos and jun in response to IL-2 binding (Satoh et al 1992). Recent studies showed that a constitutively active form of p56[lck] delays apoptotic cell death and complements either c-myc or bcl-2 in supporting IL-2-independent growth of Baf-BO3 cells. In the absence of lck activation, c-myc and bcl-2 overexpression supported IL-2-independent proliferation, suggesting functional redundancy among these three signaling molecules and their pathways (Miyazaki et al 1995).

The related kinases ZAP-70 and Syk are critical for T- or B-cell development, respectively. Syk associates with the 'S' region of the IL-2Rβ chain and with the G-CSF-R (Corey et al 1994, Minami et al 1995). IL-2 stimulation increases Syk association and kinase activity. Antibody cross-linking of a Syk chimera

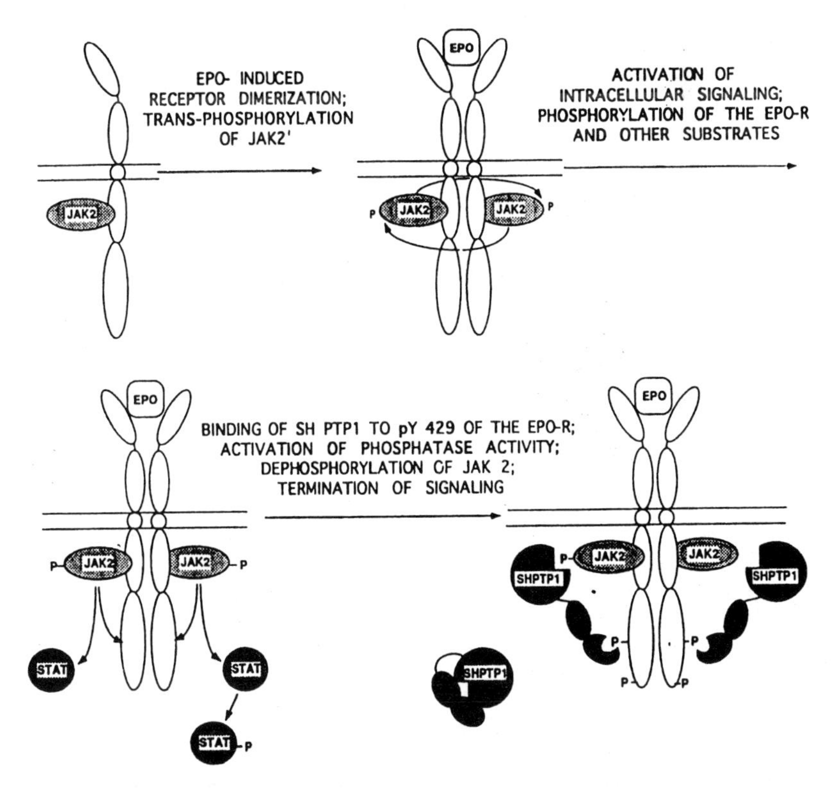

Figure 2 Model for activation and inactivation of the EpoR. Binding of EPO induces receptor dimerization and auto-or *trans*-phosphorylation of JAK2. Among the substrates of JAK2 are the EpoR and signal transduction proteins such as STATs. The protein-tyrosine phosphatase SH-PTP1 binds to a segment of the EpoR containing phosphotyrosine 429. Binding induces activation of the phosphatase and removal of the activating phosphate from JAK2 and leads to termination of signal transduction. Although specific for the EpoR, this model also illustrates some general aspects of cytokine receptor signaling; it is expected that receptor-specific signal transduction pathways will emerge with further study. This figure is modified from one in Klingmuller et al 1995.

leads to c-myc induction, suggesting that Syk activation through IL-2Rβ may be responsible for this effect of IL-2-stimulation (Minami et al 1995). Of great interest is how these different IL-2-stimulated signaling pathways coordinate in vivo to maintain lymphoid homeostasis.

Direct Signaling from the Receptor to Nucleus: STAT Proteins

Members of the STAT family (signal transducers and activators of transcription) play a pivotal role in signaling by cytokine receptors. As with the JAKs, the first indication of the importance of this family came from studies on interferon signaling, where the STATs were identified as interferon-responsive transcription factors (Levy et al 1988, 1989, Dale et al 1989, Decker et al 1991, Lew et al 1991). STATs are present in the cytoplasm of unstimulated cells; ligand binding stimulates STAT tyrosine phosphorylation, homo- or heterodimerization, and nuclear translocation (Shuai et al 1992, 1994, Improta et al 1994). In addition, the activity of some STATs is further enhanced by serine phosphorylation, possibly via MAPK, because in vitro MAPK is capable of phosphorylating the specific serine residue required for STAT activation (Boulton et al 1995, David et al 1995, Wen et al 1995, Zhang et al 1995). In the nucleus, STATs bind specific promoter elements and activate transcription of ligand-responsive genes (Pellegrini & Schindler 1993).

Conserved regions in the STAT proteins include a single carboxy-terminal tyrosine residue, which is the site of ligand-induced tyrosine phosphorylation, an SH2 domain, and an SH3-like domain. The SH2 domain plays a key role both in association with cytokine receptors and in homodimerization (Shuai et al 1994, Heim et al 1995). Activation of JAK kinase activity is required for ligand-dependent STAT phosphorylation, which leads to the hypothesis that JAKs directly phosphorylate STATs. Although there is no in vivo evidence for this, in vitro studies have shown that JAKs can phosphorylate certain STATs (Quelle et al 1995).

The Source of Specificity in STAT Activation

To date, numerically fewer JAKs have been identified than STATs, and individual JAKs can phosphorylate multiple STATs. This fact, together with the observation that for some receptors the identity of the activated JAKs, but not the STATs, varies with the cell type (Boulton et al 1994, Stahl et al 1994), led to the hypothesis that specificity is determined by the STATs (for review, see Ivashkiv 1995). The proposed model is that STAT SH2 domains bind either directly or through an adapter protein to specific phosphorylated tyrosines in the receptor cytosolic domains. This would bring them into proximity with the associated, activated JAK kinases, which would then phosphorylate and activate the associated STAT proteins (Heim et al 1995, Stahl et al 1995).

Much evidence supports this model. First, domain-swapping experiments indicate that the STAT SH2 domain determines the ability of the STAT to bind to and be activated by a specific receptor (Heim et al 1995, Stahl et al 1995). Second, phosphorylated peptides containing Y440 and surrounding residues of the IFN-γ receptor blocked IFN-γ activation of STAT1, indicating that phosphorylated Y440 in the receptor is required for STAT1 activation and presumably for binding of the STAT protein to the receptor, where it can be phosphorylated by an associated, activated, JAK kinase (Greenlund et al 1994). Both STAT1 and STAT2 are activated by interferon-α (IFN-α). IFN-α induces normal activation of STAT2 in STAT1-defective cells (Improta et al 1994), but phosphorylation and activation of STAT1 require prior activation of STAT2 (Leung et al 1995). This suggests that STAT2 "docks" directly to the activated IFN-α receptor and becomes phosphorylated. STAT1 then binds to phosphorylated STAT2 and is itself phosphorylated, leading to activation of the transcription complex (Leung et al 1995).

Mutants of gp130 or LIFRα demonstrated that specific tyrosines within a YXXQ motif in the cytoplasmic tail are required for STAT3 activation (Stahl et al 1995). Work on other receptors, including PrR, EpoR, and the IL-2Rβ chain (Damen et al 1995a, Fujii et al 1995, Lebrun et al 1995b) also supports the hypothesis that receptor tyrosines are required for specific STAT activation. However, the specific mechanisms of association between receptor phosphotyrosine residues and STAT proteins remain unclear because coimmunoprecipitation experiments or binding studies with bacterially expressed GST-fusion proteins have been unsuccessful so far (Damen et al 1995a). In addition, evidence indicates that tyrosine residues in the GH and Epo receptors are not required for STAT activation (Damen et al 1995a, Hackett et al 1995). Presumably, some other type of association between the STAT and receptor, possibly involving an accessory molecule, is important. Intriguingly, a chimeric CD16/JAK2 receptor protein, which completely lacks receptor cytoplasmic sequences, induced STAT DNA-binding activity when activated by cross-linking (Sakai et al 1995). Thus under certain circumstances, no receptor sequences may be required for STAT activation. One possibility is that STAT SH2 domains can bind to phosphorylated tyrosine residues on the JAKs themselves, leading to subsequent phosphorylation and activation of the STATs.

Biological Functions of STATs

Some examples of STAT-induced transcription are well established, including the role of STAT1 and 2 in induction of IFN-regulated genes, of STAT3 in induction of acute-phase response genes, and of STAT5 in prolactin-responsive induction of milk protein genes (Pellegrini & Schindler 1993, Akira et al 1994,

Wakao et al 1994, Wegenka et al 1994, Welte et al 1994). However, most of the genes induced by activated STAT proteins are unknown. Because over 25 ligands are known to activate one or more of the 6 known STATs, there must be factors other than the STATs themselves that govern response specificity to a ligand. There may be other tissue-specific proteins, as yet unidentified, that interact with STATs and are required for efficient and specific transcriptional responses. STAT modification by other signaling pathways (see above) may also provide specificity.

PI 3-kinase, IRS-1 and IRS-2 Proteins

Phosphatidylinositol 3-kinase (PI3K) activity is increased following stimulation by many cytokines (Truitt et al 1994, Damen et al 1995b, Jucker & Feldman 1995) and its specific inhibitor, wortmannin, can impair cytokine-dependent proliferation (Damen et al 1995b). Stimulation of PI3K can occur through direct binding of its regulatory p85 subunit to a receptor via an interaction between the SH2 group of p85 and a receptor phosphotyrosine residue, or through an indirect interaction (Truitt et al 1994, Damen et al 1995b; see below). However, the function and target molecules of PI3K are not clear. Proposed downstream targets of PI3K include p70 S6 kinase, members of the protein kinase C family, and MEK (MAP kinase kinase). The latter would represent a novel pathway for activation of the MAP kinase signaling cascade, independent of the ras/raf activation pathway (Karnitz et al 1995).

IRS-1 was initially detected as a substrate of the insulin receptor. It contains many potential serine/threonine phosphorylation sites and at least 20 potential tyrosine phosphorylation sites; specific phosphotyrosine residues dock the SH2 domains of many regulatory proteins (White 1994). The first indication that IRS-1 or IRS-2 may be involved in hematopoietic cytokine receptor signaling came from studies of the IL-4 receptor. Neither insulin nor IL-4 signal in the IL-3-dependent cell line 32D, which lacks both IRS-1 and IRS-2. Transfection of IRS-1 into 32D cells restored mitogenic signaling for both growth factors (Wang et al 1993). Similar to the insulin receptor, an NPXY motif in the IL-4R associates with IRS-1 (Keegan et al 1994). However, phosphotyrosines in other cytokine receptors may not be required for association with IRS-1 or IRS-2 proteins (Argetsinger et al 1995). JAKs coimmunoprecipitate with IRS-1 or IRS-2 after stimulation by several cytokines, including IL-2, IL-4, IL-7, and IL-15, and may be responsible for their phosphorylation (Johnston et al 1995, Ridderstrale et al 1995). In any case, IRS-1 or IRS-2 may act as the docking site for the p85 subunit of PI3K and thus induce its activation after cytokine stimulation. Much work is needed to identify all components of the cytokine signal transduction cascades and to understand their specific roles.

Phosphatases and Signal Termination

An interesting and often overlooked question is what regulates the termination of receptor signal transduction? Several lines of evidence have shown that the phosphatase SH-PTP1 plays an important role in down-regulating the proliferative signal of the EpoR through JAK2. Mice with defects in the *motheaten* locus, encoding SH-PTP1, have increased numbers of splenic CFU-Es in vivo, and erythroid progenitor cells isolated from these mice are hypersensitive to Epo in vitro (van Zant & Schultz 1989). This suggests that wild-type SH-PTP1 down-modulates signaling through the EpoR (Figure 2). As mentioned above, a negative regulatory domain was identified in the EpoR cytoplasmic tail. Individuals with mutations in this region have benign erythrocytosis, and their erythroid progenitor cells are hypersensitive to Epo. Further studies show that SH-PTP1 binds directly to the EpoR via a phosphorylated tyrosine residue (Y429) in the negative regulatory domain. Mutation of this tyrosine to phenylalanine results in a receptor which, when expressed in hematopoietic cells, allows them to grow in one fifth the normal concentration of Epo (Klingmuller et al 1995). SH-PTP1 appears to terminate EpoR signaling by dephosphorylation of JAK2 (Klingmuller et al 1995) (Figure 2). These experiments provide a molecular explanation for the negative regulatory activity associated with SH-PTP1 and the membrane distal portion of the EpoR cytoplasmic tail. Because SH-PTP1 also interacts with the IL-3Rβ chain (Yi et al 1993), it may be involved in downstream signaling pathways of other cytokine receptors.

PATHOLOGICAL EFFECTS OF ABERRANT CYTOKINE RECEPTOR FUNCTION

A variety of pathologies, including leukemia, neutropenia, and dwarfism, are due to alterations in cytokine receptor function. To date, the most common cause of altered function is a mutation in the receptor, although functional alterations can also arise from direct interaction between receptors and viral gene products. Defective signaling proteins, such as SH-PTP1 and JAK3, have been implicated in disease pathogenesis, and it is likely that the molecular basis of other pathologies will be revealed by the continued study of cytokine receptor signaling.

Benign Erythrocytosis

Members of a large Finnish family with autosomal-dominant benign erythrocytosis have a mutation in one allele of the EpoR, which introduces a premature stop codon, resulting in a 70 amino acid (aa) carboxy-terminal truncation (de la Chapelle et al 1993a,b). This mutation removes the negative regulatory domain of the EpoR (Yoshimura et al 1990a, D'Andrea et al 1991) including Tyr 429, the binding site of SH-PTP1 (Klingmuller et al 1995). Affected individuals have a normal life span and no apparent signs of illness, despite elevated hematocrits and blood hemoglobin (de la Chapelle et al 1993a). However, erythroid progenitors from these individuals are hypersensitive to Epo in vitro (de la Chapelle et al 1993a). In a second family with dominant erythrocytosis, a frame-shift mutation in the EpoR also results in a carboxy-terminal truncation, missing 64 aa from the negative regulatory domain. Erythroid progenitors from these individuals are also hypersensitive to Epo (Sokol et al 1994, 1995). Affected individuals are not prone to blood cell malignancies, indicating that erythroid progenitor hyper-proliferation in these cases is not contributing to the development of leukemia. Because all affected individuals thus far identified are heterozygous, it is not clear if homozygotes would be viable and, if so, more prone to malignancies.

GH Resistance and Laron Syndrome

Laron syndrome is an autosomal-recessive condition consisting of dwarfism, obesity, hypoglycemia, and high circulating free fatty acids, all due to resistance to GH action (Laron 1995). The disease is genetically linked to the GH-R gene (Amselem et al 1989, 1991, Duquesnoy et al 1991, Berg et al 1993, Laron 1995). Defects include GH-R gene deletions, as well as nonsense and frame-shift mutations, which interfere with GH-R function (Meacham et al 1993, Counts & Cutler 1995). One revealing point mutation in the extracellular region of the GH-R blocks the transport of the protein to the cell surface (Duquesnoy et al 1991). Another, which lies in the dimer interface region of the receptor (De Vos et al 1992), confers GH resistance by interfering with receptor dimerization and activation, even though the mutant receptors can bind GH (Duquesnoy et al 1994). Sex-linked dwarfism has been found in chickens, providing an animal model of Laron syndrome. In these cases, defects have also been linked to the GH-R gene (Duriez et al 1993, Agarwal et al 1994).

Severe Combined Immunodeficiency (SCID) Syndromes

Patients with SCID are deficient in T- and/or B-cell functions and have persistent, life-threatening infections from infancy. X-linked SCID (XSCID), which is responsible for $\approx 50\%$ of SCID cases, is characterized by severely low T-cell numbers. Although B-cell numbers are normal, the B cells are non functional. Mutations in the γ chain of the IL-2R result in XSCID (Noguchi et al 1993b), which establishes the critical role of the γ chain in immune cell function. The majority of these mutations are predicted to generate truncated proteins, and two mutational hotspots in the γ chain are associated with $\approx 20\%$ of XSCID cases (Pepper et al 1995). Also, a mutation of the γ chain, which affects the association between JAK3 and the γ chain, causes a mild form of XSCID (Russell et al 1994).

Recently, JAK3 was also implicated in SCID. In one study, JAK3 mutations were identified in two patients with autosomal-recessive SCID; interestingly these patients also lacked NK ($CD16^+$) cells. One patient carries a Tyr to Cys mutation in a conserved Tyr residue of the JH7 domain, whereas the second patient has a deletion of 151 base pairs in the kinase-like domain of JAK3; both mutations resulted in markedly reduced levels of JAK3 (Macchi et al 1995). A second study found premature termination codons in both alleles of JAK3 in a female patient with a SCID-like phenotype (Russell et al 1995). By comparing the phenotypes of patients carrying mutant alleles with those of mice with targeted gene deletions, the critical roles of JAK3 and the IL-2Rγ chain in lymphoid development have been revealed, with interesting similarities and differences (Noguchi et al 1993b; DiSanto et al 1995, Macchi et al 1995, Nosaka et al 1995, Park et al 1995, Russell et al 1995, Thomis et al 1995). Molecular defects in other crucial signaling proteins may be found to contribute to sporadic cases of SCID.

Severe Congenital Neutropenia

Severe congenital neutropenia (Kostmann syndrome) is characterized by absolute neutropenia, due to maturation arrest of myeloid progenitor cells at the promyelocyte-myelocyte stage. Some patients develop myelodysplastic syndromes or acute myeloid leukemia. In some patients with Kostmann syndrome, mutations have been found in the G-CSF-R. The mutant receptors contain premature stop codons and are missing portions of the cytoplasmic domain required for neutrophil-specific gene induction (Dong et al 1994, 1995). When coexpressed with the wild-type receptor in cultured cells, these truncated receptors support G-CSF-stimulated proliferation but dominantly inhibit differentiation, which suggests that the defective receptors block terminal neutrophil differentiation in vivo in a similar manner (Dong et al 1994, 1995).

Cytokine receptor gene mutations are likely to be identified in other pathological conditions. For example, a newly identified member of the IL-6 family, cardiotrophin 1, induces cardiac myocyte hypertrophy, apparently through activation of the gp130 signaling receptor subunit (Pennica et al 1995). Continuous activation of gp130 leads to myocardial hypertrophy in mice because transgenic mice expressing both IL-6 and IL-6R showed constitutive tyrosine phosphorylation of gp130 and STAT 3 (APRF) and had ventricular myocardial hypertrophy (Hirota et al 1995). Therefore, it is likely that aberrant gp130 function could contribute to cardiomyopathies; this expectation is supported by the gp130 knock-out mice, which exhibit cardiac defects (Yoshida et al 1996; Table 1).

Virus-Receptor Interactions: Viral env Proteins

The Friend virus complex contains the defective murine spleen focus-forming virus (SFFV) and a helper virus, the Friend murine leukemia virus (F-MuLV)

(Kabat 1989, Ben-David & Bernstein 1991). In mice, two strains of SFFV induce erythroleukemia (EL): The anemic strain (SFFV-A) induces EL and anemia (Friend 1957), whereas the polycythemic strain (SFFV-P) induces EL and polycythemia (Kabat 1989). After infection, the levels of erythroid progenitors (BFU-E and CFU-E) increase dramatically, followed by the emergence of malignant clones at 4–6 weeks post-infection (Kabat 1989, Ben-David & Bernstein 1991). Erythroid cells from mice infected with SFFV-P can proliferate and differentiate in vitro in the absence of Epo, whereas erythroid precursors derived from SFFV-A infected mice still require Epo for differentiation (Hankins & Troxler 1980).

The SFFV envelope protein (gp55) is responsible for oncogenicity in adult mice and for induction of erythroblastosis in vivo (Wolff & Ruscetti 1985). Coexpression of gp55 with the EpoR in growth factor–dependent cells results in growth factor–independent proliferation (Hoatlin et al 1990, Li et al 1990, Ruscetti et al 1990). Interactions between the EpoR and gp55 were detected by coimmunoprecipitation experiments (Li et al 1990). This association is likely to be the principal cause of the initial stage of Friend disease—Epo-independent polyclonal erythroblastosis—while subsequent mutations in these aberrantly cycling cells give rise to malignant cells (Lane & Benchimol 1990, Moreau-Gachelin et al 1990).

Association between the EpoR and gp55 was detected mainly between the intracellular forms of the proteins (Li et al 1990, Yoshimura et al 1990b). However, only the cell surface forms of gp55 induce growth factor independence (Ferro et al 1993). Binding and cross-linking experiments demonstrated that cell surface forms of gp55, competent for signaling through the EpoR, are in a complex with the EpoR and Epo (Ferro et al 1993). Other studies showed that ligation of an ER-retention signal to gp55 abolishes cell surface expression and oncogenic activity (Li JP et al 1995). Thus the bulk of experimental evidence supports the idea that gp55 activates the EpoR at the cell surface, although the mechanism is unclear.

Evidence indicates that the transmembrane region of gp55-P is critical for EpoR binding and transforming activities (Chung et al 1989, Showers et al 1993). Studies using chimeras between the EpoR and IL-3R demonstrated that the membrane-spanning domain of EpoR is required for mitogenic activation by gp55 (Zon et al 1992). However, interaction between gp55 and the EpoR is not sufficient for Epo-independent proliferation because the EpoR can also interact with R-gp55, a naturally occurring variant of gp55 which, like gp55-A, is unable to stimulate Epo-independent growth (Ruscetti et al 1990, Showers et al 1993).

Coexpression of gp55-P and the EpoR induces constitutive tyrosine phosphorylation on a number of proteins (Showers et al 1992). The DNA-binding activity of STAT-like proteins is constitutively activated in Epo-dependent cells rendered growth factor–independent by gp55 (Ohashi et al 1995). Interestingly, JAK2 is tyrosine phosphorylated in a cell line obtained from SFFV-P infected mice (Yamamura et al 1994). Finally, a dominant negative form of the EpoR blocks not only Epo signaling, but also gp55 activation of the wild-type EpoR (Barber et al 1994). Together, these results suggest that gp55-P may activate some of the same signaling pathways that are activated by the Epo-liganded Epo receptor. However, evidence indicates that gp55-P favors erythroid proliferation whereas Epo favors differentiation (Ahlers et al 1994). Thus while some of the same signals may be elicited by Epo or gp55 interaction with the EpoR, further work is required to establish the extent of overlap and to identify any differences.

The Friend or Moloney mink cell focus-forming viruses (MCF) are dual- or polytropic due to the env protein (gp70) but cannot infect laboratory strains of mice (Kabat 1989). Because of their LTR specificity, MCF viruses induce tissue-specific neoplasms in susceptible newborn mice. Although gp55, which is related to the MCF env gene, activates the EpoR, MCF gp70 activates both the EpoR and the β chain of the IL-2R (Li & Baltimore 1991). In addition, MCF gp70 coimmunoprecipitated with the EpoR. In contrast, ecotropic gp70, derived from Friend MuLV, is not able to activate these receptors, nor is the related IL-3R activated by MCF gp70 or by SFFV gp55. Leukemogenesis appears to be initiated by MCF or SFFV through activation of specific cytokine receptors, which induce sustained growth factor–independent progenitor cell proliferation. Additionally, Friend MuLV may be oncogenic in newborn animals because of recombination with endogenous MCF-like sequences, which generates gp55-like proteins (Ben-David & Bernstein 1991, Li & Baltimore 1991).

v-mpl, a Transduced Truncated Thrombopoietin (Tpo) Receptor

Myeloproliferative leukemia virus (MPLV) is a defective, acute oncogenic murine retrovirus that was recently discovered to have transduced a novel cellular proto-oncogene (Souyri et al 1990). The MPLV complex consists of a replication-competent Friend MuLV (F-MuLV) and a replication-defective virus, MPLV. Molecular analysis of MPLV revealed that the env gene *v-mpl* contains sequences related to members of the cytokine receptor superfamily (Souyri et al 1990). The cellular homologue of *v-mpl, c-mpl* (Vigon et al 1992), was subsequently shown to encode the receptor for thrombopoietin (TpoR) (Bartley et al 1994, de Sauvage et al 1994, Kaushansky et al 1994, Lok et al 1994, Wendling et al 1994, Choi et al 1995).

The *v-mpl* gene is a chimera, containing 64 amino acids derived from the NH2-terminal region of the F-MuLV env gene, including the signal peptide, followed by 36 amino acids from the middle portion of the F-MuLV env sequence, fused to a 184-aa MPLV-specific region. The *v-mpl* sequence predicts a 31-kDa fusion protein, with an extracellular domain of 109 aa, a 22-aa membrane-spanning domain and a 119-aa cytosolic region (Souyri et al 1990). Expression of v-mpl results in growth factor independence in vitro, and infection of adult mice with MPLV leads to splenomegaly, leukocytosis, polycythemia, and acute leukemia (Wendling et al 1986, Souyri et al 1990). Although the v-mpl protein is not a disulfide-linked homodimer, it is probable that non covalent constitutive dimerization explains its oncogenic properties because the normal TpoR is activated for growth and differentiation by Tpo-triggered homodimerization (Alexander et al 1995).

Signaling of Tpo through TpoR involves activation of JAK2 and Tyk2 (Sattler et al 1995), as well as activation of STAT5 (Pallard et al 1995). Clearly TpoR transduces a proliferative signal (Skoda et al 1993) and upon ligand binding becomes tyrosine phosphorylated (Drachman et al 1995). Activation of TpoR also results in tyrosine phosphorylation of Shc (Drachman et al 1995), and similar to other cytokine receptors, distinct regions of the TpoR cytoplasmic domain are coupled to the JAK-STAT pathway and Shc activation (Gurney et al 1995). The extent of overlapping and distinct signals elicited by ligand-occupied TpoR versus v-mpl will help elucidate the molecular basis of MPLV pathogenicity.

SUMMARY AND PERSPECTIVES

Since the cloning of cDNAs encoding the first cytokine receptors in the late 1980s, our knowledge of the complexity and functions of this ever-growing gene superfamily has been expanding at a remarkable pace. These receptors share many functional and structural features. Although cytokine receptors activate similar intracellular signal transduction pathways, different receptors support the proliferation and differentiation of distinct hematopoietic lineages. Based on available evidence, it is unlikely that all cytokine receptors are functionally equivalent—that is, they cannot functionally replace the normal receptor when expressed ectopically in a hematopoietic progenitor cell. However, we do not know the extent to which apparently different receptor functions are cell-type-specific or receptor-specific. Thus it will be of great importance to determine all of the signal transduction proteins activated by a particular cytokine receptor; to determine the function of each of these proteins in inducing the diverse proliferation, differentiation and maturation events triggered by the particular cytokine receptor; and then to determine the extent to which these signal transduction pathways are induced by other cytokine receptors in the same cell type. In addition, it is important to understand how signaling pathways from two or more receptors in the same cell might interact and contribute to the proliferation and differentiation programs. Such studies promise to yield exciting insights into the molecular events controlling normal hematopoietic cell development and differentiation, as well as into the pathophysiological origins of many hematologic disorders.

ACKNOWLEDGMENTS

The authors gratefully acknowledge Drs. Ronald Palacios, Peter J Murray, and Gregory D Longmore for their suggestions and review of the manuscript. Because of space limitations, we regret that we have been unable to cite many relevant primary references. HW was supported by a postdoctoral fellowship from the Damon Runyon-Walter Winchell Cancer Research Fund. MS is the recipient of a fellowship from the Howard Hughes Medical Institute and SNC holds a fellowship from the Anna Fuller Fund. UK is supported by a fellowship from the Deutsche Forschungsgemeinschaft. Research on the erythropoietin receptor was supported by grant HL32262 from the National Institutes of Health and by a grant from the Arris Pharmaceutical Corporation to HFL.

Literature Cited

Agarwal SK, Cogburn LA, Burnside J. 1994. Dysfunctional growth hormone receptor in a strain of sex-linked dwarf chicken: evidence for a mutation in the intracellular domain. *J. Endocrinol.* 142:427–34

Ahlers N, Hunt N, Just U, Laker C, Ostertag W, Nowock J. 1994. Selectable retrovirus vectors encoding Friend virus gp55 or erythropoietin induce polycythemia with different phenotypic expression and disease progression. *J. Virol.* 68:7235–43

Akira S, Nishio Y, Inoue M, Wang X-J, Wei S, et al. 1994. Molecular cloning of APRF, a novel IFN-stimulated gene factor 3 p91-related transcription factor involved in the gp130-mediated signaling pathway. *Cell* 77:63–71

Alexander WS, Metcalf D, Dunn AR. 1995. Point mutations within a dimer interface homology domain of c-Mpl induce constitutive receptor activity and tumorigenicity. *EMBO J.* 14:5569–78

Amselem S, Duquesnoy P, Attree O, Novelli G, Bousnina S, et al. 1989. Laron dwarfism and mutations of the growth hormone-receptor gene. *N. Engl. J. Med.* 321:989–95

Amselem S, Sobrier ML, Duquesnoy P, Rappaport R, Postel-Vinay MC, et al. 1991. Recurrent nonsense mutations in the growth hormone receptor from patients with Laron dwarfism. *J. Clin. Invest.* 87:1098–102

Arai K-I, Lee F, Miyajima A, Miyatake S, Arai N, Yokota T. 1990. Cytokines: coordinators of immune and inflammatory responses. *Annu. Rev. Biochem.* 59:783–836

Argetsinger LS, Campbell GS, Yang X, Witthuhn BA, Silvennoinen O, et al. 1993. Identification of JAK2 as a growth hormone receptor-associated tyrosine kinase. *Cell* 74:237–44

Argetsinger LS, Hsu GW, Myers MG Jr, Billestrup N, White MF, Carter-Su C. 1995. Growth hormone, interferon-γ and leukemia inhibitory factor promoted tyrosyl phosphorylation of insulin receptor substrate-1. *J. Biol. Chem.* 270:14685–92

Barber DL, DeMartino JC, Showers MO, D'Andrea AD. 1994. A dominant negative erythropoietin (EPO) receptor inhibits EPO-dependent growth and blocks F-gp55-dependent transformation. *Mol. Cell. Biol.* 14:2257–65

Bartley TD, Bogenberger J, Hunt P, Li YS, Lu HS, et al. 1994. Identification and cloning of a megakaryocyte growth and development factor that is a ligand for the cytokine receptor Mpl. *Cell* 77:1117–24

Bazan JF. 1990. Structural design and molecular evolution of a cytokine receptor superfamily. *Proc. Natl. Acad. Sci. USA* 87:6934–38

Bazan JF. 1995. Postmodern complexes. *Nature* 376:217–18

Ben-David Y, Bernstein A. 1991. Friend virus-induced erythroleukemia and the multistage nature of cancer. *Cell* 66:831–34

Berg MA, Argente J, Chernausek S, Gracia R, Guevara-Aguirre J, et al. 1993. Diverse growth hormone receptor gene mutations in Laron syndrome. *Am. J. Hum. Genet.* 52:998–1005

Bernad A, Kopf M, Kulbacki R, Weich N, Koehler G, Gutierrez-Ramos JC. 1994. Interleukin-6 is required in vivo for the regulation of stem cells and committed progenitors of the hematopoietic system. *Immunity* 1:725–31

Boulton TG, Stahl N, Yancopoulos GD. 1994. Ciliary neurotrophic factor/leukemia inhibitory factor/interleukin 6/oncostatin M family of cytokines induces tyrosine phosphorylation of a common set of proteins overlapping those induced by other cytokines and growth factors. *J. Biol. Chem.* 269:11648–55

Boulton TG, Zhong Z, Wen Z, Darnell JE, Stahl N, Yancopoulos GD. 1995. STAT3 activation by cytokines utilizing gp130 and related transducers involves a secondary modification requiring an H7-sensitive kinase. *Proc. Natl. Acad. Sci. USA* 92:6915–19

Boutin JM, Jolicoeur C, Okamura H, Gagnon J, Edery M, et al. 1988. Cloning and expression of the rat prolactin receptor, a member of the growth hormone/prolactin receptor gene family. *Cell* 53:69–77

Campbell GS, Argetsinger LS, Ihle JN, Kelly PA, Rillema JA, Carter-Su C. 1994. Activation of Jak2 tyrosine kinase by prolactin receptors in Nb2 cells and mouse mammary gland explants. *Proc. Natl. Acad. Sci. USA* 91:5232–36

Choi ES, Hokom M, Bartley T, Li YS, Ohashi H, et al. 1995. Recombinant human megakaryocyte growth and development factor (rHuMGDF), a ligand for c-Mpl, produces functional human platelets in vitro. *Stem Cells* 13:317–22

Chung SW, Wolff L, Ruscetti SK. 1989. Transmembrane domain of the envelope gene of a polycythemia-inducing retrovirus determines erythropoietin-independent growth. *Proc. Natl. Acad. Sci. USA* 86:7957–60

Corey SJ, Burkhardt AL, Bolen JB, Geahlem RL, Tkatch LS, Tweardy DJ. 1994. Granulocyte colony-stimulating factor receptor signaling involves the formation of a three-component complex with lyn and syk protein-tyrosine kinases. *Proc. Natl. Acad. Sci. USA* 91:4683–87

Cosman D, Lyman SD, Idzerda RL, Beckmann MP, Park LS, et al. 1990. A new cytokine receptor superfamily. *Trends Biochem. Sci.* 15:265–70

Counts DR, Cutler G Jr. 1995. Growth hormone insensitivity syndrome due to point deletion and frame shift in the growth hormone receptor. *J. Clin. Endocrinol. Metab.* 80:1978–81

Cunningham BC, Ultsch M, De Vos AM, Mulkerrin MG, Clauser KR, Wells JA. 1991. Dimerization of the extracellular domain of the human growth hormone receptor by a single hormone molecule. *Science* 254:821–25

Dale TC, Imam AM, Kerr IM, Stark GR. 1989. Rapid activation by interferon alpha of a latent DNA-binding protein present in the cytoplasm of untreated cells. *Proc. Natl. Acad. Sci. USA* 86:1203–7

Damen JE, Cutler RL, Jiao H, Yi T, Krystal G. 1995b. Phosphorylation of tyrosine 503 in the erythropoietin receptor (EpR) is essential for binding the P85 subunit of phosphatidylinositol (PI) 3-kinase and for EpR-associated PI 3-kinase activity. *J. Biol. Chem.* 270:23402–8

Damen JE, Wakao H, Miyajima A, Krosl J, Humphries RK, et al. 1995a. Tyrosine 343 in the erythropoietin receptor positively regulates erythropoietin-induced cell proliferation and Stat5 activation. *EMBO J.* 14:5557–68

D'Andrea AD, Fasman GD, Lodish HF. 1989a. Erythropoietin receptor and interleukin-2 receptor β chain: a new receptor family. *Cell* 58:1023–24

D'Andrea AD, Lodish HF, Wong GG. 1989b. Expression cloning of the murine erythropoietin receptor. *Cell* 57:277–85

D'Andrea AD, Yoshimura A, Youssoufian H, Zon LI, Koo JW, Lodish HF. 1991. The cytoplasmic region of the erythropoietin receptor contains nonoverlapping positive and negative growth-regulatory domains. *Mol. Cell. Biol.* 11:1980–87

DaSilva L, Howard OMZ, Rui H, Kirken RA, Farrar WL. 1994. Growth signaling and Jak2 association mediated by membrane-proximal cytoplasmic regions of the prolactin receptor. *J. Biol. Chem.* 269:18267–70

David M, Petricoin E III, Benjamin C, Pine R, Weber MJ, Larner AC. 1995. Requirement for MAP kinase (ERK2) activity in interferon α- and interferon β-stimulated gene expression through STAT proteins. *Science* 269:1721–23

Davis JA, Linzer DIH. 1989. Expression of multiple forms of the prolactin receptor in mouse liver. *Mol. Endocrinol.* 3:674–80

Davis S, Aldrich TH, Ip NY, Stahl N, Scherer S, et al. 1993a. Released form of CNTF receptor α component as a soluble mediator of CNTF responses. *Science* 259:1736–39

Davis S, Aldrich TH, Stahl N, Pan L, Taga T, et al. 1993b. LIFR beta and gp130 as heterodimerizing signal transducers of the tripartite CNTF receptor. *Science* 260:1805–8

Davis S, Aldrich TH, Valenzuela DM, Wong V, Furth ME, et al. 1991. The receptor for ciliary neurotrophic factor. *Science* 253:59–63

DeChiara TM, Vejsada R, Poueymirou WT, Acheson A, Suri C, et al. 1995. Mice lacking the CNTF receptor, unlike mice lacking CNTF, exhibit profound motor neuron deficits at birth. *Cell* 83:313–22

Decker T, Lew DJ, Mirkovitch J, Darnell JE. 1991. Cytoplasmic activation of GAF, an IFN-gamma-regulated DNA-binding factor. *EMBO J.* 10:927–32

de la Chapelle A, Sistonen P, Lehvaslaiho H, Ikkala E, Juvonen E. 1993a. Familial erythrocytosis genetically linked to erythropoietin receptor gene. *Lancet* 341:82–84

de la Chapelle A, Traskelin A-L, Juvonen E. 1993b. Truncated erythropoietin receptor causes dominantly inherited benign human erythrocytosis. *Proc. Natl. Acad. Sci. USA* 90:4495–99

de Sauvage FJ, Hass PE, Spencer SD, Malloy BE, Gurney AL, et al. 1994. Stimulation of megakaryocytopoiesis and thrombopoiesis by the c-Mpl ligand. *Nature* 369:533–38

De Vos AM, Ultsch M, Kossiakoff AA. 1992. Human growth hormone and extracellular domain of its receptor: crystal structure of the complex. *Science* 255:306–11

Devos R, Plaetinck G, Van der Heyden J, Cornelis S, Vandekerckhove J, et al. 1991. Molecular basis of a high affinity murine interleukin-5 receptor. *EMBO J.* 10:2133–37

Dexter TM, Spooncer E. 1987. Growth and differentiation in the hemopoietic system. *Annu. Rev. Cell Biol.* 3:423–41

DiSanto JP, Muller W, Guy-Grand D, Fischer A, Rajewsky K. 1995. Lymphoid development in mice with a targeted deletion of the interleukin 2 receptor γ. *Proc. Natl. Acad. Sci. USA* 92:377–81

Dong F, Brynes RK, Tidow N, Welte K, Lowenberg B, Touw IP. 1995. Mutations in the gene for the granulocyte colony-stimulating-factor receptor in patients with acute myeloid leukemia preceded by severe congenital neutropenia. *N. Engl. J. Med.* 333:487–93

Dong F, Hoefsloot LH, Schelen AM, Broeders LCAM, Meijer Y, et al. 1994. Identification of a nonsense mutation in the granulocyte-colony-stimulating factor receptor in severe congenital neutropenia. *Proc. Natl. Acad. Sci. USA* 91:4480–84

Doshi PD, DiPersio JF. 1994. Three conserved motifs in the extracellular domain of the human granulocyte-macrophage colony-stimulating factor receptor subunit are essential for ligand binding and surface expression. *Blood* 84:2539–53

Drachman JG, Griffin JD, Kaushansky K. 1995. The c-Mpl ligand (thrombopoietin) stimulates tyrosine phosphorylation of Jak2, Shc, and c-Mpl. *J. Biol. Chem.* 270:4979–82

Dranoff G, Crawford AD, Sadelain M, Ream B, Rashid A, et al. 1994. Involvement of granulocyte-macrophage colony-stimulating factor in pulmonary homeostasis. *Science* 264:713–16

Duquesnoy P, Sobrier ML, Amselem S, Goossens M. 1991. Defective membrane expression of human growth hormone (GH) receptor causes Laron-type GH insensitivity syndrome. *Proc. Natl. Acad. Sci. USA* 88:10272–76

Duquesnoy P, Sobrier ML, Duriez B, Dastot F, Buchanan CR, et al. 1994. A single amino acid substitution in the exoplasmic domain of the human growth hormone (GH) receptor confers familial GH resistance (Laron syndrome) with positive GH-binding activity by abolishing receptor homodimerization. *EMBO J.* 13:1386–95

Duriez B, Sobrier ML, Duquesnoy P, Tixier-Boichard M, Decuypere E, et al. 1993. A naturally occurring growth hormone receptor mutation: in vivo and in vitro evidence for the functional importance of the WS motif common to all members of the cytokine receptor superfamily. *Mol. Endocrinol.* 7:806–14

Dusanter-Fourt I, Muller O, Ziemiecke A, Mayeux P, Drucker B, et al. 1994. Identification of jak protein tyrosine kinases as signaling molecules for prolactin. Functional analysis of prolactin receptor and prolactin-erythropoietin receptor chimeras expressed in lymphoid cells. *EMBO J.* 13:2583–91

Elberg G, Kelly PA, Djiane J, Binder L, Gertler A. 1990. Mitogenic and binding properties of monoclonal antibodies to the prolactin receptor in Nb2 rat lymphoma cells. Selective enhancement by anti-mouse IgG. *J. Biol. Chem.* 265:14770–76

Elliot S, Lorenzini T, Yanagihara D, Elliot G,

Chang D. 1995. Isolation and characterization of anti rHuEPO receptor monoclonal antibodies that activate the EPO receptor. *Exp. Hematol.* 23(8):828

Ernst M, Gearing DP, Dunn AR. 1994. Functional and biochemical association of Hck with the LIF/IL-6 receptor signal transduction subunit gp130 in embryonic stem cells. *EMBO J.* 13:1574–84

Escary JL, Perreau J, Dum'enil D, Ezine S, Brulet P. 1993. Leukaemia inhibitory factor is necessary for maintenance of hematopoietic stem cells and thymocyte stimulation. *Nature* 363:361–64

Fairbairn LJ, Cowling GJ, Reipert BM, Dexter TM. 1993. Suppression of apoptosis allows differentiation and development of a multipotent hemopoietic cell line in the absence of added growth factors. *Cell* 74:823–32

Ferro F Jr, Kozak SL, Hoatlin ME, Kabat D. 1993. Cell surface site for mitogenic interaction of erythropoietin receptors with the membrane glycoprotein encoded by Friend erythroleukemia virus. *J. Biol. Chem.* 268:5741–47

Franks SJ, Yi W, Goldsmith JF, Gilliland G, Jiang J, et al. 1995. Regions of the jak2 tyrosine kinase required for coupling to the growth hormone receptor. *J. Biol. Chem.* 270:14776–85

Friend C. 1957. Cell-free transmission in adult Swiss mice of a disease having the character of a leukemia. *J. Exp. Med.* 105:307–18

Fuh G, Cunningham BC, Fukunaga R, Nagata S, Goeddel DV, Wells JA. 1992. Rational design of potent antagonists to the human growth hormone receptor. *Science* 256:1677–80

Fujii H, Nakagawa Y, Schindler U, Kawahara A, Mori H, et al. 1995. Activation of Stat5 by interleukin 2 requires a carboxyl-terminal region of the interleukin 2 receptor β chain but is not essential for the proliferative signal transmission. *Proc. Natl. Acad. Sci. USA* 92:5482–86

Fukunaga R, Ishizaka-Ikeda E, Nagata S. 1990a. Purification and characterization of the receptor for murine granulocyte colony-stimulating factor. *J. Biol. Chem.* 265:14008–15

Fukunaga R, Ishizaka-Ikeda E, Nagata S. 1993. Growth and differentiation signals mediated by different regions in the cytoplasmic domain of granulocyte colony-stimulating factor receptor. *Cell* 74:1079–87

Fukunaga R, Ishizaka-Ikeda E, Pan C-X, Seto Y, Nagata S. 1991. Functional domains of the granulocyte colony-stimulating factor receptor. *EMBO J.* 10:2855–65

Fukunaga R, Ishizaka-Ikeda E, Seto Y, Nagata S. 1990b. Expression cloning of a receptor for murine granulocyte colony-stimulating factor. *Cell* 61:341–50

Gearing DP, Comeau MR, Friend DJ, Gimpel SD, Thut CJ, et al. 1992. The IL-6 signal transducer, gp130: an oncostatin M receptor and affinity converter for the LIF receptor. *Science* 255:1434–37

Gearing DP, King JA, Gough NM, Nicola NA. 1989. Expression cloning of a receptor for human granulocyte-macrophage colony-stimulating factor. *EMBO J.* 8:3667–76

Gearing DP, Thut CJ, VandenBos T, Gimpel SD, Delaney PB, et al. 1991. Leukemia inhibitory factor receptor is structurally related to the IL-6 signal transducer, gp130. *EMBO J.* 10:2839–48

Giri JG, Ahdieh M, Eisenman J, Shanebeck K, Grabstein K, et al. 1994. Utilization of the beta and gamma chains of the IL-2 receptor by the novel cytokine IL-15. *EMBO J.* 13:2822–30

Giri JG, Kumaki S, Ahdieh M, Friend DJ, Loomis A, et al. 1995. Identification and cloning of a novel IL-15 binding protein that is structurally related to the alpha chain of the IL-2 receptor. *EMBO J.* 14:3654–63

Goodwin RG, Friend D, Ziegler SF, Jerzy R, Falk BA, et al. 1990. Cloning of the human and murine interleukin-7 receptors: demonstration of a soluble form and homology to a new receptor superfamily. *Cell* 60:941–51

Grabstein KH, Eisenman J, Shanebeck K, Rauch C, Srinivasan S, et al. 1994. Cloning of a T cell growth factor that interacts with the beta chain of the interleukin-2 receptor. *Science* 264:965–68

Greenlund AC, Farrar MA, Viviano BL, Schreiber RD. 1994. Ligand-induced IFNγ receptor tyrosine phosphorylation couples the receptor to its signal transduction system (p91). *EMBO J.* 13:1591–600

Gurney AL, Carver-Moore K, de Sauvage FJ, Moore MW. 1994. Thrombocytopenia in c-mpl-deficient mice. *Science* 265:1445–47

Gurney AL, Wong SC, Henzel WJ, de Sauvage FJ. 1995. Distinct regions of c-mpl cytoplasmic domain are coupled to the JAK-STAT signal transduction pathway and Shc phosphorylation. *Proc. Natl. Acad. Sci. USA* 92:5292–96

Hackett RH, Wang YD, Larner AC. 1995. Mapping of the cytoplasmic domain of the human growth hormone receptor required for the activation of Jak2 and Stat proteins. *J. Biol. Chem.* 270:21326–30

Hankins WD, Troxler D. 1980. Polycythemia- and anemia-inducing erythroleukemia viruses exhibit differential erythroid transforming effects in vitro. *Cell* 22:693–99

Hara T, Miyajima A. 1992. Two distinct func-

tional high affinity receptors for mouse interleukin-3 (IL-3). *EMBO J.* 11:1875–84
Harrison DA, Binar R, Stines T, Gilman M, Perrimon N. 1995. Activation of a *Drosophila* Janus kinase (JAK) causes hematopoietic neoplasia and developmental defects. *EMBO J.* 14:2857–65
Hatakeyama M, Kono T, Kobayashi N, Kawahara A, Levin SD, et al. 1991. Interaction of the IL-2 receptor with the src-family kinase p56lck: identification of novel intermolecular association. *Science* 252:1523–28
Hatakeyama M, Mori H, Doi T, Taniguchi T. 1989a. A restricted cytoplasmic region of IL-2 receptor β chain is essential for growth signal transduction but not for ligand binding and internalization. *Cell* 59:837–45
Hatakeyama M, Tsudo M, Minamoto S, Kono T, Doi T, et al. 1989b. Interleukin-2 receptor β chain gene: generation of three receptor forms by cloned human α and β chain cDNAs. *Science* 244:551–56
Hayashida K, Kitamura T, Gorman DM, Arai K-I, Yokota T, Miyajima A. 1990. Molecular cloning of a second subunit of the receptor for human granulocyte-macrophage colony-stimulating factor (GM-CSF): reconstitution of a high-affinity GM-CSF receptor. *Proc. Natl. Acad. Sci. USA* 87:9655–59
He TC, Jiang N, Zhuang H, Wojchowski DM. 1995. Erythropoietin recruitment of shc via a receptor phosphotyrosine-independent, Jak2-associated pathway. *J. Biol. Chem.* 270:11055–61
Heim MH, Kerr IM, Stark GR, Darnell JE. 1995. Contribution of STAT SH2 groups to specific interferon signaling by the Jak-STAT pathway. *Science* 267:1347–49
Heldin C-H. 1995. Dimerization of cell surface receptors in signal transduction. *Cell* 80:213–23
Hibi M, Murakami M, Saito M, Hirano T, Taga T, Kishimoto T. 1990. Molecular cloning and expression of an IL-6 signal transducer, gp130. *Cell* 63:1149–57
Hilton DJ, Hilton AA, Raicevic A, Raker S, Harrison-Smith M, et al. 1994. Cloning of a murine IL-11 receptor α-chain; requirement for gp130 for high affinity binding and signal transduction. *EMBO J.* 13:4765–75
Hilton DJ, Watowich SS, Katz L, Lodish HF. 1996a. Saturation mutagenesis of the WSXWS motif of the erythropoietin receptor. *J. Biol. Chem.* 271:4699–4708
Hilton DJ, Watowich SS, Murray PJ, Lodish HF. 1995. Increased cell surface expression and enhanced folding in the endoplasmic reticulum of a mutant erythropoietin receptor. *Proc. Natl. Acad. Sci. USA* 92:190–94
Hilton DJ, Zhang J-G, Metcalf D, Alexander WS, Nicola NA, Willson TA. 1996b. Cloning and characterization of a binding subunit of the interleukin 13 receptor that is also a component of the interleukin 4 receptor. *Proc. Natl. Acad. Sci. USA* 93:497–501
Hirota H, Yoshida K, Kishimoto T, Taga T. 1995. Continuous activation of gp130, a signal-transducing receptor component for interleukin 6-related cytokines, causes myocardial hypertrophy in mice. *Proc. Natl. Acad. Sci. USA* 92:4862–66
Hoatlin ME, Kozak SL, Lilly F, Chakraborti A, Kozak CA, Kabat D. 1990. Activation of the erythropoietin receptor by Friend viral gp55 and by erythropoietin and down-modulation by the murine Fv-2r resistance gene. *Proc. Natl. Acad. Sci. USA* 87:9985–89
Hooper KP, Padmanabhan R, Ebner KE. 1993. Expression of the extracellular domain of the rat liver prolactin receptor and its interaction with ovine prolactin. *J. Biol. Chem.* 268:22347–52
Horak ID, Gross RE, Lucas PJ, Horak EM, Waldmann TA, Bolen JB. 1991. T-lymphocyte interleukin-2 dependent tyrosine protein kinase signal transduction involves the activation of p56lck. *Proc. Natl. Acad. Sci. USA* 88:1996–2000
Hubbard SR, Wei L, Ellis L, Hendrickson WA. 1994. Crystal structure of the tyrosine kinase domain of the human insulin receptor. *Nature* 372:746–54
Ihle JN. 1995. Cytokine receptor signalling. *Nature* 377:591–94
Improta T, Schindler C, Horvath CM, Kerr IM, Stark GR, Darnell JE. 1994. Transcription factor ISGF-3 formation requires phosphorylated Stat91 protein, but Stat113 protein is phosphorylated independently of Stat91 protein. *Proc. Natl. Acad. Sci. USA* 91:4776–80
Ip NY, Nye SH, Boulton TG, Davis S, Taga T, et al. 1992. CNTF and LIF act on neuronal cells via shared signaling pathways that involve the IL-6 signal transducing receptor component gp130. *Cell* 69:1121–32
Ivashkiv LB. 1995. Cytokines and STATs: how can signals achieve specificity? *Immunity* 3:1–4
Johnston JA, Kawamura M, Kirken R, Chen Y, Blake TB, et al. 1994. Phosphorylation and activation of the jak2 Janus kinase in response to IL-2. *Nature* 370:151–53
Johnston JA, Wang L-M, Hanslon EP, Sun X-J, White MF, et al. 1995. Interleukins 2, 4, 7, and 15 stimulate tyrosine phosphorylation of insulin receptor substrates 1 and 2 in T cells. *J. Biol. Chem.* 270:28527–30
Jucker M, Feldman RA. 1995. Identification of a new adapter protein that may link the common β subunit of the receptor for gran-

ulocyte/macrophage colony stimulating factor, interleukin-3 (IL-3), and IL-5 to phosphatidylinositol 3-kinase. *J. Biol. Chem.* 270:27817–22

Kabat D. 1989. Molecular biology of Friend viral erythroleukemia. *Curr. Top. Microbiol. Immunol.* 148:1–42

Karnitz LM, Burns LA, Sutor SL, Blenis J, Abrahams RT. 1995. Interleukin-2 triggers a novel phosphatidylinositol 3-kinase-dependent MEK activation pathway. *Mol. Cell. Biol.* 15:3049–57

Kaushansky K. 1995. Thrombopoietin: the primary regulator of platelet production. *Blood* 86:419–31

Kaushansky K, Karplus PA. 1993. Hematopoietic growth factors: understanding functional diversity in structural terms. *Blood* 82:3229–40

Kaushansky K, Lok S, Holly RD, Broudy VC, Lin N, et al. 1994. Promotion of megakaryocyte progenitor expansion and differentiation by the c-Mpl ligand thrombopoietin. *Nature* 369:568–71

Kawahara A, Minami Y, Ihle JN, Taniguchi T. 1995. Critical role of the interleukin 2 (IL-2) receptor γ-chain-associated Jak3 in the IL-2-induced c-fos and c-myc but not bcl-2 induction. *Proc. Natl. Acad. Sci. USA* 92:8724–28

Kawahara A, Minami Y, Taniguchi T. 1994. Evidence for a critical role for the cytoplasmic region of the interleukin 2 (IL-2) receptor γ chain in IL-2, IL-4, and IL-7 signaling. *Mol. Cell. Biol.* 14:5433–40

Kawase I, Brooks CG, Kuribayashi K, Olabuenaga S, Newman W, et al. 1983. Interleukin 2 induces gamma-interferon production: participation of macrophages and NK-like cells. *J. Immunol.* 131:288–92

Keegan AD, Nelms K, White M, Wang L-M, Pierce JH, Paul WE. 1994. An IL-4 receptor region containing an insulin receptor motif is important for IL-4 mediated IRS-1 phosphorylation and cell growth. *Cell* 76:811–20

Kishimoto T, Akira S, Narazaki M, Taga T. 1995. Interleukin-6 family of cytokines and gp130. *Blood* 86:1243–54

Kishimoto T, Taga T, Akira S. 1994. Cytokine signal transduction. *Cell* 76:253–62

Kitamura T, Hayashida K, Sakamaki K, Yokota T, Arai K-I, Miyajima A. 1991b. Reconstitution of functional receptors for human granulocyte/macrophage colony-stimulating factor (GM-CSF): evidence that the protein encoded by the AIC2B cDNA is a subunit of the murine GM-CSF receptor. *Proc. Natl. Acad. Sci. USA* 88:5082–86

Kitamura T, Miyajima A. 1992. Functional reconstitution of the human interleukin-3 receptor. *Blood* 80:84–90

Kitamura T, Sato N, Arai K-I, Miyajima A. 1991a. Expression cloning of the human IL-3 receptor cDNA reveals a shared beta subunit for the human IL-3 and GM-CSF receptors. *Cell* 66:1165–74

Klingmuller U, Lorenz U, Cantley LC, Neel BG, Lodish HF. 1995. Specific recruitment of SH-PTP1 to the erythropoietin receptor causes inactivation of JAK2 and termination of proliferative signals. *Cell* 80:729–38

Kobayashi N, Kono T, Hatakeyama M, Minami Y, Miyazaki T, et al. 1993. Functional coupling of the src-family protein tyrosine kinases p59fyn and p53/56lyn with the interleukin-2 receptor: implications for redundancy and pleiotropism in cytokine signal transduction. *Proc. Natl. Acad. Sci. USA* 90:4201–5

Kondo M, Takeshita T, Higuchi M, Nakamura M, Sudo T, et al. 1994. Functional participation of the IL-2 receptor γ chain in IL-7 receptor complexes. *Science* 263:1453–54

Kondo M, Takeshita T, Ishii N, Nakamura M, Watanabe S, et al. 1993. Sharing of the interleukin-2 (IL-2) receptor gamma chain between receptors for IL-2 and IL-4. *Science* 262:1874–77

Kopf M, Baumann H, Freer G, Freudenberg M, Lamers M, et al. 1994. Impaired immune and acute-phase responses in interleukin-6-deficient mice. *Nature* 368:339–42

Kopf M, Brombacher F, Hodgkin PD, Ramsay AJ, Milbourne EA, et al. 1996. IL-5-deficient mice have a developmental defect in CD5+ B-1 cells and lack eosinophilia but have normal antibody and cytotoxic T cell responses. *Immunity* 4:15–24

Kopf M, Le Gros G, Bachmann M, Lamers MC, Bluethmann H, Kohler G. 1993. Disruption of the murine IL-4 gene blocks Th2 cytokine responses. *Nature* 362:245–48

Kopf M, Ramsay A, Brombacher F, Baumann H, Freer G, et al. 1995. Pleiotropic defects of IL-6-deficient mice including early hematopoiesis, T and B cell function, and acute phase responses. *Ann. NY Acad. Sci.* 762:308–18

Krantz SB. 1991. Erythropoietin. *Blood* 77:419–34

Kundig TM, Schorle H, Bachmann MF, Hengartner H, Zinkernagel RM, Horak I. 1993. Immune responses in interleukin-2-deficient mice. *Science* 262:1059–61

Lane DP, Benchimol S. 1990. p53: oncogene or anti-oncogene? *Genes Dev.* 4:1–8

Laron Z. 1995. Prismatic cases: Laron syndrome (primary growth hormone resistance) from patient to laboratory to patient. *J. Clin. Endocrinol. Metab.* 80:1526–31

Lebrun J-J, Ali S, Goffin V, Ullrich A, Kelly PA.

1995b. A single phosphotyrosine residue of the prolactin receptor is responsible for activation of gene transcription. *Proc. Natl. Acad. Sci. USA* 92:4031–35

Lebrun J-J, Ali S, Ullrich A, Kelly PA. 1995a. Proline-rich sequence-mediated Jak2 association to the prolactin receptor is required but not sufficient for signal transduction. *J. Biol. Chem.* 270:10664–70

Leung DW, Spencer SA, Cachianes G, Hammonds RG, Collins C, et al. 1987. Growth hormone receptor and serum binding protein: purification, cloning and expression. *Nature* 330:537–43

Leung S, Qureshi SA, Kerr IM, Darnell JE, Stark GR. 1995. Role of STAT2 in the alpha interferon signaling pathway. *Mol. Cell. Biol.* 15:1312–17

Levy DE, Kessler DS, Pine R, Darnell JE. 1989. Cytoplasmic activation of ISGF3, the positive regulator of interferon-alpha-stimulated transcription, reconstituted in vitro. *Genes Dev.* 3:1362–71

Levy DE, Kessler DS, Pine R, Reich N, Darnell JE. 1988. Interferon-induced nuclear factors that bind a shared promoter element correlate with positive and negative transcriptional control. *Genes Dev.* 2:383–93

Lew DJ, Decker T, Strehlow I, Darnell JE. 1991. Overlapping elements in the guanylate-binding protein gene promoter mediate transcriptional induction by alpha and gamma interferons. *Mol. Cell. Biol.* 11:182–91

Li J-P, Baltimore D. 1991. Mechanism of leukemogenesis induced by mink cell focus-forming murine leukemia viruses. *J. Virol.* 65:2408–14

Li J-P, D'Andrea AD, Lodish HF, Baltimore D. 1990. Activation of cell growth by binding of Friend spleen focus-forming virus gp55 glycoprotein to the erythropoietin receptor. *Nature* 343:762–64

Li J-P, Hu HO, Niu QT, Fang C. 1995. Cell surface activation of the erythropoietin receptor by Friend spleen focus-forming virus gp55. *J. Virol.* 69:1714–19

Li M, Sendtner M, Smith A. 1995. Essential function of LIF receptor in motor neurons. *Nature* 378:724–27

Lieschke GJ, Grail D, Hodgson G, Metcalf D, Stanley E, et al. 1994. Mice lacking granulocyte colony-stimulating factor have chronic neutropenia, granulocyte and macrophage progenitor cell deficiency, and impaired neutrophil mobilization. *Blood* 84:1737–46

Lin CS, Lim SK, D'Agati V, Costantini F. 1996. Differential effects of an erythropoietin receptor gene disruption on primitive and definitive erythropoiesis. *Genes Dev.* 10:154–64

Lok S, Kaushansky K, Holly RD, Kuijper JL, Lofton-Day CE, et al. 1994. Cloning and expression of murine thrombopoietin cDNA and stimulation of platelet production in vivo. *Nature* 369:565–68

Luo H, Hanratty WP, Dearolf CR. 1995. An amino acid substitution in the *Drosophila* *hop*$^{\mathrm{Tum-1}}$ Jak kinase causes leukemia-like hematopoietic defects. *EMBO J.* 14:1412–20

Macchi P, Villa A, Gillani S, Sacco MG, Frattini A, et al. 1995. Mutations of Jak-3 gene in patients with autosomal severe combined immune deficiency (SCID). *Nature* 376:65–68

Malkovsky M, Loveland B, North M, Asherson GL, Gao L, et al. 1987. Recombinant interleukin-2 directly augments the cytotoxicity of human monocytes. *Nature* 325:262–65

Masu Y, Wolf E, Holtmann B, Sendtner M, Brem G, Thoenen H. 1993. Disruption of the CNTF gene results in motor neuron degeneration. *Nature* 365:27–32

Matsuda T, Fukada T, Takahashi-Tezuka M, Okuyama Y, Fujitani Y, et al. 1995a. Activation of fes tyrosine kinase by gp130, an interleukin-6 family cytokine signal transducer, and their association. *J. Biol. Chem.* 270:11037–39

Matusda T, Takahashi-Tezuka M, Fukada T, Okuyama Y, Fujitani Y, et al. 1995b. Association and activation of Btk and Tec tyrosine kinases by gp130, a signal transducer of the interleukin-6 family of cytokines. *Blood* 85:627–33

Meacham LR, Brown MR, Murphy TL, Keret R, Silbergeld A, et al. 1993. Characterization of a noncontiguous gene deletion of the growth hormone receptor in Laron's syndrome. *J. Clin. Endocrinol. Metab.* 77:1379–83

Mertelsmann R. 1993. Hematopoietic cytokines: from biology and pathophysiology to clinical application. *Leukemia* 7:S168–77

Metcalf D. 1984. *The Hematopoietic Colony Stimulating Factors.* Amsterdam/New York/Oxford: Elsevier

Metcalf D. 1989. The molecular control of cell division, differentiation commitment and maturation in haemopoietic cells. *Nature* 339:27–30

Metcalf D. 1993. Hematopoietic regulators: redundancy or subtlety? *Blood* 82:3515–23

Minami Y, Kono T, Yamada K, Kobayashi N, Kawahara A, et al. 1993. Association of p56lck with IL-2 receptor β chain is critical for the IL-2-induced activation of p56lck. *EMBO J.* 12:759–68

Minami Y, Nakagawa Y, Kawahara A, Miyazaki T, Sada K, et al. 1995. Protein tyrosine kinase syk is associated with and activated by the

IL-2 receptor: possible link with the c-myc induction pathway. *Immunity* 2:89–100

Miura O, Cleveland JL, Ihle JN. 1993. Inactivation of erythropoietin receptor function by point mutations in a region having homology with other cytokine receptors. *Mol. Cell. Biol.* 13:1788–95

Miyajima A, Hara T, Kitamura T. 1992. Common subunits of cytokine receptors and the functional redundancy of cytokines. *Trends Biochem. Sci.* 17:378–82

Miyajima A, Mui AL-F, Ogorochi T, Sakamaki K. 1993. Receptors for granulocyte-macrophage colony-stimulating factor, interleukin-3, and interleukin-5. *Blood* 82:1960–74

Miyazaki T, Liu Z-J, Kawahara A, Minami Y, Yamada K, et al. 1995. Three distinct IL-2 signaling pathways mediated by bcl-2, c-myc and lck cooperate in hematopoietic cell proliferation. *Cell* 81:223–31

Moreau-Gachelin F, Ray D, de Both NJ, van der Feltz MJ, Tambourin P, Tavitian A. 1990. Spi-1 oncogene activation in Rauscher and Friend murine virus-induced acute erythroleukemias. *Leukemia* 4:20–23

Morrison SJ, Uchida N, Weissman IL. 1995. The biology of hematopoietic stem cells. *Annu. Rev. Cell Dev. Biol.* 11:35–71

Mosley B, Beckmann MP, March CJ, Idzerda RL, Gimpel SD, et al. 1989. The murine interleukin-4 receptor: molecular cloning and characterization of secreted and membrane bound forms. *Cell* 59:335–48

Muller M, Briscoe J, Laxton C, Guschin D, Ziemiecki A, et al. 1993. The protein tyrosine kinase Jak1 complements defects in interferon-α/β and -γ signal transduction. *Nature* 366:129–35

Murakami M, Hibi M, Nakagawa N, Nakagawa T, Yasukawa K, et al. 1993. IL-6-induced homodimerization of gp130 and associated activation of a tyrosine kinase. *Science* 260:1808–10

Murakami M, Narazaki M, Hibi M, Yawata H, Yasukawa K, et al. 1991. Critical cytoplasmic region of the interleukin 6 signal transducer gp130 is conserved in the cytokine receptor family. *Proc. Natl. Acad. Sci. USA* 88:11349–53

Narazaki M, Witthuhn BA, Yoshida K, Kishimoto T, Taga T. 1994. Activation of Jak2 kinase mediated by the IL-6 signal transducer, gp130. *Proc. Natl. Acad. Sci. USA* 91:2285–89

Nicola NA. 1989. Hemopoietic cell growth factors and their receptors. *Annu. Rev. Biochem.* 58:45–77

Nicolson SE, Oates AC, Harpur AG, Ziemiecke A, Wilks AF, Layton JE. 1994. Tyrosine kinase jak1 is associated with the granulocyte-colony stimulating factor receptor and both become tyrosine-phosphorylated after receptor activation. *Proc. Natl. Acad. Sci. USA* 91:2985–88

Nikaido T, Shimizu A, Ishida N, Sabe H, Teshigawara K, et al. 1984. Molecular cloning of cDNA encoding human interleukin-2 receptor. *Nature* 311:631–35

Nishinakamura R, Nakayama N, Hirabayashi Y, Inoue T, Aud D, et al. 1995. Mice deficient for the IL-3/GM-CSF/IL-5 beta c receptor exhibit lung pathology and impaired immune response, while beta IL3 receptor-deficient mice are normal. *Immunity* 2:211–22

Noguchi M, Nakamura Y, Russell SM, Ziegler SF, Tsang M, et al. 1993a. Interleukin-2 receptor gamma chain: a functional component of the interleukin-7 receptor. *Science* 262:1877–80

Noguchi M, Yi H, Rosenblatt HM, Filipovich AH, Adelstein S, et al. 1993b. Interleukin-2 receptor gamma chain mutation results in X-linked severe combined immunodeficiency in humans. *Cell* 73:147–57

Nosaka T, van Deursen JMA, Tripp RA, Thierfelder WE, Witthuhn BA, et al. 1995. Defective lymphoid development in mice lacking Jak3. *Science* 270:800–2

Ogawa M. 1993. Differentiation and proliferation of hematopoietic stem cells. *Blood* 81:2844–53

Ohashi T, Masuda M, Ruscetti SK. 1995. Induction of sequence-specific DNA-binding factors by erythropoietin and the spleen focus-forming virus. *Blood* 85:1454–62

Pallard C, Gouilleux F, Benit L, Cocault L, Souyri M, et al. 1995. Thrombopoietin activates a STAT5-like factor in hematopoietic cells. *EMBO J.* 14:2847–56

Park SY, Saijo K, Takahashi T, Osawa M, Arase H, et al. 1995. Developmental defects of lymphoid cells in Jak3 kinase-deficient mice. *Immunity* 3:771–82

Pellegrini S, Schindler C. 1993. Early events in signaling by interferons. *Trends Biochem. Sci.* 18:338–42

Pennica D, King KL, Shaw KJ, Luis E, Rullamas J, et al. 1995. Expression cloning of cardiotrophin 1, a cytokine that induces cardiac myocyte hypertrophy. *Proc. Natl. Acad. Sci. USA* 92:1142–46

Pepper AE, Buckley RH, Small TN, Puck JM. 1995. Two mutational hotspots in the interleukin-2 receptor gamma chain gene causing human X-linked severe combined immunodeficiency. *Am. J. Hum. Genet.* 57:564–71

Peschon JJ, Morrissey PJ, Grabstein KH, Ramsdell FJ, Maraskovsky E, et al. 1994. Early lymphocyte expansion is severely impaired in

interleukin 7 receptor-deficient mice. *J. Exp. Med.* 180:1955–60
Pharr PN, Hankins D, Hofbauer A, Lodish HF, Longmore GD. 1993. Expression of a constitutively active erythropoietin receptor in primary hematopoietic progenitors abrogates erythropoietin dependence and enhances erythroid colony-forming unit, erythroid burst-forming unit, and granulocyte/macrophage progenitor growth. *Proc. Natl. Acad. Sci. USA* 90:938–42
Philo JS, Aoki KH, Arakawa T, Owers Narhi L, Wen J. 1996. Dimerization of the extracellular domain of the erythropoietin (EPO) receptor by EPO: one high-affinity and one low-affinity interaction. *Biochemistry.* In press
Quelle FW, Thierfelder W, Witthuhn BA, Tang B, Cohen S, Ihle JN. 1995. Phosphorylation and activation of the DNA binding activity of purified Stat1 by the Janus protein-tyrosine kinases and the epidermal growth factor receptor. *J. Biol. Chem.* 270:20775–80
Rao MS, Sun Y, Escary JL, Perreau J, Tresser S. 1993. Leukaemia inhibitory factor mediates an injury response but not a target-directed developmental transmitter switch in sympathetic neurons. *Neuron* 11:1175–85
Rao P, Mufson RA. 1995. A membrane proximal domain of the human interleukin-3 receptor βc subunit that signals DNA synthesis in NIH 3T3 cells specifically binds a complex of src and janus family tyrosine kinases and phosphatidylinositol 3-kinase. *J. Biol. Chem.* 270:6886–93
Renauld J-C, Druez C, Kermouni A, Houssiau F, Uyttenhove C, et al. 1992. Expression cloning of the murine and human interleukin 9 receptor cDNAs. *Proc. Natl. Acad. Sci. USA* 89:5690–94
Ridderstrale M, Degerman E, Tornqvist H. 1995. Growth hormone stimulates the tyrosine phosphorylation of the insulin receptor substrate-1 and its association with phosphatidylinositol 3-kinase in primary adipocytes. *J. Biol. Chem.* 270:3471–74
Ronco LV, Silverman SL, Wong SG, Slamon DJ, Park LS, Gasson JC. 1994. Identification of conserved amino acids in the human granulocyte-macrophage colony-stimulating factor receptor alpha subunit critical for function. Evidence for formation of a heterodimeric receptor complex prior to ligand binding. *J. Biol. Chem.* 269:277–83
Rozakis-Adcock M, Kelly PA. 1991. Mutational analysis of the ligand-binding domain of the prolactin receptor. *J. Biol. Chem.* 266:16472–77
Rui H, Kirken RA, Farrar WL. 1994. Activation of receptor-associated tyrosine kinase Jak2 by prolactin. *J. Biol. Chem.* 269:5364–68
Ruscetti SK, Janesch NJ, Chakraborti A, Sawyer ST, Hankins WD. 1990. Friend spleen focus-forming virus induces factor independence in an erythropoietin-dependent erythroleukemia cell line. *J. Virol.* 64:1057–64
Russell SM, Johnston JA, Noguchi M, Kawamura M, Bacon CM, et al. 1994. Interaction of IL-2R beta and gamma c chains with Jak1 and Jak3: implications for XSCID and XCID. *Science* 266:1042–45
Russell SM, Keegan AD, Harada N, Nakamura Y, Noguchi M, et al. 1993. Interleukin-2 receptor gamma chain: a functional component of the interleukin-4 receptor. *Science* 262:1880–83
Russell SM, Tayebi N, Nakajima H, Riedy MC, Roberts JL, et al. 1995. Mutation of Jak3 in a patient with SCID: essential role of Jak3 in lymphoid development. *Science* 270: 797–800
Sadlack B, Merz H, Schorle H, Schimpl A, Feller AC, Horak I. 1993. Ulcerative colitis-like disease in mice with a disrupted interleukin-2 gene. *Cell* 75:253–61
Sakai I, Nabell L, Kraft AS. 1995. Signal transduction by a CD16/CD7/Jak2 fusion protein. *J. Biol. Chem.* 270:18420–27
Sakamaki K, Miyajima I, Kitamura T, Miyajima A. 1992. Critical cytoplasmic domains of the common β subunit of the human GM-CSF, IL-3, and IL-5 receptors for growth signal transduction and tyrosine phosphorylation. *EMBO J.* 11:3541–49
Satoh T, Minami T, Kono T, Yamada K, Kawahara A, et al. 1992. Interleukin 2-induced activation of ras requires two domains of interleukin 2 receptor β subunit, the essential region for growth stimulation and lck-binding domain. *J. Biol. Chem.* 267:25423–27
Sattler M, Durstin MA, Frank DA, Okuda K, Kaushansky K, et al. 1995. The thrombopoietin receptor c-MPL activates JAK2 and TYK2 tyrosine kinases. *Exp. Hematol.* 23:1040–48
Schorle H, Holtschke T, Hunig T, Schimpl A, Horak I. 1991. Development and function of T cells in mice rendered interleukin-2 deficient by gene targeting. *Nature* 352:621–24
Showers MO, De Martino JC, Saito Y, D'Andrea AD. 1993. Fusion of the erythropoietin receptor and the Friend spleen focus-forming virus gp55 glycoprotein transforms a factor-dependent hematopoietic cell line. *Mol. Cell. Biol.* 13:739–48
Showers MO, Moreau JF, Linnekin D, Druker B, D'Andrea AD. 1992. Activation of the erythropoietin receptor by the Friend spleen focus-forming virus gp55 glycoprotein induces constitutive protein tyrosine phospho-

rylation. *Blood* 80:3070–78

Shuai K, Horvath CM, Huang LHT, Qureshi SA, Cowburn D, Darnell JE. 1994. Interferon activation of the transcription factor Stat91 involves dimerization through SH2-phosphotyrosyl peptide interactions. *Cell* 76:821–28

Shuai K, Schindler C, Prezioso VR, Darnell JE. 1992. Activation of transcription by IFN-gamma: tyrosine phosphorylation of a 91-kD DNA binding protein. *Science* 258:1808–12

Sidman CL, Marshall JD, Shultz LD, Gray PW, Johnson HM. 1984. Gamma-interferon is one of several direct B cell-maturing lymphokines. *Nature* 309:801–4

Silvennoinen O, Witthuhn BA, Quelle FW, Cleveland JL, Yi T, Ihle JN. 1993. Structure of the murine Jak2 protein-tyrosine kinase and its role in interleukin 3 signal transduction. *Proc. Natl. Acad. Sci. USA* 90:8429–33

Skoda RC, Seldin DC, Chiang MK, Peichel CL, Vogt TF, Leder P. 1993. Murine c-mpl: a member of the hematopoietic growth factor receptor superfamily that transduces a proliferative signal. *EMBO J.* 12:2645–53

Smith KA. 1988. Interleukin 2: inception, impact, and implications. *Science* 240:1169–76

Sokol L, Luhovy M, Guan Y, Prchal J, Semenza G, Prchal J. 1995. Primary familial polycythemia: a frameshift mutation in the erythropoietin receptor gene and increased sensitivity of erythroid progenitors to erythropoietin. *Blood* 86:15–22

Sokol L, Prchal JF, D'Andrea AD, Rado TA, Prchal JT. 1994. Mutation in the negative regulatory element of the erythropoietin receptor gene in a case of sporadic primary polycythemia. *Exp. Hematol.* 22:447–53

Somers W, Ultsch M, De Vos AM, Kossiakoff AA. 1994. The X-ray structure of a growth hormone-prolactin receptor complex. *Nature* 372:478–81

Souyri M, Vigon I, Penciolelli JF, Heard JM, Tambourin P, Wendling F. 1990. A putative truncated cytokine receptor gene transduced by the myeloproliferative leukemia virus immortalizes hematopoietic progenitors. *Cell* 63:1137–47

Stahl N, Boulton TG, Farruggella T, Ip NY, Davis S, et al. 1994. Association and activation of Jak-Tyk kinases by CNTF-LIF-OM-IL-6 β receptor components. *Science* 263:92–95

Stahl N, Farruggella TJ, Boulton TG, Zhong Z, Darnell JE, Yancopoulos GD. 1995. Choice of STATs and other substrates specified by modular tyrosine-based motifs in cytokine receptors. *Science* 267:1349–53

Stahl N, Yancopoulos GD. 1993. The alphas, betas, and kinases of cytokine receptor complexes. *Cell* 74:587–90

Stanley E, Lieschke GJ, Grail D, Metcalf D, Hodgson G, et al. 1994. Granulocyte/macrophage colony-stimulating factor-deficient mice show no major perturbation of hematopoiesis but develop a characteristic pulmonary pathology. *Proc. Natl. Acad. Sci. USA* 91:5592–96

Stewart CL, Kaspar P, Brunet LJ, Bhatt H, Gadi I, et al. 1992. Blastocyst implantation depends on maternal expression of leukaemia inhibitory factor. *Nature* 359:76–79

Suzuki H, Kundig TM, Furlonger C, Wakeham A, Timms E, et al. 1995. Deregulated T cell activation and autoimmunity in mice lacking interleukin-2 receptor β. *Science* 268:1472–76

Taga T, Hibi M, Hirata Y, Yamasaki K, Yasukawa K, et al. 1989. Interleukin-6 triggers the association of its receptor with a possible signal transducer, gp130. *Cell* 58:573–81

Takaki S, Mita S, Kitamura T, Yonehara S, Yamaguchi N, et al. 1991. Identification of the second subunit of the murine interleukin-5 receptor: interleukin-3 receptor-like protein, AIC2B is a component of the high affinity interleukin-5 receptor. *EMBO J.* 10:2833–38

Takaki S, Murata Y, Kitamura T, Miyajima A, Tominaga A, Takatsu K. 1993. Reconstitution of the functional receptors for murine and human interleukin-5. *J. Exp. Med.* 177:1523–29

Takaki S, Tominaga A, Hitoshi Y, Mita S, Sonada E, et al. 1990. Molecular cloning and expression of the murine interleukin-5 receptor. *EMBO J.* 9:4367–74

Taniguchi T, Minami Y. 1993. The IL-2/IL-2 receptor system: a current overview. *Cell* 73:5–8

Tanner JW, Chen W, Young RL, Longmore GD, Shaw AS. 1995. The conserved Box1 motif of cytokine receptors is required for association with Jak kinases. *J. Biol. Chem.* 270:6523–30

Tartaglia LA, Dembski M, Weng X, Deng N, Culpepper J, et al. 1995. Identification and expression cloning of a leptin receptor, OB-R. *Cell* 83:1263–71

Tavernier J, Devos R, Cornelis S, Tuypens T, Van der Heyden J, et al. 1991. A human high affinity interleukin-5 receptor (IL5R) is composed of an IL5-specific alpha chain and a beta chain shared with the receptor for GM-CSF. *Cell* 66:1175–84

Thomis DC, Gurniak CB, Tivol E, Sharpe AH, Berg LJ. 1995. Defects in B lymphocyte maturation and T lymphocyte activation in mice lacking Jak3. *Science* 270:794–97

Torigo T, Saragovi HU, Reed JC. 1992. Interleukin-2 regulates the activity of the lyn protein tyrosine kinase in a B-cell line. *Proc.*

Natl. Acad. Sci. USA 89:2674–78
Truitt KE, Mills GB, Turck CW, Imboden JB. 1994. SH2-dependent association of phosphatidylinositol 3-kinase 85-kDa regulatory subunit with the interleukin-2 receptor beta chain. *J. Biol. Chem.* 269:5937–43
Vajdy M, Kosco-Vilbois MH, Kopf M, Kohler G, Lycke N. 1995. Impaired mucosal immune responses in interleukin 4-targeted mice. *J. Exp. Med.* 181:41–53
van Zant G, Schultz L. 1989. Hematologic abnormalities of the immunodeficient mouse mutant, viable motheaten (me^v). *Exp. Hematol.* 17:81–87
Vaux DL, Cory S, Adam JM. 1988. Bcl-2 gene promotes haemopoietic cell survival and cooperates with c-myc to immortalize pre-B cells. *Nature* 335:440–42
Velazquez L, Fellous M, Stark GR, Pellegrini S. 1992. A protein tyrosine kinase in the interferon α/β signaling pathway. *Cell* 70:313–22
Vigon I, Mornon J-P, Cocault L, Mitjavila M-T, Tambourin P, et al. 1992. Molecular cloning and characterization of MPL, the human homolog of the v-mpl oncogene: identification of a member of the hematopoietic growth factor receptor superfamily. *Proc. Natl. Acad. Sci. USA* 89:5640–44
von Freeden-Jeffry U, Vieira P, Lucian LA, McNeil T, Burdach SE, Murray R. 1995. Lymphopenia in interleukin (IL)-7 gene-deleted mice identifies IL-7 as a nonredundant cytokine. *J. Exp. Med.* 181:1519–26
Wakao H, Gouilleux F, Groner B. 1994. Mammary gland factor (MGF) is a novel member of the cytokine regulated transcription factor gene family and confers the prolactin response. *EMBO J.* 13:2182–91
Walter MR, Windsor WT, Nagabhushan TL, Lundell DJ, Lunn CA, et al. 1995. Crystal structure of a complex between interferon-γ and its soluble high-affinity receptor. *Nature* 376:230–35
Wang L-M, Myers MG Jr, Sun X-J, Aaronson WA, White MF, Pierce JH. 1993. IRS-1: essential for insulin- and IL-4-stimulated mitogenesis in hematopoietic cells. *Science* 261:1591–94
Ward LD, Howlett GJ, Discolo G, Yasukawa K, Hammacher A, et al. 1994. High affinity interleukin-6 receptor is a hexameric complex consisting of two molecules each of interleukin-6, interleukin-6 receptor, and gp130. *J. Biol. Chem.* 269:23286–89
Ware CB, Horowitz MC, Renshaw BR, Hunt JS, Liggitt D. 1995. Targeted disruption of the low-affinity leukemia inhibitory factor receptor gene causes placental, skeletal, neural and metabolic defects and results in perinatal death. *Development* 121:1283–99
Watling D, Guschin D, Muller M, Silvennoinen O, Witthuhn BA, et al. 1993. Complementation of a mutant cell line defective in the interferon-γ signal transduction pathway. *Nature* 366:166–70
Watowich SS, Hilton DJ, Lodish HF. 1994. Activation and inhibition of erythropoietin receptor function: role of receptor dimerization. *Mol. Cell. Biol.* 14:3535–49
Watowich SS, Yoshimura A, Longmore GD, Hilton DJ, Yoshimura Y, Lodish HF. 1992. Homodimerization and constitutive activation of the erythropoietin receptor. *Proc. Natl. Acad. Sci. USA* 89:2140–44
Wegenka UM, Lutticken C, Buschmann J, Yuan J, Lottspeich F, et al. 1994. The interleukin-6-activated acute-phase response factor is antigenically and functionally related to members of the signal transducer and activator of transcription (STAT) family. *Mol. Cell. Biol.* 14:3186–96
Weiss M, Yokoyama C, Shikama Y, Naugle C, Druker B, Sieff CA. 1993. Human granulocyte-macrophage colony-stimulating factor receptor signal transduction requires the proximal cytoplasmic domains of the α and β subunits. *Blood* 82:3298–306
Welte T, Garimorth K, Philipp S, Doppler W. 1994. Prolactin-dependent activation of a tyrosine phosphorylated DNA binding factor in mouse mammary epithelial cells. *Mol. Endocrinol.* 8:1091–102
Wen Z, Zhong Z, Darnell JE. 1995. Maximal activation of transcription by Stat1 and Stat3 requires both tyrosine and serine phosphorylation. *Cell* 82:241–50
Wendling F, Maraskovsky E, Debili N, Florindo C, Teepe M, et al. 1994. cMpl ligand is a humoral regulator of megakaryocytopoiesis. *Nature* 369:571–74
Wendling F, Varlet P, Charon M, Tambourin P. 1986. MPLV: a retrovirus complex inducing an acute myeloproliferative leukemic disorder in adult mice. *Virology* 149:242–46
White MF. 1994. The IRS-1 signaling system. *Curr. Opin. Genet. Dev.* 4:47–54
Willerford DM, Chen J, Ferry JA, Davidson L, Ma A, Alt FW. 1995. Interleukin-2 receptor α chain regulates the size and content of the peripheral lymphoid compartment. *Immunity* 3:521–30
Williams GT, Smith CA, Spooncer E, Dexter TM, Taylor DR. 1990. Haemopoietic colony stimulating factors promote cell survival by suppressing apoptosis. *Nature* 343:76–79
Witthuhn BA, Quelle FW, Silvennoinen O, Yi T, Tang B, et al. 1993. JAK2 associates with the erythropoietin receptor and is tyrosine phosphorylated and activated following stimulation with erythropoietin. *Cell* 74:227–36

Witthuhn BA, Silvennoinen O, Miura O, Lai KS, Cwik C, et al. 1994. Involvement of the jak3 Janus kinase in IL-2 and IL-4 signaling in lymphoid and myeloid cells. *Nature* 370:153–57

Wolff L, Ruscetti S. 1985. Malignant transformation of erythroid cells in vivo by introduction of a nonreplicating retrovirus vector. *Science* 228:1549–52

Wu H, Klingmuller U, Besmer P, Lodish HF. 1995a. Interaction of the erythropoietin and stem-cell-factor receptors. *Nature* 377:242–46

Wu H, Liu X, Jaenisch R, Lodish HF. 1995b. Generation of committed erythroid BFU-E and CFU-E progenitors does not require erythropoietin or the erythropoietin receptor. *Cell* 83:59–67

Yamamura Y, Noda M, Ikawa Y. 1994. Activated Ki-Ras complements erythropoietin signaling in CTLL-2 cells, inducing tyrosine phosphorylation of a 160-kDa protein. *Proc. Natl. Acad. Sci. USA* 91:8866–70

Yamasaki K, Taga T, Hirata Y, Yawata H, Kawanishi Y, et al. 1988. Cloning and expression of the human interleukin-6 (BSF-2/IFNβ 2) receptor. *Science* 241:825–28

Yi T, Mui AL-F, Krystal G, Ihle JN. 1993. Hematopoietic cell phosphatase associates with the interleukin-3 (IL-3) receptor β chain and down-regulates IL-3-induced tyrosine phosphorylation and mitogenesis. *Mol. Cell. Biol.* 13:7577–86

Yoshida K, Taga T, Saito M, Suematsu S, Kumanogoh A, et al. 1996. Targeted disruption of gp130, a common signal transducer for the interleukin 6 family of cytokines, leads to myocardial and hematological disorders. *Proc. Natl. Acad. Sci. USA* 93:407–11

Yoshida Y. 1994. Overview of the hematopoietic growth factors and cytokines—past, present and future. *Int. J. Hematol.* 60:177–83

Yoshimura A, D'Andrea AD, Lodish HF. 1990b. Friend spleen focus-forming virus glycoprotein gp55 interacts with the erythropoietin receptor in the endoplasmic reticulum and affects receptor metabolism. *Proc. Natl. Acad. Sci. USA* 87:4139–43

Yoshimura A, Longmore G, Lodish HF. 1990a. Point mutation in the exoplasmic domain of the erythropoietin receptor resulting in hormone-independent activation and tumorigenicity. *Nature* 348:647–49

Yoshimura A, Zimmers T, Neumann D, Longmore G, Yoshimura Y, Lodish HF. 1992. Mutations in the WSXWS motif of the erythropoietin receptor abolish processing, ligand-binding, and activation of the receptor. *J. Biol. Chem.* 267:11619–25

Zhang X, Blenis J, Li H-C, Schindler C, Chen-Kiang S. 1995. Requirement of serine phosphorylation for formation of STAT-promoter complexes. *Science* 267:1990–94

Zhao Y, Wagner F, Frank SJ, Kraft AS. 1995. The amino-terminal portion of the jak2 protein kinase is necessary for binding and phosphorylation of the granulocyte-macrophage colony-stimulating factor receptor β_c chain. *J. Biol. Chem.* 270:13814–18

Zhuang H, Patil SV, He T-C, Sonsteby SK, Niu Z, Wojchowski DM. 1994. Inhibition of erythropoietin-induced mitogenesis by a kinase-deficient form of Jak2. *J. Biol. Chem.* 269:21411–14

Ziegler SF, Bird TA, Morella KK, Mosley B, Gearing DP, Baumann H. 1993. Distinct regions of the human granulocyte-colony-stimulating factor receptor cytoplasmic domain are required for proliferation and gene induction. *Mol. Cell. Biol.* 13:2384–90

Ziemiecki A, Harpur AG, Wilks AF. 1994. Jak protein tyrosine kinases: their role in cytokine signaling. *Trends Cell Biol.* 4:207–12

Zon LI, Moreau JF, Koo JW, Mathey-Prevot B, D'Andrea AD. 1992. The erythropoietin receptor transmembrane region is necessary for activation by the Friend spleen focus-forming virus gp55 glycoprotein. *Mol. Cell. Biol.* 12:2949–57

Zurawski SM, Chormarat P, Djossou O, Bidaud C, McKenzie ANJ, et al. 1995. The primary binding subunit of the human interleukin-4 receptor is also a component of the interleukin-13 receptor. *J. Biol. Chem.* 270:13869–78

Zurawski SM, Vega F, Huyghe B, Zurawski G. 1993. Receptors for interleukin-13 and interleukin-4 are complex and share a novel component that functions in signal transduction. *EMBO J.* 12:2663–70

Annu. Rev. Cell Dev. Biol. 1996. 12:129–60

ACTIN: General Principles from Studies in Yeast

Kathryn R. Ayscough and David G. Drubin
Department of Molecular and Cell Biology, 401 H.A. Barker Hall, University of California, Berkeley, California 94720-3202

KEY WORDS: microfilament, structure-function, genetics, myosin, cdc42

ABSTRACT

Three of the most important questions concerning actin function are: (*a*) How does actin structure relate to actin function? (*b*) How does each of the numerous proteins that interact with actin contribute to actin cytoskeleton function in vivo? (*c*) How are the activities of these proteins regulated? Powerful molecular genetics combined with well-established biochemical techniques make the yeast *Saccharomyces cerevisiae* an ideal organism for studies aimed at answering these questions.

The protein sequences and biochemical properties of actin and its interacting proteins and the pathways that regulate these interactions all appear to be conserved, indicating that principles elucidated from studies in yeast will apply to all eukaryotes. In this review, we highlight advances in our general understanding of actin properties, interactions with other proteins, and regulation of the actin cytoskeleton, derived from studies in the budding yeast *S. cerevisiae*.

CONTENTS

1081-0706/96/1115-0129$08.00

STRUCTURE-FUNCTION RELATIONSHIPS

Overview

Elucidation of any atomic structure allows us to understand how the various elements of primary and secondary structure come together to produce a functional molecule. The atomic structure gives us an idea of which surfaces of a protein might be involved in the intra- and intermolecular contacts needed for function. However, spatial information alone cannot prove the various hypotheses elicited by such a structure. To probe the functional aspects of a structure, genetics and mutagenesis have proved extremely powerful tools, with yeast being an ideal organism for such studies. The lessons learned from yeast reveal the great complementarity of structural analysis and genetics.

Structural Analysis of Actin

Determination of the atomic structure of actin in its monomeric form (complexed with DNase I) (Kabsch et al 1990) has provided the basis for much of our current understanding of the protein. The structure reveals the 375 residue polypeptide to be folded into two large domains, each comprised of two subdomains (the subdomains are numbered 1–4 according to the nomenclature of Kabsch 1990; Figure 1). The large domains are organized to form a hinged molecule with a deep cleft. Actin's essential cofactors—an adenine nucleotide and a divalent metal ion—reside within the cleft. The metal ion and the nucleotide are predicted to make extensive contacts with domains on either side of the cleft, thus increasing connectivity between them. The solution of the atomic structure has made possible theoretical analysis of the molecular dynamics of actin and has also provided a framework for interpretation of numerous biochemical and biophysical studies concerning ligand binding, actin filament formation, and force generation.

Unlike monomeric actin (G-actin), filamentous or F-actin does not form crystalline arrays amenable to high resolution X-ray crystallographic analysis. However, F-actin can be induced to form oriented arrays appropriate for fiber diffraction that yield low-resolution structural information. When coupled with electron-microscopy data and using a range of mathematical modeling approaches, the structure of actin filaments has been proposed and subsequently refined to fit much of the data available (Holmes et al 1990, Lorenz et al 1993, Tirion et al 1995).

Subsequent to the determination of the atomic structure of actin with DNase I (Kabsch et al 1990), actin has now been crystallized with various physiological binding partners such as gelsolin segment-1 (McLaughlin et al 1993) and profilin (Schutt et al 1993). In addition, image reconstruction studies have led to proposed structures for the association of F-actin with myosin (Rayment et al 1993a, Schröder et al 1993) and with tropomyosin (Lorenz et al 1995). These molecular structures demonstrate which domains of the protein are involved in interactions and provide a framework for understanding how such contacts might influence the dynamic properties of actin.

Prior to the elucidation of the atomic structure of actin, the major approach to gain structural information about actin was chemical covalent derivatization. This technique can evaluate the proximity of certain amino acids within the actin monomer and can be used to determine which residues are at the interface between actin and a specific ligand (for a thorough review, see Sheterline & Sparrow 1994). These approaches do, however, have inherent limitations. In particular, they are limited by the chemical reactivity of each amino acid with only certain residues in the actin monomer sufficiently reactive to allow modification or cross-linking.

The structural information gained from chemical derivatization and the atomic structure allows prediction of the roles of various actin domains. Residues surrounding the cleft region are likely to be involved in binding or hydrolyzing nucleotide and possibly in stabilizing monomer structure. Other regions of the protein are likely to be involved in making contacts essential for filament formation or for interactions with a plethora of binding proteins. Some of the predictions from these studies have been tested using yeast. The following sections describe the in vivo and in vitro results of these investigations.

Using Yeast to Study Structure-Function Relationships for Actin

There are a number of unique advantages inherent in the yeast *S. cerevisiae* making it a particularly powerful organism with which to study the structure-function relationships of actin. For example, specific codons of *ACT1*, the yeast actin gene, can be changed in vitro and then, because of the high frequency of homologous recombination in this organism, the single wild-type copy of the gene can be precisely replaced by the mutant gene. Because *ACT1* encodes the only source of conventional actin in budding yeast, its replacement with a mutant actin gene allows in vivo analysis of a mutant actin in the absence of wild-type actin (Shortle et al 1984). Cells expressing the mutant actin can then be compared with control cells identical in genetic background apart from the single mutation that has been engineered in the actin gene. In most other eukaryotes, the replacement of conventional actin is not as straightforward because of low frequencies of homologous recombination and the presence of multiple actin genes that encode isoforms of the protein. Secondly, actin can be purified from yeast in sufficient quantities for biochemical analysis, allowing the role of the residue of interest to be studied by a variety of well-characterized assays (e.g. filament assembly, actin-myosin interactions). Finally, the generation of mutations in actin can act as a starting point for genetic studies to find interacting proteins.

The act1 Alleles in S. cerevisiae

Mutations in actin have been generated largely by in vitro mutagenesis of the actin gene. The first alleles of *ACT1* were generated by random in vitro mutagenesis of *ACT1* (Shortle et al 1984, Dunn & Shortle 1990). However, this approach yielded only three *act1* alleles. In other studies, sites for mutagenesis were selected to test ideas about actin based on chemical cross-linking data and the atomic structure. Implicit in this approach are assumptions about the role of residues, which the investigators were attempting to verify by mutagenesis. In many cases, alteration of residues proposed to be involved in certain binding interactions had no effect in vivo or in vitro (Johannes & Gallwitz 1991) and thus failed to map functions to sites on the surface of actin.

In an attempt to generate a collection of actin mutants that would be more generally useful in the analysis of actin structure-function relationships, and one that was not biased by previous studies, a systematic mutational scan of *ACT1*, in which charged residues were replaced by alanine, was performed (Wertman et al 1992). The mutagenesis was undertaken prior to the elucidation of the atomic structure of actin; thus, based on the observation that clusters of charged residues are most often on the surface of proteins (Bennett et al 1991), the intent was to change the surface residues and thereby maximize the probability of disrupting specific interactions with actin-associating proteins. In addition, surface residues are less likely to be involved in intramolecular interactions necessary for correct protein folding so it was postulated that these mutations would produce minimal perturbation to monomer structure.

The alleles generated by the alanine scan have been particularly valuable because their characterization has provided insights into the functions and interactions of actin and have been used to identify binding sites for actin-binding proteins and actin-associating drugs (Drubin et al 1993, Holtzman et al 1994, Honts et al 1994, Amberg et al 1995, Karpova et al 1995, Smith et al 1995). A full range of phenotypes was observed for the alleles generated (Figure 1*a*). Of those 36[1] alleles, 13 cause lethality when expressed as

[1]The original alanine scan of Wertman et al (1992) reported 36 actin alleles. However, one of the alleles, *act1-118*, has more mutations than initally thought (D Amberg, personal communication), and it is possible that the lethality of the mutation is not caused by the pairwise charged for alanine substitution as indicated.

Table 1 Polar residues of actin proposed to make electrostatic interactions with tropomyosin. Comparison to appropriate actin allele phenotypes[a]

Actin residue	Electrostatic energy (cal/mol)	*act1* allele	Phenotype in yeast[b]
Asp211	−370	*act1–113*	weak Ts$^-$
Glu215	−550	*act1–112*	Ts$^-$, Cs$^-$
Asp222	−300	*act1–111*	Ts$^-$, Cs$^-$
Lys226	−390	*act1–111*	Ts$^-$, Cs$^-$
Lys238	−470	*act1–110*	lethal
Glu241	−320	*act1–128*	lethal
Arg254	−930	*act1–109*	lethal
Asp288	−300	*act1–107*	lethal
Lys291	−370	*act1–106*	lethal
Asp311	−850	*act1–105*	Ts$^-$, Cs$^-$
Asp315	−330	*act1–104*	wt
Lys326	−500	*act1–126*	lethal
Lys328	−630	*act1–126*	lethal
Glu334	−400	*act1–103*	lethal
Arg335	−430	*act1–103*	lethal
Lys336	−530	*act1–103*	lethal

[a]Adapted from Lorenz et al 1995.
[b]wt = growth at all temperatures resembles that of wild type; Ts$^-$ = heat sensitive; Cs$^-$ = cold sensitive.

the sole source of actin in the cell, 16 cause temperature-sensitive defects, and 7 have no readily observable phenotype. Interestingly, 7 of the 13 lethal actin alleles have mutations in residues proposed (from structural data) to be involved in tropomyosin binding (see Table 1; Lorenz et al 1995). Furthermore, mutation of the actin residues postulated to contribute the majority of the electrostatic energy to the actin-tropomyosin interaction (see Table 1) results in lethality. If these mutations do in fact impair tropomyosin binding, the data could be explained if tropomyosin binding is essential for viability. In support of this possibility, deletion of the two genes that encode tropomyosins in *S. cerevisiae* is lethal (Drees et al 1995). It would be pertinent to investigate genetically and biochemically the interactions between these actin alleles and the tropomyosins to more definitively define the residues involved in this interaction.

In addition to site-directed mutagenesis, several actin mutants have been identified through genetic screens (Adams & Botstein 1989, Welch et al 1993, Karpova et al 1995). As explained below, such mutations can be highly indicative of which amino acids are involved in direct interaction of actin with specific ligands.

Table 2 lists the 82 published *ACT1* alleles generated by the various approaches described above. An immediate observation is that substitution of different amino acids in actin results in many phenotypes that vary considerably in their severity. On the first level of analysis, phenotypes include dominant and recessive lethality and restricted temperatures for growth. These mutants highlight certain residues involved in the essential functions of actin. In several cases, mutations did not have any readily detectable phenotypes when expressed in cells. These residues may not be involved in interactions made by actin. Alternatively, the interactions disrupted by these mutations may not be essential under normal growth conditions (see Functional Complexity below for discussion of essential and nonessential interactions).

The mutations listed in Table 2 represent changes in 88 of the 375 amino acids of actin. Some of the *act1* mutants studied contain a single amino acid substitution, whereas others contain a combination of two or three altered amino acids. In some cases, a number of amino acid substitutions have been made for a single residue. For example, the aspartate at position 11 is extremely conserved and is found in all actins sequenced to date. The position of this residue in the structure of actin indicates that it could make electrostatic interactions with the bound divalent cation, which itself is thought to play an important role in stabilizing the monomer (Kabsch et al 1990). This residue has been mutagenized to a variety of other residues including asparagine, which changes a charged side chain to a polar side chain, and to lysine, which results in a charge reversal (Cook & Rubenstein 1991, Johannes & Gallwitz 1991). Only two, relatively conservative substitutions (D11N and D11E)[2] supported growth and only the D11E mutation yielded cells capable of producing viable spores (Cook & Rubenstein 1991). This result adds experimental support to the proposed interaction of D11 with the bound cation and highlights the importance of considering which amino acid to substitute for a given residue when designing a mutagenesis strategy.

Actin Nucleotide Binding and Filament Formation

An actin filament is formed by the noncovalent association of actin monomers carrying ATP in their nucleotide-binding cleft. Following incorporation of a monomer in the filament, the ATP bound to the assembling monomers is hydrolyzed to yield ADP-actin. This hydrolysis is thought to be accompanied by a conformational change such that the cleft is narrowed (Lorenz et al 1993,

[2]This residue is designated D10 in the cited publication. At the time of mutant construction it was thought that, as with other eukaryotic actins, the N-terminal methionine was cleaved, thus numbering in this paper was from the predicted start of the protein. In fact, the methionine is not cleaved from yeast actin (Cook et al 1991), so for consistency with all the other *act1* alleles, we have designated the substitution D11N and D11E, numbering from the initial methionine.

Tirion et al 1995). A number of studies using yeast have been directed at increasing understanding of actin nucleotide binding and filament formation.

Residues important for the binding of adenine nucleotide were proposed, based on atomic structure (Kabsch et al 1990). To test the functional importance of one of these residues, Chen and co-workers (Chen et al 1995, Chen & Rubenstein 1995) substituted the Ser14 residue with a number of alternative residues. The amide proton of this residue is thought to form hydrogen bonds with the γ-phosphate of ATP, thus stabilizing the nucleotide protein complex. Cells were viable if this residue was substituted with Thr or Ala but not with Cys or Gly. In vitro analysis of actin made from cells expressing the S14A or the S14T mutants indicated that the latter mutation has no detrimental effect. The S14A mutation had no effect on the critical concentration for polymerization, but it did increase the rate of polymerization, decrease assembly-stimulated ATPase activity, and cause altered protease sensitivity. These results suggest that the hydroxyl group of serine (or threonine in the S14T mutant) is likely to be involved in interactions that stabilize the monomer and thereby its ability to polymerize. However, the altered protease sensitivity indicates more widespread changes in protein structure than just at the nucleotide-binding site. In particular, changes in protease sensitivity have been correlated to alterations in subdomain-2 (Mornet & Ue 1984, Strzelecka-Golaszewska et al 1993). From the structure of actin it is apparent that Ser14 might not interact only with the nucleotide but also with the amide nitrogen of Gly74 in subdomain-2. Removal of the residue 14 hydroxyl would abrogate this interaction and, as observed through protease sensitivity changes, result in an altered structure for this subdomain (Chen et al 1995).

These data demonstrate that even changing a single residue can affect multiple interactions within actin. Therefore, caution is needed in the interpretation of such results because determining the mechanistic basis for the observed effects is not necessarily straightforward. Mutations in other residues surrounding the adenine nucleotide-binding cleft may not cause such widespread changes in actin structure, and these could be used to further test the effects of alteration in nucleotide binding. Such studies may provide avenues toward developing a better understanding of the ATP- and ADP-bound states of actin, and how interconversion between these states is regulated.

The original model of Holmes and co-workers for F-actin (1990) predicted a conformational change in actin structure on polymerization such that a hydrophobic region (residues 265–268) could project from a monomer. This hydrophobic plug could then insert into a pocket formed by the interface of two adjacent subunits and would result in filament stabilization. The hypothesis of a hydrophobic plug was tested by Chen and co-workers (1993). A mutant, L267D[3], was created and expressed in cells. In vivo, cells showed little effect

[3]This allele is designated L266D in the cited publication. See footnote 2.

Table 2 Yeast actin mutations and their effects

Substitution position	Actin allele	Results[a]	References
DSE 2–4 DEDE (P)[b]	—	>S1-ATPase	Cook et al 1993
DSE 2–4 NSQ	—	(+) Growth	Cook et al 1992
DSE 2–4 NSQ (P)	—	<S1-ATPase	
DSE 2–4 deleted	—	(+) Growth	Cook et al 1992
DSE 2–4 deleted	—	<S1-ATPase	Cook et al 1992
D2V	—	wt	Johannes & Gallwitz 1991
D2A	*act1–136*	Cs$^-$, Ts$^-$, low spore viability	Wertman et al 1992
D2V/E4V	—	wt	Johannes & Gallwitz 1991
E4V	*act1–135*	wt	Johannes & Gallwitz 1991
E4A	—	wt	Wertman et al 1992
D11Q	—	Dominant lethal	Johannes & Gallwitz 1991
D11K	—	Dominant lethal	Johannes & Gallwitz 1991
D11A	—	Partial dominant lethal	Wertman et al 1992
D11E[c]	—	(+) Growth, viable spores	Cook & Rubenstein 1991
D11N[c]	—	(+) Growth, no viable spores	Cook & Rubenstein 1991
S14A	—	Ts$^-$	Chen & Rubenstein 1995
S14A (P)	—	< filament ATPase,	Chen & Rubenstein 1995
	—	> rate of polymerization	
S14G	—	Dominant lethal	Chen & Rubenstein 1995
S14T	—	wt	Chen & Rubenstein 1995
S14C	—	Dominant lethal	Chen & Rubenstein 1995
M161	*act1–66*	Ts$^-$	Karpova et al 1995
D24A/D25A	*act1–133*	Recessive Cs$^-$, Ts$^-$	Wertman et al 1992
P32L	*act1–1act1–3*[d]	Ts$^-$	Shortle et al 1984
R37A/R39A	*act1–132*	Recessive Cs$^-$, Ts$^-$	Wertman et al 1992
K50AD51A	*act1–125*	Recessive Cs$^-$, Ts$^-$	Wertman et al 1992
D56A	*act1–9*	Ts$^-$	Adams & Bostein 1989
D56A/E57A	*act1–124*	Recessive Ts$^-$	Wertman et al 1992
A58T	*act1–2*	Ts$^-$	Shortle et al 1984
K61N	*act1–7*	Ts$^-$	Adams & Botstein 1989
K61A/R62A	*act1–131*	Partial dominant lethal	Wertman et al 1992
R68A/E72A	*act1–123*	wt	Wertman et al 1992
D80A/D81A	*act1–122*	Recessive Cs$^-$, Ts$^-$	Wertman et al 1992
E83A/K84A	*act1–121*	Recessive Cs$^-$, Ts$^-$	Wertman et al 1992
H88Y	*act1–8*	Ts$^-$	Adams & Botstein 1989
T89I	*act1–10*	wt	Adams & Botstein 1989
E93A/R95A	*act1–130*	Recessive lethal	Wertman et al 1992
E99/E100	*act1–120*	Recessive Ts$^-$	Wertman et al 1992
R116A/E117A/K118A	*act1–119*	Recessive Ts$^-$	Wertman et al 1992
D154A/D157A	*act1–118*[e]	Partial dominant lethal	Wertman et al 1992
R177A/D179A	*act1–129*	Recessive Ts$^-$	Wertman et al 1992
R183A/D184A	*act1–117*	wt	Wertman et al 1992
D187A/K191A	*act1–116*	wt	Wertman et al 1992
K191M	—	wt	Johannes & Gallwitz 1991
K191M/C374A	—	wt	Johannes & Gallwitz 1991
A195A/R196A	*act1–115*	wt	Wertman et al 1992

(*Continues*)

Table 2 (*Continued*)

Substitution position	Actin allele	Results[a]	References
E205A/R206A/E207A	*act1–114*	Possible dominant lethal	Wertman et al 1992
R210A/D211A	*act1–113*	Recessive weak Ts^-	Wertman et al 1992
K213A/E214A/K215A	*act1–112*	Recessive, Cs^-, Ts^-	Wertman et al 1992
D222A/E224A/E226A	*act1–111*	Recessive, Ts^-	Wertman et al 1992
E237A/K238A	*act1–110*	Partial dominant lethal	Wertman et al 1992
E241A/D244A	*act1–128*	Partial dominant lethal	Wertman et al 1992
E253A/R254A	*act1–109*	Partial dominant lethal	Wertman et al 1992
R256A/E259A	*act1–108*	Recessive, Cs^-, Ts^-	Wertman et al 1992
E259V	*act1–4*	Recessive, Ts^-	Dunn & Shortle 1990
L267D[f]	—	Cs^-	Chen et al 1993
L267D (P)	—	Cs^- for polymerization	Chen et al 1993
E270A/D275A	*act1–127*	Recessive lethal	Wertman et al 1992
D286A/D288A	*act1–107*	Recessive lethal	Wertman et al 1992
R290A/K291A/E292A	*act1–106*	Recessive lethal	Wertman et al 1992
E292K	*act1–87*	Ts^-	Karpova et al 1995
E311A/R312A	*act1–105*	Recessive, Cs^-, Ts^-	Wertman et al 1992
K315A, E316A	*act1–104*	wt	Wertman et al 1992
K326A/K328A	*act1–126*	Possibly dominant lethal	Wertman et al 1992
E334A/R335A/K336A	*act1–103*	Recessive lethal	Wertman et al 1992
K336M	—	wt	Johannes & Gallwitz 1991
K336Q	—	wt	Johannes & Gallwitz 1991
I341A	*act1–45*	wt	Miller et al 1996
I341A (P)	—	<strong binding to myosin	Miller et al 1996
I345A	*act1–53*	wt	Miller et al 1996
I341A/I345A	*act1–47*	Recessive lethal	Miller et al 1996
I341K/I345K	*act1–60*	Dominant lethal	Miller et al 1996
I341F	*act1–65*	Dominant lethal	Miller et al 1996
I345F	*act1–69*	Recessive lethal	Miller et al 1996
W356Y	—	wt	Johannes & Gallwitz 1991
W356H	—	wt	Johannes & Gallwitz 1991
K359A/E361A	*act1–102*	wt	Wertman et al 1992
D363A/E364A	*act1–101*	Recessive, Ts^-	Wertman et al 1992
K373Term	—	Lethal	Johannes & Gallwitz 1991
C374A	—	wt	Johannes & Gallwitz 1991
C374Term	—	Ts^-	Johannes & Gallwitz 1991
F375Term	—	Ts^-	Johannes & Gallwitz 1991

[a]Phenotypes are as reported in the paper cited: wt = growth at all temperatures resembles that of wild type; Cs^- = cold sensitive; Ts^- = heat sensitive; (+) Growth = cells are viable but no information was reported regarding growth for a full range of temperatures.

[b](P) = Yeast actin was purified for in vitro assay.

[c]This residue is designed D10 in the cited publication. At the time of mutant construction, it was thought that, as with other eukaryotic actins, the N-terminal methionine was cleaved, so numbering in this paper was from the predicted start of the protein, In fact, the methionine is not cleaved from yeast actin (Cook et al 1991); thus for consistency with all the other *act1* alleles, we have designated the substitutions D11N and D11E.

[d]*act1-1* and *act1-3* were found independently, subsequently sequenced, and found to be the same mutant actin allele.

[e]*act1–118* was found to have multiple mutations, not just those indicated (D Amberg, personal communication). Thus it is possible that the lethality of the mutation is not caused by the pairwise charged for alanine substitution originally reported by Wertman et al 1992.

[f]This allele is designated L266D in the cited publication. See footnote c.

other than a slight cold sensitivity. In vitro tests were more dramatic, however, with mutant actin having a 10-fold increase in the critical concentration required for polymerization at 4°C compared with wild-type actin. In addition, nucleation of actin assembly was barely detectable at this temperature. Other properties, such as binding to myosin fragment S1 and activation of S1-ATPase, were not affected by the mutation. These results are consistent with this residue being involved in polymer stabilization. However, two structural studies, one by Schutt and co-workers on β-actin (1993) and one by Tirion and co-workers on the α-actin filament (1995), have indicated that it is unlikely, from energy considerations, that the conformational change to produce a hydrophobic plug occurs. Thus although residues important for polymer stability can be identified by mutant analysis, it is more difficult to understand how the residues contribute to stability.

Using Genetics to Map Surfaces of Actin Involved in Interactions In Vivo

The yeast actin mutants have been used in a number of genetic studies to analyze the association of actin with interacting proteins. Four powerful approaches have been applied: suppression, synthetic lethality, two-hybrid analysis, and intragenic complementation. The conceptual basis for each of these approaches is shown schematically in Figures 2–6. The goal of using these methods to identify residues involved in the physical association between actin and its binding proteins is to generate a functional map of the actin surface. This will allow us to better understand how the proteins can act alone or in concert to influence assembly, disassembly, and formation of higher order structures such as bundles and networks.

SUPPRESSOR ANALYSIS AND THE BINDING SITE ON ACTIN FOR YEAST FIMBRIN

Yeast fimbrin (Sac6p) was identified as an actin-binding protein, by affinity chromatography with actin filaments (Drubin et al 1988), and as a suppressor of an actin mutation, P32L, by a genetic screen (Adams et al 1989). To more fully understand the interaction between fimbrin and actin, Honts and co-workers (1994) looked at mutations in actin that suppress (or were suppressed by) mutations in *SAC6*.

The rationale behind suppression analysis is illustrated in Figure 2. Actin can be mutated such that binding to an interacting protein, in this case fimbrin, is impaired (Figure 2*b*). This *act1* mutation can be suppressed if a mutation in fimbrin compensates for the initial mutation in actin thus restoring the full binding interaction (Figure 2*c*). In this case, the mutated fimbrin residues are likely to be involved in the interaction with actin although, in principle, mutations that change the conformation of fimbrin might have the same effect.

When the fimbrin mutation is present in a background with wild-type actin, it will give a similar phenotype to the actin mutation alone because the interaction is again impaired (Figure 2*d*). Similarly, suppression of the resulting fimbrin mutations is likely to identify residues of actin important for fimbrin binding.

Eight different *act1* mutations that suppress *sac6* (fimbrin) mutations were sequenced. All these mutations lie in a discrete region of the actin surface (Figure 1*b*). The data indicate that the main binding site for fimbrin is on subdomain-2 and the part of subdomain-1 adjacent to subdomain-2. Cross-linking studies with α-actinin, a protein whose actin-binding domain is structurally related to that of fimbrin, has shown binding to the same region of actin as marked by these mutations (Mimura & Asano 1987, Lebart et al 1993). In addition, electron microscopy and image enhancement of α-actinin binding to F-actin also indicate a binding site in this region of subdomain-1 and -2 (McGough et al 1994). The genetic data were supported by biochemical analysis, which demonstrated the importance of the identifed residues for co-pelleting of actin and fimbrin (Holtzman et al 1994, Honts et al 1994, Karpova et al 1995).

SYNTHETIC LETHAL ANALYSIS AND THE FIMBRIN-BINDING SITE A second genetic approach also identified the same region on actin as a binding site for fimbrin. Holtzman and co-workers (1994) also used an in vivo strategy to identify amino acids of actin required for functional interactions. The approach is based on the assumption that an actin mutation that specifically impairs the interaction with an actin-binding protein will cause a phenotype similar to a null mutation in the gene that encodes the actin-binding protein itself (Figure 3). For example, deletion of the *SAC6* gene causes cells to become temperature sensitive (Figure 3*a*). In addition, these cells show synthetic lethality with mutations in two other genes, *ABP1* and *SLA2*, that also encode components of

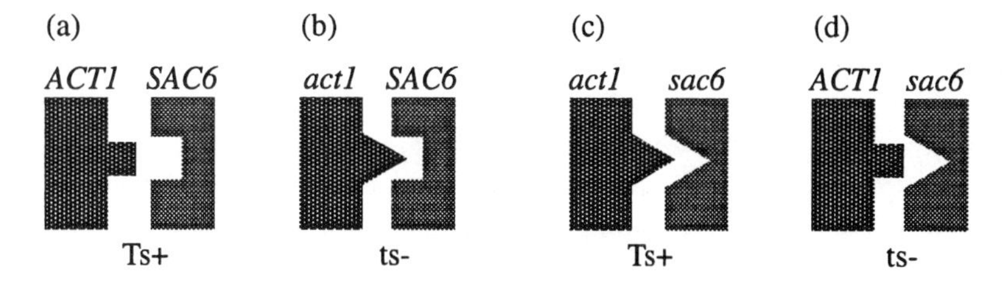

Figure 2 Suppression of mutations in the actin gene *ACT1* by mutations in the fimbrin gene *SAC6*. (*a*) Normal interaction of actin and fimbrin; (*b*) a mutation in actin at the fimbrin-binding site impairs the interaction and cells are temperature sensitive for growth (ts^-); (*c*) a mutation in fimbrin can compensate for the original actin mutation, restoring the interaction; (*d*) the fimbrin mutation alone impairs actin binding and cells are ts^-. (Adapted from Adams et al 1989.)

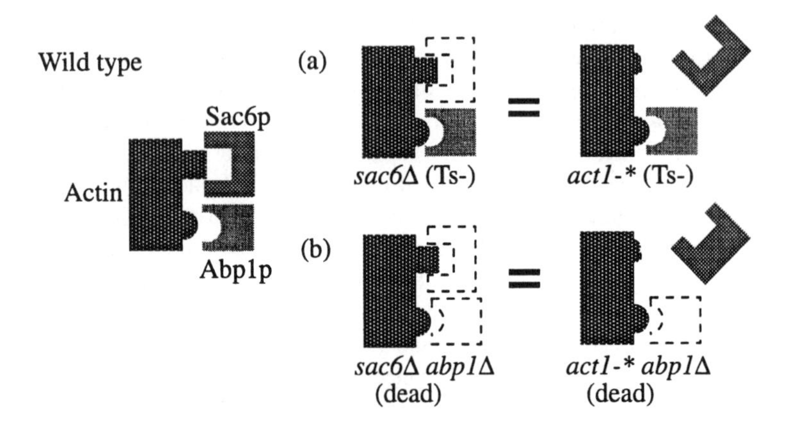

Figure 3 Use of synthetic lethality to identify functional domains of actin. (*a*) Deletion of *sac6* causes cells to be temperature sensitive. This deletion is equivalent to a mutation in actin that cannot bind fimbrin (Sac6p). (*b*) A pairwise combination of null alleles of *sac6* and *abp1* causes synthetic lethality. This is equivalent to a combination of an *abp1* null allele and an actin allele that does not bind fimbrin. (Adapted from Holtzman 1994.)

the actin cytoskeleton (Figure 3*b*). By synthetic lethality, we mean that each single mutant is viable, but the combination of mutations causes inviability. Thus a mutation in the actin gene that impairs the interaction with fimbrin could cause cells to become temperature sensitive and show the same synthetic lethal interaction as the *SAC6* null allele. The charged-to-alanine collection of actin mutants was tested for interactions with *SAC6, ABP1,* and *SLA2*. One mutant, E99A/E100A, is temperature sensitive and showed the same synthetic lethal interactions as cells deleted for *SAC6*, which indicates that it could form part of the binding site for fimbrin. This conclusion was further tested by in vitro analysis showing that Sac6p would not co-pellet with the mutant actin E99A/E100A, whereas co-pelleting was observed with wild-type actin or with actin with a mutation in a different part of the protein. In a similar study, Karpova and co-workers (1995) looked at synthetic lethal interactions between actin, capping protein, and fimbrin. Their work added further support to the conclusion that subdomain-1 and -2 contain the sites for binding fimbrin.

ANALYSIS USING THE TWO-HYBRID SYSTEM To identify and analyze proteins that bind actin in vivo, the two-hybrid assay of Fields & Song (1989) was employed. The strategy is based on the fact that the Gal4 protein needs its two domains to be brought into close proximity to function as a transcription activator. The upstream activating sequence (UAS) to which the Gal4 DNA-binding domain binds can be cloned upstream of a reporter gene such as *lacZ* to allow detection when an active complex is formed. To identify proteins that bind actin in vivo, Amberg and co-workers (Amberg et al 1995) fused the Gal4 DNA-binding domain to actin and fused the transcription-activating domain to a library of yeast cDNAs (Figure 4). In this way, several new proteins that interact with actin in vivo have been identified (Amberg et al 1995).

After demonstrating interactions with wild-type actin, the interactions of various actin-binding proteins with the alanine scan mutant actins were assessed. Mutations in actin that destroy binding to a particular protein can be located on the molecular model of actin as an approach to identifying binding sites for these proteins on the actin surface. The most striking findings described in this work were for an actin-binding protein called Aip1p (for actin interacting protein). All the actin mutants showing impaired association of the Gal4p fusions were localized to a distinct region of the actin surface lying largely on the front face of subdomain-3 (Amberg et al 1995). However, one caveat of this approach is that because the interactions are scored in vivo, it is possible that the association between actin and a protein of interest might not be direct. An intermediate linker protein that itself binds to actin could mediate a binding interaction, and it is the binding site of this second protein that is actually being mapped. This two-hybrid method promises to be a powerful approach for future studies on the interaction of actin with its plethora of associated proteins.

INTRAGENIC COMPLEMENTATION A powerful tool for identification of separable functional domains uses intragenic complementation. This approach has been used in studies on calmodulin and has demonstrated the existence of four functionally distinct regions within this small protein (Ohya & Botstein 1994). The existence of many alleles of actin with a range of phenotypes makes actin an ideal candidate for such a study. The molecular basis for intragenic complementation is illustrated in Figure 5. In wild-type cells, actin associates with a variety of binding proteins. If actin is mutated such that it can no longer bind a protein X, the cells may exhibit a phenotype, in this case temperature sensitivity. Similarly, another allele cannot bind protein Y, which causes cells to be inviable at elevated temperatures. However, if the two mutant strains are crossed to produce a diploid cell expressing both actin alleles, the binding site for protein X will be preserved on actin, which is defective in binding Y, and the binding site involved in binding Y will be preserved on the actin mutant defective in binding X. In this case, functional actin filaments will be formed, and cells will be viable at elevated temperatures. Thus intragenic complementation allows actin alleles that affect specific, non-overlapping functional domains to be identified. This approach is currently being used to identify functional domains on actin (D Botstein, personal communication).

a) Native Gal4p -
Reporter gene turned on

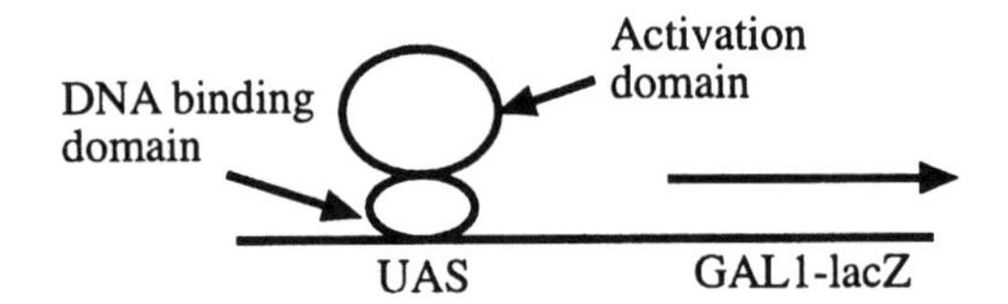

b) Individual hybrid proteins with Gal4 domains -
No transcription.

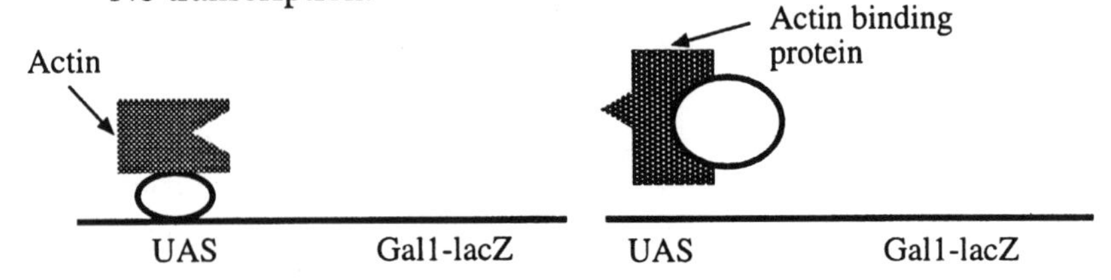

c) Interaction between hybrids reconstitutes Gal4 activity
Reporter gene turned on.

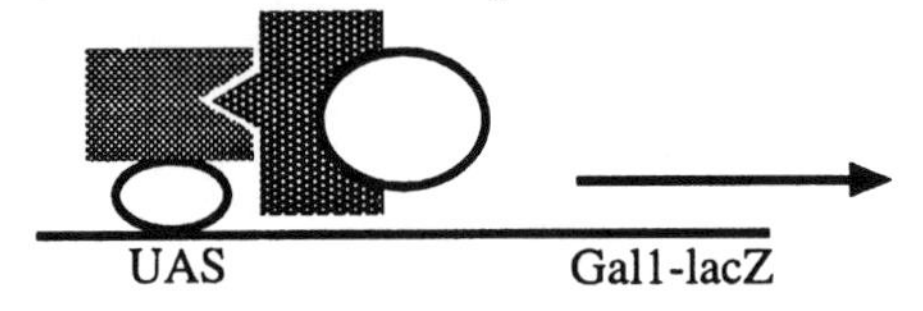

d) *act1* mutant that disrupts binding site for protein of interest.
No transcription

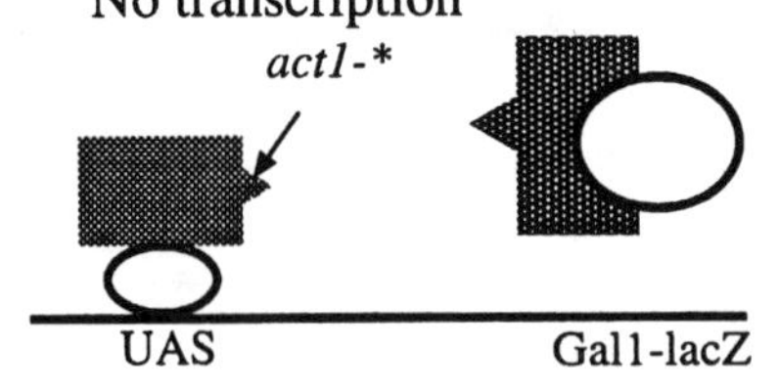

Figure 4 Use of the two-hybrid technique to identify actin binding proteins and to map functional domains of actin. (Panels a–c adapted from Fields 1989.)

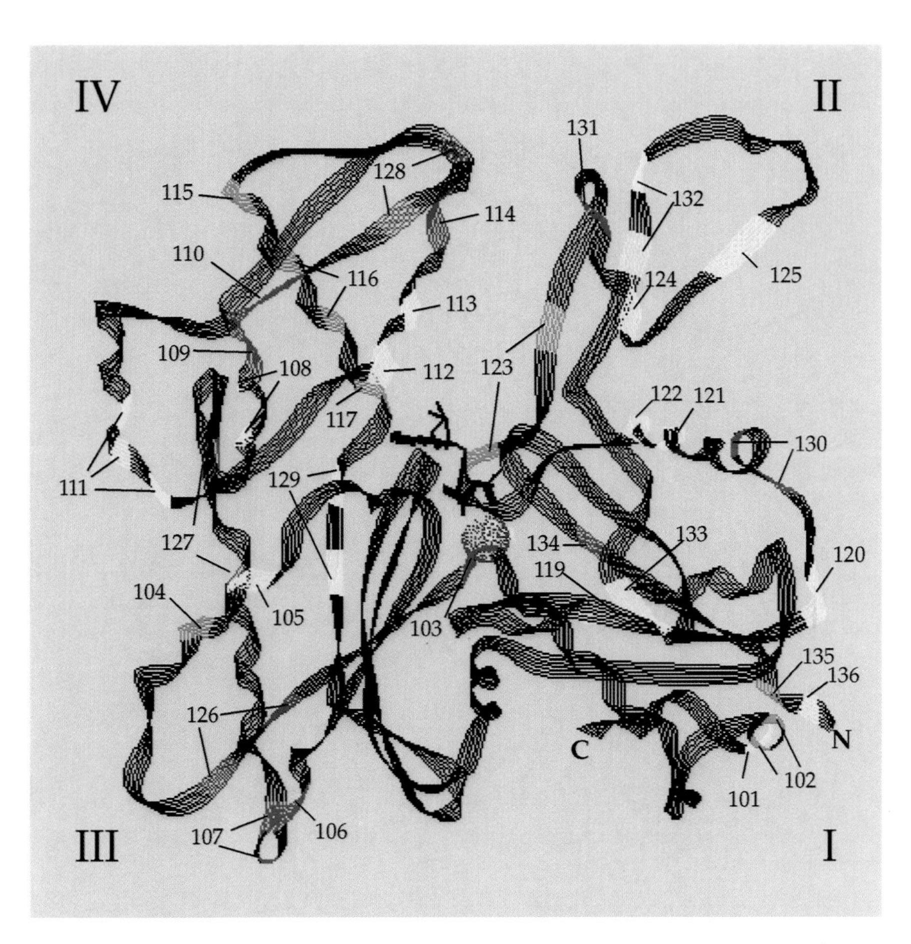

Figure 1a Ribbon diagram of the actin monomer from the coordinates of rabbit muscle actin as determined by Lorenz and co-workers (1993). Subdomains are marked I to IV; the amino and carboxy terminii are marked N and C, respectively. The positions of the backbone residues substituted in the charged for alanine scan are color coded and numbered by their allele designation (Wertman et al 1992; see Table 2). Alleles with no readily observable phenotype, *green;* temperature-sensitive alleles, *yellow;* lethal alleles, *red;* putative dominant lethal alleles, *blue.* The adenine nucleotide is shown in the prominent cleft as a simple stick model and the divalent cation as a van der Waal's sphere.

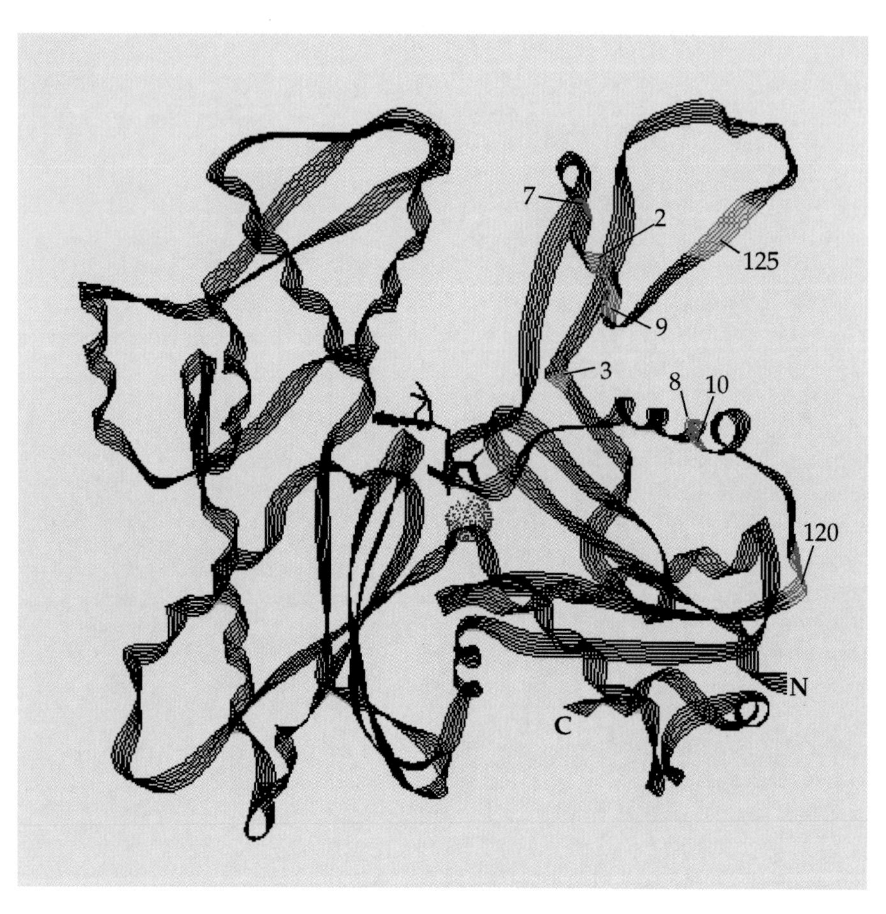

Figure 1b Ribbon diagram of the actin monomer showing the positions of residues corresponding to the mutations in yeast actin that suppress or are suppressed by *sac6* mutant alleles. (Adapted from Honts et al 1994.)

Using Mutant Actins for Analysis of the Actomyosin Interface

Elucidation of the molecular contacts between actin and myosin is central to understanding how the interaction of these two proteins results in force generation. Yeast contain both conventional (Myo1) (Rodriguez & Paterson 1990) and unconventional myosins (Myo3, Myo5—class I myosins) (Goodson & Spudich 1995; H Goodson & J Spudich, personal communication) (Myo2, Myo4—class V myosins) (Haarer et al 1994, Johnston et al 1991) that are postulated to be involved in cytokinesis (Rodriguez & Paterson 1990), mitochondrial organization (Drubin et al 1993, V Simon et al 1995, Smith et al 1995), vesicle trafficking (Johnston et al 1991, Lillie & Brown 1994, Govindan et al 1995), and the asymmetric inheritance of a transcriptional repressor (Bobola et al 1996). To study the interaction between actin and myosin, researchers have turned to yeast as a source of genetically engineered actin. Purified yeast actin activates rabbit muscle myosin ATPase activity (Greer & Schekman 1982), and it displays ATP-dependent movement over surfaces coated with rabbit muscle myosin (Kron et al 1992), which demonstrates the conservation of the residues involved in interactions at the actomyosin interface. Most of this work has been performed in vitro. The potential to determine the in vivo consequences for cell physiology of mutating the actomyosin interface has not been fully exploited. Studies on mitochondria do, however, highlight the importance of the residues encompassed by the myosin footprint for organization and inheritance of this organelle (Drubin et al 1993, Lazzarino et al 1994).

The first component of the actomyosin interface to be investigated was the extreme N-terminus of actin. A role for this domain in the actomyosin interaction was implicated from chemical cross-linking and biochemical and

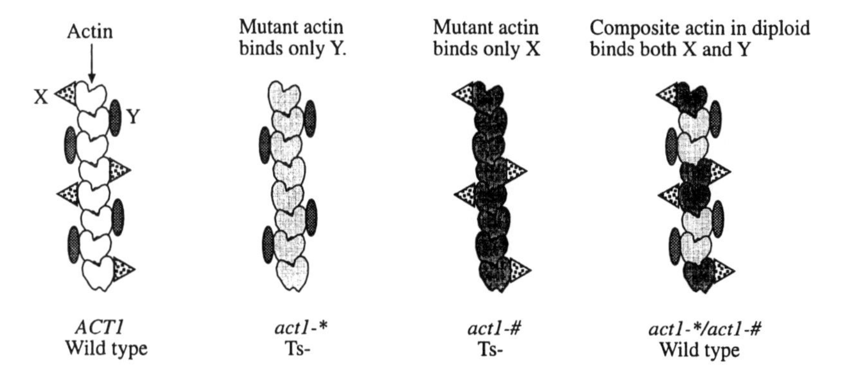

Figure 5 Use of intragenic complementation to identify functionally distinct domains of actin. This model assumes that proteins X and Y must both bind to actin filaments for wild-type growth.

immunochemical approaches (Sutoh, 1982, DasGupta & Reisler 1989, 1991). In vitro studies using vertebrate β-actins expressed in yeast revealed that if acidic residues at the N-terminus were substituted with basic residues, the resulting actins could not be decorated with myosin S1. Mutations less extreme than charge reversal, such as substitution with alanine, produced only subtle defects in myosin S1 association with actin, as assessed by electron microscopy (Aspenström & Karlsson 1991). Using yeast actin mutants, Cook and co-workers showed that charged residues at positions 2 and 4 of actin are needed to give full activation of the S1 ATPase in vitro (Cook et al 1992). Furthermore, if additional negative charges are added at the N-terminus, an increase in S1 ATPase activity is seen (Cook et al 1993). Thus the mutagenesis data support the initial cross-linking and biochemical and immunochemical approaches, indicating a role for the N-terminus in the actomyosin interaction. However, it was also clear from the mutagenesis approach that the deletion of charge did not completely abrogate binding or inhibit the S1 ATPase activity (Aspenström & Karlsson 1991, Cook et al 1992). This would indicate that the N-terminus represents only one part of the actomyosin interface.

These studies also highlight the dramatic differences in behavior associated with a particular substitution made and indicate that the amino acid selected to replace a wild-type residue should be chosen carefully so as to allow the role of a specific residue to be addressed. A substitution causing charge reversal can introduce repulsive forces at an interface of interest. In such a case, interpretation of results is likely to be difficult. A change in the binding of two proteins such as actin and myosin could be attributable to alteration of the residue that actually participates in the interaction or to alteration of a residue proximal to the site of interaction, which nonspecifically interferes with the association.

Other regions of the protein were also implicated in the interaction of actin and myosin. Charge reversal of acidic residues at positions 24/25 and 99/100 severely impaired actin sliding in in vitro motility assays using *Dictyostelium* actin (Johara et al 1993). As mentioned above, however, a charge reversal could simply be causing a relatively nonspecific repulsive effect and not addressing the specific role of residues in an interaction.

The publication of the atomic structure of myosin subfragment 1 (Rayment et al 1993b) and of a modeled structure of the actomyosin complex (Rayment et al 1993a, Schröder et al 1993) has allowed for a much better understanding of which residues are likely to be involved at the actomyosin interface. By substituting actin residues with the small neutral amino acid alanine, Miller and co-workers (1995) have shown that the elimination of the D24/D25 and E99/E100 charged pairs decreases the weak binding of actin to myosin in the presence of ATP but does not abolish force generation. More recently, the role of hydrophobic forces in the actomyosin interaction have been investigated. Mutagenesis of actin residues in the α-helix 338–348 indicates that I341, but not I345, has a major role in the strong binding of myosin to actin (Miller et al 1996). These data are in agreement with the structural data showing that the major interface for interaction in the strongly bound actomyosin complex encompasses the α-helix (338–348) on actin subdomain-1.

The use of yeast for the production of genetically engineered actin variants has allowed critical testing of the contributions of the specific actin amino acids to the actomyosin interactions. Further studies should allow increased understanding of the complete cycle of actin-myosin interactions and how this leads to force generation.

Identification of Binding Sites for Actin-Interacting Drugs

A number of drugs are known to interact with actin in vivo and in vitro that affect its ability to function. These drugs can mimic effects of actin-binding proteins in that they stabilize actin filaments (phalloidin) (Estes et al 1981), cap filaments (cytochalasin D) (Brown & Spudich 1979, Flanagan & Lin 1980, Godette & Frieden 1986a,b), sever filaments [(mycalolide B) (Saito et al 1994) (swinholideA) (Bubb et al 1995)], and bind to actin monomers (latrunculin-A) (Coué et al 1987). The use of drugs can help us to understand the inherent properties of actin as well as its physiological functions. As such, knowing where these drugs bind to actin provides a framework for developing a better understanding of their actions. The charged-to-alanine actin mutants can be used to map the binding sites for these drugs on the surface of actin.

Phalloidin is a drug commonly used to stabilize actin filaments in vitro. It can also be conjugated to fluorophors and used to visualize filamentous actin structures within cells. Drubin and co-workers (1993) observed that one actin mutant (R177/D179) was unable to bind rhodamine-phalloidin and postulated that these residues identified the binding site for phalloidin on actin. This hypothesis was supported by a refinement of an actin filament structure in which the filament was polymerized in the presence and absence of phalloidin (Lorenz et al 1993). Residues 177 and 179 lie proximal to the interface between three actin monomers in the filament, suggesting that phalloidin stabilizes filaments by binding to two or three monomers at this interface. It is now important to determine if any actin-binding proteins bind at this site to modulate filament stability.

The drug latrunculin-A has been shown to bind to actin monomers (Coué et al 1987). Residues important for latrunculin-A action were identified by exposing all of the charged-to-alanine actin mutants to varying concentrations of the drug (K Ayscough, unpublished observations). Most mutant strains, including those expressing wild-type actin, were sensitive or super-sensitive to the drug. However, three mutants showed resistance at all levels of the drug tested. Although not contiguous in the primary sequence of actin, the mutations are close when mapped to the atomic structure, which implicates this surface of the protein as the binding site for latrunculin-A, a possibility currently being tested (K Ayscough, unpublished data).

FUNCTIONAL COMPLEXITY

Overview

Another important application of yeast genetics is elucidation of the functional relationships among cytoskeletal proteins. In a system as integrated and complex as the eukaryotic cytoskeleton, it is extremely difficult to determine the contributions of individual proteins within the network. In yeast, as in other eukaryotes, it appears that many components of the actin cytoskeleton have multiple overlapping functions. This gives rise to the concept of functional redundancy in which a single essential function can be mediated by a number of proteins.

Many proteins that associate in complexes with the actin cytoskeleton have been purified for biochemical analysis and dozens of these proteins are expressed in a given eukaryotic cell. In vitro assays have revealed a variety of activities for many of these proteins, ranging from their effects on actin dynamics, increasing or decreasing the rate or extent of actin polymerization, to promoting nucleotide exchange, or arranging actin filaments into higher order structures. However, even if a thorough characterization of the in vitro activities of every cytoskeletal protein could be achieved, it is unlikely that this would provide a complete understanding of how the actin cytoskeleton influences cell behavior. Many of these studies look at effects of individual proteins in in vitro systems. Although the binding of a protein to actin and identification of domains of a protein involved in binding can be investigated, it is much harder to assess the relevance of such an interaction. This is because there is much functional redundancy among cytoskeletal proteins and because many regulatory as well as competitive and cooperative interactions are difficult to model in vitro.

Approaches such as microinjection of a cytoskeletal protein, or an antibody against the cytoskeletal protein, into a mammalian cell have inherent limitations. Injection of a cytoskeletal protein at greater than physiological levels can saturate normal binding partners and force nonphysiological interactions. In addition, nonspecific defects can be encountered when injected antibodies sterically hinder molecular interactions and when they recognize proteins other than those intended because of the existence of common epitopes. The effects of a null mutation in a cytoskeletal protein gene, on the other hand, are more likely to be specific because there are fewer ways in which the absence of a single protein can interfere with an unrelated process.

The ability in yeast, *Dictyostelium*, and mice to isolate null mutations in cytoskeletal protein genes by molecular genetic techniques has often shown that the in vivo contributions of individual cytoskeletal proteins are more subtle than predicted (DeLozanne & Spudich 1987, Witke et al 1992, Adams et al 1993, Colucci-Guyon 1994). It would appear that rather than carry out essential functions, many cytoskeletal proteins are present for fine-tuning of cytoskeletal function.

The subtlety of phenotypes resulting from null mutations is thought to be the result of functional redundancy, with subsets of gene products able to mediate the same process. Functional redundancy can be demonstrated when pairwise combinations of null mutations leads to a synergistic increase in the severity of a phenotype. When the combination of the two mutants results in cell death, the interaction is referred to as synthetically lethal. In some cases, redundancy might occur between genes encoding structurally related proteins with similar biochemical activities. Examples in yeast include *TPM1, TPM2* (tropmyosin; Drees et al 1995) and *MYO3, MYO5* (type I myosins; H Goodson & J Spudich, personal communication). Much more common, however, is the occurrence of strong negative synergism (synthetic lethality) between genes encoding nonhomologous proteins. In some cases such as the synthetic lethality of *TPM1* and *MYO2*, the synthetic lethal interaction provides in vivo support for a suspected functional interaction in vivo (Liu & Bretscher 1992). Here, the interaction of partial loss-of-function mutations provides support for the possibility that tropomyosin regulates the interaction of myosin with actin in yeast. In other cases described below, the identification of synthetic lethal interactions provides an avenue toward the identification of unsuspected functional relationships.

Functional redundancy has been investigated at a cellular level, using the powerful Mendelian genetics available in yeast, to begin to determine the roles of individual proteins within the actin cytoskeleton and to gain a deeper understanding of the extent to which the functions of cytoskeletal proteins are integrated in vivo.

To find components involved in common processes, synthetic lethal screens have been particularly successful (Bender & Pringle 1991). In this approach, a nonessential gene of interest is deleted from the genome but is present in the cells on a plasmid. Cells are then mutagenized, and selection is made for cells that cannot lose the plasmid. Hence, the gene of interest has become essential through mutation of a second gene. This mutated gene, which requires the original gene for function, can then be isolated, cloned, and sequenced. Synthetic lethal interactions have also been found by constructing double mutants from two single mutants, which alone do not have lethal phenotypes. Significantly, only mutations in specific combinations of genes result in synthetic lethality (Adams et al 1993).

Synthetic lethal screens have identified novel cytoskeletal proteins in yeast and have highlighted a plethora of genetic interactions. Investigations of these interactions promise to greatly increase our understanding of functional relationships between cytoskeletal proteins inside the cell.

Synthetic Lethal Interactions Identify New Components and Functional Relationships in the Actin Cytoskeleton

One of the first proteins found to bind to yeast actin was actin-binding protein-1 (Abp1) (Drubin et al 1988). It was found to bind to filamentous actin on an affinity column and subsequently was immunolocalized to the cortical actin patches of yeast. Abp1 has an N-terminal domain homologous with an actin monomer-binding and filament-severing protein, cofilin (Moon et al 1993), and a C-terminal SH3 domain. SH3 domains have been found in many proteins associated with cell signaling, especially those localized to the actin cytoskeleton (Drubin et al 1990). However, when the *ABP1* gene was deleted, cells showed no readily observable phenotypes (Drubin 1990). No sequence homologues of the protein could be identified; thus it was supposed that cells contain nonhomologous proteins with functions overlapping those of Abp1. In order to test this possibility, Holtzman and co-workers (1993) performed a synthetic lethal screen to identify genes that when mutated caused cells to require expression of *ABP1*. Three genes were isolated in this screen: *SAC6*, encoding yeast fimbrin; and two previously unknown genes called *SLA1* and *SLA2* (synthetically lethal with *ABP1*). Proteins encoded by both genes have subsequently been localized to the cortical actin cytoskeleton, thus demonstrating the efficacy of this approach for identifying new components of large protein complexes. Moreover, mutations in *SLA1, SLA2,* and *SAC6* all cause defects in the cortical actin cytoskeleton.

Additional interactions involving these and other cytoskeletal proteins have been demonstrated by the generation of double mutants from combinations of various single mutants (Adams et al 1993, Botstein et al 1996; T Lila, personal communication). These interactions add to those previously identified and described above. As shown in Figure 6*a*, there are complex patterns of synthetic lethality among null alleles of cytoskeletal genes. In addition to the synthetic lethal interactions of *ABP1* with *SLA1, SLA2,* and *SAC6,* another gene implicated in yeast actin cytoskeleton function, *RVS167* (Bauer et al 1993), also shows these interactions and an additional synthetic lethal interaction with *SRV2* (also known as cyclase-associated protein, CAP) (Fedor-Chaiken 1990, Field et al 1990). The similarity of *ABP1* and *RVS167* genetic interactions suggests that their gene products might form a complex, the integrity of which must be maintained in the absence of any one of the identified interacting genes. Biochemical data (T Lila & D Drubin, unpublished observation) support this possibility.

The model in Figure 6*b* explains how functional redundancy might occur in cells and how it is that each gene might exhibit a unique constellation of genetic interactions. In this model, each protein has one or more functions in the cell. Indeed, many cytoskeletal proteins are multifunctional, having combinations of activities such as severing, capping, and nucleating. However, each function can be performed by more than one protein. For example, essential protein function C can be performed by proteins 2 or 3. Deletion of either of these proteins alone will not lead to lethality because all their functions are covered by other proteins. However, deletion of proteins 2 and 3 together leads to the loss of essential function C and thus the cells will not be viable.

Using Synthetic Lethality to Understand Complex Functional Relationships

Although it is not difficult to comprehend the reasons for synthetic lethality between genes encoding two myosins or a myosin and its regulator tropomyosin, the basis for synthetic lethality when the proteins bear little or no resemblance in structure or biochemical activity (Figure 6*a*) is harder to understand.

Determination of the basis for one synthetic lethal interaction has shown the value of identifying such interactions and investigating their molecular premise. As shown in Figure 6*a*, the generation of combinations of null mutations in genes encoding components of the actin cytoskeleton reveal synthetic lethality between mutations in the genes that encode yeast fimbrin (*SAC6*) and the genes that encode capping protein (either *CAP1* or *CAP2*). Combinations with many other actin-binding protein mutations did not show negative synergism (Adams et al 1993). In this work, Adams and co-workers postulated that capping protein and fimbrin might share some regulatory function. For example, even though fimbrin acts by binding to the sides of actin filaments (Bretscher 1981, Glenney et al 1981) and capping protein binds to the ends of actin filaments (Caldwell et al 1989, Amatruda et al 1990), both types of interaction may prevent actin depolymerization and thereby stabilize filaments in vivo. To test this hypothesis, Karpova and co-workers (1995) performed experiments investigating the cellular basis for functional redundancy between *SAC6* and *CAP2*. Using complementary approaches, the levels of G-actin and F-actin in cells were measured. For both *sac6* and *cap2* null mutants, the level of F-actin was reduced compared with wild type, and the level of G-actin was elevated.

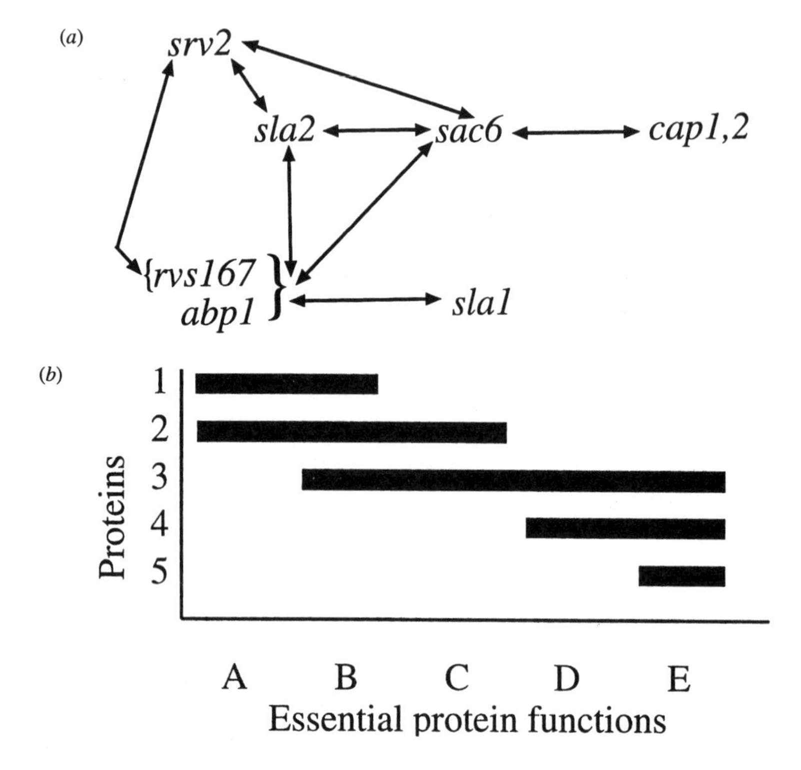

Figure 6 (*a*) Synthetic lethal interactions between several components of the actin cytoskeleton. Illustrated are the known synthetic lethal interactions between these genes. Interactions with other genes encoding cytoskeletal and regulatory proteins have also been demonstrated for some of these genes. (*b*) A model to explain the concept of functional redundancy in cells (see text).

This result supports the hypothesis that both proteins, by different mechanisms, act to stabilize actin polymers.

Taking the investigation further, Karpova and co-workers looked for synthetic lethal interactions between *CAP2* and a collection of actin mutants. Seven actin mutants showed some effect, and these were also measured for their F-actin and G-actin content. Most of these alleles showed either decreased F-actin or increased G-actin. These effects were then investigated more fully using in vitro assays for actin polymerization. All of the actin mutants showing negative synergism with the null alleles of *CAP2* also showed defects in the ability to polymerize. In addition, a final test showed that several of the actin mutants were unable to bind Sac6p, as well as wild-type actin. Therefore, cells expressing these actins would be expected to behave in a fashion similar to cells that were deleted for *SAC6* itself.

These studies have provided unique insights into the basis for the *SAC6* (fimbrin) *CAP1,2* (capping protein) genetic interaction and the functional relationship of these two proteins. However, although important in the final interpretation of results, prior biochemical elucidation of the activities of these proteins was of limited predictive value. As a result of combined genetical and biochemical analysis, it was concluded that fimbrin, which bundles actin filaments, and capping protein, which binds the ends of filaments, both stabilize filaments in vivo. However, tropomyosin has a well-characterized filament-stabilizing activity. Thus, in light of these results, it was surprising that null mutations in the tropomyosin gene *TPM1* do not show synthetic lethality with *sac6, cap1,* or *cap2* mutations. The absence of these interactions might be the result of compartmentalization; tropomyosin is present on cytoplasmic actin cables and capping protein is associated with cortical actin. Fimbrin is associated with both structures but might only be required for filament stabilization at the cell cortex.

Demonstration of the basis for the *sac6-cap1,2* interaction reveals the potential benefits likely to be gained by determining the basis for other genetic interactions identified among yeast cytoskeletal mutants (Figure 6*a*; Holtzman et al 1993, Haarer et al 1994, Wang & Bretscher 1995). These studies promise to provide a deeper understanding of the roles of cytoskeletal proteins and how their activitites are integrated and regulated. Moreover, because homologues of Srv2p (Vojtek & Cooper 1993), fimbrin (Bretscher 1981, Glenney et al 1981), Sla2p (Holtzman 1993, Wilson et al 1994), Rvs167 (David et al 1994, Sivadon et al 1995), and Cap1,2 (Caldwell et al 1989) exist in a variety of other eukaryotes, it is likely that much of what is learned from these studies in yeast will be more generally applicable.

Finally, the model we present in Figure 6*b* indicates that there is much complexity and redundancy in the actin cytoskeleton. It is worth considering what advantages such a network might impart to the cell. Possession of two fully functional redundant genes should be, on an evolutionary time scale, an unstable condition. However, redundancy among genes is observed in a wide range of organisms, indicating that it must impart a selective advantage to the organism. Several mechanisms have been proposed to explain how selection might act to maintain genetic redundancy (reviewed by Thomas 1993). In the case of the cytoskeletal interactions discussed herein (Figure 6*a*), it is important to note that although each individual gene is nonessential, each (with perhaps the exception of *ABP1*) seems to impart some unique advantage to the cell, as reflected in better growth over a broader temperature range. Thus it is possible that having multiple multifunctional proteins with partially overlapping functions allows fine-tuning of cytoskeletal properties and provides for more regulatory possibilities.

REGULATION

Overview

One of the hallmarks of genetic analysis has been the ability to delineate regulatory pathways. While the great promise of determining the mechanisms regulating the assembly and disassembly of actin structures is only beginning to be realized in yeast, some important progress has been made. Because cytoskeletal proteins are highly conserved, it is perhaps not surprising that regulatory mechanisms revealed by studies in yeast appear similar to those implicated in regulation of the mammalian cytoskeleton.

In all eukaryotic cells, the actin cytoskeleton can be induced to undergo rearrangements in response to extracellular or intracellular cues. Both yeast and mammalian cells show dramatic reorganization of the actin cytoskeleton during mitosis and cell division, firstly, bringing about the shape changes involved and, secondly, forming a ring structure at the plane of cell division. Both cell types also respond to external signals. For yeast, changes have been found in response to a variety of stimuli, including changes in osmolarity or nutrient availability. The best characterized stimulus, however, is pheromone which, in haploid cells of an appropriate mating type, induces formation of a mating projection. Extension of a mating projection, oriented along a pheromone gradient, depends on assembly of an asymmetrically organized actin cytoskeleton (Read et al 1992). In mammalian cells, the addition of growth factors can induce a variety of actin rearrangements depending on the growth factor added and the cell type, and chemotactic agents can induce assembly of cortical actin structures at the cell margin facing the source of chemoattractant (Ridley & Hall 1992, Ridley et al 1992, Norman et al 1994, Stowers et al 1995).

It is clear from these and other examples that the spatio-temporal organization of the actin cytoskeleton is under tight regulatory control. Many of the actin-binding proteins characterized are likely to be targets of signaling pathways leading to actin reorganization.

Using Yeast to Elucidate Signaling Pathways

In studies of yeast morphogenesis, many mutations affecting bud formation have been isolated. Actin is asymmetrically organized in growing yeast and has been demonstrated to be a key element of bud formation (Adams & Pringle 1984, Kilmartin & Adams 1984, Novick & Botstein 1985). Cells with temperature-sensitive mutations in actin, when shifted to the nonpermissive temperature, can no longer direct their growth and cannot form buds (Novick & Botstein 1985). Instead they grow isotropically. Because actin plays a central role in bud formation, other mutations that cause defects in bud formation are likely to occur in genes that regulate actin. Mutations in *CDC24, CDC42,* and *BEM1* cause severe defects in bud formation (reviewed in Drubin & Nelson 1996). A combination of genetics and biochemistry has identified the proteins encoded by these and other genes as components of a signaling complex that assembles at a particular site on the cell surface and generates cellular asymmetry by regulating actin and other molecules involved in cell polarity development (Figure 7). Briefly, activation of Rsr1 GTP-binding protein at a particular site on the cell cortex recruits Cdc24, which itself is a guanine-nucleotide exchange factor (GEF) for Cdc42, a small GTP-binding protein (Zheng et al 1994, 1995; reviewed in Herskowitz et al 1995). GTP-bound Cdc42 in the complex can then bind and activate Cla4/Ste20 protein kinases (Cvrckova et al 1995). Bem1 links several of these proteins together as shown in Figure 7 (Peterson et al 1994, Zheng et al 1995; reviewed in Pringle et al 1995, Herskowitz et al 1995, Drubin & Nelson 1996). It is also important to note that yeast can develop polarity in a spatial gradient of pheromone through activation of a pheromone receptor-coupled heterotrimeric G-protein complex (reviewed in Chenevert 1994). Orientation

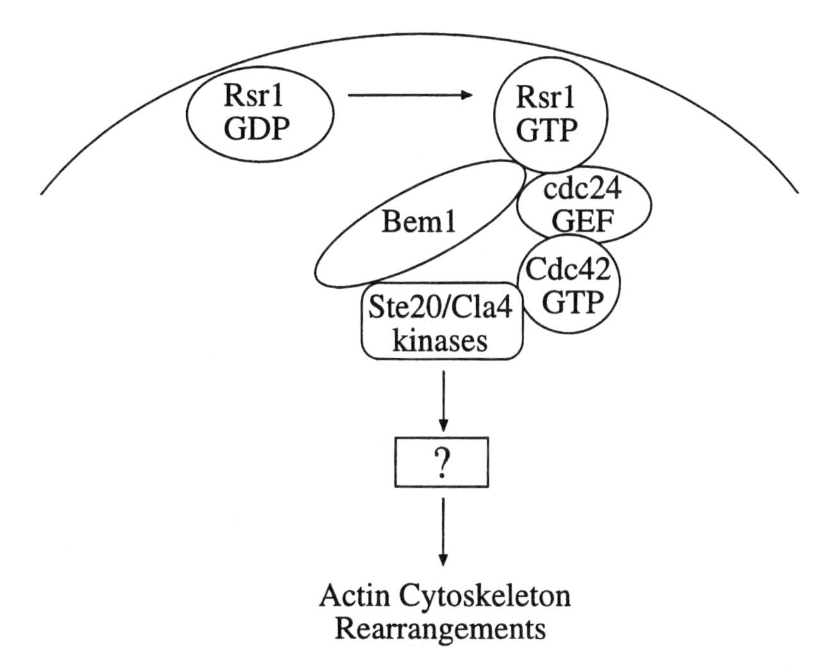

Figure 7 Proteins involved in polarity establishment in budding yeast. A complex of proteins at the presumpive bud site, including Rsr1, Cdc24, Cdc42, Bem1, and Ste20/Cla4 kinases, is able to trigger rearrangements in the actin cytoskeleton to allow polarized cell growth and the formation of a new bud.

in pheromone gradients requires asymmetric polarization of actin organization via the same signaling proteins involved in bud formation, minus Rsr1 (MN Simon et al 1995, Zhao et al 1995; reviewed in Herskowitz et al 1995).

Homologues for most of the signaling proteins shown in Figure 7 have been identified in mammalian cells and, significantly, in several cases the proteins appear to be involved in inducing rearrangements of the actin cytoskeleton. Members of the rho family of small GTP-binding proteins, particularly rho and cdc42HS, are important in bringing about cytoskeletal changes (Ridley & Hall 1992, Ridley et al 1992, Norman et al 1994, Kozma et al 1995, Nobes & Hall 1995). Current work is aimed at identifying the downstream effectors of this family of proteins. One protein shown to associate with cdc42HS is a mammalian brain protein PAK65, which is a homologue of Ste20p and Cla4 in yeast (Manser et al 1994). Other mammalian homologues of these proteins have now been identified, and in both yeast and mammals, the kinases are activated by GTP-bound Cdc42 (Bagrodia et al 1995, Creasy & Chernoff 1995, Martin et al 1995).

Although much investigation has concentrated on actin and its associated proteins, other research has focused on the complex of signaling proteins required for the selection of a bud site and formation of the bud (Figure 7). It is now clear that the gap separating the two areas needs to be bridged. What is learned is likely to apply to eukaryotes generally owing to the conservation of the GTP-binding protein/kinase signaling module. The powerful genetics available in yeast can be exploited to identify the critical regulatory interactions. In addition, epistasis tests can elucidate the order in which various components of these pathways are activated. Biochemical assays and molecular genetic approaches such as the two-hybrid assay will be able to determine which genetic interactions also represent physical interactions.

A recently developed assay allows regulation of the cortical actin cytoskeleton to be studied in permeabilized yeast cells (Li et al 1995). Rhodamine-labeled actin added to the permeabilized cells is specifically incorporated into structures at the cell cortex that resemble the cortical actin patches seen in vivo. This incorporation requires actin assembly, apparently from nucleation sites other than the ends of actin filaments. Studies on permeabilized yeast carrying mutations in several genes known to encode proteins associated with the actin cytoskeleton, including *CAP2, SLA1,* and *SLA2*, are helping to elucidate the roles of these proteins. Of particular interest, with respect to regulation, is the observation that the nucleated assembly of the exogenously added actin is impaired by mutations in *CDC42* but can be rescued by addition of exogenous Cdc42-GTP (Li et al 1995). The isolation and biochemical characterization of proteins required for nucleated actin assembly in these permeabilized cells, along with continued genetic analysis, should allow the targets of the signaling pathways that regulate the actin cytoskeleton to be identified.

An additional level of regulation imposed on the cytoskeleton is that of cell cycle control. Both yeast and other eukaryotes undergo cytoskeletal rearrangements at specific points in the cell cycle. In yeast, these changes are thought to be regulated by the Cdc28 cell cycle control protein kinase and its associated cyclins. Experimental manipulation of the timing of activation or inactivation of these cdc28/cyclin complexes results in acceleration or delay in morphological changes involving the actin cytoskeleton (Lew et al 1992, Lew & Reed 1993; reviewed in Lew & Reed 1995), demonstrating the coupling of cell cycle state to specific programs of actin assembly and disassembly. The continued application of genetic analysis and the use of biochemical studies in permeabilized cells provides a unique opportunity to elucidate the molecular mechanisms by which the cell cycle controls spatial programs of morphogenesis via rearrangements of cytoskeletal proteins.

CONCLUDING REMARKS

In this review we have illustrated the ways in which the yeast *S. cerevisiae* is being used to study the actin cytoskeleton. In several areas, this research has been particularly informative. For analyses of structure-function relationships, the ability to mutagenize the single conventional actin gene has provided a greater understanding of the role of individual residues in the actin monomer. In combination with a range of genetic approaches, strides are being made in constructing a functional map of the surface of actin. Such studies reveal how the powerful genetics available in yeast are complementary to the structural approach afforded by the generation of X-ray diffraction data and the modeling of atomic structures.

Genetic analysis of components of the actin cytoskeleton reveals a great deal of complexity and functional redundancy. The physiological basis for redundancy between nonhomologous proteins has been particularly difficult to comprehend. With use of a combination of genetics and biochemistry, we now understand how one pair of proteins in yeast—fimbrin and capping protein—shows functional redundancy. Further work should allow the basis for other redundancies to be better understood, thus increasing understanding of how the activities of many proteins are integrated to control the cytoskeleton.

Finally, the actin cytoskeleton is known to be under finely tuned and highly responsive regulatory control. In all eukaryotes, morphological changes dependent on the actin cytoskeleton are observed in response to intracellular and extracellular cues. How such cues are transmitted to cause actin rearrangements is not well understood. However, one of the hallmarks of genetic analysis has been the ability to elucidate regulatory pathways, and in this regard several of the proteins involved in regulation of the yeast actin cytoskeleton have been identified and shown to have homologues in other eukaryotes.

We have tried to address the main areas in which progress is being made in understanding the actin cytoskeleton in yeast. The actin cytoskeleton impinges on many aspects of cell physiology, so the recent advances reported here are likely to provide insight into a wide range of cellular processes from cell morphogenesis to endocytosis, vesicle trafficking, and organelle inheritance. With all of the yeast genome sequence known and with its experimental advantages, actin regulation and the integration of the actin cytoskeleton with numerous physiological processes are likely to be understood better in yeast than in any other organism. This should provide a conceptual framework for studies on all other eukaryotes.

ACKNOWLEDGMENTS

We thank David Botstein, Emil Reisler, Peter Rubenstein, and John Cooper for sending reprints and preprints and for communicating unpublished results, and Keith Kozminski and Tom Lila for helpful comments and for critical reading of the manuscript. We are also grateful to Morgan Conn for considerable assistance in preparing figures for the molecular structure of actin. This work was supported by grants to David Drubin from the National Institute of General Medicine Sciences (GM-42759) and the American Cancer Society (CB-106, FRA-442). Kathryn Ayscough is an International Prize Travelling Fellow of the Wellcome Trust (038110/Z/93/Z).

Literature Cited

Adams AEM, Botstein D. 1989. Dominant suppressors of yeast actin mutations that are reciprocally suppressed. *Genetics* 121:675–83

Adams AEM, Botstein D, Drubin DG. 1989. A yeast actin-binding protein is encoded by *SAC6*, a gene found by suppression of an actin mutation. *Science* 243:231–33

Adams AEM, Cooper JA, Drubin DG. 1993. Unexpected combinations of null mutations in genes encoding the actin cytoskeleton are lethal in yeast. *Mol. Biol. Cell* 4:459–68

Adams AEM, Pringle JR. 1984. Relationship to actin and tubulin distribution to bud growth in wild-type and morphogenetic mutant *Saccharomyces cerevisiae. J. Cell Biol.* 98:934–45

Amatruda J, Cannon J, Tatchell K, Hug C, Cooper J. 1990. Disruption of the actin cytoskeleton in yeast capping protein mutants. *Nature* 344:352–54

Amberg D, Basart E, Botstein D. 1995. Defining protein interactions with yeast actin in vivo. *Nature* 2:28–35

Aspenström P, Karlsson R. 1991. Interference with myosin subfragment-1 binding by site directed mutagenesis of actin. *Eur. J. Biochem.* 200:35–41

Bagrodia S, Taylor S, Creasy C, Chernoff J, Cerione R. 1995. Identification of a mouse p21$^{Cdc42/Rac}$ activated kinase. *J. Biol. Chem.* 270:22731–37

Bauer F, Urdachi M, Aigle M, Crouzet M. 1993. Alteration of a yeast SH3 protein leads to conditional viability with defects in cytoskeletal and budding patterns. *Mol. Cell Biol.* 13:5070–84

Bender A, Pringle JR. 1991. Use of a screen for synthetic lethal and multicopy suppressee mutants to identify two new genes involved in morphogenesis in *Saccharomyces cerevisiae. Mol. Cell Biol.* 11:1295–305

Bennett W, Paoni N, Keyt B, Botstein D, Jones A, et al. 1991. High resolution analysis of functional determinants on human tissue-type plasminogen activator. *J. Biol. Chem.* 266:5191–201

Bobola N, Jansen RP, Shin TH, Nasmyth K. 1996. Asymmetric accumulation of Ash1p in post anaphase nuclei depends on myosin and restricts yeast mating type switching to mother cells. *Cell* 84:699–709

Botstein D, Adams A, Amberg D, Drubin D, Huffaker T, et al. 1996. The yeast cytoskeleton. In *The Molecular and Cellular Biology of the Yeast Saccharomyces*, ed. JR Pringle, JR Broach, EW Jones. New York: Cold Spring Harbor Lab. Press. Vol. 3. In press

Bretscher A. 1981. Fimbrin is a cytoskeletal protein that crosslinks F-actin in vitro. *Proc. Natl. Acad. Sci. USA* 78:6849–53

Brown S, Spudich J. 1979. Cytochalasin inhibits the rate of elongation of actin filament fragments. *J. Cell Biol.* 83:657–62

Bubb M, Spector I, Bershadsky A, Korn E. 1995. Swinholide A is a microfilament disrupting marine toxin that stabilizes actin dimers and severs actin filaments. *J. Biol. Chem.* 270:3463–66

Caldwell J, Heiss S, Mermall V, Cooper J. 1989. Effects of CapZ, an actin capping protein of muscle, on the polymerization of actin. *Biochemistry* 28:8506–14

Chen X, Cook R, Rubenstein P. 1993. Yeast actin with a mutation in the "hydrophobic plug" between subdomains 3 and 4 ($L_{266}D$) displays a cold-sensitive polymerization defect. *J. Cell Biol.* 123:1185–95

Chen X, Peng J, Pedram M, Swenson C, Rubenstein P. 1995. The effect of the S14A mutation on the conformation and thermostability of *Saccharomyces cerevisiae* G-actin and its interaction with adenine nucleotides. *J. Biol. Chem.* 270:11415–23

Chen X, Rubenstein P. 1995. A mutation in an ATP-binding loop of *Saccharomyces cerevisiae* actin (S14A) causes a temperature-sensitive phenotype. *J. Biol. Chem.* 270: 11406–14

Chenevert J. 1994. Cell polarization directed by extracellular cues in yeast. *Mol. Biol. Cell.* 5:1169–75

Colucci-Guyon E, Portier M, Dunia I, Paulin D, Pournin S, Babinet C. 1994. Mice lacking vimentin develop and reproduce without an obvious phenotype. *Cell* 79:679–94

Cook R, Blake W, Rubenstein P. 1992. Removal of the amino-terminal acidic residue of yeast actin. *J. Biol. Chem.* 267:9430–36

Cook R, Root D, Miller C, Reisler E, Rubenstein P. 1993. Enhanced stimulation of myosin subfragment-1 ATPase activity by addition of negatively charged residues to the yeast actin NH_2 terminus. *J. Biol. Chem.* 268:2410–15

Cook R, Rubenstein P. 1991. The effects of actin Asp_{10} mutations on yeast viability. *J. Cell Biol.* 109:271a

Cook R, Sheff D, Rubenstein P. 1991. Unusual metabolism of the yeast actin amino teminus. *J. Biol. Chem.* 266:16825–33

Coué M, Brenner S, Spector I, Korn E. 1987. Inhibition of actin polymerization by latrunculin A. *FEBS Lett.* 213:316–18

Creasy C, Chernoff J. 1995. Cloning and characterization of a human protein kinase with homology to Ste20. *J. Biol. Chem.* 270:21695–700

Cvrckova F, DeVirgilio C, Manser E, Pringle JR, Nasmyth K. 1995. Ste20-like protein kinases are required for normal localization of cell growth and for cytokinesis in budding yeast. *Genes Dev.* 9:1817–30

DasGupta G, Reisler E. 1989. Antibody against the amino terminus of α-actin inhibits actomyosin interactions in the presence of ATP. *J. Mol. Biol.* 207:833–36

DasGupta G, Reisler E. 1991. Nucleotide-induced changes in the interaction of myosin subfragment-1 with actin: detection by antibodies against the N-terminal segment of actin. *Biochemistry* 30:9961–66

David C, Solimena M, DeCamilli P. 1994. Autoimmunity in stiff-man syndrome with breast cancer is targeted to the C-terminal region of human amphiphysin, a protein similar to the yeast proteins, Rvs167 and Rvs161. *FEBS Lett.* 351:73–79

DeLozanne A, Spudich J. 1987. Disruption of the *Dictyostelium* myosin heavy chain gene by homologous recombination. *Science* 236:1086–91

Drees B, Brown C, Barrell B, Bretscher A. 1995. Tropomyosin is essential in yeast, yet the TPM1 and TPM2 products perform distinct functions. *J. Cell. Biol.* 128:383–92

Drubin D, Jones H, Wertman K. 1993. Actin structure and function: roles in mitochondrial organization and morphogenesis in budding yeast and identification of the phalloidin-binding site. *Mol. Biol. Cell* 4:1277–94

Drubin DG, Miller KG, Botstein D. 1988. Yeast actin binding proteins: evidence for a role in morphogenesis. *J. Cell Biol.* 107:2551–61

Drubin DG, Mulholland J, Zhu Z, Botstein D. 1990. Homology of a yeast actin binding protein to signal transduction proteins and myosin-1. *Nature* 343:288–90

Drubin DG, Nelson WJ. 1996. Origins of cell polarity. *Cell* 84:335–44

Dunn T, Shortle D. 1990. Null alleles of *SAC7* suppress temperature sensitive actin mutations in *Saccharomyces cerevisiae. Mol. Cell Biol.* 10:2308–14

Estes J, Selden L, Gershman L. 1981. Mechanism of action of phalloidin on the polymerization of muscle actin. *Biochemistry* 20:708–12

Fedor-Chaiken M, Deschenes R, Broach J. 1990. *SRV2*, a gene required for RAS activation of adenylate cyclase in yeast. *Cell* 61:329–40

Field J, Vojtek A, Ballester R, Bolger G, Colicelli J, Ferguson K, et al. 1990. Cloning and characterization of *CAP*, the S. cerevisiae gene encoding the 70 kD adenylyl cyclase-associated protein. *Cell* 61:319–27

Fields S, Song O. 1989. A novel genetic system to detect protein-protein interactions. *Nature* 340:245–46

Flanagan M, Lin S. 1980. Cytochalasins block actin filament elongation by binding to high affinity sites associated with F-actin. *J. Biol. Chem.* 255:835–38

Glenney J, Kaulfus P, Matsudaira P, Weber K. 1981. F-actin binding and bundling properties of fimbrin, a major cytoskeletal protein of microvillus core filaments. *J. Biol. Chem.* 256:9283–88

Godette D, Frieden C. 1986a. The kinetics of cytochalasin D binding to monomeric actin. *J. Biol. Chem.* 261:15970–73

Godette D, Frieden C. 1986b. Actin polymerization. *J. Biol. Chem.* 261:15974–80

Goodson HV, Spudich JA. 1995. Identification and molecular characterization of a yeast myosin I. *Cell Motil. Cytoskelet.* 30:73–84

Govindan B, Bowser R, Novick P. 1995. The role of Myo2, a yeast class V myosin, in vesicular transport. *J. Cell Biol.* 128:1055–68

Greer C, Schekman R. 1982. Actin from *Saccharomyces cerevisiae. Mol. Cell Biol.* 2:1270–78

Haarer B, Petzold A, Lillie S, Brown S. 1994. Identification of *MYO4*, a second class V myosin gene in yeast. *J. Cell Sci.* 107:1055–64

Herskowitz I, Park H, Sanders S, Valtz N, Pstar M. 1996. Programming of cell polarity in budding yeast by endogenous and exogenous signals. *Cold Spring Harbor Symp. Quant. Biol.* 60:In press

Holmes K, Popp D, Gebhard W, Kabsch W. 1990. Atomic model of the actin filament. *Nature* 347:44–49

Holtzman DA, Wertman K, Drubin DG. 1994. Mapping actin surfaces required for functional interactions in vivo. *J. Cell Biol.* 126:423–32

Holtzman DA, Yang S, Drubin DG. 1993. Synthetic-lethal interactions identify two novel genes, *SLA1* and *SLA2*, that control membrane cytoskeleton assembly in *Saccharomyces cerevisiae. J. Cell Biol.* 122:635–44

Honts J, Sandrock T, Brower S, O'Dell J, Adams AEM. 1994. Actin mutatins that show suppression with fimbrin mutations identify a likely fimbrin binding site on actin. *J. Cell Biol.* 126:413–22

Johannes F, Gallwitz D. 1991. Site directed mutagenesis of the yeast actin gene: a test for actin function in vivo. *EMBO J.* 10:3951–58

Johara M, Toyoshima Y, Ishijima A, Kojima H, Yanagida T, Sutoh K. 1993. Charge-reversion mutagenesis of *Dictyostelium* actin to map the surface recognized by myosin during ATP-driven sliding motion. *Proc. Natl. Acad. Sci. USA* 90:2127–31

Johnston G, Prendergast J, Singer R. 1991. The *Saccharomyces cerevisiae MYO2* gene encodes an essential myosin for vectorial transport of vesicles. *J. Cell Biol.* 113:539–51

Kabsch W, Mannherz H, Suck D, Pai E, Holmes K. 1990. Atomic structure of the actin:DNase1 complex. *Nature* 347:37–44

Karpova T, Tatchell K, Cooper J. 1995. Actin filaments in yeast are unstable in the absence of capping protein. *J. Cell Biol.* 131:1483–93

Kilmartin JV, Adams AEM. 1984. Structural rearrangements of tubulin and actin during the cell cycle of the yeast, *Saccharomyces. J. Cell Biol.* 98:922–39

Kozma R, Ahmed S, Best A, Lim L. 1995. The Ras related protein Cdc42hs and bradykinin promote formation of peripheral actin microspikes and filopodia in Swiss 3T3 fibroblasts. *Mol. Cell Biol.* 15:1942–52

Kron SJ, Drubin DG, Botstein D, Spudich JA. 1992. Yeast actin filaments display ATP-dependent sliding movement over surfaces coated with rabbit muscle myosin. *Proc. Natl. Acad. Sci. USA* 89:4466–70

Lazzarino D, Boldogh I, Smith M, Rosand J, Pon L. 1994. Yeast mitochondria contain ATP-sensitive, reversible actin-binding activ-

ity. *Mol. Biol. Cell* 5:807–18
Lebart M, Mejéan C, Roustan C, Benyamin Y. 1993. Further characterization of the α-actinin-actin interface and comparison with filamin-binding sites on actin. *J. Biol. Chem.* 268:5642–48
Lew D, Marini N, Reed SI. 1992. Different G_1 cyclins control the timing of cell cycle commitment in mother and daughter cells of the budding yeast S. cerevisiae. *Cell* 69:317–27
Lew D, Reed SI. 1993. Morphogenesis in the yeast cell cycle: regulation by Cdc28 and cyclins. *J. Cell Biol.* 120:1305–20
Lew D, Reed SI. 1995. Cell cycle control of morphogenesis in budding yeast. *Curr. Opin. Genet. Dev.* 5:17–23
Li R, Zheng Y, Drubin DG. 1995. Regulation of cortical actin cytoskeleton assembly during polarized cell growth in budding yeast. *J. Cell Biol.* 128:599–615
Lillie S, Brown S. 1994. Immunofluorescence localization of the unconventional myosin, Myo2p, and the putative kinesin-related protein, Smy1p to the same regions of polarized growth in *Saccharomyces cerevisiae. J. Cell Biol.* 125:825–42
Liu H, Bretscher A. 1992. Characterization of *TPM1* disrupted yeast cells indicates an involvement of tropomyosin in directed vesicular transport. *J. Cell Biol.* 118:285–99
Lorenz M, Poole K, Popp D, Rosenbaum G, Holmes, K. 1995. An atomic model of the unregulated thin filament obtained by X-ray fiber diffraction on oriented actin-tropomyosin gels. *J. Mol. Biol.* 246:108–19
Lorenz M, Popp D, Holmes K. 1993. Refinement of the F-actin model against X-ray fiber diffraction data by the use of a directed mutation algorithm. *J. Mol. Biol.* 234:826–36
Manser E, Leung T, Salihuddin H, Zhao Z, Lim L. 1994. A brain serine/threonine protein kinase activated by Cdc42 and Rac1. *Nature* 367:40–46
Martin G, Bollag G, McCormick F, Abo A. 1995. A novel serine kinase activated by rac1/CDC42Hs-dependent autophosphorylation is related to PAK65 and STE20. *EMBO J.* 14:1970–78
McGough A, Way M, DeRosier D. 1994. Determination of the α-actinin-binding site on actin filaments by cryoelectron microscopy and image analysis. *J. Cell Biol.* 126:433–43
McLaughlin P, Gooch J, Mannherz H, Weeds A. 1993. Structure of gelsolin segment 1-actin complex and the mechanism of filament severing. *Nature* 364:685–92
Miller C, Doyle T, Bobkova E, Botstein D, Reisler E. 1996. Mutational analysis of the role of hydrophobic residues in the 338–348 helix in actomyosin interactions. *Biochemistry* 35:3670–76
Miller C, Reisler E. 1995. Role of charged amino acid pairs in subdomain-1 of actin in interaction with myosin. *Biochemistry* 34:2694–700
Mimura N, Asano A. 1987. Further characterization of a conserved actin-binding 27 kDa fragment of actinogelin and α-actinins and mapping of their binding sites on the actin molecule by chemical cross-linking. *J. Biol. Chem.* 262:4717–23
Moon AL, Janmey PA, Louie A, Drubin DG. 1993. Cofilin is an essential component of the yeast cortical actin cytoskeleton. *J. Cell Biol.* 120:421–35
Mornet D, Ue K. 1984. Proteolysis and structure of skeletal muscle actin. *Proc. Natl. Acad. Sci. USA* 81:3680–84
Nobes C, Hall A. 1995. Rho, Rac, and Cdc42 GTPases regulate the assembly of multimolecular focal complexes associated with actin stress fibers, lamellipodia and filopodia. *Cell* 81:53–62
Norman J, Price L, Ridley A, Hall A, Koffer A. 1994. Actin filament organization in activated mast cells is regulated by heterotrimeric and small-GTP binding proteins. *J. Cell Biol.* 126:1005–15
Novick P, Botstein D. 1985. Phenotypic analysis of temperature-sensitive yeast actin mutants. *Cell* 40:405–16
Ohya Y, Botstein D. 1994. Diverse essential functions revealed by complementing yeast calmodulin mutants. *Science* 263:963–66
Peterson J, Zheng Y, Bender L, Myers A, Cerione R, Bender, A. 1994. Interactions between the bud emergence proteins Bem1p and Bem2p and Rho-type GTPases in yeast. *J. Cell Biol.* 127:1395–406
Pringle JR, Bi E, Harkin H, Zahner J, DeVirgilio C, et al. 1996. Establishment of cell polarity in yeast. *Cold Spring Harbor Symp. Quant. Biol.* 60:In press
Rayment I, Holden H, Whittaker M, Yohn C, Lorenz M, et al. 1993a. Structure of the actin-myosin complex and its implications for muscle contraction. *Science* 261:58–65
Rayment I, Rypniewski W, Schmidt-Base K, Smith R, Tomchick D, et al. 1993b. Three-dimensional structure of myosin subfragment-1: a molecular motor. *Science* 261:50–58
Read EB, Okamura HH, Drubin DG. 1992. Actin- and tubulin-dependent functions during *Saccharomyces cerevisiae* mating projection. *Mol. Biol. Cell* 3:429–44
Ridley A, Hall A. 1992. The small GTP-binding protein rho regulates the assembly of focal adhesion and actin stress fibres in response to growth factors. *Cell* 70:389–99

Ridley AJ, Paterson HF, Johnston CL, Diekmann D, Hall A. 1992. The small GTP-binding protein rac regulates growth factor induced membrane ruffling. *Cell* 70:401–10
Rodriguez J, Paterson B. 1990. Yeast myosin heavy chain mutant: maintenance of the cell type specific budding pattern and the normal deposition of chitin and cell wall components requires an intact heavy chain gene. *Cell Motil. Cytoskelet.* 17:301–8
Saito S, Watabe S, Ozaki H, Fusetani N, Karaki H. 1994. Mycalolide B, a novel actin depolymerizing agent. *J. Biol. Chem.* 269:29710–14
Schröder R, Manstein D, Jahn W, Holden H, Rayment I, et al. 1993. Three dimensional atomic model of F-actin decorated with *Dictyostelium* myosin S1. *Nature* 364:171–74
Schutt C, Myslik J, Rozycki M, Goonesekere N, Lindberg U. 1993. The structure of crystalline profilin-β-actin. *Nature* 365:810–16
Sheterline P, Sparrow J. 1994. Actin. *Protein Profile* 1:1–121
Shortle D, Novick P, Botstein D. 1984. Construction and genetic characterization of temperature-sensitive mutant alleles of the yeast actin gene. *Proc. Natl. Acad. Sci. USA* 81:4889–93
Simon MN, DeVirgilio B, Souza B, Pringle JR, Abo A, Reed SI. 1995. Role for the Rho-family GTPase Cdc42 in yeast mating pheromone signal pathway. *Nature* 376:702–5
Simon V, Swayne T, Pon L. 1995. Actin-dependent mitochondrial motility in mitotic yeast and cell free systems: identification of a motor activity on the mitochondrial surface. *J. Cell Biol.* 130:345–54
Sivadon P, Bauer F, Aigle M, Crouzet M. 1995. Actin cytoskeleton and budding pattern are altered in the yeast *rvs161* mutant: the Rvs161 protein shares common domains with the brain protein amphiphysin. *Mol. Gen. Genet.* 246:485–95
Smith M, Simo V, O'Sullivan H, Pon L. 1995. Organelle-cytoskeletal interactions: actin mutations inhibit meiosis-dependent mitochondrial rearrangement in the budding yeast, *Saccharomyces cerevisiae. Mol Biol. Cell* 6:1381–96
Stowers L, Yelon D, Berg L, Chant J. 1995. Regulation of the polarization of T-cells toward antigen presenting cells by Ras-related GTPase CDC42. *Proc. Natl. Acad. Sci. USA* 92:5027–31
Strzelecka-Golaszewska H, Moraczewska J, Khaitlina S, Mossakowska M. 1993. Localization of the tightly bound divalent cation-dependent and nucleotide-dependent conformation changes in G-actin using limited proteolytic digestion. *Eur. J. Biochem.* 211:731–42
Sutoh K. 1982. Identification of myosin binding sites on the actin sequence. *Biochemistry* 21:3654–61
Tirion M, ben-Avraham D, Lorenz M, Holmes K. 1995. Normal modes as refinement parameters for the F-actin model. *Biophys. J.* 68:5–12
Thomas J. 1993. Thinking about genetic redundancy. *Trends Genet.* 9:395–98
Vojtek A, Cooper J. 1993. Identification and characterization of a cDNA encoding mouse CAP: a homolog of the yeast adenylyl cyclase associated protein. *J. Cell Sci.* 105:777–85
Wang T, Bretscher A. 1995. The rho-GAP encoded by *BEM2* regulates cytoskeletal structure in budding yeast. *Mol. Biol. Cell* 6:1011–24
Welch MD, Vinh DBN, Okamura HH, Drubin DG. 1993. Screens for extragenic mutations that fail to complement *act1* alleles identify genes that are important for actin function in *Saccharomyces cerevisiae. Genetics* 135:265–74
Wertman KF, Drubin DG, Botstein D. 1992. Systematic mutational analysis of the yeast *ACT1* gene. *Genetics* 132:337–50
Wilson R, Ainscough R, Anderson K, Baynes C, Berks M, et al. 1994. 2.2Mb of contiguous nucleotide sequence form chromosome III of *C. elegans. Nature* 368:32–38
Witke W, Schleicher M, Noegel A. 1992. Redundancy in the microfilament system: abnormal development of Dictyostelium cells lacking two F-actin cross-linking proteins. *Cell* 68:53–62
Zhao Z, Leung T, Manser E, Lim L. 1995. Pheromone signalling in *Saccharomyces cerevisiae* requires the small GTP-binding protein Cdc42p and its activator CDC24. *Mol. Cell Biol.* 15:5246–57
Zheng Y, Bender A, Cerione R. 1995. Interactions among proteins involved in bud-site selection and bud-site assembly in *Saccharomyces cerevisiae. J. Biol. Chem.* 270:626–30
Zheng Y, Cerione R, Bender A. 1994. Control of the yeast bud-site assembly GTPase Cdc42. *J. Biol. Chem.* 269:2369–72

Annu. Rev. Cell Dev. Biol. 1996. 12:161–80

ORGANIZING SPATIAL PATTERN IN LIMB DEVELOPMENT

William J. Brook, Fernando J. Diaz-Benjumea, and Stephen M. Cohen
European Molecular Biology Laboratory, Postfach 10.2209, 69012 Heidelberg, Germany

KEY WORDS: pattern formation, *Drosophila*, cell interaction, signal transduction

ABSTRACT

Recent studies on the development of the legs and wings of *Drosophila* have led to the conclusion that insect limb development is controlled by localized pattern organizing centers, analogous to those identified in vertebrate embryos. Genetic analysis has defined the events that lead to the formation of these organizing centers and has led to the identification of gene products that mediate organizer function. The possibility of homology between vertebrate and insect limbs is considered in light of recently reported similarities in patterns of gene expression and function.

CONTENTS

1081-0706/96/1115-0161$08.00

INTRODUCTION

The idea that cell interactions play an important role in pattern formation is widely accepted among developmental biologists. The pioneering experiments of Spemann & Mangold first established that a specific group of cells can behave as an organizing center that specifies the developmental fates of nearby cells in a non-autonomous manner (Spemann & Mangold 1924). They demonstrated that cells located in the dorsal lip of the blastopore of an amphibian embryo were able to organize a complete secondary dorsal-ventral body axis when transplanted to an ectopic ventral location. The secondary axis consisted of both host and graft tissues, indicating that cells from the organizer region were able to exert a long-range influence on the surrounding host cells so as to completely respecify axial pattern in the embryo. Subsequently, other organizing centers have been identified in vertebrate embryos by grafting experiments. Cells from the posterior margin of the chick limb bud are able to induce a secondary axis when transplanted to an ectopic anterior position in a host limb (Saunders & Gasseling 1968, Tickle et al 1975). Similarly, the notochord behaves as an inducer of the floorplate, which in turn organizes dorsal-ventral pattern in the neural tube (e.g. Yamada et al 1991).

How can a small group of cells organize the spatial pattern of an entire developmental field? One possibility is that these cells serve as a localized source of a secreted signaling molecule that instructs nearby cells about their prospective fate. A number of different signaling molecules have been proposed as candidates to mediate organizer function. For example, microinjection of mRNA encoding *Wnt* genes, *activin* or *noggin*, into blastomeres of amphibian embryos mimics the effects of organizer grafts in generating a secondary body axis (reviewed in Harland 1994, Slack 1994). Similarly, beads soaked in retinoic acid mimic the activity of the polarizing region when implanted into the anterior of a host limb bud (Tickle et al 1982). More recently, vertebrate homologues of the *Drosophila* segmentation gene *hedgehog* have been implicated as mediators of organizer activity in the dorsal blastopore lip, the polarizing region of the chick limb, and in the notochord (reviewed in Fietz et al 1994, Tickle & Eichele 1994).

GENETIC APPROACHES TO ANALYZING ORGANIZER FUNCTION

Experiments designed to assess organizer function in vertebrate embryos have traditionally involved transplantation of cells from the organizer region into a suitable host. Three strategies have been developed to assess the activity of candidate organizer molecules in the chicken limb bud: implantation of beads soaked in a candidate protein (or retinoid); implantation of heterologous cells that express a candidate gene; infection of the tissue with a recombinant retrovirus expressing a candidate gene (e.g. Niswander et al 1993, Riddle et al 1993, 1995, Vogel et al 1995, Yang & Niswander 1995). Groups of cells with organizer activity can be identified in the developing appendages of *Drosophila* by producing genetic mosaics that alter patterns of gene expression in situ. Two types of genetic mosaic experiments have been used to address the properties of organizers (Figure 1). One type of experiment is based on mis-expression of a candidate gene, similar in principle to the transplantation experiments performed in the chick limb bud. The gene encoding a putative organizer molecule is mis-expressed in randomly positioned clones of cells in the developing limb (Struhl & Basler 1993). If the product of the gene contributes to organizer function, the clone of cells expressing it will exert a non-autonomous influence on the surrounding tissue to reorganize pattern. The second type of experiment involves generating clones of cells that lack the function of a particular gene. In this case the experiment asks whether localized activity of the gene is required to establish the organizer (Diaz-Benjumea & Cohen 1993).

CELL INTERACTIONS INDUCE THE ORGANIZER

In the early 1970s, García-Bellido and colleagues showed that the developing appendages of *Drosophila* are subdivided into anterior-posterior and dorsal-ventral compartments (García-Bellido et al 1973, Wieschaus & Gehring 1976, Steiner 1976). Compartments were first recognized by virtue of the boundaries of cell lineage restriction that separate them, and it was proposed that these boundaries might serve as organizing centers responsible for generating spatial pattern in the developing appendages (Crick & Lawrence 1975). Recent studies have shown that the anterior-posterior and dorsal-ventral compartment boundaries each produce an organizing center required for wing patterning (Diaz-Benjumea & Cohen 1993, Basler & Struhl 1994). The formation of these organizers can be separated into three distinct steps: (*a*) cell fate specification to produce two differently determined populations of cells in adjacent compartments, (*b*) asymmetric cell signaling at the interface between the two cell populations, leading to (*c*) a symmetric organizing signal emanating from cells near the boundary of cell determination (Figures 2 and 3).

ORGANIZING THE ANTERIOR-POSTERIOR AXIS OF THE WING

Step 1: Establishing Asymmetry

García-Bellido proposed that cells would acquire their compartment-specific identity through localized expression of regulatory genes, which he called

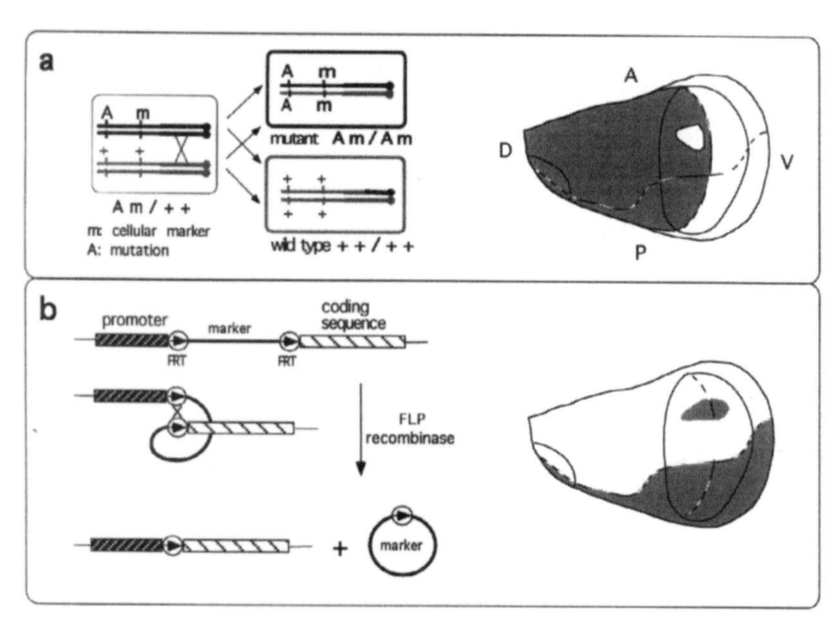

Figure 1 Tools for making genetic mosaics in *Drosophila.* (*a*) Scheme for producing clones of mutant cells by X-ray-induced somatic recombination. Exchange between homologous chromosomes heterozygous for a mutation (A) and a cell marker (m) leads to the production of a daughter cell homozygous for both the mutant and the marker (A m/A m) and a twin homozygous for both wild-type alleles (+ +/+ +). The homozygous mutant cell will proliferate to produce a patch of mutant tissue surrounded by wild-type cells. The illustration shows how a clone of *apterous*-minus cells (white triangular area) would appear in a field of *apterous*-expressing cells in the dorsal compartment of the wing imaginal disc. The phenotypic consequences of producing a dorsal *apterous* clone are shown in Figure 3*a*. (*b*) Scheme for producing clones of cells expressing a gene of interest (after Struhl & Basler 1993). The transgene carries a cell marker flanked by FRT sites separating a constitutive promoter and a protein coding sequence of interest (e.g. *hh* or *wg* cDNA). Excision of the "flip-out" cassette catalyzed by the yeast FLP recombinase allows the activation of the target gene in any cell in the imaginal disc. The result is a clone of cells expressing the gene of interest and not the cell marker. This clone is surrounded by non-expressing, marked cells. The diagram represents a clone of *hh*-expressing cells in the anterior compartment, corresponding to the wing shown in Figure 3*c*.

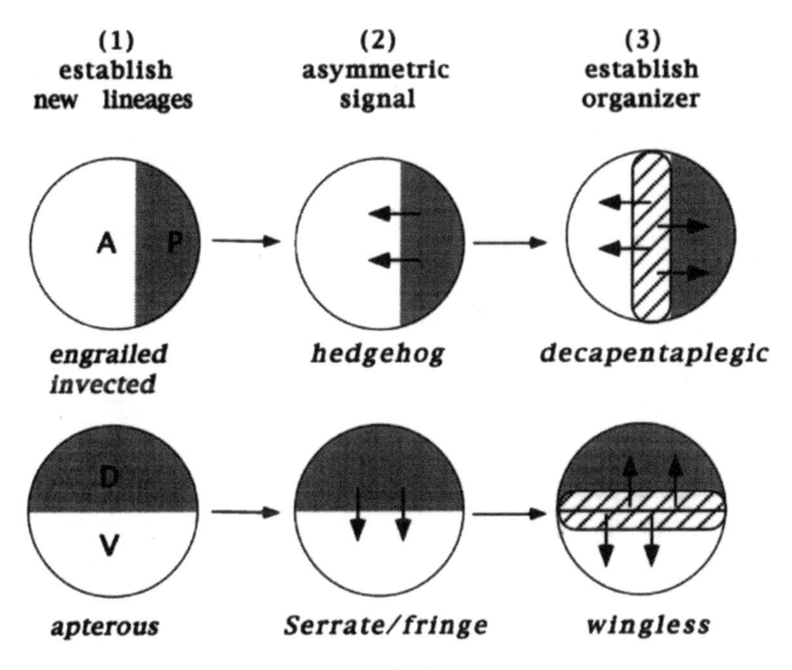

Figure 2 Genes implicated at the three stages of A/P and D/V organizer activity: (Step 1) Compartments are defined by localized expression of a selector gene: *engrailed/invected* = posterior, *apterous* = dorsal. (Step 2) Asymmetric signaling from posterior to anterior, mediated by *hedgehog* and from dorsal to ventral mediated by *fringe* and *Serrate*. (Step 3) Localized expression of a second signaling molecule in turn mediates the organizing activity of the compartment boundary. The organizing signal acts symmetrically to pattern cells in both compartments. In the A/P system, cells comprising the organizer are located exclusively on the anterior side of the compartment boundary and express *decapentaplegic*. This population lies in the center of the disc because the compartment boundary is positioned asymmetrically within the disc primordium (see Cohen 1993 for more detailed explanation). *wingless* mediates activity of the D/V organizer. The D/V organizer differs from the A/P in that *wg* is expressed symmetrically on both sides of the compartment boundary. The mechanism by which this symmetric arrangement arises is not fully understood (but see Figure 3).

selector genes (García-Bellido 1975). The homeobox gene *engrailed* (*en*) and its homologue *invected* (*inv*) function together as a selector gene for the posterior compartment of the appendages. *en* and *inv* are expressed in all cells of the posterior compartment, where they are required to specify the fates of posterior cells (reviewed in Lawrence & Morata 1994, Blair 1995; see also Hidalgo 1994, Guillen et al 1995, Sanicola et al 1995, Simmonds et al 1995, Tabata et al 1995).

Cell lineage analysis has shown that the appendage primordia are subdivided into anterior and posterior compartments in the embryo (Steiner 1976,

Wieschaus & Gehring 1976). The imaginal disc primordia originate straddling the parasegment boundary in the embryonic ectoderm in response to a positional signal from the Wingless protein (Cohen 1990, Cohen et al 1993). Consequently, the group of founder cells that comprises the disc primordium consists of cells from two adjacent parasegments (reviewed in Cohen 1993). The founder cells that comprise the posterior compartment of the appendage primordium inherit *en* expression from the embryonic segment. Thus the boundary of cell lineage restriction separating the compartments of the appendages corresponds to the boundary of cell lineage restriction between the parasegments of embryonic ectoderm (Vincent & O'Farrell 1992).

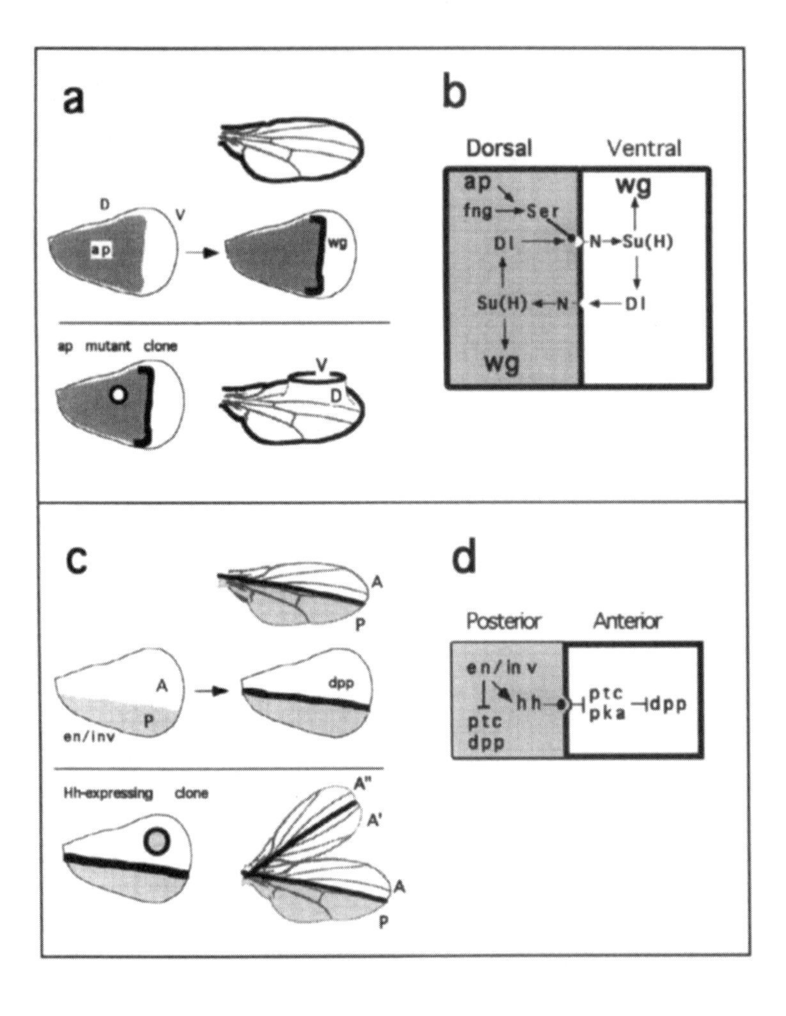

Step 2: Interaction Between Distinctly Specified Cells

Interaction between *en*-expressing cells in the posterior compartment and the adjacent cells of the anterior compartment is required for patterning in the embryo and in the limb primordia. Cells in the posterior compartment communicate with nearby cells in the anterior compartment through localized expression of the secreted signaling molecule *hedgehog* (*hh*; Lee et al 1992, Mohler & Vani 1992, Tabata et al 1992). *hh* activity in posterior cells is required for development of both compartments in the imaginal discs (Mohler 1988, Basler & Struhl 1994). To determine whether *hh* transmits a signal between posterior and anterior cells, Basler & Struhl produced clones of cells that express *hh* (Basler & Struhl 1994). *hh*-expressing cells in the anterior compartment exert a non-autonomous influence on the developmental fates of the surrounding wild-type cells that can lead to axis duplication (Figure 3). Appropriately positioned clones can lead to bifurcation of the wing or the leg. (The supernumerary limb consists of two complete, symmetrically arranged anterior compartments flanking the clone.) These experiments provide a compelling demonstration that clones of *hh*-expressing cells are able to direct axis formation.

A similar result has been obtained by grafting cells expressing *Sonic Hedgehog* (*Shh*, a vertebrate homologue of *hh*) into the anterior of the chick limb bud

←

Figure 3 Cell interactions establish the D/V and A/P organizers. (*a*) Interaction between dorsal (D) *apterous*-expressing cells (*shaded*) and ventral (V) non-expressing cells (*white*) leads to the localized expression of the secreted protein Wingless (*heavy black line*) at the wing margin. A dorsal clone mutant for *apterous* (*white circle, bottom left*) leads to the ectopic *wg* expression at the clone interface and causes an axis duplication (*bottom right*). Cells inside the clone are ventral and cause duplication of surrounding dorsal tissue. (*b*) Model for D/V cell interactions: *fng* expression in dorsal cells activates *Serrate*, which signals through *Notch* to adjacent ventral cells. *Notch* signals through Su(H) to activate *wg*. The initial activation of *wg* at the D/V boundary is ventral (Ng et al 1996). To explain the final symmetry of *wg* expression, it is necessary to invoke a secondary ventral-to-dorsal signal. This cannot be *wg* because ventral *wg* is not needed for expression of *wg* dorsally (Diaz-Benjumea & Cohen 1995). *Delta* is a good candidate to be the ventral signal (De Celis et al 1996, Doherty et al 1996). *Delta* may signal back through *Notch* to activate *wg* dorsally and may contribute to later signaling from dorsal to ventral. (*c*) Interaction between posterior (P) *engrailed*-expressing cells (*shaded*) and anterior (A) non-expressing cells (*white*) leads to localized expression of the secreted protein Decapentaplegic (*heavy black line*) in anterior cells adjacent to the A/P compartment boundary. An *hh*-expressing clone (*shaded circle, bottom left*) leads to the ectopic *Dpp* expression in and around the clone (*black ring around shaded circle*; for clarity, *Dpp* expression inside the clone is not shown). Ectopic *Dpp* causes axis duplication, as shown in the bottom right diagram. The clone runs down the center of the supernumerary wing and has organized a complete anterior compartment on each side (A′, A″). Other types of clones leading to A/P axis bifurcations are described in the text. (*d*) Model for A/P cell interactions: *engrailed/invected* (*en/inv*) repress *patched* (*ptc*) and *decapentaplegic* (*dpp*) and activate *hedgehog* (*hh*) in the posterior compartment. Hh is secreted and activates *dpp* in anterior cells by alleviating repression by Protein-kinase-A (Pka) and Ptc. Asymmetry results from the repression of *dpp* by by *en/inv*.

(Riddle et al 1993). *Shh*-expressing cells mimic the organizing activity of the polarizing region and produce a duplication of the anterior-posterior (A/P) axis of the wing. *Shh* is expressed in the posterior region of the chick limb bud, a region known to have polarizing activity.

Step 3: Transmitting the Organizing Signal

Although *hh* is expressed throughout the posterior compartment, clones of *hh*-expressing cells can only organize the pattern in the anterior compartment. Hh induces localized expression of the TGF-β homologue *decapentaplegic* (*dpp*) in a stripe of cells near the compartment boundary (Basler & Struhl 1994). Posterior cells are rendered insensitive to the *hh* signal by *en* and *inv* (Sanicola et al 1995, Tabata et al 1995, Zecca et al 1995). Expression of *dpp* is repressed by *patched* (*ptc*) and by *protein kinase A-C1* (*pka-C1*), except in cells near the compartment boundary, where Hh antagonizes *pka-C1* and *ptc* activity and thereby alleviates repression of *dpp* (Capdevila et al 1994, Jiang & Struhl 1995, Lepage et al 1995, Li et al 1995, Pan & Rubin 1995). Clones of cells that lack these repressors repattern the anterior compartment, and in the case of *pka-C1*, the repatterning has been shown to be *dpp* dependent (Jiang & Struhl 1995, Li et al 1995). Ectopic expression of *dpp* in either compartment of the wing is sufficient to organize an ectopic A/P axis, and the ectopic axes are centered around the clones of *dpp*-expressing cells (Capdevila & Guerrero 1994, Zecca et al 1995). Thus the role of the Hh signal in A/P patterning is to induce localized expression of *dpp*, which in turn organizes pattern and controls growth symmetrically in both compartments (Figure 3). It has been suggested that the secreted Dpp protein may specify cell fates along the A/P axis in a concentration-dependent manner (Zecca et al 1995, Nellen et al 1996), although this model is controversial (Lecuit et al 1996).

Some aspects of the *hh-dpp* signaling cascade are evolutionarily conserved. A vertebrate homologue of *ptc* acts downstream of *Shh* in the limb mesenchyme (Marigo et al 1996). Vertebrate *dpp* homologues (BMPs) are expressed in the limb bud and are induced by polarizing region grafts (Francis et al 1994). In spite of the fact that these genes appear to be regulated by *Shh* expression, there is as yet little or no evidence that BMPs have organizing activity in the A/P axis of the vertebrate limb, analogous to that of *dpp* in *Drosophila* (Francis-West et al 1995).

ORGANIZING THE DORSAL-VENTRAL AXIS OF THE WING

The primary subdivision of the limb primordia into anterior and posterior compartments depends on earlier patterning activity in the embryonic ectoderm. Later in larval development, the wing disc is subdivided a second time into dorsal and ventral compartments (García-Bellido 1975). Similar to the A/P boundary, the dorsal-ventral (D/V) compartment boundary serves as an organizing center that controls growth and specifies spatial pattern with reference to the D/V axis (Diaz-Benjumea & Cohen 1993). As described for the A/P system, production of an ectopic D/V organizing center leads to formation of a secondary axis in the wing. In the case of the A/P system, it is easy to see the secondary axis because the supernumerary wing lies in the same plane as the normal wing (Figure 3). By contrast, a secondary D/V axis is oriented perpendicular to the flat surface of the wing, so the supernumerary wing appears as an outgrowth, projecting dorsally from the surface of the wing blade (see Figure 3).

Step 1: Establishing Asymmetry

The dorsal compartment is specified by localized expression of the selector gene *apterous* (Diaz-Benjumea & Cohen 1993, Blair 1993). *apterous* encodes a LIM/homeodomain protein that is expressed in dorsal cells and specifies their fate as dorsal. Using genetic mosaics, it is possible to generate clones of cells in the dorsal compartment that lack *apterous* function. *apterous* mutant cells in the clone change their fate from dorsal to ventral, which results in a new interface between dorsal and ventral cells at the perimeter of the clone (Figure 1). Juxtaposition of dorsal and ventral cells mimics the normal situation at the D/V compartment boundary and leads to the formation of an ectopic wing margin encircling the clone and to a secondary D/V organizer (Diaz-Benjumea & Cohen 1993, Blair et al 1994, Williams et al 1994; see Figure 3). An analogous confrontation can be generated by ectopic expression of *apterous* in the ventral compartment. In this case, ventral cells adopt dorsal fate (Blair et al 1994), and the juxtaposition produces an ectopic organizing center on the ventral side of the wing (CJ Neumann & SM Cohen, unpublished data).

A LIM/homeodomain protein has recently been implicated as a determinant of dorsal fate in vertebrate limbs. *c-LMX-1* is induced in dorsal mesenchyme by *WNT-7a* expression in the overlying dorsal ectoderm (Riddle et al 1995, Vogel et al 1995). Loss of *WNT-7a* function causes ventralization of dorsal limb structures (Parr & McMahon 1995). Ectopic expression of *c-LMX-1* in the ventral mesenchyme, induced by ectopic *WNT-7a* expression or by *c-LMX-1* retrovirus infection, causes them to give rise to dorsal structures (Riddle et al 1995, Vogel et al 1995). These observations provide a molecular explanation for the transplantation studies suggesting that the ectoderm may be the primary determinant of D/V fate (reviewed in Tickle & Eichele 1994).

Step 2: Interaction Between Distinctly Specified Cells

Interaction between dorsal and ventral cells is needed to establish the organizing center at the D/V boundary. *fringe* and *Serrate* have been identified as candidates to mediate the signal from dorsal to ventral cells. *Serrate* encodes a predicted membrane-spanning protein with extracellular EGF repeats (Thomas et al 1991). *fringe* encodes a protein predicted to be secreted, and one that does not belong to a previously defined family (Irvine & Wieschaus 1994). Expression of both genes is restricted to the dorsal compartment during the second and early third larval instars, when cell interactions establish the D/V organizer (Irvine & Wieschaus 1994, Speicher et al 1994, Couso et al 1995, Diaz-Benjumea & Cohen 1995).

Evidence that *Serrate* mediates the signal derives from lack-of-function mosaics and from ectopic expression. Although *Serrate* is expressed throughout the dorsal compartment, mosaic analysis has shown that its activity is only required in dorsal cells that abut the compartment boundary (Couso et al 1995, Diaz-Benjumea & Cohen 1995, De Celis et al 1996). Clones of *Serrate* mutant cells that touch the margin from the dorsal side cause large nicks in the wing blade that result from loss of both dorsal and ventral wing tissue. Comparably positioned ventral clones have no effect. Mis-expression of *Serrate* in the ventral compartment leads to the formation of ectopic wing margin structures and to outgrowths of the wing reminiscent of those produced by ectopic ventral expression of *apterous* (Speicher et al 1994, Diaz-Benjumea & Cohen 1995; CJ Neumann & SM Cohen, unpublished data) and results in the expression of downstream genes such as *wingless* and *vestigial* (Couso et al 1995, Diaz-Benjumea & Cohen 1995, Kim et al 1995). Thus the properties of *Serrate* in the D/V signaling system are analogous to those of *hedgehog* in the A/P system (Figure 2).

As with *Serrate,* ectopic expression of *fringe* in the ventral compartment leads to formation of an ectopic margin, to outgrowths, and to the ectopic expression of downstream genes (Irvine & Wieschaus 1994, Kim et al 1995). Notably, one of the genes induced by *fringe* is *Serrate*. *Serrate*, however, cannot induce *fringe* in ventral cells, which suggests that *fringe* exerts its activity through *Serrate* (Kim et al 1995). Mosaic analysis shows that *fringe* activity is required throughout the dorsal compartment (Irvine & Wieschaus 1994). *fringe* mutant clones on the dorsal surface lead to the production of an ectopic organizer that expresses high levels of *Serrate, wg,* and *vestigial* (Irvine & Wieschaus 1994, Kim et al 1995). Thus *fringe* mutant cells behave like *apterous* mutant cells (except that *fringe* mutant cells retain their dorsal identity). Taken together, these observations suggest that *fringe* mediates *apterous* function to establish the D/V organizer by activating *Serrate* expression (Figure 3*b*). Some other *apterous*-dependent function such as *Dorsal-wing* (Tiong et al 1995) may control the differentiation of dorsal cell identity. It should be noted that no gene with a function analogous to that of *fringe* has been identified in the A/P system.

Notch encodes a membrane-spanning protein that serves as a receptor for signals mediated by *Serrate* and *Delta* (Rebay et al 1991). Several lines of evidence suggest that *Notch* serves as the receptor for the dorsal-to-ventral *Serrate* signal. Mosaic analysis has shown that *Notch* activity is required at the wing margin to promote wing formation (De Celis & García-Bellido 1994), and studies using hypomorphic *Notch* alleles show a greater requirement for *Notch* activity on the ventral margin (De Celis et al 1996). *Notch* has been implicated in *wg* function at the wing margin and is suggested to serve as a receptor for Wingless protein (Couso & Martinez Arias 1994, Hing et al 1994, Couso et al 1995). An alternative view of the relationship between *wg* and *Notch* comes from the observation that *Notch* activity is required in a cell-autonomous manner for activation of *wg* at the wing margin (Rulifson & Blair 1995). The notion that *Notch* acts upstream of *wg* gains additional support from the observation that a ligand-independent, constitutively active form of *Notch* is sufficient to activate *wg* throughout the wing disc (Diaz-Benjumea & Cohen 1995) and that activation of genes downstream of *wg* signaling can occur in the absence of *Notch* function (Rulifson & Blair 1995). Thus *Notch* is an excellent candidate to serve as the receptor for the *Serrate* signal and apparently acts to activate *wg* expression at the wing margin (Figure 3*b*). However, these results do not exclude the possibility that *Notch* may also have a function downstream of *wg*. *Delta* has also been implicated in signaling between ventral and dorsal cells (De Celis et al 1996, Doherty et al 1996).

Step 3: Transmitting the Organizing Signal

The secreted signaling molecule Wingless is the best candidate to serve as the mediator of the organizing activity of the D/V boundary. *wg* activity is required at two distinct stages for development of the wing. In the early stages, before establishment of the D/V boundary, removal of *wg* activity leads to loss of the entire wing (Couso et al 1993, Williams et al 1993). Early ectopic expression of *wg* is sufficient to induce an ectopic wing, which suggests a role for *wg* in establishing the wing field (Ng et al 1996). Subsequently *wg* is expressed in a stripe of cells that straddles the D/V boundary (Phillips & Whittle 1993, Williams et al 1993, Couso et al 1994) as a result of interaction between dorsal and ventral cells (Diaz-Benjumea & Cohen 1995). By producing clones of cells that express Wg protein at ectopic positions in the wing disc, we have shown that *wg* activity is sufficient to mediate the organizing activity of the wing margin (Diaz-Benjumea & Cohen 1995). Clones of *wg*-expressing cells are able to induce the formation of wing margin structures and pattern duplications, providing direct evidence that the organizing activity of the dorsal-ventral compartment boundary is mediated by the Wg protein. *vestigial*, a novel nuclear protein, and *scalloped*, a predicted TEA transcription factor, are also activated by cell interactions at the D/V boundary and play a role in growth of the wing (Williams & Carroll 1993, Williams et al 1994, Couso et al 1995). Thus *wg* serves a role analogous to *dpp* in the A/P system (Figure 2).

AXIS FORMATION IN THE LEG

The first two steps in A/P axis formation in the leg are identical to those in wing formation. Expression of *engrailed* and *invected* specifies the posterior compartment and directs localized expression of Hedgehog (Sanicola et al 1995). *hh* in turn signals to induce the organizer in nearby cells of the anterior compartment (Basler & Struhl 1994). The principal difference from the wing lies in the fact that the targets for *hh* activity differ in the dorsal and ventral regions of the leg disc (Figure 4). As in the wing, *dpp* is the target for *hh* activity, but it is expressed at higher levels on the dorsal side of the disc than on the ventral side (Masucci et al 1990), whereas *wg* is the target for *hh* activity on the ventral side (Basler & Struhl 1994). A/P axis bifurcations can be induced by clones of *hh*-expressing cells in the leg disc, but only when such clones occupy both the dorsal and ventral quadrants, so that they cause mis-expression of both *wg* and *dpp* (Basler & Struhl 1994). Mis-expression of *wg* in the ventral quadrant or *dpp* in the dorsal quadrant causes local repatterning but is not sufficient to lead to complete axis duplication.

The differential readout of the *hh* signal in dorsal and ventral leg regions reflects an underlying subdivision of the leg disc into dorsal and ventral quadrants. This is clearly distinct from the D/V subdivision of the wing, in that it is not associated with a subdivision into dorsal and ventral compartments through expression of *apterous* (Cohen et al 1992). Nonetheless this difference indicates that the ventral anterior and dorsal anterior quadrants have distinct identities. Although the basis for this difference is not well understood, one attractive possibility is that the limb primordia are established at sites overlapping the intersection point between *wg*-expressing and *dpp*-expressing cells in the embryo, such that ventral cells in the nascent primordium express *wg* whereas dorsal cells express *dpp* (Cohen et al 1993).

wg and *dpp* play central roles in D/V patterning of the leg (reviewed in Blair 1995, Campbell & Tomlinson 1995, Held 1995). *wg* activity is required for specification of ventral fate in the leg. In larvae mutant for *wg* or other genes in the *wg* signal transduction pathway, ventral structures are lost and are replaced by a symmetric duplication of dorsal structures (Baker 1988, Couso et al 1993, Held et al 1994, Klingensmith et al 1994, Theisen et al 1994). Furthermore, ectopic expression of *wg*, or activation of the *wg* signal transduction pathway on the dorsal side of the leg, re-specifies dorsal cells to ventral fate and causes

D/V axis duplication (Struhl & Basler 1993, Diaz-Benjumea & Cohen 1994, Wilder & Perrimon 1995) (see Figure 4).

Likewise, *dpp* activity is required to specify dorsal cell fate in the leg. In weak mutants for *dpp*, some dorsal structures are replaced by a symmetric duplication of ventral structures in mutant legs (Held et al 1994). Using molecular markers for ventral fate (such as Wg expression), we have observed complete transformation from dorsal to ventral leg fate in stronger mutant combinations (WJ Brook & SM Cohen, unpublished data). Ectopic expression of *dpp* in the ventral leg re-specifies cell fate from ventral to dorsal and causes D/V axis duplication (Diaz-Benjumea et al 1994, WJ Brook & SM Cohen, unpublished data). Although *dpp* is expressed at low levels in ventral cells, its patterning activity appears to be restricted to the dorsal quadrant where *wg* is not expressed. Furthermore, we have observed mutually antagonistic activities of *wg* and *dpp* (WJ Brook & SM Cohen, unpublished data). *wg* represses *dpp* (incompletely) on the ventral side of the disc, whereas *dpp* represses *wg* on the dorsal side. These antagonistic interactions appear to be important for establishing the distinct domains in which these genes act for D/V axis formation.

In addition to their roles as mediators of D/V patterning, *wg* and *dpp* are required for formation of the proximal/distal (P/D) axis of the leg (Figure 4). Based on their reciprocal expression patterns, it has been suggested that the P/D axis of the leg might be specified by interaction between *wg* and *dpp* in the center of the leg disc (Meinhardt 1983, Gelbart 1989, Campbell et al 1993). In support of this view, strong reduction in the activity of either gene leads to truncation of the limb along the P/D axis (Spencer et al 1982, Baker 1988, Couso et al 1993, Diaz-Benjumea et al 1994). The combined activity of *wg* and *dpp* is required for expression of two homeobox genes, *aristaless* and *Distal-less*, in the center of the leg disc (Campbell et al 1993, Diaz-Benjumea et al 1994). Mutants with reduced *Distal-less* activity show truncation of the P/D axis of the leg (Cohen et al 1989). In strong *Distal-less* mutants, the entire limb is deleted (Cohen & Jürgens 1989a,b). *Distal-less* mutants show synergistic genetic interaction with mutants of *wg* and *dpp*, supporting the suggestion that these genes work together in a common pathway (Held et al 1994). Based on its expression pattern, *aristaless* has been suggested to play a role similar to that of *Distal-less* in P/D axis formation (Campbell et al 1993, Campbell & Tomlinson 1995). However, analysis of *aristaless* mutant phenotypes has not yet provided strong support for this proposal.

Comparative studies suggest that *Distal-less* may also play an important role in P/D axis formation in other insects and arthropods (Panganiban et al 1994, 1995). Vertebrate homologues of *Distal-less* have been identified and shown to be expressed in the apical ectodermal ridge (AER) and ventral ectoderm of the limb bud (e.g. Bulfone et al 1993). However, no information is available to date concerning their function.

DIFFERENCES IN WING AND LEG DEVELOPMENT

Although *wg* and *dpp* play important roles in patterning the wing and leg discs, the way in which they do so is different. Formation of the A/P organizer is quite similar in leg and wing. The major mechanistic differences lie in dorsal-ventral and proximal/distal axis formation. *wg* and *dpp* are expressed in essentially reciprocal domains in the leg, which are responsible for specification of ventral and dorsal fate and for D/V axis formation. As outlined above, D/V patterning of the wing depends on localized expression of *apterous* in dorsal cells, for which there is no parallel in the leg. The D/V boundary corresponds to the wing margin (a line that runs along the outer edge of the wing). *wg* is expressed along the D/V boundary of the wing, from which it controls growth and patterning of the wing. Although outgrowth of the wing is often referred to in terms of a proximal-distal axis (e.g. Williams & Carroll 1993, Couso et al 1995), it is really a function of the D/V patterning system, and growth of the wing is directed toward the periphery. It is questionable whether this can be considered analogous to growth of the leg toward a distal center. Although both are responsible for outgrowth of the appendages, the D/V system in the wing and the P/D system in the leg do not appear to be mechanistically comparable.

HOW SIMILAR IS LIMB DEVELOPMENT IN *DROSOPHILA* AND VERTEBRATES?

As outlined above, intriguing similarities between the molecules controlling *Drosophila* and vertebrate limb development raise the question of whether the mechanisms underlying growth and patterning in the two systems are evolutionarily conserved. However, vertebrate and arthropod limbs are generally believed to have independent origins. With this in mind, it is worth considering whether the recent molecular studies provide a substantive basis for reconsidering this view.

A number of differences between *Drosophila* and vertebrate limb development are apparent. The cell interactions that give rise to *Drosophila* limbs occur within a single epithelial cell layer, whereas vertebrate limb development is controlled by interactions between ectodermal and mesenchymal cell layers. Unlike signaling in the *Drosophila* wing, anterior-posterior and dorsal-ventral signaling are tightly coupled in vertebrate limb development through the interdependent signaling between Shh, FGF-4, FGF-8, and Wnt-7 (Laufer et al

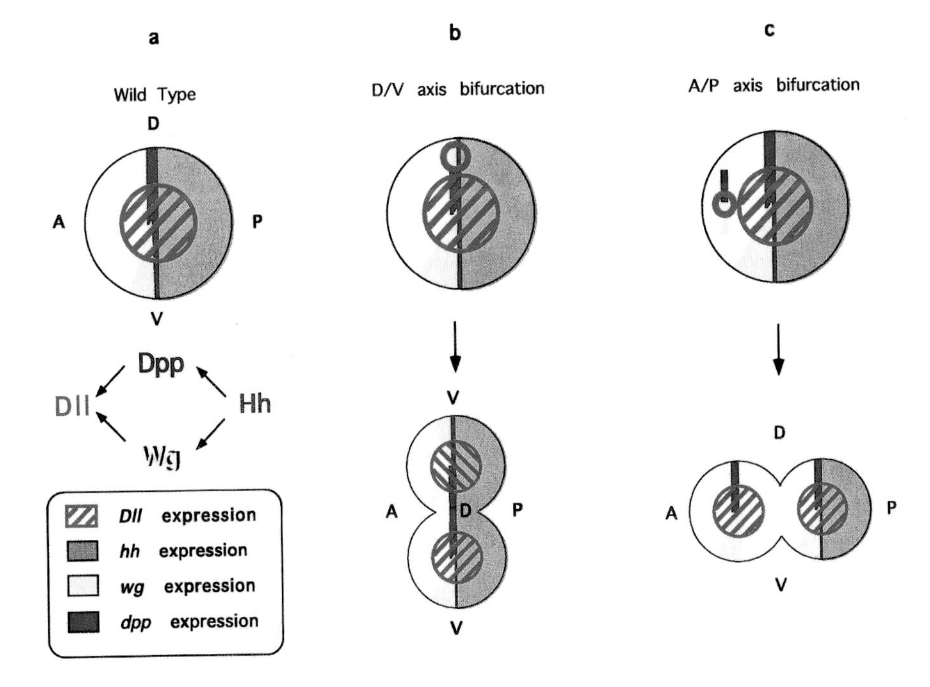

Figure 4 A/P, D/V and P/D axes of the leg. *(a)* Patterns of expressions of *hh, wg, dpp,* and *Dll* in the wild-type leg. *hh* signals from posterior to anterior to direct expression of *dpp* in anterior cells and *wg* in ventral anterior cells. *wg* partially represses *dpp* ventrally. *dpp* represses *wg* dorsally. *wg* and *dpp* cooperate to direct *Dll* expression in the center of the disc, thereby initiating P/D axis formation. *(b)* D/V axis bifurcation can be caused by mis-expressing *wg* dorsally. This leads to local repression of *dpp* and activation of *Dll*. The resulting bifurcated leg has V/D/D/V symmetry. Expression of *dpp* ventrally produces the reciprocal duplication (not illustrated). *(c)* A/P axis bifurcation can be caused by mis-expressing *hh* anteriorly. The clone of *hh*-expressing cells must activate *dpp* dorsally and *wg* ventrally to in turn activate *Dll*. Clones restricted to either dorsal or ventral cause local repatterning but do not cause axis bifurcation.

1994, Niswander et al 1994, Yang & Niswander 1995, Crossley et al 1996). There is no evidence for such coupling in *Drosophila.*

The D/V margin of the *Drosophila* wing shares some characteristics with the AER of the vertebrate limb (reviewed in Tickle & Eichele 1994). Both structures lie at the interface of the dorsal and ventral halves of the limb and generate signals for outgrowth and patterning of the limb field. The wing margin is established by interactions between *apterous*-expressing and non-expressing cells. As noted above, localized expression *Wnt7a* in the dorsal ectoderm directs expression of *c-LMX-1* in the dorsal mesenchyme. It is therefore likely that D/V subdivision of the ectoderm is the primary determinant, and it will be interesting to see whether the AER is specified by local interaction between dorsal and ventral ectodermal cells. Although it is remarkable that a LIM/homeodomain protein specifies dorsal cell identity mesenchyme of the vertebrate limb, it should be recalled that the dorsal-ventral body axes are thought to be reversed in vertebrates and insects (reviewed in De Robertis & Sasai 1996); i.e. dorsal in the fly wing is compared with ventral in the vertebrate limb. Also, sequence comparison of LIM/homeobox genes suggests that *c-LMX-1* is not the closest vertebrate homologue of *apterous* (M Averof, personal communication).

The patterning activity of the wing margin is mediated by the Wingless protein. The secreted growth factors FGF-4 and FGF-8 are expressed by the AER, and much of AER function can be substituted by FGF-beads (Niswander et al 1993). FGF-8 has also been suggested to induce formation of the limb field (Cohn et al 1995, Crossley et al 1996), which again is a role for Wingless in *Drosophila* (Cohen 1990). So far, no *Wnt* gene has been demonstrated to have a function in the AER similar to that of *wingless* in the wing margin, and there are no identified members of the FGF family in *Drosophila.* In this context it is worth pointing out that growth of the vertebrate limb bud occurs principally in a localized region under the AER called the progress zone (reviewed in Tickle & Eichele 1994). There is no comparable growth zone in the insect wing (González-Gaitán et al 1994).

In view of the many dissimilarities outlined above, we question whether the molecular similarities might be fortuitous. One example that cannot be easily dismissed is the similarity in the roles of Hedgehog and Sonic Hedgehog in A/P patterning of vertebrate limbs, including the conserved aspects of the *hh-dpp* signaling cascade (Marigo et al 1996). Nonetheless, BMPs have not yet been shown to have a function equivalent to that of *dpp* in A/P patterning. Until more compelling evidence can be found to support the conservation of mechanism in this and other examples, we suggest that the case for homology between vertebrate and insect limbs should be considered with caution.

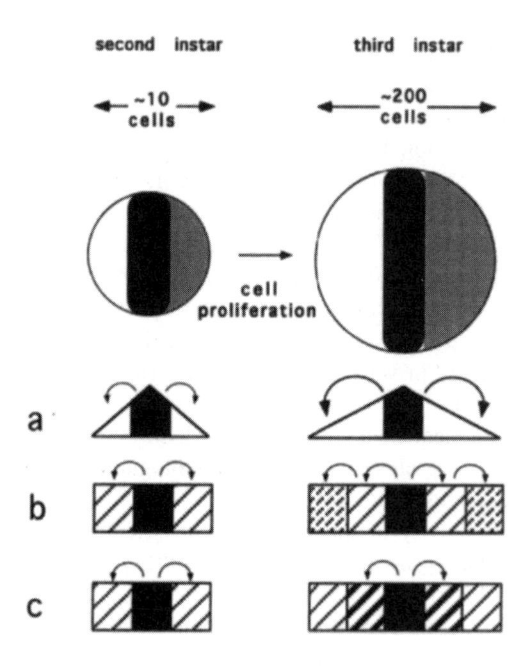

Figure 5 Imaginal disc cell patterning by the organizing signal. Organizing signals are present when the imaginal disc is only a few cells in diameter in second instar. As the disc grows, the width of the organizer remains constant, but the diameter of the field it must pattern increases dramatically. Three alternative ways in which local expression of signaling molecules could specify cells' fates at a distance are presented. (*a*) Morphogen: The signaling molecule could serve as a concentration-dependent morphogen to directly specify cell fate over long distances. (*b*) Induction/signal relay: The signaling molecule could act at a short range to induce a fate change in nearby cells (as Hh induces *dpp*). Cells near the boundary get the same information at all stages, and more distant fates could be specified by relay of a second signal. (*c*) Temporal integration: The signal could act at a short range and specify cell fates. As the disc grows, cells that remain within range continue to receive information; however, other cells will be moved out of the range of the signal. The duration of signaling might lead cells to adopt different final identities. It is also possible that cells moving out of range of the signal may retain some information about their state of specification. In this proposal, cells would integrate the effects of the organizer over time rather than reading out their position from a concentration gradient in space. This model should be distinguished from the progress zone of the vertebrate limb because it does not involve a localized growth zone.

PROSPECTS

Progress in understanding organizer function has been very rapid in the past few years. We expect that the near future will show equally rapid progress in the genetic dissection of the signal transduction pathways through which *hh, wg,* and *dpp* specify cell fate. Equally important will be efforts to understand how cells interpret the positional information generated by the organizers. It has been suggested that the secreted signaling molecules mediating organizer activity may specify cell fate as concentration-dependent morphogens. Alternatively, the genes may act as local inducers triggering cascades of cell interactions, or through integration of signaling over time (progress zone, Figure 5). Further work on the mechanisms by which specific cell fates are specified along the limb axes and on how this information is used to control cell proliferation should help to distinguish between these models.

Literature Cited

Baker NE. 1988. Embryonic and imaginal requirements for wingless, a segment-polarity gene in *Drosophila. Dev. Biol.* 125:96–108

Basler K, Struhl G. 1994. Compartment boundaries and the control of *Drosophila* limb pattern by *hedgehog* protein. *Nature* 368:208–14

Blair SS. 1993. Mechanisms of compartment formation: evidence that non-proliferating cells do not play a critical role in defining the D/V lineage restriction in the developing wing of *Drosophila. Development* 119:339–51

Blair SS. 1995. Compartments and appendage development in *Drosophila. BioEssays* 17:299–309

Blair SS, Brower DL, Thomas JB, Zavortink M. 1994. The role of *apterous* in the control of dorsoventral compartmentalization and PS integrin gene expression in the developing wing of *Drosophila. Development* 120: 1805–15

Bulfone A, Kim H-J, Puelles L, Porteus MH, Grippo JF, Rubenstein JLR. 1993. The mouse *Dlx-2* (*Tes-1*) gene is expressed in spatially restricted domains of the forebrain, face and limbs in midgestation embryos. *Mech. Dev.* 40:129–40

Campbell G, Tomlinson A. 1995. Initiation of the proximodistal axis in insect legs. *Development* 121:619–28

Campbell G, Weaver T, Tomlinson A. 1993. Axis specification in the developing Drosophila appendage: the role of *wingless, decapentaplegic,* and the homeobox gene *aristaless. Cell* 74:1113–23

Capdevila J, Estrada MP, Sánchez-Herrero E, Guerrero I. 1994. The *Drosophila* segment polarity gene patched interacts with *decapentaplegic* in wing development. *EMBO J.* 13:71–82

Capdevila J, Guerrero I. 1994. Targeted expression of the signalling molecule decapentaplegic induces pattern duplications and growth alerations in *Drosophila* wings. *EMBO J.* 13:4459–68

Cohen B, McGuffin ME, Pfeifle C, Segal D, Cohen SM. 1992. *apterous*: a gene required for imaginal disc development in *Drosophila* encodes a member of the LIM family of developmental regulatory proteins. *Genes Dev.* 6:715–29

Cohen B, Simcox AA, Cohen SM. 1993. Allocation of the thoracic imaginal disc primordia in the *Drosophila* embryo. *Development* 117:597–608

Cohen SM. 1990. Specification of limb development in the *Drosophila* embryo by positional cues from segmentation genes. *Nature* 343:173–77

Cohen SM. 1993. Imaginal disc development. In *Drosophila Development,* ed. A Martinez-Arias, M Bate, pp. 747–841. Cold Spring Harbor, NY: Cold Spring Harbor Press

Cohen SM, Brönner G, Küttner F, Jürgens G, Jäckle H. 1989. *Distal-less* encodes a homoeodomain protein required for limb development in *Drosophila. Nature* 338:432–34

Cohen SM, Jürgens G. 1989a. Proximal-distal pattern formation in *Drosophila*: cell autonomous requirement for *Distal-less* gene activity in limb development. *EMBO J.* 8:2045–55

Cohen SM, Jürgens G. 1989b. Proximal-distal pattern formation in *Drosophila*: graded requirement for *Distal-less* gene activity during limb development in *Drosophila. Roux's Arch. Dev. Biol.* 198:157–69

Cohn MJ, Izpisua-Belmonte JC, Abud H, Heath JK, Tickle C. 1995. Fibroblast growth factors induce additional limb development from the flank of chick embryos. *Cell* 80:739–46

Couso JP, Bate M, Martinez Arias A. 1993. A *wingless*-dependent polar coordinate system in the imaginal discs of *Drosophila. Science* 259:484–89

Couso JP, Bishop S, Martinez Arias A. 1994. The wingless signalling pathway and the patterning of the wing margin in *Drosophila. Development* 120:621–36

Couso JP, Knust E, Martinez Arias A. 1995. *Serrate* and *wingless* in *Drosophila* wing development. *Curr. Biol.* 5:1437–48

Couso JP, Martinez Arias A. 1994. Notch is required for *wingless* signalling in the epidermis of Drosophila. *Cell* 79:259–72

Crick FH, Lawrence PA. 1975. Compartments and polyclones in insect development. *Science* 189:340–47

Crossley PH, Minowada G, MacArthur CA, Martin GR. 1996. Roles for FGF-8 in the induction, initiation and maintenance of chick limb development. *Cell* 84:127–36

De Celis JF, García-Bellido A. 1994. Roles of the *Notch* gene in *Drosophila* wing morphogenesis. *Mech. Dev.* 46:109–22

De Celis JF, García-Bellido A, Bray SJ. 1996. Activation and function of *Notch* at the dorsal-ventral boundary of the wing imaginal disc. *Development* 122:359–69

De Robertis EM, Sasai Y. 1996. A unity of plan for dorso-ventral patterning in the development of animal species. *Nature* 380:37–40

Diaz-Benjumea FJ, Cohen B, Cohen SM. 1994. Cell interaction between compartments establishes the proximal-distal axis of *Drosophila* legs. *Nature* 372:175–79

Diaz-Benjumea FJ, Cohen SM. 1993. Interaction between dorsal and ventral cells in the imaginal disc directs wing development in Drosophila. *Cell* 75:741–52

Diaz-Benjumea FJ, Cohen SM. 1994. *wingless* acts through the *shaggy/zeste-white 3* kinase to direct dorsal-ventral axis formation in the *Drosophila* leg. *Development* 120:1661–70

Diaz-Benjumea FJ, Cohen SM. 1995. Serrate signals through Notch to establish a Wingless-dependent organizer at the dorsal/ventral compartment boundary of the *Drosophila* wing. *Development* 121:4215–25

Doherty D, Fenger G, Younger-Shepherd S, Jan L-Y, Jan Y-N. 1996. Dorsal and ventral cells respond differently to the *Notch* ligands *Delta* and *Serrate* during *Drosophila* wing development. *Genes Dev.* 10:421–34

Fietz MJ, Concordet J-P, Barbosa R, Johnson R, Krauss S, et al. 1994. The *hedgehog* gene family in *Drosophila* and vertebrate development. *Development Suppl.* pp. 43–51

Francis PH, Richardson MK, Brickell PM, Tickle C. 1994. Bone morphogenetic proteins and a signalling pathway that controls patterning in the developing chick limb bud. *Development* 120:209–18

Francis-West PH, Robertson KE, Ede DA, Rodriguez C, Izpisua-Belmonte JC, et al. 1995. Expression of genes encoding bone morphogenetic proteins and sonic hedgehog in talpid(3) limb buds: their relationships in the signaling cascade involved in limb patterning. *Dev. Dyn.* 203:187–97

García-Bellido A. 1975. Genetic control of wing disc development in *Drosophila.* In *Cell Patterning, Ciba Found Symp.,* ed. S Brenner, pp. 161–82. New York: Assoc. Sci.

García-Bellido A, Ripoll P, Morata G. 1973. Developmental compartmentalisation of the wing disk of *Drosophila. Nature New Biol.* 245:251–53

Gelbart WM. 1989. The *decapentaplegic* gene: a TGFβ homologue controlling pattern formation in *Drosophila. Development* 107:65–74 (Suppl.)

González-Gaitán M, Paz Capdevila M, García-Bellido A. 1994. Cell proliferation patterns in the wing imaginal disc of *Drosophila. Mech. Dev.* 40:183–200

Guillen I, Mullor JL, Capdevila J, Sanchez-Herrero E, Morata G, Guerrero I. 1995. The function of *engrailed* and the specification of *Drosophila* wing pattern. *Development* 121:3447–56

Harland R. 1994. Neural induction in *Xenopus. Curr. Opin. Genet. Dev.* 4:534–49

Held LI Jr. 1995. Axes, boundaries and coordinates: the ABCs of fly leg development. *BioEssays* 17:721–32

Held LI Jr, Heup MA, Sappington JM, Peters SD. 1994. Interactions of *decapentaplegic,*

wingless and *Distal-less* in the *Drosophila* leg. *Roux's Arch. Dev. Biol.* 203:310–19
Hidalgo A. 1994. Three distinct roles for the *engrailed* gene in *Drosophila* wing development. *Curr. Biol.* 4:1087–98
Hing HK, Sun X, Artavanis-Tsakonas S. 1994. Modulation of wingless signaling by Notch in *Drosophila. Mech. Dev.* 47:261–68
Irvine K, Wieschaus E. 1994. *fringe,* a boundary specific signalling molecule, mediates interactions between dorsal and ventral cells during Drosophila wing development. *Cell* 79:595–606
Jiang J, Struhl G. 1995. Protein kinase A and Hedgehog signaling in Drosophila limb development. *Cell* 80:563–72
Kim J, Irvine KD, Carroll SB. 1995. Cell recognition, signal induction and symmetrical gene activation at the dorsal/ventral boundary of the developing Drosophila wing. *Cell* 82:795–802
Klingensmith J, Nusse R, Perrimon N. 1994. The *Drosophila* segment polarity gene *dishevelled* encodes a novel protein required for response to the *wingless* signal. *Genes Dev.* 8:118–30
Laufer E, Nelson CE, Johnson RL, Morgan BA, Tabin C. 1994. *Sonic hedgehog* and *Fgf-4* act through a signalling cascade and feedback loop to integrate growth and patterning of the developing limb bud. *Cell* 79:993–1003
Lawrence PA, Morata G. 1994. Homeobox genes: their function in Drosophila segmentation and pattern formation. *Cell* 78:181–89
Lecuit T, Brook WJ, Ng M, Calleja M, Sun H, Cohen SM. 1996. Two distinct mechanisms for long-range patterning by *Dpp* in the *Drosophila* wing. *Nature* 381:387–93
Lee JJ, von Kessler DP, Parks S, Beachy PA. 1992. Secretion and localized transcript suggest a role in positional signalling for products of the segmentation gene *hedgehog. Cell* 71:33–50
Lepage T, Cohen SM, Diaz-Benjumea FJ, Parkhurst SM. 1995. Signal transduction by cAMP-dependent protein kinase A in *Drosophila* limb patterning. *Nature* 373:711–15
Li W, Ohlmeyer JT, Lane ME, Kalderon D. 1995. Function of protein kinase A in Hedgehog signal transduction and Drosophila imaginal disc development. *Cell* 80:553–62
Marigo V, Scott MP, Johnson RL, Goodrich LV, Tabin CJ. 1996. Conservation of *hedgehog* signaling: induction of a chicken *patched* homolog by *sonic hedgehog* in the developing limb. *Development* 122:1225–33
Masucci JD, Miltenberger RJ, Hoffman FM. 1990. Pattern specific expression of the *Drosophila decapentaplegic* gene in imaginal discs is regulated by 3′ *cis* regulatory elements. *Genes Dev.* 4:2011–23
Meinhardt H. 1983. Cell determination boundaries as organizing regions for secondary embryonic fields. *Dev. Biol.* 96:375–85
Mohler J. 1988. Requirements for *hedgehog,* a segment polarity gene, in patterning larval and adult cuticle of *Drosophila. Genetics* 120:1061–72
Mohler J, Vani K. 1992. Molecular organisation and embryonic expression of the *hedgehog* gene involved in cell-cell communication in segmental patterning of *Drosophila. Development* 115:957–71
Nellen D, Burke R, Struhl G, Basler K. 1996. Direct and long-range action of a *Dpp* morphogen gradient. *Cell* 85:357–68
Niswander L, Jeffrey S, Martin GR, Tickle C. 1994. A positive feedback loop coordinates growth and patterning of the vertebrate limb. *Nature* 371:609–12
Niswander L, Tickle C, Vogel A, Booth I, Martin GR. 1993. FGF-4 replaces the apical ectodermal ridge and directs outgrowth and patterning of the limb. *Cell* 75:579–87
Ng M, Diaz-Benjumea FJ, Vincent J-P, Wu J, Cohen SM. 1996. Specification of the wing primordium in *Drosophila. Nature* 381:316–18
Pan DJ, Rubin G. 1995. Protein kinase and hedgehog act antagonistically in regulating *decapentaplegic* transcription in Drosophila imaginal discs. *Cell* 80:543–52
Panganiban G, Nagy L, Carroll SB. 1994. The role of the *Distal-less* gene in the development and evolution of insect limbs. *Curr. Biol.* 4:671–75
Panganiban G, Sebring A, Nagy L, Carroll SB. 1995. The development of crustacean limbs and the evolution of arthropods. *Science* 270:1363–66
Parr BA, McMahon A. 1995. Dorsalizing signal *Wnt-7a* required for normal polarity of A/P and D/V axes on mouse limb. *Nature* 374:350–53
Phillips R, Whittle JRS. 1993. *wingless* expression mediates determination of peripheral nervous system elements in late stages of *Drosophila* wing disc development. *Development* 118:427–38
Rebay I, Fleming RJ, Fehon RG, Cherbas L, Cherbas P, Artavanis-Tsakonas S. 1991. Specific EGF repeats of Notch mediate interactions with Delta and Serrate: implications for Notch as a multifunctional receptor. *Cell* 67:687–99
Riddle RD, Ensini M, Nelson C, Tsuchida T, Jessell TM, Tabin C. 1995. Induction of LIM homeobox gene *Lmx1* by *WNT7a* establishes dorsoventral pattern in the vertebrate limb. *Cell* 83:631–40
Riddle RD, Johnson RL, Laufer E, Tabin C.

1993. *sonic hedgehog* mediates the polarizing activity of the ZPA. *Cell* 75:1401–16
Rulifson EJ, Blair SS. 1995. *Notch* regulates *wingless* expression and is not required for reception of the paracrine *wingless* signal during wing margin neurogenesis in *Drosophila*. *Development* 121:2813–24
Sanicola M, Sekelsky JSE, Gelbart WM. 1995. Drawing a stripe in *Drosophila* imaginal disks: negative regulation of *decapentaplegic* and *patched* expression by *engrailed*. *Genetics* 139:745–56
Saunders JW, Gasseling MT. 1968. Ectodermal-mesenchymal interaction in the origin of limb symmetry. In *Epithelial Mesenchymal Interactions,* ed. R Fleischmeyer, RF Billingham, pp. 78–97. Baltimore: Williams & Wilkins
Simmonds AJ, Brook WJ, Cohen SM, Bell JB. 1995. Distinguishable functions for *engrailed* and *invected* in anterior-posterior patterning in the *Drosophila* wing. *Nature* 376:424–27
Slack J. 1994. Inducing factors in *Xenopus* early embryos. *Curr. Biol.* 4:116–25
Speicher SA, Thomas U, Hinz U, Knust E. 1994. The *Serrate* locus of *Drosophila* and its role in morphogenesis of the wing imaginal discs: control of cell proliferation. *Development* 120:535–44
Spemann H, Mangold H. 1924. Uber Induktion von Embryonenanlagen durch Implantation artfremder Organisatoren. *Arch. Mikrosk. Anat. EntwMech.* 100:599–638
Spencer F, Hoffman M, Gelbart WM. 1982. Decapentaplegic: a gene complex affecting morphogenesis in Drosophila melanogaster. *Cell* 28:451–61
Steiner E. 1976. Establishment of compartments in the developing leg imaginal discs of *Drosophila melanogaster. Roux's Arch. Dev. Biol.* 180:9–30
Struhl G, Basler K. 1993. Organizing activity of *wingless* protein in Drosophila. *Cell* 72:527–40
Tabata T, Eaton S, Kornberg TB. 1992. The *Drosophila hedgehog* gene is expressed specifically in posterior compartment cells and is a target of *engrailed* regulation. *Genes Dev.* 6:2635–45
Tabata T, Schwartz C, Gustavson E, Ali Z, Kornberg TB. 1995. Creating a *Drosophila* wing de novo: the role of *engrailed* and the compartment border hypothesis. *Development* 121:3359–69
Theisen H, Purcell J, Bennett M, Kansagara D, Syed A, Marsh JL. 1994. *dishevelled* is required during *wingless* signalling to establish both cell polarity and cell identity. *Development* 120:347–60
Thomas U, Speicher SA, Knust E. 1991. The *Drosophila* gene *Serrate* encodes an EGF-like transmembrane protein with complex expression patterns in embryos and wing discs. *Development* 111:749–61
Tickle C, Alberts B, Wolpert L, Lee J. 1982. Local application of retinoic acid to the limb bud mimics action of the polarizing region. *Nature* 296:564–65
Tickle C, Eichele G. 1994. Vertebrate limb development. *Annu. Rev. Cell Biol.* 10:121–52
Tickle C, Summerbell D, Wolpert L. 1975. Positional signaling and specification of digits in chick limb morphogenesis. *Nature* 254:199–202
Tiong SYK, Nash D, Bender W. 1995. *Dorsal wing,* a locus that affects dorsoventral wing patterning in *Drosophila. Development* 121:1649–56
Vincent JP, O'Farrell PH. 1992. The state of *engrailed* expression is not clonally transmitted during early Drosophila development. *Cell* 68:923–31
Vogel A, Rodriguez C, Warnken W, Ispizua Belmonte JC. 1995. Dorsal cell fate specified by chick *lmx-1* during vertebrate limb development. *Nature* 378:716–20
Wieschaus E, Gehring W. 1976. Clonal analysis of primordial disc cells in the early embryo of *Drosophila melanogaster. Dev. Biol.* 50:249–63
Wilder EL, Perrimon N. 1995. Dual functions of *wingless* in the *Drosophila* leg imaginal disc. *Development* 121:477–88
Williams JA, Carroll SB. 1993. The origin, patterning and evolution of insect appendages. *BioEssays* 15:567–77
Williams JA, Paddock SW, Carroll SB. 1993. Pattern formation in a secondary field: a hierarchy of regulatory genes subdivides the developing *Drosophila* wing disc into discrete subregions. *Development* 117:571–84
Williams JA, Paddock SW, Vorwerk K, Carroll SB. 1994. Organization of wing formation and induction of a wing-patterning gene at the dorsal/ventral compartment boundary. *Nature* 368:299–305
Yamada T, Placzek M, Tanaka H, Dodd J, Jessell TM. 1991. Control of cell pattern in the developing nervous system: polarizing activity of the floor plate and notochord. *Cell* 64:635–47
Yang Y, Niswander L. 1995. Interaction between the signalling molecules WNT7a and SHH during vertebrate limb development: dorsal signals regulate anteroposterior patterning. *Cell* 80:939–47
Zecca M, Basler K, Struhl G. 1995. Sequential organizing activities of engrailed hedgehog and decapentaplegic in the *Drosophila* wing. *Development* 121:2265–78

Annu. Rev. Cell Dev. Biol. 1996. 12:181–220

Fc RECEPTORS AND THEIR INTERACTIONS WITH IMMUNOGLOBULINS

*Malini Raghavan and Pamela J. Bjorkman**
Division of Biology 156-29, and *Howard Hughes Medical Institute, California Institute of Technology, Pasadena, California 91125

KEY WORDS: immunoglobulin gene superfamily, structure, binding, transcytosis, effector functions

ABSTRACT

Receptors for the Fc domain of immunoglobulins play an important role in immune defense. There are two well-defined functional classes of mammalian receptors. One class of receptors transports immunoglobulins across epithelial tissues to their main sites of action. This class includes the neonatal Fc receptor (FcRn), which transports immunoglobulin G (IgG), and the polymeric immunoglobulin receptor (pIgR), which transports immunoglobulin A (IgA) and immunoglobulin M (IgM). Another class of receptors present on the surfaces of effector cells triggers various biological responses upon binding antibody-antigen complexes. Of these, the IgG receptors (FcγR) and immunoglobulin E (IgE) receptors (FcϵR) are the best characterized. The biological responses elicited include antibody-dependent, cell-mediated cytotoxicity, phagocytosis, release of inflammatory mediators, and regulation of lymphocyte proliferation and differentiation. We summarize the current knowledge of the structures and functions of FcRn, pIgR, and the FcγR and FcϵRI proteins, concentrating on the interactions of the extracellular portions of these receptors with immunoglobulins.

CONTENTS

1081-0706/96/1115-0181$08.00

INTRODUCTION

Immunoglobulin (Ig) molecules consist of two copies of a variable Fab region that contains the binding site for antigen, and a relatively constant Fc region that interacts with effector molecules such as complement proteins and Fc receptors (FcRs) (Figure 1). Receptors for the Fc domain of Ig molecules form crucial components of immune defense by (*a*) facilitating specialized transport of antibody molecules to regions of the body where they are needed, (*b*) forming the crucial link between antigen binding by Igs and effector responses such as inflammation and the regulation of antibody production, and (*c*) controlling the lifetime of Ig molecules in serum.

There are five isotypes of Igs in mammals: IgA, IgD, IgE, IgG, and IgM (Figure 1*a*). Particular Fc receptors are usually specific for only one or two of the Ig isotypes. FcγR and FcRn proteins are specific for IgG, and FcεR proteins are specific for IgE. The polymeric immunoglobulin receptor (pIgR) recognizes dimeric[1] IgA and pentameric IgM. Other Fc receptors specific for

[1]Throughout this review, monomeric Ig refers to an Ig molecule consisting of two identical heavy chains and two identical light chains arranged into two Fab arms and a single Fc region (Figure 1). Thus dimeric IgA contains the four Fab arms and two Fc regions contributed from two IgA molecules.

Figure 1 Antibody structures. (*a*) Schematic representations of the structures of the five Ig isotypes. Variable and constant domains are represented by ovals. The Fc regions of the IgE and IgM isotypes contain an extra constant domain ($C_\epsilon 2$ and $C_\mu 2$, respectively) that replaces the hinge of the other Ig isotypes. The dimeric form of IgA and the pentameric form of IgM include an additional 15-kDa polypeptide called the J chain. However, some antibody-producing cells secrete a hexameric IgM that lacks the J chain (Randall et al 1992). (*b*) Three-dimensional structures of IgG and the Fc fragment. *Left:* ribbon diagram of the structure of an intact IgG (Harris et al 1992). The hinge region is disordered in the crystals, thus its location and the location of the disulfide bonds linking the two heavy chains (indicated as horizontal lines) are approximate. *Right:* carbon-α trace from the crystal structure of the Fc fragment (Deisenhofer 1981). The space between the two Cγ2 domains is filled with carbohydrate residues. The residues within the lower hinge region that define the binding site for the FcγR receptors are disordered in this structure. Fc histidine residues implicated in the pH-dependent interaction with FcRn (Burmeister et al 1994a, Kim et al 1994, Raghavan et al 1994, 1995a) are highlighted as sidechains with the residue name and number. Figures were prepared from coordinates available from the PDB (1FC2 for Fc) or provided by the authors (intact Ig coordinates obtained from A McPherson).

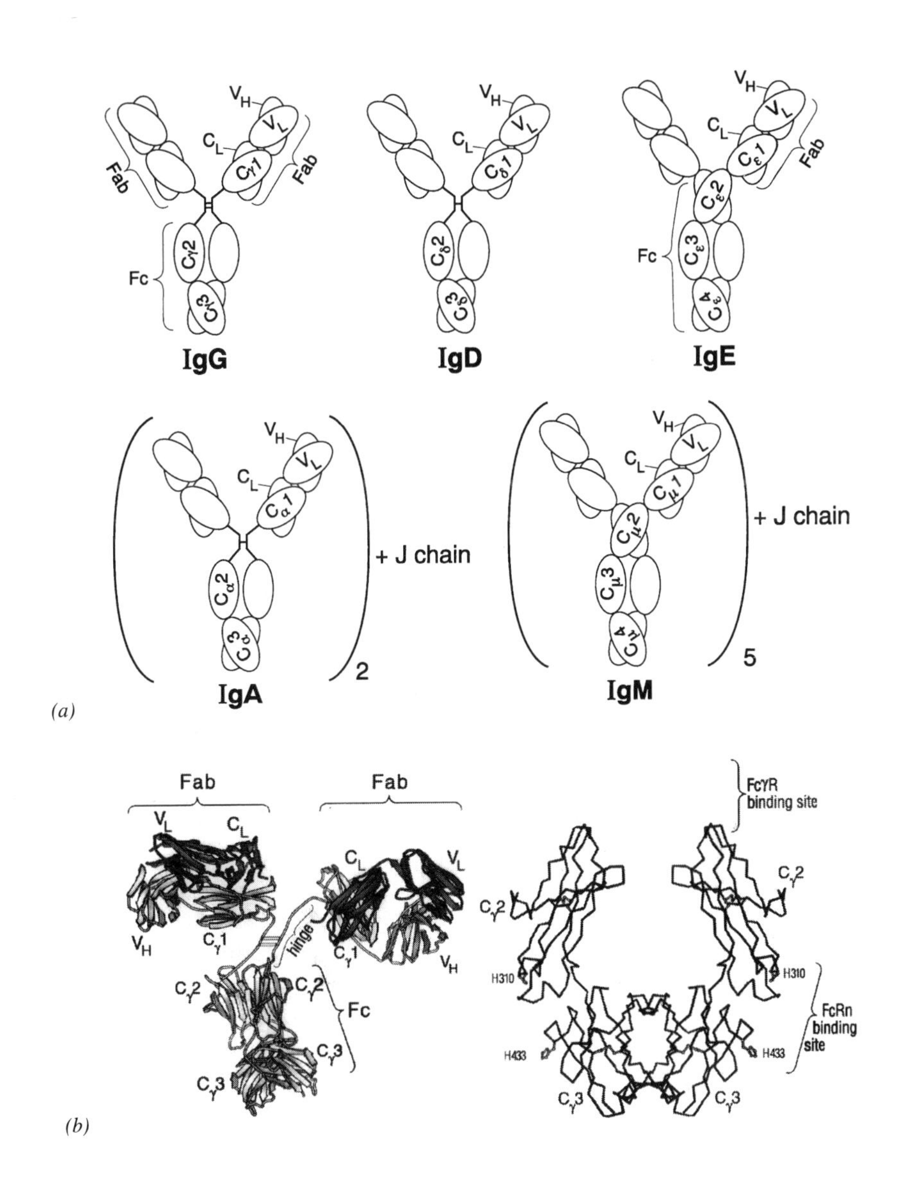

Fab
Fab
Fc
IgG
IgD
Fc
Fab
IgE
+ J chain
2
IgA
+ J chain
5
IgM
(a)
Fab
Fab
hinge
Fc
FcγR binding site
H310
H310
FcRn binding site
H433
H433
(b)

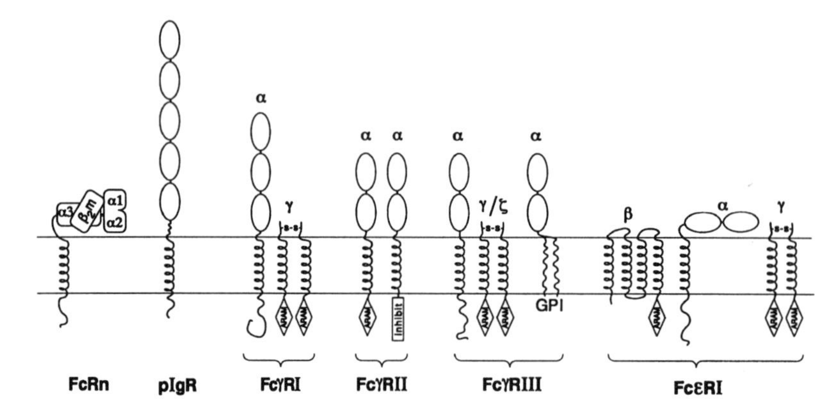

Figure 2 Schematic representation of the structures of the Ig superfamily FcRs. The Ig C1-set domains in FcRn are indicated by rectangles; V-like domains in the other FcRs are depicted as ovals. Greek letter names are written above the subunits. The Ig-binding domains of the FcγRs and FcϵRI are the α domains. Protein transmembrane regions are indicated as helices, and the glycophosphatidyl inositol anchor of FcγRIIIB is labeled GPI. ARAMs located in the cytoplasmic domains are indicated as diamonds, and the inhibitory motif of FcγRIIB is indicated as a rectangle. FcRn and FcϵRI are depicted in their lying-down orientations relative to the membrane, based on experimental data described in the text. The orientations of the other receptors with respect to the membrane are unknown (adapted from Figure 1 in Ravetch 1994).

IgA (Maliszewski 1990), IgM (Ohno et al 1990), and IgD (Sjoberg 1980) have been reported, but these receptors have not been extensively investigated and are not discussed in this review. Instead, we focus on the well-characterized Fc receptors that are members of the Ig superfamily: FcRn, pIgR, the FcγRs, and FcϵRI (Figure 2), concentrating on the extracellular regions of these receptors that are involved in ligand binding.

Ig superfamily members are defined as molecules that contain domains with sequence similarity to the variable or constant domains of antibodies (Williams & Barclay 1988). On the basis of sequence and structural similarities, Ig superfamily member domains are divided into three sets: C1, C2, and V-like (Williams & Barclay 1988). The C1 set includes antibody constant and topologically equivalent domains. FcRn is the only FcR that is a member of the Ig superfamily by virtue of containing constant or C1-set domains. C2 and V-like, the other Ig superfamily domains, are common building blocks of cell adhesion molecules (Wagner & Wyss 1994, Vaughn & Bjorkman 1996), which are structurally similar to the other FcRs discussed in this review. The V-like set includes domains that have a folding topology closely related to Ig variable domains. The C2 set has a slightly different organization of β-strands compared with the C1 and V-like sets (Figure 3). Many Ig superfamily domains were initially classified as C2 based primarily upon the number of residues separating the cysteines that form a disulfide bond in the folded domain structure. Indeed, the Ig-like domains in the FcγRs and in FcϵRI were classified as C2 and, in some cases, modeled as such (Sutton & Gould 1993, Hibbs et al 1994, Hulett & Hogarth 1994). Recent studies by Chothia and colleagues (Harpaz & Chothia 1994) suggest that some of the original classifications of Ig superfamily domains need to be reconsidered in light of recent crystal structures of Ig superfamily members (for reviews, see Harpaz & Chothia 1994, Wagner & Wyss 1994, Vaughn & Bjorkman 1996). We have therefore examined the sequences of the FcRs, taking the new structural information into account. We used consensus sequences for V-like and C2 domains obtained from structure-based sequence alignments to classify the FcR domains (Vaughn & Bjorkman 1996). Our analysis indicates that the Ig-like domains within pIgR, the FcγRs, and FcϵRI share features more in common with the V-like set than with the C2 set (Table 1).

Although crystal structures of the Fc binding regions of any of the receptors except FcRn (Burmeister et al 1994a) are not available, the structures of V-like and C2 domains in cell adhesion molecules (e.g. CD4, CD2, VCAM-1; Ryu et al 1990, Wang et al 1990, Jones et al 1992, 1995) can be used as approximate models for the three-dimensional structures of pIgR, the FcγRs, and FcϵRI. The extracellular domains of these FcRs consist of two or more Ig-like domains arranged in tandem, a domain organization in common with many cell adhesion molecules including those for which three-dimensional structures are available. As a framework for mutagenesis studies to map the interaction site on the receptors, we present topology and ribbon diagrams for the structures of Ig superfamily domains (Figure 3). Structural information about the immunoglobulin ligands of the Fc receptors is more complete: crystal structures of the IgG Fc fragment (Deisenhofer 1981) and of an intact IgG (Harris et al 1992) are available (Figure 1*b*) and can be used for interpretation of interaction site mapping studies.

Using these structural models as guides, we explore both the common features and the differences in the way that the FcRs interact with their ligands and discuss evidence showing that some of the Fc regions are less symmetric in solution that one might assume from the IgG Fc crystal structure. We also refer to the many excellent reviews covering topics not explored in detail, including the heterogeneity of the FcγR and FcϵR transmembrane and intracellular regions, the functional roles of different cytoplasmic domains, and FcR-mediated signaling.

Table 1 Alignment of FcR sequences with structure-based sequence alignments of V-like and C2 domains

```
strands                          A'     B                 C        D          E             F               G
V-like consensus                    Gxx*x*xC            *xW         +*      Lx*xx*xxxDx#xYxC               *x*x*
FcγRI       (domain 1)  ISLQPPWVSVFQEETVTLHCEVLHL--PGSSSTQWFLNGTATQTST----PSYRITSASVNDSGEYRCQRGL-----SGRSDPIQLEIHRGW
FcγRII      (domain 1)  LKLEPPWINVLQEDSVTLTCQGARS--PESDSIQWFHNGNLIPTHTQ---PSYRFK-ANNNDSGEYTCQTGQ-----TSLSDPVHLTVLSEW
FcγRIII     (domain 1)  VFLEPQWYSVLEKDSVTLKCQGAYS--PEDNSTQWFHNESLISSQA----SSYFIDAATVNDSGEYRCQTNL-----STLSDPVQLEVHIGW
FcεRI       (domain 1)  VSLDPPWNRIFKGENVTLTCNGNNF--FEVSSTKWFHNGSLSEETN----SSLNIVNAKFEDSGETKCQHQQ-----VNESEPVTLEVFSDW
C2 consensus                             *xCx*          *x*               +x*x*x         *xCx*
strands                       A            B              C     C'            E             F               G

strands                          A'     B                 C        D          E             F               G
V-like consensus                    Gxx*x*xC            *xW         +*      Lx*xx*xxxDx#xYxC               *x*x*
FcγRI       (domain 2)  LLLQVSSRVFTEGEPLALRCHAWKD--KLVYNVLYYRNG--KAFKFFHWNSNLTILKTNISHNGTYHCSGM---GKHRYTSAGISVTVK
FcγRII      (domain 2)  LVLQTPHLEFQEGETIMLRCHSWKD--KPLVKVTFFQNG--KSQKFSRLDPTFSIPQANHSHSGDYHCTGN--IGYTLFSSKPVTITVQ
FcγRIII     (domain 2)  LLLQAPRWVFKEEDPIHLRCHSWKN--TALHKVTYLQNG--KDRKYFHHNSDFHIPKATLKDSGSYFCRGL--VGSKNVSSETVNITIT
FcεRI       (domain 2)  LLLQASAEVVMEGQPLFLRCHGWRN--WDVYKVIYYKDG--EALKYWYENHNISITNATVEDSGTYYCTGK--VWQLDYESEPLNITVI//
C2 consensus                             *xCx*          *x*               +x*x*x         *xCx*
strands                       A            B              C     C'            E             F               G

strands                          A'     B                 C        D          E             F               G
V-like consensus                    Gxx*x*xC            *xW         +*      Lx*xx*xxxDx#xYxC               *x*x*
FcγRI       (domain 3)    //SVTSPLLEGNLVTLSCETKLLLQRPGLQLYFSFYMGSKTLRGRNTSSEYQILTARREDSGLYWCEAATEDGNVLKRSPELELQVL
C2 consensus                             *xCx*          *x*               +x*x*x         *xCx*
strands                       A            B              C     C'            E             F               G
```

Consensus sequences derived from a structure-based sequences alignment (Vaughn & Bjorkman 1996) are presented above (V-like) or below (C2) each domain sequence (adapted from the human sequences presented in Kochan et al 1988, Allen & Seed 1989). The alignment for the V-like consensus sequences was obtained after superposition of the available three-dimensional structures of V-like domains [CD2 domain 1, CD4 domains 1 and 3, telokin, V_H and V_L from a Fab, CD8, and VCAM-1 domain1; (Davis & Metzger 1983, Ryu et al 1990, Wang et al 1990, Holden et al 1992, Jones et al 1992, 1995, Leahy et al 1992, Brady et al 1993, Bodian et al 1994)] were superimposed to identify structurally corresponding residues. The alignment for the C2 consensus sequences was obtained after superposition of CD4 domains 2 and 4, CD2 domain 2, and VCAM-1 domain 2 (Ryu et al 1990, Wang et al 1990, Jones et al 1992, 1995, Brady et al 1993) (see Vaughn & Bjorkman 1996 for details). The consensus sequences are meant for comparison with FcR sequences and are not derived from the FcR sequences themselves. Consensus primary sequence patterns are identified by the one-letter code or symbols; *indicates a hydrophobic amino acid; + represents a basic amino acid, # indicates a Gly, Ala, or Asp; and an x indicates any amino acid. The approximate locations of the centers of β-strands are indicated by the letter name of the strand above or below the sequences. Gaps in one sequence compared with the others are represented by dashes and slashes indicate a missing portion of sequence. Residues implicated in Ig binding are underlined (mapping studies for FcγRII: Hulett et al 1995; mapping studies for FcγRIII: Hibbs et al 1994; FcϵRI mapping studies reviewed in Hulett & Hogarth 1994). The individual FcR domain sequences do not exactly match either V-like or C2 domain consensus sequences, but all the domains share some common features with the V-like domains. Some of the features that distinguish V-like domains from C2 domains are a β-turn connecting strand A′ to B, usually identifiable at the primary sequence level by a glycine seven residues before the first of the cysteines in the characteristic disulfide bond, and a distinguishing sequence motif in the vicinity of the E to F loop (the tyrosine corner) that includes a tyrosine two residues prior to the second cysteine (Harpaz & Chothia 1994).

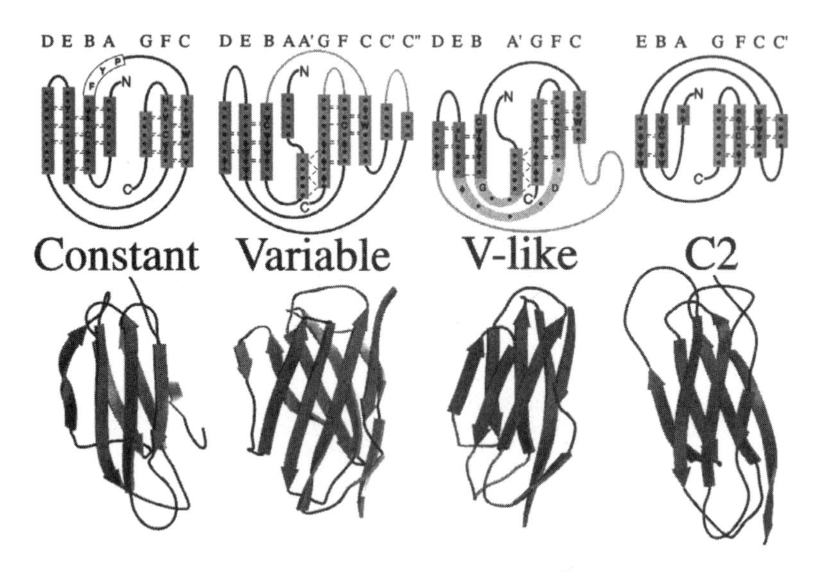

Figure 3 Structures of the Ig superfamily domains. Topology (*top*) and ribbon (*bottom*) diagrams are presented for each fold. β-strands are identified by letters in the topology diagrams. In the topology diagrams, the β-sheet containing strands A, B, and E is dark grey and the sheet containing strands G, F, and C is light grey. In the ribbon diagrams, the A-B-E containing β-sheet is in front of the G-F-C containing sheet. Amino acids that have equivalent positions in all structures of the domain type are indicated by a solid circle; by the one-letter code if the identity of the amino acid is conserved; by ϕ for hydrophobic residues; or by the symbol ψ for hydrophilic residues. β-sheet hydrogen bonding is indicated by dashed lines. Regions of irregular secondary structure are indicated by open rectangles. The antigen-binding loops in the Ig variable domain are B to C (CDR1), C′ to C″ (CDR2), and F to G (CDR3). Loops that are structurally conserved in V-like domains (A′ to B and E to F loops; see consensus sequence in Table 1) are thick. Ribbon diagrams were prepared from coordinates available from the PDB (7fab for Ig constant and variable domains) or provided by the authors (VCAM-1 coordinates from EY Jones for V-like and C2 domains) (adapted from Figure 3 in Vaughn & Bjorkman 1996).

FcRs IN ANTIBODY TRANSPORT

Transfer of maternal IgG molecules from the mother to the fetus or infant is a mechanism by which mammalian neonates acquire humoral immunity to antigens encountered by the mother. Passive acquisition of antibody is important to the neonate during the immediate time after birth when its immune system is not yet fully functional. The protein responsible for the transfer of IgG is called FcRn, for FcR neonatal (Rodewald & Kraehenbuhl 1984, Simister & Rees 1985). Here the word neonatal derives from the initial discovery of the receptor in the gut of suckling rats on the apical surface of intestinal epithelial cells (Jones & Waldman 1972). Maternal IgG in ingested milk is bound by the receptor, transported across the gut epithelium in a process called transcytosis (Mostov & Simister 1985), and then released into the bloodstream from the basolateral surface (Rodewald & Kraehenbuhl 1984). There is a net pH difference between the apical (pH 6.0–6.5) and basolateral (pH 7.0–7.5) sides of intestinal epithelial cells (Rodewald & Kraehenbuhl 1984). This pH difference facilitates the efficient unidirectional transport of IgG, because FcRn binds IgG at pH 6.0–6.5 but not at neutral or higher pH (Rodewald & Kraehenbuhl 1984, Simister & Mostov 1989b, Raghavan et al 1995a).

Recent studies have shown that FcRn is also expressed in the fetal yolk sac of rats and mice (Roberts et al 1990, Ahouse et al 1993) and in the human placenta, where it is thought to transfer maternal IgG to the fetus (Story et al 1994, Simister & Story 1996). In addition, FcRn is expressed on the canalicular cell surface of adult rat hepatocytes, where it could bind IgG in bile and potentially function to provide communication between bile and parenchymal immune cells (Blumberg et al 1995). In this case, the proposed model for FcRn function is to transport antibody-antigen complexes from bile to the parenchyma. Antigen-presenting cells in parenchyma would process the antigens and present them to T cells, which would then stimulate local B cells to secrete antigen-specific antibodies that are transported back into the bile by unknown mechanisms (Blumberg et al 1995). Thus FcRn appears to be used for transport of IgG in sites other than the intestine, and in adult animals as well as neonates. Human placental FcRn, FcRn from rat fetal yolk sac, and FcRn from rat hepatocytes all show the same pH-dependent interaction with IgG that is observed with the intestinal FcRn, i.e. IgG binding at pH 6.0–6.5 but not at neutral or basic pH (Roberts et al 1990, Story et al 1994, Blumberg et al 1995). This feature of the FcRn molecule may have evolved to facilitate IgG release from the receptor at the slightly basic pH of blood. Because a pH gradient does not exist across the placenta or the mammalian yolk sac, it is thought that FcRn binds IgG intracellularly, in acidic vesicles that are targeted to the cell surface where IgG is released upon exposure to the pH environment of blood (Roberts et al 1990, Story et al 1994, Simister & Story 1996).

In addition to these transport functions, current evidence suggests that FcRn is the catabolic receptor that controls the lifetime of Igs in serum. For example, mutant Fc fragments that show impaired FcRn binding in vitro and that are deficient for transcytosis in neonatal mice also have abnormally short serum half lives (Kim et al 1994, Popov et al 1996). Furthermore, murine IgG1 is degraded significantly faster in mice that are deficient in the FcRn light chain (β2-microglobulin; β2 m) (Ghetie et al 1996, Junghans & Anderson 1996, EJ Israel, D Wilsker & NE Simister, manuscript in preparation). Because the pH of serum is not permissive for IgG binding by FcRn, the postulated model for the catabolic receptor function of FcRn is that IgG is sequestered by FcRn in acidic intracellular vesicles and subsequently rereleased into serum.

Transport of other classes of Ig in the adult mammal occurs using different mechanisms. In order to confer specific protective functions at diverse sites, the Ig isotypes are distributed differently throughout the body, necessitating specific transport in some instances. The transport of IgA from the circulation into secretions has been well studied. IgA is found in epithelial secretions such as the lumen of the gut, the salivary and tear glands, respiratory secretions, and breast milk, where the neutralizing capacity of IgA molecules forms the first line of defense against entering pathogens (Underdown & Schiff 1986, Childers et al 1989, Kerr 1990, Hanson & Brandtzaeg 1993). IgA is synthesized by plasma cells beneath the basement membranes of surface epithelia. Subsequent to synthesis, the IgA molecules must cross the epithelial cell barrier to enter into secretions. Transcytosis of IgA across polarized epithelial cells first involves the binding of IgA to pIgR proteins on the basolateral surface (reviewed in Kraehenbuhl & Neutra 1992, Mostov 1994). pIgR-IgA complexes are internalized and sorted into endosomes destined for the apical surface. At the apical surface, pIgR is cleaved and the extracellular portion of the receptor is released into secretions as a complex with IgA. This complex, called secretory IgA, protects against pathogens in the digestive, respiratory, and genital tracts.

The structures of FcRn and pIgR and how they interact with their respective Ig molecules are now discussed.

FcRn

FcRn is a type I membrane glycoprotein that acts as a specific receptor for the IgG isotype (Jones & Waldman 1972). Reported values of the equilibrium association constant (K_A) range from $\approx 2 \times 10^7$ M^{-1} for the interaction of monomeric IgG with isolated microvilli membranes from neonatal rat intestine (Wallace & Rees 1980) to $\approx 5 \times 10^7$ M^{-1} for the interaction between purified soluble rat FcRn and monomeric IgG (Raghavan et al 1995b). This rather high affinity ensures that FcRn can efficiently transport unliganded IgG, which is the most useful form of Ig for the newborn, although transport of antigen-antibody complexes has also been observed (Abrahamson et al 1979). Unlike the other known FcRs, which are presumed to have monomeric extracellular Ig-binding regions, FcRn is a heterodimer. The FcRn light chain is β2m (Simister & Rees 1985), the same light chain that is associated with class I major histocompatibility complex (MHC) molecules (Figure 4*a*). Molecular cloning of the FcRn heavy chain revealed additional similarity to class I MHC molecules rather than to any of the other FcRs (Simister & Mostov 1989a). The heavy chains of both FcRn and class I MHC molecules consist of three extracellular domains, α1, α2, and α3, followed by a transmembrane region and a short cytoplasmic sequence

(Simister & Mostov 1989a, Bjorkman & Parham 1990) (Figure 4*a*). Although the extracellular region of FcRn and class I MHC molecules exhibits low but significant sequence similarity (22–30% identity for the $\alpha 1$ and $\alpha 2$ domains; 35–37% for the $\alpha 3$ domain), the transmembrane and cytoplasmic regions of the two types of proteins show no detectable sequence similarity (Simister & Mostov 1989a). MHC class I molecules have no reported function as Ig receptors; instead they bind and present short peptides to T cells (Townsend & Bodmer 1989). The similarity at the primary, tertiary, and quaternary structure levels between FcRn and class I MHC molecules suggests that they share a common ancestor, in spite of their apparently unrelated immunological functions.

THREE-DIMENSIONAL STRUCTURE OF FcRn The crystal structure of rat FcRn confirmed that the overall fold is very similar to that of class I MHC molecules, such that the $\alpha 1$ and $\alpha 2$ domains form a platform of eight anti-parallel β-strands topped by two long α-helices, and the $\alpha 3$ and $\beta 2$m domains are β-sandwich structures similar to antibody constant or C1-set domains (Figure 4*a*) (Burmeister et al 1994a). Although the two α-helices that form the sides of the MHC peptide-binding groove are present in FcRn, they are closer together than their MHC counterparts, rendering the FcRn groove incapable of binding peptides, consistent with earlier biochemical studies (Raghavan et al 1993). To date, the FcRn counterpart of the MHC peptide-binding groove has not been implicated in any FcRn binding function.

FcRn DIMERIZATION A dimer of FcRn heterodimers (hereafter referred to as the FcRn dimer) mediated by contacts between the $\alpha 3$ and $\beta 2$m domains was observed in three different crystal forms of FcRn (Burmeister et al 1994a) (Figure 4*b*). Because of the 2:1 binding stoichiometry between FcRn and IgG (Huber et al 1993), it was suggested that the FcRn dimer could represent a receptor dimer induced by IgG binding (Burmeister et al 1994a,b). In the absence of IgG, FcRn is predominantly monomeric in solution at μM concentrations (Gastinel et al 1992); thus it was assumed that the high protein concentrations required for crystallization could induce formation of the FcRn dimer in the absence of IgG. If the observed dimer functions in IgG binding at the cell surface, each FcRn heterodimer would be aligned with its longest dimension roughly parallel to the plane of the membrane ("lying down") (Figure 5*b*), in contrast to the "standing up" orientation generally assumed for class I MHC molecules (as shown in Figure 4*a*).

The biological relevance of crystallographically observed dimers or oligomers is a question often encountered in structural biology, because crystal packing can induce protein multimers that are not found under physiological conditions. Several lines of evidence, including the observation that the

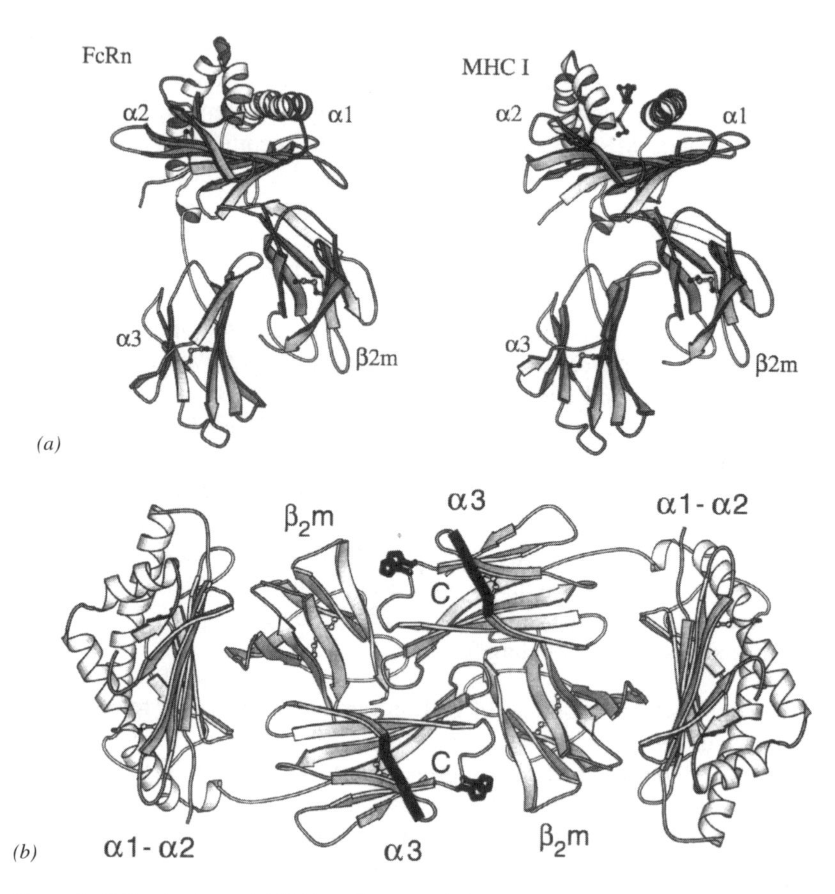

Figure 4 FcRn structure. (*a*) Ribbon diagrams of the structure of FcRn and a class I MHC molecule. Both are shown in the standing-up orientation believed to be relevant for the interactions of class I molecules with T-cell receptors. (*b*) Structure of the FcRn dimer observed in crystals of FcRn alone (Burmeister et al 1994a) and crystals of a 2:1 complex between FcRn and Fc (Burmeister et al 1994b). FcRn residues identified by site directed mutagenesis to affect the affinity of the FcRn interaction with IgG are highlighted (His 250 and His 251, dark sidechains; residues 219–224, dark loop; Raghavan et al 1994). The C-termini of the truncated FcRn heavy chains are labeled C. The FcRn dimer is believed to be oriented on the cell membrane as shown in Figure 5*b* (a 90° rotation about the horizontal axis from this view), so that in this orientation, the membrane is presumed to be parallel to and below the plane of the paper. This view of the dimer corresponds to the view in Figure 6*a*.

same dimer was observed in cocrystals of a complex between rat FcRn and Fc (Burmeister et al 1994b), indicate that the crystallographically observed FcRn dimer is not a crystal packing artifact. The crystal structure of the FcRn-Fc complex was solved at low resolution (4.5–6 Å) because the crystals diffracted poorly. However, the cocrystal structure confirmed the 2:1 FcRn-Fc binding stoichiometry, and the high resolution structures of FcRn (Burmeister et al 1994a) and Fc (Deisenhofer 1981) were used to deduce the mode of interaction between the two proteins. The cocrystal structure showed that the edge of the $\alpha 1$-$\alpha 2$ domain platform and the N-terminal portion of $\beta 2$m contact Fc at the interface between the C$\gamma 2$ and C$\gamma 3$ domains. The FcRn-binding site on Fc overlaps with the binding site for fragment B of protein A (Deisenhofer 1981), as predicted from previous mutagenesis and inhibition studies (Kim et al 1994, Raghavan et al 1994). Because FcRn is structurally distinct from the other FcRs, parallels between the FcRn-IgG interaction and the interaction of Igs with other FcRs are unlikely. Indeed, the binding site on Fc for most other FcRs (lower hinge region and top of the C$\gamma 2$ domain, see below; Figure 1*b*) does not overlap with the binding site for FcRn. The finding that FcRn interacts with a different portion of the Fc region than that contacted by the other FcRs can be rationalized by the observation that at least some of the pH dependence of the FcRn interaction with IgG arises from titration of Fc histidines at the FcRn binding site (Figure 1*b*; see below).

TWO DIFFERENT 2:1 FcRn-Fc COMPLEXES The interpretation of the cocrystal structure is complicated by the packing in the cocrystal, which creates two distinct 2:1 FcRn-Fc complexes, only one of which incorporates the FcRn dimer. The first 2:1 complex (the lying-down complex; Figure 5*a,b*) incorporates the same FcRn dimer observed in the crystals of FcRn alone. In this complex, the receptor dimer binds a single Fc molecule asymmetrically, with most of the contacts involving one molecule of the dimer (Burmeister et al 1994b). The second 2:1 FcRn-Fc complex (the standing-up complex; Figure 5*c*) does not involve the receptor dimer. Instead, Fc binds symmetrically between two FcRn molecules that are oriented with their long axes perpendicular to the membrane. Both 2:1 complexes are built from the same 1:1 FcRn-Fc building block (compare Figures 5*a* and *c*).

Information from the crystal structure alone could not resolve the issue of which of these two 2:1 complexes is physiologically relevant: the standing-up complex, the lying-down complex, or both. However, it was argued that intact IgG would be sterically prohibited from binding to cell surface FcRn in the standing-up complex because of collisions between the Fab arms and the membrane surface (Burmeister et al 1994b). In addition, biochemical studies established that the lying-down complex involving the FcRn dimer is required

(a)

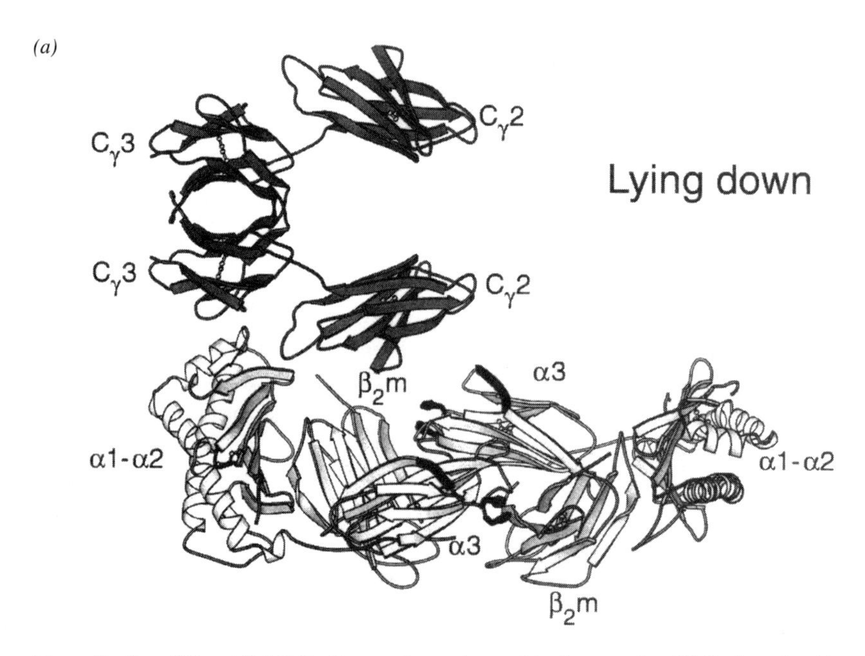

Figure 5 Two different 2:1 FcRn-Fc complexes observed in the crystals of FcRn bound to Fc (Burmeister et al 1994b). FcRn residues identified by site-directed mutagenesis to affect the affinity of the FcRn interaction with IgG are highlighted (His 250 and His 251, dark sidechains; residues 219–224, dark loop; Raghavan et al 1994). (*a*) Lying-down 2:1 complex involving the FcRn dimer. The Fc molecule interacts asymmetrically with the receptor dimer so that most of the contacts involve one of the FcRn molecules. Contacts are predicted between the N-terminal portion of the $C\gamma 2$ domain and the partner FcRn molecule, in the vicinity of FcRn heavy chain residues 219–224 (loop, shown in bold). His 250 and His 251 (highlighted) are predicted to exert their effect on the interaction with IgG by modulating the formation of the FcRn dimer, which is required for high-affinity binding of IgG (Raghavan et al 1995b). (*b*) View of the lying-down 2:1 complex rotated by 30° about the horizontal axis with respect to the orientation shown in *a*. The plane of the membrane for cell surface FcRn is horizontal. No steric hindrance between intact IgG and the membrane is expected in this orientation, as the two Fab arms of IgG could project out of the plane of the paper and into the plane of the paper. (*c*) The standing-up 2:1 complex oriented so that the plane of the membrane for cell surface FcRn is horizontal. Steric hindrance between the Fab arms of intact IgG and the membrane is predicted for this complex. In addition, neither of the regions indicated by site-directed mutagenesis to affect IgG binding affinity (histidines 250 and 250, highlighted; loop, comprising residues 219–224, bold) (Raghavan et al 1994) are at the Fc interface or any other protein-protein interface.

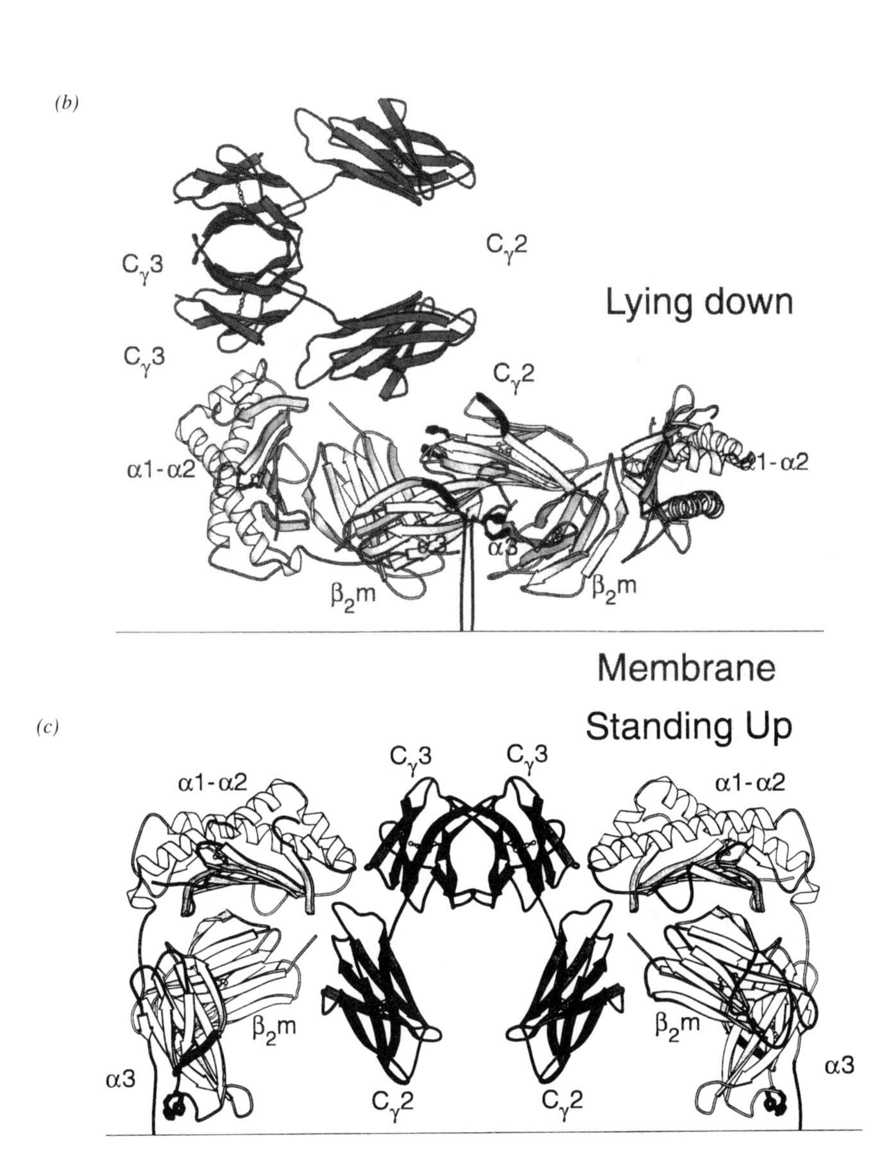

Figure 5 (*Continued*)

for high-affinity binding of IgG. The mutation of two histidine residues located at the FcRn dimer interface (Figures 4*b*, 5*a,b*) reduced the binding affinity for IgG (Raghavan et al 1994). These histidines are distant from the IgG-binding site and cannot directly contact Fc; thus their effect on IgG affinity is hypothesized to be an indirect result of modulating the formation of the FcRn dimer, which is predicted to have a higher affinity for IgG than an FcRn monomer. The prediction that the FcRn dimer has a higher IgG-binding affinity than an FcRn monomer was directly tested in biosensor studies using oriented coupling of FcRn molecules. These studies showed high-affinity IgG binding when FcRn was immobilized on a biosensor chip in an orientation facilitating dimerization, but an affinity reduction of over 100-fold when FcRn dimer formation was hindered (Raghavan et al 1995b). Examination of the predicted contacts between Fc and the two FcRn molecules in the lying-down 2:1 complex lends support to the idea that dimerization of FcRn results in an increased affinity for IgG, in that contacts between Fc and the second FcRn molecule are predicted in the regions near residues 219 and 245 of the $\alpha 3$ domain of FcRn and residues 272 and 285 in the $C\gamma 2$ domain of Fc (Burmeister et al 1994b) (Figure 5*a*). Site-directed mutagenesis confirmed that alteration of residues in the region of residue 219 of the FcRn heavy chain affected IgG binding affinity (Raghavan et al 1994) (Figure 5*a,b*).

Thus the available evidence suggests that FcRn dimerization is indeed involved in IgG binding and that the lying-down 2:1 complex is physiologically relevant for the binding of intact IgG by FcRn molecules at the cell surface. The crystallographic and biochemical results raise the possibility that FcRn functions like many other cell surface receptors in which ligand-induced dimerization is required for downstream signal transduction (Heldin 1995) which, in this case, would initiate the endocytosis of IgG-FcRn complexes. However, the lying-down 2:1 FcRn-Fc complex does not take advantage of the twofold symmetry of the Fc homodimer, because the FcRn binding site on one of the Fc polypeptide chains is not used (Figure 5*a,b*). Both binding sites are utilized in the cocrystals because the standing-up and lying-down complexes coexist to form an oligomeric ribbon with an overall stoichiometry of two receptors per Fc (Figure 6). If both complexes exist in the cocrystals, could both complexes form in vivo?

We recently proposed a model in which IgG-induced dimerization of FcRn followed by the creation of networks that include both the standing-up and lying-down 2:1 FcRn-IgG complexes (Figure 6) signals the initiation of transcytosis in vivo (Burmeister et al 1994b). This model requires that networks of FcRn-IgG complexes are formed between parallel adjacent membranes separated by $\approx$170 Å (Figure 6*b*). The requirement for closely spaced parallel adjacent membranes

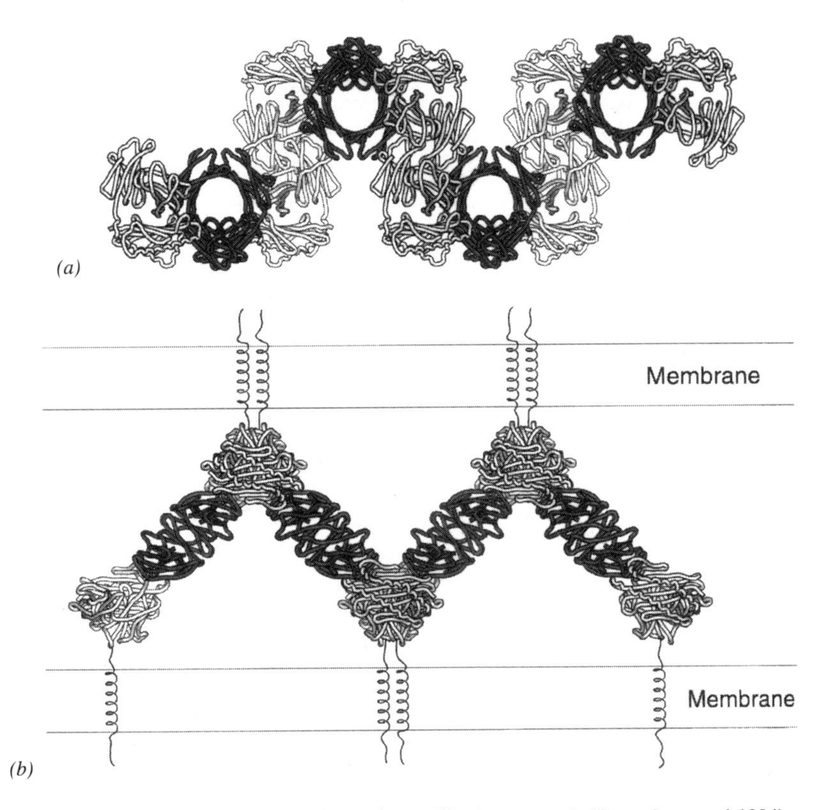

Figure 6 Network of FcRn-Fc complexes observed in the cocrystals (Burmeister et al 1994) as a model for the formation of higher order oligomers upon binding of IgG to cell surface FcRn. FcRn is grey and Fc is black. (*a*) Top view illustrating the simultaneous formation of lying-down and standing-up FcRn-Fc complexes. If this network is formed under physiological conditions, each FcRn dimer would be associated with a membrane parallel to the plane of the paper: the left-most dimer is associated with a membrane below the plane of the paper, the central dimer with a membrane above the plane of the paper, and the right-most dimer is again associated with a membrane below the plane of the paper. (*b*) Side view, rotated by 90° about the horizontal axis from the view in *a*. The FcRn dimers are seen looking down their long axes (vertical in *a*). Existence of this network under physiological conditions would require two parallel membranes separated by 160 to 170 Å (Burmeister et al 1994b). Adjacent microvilli membranes in the brush borders of neonatal rat epithelial cells can be separated by as little as 200 Å (Rodewald 1973), consistent with the possibility that networks of 2:1 FcRn-IgG complexes form between adjacent microvilli membranes.

is fulfilled in the microvilli of the brush border epithelial cells (Rodewald 1973) where FcRn is expressed and functions, but the bridging of adjacent membranes by IgG bound to FcRn dimers remains to be demonstrated. However, recent evidence suggests that both Fc binding sites are required for FcRn function in vivo, in that a hybrid Fc heterodimer containing one functional and one disrupted FcRn binding site is not transcytosed efficiently into the bloodstream of newborn mice (Kim et al 1994). The same hybrid Fc heterodimer was found to bind with normal affinity to mouse FcRn immobilized on a biosensor chip (Popov et al 1996). This latter observation was interpreted as support for a 1:1 binding stoichiometry between mouse FcRn and IgG (Popov et al 1996) but is also consistent with a 2:1 stoichiometry involving the FcRn dimer and the utilization of only one FcRn binding site on the Fc molecule (Figure 5*a,b*). In combination with the previously described results, these experiments suggest that the lying down FcRn-Fc 2:1 complex is necessary, but not sufficient, for transcytosis of IgG, and that both 2:1 complexes are relevant for FcRn function in vivo.

We have argued that the bridging of FcRn dimers on two adjacent membranes by IgG molecules is a prerequisite for transcytosis. However, when we prepare FcRn-Fc complexes at μM concentrations in solution, size exclusion chromatography suggests that the predominant species is a 2:1 receptor-Fc complex, rather than a higher-order multimer (AH Huber, L Sanchez & PJ Bjorkman, unpublished data). It is possible that higher-order multimers appear only under special circumstances of high local protein concentrations, i.e. in the cocrystal (mM concentration) or when FcRn dimers are constrained to adjacent membranes separated by the optimal distance for bridging by IgGs. One explanation for this involves the idea that the Fc region of IgG is asymmetric, such that only one of the FcRn-binding sites is in an optimal conformation for receptor binding. As a precedent for this idea, the Fc region of IgE is strongly bent both in solution and when bound to its receptor (reviewed in Baird et al 1993) and thus only one of the FcϵRI binding sites on the IgE Fc appears to be accessible (discussed in FcϵRI section; reviewed by Sutton & Gould 1993). There is evidence that IgG is also bent in solution (Baird et al 1993), and the FcRn binding sites might therefore be in slightly different conformations on both polypeptide chains. The low resolution of the cocrystal structure and the disorder of the N-terminal portion of the Cγ2 Fc domain (Burmeister et al 1994b) prevent the use of this crystal structure to determine whether Fc is bent when bound to FcRn. However, it is clear that in the crystals both FcRn-binding sites on Fc interact with an FcRn molecule, so one must assume that any Fc bending that occurs does not completely occlude either of the FcRn binding sites under the conditions of high protein concentration in the cocrystal.

pH DEPENDENCE OF THE FcRn-Fc INTERACTION To function as an efficient unidirectional transporter of IgG molecules, FcRn must bind IgG with high affinity prior to and during transcytosis and release intact IgG molecules at the end point of transport. As mentioned above, release of IgG from FcRn occurs rapidly upon exposure to environments of neutral pH such as the blood. Indeed, the affinity of FcRn for IgG is relatively stable between pH 5 and 6, then drops steeply by over two orders of magnitude as the pH is raised from 6.0 to 7.0 (Raghavan et al 1995a). The dramatic pH dependence of FcRn binding to IgG is unusual for protein-protein interactions. Histidine residues are likely candidates for effecting such pH-dependent affinity changes because the imidazole side chain usually deprotonates over the pH range of 6.0 to 7.0. Mutagenesis (Kim et al 1994), inhibition studies (Raghavan et al 1994), and the FcRn-Fc cocrystal structure (Burmeister et al 1994b) implicate conserved histidines at the interface of the $C\gamma2$ and $C\gamma3$ domains. A biosensor pH dependence assay was used to analyze the role of histidine residues in Fc and FcRn in the affinity transition (Raghavan et al 1995a). These studies suggest that the lying-down 2:1 FcRn-IgG complex is destabilized at neutral or basic pH values by titration of IgG histidine residues located at the IgG-FcRn interface (IgG His 310 and His 433; Figure 1*b*) as well as FcRn histidines located at the FcRn-FcRn dimer interface (FcRn His 250 and His 251; Figures 4*b*, 5*a,b*) (Raghavan et al 1995a). Thus an intricate molecular network of interactions has evolved to facilitate the binding and release of IgG by FcRn, a situation quite different from pIgR-mediated transport of IgA, where a proteolytic cleavage step is utilized to release IgA into secretions (see below). The advantage conferred upon the FcRn transporter system is that the same receptor molecule can be used for multiple rounds of transport.

pIgR

pIgR is so called because of its ability to transport polymeric Ig molecules: IgA and IgM (Brandtzaeg 1981). Serum IgA is produced in the bone marrow in a primarily monomeric form, whereas the IgA in secretions, the product of local synthesis at mucosal surfaces, is associated with a protein called the J chain and is primarily dimeric (Kerr 1990) (Figure 1*a*). The J chain is not required for IgA dimerization, but it appears to have a role in maintaining dimer stability (Hendrickson et al 1995). IgM is found in both pentameric and hexameric forms; only the pentameric form includes the J chain (Randall et al 1992). pIgR transports the dimeric form of IgA and the pentameric form of IgM (Brandtzaeg 1981). It is not known if IgM hexamers are transported by pIgR.

pIgR TRANSPORT After synthesis, pIgR transits from the endoplasmic reticulum to the Golgi and then to the *trans*-Golgi network, where the receptor is specifically targeted to the basolateral membrane of epithelial cells (reviewed in Mostov 1994). The basolateral targeting signal is contained in a 17-amino acid sequence of the cytoplasmic domain immediately C-terminal to the predicted transmembrane region (Casanova et al 1991). IgA binding occurs at the basolateral surface. Binding of IgA to pIgR stimulates transcytosis to the apical side of the cell, but some transcytosis of unliganded pIgR occurs constitutively (Song et al 1994). The direction of pIgR transcytosis, from the basolateral to apical side of the cell, is opposite to the direction of FcRn transcytosis, consistent with the opposite functions of the two receptors: pIgR to deposit Ig into secretions such as milk and FcRn to retrieve Ig from ingested milk.

In the transcytotic pathway, pIgR and pIgR-Ig complexes are first internalized into endosomes. Targeting of endosomes containing pIgR and pIgR-Ig complexes to the apical surface is thought to involve the same sorting signals and proteins required for basolateral targeting from the *trans*-Golgi network (Aroeti & Mostov 1994). At the apical surface, the extracellular portion of pIgR is cleaved at a site near the transmembrane sequence, and the extracellular portion of pIgR (called secretory component; SC) is released either alone or as a complex with IgA. The protease(s) involved in cleavage have not been characterized, but multiple C-terminal residues have been identified on SC, implying cleavage by multi- or nonspecific protease(s) (Eiffert et al 1984). Because pIgR is cleaved during the transcytosis process, it is used for only a single round of transport.

The cleavage of pIgR and formation of the SC-IgA complex confer enhanced resistance to proteolysis in the protease-rich mucosal environment (Underdown & Dorrington 1974). Once SC-IgA complexes are released into secretions, target microorganisms are efficiently cross-linked because dimeric IgA contains a total of four antigen-binding sites. The resulting large particles show retarded movement to the mucosal surface and can therefore be trapped in the mucus and cleared (Underdown & Schiff 1986, Kraehenbuhl & Neutra 1992). SC-IgA may also function by binding to microorganisms so as to prevent their interactions with the epithelia. In addition, SC-IgA can participate in direct killing of pathogens by facilitating the interaction of an antibody-bound microorganism with Fcα receptor-bearing macrophages and neutrophils in the mucosal surfaces (Underdown & Schiff 1986).

pIgR STRUCTURE pIgR is a glycosylated type I membrane protein containing an extracellular domain of about 629 amino acids, a single transmembrane region, and a cytoplasmic domain of 103 amino acids, with a total molecular weight of 100,000–105,000 (Mostov et al 1984). The extracellular portion is organized into five homologous domains that most closely resemble V-like domain members of the Ig superfamily (Mostov et al 1984) (Figure 2). The structure of pIgR is therefore very different from that of FcRn, the previously discussed FcR involved in Ig transport. As is the case for most FcRs, the extracellular portion of pIgR consists of multiple Ig-like domains arranged in tandem (Figure 2).

THE pIgR-Ig INTERACTION The derived affinity of interaction between SC and polymeric IgM is high: reported values for the K_A are $\approx 6 \times 10^8$ M^{-1} to 2×10^9 M^{-1} for the human SC-IgM interaction (Goto & Aki 1984). The binding of IgA to pIgR involves noncovalent interactions as well as covalent interactions thought to stabilize the complex (Mestecky & McGhee 1987); thus SC remains complexed to IgA in the mucosal environment to provide protection from proteolysis. By contrast to human SC-Ig complexes, some rabbit SC-Ig complexes are noncovalent (Knight et al 1975). The reported K_A values for the rabbit SC-IgA interaction are $\approx 1 \times 10^8$ M^{-1} (Kühn & Kraehenbuhl 1979), indicating tight binding even in the absence of covalent interactions.

Several pieces of evidence suggest that the N-terminal (membrane distal) Ig domain of pIgR is the primary determinant of the noncovalent interaction of pIgR with IgA (Frutiger et al 1986, Bakos et al 1991, Coyne et al 1994). It was originally demonstrated that a proteolytic fragment containing pIgR domain 1 could bind to IgA (Frutiger et al 1986). Site-directed mutagenesis of sequences in the regions predicted to encompass the B to C, C′ to C″, and F to G loops of domain 1 resulted in a significant reduction or nearly complete abrogation of IgA binding, suggesting that the binding site for IgA includes the counterparts of the antigen-binding loops in Ig variable domains (Coyne et al 1994) (Figure 3). In addition, the human SC-pIgR interaction involves a covalent interaction between Cys 311 of the Cα2 domain of one IgA heavy chain and Cys 467 of domain 5 of SC (Fallgreen-Gebauer et al 1993). Thus the binding interactions of pIgR involve the tip of domain 1, as well as domain 5, which are predicted to be separated by 150 to 170 Å if the five Ig-like domains in pIgR are aligned in a strictly end-to-end fashion with no significant kinks between domains (based on measurements of the Ig-like domains in crystal structures of CD4, CD2, and VCAM-1; Ryu et al 1990, Wang et al 1990, Jones et al 1992, 1995, Brady et al 1993). By contrast, the Fc region of IgA would be no more than 60 Å in length. This information suggests either that pIgR contacts both Fc regions in the IgA dimer or that it is significantly bent, or a combination of both possibilities.

There are little available data concerning the location of the sites on the IgA dimer or the IgM pentamer that determine their interactions with pIgR. Inhibition studies using monoclonal antibodies have suggested that regions of the Cα2 and Cα3 domains of IgA are involved in interactions with SC (Geneste et al 1986), although the contribution of residues 381–411 of the Cα3 domain was ruled out in studies of pIgR binding by an IgA molecule lacking these residues (Switzer et al 1992). It has been suggested that the pIgR binding site on IgA and IgM involves the J chain (Brandtzaeg & Pyrdz 1984), but the involvement of the J chain in pIgR binding is controversial (see Mestecky & McGhee 1987). Recent work using transport assays with pIgR-transfected Madin-Darby canine kidney (MDCK) cells showed impaired transport of IgA isolated from mice lacking the J chain (Hendrickson et al 1995), but this result does not distinguish between the possibility that pIgR directly contacts the J chain or that the IgA structure is altered in the absence of the J chain. The reported 1:1 stoichiometry of the SC-IgA interaction [one molecule of SC associated with one $(IgA)_2$-J chain complex] (Goto & Aki 1984, Mestecky & McGhee 1987) could be interpreted as consistent with the idea that SC interacts with the J chain. However, more structural and biochemical information will be required to understand the complete molecular details of the pIgR-Ig complexes.

FcRs IN ANTIBODY-MEDIATED EFFECTOR RESPONSES

FcRs for IgG and IgE are present on the surfaces of several accessory cells of the immune system. These receptors, designated FcγRs for those that bind IgG, and FcϵRs for those that bind IgE, interact with antibody-antigen complexes to activate various biological responses. Here we review the functions of the FcγR and FcϵR proteins and discuss the various forms of each type of receptor and how they are thought to interact with their Ig ligands.

Effector responses mediated by FcγRs include phagocytosis, endocytosis, antibody-dependent cell-mediated cytotoxicity (ADCC), the release of mediators of inflammation, and the regulation of B cell activation and antibody production (reviewed in Unkeless et al 1988, van de Winkel & Anderson 1991). Phagocytosis involves the engulfing of microbial particles, internalization into acidified cytoplasmic vesicles called phagosomes, and fusion of phagosomes with lysosomes, upon which lysosomal enzymes destroy the microbe (Silverstein et al 1989, CL Anderson et al 1990). The process of phagocytosis is distinguished from endocytosis by the size of the particle that is internalized and degraded, with phagocytosis referring to ingestion of particles of 1 μM or greater in diameter (Silverstein et al 1977). FcγRs mediate the internalization of smaller IgG-antigen complexes via endocytosis, which enhances the efficiency of antigen presentation by class II MHC molecules (Amigorena et al 1992b). In addition, FcγRs present on cells such as natural killer cells mediate interactions with antibody-coated target cells, resulting in the destruction of target cells by ADCC (reviewed in Unkeless et al 1988, van de Winkel & Anderson 1991).

On mast cells, the receptor for IgE, FcϵRI, binds monomeric IgE with high affinity. Upon cross-linking of the IgE by interaction with multivalent antigens, inflammatory responses are activated. These responses include the release of histamines, serotonin, and leukotrienes, resulting in fluid accumulation and the influx of cells and proteins to contain infection (Beaven & Metzger 1993, Sutton & Gould 1993, Ravetch 1994).

Molecular cloning of the genes encoding the FcγR and FcϵR proteins revealed the existence of three major classes of FcγRs (FcγRI, FcγRII, FcγRIII) and two classes of FcϵRs (FcϵRI and FcϵRII). The Ig-binding portions of FcγRI, FcγRII, FcγRIII, and FcϵRI (the α chains) are members of the Ig gene superfamily, all type I transmembrane proteins containing an extracellular region with two or more Ig-like domains and a polypeptide or lipid anchor in the membrane (Figure 2) (reviewed in Ravetch & Kinet 1991, Burton & Woof 1992, Hulett & Hogarth 1994). The extracellular regions of the FcγR and FcϵRI receptors show significant sequence similarity to each other: 70–98% sequence identity within the FcγRs and about 40% sequence identity between the FcγRs and FcϵRI (reviewed in Ravetch & Kinet 1991, Ravetch 1994). FcϵRII, a type II integral membrane protein containing a C-terminal extracellular region that includes a C-type lectin domain (Kikutani et al 1986), is not a member of the Ig superfamily and is not included in this discussion of the Ig superfamily FcRs.

The α subunits of many of the Ig superfamily FcRs are found in multiple forms (Ravetch et al 1986, Stuart et al 1989, Qiu et al 1990, Ernst et al 1992; reviewed in Ravetch & Kinet 1991, Fridman et al 1992, Hulett & Hogarth 1994). In the mouse, single genes encode each of the three classes of FcγR, whereas in humans, three FcγRI, three FcγRII, and two FcγRIII genes have been identified (reviewed in Ravetch 1994, Hulett & Hogarth 1994). Figure 2 shows the isoforms of the FcγRs discussed in this review. Differences in the cytoplasmic domains of the two FcγRII isoforms result in important functional differences (see below). The two FcγRIII isoforms differ in their membrane anchorage; one isoform is anchored to the membrane with a polypeptide chain, and the second isoform is anchored to the cell surface with a glycophosphatidyl inositol (GPI) linkage (Ravetch & Perussia 1989) (Figure 2). In addition to the membrane-bound form of FcRs, soluble versions generated by alternative splicing of the transmembrane exon or by proteolysis of membrane-bound forms are found in the circulation (Fridman 1991, Galon et al 1995). The soluble forms of FcRs may play an immunoregulatory role by interfering with the functions of their membrane bound counterparts (Fridman 1991, Galon et al 1995).

The ligand-binding domains of some FcRs are associated in the membrane with other proteins that are required for receptor assembly and signaling. The

α chains of FcϵRI and FcγRI associate with an integral membrane protein called the γ chain (Perez-Montfort et al 1983, Ra et al 1989, Ernst et al 1993, Scholl & Geha 1993). The γ chain is homologous to the ζ chain, a protein originally identified as essential for the assembly and signaling of the T-cell receptor-CD3 complex (Weissman et al 1989). Both γ and ζ chains associate with FcγRIII (Hibbs et al 1989, Lanier et al 1989). The cytoplasmic domains of γ and ζ chains share a common tyrosine-containing sequence motif called the antigen receptor activation motif (ARAM) (Reth 1989, Weiss & Littman 1994), ITAM (Cambier 1995), TAM (Samelson & Klausner 1992), or the ARH1 motif (Cambier 1992). In FcR signaling, as observed for signaling via the T and B cell receptors (Weiss & Littman 1994), the ARAM motifs associate with Src family protein tyrosine kinases that are activated upon receptor cross-linking to initiate a cascade of signal transduction events (Keegan & Paul 1992, Lin et al 1994).

We now review current knowledge about the structures of FcγRI, FcγRII, FcγRIII, and FcϵRI; their affinities and specificities for Ig; and binding domains interactions, and we discuss recent insights into the biological responses elicited by each receptor.

FcγRI

FcγRI is expressed on the surfaces of neutrophils, monocytes, granulocytes and macrophages (van de Winkel & Anderson 1991, Hulett & Hogarth 1994). Three human FcγRI genes have been identified, one encoding a transmembrane receptor with three Ig-like domains (Figure 2), and two encoding soluble receptors with three Ig-like domains. Alternative splicing of one of the soluble receptor genes results in an mRNA for a transmembrane receptor with two Ig-like domains (Ernst et al 1992). Of these isoforms, expression of only the transmembrane receptor with three Ig-like domains (FcγRIA) has been demonstrated. This receptor binds monomeric IgG with high affinity with reported K_A values ranging from 2×10^9 M^{-1} (for the binding of human IgG1 to human monocyte-like U937 cells expressing endogenous FcγRI; Shopes et al 1990) to 5×10^9 M^{-1} (for the binding of human IgG1 to COS cells transfected with human FcγRI; Allen & Seed 1989). Human IgG3 binds with an affinity comparable to human IgG1, but human IgG4 and human IgG2 bind more weakly (Allen & Seed 1989). The unique role of FcγRI compared with the low-affinity receptors FcγRII and FcγRIII may lie in the capability of FcγRI to trigger effector responses at low IgG concentrations, which are typical of early immune responses in vivo (Shen et al 1987).

The α chains of human and murine FcγRI consist of three extracellular Ig-like domains that show features most closely resembling the V-like set (Figure 2, Table 1), a transmembrane region, and a cytoplasmic domain (Allen & Seed

1989, Sears et al 1990, Ernst et al 1992). The α chain associates on cell surfaces with a γ homodimer (Ernst et al 1993) (Figure 2). Although transfection studies in COS cells show that the γ chain is not required for stable cell surface expression of FcγRI (Ernst et al 1993), macrophages derived from γ chain knockout mice do not bind murine IgG2a, which suggests that γ is required for FcγRI function in vivo (Takai et al 1994).

FcγRI FUNCTIONS FcγRI mediates ADCC, endocytosis, and phagocytosis in vitro (Shen et al 1987, CL Anderson et al 1990, Davis et al 1995). Despite the ability of FcγRI to bind monomeric IgG with high affinity, signals for endocytosis/degradation and phagocytosis are transduced by FcγRI only upon receptor cross-linking (Davis et al 1995). Although association of FcγRI with monomeric IgG results in internalization, the receptor-IgG complexes are rapidly recycled to the cell surface (Harrison et al 1994). Cross-linking of receptors at the cell surface also promotes internalization, but instead of being recycled, the receptor-ligand complexes are retained in intracellular compartments and subsequently degraded (Mellman & Plutner 1984, Harrison et al 1994). Internalization and degradation of cross-linked FcγRI-IgG-antigen complexes could lead to the enhanced presentation of peptide antigens on class II MHC molecules, as described for FcγRIII-immune complex interactions (Amigorena et al 1992b). However, because antigens associated with monomeric IgG bound to FcγRI are not degraded, FcγRI may not have a role in the enhancement of presentation of antigens with single epitopes.

FcγRI-IgG INTERACTION FcγRI contains an extra Ig-like domain in its extracellular portion compared with the lower-affinity receptors FcγRII and FcγRIII (Figure 2). The first two extracellular domains of FcγRI share greater sequence similarity with the two extracellular domains of FcγRII and FcγRIII than does the third domain, which suggests that the third domain is responsible for some of the interactions that confer high-affinity binding upon FcγRI (Allen & Seed 1989). Studies with mutant and chimeric FcRs demonstrate that removal of the third domain of murine FcγRI abrogates high-affinity binding to monomeric IgG, although domains 1 and 2 on their own retain a weak affinity for IgG. Linking FcγRI domain 3 to domains 1 and 2 of FcγRII does not confer high-affinity binding of monomeric IgG to FcγRII (Hulett et al 1991). Thus regions of FcγRI in addition to domain 3 are required for high-affinity binding of monomeric IgG (Hulett & Hogarth 1994).

The interaction sites on IgG for human FcγRI have been mapped in several studies that measured the binding of IgG and chimeric Igs to FcγRI expressed on U937 cells. Domain swap experiments involving chimeric IgGs constructed from different subtypes and chimeric IgE-IgG molecules identified the Cγ2 domain of IgG as an interaction site for FcγRI (Shopes et al 1990, Canfield & Morrison 1991). Mutagenesis studies showed that residues 234–237 of the lower hinge region as well as residue 331 in the Cγ2 domain of IgG are important (Duncan et al 1988, Canfield & Morrison 1991, Lund et al 1991) (Figure 1*b*). The binding sites for the closely related low affinity FcγRs (FcγRII and FcγRIII) have been mapped to a similar region of IgG, and FcϵRI appears to bind to an analogous place on the IgE Fc region (Hulett & Hogarth 1994; see below). Asn 297 of the Cγ2 domain is a conserved N-linked glycosylation site in IgG molecules. Alteration of the carbohydrate structure reduced the binding affinity of human FcγRI for IgG1 by four- to sixfold (Wright & Morrison 1994), implying involvement of carbohydrate in the receptor binding, either by direct interaction with the receptor, or by indirectly affecting the Fc conformation.

The Ig-binding domains of the FcγRs and FcϵRI proteins are generally assumed to be monomeric, based on analogy to similar Ig superfamily structures including CD4, CD2, and VCAM-1 (Ryu et al 1990, Wang et al 1990, Jones et al 1992, 1995). Indeed, FcγRI has been shown to bind only a single IgG molecule (O'Grady et al 1986), consistent with, but not conclusively demonstrating, the supposition that it is a monomer. However, all Ig Fc regions are dimers composed of two identical polypeptide chains (Figure 1*b*), each of which could theoretically interact with a single FcR to produce a 2:1 receptor-Ig stoichiometry. However, only one of two binding sites on IgG is utilized in the interaction with FcγRI (Koolwijk et al 1989), implying a 1:1 receptor-IgG stoichiometry. This issue is explored in more detail in the sections on FcγRIII and FcϵRI when other measurements of receptor-Ig stoichiometry are discussed.

FcγRII

FcγRII is very widely distributed among cells of the immune system (van de Winkel & Anderson 1991, Hulett & Hogarth 1994). Unlike FcγRI, which binds unaggregated IgG with high affinity, FcγRII binds monomeric IgG with low to undetectable affinity (estimated $K_A < 1 \times 10^7\ M^{-1}$ for human FcγRII binding to IgG; Hulett & Hogarth 1994). Under physiological conditions, the low affinity of FcγRII for monomeric IgG ensures that this receptor (and FcγRIII) interact only with IgG that has been aggregated by binding to multivalent antigens.

Although the γ chain is not required for FcγRII expression, γ has been observed in association with the α chain of human FcγRII (Masuda & Roos 1993). In mice, γ chain knockouts are impaired for the FcγRII-mediated phagocytic function of macrophages, although the expression levels of FcγRII are normal (Takai et al 1994). These results suggest that γ may be required to elicit biological responses associated with murine FcγRII (Takai et al 1994), consistent with the absence of an ARAM motif in murine FcγRII.

FcγRII ISOFORMS AND FUNCTION The α chains of human and murine FcγRII contain two extracellular Ig-like domains connected to a transmembrane and cytoplasmic region (Figure 2). The sequences of the two domains show characteristics of Ig V-like domains (Table 1), although they have been previously classified and modeled as C2 domains (Hulett & Hogarth 1994). Three human genes and one murine FcγRII gene have been identified (Ravetch et al 1986, Brooks et al 1989, Stuart et al 1989) encoding multiple transcripts that differ primarily in their cytoplasmic tails (Ravetch & Kinet 1991).

The FcγRII proteins exemplify how distinct biological responses can be elicited by differences in the cytoplasmic domains of receptors that have common extracellular ligand-binding domains (Ravetch 1994). The murine FcγRII gene has two transcripts, $\beta 1$ and $\beta 2$ (or b1 and b2). $\beta 1$ contains a 46-amino acid insertion in the cytoplasmic region (Ravetch et al 1986) that prevents efficient FcγRII$\beta 1$-mediated endocytosis (Miettinen et al 1989). B lymphocytes preferentially express the $\beta 1$ isoform of FcγRII and are thus deficient in FcγRII-mediated endocytosis and enhancement of peptide antigen presentation by class II MHC molecules (Amigorena et al 1992a). This is thought to be important in order to limit B cell stimulation by T cells to antigens for which the B cell is specific; i.e. those internalized because of binding to the B cell receptor, rather than those internalized via IgG binding to FcγRII (Amigorena et al 1992a).

Both in vitro and in vivo studies have shown that FcγRII acts as a negative regulator of immune complex–triggered activation (Muta et al 1994, Daeron 1995, Takai et al 1996). On B lymphocytes, cross-linking of FcγRII$\beta 1$ and membrane bound Ig by complexes of antigen and soluble antibody modulates B cell activation (Amigorena et al 1992a, Muta et al 1994), providing a feedback mechanism for regulation of B cell stimulation at high soluble antibody concentrations. In vitro reconstitution studies suggest that FcγRII can also inhibit FcϵRI-triggered activation of mast cells (Daeron 1995). Recent studies of FcγRII-deficient mice confirmed that FcγRII functions in vivo as an inhibitory receptor for both B cells and mast cells (Takai et al 1996). FcγRII-deficient mice display elevated Ig levels and demonstrate decreased thresholds for mast cell activation induced by FcγRIII (Takai et al 1996). The FcγRII isoforms that mediate these inhibitory functions do not contain ARAM motifs, and their cytoplasmic regions are indicated as "inhibit" in Figure 2. Both the murine $\beta 1$ and $\beta 2$ isoforms inhibit B-lymphocyte activation (Amigorena et al 1992a), as does the human homologue FcγRIIB (Ravetch 1994) [this protein has also been referred to as FcγRIIb (Brooks et al 1989) or FcγRIIC (Stuart et al 1989, Ravetch 1994)]. Other human FcγRII proteins contain the ARAM motifs in their cytoplasmic domains, which allows these receptors to mediate conventional cellular activation upon receptor cross-linking (indicated as ARAM in Figure 2).

FcγRII-IgG INTERACTION Unlike pIgR, which mainly uses its N-terminal Ig-like domain for interactions with IgA, most of the direct binding interactions between FcγRII and IgG involve the second domain of this receptor (Hulett & Hogarth 1994). Residues 154–161, 109–116, and 130–135 of the second Ig-like domain have been implicated in ligand binding (Hulett et al 1995) (*underlined* in Table 1). When these regions were modeled onto the structure of a Ig superfamily C2 domain (Hulett et al 1995), they corresponded to the F to G, B to C, and C′ to E loops (Figure 3). All three loops are predicted to be on the upper portion of the second domain, at the interface with the bottom of the first domain (Hulett et al 1995). The IgG binding site is therefore thought to involve the domain 1-domain 2 interface, with the first domain being required for stabilization of the regions of the second domain that interact with IgG (Hulett et al 1995).

The binding site on IgG for human FcγRII involves the same region of the IgG molecule as does the binding site for FcγRI; mutagenesis results implicate residues 234 to 237 of the lower hinge region adjacent to the Cγ2 domains (Lund et al 1991). Removal of the carbohydrate residues between the two Cγ2 domains also affects the ability of FcγRII to bind IgG (Walker et al 1989). Taken together with results mapping the FcεRI-IgE interface to analogous regions on both the receptor and Ig ligand, the idea that the related IgG and IgE FcRs employ similar modes of interaction seems reasonable (Hulett & Hogarth 1994; see below).

FcγRIII

FcγRIII is expressed on macrophages, neutrophils, and mast cells and is the only FcR found on natural killer cells (van de Winkel & Anderson 1991, Hulett & Hogarth 1994). Like FcγRII, FcγRIII is classified as a low-affinity receptor for IgG. The Fc fragment of IgG binds human FcγRIII with a K_A value of 1.7 $\times 10^5$ M^{-1} (Ghirlando et al 1995).

FcγRIII ISOFORMS AND FUNCTION Two human FcγRIII α chain genes have been identified, FcγRIIIA and FcγRIIIB, both encoding proteins containing an extracellular portion of 180 amino acids with two Ig-like domains (Ravetch & Perussia 1989). The Ig-like domains show sequence characteristics of Ig V-like domains (Table 1) although they were previously classified and modeled as C2 domains (Hibbs et al 1994). The most significant difference between FcγRIIIA and FcγRIIIB is that FcγRIIIA encodes a protein with a polypeptide transmembrane region and a cytoplasmic domain of 25 amino acids, whereas

FcγRIIIB encodes a protein that is anchored to the membrane by a glycophosphatidyl inositol linkage (Ravetch & Perussia 1989) (Figure 2). The FcγRIIIA and FcγRIIIB receptors have different cellular distributions, with FcγRIIIA found on macrophages, natural killer cells, and mast cells, whereas FcγRIIIB is expressed mainly on neutrophils (van de Winkel & Anderson 1991, Hulett & Hogarth 1994). By contrast to the human system, a single murine FcγRIII isoform expressed on macrophages and natural killer cells consists of two extracellular Ig domains and a polypeptide membrane linkage (previously called FcγRIIa; Ravetch et al 1986).

The α chain of murine FcγRIII associates with the γ chain (Kurosaki & Ravetch 1989), whereas human FcγRIIIA α associates with γ (Hibbs et al 1989) as well as ζ chains (in natural killer cells) (P Anderson et al 1990, Lanier et al 1989). The γ and ζ chains protect the α chain from degradation in the endoplasmic reticulum, and the absence of associated γ or ζ chains results in a reduction in cell surface expression of the transmembrane FcγRIII (Kurosaki & Ravetch 1989, Ra et al 1989). The in vivo role of FcγRIII in mediating ADCC was examined by studying γ-deficient mice. Natural killer cells from γ-deficient mice cannot mediate ADCC because of the absence of cell surface FcγRIII (Takai et al 1994).

In addition to its role in ADCC, in vitro experiments demonstrated that FcγRIII functions in endocytosis and phagocytosis (van de Winkel & Anderson 1991, Amigorena et al 1992b, Daeron et al 1994, Nagarajan et al 1995). Murine FcγRIIIA mediates rapid internalization of antibody-antigen complexes and enhances the efficiency of antigen presentation. Tyrosines in the ARAM motif of the γ subunit are required for signaling the internalization (Amigorena et al 1992b) and phagocytosis of antibody-coated erythrocytes (Daeron et al 1994). Transfection experiments in Chinese hamster ovary (CHO) cells showed that FcγRIIIA, when co-expressed with the γ chain, mediates phagocytosis of IgG-coated erythrocytes (Nagarajan et al 1995). By contrast, CHO cells expressing FcγRIIIB alone can bind IgG-coated erythrocytes, but not mediate their phagocytosis, suggesting that the lipid-linked FcγRIIIB does not deliver a phagocytic signal in CHO cells (Nagarajan et al 1995). However, it has been suggested that FcγRIIIB can work synergistically with FcγRII to enhance responses such as phagocytosis (Edberg et al 1992, Edberg & Kimberly 1994).

FcγRIII-IgG INTERACTION Similar to results described for FcγRI and FcγRII, the IgG-binding site of FcγRIII is thought to primarily involve the second Ig domain (the membrane proximal domain) (Hibbs et al 1994). When the results of mutagenesis and binding studies were mapped onto a model of this domain as a C2-set domain, the C to C′ loop was identified as the major site of IgG interaction, with contributions from the B to C and E to F loops (Hibbs et al 1994) (*underlined* in Table 1). The authors concluded that the FcγRIII interaction site lies on the portion of domain 2 that is furthest from domain 1 because of the positions of two of the three implicated loops (C to C′ and E to F, at the bottom of a C2 domain; Figure 3). By contrast, mutagenesis studies implicate the opposite portion of domain 2 of FcγRII in the interaction with IgG (Hulett & Hogarth 1994) (see above). Further work will be required to resolve whether the modes of FcγRIII and FcγRII interactions with IgG are different.

The interaction site on IgG for FcγRIII involves the same region of the IgG molecule as does the binding site for FcγRI and FcγRII, the lower hinge region just N-terminal to the Cγ2 domain (Jefferis et al 1990, Morgan et al 1995) (Figure 1*b*), but regions within the Cγ3 domain might also be involved (Gergely & Sarmay 1990).

Recent analytical ultracentrifugation experiments demonstrated that the stoichiometry of the FcγRIII-IgG interaction is 1:1 (Ghirlando et al 1995). A steric hindrance mechanism is one explanation for the observation that only one receptor binds per homodimeric Fc region, i.e. the binding of the first receptor to IgG hinders binding of the second. The location of the FcγR binding site on the lower hinge, where the two Cγ2 domains are in close spatial proximity, is consistent with this idea. The authors use thermodynamic data in support of a conformational change mechanism to explain the occupancy of only one of the two available sites on the Fc homodimer, whereby the binding of the first receptor molecule induces a conformational change that renders the second binding site on Fc nonfunctional (Ghirlando et al 1995). The exact details of the interaction of FcγRIII and the other FcγRs with IgG await the appropriate crystal structures. However, the currently available studies of the interactions of IgGs with the FcγRs and FcRn, and IgE with FcϵRI (see below), hint that these interactions are not as symmetric as might be initially assumed given the twofold symmetry of the IgGFc fragment in the crystal structure (Deisenhofer 1981) (Figure 1*b*).

FcϵRI

FcϵRI is a high-affinity receptor, binding to monomeric IgE with an equilibrium constant of $\approx 1 \times 10^{10}$ M^{-1} (Beavil et al 1993). IgE is present in very low concentrations in human serum (5–300 ng/ml), and the high-affinity binding permits a dramatic amplification of IgE-antigen interactions (Sutton & Gould 1993). Upon being cross-linked by the binding of IgE to a multivalent antigen, FcϵRI mediates a variety of allergic and inflammatory responses (Metzger 1991, Sutton & Gould 1993). Mast cells in tissues and basophils in blood express FcϵRI and are thought to be the cells principally responsible for an allergic response. Allergic reactions commonly begin with the production of IgE in response to pollen, cat dander, or other environmental antigens.

Receptor cross-linking triggered by the binding of these multivalent allergens to cell surface IgE-FcϵRI complexes results in the degranulation of cells and the release of stored mediators of inflammation, including histamine, serotonin, and leukotrienes (Metzger 1991, Sutton & Gould 1993). Recent in vivo studies confirm the importance of FcϵRI in the allergic response, in that the disruption of the FcϵRI α chain gene in mice resulted in a failure of the mice to mount anaphylactic responses (Dombrowicz et al 1994). The more beneficial role of FcϵRI is thought to involve the immune defense against parasites (Janeway & Travers 1994). Mast cell release of inflammatory mediators caused by cross-linking of FcϵRI through IgE binding of multivalent antigen induces smooth muscle contractions, with the result that mast cell degranulation results in protective responses such as vomiting, coughing, and sneezing that can help in expelling an invading organism (Janeway & Travers 1994).

FCϵRI STRUCTURE FcϵRI occurs on cell surfaces as a complex of four polypeptide chains: a ligand binding or α chain, a β chain, and a disulfide-linked γ chain homodimer (Blank et al 1989) (Figure 2). The α chain consists of two extracellular Ig-like domains, a single transmembrane sequence, and a cytoplasmic domain of about 20 to 31 amino acids (Kinet et al 1987, Metzger 1991). Although the Ig-like domains of FcϵRI were classified and modeled as C2-type domains (Sutton & Gould 1993; reviewed in Hulett & Hogarth 1994), we believe they share more similarity with V-like domains (Table 1). The β chain is predicted to contain four membrane-spanning domains, with the amino and carboxy termini being located on the cytoplasmic side (Kinet et al 1988).

FCϵRI-IgE INTERACTION Chimeric FcϵRI-FcγRIIIA and chimeric rodent-human FcϵRI were used to demonstrate that substitution of the second FcϵRI α chain extracellular domain (the membrane proximal domain) resulted in a loss of IgE binding (Mallamaci et al 1993, Robertson 1993). Further mutagenesis studies identified four regions in domain 2 as being important for interaction with IgE: residues 93–104, 111–125 and 129–137, and 154–161 (Mallamaci et al 1993, Hulett & Hogarth 1994) (*underlined* in Table 1). When the results of these studies were mapped onto a C2 set-based model of domain 2, the IgE interaction site was situated primarily in the portion of domain 2 near the interface with domain 1: the F to G, C′ to E, and B to C loops, and contributions from the B and C strands (Table 1) (Sutton & Gould 1993, Hulett & Hogarth 1994). Although the main site of IgE interaction involves the second FcϵRI domain, the observation that a mutant FcϵRI α chain containing only the second extracellular domain binds with lower affinity than the two-domain protein suggests that regions in the first domain are required for high-affinity binding (Robertson 1993).

The binding site on IgE for FcϵRI has been mapped to the interface between the $C_\epsilon 2$ and $C_\epsilon 3$ domains (Helm et al 1988, Weetall et al 1990, Nissim et al 1993), primarily involving residues of the $C_\epsilon 3$ domain, based upon the studies of the binding of chimeric rodent-human IgE molecules to rodent and human FcϵRI (Nissim et al 1993), and numerous other studies (Beavil et al 1993, Hulett & Hogarth 1994). Because the IgE heavy chain contains an extra domain, $C_\epsilon 2$, in the region corresponding to the IgG hinge (Figure 1*a*), the $C_\epsilon 2$-$C_\epsilon 3$ interface may be analogous to the IgG hinge-Cγ2 interface identified as the binding site of the FcγRs.

Fluorescence resonance energy transfer measurements demonstrated that IgE in solution is bent, such that the average end-to-end distance between the Fab and Fc segments is only 75 Å rather than the 180 Å predicted for a planar Y-shaped IgE structure (Baird et al 1993). Fluorescence energy transfer studies were also used to compare the conformation of IgE alone and IgE when bound to FcϵRI. The receptor-bound IgE was also bent, with an average end-to-end distance between the Fab and Fc segments of about 71 Å (Baird et al 1993). The bending of IgE should result in an asymmetric structure of the homodimeric IgE Fc region, such that only one of the two polypeptide chains of the Fc provides a binding site for FcϵRI. Indeed, the observed stoichiometry of the IgE-FcϵRI interaction is 1:1 (Robertson 1993), consistent with the idea that the bent region of IgE presents one concave and one convex surface, with the receptor site occluded on the concave surface (Baird et al 1993, Sutton & Gould 1993). Further analysis of the spectroscopic studies suggests that the Fc region of FcϵRI-bound IgE lies parallel to the membrane, based on measurements of the distances between the fluorophores on IgE and a fluorophore in the membrane (Baird et al 1993). This information implies a lying-down orientation for cell surface FcϵRI (Figure 2), reminiscent of the lying-down orientation proposed for the FcRn dimer (Burmeister et al 1994a,b) (Figure 5*a,b*). Because of the similarities in the predicted interfaces of the FcγR-IgG and the FcϵRI-IgE complexes, and the sequence similarities between the FcγRs and FcϵRI, a lying-down orientation on the membrane may be a common feature of many FcRs.

CONCLUSIONS

Most FcRs that are members of the Ig superfamily have related structures consisting of two or more Ig-like domains arranged in tandem, a domain organization shared by many cell adhesion molecules. FcRn, a heterodimer with an overall structural similarity to class I MHC molecules, is structurally distinct from the other Ig superfamily FcRs reviewed here (pIgR, the FcγRs, and FcϵRI). There are several common elements in the interactions of the adhesion molecule-like FcRs with Ig, which usually are not features of the interaction of FcRn with IgG.

First, the binding site on IgG for all the FcγRs involves the hinge and the hinge-proximal portion of the Cγ2 domain. The binding site for FcϵRI is at the $C_\epsilon 2$-$C_\epsilon 3$ interface, an analogous location on the IgE molecule. By contrast, FcRn binds to IgG at a different site, the interface between the Cγ2 and Cγ3 domains, where Fc histidine residues mediate part of the pH-dependent interaction between FcRn and IgG.

A second common feature of the FcγRs and FcϵRI is a 1:1 receptor-Ig stoichiometry in the case of FcγRI, FcγRIII, and FcϵRI, which suggests asymmetry in the Fc regions of receptor-bound Igs, such that one receptor binding site on the Fc homodimer is inaccessible. Strong evidence suggests that the IgE Fc region is bent, which may contribute to the inaccessibility of the second receptor binding site on IgE molecules. Biophysical studies also suggest that the Fc region of IgG is bent and can therefore bind asymmetrically to the FcγR receptors, which may explain the observation that only one of the two sites at the hinge-proximal end of the Cγ2 domain is available for interaction with FcγRI and FcγRIII. By contrast, both sites at the Cγ2-Cγ3 domain interface of IgG are accessible for interaction with FcRn in the crystals of FcRn bound to Fc; thus bending of IgG does not occlude either of the sites in this region of Fc at the high protein concentrations found in the crystals.

Another common feature of some of the adhesion molecule-like FcRs is the use of the membrane-proximal domain (domain 2 of FcγRII, FcγRIII, and FcϵRI; domain 3 of FcγRI) as the primary Ig-binding site, with possible contributions from the membrane distal domain(s). By contrast, pIgR binds to IgA primarily with its N-terminal domain, which is more typical of the interactions of adhesion molecules such as CD2, CD4, and ICAM-1 with their ligands. Because FcRn is so different structurally from the other FcRs, its interactions with IgG bear no resemblance to the Ig interactions of either pIgR or the other FcRs. The IgG-binding site is on the edge of the $\alpha 1$ and $\alpha 2$ domain platform, with contributions from the $\beta 2$m subunit. This mode of interaction also does not resemble the ways that MHC proteins interact with any other molecules, including peptides, T-cell receptors and coreceptors, or superantigens.

Finally, in what appears to be the only common feature of Ig interaction between FcRn and the other Ig superfamily FcRs, a lying-down orientation was proposed for cell surface FcϵRI and FcRn. The orientations of the remaining receptors on the membrane are unknown, but the similarities in sequence and domain organization between FcϵRI and the FcγRs suggests this orientation may be relevant for the other receptors as well.

A more complete understanding of FcR-Ig interaction awaits high-resolution crystal structures of the adhesion molecule-like receptors in complex with their Fc ligands. In the absence of these structures, models of the FcγR and FcϵRI domains were constructed using the available Ig-like domain structures from cell adhesion molecules. Our analysis of the sequences of the FcγR and FcϵRI domains suggests that they have many features in common with V-like domains, although they had been classified as members of the C2 set. V-like and C2 domains differ primarily in the region between the C and E strands, which includes a region mapped as an Ig interaction site for some of these receptors. Further studies are required to resolve this issue, which promise to be of great interest owing to the recent advances in the understanding of the physiological functions of these receptors.

ACKNOWLEDGMENTS

We thank Dan Vaughn and Bob Turring for help with figures, Neil Simister and Sally Ward for communicating results prior to publication and for helpful discussions, Sherie Morrison and Dan Vaughn for helpful discussions, and Dan Vaughn and Luis Sanchez for critical reading of the manuscript. MR was supported by a fellowship from the Cancer Research Institute. PJB is supported by the Howard Hughes Medical Institute and a Camille and Henry Dreyfuss Teacher Scholar Award. Ribbon diagrams were prepared using MOLSCRIPT (Kraulis 1991).

Literature Cited

Abrahamson DR, Powers A, Rodewald R. 1979. Intestinal absorption of immune complexes by neonatal rats: a route of antigen transfer from mother to young. *Science* 206:567–69

Ahouse JJ, Hagerman CL, Mittal P, Gilbert DJ, Copeland NG, et al. 1993. Mouse MHC class-I like Fc receptor encoded outside the MHC. *J. Immunol.* 151:6076–88

Allen JM, Seed B. 1989. Isolation and expression of functional high affinity Fc receptor cDNAs. *Science* 243:378–80

Amigorena S, Bonnerot C, Drake JR, Choquet D, Hunziker W, et al. 1992a. Cytoplasmic domain heterogeneity and functions of IgG Fc receptors in B lymphocytes. *Science* 256:1808–12

Amigorena S, Salamero J, Davoust J, Fridman WH, Bonnerot C. 1992b. Tyrosine-containing motif that transduces cell activation signals also determines internalization and antigen presentation via type III receptors for IgG. *Nature* 358:337–41

Anderson CL, Shen L, Eicher DM, Wewers MD, Gill JK. 1990. Phagocytosis mediated by three distinct Fcγ receptor classes on human leukocytes. *J. Exp. Med.* 171:1333–45

Anderson P, Caligiuri M, O'Brien C, Manley T, Ritz J, Schlossman SF. 1990. Fcγ receptor type III (CD16) is included in the ζNK receptor complex expressed by human nat-

ural killer cells. *Proc. Natl. Acad. Sci. USA* 87:2274–78
Aroeti B, Mostov KE. 1994. Polarized sorting of the polymeric immunoglobulin receptor in the exocytic and endocytic pathways is controlled by the same amino acids. *EMBO J.* 13:2297–304
Baird B, Zheng Y, Holokwa D. 1993. Structural mapping of IgE-FcϵRI, an immunoreceptor complex. *Acc. Chem. Res.* 26:428–34
Bakos M-A, Kurosky A, Goldblum RM. 1991. Characterization of a critical binding site for human polymeric Ig on secretory component. *J. Immunol.* 147:3419–26
Beaven MA, Metzger H. 1993. Signal transduction by Fc receptors: the FcϵRI case. *Immunol. Today* 14:222–26
Beavil AJ, Beavil RL, Chan CMW, Cook JPD, Gould HJ, et al. 1993. Structural basis of the IgE-FcϵRI interaction. *Biochem. Soc. Trans.* 21:968–72
Bjorkman PJ, Parham P. 1990. Structure, function and diversity of class I major histocompatibility complex molecules. *Annu. Rev. Biochem.* 90:253–88
Blank U, Ra C, Miller L, Metzger H, Kinet JP. 1989. Complete structure and expression in transfected cells of high affinity IgE receptor. *Nature* 337:187–89
Blumberg RS, Koss T, Story CM, Barisani D, Polischuk J, et al. 1995. A major histocompatibility complex class I-related Fc receptor for IgG on rat hepatocytes. *J. Clin. Invest.* 95:2397–402
Bodian DL, Jones EY, Harlos K, Stuart DI, Davis SJ. 1994. Crystal structure of the extracellular region of the human cell adhesion molecule CD2 at 2.5 Å resolution. *Structure* 2:755–66
Brady RL, Dodson EJ, Dodson GG, Lange G, Davis SJ, et al. 1993. Crystal structure of domains 3 and 4 of rat CD4: relation to the NH_2-terminal domains. *Science* 260:979–83
Brandtzaeg P. 1981. Transport models for secretory IgA and secretory IgM. *Clin. Exp. Immunol.* 44:221–32
Brandtzaeg P, Pyrdz H. 1984. Direct evidence for an integrated function of J chain and secretory component in epithelial transport of immunoglobulins. *Nature* 311:71–73
Brooks DG, Qiu WQ, Luster AD, Ravetch JV. 1989. Structure and expression of human IgG FcγRII(CD32). *J. Exp. Med.* 170:1369–85
Burmeister WP, Gastinel LN, Simister NE, Blum ML, Bjorkman PJ. 1994a. The 2.2 Å crystal structure of the MHC-related neonatal Fc receptor. *Nature* 372:336–43
Burmeister WP, Huber AH, Bjorkman PJ. 1994b. Crystal structure of the complex between the rat neonatal Fc receptor and Fc. *Nature* 372:379–83
Burton DR, Woof JM. 1992. Human antibody effector functions. *Adv. Immunol.* 51:1–85
Cambier JC. 1992. Signal transduction by T- and B-cell antigen receptors: converging structures and concepts. *Curr. Opin. Immunol.* 4:257–64
Cambier JC. 1995. Antigen and FcR signaling—the awesome power of the immunoreceptor tyrosine-based activation motif (ITAM). *J. Immunol.* 155:3281–85
Canfield SM, Morrison SL. 1991. The binding affinity of human IgG for its high affinity Fc receptor is determined by multiple amino acids in the CH2 domain and is modulated by the hinge region. *J. Exp. Med.* 173:1483–91
Casanova JE, Apodaca G, Mostov KE. 1991. An autonomous signal for basolateral sorting in the cytoplasmic domain of the polymeric immunoglobulin receptor. *Cell* 66:65–75
Childers NK, Bruce MG, McGhee JR. 1989. Molecular mechanisms of immunoglobulin A defense. *Annu. Rev. Microbiol.* 43:503–36
Coyne RS, Siebrecht M, Peitsch MC, Casanova JE. 1994. Mutational analysis of polymeric immunoglobulin receptor/ligand interactions. *J. Biol. Chem.* 269:31620–25
Daeron M. 1995. Regulation of high affinity IgE receptor-mediated mast cell activation by murine low affinity IgG receptors. *J. Clin. Invest.* 95:577–85
Daeron M, Malbec O, Bonnerot C, Latour S, Segal DM, Fridman W. 1994. Tyrosine-containing activation motif-dependent phagocytosis in mast cells. *J. Immunol.* 152:783–92
Davies DR, Metzger H. 1983. Structural basis of antibody function. *Annu. Rev. Immunol.* 1:87–117
Davis W, Harrison PT, Hutchinson MJ, Allen JM. 1995. Two distinct regions of FcγRI initiate separate signaling pathways involved in endocytosis and phagocytosis. *EMBO J.* 14:432–41
Deisenhofer J. 1981. Crystallographic refinement and atomic models of a human Fc fragment and its complex with fragment B of protein A from *Staphylococcus aureus* at 2.9- and 2.8-Å resolution. *Biochemistry* 20:2361–70
Dombrowicz D, Flamand V, Brigman KK, Koller BH, Kinet J-P. 1994. Abolition of anaphylaxis by targeted disruption of the high affinity immunoglobulin E receptor α chain gene. *Cell* 75:969–76
Duncan AR, Woof JM, Partridge LJ, Burton DR, Winter G. 1988. Localization of the binding site for the human high-affinity Fc receptor on IgG. *Nature* 332:563–64
Edberg JC, Kimberly RP. 1994. Modulation of

Fcγ and complement receptor function by the glycosyl-phosphatidylinositol-anchored form of FcγRIII. *J. Immunol.* 1994:5826–35

Edberg JC, Salmon JE, Kimberly RP. 1992. Functional capacity of Fcγ receptor III (CD16) on human neutrophils. *Immunol. Res.* 11:239–51

Eiffert H, Quentin E, Decker J, Hillemeir S, Hufschmidt M, et al. 1984. Die Primärstruktur der menschlichen freien Sekretkomponente und die Anordnung der Disulfidbrucken. *Hoppe-Seyler's Z. Physiol. Chem.* 365:1489

Ernst LK, Duchemin A-M, Anderson CL. 1993. Association of the high affinity receptor for IgG (FcγRI) with the γ subunit of the IgE receptor. *Proc. Natl. Acad. Sci. USA* 90:6023–27

Ernst LK, van de Winkel JGJ, Chiu I, Anderson CL. 1992. Three genes for the human high affinity Fc receptor encode four different transcription products. *J. Biol. Chem.* 267:15692–700

Fallgreen-Gebauer E, Gebauer W, Bastian A, Kratzin HD, Eiffert H, et al. 1993. The covalent linkage of secretory component to IgA-structure of IgA. *Hoppe-Seyler's Z. Biol. Chem.* 374:1023–28

Fridman W. 1991. Fc receptors and immunoglobulin binding factors. *FASEB J.* 5:2684–90

Fridman WH, Bonnerot C, Daeron M, Amigorena S, Teillaud J-L, Sautes C. 1992. Structural bases of Fcγ receptor function. *Immunol. Rev.* 125:49–76

Frutiger S, Hughes GJ, Hanly WC, Kingzette M, Jaton JC. 1986. The amino-terminal domain of rabbit secretory component is responsible for noncovalent binding to immunoglobulin A dimers. *J. Biol. Chem.* 261:16673–81

Galon J, Bouchard C, Fridman WH, Sautes C. 1995. Ligands and biological activities of soluble Fcγ receptors. *Immunol. Lett.* 44:175–81

Gastinel LN, Simister NE, Bjorkman PJ. 1992. Expression and crystallization of a soluble and functional form of an Fc receptor related to class I histocompatibility molecules. *Proc. Natl. Acad. Sci. USA* 89:638–42

Geneste C, Iscaki S, Mangalo R, Pillot J. 1986. Both Fcα domains of human IgA are involved in in vitro interaction between secretory component and dimeric IgA. *Immunol. Lett.* 13:221–26

Gergely J, Sarmay G. 1990. The two binding site models of human IgG binding Fcγ receptors. *FASEB J.* 4:3275–83

Ghetie V, Hubbard JG, Kim J-K, Lee Y, Ward ES. 1996. Abnormally short serum half-lives of IgG in β2-microglobulin-deficient mice. *Eur. J. Immunol.* 26:690–96

Ghirlando R, Keown MB, Mackay GA, Lewis MS, Unkeless MS, Gould HJ. 1995. Stoichiometry and thermodynamics of the interaction between the Fc fragment of human IgG1 and its low-affinity receptor FcγRIII. *Biochemistry* 34:13320–27

Goto Y, Aki K. 1984. Interaction of the fluorescence-labeled secretory component with human polymeric immunoglobulins. *Biochemistry* 23:6736–44

Hanson A, Brandtzaeg P. 1993. The discovery of secretory IgA and the mucosal immune system. *Immunol. Today* 14:416–17

Harpaz Y, Chothia C. 1994. Many of the immunoglobulin superfamily domains in cell adhesion molecules and surface receptors belong to a new structural set which is close to that containing variable domains. *J. Mol. Biol.* 238:528–39

Harris LJ, Larson SB, Hasel KW, Day J, Greenwood A, McPherson A. 1992. The three-dimensional structure of an intact monoclonal antibody for canine lymphoma. *Nature* 360:369–72

Harrison PT, Davis W, Norman JC, Hockaday AR, Allen JA. 1994. Binding of monomeric immunoglobulin G triggers FcγRI-mediated endocytosis. *J. Biol. Chem.* 269:24396–402

Heldin C-H. 1995. Dimerization of cell surface receptors in signal transduction. *Cell* 80:213–23

Helm B, Marsh P, Vercelli D, Padlin E, Gould H, Geha R. 1988. The mast cell binding site on human immunoglobulin E. *Nature* 331:180–83

Hendrickson BA, Conner DA, Ladd DJ. 1995. Altered hepatic transport of immunoglobulin A in mice lacking the J chain. *J. Exp. Med.* 182:1905–11

Hibbs ML, Selvaraj P, Carpen O, Springer TA, Kuster H, et al. 1989. Mechanisms for regulating expression of membrane isoforms of FcγRIII (CD16). *Science* 246:1608–11

Hibbs ML, Tolvanen M, Carpen O. 1994. Membrane-proximal Ig-like domain of FcγRIII (CD16) contains residues critical for ligand binding. *J. Immunol.* 152:4466–74

Holden HM, Ito M, Hartshorne DJ, Rayment I. 1992. X-ray structure determination of telokin, the C-terminal domain of myosin light chain kinase, at 2.8 Å resolution. *J. Mol. Biol.* 277:840–51

Huber AH, Kelley RF, Gastinel LN, Bjorkman PJ. 1993. Crystallization and stoichiometry of binding of a complex between a rat intestinal Fc receptor and Fc. *J. Mol. Biol.* 230:1077–83

Hulett MD, Hogarth PM. 1994. Molecular basis

of Fc receptor function. *Adv. Immunol.* 57:1–127
Hulett MD, Osman N, McKenzie IFC, Hogarth PM. 1991. Chimeric Fc receptors identify functional domains of murine FcγRI. *J. Immunol.* 142:1863–68
Hulett MD, Witort E, Brinkworth RI, McKenzie IFC, Hogarth PM. 1995. Multiple regions of human FcγRII (CD32) contribute to the binding of IgG. *J. Biol. Chem.* 270:21188–94
Janeway CA, Travers P. 1994. The immune system in health and disease. In *Current Biology* 8:33–34. London: Current Biology/New York: Garland
Jefferis R, Lund JJP. 1990. Molecular definition of interaction sites on human IgG for Fc receptors (huFcγR). *Mol. Immunol.* 27:1237–40
Jones EA, Waldman TA. 1972. The mechanism of intestinal uptake and transcellular transport of IgG in the neonatal rat. *J. Clin. Invest.* 51:2916–27
Jones EY, Davis SJ, Williams AF, Harlos K, Stuart DI. 1992. Crystal structure at 2.8 Å resolution of a soluble form of the cell adhesion molecule CD2. *Nature* 360:232–39
Jones EY, Harlos K, Bottomley MJ, Robinson RC, Driscoll PC, et al. 1995. Crystal structure of an integrin-binding fragment of vascular cell adhesion molecule-1 at 1.8 Å resolution. *Nature* 373:539–44
Junghans RP, Anderson CL. 1996. The protection receptor for IgG catabolism is the β2-microglobulin-containing neonatal intestinal transport receptor. *Proc. Natl. Acad. Sci. USA* 93:5512–16
Keegan A, Paul WE. 1992. Multichain immune recognition receptors: similarities in structure and signaling pathways. *Immunol. Today* 13:63–68
Kerr MA. 1990. The structure and function of human IgA. *Biochem. J.* 271:285–96
Kikutani H, Inui S, Sato R, Barsumain EL, Owaki H, et al. 1986. Molecular structure of human lymphocyte receptor for immunoglobulin E. *Cell* 47:657–65
Kim J-K, Tsen M-F, Ghetie V, Ward ES. 1994. Localization of the site of the murine IgG1 molecule that is involved in binding to the murine intestinal Fc receptor. *Eur. J. Immunol.* 24:2429–34
Kinet JP, Blank U, Ra C, Metzger H, Kochan J. 1988. Isolation and characterization of cDNAs coding for the beta subunit of the high affinity receptor for immunoglobulin E. *Proc. Natl. Acad. Sci. USA* 85:6483–87
Kinet JP, Metzger H, Hakimi J, Kochan J. 1987. A cDNA presumptively coding for the alpha subunit of the receptor with high affinity for immunoglobulin E. *Biochemistry* 26:4605–10
Knight KL, Vetter ML, Malek TR. 1975. Distribution of covalently-bound and non-covalently bound secretory component on subclasses of rabbit secretory IgA. *J. Immunol.* 115:595–98
Kochan J, Pettine LF, Hakimi J, Kishi K, Kinet JP. 1988. Isolation of the gene coding for the α subunit of the human high affinity IgE receptor. *Nucleic Acid Res.* 16:3584
Koolwijk P, Spierenburg GT, Frasa H, Boot JHA, van de Winkel JGJ, Bast BJEG. 1989. Interaction between hybrid mouse monoclonal antibodies and the human high-affinity IgG FcR, huFcγRI, on U937: involvement of only one of the mIgG heavy chains in receptor binding. *J. Immunol.* 143:1656–62
Kraehenbuhl J-P, Neutra MR. 1992. Transepithelial transport and mucosal defense II: secretion of IgA. *Trends Cell Biol.* 2:134–38
Kraulis PJ. 1991. MOLSCRIPT: a program to produce both detailed and schematic plots of protein structures. *J. Appl. Crystallogr.* 24:946–50
Kühn L, Kraehenbuhl J-P. 1979. Interaction of rabbit secretory component with rabbit IgA dimer. *J. Biol. Chem.* 254:11066–71
Kurosaki T, Ravetch JV. 1989. A single amino acid in glycosyl phosphatidylinositol attachment domain determines the membrane topology of FcγRIII. *Nature* 326:292–95
Lanier LL, Yu G, Phillips JH. 1989. Co-association of CD3ζ with a receptor (CD16) for IgG Fc on human natural killer cells. *Nature* 342:803–5
Leahy DJ, Axel R, Hendrickson WA. 1992. Crystal structure of a soluble form of the human T cell coreceptor CD8 at 2.6 Å resolution. *Cell* 68:1145–62
Lin C, Shen Z, Boros P, Unkeless JC. 1994. Fc receptor mediated signal transduction. *J. Clin. Immunol.* 14:1–13
Lund J, Winter G, Jones PT, Pound JD, Tanaka T, et al. 1991. Human FcγRI and FcγRII interact with distinct but overlapping sites on human IgG. *J. Immunol.* 147:2657–62
Maliszewski CR. 1990. Expression cloning of a human Fc receptor for IgA. *J. Exp. Med.* 172:1665–72
Mallamaci MA, Chizzonite R, Griffin M, Nettleton M, Hakimi J, et al. 1993. Identification of sites on the human FcϵR1α subunit which are involved in binding human and rat IgE. *J. Biol. Chem.* 268:22076–83
Masuda M, Roos D. 1993. Association of all three types of FcγR (CD64, CD32 and CD16) with a γ chain homodimer in cultured human monocytes. *J. Immunol.* 151:6382–88
Mellman I, Plutner H. 1984. Internalization

and degradation of macrophage Fc receptors bound to polyvalent immune complexes. *J. Cell Biol.* 98:1170–77

Mestecky J, McGhee JR. 1987. Immunoglobulin A (IgA): molecular and cellular interactions involved in IgA biosynthesis and immune response. *Adv. Immunol.* 40:153–245

Metzger H. 1991. The high affinity receptor for IgE on mast cells. *Clin. Exp. Allergy* 21:269–79

Miettinen HM, Rose JK, Mellman I. 1989. Fc receptor isoforms exhibit distinct abilities for coated pit localization as a result of cytoplasmic domain heterogeneity. *Cell* 58:317–27

Morgan A, Jones ND, Nesbitt AM, Chaplin L, Bodmer MW, Emtage JS. 1995. The N-terminal end of the CH2 domain of chimeric human IgG1 and anti-HLA-DR is necessary for C1q, FcγRI and FcγRIII binding. *Immunology* 86:319–24

Mostov KE. 1994. Transepithelial transport of imunoglobulins. *Annu. Rev. Immunol.* 12:63–84

Mostov KE, Friedlander M, Blobel G. 1984. The receptor for trans-epithelial transport of IgA and IgM contains multiple immunoglobulin-like domains. *Nature* 308:37–43

Mostov KE, Simister NE. 1985. Transcytosis. *Cell* 43:389–90

Muta T, Kurosaki T, Misulovin Z, Sanchez M, Nussenzweig MC, Ravetch JV. 1994. A 13-amino acid motif in the cytoplasmic domain of FcγRIIB modulates B-cell signalling. *Nature* 368:70–73

Nagarajan S, Chesla S, Cobern L, Anderson P, Zhu C, Selvaraj P. 1995. Ligand binding and phagocytosis by CD16 (Fcγ receptor III) isoforms. *J. Biol. Chem.* 270:25762–70

Nissim A, Schwarzbaum S, Siraganian R, Eshhar Z. 1993. Structural mapping of IgE-FcϵRI, an immunoreceptor complex. *J. Immunol.* 150:428–34

O'Grady JH, Looney RJ, Anderson CL. 1986. The valence for ligand of the human mononuclear phagocyte 72 kD high-affinity IgG Fc receptor is one. *J. Immunol.* 137:2307–10

Ohno T, Kubagawa H, Sanders SK, Cooper MD. 1990. Biochemical nature of an Fcμ receptor on human B-lineage cells. *J. Exp. Med.* 172:1165–75

Perez-Montfort R, Kinet JP, Metzger H. 1983. A previously unrecognized subunit of the receptor for immunoglobulin E. *Biochemistry* 22:5722–28

Popov S, Hubbard JG, Kim J-K, Ober B, Ghetie V, Ward ES. 1996. The stoichiometry and affinity of murine Fc fragments with the MHC class I-related receptor, FcRn. *Mol. Immunol.* In press

Qiu WQ, De Bruin D, Brownstein BH, Pearse R, Ravetch JV. 1990. Organization of the human and mouse low affinity FcγR genes: duplication and recombination. *Science* 248:732–35

Ra C, Jouvin M-H, Blank U, Kinet JP. 1989. A macrophage Fcγ receptor and the mast cell receptor for IgE share an identical subunit. *Nature* 341:752–54

Raghavan M, Bonagura VR, Morrison SL, Bjorkman PJ. 1995a. Analysis of the pH dependence of the neonatal Fc receptor/immunoglobulin G interaction using antibody and receptor variants. *Biochemistry* 34:14649–57

Raghavan M, Chen MY, Gastinel LN, Bjorkman PJ. 1994. Investigation of the interaction between the class I MHC-related Fc receptor and its immunoglobulin G ligand. *Immunity* 1:303–15

Raghavan M, Gastinel LN, Bjorkman PJ. 1993. The Class I MHC-related Fc receptor shows pH-dependent stability differences correlating with immunoglobulin binding and release. *Biochemistry* 32:8654–60

Raghavan M, Wang Y, Bjorkman PJ. 1995b. Effects of receptor dimerization on the interaction between the class I MHC related Fc receptor and immunoglobulin G. *Proc. Natl. Acad. Sci. USA* 92:11200–4

Randall TD, Brewer JW, Corley RB. 1992. Direct evidence that J-chain regulates the polymeric structure of IgM in antibody-secreting B-cells. *J. Biol. Chem.* 267:18002–7

Ravetch JV. 1994. Fc receptors: rubor redux. *Cell* 78:553–60

Ravetch JV, Kinet J-P. 1991. Fc receptors. *Annu. Rev. Immunol.* 9:457–92

Ravetch JV, Luster AD, Weinshank R, Kochan J, Pavlovec A, et al. 1986. Structural heterogeneity and functional domains of murine immunoglobulin G Fc receptors. *Science* 234:718–25

Ravetch JV, Perussia B. 1989. Alternative membrane forms of FcγRIII(CD16) on human natural killer cells and neutrophils. *J. Exp. Med.* 170:481–97

Reth M. 1989. Antigen receptor tail clue. *Nature* 338:383–84

Roberts DM, Guenthert M, Rodewald R. 1990. Isolation and characterization of the Fc receptor from fetal yolk sac of the rat. *J. Cell Biol.* 111:1867–76

Robertson MW. 1993. Phage and *Escherichia coli* expression of the human high affinity immunoglobulin E receptor α-subunit ectodomain. *J. Biol. Chem.* 268:12736–43

Rodewald R. 1973. Intestinal transport of antibodies in the newborn rat. *J. Cell Biol.* 58:189–211

Rodewald R, Kraehenbuhl J-P. 1984. Receptor-mediated transport of IgG. *J. Cell Biol.* 99:159-64s

Ryu S-E, Kwong PD, Truneh A, Porter TG, Arthos J, et al. 1990. Crystal structure of an HIV-binding recombinant fragment of human CD4. *Nature* 348:419–26

Samelson LE, Klausner RD. 1992. Tyrosine kinases and tyrosine-based activation motifs. Current research on activation via the T cell antigen receptor. *J. Biol. Chem.* 267:24913–16

Scholl PR, Geha RS. 1993. Physical association between the high affinity IgG receptor (FcγRI) and the gamma subunit of the high affinity IgE receptor (FcϵRI γ). *Proc. Natl. Acad. Sci. USA* 90:8847–50

Sears DW, Osman N, Tate B, McKenzie IFC, Hogarth PM. 1990. Molecular cloning and expression of the mouse high affinity receptor for IgG. *J. Immunol.* 144:371–78

Shen L, Guyre PM, Fanger MW. 1987. Polymorphonuclear leukocyte functions triggered through the high affinity Fc receptor for monomeric IgG. *J. Immunol.* 139:534–38

Shopes B, Weetall M, Holowka D, Baird B. 1990. Recombinant human IgG1-murine IgE chimeric Ig. Construction, expression and binding to Fcγ receptors. *J. Immunol.* 145:3842–48

Silverstein SC, Greenberg S, DiVirgillo F, Steinberg TH. 1989. Phagocytosis. In *Fundamental Immunology,* ed. WE Paul, 703–720. New York: Raven

Silverstein SC, Steinman RM, Cohn ZA. 1977. Endocytosis. *Annu. Rev. Biochem.* 46:669–722

Simister NE, Mostov KE. 1989a. Cloning and expression of the neonatal rat intestinal Fc receptor, a major histocompatibility complex class I antigen homolog. *Cold Spring Harbor Symp. Quant. Biol.* LIV:571–80

Simister NE, Mostov KE. 1989b. An Fc receptor structurally related to MHC class I antigens. *Nature* 337:184–87

Simister NE, Rees AR. 1985. Isolation and characterization of an Fc receptor from neonatal rat small intestine. *Eur. J. Immunol.* 15:733–38

Simister NE, Story CM. 1996. Fcγ receptors in human placenta. In *Human IgG Fc Receptors,* ed. JGJ van de Winkel, PJA Capel, pp. 25–38. Landes

Sjoberg O. 1980. Presence of receptors for IgD on human T and non-T lymphocytes. *Scand. J. Immunol.* 11:377–82

Song WX, Bomsel M, Casanova J, Vareman JP, Mostov K. 1994. Stimulation of transcytosis of the polymeric immunoglobulin receptor by dimeric IgA. *Proc. Natl. Acad. Sci. USA* 91:163–66

Story CM, Mikulska JE, Simister NE. 1994. MHC class I-like Fc receptor cloned from human placenta. *J. Exp. Med.* 180:2377–81

Stuart S, Simister NE, Clarkson SB, Kacinski BK, Shapiro M, Mellman I. 1989. Human IgG Fc receptor (hFcγRII; CD32) exists as multiple isoforms in macrophages, lymphocytes and IgG-transporting placental epithelium. *EMBO J.* 8:3657–66

Sutton BJ, Gould HJ. 1993. The human IgE network. *Nature* 366:421–28

Switzer IC, Loney GM, Yang DS, Underdown BJ. 1992. Binding of secretory component to protein 511, a pIgA mouse protein lacking 36 amino acid residues of the Cα3 domain. *Mol. Immunol.* 29:31–35

Takai T, Li M, Sylvestre D, Clynes R, Ravetch JV. 1994. FcRγ deletion results in pleiotrophic effector cell defects. *Cell* 76:519–29

Takai T, Ono M, Hikida M, Ohmori H, Ravetch JV. 1996. Augmented humoral and anaphylactic responses in FcγRII-deficient mice. *Nature* 379:346–49

Townsend A, Bodmer H. 1989. Antigen recognition by class I-restricted T lymphocytes. *Annu. Rev. Immunol.* 7:601–24

Underdown BJ, Dorrington KJ. 1974. Studies on the structural and conformational basis for the relative resistance of serum and secretory IgA to proteolysis. *J. Immunol.* 112:949–59

Underdown BJ, Schiff JM. 1986. Immunoglobulin A: strategic defense at the mucosal surface. *Annu. Rev. Immunol.* 4:389–417

Unkeless JC, Scigliano E, Freedman JH. 1988. Structure and function of human and murine receptors for IgG. *Annu. Rev. Immunol.* 6:251–81

van de Winkel JGJ, Anderson CL. 1991. Biology of human immunoglobulin G Fc receptors. *J. Leukocyte Biol.* 49:511–24

Vaughn DE, Bjorkman PJ. 1996. The (Greek) key to structures of neural adhesion molecules. *Neuron* 16:261–73

Wagner G, Wyss DF. 1994. Cell surface adhesion receptors. *Curr. Opin. Struct. Biol.* 4:841–51

Walker MR, Lund J, Thompson KM, Jefferies R. 1989. A glycosylation of human IgG1 and IgG3 monoclonal antibodies can eliminate recognition by human cells expressing FcγRI and/or FcγRII receptors. *Biochem. J.* 259:1347–53

Wallace KH, Rees AR. 1980. Studies on the immunoglobulin G Fc fragment receptor from neonatal rat small intestine. *Biochem. J.* 188:9–16

Wang J, Yan Y, Garrett TPJ, Liu J, Rodgers DW, et al. 1990. Atomic structure of a fragment of human CD4 containing two immunoglobulin-like domains. *Nature* 348:411–19

Weetall M, Shopes B, Holowka D, Baird B. 1990. Mapping the site of interaction between murine IgE and its high affinity receptor with chimeric Ig. *J. Immunol.* 145:3849–54

Weiss A, Littman DR. 1994. Signal transduction by lymphocyte antigen receptors. *Cell* 76:263–74

Weissman AM, Baniyash M, Hou D, Samuelson LE, Burgess WH, Klausner RD. 1989. Molecular cloning of the ζ chain of the T cell antigen receptor. *Science* 239:1018–21

Williams AF, Barclay AN. 1988. The immunoglobulin superfamily-domains for cell surface recognition. *Annu. Rev. Immunol.* 6:381–405

Wright A, Morrison SL. 1994. Effect of altered CH2-associated carbohydrate structure on the functional properties and in vivo fate of chimeric mouse-human immunoglobulin G1. *J. Exp. Med.* 180:1087–96

Annu. Rev. Cell Dev. Biol. 1996. 12:221–55

CROSS-TALK BETWEEN BACTERIAL PATHOGENS AND THEIR HOST CELLS

Jorge E. Galán and James B. Bliska
Department of Molecular Genetics and Microbiology, State University of New York at Stony Brook, Stony Brook, New York 11794-5222

KEY WORDS: bacterial pathogenicity, signal transduction, protein secretion

ABSTRACT

A taxonomically diverse group of bacterial pathogens have evolved a variety of strategies to subvert host-cellular functions to their advantage. This often involves two-way biochemical interactions leading to responses in both the pathogen and host cell. Central to this interaction is the function of a specialized protein secretion system that directs the export and/or translocation into the host cells of a number of bacterial proteins that can induce or interfere with host-cell signal transduction pathways. The understanding of these bacterial/host-cell interactions will not only lead to novel therapeutic approaches but will also result in a better understanding of a variety of basic aspects of cell physiology and immunology.

CONTENTS

1081-0706/96/1115-00221$08.00

INTRODUCTION

Microbial pathogens have evolved unique ways to interact with their hosts. In many instances the terms of this interaction reflect the co-evolutionary balance that the host and pathogen must reach in order to secure their survival. It is therefore not surprising that bacterial pathogens have evolved a great array of virulence factors well-suited to interfere with or stimulate a variety of host-cell physiological responses (Bliska et al 1993b). The identification and characterization of these virulence factors is proving to be a fruitful area of research in more ways than expected. Understanding how pathogens interact with their hosts is providing the basis for the development of novel therapeutic approaches, as well as a number of very sophisticated tools for probing basic aspects of cellular physiology and immunology.

The interaction of bacterial pathogens with host cells is by no means unidirectional. Indeed, it has become clear that bacterial pathogens and host cells engage in a true biochemical cross-talk that often leads to overt responses from both parties. At the center of the two-way interaction between a distinct group of pathogenic bacteria and their host cells lies a dedicated protein secretion apparatus designed to export to and/or deliver into the host cell a variety of bacterial effector molecules (reviewed in Salmond & Reeves 1993, Van Gijsegem et al 1993). Remarkably, this secretion apparatus is found not only in animal pathogens such as *Salmonella* spp, *Shigella* spp, *Yersinia* spp, and *Escherichia coli* spp, but also in plant pathogenic bacteria such as *Pseudomonas* spp, *Erwinia* spp, and *Xanthomonas* spp (Van Gijsegem et al 1993). Herein we describe the most salient features of this protein secretion system, as well as the host-cell responses induced by this group of bacterial pathogens.

HOST-CELL CONTACT-DEPENDENT PROTEIN SECRETION BY BACTERIAL PATHOGENS

Gram-positive and gram-negative bacteria secrete to the extracellular environment numerous proteins including a variety of degradative enzymes required for the metabolism of high-molecular weight compounds, as well as a wide array of virulence factors and toxins. Secretion of proteins by gram-positive organisms, whose cell wall possesses only one membrane, resembles the targeting of proteins through the endoplasmic reticulum of eukaryotic cells. This process usually involves an amino-terminal signal sequence that is cleaved upon translocation, and a variety of cytoplasmic chaperones (Simonen & Palva 1993). The well-characterized *sec*-dependent export apparatus can only direct the export of proteins through the inner membrane. In gram-negative bacteria, however, proteins secreted to the extracellular environment have to traverse two membranes. Therefore, these microorganisms have evolved additional strategies to secrete proteins to the extracellular milieu (reviewed in Pugsley 1993, Salmond & Reeves 1993). The simplest mechanism of all is that employed by *Neisseria gonorrheae* for the export of the IgA protease, which has within its amino acid sequence all the information required for secretion and therefore can exit the bacterial cell without further help from accessory proteins (Klauser et al 1993). Other mechanisms involve the use of the *sec*-dependent pathway to export the target proteins through the inner membrane and additional accessory proteins to direct the secretion through the outer membrane. In general, the exported proteins in this category are said to be secreted via a Type II secretion system, also known as the general secretory pathway, and are best exemplified by the *Klebsiella oxytoca* pullulanase enzyme (Pugsley et al 1991). Yet other target proteins are exported independently of the *sec* machinery and require the function of accessory proteins for secretion. Among these proteins are several toxins and/or extracellular products that are secreted through what is known as the Type I secretion pathway (Holland et al 1990). These proteins, best exemplified by the *E. coli* hemolysin, possess a signal sequence located at the carboxy terminus and require the function of at least three accessory proteins for their secretion (Holland et al 1990). The secretion through the two bacterial membranes is thought to occur in one step.

Recently, a novel protein secretion pathway apparently dedicated to functions related to the interaction of pathogenic bacteria with host cells has been identified in a variety of gram-negative bacteria that are pathogenic for animals and plants (Salmond & Reeves 1993, Van Gijsegem et al 1993). This secretion system also has some similarity to the flagellar assembly apparatus (Dreyfus et al 1993). Characteristic features of this protein secretion system, termed Type III or contact-dependent, are (*a*) the absence of a typical, cleavable, *sec*-dependent signal sequence in the secreted targets; (*b*) the requirement of a large number of accessory proteins for the export process; and (*c*) the export of the target proteins through both the inner and outer membranes.

Perhaps the most striking feature of this protein secretion system is the requirement of activating extracellular signals for its full function. As discussed below, it is now generally accepted that the biologically relevant activating signals result from the interaction between this group of bacteria with their host cells.

Structural Components of the Type III Secretion Systems

Molecular genetic analyses of the Type III secretion systems in various microorganisms have revealed that the structural components of this protein translocon are usually encoded either in extrachromosomal elements, such as large virulence-associated plasmids in *Shigella, Yersinia,* and *Pseudomonas solanacearum,* or in discrete chromosomal regions known as pathogenicity islands, such as those in *Salmonella* and enteropathogenic *E. coli* (EPEC). These regions are thought to have been acquired as a block by horizontal gene transfer because their G + C content is significantly different from that of the genomes of the host microorganisms. This suggests a common ancestral origin for all these homologous systems.

The general organization of the Type III secretion apparatus in the different organisms appears to be very similar. A group of nine proteins, which constitute the core components of this apparatus, are highly conserved in all microorganisms (see Table 1). In some instances, functional complementation by homologous genes from heterologous bacteria has been observed (Groisman & Ochman 1993, Ginocchio & Galán 1995). Furthermore, the secretion of heterologous targets has also been demonstrated (Rosqvist et al 1995, Hermant et al 1995); e.g. *Yersinia pseudotuberculosis* secreted the *Shigella* IpaB protein and *Salmonella typhimurium* translocated the *Y. pseudotuberculosis* toxin YopE into HeLa cells (Rosqvist et al 1995). Furthermore, *sipB* from *S. typhimurium* complemented an *ipaB* mutation in *Shigella*, implying that this organism could secrete the *S. typhimurium* SipB protein (Hermant et al 1995). YopE, SipB, and IpaB are targets of homologous Type III systems in the respective microorganisms. In contrast, some proteins associated with this system appear to be unique to a bacterial species or to either animal or phytopathogenic bacteria. This may reflect the adaptation of the system to secrete specific targets or to operate in different hosts.

Although the actual mechanisms by which the different target proteins are exported via this secretion pathway are poorly understood, studies of these systems in various microorganisms have identified most, if not all, the components of the secretion machinery (Andrews et al 1991, Michiels et al 1991, Plano et al 1991, Allaoui et al 1992, 1993a,b, 1994, Andrews & Maurelli 1992, Fenselau et al 1992, Galán et al 1992a, Ginocchio et al 1992, Gough et al 1992,

Table 1 Components of the Type III secretion systems with homologues in one or more bacterial pathogens[a]

Salmonella spp.	*Shigella* spp.	*Yersinia* spp.	*P. solanacearum*	*P. syringae*	*X. compestris*	Flagellar assembly	Localization[b]
InvA	MxiA	LcrD	HrpO	HrpI	HrpC2	FlhA	inner membrane
SpaP	Spa24	YscR	HrpT	HrpW	ORF2	FliP	inner membrane
SpaQ	Spa9	YscS	HrpU	HrpO	—	FliQ	inner membrane
SpaR	Spa29	YscT	HrpC	HrpX	HrpB8	FliR	inner membrane
SpaS	Spa40	YscU	HrpN	HrpY	—	FlhB	inner membrane
InvG	MxiD	YscC	HrpA	HrpH	HrpAl	—	outer membrane
PrgH	MxiG	—	—	—	—	—	outer membrane
PrgK	MxiJ	YscJ	HrpI	HrpC	HrpB3	FliF	outer membrane
InvE	MxiC	LcrE/YopN	—	—	—	—	unknown
InvB	Spal5	—	—	—	—	FliH	unknown
InvC	Spa47	YscN	HrpE	HrpJ4	HrpB6	FliI	unknown
SpaO	Spa33	YscQ	HrpQ	HrpU	—	FliN/FliY	extracellular
IagB	IpgF	—	—	—	—	—	unknown
PrgI	MxiH	YscF	—	—	—	—	unknown
PrgJ	MxiI	—	—	—	—	—	unknown
OrgA	MxiK	—	—	—	—	—	unknown
SicA	IpgC	SycD/LcrH	—	—	—	—	cytoplasmic
IacP	OrfX	—	—	—	—	—	unknown
—	—	YscL	HrpF	HrpE	HrpB5	—	unknown
—	—	—	HrpW	HrpJ3	—	—	unknown
—	—	YscI	—	HrpB	—	—	unknown
—	—	—	—	HrpJ5	—	—	unknown
—	—	—	HrpK	—	HrpB1	—	unknown
—	—	—	HrpJ	—	HrpB2	—	unknown
—	—	—	HrpH	—	HrpB4	—	unknown

[a]Searches were conducted at the NCBI server using the Blast program. Sequences were from Genbank release 92.0.
[b]Localization based on experimental data obtained in at least one member of the family.

1993, Haddix & Straley 1992, Huang et al 1992, 1993, 1995, Venkatesan et al 1992, Groisman & Ochman 1993, Sasakawa et al 1993, Van Gijsegem et al 1993, 1995, Wattiau & Cornelis 1993, Bergman et al 1994, Eichelberg et al 1994, Genin & Boucher 1994, Kaniga et al 1994, 1995a,b, Lidell & Hutcheson 1994, Wattiau et al 1994, Woestyn et al 1994, Collazo et al 1995, Pegues et al 1995, Plano & Straley 1995, Collazo & Galán 1996, Jarvis et al 1995). Protein sequence and biochemical analysis have assigned putative functions to some of these components. For example, several proteins are predicted to be localized in the inner membrane where they form a platform for the secretion targets to be engaged for translocation. One of these inner membrane proteins exemplified by *Yersinia* spp LcrD or *Salmonella* spp InvA, has a topology consistent with that of a polytopic membrane protein, with seven transmembrane domains located in the amino terminal half and a largely hydrophilic carboxy terminus likely located in the cytoplasm (Plano et al 1991, Andrews & Maurelli 1992, Fenselau et al 1992, Galán et al 1992a, Gough et al 1993, Huang et al 1993). This configuration, which is similar to that of the protein translocator SecY, suggests that this family of proteins serves as a channel for the secreted proteins to move through the inner membrane. However, this hypothesis has yet to be experimentally substantiated. In addition to the InvA/LcrD protein family, at least four other proteins, highly conserved among Type III systems in all species of bacteria, are predicted to be localized to the inner membrane (Venkatesan et al 1992, Groisman & Ochman 1993, Bergman et al 1994, Huang et al 1995, Van Gijsegem et al 1995).

At least three components of the secretion machinery either have been predicted or have been shown to be localized in the outer membrane where they may assist the translocation of the target proteins through this membrane. Two of these components, YscC of *Yersinia* and InvG of *Salmonella*, share sequence similarity with members of the PulD family of protein translocators (Michiels et al 1991, Allaoui et al 1993b, Genin & Boucher 1994, Kaniga et al 1994). Members of this protein family play a critical role in filamentous phage and type IV pilus assembly (Russel 1995) and in the translocation of proteins through the outer membrane in the Type II secretion systems (Pugsley et al 1991). Members of this protein family associated with Type III secretion systems share sequence similarity throughout their entire lengths, whereas their homology to other members of the PulD superfamily is restricted to a C-terminal domain. Thus these proteins appear to consist of at least two domains: a highly conserved C-terminal domain shared by all members, which may be involved in protein localization, and an amino-terminal domain specifically tailored to Type II or Type III systems. Two other components of the secretion machinery putatively located in the outer membrane have sequence motifs characteristic of bacterial lipoproteins, and those that have been investigated are lipidated (Michiels et al 1991, Allaoui et al 1992, Fenselau et al 1992, Gough et al 1992, Huang et al 1995, Pegues et al 1995).

Another highly conserved component of this apparatus is an ATPase, which shares sequence similarity to the β subunit of the F_0F_1 ATPase family of proteins (Venkatesan et al 1992, Eichelberg et al 1994, Lidell & Hutcheson 1994, Woestyn et al 1994, Van Gijsegem et al 1995). The conservation of these ATPases in the Type III secretion systems suggests that these proteins play an important role in providing energy to the translocation machinery. One of these proteins, *S. typhimurium* InvC, was shown to be capable of hydrolizing ATP (Eichelberg et al 1994). The change of a single residue in the predicted nucleotide-binding region of this protein abolished its ATPase activity as well as its ability to complement an *invC* null mutation. Further experiments will be required to understand the role of these ATPases in energizing the secretion process.

Accessory Proteins

In addition to the putative structural components of the translocation apparatus, there are numerous accessory proteins required for efficient export and/or function of the secreted proteins. For example, for export several target proteins require the function of specific cytoplasmic chaperones (Wattiau & Cornelis 1993, Ménard et al 1994b, Wattiau et al 1994, Kaniga et al 1995b; reviewed in Wattiau et al 1996). These chaperones bind to their cognate target proteins and thereby assist in their secretion. Unlike classical chaperones such as GroEL or Hsp70, these chaperones do not hydrolyze ATP and usually associate with only one protein, although some may bind more than one protein target. Although the function of these chaperones may vary, it is thought that they stabilize the target proteins in the bacterial cytoplasm and prevent their premature association and subsequent degradation. In some cases, they may also aid the targeting of the secreted proteins to the inner membrane translocon. Although the cytoplasmic chaperones do not share significant protein sequence similarity, they do share certain structural features such as small size, acidic pI, and regions predicted to form amphipathic α-helixes. Additionally, these chaperones tend to be encoded immediately adjacent to their cognate target proteins.

Another accessory protein associated with these systems in at least two different microorganisms, *Salmonella* and *Shigella*, is an acyl carrier protein (ACP) (Yao & Palchauduri 1992, Kaniga et al 1995a). Although acyl carrier proteins are usually associated with the synthesis of essential lipids (Magnuson et al 1993), it is unlikely that the ACPs associated with Type III secretion systems function in this capacity. More likely, these proteins may be involved in the lipid modification of some yet-unidentified secreted target protein. In fact, ACPs are involved in the fatty acylation of several membrane-targeted toxins, including the *E. coli* hemolysin (Issartel et al 1991, Stanley et al 1994), *Pasteurella* (Lo et al 1987), *Actinobacillus* leukotoxins (Chang et al 1989), and the adenylate cyclase-hemolysin of *Bordetella pertussis* (Hackett et al 1994). In these cases, modification is essential for activation, perhaps by increasing affinity of the protein for the plasma membrane of the host cell. Interestingly, secreted targets in at least three of these microorganisms, *Salmonella* (SipB), *Shigella* (IpaB), and *Yersinia* (YopB) (see below), have a hydrophobic domain that is distantly related to a similar domain in a family of pore-forming toxins (Haåkansson et al 1993). The ACPs encoded adjacent to the secretion apparatus of these bacteria may be involved in fatty acylation of a secretory protein. However, this has not been demonstrated.

Regulatory Proteins

The expression of the genes encoding the components and targets of Type III secretion systems is carefully regulated at both the transcriptional and post-transcriptional levels. Several regulatory genes have been identified that are encoded adjacent to genes associated with these secretion systems (Table 2). These regulatory genes are members of either the AraC (e.g. *Salmonella* InvF, *Shigella*, *Yersinia* VirF, and *Pseudomonas solanacearum* HrpB) (Sakai et al 1986, Cornelis et al 1989, Genin et al 1992, Kaniga et al 1994), NtrC (Xiao et al 1994), or Par B (Adler et al 1989) family of proteins, or the two-component regulatory family of proteins (e.g. HrpB from *Pseudomonas syringae* pv *syringeae*, HrpR/HrpS from *Pseudomonas syringae* pv *phaseolicola* and HilA from *Salmonella*) (Bajaj et al 1995, Grimm et al 1995, Miras et al 1995). Alternative sigma factors involved in the regulation of the *Pseudomonas syringae* pv *syringae* (Xiao et al 1994) and *Erwinia amylovora* (Wei & Beer 1995) *hrp* loci have also been identified. A list of the regulatory genes associated with Type III secretion systems is presented in Table 2. In addition, it appears that the expression of these systems is also influenced by other complex regulatory loops such as those that control flagellar assembly (Mulholland et al 1993, Eichelberg et al 1995) or intracellular survival (Pegues et al 1995), or by a variety of environmental cues such as growth temperature (Maurelli et al 1984, Kenny & Finlay 1995), osmolarity (Galán & Curtiss 1990, Rahme et al 1992), oxygen concentration (Ernst et al 1990, Lee & Falkow 1990), pH (Rahme et al 1992), calcium or nucleotide concentration (Straley 1991), growth phase (Lee & Falkow 1990), or a variety of nutrients (Huynh et al 1989, Rahme et al 1992, Wei et al 1992b). A discussion of the regulatory mechanisms that control the expression of Type III secretory systems is beyond the scope of this review.

TARGETS OF THE TYPE III PROTEIN SECRETION SYSTEMS

A number of proteins whose secretion is directed by the Type III secretion apparatus have been identified in different microorganisms (Table 3). Some of these proteins can induce signaling events in the host cells and are therefore directly

Table 2 Regulatory proteins specifically associated with the Type III protein secretion systems

Salmonella spp.	*Shigella* spp.	*Yersinia* spp.	*P. solanacearum*	*X. campestris*	*P. syringae*
InvF	MxiE	VirF	HrpB	HrpX	HrpL
HilA	VirF	—	—	—	HrpR/HrpS
—	VirB	—	—	—	—

Table 3 Secreted target proteins of the Type III secretion systems[a]

Salmonella spp.	*Shigella* spp.	*Yersinia* spp.	*P. solanacearum*	*P. syringae*	EPEC[b]
SipA	IpaA	YopE	PopA1	HrpZ	EspA
SipB	IpaB	YopH (Yop51)	—	—	EspB (EaeB)
SipC	IpaC	YopQ (YopK)	—	—	—
SipD	IpaD	YopB	—	—	—
InvJ	VirA	YopD	—	—	—
SpaO	IpgF	YopM	—	—	—
SptP	—	YpkA (YopO)	—	—	—
—	—	LcrV (V antigen)	—	—	—
—	—	LcrE (YopN)	—	—	—
—	—	YopJ	—	—	—
—	—	YscM (LcrQ)	—	—	—
—	—	LcrG	—	—	—

[a]Alternative names are listed in parentheses.
[b]EPEC: enteropathogenic *Escherichia coli.*

responsible for some host-cell responses triggered by these microorganisms. Others are involved in the secretion process itself, its regulation, or the delivery of secreted proteins to the target cell.

Secreted Target Proteins in Yersinia spp

Twelve proteins have been identified in pathogenic *Yersinia* species that are secreted by the Type III pathway (Straley et al 1993, Bliska 1994; Table 3). Most of these protein are essential for *Yersinia* pathogenesis, although an essential function for YopJ has not been identified. Several of these secreted proteins are thought to be translocated into host cells (see below). The best-characterized is the 51-kDa YopH (also known as Yop51). YopH is involved in the antiphagocytic activity of *Yersinia* (Rosqvist et al 1988, Bliska et al 1993a, Bliska & Black 1995, Fällman et al 1995) and is composed of several modular domains (Figure 1), which include two N-terminal domains important for secretion and translocation (Persson et al 1995, Sory et al 1995), a central proline-rich sequence that binds host-cell Src homology 3 (SH3) domains (SS Black et al, manuscript in preparation), and a C-terminal protein tyrosine phosphatase catalytic domain that is responsible for dephosphorylation of host proteins (Guan & Dixon 1990, Bliska et al 1991) (Figure 1). YopE is also involved in antiphagocytosis and is the primary factor involved in cytotoxicity (see below) (Rosqvist et al 1990). Although no specific function has been established for YopE, it is also organized in a modular fashion. YopE contains N-terminal secretion and translocation domains (Sory et al 1995) and a C-terminal domain related to the ADP ribosyltransferase exoenzyme of *Pseudomonas aeruginosa* (Kulich et al

1994) and SptP of *Salmonella* (Kaniga et al 1996) (Figure 1). A third protein with a putative intracellular targeting function is YpkA (*Yersinia* protein kinase A) (Galyov et al 1993, 1994) (Figure 1). YpkA displays extensive homology to eukaryotic serine/threonine kinases and catalyzes autophosphorylation of a serine residue in vitro (Galyov et al 1993). No function has been identified for YpkA, although *Y. pseudotuberculosis ypkA* mutants have a significant defect in pathogenesis (Galyov et al 1993, 1994).

YopM is distinct from the previously described Yop proteins in that it appears to function outside the host cell. YopM contains a segment of homology to an external domain of the platelet-specific receptor GP1ba, a seven-transmembrane G protein-linked receptor for thrombin and for the von Willebrand factor (Leung & Straley 1989). Purified YopM has thrombin-binding activity and competitively inhibits thrombin-induced platelet activation in vitro (Leung et al 1990, Reisner & Straley 1992). YopM may sequester thrombin during infection in order to subvert host inflammatory responses mediated by activated platelets.

A third class of proteins includes YopB and YopD, which are secreted by the Type III system in *Yersinia* and are required for the efficient translocation of YopE and YopH into mammalian cells. YopB and YopD are encoded in the same operon, and both have structures compatible with computer-generated models of transmembrane proteins (Haåkansson et al 1993). YopB has a

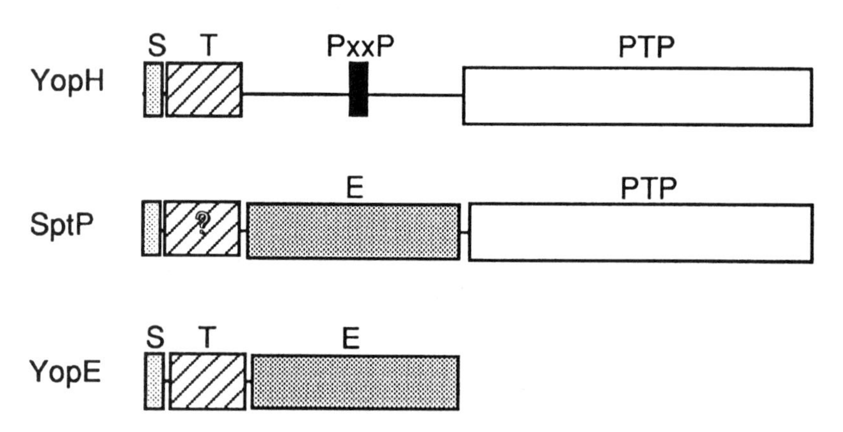

Figure 1 Modular organization of proteins secreted by Type III systems. YopH and YopE are secreted by the Type III system in *Yersinia*. SptP is secreted by the Type III system in *Salmonella*. Domains that have been identified on the basis of functionality or homology are indicated by boxes. S and T, secretion domain and translocation domains, respectively (Sory et al 1995); PxxP, proline-rich SH3-binding site (SS Black et al, manuscript in preparation); PTP, protein tyrosine phosphatase catalytic domain (Guan & Dixon 1990, Kaniga et al 1996); E, YopE homology domain (Kulich et al 1994, Kaniga et al 1996). The secretion and translocation domains in SptP are putative.

hydrophobic region that shares homology with IpaB of *Shigella flexneri*, SipB of *Salmonella* spp, and the RTX family of hemolysins and leukotoxins, whereas YopD contains an amphipathic helix in the carboxy-terminal end (Haåkansson et al 1993). These features suggest a possible role for YopB and YopD in the formation of a membrane pore to allow translocation of YopH and YopE into host cells (Haåkansson et al 1993). YopB has also been implicated in suppressing the production of TNFα by macrophages infected with *Y. pseudotuberculosis* (Beuscher et al 1995). This effect seems to be distinct from its role in translocating other Yops because purified YopB was shown to suppress production of TNFα.

Several of the proteins secreted by the Type III pathway in *Yersinia* regulate the secretion process itself. These include YopN (LcrE) (Forsberg et al 1991), LcrV (V antigen) (Price et al 1989), LcrQ (YscM) (Rimpilainen et al 1992), and LcrG (Skrzypek & Straley 1995). YopN negatively regulates the Type III pathway in *Yersinia* because *yopN* mutants are derepressed for Yop secretion in the presence of calcium (Forsberg et al 1991). In addition, YopN plays a role in the translocation of Yops into host cells because a *Y. pseudotuberculosis yopN* mutant showed reduced levels of YopE and YopH translocation into HeLa cells (Rosqvist et al 1994, Persson et al 1995). It has been suggested that YopN functions in a gating mechanism at the bacterial surface to regulate secretion and to focus Yop translocation across the host-cell membrane (Rosqvist et al 1994, Persson et al 1995).

LcrV positively regulates Yop secretion, and genetic data suggest that it functions as an inhibitor of a negative regulator (Skrzypek & Straley 1995). Antibodies directed against LcrV are passively protective against experimental plague in mice (Motin et al 1994). LcrV is proposed to have a direct anti-host function, indicating that the protein may be bifunctional; however, this anti-host function remains poorly understood. By deleting small internal regions in the N-terminus of LcrV, Skrzypek & Straley have produced mutant LcrV proteins that are not secreted but retain wild-type regulatory activity in the Type III pathway (Skrzypek & Straley 1995). *Yersinia pestis* strains that encode these mutant proteins are avirulent, which suggests that secretion of LcrV is necessary for its putative anti-host function but is not necessary for its regulatory function (Skrzypek & Straley 1995).

Secreted Target Proteins in Shigella spp

Several proteins whose secretion requires an intact Type III secretion apparatus have been identified in *Shigella*. The best-characterized are the Ipa proteins (IpaA, IpaB, IpaC, and IpaD). The identification of these proteins was facilitated by their strong reactivity to the sera of convalescent patients (Hale 1991). Subsequent molecular genetic analysis demonstrated that IpaB, IpaC, and IpaD, but not IpaA, were required for *Shigella* entry into cultured epithelial cells (Ménard et al 1993). The Ipa proteins assemble into multiprotein complexes that appear to be essential for their function (Ménard et al 1994b, Parsot et al 1995). Complexes of IpaB and IpaC assembled on microcarrier beads were able to trigger membrane ruffling and subsequent internalization of the coated beads (Ménard et al 1996) (see below). Although these results strongly suggest that the two proteins are sufficient to trigger these responses, the experimental design could not rule out the possibility of smaller amounts of other *Shigella* proteins associated with IpaB and IpaC. In fact, it has been shown that in addition to the Ipas, several other proteins are also secreted through the Type III secretion system (Alloui et al 1993, Parsot et al 1995, Uchiya et al 1995). One of this proteins, VirA, is required for both efficient invasion and intercellular spread (Uchiya et al 1995).

IpaB is involved in the lysis of the phagocytic vesicle subsequent to bacterial entry into host cells (High et al 1992), as well as in the induction of apoptosis in infected macrophages (Zychlinsky et al 1994b). IpaB contains two hydrophobic domains similar to those found in the RTX family of pore-forming proteins (Haåkansson et al 1993) and is required for the contact hemolytic activity observed in *Shigella* (Ménard et al 1993). It is not clear how the putative pore-forming ability of IpaB is related to its ability to mediate all these phenotypes. One possibility is that it forms a channel to allow entrance of other *Shigella* protein(s) into the host cell. IpaB and IpaD appear to be involved in the modulation of the secretion process itself because mutations in these genes result in increased secretion of other targets of the Type III secretion system (Parsot et al 1995). This observation, coupled with the fact that both proteins are associated in the bacterial membrane (Ménard et al 1994b), has led to the hypothesis that this complex plays an anti-secretion function to plug the translocon in the absence of activation signals. This complex may not only prevent secretion in the absence of the activating signal (see below) but may also control the secretion process in the presence of the stimulating signal.

Secreted Target Proteins in Salmonella spp

Several proteins secreted via the specialized Type III secretion system have been identified in *Salmonella* (Collazo et al 1995, Kaniga et al 1995a,b, 1996, Li et al 1995, Pegues et al 1995). These proteins can be classified in at least two functional groups (Collazo & Galán 1996): (*a*) proteins that are required for secretion through this specialized system (InvJ and SpaO) or that can modulate the secretion process itself (SipD) and (*b*) proteins that are candidates for effector molecules (SipA, SipB, SipC, and SptP).

Mutations in either *invJ* or *spaO* prevent the export of all proteins secreted via the Type III secretion system (Collazo & Galán 1996). This is in contrast to null mutations in the genes encoding other secreted proteins such as the Sip proteins, which do not prevent the secretion of any other target proteins (Kaniga et al 1995a,b). There appears to be a hierarchy in the secretion process in which the export of certain proteins (i.e. InvJ and SpaO) is essential for the secretion of others (Sips and SptP). Whether this observation is a general feature of all Type III secretion systems or is unique to *Salmonella* spp remains to be determined. InvJ and SpaO are encoded between two clusters of genes that are highly conserved among microorganisms possessing a Type III secretion apparatus (Collazo et al 1995). However, these two proteins share limited sequence similarity with their cognate proteins in other microorganisms, which suggests that they may be involved in functions that are specific for *Salmonella.* For example, InvJ shares only 18% identical residues with its putative *Shigella* homologue Spa32, which is statistically insignificant. Furthermore, Spa32 is not secreted to the culture supernatant of *Shigella* and has been implicated in the release of Ipa proteins from the bacterial surface (Watarai et al 1995a,b). Thus mutations in *spa32* prevent the release of the Ipa proteins but not their surface presentation. In contrast, mutations in *invJ* prevent both the secretion and the surface presentation of the Sip proteins (Collazo et al 1995). Therefore, it appears that InvJ and Spa32 do not perform equivalent functions. This perhaps underscores significant functional differences between the Type III secretion systems of these two microorganisms. InvJ has homology, albeit low, to EaeB, a protein of enteropathogenic *E. coli* that is secreted through a Type III system (see below). The functional significance of this similarity is unknown.

Four other secreted proteins, SipA, SipB, SipC, and SipD, exhibit significant sequence similarity to the *Shigella* IpaA, IpaB, IpaC, and IpaD proteins, respectively (Kaniga et al 1995a,b). SipB, SipC, and SipD are required for bacterial entry into host cells. SipA, however, does not seem to play a role in either bacterial entry into cultured cells or virulence in a mouse model of infection. A null mutation in *sipD* leads to increased secretion of a subset of invasion proteins, which suggests a role for SipD in modulating protein export in a manner similar to that of its homologue in *Shigella* (Kaniga et al 1995a). Although the homology between these proteins varies considerably, certain regions are remarkably conserved. For example, a 150 amino-acid hydrophobic domain of SipB shares 65% identity with that of IpaB. This domain is also present in the distantly related RTX family of pore-forming toxins. The putative pore-forming function of these proteins may be implicated in the induction of signaling events in the host cell, perhaps by creating an ion channel in the host cell plasma membrane. However, no direct evidence has been presented that demonstrates actual signaling capacity for any of the Sip proteins or their *Shigella* counterparts. Equally likely is the possibility that any one of the Sip proteins, particularly SipB, may be involved in the delivery of effector molecules to the host cell. Recently, a tyrosine phosphatase with modular effector domains, whose secretion also occurs via the Type III secretion apparatus, was identified in *S. typhimurium* (Kaniga et al 1996). This protein, termed SptP, is composed of two distinct domains: an amino-terminal domain with homology to the *Yersinia* spp YopE and the *P. aeruginosa* ExoS toxins, and a carboxy-terminal domain with sequence similarity to protein tyrosine phosphatases (Figure 1). Consistent with this finding, purified SptP, but not a mutant protein in which a critical cysteine residue in the catalytic site had been changed to alanine, exhibited protein tyrosine phosphatase activity. Although this protein is not involved in bacterial entry into epithelial cells, as measured in an in vitro invasion assay, a *Salmonella* strain carrying a null mutation in *sptP* was less virulent in a mouse model of infection, thus indicating that it is important for pathogenesis.

Secreted Target Proteins in Enteropathogenic E. coli

A set of polypeptides secreted via the *sep*-encoded Type III system has been identified in enteropathogenic *E. coli* (EPEC) (Jarvis et al 1995, Kenny & Finlay 1995). These proteins appear to be involved in the formation of attaching and effacing lesions, a pronounced rearrangement of the host-cell cytoskeleton in response to bacterial attachment (see below). Proteins of 40, 39, 37, 25, and 19 kDa are secreted when EPEC are grown in tissue culture medium, but not in typical bacterial culture media. A sixth EPEC protein with a size of 110 kDa is also released into tissue culture medium, although secretion of this protein does not appear to require the Type III pathway (Jarvis et al 1995, Kenny & Finlay 1995). Amino-terminal protein sequence analysis has been performed on each of these secreted proteins (Kenny & Finlay 1995) except for the 19-kDa product. The 37-kDa polypeptide was found to be identical to EspB (formerly EaeB), a protein required for signaling and rearrangement of the host-cell cytoskeleton during EPEC attachment (Kenny & Finlay 1995). The amino-terminal sequence of the 25-kDa protein, termed EspA, matched an open reading frame that lies adjacent to *espB* in a region of chromosomal DNA required for EPEC pathogenesis (Kenny & Finlay 1995, Kenny et al 1996). The amino-terminus of the 39-kDa secreted protein was found to be homologous to that of glyceraldehyde-3-phosphate dehydrogenase (GAPDH), a housekeeping enzyme of *E. coli.* This observation is intriguing because a surface-localized GAPDH is suggested to contribute to the virulence of a number of microorganisms by virtue of its ability to bind plasmin (Lottemberg et al 1994). Whether the GAPDH secreted by EPEC has a similar function is unknown.

Secreted Target Proteins in Plant Pathogenic Bacteria

Only three targets of the Type III secretion systems have been identified in plant pathogenic bacteria: the harpins of *Erwinia amylovora* and *Pseudomonas syringae* and the PopA1 protein of *Pseudomonas solanacearum* (Wei et al 1992a, He et al 1993, Arlat et al 1994). Although there is no overall sequence similarity among these different target proteins, they all share some structural features such as size, high glycine content, and heat stability. The PopA1 protein of *P. solanacearum* is proteolytically processed upon secretion by the removal of the 93 amino-terminal amino acids, which results in a biologically active form of this protein known as PopA3 (Arlat et al 1994).

The three secreted target proteins induce plant cell death and tissue necrosis when artificially infiltrated into nonsusceptible hosts (Wei et al 1992a, He et al 1993, Arlat et al 1994), which suggests that these proteins are sufficient for eliciting the hypersensitivity response (HR) in nonsusceptible hosts. However, a *Pseudomonas syringae* pv *syringae* strain carrying a deletion mutation in the *hrpZ* gene was still capable of inducing a typical HR (Alfano et al 1996). Furthermore, a mutation in another gene in the *hrp* cluster, *hrmA*, resulted in a strain that was capable of exporting wild-type levels of HrpZ, but was unable to induce the HR response. These results indicate that HrpZ is required but not sufficient for HR elicitation. Most likely there are other proteins that can substitute for HrpZ. A more detailed analysis of HrpZ established that the eliciting activity resides in multiple regions of this protein (Alfano et al 1996).

The Type III secretion system encoded in the *hrp* locus is required for the hypersensitivity response in heterologous hosts and pathogenicity in homologous hosts. With the exception of the *Erwinia amylovora* HrpN protein (Wei et al 1992a), the secreted targets identified so far do not act in a host-specific manner. Therefore, at least for several of the plant pathogens, some unidentified secreted factor(s) must exist that determine whether the bacterial/host-cell interaction leads to disease or defense (see section on host responses to plant pathogenic bacteria).

ACTIVATION OF THE TYPE III PROTEIN SECRETION SYSTEMS

A distinct feature of the Type III protein secretion systems is the requirement of environmental signals for their activation. A variety of inducers of secretion have been identified in different microorganisms. For example, the presence of serum (Ménard et al 1994a, Zierler & Galán 1995), Congo red (Parsot et al 1995), or low levels of calcium (Brubaker 1986) or carbon dioxide, (Haigh et al 1995) as well as growth in tissue culture (Kenny & Finlay 1995) or other

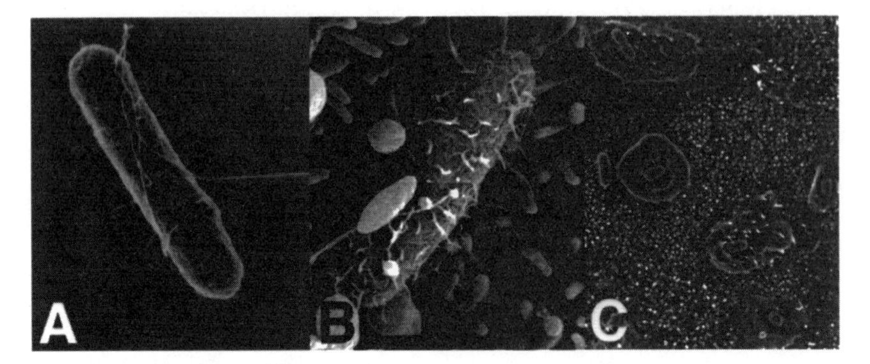

Figure 2 Low-voltage scanning electron micrograph illustrating the interaction of *S. typhimurium* with cultured epithelial cells and the assembly of the surface invasomes. Note that microorganisms not exposed to cultured cells do not exhibit the surface invasomes (*A*). In contrast, microorganisms that have been exposed for 15 min to cultured epithelial cells show the invasome structures on their surfaces (*B*). At later times (30 min), bacteria are seen associated with host-cell membrane ruffles (*C*) (Ginocchio et al 1994).

complex media (Wei et al 1992b), stimulate secretion through Type III systems. However, it is now largely accepted that the biologically relevant activating signals derive from the interaction of this group of bacteria with their host cells. The nature of any of these signals remains unknown.

Induction of Protein Secretion and Invasome Assembly by Salmonella spp

Contact with host cells leads to the stimulation of the secretion of *Salmonella* invasion proteins through the Type III secretion apparatus (Zierler & Galán 1995). Stimulation does not require de novo protein synthesis and can also be induced by a factor present in bovine calf serum. Host cells grown in the absence of serum were also able to induce secretion, which indicates that an inducer must be produced by the host cell. Whether the serum and host-cell factor are identical is not known. Contact of *Salmonella* with cultured cells also results in the assembly of surface structures called invasomes (Ginocchio et al 1994) (Figure 2). These structures are transient and presumed to be rapidly shed from the bacterial surface (Ginocchio et al 1994). The assembly and subsequent shedding of these structures is dependent on the Type III secretion system and does not require de novo protein synthesis (Ginocchio et al 1994). Although the actual components of the invasome have not been identified, its dependence on the Type III secretion system strongly suggests that it is made of targets of this secretion apparatus. The role of the invasome in *Salmonella* interaction with host cells is not known. Although there is a close correlation between invasome assembly and bacterial entry, a direct role for this structure has not been demonstrated. The transient nature of the invasomes rules out any role in the physical attachment of bacteria to host cells, a function more commonly associated with other structures that are firmly anchored to the bacterial cell surface such as pili or fimbriae. It is possible that invasomes represent the association of secreted proteins as they exit the bacterial cell through discrete exit points. Such association may be required for these proteins to acquire signaling competency. Alternatively, invasomes may represent an organelle involved in the delivery of bacterial proteins to the host cell.

Host-Cell-Dependent Stimulation of Protein Secretion in Shigella spp

Contact with epithelial cells stimulates the secretion of the Ipa proteins of *Shigella* (Ménard et al 1994a), as it does in *Salmonella.* This stimulation does not result from increased gene expression of either the genes encoding components of the secretion apparatus or the target proteins themselves. Watarai et al (1995a) report that release of the Ipa proteins upon contact with host cells occurs from an intermediate stage in which these proteins are bound to the cell surface, a process mediated by the protein Spa32. However, Ménard et al (1994b) reported that secretion upon contact occurs directly from a cytoplasmic pool without a bacterial surface intermediate. Although the reasons for these inconsistencies are not clear, they may reflect differences in either the growth conditions or the bacterial strains used in the different studies. Stimulation of secretion can also be mediated by a soluble factor in fetal bovine serum (Ménard et al 1994a) or by interaction of *Shigella* with extracellular matrix proteins such as fibronectin, laminin, or collagen type IV (Watarai et al 1995a), which suggests that stimulation of Ipa protein secretion in vivo may also occur at a distance from the host-cell surface. Interestingly, when *Shigella* are grown under inducing conditions, the secreted proteins also associate into a supramolecular structure in the form of extended sheets (Parsot et al 1995). These structures are composed of several proteins, such as IpaA, IpaB, and IpaC, whose secretion is known to occur through the Type III system. The functional significance of these structures and their relationship, if any, to the *Salmonella* invasomes are unknown.

Host-Cell Contact-Dependent Translocation of Yop Proteins by Yersinia spp

Secretion of proteins by the Type III system in *Yersinia* shares a number of features in common with those described for *Salmonella* and *Shigella.* For example, secretion of these proteins is stimulated upon contact of *Yersinia* with epithelial cells (Rosqvist et al 1990, Forsberg et al 1994), and these systems share the ability to translocate proteins into eukaryotic cells (Figure 3) (Rosqvist et al 1995). Little is known about the molecular basis of this translocation process, although YopD, and possibly YopB, are required (see below). Translocation of YopH or YopE into macrophages does not appear to require de novo bacterial protein synthesis (Goguen et al 1986; SS Black et al, manuscript in preparation). However, continuous protein synthesis is necessary under experimental conditions that require high levels of Yops to be translocated into host cells (i.e. detection of YopH or YopE by immunofluorescence).

Following the discovery of the tyrosine phosphatase activity of YopH, it was proposed that this enzyme entered into the mammalian host cell to perform

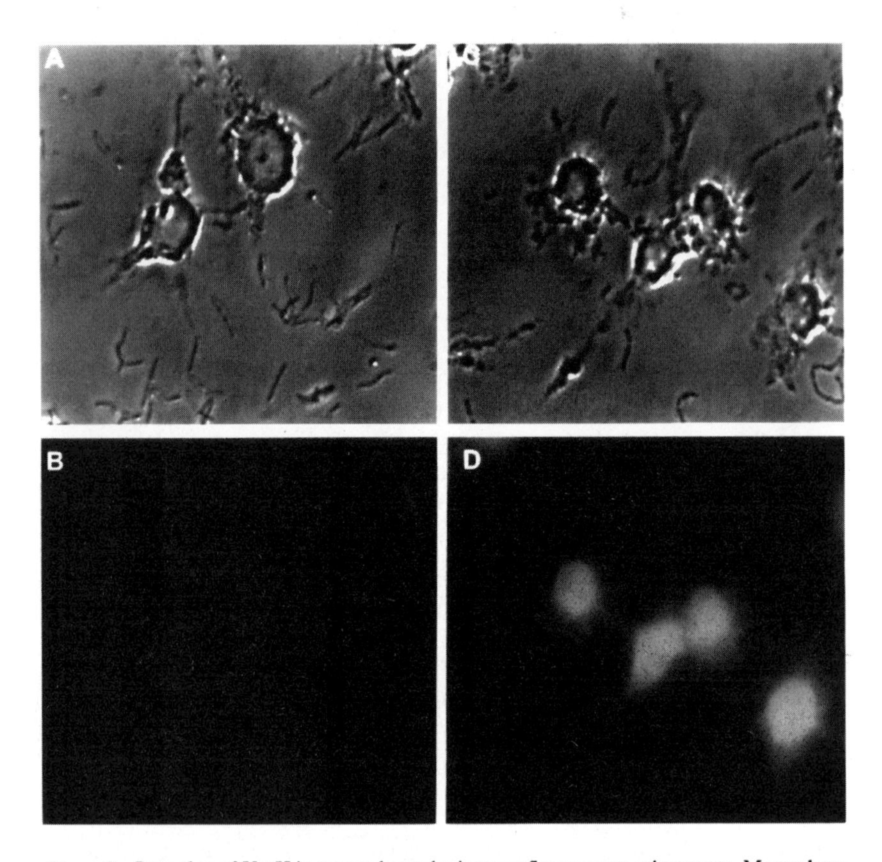

Figure 3 Detection of YopH in macrophages by immunofluorescence microscopy. Macrophage were infected with wild-type *Y. pseudotuberculosis* (*C, D*) or a *yopH* deletion mutant (*A, B*) for 2 h. Cells were fixed, permeabilized and stained with an affinity-purified rabbit antibody specific for YopH, followed by anti-rabbit antibody conjugated to FITC. Cells were photographed by phase contrast (*A, C*) or epifluorescence microscopy (*B, D*). Figure provided courtesy of D Black.

its function (Guan & Dixon 1990, Bliska et al 1991). Several experimental approaches examined the possibility that YopH, as well as YopE, enters mammalian cells. Rosqvist et al utilized a microinjection technique with glass beads to show that crude protein fractions containing YopE had to be introduced into HeLa cells to cause cytotoxicity (Rosqvist et al 1991). The same study showed that a *yopD* mutant of *Y. pseudotuberculosis* secreted YopE in a normal fashion but was defective for cytotoxicity. These results suggest that YopE is translocated across the host-cell membrane and that YopD is required for this process (Rosqvist et al 1991).

YopH was shown to dephosphorylate cytoplasmic proteins in macrophages or HEp-2 cells infected with *Y. pseudotuberculosis* or *Yersinia enterocolitica* (Bliska et al 1991), which suggests that the enzyme reaches the cytoplasm of the host cell. The use of a strain defective for entry (an invasin mutant) indicated that dephosphorylation of host proteins did not require efficient internalization of *Y. pseudotuberculosis* into HEp-2 cells. This led to the suggestion that YopH was translocated across the eukaryotic plasma membrane (Bliska et al 1991). A *Y. enterocolitica yopD* mutant secreted YopH normally, although it was defective for dephosphorylation of macrophage proteins, which suggests a role for YopD in translocation of YopH through the plasma membrane (Hartland et al 1994). In these experiments however, it is difficult to rule out the possibility that host protein dephosphorylation may have occurred after lysis of infected cells. To circumvent this problem, Sory & Cornelis developed a reporter enzyme approach based on the fusion of YopH or YopE to the catalytic domain of the calmodulin-dependent adenylate cyclase (Cya) (Sory & Cornelis 1994, Sory et al 1995). Because bacteria lack calmodulin and because the catalytic domain of Cya is unable to enter mammalian cells on its own, the accumulation of cAMP in HeLa cells or macrophages infected with *Y. enterocolitica* can be used as a measure of internalization of Yop-Cya fusion proteins.

Translocation of the YopE- or YopH-Cya fusion proteins requires an active Type III secretion pathway, as well as the function of YopB and/or YopD (Sory & Cornelis 1994, Sory et al 1995). The Yop-Cya fusion proteins appear to be translocated primarily by extracellular bacteria, because reduction of bacterial internalization through inactivation of invasin, or by treatment of mammalian cells with cytochalasin D, causes only a marginal decrease in cAMP production (Sory & Cornelis 1994, Sory et al 1995). The Yop-Cya fusion approach has also been used to map the domains of YopH and YopE that are required for secretion and translocation (Sory & Cornelis 1994). The N-terminal residues 15 and 17 were sufficient for secretion of YopE- and YopH-Cya fusions, respectively, which indicates that a relatively small domain is sufficient to direct the secretion of hybrid proteins by the Type III pathway in *Yersinia*. The N-terminal 50 and 71 amino acids were sufficent to mediate translocation of the YopE- and YopH-Cya fusion proteins, respectively. These results suggest that secretion and translocation functions of Yops are separable and that relatively few amino acids are able to direct the movement of Yops from the bacterium into the host cell.

Immunofluorescence microscopy has been used to directly examine the location of YopE and YopH in HeLa cells infected with *Y. pseudotuberculosis* (Rosqvist et al 1994, Persson et al 1995). These studies utilized relatively long infections (2–3 h) in the absence of bacterial protein synthesis inhibitors because a high level of Yop translocation is required to detect these proteins by immunofluorescence techniques. Under these conditions, YopE was localized in the perinuclear region of HeLa cells infected with a wild-type strain, whereas the bacteria were devoid of the protein (Rosqvist et al 1994). In a similar fashion, YopH was largely localized to the cytoplasm of HeLa cells, although a small fraction colocalized with the plasma membrane. Assays performed with an invasin mutant, or in the presence of gentamicin or cytochalasin D, indicated that YopH and YopE were translocated primarily from extracellular bacteria. A *Y. pseudotuberculosis yopN* mutant translocated YopE and YopH into HeLa cells with reduced kinetics, and in this case, YopE was detected as a halo around the bacterium (Rosqvist et al 1994, Persson et al 1995). A similar localization pattern was observed for a strain expressing a mutant form of YopH with a deletion in the translocation domain. In this case, the protein appeared to have reached the bacterial surface but was defective for translocation into the HeLa cell. A *Y. pseudotuberculosis* strain defective in *yopD* and or *yopB* was incapable of translocating either YopH or YopE, and the respective proteins were localized in discreet spots on the surface of the bacteria. These discrete areas of protein localization may reflect zones at the interface of bacteria and host-cell membranes where the Yops are translocated (Rosqvist et al 1994, Persson et al 1995).

HOST-CELL RESPONSES DEPENDENT ON THE FUNCTION OF THE TYPE III PROTEIN SECRETION SYSTEMS

Despite the significant similarity in the structural components of the Type III secretion systems in the different microorganisms, the type of responses that they induce in their host cells vary greatly, which reflects the variety of putative signaling proteins secreted through this system by these microorganisms. Indeed, with a few notable exceptions (e.g. *Shigella* and *Salmonella* Ipa and Sip proteins), there appears to be limited similarity among the targets secreted by the different microorganisms. Thus although all these microorganisms generally use this specialized protein secretion system to induce host-cell responses through the export of specific signaling molecules, the type of response they elicit has been evolutionarily tailored for their specific needs.

Host-Cell Responses Induced by Shigella spp

The Type III secretion system of *Shigella* spp encoded in the *mxi* and *spa* loci is involved in triggering host-cell signal transduction pathways that result in several responses including bacterial internalization. Subsequent to bacterial internalization, *Shigella* gains access to the cell cytosol where it subverts the host-cell actin cytoskeleton in order to spread to neighboring cells. The molecular and cellular details of this fascinating form of subversion of host-cell functions have been reviewed elsewhere (Goldberg & Sansonetti 1993, Theriot 1995) and are not dealt with here. *Shigella* internalization into host cells occurs after the induction of dramatic cytoskeletal rearrangements characterized by the formation of membrane projections immediately adjacent to the point of contact of the bacterium with the host cell (Adam et al 1995). The morphological changes, which resemble membrane ruffles, are similar to those induced by other stimuli such as growth factors (Bar-Sagi & Ferasmico 1986) or even the bacterial pathogen *Salmonella* (see below). The accompanying cytoskeletal rearrangements are characterized by the presence of long actin filaments, polymerized with their positive ends towards the tip of the projections, in the proximity of the incoming bacterium (Adam et al 1995). The actin structures are tightly cross-linked at least in part by plastin, which has been shown to be associated with these structures. Expression of dominant-negative mutants of plastin in infected cells interfered with bacterial entry, which supports the involvement of this protein in the entry process. In addition to plastin, other cytoskeletal proteins, including α-actinin, vinculin, paxilin and talin, have been localized to the point of bacterial entry. The signaling events leading to bacterial internalization are not well understood. A role for the small GTP-binding protein Rho has been suggested because a Rho-inhibitor blocked bacterial entry into cultured cells (Adam et al 1996). Another signaling molecule implicated in the entry process is the protein tyrosine kinase $pp60^{c-src}$ (Dehio et al 1995). The interaction of *Shigella* with cultured cells resulted in the tyrosine phosphorylation of cortactin, one of the major substrates of $pp60^{c-src}$. In addition, overexpression of $pp60^{c-src}$ induced cytoskeletal and plasma membrane changes similar to those induced by *Shigella* and rescued the invasion phenotype of an invasion-defective strain of *Shigella*. Although these experiments do not demonstrate a direct role for either $pp60^{c-src}$ or cortactin in the entry mechanisms, the correlation suggests a possible connection between these signaling events and the cytoskeletal remodeling induced by *Shigella*.

Unlike the responses triggered in non-phagocytic cells, *Shigella* induces programmed cell death in infected macrophages (Zychlinsky et al 1992). If macrophages have been previously activated, the *Shigella*-induced apoptotic death is preceded by a massive release of interleukin-1β, a mechanism that may contribute to the profound inflammatory response induced by these microorganisms in the infected host (Zychlinsky et al 1994a). Little is known about the mechanisms by which *Shigella* induces apoptosis in infected macrophages. The apoptotic event is intimately linked to the expression of IpaB and the ability of these microorganisms to exit the phagocytic vesicle (Zychlinsky et al 1994a). Addition of cytochalasin, a drug that prevents bacterial entry into macrophages, effectively prevented the *Shigella*-induced apoptosis of macrophages (Clerc et al 1987), which is consistent with the notion that *Shigella* induces programmed cell death from the cell cytosol and not via transmembrane signaling.

Host-Cell Responses Induced by Salmonella spp

The Type III secretion system encoded in centisome 63 of the *Salmonella* chromosome is involved in triggering a variety of host-cell responses. Perhaps the most overt response is the induction of profound changes in the plasma membrane shortly after the bacteria come in contact with the host cell (Takeuchi 1967, Finlay & Falkow 1990) (Figure 2). These morphological changes, which are similar to those induced by *Shigella*, resemble the membrane ruffles induced by a variety of agonists such as growth factors or oncogene activation (Bar-Sagi & Ferasmico 1986, Kadowaki et al 1986). The disruption of the host-cell plasma membrane requires an intact cytoskeleton (Kihlstrom & Nilsson 1977, Buckholm 1984) and is intimately linked to the induction of macropinocytosis (Garcia-del Portillo & Finlay 1994). The macropinocytotic activity induced by *Salmonella* ultimately leads to its own internalization. A number of cytoskeletal-associated proteins such as actin, α-actinin, talin, and ezrin have been observed at the point of bacteria/host-cell contact (Finlay & Ruschkowski 1991). However, the individual contribution of these proteins to the cytoskeletal rearrangements has not been investigated.

Although it is clear that the morphological changes induced by *Salmonella* are the result of complex signaling events, the nature of these signaling events and how *Salmonella* triggers them are not well understood. Contact of *Salmonella* with the host cell is accompanied by rapid calcium fluxes (Ginocchio et al 1992, Pace et al 1993). Induction of this response does not require bacterial internalization because addition of cytochalasin, which effectively prevents bacterial entry by disrupting the host-cell cytoskeleton, does not prevent calcium fluxes. In fact, addition of intracellular calcium chelators prevents bacterial entry (Ruschkowski et al 1992), which suggests that calcium mobilization may be a requirement for bacterial internalization. Consistent with this requirement, invasion-defective mutants of *Salmonella* failed to stimulate calcium mobilization (Ginocchio et al 1992, Pace et al 1993). The signaling events leading to calcium mobilization appear to be dependent upon the particular cell. In certain cell lines, *Salmonella* infection results in the production of inositol phospholipids (most likely inositol (1,4,5)trisphosphate) which may ultimately mediate calcium fluxes in these cells by inducing its release from intracellular stores (Ruschkowski et al 1992). In other cell lines such as intestinal Henle-407, however, calcium fluxes are the result of an influx from the extracellular milieu (Pace et al 1993). In this cell line, *Salmonella* infection results in the phosphorylation of the EGF receptor and a number of responses that are characteristic of receptor stimulation (Galán et al 1992b, Pace et al 1993). These responses include not only calcium fluxes but also the activation of MAP (ERK) kinase and the production of arachidonic acid metabolites such as peptidoleukotrienes as a consequence of phospholipase A_2 (PLA_2) and 5-lipooxygenase (5-LO) activation (Pace et al 1993). These eicosanoids, in particular leukotriene D4, are likely involved in the modulation of the calcium fluxes because addition of this compound to Henle-407 cells induced a calcium flux similar to that triggered by *Salmonella*. Pharmacological interference with this pathway by blocking the activities of PLA_2 or 5-LO prevented *Salmonella* entry, which implicates these events in bacterial internalization (Pace et al 1993). Furthermore, the invasion phenotype of an entry-defective *Salmonella* mutant that was unable to trigger this signaling pathway was rescued by supplementing the infection assay with EGF, the natural ligand of the EGFR or LTD_4 (Pace et al 1993). It should be noted that Rosenshine et al failed to detect tyrosine phosphorylation of the EGF receptor in *Salmonella*-infected Henle-407 cells (Rosenshine et al 1994). The reasons for these apparently contradictory results, however, are not clear.

How these different second messengers are involved in mediating the cytoskeletal rearrangements is not understood. Cytoskeletal dynamics are influenced by Ca^{2+} (Mooseker 1985) and arachidonic acid metabolites (Peppelenbosch et al 1993, 1995). The latter regulate the activity of actin-organizing small GTP-binding proteins and may, therefore, participate in the events leading to the *Salmonella*-induced cytoskeletal rearrangements (Peppelenbosch et al 1995, Tsai et al 1990, Han et al 1991, Chuan et al 1993). Microinjection of dominant interfering mutants of the small G proteins ras and rac or the rho-inhibitor C3 exotoxin from *Clostridium botulinum* did not prevent *Salmonella*-induced cytoskeletal rearrangements (Jones et al 1993). However, expression of a dominant interfering mutant of CDC42 effectively prevented the *Salmonella*-induced rearrangement of the host-cell cytoskeleton that leads to macropinocytosis (L Chen & M Galán, manuscript in preparation). Additionally, expression of an activated form of CDC42 rescued the phenotype of an invasion-defective mutant of *Salmonella*. These results implicate CDC42 in the signaling events leading to *Salmonella* entry.

In addition to membrane ruffling, *Salmonella* induces a variety of nuclear responses leading to the production of pro-inflammatory cytokines, particularly IL-8 (Eckmann et al 1993; S Hobbie et al, submitted). These responses are partially mediated by the transcription factors AP1 and NFκB, both of which become activated after *Salmonella* infection of cultured epithelial cells (S Hobbie et al, submitted). Although the signal transduction pathways that lead to these responses are not understood, the activation of MAP (ERK) kinase induced by *Salmonella* is likely to be involved. In support of this hypothesis, a tyrosine kinase inhibitor with selectivity for the EGFR tyrosine kinase effectively prevented *Salmonella*-induced nuclear responses in Henle-407 cells (S Hobbie et al, submitted). More studies will be required to unravel the complexities of the host-cell responses to *Salmonella*, and in particular, to determine whether the signaling pathways that lead to either the morphological or nuclear cell responses share some features in common.

Host-Cell Responses Induced by Yersinia spp

The Type III secretion system encoded on the virulence plasmids of *Yersinia* spp is primarily associated with alterations in host-cell morphology and negative effects on host-cell processes involving phagocytosis of bacteria. Infection of epithelial cells or macrophages with *Yersinia* leads to dramatic alterations in the microfilament structure of the host cell. Typically, within minutes of infection, the host cells round up and contract away from the extracellular matrix, a phenomenon referred to as cytotoxicity (Rosqvist et al 1990). Visualization of the actin framework of cells with FTIC-labeled phalloidin indicates that as the infection progresses, ordered actin filaments are reduced to disordered structures that give rise to a granular appearance (Rosqvist et al 1991). YopE is largely responsible for this alteration in cell morphology, although in the absence of YopE, a weak cytotoxic effect due to YopH has been detected (Rosqvist et al 1990, Bliska et al 1993a).

YopE and YopH also antagonize internalization of *Yersinia*, a process referred to as antiphagocytosis (Rosqvist et al 1988, 1990). Several pathways are inhibited: YadA-mediated uptake (Fällman et al 1995, Bliska et al 1993a) and Fc receptor-mediated uptake of IgG-opsonized bacteria (Fällman et al 1995). YopE and YopH each contribute to the antiphagocytic effect, depending upon the type of host cell used (Rosqvist et al 1988, Bliska et al 1993a, Fällman et al 1995). Paradoxically, the antiphagocytic effect of YopE and YopH is more pronounced in epithelial cells than in macrophages (Rosqvist et al 1988, Bliska et al 1993a, Fällman et al 1995). In macrophages, the tyrosine phosphatase activity of YopH is also responsible for inhibiting the oxidative burst mediated by ligation of Fc-receptors (Bliska & Black 1995). Hartland et al have shown that wild-type *Y. enterocolitica* inhibits the oxidative burst in macrophages triggered by unopsonized zymosan (Hartland et al 1994). This process requires the function of YopD, but not YopH, which suggests that another Yop(s) translocated into macrophages by the Type III system is responsible for inhibition of zymosan-induced signaling events.

Inhibition of host-cell signaling processes by YopH is likely related to its ability to dephosphorylate host proteins (Bliska et al 1991, 1992, Bliska & Black 1995) and bind host-cell SH3 domains (SS Black et al, manuscript in preparation). As discussed above, YopE is homologous to a region of the *P. aeruginosa* exoenzyme S (ExoS) (Kulich et al 1994). Also, both proteins are capable of disrupting host-cell actin microfilaments, which suggests that they have a common mechanism of action. Although the target of ExoS in vivo is not known, it has been shown to ribosylate Ras and other low-molecular weight GTPases in vitro (Kulich et al 1994). In addition, an accessory host factor for ExoS has been identified as a member of the 14-3-3 family of proteins (Kulich et al 1994). Because GTPases regulate the polymerization of actin in eukaryotic cells, it is tempting to speculate that YopE modulates host-cell responses through interaction with GTPases. This concept is tentative because it is uncertain whether the region of shared homology is required for modification of GTPases by ExoS.

Host-Cell Responses Induced by Enteropathogenic E. coli

Proteins secreted by the Type III pathway in EPEC (*sep*) appear to be required for the formation of attaching and effacing lesions. These lesions occur at the sites of bacterial attachment and represent localized areas of denuded brush border microvilli (Donnenberg & Kaper 1992). Several host-cell responses are associated with the formation of attaching and effacing lesions. A pedestal-like structure forms under the site of bacteria attachment. This structure is enriched in filamentous actin, several actin-associated proteins, and one or more tyrosine phosphorylated proteins (Finlay et al 1992). Other responses include elevated levels of intracellular Ca^{2+} (Baldwin et al 1991, 1993), fluxes in inositol trisphosphate levels (Foubister et al 1994), and phosphorylation of myosin light chain (Manjarrez-Hernandez et al 1991) and a 90-kDa protein (Hp90) (Rosenshine et al 1992). A mutant defective in one of the secretory proteins (*sepB*) was unable to form attaching and effacing lesions, indicating that secretion of proteins by the Type III secretion system is required for inducing host-cell responses (Jarvis et al 1995). In addition, EaeB, one of the proteins secreted by this pathway, appears to be required for inducing pedestal formation and phosphorylation of host proteins (Kenny & Finlay 1995). These results suggest that secretion of EaeB, and possibly other proteins, is required for EPEC to induce attaching and effacing lesions.

hrp-Dependent Host-Cell Responses Induced by Phytopathogenic Bacteria

At least two types of interactions can occur between pathogenic bacteria and their plant hosts (reviewed in Lindsay et al 1993, Long & Staskawicz 1993, Staskawicz et al 1995). Interactions between phytopathogenic bacteria and susceptible hosts are said to be compatible and lead to disease that is usually characterized by the generalized collapse of the plant. The production of disease requires the function of virulence genes that include a variety of extracellular enzymes, polysaccharides, hormones, toxins, etc. In contrast, interactions of bacteria with resistant hosts are said to be incompatible and lead to a plant defense response known as the hypersensitivity response, HR. This response is characterized by rapid but localized tissue necrosis resulting in little or no disease and requires the function of specific bacterial avirulence genes (*avr*) acting in concert with paired host genes. Therefore, this gene-for-gene model of plant-pathogen interaction establishes that the induction of the HR is initiated by the plant's recognition of a specific bacterial elicitor molecule, which in turn triggers host-cell responses that drastically limit pathogen growth. These responses include a rapid oxidative burst, ion fluxes, and production of antimicrobial compounds such as phytoalexins, or a variety of enzymes such chitinases and glucanases. Although the different signaling events triggered in the plant cell by these interactions are beginning to be unraveled and several plant resistance genes are beginning to be identified, their description is beyond the scope of this review (for reviews see Lindsay et al 1993, Long & Staskawicz 1993, Staskawicz et al 1995).

The Type III protein secretion system encoded in the *hrp* locus is involved in both types of responses induced by phytopathogenic bacteria: disease production in susceptible hosts and elicitation of the HR in resistant hosts (Niepold et al 1985, Lindgren et al 1986). This implies a close relationship between the protein secretion system and the *avr* gene products and the possibility that some Avr proteins may be secreted and/or translocated to the plant cell via this system. *hrp*-dependent secretion of Avr proteins, however, has not yet been demonstrated. Nevertheless, *avrB* and other *avr* genes in *P. syringae* were shown to be members of the *hrp* regulon (Huynh et al 1989), and the *hrp* gene cluster of *P. syringae* was capable of delivering *avr* signals when cloned in an heterologous host, which suggests a close functional relationship between the *hrp* and the *avr* gene products (Collmer et al 1996).

The mechanisms by which the Harpin proteins elicit these responses are poorly understood. The elicitation of the HR by the artificial application of the HrpZ protein in nonsusceptible hosts involves an active response of the plant (He et al 1993) because addition of a number of pharmacological inhibitors such as cycloheximide, α-amanitin, vanadate, and lanthanum effectively prevented such a response. However, the specific role of the inhibited processes is not understood.

SUMMARY AND PERSPECTIVE

It is now clear that microbial pathogens have evolved sophisticated ways to interact with their hosts. The last few years have seen remarkable progress in our understanding of these interactions. Herein we have dealt with a selected group of pathogens that utilize a specialized protein secretion system to interact with host cells. This protein secretion pathway, termed Type III, is activated when these bacteria come in contact with or in close proximity to the host cell, which results in the secretion and/or translocation into the host cell of a number of putative signaling proteins. These signaling proteins, in turn, interfere with or stimulate a variety of host responses. Common themes among these interactions include the manipulation of the host-cell actin cytoskeleton, either to gain access to non-phagocytic cells or to avoid ingestion by professional phagocytes. In many cases this involves interference with host-cell signal transduction pathways that utilize tyrosine phosphorylation. Induction of nuclear responses and the subsequent activation of eukaryotic gene transcription are also common themes in the interaction of animal and plant pathogenic bacteria with their host cells. Furthermore, a number of proteins secreted by the Type III system contain multiple modular domains. The presence of modular domains in proteins is a recurring theme in the field of signal transduction. It is likely that other signaling proteins secreted by Type III systems will have a similar modular organization.

At first glance it seems surprising that such a taxonomically diverse group of microorganisms has evolved similar mechanisms to engage a diverse group of host cells in complex two-way interactions. However, it is clear that there is much in common between plant and animal pathogens. Despite the large evolutionary gap, it is now well documented that certain bacteria can infect both plants and animals and make use of the same virulence genes to cause disease in both hosts (Rahme et al 1995). Phytopathogenic bacteria engage the plant hosts in a gene-for-gene interaction known as the HR. This response, which requires a functional Type III secretion system, is elicited by the interaction of specific *avr* (avirulence) genes encoded by a noncompatible (avirulent) bacteria, with the corresponding plant resistance (*R*) gene products. This interaction leads to a cascade of events that severely limits the growth of the invading pathogen. It is possible that animal pathogens with functional Type III systems may participate in some form of gene-for-gene interaction with their hosts similar to that seen in plant pathogenic bacteria. Although no *avr* genes have been identified in this group of animal pathogens, it is interesting to note that no function has yet been assigned to a number of proteins secreted via the Type III secretion system in these pathogenic bacteria. The investigation of this hypothesis is undoubtedly warranted.

Recent research has defined, in at least a few bacterial species, most of the bacterial genes required for their interaction with host cells through the Type III secretion systems. The future challenge will be to understand the mechanisms by which this protein secretion apparatus delivers bacterial proteins to the host cell and how these different effector proteins trigger host responses. Understanding these mechanisms at the molecular level will undoubtedly provide insight into a variety of animal and plant host-cell processes and open new avenues to explore the development of novel therapeutic strategies. The use of these natural interactions, which have been shaped by evolution, will continue to teach us perhaps unexpected but undoubtedly relevant aspects of host-cellular processes.

ACKNOWLEDGMENTS

We thank Debbie Black, Katrin Eichelberg, Wolf-Dietrich Hardt, and Stephanie Tucker for critical review of this manuscript and Carmen Collazo for help in assembling Tables 1 through 3. Work in our laboratories is supported by grants from the National Institutes of Health, the Pew charitable trust, and the American Heart Association.

Literature Cited

Adam T, Arpin M, Prévost M-C, Gounon P, Sansonetti PJ. 1995. Cytoskeletal rearrangements and the functional role of T-plastin during entry of *Shigella flexneri* into HeLa cells. *J. Cell. Biol.* 129:367–81

Adam T, Arpin M, Prévost M-C, Gounon P, Sansonetti PJ. 1996. *EMBO J.* In press

Adler B, Sasakawa C, Tobe T, Makino S, Komatsu K, Yoshikawa M. 1989. A dual transcriptional activation system for the 230 kilobase plasmid genes coding for virulence associated antigens of *Shigella flexneri*. *Mol. Microbiol.* 3:627–35

Alfano JR, Bauer DW, Milos TM, Collmer A. 1996. Analysis of the role of the *Pseudomonas syringae* pv. *syringae* HrpZ harpin in elicitation of the hypersensitive response in tobacco using functionally nonpolar deletion mutations, truncated HrpZ fragments, and *hrmA* mutations. *Mol. Microbiol.* 19:715–24

Allaoui A, Ménard R, Sansonetti PJ, Parsot C. 1993a. Characterization of the *Shigella flexneri ipgD* and *ipgF* genes, which are located in the proximal part of the *mxi* locus. *Infect. Immun.* 61:1707–14

Allaoui A, Sansonetti PJ, Parsot C. 1992. MxiJ, a lipoprotein involved in secretion of *Shigella* Ipa invasins, is homologous to YscJ, a secretion factor of the *Yersinia* Yop proteins. *J.*

Bacteriol. 174:7661–69

Allaoui A, Sansonetti PJ, Parsot C. 1993b. MxiD, an outer membrane protein necessary for the secretion of the *Shigella flexneri* Ipa invasins. *Mol. Microbiol.* 7:59–68

Allaoui A, Woestyn S, Sluiters C, Cornelis GR. 1994. YscU, a *Yersinia enterocolitica* inner membrane protein involved in Yop secretion. *J. Bacteriol.* 176:4534–42

Andrews GP, Hromockyj AE, Coker C, Maurelli AT. 1991. Two novel virulence loci, *mxi*A and *mxi*B, in *Shigella flexneri* 2a facilitate excretion of invasion plasmid antigen. *Infect. Immun.* 59:1997–2005

Andrews GP, Maurelli AT. 1992. *mxi*A of *Shigella flexneri* 2a, which facilitates export of invasion plasmid antigens, encodes a homologue of the low-calcium-response protein, LcrD, of *Yersinia pestis. Infect. Immun.* 60:3287–95

Arlat M, Van Gijsegem F, Pernollet JC, Boucher CA. 1994. PopA1, a protein which induces a hypersensitivity-like response on specific petunia genotypes, is secreted via the Hrp pathway of *Pseudomonas solanacearum. EMBO J.* 13:543–53

Bajaj V, Hwang C, Lee CA. 1995. *hil*A is a novel *ompR.toxR* family member that activates the expression of *Salmonella typhimurium* expression genes. *Mol. Microbiol.* 18:715–27

Baldwin TJ, Lee-Delaunay MB, Knutton S, Williams PH. 1993. Calcium-calmodulin dependence of actin accretion and lethality in cultured HEp-2 cells infected with enteropathogenic *Escherichia coli. Infect. Immun.* 61:760–63

Baldwin TJ, Ward W, Aitken A, Knutton S, Williams PH. 1991. Elevation of intracellular free calcium levels in HEp-2 cells infected with enteropathogenic *Escherichia coli. Infect. Immun.* 59:1599–604

Bar-Sagi D, Ferasmico JR. 1986. Induction of membrane ruffling and fluid-phase pinocytosis in quiescent fibroblasts by the ras proteins. *Science* 233:1061–68

Bergman T, Erickson K, Galyov E, Persson C, Wolf-Watz H. 1994. The *lcr*B (*ysc*N/U) gene cluster of *Yersinia pseudotuberculosis* is involved in Yop secretion and shows high homology to the *spa* gene clusters of *Shigella flexneri* and *Salmonella typhimurium. J. Bacteriol.* 176:2619–26

Beuscher HU, Rodel F, Forsberg A, Rollinghoff M. 1995. Bacterial evasion of host immune defense: *Yersinia enterocolitica* encodes a suppressor for tumor necrosis factor alpha expression. *Infect. Immun.* 63:1270–77

Bliska JB, Black DS. 1995. Inhibition of the Fc receptor-mediated oxidative burst in macrophages by the *Yersinia pseudotuberculosis* tyrosinephosphatase. *Infect. Immun.* 63:681–85

Bliska JB. 1994. Yops of the pathogenic *Yersinia* spp. In *Molecular Genetics of Bacterial Pathogenesis,* ed. VL Miller, JB Kasper, DA Portnoy, and RR Isberg. Washington DC: Am. Soc. Microbiol.

Bliska JB, Clemens JC, Dixon JE, Falkow S. 1992. The *Yersinia* tyrosine phosphatase: specificity of a bacterial virulence determinant for phosphoproteins in the J774A.1 macrophage. *J. Exp. Med.* 176:1625–30

Bliska JB, Copass MC, Falkow S. 1993a. The *Yersinia pseudotuberculosis* adhesin YadA mediates intimate bacterial attachment to and entry into HEp-2 cells. *Infect. Immun.* 61:3914–21

Bliska JB, Galán JE, Falkow S. 1993b. Signal transduction in the mammalian cell during bacterial attachment and entry. *Cell* 73:903–20

Bliska JB, Guan KL, Dixon JE, Falkow S. 1991. Tyrosine phosphate hydrolysis of host proteins by an essential *Yersinia* virulence determinant. *Proc. Natl. Acad. Sci. USA* 88:1187–91

Buckholm G. 1984. Effect of cytochalasin B and dihydrocytochalasin B on invasiveness of enteroinvasive bacteria in HEp-2 cell cultures. *Acta Pathol. Microbiol. Immun. Scand.* 92:145–49

Chang Y, Young R, Struck DK. 1989. Cloning and characterization of a hemolysin gene from *Actinobacillus (Haemphilus) pleuropneumoniae. DNA* 8:635–47

Chuan T-H, Bohl BP, Bokoch GM. 1993. Biologically active lipids are regulators of Rac-GDI complexation. *J. Biol. Chem.* 268:26206–11

Clerc P, Ryter A, Mounier J, Sansonetti PJ. 1987. Plasmid-mediated early killing of eukaryotic cells by *Shigella flexneri* as studied by infection of J774 macrophages. *Infect. Immun.* 55:521–27

Collazo C, Galán JE. 1996. Requirement of exported proteins for secretion through the invasion-associated Type III system in *Salmonella typhimurium. Infect. Immun.* In press

Collazo CM, Zierler MK, Galán JE. 1995. Functional analysis of the *Salmonella typhimurium* invasion genes *inv*I and *inv*J and identification of a target of the protein secretion apparatus encoded in the *inv* locus. *Mol. Microbiol.* 15:25–38

Collmer A, Bauer DW, Alfano JR, Preston G, Loniello AO, Milos TM. 1996. Extracellular proteins as determinants of pathogenicity in *Pseudomonas syringae.* In *Proc. 5th Int. Conf. Pseudomonas syringae. Pathovars and Related Pathogens, Berlin, 1995*. ed. K Rudolph, In press

Cornelis G, Sluiters C, de Rouvroit CL, Michiels T. 1989. Homology between VirF, the transcriptional activator of the *Yersinia* virulence regulon, and AraC, the *Escherichia coli* arabinose operon regulator. *J. Bacteriol.* 171:254–62

Dehio C, Prévost M-C, Sansonetti PJ. 1995. Invasion of epithelial cells by *Shigella flexneri* induces tyrosine phosphorylation of cortactin by a $pp60^{c\text{-}src}$-mediated signalling pathway. *EMBO J.* 14:2471–82

Donnenberg MS, Kaper JB. 1992. Enteropathogenic *Escherichia coli*. *Infect. Immun.* 60:3953–61

Dreyfus G, Williams AW, Kawagishi I, Macnab RM. 1993. Genetic and biochemical analysis of *Salmonella typhimurium* FliI, a flagellar protein related to the catalytic subunit of the F_0F_1 ATPase and to virulence proteins of mammalian and plant pathogens. *J. Bacteriol.* 175:3131–38

Eckmann L, Kagnoff MF, Fierer J. 1993. Epithelial cells secrete the chemokine interleukin-8 in response to bacterial entry. *Infect. Immun.* 61:4569–74

Eichelberg K, Ginocchio C, Galán JE. 1994. Molecular and functional characterization of the *Salmonella typhimurium* invasion genes *invB* and *invC*: Homology of *InvC* to the F_0F_1 ATPase family of proteins. *J. Bacteriol.* 176:4501–10

Eichelberg K, Kaniga K, Galán JE. 1995. Regulation of *Salmonella inv* gene expression by the flagellar sigma factor FliA (s^{28}). *Annu. Meet. Am. Soc. Microbiol.* Abstr.

Ernst RK, Domboski DM, Merrick JM. 1990. Anaerobiosis, Type 1 fimbriae, and growth phase are factors that affect invasion of HEp-2 cells by *Salmonella typhimurium*. *Infect. Immun.* 58:2014–16

Fällman M, Andersson K, Haåkansson S, Magnusson K-E, Stendahl O, Wolf-Watz H. 1995. *Yersinia pseudotuberculosis* inhibits Fc receptor-mediated phagocytosis in J774 cells. *Infect. Immun.* 63:3117–24

Fenselau S, Balbo I, Bonas U. 1992. Determinants of pathogenicity in *Xanthomonas campestris* pv. *vesicatoria* are related to proteins involved in the secretion in bacterial pathogens of animals. *Mol. Plant-Microbe Interact.* 5:390–96

Finlay BB, Falkow S. 1990. *Salmonella* interactions with polarized human intestinal Caco-2 epithelial cells. *J. Infect. Dis.* 162:1096–1106

Finlay BB, Rosenshine I, Donnenberg MS, Kaper JB. 1992. Cytoskeletal composition of attaching and effacing lesions associated with enteropathogenic *Escherichia coli* adherence to HeLa cells. *Infect. Immun.* 60:2541–43

Finlay BB, Ruschkowski S. 1991. Cytoskeletal rearrangements accompanying *Salmonella* entry into epithelial cells. *J. Cell Sci.* 99:283–96

Forsberg A, Rosqvist R, Wolf-Watz H. 1994. Regulation and polarized transfer of the *Yersinia* outer proteins (Yops) involved in antiphagocytosis. *Trends Microbiol.* 2:14–19

Forsberg A, Vitanen AM, Skurnik M, Wolf-Watz H. 1991. The surface-located YpoN protein is involved in calcium signal transduction in *Yersinia pseudotuberculosis*. *Mol. Microbiol.* 5:977–86

Foubister V, Rosenshine I, Finlay BB. 1994. A diarrheal pathogen, enteropathogenic *Escherichia coli* (EPEC), triggers a flux of inositol phosphates in infected epithelial cells. *J. Exp. Med.* 179:993–98

Galán JE, Curtiss R III. 1990. Expression of *Salmonella typhimurium* genes required for invasion is regulated by changes in DNA supercoiling. *Infect. Immun.* 58:1879–85

Galán JE, Ginocchio C, Costeas P. 1992a. Molecular and functional characterization of the *Salmonella typhimurium* invasion gene *inv*A: homology of InvA to members of a new protein family. *J. Bacteriol.* 17:4338–49

Galán JE, Pace J, Hayman MJ. 1992b. Involvement of the epidermal growth factor receptor in the invasion of the epithelial cells by *Salmonella typhimurium*. *Nature* 357:588–89

Galyov EE, Haåkansson S, Forsberg A, Wolf-Watz H. 1993. A secreted protein kinase of *Yersinia pseudotuberculosis* is an indispensable virulence determinant. *Nature* 361:730–32

Galyov EE, Haåkansson S, Wolf-Watz H. 1994. Characterization of the operon encoding the YpkA Ser/Thr protein kinase and the YopJ protein of *Yersinia*. *J. Bacteriol.* 176:4543–48

Garcia-del Portillo F, Finlay BB. 1994. *Salmonella* invasion of nonphagocytic cells induces formation of macropinosomes in the host cell. *Infect. Immun.* 62:4641–45

Genin S, Boucher C. 1994. A superfamily of proteins involved in different secretion pathways in gram-negative bacteria: modular structure and specificity of the N-terminal domain. *Mol. Gen. Genet.* 243:112–18

Genin S, Gough CL, Zischek C, Boucher CA. 1992. Evidence that the *hrpB* gene encodes a positive regulator of pathogenicity genes from *Pseudomonas solanacearum*. *Mol. Microbiol.* 6:3065–76

Ginocchio C, Galán JE. 1995. Functional conservation among members of the *Salmonella typhimurium* InvA family of proteins. *Infect.*

Immun. 63:729–32

Ginocchio C, Olmsted SB, Wells CL, Galán JE. 1994. Contact with epithelial cells induces the formation of surface appendages on *Salmonella typhimurium. Cell* 76:717–24

Ginocchio C, Pace J, Galán JE. 1992. Identification and molecular characterization of a *Salmonella typhimurium* gene involved in triggering the internalization of Salmonellae into cultured epithelial cells. *Proc. Natl. Acad. Sci. USA* 89:5976–80

Goguen JD, Walker WS, Hatch TP, Yother J. 1986. Plasmid-determined cytotoxicity in *Yersinia pestis* and *Yersinia pseudotuberculosis. Infect. Immun.* 51:788–94

Goldberg MB, Sansonetti PJ. 1993. *Shigella* subversion of the cellular cytoskeleton: a strategy for epithelial colonization. *Infect. Immun.* 61:4941–46

Gough CL, Genin S, Lopes V, Boucher CA. 1993. Homology between the HrpO protein of *Pseudomonas solanacearum* and bacterial proteins implicated in a signal peptide-independent secretion mechanism. *Mol. Gen. Genet.* 239:378–92

Gough CL, Genin S, Zischek C, Boucher C. 1992. *hrp* genes of *Pseudomonas solanacearum* are homologous to pathogenicity determinants of animal pathogenic bacteria and are conserved among plant pathogenic bacteria. *Mol. Plant-Microbe Interact.* 5:384–89

Grimm C, Aufsatz W, Panopoulos NJ. 1995. The hrpRS locus of *Pseudomonas syringae* pv. *phaseolicola* constitutes a complex regulatory unit. *Mol. Microbiol.* 15:155–65

Groisman EA, Ochman H. 1993. Cognate gene clusters govern invasion of host epithelial cells by *Salmonella typhimurium* and *Shigella flexneri. EMBO J.* 12:3779–87

Guan K, Dixon JE. 1990. Protein tyrosine phosphatase activity of an essential virulence determinant in *Yersinia. Science* 249:553–56

Hackett M, Guo L, Shabanowitz J, Hunt DF, Hewlett EL. 1994. Internal lysine palmitoylation in adenylate cyclase toxin from *Bordetella pertussis. Science* 266:433–35

Haddix PL, Straley SC. 1992. Structure and regulation of the *Yersinia pestis yscBCDEF* operon. *J. Bacteriol.* 174:4820–28

Haigh R, Baldwin T, Knutton S, Williams PH. 1995. Carbon dioxide regulated secretion of the EaeB protein of enteropathogenic *Escherichia coli. FEMS Microbiol Lett.* 129:63–67

Haåkansson S, Bergman T, Vanooteghem JC, Cornelis G, Wolf-Watz H. 1993. YopB and YopD constitute a novel class of *Yersinia* Yop proteins. *Infect. Immun.* 61:71–80

Hale TL. 1991. Genetic basis of virulence in *Shigella* species. *Microbiol. Rev.* 55:206–24

Han JW, McCormick F, Macara IG. 1991. Regulation of Ras-Gap and neurofibromatosis-1 gene product by eicosanoids. *Science* 252:576–79

Hartland EL, Green SP, Phillips WA, Robins-Browne RM. 1994. Essential role of YopD in inhibition of the respiratory burst of macrophages by *Yersinia enterocolitica. Infect. Immun.* 62:4445–53

He SY, Huang H-C, Collmer A. 1993. *Pseudomonas syringae* pv. *syringae* Harpin^{Pss}: a protein that is secreted via the Hrp pathway and elicits the hypersensitive response in plants. *Cell* 73:1255–66

Hermant D, Ménard R, Arricau N, Parsot C, Popoff MY. 1995. Functional conservation of the *Salmonella* and *Shigella* effectors of entry into epithelial cells. *Mol. Microbiol.* 17:781–89

High N, Mounier J, Prévost M-C, Sansonetti PJ. 1992. IpaB of *Shigella flexneri* causes entry into epithelial cells and escape from the phagocytic vacuole. *EMBO J.* 11:1991–99

Holland IB, Blight MA, Kenny B. 1990. The mechanism of secretion of hemolysin and other polypeptides from gram-negative bacteria. *J. Bioenerg. Biomembr.* 22:473–91

Huang H-C, He SY, Bauer DW, Collmer A. 1992. The *Pseudomonas syringae* pv. *syringae* 61 hrpH product, an envelope protein required for elicitation of the hypersensitive response in plants. *J. Bacteriol.* 174:6878–85

Huang H-C, Lin R-H, Chang C-J, Collmer A, Deng W-L. 1995. The complete *hrp* gene cluster of *Pseudomonas syringae* pv. *syringae* 61 includes two blocks of genes required for Harpin Pss secretion that are arranged colinearly with *Yersinia ysc* homologs. *Mol. Plant-Microbe Interact.* 8:733–46

Huang H-C, Xiao Y, Lin R-H, Lu Y, Hutcheson SW, Collmer A. 1993. Characterization of the *Pseudomonas syringae* pv. *syringae* 61 *hrpJ* and *hrpI* genes: homology of HrpI to a superfamily of proteins associated with protein translocation. *Mol. Plant Microbe Interact.* 6:515–20

Huynh TV, Dahlbeck D, Staskawicz BJ. 1989. Bacterial blight of soybean: regulation of a pathogen gene determining host cultivar specificity. *Science* 245:1374–77

Issartel J-P, Koronakis V, Hughes C. 1991. Activation of *Escherichia coli* prohaemolysin to the mature toxin by acyl carrier protein-dependent fatty acylation. *Nature* 351:759–61

Jarvis KG, Giron JA, Jerse AE, McDaniel TK, Donnenberg MS, Kaper JB. 1995. Enteropathogenic *Escherichia coli* contains a putative type III secretion system necessary for export of proteins involved in attaching and effacing lesion formation. *Proc. Natl. Acad. Sci. USA* 92:7996–8000

Jones BD, Paterson HF, Hall A, Falkow S. 1993. *Salmonella typhimurium* induces membrane ruffling by a growth factor-receptor-independent mechanism. *Proc. Natl. Acad. Sci. USA* 90:10390–94

Kadowaki T, Koyasu S, Nishida E, Sakai H, Takaku F, et al. 1986. Insulin-like growth factors, insulin, and epidermal growth factor cause rapid cytoskeletal reorganization in KB cells. *J. Biol. Chem.* 261:16141–47

Kaniga K, Bossio JC, Galán JE. 1994. The *Salmonella typhimurium* invasion genes *inv*F and *inv*G encode homologues to the PulD and AraC family of proteins. *Mol. Microbiol.* 13:555–68

Kaniga K, Trollinger D, Galán JE. 1995a. Identification of two targets of the type III secretion system encoded in the *inv* and *spa* loci of *Salmonella typhimurium* that share homology to IpaD and IpaA proteins. *J. Bacteriol.* 177:7078–85

Kaniga K, Tucker SC, Trollinger D, Galán JE. 1995b. Homologues of the *Shigella* invasins IpaB and IpaC are required for *Salmonella typhimurium* entry into cultured cells. *J. Bacteriol.* 177:3965–71

Kaniga K, Uralil J, Bliska JB, Galán JE. 1996. A secreted tyrosine phosphatase with modular effector domains encoded by the bacterial pathogen *Salmonella typhimurium*. *Mol. Microbiol.* In press

Kenny B, Finlay BB. 1995. Protein secretion by enteropathogenic *Escherichia coli* is essential for transducing signals to epithelial cells. *Proc. Natl. Acad. Sci. USA* 92:7991–95

Kenny B, Lai L-C, Finlay BB, Donnemberg MS. 1996. EspA, a protein secreted by enteropathogenic *Escherichia coli* is required to induce signals in epithelial cells. *Mol. Microbiol.* 20:313–23

Kihlstrom E, Nilsson L. 1977. Endocytosis of *Salmonella typhimurium* 395 MS and MR10 by HeLa cells. *Acta Pathol. Microbiol. Scand.* 85:322–28

Klauser T, Pohlner J, Meyer TF. 1993. The secretion pathway of IgA protease-type proteins in gram-negative bacteria. *BioEssays* 15:799–805

Kulich SM, Yahr TL, Mende-Mueller LM, Barbieri JT, Frank DW. 1994. Cloning the structural gene for the 49-kDa form of exoenzyme S (*exoS*) from *Pseudomonas aeruginosa* strain 388. *J. Biol. Chem.* 269:10431–37

Lee CA, Falkow S. 1990. The ability of *Salmonella* to enter mammalian cells is affected by bacterial growth state. *Proc. Natl. Acad. Sci. USA* 87:4304–8

Leung KY, Reisner BS, Straley S. 1990. YopM inhibits platelet aggregation and is necessary for virulence of *Yersinia pestis* in mice. *Infect. Immun.* 58:3262–71

Leung KY, Straley SC. 1989. The *yop*M gene of *Yersinia pestis* encodes a released protein having homology with the platelet surface protein GP1b. *J. Bacteriol.* 171:4623–32

Li J, Ochman H, Groisman EA, Boyd EF, Salomon F, et al. 1995. Relationship between evolutionary rate and cellular location among the Inv/Spa invasion proteins of *Salmonella enterica*. *Proc. Natl. Acad. Sci. USA* 92:7252–56

Lidell MC, Hutcheson SW. 1994. Characterization of the *hrpJ* and *hrpU* operons of *Pseudomonas syringae* pv. *syringae* Pss61: similarity with components of enteric bacteria involved in flagellar biogenesis and demonstration of their role in Harpin Pss secretion. *Mol. Plant-Microbe Interact.* 7:488–97

Lindgren PB, Peet RC, Panopoulos NJ. 1986. Gene cluster in *Pseudomonas syringae* pv. *phaseolicola* controls pathogenicity of bean plants and hypersensitivity on non-host plants. *J. Bacteriol.* 168:512–22

Lindsay WP, Lamb CJ, Dixon RA. 1993. Microbial recognition and activation of plant defense systems. *Trends Microbiol.* 1:181–87

Lo RYC, Strathdee CA, Shewen PE. 1987. Nucleotide sequence of the leukotoxin genes of *Pasterurella haemolytica* A1. *Infect. Immun.* 55:1987–96

Long SR, Staskawicz BJ. 1993. Prokaryotic plant parasites. *Cell* 73:921–35

Lottemberg R, Minning-Wenz D, Boyle DP. 1994. Capturing host plasmin(ogen): a common mechanism for invasive pathogens? *Trends Microbiol.* 2:20–24

Magnuson K, Jackowski S, Rock CO, Cronan JE. 1993. Regulation of fatty acid biosynthesis in *Escherichia coli*. *Microbiol Rev.* 57:522–42

Manjarrez-Hernandez HA, Amess B, Sellers L, Baldwin TJ, Knutton S, et al. 1991. Purification of a 20 kDa phosphoprotein from epithelial cells and identification as a myosin light chain. Phosphorylation induced by enteropathogenic *Escherichia coli* and phorbol ester. *FEBS Lett.* 292:121–27

Maurelli AT, Blackman B, Curtiss R III. 1984. Temperature dependent expression of virulence genes in *Shigella* species. *Infect. Immun.* 43:195–201

Ménard R, Prévost M-C, Gounon P, Sansonetti PJ, Dehio C. 1996. The secreted Ipa complex of *Shigella flexneri* promotes entry into mammalian cells. *Proc. Natl. Acad. Sci. USA*

93:1254–58

Ménard R, Sansonetti PJ, Parsot C. 1993. Non-polar mutagenesis of the *ipa* genes defines IpaB, IpaC, and IpaD as effectors of *Shigella flexneri* entry into epithelial cells. *J. Bacteriol.* 175:5899–906

Ménard R, Sansonetti PJ, Parsot C. 1994a. The secretion of the *Shigella flexneri* Ipa invasins is induced by the epithelial cell and controlled by IpaB and IpaD. *EMBO J.* 13:5293–302

Ménard R, Sansonetti PJ, Parsot C, Vasselon T. 1994b. The IpaB and IpaC invasins of *Shigella flexneri* associate in the extracellular medium and are partitioned in the cytoplasm by a specific chaperon. *Cell* 79:515–29

Michiels T, Vanooteghem JC, Lambert de Rouvroit C, China B, Gustin A, et al. 1991. Analysis of *virC*, an operon involved in the secretion of Yop proteins by *Yersinia enterocolitica*. *J. Bacteriol.* 173:4994–5009

Miras I, Hermant D, Arricau N, Popoff MY. 1995. Nucleotide sequence of *iagA* and *iagB* genes involved in invasion of HeLa cells by *Salmonella enterica* ssp. *enterica* ser. Typhi. *Res. Microbiol.* 146:17–20

Mooseker MS. 1985. Organization, chemistry and assembly of the cytoskeletal apparatus of the intestinal brush border. *Annu. Rev. Cell. Biol.* 1:209–41

Motin VL, Nakajima R, Smirnov GB, Brubaker RR. 1994. Passive immunity to Yersiniae mediated by anti-recombinant V antigen and protein A-V fusion peptide. *Infect. Immun.* 62:4192–201

Mulholland V, Hinton JCD, Sidebotham J, Toth IK, Hyman LJ, et al. 1993. A pleiotropic reduced virulence (Rvi$^-$) mutant of *Erwinia carotovora* subspecies *atroseptica* is defective in flagella assembly proteins that are conserved in plant and animal bacterial pathogens. *Mol. Microbiol.* 9:343–56

Niepold F, Anderson D, Mills D. 1985. Cloning determinants of pathogenesis from *Pseudomonas syringae* pv. *syringae*. *Proc. Natl. Acad. Sci. USA* 82:406–10

Pace J, Hayman MJ, Galán JE. 1993. Signal transduction and invasion of epithelial cells by *Salmonella typhimurium*. *Cell* 72:505–14

Parsot C, Ménard R, Gounon P, Sansonetti PJ. 1995. Enhanced secretion through the *Shigella flexneri* Mxi-Spa translocon leads to assembly of extracellular proteins into macromolecular structures. *Mol. Microbiol.* 16:291–300

Pegues DA, Hantman MJ, Behlau I, Miller SI. 1995. PhoP/PhoQ transcriptional repression of *Salmonella typhimurium* invasion genes: evidence for a role in protein secretion. *Mol. Microbiol.* 17:169–81

Peppelenbosch MP, Qiu RG, de Vries-Smits AM, Tertoolen LG, de Laat SW, et al. 1995. Rac mediates growth factor-induced arachidonic acid release. *Cell* 81:849–56

Peppelenbosch MP, Tertoolen LG, Hage WJ, de Laat SW. 1993. Epidermal growth factor-induced actin remodeling is regulated by 5-lipoxygenase products. *Cell* 74:565–75

Persson C, Nordfelth R, Holmström A, Haåkansson S, Rosqvist R, Wolf-Watz H. 1995. Cell surface-bound *Yersinia* translocate the protein tyrosine phosphatase YopH by a polarized mechanism into the target cell. *Mol. Microbiol.* 18:135–50

Plano GV, Barve SS, Straley SC. 1991. LcrD, a membrane-bound regulator of the *Yersinia pestis* low-calcium response. *J. Bacteriol.* 173:7293–303

Plano GV, Straley SC. 1995. Mutations in yscC, yscD, and yscG prevent high-level expression and secretion of V antigen and Yops in *Yersinia pestis*. *J. Bacteriol.* 177:3843–54

Price SB, Leung KY, Barve SS, Straley SC. 1989. Molecular analysis of lcrGVH, the V antigen operon of *Yersinia pestis*. *J. Bacteriol.* 171:5646–53

Pugsley AP. 1993. The complete general secretory pathway in gram-negative bacteria. *Microbiol. Rev.* 57:50–108

Pugsley AP, Kornacker MG, Poquet I. 1991. The general protein-export pathway is directly required for extracellular pullulanase secretion in *Escherichia coli* K12. *Mol. Microbiol.* 5:343–52

Rahme LG, Mindrinos MN, Panopoulos NJ. 1992. Plant and environmental sensory signals control the expression of *hrp* genes in *Pseudomonas syringae* pv. *phaseolicola*. *J. Bacteriol.* 174:3499–507

Rahme LG, Stevens EJ, Wolfort SF, Shao J, Tompkins RG, Ausubel FM. 1995. Common virulence factors for bacterial pathogenicity in plants and animals. *Science* 268:1899–902

Reisner BS, Straley SC. 1992. *Yersinia pestis* YopM: thrombin binding and overexpression. *Infect. Immun.* 60:5242–52

Rimpilainen M, Forsberg A, Wolf-Watz H. 1992. A novel protein, LcrQ, involved in the low-calcium response of *Yersinia pseudotuberculosis* shows extensive homology to YopHJ. *Bacteriol.* 174:3355–63

Rosenshine I, Donnenberg MS, Kaper JB, Finlay BB. 1992. Signal transduction between enteropathogenic *Escherichia coli* (EPEC) and epithelial cells: EPEC induces tyrosine phosphorylation of host cell proteins to initiate cytoskeletal rearrangement and bacterial uptake. *EMBO J.* 11:3551–60

Rosenshine I, Ruschkowski S, Foubister V, Finlay BB. 1994. *Salmonella typhimurium* invasion of epithelial cells: role of the induced host cell tyrosine protein phosphorylation. *Infect. Immun.* 62:4969–74

Rosqvist R, Bolin I, Wolf-Watz H. 1988. Inhibition of phagocytosis in *Yersinia pseudotuberculosis*: a virulence plasmid-encoded ability involving the Yop2b protein. *Infect. Immun.* 56:2139–43

Rosqvist R, Forsberg A, Wolf-Watz H. 1990. The cytotoxic protein YopE of *Yersinia* obstructs the primary host defence. *Mol. Microbiol.* 4:657–67

Rosqvist R, Forsberg A, Wolf-Watz H. 1991. Intracellular targeting of the *Yersinia* YopE cytotoxin in mammalian cells induces actin microfilament disruption. *Infect. Immun.* 59:4562–69

Rosqvist R, Haåkansson S, Forsberg A, Wolf-Watz H. 1995. Functional conservation of the secretion and translocation machinery for virulence proteins of Yersiniae, Salmonellae and Shegellae. *EMBO J.* 14:4187–95

Rosqvist R, Magnusson KE, Wolf-Watz H. 1994. Target cell contact triggers expression and polarized transfer of *Yersinia* YopE cytotoxin into mammalian cells. *EMBO J.* 13:964–72

Ruschkowski S, Rosenshine I, Finlay BB. 1992. *Salmonella typhimurium* induces an inositol phosphate flux in infected epithelial cells. *FEMS Lett.* 74:121–26

Russel M. 1995. Moving through the membrane with filamentous phages. *Trends Microbiol.* 3:223–28

Sakai T, Sasakawa C, Makino S, Yoshikawa M. 1986. DNA sequence and product analysis of the *vir*F locus responsible for Congo red binding and cell invasion in *Shigella flexneri* 2a. *Infect. Immun.* 54:395–402

Salmond GPC, Reeves PJ. 1993. Membrane traffic wardens and protein secretion in gram-negative bacteria. *Trends Biochem. Sci.* 18:7–12

Sasakawa C, Komatsu K, Tobe T, Suzuki T, Yoshikawa M. 1993. Eight genes in region 5 that form an operon are essential for invasion of epithelial cells by *Shigella flexneri* 2a. *J. Bacteriol.* 175:2334–46

Simonen M, Palva I. 1993. Protein secretion in *Bacillus* species. *Microbiol. Rev.* 57:109–37

Skrzypek E, Straley S. 1995. Differential effects of deletions in *lcrV* on secretion of V antigen, regulation of the low-Ca^{2+} response, and virulence of *Yersinia pestis*. *J. Bacteriol.* 177:2530–42

Sory M-P, Boland A, Lambermount I, Cornelis G. 1995. Identification of the YopE and YopH domains required for secretion and internalization into the cytosol of macrophages, using the *cyaA* gene fusion approach. *Proc. Natl. Acad. Sci. USA* 92:11998–2002

Sory M-P, Cornelis GR. 1994. Translocation of a hybrid YopE-adenylate cyclase from *Yersinia enterocolitica* into HeLa cells. *Mol. Microbiol.* 14:583–94

Stanley P, Packman LC, Koronakis V, Hughes C. 1994. Fatty acylation of two internal lysine residues required for the toxic activity of *Escherichia coli* hemolysin. *Science* 266:1992–96

Staskawicz BJ, Ausubel FM, Baker BJ, Ellis JG, Jones JDG. 1995. Molecular genetics of plant disease resistance. *Science* 268:661–67

Straley SC. 1991. The low-Ca^{2+} response virulence regulon of human-pathogenic Yersiniae. *Microb. Pathog.* 10:87–91

Straley SC, Skrzypek E, Plano GV, Bliska JB. 1993. Yops of *Yersinia* spp. pathogenic for humans. *Infect. Immun.* 61:3105–10

Takeuchi A. 1967. Electron microscopic studies of experimental *Salmonella* infection. 1. Penetration into the intestinal epithelium by *Salmonella typhimurium*. *Am. J. Pathol.* 50:109–36

Theriot J. 1995. The cell biology of infection by intracellular bacterial pathogens. *Annu. Rev. Cell Dev. Biol.* 11:213–39

Tsai MH, Yu CL, Stacey DW. 1990. A cytosplasmic protein inhibits the GTPase activity of H-Ras in a phospholipid dependent manner. *Science* 250:962–65

Uchiya K, Tobe T, Komatsu K, Suzuki T, Watari M, et al. 1995. Identification of a novel virulence gene, *virA*, on the large plasmid of *Shigella*, involved in invasion and intercellular spreading. *Mol. Microbiol.* 17:241–50

Van Gijsegem F, Genin S, Boucher C. 1993. Conservation of secretion pathways for pathogenicity determinants of plant and animal bacteria. *Trends Microbiol.* 1:175–80

Van Gijsegem F, Gough C, Zischek C, Niqueux E, Arlat M, et al. 1995. The *hrp* gene locus of *Pseudomonas solanacearum*, which controls the production of a type III secretion system, encodes eight proteins related to components of the bacterial flagellar biogenesis complex. *Mol. Microbiol.* 15:1095–114

Venkatesan MM, Buysse JM, Oaks EV. 1992. Surface presentation of *Shigella flexneri* invasion plasmid antigens requires the products of the *spa* locus. *J. Bacteriol.* 174:1990–2001

Watarai M, Tobe T, Yoshikawa M, Sasakawa C. 1995a. Contact of *Shigella* with host cells triggers release of Ipa invasins and is an essential function of invasiveness. *EMBO J.* 14:2461–70

Watarai M, Tobe T, Yoshikawa M, Sasakawa C. 1995b. Disulfide oxireductase activity of *Shigella flexneri* is required for release of Ipa proteins and invasion of epithelial cells. *Proc.*

Natl. Acad. Sci. USA 92:4927–31

Wattiau P, Bernier B, Deslée P, Michiels T, Cornelis GR. 1994. Individual chaperones required for Yop secretion by *Yersinia. Proc. Natl. Acad. Sci. USA* 91:10493–97

Wattiau P, Cornelis GR. 1993. SycE, a chaperone-like protein of *Yersinia enterocolitica* involved in the secretion of YopE. *Mol. Microbiol.* 8:123–31

Wattiau P, Woestyn S, Cornelis GR. 1996. Customized secretion chaperones in pathogenic bacteria. *Mol. Microbiol.* In press

Wei ZM, Beer SV. 1995. hrpL activates *Erwinia amylovora hrp* gene transcription and is a member of the ECF subfamily of sigma factors. *J. Bacteriol.* 177:6201–10

Wei ZM, Laby RJ, Zumoff CH, Bauer DW, He SY, et al. 1992a. Harpin, elicitor of the hypersensitive response produced by the plant pathogen *Erwinia amylovora. Science* 257:85–88

Wei ZM, Sneath BJ, Beer SV. 1992b. Expression of *Erwinia amylovora hrp* genes in response to environmental stimuli. *J. Bacteriol.* 174:1875–82

Woestyn S, Allaoui A, Wattiau P, Cornelis GR. 1994. YscN, the putative energizer of the *Yersinia* Yop secretion machinery. *J. Bacteriol.* 176:1561–69

Xiao Y, Heu S, Yi J, Lu Y, Hutcheson SW. 1994. Identification of a putative alternate sigma factor and characterization of a multicomponent regulatory cascade controlling the expression of *Pseudomonas syringae* pv. *syringae* Pss61 *hrp* and *hrmA* genes. *J. Bacteriol.* 176:1025–36

Yao R, Palchauduri S. 1992. Nucleotide sequence and transcriptional regulation of a positive regulatory gene of *Shigella dysenteriae. Infect. Immun.* 60:1163–69

Zierler M, Galán JE. 1995. Contact with cultured epithelial cells induces the secretion of the *Salmonella typhimurium* invasion protein InvJ. *Infect. Immun.* 63:4024–28

Zychlinsky A, Fitting C, Cavaillon JM, Sansonetti PJ. 1994a. Interleukin 1 is released by macrophages during apoptosis induced by *Shigella flexneri. J. Clin. Invest.* 94:1328–32

Zychlinsky A, Kenny B, Ménard R, Prévost M-C, Holland IB, Sansonetti PJ. 1994b. IpaB mediates macrophage apoptosis induced by *Shigella flexneri. Mol. Microbiol.* 11:619–27

Zychlinsky A, Prévost MC, Sansonetti PJ. 1992. *Shigella flexneri* induces apoptosis in infected macrophages. *Nature* 358:167–69

Annu. Rev. Cell Dev. Biol. 1996. 12:257–304

ACQUISITION OF IDENTITY IN THE DEVELOPING LEAF

Anne W. Sylvester
Department of Biological Sciences, University of Idaho, Moscow Idaho 83844

Laurie Smith
Department of Biology, University of North Carolina, Chapel Hill, North Carolina 27599

Michael Freeling
Department of Plant Biology, University of California, Berkeley, California 94720

KEY WORDS: maize, tobacco, position-dependent, cell lineage, mutant

ABSTRACT

Leaves are produced repeatedly from the shoot apical meristem of plants. Molecular and cellular evidence show that identity of the leaf and its parts is acquired progressively and that the underlying process changes as the leaf matures. The relative importance of cell lineage compared with a position-dependent model for specifying cell fates is discussed.

CONTENTS

1081-0706/96/1115-00257$08.00

INTRODUCTION

Leaves are distinctive plant organs that arise from an indistinct group of cells in the shoot apical meristem. For this reason, leaves are developmentally intriguing organs. Not only is the unique shape and function of each leaf acquired gradually from an otherwise uniform set of cells, but the partitioning of a new leaf from the shoot apical meristem occurs repeatedly. Plant developmental biologists face a myriad of questions regarding how leaves develop. How do the undifferentiated cells of the meristem give rise to cells destined to become leaf initials? Once emerged, how is regional and cellular identity acquired within the developing leaf? Underlying these questions, and at the focus of this review, is the issue of commitment; that is, when and how does a plant cell or field of cells become committed to a particular fate?

Plant cells have certain unique characteristics that influence how identity is acquired. For instance, plant cells are flexible in their developmental programming (Waring 1978, Henshaw et al 1982, Steeves & Sussex 1989, Fosket 1994). Generally the outcome of cellular commitment is differentiation; yet it is well known that living differentiated cells in the plant body, once excised and given supplemental hormones and nutrients, are capable of transdifferentiation (reviewed in Halperin 1969, Steeves & Sussex 1989, Sugiyama & Komamine 1990). The ease with which the identity of plant cells can be manipulated suggests that determination is not strictly an irreversible process. Any understanding of the process of acquiring identity must take into account this element of totipotency.

Another unique aspect of plant cells is the presence of a cell wall, which imposes certain structural constraints on cell activity and morphology. Plant cells do not migrate to acquire developmental information but instead remain relatively immobile. Based on this fact, the orientation of cell division is deemed important because it establishes the placement of daughter cells in the context of the developing organ. It is frequently argued that there is concordance between cell orientation and organ identity: A precise series of cell divisions and cell expansions is a prerequisite for the proper acquisition of identity (see Alberts et al 1994). Acquisition of identity is not a single process or a one-time event, however, and the impact of individual cell divisions may vary depending on the stage of development. Early growth of the primordium, for example, when shape is being established, may be less dependent on specific orientations of

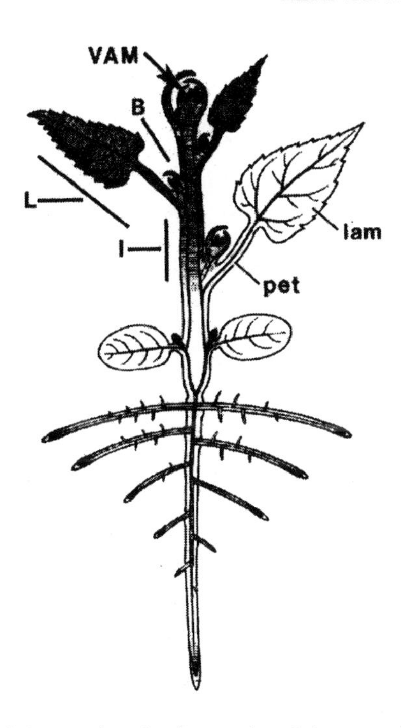

Figure 1 Typical dicotyledonous plant showing a series of phytomers. Each phytomer produced by the vegetative apical meristem (VAM) consists of three parts: a leaf (L), internode (I), and axillary bud (B). The leaf is subdivided into the flattened photosynthetic lamina (lam), petiole (pet), and adjoining basal structures (not shown). Shading represents areas of cell division and expansion in the shoot. (Redrawn from Foster & Gifford 1974.)

individual cells than later aspects of differentiation, when more precise cell positioning is required to maintain cell relationships in a functioning tissue.

Basic organ identity is first acquired when the apical meristem begins to produce a reiterated segment of the shoot system, frequently termed a phytomer (e.g. Poethig & Szymkowiak 1995). Although not a precisely defined unit (Poethig 1994), the phytomer generally consists of several parts: a foliage organ, or leaf, subtended by an internode, and an axillary bud (Figures 1, 2). Phytomers are present in both dicotyledons (Figure 1) and monocotyledons (Figure 2), but variations in the pattern and timing of growth of the phytomer parts provide the range of architectural patterns and diversity of form in plants (e.g. Kaplan 1970, Steeves & Sussex 1989). Herein, we consider development of the leaf organ only and specifically discuss how identity of the leaf and its parts is acquired.

Several overlapping but distinctive developmental events occur during emergence and subsequent growth of the leaf: The organ is first partitioned from the shoot apical meristem, at which time morphological continuity with the shoot apex is maintained while leaf shape unfolds. Next, cell types and tissues of a given function begin to differentiate within the organ. It is likely that cell position within the context of the developing organ impacts future fate more commonly than does the precise lineage of a cell. However, cells and fields of cells may acquire identity in different ways, in part depending on the stage of development of the leaf. To accomodate this general trend, we define three stages of leaf development. The first stage is organogenesis, the time when future leaf cells are set aside from the meristem and the leaf primordium emerges.

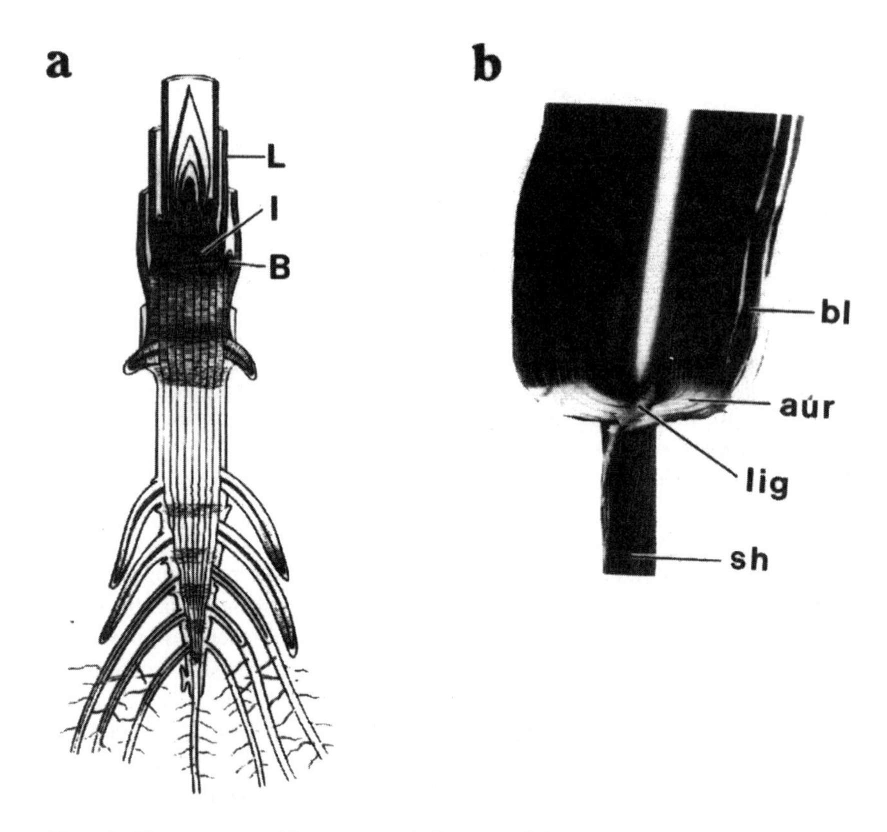

Figure 2 Shoot structure of *Zea mays*, a typical monocotyledon. (*a*) Each phytomer, derived from the vegetative apical meristem consists of an encircling leaf (L) and subtending internode (I) and axillary bud (B). (*b*) The mature leaf is subdivided into a distal blade (bl), a proximal sheath (sh), separated by the ligule (lig) and auricle (aur). (Redrawn from Foster & Gifford 1974; photograph courtesy of R Frey.)

Parameters defining how this occurs remain largely unknown but are considered briefly herein (for reviews, see Sussex 1989, Steeves a & Sussex 1989, Poethig 1989, Medford 1992, Smith & Hake 1992, Harper & Freeling 1996). The second stage, referred to as primary morphogenesis by D Kaplan (manuscript in preparation), encompasses early growth of the primordium, when regions of the developing leaf are established. We review how and when regional identity is acquired, using maize as a model analytical system (Freeling 1992, Smith & Hake 1992, Sinha et al 1993, Harper & Freeling 1996). The third stage is differentiation of the leaf, when cells and tissue acquire final identity and function. Although oriented cell divisions are frequently required during this third stage, questions of the spatial control of cytokinesis, particularly via the plant cytoskeleton, though pertinent, are not considered here (see Wick 1991, Lloyd 1991, 1995, Staiger & Lloyd 1991, Traas et al 1995). Instead, the epidermis is presented as a model for understanding how individual cell types assume the unique distribution and form characteristic of a differentiated tissue.

These three stages should not be viewed as discrete and isolated steps, but rather as landmarks in a continuous developmental process: Leaf shape continues to unfold even later during stage three of differentiation. Similarly, the differentiation of particular tissues can begin early in localized regions of a developing primordium. The stages discussed here are tools with which to appreciate subtle differences in how cells and fields of cells acquire identity. In the following, we present a descriptive account of each of the pertinent stages, followed by a discussion of general concepts of how identity is acquired. Finally, we consider the details of how identity is acquired at each of these three stages.

DESCRIPTION OF STAGES OF LEAF DEVELOPMENT

The leaf pattern first emerges during embryogenesis, when the shoot axis is partitioned into an apical meristem flanked by cotyledons. Mutants arrested very early in the formation of cotyledons do not produce normal leaf organs in *Arabidopsis* (Vernon & Meinke 1995) or maize (Clark & Sheridan 1991, Sheridan 1995). Limited histogenesis can still proceed in mutant *Arabidopsis* embryos arrested at the earliest globular stage of embryogenesis (Yadegari et al 1994, Meinke 1991), but neither cotyledons nor leaves are produced in these mutants. The basic pattern of a repetitively produced leaf appears to be associated with the proper formation of a shoot apical meristem: A mutant, *shootmeristemless*, develops normal cotyledons but lacks an embryonic shoot meristem (Barton & Poethig 1993). The molecular mechanism by which cells commit to producing leaves in the embryonic shoot axis is currently being investigated and is considered elsewhere (Meinke 1991, 1995, West & Harada 1993, Jurgens et al 1994, Clark 1996, Sheridan 1995, Long et al 1996).

The shoot pattern established in the embryo persists throughout the vegetative life of the plant. Such a pre-established design is not absolutely required for the production of leaves, however. Many plants retain the ability to generate vegetative meristems out of previously unpatterned or unique tissue (Mauseth 1988; D Kaplan, manuscript in preparation). For example, primordia do not have to be present for an isolated shoot meristem to continue to generate new leaves (Ball 1960, Smith & Murashige 1970). An isolated mature leaf of *Graptopetalum* can give rise to a new meristem, new leaves, and ultimately a normal shoot (Green & Brooks 1978). New shoot-forming meristems emerge naturally from mature leaf margins in some plant species such as *Kalanchoe* (D Kaplan, manuscript in preparation). Finally, ectopic expression of *knotted 1* (*kn1*), a gene expressed in shoot apical meristems (see below), will induce new shoots to form on already mature leaves in transgenic tobacco plants (Sinha et al 1993). Regardless of whether the shoot meristem arises from embryogenic tissue, the subsequent steps in the development of the foliage organ are relatively similar.

Stage One: Organogenesis, When the Leaf Is Partitioned and First Emerges from the Meristem

Anatomical changes predict the impending emergence of a leaf primordium from the flanks of the shoot apical meristem (Figure 3*b*; Esau 1977, Steeves & Sussex 1989, Medford 1992). Although not a proven prerequisite, histologically distinct zones are visible in apical meristems of dicotyledons and gymnosperms (first described by Foster 1938, reviewed in Gifford & Corson 1971, Lyndon 1990, Medford 1992). These zones reflect differences in cell cycle rate. Larger, more vacuolate cells at the summit of the meristem divide more slowly than the small cytoplasmically dense cells at the meristem flank, as confirmed by calculating mitotic indices (summarized by Lyndon 1976), by autoradiography (Davis et al 1979), and by expression patterns of cell cycle genes (Fleming et al 1993, Ferreira et al 1994, Brandstadter et al 1994, Hemerly et al 1995). The more rapidly cycling cells at the flanks coincide with the portion of the meristem that is ultimately partitioned off into a newly emerging leaf.

A change in meristem shape is another indication of impending leaf emergence. In some dicotyledons, the apical dome of the meristem reaches a maximum size just prior to the initiation of a new leaf (Esau 1977, Mauseth 1988) and the flanks of the meristem then swell (Figure 3). Anatomically, swelling of the flank is also associated with changes in cell activity. The rate of cell division, numbers of periclinal divisions (parallel to the surface), and degree of cell elongation all increase in the region of swelling (Esau 1977). The resulting protruberance, called a foliage buttress (Foster 1936), is depicted in Figure 3*b*

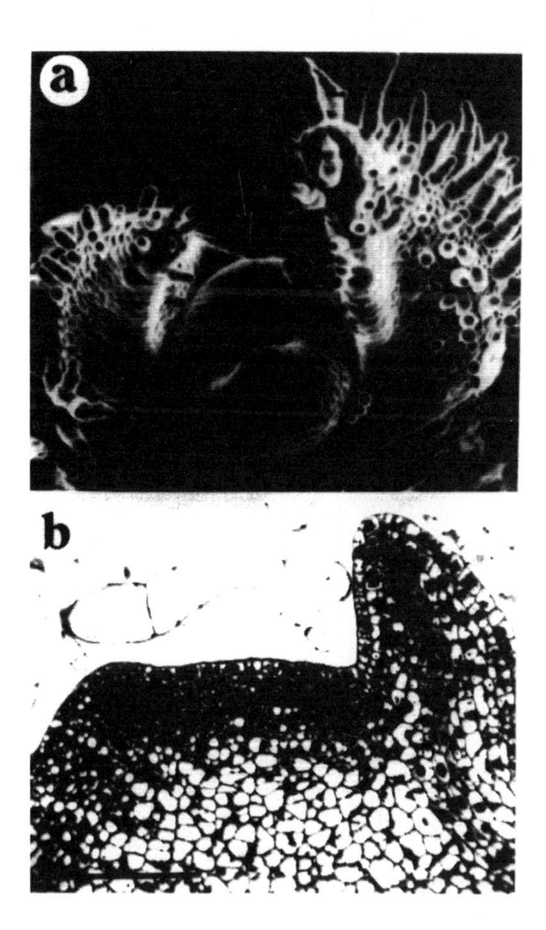

Figure 3 Stages of development of dicotyledonous leaves. The shoot apical meristem of *Nicotiana tabacum* (tobacco), ≈ 100 μm in diameter, produces leaves in a spiral arrangement. (*a*) An incipient primordium is indicated by the arrow. P1-P3 are the next oldest primordia present (reprinted from Poethig 1989). (*b*) A median longitudinal section of a similar stage is shown (reprinted from Poethig & Sussex 1985a,b).

for tobacco. Figure 4*a* depicts the appearance of the foliage buttress in maize, sometimes referred to as a disc of insertion, which resembles a protruding crescent ensheathing the meristem base.

The group of cells giving rise to the buttress are called founder cells, to define the population of future leaf cells within the meristem (Poethig 1984). Leaf founder cells may also be a subset of cells within the apical meristem that gives rise to all organs of the phytomer, including internode, bud, and leaf (Galinat 1959, Scanlon et al 1996). Founder cells occupy a distinct region within the meristem of both tobacco and maize, as determined by clonal analysis (Poethig 1984, Poethig & Sussex 1985b). Numbers of founder cells in the meristem of maize and tobacco are estimated to be from 100 to 200 cells for the large encircling leaf base of maize (Poethig 1984, Poethig & Szymkowiak 1995) and approximately 100 for the more confined leaf base of tobacco (Poethig & Sussex 1985b).

The diversity of plant form suggests that the details of patterning and even the mechanism for gaining identity may be variable. To simplify discussion, most further description is based on documentation of leaf development in maize as a model monocotyledon (Sharman 1942, Poethig 1984, Sylvester et al 1990, Poethig & Szymkowiak 1995) and tobacco, whenever possible, as a model dicotyledon (Avery 1933, Poethig & Sussex 1985a,b).

Stage Two: Early Growth of the Primordium When Leaf Regions Are Initiated

The foliage buttress grows out to produce a leaf primordium that is peg-shaped in tobacco (Figure 3*a*) and hood-shaped in maize (Figure 4*a*). Plastochron number is frequently used to identify primordial stages of development. Originally defined as a time interval between similar stages of leaf development (Hill & Lord 1990), the term plastochron (P) is often used to identify the position of a primordium relative to the apical meristem, such that P0 represents the site of the incipient leaf, P1 the primordium closest to the meristem, P2 the next oldest primordium, etc (see Figures 3, 4). Thus stages of leaf development may be classified according to P number. In maize, for example, the hood-shaped stage corresponds to plastochron 2 in Figure 4*a*. In tobacco, the peg-shaped stage is evident in P1, as shown in Figure 3*a*. It is possible, however, that morphological criteria rather than P number better suit a discussion of when identity is acquired. For maize, varied growth rates, different genetic backgrounds, and particularly the age of the plant produce an apex with varied numbers and sizes of leaves (Greyson et al 1982, Sinha et al 1993; A Sylvester, unpublished data). To retain consistency with previous terminology, however, plastochron number will be used here in conjunction with the morphological staging.

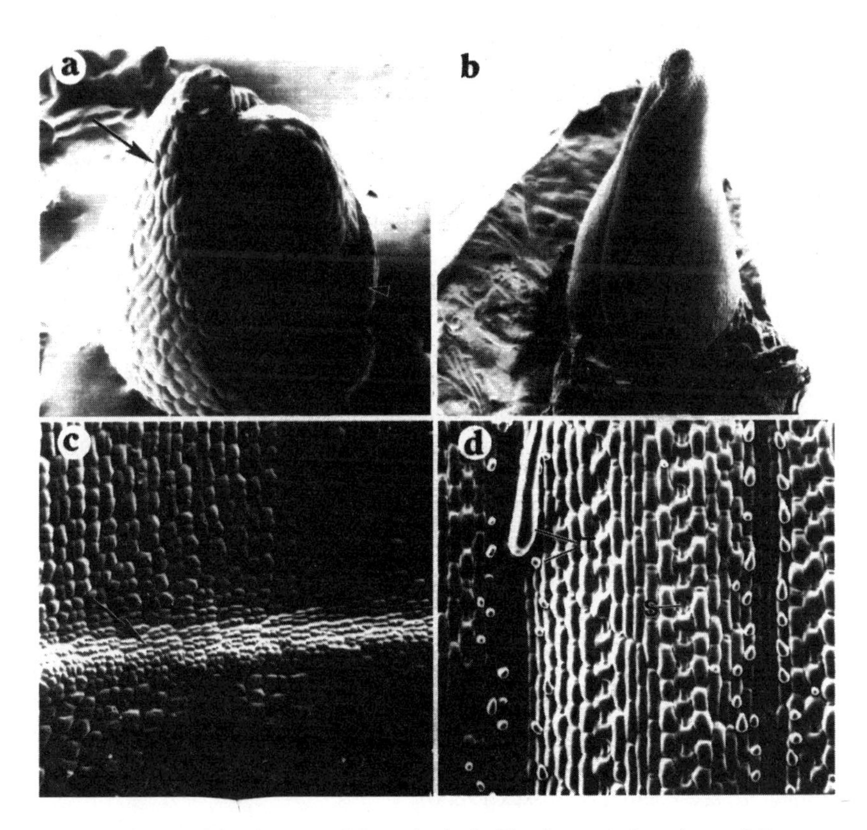

Figure 4 Stages of development of the maize leaf. The shoot apical meristem of *Zea mays* (maize), ≈ 250 μm in height, produces leaves in an opposite (distichous) arrangement. (*a*) A newly emerging primordium, P1, appears as the foliage buttress, or disc of insertion (*arrowhead*). The next oldest leaf is a P2 primordium at the hood-shaped stage (*arrow*). (*b*) Stage two shows that the cone-shaped leaf establishes blade and sheath regions. (*c*) The boundary of the emerging ligule is visible in an adaxial view of a stage two leaf (*arrow*). (*d*) Specific cells, such as trichomes (T) and stomata (S), are differentiated in distinct linear alignment in a stage three leaf. (Figures *b, c, d* reprinted from Sylvester et al 1990.)

Outgrowth of the peg or hood is usually accomplished by uniform cell divisions throughout the primordium in both maize and tobacco (Poethig 1984, Poethig & Sussex 1985a, Sylvester et al 1990, Poethig & Szymkowiak 1995). Dorsiventrality, or flattened appearance of the primordium, is evident from the outset; the curvature and growth pattern of the meristem flank is continuous with the base of the primordium. Leaf parts are then initiated as cell division is restricted and localized within the primordium. The distal lamina emerges in the dicotyledon primordium, accompanied by localized and oriented cell divisions at the primordium flank. In the case of monocotyledons such as maize, a distal blade region and a proximal sheath region are distinct in the hood-shaped primordium when cell division is restricted at the tip of the leaf (Sylvester et al 1990).

Formation of the proximal-distal regions in maize have been documented (Figure 4*c*). A visible boundary between blade and sheath is made evident by a linear band of uniquely shaped square cells, termed a preligular band (Sylvester et al 1990). The appearance of the preligular band is achieved by an increase in anticlinal divisions (perpendicular to the leaf surface) at the boundary between blade and sheath, when the hood is surpassing the tip of the meristem (P2 in Sylvester et al 1990). Cells in the preligule band subsequently divide periclinally and elongate, producing the distinct structure of the ligule proper. The auricle is a wedge-shaped group of cells that also differentiates from the preligular band of cells and suprajacent cells, just above the emerging ligule (Figure 2*b*).

The tobacco leaf is also visibly subdivided into parts during this second stage. The dorsiventral peg (P1: Figure 3*a*) becomes more obviously flattened and curved distally due to increased growth of the abaxial compared with the adaxial side (P3: Figure 3*a*). Distinction of lamina from petiole is achieved by more rapid divisions near the flanks of the developing peg (approximately P3; Figure 3*a*; see Poethig & Sussex 1985b).

Stage Three: Leaf Differentiation, When Tissues and Cell Types Become Well-Established

Stage three is the stage during which functional tissues are established within the framework of basic leaf parts previously specified. Stage three is the least well defined, overlaps with, and is the most prolonged, of the three described stages. For maize, traditional terminology places stage three as beginning approximately with plastochron 3 or 6. Morphologically the stage begins when the leaf appears as a cone that encloses the apex entirely (Figure 4*b*). At some time during the maize cone stage, the ligule at the blade-sheath boundary begins to differentiate further, as a periclinal elaboration of the epidermis (Figure 4*c*).

In maize, unique cell growth patterns produce functionally distinct cell and tissue types during this third stage of primordial development. Particularly apparent is the linear alignment of tissues, characteristic of the monocotyledonous leaf. Lateral veins differentiating acropetally are separated from one another by intermediate veins, which differentiate basipetally (Sharman 1942, Bosabalidis et al 1994). The distribution of epidermal cells reflects the pattern of differentiation in the underlying tissues (Freeling & Lane 1994). For example, stomata first arise when an intercostal epidermal cell undergoes an asymmetrical cell division (Figure 4*d*), as described in more detail below. A given maize leaf during stage three of development thus exhibits a typical gradient: The base or least developed part of the leaf is still acquiring cellular identity of unique cell types such as stomata, whereas the tip of the leaf is already dividing to produce the specified cell type. Figure 4*d* depicts the tip of the cone-shaped maize leaf that is undergoing divisions characteristic of stomata. The base of the same leaf would be engaged in proliferative cell divisions, not the formative divisions that precede differentiation.

Considerable overlap exists between stage two and three in tobacco. For example, epidermal trichomes, considered a typical example of stage three tissue differentiation, first emerge prior to elaboration of the lamina in tobacco (Poethig & Sussex 1985b). Despite these ambiguities, the usefulness of staging in this manner is made evident in subsequent sections of the review.

CONCEPTS OF POSITION DEPENDENCE VERSUS CELL HISTORY

It is well established that cellular identity is acquired based on position of a cell rather than its specific lineage. Identity of the leaf is thus accumulated gradually, while the leaf grows, rather than at a given time when a lineage of cells is first established. Support for this view comes from various observations and experiments. For example, plant cells are flexible in their developmental programming. Such plasticity suggests that differentiation is not a terminal condition and the ease of reversing it may reflect the ease of acquiring it. More direct evidence comes from at least two other sources. An example of the first type of evidence is typified by microsurgical experiments conducted by Sachs (1969), in which regeneration was examined in operated leaf primordia of *Pisum*. Young bisected primordia (equivalent to stage two described here) were capable of regenerating two complete leaves, whereas the extent of regeneration was restricted in increasingly older leaves. Stage three leaves, when bisected, would no longer regenerate all leaf parts and regions (Sachs 1969). This study, and other such surgical experiments (Cutter 1958), support the view that leaf identity is flexible for a time, in this case up to stage three in dicotyledonous plants.

A second source of evidence comes from cell lineage studies. Analysis of periclinal chimeras and clonal sectors, induced by ionizing radiation, provides strong support for a position-dependent model for how cells are determined (Stewart & Dermen 1970, Stewart 1978, McDaniel & Poethig 1988, Jegla & Sussex 1989, Furner & Pumphrey 1992, Irish & Sussex 1992). For example, a cell can divide incongruously, thereby changing the relative positions of its daughter cells within a tissue. If such an event occurs in genetically marked cells, the sectors can be identified and analyzed (Dermen 1953, Poethig 1987). Such studies reveal that clones of cells resulting from the original incongruous division acquire the identity of new neighboring tissues: Lineage history is not carried over from one tissue to another. Marcotrigiano & Bernatzky (1995) argue that the patterns of deviation from a given lineage in a developing leaf are not random. However, it is clear that cell lineage does not influence many examples of histogenesis. An example is the differentiation of photosynthetic cell types in C4 plants such as maize (Langdale et al 1988, Nelson & Langdale 1989) and in *Atriplex* (Dengler et al 1995), in which C4 photosynthetic enzymes are localized to developing cell type. Cell position, rather than derived cell lineage, determines the identity of photosynthetic cells in the maturing leaf (Langdale et al 1988, Nelson & Langdale 1989).

There is evidence that precisely oriented cell divisions do not contribute to the positioning of cells within the growing primordium. For example, morphogenesis, the acquisition of form, does not depend on a precise distribution of individual cells, as shown by studies of comparative developmental morphology (Kaplan & Hagemann 1991, Cooke & Lu 1992, Kaplan 1992, Hageman 1992). Supporting evidence comes from experimental studies (Foard 1971), clonal analysis (Poethig 1987), and genetic analysis (Hemerley et al 1995, Smith et al 1996). Neither cell orientation (Smith et al 1996) nor rate of cell division (Hemerley et al 1995) appear to be major determinants of leaf form. For example, the mutant maize plant, *tangled 1* (*tan1*), shows defects in the orientation of cell division but still acquires normal leaf parts (Smith et al 1996). Tissue anatomy is altered in *tan1* mutants, however, producing leaves of normal shape, but with abnormally distributed cell types. In this case, regional changes in leaf shape and the production of basic leaf parts, which all occur during early primordial growth, do not require precisely oriented cells. Similarly, Hemerley et al (1995) show that a given rate of cell division is not a prerequisite for proper morphological development. In their experiments, transgenic *Arabidopsis* plants that contain a defective CDC2 kinase, a regulator of the cell cycle, undergo normal morphogenesis despite a significant decrease in the rate of cell division.

Despite evidence favoring the position-based model for identity acquisition, lineage-based models still retain some validity. There are examples in which precise cellular patterns are required for proper tissue differentiation, particularly during later stages of leaf differentiation and during specification of some organs. For example, clonal analysis reveals that the orientation of individual cells may be fundamentally important in defining the axes and structure of the developing anther in maize (Dawe & Freeling 1990, 1991). Similarly, cells divide in a precise pattern to produce functional stomata, and this pattern is usually invariant in a given species (e.g. Chin et al 1995; see below).

Another example of how cell pattern, as established by lineage history, may contribute to identity of the developing leaf is in clonal analysis of the maize epidermis (Bossinger et al 1992, Cerioli et al 1994). These experiments show that the epidermis of the maize leaf consists of developmental modules, or clonal sectors, that are inferred to originate from precise anticlinal divisions. The modules are evident on examination of the distribution and frequency of sectors of wild-type epidermis within leaves mutant for a gene that conditions a wax-less phenotype. The sectors originate early and are generated because the mutant allele is unstable and reverts at a given frequency. Consistent with prior observations in maize (Steffenson 1968, Coe & Neuffer 1978, Bossinger et al 1992), sectors were observed to start and end at predictable positions within the module (Cerioli et al 1994). The most frequent sectors found were those that occupy at least half a module: Within this class, the highest numbers of sectors start at the module border closest to the midrib. Sectors in the margin-half of the module appear to be derived from clones originating in the midrib-half of the module. Based on the frequency and distribution of the sectors, Cerioli et al (1994) argue that a module initiated by at least four cells follows a typical clonal pattern of development, and therefore the orientation of cell division at an early stage does influence the organization of the maize epidermis. In this case, however, basic leaf shape is presumed to be well established by the time the required oriented divisions give rise to epidermal modules.

Given the observations cited above, that cell lineage, cell orientation, and relative cell position variably influence cell fate, an increasingly complex view is emerging of how leaf identity is established. It is likely that identity is not acquired either by a single mechanism or at a single time. In particular, the cellular parameters governing early development of the primordium and later differentiation of the leaf may be different (Poethig 1987, Kaplan & Hagemann 1991, Cooke & Lu 1992, Kaplan 1992, Smith et al 1996). As a leaf grows and changes in shape and in function, identity may be acquired accordingly. Parts of a leaf may be roughly established in founder cells in the meristem; the parts elaborate a given shape with attendent subparts and then specific cells divide and differentiate in a more precisely oriented pattern. The next sections consider how identity is acquired in each of these stages separately and in more detail.

BASIC ORGAN IDENTITY: STAGE ONE

Leaf organogenesis is preceded by a partitioning of the meristem into a region that contains the future leaf founder cells. A question to consider is how and when the future leaf founder cells are set aside from the rest of the shoot apical meristem. Neither the molecular nor the cellular basis for how this identity is first acquired is understood. It is clear that a predictable series of anatomical changes precede the emergence of a leaf. However, plant biologists have long appreciated the fact that founder cell identity must be acquired at some time before anatomical change heralds the event. Consequently, a driving area of research is to understand the first steps of organogenesis—how founder cells acquire identity—as well as how the leaf buttress is then specified from the founder cells.

Establishing Founder Cell Identity

A primary anatomical feature of the meristem is its separation into cytologically distinct zones (Esau 1977, Mauseth 1988, Steeves & Sussex 1989). There is currently little data to indicate that meristem zonation is actually required for the emergence of a primordium; however, many descriptive studies, and also surgical experiments, confirm that the two events are coincident (Steeves & Sussex 1989). Meristems that are isolated and surgically divided will subsequently regain a leaf-generating function. Documentation of this process reveals that a variation in cell division rate, characteristic of zonation, coincides with meristem recovery and leaf production. In the case of *Helianthus*, Davis & Steeves (1977) showed that division slows in cells at the meristem summit at the time of leaf initiation.

The significance of these gradients of cell cycle rate to the partitioning of leaves remains unclear, despite the fact that zonation is evident only in vegetative meristems: Inflorescence meristems do not exhibit zonation (Huala & Sussex 1993). One hypothesis suggests that cells in the flanks of the meristem are destined for the vegetative organs, whereas the central slowly dividing cells are destined to give rise to the reproductive organs (Buvat 1952, Nougarede 1967). In part, the hypothesis explains the observed phenomenon of zonation: The summit cells are quiescent until reproductive signals start their differentiation as floral organs. Zonation is then thought to be absent in inflorescence meristems because summit cells divide more rapidly as they are recruited for floral development. Reminiscent of the mosaic type of development described for animals, this hypothesis states that a population of meristem cells are determined to produce a specific part of the plant. The hypothesis is not well supported. Fate maps of the shoot apical meristem show that there are not highly specific lineages from meristem zones that give rise only to leaves (McDaniel & Poethig 1988, Jegla & Sussex 1989). Clonal sectors of genetically marked cells, generated by ionizing radiation, are not restricted to either vegetative or floral organs in maize (McDaniel & Poethig 1988) or sunflower (Jegla & Sussex 1989). Rather, meristem-derived sectors traverse several vegetative phytomers and even extend up into the inflorescence, proving that a reproductive cell line is not sequestered from a vegetative cell line. In fact, such fate maps of apical meristems reveal only partial correspondence with phytomer production. Basal parts of the meristem give rise to basal parts of the shoot system, apical parts contibute to the upper and reproductive part of the maize plant (Steffenson 1968, Coe & Neuffer 1978, McDaniel & Poethig 1988). The relevant questions remain: How and when do specific cells within these generalized zones become determined to produce leaf founder cells, and what is the impact of differentially controlling cell cycle rate on the identity of a leaf-generating meristem?

A molecular or genetic approach to the question of what precedes the appearance of founder cells should be informative. Current effort is directed toward identifying genes expressed uniquely in the shoot apical meristem (Medford 1992), with some attention toward finding expression patterns that correlate with organ initiation. As yet, there are few convincing candidate genes that are expressed exclusively in the meristem or the founder cells. Several genes that are expressed in layered or zonate patterns in the shoot apical meristem are also expressed in young developing leaves (Fleming et al 1993, Pri-Hadash et al 1992, Shahar et al 1992) and/or in developing floral primordia (Brandstadter et al 1994, Kelly & Meeks-Wagner 1995), which suggests their general association with rapid cell division.

One class of genes proving to be informative are those belonging to the *knox* (*knotted 1*-like homeobox) genes of class I (Kerstetter et al 1994). These genes are expressed in various portions of the shoot apex but not in developing leaves or in leaf founder cells (Figure 5; Smith et al 1992, Jackson et al 1994, Schneeberger et al 1995; J Fowler, unpublished data). *knox* genes represent a group of genes first recognized by studying the effects of their dominant mutant alleles on leaf development (Freeling & Hake 1985, Freeling 1992). Transposon-tagging and cloning of *kn1* (Hake et al 1989, Volbrecht et al 1991) revealed that *kn1* contains a homeobox sequence, which implicates the gene as a transcription factor that may contribute to the control of cell fate during development. Twelve related sequences have now been identified and characterized (Kerstetter et al 1994).

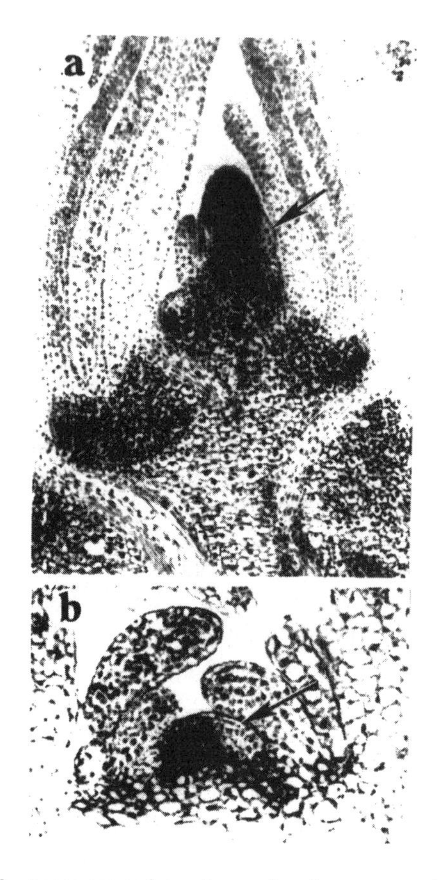

Figure 5 *knox* and *knat* gene expression patterns. *knox1* gene expression in the maize shoot apex as detected by hybridization to digoxigenin-labeled antisense probes. (*a*) *knox* expression is absent from a region of the meristem corresponding to the position of founder cells (*arrow*) (figure courtesy of S Hake). (*b*) Similarly, *knat1* is not expressed in the presumed founder cells of the *Arabidopsis* vegetative apical meristem (reprinted from Lincoln et al 1994).

Three examples of *knox* genes (*kn1*; *rough-sheath 1*, *rs1*; *liguleless 3*, *lg3*) are expressed in the shoot apex (Smith et al 1992, Jackson et al 1994, Schneeberger et al 1995, Fowler et al 1996). In these cases, the genes are down-regulated in a region of cells presumed to be leaf founder cells (Figure 5). There is strong support for the assumption that these genes are down-regulated specifically in founder cells. First, the location of lowered gene expression is consistent with where incipient leaf primordia should form (Figure 5, *arrows*). More convincing is the fact that there is good correspondence between numbers of founder cells estimated previously by clonal analysis and numbers of cells lacking *knox* expression (Smith et al 1992, Poethig 1984, Poethig & Sussex 1985b, Poethig & Szymkowiak 1995).

Homologous sequences to the *kn1* gene have been identified and cloned in *Arabidopsis* (Lincoln et al 1994). These *knat* genes (*knotted*-like from *Arabidopsis thaliana*) show patterns of expression similar to those characterized for *knox* (Figure 4). Down-regulation of the homeobox genes in leaf founder cells are thus evident in both monocotyledons and dicotyledons. This class of genes is currently the earliest known molecular marker for founder cell identity, but it is not yet well understood how the *knox* genes control cell fate. As proposed by Smith et al (1992), *kn1* and, by extension, other *knox* or *knox*-like genes may be involved in prolonging an undetermined state characteristic of a meristem (Long et al 1996). Shutting off *knox* expression could release the cells from the undetermined state, thereby establishing the founder cell position. It is possible that down-regulation of *knox* and *knat* is coincident with, but not necessarily responsible for, releasing cells from the undetermined state (Klinge & Werr 1995, Smith et al 1995). For example, Smith et al (1995) show that *kn1* expression occurs in conjunction with the first appearance of the embryonic shoot meristem, not necessarily prior to it. Although recognizing antecedent events may be difficult in histological sections, the observations suggest that *kn1* may not be a precursor marker for meristem initiation in the maize embryo. Similar expression studies using other members of the class I *knox* genes or those isolated from other plants (Lincoln et al 1994, Long et al 1996) may prove informative. It would be intriguing to characterize the timing and spatial pattern of expression of homologous *knox* genes in non-traditionally formed meristems, such as those that arise de novo from mature or isolated tissue.

Fates of the Founder Cells in the Maize Leaf

Clonal analysis confirms that individual phytomers are not strictly derived from given sets of cells in the shoot apical meristem; precise fate maps cannot be produced, as discussed above. The next question is when and if founder cells are fated to produce distinct parts of an individual leaf. Maize has been informative as a model system to answer this question, in part because the maize leaf is subdivided into a hierarchy of regions and domains. The distal blade and proximal sheath regions may acquire identity later in leaf development than the transverse domains, which span the distance from margin to margin of the leaf. Several studies support this idea. For example, the position of a founder cell around the circumference of the meristem flank contributes to its future destination in the transverse dimension of the leaf (Steffenson 1968, Poethig & Szymkowiak 1995, Scanlon et al 1996). Scanlon et al (1996) present molecular evidence from analysis of leaves of the *narrow sheath* mutant (double mutant homozygote *ns1-R;ns2-R*) that the transverse ring of founder cells in maize is subdivided to give rise to specific leaf domains in the transverse dimension. *ns* mutants are distinctive because they lack leaf margins, which indicates the presence of a genetically distinct margin domain in the leaf. Using the *knox* gene as a molecular marker for leaf founder cells by its absence of expression, Scanlon et al (1996) show that cells at the margin of the ring of founder cells are present but do not exhibit leaf identity. Thus the margin domain of the leaf is proposed to be pre-established in the margin region of the founder cells. Furthermore, because a region of internode is also affected in *narrow sheath* plants, it is likely that the leaf marginal domain also reflects an entire phytomer marginal domain.

Sector analysis in maize was also used to consider the relationship between founder cells and future leaf parts, both of transverse leaf domains and proximal-distal regions (Poethig & Szymkowiak 1995). Sectors were generated by irradiating maize plants heterozygous for a recessive mutation that produces an albino condition. Seedlings were irradiated when a given set of leaves were just emerging from the meristem and the position and extent of sectors in the same leaves was characterized at the time of maturity. To align terminology used here with that of Poethig & Szymkowiak (1995), pre-initiation stage represents founder cells or P0, and 0-buttress stage represents the disc of insertion or P1, both of which are defined herein as stage one of organogenesis. The 75–150 μm stage represents the emerging hood or P2, defined here as stage two, growth of the primordium. The clonal analysis results of Poethig & Szymkowiak (1995) show that the proximal-distal regions of blade and sheath are not yet determined in founder cells, based on the fact that few blade and sheath only sectors are produced when the pre-initiation stage is irradiated. At the leaf buttress stage, when the disc of insertion is actually initiated, sectors are more evenly distributed within the phytomer, but it is not until the next stage, when the primordium is fully emerged, that sectors predominate in the blade alone. These results suggest that the blade as the distal region of the leaf is not specified within founder cells but is specified sometime between the formation of the disc of insertion and the end of stage two.

The same study shows that the situation is different for cells in the transverse dimension of the leaf (Poethig & Szymkowiak 1995). These cells partly gain identity from their position relative to the crescent of founder cells. Sector analysis shows that later aspects of cell division in the leaf, including duration, rate, and orientation, influence the precise relationship between distribution of the transverse domains in the mature leaf and the cells from which the sectors were derived. However, the result confirms that there is a generalized fate of founder cells in the transverse span: cells around the circumference of the meristem flanks contribute to a corresponding portion of the transverse domains of the leaf. The specific fate of a founder cell is thus influenced by its starting position as well as its future pattern of cell division (Poethig & Szymkowiak 1995).

Identity of the Foliage Buttress

As mentioned earlier and shown in Figures 3 and 4, swelling of the flank of the meristem indicates the emergence of the leaf buttress from founder cells. Evidence cited above shows that the position of a cell along the ring of founder cells influences its future position in the transverse dimension of the leaf in a general way. However, it is unlikely that cells arrive at this general position due to precisely organized cell divisions. Foard (1971) showed that swelling of the flank will still occur even if division has been arrested by gamma irradiation of the wheat apical meristem. Ultimately, cell numbers must be increased to maintain growth of the primoridum, but division is evidently not a triggering factor. The nature of the trigger remains a fundamental mystery of the process.

While a molecular explanation of the triggering process awaits clarification, a variety of other influences on the process are postulated. Green (1994) proposes to connect biochemistry (i.e. molecular triggers) to solid tissue geometry, that is, biophysical considerations of the tissue itself contribute to morphogenesis (reviewed in Green 1994, Green & Selker 1991). Morphogenesis is viewed as a transduction chain with the physical tissue at one end of the chain responding to molecular triggers at the starting point of the chain. Considering that leaf geometry is indeed an aspect of identity, it is reasonable to consider that the nature of the responding tissue could impact the fate of the organ. A relevant question is how leaf position, or phyllotaxy, is established. By observing alignment of cellulose microfibrils in cell walls of formative tissue such as an apical meristem, Green & coworkers show that the structure of a tissue changes in conjunction with (Jesuthasan & Green 1989) or even prior to a morphogenetic event such as the emergence of a leaf primordium (Green & Brooks 1978, Tiwari & Green 1991). These studies are based on the observation that alignment of cellulose microfibrils in the cell wall contributes to directional expansion. Cell wall alignment reinforces the expanding cell, further encouraging directional growth. Shared alignment of microfibrils among cells serves to reinforce the tissue itself, thereby encouraging a given directional expansion of the organ. Any site of alignment disjunction between microfibrils of neighboring cells represents a weak point and is consequently the least structurally stable site of the tissue (Green 1994). At these sites, leaves appear to emerge. A crucial element of the theory still relies on the presence of triggering factors. The unidentified molecular signals that impart identity to the group of cells in question do so at some time during the chain of events leading to organogenesis. Green (1994) points out that, as in any transduction chain, there is likely to be considerable and subtle feedback between all steps in the chain from the generating molecules to the responding tissue. When and to what extent the signals must be maintained in order to initiate the physical responses remain in question. An exciting area for future research concerns the nature of the triggering signals, particularly those that may change the physical condition of the tissue.

IDENTITY OF LEAF PARTS: STAGE TWO

With the basic leaf identity acquired and the primordium well-established, the next stage in the process is the elaboration of basic leaf parts as the primordium grows. Several experimental approaches have been used to determine when and where commitment occurs in the growing primordium. Surgical experiments, such as those of Sachs (1969), provide some evidence that commitment is gradually restricted as the leaf ages. A genetic approach to the question is particularly revealing because it considers the nature of the triggering signals and molecules. Maize has been informative in this regard due to a group of mutants that affect different aspects of leaf structure, as discussed below (Hake 1992, Freeling 1992).

Identity of Blade and Sheath in the Maize Leaf

Subdivision of the maize leaf into blade and sheath is first recognized by cellular changes in the future boundary between the two regions (see Figure 4*c*). Cells in such a concise boundary must be determined at or before the cell growth patterns visibly change. Mutations that remove the ligule and auricles, or alter the shape of this boundary region, are of particular importance when considering how the regions gain identity.

Mutations at 15 different genes alter the blade-sheath boundary or remove organs that initiate at the boundary, and so it is presumed that the mutations must act early, at or before stage two of leaf development. Figure 6 diagrams the adaxial surfaces of idealized young leaves, each showing a typical mutant phenotype. The genic designations refer to specific mutant alleles expressing in

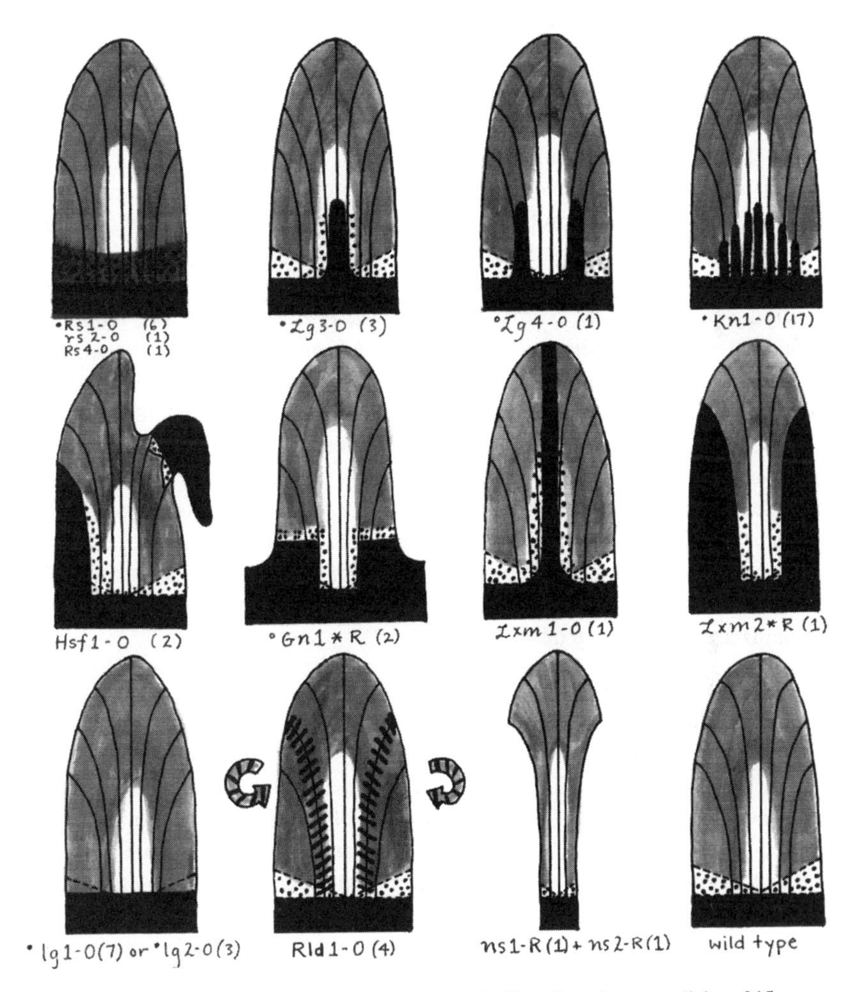

Figure 6 Schematic leaf primordial phenotypes specified by selected mutant alleles of 15 genes. The alleles are denoted as either O for original or R for reference and begin with a capital letter if the mutant allele is dominant. The minimum number of unselected independent alleles for each gene is in parentheses. A solid circle indicates that DNA sequence is available; an open circle indicates that the gene is probably cloned; solid black is sheath; light gray is blade; dots denote ligule-auricle; cross-hatches on the *Rld-O* schematic indicate a switching of abaxial-adaxial epidermal identities; the dark gray auricle of *Rs1-O* denotes simultaneous auricle and sheath identity.

a permissive genetic background, as detailed in the legend, and the numbers in parentheses are the total number of independent alleles. New recessive mutants screened against a reference mutant tester are not included in this parenthetical number because only mutants in this one gene are recovered. Although specific phenotypes, such as liguleless (*lg1* and *lg2*), knotted (*kn1*), and rolled (*rld1*) may be saturated for possible mutants, the more general phenotype of a defective blade-sheath boundary is not saturated.

The 15 different mutant leaves diagrammed in Figure 6 represent three basically different classes of mutants, termed ligule-polarity, deletion, and upside-down mutants. The ligule-polarity mutants are so called because they shift the direction of the ligule from a basal to a more apical position. Three of these ten ligule-polarity genes (*Hairy sheath frayed 1*, *Hsf1*; *Laxmidrib*, *Lxm1* and *Lxm2*) are pleiotropic, display multiple heterochronic effects (Bertrand-Garcia & Freeling 1991, D Schichness & M Freeling, unpublished data), and will not discussed be further here. The deletion class of genes is represented by mutants in which either the margin domain of the leaf (double mutant *ns1;ns2*) or the ligule-auricle boundary (*lg1* or *lg2*) is absent. The upside-down class of genes includes mutants (*Rld1*) with traits exclusive to the abaxial (down) side of the leaf that are often found on the adaxial (upper) side (B Lane & M Freeling, personal observation). Most of the mutants in these three classes affect particular domains in the leaf (Figure 6), permitting inferences regarding how and when domains are fated during development.

LIGULE-POLARITY GENES Mutations in the ten ligule-polarity genes share certain similar features that have been explained by a current model of leaf development (Freeling 1992). These mutations all change the shape of the blade-sheath boundary in the same direction; blade cells transform to sheath, but not vice versa (Freeling 1992). The position of the blade-sheath boundary is altered in these mutants, but normal ligules and auricles differentiate along the new boundaries. Even if the boundary extends in a new direction, such as up rather than across the leaf, as in *Liguleless 3-Original* (*Lg3-O*), a morphologically normal ligule and auricle often form (Fowler & Freeling 1996). As a minor exception, the *Rs* mutations transform the ligular region to a region that is anatomically similar to both sheath and auricle (Becraft & Freeling 1994).

Three of ten ligule-polarity genes have been cloned and characterized (*kn1, rs1, lg3*). Analysis of independent alleles confirms that the clones correspond to the identified genes and their corresponding cDNAs have been sequenced (Hake et al 1989, Vollbrecht et al 1991, Freeling 1992, Becraft & Freeling 1994, Schneeberger et al 1995). They all share regions of sequence homology, and, in particular, share the homeobox sequence similar to the class I *knox* genes described above. Two other genes in the group also defined by dominant mutants (*Lg4-O* and *Gn1-R*; see Figure 6 for details) map close to class I *knox* sequences, but genetic confirmation is still in progress (R Schneeberger & T Foster, unpublished data).

The dominant mutants of the three cloned and best-understood *knox* genes share similar patterns of gene expression. In wild-type plants, these genes are all expressed in the shoot apical meristem but not in the leaves or founder cells, as described above (Jackson et al 1994, Schneeberger et al 1995, Fowler et al 1996). The mutant phenotype in each case may be explained by ectopic expression of the KNOX protein in mutant leaves. Furthermore, transposon insertions into specific regions of *kn1* (Smith et al 1992) and *rs1* (Schneeberger et al 1995) are responsible for the ectopic expression. Apparently, *knox* genes could express in leaves were it not for specific regulatory sequences that, when rendered functionless by DNA insertion, result in ectopic leaf expression.

The maize leaf is subdivided into domains The fact that the ligule-polarity mutants share molecular and genetic characteristics, but unique and different phenotypes, is relevant to the question of how identity is acquired in different parts of the leaf. Figure 6 reveals that the proximal-distal regions of blade and sheath are subdivided further into domains of distinct gene expression, encompassing the transverse domains of the leaf. For example, the leaf margin is viewed as a domain because it is genetically distinct from the rest of the blade and sheath: Deletion of the domain is possible in mutant leaves (Scanlon et al 1996). Much as the *Drosophila* wing is a patchwork of semi-independent domains (Williams et al 1994), the maize leaf may be viewed as a patchwork of independently derived parts.

It is interesting to consider how shared molecular characteristics of the *knox* genes can give rise to divergent phenotypes in mutant leaves. One hypothesis is that the different phenotypic patterns specified by the dominant *knox* mutants are due to region-specific gene expression. Each domain is regulated differently, leading to different patterns of ectopic *knox* expression and consequently to different phenotypic patterns. An alternative to this regulatory hypothesis is a competence hypothesis, which contends that cells in different domains are at different competency states (Freeling 1992) and thus respond differently to signal(s) to make a leaf part. A competency state is the ability of a cell to acquire a given fate: Therefore, the hypothesis argues that cells in each domain have a different inherent ability to respond to the same set of signals. Results of analysis of *rs1* (Schneeberger et al 1995) support but do not prove the competence alternative because the region of ectopic RS1 expression is greater than the region of transformation. Distinguishing these and other hypotheses are currently active areas of research.

How is identity of the leaf domains acquired? Cell-to-cell signaling may be an important component of how domains acquire identity during stage two. Mosaic analysis of *Kn1-0* shows that ectopic expression in the middle layers of the leaf induces a mutant *Kn1* phenotype in the overlying tissue, even when the epidermal genotype is homozygous wild-type (Hake & Freeling 1986, Sinha & Hake 1993). The period during which tissues are primarily differentiating (Sharman 1942; R Frey & A Sylvester, unpublished data) coincides with the time when epidermal traits of the *Kn1* phenotype are first recognized (Hake et al 1985, Hake & Freeling 1986). Another study reveals that a particular KNOX protein, LG3, acts only in the internal tissues of the leaf (Fowler et al 1996), despite the fact that the epidermal tissues of the ligule in the transverse domain bordering the midrib are primarily affected in the mutant *Lg3-0* leaf.

The signaling mechanism may be in the transport of the protein itself: KN1 protein is detected immunologically in all cells of a meristem (Smith et al 1992, Jackson et al 1994), whereas mRNA is restricted from the outermost layer of cells (Jackson et al 1994). Furthermore, transport experiments confirm that KN1 and its mRNA are capable of migrating from cell to cell via the plasmodesmata (Lucas et al 1995). In these studies, labeled KN1, when injected into the cytoplasm of mesophyll leaf cells, migrated into immediately neighboring cells, both of the same and different cell type. Similar to many viral movement proteins, KN1 appears to mediate its own movement by altering the size exclusion limit for the plasmodesmata. Molecules of greater molecular weight can be transported through the plasmodesmata in the presence of KN1. The protein domain responsible for the transport phenomenon was investigated, revealing a potentially effective sequence in the NH_2 terminal region of the homeodomain (Lucas et al 1995). Although these experiments do not prove that KN1 gene products actually traffic from cell to cell during development, they are suggestive results. For example, the cell types or plasmodesmata characteristic of a given leaf domain may respond differently to the same movement signal. Alternatively, the protein domain responsible for movement may be specific for the leaf domain it acts in.

Timing of ectopic expression is important for fate to be altered Mosaic analysis has been useful for relating the timing of *knox* gene expression with a specific cell transformation phenotype (Fowler et al 1996). These experiments use transposons and transposon-insertion alleles to generate mosaics of genetically marked sectors expressing LG3 KNOX protein in leaves that otherwise do not express LG3 (Figure 7). The results suggest that the blade-to-sheath transformation phenotype of LG3 is a consequence of LG3 expression early in leaf development, but that the LG3 protein itself does not instruct the cells to "make sheath" specifically. Leaf sectors in this experiment represent LG3 on tissue surrounded by normal off tissue, as depicted in Figure 7. Sector size corresponds to the timing of gene expression: Larger sectors of LG3 expression, generated by an early event, show the expected blade-to-sheath transformation in the region indicated in Figure 7. Sector phenotypes that reflect later LG3 expression are particularly informative. Medium-sized sectors transform blade to auricle. Very small sectors transform blade to modified blade or not at all. Unusual transformations were observed; the region of the large sector in Figure 7 marked with an asterisk remains blade-like but is not anatomically identical to adjacent wild-type cells. Clearly, the presence of LG3 in leaf cells

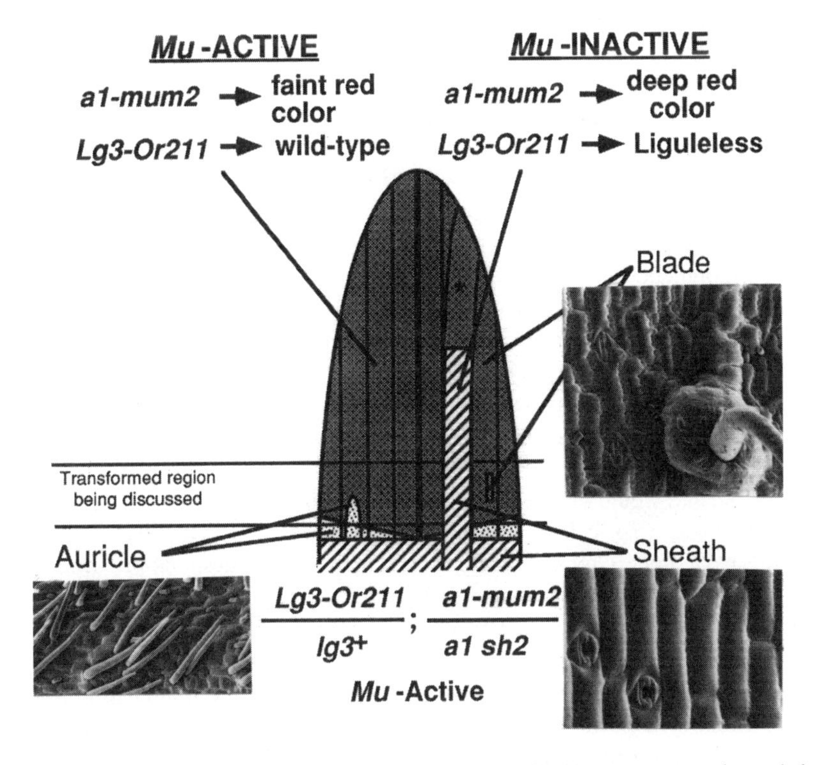

Figure 7 Schematic of how a mutant homeobox gene, *Lg3-Or211*, was constructed to switch on in leaf sectors marked with red pigment (encoded by *a1-mum2*). The result is a green leaf without LG3, except in the red sectors. Depending on the size of the sector, including cells in the transformed region, this ectopic expression of LG3 homeodomain protein has different cellular transformation phenotypes, evidenced by the three SEMs of distinctive mature leaf, adaxial, and epidermal surfaces. The * denotes a sector of immature blade surrounded by more mature blade. The switch that simultaneously turns *Lg3-Or211* and *a1-mum2* on is the spontaneous autodeletion of the regulator *Mu* family transposon MuDR-1(p1) (from Fowler et al 1996).

does not result in a single type of regional transformation, but is dependent on when, rather than where, LG3 is expressed during early primordium growth.

Analysis of ligule-polarity mutants supports the maturation schedule hypothesis Freeling (1992) proposed that the proximal-distal regions of the leaf (blade and sheath) have a hierarchy of competencies, as defined above. According to the hypothesis, founder cells in the meristem are all competent only to be sheath cells. As the leaf develops, a schedule of maturation unfolds, and regions of the leaf acquire new competencies, such that cells become competent to differentiate as blade or auricle. Clonal analysis studies described earlier are consistent with aspects of the hypothesis (Poethig & Szymkowiak 1995): Albino sectors are not generated in the blade until after the primoridum is initiated (75–150 μm stage leaf, P1 leaf in their study), suggesting that blade identity is not acquired until the primordium is well established.

There are three unique and relevant features to the maturation hypothesis. First, the progress of maturation occurs in only one direction, the upward growth direction. Unidirectional progression is supported by the fact that ligule-polarity mutants all transform tissue in this one direction. Second, maturation begins in founder cells and is complete near the end of stage two of primoridal development. Such developmental timing is supported by the fact that all ligule-polarity mutants show defects in the blade-sheath boundary, known to be well established by the end of stage two. Third, the ligule-polarity mutants can each retard the maturation schedule in their unique domains of expression. If progress through the maturation schedule is slowed as a consequence of the ectopic expression of a *knox* gene, then the domain would not yet have reached its appropriate competency state. Cells would thus be immature relative to neighbors and would differentiate according to the characteristics of that more immature state (e.g. cells would appear more sheath-like than blade-like).

The best support for the maturation schedule hypothesis comes from the LG3 on sector data diagrammed in Figure 7. The *lg3 knox* gene expressed very early in leaf development produced the highest degree of blade-to-sheath transformation because the cells were at the earliest competency state, that of sheath. If the same *knox* gene is ectopically turned on at a later time in the same cells, then an intermediate type of transformation would occur because cells had progressed to the next level of competency, that of auricle. This was in fact observed. Increasingly smaller sectors show a less advanced degree of transformation. The maturation schedule hypothesis is compatible with the suggestion that ectopic expression of class I *knox* genes such as *kn1* promotes the meristematic state (see Smith et al 1992). Using the terminology of the hypothesis, the meristem might be thought of as the least mature leaf competency state.

LEAF DELETION MUTANTS Another class of leaf mutants in maize are those in which a given domain of the leaf is entirely absent, such as *ns* and *lg1* mutants. Analysis of the ligule-polarity mutants suggests that the maize leaf is composed of domains, and it was further proposed that these domains were first initiated in founder cells of the meristem (Freeling 1992). Subsequent support came from analysis of the *ns* mutants, in which the meristems of *ns* mutant plants show a failure to down-regulate KNOX at the site corresponding to future margins of the leaf (Scanlon et al 1996).

Other well-characterized deletion mutants are those that remove ligule and auricle (*lg1-0* and *lg2-0*). Both phenotypes reflect loss-of-function mutations. Molecular analysis by cloning and sequencing shows *lg1* and *lg2* to be deletions (M Moreno et al, unpublished data; J Walsh & M Freeling, unpublished data). The LG1 product acts in a cell-autonomous manner in the epidermis to promote the anticlinal divisions of the ligule and the epidermal characteristics of the auricle, and acts in internal tissues less autonomously to specify mesophyll-vascular characteristics of the auricle (Becraft et al 1990). LG2 acts earlier than LG1 and is not cell autonomous, as if it acted as part of the "make ligule" signaling pathway (L Harper & M Freeling, unpublished data). Studies of ligule interruption and reinitiation using genetic mosaic sectors have been used to trace the direction of this "make ligule" signal and define cells competent to be induced (Becraft & Freeling 1991). The results show that the signal progresses directionally, from midrib toward leaf margin. Unlike the leaf itself, the ligule signaling pathway does appear to be saturated with mutants.

Recently, other deletion mutants have been identified in maize, including a midribless (*mrl*) mutant (J Paxton & T Nelson, unpublished data). Preliminary investigations of the mutant phenotype support the idea that the midrib may well represent a distinct domain of the leaf. In this mutant, the central midrib, as identified by its characteristic distribution of lignified cells and elaborated mesophyll, is absent, and the midvein is reduced in size. Similar to the vasculature of a similar mutant, *mrl*, in *Panicum* (Fladung et al 1991), mutant vasculature in the maize *mrl* is altered in what should be the midrib domain. Vein number and distribution is more similar to that present in the flattened lamina of the blade. Preliminary analysis suggests that midrib identity can be acquired independently from that of the other leaf domains. Further analysis of this mutant will contribute to an understanding of another important domain of the patchwork of the maize leaf and may shed light on how development of domains is coordinated.

Identity of Lamina and Petiole in Dicotyledons

The parts of a dicotyledon leaf also arise during stage two of primordium growth. Two identity changes occur at this time: The first involves the emergence of a photosynthetic lamina followed by outgrowth of the peg-shaped supporting petiole. The second event is the establishment of planar leaf axiality. The lamina differentiates as a flattened, dorsiventral part of the leaf with the adaxial side (the upper side, adjacent to the stem axis) distinct anatomically from the abaxial side (the lower side, away from the stem axis). Lateral flattening accentuates the dorsiventrality already present in the peg, and is also first evident when distinct cell types and tissues begin to differentiate.

The question of when lamina and petiole identity is acquired has been considered in tobacco (Poethig & Sussex 1985a,b). Clonal analysis suggests that lamina and petiole are determined after the tobacco primordium has emerged and during stage two (Poethig & Sussex 1985b). When induced prior to leaf initiation, clonal sectors are not confined to either lamina or petiole but span both regions (Poethig & Sussex 1985b). Thus leaf founder cells are not sequestered as future lamina or petiole. These results are similar to those found for maize, in that identity of the maize blade is also not acquired until the emergence of the primordium (Poethig & Szymkowiak 1995). Furthermore, these results confirm the idea that there is gradual and position-dependent acquisition of identity during the early stages of leaf development.

Lamina initiation may be arrested or altered by mutation with little effect on plant function (McHale 1993, Waites & Hudson 1995), permitting genetic dissection of the process, just as in the maize leaf. For example, leaves of a *Nicotiana* mutant *lam-1* lack laminae, but the plant otherwise grows to maturity. Analysis shows that the lamina initiates at the right time and place in the emerging primordium, but subsequent outgrowth, which is usually associated with periclinal divisions of inner tissues, is arrested (McHale 1993). Another mutation *fat1* shows excessive numbers of periclinal divisions in inner tissues of mutant leaves. *fat1* can only partially free the *lam1* arrest. *fat1;lam1* double mutants show proliferation of cells in the proper position, but this vestigial lamina lacks the characteristic anatomy of layers and organized venation of wild-type. The double-mutant phenotype suggests the two genes are additive in effect. That mutant *lam1* leaves grow to a normal length, with anatomically normal petiole and blade midrib, suggests that *lam1* could be required late in the establishment of the dorsiventral lamina. Alternatively, continued growth of the leaf midrib and petiole could be independent events not part of a temporal sequence that specifies one region of the leaf over another. It is pertinent, however, that new information is required to trigger the cellular changes associated with lamina outgrowth, whereas midrib and petiole growth follow the developmental pattern already present in the primordium. Further analysis of the functions of these genes awaits molecular analysis and additional genetic characterization.

The question of how the primordium partitions into regions that give rise to a dorsiventral lamina is of continuing interest. Although dorsiventrality is established at the outset (even predicted by the arrangement of founder cells in the meristem) the lamina still acquires an anatomically distinct planar anatomy in which an adaxial side is distinct from the abaxial. A series of mutant alleles at the *phantastica* locus (*phan*) affect the emergence of the lamina of *Antirrhinum* (Waites & Hudson 1995). Although several pleiotropic effects may also be present, it appears that *phan* mutants lack a coherent adaxial-abaxial (ad-ab) axis in the leaf (referred to as dorsal-ventral, respectively, by Waites & Hudson 1995). In the most severe case, needle-like mutant leaves may resemble petioles in their anatomy, although wild-type petioles are not depicted for direct comparison (Waites & Hudson 1995). A more mild phenotype is one in which a narrow lamina is produced but is improperly positioned in the proximal-distal axis of the leaf. Ad-ab anatomy can also be reversed: Stomata characteristic of the abaxial epidermis are found in patches on the adaxial epidermis. To explain these observations, Waites & Hudson (1995) propose that *phan* is required for adaxial cell identity. Varying degrees of the adaxializing function of *phan* determine the fate of the lamina. Inherent to the model is the assumption that the lamina represents an adaxial character of the primordium. A wild-type leaf is thus viewed as having equal distribution of adaxializing function in the primordium. Absence of adaxializing function in the most extreme case prevents any expression of adaxial characters and a needle-like leaf lacking lamina results. This model is based on changes in the distribution of *phan* expression in the primordium to explain the mutant phenotypes.

Additional models can be formulated to explain the observed *phan* mutant phenotypes. For example, the timing of the response of *phan* to lamina initiation signals, rather than its distribution in the primordium, could be affected, or its control over the distribution of an adaxializing function could change. If *phan* does not respond early enough, the primordium could continue growing in its given trajectory, as a needle-like peg. As an extension of this idea, a given distance from the apical meristem may be required for lamina initiation to be successful. The farther the primordium grows, the less likely normal lamina initiation signals can evoke the appropriate response. In this case, *phan* could normally arrest the continued outgrowth of the primordium, whereas mutant *phan* could prolong the outgrowth such that the primordium loses competence to initiate a lamina. It is interesting that the loss of *phan* function, or the change of its expression pattern in the primordium, must occur after the primordium is initiated. There is no evidence that altered *phan* function is present earlier, when the primordium is first growing. These speculations will likely be subject to rigorous testing. One of the *phan* mutant alleles shows evidence of germinal instability at a frequency characteristic of transposon insertion. Thus the *phan* locus may be isolated by transposon-tagging and expression pattern of the gene identified. These mutants will continue to be important tools in understanding how lamina identity is established in the dicotyledons.

IDENTITY DURING STAGE THREE OF LEAF DIFFERENTIATION

After the leaf and its given parts become well-established during stages one and two, a more individualized differentiation of tissues and cell types occurs. This third stage of leaf differentiation is one in which more precisely oriented patterns of cell division and expansion predominate. Position dependence still remains an important aspect of tissue differentiation (see above discussion of tissue differentiation of C4 vascular tissue), but the relative importance of cellular orientation increases during this stage. As shown clearly by Poethig & Szymkowiak (1995), cell position early in development and subsequent cell division patterns, generated by changes in the duration, rate, and orientation of cell division interact to produce the linear structure of the maize leaf.

Differentiation of the epidermis is a good example of the increasing importance of cell division and expansion in the process of cellular determination. Two types of observations support the view that individual cells are relatively more important during differentiation of the epidermis. First, one of the distinguishing features of the leaf epidermis is the presence of a limited number of dispersed and specialized cell types, most of which arise by oriented cell divisions and expansions. Second, clonal analysis of leaf sectors in maize shows that related cells are generally confined along vascular boundaries (Steffensen 1968, Coe & Neuffer 1978, Bossinger et al 1992, Cerioli et al 1994). These studies argue for the importance of cell lineage in the establishment of the maize epidermis. The epidermis is thus a particularly good model for understanding how specific cell types acquire identity during stage three of leaf development. Relevant questions include the following: How are particular cells selected as the precursors to specialized cell types such as stomata or trichomes? What contributes to the patterning of these precursors within the epidermis and when is the identity of these precursor cells acquired? These questions are considered using two well-studied examples of cell differentiation in the epidermis: the formation of stomatal complexes and of trichomes.

Identity and Patterning of Stomata

Stomata are specialized cell complexes, consisting of a guard cell pair and the surrounding epidermal cells, that regulate gas exchange between the leaf and the environment (Stevens & Martin 1978). Guard cell pairs are often, although not always, flanked by subsidiary cells of size and shape distinct from that of other epidermal cells. However, epidermal cells adjacent to guard cell pairs are thought to be involved in stomatal regulation whether or not they are distinguishable as subsidiaries (Raschke 1975).

The formation of stomata begins with an asymmetric cell division, the smaller product of which is defined as a stomatal initial. The sequence of events leading from this first asymmetric division to the formation of the mature stomatal complex is consistent within a species, but varies from species to species (Figure 4*d*). Division pattern is also dependent on the type of stomatal complex produced (Stebbins & Khush 1961, Pant 1965, Tomlinson 1974, Stevens & Martin 1978). For example, stomatal complexes lacking subsidiary cells (Figure 8*a,b*) may achieve this condition in several ways: A stomatal initial may itself function as a guard mother cell (GMC), defined as the cell that will give rise to guard cell pairs. As in *Anemarrhena* (Figure 8*a*), the stomatal initial divides symmetrically, giving rise to a guard cell pair directly. Alternatively, the initial may undergo additional divisions to produce one or more ordinary epidermal cells before forming an intermediate cell, the GMC, which then divides symmetrically to form a guard cell pair. An example of this type of stomatal development is found in *Pisum* (Figure 8*b*).

The pattern of cell division is different in stomatal complexes that have distinct subsidiary cells associated with the guard cells (Figure 8*c,d*). In some cases, all of the cells in the complex are clonally related because the stomatal initial gives rise to one or more subsidiaries before finally forming the GMC. The stomatal complex thus represents a lineage unit in which all cells are derived from the same initial cell. This type of stomatal development is common among dicotyledons and is illustrated in Figure 8*c* for *Thunbergia erecta*. Alternatively, subsidiary cells may not be immediately related to guard cells because the initial cell gives rise to only the guard cells, whereas neighboring epidermal cells undergo coordinated divisions to give rise to adjacent subsidiary cells. This type of stomatal complex, as exemplified by *Tradescantia* (Figure 8*d*), is thus not composed of a single lineage unit but is multiclonal.

Regardless of the pattern of division, little is known about how cells of the stomatal complex acquire their unique fates in a coordinately controlled manner. How and when are individual cells within the developing epidermis selected as stomatal initials? Once selected, how is their subsequent development controlled so that a complex of differentiated cells is formed in the proper arrangement either from a single stomatal initial or by means of the local coordination of multiple lineages?

STOMATAL DISTRIBUTION IS NONRANDOM Statistical analyses show that the arrangement of stomata within the epidermis, although not perfectly regular,

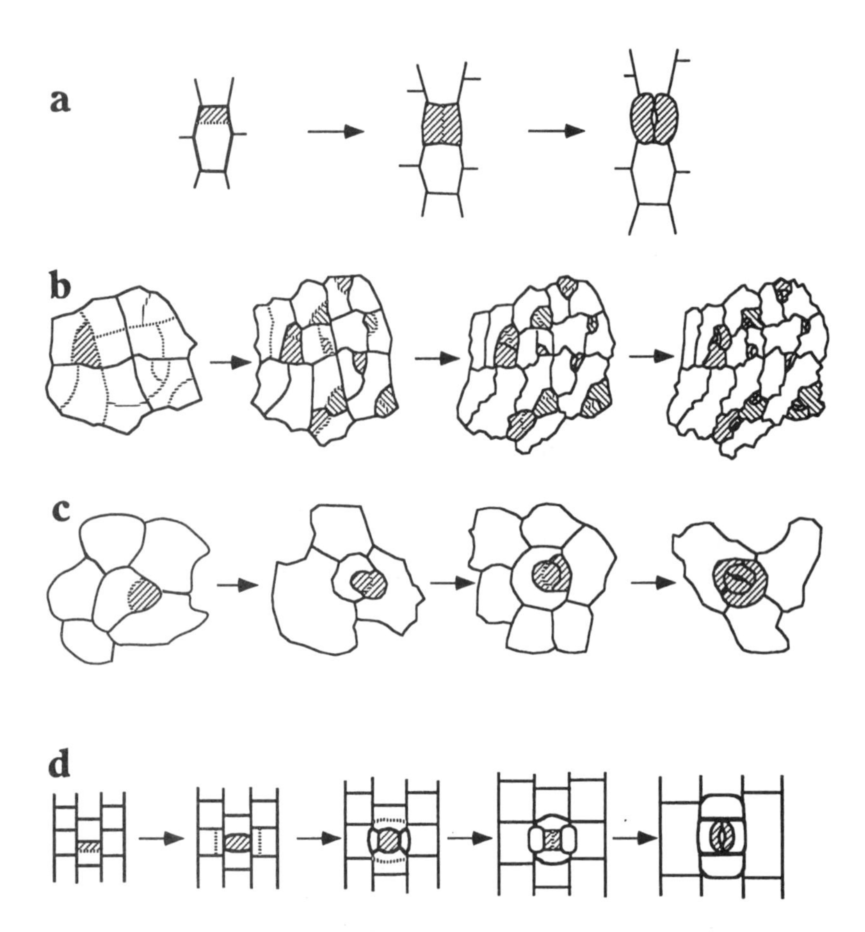

Figure 8 Four different patterns of cell division giving rise to stomatal complexes. Stomata that lack conspicuous subsidiary cells are shown for (*a*) *Anemarrhena asphodeloides* and (*b*) *Pisum*. Two patterns of deriving stomatal complexes with subsidiary cells are typified by (*c*) *Thunbergia erecta* and (*d*) *Tradescantia* [drawings adapted from Rasmussen (1986), Kagan et al (1992), Paliwal (1966), and Croxdale et al (1992)]. Broken lines indicate recently formed cell walls; diagonal shading indicates groups of cells derived from a common precursor.

is ordered to some degree. In all observed cases, adjacent stomata are found less frequently than would be expected by chance (Korn 1972, Sachs 1974, 1978, Marx & Sachs 1977, Rasmussen 1986, Charlton 1988, Croxdale et al 1992, Kagan & Sachs 1991, Sachs et al 1993). A mechanism that ensures some degree of interstomatal spacing presumably allows for optimal stomatal functioning at a given stomatal density. The nonrandom arrangement of stomata within the epidermis implies that their development is guided by a patterning process, which must be linked to some degree with how the initials gain identity.

CELL DIVISION AND LINEAGE-BASED MODELS TO EXPLAIN STOMATAL PATTERNING Several models have been proposed to account for stomatal patterning (Sachs 1978, 1991). One theory is that oriented cell divisions give rise to stomata that are nonrandomly distributed in the mature epidermis (Sachs 1978, Charlton 1990). In *Pisum* and *Sedum*, for example, stomatal initials are distributed randomly in the epidermis at the outset. A nonrandom distribution pattern is generated during subsequent cell divisions when the initial divides to give rise to a lineage group, consisting of a guard cell pair and one or more adjacent epidermal cells (Korn 1972, Sachs 1978). In other plants, such as in many monocotyledons, stomatal initials are distributed nonrandomly at the outset and all subsequent divisions contribute cells only to the stomatal complex itself (Tomlinson 1974, Charlton 1988). The linear alignment of stomata in monocotyledons thus arises from a linear alignment of stomatal initials, and subsequent divisions of the initials have no impact on the final relative positions of stomatal complexes (Stebbins & Shah 1960, Stebbins & Jain 1960, Sachs 1974, Charlton 1988, Croxdale et al 1992). Nevertheless, stomatal spacing within a linear series is assured by the pattern of cell division. Asymmetric divisions are oriented so that the small apical daughter cell becomes the stomatal initial and the larger basal daughter cell becomes an ordinary epidermal cell. Thus a minimum distance of one cell between neighboring stomata within a linear series is maintained.

As an extension of the cell division theory described above, Charlton (1990) proposed that the linear patterning observed in several monocotyledons (Charlton 1988) is influenced by the stage of the cell cycle for a particular group of initials. He proposed that there are distinct zones in the monocotyledon leaf that influence the patterning of stomata in the linear series. The basal-most zone is one in which proliferative cell divisions occur followed by a formative zone, in which stomatal initials are first established. Charlton (1990) argues that a cell's fate is determined by its cell cycle stage upon entry into the formative zone. This model predicts (*a*) that stomata belonging to one linear series would be more closely related to each other by lineage than they are to stomata of another series in the vicinity and (*b*) that all the cells within the series should be at the same stage of the cell cycle. Although direct clonal relationships within a linear series have not yet been proven, Chin et al (1995) show that the theory is supported by observations of stomatal development in *Tradescantia.* For example, cells within a linear group (referred to as strings) appear to be at similar stages of the cell cycle, based on the fact that stomatal development within a string is synchronous and that DAPI-stained nuclei appear to be at similar stages of DNA synthesis. In addition, formative zones are present in individual cell files as detected by mapping entire leaf surfaces (Chin et al 1995). Strings within linear files of cells were classified according to the stage of stomatal development, and the extent of the leaf occupied by each stage was recorded. Results show that the formative zone does not span the entire leaf width uniformly when viewed for individual cell files. Rather, longitudinal files of cells exhibit individualized separation into proliferative and formative zones. The suggestion was made that a patterning signal travels down files of cells rather than across the width of the leaf (Chin et al 1995).

CELL INTERACTIONS CONTRIBUTE TO STOMATAL PATTERNING A second theory to explain stomatal patterning is that developing stomata or initials inhibit the formation of neighboring stomata (Korn 1972, 1981, Kagan & Sachs 1991, Croxdale et al 1992, Sachs et al 1993). Mathematical models demonstrate that the inhibition theory could account for stomatal patterning in some dicotyledons (Korn 1972, 1981, 1994). The theory was not supported, however, by ablation experiments in the monocotyledon *Tradescantia*, in which the presumed inhibitory action was removed by ablating stomatal initals: Additional stomata did not form in adjacent cells (Croxdale et al 1992). On the other hand, some observations do support the idea that inhibition over longer distances may suppress stomatal development. For example, in three distantly related monocotyledons, cells resembling arrested stomata are observed closer to mature stomata than mature stomata are to each other. These observations imply that stomata already initiated may suppress the final differentiation of initials nearby (Kagan & Sachs 1991, Croxdale et al 1992, Sachs et al 1993). Furthermore, Boetsch et al (1995) show that initials are arrested in *Tradescantia* based only on the presence of stomata in immediately adjacent cell files; thus arrest of stomatal development is distance dependent. Cellular analysis suggests the arrested initials acquire epidermal cell identity following the original assumption and arrest of the stomatal pathway (Boetsch et al 1995). Arrested cells share features of both stomatal initials (small size and proper position) and ordinary epidermal cells (transverse cell profile and accumulation of anthocyanin). These results suggest that the initials retain responsiveness to the patterning signal. The initials receive the signal to differentiate but remain responsive to inhibitory factors and can then redifferentiate as ordinary epidermal cells.

PREPATTERNING ESTABLISHES IDENTITY OF STOMATAL INITIALS A third theory is that a prepattern is responsible for the distribution of stomatal initials. For example, stomatal position might be imposed on the epidermis by a morphogen that is distributed non-uniformly (Meinhardt 1984), or by patterns that have already been established within the internal tissue layers. Although evidence for a morphogen is lacking, in some species stomata do not form in epidermal cells directly (Stebbins & Shah 1960, Sachs 1978, Kaufman et al 1985, Rasmussen 1986, Croxdale et al 1992). This is also apparent in C4 grasses such as maize, in which the veins are closely spaced and stomatal files are parallel and aligned with the underlying vascular pattern (Freeling & Lane 1994). The observed correlation suggests that veins could have an inhibitory influence on stomatal development because the vascular pattern is established before stomatal initials are formed. Such an inhibitory relationship could be purely structural in nature; a biochemical morphogen is not necessary to invoke inhibitory effects. There is considerable evidence, however, that cell layer interactions are an intrinsic part of leaf development, particularly for specification of epidermal identity (Hake & Freeling 1986, Sinha & Hake 1990, Fowler et al 1996). Clearly, the mechanism of intervention by veins on the acquisition of stomatal identity remains a question. Further inquiry into the relationship between vascular and stomatal pattern in dicotyledons is warranted.

BOTH CELL LINEAGE AND CELL INTERACTIONS CONTRIBUTE TO STOMATAL PATTERNING Prepatterning is the least supported and also the least thoroughly studied of the theories presented to explain how stomatal initials acquire identity in particular patterns. Observations of different plant species do support the view that effects of cell lineage and intercellular interaction, acting together, contribute to stomatal patterning. Both processes appear to influence the final stomatal distribution in *Pisum.* As described earlier, stomatal initials in this species arise in a random pattern; the final nonrandom distribution of stomata in the mature epidermis emerges as the initials divide further and is thus dependent on cell division pattern. However, Kagan & Sachs (1991) show that the pattern of divisions producing stomatal initial cells is variable, thereby demonstrating that spacing in the mature epidermis is not the result of a rigidly programmed sequence. Instead, the division patterns and fates of stomatal initials appear to be influenced by cell-cell interactions. For example, when two stomatal initials form side by side, additional divisions result in the separation of these initials by an ordinary epidermal cell. Alternatively, one or both of the adjacent stomatal initials do not produce guard cells (Kagan & Sachs 1991, Kagan et al 1992).

Genetics will ultimately contribute to our understanding of how stomatal cells acquire identity. Zeiger & Stebbins (1972) described a mutant of barley,

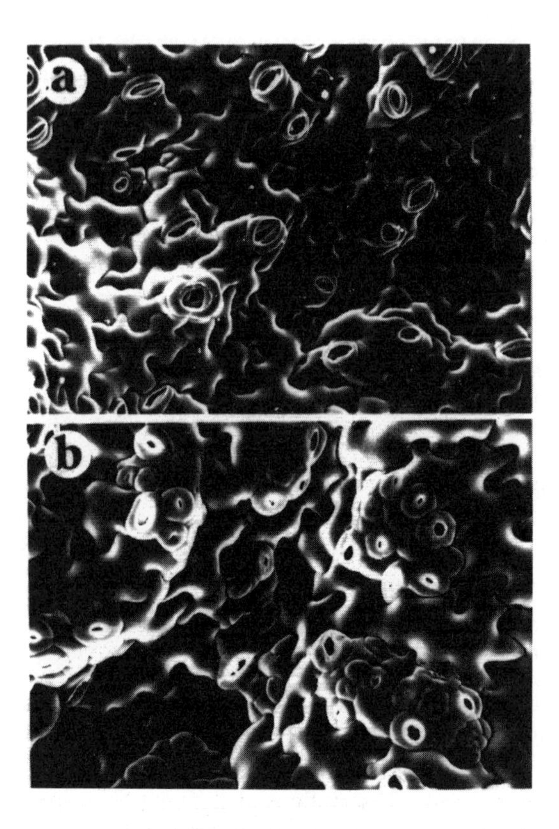

Figure 9 Wild-type leaf and stomatal mutant *tmm* in *Arabidopsis*. (*a*) Stomata are evenly distributed in the wild-type epidermis. (*b*) The mutant *tmm* epidermis shows altered spacing, with many clustered stomata in direct contact. Stomatal divisions are also defective in some cases. (Figure reprinted from Yang & Sack 1995).

cer-g, with alterations in both the pattern of wax biosynthesis and stomatal development. Intriguing patterns of guard cell divisions are described in this mutant, and predictions of wild-type gene function in contributing to cell identity are forthcoming. Yang & Sack (1995) describe two new stomatal mutants in *Arabidopsis* (Figure 9). In the mutant *too many mouths* (*tmm*), a high proportion of stomata are found in clusters containing two or more adjacent pairs of guard cells. In another mutant, *four lips* (*flp*), guard cells are frequently found unpaired or in odd combinations. Guard cells in these mutants appear to differentiate normally. Analysis of the frequency and distribution of guard cell clusters suggest that they result from the overproduction of GMCs by stomatal initials that formed in normal positions. TMM and FLP may thus function to restrict the guard mother cell fate to one cell within a lineage founded by a single stomatal initial. The genes could also be required for the proper timing of cell division following specification of guard mother cell fate. Analysis of the process of stomatal development in *tmm* and *flp* mutants will most likely provide further insight into the roles of these genes in the regulation of stomatal development. Identification of stomatal mutants in monocotyledons such as maize would also help round out the inquiry into the molecular control of stomatal patterning.

Identity and Patterning of Trichomes

Trichomes are similar to stomata in that they are unique dispersed cell types appearing during differentiation of the leaf epidermis. A trichome can be defined as any structure that originates as a local outgrowth of the epidermis, most often of a single epidermal cell, including hairs, scales, bladders, spines and glands (Mauseth 1988). Trichomes are diverse structures that may be unicellular or multicellular; unbranched, branched, or peltate (Figure 10). Evidence for trichome function is circumstantial, but they probably serve a range of functions including protection against herbivore damage, reduction of water loss, and protection from UV light damage by scattering UV irradiation. Trichomes, like stomata, are regularly spaced throughout the leaf epidermis and initiate from the smaller daughter of an asymmetric cell division, termed a trichome precursor cell (Rasmussen 1986, Sachs et al 1993). In *Arabidopsis*, the trichome precursor cell first undergoes several rounds of DNA replication without intervening mitoses (endoreduplication). The precursor then expands out from the leaf surface with the enlarged nucleus migrating toward the tip of the expanding trichome (Hulskamp et al 1994; reviewed in Marks 1994). Frequently, cells near the base of the trichome, termed accessory cells, differentiate further as part of the trichome unit. Patterning mechanisms are most likely linked to cellular determination. The development of trichomes, therefore, raises many of the same questions that have already been discussed in connection with stomatal development. When and how are individual epidermal cells selected as trichome precursors? How is their subsequent differentiation controlled?

In contrast to stomata, trichomes have been particularly amenable to genetic methods of analysis. Mutations in approximately 24 genes have been shown to alter various aspects of trichome development in *Arabidopsis* (summarized by Marks & Esch 1994, Hulskamp et al 1994, Marks 1994). Mutant phenotypes indicate that most of these genes function during trichome morphogenesis, well after trichome cell fates have been determined. However, the mutant phenotypes associated with four of these genes imply that they play important roles in the specification and patterning of trichomes. Three of the four genes, *GLABROUS1* (*GL1*), *TRANSPARENT TEST A GLABROUS* (*TTG*), and

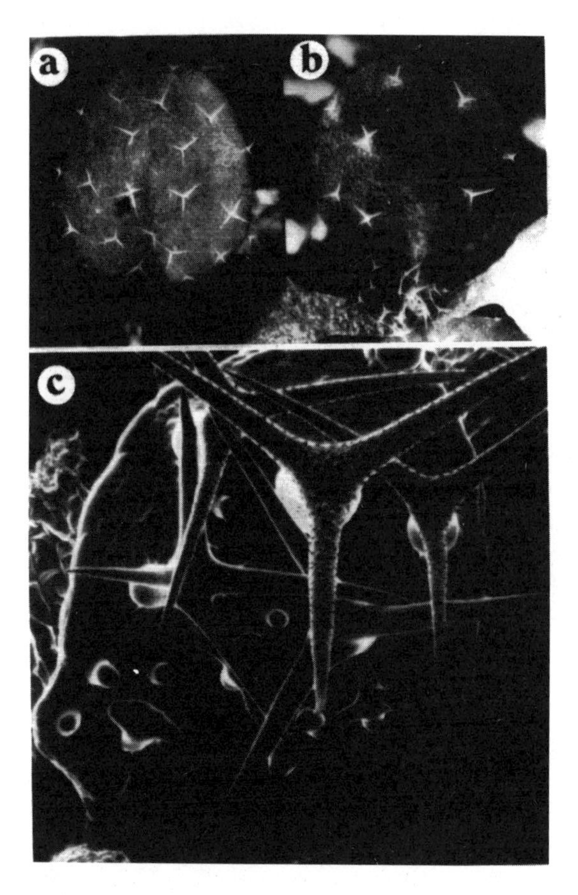

Figure 10 Trichome spacing and development in *Arabidopsis*. (*a,b*) Trichome spacing can vary in ecotypes but shows a nonrandom distribution. Trichomes mature in a roughly basipetal pattern (reprinted from Larkin et al 1996). (*c*) The least mature trichome is marked with an arrow near the base of the leaf. An arrowhead marks the mature branched trichome near the leaf tip (reprinted from Marks 1994).

TRYPTICHON (*TRY*), are discussed below with reference to initiation and patterning of trichome placement. The fourth recently described gene, *REDUCED TRICHOME NUMBER* (*RTN*), appears to be associated with an arrest of trichome initiation and is not discussed here (Larkin et al 1996).

AT LEAST TWO GENES ARE REQUIRED FOR TRICHOME INITIATION Recessive mutations in the *GL1* and *TTG* genes block the earliest recognizable events in trichome development: expansion of prospective trichome-forming cells in the plane of the leaf surface, and the nuclear enlargement associated with endoreduplication (Haughn & Somerville 1988, Marks 1994, Hulskamp et al 1994). GL1 and TTG are thus required for trichome initiation. *ttg* mutations also block the production of mucilage and anthocyanin in the seed coat, thus indicating that *TTG* has functions in addition to its role in trichome initiation. Molecular cloning and sequencing of the *GL1* gene reveals that its product has homology to the *myb* class of transcription factors; therefore, *GL1* probably functions to regulate the transcription of other genes required for trichome development (Oppenheimer et al 1991). *GL1* transcripts are present at high levels in trichome precursors and in developing trichomes and additionally are found at lower levels in other epidermal cells (Larkin et al 1993). Thus an increase in *GL1* mRNA appears to be an essential and early event in the commitment of an epidermal cell to the trichome developmental pathway. Genetic mosaics, in which patches of trichomeless *gl1* mutant tissue were identified within trichome-bearing wild-type plants, also support this conclusion. The presence of patches of phenotypically mutant tissue suggest that the *GL1* gene acts locally, rather than over a long distance, to direct trichome initiation (Hulskamp et al 1994).

The *TTG* gene may be a homologue of the *R* gene of maize, a transcription factor of the *myc* class that is involved in the regulation of anthocyanin biosynthesis (Ludwig et al 1989). Lloyd et al (1992) showed that the *ttg* mutation could be complemented by expression of the maize *R* gene in transgenic *Arabidopsis* under the control of the CaMV35S promoter. These results suggest that either the maize *R* gene is functionally homologous to the *TTG* gene, or the expression of *R* in *Arabidopsis* bypasses the requirement for the *TTG* gene in trichome development. *R* is also known to cooperate with a second gene, *C1*, which encodes a transcription factor belonging to the *myb* family, in the regulation of anthocyanin biosynthesis in maize (Cone et al 1986, Paz-Ares et al 1987, Dooner et al 1991, Roth et al 1991). By analogy, *TTG* could encode a *myc*-like transcription factor that cooperates with the *myb*-like product of the *GL1* gene in the control of trichome development. Preliminary evidence indicates that R and GL1 proteins do interact directly in the yeast two-hybrid system, further supporting the hypothesis that *TTG* is a homologue of *R* (A Lloyd, unpublished data). Proof of this hypothesis clearly relies on the molecular analysis of the *TTG* gene itself.

Experiments using the CaMV35S-driven expression of the maize *R* gene and the *GL1* gene in transgenic *Arabidopsis* (Larkin et al 1994) showed that constitutive expression of *GL1* does not cause excess trichome production in a wild-type background, and also does not functionally complement the *ttg* mutation. Conversely, overexpression of *R* does not functionally complement the *gl1* mutation. However, 35S-driven expression of both *R* and *GL1* causes overproduction of hairs on both stems and leaves. These observations suggest that *GL1* and *TTG* function together to promote trichome initiation, a situation that is analogous to the coordinate regulation of anthocyanin biosynthesis by *R* and *C1* in maize.

CONTROL OF TRICHOME PATTERNING BY LINEAGE MODELS OR BY CELL INTERACTION? What kind of patterning process(es) might be responsible for the selection of dispersed, individual epidermal cells as trichome precursor cells? For stomatal complexes, there is evidence suppporting both a lineage-based model, in which cell division is important, and a cell interaction model, in which neighboring initials influence the fate of neighboring epidermal cells. There is mounting evidence that cell interactions, rather than cell lineage, influence trichome identity (Lloyd et al 1994, Larkin et al 1994, 1996). In one experiment, Lloyd et al (1994) produced transgenic *Arabidopsis* plants, in which trichome initiation could be modulated experimentally. The results show that any epidermal cell in a stage three *Arabidopsis* leaf is competent to produce a trichome. The experimental system made use of the fact that *ttg* defects in *Arabidopsis* can be reversed by expression of the maize *R* gene. A plasmid construct combining the rat glucocorticoid receptor (GR), which has a steroid-binding domain, and the maize *R* gene was placed under the control of the 35S promoter (35SR-GR) and was introduced into wild-type and *ttg* mutants of *Arabidopsis*. The resulting transformants show hormone-dependent induction of trichomes, thus providing a means to modulate trichome initiation. A transformed leaf (*ttg* 35SR-GR) at stage three of development showed an increased number and clustering of trichomes in response to hormone induction. Thus it appears that individual epidermal cells are equally competent to produce trichomes. The stage at which epidermal cells are committed, based on an absence of response to the hormone, could be determined by varying the time of hormone treatment. These results confirm the basipetal differentiation pattern of the *Arabidopsis* leaf. Although all epidermal cells appear competent to differentiate as trichomes, the fact that only select cells actually do so was explained by a cell interaction model; e.g. inhibition by emerging trichomes prevents neighboring epidermal cells from differentiating further (Lloyd et al 1994).

Another set of experiments was conducted to find out whether trichome patterning results from a lineage-based process or from cell interactions such as lateral inhibition (Larkin et al 1996). First, it was shown that trichomes in

Arabidopsis are nonrandomly spaced from the earliest visible stages of differentiation, when the trichome precursor is first expanding. Second, clonal sectors were examined to determine the relationship between sector boundaries and the position of a trichome unit. If a lineage-based model is correct, then genetically marked sectors (produced by early excision of the transposable element *Ac* from a GUS transgene and made visible by histochemical staining) should include either all or none of the cells in the trichome unit. Alternatively, if cells of the trichome unit form regardless of cell lineage, then sector boundaries should not align with the cells of the trichome unit. The analysis confirms the latter: Sector boundaries do not show a defined relationship relative to the trichome unit. Rather, individual accessory cells within the trichome unit were stained in several cases with no apparently consistent pattern (Larkin et al 1996). These results demonstrate that trichome spacing cannot be explained on the basis of cell lineage because "related" cells were not similar in staining patterns. Additional observations of arrested trichome initials support the view that lateral inhibition is more effective in establishing trichome distribution than cell division pattern, at least in *Arabidopsis*. Similar analyses in monocotyledons, which show a linear alignment of trichomes, would be informative. As is the case for stomatal positioning, the control of cell fate may be quite different in monocotyledons compared with dicotyledons.

The molecular basis for establishing trichome spacing remains to be clarified. Of interest is the suggested role for *TTG* in trichome patterning, as proposed by Larkin et al (1994). Plants heterozygous for the *ttg* mutation and carrying the 35S-*GL1* transgene show unusual trichome distribution: one third of the trichomes produced were found in clusters. Trichome clusters are uncommon both in wild-type and in plants carrying the 35S-*GL1* transgene in a fully ttg^{+} background. Furthermore, another *Arabidopsis* trichome mutant, *TRY*, also shows trichome clustering (Hulskamp et al 1994). A possible explanation of these effects is that *TRY* and *TTG* may be required for a lateral signaling process that inhibits trichome initiation in the immediate vicinity of a trichome precursor cell. Another possibility is that clustering could result if precursor cells undergo further cell divisions before giving rise to trichomes (Larkin et al 1994). In this case, the time of commitment of precursor cells is premature relative to the timing of cell division. Another idea proposed by Larkin et al (1994) is that *TRY* may be a target of transcriptional regulation by *TTG*. Clearly, there are several hypotheses to test regarding the interaction of these genes once they are cloned.

CONCLUSIONS AND PROSPECTS

A leaf is a progressively determined organ, the identity of which is acquired in different ways at each stage of development. Individual cells participate little in the assigning of identity, particularly during early stages of development when leaf shape is emerging. Even though a position-dependent model requires that cells occupy a given position, it is apparent that this is a regional phenomenon early in development. Later, however, precisely oriented cells participate in tissue differentiation, such as described here for stomata and trichomes. Indeed, many of the final details of anatomy and the preservation of a functional tissue are intimately dependent on the behavior of individual cells. The trend is thus that the progressive acquisition of identity involves a refining of the behavior of individual cells.

Early events of stage one are associated with partitioning the meristem and segregating founder cells from the general meristem population. Specific cell divisions do not appear to be influential, yet in some cases, relative positions of cells in the meristem affect future identity in a generalized way. Two important questions remain: The first question concerns how founder cells are established at the flanks of the meristem and what the relationship is between founder cell domains and phytomer domains. The second question concerns the fate of the founder cell domains in the leaf. It is intriguing that the constant feature of organogenesis is the presence of the flanking zone in the meristem. The changing features are the cells themselves, which enter the zone without identity and exit the zone with the identity of leaf initials. It is postulated that there is a hierarchy of transcription factors, each regulating the maintenance or proliferation of the founder cell zone. To date, the *knox* genes are the only indicators of such a hierarchy and future work, separating the "meristem bureaucracy" into its attendant molecular parts will be enlightening.

Another question with regard to organogenesis concerns how the tissue reponds to the established founder cells; that is, how and when is the foliage buttress specified? Evidence cited above suggests that tissue response to founder cell identity is rapid. For example, the number of visible cells in the foliage buttress is similar to the number of founder cells estimated by clonal analysis (Poethig & Sussex 1985a,b), suggesting the buttress emerges directly from the founder cells with no intervening cell division. The nature of the response mechanism is not understood. The intriguing fact is that founder cells are poised to expand in the direction of a primordium, and they will do so even if division is prevented.

Organogenesis may not be particularly amenable to genetic dissection, if only because mutations that affect such fundamental aspects of development are difficult to identify and study. This is particularly true if such fundamental control functions are redundant. The next stage, that of the basic elaboration of leaf parts, is quite different in this regard. Once the leaf has been initiated, it appears that many plants can support normal function with a high degree of altered structure. Thus a genetic approach is promising, particularly in conjunction with morphological and anatomical analyses. Maize, a monocotyledon, and several dicotyledonous species (tobacco, *Arabidopsis*, and *Antirrhinum*) are at the forefront of such analyses. It is interesting that these species share a common feature, namely, the ultimate identity of the distal leaf part (blade or lamina) is not acquired until after the primordium is established during stage two, even though those distal parts arise from distinctly different parts of the primordium (D Kaplan, manuscript in preparation). These results align with the position-dependent model. Cells in the proper position (the distal area) will acquire the appropriate identity, but precisely oriented divisions are not required except in special circumstances, to confer identity. The final stage, differentiation, relies more heavily on the behavior of individual cells. In many cases, oriented cell division and specific cell expansions are required to realize the identity of a tissue. The future promise in this area rests in cell biological investigations to understand how cells switch from acting as groups and acquiring identity regionally to more individualized actions. Leaf development is not disjunct, but is a continuous process, with relative variation in the extent to which regions and cells participate in the process of identity acquisition. Understanding the subtle relationship between the molecular and cellular control of the continuum of leaf development is a current goal of research.

Literature Cited

Alberts B, Bray D, Lewis J, Raff M, Roberts K, Watson JD. 1994. *Molecular Biology of the Cell.* New York: Garland. 2nd ed.

Avery GSJ. 1933. Structure and development of the tobacco leaf. *Am. J. Bot.* 20:565–91

Ball E. 1960. Sterile culture of the shoot apex of *Lupinus albus. Growth* 24:91–110

Barton MK, Poethig RS. 1993. Formation of the shoot apical meristem in *Arabidopsis thaliana*: an analysis of development in the wild type and in the shoot meristemless mutant. *Development* 119:823–31

Becraft P, Bongard-Pierce DK, Sylvester AW, Poethig RS, Freeling M. 1990. The *liguleless-1* gene acts tissue-specifically in maize leaf development. *Dev. Biol.* 141:220–32

Becraft PW, Freeling M. 1991. Sectors of *liguleless-1* tissue interrupt an inductive signal during maize leaf development. *Plant Cell* 3:801–7

Becraft PW, Freeling M. 1994. Genetic analysis of *Rough sheath1* developmental mutants of maize. *Genetics* 136:295–311

Bertrand-Garcia R, Freeling M. 1991. *Hairy-sheath frayed 1–0*: a systemic, heterochronic mutant of maize that specifies slow developmental stage transitions. *Am. J. Bot.* 78:747–65

Boetsch J, Chin J, Croxdale J. 1995. Arrest of stomatal initials in *Tradescantia* is linked to the proximity of neighboring stomata and results in the arrested initials acquiring properties of epidermal cells. *Dev. Biol.* 168:28–38

Bosabalidis, AM, Evert RF, Russin, WA. 1994. Ontogeny of the vascular bundles and contiguous tissues in the maize leaf blade. *Am. J. Bot.* 81:745–52

Bossinger G, Maddaloni M, Motto M, Salamini

F. 1992. Formation and cell lineage patterns of the shoot apex of maize. *Plant J.* 2:311–20

Brandstadter J, RoBback C, Theres K. 1994. The pattern of histone H4 expression in the tomato shoot apex changes during development. *Planta* 192:69–74

Buvat R. 1952. Structure, evolution et functionnement du meristeme apical de quelques dicotyledones. *Ann. Sci. Nat. Bot.* 13:199–300

Cerioli S, Marocco A, Maddaloni M, Motto M, Salamini F. 1994. Early event in maize leaf epidermis formation as revealed by cell lineage studies. *Development* 120:2113–20

Charlton WA. 1988. Stomatal pattern in four species of monocotyledons. *Ann. Bot.* 61:611–21

Charlton WA. 1990. Differentiation in leaf epidermis of *Chlorophytum comosum* Baker. *Ann. Bot.* 66:567–78

Chin J, Wan Y, Smith J, Croxdale J. 1995. Linear aggregations of stomata and epidermal cells in *Tradescantia* leaves: evidence for their group patterning as a function of the cell cycle. *Dev. Biol.* 168:39–46

Clark JK, Sheridan WF. 1991. Isolation and characterization of 51 embryo-specific mutations of maize. *Plant Cell* 3:935–51

Clark JK. 1996. Maize embryogenesis mutants. In *Embryogenesis: The Generation of Plant Form,* ed. TL Wang, A Cuming. Oxford: Bios Scientific. In press

Coe EH, Neuffer MG. 1978. Embryo cells and their destinies in the corn plant. In *The Clonal Basis of Development,* ed. S Subtelny, I Sussex. pp.113–29. New York: Academic

Cone KC, Burr FA, Burr B. 1986. Molecular analysis of the maize anthocyanin regulatory locus *Cl. Proc. Natl. Acad. Sci. USA* 83:9631–35

Cooke TJ, Lu B. 1992. The independence of cell shape and overall form in multicellular algae and land plants: cells do not act as building blocks for constructing plant organs. *Int. J. Plant Sci.* 153:S7–27

Croxdale J, Smith J, Yandell B, Johnson B. 1992. Stomatal patterning in *Tradescantia*: evaluation of the cell lineage theory. *Dev. Biol.* 149:158–67

Cutter EG. 1958. Studies of morphogenesis in the Nymphaeaceae III. Surgical experiments on leaf and bud formation. *Phytomorphology* 8:74–95

Davis EL, Rennie P, Steeves TA. 1979. Further analytical and experimental studies on the shoot apex of *Helianthus annuus*: variable activity in the central zone. *Can. J. Bot.* 57:971–80

Davis EL, Steeves TA. 1977. Experimental studies on the shoot apex of *Helianthus annuus*: the effect of surgical bisection on quiescent cells in the apex. *Can. J. Bot.* 55:606–14

Dawe RK, Freeling M. 1990. Clonal analysis of the cell lineages in the male flower of maize. *Dev. Biol.* 142:233–45

Dawe RK, Freeling M. 1991. Cell lineage and its consequences in higher plants. *Plant J.* 1:3–7

Dengler NG, Dengler RE, Donnelly PM, Filosa MF. 1995. Expression of the C4 pattern of photosynthetic enzyme accumulation during leaf development in *Atriplex rosea* (Chenopodaceae). *Am. J. Bot.* 82:318–27

Dermen H. 1953. Periclinal cytochimeras and origin of tissues in stem and leaf of peach. *Am. J. Bot.* 40:154–68

Dooner HK, Robbins TP, Jorensen RA. 1991. Genetic and developmental control of anthocyanin biosynthesis. *Annu. Rev. Genet.* 25:173–99

Esau K. 1977. *Anatomy of Seed Plants.* New York: Wiley & Sons. 550 pp.

Ferreira PCG, Hemerly AS, Engler JD, Van Montagu M, Engler G, Inze D. 1994. Developmental expression of the *Arabidopsis* cyclin gene *cyc1A. Plant Cell* 6:1763–74

Fladung M, Bossinger G, Roeb GW, Salamini F. 1991. Correlated alterations in leaf and flower morphology and rate of leaf photosynthesis in a midribless (mbl) mutant of *Panicum maximum* Jacq. *Planta* 184:356–61

Fleming AJ, Mandel T, Roth I, Kuhlemeier C. 1993. The patterns of gene expression in the tomato shoot apical meristem. *Plant Cell* 5:297–309

Foard DE. 1971. The initial protrusion of a leaf primordium can form without concurrent periclinal cell divisions. *Can. J. Bot.* 49:1601–3

Fosket DE. 1994. *Plant Growth and Development: A Molecular Approach.* San Diego: Academic

Foster AS. 1936. Leaf differentiation in angiosperms. *Bot. Rev.* 2:349–72

Foster AS. 1938. Structure and growth of the shoot apex in *Ginkgo biloba. Bull. Torrey Botan. Club* 65:531–56

Foster AS, Gifford EMJ. 1974. *Comparative Morphology of Vascular Plants.* San Francisco: WH Freeman

Fowler JE, Freeling M. 1996. Genetic analysis of mutations that alter cell fates in maize leaves: dominant *liguleless* mutations. *Dev. Genet.* 18:198–222

Fowler JE, Muehlbauer GJ, Freeling M. 1996. Mosaic analysis of the *liguleless 3* mutant phenotype in maize by coordinate suppression of mutator–insertion alleles. *Genetics* 143:489–503

Freeling M. 1992. A conceptual framework for maize leaf development. *Dev. Biol.* 153:44–58

Freeling M, Hake S. 1985. Developmental genetics of mutants that specify knotted leaves in maize. *Genetics* 111:617–34

Freeling M, Lane B. 1994. The maize leaf. In *The Maize Handbook,* ed. M Freeling, V. Walbot. pp. 17–28 New York: Springer-Verlag

Furner IJ, Pumphrey JE. 1992. Cell fate in the shoot apical meristem of *Arabidopsis thaliana. Development* 115:755–64

Galinat WC. 1959. The phytomer in relation to floral homologies in the American maydeae. *Bot. Mus. Leafl.* 19:1–32

Gifford EMJ, Corson GEJ. 1971. The shoot apex in seed plants. *Bot. Rev.* 37:143–229

Green PB. 1994. Connecting gene and hormone action to form, pattern and organogenesis: biophysical transductions. *J. Exp. Bot.* 45:1775–88

Green PB, Brooks KE. 1978. Stem formation from a succulent leaf: its bearing on theories of axiation. *Am. J. Bot.* 65:13–26

Green PB, Selker JML. 1991. Mutual alignments of cell walls, cellulose, and cytoskeletons: their role in meristems. In *The Cytoskeletal Basis of Plant Growth and Form,* ed. C Lloyd. pp. 303–22. London: Academic

Greyson RI, Walden DB, Smith WJ. 1982. Leaf and stem heteroblasty in *Zea. Bot. Gaz.* 143:73–78

Hagemann W. 1992. The relationship of anatomy to morphology in plants: a new theoretical perspective. *Int. J. Plant Sci.* 153:S38–48

Hake S. 1992. Unraveling the knots in plant development. *Trends Genet.* 8:109–14

Hake S, Bird RM, Neuffer MG, Freeling M. 1985. Development of the maize ligule and mutants that affect it. In *Plant Genetics,* ed. M Freeling, pp. 61–71. New York: Liss

Hake S, Freeling M. 1986. Analysis of genetic mosaics shows that the extra epidermal cell divisions in *Knotted* mutant maize plants are induced by adjacent mesophyll cells. *Nature* 320:621–23

Hake SE, Vollbrecht E, Freeling M. 1989. Cloning *Knotted*, the dominant morphological mutant in maize using *Ds2* as a transposon tag. *EMBO J.* 8:15–22

Halperin W. 1969. Morphogenesis in cell cultures. *Annu. Rev. Plant Physiol.* 20:395–418

Harper L, Freeling M. 1996. Recent studies on early leaf development. *Curr. Opin. Biotechnol.* 7:139–44

Haughn GW, Somerville CR. 1988. Genetic control of morphogenesis in *Arabidopsis. Dev. Genet.* 9:73–89

Hemerly A, Engler JDA, Bergounioux C, Van Montagu M, Engler G, et al. 1995. Dominant negative mutants of the Cdc2 kinase uncouple cell division from iterative plant development. *EMBO J.* 14:3925–36

Henshaw GG, O'Hara JF, Webb KJ. 1982. Morphogenetic studies in plant tissue cultures. In *Differentiation In Vitro,* ed. MM Yeoman, DES Truman. Cambridge: Cambridge Univ. Press

Hill JP, Lord EM. 1990. A method for determining plastochron indices during heteroblastic shoot growth. *Am. J. Bot.* 77:1491–97

Huala E, Sussex IM. 1993. Determination and cell interactions in reproductive meristems. *Plant Cell* 5:1157–65

Hulskamp M, Misera S, Jurgens G. 1994. Genetic dissection of trichome cell development in Arabidopsis. *Cell* 76:555–66

Irish VF, Sussex IM. 1992. A fate map of the *Arabidopsis* embryo shoot apical meristem. *Development* 115:745–53

Jackson D, Veit B, Hake S. 1994. Expression of maize *KNOTTED1* related homeobox genes in the shoot apical meristem predicts patterns of morphogenesis in the vegetative shoot. *Development* 120:405–13

Jegla DE, Sussex IM. 1989. Cell lineage patterns in the shoot meristem of the sunflower embryo in the dry seed. *Dev. Biol.* 131:215–25

Jesuthasan S, Green PB. 1989. On the mechanism of decussate phyllotaxis: biophysical studies on the tunica of *Vinca major. Am. J Bot.* 76:1152–66

Jurgens G, Torres RA, Berleth T. 1994. Embryonic pattern formation in flowering plants. *Annu. Rev. Genet.* 28:351–71

Kagan ML, Novoplansky N, Sachs T. 1992. Variable cell lineages form the functional pea epidermis. *Ann. Bot.* 69:303–12

Kagan ML, Sachs T. 1991. Development of immature stomata: evidence for epigenetic selection of a spacing pattern. *Dev. Biol.* 146:100–5

Kaplan DR. 1970. Comparative foliar histogenesis in *Acorus calamus* and its bearing on the phyllode theory of monocotyledonous leaves. *Am. J. Bot.* 57:331–61

Kaplan DR. 1992. The relationship of cells to organisms in plants: problem and implications of an organismal perspective. *Int. J. Plant Sci.* 153:S28–37

Kaplan DR, Hagemann W. 1991. The relationship of cell and organism in vascular plants. *Bioscience* 41:693–703

Kaufman PB, Dayanandan P, Franklin CI. 1985. Structure and function of silica bodies in the epidermal system of grass shoots. *Ann. Bot.* 55:487–507

Kelly AJ, Meeks-Wagner DR. 1995. Characterization of a gene transcribed in the L2 and L3 layers of the tobacco shoot apical meristem. *Plant J.* 8:147–53

Kerstetter R, Vollbrecht E, Lowe B, Veit B, Yamaguchi J, Hake S. 1994. Sequence analysis and expression patterns divide the maize *knotted1*-like homeobox genes into two classes. *Plant Cell* 6:1877–87

Klinge B, Werr W. 1995. Transcription of the *Zea mays* homeobox (*ZmHox*) genes is activated early in embryogenesis and restricted to meristems of the maize plant. *Dev. Genet.* 16:349–57

Korn RW. 1972. Arrangement of stomata on leaves of *Pelargonium zonale* and *Sedum stahli*. *Ann. Bot.* 36:325–33

Korn RW. 1981. A neighboring-inhibition model for stomate patterning. *Dev. Biol.* 88:115–20

Korn RW. 1994. Pattern formation in plant epidermis through inhibition of immediately adjacent cells by pattern elements. *Protoplasma* 180:145–52

Langdale JA, Zelitch I, Miller E, Nelson T. 1988. Cell position and light influence C4 versus C3 patterns of photosynthetic gene expression in maize. *EMBO J.* 7:3643–51

Larkin JC, Oppenheimer DG, Lloyd AM, Paparozzi ET, Marks MD. 1994. Roles of the *GLABROUS1* and *TRANSPARENT TESTA GLABRA* genes in *Arabidopsis* trichome development. *Plant Cell* 6:1065–76

Larkin JC, Oppenheimer DG, Pollock S, Marks MD. 1993. *Arabidopsis GLABROUS1* gene requires downstream sequences for function. *Plant Cell* 5:1739–48

Larkin JC, Young N, Prigge M, Marks MD. 1996. The control of trichome spacing and number in *Arabidopsis*. *Development* 122:997–1005

Lincoln C, Long J, Yamaguchi J, Serikawa K, Hake S. 1994. A *knotted1*-like homeobox gene in *Arabidopsis* is expressed in the vegetative meristem and dramatically alters leaf morphology when overexpressed in transgenic plants. *Plant Cell* 6:1859–76

Lloyd AM, Schena M, Walbot V, Davis RW. 1994. Epidermal cell fate determination in *Arabidopsis*: patterns defined by a steroid-inducible regulator. *Science* 266:436–39

Lloyd AM, Walbot V, Davis RW. 1992. *Arabidopsis* and *Nicotiana* anthocyanin production activated by maize regulators *R* and *Cl*. *Science* 258:1173–75

Lloyd C. 1991. *The Cytoskeletal Basis of Plant Growth and Form*. London: Academic

Lloyd C. 1995. Life on a different plane. *Curr. Biol.* 5:1085–87

Long JA, Moan EI, Medford JI, Barton MK. 1996. A member of the KNOTTED class of homeodomain proteins encoded by the *STM* gene of *Arabidopsis*. *Nature* 379:66–72

Lucas WJ, Bouche-Pillon S, Jackson DP, Nguyen L, Baker L, et al. 1995. Selective trafficking of *KNOTTED1* homeodomain protein and its mRNA through plasmodesmata. *Science* 270:1980–83

Ludwig SR, Habera LF, Dellaporta SL, Wessler SR. 1989. *Lc*, a member of the maize *R* gene family responsible for tissue-specific anthocyanin production, encodes a protein similar to transcriptional activators and contains the *myc*-homology region. *Proc. Natl. Acad. Sci. USA* 86:7092–96

Lyndon RF. 1976. The shoot apex. In *Cell Division in Higher Plants*, ed. M Yeoman, pp. 285–314. New York: Academic

Lyndon RF. 1990. *Plant Development. The Cellular Basis*. London: Unwin Hyman

Marcotrigiano M, Bernatzky R. 1995. Arrangement of cell layers in the shoot apical meristem of periclinal chimeras influences cell fate. *Plant J.* 7:193–202

Marks MD. 1994. The making of a plant hair. *Curr. Biol.* 4:621–23

Marks MD, Esch JJ. 1994. Trichomes. In *Arabidopsis: An Atlas of Morphology and Development*, ed. J Bowman, pp. 56–73. New York: Springer-Verlag

Marx A, Sachs T. 1977. The determination of stomata pattern and frequency in *Anagallis*. *Bot. Gaz.* 138(4):385–92

Mauseth JD. 1988. *Plant Anatomy*. Menlo Park: Benjamin/Cummings

McDaniel CN, Poethig RS. 1988. Cell-lineage patterns in the shoot apical meristem of the germinating maize embryo. *Planta* 175:13–22

McHale NA. 1993. *LAM-1* and *FAT* genes control development of the leaf blade in *Nicotiana sylvestris*. *Plant Cell* 5:1029–38

Medford JI. 1992. Vegetative apical meristems. *Plant Cell* 4:1029–39

Meinhardt H. 1984. Models of pattern formation and their application to plant development. In *Positional Controls in Plant Development*, ed. PW Barlow, DJ Carr. pp. 1–32 Cambridge: Cambridge Univ. Press

Meinke DW. 1991. Embryonic mutants of Arabidopsis thaliana. *Dev. Genet.* 12:382–92

Meinke DW. 1995. Molecular genetics of plant embryogenesis. *Annu. Rev. Plant Physiol. Plant Mol. Biol.* 46:369–94

Nelson T, Langdale JA. 1989. Patterns of leaf development in C4 plants. *Plant Cell* 1:3–13

Nougarede A. 1967. Experimental cytology of the shoot apical cells during vegetative growth and flowering. *Int. Rev. Cytol.* 21:203–351

Oppenheimer DG, Herman PL, Sivakumaran S, Esch J, Marks MD. 1991. A *myb* gene required for leaf trichome differentiation in Arabidopsis is expressed in stipules. *Cell*

67:483–93
Paliwal GS. 1966. Structure and ontogeny of stomata in some Acanthaceae. *Phytomorphology* 16:527–39
Pant DD. 1965. On the ontogeny of stomata and other homologous structures. *Plant Sci. Ser.* 1:1–24
Paz-Ares J, Ghosal D, Wienand U, Peterson PA, Saedler H. 1987. The regulatory *c1* locus of *Zea mays* encodes a protein with homology to *myb* proto-oncogene products and with structural similarities to transcriptional activators. *EMBO J.* 6:3553–58
Poethig S. 1984. Cellular parameters of leaf morphogenesis in maize and tobacco. In *Contemporary Problems in Plant Anatomy,* ed. RA White, WC Dickinson, pp. 235–59. New York: Academic
Poethig RS. 1987. Clonal analysis of cell lineage patterns in plant development. *Am. J. Bot* 74:581–94
Poethig S. 1989. Genetic mosaics and cell lineage analysis in plants. *Trends Genet.* 5:272–77
Poethig RS. 1994. The maize shoot. In *The Maize Handbook,* ed. M Freeling, V Walbot, pp. 11–17. New York: Springer-Verlag
Poethig RS, Sussex IM. 1985a. The developmental morphology and growth dynamics of the tobacco leaf. *Planta* 165:158–69
Poethig RS, Sussex IM. 1985b. The cellular parameters of leaf development in tobacco: a clonal analysis. *Planta* 165:170–84
Poethig RS, Szymkowiak EJ. 1995. Clonal analysis of leaf development in maize. *Maydica* 40:67–76
Pri-Hadash A, Hareven D, Lifschitz E. 1992. A meristem-related gene from tomato encodes a *dUTPase*: analysis of expression in vegetative and floral meristems. *Plant Cell* 4:149–59
Raschke K. 1975. Stomata action. *Annu. Rev. Plant Physiol.* 26:306–40
Rasmussen H. 1986. Pattern formation and cell interactions in epidermal development of *Anemarrhena asphodeloides* (Liliaceae). *Nord. J. Bot.* 6:467–77
Roth BA, Goff SA, Klein TM, Fromm ME. 1991. *C1* and *R*-dependent expression of the maize *Bz1* gene requires sequences with homology to mammalian *myb* and *myc* binding sites. *Plant Cell* 3:317–25
Sachs T. 1969. Regeneration experiments on the determination of the form of leaves. *Isr. J. Bot.* 18:21–30
Sachs T. 1974. The developmental origin of stomata pattern. *Crinum. Bot. Gaz.* 135:314–18
Sachs T. 1978. The development of spacing patterns in the leaf epidermis. In *The Clonal Basis of Development,* ed. S Subtelny, I Sussex. Cambridge: Cambridge Univ. Press
Sachs T. 1991. *Pattern Formation in Plant Tissues.* Cambridge: Cambridge Univ. Press
Sachs T, Novoplansky N, Kagan ML. 1993. Variable development and cellular patterning in the epidermis of *Ruscus hypoglossum. Ann. Bot.* 71:237–43
Scanlon MJ, Schneeberger RG, Freeling M. 1996. The maize mutant *narrow sheath* fails to establish leaf margin identity in a meristematic domain. *Development.* In press
Schneeberger RG, Becraft PW, Hake S, Freeling M. 1995. Ectopic expression of the *knox* homeobox gene *rough sheath1* alters cell fate in the maize leaf. *Genes Dev.* 9:2292–304
Shahar T, Hennig N, Gutfinger T, Hareven D, Lifschitz E. 1992. The tomato 66.3 kD *polyphenoloxidase* gene: molecular identification and developmental expression. *Plant Cell* 4:135–47
Sharman BC. 1942. Developmental anatomy of the shoot of *Zea mays* L. *Ann. Bot.* 6:245–82
Sheridan WF. 1995. Genes and embryo morphogenesis in angiosperms. *Dev. Genet.* 16:291–97
Sinha N, Hake S. 1990. Mutant characters of *Knotted* maize leaves are determined in the innermost tissue layers. *Dev. Biol.* 141:203–10
Sinha N, Hake S, Freeling M. 1993. Genetic and molecular analysis of leaf development. *Curr. Top. Dev. Biol.* 28:47–80
Smith LG, Greene B, Veit B, Hake S. 1992. A dominant mutation in the maize homeobox gene, *knotted1*, causes its ectopic expression in leaf cells with altered fates. *Development* 116:21–30
Smith LG, Hake S. 1992. The initiation and determination of leaves. *Plant Cell* 4:1017–27
Smith LG, Hake S, Sylvester AW. 1996. The *tangled1* mutation alters cell division orientations throughout maize leaf development without altering leaf shape. *Development* 122:481–89
Smith LG, Jackson D, Hake S. 1995. Expression of *knotted1* marks shoot meristem formation during maize embryogenesis. *Dev. Genet.* 16:344–48
Smith RH, Murashige T. 1970. In vitro development of the isolated shoot apical meristem of angiosperms. *Am. J. Bot.* 57:562–68
Staiger CJ, Loyd CW. 1991. The plant cytoskeleton. *Curr. Opin. Cell Biol.* 3:33–42
Stebbins GL, Jain SK. 1960. Developmental studies of cell differentiation in the epidermis of monocotyledons. I. *Allium, Rhea,* and *Commelina. Dev. Biol.* 2:409–26
Stebbins GL, Khush GS. 1961. Variation in

the organization of the stomatal complex in the leaf epidermis of monocotyledons and its bearing on their phylogeny. *Am. J. Bot.* 48:51–9

Stebbins GL, Shah SS. 1960. Developmental studies of cell differentiation in the epidermis of monocotyledons. II. Cytological features of stomatal development in the Gramineae. *Dev. Biol.* 2:477–500

Steeves TA, Sussex IA. 1989. *Patterns in Plant Development.* New York: Cambridge Univ. Press. 2nd ed.

Steffenson DM. 1968. A reconstruction of cell development in the shoot apex of maize. *Am. J. Bot.* 55:354–69

Stevens RA, Martin ES. 1978. A new ontogenetic classification of stomatal types. *Bot. J. Linn. Soc.* 77:53–64

Stewart RN. 1978. Ontogeny of the primary body in chimeral forms of higher plants. *Bot. J. Linn. Soc.* 77:131–60

Stewart RN, Dermen H. 1970. Determination of number and mitotic activity of shoot apical initial cells by analysis of mericlinal chimeras. *Am. J. Bot.* 57:816–26

Sugiyama M, Komamine A. 1990. Transdifferentiation of quiescent parenchymatous cells into tracheary elements. *Cell Diff. Dev.* 31:77–87

Sussex IM. 1989. Developmental programming of the shoot meristem. *Cell* 56:225–29

Sylvester AW, Cande WZ, Freeling M. 1990. Division and differentiation during normal and *liguleless1* maize leaf development. *Development* 110:985–1000

Tiwari SC, Green PB. 1991. Shoot initiation on a *Graptopetalum* leaf: sequential scanning electron microscopic analysis for epidermal division patterns and quantitation of surface growth (kinematics). *Can. J. Bot.* 69:2302–19

Tomlinson PB. 1974. Development of the stomatal complex as a taxonomic character in the monocotyledons. *Taxon* 23:109–28

Traas J, Bellini C, Nacry P, Kronenberger J, Bouchez D, Caboche M. 1995. Normal differentiation patterns in plants lacking microtubular preprophase bands. *Nature* 375:676–77

Vernon DM, Meinke DW. 1995. Late embryo-defective mutants of *Arabidopsis*. *Dev. Genet.* 16:311–20

Vollbrecht E, Veit B, Sinha N, Hake S. 1991. The developmental gene *knotted-1* is a member of a maize homeobox gene family. *Nature* 350:241–43

Waites R, Hudson A. 1995. *phantastica*: a gene required for dorsoventrality of leaves in *Antirrhinum majus*. *Development* 121:2143–54

Waring PF. 1978. Determination in plant development. *Bot. Mag.* 1:3–17

West MAL, Harada JJ. 1993. Embryogenesis in higher plants: an overview. *Plant Cell* 5:1361–69

Wick SM. 1991. Spatial aspects of cytokinesis in plant cells. *Curr. Biol.* 3:253–60

Williams JA, Paddock SW, Vorwerk K, Carrol SB. 1994. Organization of wing formation and induction of a wing-patterning gene at the dorsal/ventral compartment boundary. *Nature* 368:299–305

Yadegari R, Depaiva GR, Laux T, Koltunow AM, Apuya N, et al. 1994. Cell differentiation and morphogenesis are uncoupled in *Arabidopsis raspberry* embryos. *Plant Cell* 6:1713–29

Yang M, Sack F. 1995 The *too many mouths* and *four lips* mutations affect stomatal production in *Arabidopsis*. *Plant Cell* 7:2227–39

Zeiger E, Stebbins GL. 1972. Developmental genetics in barley: a mutant for stomatal development. *Am. J. Bot.* 59:143–48

Annu. Rev. Cell Dev. Biol. 1996. 12:305–33

MITOTIC CHROMOSOME CONDENSATION

Douglas Koshland and Alexander Strunnikov

Department of Embryology, Carnegie Institution of Washington, 115 W. University Parkway, Baltimore, Maryland 21210

KEY WORDS: chromosome structure, mitosis, topoisomerase II, SMC proteins

ABSTRACT

In this chapter, we review the structure and composition of interphase and mitotic chromosomes. We discuss how these observations support the model that mitotic condensation is a deterministic process leading to the invariant folding of a given chromosome. The structural studies have also placed constraints on the mechanism of condensation and defined several activities needed to mediate condensation. In the context of these activities and structural information, we present our current understanding of the role of *cis* sites, histones, topoisomerase II, and SMC proteins in condensation. We conclude by using our current knowledge of mitotic condensation to address the differences in chromosome condensation observed from bacteria to humans and to explore the relevance of this process to other processes such as gene expression.

CONTENTS

1081-0706/96/1115-0305$08.00

INTRODUCTION

Mitotic chromosomes were first observed by Flemming in the late 1800s. Over the next decade additional studies revealed that chromosomes became invisible upon the completion of mitosis. They remained invisible in the newly divided cell during interphase, the resting phase of the cell cycle, and then reappeared when the this cell entered mitosis. Careful examination of the mitotic chromosomes as they formed suggested that they were composed of thin thread-like material or chromatin. This observation led to the idea that a chromosome was invisible in interphase because its chromatin was unraveled. In contrast, a chromosome became visible in mitosis because its chromatin bundled to a more condensed state. Hence over 90 years ago the phenomenon of mitotic chromosome condensation was defined.

Mitotic chromosome condensation is thought to help resolve two topological problems of chromosome segregation. First, tangles between replicated chromosomes (sister chromatids) arise during DNA replication (Sundin & Varshavsky 1981, Weaver et al 1985). Additional tangles between sister chromatids and nonhomologous chromosomes presumably result from simple diffusion of interphase chromosomes. If these tangles were to persist, they would impair the movement of neighboring chromosomes on the mitotic spindle and the eventual segregation of sister chromatids. Chromosome condensation apparently helps to eliminate tangles between sister chromatids or between neighboring nonhomologous chromosomes (see below). A second potential topological problem for chromosome segregation is that interphase chromosomes are generally much longer than the length of the dividing cell. These uncondensed chromosomes would be impossible to segregate because portions of the chromosomes would often cross the plane of cell division and be cleaved or entrapped by cytokinesis. Chromosome condensation compacts chromosomes sufficiently so that they are always displaced from the cytokinetic furrow at the end of mitosis (Guacci et al 1994).

Despite the early description of chromosome condensation and an appreciation of its biological significance, the molecular analysis of condensation has lagged behind the molecular studies of other forms of DNA metabolism such as transcription and DNA replication. The study of condensation has been confounded by several problems. For example, researchers were unable to generate the appropriate in vitro substrates. In particular, the interphase chromosome was difficult to isolate or reconstitute because of its complex composition and unwieldy size. Genetic approaches were also hindered because the isolation of relevant condensation mutants was elusive. The signature phenotype of these types of mutants was obscured in many unicellular organisms like fungi because of poor cytology, and in multicellular organisms such as *Drosophila* because of the inheritance of large pools of wild-type products in the mutant egg.

However, in the past few years several new cytological, biochemical, and genetic tools have been generated for the study of chromosome condensation in various eukaryotic model systems. For example, an in vitro system for condensation has been developed using demembranated sperm DNA and mitotic extracts from *Xenopus* eggs (Newport & Spann 1987, Hirano & Mitchison 1993). The improvements in cytological methods such as fluorescent in situ hybridization (FISH) and digital-enhanced light microscopy have allowed new structural studies of chromosomes in both fixed and live cells (Mathog et al 1984, Hiraoka et al 1989, Branddff et al 1991, Selig et al 1992). In addition, the application of these cytological methods to fungi has opened the door to sophisticated genetic dissection of condensation in these organisms (Funabiki et al 1993, Guacci et al 1994). These new tools have already led to exciting observations and have generated hope that a study of the molecular basis of mitotic condensation is now feasible.

In this chapter we review what is known about the structure and composition of interphase and mitotic chromosomes. We discuss how these observations support the model that mitotic condensation is a deterministic process leading to the invariant folding of a given chromosome. These structural studies have also placed constraints on the mechanism of condensation and defined several activities needed to mediate condensation. In the context of these activities and structural information, we present our current understanding of the role of *cis* sites, histones, topoisomerase II, and SMC (*s*table *m*aintenance of *c*hromosomes) proteins in condensation. We conclude by using our current understanding of mitotic condensation to address the differences in chromosome condensation observed from bacteria to human and to explore the relevance of this process to other processes such as gene expression. The cell cycle regulation of mitotic condensation is not covered.

INTERPHASE CHROMOSOMES

Analysis of the composition and structure of interphase chromosomes is important for the elucidation of mitotic chromosome condensation because interphase chromosomes are the initial substrates in this process. A major breakthrough in our understanding of interphase chromatin came with the isolation and characterization of histones, the major protein component of chromatin (reviewed by Smith 1991, Moudrianakis & Arents 1993). Four histones (H2A, H2B, H3, and H4) assemble into an octamer in the presence of DNA to form the nucleosome. Each nucleosome has approximately two complete turns of DNA wrapped around the exterior of the histone octamer. Nucleosomes co-assemble concomitant with DNA replication in vivo although replication-independent assembly is observed in vitro. The packaging of long DNA molecules into nucleosomes gives rise to a 10-nm fiber. The binding of H1 (a fifth histone) to the nucleosome can give rise to a 30-nm fiber, at least in vitro. The relevance of the 30-nm fiber to the structure of in vivo chromatin is controversial (Van Holde & Zlatanoval 1995). Besides histones, interphase chromatin consists of a complex profile of proteins including transcription factors, nonhistone high-mobility group proteins, topoisomerases, and RNA and DNA polymerases.

The structure of interphase chromosomes has been difficult to determine for two reasons. First, conventional light and electron microscopy have insufficient resolution to follow the path of individual chromatin strands in the unperturbed interphase nucleus. Second, the structure of interphase chromosomes may vary because interphase cells are a mixture of cells in G_1, S, or G_2. In particular, it has been suggested that chromosome condensation initiates at the end of S phase and culminates in mitosis (Rao & Adlakha 1984). If this is the case, interphase chromosomes may not have a uniform structure, but may be a mixture of completely decondensed chromosomes as well as condensation intermediates. The visible appearance of chromosomes at the beginning of mitosis may reflect only the terminal stages of condensation. With these caveats in mind, we discuss relevant observations about interphase chromosome structure.

Insights into the global structure of interphase chromosomes has come from several recent cytological studies. The position of entire chromosomes in interphase nuclei was analyzed in mammalian cells (Lichter et al 1988, Pinkel et al 1988) by FISH and in *Drosophila* (Mathog et al 1984) by computer-enhanced light microscopy. These studies show that interphase chromosomes occupy relatively distinct domains within the nucleus. This partial sorting may help to eliminate steric problems that would arise from trying to condense completely entangled chromosomes. In another study, FISH was used to assess the physical distance between pairs of sequences on a mammalian chromosome as a function of the number of nucleotides between these sequences (Yakoto et al 1995). The studies reveal that these chromosomes form large ($\approx$2 megabases) loop-like structures. Using standard cytological dyes, interphase chromosomes also appear to have both heterochromatin and euchromatin, regions with potentially different levels of condensation. Taken together, these structural studies suggest that interphase chromosomes have at least some global higher-order structure.

Interphase chromosomes may have even more structure than suggested by these relatively crude structural studies. First, the superhelicity of chromosomes was analyzed by sedimentation studies in the presence or absence of intercalating or nicking agents (Cook & Brazen 1975, Benyajati & Worcel 1976). By these methods, a significant fraction of interphase chromatin is in superhelical domains estimated to be 75 to 220 kb in size. Second, when nuclei are isolated under mild nondenaturing conditions, chromatin fragments of only 75 kb or smaller could be solubilized by limited digestion with nucleases (Igó-Kemens & Zachau 1977). Finally, loops have been directly observed for the lampbrush chromosomes in the ovarial eggs (oocytes) of amphibia and other organisms (Callan 1986). While these oocytes are in the first meiotic prophase, they are transcriptionally active, similar to interphase chromosomes. Therefore, interphase chromosomes may be organized into 10- to 200-kb loops, as well as into two megabase loops.

However, it is important to realize the limitations of these data in their support of 50–200-kb loops in interphase chromosomes. One limitation is that loops are not the only means to form superhelical domains or chromatin with limited solubility. Superhelical domains can be formed by any mechanism that restricts the rotation of the chromosome. For example, after DNA replication, replicated chromosomes are paired. If pairing sites exist at regular intervals, then pairing at these sites could restrict the rotation of the chromatids and generate the apparent superhelical domains. Alternatively, chromosomes could be tethered every 70 kb to some nuclear structure such as lamins. The size range of soluble chromatin could simply reflect the distribution of an insoluble protein-DNA complex along the chromosome. A second limitation is that the superhelical or insoluble fraction of chromatin may be derived from a subset of interphase cells whose chromosomes have begun initial steps in chromosome condensation (Rao & Adlakha 1984). In this light, the lampbrush chromosomes observed in prophase cells are likely intermediates in condensation, and their lateral loops may be formed after the initiation of condensation rather than arising from preexisting interphase loops. Hence, we conclude that the process of chromosome condensation is initiated on interphase chromosomes that have higher order structure. However, it is not clear whether the 50–200-kb loops exist in interphase chromosomes, and if they do, whether they are present on all interphase chromosomes or on only those that have initiated condensation.

MITOTIC CHROMOSOMES

Studies of the composition of mitotic chromosomes have revealed three major classes of proteins. One class includes the basic structural components of interphase chromatin such as core histones and histone H1. The second class includes molecules that are present on both interphase and mitotic chromosomes but whose activities are used for different purposes, e.g. topoisomerase II (see below). The third class consists of the recently identified SMC family of proteins. These chromosomal proteins are representatives of factors likely to be specific for condensation (see below). Taken together, these results suggest that condensation occurs by modifying interphase chromatin through the action of topoisomerase II, SMC proteins, and other unidentified factors rather than by completely remodeling the DNA de novo. This may not always be the case. For example, condensation of sperm chromatin in many species involves the complete replacement of core histones by protamines.

Structural analyses of mitotic chromosomes have also provided insights into mitotic condensation. One insight came from a comparison of the length of a mitotic chromosome to its interphase length. This simple ratio provides a measure of the magnitude of compaction, which can be obtained by FISH. Chromosomes from interphase and mitotic cells are hybridized with probes to two distinct regions of a chromosome, and the distance between the hybridization signals is measured. By this approach, the compaction of mitotic chromosomes relative to interphase chromosomes was estimated to be five to tenfold in mammalian cells (Lawrence et al 1988). Using an identical approach for budding yeast, the magnitude of compaction was only twofold (Guacci et al 1994). This apparent reduction in compaction is in agreement with the observation that budding-yeast mitotic chromosomes are not visible by stains that visualize the more condensed chromosomes of other eukaryotic cells. The difference in the magnitude of compaction between yeast and mammalian chromosomes suggests that not all aspects of chromosome condensation are conserved between different eukaryotic cells (see below).

In order to compact chromatin, it must be folded. The first insight into how mitotic chromosomes might be folded came from the study of lampbrush chromosomes in the ovarial eggs (oocytes) of amphibia and fish (Callan 1986). These cells are in the first meiotic prophase in which chromosomes have replicated to form paired sister chromatids, but homologues have not yet segregated. When the chromosomes in these oocytes are stained, they exhibit a characteristic structure of a central axis containing highly condensed regions of the paired sister chromatids. Lateral loops extend from this axis and are visualized indirectly because of actively ongoing localized transcription. When transcription is blocked, the extended loops collapse, and the resulting chromosome forms a mitotic-like structure. These results showed that chromosomes could fold in an axial loop structure and that further condensation could occur by folding of the loops (Figure 1*A*).

The relevance of the structure of lampbrush chromosomes to mitotic chromosomes became apparent from two studies of mitotic chromosomes. First, when mitotic chromosomes are depleted of histones and other proteins by treatment with either high salt or polyanions, the chromosomes swell (Adolph et al 1977, Paulson & Laemmli 1977). An examination of these chromosomes in the electron microscope revealed approx 50–200-kb loops of chromatin extending from a central core. Second, scanning electron microscopy of native mitotic chromosomes revealed a bar-like structure covered with little knobs. It is reasonable to assume that the knobs are folded forms of the loops observed upon histone depletion (Marsden & Laemmli 1979). These results coupled with the lampbrush chromosome studies suggest that the compaction of mitotic chromosomes occurs at least in part by the formation of an axial loop structure followed by further folding of the loops.

A final level of compaction may occur by additional coiling of the axis. Lampbrush chromosomes are significantly longer than their mitotic counterparts, suggesting that looping by itself may not account for the shortening of the chromosome. This conclusion, of course, is subject to the caveat that the composition or looping of mitotic and meiotic chromosomes might be different. However, light and electron micrographs of human mitotic chromosomes also revealed a zigzag pattern consistent with a coiling of the chromosome into 200-nm helical fiber (Rattner & Lin 1985, Boy de la Tour & Laemmli 1988).

In principle, the extent of coiling of the axis, the folding of the loops, or the number and position of the loops in a given mitotic chromosome could be variable, reflecting a flexible process of condensation (Figure 1*B*). Indeed, the variable structure of the chromatin in the sperm head provides precedent for flexible packaging (Ward et al 1995). However, several features of mitotic chromosomes suggest that they fold to an invariant structure. First, each mitotic chromosome in a given cell-type has a characteristic and reproducible length. Second, each mitotic chromosome has a signature pattern of bands after staining with specific dyes like Giemsa. Although the exact molecular basis of bands is still unclear, they presumably reflect some aspect of the underlying structure that results from invariant folding. The reproducible length and the reproducible banding pattern of mitotic chromosomes imply that a given interphase chromosome must fold to an invariant mitotic structure. Additional confirmation of invariant folding has come from recent experiments where the position of specific DNA sequences on mitotic chromosomes were determined by FISH (Baumgartner et al 1991). Each sequence analyzed occupied a reproducible position along the long and transverse axes of the chromosome.

Furthermore, with respect to the transverse position of any sequence, sister chromatids exhibited a mirror symmetry. This conclusion was foreshadowed by the observation that the helical configuration of the 200-nm fiber for the two sister chromatids was apparently of opposite handedness (Boy de la Tour & Laemmli 1988). It is possible that this mirror symmetry facilitates the cohesion of sister chromatids by allowing the spatial juxtaposition of the homologous sequences that lie immediately adjacent to the axis of symmetry (Baumgartner

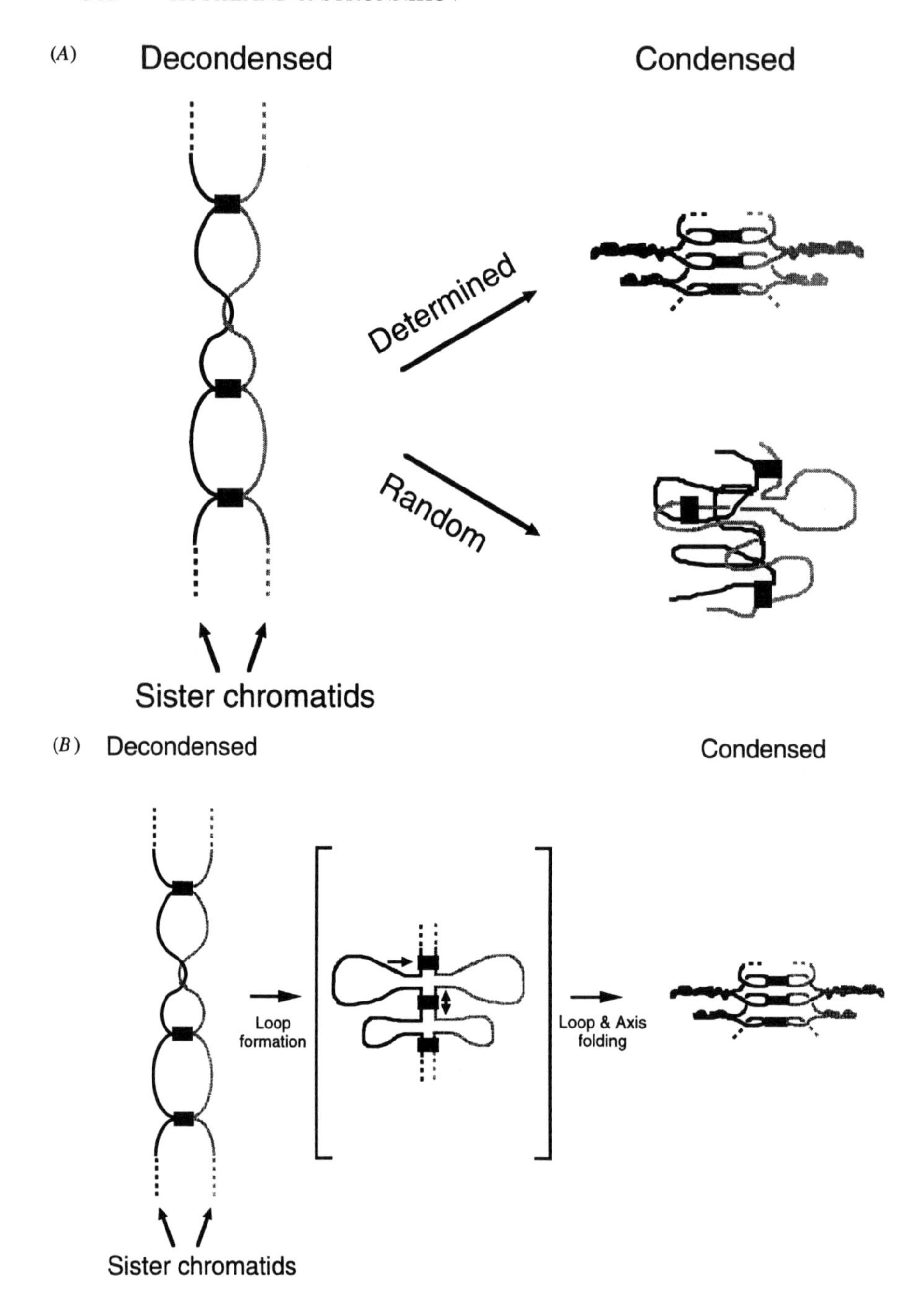
(A)
Decondensed
Condensed
Determined
Random
Sister chromatids
(B)
Decondensed
Condensed
Loop formation
Loop & Axis folding
Sister chromatids

et al 1991). However, an alternative interpretation stems from the observation that sister chromatids are paired from the time of replication, long before condensation (Selig et al 1992, Guacci et al 1994). Thus it is likely that the pairing of sister chromatids generates a steric constraint, forcing sister chromatids to fold with opposite handedness. In conclusion, the fixed length, banding pattern, and position of specific sequences for any given mitotic chromosome demonstrate that it has an invariantly folded structure. This regularity in structure suggests that mitotic chromosome condensation must occur by a deterministic process.

A comparison of the structure of interphase and mitotic chromosomes suggests that mitotic condensation occurs probably by the formation of three types of folds: those that form the chromatin loops (although a subset may preexist in interphase); those that compact the loops; and those that compact the axis (putative coiling). The stable maintenance of each of these folds requires the existence of adhesive activities that lock chromatin to itself or to some other structure. For example, adhesive molecules would presumably exist at the base of a loop.

It is unlikely that these adhesive activities by themselves provide the specificity for the deterministic process of condensation that results in the invariant folding of mitotic chromosomes. If this were the case, then every fold would require its own set of unique adhesive activities. It is more likely that a particular type of fold is mediated by a class of common adhesive activities. In

←

Figure 1 (*A*) Model for the compaction of chromosomes during mitotic chromosome condensation. At the completion of DNA replication, sister chromatids (*black, grey lines*) are paired and catenated. The pairing sites are indicated by the *black boxes*, which reflect the fact that the molecular nature of the pairing mechanism remains obscure. Loops are formed as a first step in condensation, although some loops may preexist (see text). This intermediate resembles the lampbrush chromosomes observed in meiotic prophase. Different regions of the chromosome form loops of variable size, but a given region of a chromosome will form a loop of a specific size. This results in variable loop size along the chromosome axis (compare the top and bottom loops), but identical loop size at a particular position (compare loops at the same position on sister chromatids). Further compaction is achieved by folding of the loops and the axis. These different types of folds presumably are locked in place by adhesive activities. Upon completion of condensation, the paired sister chromatids exhibit mirror symmetry. (*B*) Determined versus random condensation of chromatin. Chromatin condenses by a deterministic process leading to an invariantly folded chromosome or by a random process leading to a variably folded chromosome. The invariant structure of mitotic chromosomes suggests that they undergo determined condensation. This process requires specific *cis* sites, condensation-directing activities that use these sites, and adhesive activities to generate and maintain specific folds. Determined condensation ensures that a sister chromatid condenses only with itself and not with its neighboring sister chromatid. As a result, sister chromatids form two distinct domains rather than one entangled domain.

order for these adhesive molecules to make specific folds rather than nonspecific wads, we suggest that the interaction of these adhesive molecules is regulated by a second class of condensation-directing activities. A remarkable feature of the condensation-directing activities is evident from a particular structural feature of mitotic chromosomes—the juxtaposition of distinctly condensed sister chromatids. As mentioned above, sister chromatids become paired during interphase, concomitant with or immediately following DNA replication (Selig et al 1992, Guacci et al 1994). Therefore, these condensation-directing activities must ensure that the chromatin of one sister chromatid preferentially condenses upon itself and not the chromatin of the identical sister chromatid in the immediate vicinity. This strand preference could be achieved if the condensation-directing activities work by a processive mechanism.

POTENTIAL *CIS* SITES FOR CONDENSATION

It is difficult to imagine how the invariant condensation of chromosomes could be achieved without the existence of *cis*-acting chromosomal sites that serve as landmarks to direct condensation. Indeed, in vivo analysis of the condensation of *Drosophila* chromosomes showed that condensation was initiated at specific chromosomal foci (Hiraoka et al 1989). Further support for the existence of these *cis* sites comes from analysis of the loops of lampbrush chromosomes (Callan 1986).

Loops at different positions along the chromosome vary in size. However, a loop at a specific locus is always relatively the same size, as evidenced by the comparison of different lampbrush preparations and symmetrical patterns of paired loops present on paired sister chromatids. These results support the idea that the size and position of each loop must be dictated by *cis* sites.

One possibility is that *cis* sites for condensation are coincident with known chromosomal landmarks. Because heterochromatin is thought to be condensed interphase chromatin, heterochromatic regions are potential *cis* sites that may direct mitotic condensation. However, when nonheterochromatic sequences are rearranged relative to centromere or telomere heterochromatin, they retain the banding pattern of their unrearranged counterpart. Furthermore, the in vivo analysis of the condensation of *Drosophila* chromosomes identifies sites for the initiation of condensation that are not coincident with heterochromatin (Hiraoka et al 1989). Therefore, even if heterochromatic regions are sites of mitotic condensation, other heterochromatin-independent sites must exist. Origins of replication are another landmark that could also act as *cis* sites for condensation. Indeed, the 50–100-kb spacing of origins of replication is similar to the apparent loop size of mitotic chromosomes. However, no evidence exists that demonstrates a direct role of origins in condensation.

Another possibility is that condensation involves novel sites other than established landmarks. A structural approach to identify condensation sites comes from the analysis of mitotic chromosome scaffolds. The proteins that remained bound to chromosomal DNA after the depletion of histones by treatment with various salts were operationally defined as the mitotic chromosomal scaffold (Adolph et al 1977, Gasser 1992). It was assumed that DNA sequences that bound with high affinity to proteins of the scaffold fraction might be important components of condensation landmarks (Mirkovitch et al 1984, 1988). Indeed, scaffold-bound sequences were found and named scaffold attachment regions (SARs). The nucleotide sequence analysis of SARs shows that they do not conform to a single consensus sequence (Gasser 1992). However, they do share certain characteristics that include a length of greater than 200 bp; an AT content of greater than 70%, including runs of homopolymeric dT; and the absence of repetitive elements. The minor groove of these sequences provides a binding site for a subset of DNA-binding proteins such as HMGI-Y.

Several observations place SARs as chromatin landmarks that help mediate chromosome condensation. First, two major scaffold proteins, topoisomerase II and SC2, a SMC family member, have been shown to be involved in condensation (see below). This fact increases the likelihood that SARs may function in condensation. Second, *cis* sites for condensation are expected to lie along the chromosomal axis in vivo. The position of SARs on the chromosomal axis has been proposed based upon the staining pattern of mitotic chromosomes with fluorescent probes specific for AT-rich sequences (Saitoh & Laemmli 1994). However, no direct link between SARs and condensation has been established.

Recently, artificial proteins (MATH) have been constructed with *m*ultiple tandem copies of the *AT h*ook, a 14 residue sequence from HMGI-Y proteins with high affinity for the minor groove of extensive poly dT stretches of DNA (Strick & Laemmli 1995). MATHs bind with high affinity to SARs and block chromosome condensation in vivo and in vitro. Although exciting, the relevance of this particular observation is not clear. In particular, the addition of exogenous SAR DNA to condensation-competent extracts blocks the inhibitory action of MATHs on condensation but does not block condensation itself. This result suggests that SARs are the targets of MATHs but not the targets of endogenous condensation factors. In other words, chromosomes may condense in a SAR-independent manner. In this light, the MATHs may inhibit condensation by an abnormal mechanism, perhaps by inappropriately cross-linking two aberrant sites on the same chromatid, on sister chromatids, or even on nonhomologous chromosomes. Additional studies will be needed to test the biological relevance of SARs to condensation and to pursue the interesting observation of MATHs. Because the identification of *cis* sites for condensation would be an important conceptual advance and a tremendous tool for future studies of condensation, the scaffold studies need to be complemented by additional functional approaches.

PROTEIN FACTORS OF CONDENSATION

The *cis* components of mitotic chromosome condensation must act in concert with transacting factors. Many transacting factors are likely to be required for condensation because at least three putative activities (adhesion, condensation-directing, and facilitating) are required for condensation, and other types of chromosome metabolism such as DNA replication or transcription invariably involve numerous factors. Nonetheless, at present only three chromosomal components, histones, topoisomerase II, and SMC family proteins, have been implicated in this process.

Histones

Initially, histone H1 was thought to be a prime candidate for an important compacting activity of mitotic chromosome condensation based on two observations. First, chromatin assembled with core histones was converted from a 11-nm fiber to a more condensed 30-nm fiber by the addition of histone H1. This led to the idea that H1 has a condensing activity. Second, H1 becomes hyperphosphorylated about the time that condensed chromosomes appear. Thus it was thought that the modified form of H1 might be a hypercondensing form. However, chromosomes can condense in the absence of H1 hyperphosphorylation (Guo et al 1995). Furthermore, when histone H1 is depleted from mitotic extracts (Ohsumi et al 1993), or when the H1 gene is disrupted in *Tetrahymena* (Shen et al 1995), mitotic condensation proceeds apparently unperturbed. Although these results cannot rule out that H1 function is being provided by another H1-like molecule, they open up the possibility that histone H1 is not required for condensation at all. In fact, an alternative model has been suggested where interphase H1 bound to chromatin blocks the binding of condensing factor(s), and H1 hyperphosphorylation reduces its affinity for DNA, which allows this factor access to the chromatin (Roth & Allis 1992).

As the likelihood of an important role for H1 in chromosome condensation has diminished, attention has been focused on the potential role of other histones in this process. Prior to metaphase, at the time of terminal steps in condensation, histone H3 is phosphorylated whereas H2A is hyperphosphorylated (Bradbury 1992). These phosphorylated forms are also found when premature chromosome condensation is induced in vivo by fostriecin, a phosphatase inhibitor (Guo et al 1995). Fostriecin-induced condensation is blocked by a kinase inhibitor staurosporine. This inhibitor blocks H3 but not H2A phosphorylation. These results suggest that a staurosporine-sensitive kinase is required for condensation, which further strengthen the correlation between H3 phosphorylation and condensation. Histone H2A and H2B are also modified by ubiquitin (Mueller et al 1985, Bradbury 1992). At metaphase, these histones lose their ubiquitin modification only to become remodified in anaphase. Thus histone de-ubiquitination also correlates with the late stages of condensation. However, as in the case of H1, these observations concerning H2A, H2B, and H3 are only correlations, and no evidence for a causal relationship has been established.

Topoisomerase II

Topoisomerase II is an excellent example of a protein that participates in chromosome condensation as well as in other cellular processes including chromosome segregation (Holm et al 1985, Uemura & Yanagida 1986, Shamu & Murray 1992, Adams et al 1992, Buchenau et al 1993, Clarke et al 1993, Gorbsky 1994, Spell & Holm 1994), transcription and DNA replication (Brill & Sternglanz 1988, Kim & Wang 1989, Schultz et al 1992, Gartenberg & Wang 1993). Its function in chromosome condensation is controversial, whereas its function in other processes is fairly well understood. Hence, we begin with a summary of topoisomerase II function in transcription, DNA replication, and chromosome segregation and use these insights as a framework to evaluate the functions of topoisomerase II in chromosome condensation.

Topoisomerase II is one of several type two topoisomerases having the unique ability to create a transient double-strand break in a DNA molecule that allows the passage of one DNA strand through another (Hsieh & Brutlag 1980, Kreuzer & Cozzarelli 1980). This activity can be detected by the resolution of catenated DNA circles in vitro (Baidi et al 1980, Goto & Wang 1982). The same activity can also remove intertwines in a superhelical DNA molecule, which reduces its superhelicity. Indeed, this activity of type II topoisomerases apparently facilitates DNA replication and transcription by removing superhelical twists that result from the progression of the DNA and RNA polymerases along a chromosome (Sundin & Varshavsky 1981, Brill & Sternglanz 1988, Kim & Wang 1989). However, although topoisomerase II is essential for the viability of eukaryotic cells and is associated with the chromosomes (Earnshaw & Heck 1985, Heller et al 1986, Moens & Earnshaw 1989, Klein et al 1992, Zini et al 1992, Hock et al 1996, Sumner 1996), it is apparently dispensable for DNA transcription and replication. Eukaryotic mutants defective in topoisomerase II do not block transcription in vivo. They also do not arrest at a characteristic DNA damage-activated checkpoint (Holm et al 1985, Uemura & Yanagida 1986, Gorbsky 1994) and do not accumulate replication intermediates (Holm et al 1989). The function of topoisomerase II is dispensable during S phase because the relaxation of superhelical coils can be achieved by the nicking-closing activity of type I topoisomerases (Kim & Wang 1989).

Although topoisomerase I can substitute for topoisomerase II in removing superhelicity within one DNA molecule, only topoisomerase II can remove the intertwines between two catenated DNA molecules. In vivo, this decatenating activity can remove the interlocks between sister DNA molecules that result, for example, from DNA replication. Mutations or drugs that block topoisomerase II catalytic function block the completion of mitosis apparently because the catenation between sister chromatids physically impedes their segregation (Holm et al 1985, 1989, Uemura et al 1987). Bacterial mutants defective in a type II topoisomerase (topoisomerase IV) also fail to decatenate replicated chromosomes and are impaired in segregation (Adams et al 1992). Together, these results suggest why the catalytic activity of type II topoisomerases is critical for chromosome segregation and explain, in part, the essential function of these enzymes for cell viability.

Several in vivo and in vitro studies reveal that topoisomerase II is also required in an early step of chromosome segregation, particularly chromosome condensation in mitotic prophase. First, mutants defective in topoisomerase II fail to condense their chromosomes in prophase (Uemura et al 1987). Second, the depletion of topoisomerase II from mitotic extracts prevents chromosome condensation in vitro (Adachi et al 1991, Hirano & Mitchison 1993). Third, specific and nonspecific inhibitors of topoisomerase II enzymatic activity block chromosome condensation in vivo and in vitro (Newport 1987, Newport & Spann 1987, Wright & Schatten 1990, Hirano & Mitchison 1991, Sumner 1992, Buchenau et al 1993, Gorbsky 1994).

What is the molecular function of topoisomerase II in chromosome condensation? The most straightforward model is that the cell uses topoisomerase II enzymatic activity for condensation the same as it does for transcription, DNA replication, and chromosome segregation. In support of this model, condensation is blocked by a class of inhibitors that interact specifically with the active site of topoisomerase II. One possibility is that the catalytic activity of topoisomerase II relaxes superhelical domains during folding of the condensing chromatin. However, because other molecules such as topoisomerase I can relax superhelical domains, this model would not account for the essential function of topoisomerase II in condensation. A second model is that decatenating activity of topoisomerase II is used to eliminate steric problems during condensation. Sister chromatids become catenated during replication, and intertwines between the neighboring chromosomes may result simply from chromatin strand diffusion. Left unresolved, these intertwines would impede the formation of correct folds. By removing catenation, topoisomerase II would facilitate the formation of proper folds needed for condensation.

This second model nicely explains several observations. First, the inactivation of topoisomerase II prior to condensation reduces compaction only threefold, as would be expected if topoisomerase II only facilitated condensation by removing tangles. In other words, occasional tangles would not be expected to impede all condensation. Second, inactivation of topoisomerase II prior to condensation leads to partially compacted chromosomes that are apparently tangled with each other (Uemura et al 1987). This observation supports the idea that topoisomerase II is necessary during condensation to remove accidental intertwines between chromosomes. Third, inactivation of topoisomerase II after chromosome condensation has been completed does not alter chromosome appearance (Uemura et al 1987, Hirano & Mitchison 1993, Swedlow et al 1993). This result suggests that topoisomerase II is needed for the establishment but not the maintenance of condensation. Indeed, we would expect a detangling activity to be required only during the formation of folds but not after they are formed.

An interesting aspect of this model is that condensation and detangling are possibly synergistic. That is, detangling not only facilitates condensation of sister chromatids, but condensation in its turn may prevent sister chromatids from retangling. It is important to remember that although topoisomerase II passes one strand through another, it does not "know" whether it is removing or adding intertwines. The removal of catenations by topoisomerase II is favored only if the concentration of the decatenated molecules is low. However, prior to condensation, the concentration of decatenated sister chromatids is very high because sister chromatids are paired. Therefore, in the absence of condensation, topoisomerase II is as likely to put in intertwines as it is to remove them, setting up a futile cycle of detangling and retangling of sister chromatids as well as chromosomes. However, during condensation, the proper packaging of the decatenated sister chromatids would lower their effective concentration, preventing topoisomerase II from retangling them. Of course, the proximity of the paired sister chromatids, even after condensation, still prevents the complete removal of intertwines. These remaining intertwines are removed at anaphase when the topoisomerase II is further deprived of the possibility of interlocking sister chromatids because of the poleward movement of chromosomes (Holm et al 1985, Holm 1994). Thus by providing directionality to topoisomerase II activity, the condensation machinery may drive sister chromatids into distinct domains and perform a critical early step in their segregation (Wood & Earnshaw 1990).

In addition to its enzymatic role in mitotic chromosome condensation, topoisomerase II has been proposed to play a structural role in the architecture of the mitotic chromosomes (Lewis & Laemmli 1982, Adachi et al 1991). This hypothesis originated from several observations, the most evident and reproducible of which is that under certain fixation conditions, eukaryotic chromosomes have a significant portion of their topoisomerase II concentrated at the chromosome core (Moens & Earnshaw 1989, Hock et al 1996). This fraction of topoisomerase II is part of an insoluble complex, also referred to as the chromosome scaffold (Lewis & Laemmli 1982, Earnshaw et al 1985). However, the structural role of topoisomerase II in condensation has been recently challenged. When condensed chromosomes were first generated by mixing sperm DNA to mitotic extracts and then subjected to complete topoisomerase II extraction, no significant changes in the degree of condensation were observed (Hirano & Mitchison 1993). Thus even if insoluble topoisomerase II is part of a chromosome core or scaffold in vivo, its presence there is not essential for maintaining mitotic condensation. No core staining of condensed mitotic chromosomes was observed when topoisomerase II localization was followed in vitro (Swedlow et al 1993). Furthermore, no domains of topoisomerase II, other than its enzymatic part, have been shown to be essential for condensation. Topoisomerase II has no established polymerization motifs, and the only data supporting the aggregation of topoisomerase II molecules with themselves have been obtained in vitro (Vassetzky et al 1994). Taken together, these observations argue against a separate structural role of topoisomerase II.

If the scaffold part of the topoisomerase II pool forming a regular structure along the chromosome core is not an essential structural element of the condensed chromosomes, why is it present and what is its possible biological role? We offer two explanations. First, we suggest that the accumulation of the chromosome core topoisomerase II molecules occurs in higher eukaryotes as a consequence of active DNA and RNA synthesis. This model is supported by the observation that scaffolds are not observed when chromosomes are condensed in the absence of DNA replication and transcription (Hirano & Mitchison 1993). We assume that topoisomerase II accumulation occurs at the regions adjacent to the points of initiation and termination of transcription and at the terminal points of DNA replication where catenation and superhelical intertwines occur (Gasser & Laemmli 1986, Udvardy & Schedl 1991, Ma et al 1993). These sites provide an excellent means of distributing topoisomerase II molecules along the chromosome arms in the form of regularly spaced foci (Hiraoka et al 1989), which likely correspond to some of the SARs (Earnshaw & Heck 1985, Iarovaia et al 1996). A second model arises from the observation that transcription and replication are mediated by complex macromolecular machines (Alberts 1984). These machines have multiple independent enzymatic activities and structural components. By analogy, condensation may also be performed by condensation machines, which localize along the chromosome and contain adhesive, condensation-directing, and condensation-facilitating activities. In this light, topoisomerase II may be tethered to the uniformly distributed condensation machines to enhance their function during condensation. However, this model does not explain the absence of core topoisomerase II fraction when chromosomes are condensed in the absence of replication or transcription.

Although eukaryotic topoisomerase II has received an overwhelming amount of attention over the last 15 years, it is evident from the preceding discussion that many questions remain about its role in mitotic chromosome condensation. We will need to understand the biological significance of the topoisomerase II isozymes and multiple intracellular pools (Zini et al 1992, Swedlow et al 1993) to determine if any of the topoisomerase II molecules have a chromosome scaffolding role and to analyze topoisomerase II regulation in mitosis. These studies should be facilitated by recent advances in the understanding of the molecular mechanics and structure of the topoisomerase II molecule (Roca & Wang 1994, Zechiedrich & Cozzarelli 1995, Berger et al 1996). A role for topoisomerase II in mitotic chromosome decondensation also has been suggested (Newport 1987). Thorough analysis of this aspect of topoisomerase II function should provide insight into its role in condensation.

SMC Family

The discovery of the SMC proteins is an important landmark in the study of chromosome structure. This is the newest group of chromosomal proteins found to be indispensable for chromosome condensation. In contrast to the histones, HMG proteins, and topoisomerase II, the function of SMC family proteins appears to be mainly restricted to mitotic chromosome condensation. Recently, the importance of this group has become evident after a number of corresponding genes had been cloned and their products characterized from a variety of organisms, including bacteria, fungi, and invertebrate and vertebrate animals (Falk & Walker 1988, Notarnicola et al 1991, Strunnikov et al 1993, Hirano & Mitchison 1994, Chuang et al 1994, Saitoh et al 1994, Saka et al 1994, Strunnikov et al 1995, Holt & May 1996). The initial description of two family members is dated much earlier and corresponds to two independent findings: the fortuitous isolation of the *smc1* mutation in baker's yeast and partial purification of SC2 protein as a component of chromosome scaffolds (Earnshaw & Laemmli 1983, Larionov et al 1985). Subsequent cloning of the *SMC1* gene revealed that it is a member of a ubiquitous protein family. We first discuss the structural features of the SMC proteins and then the particular experiments in which their functional properties have been revealed. The SMC family has already been a subject of several reviews (Peterson 1994, Hirano

1995, Hirano et al 1995, Saitoh et al 1995, Yanagida 1995); thus we emphasize recent developments and unresolved issues.

Members of the SMC family range from 115 to 160 kDa in their molecular mass, and can be subdivided into at least five structurally distinct subgroups, four eukaryotic and one prokaryotic (Figure 2). The amino acid sequence of a typical SMC protein predicts five major regions (Figure 2): an NTP-binding region; the first stretch of the coiled-coil alpha helix; a loop or hinge; the second coiled-coil region; and a COOH-terminal region harboring the DA-box, a signature motif for the SMC family (Strunnikov et al 1993). Deletion of the DA-box or NTP-binding region (Strunnikov et al 1995), as well as point mutations in the NTP-binding region (B Meyer, personal comminication), result in a nonfunctional protein, which demonstrates that these motifs are essential for the SMC proteins. A number of conditional lethal mutations have been mapped to the coiled-coil regions, thus showing the importance of these potential protein-protein interaction domains for SMC protein function. A number of less important regions have also been identified through their tolerance to the insertion of foreign epitopes (Strunnikov et al 1993, 1995). Based upon these five major regions, several hypotheses have been proposed for the structure and molecular function of SMC proteins. However, only the model of oligomerization of SMC proteins, inferred from the presence of the coiled-coil domains, has been confirmed (Hirano & Mitchison 1994, Strunnikov et al 1995).

Most of the in vivo data regarding the function of SMC proteins in mitosis has come from studies of fungi. In *Saccharomyces cerevisiae*, four structurally distinct SMC proteins are found: Smc1 (Strunnikov et al 1993), Smc2 (Strunnikov et al 1995), Smc3, and Smc4 (A Strunnikov & D Koshland, unpublished data). An *SMC4* gene homologue, *cut3*, and an *SMC2* homologue, *cut14*, have been characterized in *Schizosaccharomyces pombe* (Saka et al 1994). An *SMC3* gene homologue, *sudA*, has been isolated in *Aspergillus* (Holt & May 1996). Among these genes, *SMC1, SMC2, SMC3, cut3*, and *sudA* are essential for viability; *cut14* and *SMC4* are yet to be tested. Thus it appears that multiple types of the SMC proteins in the same cell have at least some essential nonoverlapping functions.

An insight into the functional diversity among the SMC proteins is given by the analysis of the phenotypes of the given SMC family gene impaired by a conditional lethal mutation. The first and most characteristic feature of all SMC protein-deprived cells (*smc1, smc2, cut3, cut14, sudA*) is their failure to segregate chromosomes in mitosis. This failure, coupled with cytokinesis, results in a mitotic "cut" phenotype, i.e. partial splitting of the undivided nucleus into two halves (Samejima et al 1993). The cut morphology is very reminiscent of topoisomerase II mutants in yeast (Uemura & Yanagida 1984, Holm et al 1985), and at least in budding yeast, this could be a general landmark of a structural obstacle to segregation. The block of mitosis in *smc* mutants is probably not generated by a checkpoint response. These mutant cells arrest with non-uniform morphology and quickly lose viability at the nonpermissive condition, whereas a checkpoint arrest is usually characterized by uniform cell morphology and high viability. Abnormal segregation in *smc* mutants is apparently the result of a defect in chromosome condensation. Using FISH, it has been shown that the *cut14* mutations and, to a lesser degree, *cut3* and *smc2* mutations perturb chromosome condensation (Saka et al 1994, Strunnikov et al 1995). The ability of different *smc* mutations to perturb either the establishment of condensation or the integrity of already condensed chromosomes suggests that the SMC proteins are central players in chromosome condensation. This is in contrast to topoisomerase II, whose input in condensation is limited to the establishment of condensed chromosomes. One study reported an allele-specific suppression and/or synthetic lethality between the *cut3* and *cut14* mutations of *S. pombe* and the genes for topoisomerase I and topoisomerase II (Saka et al 1994). However, these genetic interactions appear to be too weak to provide a clear indication of a direct physical interaction of SMC family proteins with topoisomerase I or II. Biochemical studies also have not revealed a physical interaction between them (Hirano 1994; A Strunnikov & D Koshland, unpublished data). Most likely the condensation activity of SMC proteins is independent of topoisomerases I and II.

Additional information about the in vivo function of SMC proteins originated from the description of the SC2 scaffold protein (Earnshaw & Laemmli 1983). The corresponding cDNA was cloned recently and was reported to contain an open reading frame for an SMC family protein (Saitoh et al 1994). The localization of SC2 in vivo is consistent with its role as an important component of chromosome condensation. The protein has been found in the nuclear region throughout the cell cycle; however, upon initiation of the chromosome condensation, it was accumulated in foci near the geometrical axis of a condensed chromosome. This pattern of localization on mitotic chromosomes was also observed for the *Xenopus laevis* SMC family proteins XCAP-E and XCAP-C (Hirano & Mitchison 1994). Interestingly, in one case SC2 was found to colocalize with topoisomerase II, consistent with the idea that condensation proteins assemble as part of a large macromolecular complex (Ma et al 1993).

Additional studies have also been conducted in vitro to clarify the role of the SMC proteins in mitotic chromosomes. These studies are based on an excellent in vitro system in which the formation of condensed chromosomes is recapitulated by the addition of sperm chromatin to the mitotic extracts from *X. laevis* eggs. One major advantage of this system is that only a few proteins

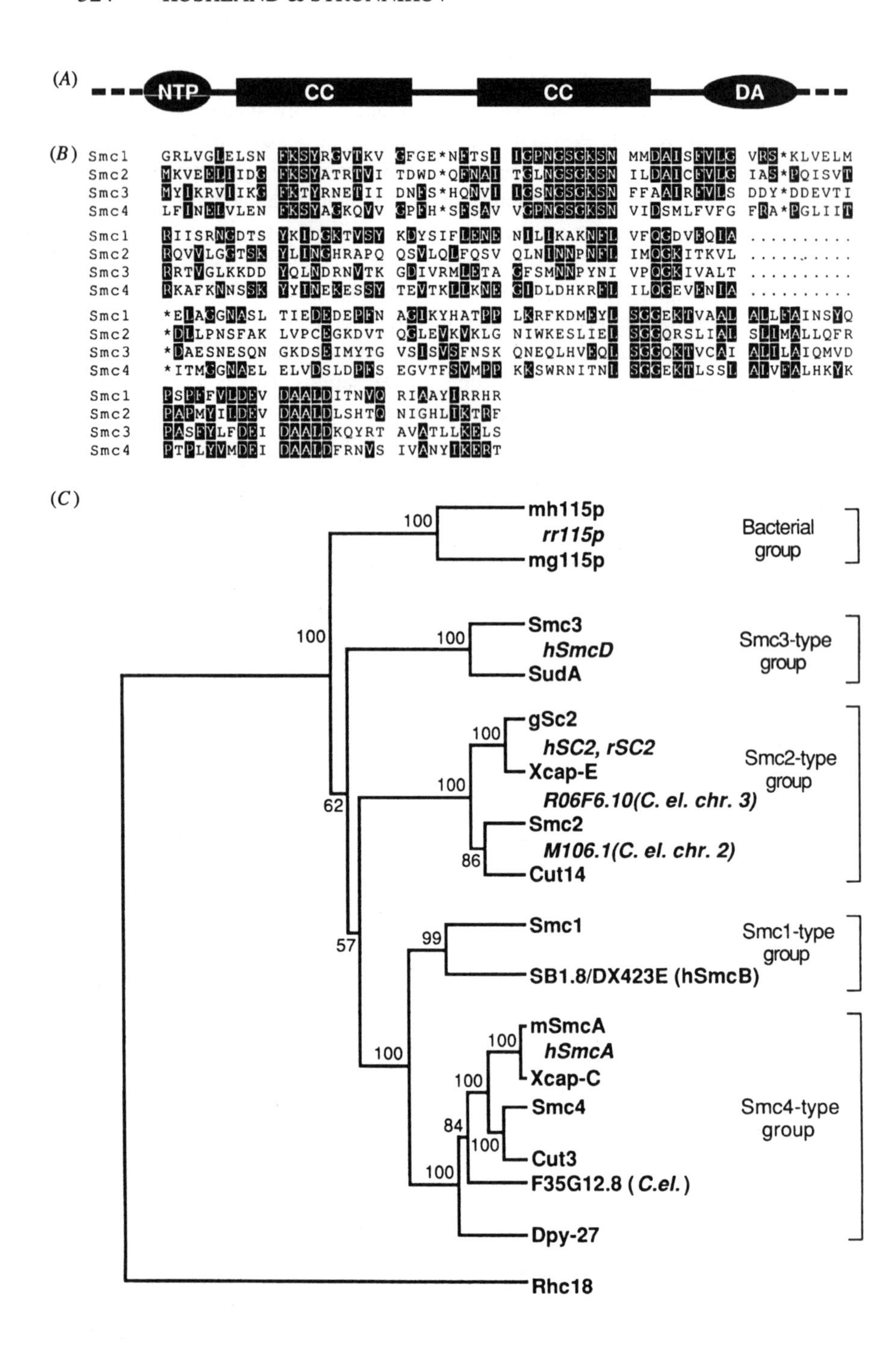
(A)
NTP
CC
CC
DA
(B)
Smc1 GRLVGLELSN FKSYRGVTKV GFGE*NFTSI IGPNGSGKSN MMDAISFVLG VRS*KLVELM
Smc2 MKVEELIIDG FKSYATRTVI TDWD*QFNAI TGLNGSGKSN ILDAICFVLG IAS*PQISVT
Smc3 MYIKRVIIKG FKTYRNETII DNFS*HQNVI IGSNGSGKSN FFAAIRFVLS DDY*DDEVTI
Smc4 LFINELVLEN FKSYAGKQVV GPFH*SFSAV VGPNGSGKSN VIDSMLFVFG FRA*PGLIIT
Smc1 RIISRNGDTS YKIDGKTVSY KDYSIFLENE NILIKAKNFL VFQGDVEQIA ....
Smc2 RQVVLGGTSK YLINGHRAPQ QSVLQLFQSV QLNINNPNFL IMQGKITKVL ....
Smc3 RRTVGLKKDD YQLNDRNVTK GDIVRMLETA GFSMNNPYNI VPQGKIVALT ....
Smc4 RKAFKNNSSK YYINEKESSY TEVTKLLKNE GIDLDHKRFL ILQGEVENIA ....
Smc1 *ELAGGNASL TIEDEDEPFN AGIKYHATPP LKRFKDMEYL SGGEKTVAAL ALLFAINSYQ
Smc2 *DLLPNSFAK LVPCEGKDVT QGLEVKVKLG NIWKESLIEL SGGQRSLIAL SLIMALLQFR
Smc3 *DAESNESQN GKDSEIMYTG VSISVSFNSK QNEQLHVEQL SGGQKTVCAI ALILAIQMVD
Smc4 *ITMGGNAEL ELVDSLDPFS EGVTFSVMPP KKSWRNITNL SGGEKTLSSL ALVFALHKYK
Smc1 PSPFFVLDEV DAALDITNVQ RIAAYIRRHR
Smc2 PAPMYILDEV DAALDLSHTQ NIGHLIKTRF
Smc3 PASFYLFDEI DAALDKQYRT AVATLLKELS
Smc4 PTPLYVMDEI DAALDFRNVS IVANYIKERT
(C)
mh115p
rr115p
mg115p
Bacterial group
Smc3
hSmcD
SudA
Smc3-type group
gSc2
hSC2, rSC2
Xcap-E
R06F6.10(C. el. chr. 3)
Smc2
M106.1(C. el. chr. 2)
Cut14
Smc2-type group
Smc1
SB1.8/DX423E (hSmcB)
Smc1-type group
mSmcA
hSmcA
Xcap-C
Smc4
Cut3
F35G12.8 (C.el.)
Dpy-27
Smc4-type group
Rhc18
100
100
62
100
100
86
57
99
100
100
100
84
100
100
100

are found to be associated with these condensed chromosomes, in contrast to hundreds of proteins associated with mitotic chromosomes in vivo. This property of the chromosomes assembled from the sperm chromatin served as a powerful purification step in identification of several new chromosomal proteins (Hirano & Mitchison 1994, Vernos et al 1995). A second advantage of this system is in its ability to serve as a potent biochemical assay for chromosome condensation.

This in vitro system has been used to characterize two proteins, XCAP-C and XCAP-E, in *X. laevis* (Hirano & Mitchison 1994). They belong to two different subgroups of the SMC family (Figure 2). The role of XCAP-C in chromosome condensation in vitro has been established through a number of direct experiments. It was found that chromosome condensation in vitro fails when anti-XCAP-C antibodies are present in the extract. Moreover, the already assembled and condensed chromosome decondenses slightly in response to treatment with the same antibodies. These results suggest that XCAP-C plays an important role in the establishment of the condensed chromosome, and it is in part responsible for the maintenance of condensation. It is possible, however, that the condensation defect seen after the antibody treatment is not attributable to the XCAP-C depletion alone. It was shown, for instance, that XCAP-C is tightly associated with XCAP-E. In addition, we now realize that the SMC proteins can interact with non-SMC subunits (A Strunnikov & D Koshland, unpublished data). The addition of the anti-XCAP-C antibody may disrupt this complex or prevent the whole complex from binding to the chromosomal DNA. In the latter case, the observed condensation defect might be a cumulative result of the depletion of several proteins involved in condensation. These possibilities

←

Figure 2 The SMC family of proteins. (*A*) Schematic depiction of the putative structure of a SMC protein. NTP, putative NTP-binding region; CC, coiled-coil regions; DA, DA-box region. (*B*) Amino and carboxyl terminal regions of the SMC proteins used to build the molecular phylogeny. As an example, an alignment is given for the Smc1, Smc2, Smc3, and Smc4 proteins. The gap between amino and carboxyl terminal regions is indicated by a dotted line. Asterisks indicate gaps, stripped or introduced for the multiple sequence alignment. (*C*) Phylogenetic tree representing clustering of the SMC proteins into subgroups. The phylogeny was established using the Fitch-Margoliash algorithm as described previously with minor modifications (Strunnikov et al 1995). The numbers at the fork are percentages indicating the probability that all species to the right of that fork fall into that group. The species shown in italics are represented only by partial sequences and were placed on the tree through the analysis of truncated alignments. Rhc18 protein was used as an outside non-SMC reference. Abbreviations: mh115p, *Mycoplasma hyorhinis* p115p; mg115p, *Mycoplasma genitalium* p115p; rr115p, *Rhodospirillum rubrum* p115; hSmcD, human SmcD; mSmcA, mouse SmcA; gSC2, *Gallus gallus* SC2; rSC2, rat SC2, M106.1; (*C. el.* chr. 2), *C. elegans* gene M106.1 from chromosome II.

could be addressed by the analysis of the protein profile of the decondensed chromatin after treatment with antibody and by the addition of purified XCAP-C to immune-depleted extracts.

These in vitro and in vivo studies suggest that the primary function of the SMC proteins is to condense chromatin in mitosis. However, two outstanding questions remain. First, what is the molecular mechanism of the SMC protein action and, second, why are there multiple species of the SMC proteins within the same cell? The putative structure of the SMC family proteins gives some clues regarding the first of these two problems. It was noticed very early that the structure of SMC proteins resembles that of mechanochemical proteins like myosin and kinesin, which are able to move along an array of macromolecules using the energy of ATP hydrolysis (Notarnicola et al 1991, Strunnikov et al 1993). Thus it is possible that SMC family molecules are condensation "motors" bringing distant chromatin regions together in a processive and energy-dependent fashion and making the energetically unfavorable condensed state possible. Another feature of the SMC proteins, their long coiled-coil regions, could serve as an extended surface of interaction between themselves and other proteins. This adhesiveness could help to solidify and stabilize the folds of the condensed chromosomes once they are formed. In support of this model, a physical interaction has been demonstrated between XCAP-C and XCAP-E proteins, between Smc1 and Smc2, and between two Smc2 molecules (Hirano & Mitchison 1994, Strunnikov et al 1995). Thus it is possible the SMC proteins can oligomerize and form a multisubunit complex.

The multitude of SMC family proteins and the diversity of phenotypes of different *smc* and *cut* mutants suggest that there is functional specialization between different SMC family molecules. This specialization could be a result of specific interactions with non-SMC subunits of the complex. On the other hand, specialization of the SMC family proteins could also be a result of their differential spatial distribution within the nucleus or along the chromosome. It is known that when some chromosomal domains, e.g. kinetochore regions, are condensed, they have a structure that is distinctly different from the rest of the chromosome (Rattner & Lin 1985). Thus it is possible that some of the SMC molecules would be restricted to a specific region. In fact, the spatial restriction of a SMC family protein is well documented. Dpy-27, a SMC family member, specifically localizes to the X chromosomes in XX *Caenorhabditis elegans* animals (Chuang et al 1994). However, it is interesting that Dpy-27 is targeted to a sex chromosome by another protein (B Meyer, personal communication).

The discovery of the SMC proteins provides an important new tool to explore eukaryotic chromosome condensation. First, the study of the molecular activity of SMC proteins should reveal the underlying mechanisms of chromosome condensation. Second, the SMC proteins can also serve as promising biochemical and genetic bait in the isolation of novel proteins involved in chromosome condensation. Among the potential candidates for molecules interacting with SMC proteins are a budding yeast protein that shares homology with the Rad 21 protein of *S. pombe*; proteins in humans, mice, and worms (Birkenbihl et al 1995; V Guacci et al, in preparation); and the bimD protein of *Aspergillus,* which also has a *S. cerevisiae* homologue, Pds5 (Denison et al 1992; V Guacci & D Koshland, unpublished data).

CONCLUSION

In this review we have summarized our current understanding of mitotic chromosome condensation, beginning with the structural information that reveals mitotic condensation as a programmed process of invariant folding. We have analyzed experiments from numerous in vivo and in vitro systems aimed at identifying and characterizing the *cis* sites and *trans* factors that are the molecular basis of this process. To conclude, we revisit some of the central themes herein by addressing two questions about the conservation of mitotic chromosome condensation.

First, how conserved is the molecular basis of mitotic chromosome condensation between evolutionary distant organisms? An assumption in our review has been that this process is highly conserved, as evident from our use of results from many different organisms to dissect a particular aspect of the process. This assumption is vindicated largely by the existence of both topoisomerase II and SMC homologues in all eukaryotes examined to date and by studies from fungi to vertebrates implicating their function in mitotic chromosome condensation. Furthermore, the presence of type II topoisomerases and SMC family proteins in bacteria suggests that at least some aspects of segregation-related condensation may be conserved, even between eukaryotes and prokaryotes. This conservation was not anticipated because of the dramatically different appearance of segregating chromosomes from mammals, fungi, and bacteria.

Although the general mechanism of segregation-related chromosome condensation is likely to be conserved, the differences in chromosome appearance in organisms such as bacteria, fungi, and vertebrates suggest that the sophisticated machinery for chromosome condensation of higher eukaryotes probably evolved through several distinct steps. For example, budding yeast chromosomes condense some three- to fivefold less than mammalian chromosomes. It is possible that yeast chromosomes have fewer chromatin folds because of a lower density of *cis* sites or a limiting number of some *trans* factors. Alternatively, one type of fold present in mammalian chromosomes may be completely missing in budding yeast. This may explain why mitotic chromosomes of budding yeast are more transcriptionally permissive than the mitotic chromosomes of larger eukaryotes. This difference cannot be attributed to the absence of H1 in budding yeast because an apparent H1 homologue has recently been identified in this organism (H Bussey et al, personal communication). Another evolutionary aspect arises from the analysis of DNA sequences from the genome sequencing projects (Fleischmann et al 1995, Fraser et al 1995). From these studies we know that some bacteria (*Mycoplasma* and purple bacteria) contain a SMC family protein (p115), whereas others do not (*Haemophilus influenza* and probably *E. coli*). An operon similar to the p115 operon in *Mycoplasma* is present in *H. influenza* and *E. coli,* but the p115 gene is replaced by *ftsE* and *ftsX*, which are known to play a role in cell division (Gill & Salmond 1987). Although ftsX and ftsE proteins are not structurally similar to p115, perhaps they are functionally related. A comparison of chromosome structure in bacteria with or without p115 would be extremely interesting.

Second, are the same principles and/or molecules utilized in chromosome condensation and other processes that modulate chromatin structure? Indeed, there is a precedent for this. For example, Dpy-27, a SMC family protein in *C. elegans,* has been implicated in transcription control as an essential component of dosage compensation. In this animal, the mechanism of dosage compensation results in the twofold suppression of transcription originating from a pair of X chromosomes in hermaphrodites (Meyer & Casson 1986). Although it still remains to be shown whether the transcriptionally repressed X chromosome is also partially condensed, this situation can be regarded as an analogue of mitotic chromosomes in budding yeast, where mitotic transcription is reduced compared with the S-phase but not shut down completely. This leads to the attractive concept that the global changes in chromatin structure mediated by mitotic condensation factors are used to regulate the transcription of large blocks of genes.

A number of other proteins structurally similar to SMC proteins, although not members of the SMC family, have been implicated in chromosome-related processes. For example, the Rhc18 and Rad18 proteins from budding and fission yeast, respectively, are part of the DNA-repair machinery (Lehmann et al 1995). The Rad5O protein of *S. cerevisiae* and bacterial RecN proteins have a major role in genetic recombination (Rostas et al 1987, Alani et al 1990). These proteins have a similar arrangement of NTP-binding sites and coiled-coil regions but lack a canonical DA-box. The presence of SMC-like molecules in repair and recombination pathways supports the idea that these processes and mitotic chromosome condensation share some common molecular events. In summary, we anticipate that further studies of mitotic chromosome condensation will not only reveal important insights into this most fundamental of biological processes but also into a broad spectrum of chromatin-related processes.

ACKNOWLEDGMENTS

We wish to acknowledge Joseph Gall and Connie Holm for helpful discussions and critical reading of this manuscript.

Literature Cited

Adachi Y, Luke M, Laemmli U. 1991. Chromosome assembly in vitro: topoisomerase II is required for condensation. *Cell* 64:137–48

Adams DE, Shekhtman EM, Sechiedrich EL, Schmid MB, Cozzarelli NR. 1992. The role of topoisomerase IV in partitioning bacterial replications and the structure of catenated intermediates in DNA replication. *Cell* 71:277–88

Adolph KW, Cheng SM, Laemmli UK. 1977. Role of nonhistone proteins in metaphase chromosome structure. *Cell* 12:805

Alani E, Padmore R, Kleckner N. 1990. Analysis of wild-type and *rad5O* mutants of yeast suggests an intimate relationship between meiotic chromosome synapsis and recombination. *Cell* 61:419–36

Alberts BM. 1984. The DNA enzymology of protein machines. *Cold Spring Harbor Symp. Quant. Biol.* 49:1–12

Baldi MI, Benedetti P, Mattoccia E, Tocchini-Valentini GP. 1980. In vitro catenation and decatenation of DNA and a novel eucaryotic ATP-dependent topoisomerase. *Cell* 20:461–67

Baumgartner M, Dutrillaux B, Lemieux N, Lillenbaum A, Paulin D, et al. 1991. Genes occupy a fixed and symmetrical position on sister chromatids. *Cell* 64:761–66

Benyajati C, Worcel A. 1976. Isolation, characterization and structure of the folded interphase genome of Drosophila melanogaster. *Cell* 9:393–407

Berger JM, Gamblin SJ, Harrison SC, Wang JC. 1996. Structure and mechanism of DNA topoisomerase II. *Nature* 379:225–32

Birkenbihl RP, Subramani S. 1995. The *rad21* gene product of *Schizosaccharomyces pombe* is a nuclear, cell cycle-regulated phosphoprotein. *J. Biol. Chem.* 270:7703–11

Boy de la Tour E, Laemmli UK. 1988. The metaphase scaffold is helically folded: sister chromatids have predominantly opposite helical handedness. *Cell* 55:937–44

Bradbury EM. 1992. Reversible histone modifications and the chromosome cell cycle. *BioEssays* 14:9–16

Brandriff B, Gordon L, Trask B. 1991. A new system for high-resolution DNA sequence mapping in interphase pronuclei. *Genomics* 10:75–82

Brill SJ, Sternglanz R. 1988. Transcription-dependent DNA supercoiling in yeast DNA topoisomerase mutants. *Cell* 54:403–11

Buchenau P, Saumweber H, Arndt-Jovin DJ. 1993. Consequences of topoisomerase II inhibition in early embryogenesis of *Drosophila* revealed by in vivo confocal laser scanning microscopy. *J. Cell Sci.* 104:1175–85

Callan HG. 1986. *Lampbrush Chromosomes.* Berlin: Springer-Verlag. 255 pp.

Chuang P-T, Albertson DG, Meyer BJ. 1994. DPY-27: a chromosome condensation protein homolog that regulates C. elegans dosage compensation through association with the X chromosome. *Cell* 79:459–74

Clarke D, Johnson R, Downes C. 1993. Topoisomerase II inhibition prevents anaphase chromatid segregation in mammalian cells independently of the generation of DNA strand breaks. *J. Cell Sci.* 105:563–69

Cook PR, Brazen IA. 1975. Supercoils in human DNA. *J. Cell Sci.* 19:261

Denison SH, Kafer E, May GS. 1992. Mutation in the *bimD* gene of *Aspergillus nidulans* confers a conditional mitotic block and sensitivity to DNA damaging agents. *Genetics* 134:1085–96

Earnshaw WC, Halligan B, Cooke CA, Heck MMS, Liu LF. 1985. Topoisomerase II is a structural component of mitotic chromosome scaffolds. *J. Cell Biol.* 100:1706–15

Earnshaw WC, Heck MMS. 1985. Localization of topoisomerase II in mitotic chromosomes. *J. Cell Biol.* 100:1716–25

Earnshaw WC, Laemmli UK. 1983. Architecture of metaphase chromosomes and chromosome scaffolds. *J. Cell Biol.* 96:84–93

Falk G, Walker J. 1988. DNA sequence of a gene cluster coding for subunits of the F-0 membrane sector of ATP synthase in *Rhodospirillum rubrum:* support for modular evolution of the F-1 and F-0 sectors. *Biochem. J.* 254:109–22

Fleischmann RD, Adams MD, White D, Clayton RA, Kirkness EF, et al. 1995. Whole-genome random sequencing and assembly of *Haemophilis influenzas* Rd. *Science* 269:496–512

Fraser CM, Gocayne JD, White D, Adams MD, Clayton RA, et al. 1995. The minimal gene complement of *Mycoplasma genitalium. Science* 270:397–403

Funabiki H, Hagan I, Uzawa S, Yanagida M. 1993. Cell cycle-dependent specific positioning and clustering of centromeres and telomeres in fission yeast. *J. Cell Biol.* 121:961–76

Gartenberg MR, Wang JC. 1993. Identification of barriers to rotation of DNA segments in yeast from the topology of DNA rings excised by an inducible site-specific recombinase. *Proc. Natl. Acad. Sci. USA* 90:10514–18

Gasser SM. 1992. Functional aspects of chromosome organization: scaffold attachment regions and their ligands. In *Advances in Molecular and Cell Biology,* pp. 75–101. Greenwich, CT: JAI Press

Gasser SM, Laemmli UK. 1986. Cohabitation of scaffold binding regions with upstream/enhancer elements of three developmentally regulated genes in D. melanogaster. *Cell* 46:521–30

Gill DR, Salmond GP. 1987. The *Escherichia coli* cell division proteins FtsY, FtsE and FtsX are inner membrane-associated. *Mol. Gen. Genet.* 210:504–8

Gorbsky G. 1994. Cell cycle progression and chromosome segregation in mammalian cell cultured in the presence of the topoisomerase II inhibitors ICRF-187 [(+)-1,2-bis(c,5-dioxopiperazinyl1-1-yl)propane; ADR-529] and ICRF-159 (Razoxane). *Cancer Res.* 54:1042–48

Goto T, Wang JC. 1982. Yeast DNA topoisomerase II. *J. Biol. Chem.* 257:5866–72

Guacci V, Hogan E, Koshland D. 1994. Chromosome condensation and sister chromatid pairing in budding yeast. *J. Cell Biol.* 125:517–23

Guo X, Thing J, Swank R, Anderson H, Tudan C, et al. 1995. Chromosome condensation induced by tostriecein does not require p34cdc2 kinase activity and histone H1 hyperphosphorylation, but is associated with enhanced histone H2A and H3 phosphorylation. *EMBO J.* 14:976–85

Heller RA, Shelton ER, Dietrich V, Elgin SCR, Brutlag DL. 1986. Multiple forms and cellular localization of *Drosophila* DNA topoisomerase II. *J. Biol. Chem.* 261:8063–69

Hirano T. 1995. Biochemical and genetic dissection of mitotic chromosome condensation. *Trends Biol. Sci.* 20:357–61

Hirano T, Mitchison TJ. 1991. Cell cycle control of higher-order chromatin assembly around naked DNA in vitro. *J. Cell Biol.* 115:1479–89

Hirano T, Mitchison TJ. 1993. Topoisomerase II does not play a scaffolding role in the organization of mitotic chromosomes assembled in *Xenopus* egg extracts. *J. Cell Biol.* 120:601–12

Hirano T, Mitchison TJ. 1994. A heterodimeric coiled-coil protein required for mitotic chromosome condensation in vitro. *Cell* 79:449–58

Hirano T, Mitchison TJ, Swedlow JR. 1995. The SMC family: from chromosome condensation to dosage compensation. *Curr. Opin. Cell Biol.* 7:329–36

Hiraoka Y, Minden JS, Swedlow JR, Sedat JW, Agard DA. 1989. Focal points for chromosome condensation and decondensation revealed by three-dimensional in vivo time-lapse microscopy. *Nature* 342:293–96

Hock R, Carl M, Lieb B, Gebauer D, Scheer U. 1996. A monoclonal antibody against DNA topoisomerase II labels the axial granules of *Pleurodeles* lampbrush chromosomes. *Chromosoma* 104:358–66

Holm C. 1994. Coming undone: how to untangle a chromosome. [Review] *Cell* 77:955–57

Holm C, Goto T, Wang JC, Botstein D. 1985. DNA topoisomerase II is required at the time of mitosis in yeast. *Cell* 41:553–63

Holm C, Stearns T, Botstein D. 1989. DNA topoisomerase II must act at mitosis to prevent nondisjunction and chromosome breakage. *Mol. Cell. Biol.* 9:159–68

Holt CL, May GS. 1996. An extragenic suppressor of the mitosis-defective *bim6* mutation of *Aspergillus nidulans* codes for a chromosome scaffold protein. *Genetics* 142:777–87

Hsieh T-S, Brutlag D. 1980. ATP-dependent DNA topoisomerase from D. melanogaster reversibly catenates duplex DNA rings. *Cell* 21:115–25

Iarovaia O, Hancock R, Lagarkova M, Miassod R, Razin SV. 1996. Mapping of genomic DNA loop organization in a 500-kilobase region of the *Drosophila* X chromosome by the topoisomerase II-mediated DNA loop excision protocol. *Mol. Cell Biol.* 16:302–8

Igó-Kemens T, Zachau HG. 1977. Domains in chromatin structure. *Cold Spring Harbor Symp. Quant. Biol.* 42:109–18

Kim RA, Wang JC. 1989. Function of DNA topoisomerases as replication swivels in *Saccharomyces cerevisiae*. *J. Mol. Biol.* 208:257–67

Klein F, Laroche T, Cardenas ME, Hofmann JF, Schweizer D, et al. 1992. Localization of RAP1 and topoisomerase II in nuclei and meiotic chromosomes of yeast. *J. Cell Biol.* 117:935–48

Kreuzer KN, Cozzarelli NR. 1980. Formation and resolution of DNA catenanes by DNA gyrase. *Cell* 20:245–54

Larionov V, Karpova T, Kouprina N, Jouravleva G. 1985. A mutant of *Saccharomyces cerevisiae* with impaired maintenance of centromeric plasmids. *Curr. Genet.* 10:15–20

Lawrence JB, Villnave CA, Singer RH. 1988. Sensitive, high-resolution chromatin and chromosome mapping in situ: presence and orientation of two closely integrated copies of EBV in a lymphoma line. *Cell* 52:51–61

Lehmann AR, Walicka M, Griffiths DJF, Murray JM, Watts FZ, et al. 1995. The *radl8* gene of *Schizosaccharomyces pombe* defines a new subgroup of the SMC superfamily involved in DNA repair. *Mol. Cell Biol.* 15:7067–80

Lewis CD, Laemmli UK. 1982. Higher order metaphase chromosome structure: evidence for metalloprotein interactions. *Cell* 29:171–81

Lichter P, Cremer T, Borden J, Manuelidis L, Ward DC. 1988. Delineation of individual human chromosomes in metaphase and interphase cells by in situ suppression hybridization using recombinant DNA libraries. *Hum. Genet.* 80:224–34

Ma X, Saitoh N, Curtis PJ. 1993. Purification and characterization of a nuclear DNA-binding factor complex containing topoisomerase II and chromosome scaffold protein 2. *J. Biol. Chem.* 268:6182–88

Marsden MP, Laemmli UK. 1979. Metaphase chromosomes: evidence for a radial loop model. *Cell* 17:849–58

Mathog D, Hochstrasser M, Gruenbaum Y, Saumweber H, Sedat J. 1984. Characteristic folding pattern of polytene chromosomes in *Drosophila* salivary gland nuclei. *Nature* 308:414–21

Meyer BJ, Casson LP. 1986. Caenorhabditis elegans compensates for the difference in X chromosome dosage between the sexes by regulating transcript levels. *Cell* 47:871–81

Mirkovitch J, Gasser SM, Laemmli UK. 1988. Scaffold attachment of DNA loops in metaphase chromosomes. *J. Mol. Biol.* 200:101–10

Mirkovitch J, Mirault M-E, Laemmli UK. 1984. Organization of the higher order chromatin loop: specific DNA attachment sites on nuclear scaffolds. *Cell* 39:223–32

Moens PB, Earnshaw WC. 1989. Anti-topoisomerase II recognizes meiotic chromosome cores. *Chromosoma* 98:317–22

Moudrianakis E, Arents G. 1993. Structure of the histone octamer core of the nucleosome and its potential interactions with DNA. *Cold Spring Harbor Symp. Quant. Biol.* 58:273–79

Mueller R, Yasuda H, Hatch C, Bonner W, Bradbury E. 1985. Identification of ubiquitinated histones 2A and 2B in *Physarum polycephalum*. Disappearance of these proteins at metaphase and reappearance at anaphase. *J. Biol. Chem.* 260:5147–53

Newport J. 1987. Nuclear reconstitution in vitro: stages of assembly around protein-free DNA. *Cell* 48:205–17

Newport J, Spann T. 1987. Disassembly of the nucleus in mitotic extracts: membrane vesicularization, lamin disassembly and chromosome condensation are independent processes. *Cell* 48:219–30

Notarnicola S, McIntosh M, Wise K. 1991. A *Mycoplasma hyorhinis* protein with sequence similarities to nucleotide-binding enzymes. *Gene* 97:77–85

Ohsumi K, Katagiri C, Kishimoto T. 1993. Chromosome condensation in *Xenopus* mitotic extracts without histone H1. *Science* 262:2033–35

Paulson JR, Laemmli UK. 1977. The structure of histone-depleted metaphase chromosomes. *Cell* 12:817

Peterson C. 1994. The SMC family: novel motor proteins for chromosome condensation? *Cell* 79:389–92

Pinkel D, Landegent J, Collins C, Fuscoe J, Seagraves R, et al. 1988. Fluorescence in situ hybridization with human chromosome-specific libraries: detection of trisomy 21 and translocations of chromosome 4. *Proc. Natl. Acad. Sci. USA* 85:9138–42

Rao P, Adlakha R. 1984. Chromosome condensation and decondensation factors in the life cycle of eukaryotic cells. *Symp. Fund. Cancer Res.* 37:45–69

Rattner JB, Lin CC. 1985. Radial loops and helical coils coexist in metaphase chromosomes. *Cell* 42:291–96

Roca J, Wang JC. 1994. DNA transport by a type II DNA topoisomerase: evidence in favor of a two-gate mechanism. *Cell* 77:609–16

Rostas K, Morton SJ, Picksley SM, Lloyd RG. 1987. Nucleotide sequence and LexA regulation of the *Escherichia coli recN* gene. *Nucleic Acids Res.* 15:5041–49

Roth SY, Allis CD. 1992. Chromatin condensation: does histone H1 dephosphorylation play a role? *Trends Biol. Sci.* 17:93–98

Saitoh N, Goldberg IG, Earnshaw WC. 1995. The SMC proteins and the coming of age of the chromosome scaffold hypothesis. *BioEssays* 17:759–66

Saitoh N, Goldberg IG, Wood ER, Earnshaw WC. 1994. ScII: an abundant chromosome scaffold protein is a member of a family of putative ATPases with an unusual predicted tertiary structure. *J. Cell Biol.* 127:303–18

Saitoh Y, Laemmli UK. 1994. Metaphase chromosome structure: bands arise from a differential folding path of the highly AT-rich scaffold. *Cell* 76:609–22

Saka Y, Sutani T, Yamashita Y, Saitoh S, Takeuchi M, et al. 1994. Fission yeast cut3 and cut14, members of the ubiquitous protein family, are required for chromosome condensation and segregation in mitosis. *EMBO J.* 13:4938–52

Samejima I, Matsumoto T, Nakaseko Y, Beach D, Yanagida M. 1993. Identification of seven new cut genes involved in *Schizosaccharomyces pombe* mitosis. *J. Cell Sci.* 105:135–43

Schultz MC, Brill SJ, Ju Q, Sternglanz R, Reeder RH. 1992. Topoisomerases and yeast rRNA transcription: negative supercoiling stimulates initiation and topoisomerase activity is required for elongation. *Genes Dev.* 6:1332–41

Selig S, Okumura K, Ward DC, Cedar H. 1992. Delineation of DNA replication time zones by fluorescence in situ hybridization. *EMBO J.* 11:1 217–25

Shamu CE, Murray AW. 1992. Sister chromatid separation in frog egg extracts required DNA topoisomerase II activity during anaphase. *J. Cell Biol.* 117:921–34

Shen X, Yu L, Weir JW, Grovosky MA. 1995. Linker histones are not essential and affect chromatin condensation in vivo. *Cell* 82:47–56

Smith MM. 1991. Histone structure and function. *Curr. Opin. Cell Biol.* 3:429–37

Spell RM, Holm C. 1994. Nature and distribution of chromosomal intertwinings in *Saccharomyces cerevisiae. Mol. Cell. Biol.* 14:1465–76

Strick R, Laemmli U. 1995. SARs are *cis* DNA elements of chromosome dynamics: synthesis of a SAR repressor protein. *Cell* 83:1137–48

Strunnikov AV, Hogan E, Koshland D. 1995. *SMC2*, a *Saccharomyces cerevisiae* gene essential for chromosome segregation and condensation defines a subgroup within the SMC-family. *Genes Dev.* 9:587–99

Strunnikov AV, Larionov VL, Koshland D. 1993. *SMC1*: an essential yeast gene encoding a putative head-rod-tail protein is required for nuclear division and defines a new ubiquitous protein family. *J. Cell Biol.* 123:1635–48

Sumner A. 1992. Inhibitors of topoisomerases do not block the passage of human lymphocyte chromosomes through mitosis. *J. Cell Sci.* 103:105–15

Sumner AT. 1996. The distribution of topoisomerase II on mammalian chromosomes. *Chromosome Res.* 4:5–14

Sundin O, Varshavsky A. 1981. Arrest of segregation leads to accumulation of highly intertwined catenated dimers: dissection of the final stages of SV40 DNA replication. *Cell* 25:659–69

Swedlow JR, Sedat JW, Agard DA. 1993. Multiple chromosomal populations of topoisomerase II detected in vivo by time-lapse, three-dimensional side-field microscopy. *Cell* 73:97–108

Udvardy A, Schedl P. 1991. Chromatin structure, not DNA sequence specificity, is the primary determinant of topoisomerase II sites of action in vivo. *Mol. Cell Biol.* 11:4973–84

Uemura T, Ohkura H, Adachi Y, Morino K, Shiozaki K, et al. 1987. DNA topoisomerase II is required for condensation and separation of mitotic chromosomes in S. pombe. *Cell* 50:917–25

Uemura T, Yanagida M. 1984. Isolation of type I and II DNA topoisomerase mutants from fission yeast: single and double mutants show different phenotypes in cell growth and chromatin organization. *EMBO J.* 3:1737–44

Uemura T, Yanagida M. 1986. Mitotic spindle pulls but fails to separate chromosomes in type II DNA topoisomerase mutants: uncoordinated mitosis. *EMBO J.* 5:1003–10

Van Holde K, Zlatanoval J. 1995. Chromatin higher order structure: chasing a mirage. *J. Biol. Chem.* 270:8373–76

Vassetzky YS, Dang O, Benedetti P, Gasser SM. 1994. Topoisomerase II forms multimers in vitro: effects of metals, beta-glycerophosphate and phosphorylation of its C-terminal domain. *Mol. Cell Biol.* 14:6962–74

Vernos I, Reats J, Hirano T, Heasman J, Karsenti E, et al. 1995. Xklp1, a chromosomal Xenopus kinesin-like protein essential for spindle organization and chromosome positioning. *Cell* 81:117–27

Ward S, McNeil J, de Lara J, Lawrence J. 1995. Localization of three genes in the asymmetric hamster sperm nucleus by fluorescent in situ hybridization. *Mol. Biol. Cell* 6:72a

Weaver DT, Fields-Berry SC, DePamphilis ML. 1985. The termination region for SV40 DNA replication directs the mode of separation for the two sibling molecules. *Cell* 41:565–75

Wood ER, Earnshaw WC. 1990. Mitotic chromatin condensation in vitro using somatic cell extracts and nuclei with variable levels of endogenous topoisomerase II. *J. Cell Biol.* 111:2839–50

Wright S, Schatten G. 1990. Teniposide, a topoisomerase II inhibitor, prevents chromosome condensation and separation but not decondensation in fertilized surf clam (*Spisula solidissima*) oocytes. *Dev. Biol.* 142:224–32

Yakoto H, van den Engh G, Hearst JE, Sachs RK, Trask BJ. 1995. Evidence for the organization of chromatin in megabase pair-sized loops arranged along a random walk path in the human GO/G1 interphase nucleus. *J. Cell Biol.* 130:1239–49

Yanagida M. 1995. Frontier questions about sister chromatid separation in anaphase. *BioEssays* 17:519–26

Zechiedrich EL, Cozzarelli NR. 1995. Roles of topoisomerase IV and DNA gyrase in DNA unlinking during replication in *Escherichia coli*. *Genes Dev.* 9:2859–69

Zini N, Martelli AM, Sabatelli P, Santi S, Negri C, et al. 1992. The 180-kDa isoform of topoisomerase II is localized in the nucleolus and belongs to the structural elements of the nucleolar remnant. *Exp. Cell Res.* 200:460–66

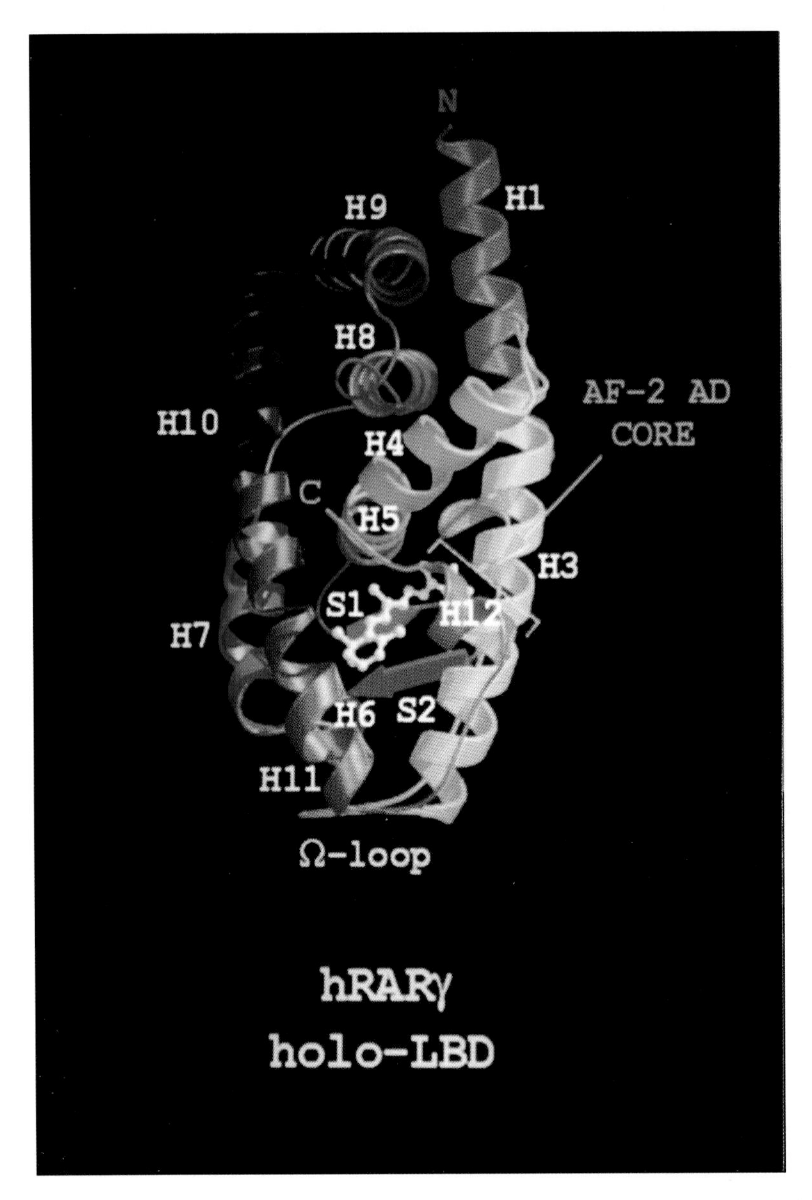

Figure 1c Ribbon drawing of the ligand-binding domain of RAR γ bound to all-*trans* retinoic acid. Helices (H1, H3 to H12) and β-strands (S1 and S2) are indicated and color-coded, starting with red at the N-terminus (H1) and progressing to purple at the C-terminus (H12). The molecule of all-*trans* retinoic acid is in white. (Communicated by D Moras; Renaud et al 1995.)

Annu. Rev. Cell Dev. Biol. 1996. 12:335–63

PEROXISOME PROLIFERATOR-ACTIVATED RECEPTORS: A Nuclear Receptor Signaling Pathway in Lipid Physiology

Thomas Lemberger, Béatrice Desvergne and Walter Wahli
Institut de Biologie Animale, Université de Lausanne, CH-1015 Lausanne, Switzerland

KEY WORDS: adipocyte differentiation, fatty acids, insulin resistance, prostaglandins, transcription

ABSTRACT

Peroxisome proliferator-activated receptors (PPARs) are lipid-activated transcription factors that belong to the steroid/thyroid/retinoic acid receptor superfamily. All their characterized target genes encode proteins that participate in lipid homeostasis. The recent finding that antidiabetic thiazolidinediones and adipogenic prostanoids are ligands of one of the PPARs reveals a novel signaling pathway that directly links these compounds to processes involved in glucose homeostasis and lipid metabolism including adipocyte differentiation. A detailed understanding of this pathway could designate PPARs as targets for the development of novel efficient treatments for several metabolic disorders.

CONTENTS

1081-0706/96/1115-0335$08.00

INTRODUCTION

For their survival, living organisms have to cope with unpredictable changes in their energetic needs, while their access to food is often irregular. Maintenance of caloric homeostasis is hence a vital task that requires an efficient regulation of fuel metabolism in response to internal and environmental stimuli. While immediate regulation of metabolic processes involves modification of the catalytic activity of individual enzymes, the modulation of gene expression provides a longer-term mechanism that adjusts the synthesis of new gene products suitable to the physiological needs of the organism. In the latter case, factors that transduce metabolic parameters, such as glucose or fatty acid concentrations, into nuclear transcriptional regulatory events are fundamental components of the control system. The identification of proteins participating in this signal transduction process is a crucial step toward the elucidation of the molecular basis underlying regulation of metabolism and its possible disorders.

Peroxisome proliferator-activated receptors (PPARs) are transducer proteins belonging to the steroid/thyroid/retinoid receptor superfamily. They were initially shown to be activated by hypolipidemic drugs of the fibrate class (Issemann & Green 1990, Dreyer et al 1992). Soon after their discovery, these receptors were demonstrated to be activated by natural fatty acids and to control a variety of target genes involved in key steps of lipid metabolism (Wahli et al 1995). Further interest arose when these receptors were found to be critical determinants of adipocyte differentiation and direct targets of antidiabetic drugs of the thiazolidinedione class (Tontonoz et al 1995b, Lehmann et al 199). The rapid progress made in the characterization of PPAR biology has unveiled new mechanisms for the regulation of lipid metabolism and provided insight into possible molecular determinants of metabolic disorders including obesity and type II diabetes. In this context, the recent finding that certain eicosanoids are natural PPAR ligands opens unexpected perspectives for investigating new regulatory routes of energy homeostasis (Kliewer et al 1995, Forman et al 1995).

PPARs AS MEMBERS OF THE NUCLEAR RECEPTOR SUPERFAMILY

Several related PPAR subtypes have been identified in different vertebrate species, e.g. *Xenopus* (Dreyer et al 1992); mouse (Issemann & Green 1990, Chen et al 1993, Zhu et al 1993, Kliewer et al 1994, Amri et al 1995); rat (Göttlicher et al 1992, Xing et al 1995); hamster (Aperlo et al 1995); and human (Schmidt et al 1992, Sher et al 1993, Greene et al 1995). Based on the analysis of the highly conserved DNA- and ligand-binding domains, the PPARs have been assigned to a subfamily of nuclear receptors that includes the retinoic acid receptors (RARs), the thyroid hormone receptors (TRs), and the revErbAα-related orphan receptors (Laudet et al 1992, Desvergne & Wahli, 1995). The PPAR subfamily comprises three distinct subtypes, PPARα, PPARβ or δ (also called NUCI or FAAR), and PPARγ. The rodent PPARδ or FAAR appears to be the homologue of the human NUCI; however, it is not yet clear whether they are the homologues of the *Xenopus* PPARβ. Because the *Xenopus* PPARβ was the first reported representative of this subtype, for the sake of simplicity, we refer to PPARδ, FAAR, and NUCI as PPARβ.

The mouse PPARα, β, and γ genes are localized to chromosome 15, 17, and 6, respectively, and the human genes for the α and γ subtypes have been mapped to the chromosome regions 22q12-q13.1 and 3p25, respectively (Sher et al 1993, Greene et al 1995, Jones et al 1995). The genomic organization of mouse PPARα and γ and *Xenopus* PPARβ genes has been characterized and consists of eight exons (exons A1, A2, and exons 1 to 6), of which the two first (A1 and A2) are noncoding (Krey et al 1993, Gearing et al 1994, Zhu et al 1995). The positions of the introns in the coding region are well conserved among the three subtypes, indicating that the three PPAR genes are derived from a common ancestral gene.

Like other members of the nuclear hormone receptors superfamily, PPARs display a modular structure consisting of four functional domains: (*a*) the poorly conserved AB domain, which has not been assigned a clear function in PPARs; (*b*) the DNA-binding C domain (DBD; Figure 1*a*); (*c*) the D domain, comprising the so-called DBD carboxy-terminal extension (CTE; Figure 1*b*), and a flexible hinge connecting the C and E domains; and (*d*) the E domain (Figure 1*b*), which binds the ligand, mediates hormone-dependent transactivation, and provides the major dimerization interface.

The DNA-binding domain is the most conserved domain of nuclear receptors. It contains two zinc fingers and in the case of glucocorticoid, estrogen, and retinoic X receptors has been resolved in molecular structure by crystallography and nuclear magnetic resonance analyses (Härd et al 1990, Schwabe et al 1990, Luisi et al 1991, Lee et al 1993). The three-dimensional structure of the DNA-binding domain consists of two perpendicular α-helices, the first of which lies in the DNA major groove (Rastinejad et al 1995). A crucial role in DNA-binding specificity is played by the P-box amino acids located at the carboxy end of the first zinc finger. The primary sequence of the PPAR P-box (CEGCKG; Figure 1*a*) is identical to that of members of the TR/RAR subfamily of nuclear receptors, and of several orphan receptors as well, and

a **DNA-binding domain**

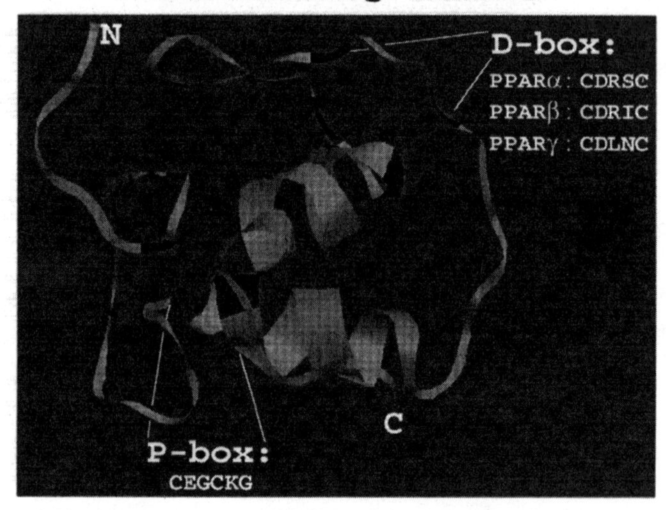

b **Ligand-binding domain**

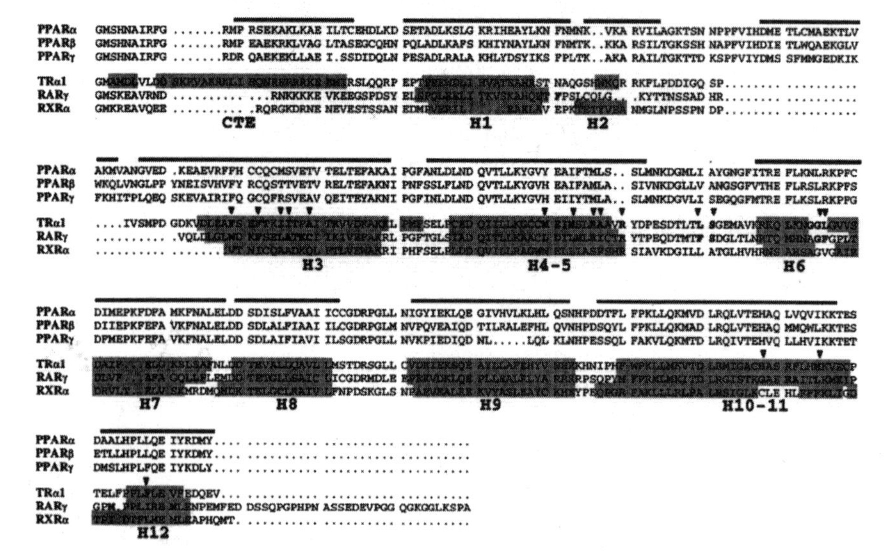

must therefore confer similar DNA-binding properties to this group of receptors (see below). The amino acids between the first and second cysteines of the second zinc finger, the so-called D-box, are involved in contacts between dimerizing receptors. In contrast to other nuclear receptors, PPARs have a unique D-box of only three amino acids instead of the five found in most of the superfamily members (Figure 1*a*). A putative functional characteristic of this unique D-box is discussed below. The DNA-binding domain is immediately followed by its carboxy-terminal extension (CTE), which is predicted to adopt an α-helical conformation in PPARs, consistent with the existing structure in TRβ (Rastinejad et al 1995; Figure 1*b*).

The ligand-binding domain (LBD) of nuclear receptors plays a pivotal role in the transduction of the hormonal signal into transcriptional activation. The three-dimensional structures of the LBDs of TRα1, RXRα, and RARγ have been solved recently and appear to be extremely well conserved among these three receptors (Bourguet et al 1995, Renaud et al 1995, Wagner et al 1995). The structure consists of an antiparallel α-helical sandwich comprising 12 major α-helices (H1-H12; Figure 1*b*, and 1*c* and at least two short β-strands (S1–S2; Figure 1*c*). The ligands of TRα1 and RARγ, tri-iodothyronine (T3) and all-*trans* retinoic acid, respectively, are buried inside the domain, and despite their structural unrelatedness, make contacts with residues occupying equivalent positions in both receptors. Upon ligand binding, the domain is thought to undergo conformational changes involving the swinging of H12, which closes the ligand pocket like a lid (Figure 1*c*). The strong resemblance of the three LBD structures suggests that they define a common fold for all nuclear receptors, including PPARs. Indeed, secondary structure prediction by computer analysis of the PPAR primary sequence proposes

Figure 1 DNA- and ligand-binding domains of PPARs. (*a*) A ribbon diagram of the PPAR DNA-binding domain was generated by knowledge-based molecular modeling (performed at GLAXO Institute for Molecular Biology, using ProMod software; Peitsch & Jongeneel 1993). The cysteine residues of the two zinc fingers are shown in black. The location and primary sequences of the P- and D-boxes are depicted. (*b*) The sequences of the D and E domains of the three murine PPAR subtypes have been aligned to rat TRα1, human RARγ, and human RXRα. The carboxy-terminal extension and ligand-binding domain α-helices (CTE and H1 to H12, respectively) are delineated as described in the crystal structure of TRα1, RARγ, and RXRα (*stippled boxes*; Bourguet et al 1995, Renaud et al 1995, Wagner et al 1995) or as predicted by computer analysis of the PPARα sequence (*bars*). Residues that directly contact the ligand in TRα1 or RARγ are in boldface and identified by an arrowhead when located at identical positions in both receptors. (*c*) Ribbon drawing of the ligand binding domain of RARγ bound to all-trans retinoic acid (communicated by D Moras; Renaud et al 1995). Helices (H1, H3 to H12) and β- strands (S1 and S2) are indicated and color-coded, starting with red at the N terminus (H1) and progressing to purple at the C terminus (H12). The molecule of all-*trans* retinoic acid is in white.

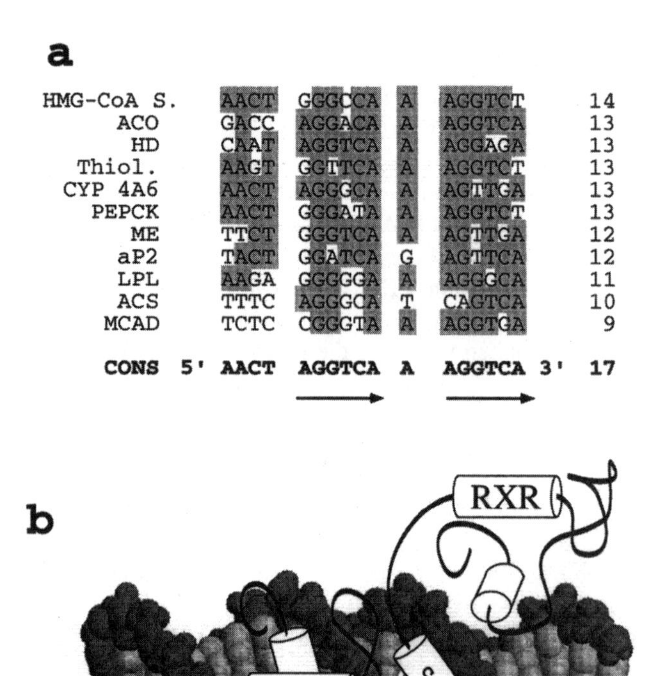

Figure 2 DNA recognition by PPARs. (*a*) Comparison of natural PPREs. The number and location of nucleotides matching the consensus PPRE (CONS) are indicated (*stippled area*). (*b*) The hypothetical arrangement of PPAR and RXR DNA-binding domains on a PPRE is schematically depicted. Note the proposed polarity of the PPAR · RXR heterodimer and the putative interactions between PPAR D-box and RXR CTE (*arrow*). ACO, acyl-CoA oxidase; ACS, acyl-CoA synthetase; aP2, adipocyte fatty acid-binding protein; CTE, carboxy-terminal extension; CYP 4A6, cytochrome P450 ω-hydroxylase; HMG-CoA S, hydroxymethylglutaryl-CoA synthase; HD, hydratase/dehydrogenase; LPL, lipoprotein lipase; MCAD, medium-chain acyl-CoA dehydrogenase; ME, malic enzyme; PEPCK, phosphoenolpyruvate carboxykinase; Thiol, keto-acyl-CoA thiolase. (Zhang et al 1992, Kliewer et al 1992, Muerhoff et al 1992, Castelein et al 1994, Gulick et al 1994, Rodriguez et al 1994, Tontonoz et al 1994a, 1995a, Schoonjans et al 1995; K Schoonjans & J Auwerx, personal communication).

α-helices at locations consistent with the determined LBD folds. Interestingly, PPARs seem to have a putative α-helix inserted before H3 that might represent an extension of H3 (Figure 1*b*). A distinctive feature of the PPAR subfamily is that the average divergence in the LBD among the various PPAR subtypes is greater than that among the subtypes of the TR or RAR subfamilies. Similarly, for a given PPAR subtype in different species, the divergence is higher than the interspecies differences seen in the TR or RAR subtypes. This suggests that the E domain in PPARs has evolved more rapidly than in TRs and RARs (Dreyer et al 1993). Consequently, one might expect the three PPAR subtypes to have divergent ligand-binding specificities and thus distinct biological functions.

DNA-BINDING PROPERTIES OF PPARs

Nuclear hormone receptors bind to DNA by recognizing target sequences, or core recognition motifs, composed of six nucleotides (reviewed in Glass 1994). The motif recognized by PPARs and other members of the TR/RAR class has the consensus sequence AGGTCA. Because receptors belonging to this class bind DNA as dimers, in which both partners contact the DNA, two copies of the core motif are necessary to constitute a functional hormone response element (HRE). The relative orientation (e.g. direct repeat, palindromic, or inverted palindromic configurations) and spacing of the two motifs dictate which particular set of receptors bind to a given HRE (Näär et al 1991, Umesono et al 1991).

Like the other members of the TR/RAR class, PPARs heterodimerize with RXRs (Kliewer et al 1992, Gearing et al 1993, Keller et al 1993a). Indeed, PPARs strictly depend on RXRs as DNA-binding partners because they do not function as homodimers or monomers. The first PPAR response element (PPRE) characterized was found in the promoter of the acyl-CoA oxidase (ACO) gene and was defined as a direct repeat of two core recognition motifs spaced by one nucleotide (DR-1; Dreyer et al 1992, Tugwood et al 1992). Systematic in vitro binding studies, using direct repeat elements (DR) with various spacer (from 0 to 5 base pairs; DR-1 to DR-5), revealed that PPAR · RXR heterodimers bind preferentially to DR-1 elements (Kliewer et al 1992). This binding specificity allows PPARs to discriminate between PPREs (DR-1) and other direct repeat response elements such as the binding sites recognized by VDR (DR-3), TR (DR-4), and RAR (DR-5).

Additional binding determinants of natural PPREs have been identified by the detailed analyses of PPREs from the CYP4A6 gene (Palmer et al 1995) and the malic enzyme (ME) gene (AI Jpenberg & B Desvergne, unpublished results). First, mutations or deletions affecting the 5′ flanking nucleotides of the DR-1 of these PPREs result in a loss of PPAR · RXR heterodimer binding, showing that the actual PPAR-binding site extends 5′ of the DR-1. Second,

PPAR · RXR heterodimers are found to display a strong preference for an A:T base pair as spacer nucleotide in the malic enzyme PPRE. Accordingly, the comparison of the natural PPREs described thus far identifies the consensus sequence 5′-AACT AGGTCA A AGGTCA-3′ (Figure 2*a*).

DR-1 elements are also recognized by RAR · RXR heterodimers and by RXR, ARP-1, HNF-4, and COUP-TF homodimers (reviewed in Mangelsdorf & Evans 1995). Hence, competition for binding to a DR-1 element provides one mechanism for interactions between the signaling pathways mediated by this group of receptors. For example, HNF-4 and COUP-TF homodimers have been interpreted to compete with PPAR signaling by displacing PPAR · RXR from its binding site (Miyata et al 1993, Winrow et al 1994, Baes et al 1995). However, subtle sequence differences can also confer preference to a given receptor complex. Indeed, the core motifs of natural PPREs often exhibit considerable divergence from the idealized consensus AGGTCA sequence (Figure 2*a*). Because RXR or ARP-1 homodimers bind poorly to nonconsensus DR-1 elements, the combination of divergent core motifs with consensus 5′-flanking sequences results in preferential binding of PPAR · RXR to natural PPREs (Palmer et al 1995).

Receptors that can bind as monomers, such as NGFI-B, ROR, and revErbAα, also require an AT-rich, 5′-extended binding site (reviewed in Mangelsdorf & Evans 1995). Interaction of these receptors with the 5′-flanks is thought to involve the carboxy-terminal extension (CTE). In TRβ, which can also bind DNA as a monomer, this region consists of an α-helix that interacts with the minor groove of DNA (Rastinejad et al 1995). Thus it is possible that the CTE region of PPARs plays a similar role and is responsible for the recognition of the 5′ flanking sequences of the DR-1 in PPREs. This would imply that PPAR interacts with the upstream core recognition motif of the DR-1, whereas RXR would occupy the downstream motif (Figure 2*b*). This represents a reversed polarity compared with RXR · VDR and RXR · TR heterodimers, where RXR occupies the upstream position (Mangelsdorf & Evans 1995). X-ray crystallographic studies of the TR · RXR DNA-binding domain heterodimers bound to a DR-4 revealed that important TR-RXR contacts are made between the D-box of the upstream receptor (RXR) and the CTE region of the downstream receptor (TR; Rastinejad et al 1995). By analogy, if PPAR binds the upstream motif, it can be postulated that the D-box of PPAR makes contact with the CTE region of RXR (see *arrow* in Figure 2*b*). Thus it is tempting to speculate that the unique PPAR D-box, which comprises only three amino acids, is responsible for the reversed binding orientation of the PPAR · RXR heterodimer.

The same orientation has been proposed for a RAR · RXR heterodimer bound to a DR-1 element, which is transcriptionally silent, independently of the presence or absence of RAR and RXR ligands (Kurokawa et al 1994). In contrast, cotransfection studies have shown that the PPAR · RXR heterodimer is responsive to both peroxisome proliferators and 9-*cis* retinoic acid (Kliewer et al 1992, Keller et al 1993a). Thus allosteric interactions between PPAR and RXR LBDs must create a permissive configuration that allows 9-*cis* retinoic acid responsiveness of RXR.

Heterodimerization with RXR appears to be a cause for reciprocal negative interactions between the PPAR and TR signaling pathways. The mechanism of interaction, as studied by transfection and in vitro binding analyses, involves a competition for RXR between PPAR and TR, leading to the sequestration of RXR by the dominant receptor (i.e. either the receptor that is present at a higher concentration or the receptor that is activated by it cognate ligand) (Chu et al 1995, Juge-Aubry et al 1995). The relative amount of PPAR with respect to TRs and their relative affinities for RXR is believed to determine both the pathway that is inhibited and the extent of inhibition. A similar competition has been proposed to occur between PPAR subtypes. For example, high levels of PPARβ efficiently down-regulate PPARα-mediated transcription (Jow & Mukherjee 1995). In this case, competition for RXR, or another common co-factor, may be coupled to competition for PPRE binding. As a consequence, if several PPAR subtypes are co-expressed in a single cell type, only one of these PPARs might be functionally active. However, it is important to note that this phenomenon is likely to occur only when the amounts of RXR are limiting. Whether this is the case in any tissue in vivo is not known, but a recent report showing an in vivo inhibition by thyroid hormone of a ciprofibrate-induced transgene (Chu et al 1995) is consistent with this model.

DIFFERENTIAL ACTIVATION OF PPARs

As discussed above, a distinctive feature of the PPAR subfamily is the poor conservation of the ligand-binding domains between the three subtypes. It was therefore not surprising that these PPAR subtypes can be distinguished pharmacologically.

In this section, the aim is not to provide a comprehensive inventory of all PPAR activators. Instead, following a short methodological introduction, focus is placed on the differential activation properties of the PPAR α, β, and γ subtypes and their sensitivities to specific ligands.

Most studies on PPAR activation have involved transfection experiments. Obviously, in these experiments, the activation of the receptor may result from a cascade of events rather than a direct binding of the tested compound. Furthermore, the potency of a given compound is dependent on the rate at which it is taken up and metabolized by the cell type used. Finally, due to structural similarity, exogenously applied ligands probably interfere with the metabolic processing of endogenous ligands, possibly fatty acids and/or eicosanoids (see below). Thus the measured PPAR response presumably results from the combined actions of direct receptor binding of the exogenous ligand and the indirect effects owing to the perturbation of endogenous ligand levels.

Different assays measuring PPAR activation consist of transfecting either the full-length receptor or chimeric receptors, in which the ligand-binding domain of PPARs has been fused to the DNA-binding domain of a steroid receptor (ER or GR; Dreyer et al 1992, Göttlicher et al 1992), the yeast GAL4 protein (Lehmann et al 1995), or the bacterial tetracycline repressor (tetR; Yu et al 1995). The activity of the chimeric receptor can in this way be assessed using reporter constructs containing the proper DNA-binding sites (PPRE, GRE, ERE, GAL4-binding site, tet operator, respectively). The chimeric receptor system abolishes possible biases arising from variable DNA-binding specificities of PPAR subtypes to different PPREs and from endogenous levels of PPARs.

The peroxisome proliferator and hypolipidemic agent WY-14,643 (WY) was among the first identified PPAR activator, and it is now experimentally widely used. Comparative studies involving tetR/PPAR, GAL4/PPAR, and PPAR full-length receptors have shown that at low concentrations (10 μM), WY specifically activates PPARα (Kliewer et al 1994, Lehmann et al 1995, Yu et al 1995). However, higher concentrations (100 μM) lead to the activation of all three PPAR subtypes. Two other peroxisome proliferators, the hypolipidemic agent clofibrate and the leukotriene D_4 antagonist LY 171883, have also been reported to preferentially activate PPARα, suggesting that the PPARα subtype is the main target of peroxisome proliferators (Kliewer et al 1994, Yu et al 1995). PPARα is consistently and predominantly expressed in liver (see below), which is the major tissue showing peroxisome proliferation in rodents. Indeed, conclusive proof for the essential role of PPARα in peroxisome proliferation has been given by the targeted disruption of the PPARα gene in mice (Lee et al 1995). These mice no longer display hepatic peroxisome proliferation upon treatment by clofibrate or WY-14,643.

That PPARs are pharmacologically distinguishable has been further illustrated by the recent identification of highly specific PPARγ activators among the class of antidiabetic drugs termed thiazolidinediones (TZDs; Lehmann et al 1995, Willson et al 1996). All TZDs tested so far, including pioglitazone, englitazone, ciglitazone, and BRL 49653, preferentially activate the PPARγ subtype when assessed using either full-length or GAL4/PPAR receptors. The half-maximal concentration for PPARγ activation of the most potent of these compounds is several orders of magnitude lower than that of peroxisome proliferators ($EC_{50} = 100$ nM for PPARγ2 activation by BRL 49653). Also, TZDs activate PPARγ at much lower concentrations that the Wyeth compound activates PPARα. Activation of PPARβ by BRL 49653 has also been reported but only with a 100-fold higher EC_{50} (10 mM; Ibrahimi et al 1994, Forman et al 1995). Ligand-binding experiments show that TZDs bind directly to PPARγ, although some discrepancies exist with regard to the measured affinities (for BRL 49653, $K_d = 43$ nM, Lehmann et al 1995; $K_d = 325$ nM, Forman et al 1995). Binding affinities of TZD to PPARα and β subtypes have not yet been determined. However, data from transactivation studies strongly suggest that TZDs are specific PPARγ agonists.

Great interest has arisen from the finding that PPARs are activated by fatty acids. So far, however, only a few studies have reported differential activation of PPARs by fatty acids. PPARα, the most extensively studied subtype, is activated by a broad array of medium- to long-chain unsaturated fatty acids (C10–C22). Unmetabolized fatty acids, such as 3-thia fatty acids derivatives or 2-bromopalmitate, are generally more efficient than the parent molecules (Göttlicher et al 1992, 1993, Keller et al 1993a). Two of the best activating fatty acids, linoleic acid (LA) and docosahexaenoic acid (DHA), have been tested in comparative studies (Yu et al 1995). The chimeric receptor tetR/PPARα is activated equally well by LA and DHA. The chimera tetR/PPARβ also responds well to DHA and less well to LA, whereas tetR/PPARγ is activated only by DHA. Similar observations were made with full-length *Xenopus* PPARα, β, and γ (Krey et al 1993). Thus all three PPAR subtypes are able to significantly respond to fatty acids with different yet overlapping specificities. Fatty acids like DHA may be used as a general activator of PPARs.

The fact that arachidonic acid can activate PPARα suggests that some of its eicosanoid metabolites are responsible for this effect. Thus, in early studies, several inhibitors of arachidonic acid metabolism, including cyclooxygenase (aspirin, indomethacin), lipoxygenase (nordihydroguarietic acid, NDGA; eicosatetraynoic acid, ETYA) and epoxygenase (metyrapone) blockers have been tested. ETYA was found, surprisingly, to be a potent activator of PPARα (Keller et al 1993a,b). In contrast, indomethacin, NDGA, and metyrapone have no effect on PPARα activation by arachidonic acid (Göttlicher et al 1993, Keller et al 1993a,b) or by DHA (Yu et al 1995), suggesting that the activation of PPARα by fatty acids does not necessarily require the production of eicosanoids metabolites. Alternatively, these compounds might have some activation properties by themselves. More recently, PGs from the A, D, and J series (Figure 3) were shown to activate the three PPAR subtypes when applied at relatively high concentrations (10 μM; Yu et al 1995). The most responsive subtype appears to be PPARγ, for which it was further shown that PGJ_2 metabolites, Δ^{12}-PGJ_2 and 15-deoxy-$\Delta^{12,14}$-PGJ_2 (15d-PGJ_2), are even more potent activators than the parent molecule (Forman et al 1995, Kliewer et al 1995). In vitro ligand-binding assays have shown that PGJ_2 and its metabolites compete for the

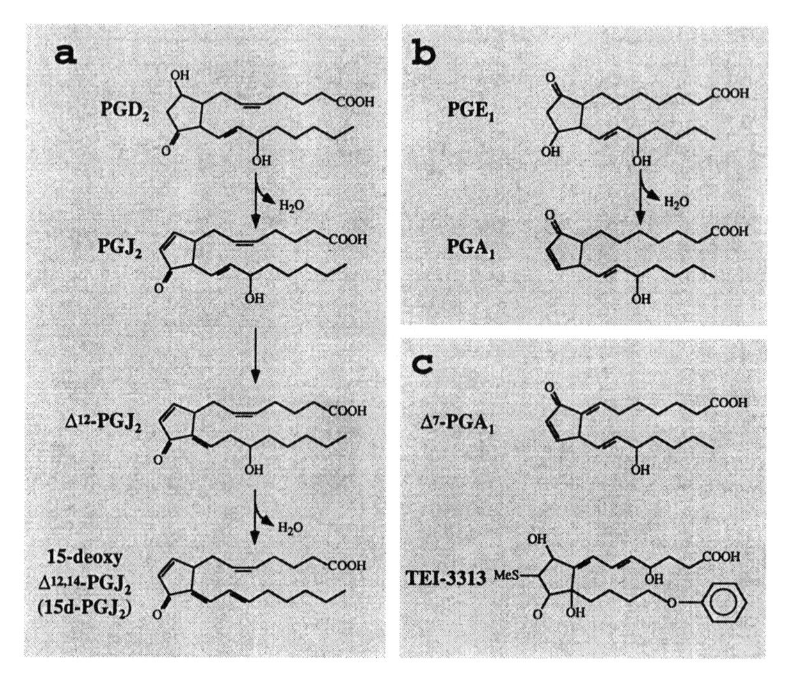

Figure 3 Structure of cyclopentenone and related prostaglandins. (*a*) PGD_2 and (*b*) PGE_1 metabolic pathways. (*c*) Δ^7-PGA_1 and TEI-3313 are Δ^{12}-PGJ_2 and PGA_1 analogues, respectively.

binding of TZDs to PPARγ, which demonstrates a direct interaction with this receptor. The 15d-PGJ_2 metabolite, which is also the most potent when tested in transactivation assays, has the highest affinity for PPARγ ($K_i = 2.5\ \mu M$ in competition assays with BRL 49653). In this direct ligand-binding assay, no binding of PGD_2 was detected, suggesting that PGD_2 activates PPARγ through one or more of its metabolites. The most probable candidates are PGJs, which arise from the non-enzymatic dehydration of PGD_2 (see below and Figure 3*a*).

Because PGJ_2 metabolites bind directly to PPARγ and because the three PPAR subtypes are activated by the same spectrum of structurally and functionally related PGs (Yu et al 1995), it is possible that some other PGs from the A, D, and J series will bind to PPARα and PPARβ as well. Other eicosanoids may also be PPAR ligands, as suggested by the specific activation of tetR/PPARα by 8-hydroxyeicosatetraenoic acid, a product of the lipoxygenase pathway (Yu et al 1995). However, it is also possible that PGs may represent a particular but not exclusive class of fatty acids that binds to PPARs. Indeed, the EC_{50}s of fatty acids and of PGs are of the same order of magnitude (about 5 μM), which is consistent with a similar mechanism of action. In this context, it would be interesting to assess whether fatty acids are able to compete PGJ_2 metabolites or TZDs for PPARγ binding.

DIFFERENTIAL EXPRESSION OF PPARs

The discovery of several subtypes of PPARs has raised the question of the biological significance of PPAR diversity. One step to address this question consists of analyzing the patterns of expression of the different PPAR subtypes, which so far has been done in *Xenopus*, rodents (rat and mouse), and to a more limited extent, humans.

In *Xenopus*, PPARα and PPARβ display a similar ubiquitous pattern of expression (Dreyer et al 1992, 1993). In contrast, PPARγ expression is more restricted and mostly prevalent in fat body and, to some extent, in mesonephros. Oocytes express high levels of PPARβ and smaller amounts of PPARα. No PPARγ transcripts of the regular size (2.2 kb) are detected in oocytes. However, a shorter transcript of unknown significance (1 kb) is expressed. During early development, zygotic transcripts of PPARβ appear earlier (neurula stage) than those of PPARα (tail bud stage), whereas PPARγ is not detectable.

The tissue distribution of PPARs in mammals has been extensively investigated in rats and mice (Kliewer et al 1994, Braissant et al 1996). Similar to the situation in *Xenopus*, PPARβ is expressed in most of the tissues analyzed; however, some exceptions are notable. First, PPARβ is only weakly expressed in liver as compared with other tissues such as kidney or lung. Second, although PPARβ is abundantly expressed in skeletal and cardiac muscle, it is below detection levels in smooth muscle cells (Braissant et al 1996). Finally, in several cell lines cultured in vitro (C_2C_{12} myoblastic cell line; ob 1771, 3T3-L1, and 3T3-F442A adipose cell lines), the level of PPARβ expression appears to depend on the growth or differentiation state of the cells (Amri et al 1995, Teboul et al 1995).

In rodents, PPARα has a more restricted pattern of expression than in *Xenopus*. Tissues that express high levels of PPARα include liver, kidney, heart, and mucosa of stomach and duodenum. Interestingly, PPARα is heavily expressed in brown adipose tissue, with three- to fourfold higher mRNA levels than in liver (Lemberger et al 1996a). Other organs expressing significant levels of PPARα include retina, adrenal gland, and skeletal muscle. Importantly, hepatic expression of PPARα is subject to positive and negative regulation by glucocorticoids and insulin, respectively (Lemberger et al 1994, Steineger et al 1994). Accordingly, PPARα expression in liver is increased by stress or fasting, and its level cycles according to the circadian rhythm of circulating glucocorticoids (Lemberger et al 1996b).

Similar to the situation first described in *Xenopus,* PPARγ has a restricted pattern of expression in rodents. Very high levels of PPARγ mRNA have been reported so far in both white and brown adipose tissue (Tontonoz et al 1994a; F Foufelle & W Wahli, unpublished results). Thus brown adipocytes represent a unique example of a cell type that co-expresses high levels of both PPARα and PPARγ. Although adipose tissue is indeed a major site of PPARγ expression, it is important to realize that several non-adipose tissues, such as spleen, the mucosa of duodenum, or the retina, also express PPARγ, albeit at lower levels (Braissant et al 1996). In spleen, PPARγ is expressed in both red and white pulp. The latter represents lymphocyte proliferating centers. Strikingly, PPARγ is also abundant in other lymphoid nodes such as the Peyer's patches in the digestive tract. In human, the expression pattern of PPARγ may differ to some extent as compared with that of rodents (Greene et al 1995). For example, PPARγ is detected neither in spleen nor in circulating T-lymphocytes, although it is expressed in several transformed human B-lymphocyte and myeloid cell lines. Intriguingly, similar to the *Xenopus* oocytes, circulating lymphocytes or polymorphonuclear cells express a short PPARγ messenger (0.65 kb) of unknown function.

PPARα AND LIVER FATTY ACID METABOLISM

Liver is a pivotal organ in determining the metabolic fate of fatty acids, directing fatty acids either to esterification into triglycerides or to oxidation. On one hand, the esterification pathway leads to secretion into the blood stream of the excess of triglycerides as VLDL, which are mainly taken up by adipose tissue for triglyceride storage. On the other hand, when plasma fatty acid levels are high, the oxidative pathway leads to the production and secretion of ketone bodies that serve as fuel for brain, muscle, and kidney. Thus liver is able to regulate the levels of the three circulating fat fuels, e.g. non-esterified fatty acids, triglycerides, and ketone bodies, by modulating the relative rates of fatty acid uptake, esterification into triglycerides, and oxidation, respectively. Based on the various PPAR target genes identified so far, there is accumulating evidence that PPARα plays an important role in fatty acid oxidation in liver (Figure 4*a*).

Fatty acids taken up by the liver originate mainly from non-esterified fatty acids carried by serum albumin and from the hydrolysis of triglycerides conveyed in chylomicrons. Triglycerides are hydrolyzed extracellularly into glycerol and fatty acids by the lipoprotein lipase (LPL), whose gene is stimulated by PPARs (K Schoonjans & J Auwerx, personal communication). Once imported into hepatocytes, fatty acids are activated into acyl-CoA thioesters by various acyl-CoA synthetases (ACS), one of which, the long-chain fatty acid ACS, is regulated by PPARα at the transcriptional level (Schoonjans et al 1995).

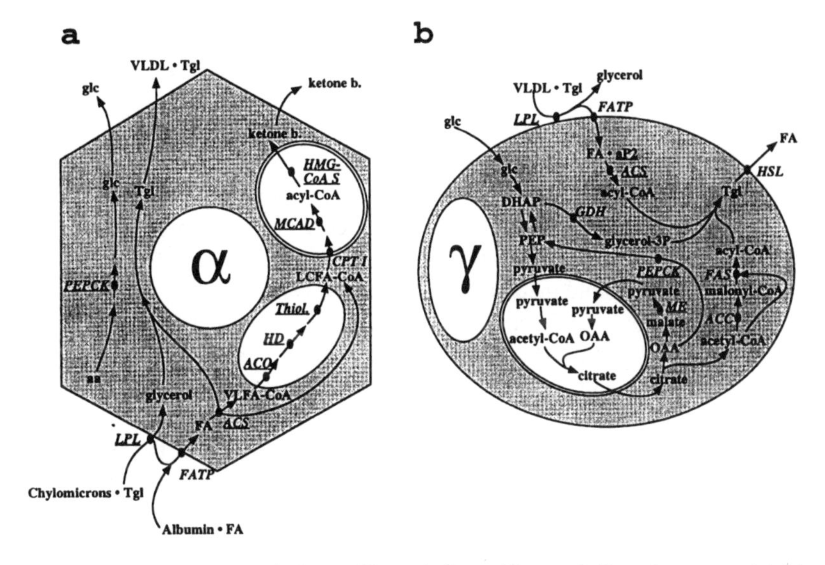

Figure 4 PPAR target genes in fatty acid metabolism. The metabolic pathways regulated by PPARα in hepatocytes (*a*) and PPARγ in adipocytes (*b*) are summarized. The α and γ labels inside the nuclei indicate the fact that hepatocytes express predominantly PPARα, whereas PPARγ is the major subtype in adipocytes. Genes so far described as direct PPAR targets are underlined. For the sake of clarity, the lipogenic pathway, the apolipoprotein, and the cytochrome P450 CYP target genes have been omitted in hepatocytes, as well as the oxidative pathway in adipocytes. ACC, acetyl-CoA carboxylase; aa, amino acids; CPT I, carnitine palmitoyl transferase I; DHAP, dihydroxyacetone phosphate; FA, fatty acid; FAS, fatty acid synthase; GDH, glycerol-phosphate dehydrogenase; glc, glucose; HSL, hormono-sensitive lipase; ketone b, ketone bodies; LCFA, long-chain fatty acid; OAA, oxaloacetate; PEP, phosphoenol pyruvate; Tgl, triglyceride; VLDL, very-low-density lipoprotein; VLFA, very long-chain fatty acid. In addition to the cell nuclei (α, γ), white areas symbolize the peroxisomal (*single line*) and mitochondrial (*double line*) compartments.

Acyl-CoA may then either be esterified or, alternatively, enter the pathway of β-oxidation. In liver, very long-chain fatty acyl-CoA ($> C_{20}$) is β-oxidized in peroxisomes. All three enzymes involved in the peroxisomal β-oxidation pathway, e.g. acyl-CoA oxidase (ACO), enoyl-CoA hydratase/dehydrogenase bifunctional enzyme (HD), and keto-acyl-CoA thiolase, are direct PPARα target genes (Dreyer et al 1992, Tugwood et al 1992, Zhang et al 1992, 1993, Marcus et al 1993, Lee et al 1995). Long- (C14 to C20) and medium-chain (C8 to C12) fatty acids can be oxidized either in peroxisomes or in mitochondria. The flux of fatty acids into mitochondria is controlled by a carnitine-dependent facilitated transport system. Interestingly, one of its critical components, carnitine palmitoyl transferase I (CPT I), is strongly induced by peroxisome proliferators and fatty acids (Brady et al 1989, Foxworthy et al 1990). Thus the gene encoding

CPT I probably represents another PPARα target gene. PPARα further directly regulates the mitochondrial β-oxidative spiral by stimulating the expression of the medium-chain acyl-CoA dehydrogenase (MCAD) gene, the promoter of which contains a functional PPRE (Gulick et al 1994). Finally, in liver, most of the acetyl-CoA derived from fatty acid β-oxidation is converted into ketone bodies (mainly acetoacetate and 3-hydroxy butyrate). The rate-limiting enzyme of this last pathway, the hydroxymethylglutaryl-CoA synthetase, is controlled by PPARα (Rodriguez et al 1994). Additional PPARα target genes involved in hepatic lipid metabolism include the gene encoding apolipoprotein A-II (Vu-Dac et al 1995), a major constituent of high-density lipoproteins, and the CYP4A1 and CYP4A6 genes that code for cytochrome P450 enzymes that ω-hydroxylate fatty acids (mainly lauric, myristic, and arachidonic acids) in microsomes (Muerhoff et al 1992, Aldridge et al 1995). In contrast to the above-mentioned target genes that are all stimulated by PPARs, the apolipoprotein A-I and C-III genes have been reported to be down-regulated by PPARs (Vu-Dac et al 1994, Hertz et al 1995). The mechanisms responsible for such inhibitions remain unclear.

The extent to which peroxisomal β-oxidation contributes to energy production is uncertain (Mannaerts & Van Veldhoven 1993, Masters & Crane 1995). Although modest in normal physiological conditions, as compared with mitochondrial β-oxidation, this contribution might be significantly increased when peroxisomal β-oxidation activity is stimulated by PPARα. An alternative function of this stimulation could be to provide a negative feedback on the PPARα signaling pathway by degrading PPARα activators such as long-chain fatty acids or prostaglandins. Peroxisomal β-oxidation would then be also implicated in the modulation of the signal transduction mechanism in addition to its metabolic function. To understand the roles of PPARα in energy homeostasis, it is necessary to determine to what extent PPARα regulates mitochondrial β-oxidation. In addition to the fact that the MCAD and the hydroxymethylglutaryl-CoA synthase are PPAR target genes, several lines of evidences point to a significant impact of PPARα on this pathway. First, several peroxisome proliferators, including clofibrate and LY 171883, induce the rate of hepatic mitochondrial β-oxidation and ketogenesis (Mannaerts et al 1979, Foxworthy et al 1990). In the liver, the stimulation of the oxidative capacity by LY 171883 is even sufficient to reverse the accumulation of newly synthesized fatty acids caused by a high-carbohydrate diet (Foxworthy & Eacho 1991). Second, PPARα mutant mice generated by the targeted disruption the PPARα gene show a progressive accumulation of lipid droplets in the liver (FJ Gonzalez, personal communication). This is indicative of an imbalance between the esterification and oxidative pathways, which may well be the result of lack of regulation of β-oxidation and ketogenesis.

What are the natural physiological conditions in which PPARα would stimulate fatty acid catabolism? At low plasma levels of fatty acids, the esterification pathway competes effectively with that of oxidation (Newsholme & Start 1979). In contrast, fasting or stress situations stimulate adipose lipolysis, which leads to an increase of the plasma levels of non-esterified fatty acids. In turn, the rates of fatty acid uptake, fatty acid β-oxidation, and ketogenesis are strongly activated in liver. Thus fasting and stress represent physiological situations in which PPARα is likely to be activated by fatty acids and to stimulate the oxidative pathway. Fasting is also associated with an increased rate of gluconeogenesis. Notably, the gene encoding PEPCK, a key regulatory enzyme of this latter pathway in liver, contains a functional PPRE (Tontonoz et al 1995a) and is thus expected to be stimulated by PPARα in liver.

The levels of PPARα mRNA and proteins are stimulated in liver by glucocorticoids, resulting in a potentiation of the efficiency of the PPARα signaling pathway (Lemberger et al 1994, 1996b). In contrast, insulin counteracts the stimulatory action of glucocorticoids (Steineger et al 1994). Accordingly, PPARα expression is stimulated during stress situations, mainly by elevated circulating glucocorticoid levels, and after fasting, probably by means of both increased glucocorticoid and decreased insulin levels (Lemberger et al 1996a). Hence, the regulatory action of PPARα appears to be reinforced precisely according to physiological conditions that involve fatty acid mobilization and, possibly, PPARα receptor activation. The importance of the PPARα contribution to the regulation of fatty acid oxidation in fasting can be judged from PPARα knock-out mice. In these animals, an overnight fast leads to the accumulation of lipid droplets in the liver (FJ Gonzalez, personal communication), consistent with an impaired overall β-oxidative capacity. This suggests that PPARα plays an important role in regulating the catabolism of fatty acids mobilized during energy-demanding physiological situations such as fasting or stress. Importantly, its regulatory action is distributed over the entire oxidative pathway, including uptake, activation, and peroxisomal and mitochondrial β-oxidation of fatty acids, as well as ketogenesis. The modulation of fatty acid oxidation by PPARα in liver represents an example of a distributive control (Srere 1994) in which fine tuning is exerted at several points of the metabolic sequence rather than tight control of a single committed step.

PPARγ AND ADIPOGENESIS

PPARγ is expressed predominantly in adipose tissue. Several adipose target genes have been described so far, including the genes encoding lipoprotein lipase (LPL), fatty acid transporter protein (FATP), adipocyte fatty acid-binding protein aP2, acyl-CoA synthetase (ACS), phosphoenolpyruvate carboxy kinase

(PEPCK), and malic enzyme (ME) (K Schoonjans & J Auwerx, personal communication; Castelein et al 1994, Tontonoz et al 1994a, 1995a, Schoonjans et al 1995,). This array of PPAR target genes suggests that PPARγ is involved in the stimulation of critical steps of lipid accumulation (Figure 4*b*) such as fatty acid uptake (LPL, FATP, aP2), NADPH synthesis for lipogenesis (ME), glyceroneogenesis (PEPCK), and fatty acid esterification (ACS).

In contrast to fully differentiated adipocytes, preadipocytes express only trace amounts of PPARγ. The appearance of PPARγ in several adipose cell lines, including 3T3-L1 and 3T3-F442A, precedes that of several markers of the adipose differentiated state, such as aP2, PEPCK and C/EBPα (Tontonoz et al 1994a,b, 1995b). These observations, together with the fact that several PPAR activators, e.g. WY-14,643, ETYA, and TZDs (Kletzien et al 1992, Sandouk et al 1993, Chawla et al 1994, Lehmann et al 1995), are able to promote the conversion of preadipocytes to adipocytes, suggest that PPARγ might be involved in the adipose differentiation process. This hypothesis was verified by forced expression of PPARγ in NIH-3T3 fibroblasts that, after treatment with several activators (ETYA, clofibrate, LY 171883 and LA), underwent adipose differentiation and accumulated lipids (Tontonoz et al 1994b). The phenotypical conversion to lipid-accumulating cells is accompanied by the induction of several specific adipose markers such as aP2, adipsin, and PPARγ itself. A similar adipogenic effect of PPARγ was observed in several fibroblast cell lines (NIH-3T3, Balb/c-3T3, Swiss-3T3). Thus ectopic expression of PPARγ in mesodermally derived cell lines is sufficient to direct undetermined cells toward the adipocyte lineage.

The C/EBP family of transcription factors, comprising the α, β, and γ subtypes, is involved in the adipocyte differentiation program as well (McDougald & Lane 1995). Upon stimulation of 3T3-L1 cells to differentiate by an appropriate external hormonal cocktail (usually fetal bovine serum, insulin, dexamethasone, and isobutyl methyl xanthine), the levels of C/EBPβ and δ increase transiently, with a rapid and slow decline of expression, respectively (Cao et al 1991). The onset of C/EBPα expression follows the C/EBPβ and δ peak of expression and occurs during the late phase of adipose differentiation. Because PPARγ expression is stimulated by the forced expression of C/EBPβ in 3T3 cells (Wu et al 1995), and because its levels increase only shortly before that of C/EBPα in 3T3-F422A cells (Tontonoz et al 1994b), the appearance of PPARγ seems to intercalate between the C/EBPβ/δ and C/EBPα waves. Besides PPARγ, PPARβ is also expressed at significant but lower levels in adipocytes. In the ob1771 cell line, its expression is turned on shortly before the cells reach confluence (Amri et al 1995).

Adipocyte differentiation seems, therefore, to involve a sequence of gene activation with the successive stimulation of PPARβ, C/EBPδ, C/EBPβ, PPARγ, and C/EBPα expression. Even if the order of appearance of these regulatory factors is still provisional, such an expression pattern suggests that several of these genes participate in an ordered activation cascade in which each factor triggers a particular phase of the differentiation program. However, several studies have shown that forced expression of either PPARγ, C/EBPα, β, or δ variously triggers the differentiation of uncommitted fibroblasts into adipocytes (Freytag et al 1994, Tontonoz et al 1994b, Yeh et al 1995). Hence, even factors that are only expressed late in the temporal sequence of gene activation, such as C/EBPα and PPARγ, activate the entire differentiation process. This may imply that late markers exert a positive feedback action on upstream adipogenic factors, ensuring the autocatalytic nature of the differentiation process. Alternatively, early forced expression of late factors might simply mimic the action of early subtypes of these same factors. Whatever the case, adipogenic factors are probably interacting with each other in a complex fashion to build an intricate regulatory network that modulates the expression of the battery of target genes involved in the adipocyte phenotype.

Adipocytes originate from mesodermal pluripotent cells. Depending on external stimuli, these progenitor cells differentiate into either myocytes, chondrocytes, or adipocytes. Although these differentiation pathways are mutually exclusive, some degree of plasticity remains. For example, the C_2C_{12} myoblastic cell line can be directed toward the osteoblastic lineage when exposed to bone morphogenetic protein-2 (Yamaguchi et al 1991, Katagiri et al 1994). Similarly, myoblasts can differentiate into adipocytes upon treatment with PPAR activators (Hu et al 1995, Teboul et al 1995). Indeed, the forced and simultaneous expression of PPARγ and C/EBPα in G8 myoblastic cells inhibits myotube formation and represses the expression of several muscle-specific markers. Further treatment of these transfected G8 cells by a permissive hormonal cocktail (15% fetal calf serum, dexamethasone, insulin, and ETYA) leads them to differentiate into adipocyte-like cells that express typical adipocyte markers (LPL, aP2, adipsin, and PEPCK). Intriguingly, nontransfected C_2C_{12} myoblasts also differentiate into adipocytes when exposed to low concentrations (100 nM) of the potent PPARγ agonist BRL 49653 two days before confluence, a stage at which PPARβ is expressed but PPARγ and C/EBPα are not detected (Teboul et al 1995). In this case, it is not clear whether BRL 49653 acts through very low PPARγ levels or whether it activates PPARβ indirectly by perturbing the metabolism of endogenous PPARβ ligands. It will be interesting to test whether the transdifferentiating effects of PPARγ and TZDs can be observed in other cell types. One particularly interesting issue could be the differentiation of monocytes into lipid-accumulating cells, which is one of the key steps in the development of atherosclerosis.

TRENDS AND MODELS

PPARα and PPARγ in Lipid Homeostasis

PPARs were already proposed to play an important role in energy homeostasis early after the identification of ACO as a target gene and fatty acids as activators. As outlined in this review, this hypothesis has been largely confirmed. Although the function of PPARβ remains obscure, a more precise picture emerges on the metabolic roles of PPARα and γ. Indeed, PPARα and γ appear to regulate the two arms of lipid homeostasis, fatty acid oxidation, and triglyceride accumulation, respectively. Specifically, PPARα stimulates fatty acid catabolism in several tissues known for their high rates of fatty acid oxidation such as liver, heart, kidney, and brown adipose tissue, whereas PPARγ favors fatty acid storage by stimulating triglyceride accumulation in adipocytes. The differential expression of both PPAR subtypes would largely explain their differential physiological actions. When co-expressed, as is the case in brown adipose tissue, the two subtypes could be expected to stimulate preferentially a different subset of target genes involved in the oxidative and lipogenic pathways, respectively. When specific activators of PPAR subtypes become available, it should be possible to test whether PPARα and PPARγ differentially regulate the two opposite metabolic pathways in a single cell type. In liver, the activity of the anabolic and catabolic branches of fatty acid metabolism are generally mutually exclusive. Because liver expresses predominantly PPARα, it should be interesting to test whether PPARα is able to preferentially activate target genes involved in fatty acid oxidation rather than in the lipogenic or esterification pathways.

Insulin Resistance

Energy homeostasis is achieved by the action of a complex hormonal network that regulates the intake, storage, mobilization, and expenditure of the various fuel sources necessary to the organism. Coordination and cross talk among the multiple components of this network are critical to ensure correct energy balance. Understanding of the physiological roles of PPARs requires an analysis of their interaction with the other important hormonal systems involved in energy homeostasis. The regulation of PPARα expression by glucocorticoids and insulin provides the first example of such cross talk. Insulin is the major regulator of glucose levels in blood. It stimulates glucose uptake and metabolism in insulin-sensitive tissues such as liver, adipose tissue, and muscle, and inhibits hepatic gluconeogenesis. A link between PPARs and the insulin-signaling pathway has been suggested by the recent finding that the TZD class of antidiabetic agents are high-affinity ligands of PPARγ. In contrast to other hypoglycemic drugs such as sulfonylurea, TZDs do not enhance insulin secretion, but rather improve the insulin sensitivity of several peripheral target tissues, including adipose tissue, liver, heart, and muscle (Hofmann & Colca 1992). The fact that TZDs are PPARγ ligands suggests that these effects could be mediated by this receptor and indicates that adipose tissue represents one of the major direct targets of TZD action. To explain the action TZDs exert in vivo on non-adipose tissues, one has to postulate that TZDs regulate, possibly through PPARγ activation, the secretion by the adipocyte of some messenger molecules that would modulate insulin sensitivity of peripheral organs. The identity of such messengers is still elusive, but several molecules are good candidates for adipose endocrine modulators of insulin action. For example, the cytokine TNFα, whose expression in adipose tissue is markedly increased in several animal models of obesity and diabetes, has been proposed to induce insulin resistance (reviewed in Spiegelman & Hotamisligil 1993, Tracey & Cerami 1993). Remarkably, treatment of diabetic KKA^y mice with the TZD pioglitazone partially corrects the increased levels of TNFα expression in adipose tissue (Hofmann et al 1994). Thus it is possible that the activation of PPARγ by TZDs directly or indirectly down-regulates TNFα expression in adipose tissue, thereby attenuating TNF-induced insulin resistance. Another major signaling molecule secreted by adipocytes has been recently identified as leptin, the product of the *ob* gene (McGarry 1995). Whether PPARγ regulates *ob* gene expression is presently unknown.

Beside accumulating triglycerides, one crucial physiological function of adipose tissue is to release fatty acids into the blood stream in a highly controlled manner. Nonesterified fatty acids can thus be envisioned as messenger molecules between adipose and other tissues. In vivo, TZDs have strong hypolipidemic effects—decreasing plasma levels of nonesterified fatty acids, glycerol, and ketone bodies—consistent with their antilipolytic property (Oakes et al 1994). Interestingly, free fatty acids are known to cause insulin resistance in muscle and liver and to increase hepatic gluconeogenesis (Björntorp 1991, McGarry 1992). A decrease in free fatty acids levels could thus account for the insulin-sensitizing action of TZDs on non-adipose tissues. One exciting hypothesis is that the induction of insulin resistance by fatty acids is mediated by PPARα, which is well expressed in insulin-sensitive tissues such as liver or muscle. Hence, while insulin down-regulates PPARα/fatty acid signaling (Steineger et al 1994), activation of PPARα by fatty acids would reciprocally impair insulin-dependent glucose uptake and metabolism. Such an antagonism between PPARα and insulin would be consistent with Randle's glucose/fatty acid theory stating a reciprocal relation between glucose and fatty acid metabolism (Newsholme & Start 1979, Randle et al 1988). Although these speculative proposals allow an integrated view of PPARα and γ functions in fatty acid and glucose metabolism, several questions remain. Why do TZDs have insulin-sensitizing effects in hepatocytes or muscle cells cultured in vitro (Blackmore et al 1993, el-Kebbi et al 1994)? Are very low levels of PPARγ able to mediate these effects? Will TZDs induce obesity by stimulating adipose differentiation and triglyceride accumulation?

Prostaglandins A, D, and J

As noted throughout this review, PPAR research has concentrated on the metabolic roles of PPARs. For example, all PPAR target genes identified thus far encode proteins involved in fuel metabolism. Whether this situation reflects the dominant role of PPARs in vivo or results from a biased focus of research is still hard to tell. Curiously, the first natural compound identified as a PPAR ligand, 15d-PGJ_2, belongs to a family of PGs including the A and J series, for which metabolic effects have not been reported. This may open the possibility for more diversified roles of PPARs or, alternatively, for unforeseen metabolic functions of this class of PGs.

Prostaglandins are arachidonic acid metabolites produced via the cyclooxygenase pathway (Goetzl et al 1995). Prostaglandin J_2 and PGAs arise from the non-enzymatic dehydration in aqueous solutions of PGD_2 and PGEs, respectively (reviewed in Fukushima 1990, 1992, Noyori & Suzuki 1993; Figure 3*a,b*). Albumin accelerates these conversions and catalyzes the subsequent isomerization of PGAs and PGJ_2 to PGBs and Δ^{12}-PGJ_2, respectively. In turn, Δ^{12}-PGJ_2 is converted to 15d-PGJ_2. A common structural characteristic of PGs from the A and J series is the cyclopentenone ring with an α, β unsaturated carbonyl that can be readily attacked by cellular nucleophilics such as glutathione and the thiol moieties of cysteines. In contrast to classical PGs such as PGEs or $PGF_{2\alpha}$, cyclopentenone PGs are thought to have no membrane receptors. Rather, they are actively taken up by cells and accumulate in the nucleus, which is consistent with their binding to a nuclear receptor. Intriguingly, Δ^{12}-PGJ_2 has been shown to remain covalently bound to nuclear proteins (Narumiya et al 1987).

Very little is known about the endogenous production and distribution of cyclopentenone PGs. In particular, it is not known whether specific enzymes catalyze the formation of this class of PGs. Δ^{12}-PGJ_2 levels in urine increase after an injection of PGD_2 and decrease after indomethacin administration, indicating that these PGs are produced in vivo (Hirata et al 1988).

Both A and J series of PGs display strong antiproliferative activities on a broad array of cell lines, including melanoma, retinoblastoma, glioma, neuroblastoma, ovarian cancer, leukemia, and HeLa S3 cell lines (reviewed in

Fukushima 1990). Much interest has therefore arisen favoring the use of cyclopentenone PGs and analogues as anti-cancer drugs. Analysis of the structure-activity relationship of PGs suggests that the chemical reactivity of anti-tumor PGs determines much of their biological activity. Strikingly, the most potent PGJs with respect to their antiproliferative action also bind with the highest affinities to PPARγ. As expected for PPARγ ligands, 15d-PGJ_2 efficiently triggers the differentiation of C3H10$T_{1/2}$ fibroblasts or PPARγ transfected NIH-3T3 fibroblasts into adipocytes (Forman et al 1995, Kliewer et al 1995). It is not known whether induction of adipose differentiation is linked to the antiproliferative property of this PG. Furthermore, it is possible that the ubiquitous expression of PPARβ, which is activated to some extent by cyclopentenone PGs (Forman et al 1995, Yu et al 1995), may account for the variety of cell lines that are susceptible to anti-tumor PGs action. However, there is no proof so far that this anti-tumor action is mediated by PPARs.

A PGA_1 synthetic analogue, TEI-3313 (Figure 3*c*), has been observed to markedly increase osteoblast mineralization in vitro and bone formation in vivo (Koshihara et al 1991, Ohta et al 1995). Therefore, stimulation of osteogenesis could be another biological function of cyclopentenone PGs. Interestingly, human PPARβ (or NUCI), which has been cloned from an osteoblastic cell line, is expressed in bone and is thought to be involved in osteogenesis by activating osteocalcin and bone sialoprotein promoters (Schmidt et al 1992, 1994). In addition, the promoter of the bone morphogenetic protein-4 gene contains several DR-1 elements (Feng et al 1995). Because PPARβ responds to PGA1 (Yu et al 1995), it is possible that some of the TEI-3313 effects on bone formation are mediated by PPARβ.

CONCLUSIONS

From their differential pharmacological profiles and discrete patterns of expression, it is clear that the three PPAR subtypes play distinct roles. Although the ultimate biological function of PPARβ remains enigmatic, one emerging picture is that of a dual role for PPARα and γ in the regulation of the catabolic (PPARα) and anabolic (PPARγ) aspects of lipid metabolism, respectively. Moreover, PPARγ is a potent stimulator of adipocyte differentiation. Because PPARs act as lipid-activated transcription factors, the new paradigm implies that lipids are not merely substrates or products of the metabolic machinery, but are also signaling molecules capable of directly regulating gene expression through nuclear receptors. Hence, we propose that the PPAR/fatty acid signaling transduction mechanism plays a fundamental role in the metabolic cross talk and coordination between organs and tissues involved in energy homeostasis.

It also appears that the regulatory pathways controlling the metabolic processing of the various fuel sources, such as glucose or fatty acids, are highly interdependent. Thus pharmacological manipulation of PPAR signaling has profound effects on overall energy balance, as illustrated by the hypoglycemic effects of PPARγ agonists such as TZDs. This also raises the intriguing possibility that endogenous levels of cyclopentenone PGs might critically affect the metabolic equilibrium of the organism. The future development of specific PPAR agonists and antagonists will be of great importance in deciphering the interconnections linking lipid and carbohydrate metabolism and should provide rational tools for therapeutic interventions in disorders like obesity, diabetes, or atherosclerosis.

ACKNOWLEDGMENTS

We are grateful to D Moras for Figure 1*c* and F Gonzalez and J Auwerx for communicating unpublished data. We also thank H Keller, P Devchand, and G Krey for critical reading of the manuscript and helpful discussions. Because of strict space limitations, numerous primary references could not be cited directly but can be found in reviews quoted in the text. The work done in the authors' laboratories has been supported by the État de Vaud and the Swiss National Science Foundation.

Literature Cited

Aldridge TC, Tugwood JD, Green S. 1995. Identification and characterization of DNA elements implicated in the regulation of CYP4A1 transcription. *Biochem. J.* 306:473–79

Amri EZ, Bonino F, Ailhaud G, Abumrad NA, Grimaldi PA. 1995. Cloning of a protein that mediates transcriptional effects of fatty acids in preadipocytes. *J. Biol. Chem.* 270:2367–71

Aperlo C, Pognonec P, Saladin R, Auwerx J, Boulukos KE. 1995. cDNA cloning and characterization of the transcriptional activities of the hamster peroxisome proliferator-activated receptor haPPAR-gamma. *Gene* 162:297–302

Baes M, Castelein H, Desmet L, Declercq PE. 1995. Antagonism of COUP-TF and PPAR-alpha/RXR-alpha on the activation of the malic enzyme gene promoter: modulation by 9-*cis* RA. *Biochem. Biophys. Res. Commun.* 215:338–45

Björntorp P. 1991. Metabolic implications of body fat distribution. *Diabetes Care* 14: 1132–43

Blackmore PF, McPherson RK, Stevenson RW. 1993. Actions of the novel antidiabetic agent englitazone in rat hepatocytes. *Metabolism* 42:1583–87

Bourguet W, Ruff M, Chambon P, Gronemeyer H, Moras D. 1995. Crystal structure of the ligand-binding domain of the human nuclear receptor RXR alpha. *Nature* 375:377–82

Brady PS, Marine KA, Brady LJ, Ramsay RR. 1989. Co-ordinate induction of hepatic mitochondrial and peroxisomal carnitine acyltransferase synthesis by diet and drugs.

Biochem. J. 260:93–100

Braissant O, Foufelle F, Scotto C, Dauça M, Wahli W. 1996. Differential expression of peroxisome proliferator-activated receptors (PPARs): tissue distribution of PPAR alpha, beta and gamma in the adult rat. *Endocrinology* 137:354–66

Cao ZD, Umek RM, McKnight SL. 1991. Regulated expression of three C/EBP isoforms during adipose conversion of 3T3-L1 cells. *Genes Dev.* 5:1538–52

Castelein H, Gulick T, Declercq PE, Mannaerts GP, Moore DD, Baes MI. 1994. The peroxisome proliferator activated receptor regulates malic enzyme gene expression. *J. Biol. Chem.* 269:26754–58

Chawla A, Schwarz EJ, Dimaculangan DD, Lazar MA. 1994. Peroxisome proliferator-activated receptor (PPAR) gamma: adipose-predominant expression and induction early in adipocyte differentiation. *Endocrinology* 135:798–800

Chen F, Law SW, O'Malley BW. 1993. Identification of two mPPAR related receptors and evidence for the existence of five subfamily members. *Biochem. Biophys. Res. Commun.* 196:671–77

Chu RY, Madison LD, Lin YL, Kopp P, Rao MS, et al. 1995. Thyroid hormone (T3) inhibits ciprofibrate-induced transcription of genes encoding beta-oxidation enzymes: cross talk between peroxisome proliferator and T3 signaling pathways. *Proc. Natl. Acad. Sci. USA* 92:11593–97

Desvergne B, Wahli W. 1995. PPAR: a key nuclear factor in nutrient/gene interactions? In *Inducible Transcriptions,* ed. P Baeuerle, 1:142–76. Boston: Birkhaüser

Dreyer C, Keller H, Mahfoudi A, Laudet V, Krey G, Wahli W. 1993. Positive regulation of the peroxisomal beta-oxidation pathway by fatty acids through activation of peroxisome proliferator-activated receptors (PPAR). *Biol. Cell* 77:67–76

Dreyer C, Krey G, Keller H, Givel F, Helftenbein G, Wahli W. 1992. Control of the peroxisomal beta-oxidation pathway by a novel family of nuclear hormone receptors. *Cell* 68:879–87

el-Kebbi IM, Roser S, Pollet RJ. 1994. Regulation of glucose transport by pioglitazone in cultured muscle cells. *Metabolism* 43:953–58

Feng JQ, Chen D, Cooney AJ, Tsai MJ, Harris MA, et al. 1995. The mouse bone morphogenetic protein-4 gene. Analysis of promoter utilization in fetal rat calvarial osteoblasts and regulation by COUP-TFI orphan receptor. *J. Biol. Chem.* 270:28364–73

Forman BM, Tontonoz P, Chen J, Brun RP, Spiegelman BM, Evans RM. 1995. 15-deoxy-delta (12,14)-prostaglandin J2 is a ligand for the adipocyte determination factor PPAR-gamma. *Cell* 83:803–12

Foxworthy PS, Eacho PI. 1991. Effect of the peroxisome proliferator LY171883 on triglyceride accumulation in rats fed a fat-free diet. *Biochem. Pharmacol.* 42:1487–91

Foxworthy PS, Perry DN, Hoover DM, Eacho PI. 1990. Changes in hepatic lipid metabolism associated with lipid accumulation and its reversal in rats given the peroxisome proliferator LY171883. *Toxicol. Appl. Pharmacol.* 106:375–83

Freytag SO, Paielli DL, Gilbert JD. 1994. Ectopic expression of the CCAAT/enhancer-binding protein alpha promotes the adipogenic program in a variety of mouse fibroblastic cells. *Genes Dev.* 8:1654–63

Fukushima M. 1990. Prostaglandin J2—anti-tumour and anti-viral activities and the mechanisms involved. *Eicosanoids* 3:189–99

Fukushima M. 1992. Biological activities and mechanisms of action of PGJ2 and related compounds: an update. *Prostaglandins Leuk. Essential Fatty Acids* 47:1–12

Gearing KL, Crickmore A, Gustafsson JÅ. 1994. Structure of the mouse peroxisome proliferator activated receptor alpha gene. *Biochem. Biophys. Res. Commun.* 199:255–63

Gearing KL, Göttlicher M, Teboul M, Widmark E, Gustafsson JÅ. 1993. Interaction of the peroxisome-proliferator-activated receptor and retinoid X receptor. *Proc. Natl. Acad. Sci. USA* 90:1440–44

Glass CK. 1994. Differential recognition of target genes by nuclear receptor monomers, dimers, and heterodimers. *Endocr. Rev.* 15:391–407

Goetzl EJ, An S, Smith WL. 1995. Specificity of expression and effects of eicosanoid mediators in normal physiology and human diseases. *FASEB J.* 9:1051–58

Göttlicher M, Demoz A, Svensson D, Tollet P, Berge RK, Gustafsson JÅ. 1993. Structural and metabolic requirements for activators of the peroxisome proliferator-activated receptor. *Biochem. Pharmacol.* 46:2177–84

Göttlicher M, Widmark E, Li Q, Gustafsson JÅ. 1992. Fatty acids activate a chimera of the clofibric acid-activated receptor and the glucocorticoid receptor. *Proc. Natl. Acad. Sci. USA* 89:4653–57

Greene ME, Blumberg B, McBride OW, Yi HF, Kronquist K, et al. 1995. Isolation of the human peroxisome proliferator activated receptor gamma cDNA: expression in hematopoietic cells and chromosomal mapping. *Gene Exp.* 4:281–99

Gulick T, Cresci S, Caira T, Moore DD,

Kelly DP. 1994. The peroxisome proliferator-activated receptor regulates mitochondrial fatty acid oxidative enzyme gene expression. *Proc. Natl. Acad. Sci. USA* 91:11012–16

Härd T, Kellenbach E, Boelens R, Maler BA, Dahlman K, et al. 1990. Solution structure of the glucocorticoid receptor DNA-binding domain. *Science* 249:157–60

Hertz R, Bishara-Shieban J, Bar-Tana J. 1995. Mode of action of peroxisome proliferators as hypolipidemic drugs. Suppression of apolipoprotein C-III. *J. Biol. Chem.* 270:13470–75

Hirata Y, Hayashi H, Ito S, Kikawa Y, Ishibashi M, et al. 1988. Occurrence of 9-deoxy-delta 9,delta 12–13,14-dihydroprostaglandin D2 in human urine. *J. Biol. Chem.* 263:16619–25

Hofmann CA, Colca JR. 1992. New oral thiazolidinedione antidiabetic agents act as insulin sensitizers. *Diabetes Care* 15:1075–78

Hofmann CA, Lorenz K, Braithwaite SS, Colca JR, Palazuk BJ, et al. 1994. Altered gene expression for tumor necrosis factor-alpha and its receptors during drug and dietary modulation of insulin resistance. *Endocrinology* 134:264–70

Hu ED, Tontonoz P, Spiegelman BM. 1995. Transdifferentiation of myoblasts by the adipogenic transcription factors PPAR-gamma and C/EBP-alpha. *Proc. Natl. Acad. Sci. USA* 92:9856–60

Ibrahimi A, Teboul L, Gaillard D, Amri EZ, Ailhaud G, et al. 1994. Evidence for a common mechanism of action for fatty acids and thiazolidinedione antidiabetic agents on gene expression in preadipose cells. *Mol. Pharmacol.* 46:1070–76

Issemann I, Green S. 1990. Activation of a member of the steroid hormone receptor superfamily by peroxisome proliferators. *Nature* 347:645–50

Jones PS, Savory R, Barratt P, Bell AR, Gray T, et al. 1995. Chromosomal localisation, inducibility, tissue-specific expression and strain differences in three murine peroxisome-proliferator-activated-receptor genes. *Eur. J. Biochem.* 233:219–26

Jow L, Mukherjee R. 1995. The human peroxisome proliferator-activated receptor (PPAR) subtype NUC1 represses the activation of hPPAR alpha and thyroid hormone receptors. *J. Biol. Chem.* 270:3836–40

Juge-Aubry CE, Gorla-Bajszczak A, Pernin A, Lemberger T, Wahli W, et al. 1995. Peroxisome proliferator-activated receptor mediates cross-talk with thyroid hormone receptor by competition for retinoid X receptor. Possible role of a leucine zipper-like heptad repeat. *J. Biol. Chem.* 270:18117–22

Katagiri T, Yamaguchi A, Komaki M, Abe E, Takahashi N, et al. 1994. Bone morphogenetic protein-2 converts the differentiation pathway of C2C12 myoblasts into the osteoblast lineage. *J. Cell Biol.* 127:1755–66

Keller H, Dreyer C, Medin J, Mahfoudi A, Ozato K, Wahli W. 1993a. Fatty acids and retinoids control lipid metabolism through activation of peroxisome proliferator-activated receptor-retinoid X receptor heterodimers. *Proc. Natl. Acad. Sci. USA* 90:2160–64

Keller H, Mahfoudi A, Dreyer C, Hihi AK, Medin J, et al. 1993b. Peroxisome proliferator-activated receptors and lipid metabolism. *Ann. NY Acad. Sci.* 684:157–73

Kletzien RF, Clarke SD, Ulrich RG. 1992. Enhancement of adipocyte differentiation by an insulin-sensitizing agent. *Mol. Pharmacol.* 41:393–98

Kliewer SA, Forman BM, Blumberg B, Ong ES, Borgmeyer U, et al. 1994. Differential expression and activation of a family of murine peroxisome proliferator-activated receptors. *Proc. Natl. Acad. Sci. USA* 91:7355–59

Kliewer SA, Lenhard JM, Willson TM, Patel I, Morris DC, Lehmann JM. 1995. A prostaglandin J2 metabolite binds peroxisome proliferator-activated receptor gamma and promotes adipocyte differentiation. *Cell* 83:813–19

Kliewer SA, Umesono K, Noonan DJ, Heyman RA, Evans RM. 1992. Convergence of 9-*cis* retinoic acid and peroxisome proliferator signalling pathways through heterodimer formation of their receptors. *Nature* 358:771–74

Koshihara Y, Takamori R, Nomura K, Sugiura S, Kurozumi S. 1991. Enhancement of in vitro mineralization in human osteoblasts by a novel prostaglandin A1 derivative TEI-3313. *J. Pharmacol. Exp. Ther.* 258:1120–26

Krey G, Keller H, Mahfoudi A, Medin J, Ozato K, et al. 1993. *Xenopus* peroxisome proliferator activated receptors: genomic organization, response element recognition, heterodimer formation with retinoid X receptor and activation by fatty acids. *J. Steroid Biochem. Mol. Biol.* 47:65–73

Kurokawa R, DiRenzo J, Boehm M, Sugarman J, Gloss B, et al. 1994. Regulation of retinoid signalling by receptor polarity and allosteric control of ligand binding. *Nature* 371:528–31

Laudet V, Hanni C, Coll J, Catzeflis F, Stehelin D. 1992. Evolution of the nuclear receptor gene superfamily. *EMBO J.* 11:1003–13

Lee MS, Kliewer SA, Provencal J, Wright PE, Evans RM. 1993. Structure of the retinoid-X receptor-alpha DNA-binding domain: a helix required for homodimeric DNA binding. *Science* 260:1117–21

Lee SS, Pineau T, Drago J, Lee EJ, Owens JW,

et al. 1995. Targeted disruption of the alpha isoform of the peroxisome proliferator-activated receptor gene in mice results in abolishment of the pleiotropic effects of peroxisome proliferators. *Mol. Cell Biol* 15:3012–22

Lehmann JM, Moore LB, Smith-Oliver TA, Wilkison WO, Willson TM, Kliewer SA. 1995. An antidiabetic thiazolidinedione is a high affinity ligand for peroxisome proliferator-activated receptor gamma (PPAR gamma). *J. Biol. Chem.* 270:12953–56

Lemberger T, Braissant O, Juge-Aubry C, Keller H, Saladin R, et al. 1996a. PPAR tissue distribution and interactions with other hormone-signaling pathways. *Ann. NY Acad. Sci.* In press

Lemberger T, Saladin R, Vázquez M, Assimacopoulos F, Staels B, et al. 1996b. Expression of the peroxisome proliferator-activated receptor alpha gene is stimulated by stress and follows a diurnal rhythm. *J. Biol. Chem.* 271:1764–69

Lemberger T, Staels B, Saladin R, Desvergne B, Auwerx J, Wahli W. 1994. Regulation of the peroxisome proliferator-activated receptor alpha gene by glucocorticoids. *J. Biol. Chem.* 269:24527–30

Luisi BF, Xu WX, Otwinowski Z, Freedman LP, Yamamoto KR, et al. 1991. Crystallographic analysis of the interaction of the glucocorticoid receptor with DNA. *Nature* 352:497–505

Mangelsdorf DJ, Evans RM. 1995. The RXR heterodimers and orphan receptors. *Cell* 83:841–50

Mannaerts GP, Debeer LJ, Thomas J, De Schepper PJ. 1979. Mitochondrial and peroxisomal fatty oxidation in liver homogenates and isolated hepatocytes from control and clofibrate-treated rats. *J. Biol. Chem.* 254:4585–95

Mannaerts GP, Van Veldhoven PP. 1993. Metabolic role of mammalian peroxisomes. In *Peroxisomes: Biology and Importance in Toxicology and Medicine,* ed. G Gibson, B Lake, pp. 19–62. London: Taylor & Francis

Marcus SL, Miyata KS, Zhang B, Subramani S, Rachubinski RA, Capone JP. 1993. Diverse peroxisome proliferator-activated receptors bind to the peroxisome proliferator-responsive elements of the rat hydratase/dehydrogenase and fatty acyl-CoA oxidase genes but differentially induce expression. *Proc. Natl. Acad. Sci. USA* 90:5723–27

Masters C, Crane D. 1995. *The Peroxisome: A Vital Organelle.* pp. 150–53. Cambridge: Cambridge Univ. Press

McDougald OA, Lane MD. 1995. Transcriptional regulation of gene expression during adipocyte differentiation. *Annu. Rev. Biochem.* 64:345–73

McGarry JD. 1992. What if Minkowski had been ageusic? An alternative angle on diabetes. *Science* 258:766–70

McGarry JD. 1995. Does leptin lighten the problem of obesity? *Curr. Biol.* 5:1342–44

Miyata KS, Zhang B, Marcus SL, Capone JP, Rachubinski RA. 1993. Chicken ovalbumin upstream promoter transcription factor (COUP-TF) binds to a peroxisome proliferator-responsive element and antagonizes peroxisome proliferator-mediated signaling. *J. Biol. Chem.* 268:19169–72

Muerhoff AS, Griffin KJ, Johnson EF. 1992. The peroxisome proliferator-activated receptor mediates the induction of CYP4A6, a cytochrome P450 fatty acid omega-hydroxylase, by clofibric acid. *J. Biol. Chem.* 267:19051–53

Näär AM, Boutin JM, Lipkin SM, Yu VC, Holloway JM, et al. 1991. The orientation and spacing core DNA-binding motifs dictate selective transcriptional responses to three nuclear receptors. *Cell* 65:1267–79

Narumiya S, Ohno K, Fukushima M, Fujiwara M. 1987. Site and mechanism of growth inhibition by prostaglandins. III. Distribution and binding of prostaglandin A2 and delta 12-prostaglandin J2 in nuclei. *J. Pharmacol. Exp. Ther.* 242:306–11

Newsholme EA, Start C. 1979. *Regulation in Metabolism.* pp. 195–328. New York: Wiley

Noyori R, Suzuki M. 1993. Organic synthesis of prostaglandins: advancing biology. *Science* 259:44–45

Oakes ND, Kennedy CJ, Jenkins AB, Laybutt DR, Chisholm DJ, Kraegen EW. 1994. A new antidiabetic agent, BRL 49653, reduces lipid availability and improves insulin action and glucoregulation in the rat. *Diabetes* 43:1203–10

Ohta T, Azuma Y, Kanatani H, Kiyoki M, Koshihara Y. 1995. TEI-3313, a novel prostaglandin A1 derivative, prevents bone loss and enhances bone formation in immobilized male rats. *J. Pharmacol. Exp. Ther.* 275:450–55

Palmer CN, Hsu MH, Griffin HJ, Johnson EF. 1995. Novel sequence determinants in peroxisome proliferator signaling. *J. Biol. Chem.* 270:16114–21

Peitsch MC, Jongeneel CV. 1993. A 3-D model for the CD40 ligand predicts that it is a compact trimer similar to the tumor necrosis factors. *Int. Immunol.* 5:233–38

Randle PJ, Kerbey AL, Espinal J. 1988. Mechanisms decreasing glucose oxidation in diabetes and starvation: role of lipid fuels and

hormones. *Diabetes Metab. Rev.* 4:623–38
Rastinejad F, Perlmann T, Evans RM, Sigler PB. 1995. Structural determinants of nuclear receptor assembly on DNA direct repeats. *Nature* 375:203–11
Renaud J-P, Rochel N, Ruff M, Vivat V, Chambon P, et al. 1995. Crystal structure of the RAR gamma ligand-binding domain bound to all-*trans* retinoic acid. *Nature* 378:681–89
Rodriguez JC, Gil-Gomez G, Hegardt FG, Haro D. 1994. Peroxisome proliferator-activated receptor mediates induction of the mitochondrial 3-hydroxy-3-methylglutaryl-CoA synthase gene by fatty acids. *J. Biol. Chem.* 269:18767–72
Sandouk T, Reda D, Hofmann C. 1993. Antidiabetic agent pioglitazone enhances adipocyte differentiation of 3T3-F442A cells. *Am. J. Physiol.* 264:C1600–8
Schmidt A, Endo N, Rutledge SJ, Vogel R, Shinar D, Rodan GA. 1992. Identification of a new member of the steroid hormone receptor superfamily that is activated by a peroxisome proliferator and fatty acids. *Mol. Endocrinol.* 6:1634–41
Schmidt A, Towler DA, Vogel R, Witherup K, Rutledge S, et al. 1994. Regulation of osteocalcin and bone sialoprotein promoters by NUCI receptor and the identification of a natural activator of NUCI. *Annu. Meet. Am. Soc. Bone Min. Res. 16th, Kansas City.* Abstr.
Schoonjans K, Watanabe M, Suzuki H, Mahfoudi A, Krey G, et al. 1995. Induction of the acyl-coenzyme A synthetase gene by fibrates and fatty acids is mediated by a peroxisome proliferator response element in the C promoter. *J. Biol. Chem.* 270:19269–76
Schwabe JWR, Neuhaus D, Rhodes D. 1990. Solution structure of the DNA-binding domain of the oestrogen receptor. *Nature* 348:458–61
Sher T, Yi HF, McBride OW, Gonzalez FJ. 1993. cDNA cloning, chromosomal mapping, and functional characterization of the human peroxisome proliferator activated receptor. *Biochemistry* 32:5598–604
Spiegelman BM, Hotamisligil GS. 1993. Through thick and thin: wasting, obesity, and TNF alpha. *Cell* 73:625–27
Srere P. 1994. Complexities of metabolic regulation. *Trends Biochem. Sci.* 19:519–20
Steineger HH, Sørensen HN, Tugwood JD, Skrede S, Spydevold Ø, Gautvik KM. 1994. Dexamethasone and insulin demonstrate marked and opposite regulation of the steady-state mRNA level of the peroxisomal proliferator-activated receptor (PPAR) in hepatic cells. Hormonal modulation of fatty-acid-induced transcription. *Eur. J. Biochem.* 225:967–74
Teboul L, Gaillard D, Staccini L, Inadera H, Amri EZ, Grimaldi PA. 1995. Thiazolidinediones and fatty acids convert myogenic cells into adipose-like cells. *J. Biol. Chem.* 270:28183–87
Tontonoz P, Hu E, Devine J, Beale EG, Spiegelman BM. 1995a. PPAR gamma 2 regulates adipose expression of the phosphoenolpyruvate carboxykinase gene. *Mol. Cell Biol.* 15:351–57
Tontonoz P, Hu E, Graves RA, Budavari AI, Spiegelman BM. 1994a. mPPAR gamma 2: tissue-specific regulator of an adipocyte enhancer. *Genes Dev.* 8:1224–34
Tontonoz P, Hu E, Spiegelman BM. 1994b. Stimulation of adipogenesis in fibroblasts by PPAR gamma 2, a lipid-activated transcription factor. *Cell* 79:1147–56
Tontonoz P, Hu E, Spiegelman BM. 1995b. Regulation of adipocyte gene expression and differentiation by peroxisome proliferator activated receptor gamma. *Curr. Opin. Gene Dev.* 5:571–76
Tracey KJ, Cerami A. 1993. Tumor necrosis factor, other cytokines and disease. *Annu. Rev. Cell Biol.* 9:317–43
Tugwood JD, Issemann I, Anderson RG, Bundell KR, McPheat WL, Green S. 1992. The mouse peroxisome proliferator activated receptor recognizes a response element in the 5′ flanking sequence of the rat acyl CoA oxidase gene. *EMBO J.* 11:433–39
Umesono K, Murakami KK, Thompson CC, Evans RM. 1991. Direct repeats as selective response elements for the thyroid hormone, retinoic acid, and vitamin-D3 receptors. *Cell* 65:1255–66
Vu-Dac N, Schoonjans K, Kosykh V, Dallongeville J, Fruchart JC, et al. 1995. Fibrates increase human apolipoprotein A-II expression through activation of the peroxisome proliferator-activated receptor. *J. Clin. Invest.* 96:741–50
Vu-Dac N, Schoonjans K, Laine B, Fruchart JC, Auwerx J, Staels B. 1994. Negative regulation of the human apolipoprotein A-I promoter by fibrates can be attenuated by the interaction of the peroxisome proliferator-activated receptor with its response element. *J. Biol. Chem.* 269:31012–18
Wagner RL, Apriletti JW, McGrath ME, West BL, Baxter JD, Fletterick RJ. 1995. A structural role for hormone in the thyroid hormone receptor. *Nature* 378:690–97
Wahli W, Braissant O, Desvergne B. 1995. Peroxisome proliferator activated receptors: transcriptional regulators of adipogenesis, lipid metabolism and more. *Chem. Biol.* 2:261–66
Willson TM, Cobb JE, Cowan DJ, Wiethe

RW, Correa ID, et al. 1996. The structure-activity relationship between peroxisome proliferator-activated receptor gamma agonism and the anti-hyperglycemic activity of thiazolidinediones. *J. Med. Chem.* 39:665–68

Winrow CJ, Marcus SL, Miyata KS, Zhang B, Capone JP, Rachubinski RA. 1994. Transactivation of the peroxisome proliferator-activated receptor is differentially modulated by hepatocyte nuclear factor-4. *Gene Exp.* 4:53–62

Wu ZD, Xie YH, Bucher N, Farmer SR. 1995. Conditional ectopic expression of C/EBP-beta in NIH-3T3 cells induces PPAR-gamma and stimulates adipogenesis. *Genes Dev.* 9:2350–63

Xing GQ, Zhang LX, Zhang L, Heynen T, Yoshikawa T, et al. 1995. Rat PPAR-delta contains a CGG triplet repeat and is prominently expressed in the thalamic nuclei. *Biochem. Biophys. Res. Commun.* 217:1015–25

Yamaguchi A, Katagiri T, Ikeda T, Wozney JM, Rosen V, et al. 1991. Recombinant human bone morphogenetic protein-2 stimulates osteoblastic maturation and inhibits myogenic differentiation in vitro. *J. Cell Biol.* 113:681–87

Yeh WC, Cao Z, Classon M, McKnight SL. 1995. Cascade regulation of terminal adipocyte differentiation by three members of the C/EBP family of leucine zipper proteins. *Genes Dev.* 9:168–81

Yu K, Bayona W, Kallen CB, Harding HP, Ravera CP, et al. 1995. Differential activation of peroxisome proliferator-activated receptors by eicosanoids. *J. Biol. Chem.* 270:23975–83

Zhang B, Marcus SL, Miyata KS, Subramani S, Capone JP, Rachubinski RA. 1993. Characterization of protein-DNA interactions within the peroxisome proliferator-responsive element of the rat hydratase-dehydrogenase gene. *J. Biol. Chem.* 268:12939–45

Zhang B, Marcus SL, Sajjadi FG, Alvares K, Reddy JK, et al. 1992. Identification of a peroxisome proliferator-responsive element upstream of the gene encoding rat peroxisomal enoyl-CoA hydratase/3-hydroxyacyl-CoA dehydrogenase. *Proc. Natl. Acad. Sci. USA* 89:7541–45

Zhu Y, Alvares K, Huang Q, Rao MS, Reddy JK. 1993. Cloning of a new member of the peroxisome proliferator-activated receptor gene family from mouse liver. *J. Biol. Chem.* 268:26817–20

Zhu Y, Qi C, Korenberg JR, Chen XN, Noya D, et al. 1995. Structural organization of mouse peroxisome proliferator-activated receptor gamma (mPPAR gamma) gene: alternative promoter use and different splicing yield two mPPAR gamma isoforms. *Proc. Natl. Acad. Sci. USA* 92:7921–25

Annu. Rev. Cell Dev. Biol. 1996. 12:365–91

GERM CELL DEVELOPMENT IN *DROSOPHILA*

Anne Williamson and Ruth Lehmann
Skirball Institute, New York University Medical Center Developmental Genetics Program, 540 First Avenue, 4th floor New York, NY 10016

KEY WORDS: pole cells, polar granules, migration, germ plasm, primordial germ cells

ABSTRACT

More than 100 years have passed since Weismann first recognized the role of germ cells in the continuity of a species. Today, it remains unclear how a germ cell is initially set aside from somatic cells and how it chooses its unique developmental path. In this review, we address various aspects of germ cell development in *Drosophila*, such as germ cell determination, germ cell migration, gonad formation, sex determination, and gametogenesis. Many aspects of germ cell development, including the morphology of germ cells, their migratory behavior, as well as the processes of gonad formation and gametogenesis, show striking similarities among organisms. Considering the conservation of factors that regulate somatic development, it is likely that some aspects of germ cell development are shared not only on a morphological but also on the molecular level between *Drosophila* and other organisms.

CONTENTS

1081-0706/96/1115-0365$08.00

PRIMORDIAL GERM CELL DETERMINATION

The *Drosophila* embryo develops from the fertilized egg through a series of synchronized, rapid nuclear divisions in a multi-nucleate syncytium. The first nuclei to become surrounded by cell membranes are those that migrate to the posterior pole of the embryo (Huettner 1923, Sonnenblick 1941). Cellularization of these pole cells depends upon a specialized cytoplasm, the posterior pole plasm, or germ plasm, that assembles at the posterior pole during oogenesis (Mahowald 1962). Two different experiments have led to the conclusion that posterior pole plasm harbors germ cell determinants that confer germ cell-precursor fate on cells containing this specialized cytoplasm. First, transplantation of posterior pole plasm to an ectopic site directs the formation of primordial germ cells at that site (Illmensee & Mahowald 1974). Second, maternal effect mutations that disrupt the assembly of posterior pole plasm prevent pole cell formation (reviewed in Rongo & Lehmann 1996).

Germ Plasm Morphology

Electron microscopic (EM) analysis has shown that pole plasm contains spherical electron-dense structures, termed polar granules (Mahowald 1962). The polar granules contain RNA and protein. In particular, Vasa, Oskar, and Tudor proteins, as well as the mitochondrial 16S large rRNA (*mtlr* RNA), have been shown to be components of the polar granules (Figure 1) (Hay et al 1988a, Bardsley et al 1993, Kobayashi et al 1993a). However, the exact structure and composition of these organelles and of the pole plasm that surrounds them remain unknown.

Immuno-EM studies suggest that germ plasm components, distinguishable as electron-dense, Vasa immuno-positive structures, are continuously present in the germ line (Mahowald 1971, Hay et al 1988a). In the mature egg cell, and immediately after egg activation, polar granules are organized in large clusters, often associated with mitochondria. During the early nuclear divisions, these clusters start to fragment (Mahowald 1968). By the time of pole cell formation,

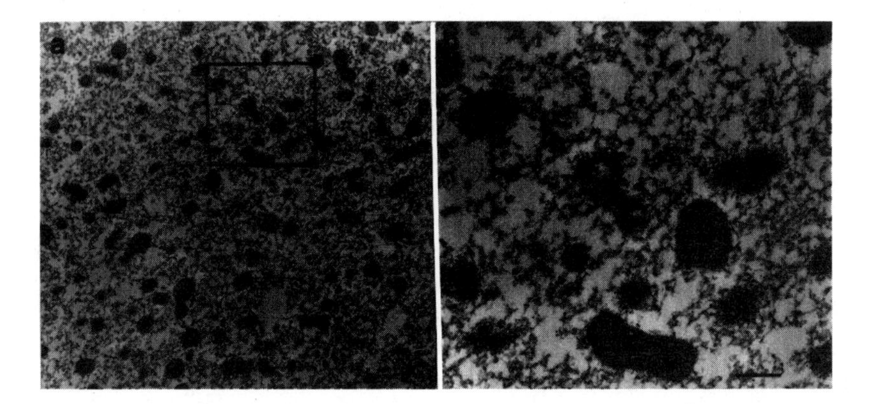

Figure 1 Ultra-structure of germ plasm. Immuno-histochemical detection of Oskar protein in early embryos using 5-nm gold particle-conjugated, anti-rat secondary antibodies. Whole embryos were fixed and stained with Oskar-specific antibodies, then sectioned and examined by electron microscopy. (*a*) Large-field view of posterior pole plasm. Polar granules of typical appearance can be seen throughout the cytoplasm (scale bar represents 0.5 microns). (*b*) Higher magnification of area boxed in (*a*), (≈ 4X, scale bar represents 0.25 microns). Gold particles can be seen associated with polar granules (p), but not with mitochondria (m).

two electron-dense, Vasa immuno-positive structures appear, called nuclear bodies and nuage: the former in the pole cell nuclei and the latter surrounding the nuclear envelope (Mahowald 1971, Hay et al 1998a). Although nuage is associated with the germ cell nuclei throughout the life cycle, nuclear bodies disappear by mid-embryogenesis (Mahowald 1971) and reappear in the ovarian nurse cell nuclei. Small, putative precursors of cytoplasmic polar granules are first detectable during oogenesis at the posterior pole in stage 9 oocytes, and by stage 10, polar granules of a characteristic appearance can be seen (Mahowald 1962, 1971).

Germ Plasm Genetics

Extensive efforts have been made to isolate polar granule components biochemically, and molecular identification of some pole plasm components has contributed to our knowledge of specific molecules included in these structures (Waring et al 1978, Kobayashi & Okada 1989, Jongens et al 1992). Molecular information about the composition of the pole plasm has come mainly from the study of maternal effect genes, whose function is required for pole plasm formation. Females mutant for any one of these genes produce embryos that lack polar granules and subsequently fail to form primordial germ cells. Because the progeny of mutant females are sterile, this class of mutations has been referred to as grandchild-less. Thus far, genetic screens have identified twelve genes required for formation of posterior pole plasm: *oskar, vasa, valois, tudor, cappuccino, spire, staufen, pipsqueak, mago nashi, orb, homeless,* and *tropomyosin II* (for review see Rongo & Lehmann 1996).

Many of the products encoded by the grandchild-less genes are localized to the posterior pole (Rongo & Lehmann 1996). Analysis of the distribution of various germ plasm components in mutant embryos has revealed that pole plasm assembly occurs stepwise. For example, localization of certain components such as Staufen protein and *oskar* RNA precedes and is a prerequisite for the localization of other components, i.e. Vasa and Tudor proteins (Hay et al 1990, Lasko & Ashburner 1990, Ephrussi et al 1991, Kim-Ha et al 1991, St Johnston et al 1991, Bardsley et al 1993). In addition, continuing interaction between germ plasm components is required for maintenance of posteriorly localized pole plasm during oogenesis and early embryogenesis (Ephrussi et al 1991, Kim-Ha et al 1991, St Johnston et al 1991, Rongo et al 1995; for review see Rongo & Lehmann, 1996).

The *oskar* gene plays a critical role in germ plasm assembly. *oskar* RNA and protein distribution foretell the appearance of polar granules at the posterior pole and their fragmentation during early embryogenesis (Figuress 1, 2) (Markussen et al 1995, Rongo et al 1995). The amount of *oskar* RNA and protein localized to the posterior pole dictates the number of pole cells formed (Ephrussi & Lehmann 1992, Smith et al 1992). Furthermore, mislocalization of *oskar* RNA to the anterior pole directs formation of polar granules and induces the formation of functional germ cells at this ectopic location. Consistent with the idea that *oskar* directs germ plasm assembly is the finding that gene products downstream of *oskar* in the germ plasm assembly pathway are mislocalized to the ectopic site and that their gene activities are required for ectopic germ cell formation (Ephrussi & Lehmann 1992, Bardsley et al 1993, Kobayashi et al 1995).

Most of the grandchild-less genes were initially identified on the basis of an abdominal segmentation phenotype in the progeny of mutant females that resembles the phenotype of the zygotic gene *knirps* (Lehmann 1988). This abdominal phenotype results from a lack of proper localization and translation of *nanos* (*nos*) RNA at the posterior pole in the absence of functional pole plasm (Gavis & Lehmann 1994, Wang et al 1994). Therefore, this group of genes has been called the grandchild-less-knirps or posterior class of genes to reflect the dual effect on germ cell formation and abdominal segmentation (Rongo & Lehmann 1996).

Germ Plasm Effector Genes

Function of *nos* is required for embryonic abdomen formation, and although it has no effect on pole cell formation, *nos* plays a role later in germ cell

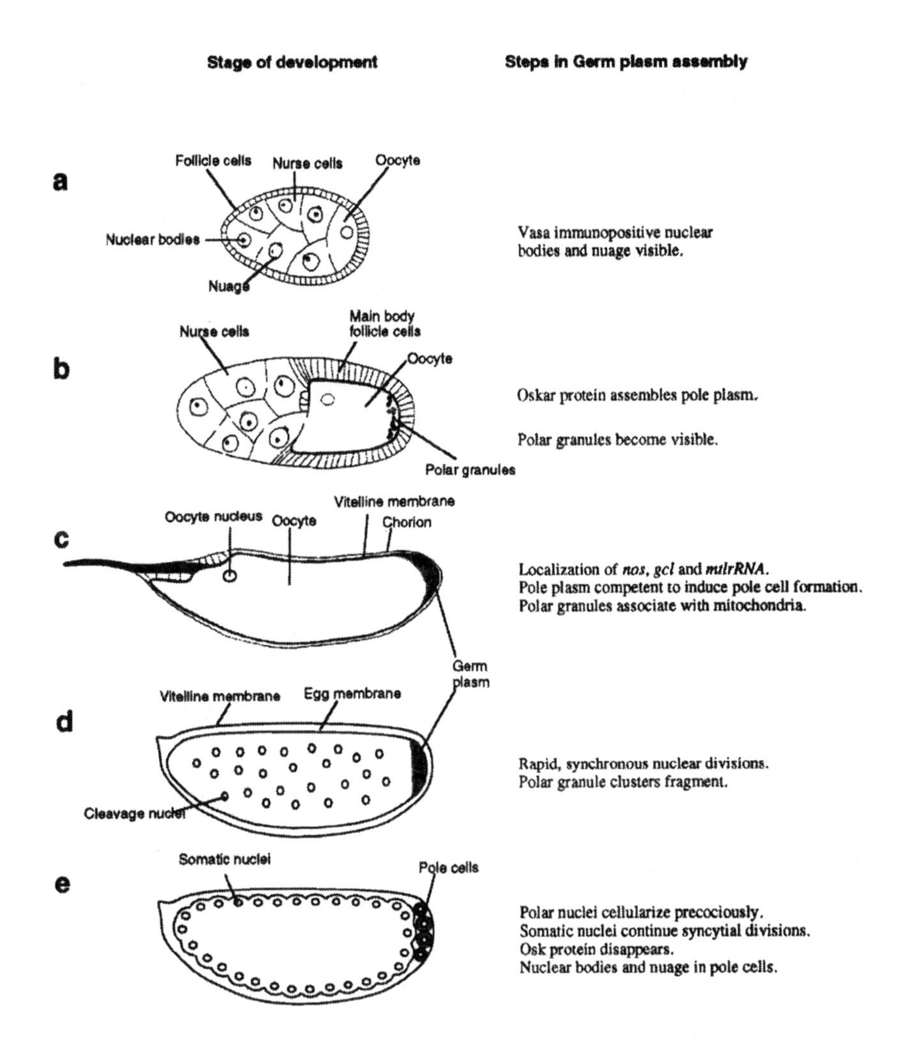

Figure 2 Germ plasm assembly during oogenesis and early embryogenesis. (*a*) Stage 6 egg chamber. Oocyte is positioned posterior to fifteen nurse cells; somatic follicle cells surround oocyte-nurse cell cluster. (*b*) Stage 9–10 egg chamber. Oocyte has increased in volume; nurse cell nuclei are polyploid. The majority of follicle cells cover the oocyte. (*c*) Mature, stage 14 egg chamber. Nurse cells have extruded their contents into oocyte and degenerated; follicle cells have secreted the egg covers, the chorion, and the vitelline membrane. (*d*) Cleavage stage embryo. Nuclei divide within cytoplasmic islands in rapid synchrony. (*e*) Syncytial blastoderm stage embryo. Nuclei have moved to the cortex. Oocytes and embryos are oriented anterior to left, dorsal to top of figure (after Rongo & Lehmann 1996).

differentiation (Lehmann & Nüsslein-Volhard 1991, Kobayashi et al 1996). *nos* RNA becomes localized to the posterior pole during oogenesis, and this localization depends on *oskar, vasa,* and *tudor*, genes known to be required for pole plasm assembly (Wang et al 1994). Nos protein is synthesized in the early embryo from the posteriorly localized RNA source.

nos RNA localization during the last stages of oogenesis coincides with the stage at which cytoplasm from the posterior pole becomes competent to induce pole cell formation when transplanted into cleavage stage host embryos (Illmensee et al 1976). Thus it is likely that genes specific for pole cell formation show developmental profiles similar to that of *nos*. Large-scale genetic screens for pole cell-specific genes have not been carried out. Thus a specific role in pole cell formation rather than in germ plasm assembly has been implied for only two genes, mitochondrial 16S large rRNA (*mtlr* RNA) and *germ cell-less* (*gcl*) (Kobayashi & Okada 1989, Jongens et al 1992). However, in contrast to *nos* RNA, which can promote abdomen formation when mislocalized, neither *mtlr* nor *gcl* RNA is sufficient to induce pole cell formation at an ectopic site (Kobayashi & Okada 1989, Jongens et al 1994). These results suggest that multiple factors may contribute to the germ cell-inducing activity of the posterior pole plasm. Deletion of a single component may impair pole cell formation, but at an ectopic site a single component alone may not be sufficient.

In support of this latter notion, Okada and coworkers showed that embryos irradiated at the posterior pole fail to form pole cells and that injection of *mtlr* RNA into such embryos restores the ability of the pole plasm to precociously cellularize posterior nuclei (Kobayashi & Okada 1989). However, the resulting pole cells are unable to give rise to germ cells. Furthermore, because injection of *mtlr* RNA into other cortical regions of the embryo fails to induce cellularization, Kobayashi et al concluded that the posterior pole plasm must contain at least two activities in addition to *mtlr* RNA: an activity that is UV resistant and is required for pole cell formation, and one that is UV sensitive and is involved in germ cell determination (Kobayashi & Okada 1989). The role of *mtlr* RNA in pole cell formation is unclear. Electron microscopy reveals an association of *mtlr* RNA with polar granules outside the mitochondria during early cleavage stages (Kobayashi et al 1993a). In this context, it is notable that the protein product of the nuclear-encoded *tudor* gene is found in polar granules and mitochondria at the posterior pole (Bardsley et al 1993). Thus it is possible that association of polar granules and mitochondria somehow relates to the assembly of polar granules and to pole cell formation.

The *germ cell-less* (*gcl*) gene was initially identified as a posteriorly localized RNA that encodes a novel protein (Jongens et al 1992). Because mutations in the *gcl* gene have not yet been isolated, Jongens et al used heat shock promoter-driven expression of antisense and sense *gcl* RNA to vary *gcl* product levels. *gcl* antisense expression during oogenesis causes a significant decrease in the number of pole cells formed, whereas an increase in the levels of wild-type *gcl* RNA and protein leads to an increase in pole cell number (Jongens et al 1992). Localization of *gcl* RNA to the anterior pole is not sufficient to induce ectopic pole cells; however, cellularization of somatic nuclei is disrupted at the anterior pole (Jongens et al 1994). These experiments strongly suggest a requirement for the *germ cell-less* gene product in pole cell formation and/or survival. Immunofluorescence antibody staining shows that Gcl protein is localized to the nuclear membrane, possibly to the nucleoplasmic side of the nuclear pores in pole cells, at the time of their formation (Jongens et al 1994). During embryogenesis, expression of *gcl* is detected in a number of different tissues, including foregut, hindgut, muscle, and nervous system (Jongens et al 1992). Thus in the absence of the true null phenotype, the specificity of *gcl* function for germ cell formation awaits further clarification.

PRIMORDIAL GERM CELL FORMATION

Upon fertilization and fusion of the male and female pronuclei, nuclear divisions in the absence of cytokinesis ensue. Approximately 8 to 10 nuclei reach the posterior pole and become surrounded by cell membranes during mitotic cycle 10. These cells undergo 0 to 2 further nonsynchronous divisions to give rise to about 40 pole cells (Sonnenblick 1950). At the blastoderm stage (3.2 h after egg deposition), pole cells mitotically arrest in G_2 of the cell cycle (TT Sue & P O'Farrell, personal communication). Transplantation of newly formed pole cells results in colonization of only the host's germ line, which demonstrates that from the time of their formation, these cells are committed to the germ cell fate (Underwood et al 1980, Technau & Campos-Ortega 1986).

In contrast to the precociously cellularized germ line, the nuclei that give rise to the somatic cells undergo four more synchronous nuclear divisions after cycle 10 and cellularize after the 13th division (Zalokar & Erk 1976). Subsequently, these somatic cells loose their synchronous mitotic behavior and initiate G_2-containing cell cycles as small groups or mitotic domains throughout the embryo (Foe et al 1993).

Precocious Cellularization

The signals that promote precocious cellularization of the pole cell nuclei are unknown. There is no evidence that the posterior pole plasm actively recruits nuclei from a distance because nuclei arrive at the posterior and anterior poles at approximately the same time, independently of germ plasm (G Schubiger, personal communication). When nuclear migration is delayed, either by a temporary ligation or by mutations that disrupt the F-actin cytoskeleton, pole cells do not form (Okada 1982, Warn et al 1985, Hatanaka & Okada 1991). This indicates that the posterior pole plasm is only competent to induce pole cell formation during the very early stages of embryogenesis. Furthermore, pole cells can form in the absence of nuclei. In aphidicoline-treated embryos, DNA replication is inhibited while centrosomes continue to replicate (Raff & Glover 1989). Those centrosomes that move to the posterior cortex are surrounded by cell membranes. This suggests that interaction of posterior pole plasm components with centrosome-associated factors may trigger cellularization of pole cells. Indeed, several posteriorly localized RNAs such as *cyclin B* and *nanos* seem to associate with the spindle poles of the pole cells (Raff et al 1990, Gavis et al 1996).

Transcriptional Quiescence

The differences between the behavior of pole cells and somatic cells are not restricted to their modes of cellularization and cell-cycle regulation. Although transcription is initiated in the somatic nuclei around cycle 10 and increases in intensity during this period, pole cells seem to be transcriptionally quiescent. Zalokar showed that while radioactively labeled uridine (^{3}H-UTP) was rapidly integrated into the newly synthesized RNA of somatic cells, little or no labeling was detected in the pole cells. However, radioactive valine (^{3}H-Val) was incorporated into newly synthesized proteins, indicating that pole cells are translationally active but transcriptionally inactive (Zalokar 1976). The mechanism that causes transcriptional quiescence in the pole cells is not known; however, it is possible that pole cells form prior to the activation of transcription in the embryo and therefore lack components of the general transcription machinery (Edgar & Schubiger 1986). Alternatively, transcriptional repression may be part of the germ cell determination process itself, e.g. by preventing the initiation of zygotic gene expression that would otherwise promote somatic differentiation.

Support for this latter notion comes from the finding that a similar phenomenon has also been observed in *Caenorhabditis elegans*. Although transcription in somatic nuclei starts in *C. elegans* at the four cell stage, there is no detectable transcription in the germ line progenitor or P cell at this stage. Upon division, only the P cell descendants remain transcriptionally repressed while the somatic daughter cells resume transcription (Seydoux & Fire 1994). Mutations in *par-4* disrupt the asymmetric pattern of P granule segregation in early blastomeres. In these embryos, germ cells do not form, and all nuclei are transcriptionally active (Seydoux & Fire 1994). These data suggest that there may be factors present in, or cosegregating with, P granules that are involved in repression of zygotic transcription in this lineage.

Recent work suggests a role for the *pie-1* gene in germ line-specific transcriptional quiescence. In embryos from *pie-1* mutant females, P granules are initially segregated correctly, but no germ cells form and the P2 cell adopts the fate of its sister, EMS (Mello et al 1992). *pie-1* encodes a nuclear protein that segregates with the germ line but not with the somatic daughter cell during each division of the P lineage (CC Mello et al, manuscript in preparation). Ectopic expression of PIE-1 in early embryos is sufficient to repress transcription in all cells (G Seydoux et al, manuscript in preparation). Thus transcriptional quiescence seems to be an active process in wild-type *C. elegans* germ cells. However, ectopic expression of PIE-1 alone does not lead to ectopic germ cell formation, which suggests that, as in *Drosophila,* regulation of transcription is just one of multiple factors contributing to germ cell determination.

In summary, the fact that germ cells are transcriptionally repressed in organisms as different as *Drosophila* and *C. elegans* suggests that transcriptional repression provides a general mechanism to prevent differentiation of primordial germ cells (PGCs) along somatic pathways. Could such a mechanism also be involved in germ cell determination in organisms such as the mouse, which lacks prelocalized germ line determinants? Although there is no direct evidence to support this idea, it is intriguing that the mouse primordial germ cell genome is not methylated during early embryogenesis, which suggests that other, methylation-independent mechanisms may control PGC-specific transcription (Grant et al 1992, Kafri et al 1992).

Translational Regulation

Regulation of translational activity in pole cells is likely to be linked to specific accumulation or localization of maternal mRNAs in the posterior pole plasm. A number of posteriorly localized RNAs such as *nanos, cyclin B,* and *gcl* appear to be regulated at the translational level (Dalby & Glover 1992, Gavis & Lehmann 1992, Jongens et al 1992). Regulatory sequences that control aspects of RNA localization as well as RNA translation are located in the 3′UTR of *cyclin B* and *nanos* mRNAs (Dalby & Glover 1992, Gavis & Lehmann 1992). *nanos* RNA is translated upon egg deposition, and its translation is linked to RNA localization such that only localized RNA becomes translated, whereas unlocalized, uniformly distributed RNA is translationally repressed (Gavis & Lehmann 1994). As in the case of *nanos* RNA, *cyclin B* RNA becomes localized to the posterior pole during late oogenesis. However, this RNA is not translated until late in embryogenesis when pole cells resume mitotic activity (Dalby & Glover 1993). Together these observations suggest that early germ cell behavior may be largely regulated by pole cell-specific translational regulation of maternal RNAs rather than by pole cell-specific transcriptional activation of zygotic genes.

PRIMORDIAL GERM CELL MIGRATION AND EMBRYONIC GONAD FORMATION

Adherence to and Migration Through Posterior Midgut Anlagen

Throughout embryogenesis, pole cells can be identified using either morphological criteria such as their size, shape, and nuclear morphology or their expression of markers such as the Nanos and Vasa proteins (Figure 3) (Campos-Ortega & Hartenstein 1985, Hay et al 1988a, Wang et al 1994). At the blastoderm stage, pole cells have a rounded morphology and are positioned just posterior to the prospective endodermal cells, which give rise to the posterior midgut (Figure 3*a*). During the onset of gastrulation, when the germ band starts to extend towards the dorsal side of the embryo, pole cells remain associated with the future posterior midgut cells (Figure 3*b*). As the midgut anlage invaginates, pole cells are carried inside the embryo (Figure 3*c*). At this point, pole cells actively migrate towards the blind end of the midgut pocket (Sonnenblick 1941, Poulson 1950, Jaglarz & Howard 1995). Blastoderm stage pole cells transplanted into the midgut pocket are able to migrate through the posterior midgut epithelium (Jaglarz & Howard 1994). This result indicates that at the blastoderm stage pole cells are already motile. Indeed, pole cells have occasionally been observed to move prematurely through the underlying blastoderm cell layer (Sonnenblick 1950).

At stage 10 of embryogenesis, pole cells migrate through the midgut epithelial layer (Figure 3*g*) (Hay et al 1990, Warrior 1994). Large pseudopodia extend from the pole cell bodies and interdigitate with protrusions formed by the endodermal epithelial cells. Coincident with the migration, this part of the endodermal epithelium reorganizes: Actin is lost from the apical side of the endodermal cells, apical cell-cell junctions are rearranged, and intercellular gaps form between the cells (Callaini et al 1995, Jaglarz & Howard 1995). Because these gaps are observed in regions where pole cells pass through the epithelium, it is possible that the gaps are a prerequisite for passage. In support of this idea is the observation that mutations in genes that affect differentiation of the endodermal epithelium such as *huckebein* and *serpent* prevent pole cell migration (Reuter 1994, Warrior 1994, Jaglarz & Howard 1995). Thus, at least in part, pole cell migration seems to require a rearrangement of endodermal cells. These changes in endodermal morphology, however, occur even in the absence of pole cells (Callaini et al 1995, Jaglarz & Howard 1995).

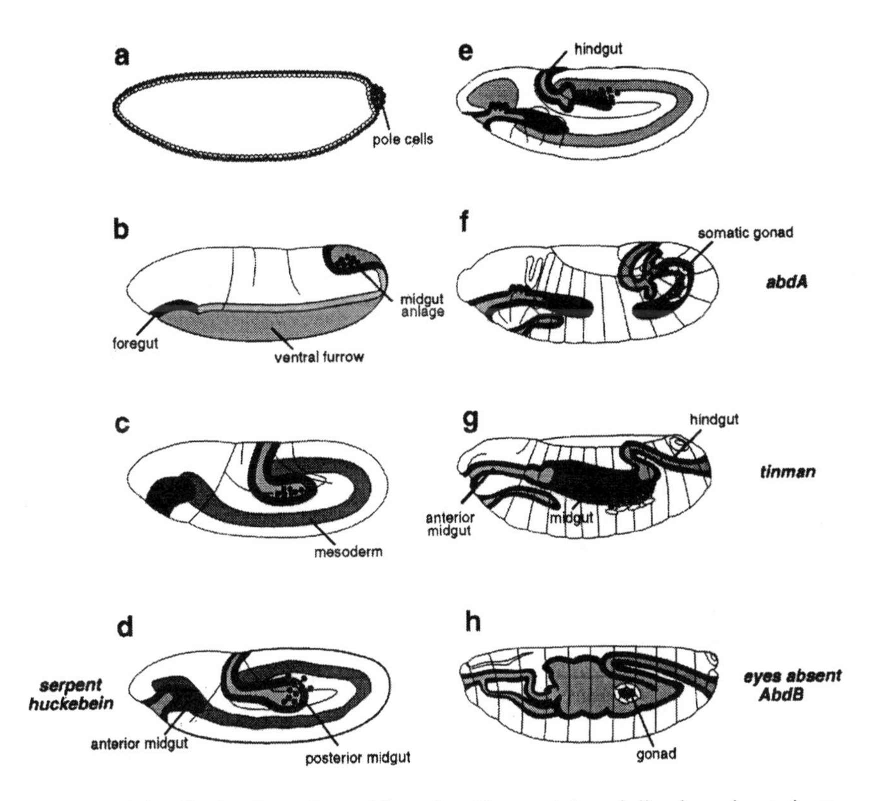

Figure 3 Pole cell migration and gonad formation. The orientation of all embryos is anterior to the left and dorsal to the top. Genes that affect particular steps in gonad formation are indicated in italics. (*a*) Stage 5 [≈3 h after egg laying (AEL), at 25°C]. Cellular blastoderm: Pole cells divide between 0 and 2 times. (*b*) Stage 7 (≈3.5 h AEL): Pole cells adhere to the prospective midgut endoderm as it moves dorsally and anteriorly during germ band extension. (*c*) Stage 8–9 (≈4–5 h AEL): Pole cells are inside the midgut pocket. (*d*) Stage 10 (≈5–7 h AEL): Pole cells migrate through the posterior midgut. (*e*) Stage 11 (≈7–9 h AEL): Pole cells make contact with and attach to the overlying mesoderm. (*f*) Stage 12 (≈9–10.5 h AEL): During germband retraction, pole cells are in close contact with gonadal mesoderm. (*g*) Stage 13 (≈10.5–11.5 h AEL): Pole cells align with somatic gonadal mesodermal cells. (*h*) Stage 15 (≈13–15 h AEL): Gonad coalescence, pole cells become encapsulated by somatic mesoderm. (Staging according to Wieschaus & Nüsslein-Volhard 1986; drawings after Hartenstein 1993.)

Attachment to Mesoderm

Once the pole cells have traversed the posterior midgut pocket, they move toward the dorsal side of the endoderm during stage 11 of embryogenesis (Figure 3*e*). At this stage, the endoderm is losing its epithelial organization and grows as a mesenchymal sheet towards the posterior. It is unlikely that chemotropic signals emanating from the overlying mesoderm cause this dorsal movement because pole cells still move to the dorsal side of the endoderm in *twist* and *snail* mutant embryos, which lack mesoderm (Jaglarz & Howard 1994). It is therefore more likely that this pole cell movement is triggered by unknown properties of the endodermal layer.

At the dorsal side of the endoderm, the pole cells come in contact with overlying mesoderm. It is still unclear how the pole cells are transferred from the endoderm to the mesoderm, but it is possible that changes in adhesive properties of the endodermal cells play an important role in this process. Adherence of the pole cells to the mesoderm occurs independently of mesoderm patterning and may therefore reflect an initial general adhesion between mesoderm and pole cells (Brookman et al 1992, Boyle & DiNardo 1995, Greig & Akam 1995; H Broihier & R Lehmann, unpublished results). Studies on PGC migration in vertebrates suggest that germ cells specifically adhere to extracellular matrix guidance factors such as fibronectin (Heasman et al 1981). Jaglarz & Howard (1995) showed that pole cells can adhere to and migrate on laminin in vitro. However, embryos mutant for the *Drosophila* laminin gene show only a late defect in gonad formation, suggesting that laminin is not required for the initial interactions of pole cells with endodermal or mesodermal cells. However, we cannot exclude the possibility that maternally supplied laminin or other extracellular matrix components are compensating for the lack of zygotic laminin in these mutants.

Association with Somatic Gonadal Precursors

Shortly after the pole cells have contacted the mesoderm, a subpopulation of mesodermal cells can be distinguished by their expression of markers such as the RNA transcript of the 412 retrotransposon (Brookman et al 1992). Although initially more broadly distributed, by stage 13 of embryogenesis 412-expressing cells are restricted to two groups that lie on either side of the ventral midline. Each group is in close contact with 10 to 15 pole cells (Figures 3*f*,*g*) (Brookman et al 1992, Warrior 1994). At this stage, pole cells and somatic gonadal cells are distributed along three segmental units between the anlagen of abdominal segments four to six [parasegments (PS) 10–12] (Boyle & DiNardo 1995). This observation corresponds well with gynandromorph studies that placed the somatic gonad primordium in a group of roughly ten cells in the prospective mesoderm of abdominal segments four and five (Szabad & Nöthiger 1992).

Association of pole cells with mesodermal cells depends on the function of the gene *tinman*, which is required for visceral mesoderm differentiation (Azpiazu & Frasch 1993, Bodmer 1993; H Broihier & R Lehmann, unpublished results), as well as homeotic genes of the bithorax complex (BX-C), which are required for region-specific mesoderm differentiation (Figures 3*f,g*) (Brookman et al 1992). In particular, mutations in *abdA* that affect specification of abdominal segments A1 to A8 (PS 7 to PS 13) inhibit gonad formation (Cumberledge 1992, Brookman 1992, Boyle & DiNardo 1995). Furthermore, ectopic expression of *abdA* in the mesoderm causes an expansion of the expression domain of the 412 retrotransposon. Consequently at stage 13, pole cells spread over a larger region of the mesoderm than in wild-type embryos (Boyle & DiNardo 1995, Greig & Akam 1995). This result indicates that *abdA* expression may be sufficient to specify somatic gonadal precursor fate. However, even in the presence of ectopic *abdA* expression and in the absence of positional information from other BX-C genes, embryonic gonads form preferentially in abdominal segment five (Boyle & DiNardo 1995, Greig & Akam 1995). This suggests that additional, as yet unknown, positional clues specify the site of embryonic gonad formation.

Embryonic Gonad Formation

At the end of embryogenesis, pole cells become encapsulated by the somatic gonadal cells to form rounded embryonic gonads located in the fifth abdominal segment on either side of the embryo (Figure 3*h*) (Sonnenblick 1950). Encapsulation requires the organized anterior movement of both pole cells and somatic gonadal precursor cells toward PS10. In wild type, *abdA* is expressed at high levels in PS 10 at this stage. This expression is controlled by the *abdA* regulatory region *iab-4*. Mutations in *iab-4* affect gonad coalescence but not the anterior movement of pole cells and gonadal mesoderm cells, suggesting that additional factors direct this movement (Boyle & DiNardo 1995). *AbdB* is not required for gonad coalescence; however, smaller gonads are formed in *AbdB* mutants (Brookman et al 1992).

Two subpopulations can be distinguished among the somatic gonadal precursors on the basis of marker gene expression: Anterior somatic gonadal cells (PS10-11) express *escargot*, whereas posterior cells (PS12) express *eyes absent* (*eya*) and *Dwnt-2* (Bonini et al 1993, Boyle & DiNardo 1995). The PS10-11 expression pattern of *escargot* depends on *abdA* function; the expression of *eya* and *Dwnt-2* in the posterior part of the gonad depends on both *abdA* and *AbdB* (Boyle & DiNardo 1995). Among these marker genes, a role in embryonic gonad formation has only been demonstrated for *eya*. In *eya* mutant embryos, pole cells sort into two lateral groups, but gonads fail to coalesce at stage 15 (H Broihier, L Moore & R Lehmann, unpublished results).

EMBRYONIC GERM CELL DIFFERENTIATION

Germ Line-Specific Splicing

One of the earliest known germ cell specific differentiation events is splicing of the third intron of the P-element transposase gene. Transposase RNA was originally shown to be spliced to produce active transposase only in the germ line-derived cells of ovaries and testes (Laski et al 1986). This tissue-specific splicing event occurs in germ cells as early as embryonic stage 9, reaching a maximum at stage 10 of embryonic development (see Figure 3 for embryonic stages) (Kobayashi et al 1993b). Not all pole cells show splicing activity. Interestingly, pole cells that productively splice the P-element third intron are more likely to populate the embryonic gonad than those that fail to remove the third intron. Thus intrinsic differences may exist among pole cells with regard to their ability to participate in further germ cell differentiation as early as embryonic stage 10. It is not clear how such a difference in developmental potential would be established. One possibility is that pole cells inherit different levels of germ line determinants at the time they form. Alternatively, interactions between somatic tissues and pole cells during their migratory phase could affect their ability to contribute to the germ line. Whatever the mechanism, it may have been co-opted by the transposition machinery to ensure maximal propagation of P-elements to the next generation (Kobayashi et al 1993b).

Germ Line-Specific Transcription

Transcription is reinitiated in pole cells just before they migrate through the midgut epithelium (Zalokar 1976). Very few genes have been shown to be specifically transcribed in the primordial germ cells; the earliest of these is *vasa*, whose zygotic transcript is detectable as early as embryonic stage 9–10, before the pole cells exit the midgut pocket (Hay et al 1988b; H Broihier, L Moore & R Lehmann unpublished). Transcription of all other genes known to be expressed in the pole cells is not observed until the gonads have reinitiated proliferation (embryonic stage 16–18) (Mével-Ninio et al 1995; A Williamson & R Lehmann, unpublished results).

In mouse, genes involved in embryonic germ cell migration and proliferation have been identified on the basis of male and female sterility. Two of the most well-studied genes are *White spotting* (*W*) and *Steel* (*Sl*) (Gomperts et al 1994; for review see Witte 1990). Strong alleles of both genes cause homozygous embryos to die of anemia before birth, whereas survivors are sterile and have white coats. *W* encodes the c-kit receptor tyrosine kinase, which is expressed in germ cells, hemopoietic stem cells, and melanoblasts. *Sl*, on the other hand, encodes the diffusible c-kit ligand and is expressed broadly along the migratory route of the three cell types, as well as in their target tissues. In vitro and in vivo studies suggest that *W* and *Sl* functions are required for survival of primordial germ cells. In contrast, among the known male and/or female sterile mutations in *Drosophila*, none affect survival of pole cells during embryogenesis. Although previous studies had suggested a role for the zygotic *ovo* product in survival of female pole cells as early as the blastoderm stage (Oliver et al 1987), recently it was shown that *ovo* RNA transcription is not detectable until late embryogenesis and that *ovo* function is not required for PGC survival until the larval stages (Mével-Ninio et al 1995, Rodesch et al 1995, Staab & Steinmann-Zwicky 1995). Thus it is possible that embryonic germ cell survival depends on maternal factors and that these genes have not yet been identified.

Embryonic Gonad Size

Estimates for the number of pole cells that populate each of the embryonic gonads and give rise to the germ line range from 5 to 20 (Wieschaus & Szabad 1979, Underwood et al 1980, Wieschaus et al 1981). Furthermore, Sonnenblick (1941) and more recently Poirié et al (1995) noted a sex-specific difference in the number of primordial germ cells incorporated into embryonic gonads, with female gonads containing, on average, fewer pole cells than male gonads (Sonnenblick 1941, Poirié et al 1995). It is not known whether the difference in the number of germ cells in late embryonic/early larval gonads is the result of cell autonomous properties of the pole cells themselves or sex-specific differences in the number of the somatic gonadal precursor cells that encapsulate them. The latter hypothesis is supported by the observations that changes in the number of somatic gonadal precursor cells, i.e. by mutation in *AbdB*, can dramatically affect the size of the gonad, whereas a significant increase or decrease in the number of pole cells, i.e. by altering *oskar* gene copy number, does not significantly affect the size of the embryonic gonad (Brookman et al 1992, Ephrussi & Lehmann 1992, R Lehmann, unpublished observations). Moreover, sex-specific differences in gonadal size have been shown to be affected by genes like *tra* and *dsx*, which act in the soma but not in the germ line (Steinmann-Zwicky 1994).

SEX DETERMINATION

Sxl is the Primary Signal for Sex Determination in the Soma

The sex of each somatic cell is determined autonomously by the X-chromosome to autosome (X:A) ratio and the primary sex regulatory gene *Sex lethal* (*Sxl*) (for recent reviews see Parkhurst & Meneely 1994, MacDougall et al 1995). In

XX animals, early transcription of *Sxl* establishes autocatalysis of *Sxl* female-specific splicing, which in turn leads to processing of *transformer* (*tra*) pre-mRNA. *tra*, together with the ubiquitous product of the *transformer-2* (*tra-2*) gene, activates production of the female-specific *doublesex* product (Dsx^F). XY animals lack *sxl* and *tra* products, and as a consequence, the male-specific *dsx* product (Dsx^M) is expressed. The X:A ratio and *sxl* also control dosage compensation (reviewed by Baker et al 1994). Lack of *sxl* product in XY animals leads to activation of the *mle* and *msl* genes and X-chromosomal hypertranscription in males. In females, X-chromosomal hypertranscription is prevented by *sxl*-dependent splicing of the *male-specific lethal-2* gene (Zhou et al 1995).

Autonomous and Inductive Signals Control Germ Line Sex Determination

In contrast to the cell-autonomous mode by which sex-specific characteristics are expressed in the soma, sex determination in germ cells is controlled by germ line-autonomous signals, as well as by inductive interactions with the soma (Figure 4). This is best demonstrated by the finding that heterogametic chimeras are sterile (Van Deusen 1976). Closer analysis of this phenomenon reveals that XX pole cells transplanted into XY host embryos at blastoderm stage become spermatogenic or degenerate, whereas XY germ cells transplanted into XX hosts develop spermatogenic characteristics. This indicates that XX germ cells require inductive signals from the soma to initiate oogenic development, whereas XY germ cells appear to initiate spermatogenesis autonomously (Steinmann-Zwicky et al 1989).

Sxl plays a critical role in germ line sex determination; however, in contrast to sex determination in the soma, *Sxl* is not the primary regulator in germ cells. In the absence of active Sxl protein, XX germ cells display morphological features characteristic of spermatogenic cells or they degenerate (Schüpbach 1985, Steinmann-Zwicky et al 1989). However, constitutively active Sxl product can promote oogenic development only in XX germ cells but not in XY germ cells (Steinmann-Zwicky et al 1989, Steinmann-Zwicky 1993). It is therefore likely that additional factors act in concert with *Sxl* to promote female germ line determination and differentiation. Identification of these additional regulatory genes may be complicated if they control sex-specific differentiation as well as proliferation. For example, mutations in the *ovo* and *ovarian tumor* (*otu*) genes, which are required for sex determination and sex-specific splicing of *Sxl* in the germ line, also affect proliferation of the female germ line (see below).

It is not known when during development sex determination occurs in female germ cells. Sxl protein has been detected in embryonic XX germ cells as early as stage 16 of embryogenesis; however, oogenesis-specific isoforms of Sxl do not appear to affect the process until egg chamber formation (Bopp et al 1993). Production of active Sxl protein in female germ cells depends upon a different set of genes from those that act in the soma. For example, pole cell transplantation experiments show that the numerator elements *sis a, sis b,* and *runt* and the maternally provided *daughterless* gene have no detectable function in the germ line (Cronmiller & Cline 1987, Granadino et al 1993, Steinmann-Zwicky 1993). Furthermore, genes known to act downstream of *Sxl* in somatic cells, including *transformer* (*tra*), *transformer-2* (*tra-2*), *doublesex* (*dsx*), and *intersex* (*ix*), have all been shown to be dispensable in the germ line for sex determination (Marsh & Wieschaus 1978, Schüpbach 1982). Thus different sets of genes act upstream and downstream of *Sxl* in the germ line sex determination pathway. Synthesis of functional Sxl product in germ cells depends on factors expressed in the germ line such as *ovo, otu, female lethal* (2)*d* (*fl*(2)*d*), *sans fille* (*snf*) (see Figure 4), as well as on inductive signals from the soma (Granadino et al 1992, Bopp et al 1993, Oliver et al 1993).

Continuous inductive interactions between the soma and the germ line throughout gametogenesis are critical to ensure differentiation of female germ cells. The somatic signals that promote female differentiation of XX germ cells are transmitted by the somatic sex determination hierarchy, as well as through a TRA-2-dependent mechanism that is independent of *tra* and *dsx* (Figure 4). One of the early targets for mediating the somatic signal is the 98-kDa product of the *otu* gene (Nagoshi et al 1995). During later stages of oogenesis, female-specific somatic signals transmitted through Dsx^F are required for the processes of cyst formation and subsequent differentiation of the female germ line (Nöthiger et al 1989, Steinmann-Zwicky et al 1989, Steinmann-Zwicky 1994).

The male germ line starts to differentiate at the end of embryogenesis. XY germ cells initiate male-specific germ line differentiation regardless of the sex of the surrounding tissue (Steinmann-Zwicky et al 1989). However, XY germ cells require a male somatic environment and Dsx^M activity to fully differentiate into motile sperm. Thus neither XX nor XY germ cells can undergo normal gametogenesis in the absence of sex-specific cues from the surrounding somatic tissues (Figure 4) (Nöthiger et al 1989, Steinmann-Zwicky et al 1989, Steinmann-Zwicky 1994).

Dosage Compensation in the Germ Line

The X:A ratio and *sxl* not only control sex differentiation but also the level of X chromosome transcription. Very little is known about dosage compensation in the germ line. It is not clear if the X chromosome is hyperactivated in XY cells and, if so, when dosage compensation occurs during gametogenesis. Finally, it is unclear whether the same genes required for dosage compensation

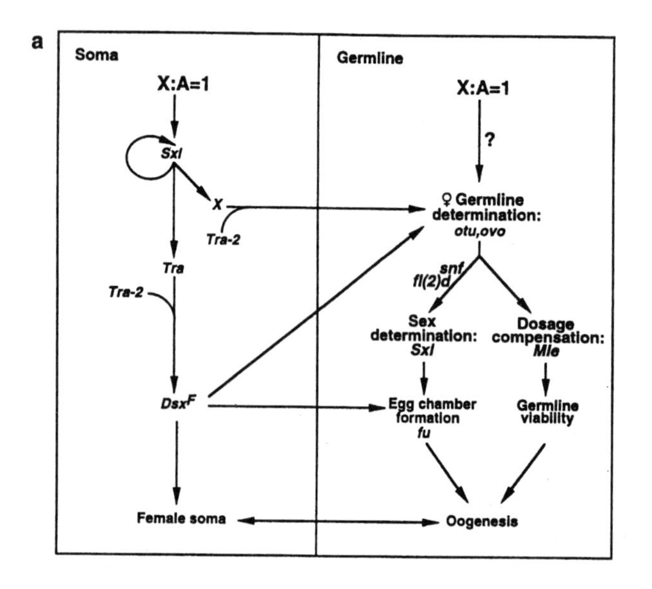
a
Soma
Germline
X:A=1
X:A=1
?
Sxl
X
Tra-2
♀ Germline determination: otu,ovo
Tra
Tra-2
snf
fl(2)d
Sex determination: Sxl
Dosage compensation: Mle
DsxF
Egg chamber formation fu
Germline viability
Female soma
Oogenesis

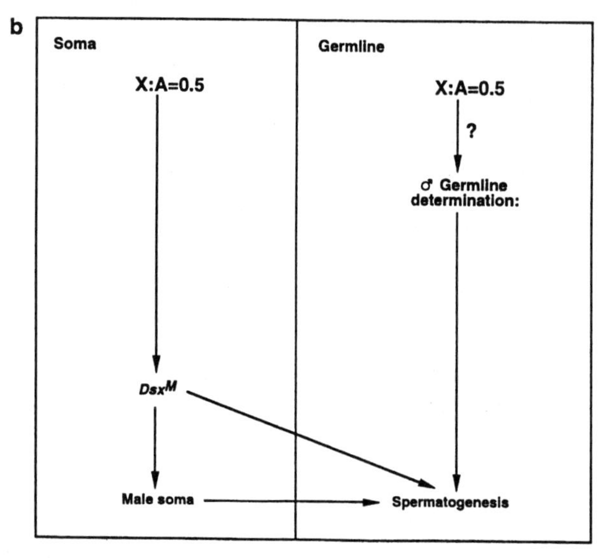
b
Soma
Germline
X:A=0.5
X:A=0.5
?
♂ Germline determination:
DsxM
Male soma
Spermatogenesis

in the soma also act in the germ line. So far, *male-specific lethal* (*mle*) is the only known gene required for somatic dosage compensation that also affects the germ line (Bachiller & Sanchez 1986). In the soma, *mle* is required for hypertranscription of X-linked genes. It has been shown that the inviablity of XX germ cells lacking functional *ovo* gene product can be partially rescued by deletion of the *mle* gene (Oliver et al 1993). This result suggests that *ovo* may be a mediator between the X:A ratio on one hand and sex differentiation and dosage compensation on the other (Figure 4).

GAMETOGENESIS

Oogenesis

Upon population of the somatic gonadal mesoderm by the pole cells during embryogenesis, female germ cells enter a mitotic phase that continues throughout larval stages. Differentiation of the ovary is initiated during the third larval instar when cells of the somatic gonadal mesoderm begin to differentiate (for reviews of oogenesis see King 1970, Mahowald & Kambysellis 1980, Spradling 1993). During metamorphosis, terminal filament cells, which are derived from the somatic gonadal precursors, have subdivided each gonad into 16–20 ovarioles. Each ovariole is a string of maturing egg chambers with terminal filament cells at the apical tip (see Figures 2 and 5).

Just distal to the terminal filament cells, in the germarial region 1 of the ovariole, 2 to 3 germ line stem cells each divide repeatedly to produce a cystoblast and another stem cell (Figure 5*a*) (King 1970, Lin & Spradling 1993). Each cystoblast undergoes four rounds of mitosis with incomplete cytokinesis to produce a 16-cell cyst of interconnected cystocytes. Germ line cysts are

←

Figure 4 Germ line sex determination. (*a*) Germ line sex determination in females. The left box summarizes the somatic sex determination hierarchy (for review see MacDougall et al 1995). The soma produces inductive signals that promote female-specific differentiation. The right box summarizes processes that occur in the germ line. *tra-2-* and *dsx*-dependent somatic signals, as well as the X:A ratio, initiate female germ line determination, which includes control of sex-specific *Sxl* splicing, intracellular distribution of Sxl protein, germ line proliferation, and possibly dosage compensation. A later *dsx*-dependent somatic signal, as well as a number of germ line-specific gene functions, i.e. the germ line product of the *fused* (*fu*) gene, control egg chamber formation. Finally reciprocal signaling between the germ line and the somatic follicle cells are required for germ cell differentiation (see Figure 2). (*b*) Germ line sex determination in males. Initiation of male germ line differentiation occurs autonomously and is presumably dependent on the X:A ratio. Later stages of spermatogenesis require somatic input through *dsx*. Terminal differentiation of sperm requires germ line expression of *tra-2*, as well as inductive signals between the soma and the germ line.

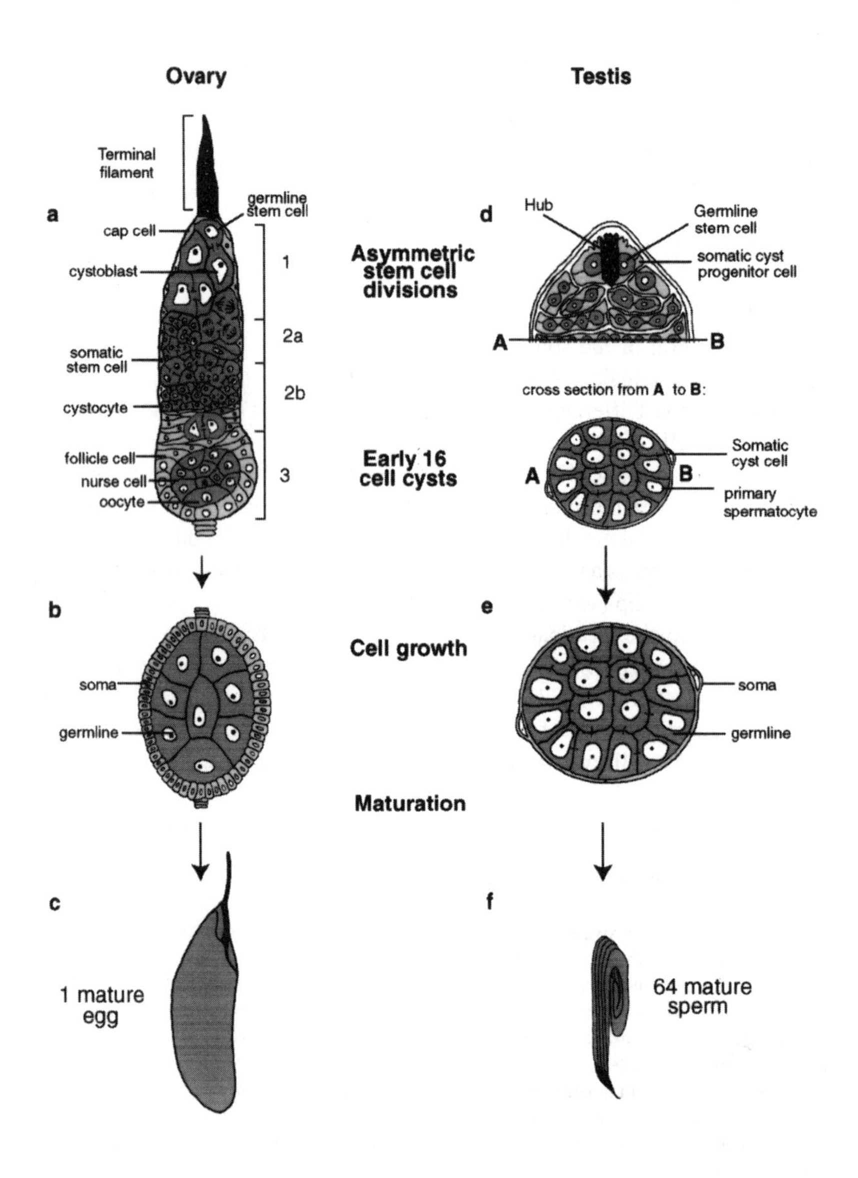
Ovary
Testis
Terminal filament
germline stem cell
a
cap cell
cystoblast
somatic stem cell
cystocyte
follicle cell
nurse cell
oocyte
1
2a
2b
3
Asymmetric stem cell divisions
Early 16 cell cysts
d
Hub
Germline stem cell
somatic cyst progenitor cell
A
B
cross section from A to B:
Somatic cyst cell
primary spermatocyte
b
soma
germline
Cell growth
e
soma
germline
Maturation
c
1 mature egg
f
64 mature sperm

displaced posteriorly along the ovariole as germ line stem cells continue to divide. Subsequently, cysts become surrounded by approximately 16 somatic follicle cells, derived from two somatic stem cells located in germarial region 2a/b (Figure 5*a*) (Margolis & Spradling 1995). The newly formed egg chamber is pinched off from the younger cysts by follicle cells, which form stalks between successive egg chambers in germarial region 3 (Figure 5*a*).

Recent results indicate that the orientation of the first cystoblast division with respect to the germ line stem cell may determine which of the 16 cystocytes will become the oocyte (Lin & Spradling 1995). Upon determination and throughout oogenesis, the oocyte is arrested in prophase I of meiosis. The remaining 15 cystocytes become polyploid nurse cells (Figure 5*b*). Connected to the oocyte by large intercellular bridges, termed ring canals, these cells act as feeder cells to the oocyte. After multiple divisions, about 1000 somatic follicle cells surround the growing oocyte at mid-oogenesis (see Figure 2*b*). Interactions between follicle cells and the oocyte establish oocyte polarity, which determines the anterior to posterior and dorsal to ventral axes of the future embryo (see review by Lehmann 1995). At the end of oogenesis, the nurse cells extrude their contents into the oocyte and degenerate. The follicle cells secrete the egg covers, the vitelline membrane, and the chorion before they degenerate (King 1970, Lin & Spradling 1993). Meiosis resumes when the egg progresses through the uterus and is completed in the freshly laid egg.

Spermatogenesis

Testis differentiation begins as soon as the first larval instar, when differentiated somatic testicular structures are visible and primordial spermatocyte-containing cysts are present (Sonnenblick 1941; for review see Lindsley & Tokuyasu 1980, Fuller 1993). Spermatogenesis is initiated in the so-called germinal proliferation center. This region, which can be viewed as the analogue of the female germarium, is comprised of an apical, cone-shaped group of small somatic cells called the hub, surrounded by six germ line stem cells and their accompanying

←

Figure 5 Oogenesis and spermatogenesis. Schematic presentation of oogenesis (*left*) and spermatogenesis (*right*). The female germarium (*a*) and male germinal proliferation center (*d*). In the ovary, germ line and somatic stem cells are located in different regions (region 1 and 2a of germarium, respectively). In region 2a/2b, somatic stem cells (follicle cells) join the dividing germ line cyst; in region 3, the cyst is surrounded by follicle cells. In the testis, germ line and somatic stem cells are organized around the hub. The female (*b*) and male (*e*) cysts. Female and male germ cells are connected by intercellular bridges and are surrounded by somatic follicle/cyst cells (about 1000 in females and 2 in males). One egg (*c*) and 64 sperm (*f*) are generated per cystoblast. Female germ line drawings after Peifer et al 1993, Lin & Spradling 1994, Rongo & Lehmann 1996. Male germ line after Lindsley & Tokuyasu 1980, Castrillon et al 1993, Fuller 1993.

somatic cyst progenitor cells (compare Figure 5*d* and *a*). Germ line stem cells divide in synchrony with somatic cyst progenitor cells, and each spermatogonial cell is encased by one pair of somatic cyst cells. The germ line and somatic stem cells remain anchored to the hub, while the spermatogonial cell enters four rounds of mitosis. As in females, cytokinesis in the germ line is incomplete and intercellular bridges connect differentiating diploid spermatocytes. In ovaries and testes, cyst formation is affected by two genes, *bag of marbles* (*bam*) and *benign gonial cell neoplasm* (*bgcn*). Therefore, these two genes are candidates for specifying the nature and number of cystocyte divisions that give rise to normal 16-cell cysts in males and females (McKearin & Spradling 1990, Spradling 1993).

The newly formed cysts are moved posteriorly along the testis and enter a growth phase (Figure 5*e*). Anywhere from 5 to 13 cysts may be progressing through the testis from the apex, where the primary gonial cell is produced, to the middle portion of the testis. During the early growth phase, cysts assume a disk-like conformation and are aligned perpendicularly to the wall of the testis. Further posterior, the growing cysts align around the circumference of the testis wall, parallel to the long axis. In the lumen of the lower testis, spermatocytes undergo meiosis, which generates 64 haploid spermatids, still interconnected by ring canals (Figure 5*f*). Subsequently, spermatids elongate, individualize, and coil their tails to generate motile sperm (Lindsley & Tokuyasu 1980).

Control of Stem Cell Development During Oogenesis and Spermatogenesis

Both male and female germ line stem cells lie in the apical region of the gonad, adjacent to a discrete set of somatic cell types that may play critical roles in controlling stem cell divisions. The putative germ line stem cells in ovaries (about 32) and testes (about 12) were originally identified by mosaic analysis as well as by morphological criteria (Schüpbach et al 1978, Wieschaus & Szabad 1979, Lindsley & Tokuyasu 1980). More recently, examination of enhancer trap lines in testes (Gönczy et al 1992) and clonal analysis in ovaries (Margolis & Spradling 1995) have allowed a more direct assessment of division patterns of the putative stem cells. Laser ablation experiments in which the three to six anterior-most germ cells in the ovariole were destroyed identified this cluster of cells as the germ line stem cells (Lin & Spradling 1993).

Because germ line stem cells make direct contact with the somatic terminal filament (TF) cells, it has been suggested that the TF cells play a role in regulating germ line stem cell mitoses similar to the somatic distal tip cell (DTC) in *C. elegans.* Laser ablation of the DTC causes germ line stem cells to enter meiosis prematurely at the expense of mitotic proliferation (Kimble & White 1981; for reviews see Ellis & Kimble 1994, Schedl 1995). In *Drosophila* ovaries, TF cells appear to regulate germ line mitosis; however, the mechanism is unclear. Although partial laser ablation of the TF leads to an increase in germ line stem cell divisions (Lin & Spradling 1993), ovaries of females mutant for the *bric-a-brac* gene, which have abnormally short TFs, show a decrease in stem cell mitoses (Sahut-Barnola et al 1995). Thus a clear determination of the role of the TF in germ cell proliferation will have to await experiments in which these cells are completely removed. It is further possible that in addition to the TF cells, other somatic cells such as the Cap and Terminal filament base cells, which lie distal to the terminal filament and in close apposition to the germ line stem cells (Figure 5*a*), also regulate germ line stem cell divisions (Margolis & Spradling 1995, Forbes et al 1996).

In testes, the obvious candidates for somatic cell types involved in regulation of germ line stem cell divisions are the hub cells and the cyst progenitor cells because of their clear physical association with germ line stem cells in the germarial proliferation center (Lindsley & Tokuyasu 1980, Gönczy et al 1992). Although it has been demonstrated that in the absence of germ cells the hub increases in size (Gönczy et al 1992), it has not been tested whether germ cell divisions are regulated by the hub.

ACKNOWLEDGMENTS

We are grateful to Laura Dickinson who prepared antibody stainings and EM sections of the embryo shown in Figure 1. We would like to thank past and present members of our laboratory for contributing results and many of the ideas summarized in this review and Mark VanDoren for critical suggestions on the manuscript. Finally, we wish to thank our colleagues for encouragement and for sending us preprints and reprints of their work and communicating unpublished results to us. Our work on germ plasm formation is supported by a grant of the National Institutes of Health. RL is a former HHMI associate investigator.

Literature Cited

Azpiazu N, Frasch M. 1993. *tinman* and *bagpipe*: two homeo box genes that determine cell fates in the dorsal mesoderm of *Drosophila. Genes Dev.* 7:1325–40

Bachiller D, Sanchez L. 1986. Mutations affecting dosage compensation in *Drosophila melanogaster.* Effects in the germ line. *Dev. Biol.* 118:379–84

Baker BS, Gorman M, Marín I. 1994. Dosage compensation in *Drosophila. Annu. Rev. Genet.* 28:491–521

Bardsley A, Macdonald K, Boswell RE. 1993. Distribution of tudor protein in the *Drosophila* embryo suggests separation of function based on site of localization. *Development* 119:207–19

Bate M, Martinez Arias A, eds. 1993. *The Development of Drosophila melanogaster.* Cold Spring Harbor, NY: Cold Spring Harbor Press. 1558 pp.

Bodmer R. 1993. The gene *tinman* is required for specification of the heart and visceral muscles in *Drosophila. Development* 118:719–29

Bonini NM, Leierson WM, Benzer S. 1993. The *eyes absent* gene: genetic control of cell survival and differentiation in the developing Drosophila eye. *Cell* 72:379–95

Bopp D, Horabin JI, Lersch RA, Cline TW, Schedl P. 1993. Expression of the *Sex-lethal* gene is controlled at multiple levels during *Drosophila* oogenesis. *Development* 118:797–812

Boyle M, DiNardo S. 1995. Specification, migration and assembly of the somatic cells of the *Drosophila* gonad. *Development* 121(6):1815–25

Brookman JJ, Toosy AT, Shashidhara LS, White RAH. 1992. The 412 retrotransposon and the development of gonadal mesoderm in *Drosophila. Development* 116:1185–92

Callaini G, Riparbelli MG, Dallai R. 1995. Pole cell migration through the gut wall of the *Drosophila* embryo: analysis of cell interactions. *Dev. Biol.* 170:365–75

Campos-Ortega JA, Hartenstein V. 1985. *The Embryonic Development of Drosophila melanogaster.* Heidelberg: Springer-Verlag

Castrillon DH, Gönczy P, Alexander S, Rawson R, Eberhart CG, et al. 1993. Toward a molecular genetic analysis of spermatogenesis in *Drosophila melanogaster:* characterization of male-sterile mutants generated by single P element mutagenesis. *Genetics* 135:489–505

Cronmiller C, Cline TW. 1987. The Drosophila sex determination gene *daughterless* has different functions in the germ line versus the soma. *Cell* 48:479–87

Cumberledge S, Szabad J, Sakonju S. 1992. Gonad formation and development requires the *abd*-A domain of the bithorax complex in *Drosophila melanogaster. Development* 115:395–402

Dalby B, Glover DM. 1992. 3′ non-translated sequences in *Drosophila* cyclin B transcripts direct posterior pole accumulation late in oogenesis and perinuclear association in syncytial embryos. *Development* 115:989–97

Dalby B, Glover DM. 1993. Discrete sequence elements control posterior pole accumulation and translational repression of maternal cyclin B RNA in *Drosophila. EMBO J.* 12:1219–27

Edgar BA, Schubiger G. 1986. Parameters controlling transcriptional activation during early Drosophila development. *Cell* 44:871–77

Ellis RE, Kimble J. 1994. Control of germ cell differentiation in *Caenorhabditis elegans. Ciba Found. Symp.* 182:179–88

Ephrussi A, Dickinson LK, Lehmann R. 1991. *Oskar* organizes the germ plasm and directs localization of the posterior determinant *nanos. Cell* 66:37–50

Ephrussi A, Lehmann R. 1992. Induction of germ cell formation by *oskar. Nature* 358:387–92

Foe V, Odell GM, Edgar BA. 1993. Mitosis and morphogenesis in the *Drosophila* embryo: point and counterpoint. See Bate & Martinez Arias 1993, pp. 149–300

Forbes AJ, Lin, H, Ingham PW, Spradling AC. 1996. *hedgehog* is required for the proliferation and specification of somatic cells during egg chamber assembly in *Drosophila* oogenesis. *Development* 122:1125–35

Fuller M. 1993. Spermatogenesis. See Bate & Martinez Arias 1993, pp. 71–147

Gavis ER, Curtis D, Lehmann R. 1996. Identification of *cis*-acting sequences that control *nanos* RNA localization. *Dev. Biol.* 176:36–50

Gavis ER, Lehmann R. 1992. Localization of *nanos* RNA controls embryonic polarity. *Cell* 71:301–13

Gavis ER, Lehmann R. 1994. Translational regulation of *nanos* by RNA localization. *Nature* 369:315–18

Gomperts M, Wylie C, Heasman J. 1994. Primordial germ cell migration. *Ciba Found. Symp.* 182:121–34

Gönczy P, Viswanathan S, DiNardo S. 1992. Probing spermatogenesis in *Drosophila* with P-element enhancer detectors. *Development* 114:89–98

Granadino B, San Juan A, Santamaria P, Sánchez L. 1992. Evidence of a dual function in *fl(2)d*, a gene needed for *Sex-lethal* expression in *Drosophila melanogaster. Genetics* 130:597–612

Granadino B, Santamaria P, Sánchez L. 1993. Sex determination in the germ line of *Drosophila melanogaster*: activation of the gene *Sex-lethal. Development* 118:813–16

Grant M, Zuccotti M, Monk M. 1992. Methylation of CpG sites of two X-linked genes coincides with X-inactivation in the female mouse embryo but not in the germ line. *Nat. Genet.* 2:161–66

Greig S, Akam M. 1995. The role of homeotic genes in the specification of the *Drosophila* gonad. *Curr. Biol.* 5:1057–62

Hartenstein V. 1993. *Atlas of Drosophila Development*. Cold Spring Harbor, NY: Cold Spring Harbor Press

Hatanaka K, Okada M. 1991. Retarded nuclear migration in *Drosophila* embryos with aberrant F-actin reorganization caused by maternal mutations and by cytochalasin treatment. *Development* 111:909–20

Hay B, Ackerman L, Barbel S, Jan LY, Jan YN. 1988a. Identification of a component of *Drosophila* polar granules. *Development* 103:625–40

Hay B, Jan LY, Jan YN. 1988b. A protein component of Drosophila polar granules is encoded by *vasa* and has extensive sequence similarity to ATP-dependent helicases. *Cell* 55:577–87

Hay B, Jan LY, Jan YN. 1990. Localization of *vasa,* a component of *Drosophila* polar granules, in maternal-effect mutants that alter embryonic anteroposterior polarity. *Development* 109:425–33

Heasman J, Hynes RO, Swan AP, Thomas V, Wylie CC. 1981. Primordial germ cells of *Xenopus* embryos: the role of fibronectin in their adhesion during migration. *Cell* 27:437–47

Huettner AF. 1923. The origin of the germ cells in *Drosophila melanogaster. J. Morphol.* 39:249–65

Illmensee K, Mahowald AP. 1974. Transplantation of posterior polar plasm in *Drosophila.* Induction of germ cells at the anterior pole of the egg. *Proc. Natl. Acad. Sci. USA* 71:1016–20

Illmensee K, Mahowald AP, Loomis MR. 1976. The ontogeny of germ plasm during oogenesis in *Drosophila. Dev. Biol.* 49:40–65

Jaglarz MK, Howard KR. 1994. Primordial germ cell migration in *Drosophila melanogaster* is controlled by somatic tissue. *Development* 120:83–89

Jaglarz MK, Howard KR. 1995. The active migration of *Drosophila* primordial germ cells. *Development* 121:3495–503

Jongens TA, Ackerman LD, Swedlow JR, Jan LY, Jan YN. 1994. *Germ cell-less* encodes a cell type-specific nuclear pore-associated protein and functions early in the germ-cell specification pathway of *Drosophila. Genes Dev.* 8:2123–36

Jongens TA, Hay B, Jan LY, Jan YN. 1992. The *germ cell-less* gene product: a posteriorly localized component necessary for germ cell development in Drosophila. *Cell* 70:569–84

Kafri T, Ariel M, Brandeis M, Shemer R, Urven L, et al. 1992. Developmental pattern of gene-specific DNA methylation in the mouse embryo and germ line. *Genes Dev.* 6:705–14

Kim-Ha J, Smith JL, Macdonald PM. 1991. *oskar* mRNA is localized to the posterior pole of the Drosophila oocyte. *Cell* 66:23–35

Kimble J, White JG. 1981. On the control of germ cell development in *Caenorhabditis elegans. Dev. Biol.* 81:208–19

King RC. 1970. *Ovarian Development in Drosophila melanogaster.* New York: Academic

Kobayashi S, Amikura R, Nakamura A, Saito H, Okada M. 1995. Mislocalization of *oskar* product in the anterior pole results in ectopic localization of mitochondrial large ribosomal RNA in *Drosophila* embryos. *Dev. Biol.* 169:384–86

Kobayashi S, Amikura R, Okada M. 1993a. Presence of mitochondrial large ribosomal RNA outside mitochondria in germ plasm of *Drosophila melanogaster. Science* 260:1521–24

Kobayashi S, Kitamura T, Sasaki H, Okada M. 1993b. Two types of pole cells are present in the *Drosophila* embryo, one with and one without splicing activity for the third P-element intron. *Development* 117:885–93

Kobayashi S, Okada M. 1989. Restoration of pole-cell-forming ability to uv-irradiated *Drosophila* embryos by injection of mitochondrial lrRNA. *Development* 107:733–42

Kobayashi S, Yamada M, Asaoka M, Kitamura T. 1996. Essential role of the posterior morphogen *nanos* for germline development in *Drosophila. Nature* 380:708–11

Laski FA, Rio DC, Rubin GM. 1986. Tissue specificity of Drosophila P element transposition is regulated at the level of mRNA splicing. *Cell* 44:7–19

Lasko PF, Ashburner M. 1990. Posterior localization of *vasa* protein correlates with, but is not sufficient for, pole cell development. *Genes Dev.* 4:905–21

Lehmann R. 1988. Phenotypic comparison between maternal and zygotic genes controlling the segmental pattern of the *Drosophila* embryo. *Development* 104:17–27 (Suppl.)

Lehmann R. 1995. Establishment of embryonic polarity during *Drosophila* oogenesis. *Sem. Dev. Biol.* 6:25–38

Lehmann R, Nüsslein-Volhard C. 1991. The maternal gene *nanos* has central role in pattern formation of the *Drosophila* embryo. *Development* 112:679–91

Lin H, Spradling AC. 1993. Germ line stem cell division and egg chamber development in transplanted *Drosophila* germaria. *Dev. Biol.* 159:140–52

Lin H, Spadling AC. 1994. The *Drosophila* fusome, a germline-specific organelle, contains membrane skeletal proteins and functions in cyst formation. *Development* 120:947–56

Lin H, Spradling AC. 1995. Fusome asymmetry and oocyte determination in *Drosophila. Dev. Genet.* 16:6–12

Lindsley D, Tokuyasu KT. 1980. Spermatogenesis. In *Genetics and Biology of Drosophila,* ed. M Ashburner, TR Wright. pp. 225–94. London: Academic

MacDougall C, Harbison D, Bownes M. 1995. The developmental consequences of alternate splicing in sex determination and differentiation in *Drosophila*. *Dev. Biol.* 172:353–76

Mahowald AP. 1962. Fine structure of pole cells and polar granules in *Drosophila melanogaster*. *J. Exp. Zool.* 151:201–5

Mahowald AP. 1968. Polar granules of *Drosophila*. II. Ultrastructural changes during early embryogenesis. *J. Exp. Zool.* 167:237–62

Mahowald AP. 1971. Polar granules of *Drosophila*. III. The continuity of polar granules during the life cycle of *Drosophila*. *J. Exp. Zool.* 176:329–44

Mahowald AP, Kambysellis MP. 1980. Oogenesis. In *The Genetics and Biology of Drosophila*, ed. M Ashburner, TRF Wright, 2:141–224. London: Academic

Margolis J, Spradling AC. 1995. Identification and behavior of epithelial stem cells in the *Drosophila* ovary. *Development* 121:3797–807

Markussen F-H, Michon A-M, Breitweiser W, Ephrussi A. 1995. Translational control of *oskar* generates Short OSK, the isoform that induces pole plasm assembly. *Development* 121:3723–32

Marsh JL, Wieschaus E. 1978. Is sex determination in germ line and soma controlled by separate genetic mechanisms? *Nature* 272:249–51

McKearin DM, Spradling AC. 1990. *bag-of-marbles*: a *Drosophila* gene required to initiate both male and female gametogenesis. *Genes Dev.* 4:2242–51

Mello CC, Draper BW, Krause M, Weintraub H, Priess JR. 1992. The *pie-1* and *mex-1* genes and maternal control of blastomere identity in early C. elegans embryos. *Cell* 70:163–76

Mével-Ninio M, Terracol R, Salles C, Vincent A, Payre F. 1995. *ovo*, a *Drosophila* gene required for ovarian development, is specifically expressed in the germ line and shares most of its coding sequences with *shavenbaby*, a gene involved in embryo patterning. *Mech. Dev.* 49:83–95

Nagoshi R, Patton S, Bae E, Geyer P. 1995. The somatic sex determines the requirement for ovarian tumor gene activity in the proliferation of the *Drosophila* germ line. *Development* 121:579–87

Nöthiger R, Jonglez M, Leuthold M, Meier-Gerschwiler P, Weber T. 1989. Sex determination in the germ line of *Drosophila* depends on genetic signals and inductive somatic factors. *Development* 107:505–18

Okada M. 1982. Loss of the ability to form pole cells in *Drosophila* embryos with artificially delayed nuclear arrival at the posterior pole. In *Embryonic Development, Part A: Genetic Aspects*. ed. M Burger, R Weber, pp. 363–72. New York: Liss

Oliver B, Kim Y-J, Baker BS. 1993. *Sex-lethal*, master and slave: a hierarchy of germ-line sex determination in *Drosophila*. *Development* 119:897–908

Oliver B, Perrimon N, Mahowald AP. 1987. The *ovo* locus is required for sex-specific germ line maintenance in *Drosophila*. *Genes Dev.* 1:913–23

Parkhurst SM, Meneely PM. 1994. Sex determination and dosage compensation: lessons form flies and worms. *Science* 264:924–32

Peifer M, Orsulic S, Sweeton D, Wieschaus E. 1993. A role for the *Drosophila* segment polarity gene *armadillo* in cell adhesion and cytoskeletal integrity during oogenesis. *Development* 118:1191–207

Poirié M, Niederer E, Steinmann-Zwicky M. 1995. A sex-specific number of germ cells in embryonic gonads of *Drosophila*. *Development* 121:1867–73

Poulson DF. 1950. Histogenesis, organogenesis, and differentiation in the embryo of *Drosophila melanogaster* (Meigen). In *Biology of Drosophila*, ed. M Demerec, pp. 168–274. New York: Wiley

Raff JW, Glover DM. 1989. Centrosomes, and not nuclei, initiate pole cell formation in Drosophila embryos. *Cell* 57:611–19

Raff JW, Whitfield WGF, Glover DM. 1990. Two distinct mechanisms localise cyclin B transcripts in syncytial *Drosophila* embryos. *Development* 110:1–13

Reuter R. 1994. The gene *serpent* has homeotic properties and specifies endoderm versus ectoderm within the *Drosophila* gut. *Development* 120:1123–35

Rodesch C, Geyer PK, Patton JS, Bae E, Nagoshi RN. 1995. Developmental analysis of the *ovarian tumor* gene during *Drosophila* oogenesis. *Genetics* 141:191–202

Rongo C, Gavis ER, Lehmann R. 1995. Localization of *oskar* RNA regulates oskar translation and requires oskar protein. *Development* 121:2737–46

Rongo C, Lehmann R. 1996. Regulated synthesis, transport and assembly of the *Drosophila* germ plasm. *Trends Genet.* 12:102–9

Sahut-Barnola I, Godt D, Laski FA, Couderc JL. 1995. *Drosophila* ovary morphogenesis: analysis of terminal filament formation and identification of a gene required for this process. *Dev. Biol.* 170:127–35

Schedl T. 1996. Developmental genetics of the germ line. In *The Nematode C. elegans II*, ed. D Riddle, T Blumental, B Meyer, J Priess. Cold Spring Harbor NY: Cold Spring Harbor Press. In press

Schüpbach T. 1982. Autosomal mutations that interfere with sex determination in somatic cells of *Drosophila* have no direct effect on the germ line. *Dev. Biol.* 89:117–27

Schüpbach T. 1985. Normal female germ cell differentiation requires the female X chromosome to autosome ratio and expression of sex-lethal in *Drosophila melanogaster. Genetics* 109:529–48

Schüpbach T, Wieschaus EF, Nöthiger R. 1978. A study of the female germ line in mosaics of *Drosophila. Roux's Arch. Dev. Biol.* 184:41–56

Seydoux G, Fire A. 1994. Soma-germ line asymmetry in the distributions of embryonic RNAs in *Caenorhabditis elegans. Development* 120:2823–34

Smith JL, Wilson JE, Macdonald PM. 1992. Overexpression of *oskar* directs ectopic activation of *nanos* and presumptive pole cell formation in Drosophila embryos. *Cell* 70:849–59

Sonnenblick BP. 1941. Germ cell movements and sex differentiation of the gonads in the *Drosophila* embryo. *Proc. Natl. Acad. Sci. USA* 26:373–81

Sonnenblick BP. 1950. The early embryology of *Drosophila melanogaster*. In *Biology of Drosophila,* ed. M Demerec, pp. 62–167. New York: Wiley

Spradling AC. 1993. Developmental genetics of oogenesis. In *The Development of Drosophila melanogaster,* ed. M Bate, A Martinez Arias, I:1–70. Cold Spring Harbor NY: Cold Spring Harbor Press

Staab S, Steinmann-Zwicky M. 1995. Female germ cells of *Drosophila* require zygotic *ovo* and *otu* product for survival in larvae and pupae respectively. *Mech. Dev.* 54:205–10

Steinmann-Zwicky M. 1993. Sex determination in *Drosophila*: *sis-b*, a major numerator element of the X:A ratio in the soma, does not contribute to the X:A ratio in the germ line. *Development* 117:763–67

Steinmann-Zwicky M. 1994. Sex determination of the *Drosophila* germ line: *tra* and *dsx* control somatic inductive signals. *Development* 120:707–16

Steinmann-Zwicky M, Schmid H, Nöthiger R. 1989. Cell-autonomous and inductive signals can determine the sex of the germ line of Drosophila by regulating the gene *Sxl. Cell* 57:157–66

St Johnston D, Beuchle D, Nüsslein-Volhard C. 1991. *Staufen*, a gene required to localize maternal RNAs in the Drosophila egg. *Cell* 66:51–63

Szabad J, Nöthiger R. 1992. Gynandromorphs of *Drosophila* suggest one common primordium for the somatic cells of the female and male gonads in the region of abdominal segments 4 and 5. *Development* 115:527–35

Technau GM, Campos-Ortega JA. 1986. Lineage analysis of transplanted individual cells in embryos of *Drosophila melanogaster.* III. Commitment and proliferative capabilities of pole cells and midgut progenitors. *Roux's Arch. Dev. Biol.* 195:489–98

Underwood EM, Caulton JH, Allis CDM, Mahowald AP. 1980. Developmental fate of pole cells in *Drosophila melanogaster. Dev. Biol.* 77:303–14

Van Deusen EB. 1976. Sex determination in germ line chimeras of *Drosophila melanogaster. J. Embryol. Exp. Morphol.* 37:173–85

Wang C, Dickinson LK, Lehmann R. 1994. The genetics of *nanos* localization in *Drosophila. Dev. Dyn.* 199:103–15

Waring GL, Allis CD, Mahowald AP. 1978. Isolation of polar granules and the identification of polar granule specific protein. *Dev. Biol.* 66:197–206

Warn RM, Smith L, Warn A. 1985. Three distinct distributions of F-actin occur during the divisions of polar surface caps to produce pole cells in *Drosophila* embryos. *J. Cell Biol.* 100:1010–15

Warrior R. 1994. Primordial germ cell migration and the assembly of the *Drosophila* embryonic gonad. *Dev. Biol.* 166:180–94

Wieschaus E, Audit C, Masson M. 1981. A clonal analysis of the roles of somatic cells and germ line during oogenesis in *Drosophila. Dev. Biol.* 88:92–103

Wieschaus E, Nüsslein-Volhard C. 1986. Looking at embryos. In *Drosophila: A Practical Approach,* ed. DB Roberts, pp. 199–227. Oxford: IRL Press

Wieschaus E, Szabad J. 1979. The development and function of the female germ line in *Drosophila melanogaster*, a cell lineage study. *Dev. Biol.* 68:29–46

Witte ON. 1990. Steel locus defines new multipotent growth factor. *Cell* 63:5–6

Zalokar M. 1976. Autoradiographic study of protein and RNA formation during early development of *Drosophila* eggs. *Dev. Biol.* 49:425–37

Zalokar M, Erk I. 1976. Division and migration of nuclei during early embryogenesis of *Drosophila melanogaster. J. Microsc. Biol. Cell.* 25:97–106

Zhou S, Yang Y, Scott JM, Pannuti A, Fehr KC, et al. 1995. *Male-specific lethal 2,* a dosage compensation gene of *Drosophila,* undergoes sex-specific regulation and encodes a protein with a RING finger and metallothionein-like cysteine cluster. *EMBO J.* 14:2884–95

Annu. Rev. Cell Dev. Biol. 1996. 12:393–416

A CONSERVED SIGNALING PATHWAY: The *Drosophila* Toll-Dorsal Pathway

Marcia P. Belvin and Kathryn V. Anderson
Genetics Division, Department of Molecular and Cell Biology, University of California, Berkeley, California 94720

KEY WORDS: Toll, Dorsal, NF-κB, IκB, dorsal-ventral patterning

ABSTRACT

The Toll-Dorsal pathway in *Drosophila* and the interleukin-1 receptor (IL-1R)-NF-κB pathway in mammals are homologous signal transduction pathways that mediate several different biological responses. In *Drosophila*, genetic analysis of dorsal-ventral patterning of the embryo has defined the series of genes that mediate the Toll-Dorsal pathway. Binding of extracellular ligand activates the transmembrane receptor Toll, which requires the novel protein Tube to activate the cytoplasmic serine/threonine kinase Pelle. Pelle activity controls the degradation of the Cactus protein, which is present in a cytoplasmic complex with the Dorsal protein. Once Cactus is degraded in response to signal, Dorsal is free to move into the nucleus where it regulates transcription of specific target genes. The *Toll, tube, pelle, cactus*, and *dorsal* genes also appear to be involved in *Drosophila* immune response. Because the IL-1R-NF-κB pathway plays a role in vertebrate innate immunity and because plant homologues of the Toll-Dorsal pathway are important in plant disease resistance, it is likely that this pathway arose before the divergence of plants and animals as a defense against pathogens.

CONTENTS

1081-0706/96/1115-0393$08.00

INTRODUCTION

Over the past few years, many examples of the evolutionary conservation of signal transduction pathways have been characterized. Through these studies it appears that although the biochemical interactions allowing signaling are frequently conserved, the same signaling pathway can mediate very different biological processes in different animals. The Toll-Dorsal pathway in *Drosophila* and the interleukin-1 receptor (IL-1R)-NF-κB pathway in mammals represent a conserved signaling pathway that is used in several, quite different biological contexts in both invertebrates and vertebrates. Dorsal and NF-κB are homologous transcription factors of the Rel family that are activated in response to binding of extracellular ligands to homologous transmembrane receptors, *Drosophila* Toll, and the mammalian IL-1R (Figure 1). A benefit of studying a signaling pathway that is conserved different species is the ability to exploit the strengths of each organism as an experimental system. In the case of the Toll-Dorsal and IL-1R-NF-κB pathways, the genetic experiments in *Drosophila* and biochemical experiments in mammalian cells contribute complementary sets of results that have accelerated the understanding of the pathway.

Because of its amenability to genetic analysis, the best characterized function of the *Drosophila* Toll-Dorsal pathway is the establishment of the dorsal-ventral pattern of the embryo (Anderson & Nüsslein-Volhard 1986, Morisato & Anderson 1995). Analysis of the effects of mutants on early embryonic patterning has defined the components of the signaling pathway and ordered the action of those components. Based primarily on expression data, the same pathway appears to be used in other processes at later developmental stages, including the *Drosophila* immune response (Ip & Levine 1994, Hoffmann 1995), morphogenetic movements (Gerttula et al 1988, Hashimoto et al 1991), and muscle development (Nose et al 1992, Halfon et al 1995).

In mammals, the function and regulation of NF-κB was first characterized in cultured cells. NF-κB is a dimer, most commonly a heterodimer composed

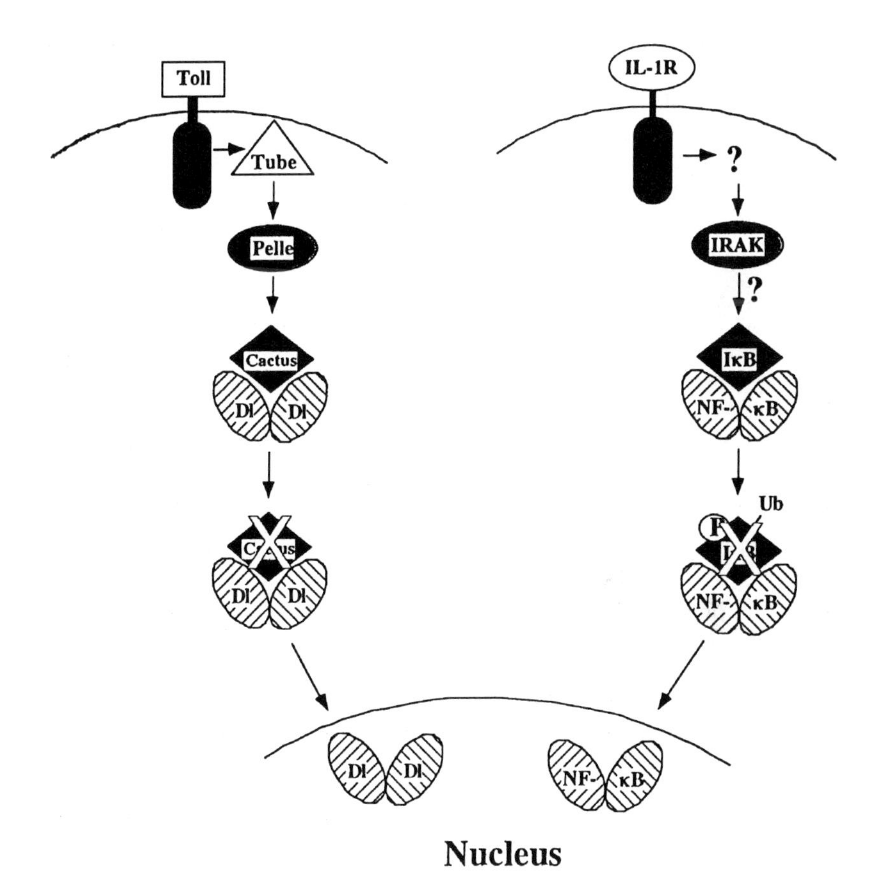

Figure 1 Homology between the Toll-Dorsal and IL-1R-NF-κB pathways. The components of the cytoplasmic portion of the dorsal-ventral signal transduction pathway are homologous to components of the IL-1R-NFκB pathway. The cytoplasmic domains of Toll and IL-1R are homologous; Pelle, Cactus, and Dorsal are homologous to IRAK, IκB and NF-κB, respectively. NF-κB is a heterodimer of p50 and p65/RelA, whereas Dorsal exists as a homodimer. It is not known if IL-1R activates IRAK directly or through another protein, potentially a homologue of Tube. IκB is phosphorylated and ubiquitinated prior to its degradation; however, this has not been demonstrated for Cactus.

of p50 and p65 (RelA) subunits; other mammalian members of the family include p52, RelB, and c-Rel, which can associate in a variety of hetero- and homodimers (Siebenlist et al 1994). Biochemical studies have been particularly important in defining the biochemical mechanisms responsible for the activation of NF-κB (Baeuerle & Baltimore 1988, Ganchi et al 1992, Beg et al 1993, Traenckner et al 1994). Studies in cultured cell lines have implicated NF-κB in a variety of functions in the immune system, including the response to inflammatory cytokines, B cell and T cell proliferation, and B cell maturation (Siebenlist et al 1994). More recently, targeted mutations in the mouse genes have shown that particular NF-κB subunits are important in B cell maturation, T cell activation, and development of the fetal liver (Beg et al 1995b, Sha et al 1995, Weih et al 1995).

Thus far the only biological process common to *Drosophila* and mammals that uses the Toll-Dorsal/IL-1R-NF-κB pathway is innate immunity—the rapid nonspecific response to pathogens that leads to the production of antimicrobial peptides and cytokines, as well as activation of macrophage-like cells. This apparent conservation of function is made even more tantalizing by data showing that resistance to pathogens in plants involves proteins that are homologous to components of the *Drosophila* pathway (Staskawicz et al 1995). Innate immunity, therefore, may be the ancestral use of this signaling pathway, predating the divergence of plants and animals. In this context, it is interesting to speculate how the immunity function of this pathway was conserved, and how and when the pathway was co-opted for other biological functions.

THE CYTOPLASMIC SIGNALING PATHWAY THAT INITIATES DORSAL-VENTRAL PATTERNING IN THE *DROSOPHILA* EMBRYO

The *Drosophila* genes that specify components of the Toll-Dorsal signaling pathway were identified because maternal effect mutations in these genes cause dorsalization of the embryo. Early in wild-type *Drosophila* development, cells adopt specific fates as a function of their dorsal-ventral position in the embryo: Ventral cells give rise to mesoderm, lateral cells give rise to the ventral nerve cord and ventral ectoderm, and dorsal cells give rise to dorsal ectoderm (Lohs-Schardin et al 1979). Twelve maternal effect genes, called the dorsal group genes, control the initial dorsal-ventral pattern of the embryo (Anderson & Nüsslein-Volhard 1984). Females that lack the activity of any one of eleven of these genes produce dorsalized embryos, in which cells at all embryonic positions give rise to dorsal ectoderm. The maternal effect phenotypes caused by these mutations are fully penetrant and constant from embryo to

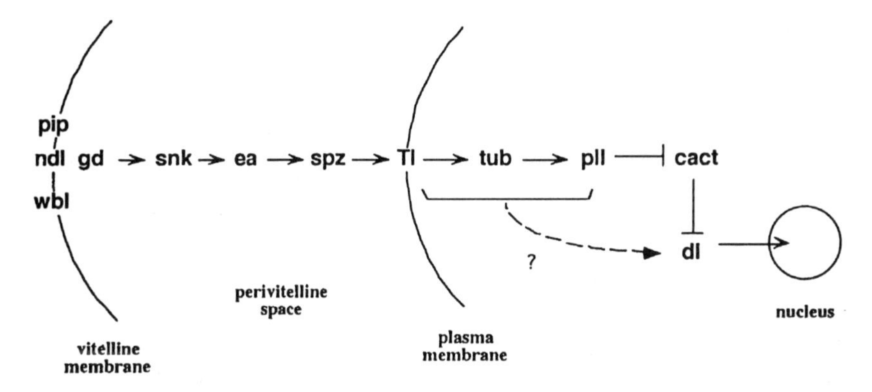

Figure 2 The dorsal-ventral signal transduction pathway. The twelve genes comprising the dorsal-ventral signal transduction pathway in *Drosophila* are listed in the order in which they are known to act. Germ-line clone analysis revealed that *pip, ndl,* and *wbl* are required in the soma. The other nine germ-line-dependent genes have been ordered with respect to each other and to the somatically required genes by genetic epistasis experiments. Abbreviations: pip, *pipe*; ndl, *nudel*; wbl, *windbeutel*; gd, *gastrulation defective*; snk, *snake*; ea, *easter*; spz, *spätzle*; Tl, *Toll*; tub, *tube*; pll, *pelle*; cact, *cactus*; dl, *dorsal*.

embryo, which has made genetic analysis of these components straightforward.

The order of action of the dorsal group genes has been defined by double-mutant analysis (Figure 2). Key in this analysis were the phenotypes of double mutants of dominant, constitutively active alleles of the *Toll* gene with loss-of-function mutations in other genes (Anderson et al 1985, Hecht & Anderson 1993, Morisato & Anderson 1994). These double mutants showed that activated Toll bypasses the requirement for seven genes; therefore these seven genes are normally required to activate Toll and lie upstream of Toll in the genetic pathway. These results also showed that three other genes, *tube, pelle,* and *dorsal,* were required for function of the activated Toll, placing these three genes downstream of Toll. The other key gene used in ordering the pathway was *cactus*. In contrast to the dorsalized phenotype caused by mutations in the other eleven dorsal group genes, loss-of-function mutations in *cactus* ventralize the embryo. Double mutants showed that *cactus* acts downstream of all the other maternal effect dorsal group genes except *dorsal.*

The molecular cloning and characterization of *Toll, tube, pelle, cactus,* and *dorsal* made it clear that these genes encode components of a signaling pathway that allows an extracellular signal to control the transcription of specific target genes in the embryo (Steward 1987, Hashimoto et al 1988, Letsou et al 1991, Geisler et al 1992, Kidd 1992, Shelton & Wasserman 1993). *Toll* encodes a receptor protein with a single transmembrane domain; *tube* encodes a novel cytoplasmic protein; *pelle* encodes a cytoplasmic serine/threonine kinase; *cactus* encodes a cytoplasmic protein with an ankyrin repeat domain; and *dorsal* encodes a transcription factor related to the rel oncogene. From their order of action and their sequences, Toll appears to be a transmembrane receptor whose activation triggers a cytoplasmic signaling pathway involving Tube, Pelle, and Cactus, ultimately controlling the activity of the transcription factor Dorsal.

Analysis of the spatial distribution of the Dorsal protein revealed that the dorsal group genes control embryonic patterning by controlling the translocation of Dorsal protein from the cytoplasm to the nuclei on the ventral side of the embryo (Roth et al 1989, Rushlow et al 1989, Steward 1989). In very young wild-type embryos, Dorsal is distributed uniformly in the cytoplasm of the embryo. At the syncytial blastoderm stages, a nuclear gradient of Dorsal arises, with the protein most concentrated in the ventral nuclei of the embryo, less concentrated in lateral nuclei, and absent from nuclei on the dorsal side of the embryo (Figure 3*a*). In mutant embryos that are missing one of the dorsal group genes, the gradient of Dorsal does not arise. For instance, in embryos laid by *Toll*$^-$ females, Dorsal protein is present, but it is uniformly distributed in the cytoplasm and is not translocated into any embryonic nuclei (Figure 3*b*). The phenotype of *Toll*$^-$ embryos is exactly the same as the phenotype of embryos that lack Dorsal altogether; both kinds of embryos are completely dorsalized, indicating that Dorsal is active only when it is present in nuclei. Thus the dorsal-ventral pattern of the embryo is generated by a ventrally active signaling pathway that leads to a gradient of translocation of Dorsal protein into ventral and lateral embryonic nuclei.

Once in the nuclei, Dorsal acts as a transcription factor to define region-specific gene expression in the embryo. Dorsal activates the transcription of particular ventrally required genes, including *twist, snail,* and *rhomboid* (Pan et al 1991, Thisse et al 1991, Ip et al 1992a,b, Jiang et al 1992). The *twist* and *snail* genes are activated by the high concentrations of nuclear Dorsal protein found near the ventral midline of the embryo, which bind low-affinity Dorsal-binding sites in the promoters of the two target genes (Jiang & Levine 1993). The *rhomboid* gene is activated in a broad ventral and lateral domain where there are lower concentrations of nuclear Dorsal; transcriptional activation of *rhomboid* in lateral regions of the embryo requires cooperation between Dorsal protein and an unidentified, but apparently ubiquitous, basic HLH protein (Jiang & Levine 1993). Dorsal also represses the transcription of other genes in ventral cells in order to limit their expression to the dorsal side of the embryo; these genes include *zerknüllt* (*zen*) and *decapentaplegic* (*dpp*) (Ip et al 1991, Schwyter et al 1995). Dorsal is converted from an activator to a repressor of transcription

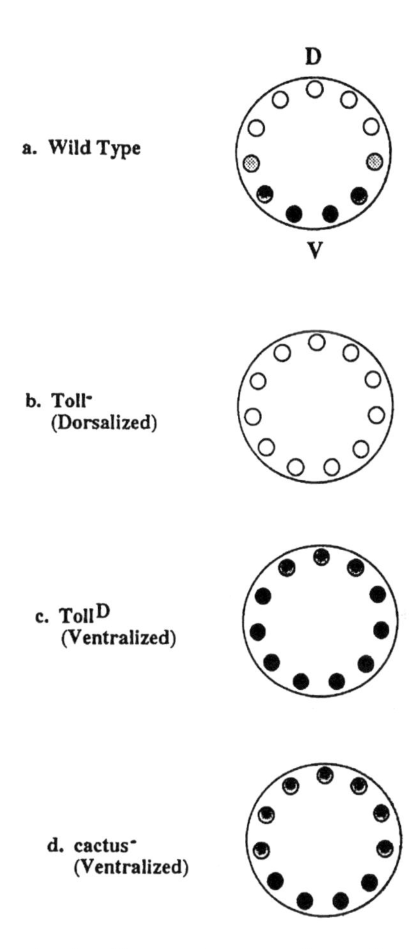

Figure 3 The Dorsal gradient in wild-type and mutant embryos. The cartoons represent a cross section of a syncytial blastoderm *Drosophila* embryo, with the nuclei at the periphery of the embryo. The dorsal side is on top (D) and the ventral side is on the bottom (V). (*a*) In wild-type embryos, nuclear Dorsal exists as a gradient, with high concentrations in ventral nuclei, and lower concentrations in lateral nuclei. On the dorsal side, all of the Dorsal protein is in the cytoplasm. (*b*) In embryos lacking Toll, no Dorsal protein enters the nuclei. (*c*) In embryos with a dominant allele of *Toll*, the Dorsal gradient is expanded. (*d*) In embryos lacking Cactus, Dorsal is nuclear all the way around the circumference, with the highest concentrations in ventral nuclei.

when it binds regulatory sites that are adjacent to binding sites for a corepressor, the HMG-like protein DSP1 (Lehming et al 1994).

Activation of Dorsal Protein Requires Disruption of the Cytoplasmic Cactus-Dorsal Complex

In the oocyte and newly-laid embryo, the Cactus and Dorsal proteins are in a complex (Kidd 1992), with two molecules of Dorsal associated with one molecule of Cactus (Isoda & Nüsslein-Volhard 1994). Cactus is a *Drosophila* homologue of mammalian IκB proteins (Haskill et al 1991), which form cytoplasmic complexes with NF-κB (Figure 4). IκB family members and Cactus each contain a similar block of five to six ankyrin repeats in the middle of the protein. The ankyrin repeats of IκB and Cactus mediate their interaction with Rel proteins (Kidd 1992, Hatada et al 1993, Jaffray et al 1995). The domains of Dorsal that interact with Cactus have been mapped (Isoda et al 1992, Kidd 1992, Norris & Manley 1992, Lehming et al 1995, Tatei & Levine 1995, Govind et al 1996). There is agreement that one region of the Dorsal Rel domain, including amino acids 233 and 234 (Lehming et al 1995), is crucial for Cactus binding. Based on the crystal structure of the Rel domain of p50, this site lies in a groove between the two p50 monomers, suggesting that Cactus binds in that groove (Kopp & Ghosh 1995, Müller & Harrison 1995). Some, but not all, assays indicate that two other regions of Dorsal are important for binding Cactus:

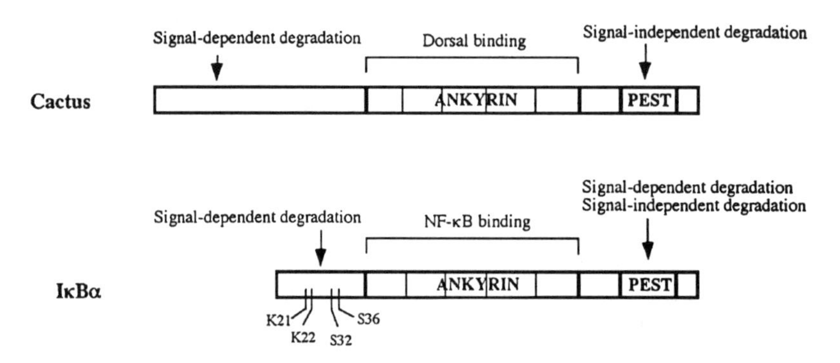

Figure 4 A comparison of the Cactus and IκBα proteins. Cactus and IκBα each contain a set of five homologous ankyrin repeats in the center of the protein and a PEST sequence at the carboxy terminus. The ankyrin repeat domain of Cactus and IκBα is the site of interaction with Rel proteins. In Cactus, signal-dependent degradation is mediated by the amino terminus, whereas signal-independent degradation is mediated by the carboxy terminus. In IκBα, signal-dependent degradation is regulated by the amino terminus, and involves the phosphorylation of serines 32 and 36 and the ubiquitination of lysines 21 and 22. The carboxy terminus of IκBα has been implicated in both signal-dependent and signal-independent degradation.

the nuclear localization signal at the C-terminus of the Rel domain (Tatei & Levine 1995) and a region outside the Rel domain at the carboxy terminus of the protein (Isoda et al 1992, Norris & Manley 1992).

NF-κB proteins contain nuclear localization sequences that are masked by the binding of IκB (Ganchi et al 1992, Henkel et al 1992), providing an explanation for the cytoplasmic localization of these complexes. Similarly, it appears that Cactus masks the nuclear localization signal of the Dorsal protein, preventing its migration into nuclei. Consistent with this, in embryos that lack Cactus, Dorsal enters nuclei around the entire dorsal-ventral circumference of the embryo. Thus Cactus holds Dorsal in the cytoplasm until ventral signaling disrupts this complex, allowing Dorsal to move into nuclei on the ventral side of the embryo.

Cactus Protein Is Degraded in Response to Signaling

Work on the mammalian homologue of Cactus raised the possibility that the signal-dependent disruption of the Cactus-Dorsal complex might involve degradation of Cactus protein. Several groups working in mammalian tissue culture cells showed that IκB is rapidly degraded after stimulation of responsive cell lines with interleukin-1 (IL-1), tumor necrosis factor alpha (TNF-α), or phorbol myristate acetate (PMA), and that IκB degradation correlates with activation of NF-κB (Beg et al 1993, Henkel et al 1993, Miyamoto et al 1994).

In *Drosophila,* it is possible to activate the signaling pathway at a controlled time by injecting the ligand for Toll, activated Spätzle, into the extracellular space of the embryo (Schneider et al 1994). To test how signaling affects Cactus, high concentrations of activated Spätzle protein were injected into the extracellular space of embryos to activate the signaling pathway in all embryonic regions at a defined time (Belvin et al 1995). By analyzing Cactus protein on Western blots after injection, it was found that more than 90% of the Cactus protein in the embryo was rapidly degraded within 15 min after Spätzle injection. In embryos that lack Toll, Tube, or Pelle, the injection of activated Spätzle fails to cause Cactus degradation. Therefore, degradation of Cactus is a consequence of the activity of the Toll-Dorsal signaling pathway in the intact organism.

In theory, two models could explain the rapid degradation of Cactus in response to signaling: either signaling directly promotes Cactus degradation, or signaling causes separation of the Cactus-Dorsal complex and free Cactus is degraded. Taking advantage of *Drosophila* mutants, it was found that Cactus is rapidly degraded in response to signal even in *dorsal*$^-$ embryos, indicating that Cactus degradation is a direct consequence of signaling even when Cactus is not complexed with Dorsal (Belvin et al 1995). It is therefore likely that in the wild-type embryo, the activity of the signaling pathway targets the Cactus protein in the Cactus-Dorsal complex for degradation, liberating Dorsal from the complex and allowing it to move into embryonic nuclei (Figure 1).

Toll, Tube, and Pelle Regulate Cactus Degradation

Genes that act upstream of Cactus in the signaling pathway somehow target Cactus for degradation. The transmembrane protein Toll is activated on the ventral side of the embryo by binding a ventrally localized ligand, apparently a proteolytically processed form of the Spätzle protein (Morisato & Anderson 1994, Schneider et al 1994). The extracellular domain of Toll is composed largely of leucine-rich repeats, whereas the cytoplasmic domain of Toll is similar to the cytoplasmic domain of the mammalian IL-1 receptor (Schneider et al 1994). The sequence homology between the cytoplasmic domains of Toll and the IL-1 receptor appears to be significant because mutations that change amino acids conserved between the two proteins destroy receptor function (Schneider et al 1991, Heguy et al 1992). However, the biochemical mechanisms that convert ligand binding into activation of either receptor are not known.

Only two genes, *tube* and *pelle,* have been identified that act between Toll and the Dorsal/Cactus complex (Anderson et al 1985, Hecht & Anderson 1993). Analyses of double mutants using activated forms of *tube* and *pelle* have shown that *pelle* acts downstream of *tube* (Grosshans et al 1994, Galindo et al 1995). Activated *tube* and *pelle* were constructed in vitro by fusing the coding region of each gene with the extracellular and transmembrane domains of an activated allele of the *torso* receptor. Both the Torso-Pelle and the Torso-Tube fusion proteins promote ventral development, even in the absence of genes that act upstream of Toll. The *torso-pelle* construct promotes ventral development in the absence of *tube*, but the *torso-tube* construct has no activity in the absence of *pelle*, which places *pelle* downstream of *tube* in the signaling pathway.

tube encodes a novel 462-amino acid protein; the only striking sequence feature of the protein is the presence of five copies of an 8-amino acid repeat in its C-terminal half (Letsou et al 1991). Although these repeats are conserved in the *Drosophila virilis tube* gene, mutant forms of *tube* that lack the C-terminal 282 amino acids are still active in signaling, indicating that the N-terminal region of the protein is sufficient for function (Letsou et al 1993). Tube protein is highly enriched in regions near the plasma membrane of the syncytial embryo (Galindo et al 1995), where it might be involved in mediating signaling from Toll.

pelle encodes a cytoplasmic serine/threonine kinase (Shelton & Wasserman 1993). As Dorsal and Cactus are both phosphoproteins (Kidd 1992, Whalen & Steward 1993, Gillespie & Wasserman 1994), they are candidate substrates for Pelle. Bacterially produced Pelle is an active kinase but does not appear to phosphorylate either Dorsal or Cactus to a high degree (Grosshans et al 1994). However, Pelle can phosphorylate Tube in vitro, which raises the possibility of a dynamic interaction between Pelle and Tube proteins.

Genetic interactions between specific *pelle* and *tube* alleles suggest that the two proteins interact in vivo (Hecht & Anderson 1993). Direct interactions between the two proteins have been identified in the yeast two-hybrid assay (Grosshans et al 1994, Galindo et al 1995), where the N-terminal noncatalytic domain of Pelle interacts with the N-terminal essential domain of Tube (Galindo et al 1995). Because Tube is localized to the membrane, it is possible that activation of Toll causes Tube to recruit Pelle to the membrane, thereby activating Pelle. However, it is not yet known where Pelle protein is localized in embryos.

The homology of the Toll-Dorsal and the IL-1R-NF-κB pathways has motivated the search for mammalian protein kinases that act downstream of the IL-1 receptor and might be homologous to Pelle. A serine/threonine kinase, the IL-1 receptor-associated kinase (IRAK), that coimmunoprecipitates with the activated interleukin-1 receptor (M Martin et al 1994, Croston et al 1995) has been identified. Among all animal kinases, IRAK is most closely related to Pelle in sequence (Cao et al 1996). Within 30 s after treatment with IL-1, IRAK becomes associated with the IL-1 receptor and is extensively phosphorylated within minutes (Cao et al 1996). Thus because of both its biochemical behavior and its homology to Pelle, IRAK is an excellent candidate to mediate signaling from the IL-1 receptor to downstream cytoplasmic targets.

Signal-Dependent Phosphorylation, Ubiquitination, and Degradation of IκB and Cactus

In mammalian cells, IκB is transiently phosphorylated prior to its degradation (Beg et al 1993, Henkel et al 1993). Although this phosphorylation event is not sufficient to release NF-κB from the complex (Henkel et al 1993, Miyamoto et al 1994, Traenckner et al 1994, Alkalay et al 1995, DiDonato et al 1995, Lin et al 1995), phosphorylation of IκB appears to be necessary to target the protein for degradation (Traenckner et al 1994, Brown et al 1995, Chen et al 1995). Signal-dependent phosphorylation of Cactus has not been observed (Whalen & Steward 1993, Belvin et al 1995). However signal-dependent phosphorylation of IκB is difficult to detect because the protein is subsequently degraded within 1 to 2 min, so it is possible that Cactus is also phosphorylated in response to the signal but is degraded so rapidly that the phosphorylated form has not been detected. At this point, an attractive model is that Pelle phosphorylates Cactus, which targets Cactus for degradation. However, there are currently no biochemical data to support a direct interaction between Pelle and Cactus.

Site-directed mutagenesis that replaces either of two serines in the amino terminus of IκB (S32 and S36) with alanine prevents phosphorylation and degradation, suggesting that phosphorylation of these particular sites targets IκB for degradation (Brown et al 1995, Chen et al 1995), apparently by the proteasome. The proteasome, a very large multiprotein structure found in all eukaryotes, is responsible for the degradation of cytoplasmic proteins that are marked by ubiquitination (Ciechanover 1994, Peters 1994). In the presence of peptides that inhibit the activity of the proteasome, ubiquitinated conjugates of IκB accumulate, with ubiquitin groups added to lysines 21 and 22 (Chen et al 1995, Scherer et al 1995). Because ubiquitinated forms of IκB do not accumulate in the S32A or S36A mutants, phosphorylation must precede ubiquitination. Therefore, the likely sequence of events in IκB degradation is signal-dependent phosphorylation of serines 32 and 36, followed by ubiquitination of lysines 21 and 22, followed by degradation by the proteasome.

Although the N-terminal domain of Cactus is much longer than that of IκB (Figure 4), there is some evidence that the N-terminal domain of Cactus is also required for signal-dependent degradation. A mutant allele of *cactus*, *cact*^{E10}, lacks the N-terminal domain of the protein (N Ito, personal communication). The Cact^{E10} product blocks nuclear localization of Dorsal, which suggests that it is not degraded in response to signal. Therefore, it is likely that the ability of Cactus to degrade in response to signaling depends on the N-terminal region of the protein.

Mutant alleles of a gene encoding one *Drosophila* proteasome subunit have been isolated (Saville & Belote 1993) and should provide a useful tool for analyzing whether the proteasome is important in the signal-dependent degradation of Cactus. Null alleles of this proteasome subunit gene are recessive lethal mutations, so partial loss-of-function alleles of the gene or genetic mosaics with mutant germ lines will be necessary to test whether the proteasome is required for Cactus degradation.

Cactus Stability Is Regulated by Both Signal-Dependent and Signal-Independent Processes

Both Cactus and IκB have PEST (Pro-Glu-Ser-Thr-rich) sequences in the region of the protein C-terminal to the ankyrin repeats (Figure 4). Because PEST sequences have been implicated in rapid protein turnover (Rogers et al 1986, Rechsteiner 1990), this region could be important for controlling the stability of these proteins. However, the PEST sequence in Cactus is not involved in signal-dependent degradation, because a mutant form of Cactus that lacks the entire C-terminus of the protein including the PEST region is degraded in response to signaling at approximately the same rate as wild-type Cactus protein (Belvin et al 1995).

Nevertheless, the PEST region is important for preventing the accumulation of free Cactus that is not complexed with Dorsal. Cactus is unstable in the absence of Dorsal protein (Kidd 1992, Whalen & Steward 1993) with the result that all the Cactus protein in the wild-type embryo is complexed with Dorsal. In contrast, the mutant form of Cactus that lacks the PEST region is much more stable than wild-type Cactus in the absence of Dorsal (Belvin et al 1995) and, if present in sufficient amounts, can effectively block signaling and dorsalize the embryo (Belvin et al 1995). This result demonstrates the importance of maintaining controlled amounts of Cactus protein. It is not clear whether free Cactus blocks signaling because it competes with Cactus in the complex for the activity of the upstream signal or whether it acts by binding Dorsal rapidly after it is freed from the complex. Altogether, these studies show that two distinct regions of the Cactus protein mediate two different types of degradation: The N-terminus mediates signal-dependent degradation, and the C-terminus mediates signal-independent degradation. Both types of degradation are important for the proper regulation of signaling.

The C-terminal region of IκB is probably also important for signal-independent degradation, because deletion of the C-terminal PEST-containing region increases the half-life of free IκBa (Miyamoto & Verma 1995). In contrast to Cactus, however, the C-terminus of IκB appears to be required for signal-dependent degradation as well (Brown et al 1995, Rodriguez et al 1995). Because the C-terminus of IκB appears to be involved in both signal-dependent and signal-independent degradation, whereas the domains of Cactus involved in the two kinds of degradation are physically separated, it may be easier to dissect these two kinds of regulation in *Drosophila* than in mammals.

Parallel Signaling Events that Control Dorsal Activity

The degradation of Cactus is the primary response to signal in the embryonic signaling pathway. However, embryos that lack all *cactus* activity still have a detectable dorsal-ventral asymmetry: These embryos have Dorsal protein in nuclei around the entire dorsal-ventral circumference, but there is more Dorsal protein in ventral than in dorsal nuclei (Roth et al 1991; A Bergmann, personal communication) (Figure 3*d*). Therefore, there must be some additional ventral signal promoting nuclear translocation of Dorsal protein that does not act through Cactus (Figure 2).

One possible parallel signaling event is the modification of Dorsal in direct response to signal. By comparing the Dorsal protein in mutant backgrounds that block or enhance signaling, it has been shown that nuclear Dorsal is more highly phosphorylated than cytoplasmic Dorsal (Whalen & Steward 1993, Gillespie & Wasserman 1994). However, it is not clear whether the increase in phosphorylation is a cause or an effect of the disruption of the Cactus-Dorsal complex. In embryos that lack Cactus and therefore have nuclear Dorsal protein around the entire embryonic circumference, a much larger fraction of the Dorsal protein is highly phosphorylated than in wild-type embryos, suggesting that free Dorsal becomes highly phosphorylated independently of signaling. This interpretation is supported by the observation that *cactus*$^-$ embryos in which signaling is blocked by a mutation in an upstream gene also have a high level of highly phosphorylated Dorsal (Gillespie & Wasserman 1994). Thus once Dorsal is free from the Dorsal-Cactus complex, it becomes phosphorylated by a kinase whose activity does not depend on the signal. It is not known what sites in Dorsal become phosphorylated, nor is it known whether phosphorylation of those sites is important for Dorsal's activity as a transcription factor. However, it is interesting to note that in cultured *Drosophila* cells, mutation of a protein kinase A consensus phosphorylation site in the Rel domain inhibits both Dorsal nuclear localization and the ability of Dorsal to activate transcription of a reporter gene (Norris & Manley 1992).

Currently there is no evidence indicating that signaling directly increases phosphorylation of Dorsal. Therefore, some other mechanism may explain how, in the absence of Cactus, there is more Dorsal protein in ventral than in dorsal nuclei. One possibility is suggested by the recent report that at least one other Drosophila gene, *relish*, encodes a Cactus/IκB-like gene (Åsling et al 1995). The *relish* gene encodes a p105-like protein with both Rel and ankyrin repeat domains and is expressed maternally as well as later in development. It is possible (for the non-Cactus-mediated component of signaling) that Cactus is partially redundant and a small fraction of Dorsal is complexed with a different IκB family member, and that this IκB protein, like Cactus, is also degraded in response to signaling. Mutants in this gene may not have been recovered either because the maternal effect phenotype caused by loss of this minor component is very subtle, or because mutations in the gene cause lethality, which would prevent the recovery of maternal effect alleles.

THE TOLL-DORSAL SIGNALING PATHWAY IN THE *DROSOPHILA* IMMUNE RESPONSE

In both insects and vertebrates, the rapid response to pathogens involves the activation of macrophages, which engulf invading pathogens, and production of antimicrobial peptides, which lyse invading pathogens (Figure 5). This process, called innate immunity, does not rely on specific recognition of pathogen and instead involves an immediate, general response to a wide range of pathogens without the subsequent development of specific long-term immunity (Sipe 1985, Hultmark 1993, Ip & Levine 1994, Hoffmann 1995). The first event in the innate immune response in mammals is the activation of macrophages, which engulf invading pathogens and secrete cytokines that activate other aspects of the immune response (Sipe 1985). Similarly, the hemocytes (blood cells) of insects

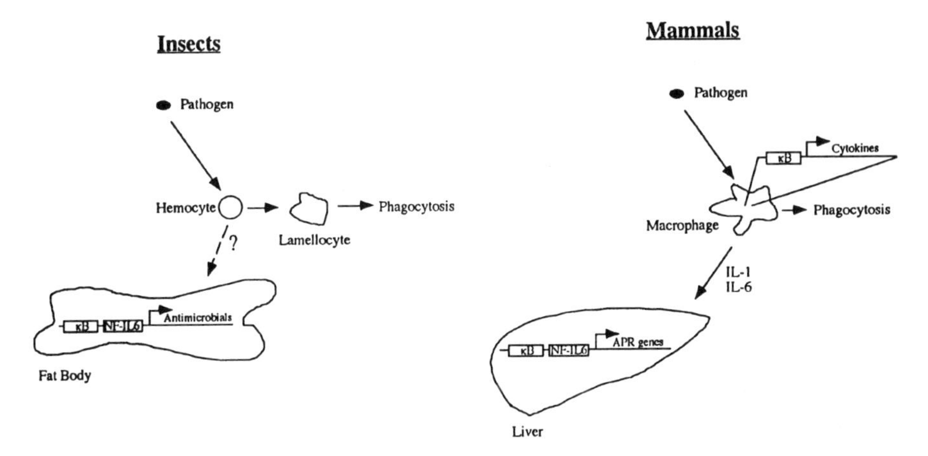

Figure 5 Innate immunity in insects and mammals. In both insects and mammals, innate immunity involves the activation of macrophage-like cells and the subsequent production of antipathogenic agents in the liver, or its insect counterpart, the fat body. In *Drosophila,* hemocytes are activated in response to infection and differentiate into lamellocytes, a macrophage-like cell that can engulf invading organisms. The mechanisms leading to the induction of antimicrobial agents in the fat body are not known. In mammals, macrophages are activated by infection to produce inflammatory cytokines and to engage in phagocytosis. Through the secretion of IL-1 and IL-6, macrophages induce the transcription of acute phase response (APR) genes in the liver.

are activated by infection and participate in phagocytosis and encapsulation of invading pathogens (Götz & Boman 1985). The second major aspect of innate immunity is the production of the acute phase proteins by the liver and their secretion into the circulation; these acute phase proteins are involved directly in killing bacteria. Analogously, the fat body (the insect counterpart to the liver) produces and secretes a set of antimicrobial peptides in response to infection.

In mammals, NF-κB has been implicated in innate immunity in the response of macrophages to pathogens and the ensuing induction of acute phase proteins in the liver. The initial activation of macrophages by bacterial lipopolysaccharide triggers the activation of NF-κB and nuclear factor NF-IL6, which act synergistically to transcribe several of the cytokines (Matsusaka et al 1993). Among the cytokines secreted by the macrophages are IL-1 and IL-6 which, in addition to their role as inflammatory cytokines, induce the acute phase response in the liver (Koj 1985, Fey & Gauldie 1990). In addition, NF-κB and NF-IL6 synergistically activate many of these acute phase response genes (Betts et al 1993, LeClair et al 1992). The observation that the promoters of the *Drosophila* antibacterial genes also contain NF-κB and NF-IL6 sites sparked the interest in comparing insect immunity and the mammalian acute phase response (Reichhart et al 1992). Data from several laboratories indicate that the Toll-Dorsal pathway is activated in the *Drosophila* immune response, but the precise function of these genes in immunity remains to be elucidated.

The Role of the Toll-Dorsal Pathway in the Humoral Immune Response

The primary humoral response of *Drosophila* to infection is the induction of genes encoding antibacterial peptides in the fat body, the functional equivalent of the mammalian liver (Cociancich et al 1994). The first evidence suggesting that *dorsal* or a related gene might be important in the humoral immune response was the discovery of κB binding motifs in the upstream regulatory regions of the genes encoding the antibacterial agents cecropin, diptericin, and attacin (Sun et al 1991, Reichhart et al 1992, Engstrom et al 1993, Kappler et al 1993). Biochemical experiments have shown that Dorsal binds κB sites in the diptericin gene and activates a reporter construct carrying these sites (Engstrom et al 1993, Reichhart et al 1993). These observations led to the search for the expression of *dorsal* and *dorsal*-related genes expressed in the fat body that might activate transcription of the target genes in response to infection.

A gene closely related in sequence to *dorsal*, called *Dif* for Dorsal-related immunity factor, was cloned by homology to *dorsal* (Ip et al 1993). *Dif* is expressed at high levels during larval, pupal, and adult stages, but is not expressed during oogenesis or early embryogenesis (Ip et al 1993). Both *Dif* and *dorsal* are expressed in the larval fat body (Reichhart et al 1993, Lemaitre et al 1995), where they could mediate transcriptional activation of the antimicrobial peptide genes. *Dif*, like *Dorsal*, can bind the κB sites in the cecropin gene and activate transcription in vitro, with *Dif* a more potent transactivator (Petersen et al 1995). At rest, both Dorsal and Dif are present in the cytoplasm of fat body cells. When larvae are infected by bacteria, Dorsal and Dif become localized to the nuclei within 30 min (Ip et al 1993, Reichhart et al 1993, Lemaitre et al 1995). However, the nuclear localization of Dorsal in response to signal is more complete: Essentially all the Dorsal moves into fat body nuclei, whereas the Dif response is more variable, and only about 30% of the fat body nuclei contain Dif after infection (Ip et al 1993, Reichhart et al 1993, Lemaitre et al 1995).

Other components of the Toll-Dorsal pathway play a role in the nuclear localization of both Dorsal and Dif in the fat body. By infecting larvae lacking components of the maternal dorsal-ventral pathway, it was determined that *Toll, tube,* and *pelle* are required for this nuclear translocation of Dorsal in the fat body (Lemaitre et al 1995). In addition, in larvae carrying *Toll* gain-of-function or *cactus* loss-of-function mutations, Dorsal and Dif are constitutively nuclear in the fat body (Ip et al 1993, Lemaitre et al 1995). In contrast, components that act upstream of Toll in the dorsal-ventral patterning pathway, *spätzle, easter, snake,* and *gastrulation defective* have no effect on fat body nuclear localization of Dorsal. Therefore, it appears that only the cytoplasmic portion of the dorsal-ventral pathway is re-utilized in the humoral immune response, raising the question of what ligand(s) activate Toll in the fat body.

Despite the clear response of the Toll-Dorsal and Toll-Dif pathways to infection, their function in the fat body is unclear. In fat bodies of larvae lacking *Toll* or *dorsal,* cecropin and diptericin are induced to wild-type levels (Lemaitre et al 1995), so Dorsal is not required for this aspect of the immune response. No mutation in *Dif* has been reported, thus it is not known whether Dif is important for induction of these peptides. It is possible that Dorsal and Dif are functionally redundant in the fat body and that only the double mutant will display a phenotype.

Is There a Role for the Pathway in the Cellular Immune Response?

There are also data suggesting that the Toll-Dorsal pathway is a component of the cellular immune response, although again, the exact function of the pathway is not clear. Larvae that lack *cactus* or that carry dominant constitutively active alleles of *Toll* have a high incidence of melanotic tumors. Melanotic tumors are not transplantable invasive growths, but instead are visible, melanized bodies that form as a normal part of the immune response and float in the larval or adult hemolymph. In the infection of a normal larva, the larval blood cells (hemocytes) change morphology from rounded cells to flattened cells (called lamellocytes) that encapsulate invading microorganisms and deposit melanin around the encapsulated pathogen. Melanotic tumor mutants have a high frequency of spontaneous melanotic tumors in the absence of infection, which apparently reflects abnormal activity of the hemocytes. Up to 50% of the hemocytes in larvae heterozygous for an activated *Toll* allele have the morphology of lamellocytes (Lemaitre et al 1995), suggesting that the spontaneous activation of blood cells may be responsible for the melanotic tumors. Because dominant *Toll* and loss-of-function *cactus* alleles share the same phenotype, it is likely that the signaling pathway controls hemocyte maturation and that constitutive activity of the pathway causes activation of unstimulated hemocytes. However, double-mutant larvae that lack both *dorsal* and *cactus* or only *dorsal* and carry a constitutively active allele of *Toll* still have melanotic tumors (Lemaitre et al 1995). Therefore, another target of the signaling pathway, besides Dorsal, must lead to activation of the blood cells.

OTHER FUNCTIONS OF THE GENES OF THE TOLL-DORSAL PATHWAY IN THE *DROSOPHILA* LIFE CYCLE

Genetic and molecular data indicate that components of the Toll-Dorsal signaling pathway are used in a variety of other processes during the *Drosophila* life cycle, although it is not clear whether the entire pathway is used as a signaling cassette in any of these processes. The clearest indication that *Toll, tube, pelle,* and *cactus* are important in other processes is that null mutations in these genes decrease the probability of homozygotes surviving to adult stages. All animals that are homozygous for null alleles of *cactus* die during the pupal stage (Roth et al 1991); nearly all (> 95%) of the animals that are homozygous for null alleles of *Toll* die as larvae (Gerttula et al 1988); and only about 70% of animals homozygous for null alleles of *tube* or *pelle* survive to adult stages (Hecht & Anderson 1993). In contrast, animals homozygous for null alleles of *dorsal* are fully viable. Because mutations in the different genes have different effects on viability, it is clear that there must be differences in how these genes are used in later development. Either these genes have other functions unrelated to the signaling pathway defined in early embryos, or there are other genes expressed at later developmental stages that can partially substitute for components of these pathways. The functions of the Toll-Dorsal pathway in later developmental processes have been difficult to define. *Toll* is expressed in a dynamic pattern in many embryonic cells in the process of undergoing morphogenetic movements, which suggests that Toll might act as a coupled adhesion and signaling receptor. This is consistent with the observation that *Toll* expressed in *Drosophila* cultured cells promotes cell adhesion (Keith & Gay 1990). However, *Toll* mutants do not have gross defects in these morphogenetic movements. Mutants that lack *Toll, tube,* or *pelle* do form abnormally short pupae (Letsou et al 1991), indicating that these genes play a role in overall body shape, although it is not known what cell types control body shape.

Toll is expressed in a specific subset of developing muscles (Nose et al 1992, Halfon et al 1995). Embryos that lack *Toll* function have a variety of defects in muscle development, including muscle deletions, muscle duplications, and inappropriate muscle insertions (Nose et al 1992, Halfon et al 1995). The muscle phenotype is both variable and incompletely penetrant, suggesting that *Toll* overlaps in function with other genes affecting development of these muscles.

The only dorsal group gene that is completely lethal is *cactus.* Because flies that lack other components of the pathway are partially or completely viable, it may be that the lack of signaling via the Toll-Dorsal pathway is less harmful to the organism than is constitutive signaling. In a provocative parallel, mice that lack IκBα die shortly after birth (Beg et al 1995a), whereas mice that lack p50 are viable. Interestingly, the lethality of IκBα mutant mice is suppressed by the concomitant loss of p50 (Beg et al 1995a), providing strong evidence that excess signaling in this pathway is more deleterious than lack of signaling. It will be interesting to test in *Drosophila* whether mutations in *dorsal* can suppress the lethality caused by *cactus* mutants, which would indicate that the lethality of *cactus* mutants results from excessive signaling through the pathway.

CONCLUDING THOUGHTS

Recent work on the resistance of plants to pathogens has uncovered a remarkable similarity of plant disease-resistance genes (*R* genes) to components of the Toll-Dorsal signaling pathway (Staskawicz et al 1995). The *R* genes mediate the response of the plant to particular pathogens (gene-for-gene plant-pathogen interactions) and are thought to encode proteins that recognize pathogen and signal the immune response.

One set of *R* genes encodes proteins that are similar to Toll. Most dramatically, the tobacco *N* gene encodes a cytoplasmic protein with a domain homologous to the cytoplasmic domains of Toll and the IL-1 receptor, as well as a leucine-rich repeat domain similar to the Toll extracellular domain (Whitham et al 1994). *R* genes from other plants encode proteins with related but distinct domain structure. For instance, the tomato *Cf-9* gene encodes a transmembrane protein with extracellular leucine-rich repeats and a short cytoplasmic tail (Jones et al 1994), and the rice *XA21* gene encodes a transmembrane protein with extracellular leucine-rich repeats and a cytoplasmic serine-threonine kinase domain (Song et al 1995).

Another set of *R* genes, including the tomato *pto, pti,* and *fen* genes (GB Martin et al 1993, 1994, Zhou et al 1995), encodes cytoplasmic kinases that are more closely related to Pelle and IRAK than other kinases. The kinase domains of the tomato kinases and Pelle and IRAK are 30% identical to each other and define a subgroup of cytoplasmic kinases that we call the serine/threonine innate immunity kinases (the SIIK group).

The remarkable fact that related proteins are used in the signaling pathways that are activated by pathogens in mammals, insects, and plants makes it likely that these components were used in an innate immune response that evolved more than a billion years ago, before the divergence of plants and animals. As the components and steps in these signaling pathways become more completely understood, it may be possible to infer what the ancient forms of these signaling molecules looked like. For instance, was the original pathogen receptor a molecule like the rice XA21, with extracellular leucine-rich repeats to bind pathogen, and a cytoplasmic kinase domain to relay information to the inside of the cell? If so, did a separate molecule like Pelle/IRAK later take over the kinase function of the pathway, allowing the receptor to diversify?

Even if we can deduce the nature of the ancient signaling pathway, the question of how this pathway was recruited to control embryonic patterning remains. Studies on early dorsal-ventral patterning in other insects and other invertebrates may shed light on how long ago this signaling pathway was adopted for use in embryonic patterning and could provide hints about the general problem of how signaling pathways are recruited for new functions.

Literature Cited

Alkalay I, Yaron A, Hatzubai A, Jung S, Avraham A, et al. 1995. In vivo stimulation of IκB phosphorylation is not sufficient to activate NF-κB. *Mol. Cell. Biol.* 15:1294–301

Anderson KV, Jurgens G, Nüsslein-Volhard C. 1985. Establishment of dorsal-ventral polarity in the Drosophila embryo: genetic studies on the role of the *Toll* gene product. *Cell* 42:779–89

Anderson KV, Nüsslein-Volhard C. 1984. Information for the dorsal-ventral pattern of the *Drosophila* embryo is stored as maternal mRNA. *Nature* 311:223–27

Anderson KV, Nüsslein-Volhard C. 1986. Dorsal-group genes of Drosophila. In *Gametogenesis and the Early Embryo,* ed. J Gall, pp. 177–94. New York: Liss

Åsling B, Dushay M, Hultmark D. 1995. *Drosophila* NF-κB-like gene isolated in a differential display screen for immune genes. *36th Annu. Drosophila Res. Conf.* 269A (Abstr.)

Baeuerle PA, Baltimore D. 1988. Activation of DNA-binding activity in an apparently cytoplasmic precursor of the NF-κB transcription factor. *Cell* 53:211–17

Beg AA, Finco TS, Nantermet PV, Baldwin AS. 1993. Tumor necrosis factor and interleukin-1 lead to phosphorylation and loss of IκB-alpha: a mechanism for NF-κB activation. *Mol. Cell. Biol.* 13:3301–10

Beg AA, Sha WC, Bronson RT, Baltimore D. 1995a. Constitutive NF-κB activation, enhanced granulopoiesis, and neonatal lethality in IκB-alpha-deficient mice. *Genes Dev.* 9:2736–46

Beg AA, Sha WC, Bronson RT, Ghosh S, Baltimore D. 1995b. Embryonic lethality and liver degeneration in mice lacking the RelA component of NF-κB. *Nature* 376:167–70

Belvin MP, Jin Y, Anderson KV. 1995. Cactus protein degradation mediates *Drosophila* dorsal-ventral signaling. *Genes Dev.* 9:783–93

Betts JC, Cheshire JK, Akira S, Kishimoto T, Woo P. 1993. The role of NF-κB and NF-IL6 transactivating factors in the synergistic activation of human serum amyloid A gene expression by interleukin-1 and interleukin-6. *J. Biol. Chem.* 268:25624–31

Brown K, Gerstberger S, Carlson L, Franzoso G, Siebenlist U. 1995. Control of IκB-alpha proteolysis by site-specific, signal-induced phosphorylation. *Science* 267:1485–88

Cao Z, Henzel WJ, Gao X. 1996. IRAK: A kinase associated with the interleukin-1 receptor. *Science* 271:1128–31

Chen Z, Hagler J, Palombella VJ, Melandri F, Scherer D, et al. 1995. Signal-induced site-specific phosphorylation targets IκB-alpha to the ubiquitin-proteasome pathway. *Genes Dev.* 9:1586–97

Ciechanover A. 1994. The ubiquitin-mediated proteolytic pathway: mechanisms of action and cellular physiology. *Biol. Chem. Hoppe-Seyler* 375:565–81

Cociancich S, Dupont A, Hegy G, Lanot R, Holder F, et al. 1994. Novel inducible antibacterial peptides from a hemipteran insect, the sap-sucking bug *Pyrrhocoris apterus*. *Biochem. J.* 300:567–75

Croston GE, Cao Z, Goeddel DV. 1995. NF-κB

activation by interleukin-1 (IL-1) requires an IL-1 receptor-associated protein kinase activity. *J. Biol. Chem.* 270:16514–17

DiDonato JA, Mercurio F, Karin M. 1995. Phosphorylation of IκB alpha precedes but is not sufficient for its dissociation from NF-κB. *Mol. Cell. Biol.* 15:1302–11

Engstrom Y, Kadalayil L, Sun SC, Samakovlis C, Hultmark D, Faye I. 1993. Kappa-B-like motifs regulate the induction of immune genes in *Drosophila. J. Mol. Biol.* 232:327–33

Fey GH, Gauldie J. 1990. In *Progress in Liver Disease,* ed. H Topper, F Schaffner, pp. 89–116. Philadelphia: Saunders

Galindo RL, Edwards DN, Gillespie SK, Wasserman SA. 1995. Interaction of the pelle kinase with the membrane-associated protein tube is required for transduction of the dorsoventral signal in *Drosophila* embryos. *Development* 121:2209–18

Ganchi PA, Sun SC, Greene WC, Ballard DW. 1992. I-κB/MAD-3 masks the nuclear localization signal of NF-κB p65 and requires the transactivation domain to inhibit NF-κB p65 DNA-binding. *Mol. Biol. Cell* 3:1339–52

Geisler R, Bergmann A, Hiromi Y, Nüsslein-Volhard C. 1992. *Cactus,* a gene involved in dorsoventral pattern formation of Drosophila, is related to the IκB gene family of vertebrates. *Cell* 71:613–21

Gerttula S, Jin YS, Anderson KV. 1988. Zygotic expression and activity of the *Drosophila Toll* gene, a gene required maternally for embryonic dorsal-ventral pattern formation. *Genetics* 119:123–33

Gillespie SK, Wasserman SA. 1994. Dorsal, a *Drosophila* Rel-like protein, is phosphorylated upon activation of the transmembrane protein Toll. *Mol. Cell. Biol.* 14:3559–68

Götz P, Boman HG. 1985. Insect immunity. In *Comprehensive Insect Physiology,* ed. G Kerkut, L Gilbert, pp. 453–85. Oxford: Pergamon

Govind S, Drier E, Huang LH, Steward R. 1996. Regulated nuclear import of the *Drosophila* Rel protein Dorsal: structure-function analysis. *Mol. Cell. Biol.* 16:1103–14

Grosshans J, Bergmann A, Haffter P, Nüsslein-Volhard C. 1994. Activation of the kinase Pelle by Tube in the dorsoventral signal transduction pathway of *Drosophila* embryo. *Nature* 372:563–66

Halfon MS, Hashimoto C, Keshishian H. 1995. The *Drosophila Toll* gene functions zygotically and is necessary for proper motoneuron and muscle development. *Dev. Biol.* 169:151–67

Hashimoto C, Gerttula S, Anderson KV. 1991. Plasma membrane localization of the Toll protein in the syncytial *Drosophila* embryo: importance of transmembrane signaling for dorsal-ventral pattern formation. *Development* 111:1021–28

Hashimoto C, Hudson KL, Anderson KV. 1988. The *Toll* gene of Drosophila, required for dorsal-ventral embryonic polarity, appears to encode a transmembrane protein. *Cell* 52:269–79

Haskill S, Beg AA, Tompkins SM, Morris JS, Yurochko AD, et al. 1991. Characterization of an immediate-early gene induced in adherent monocytes that encodes IκB-like activity. *Cell* 65:1281–89

Hatada EN, Naumann M, Scheidereit C. 1993. Common structural constituents confer I-κ-B activity to NF-κB p105 and I-κB/MAD-3. *EMBO J.* 12:2781–88

Hecht PM, Anderson KV. 1993. Genetic characterization of *tube* and *pelle,* genes required for signaling between *Toll* and *dorsal* in the specification of the dorsal-ventral pattern of the *Drosophila* embryo. *Genetics* 135:405–17

Heguy A, Baldari CT, Macchia G, Telford JL, Melli M. 1992. Amino acids conserved in interleukin-1 receptors (IL-1Rs) and the *Drosophila* toll protein are essential for IL-1R signal transduction. *J. Biol. Chem.* 267:2605–9

Henkel T, Machleidt T, Alkalay I, Kronke M, Ben-Neriah Y, Baeuerle PA. 1993. Rapid proteolysis of I-κB-alpha is necessary for activation of transcription factor NF-κB. *Nature* 365:182–85

Henkel T, Zabel U, van Zee K, Muller JM, Fanning E, Baeuerle PA. 1992. Intramolecular masking of the nuclear location signal and dimerization domain in the precursor for the p50 NF-κB subunit. *Cell* 68:1121–33

Hoffmann JA. 1995. Innate immunity of insects. *Curr. Opin. Immunol.* 7:4–10

Hultmark D. 1993. Immune reactions in *Drosophila* and other insects: a model for innate immunity. *Trends Genet.* 9:178–83

Ip YT, Kraut R, Levine M, Rushlow CA. 1991. The dorsal morphogen is a sequence-specific DNA-binding protein that interacts with a long-range repression element in Drosophila. *Cell* 64:439–46

Ip YT, Levine M. 1994. Molecular genetics of *Drosophila* immunity. *Curr. Opin. Genet. Dev.* 4:672–77

Ip YT, Park RE, Kosman D, Bier E, Levine M. 1992a. The dorsal gradient morphogen regulates stripes of *rhomboid* expression in the presumptive neuroectoderm of the *Drosophila* embryo. *Genes Dev.* 6:1728–39

Ip YT, Park RE, Kosman D, Yazdanbakhsh K, Levine M. 1992b. Dorsal-twist interactions establish *snail* expression in the presumptive mesoderm of the *Drosophila* embryo. *Genes Dev.* 6:1518–30

Ip YT, Reach M, Engstrom Y, Kadalayil L, Cai H, et al. 1993. *Dif*, a dorsal-related gene that mediates an immune response in Drosophila. *Cell* 75:753–63

Isoda K, Nüsslein-Volhard C. 1994. Disulfide cross-linking in crude embryonic lysates reveals three complexes of the *Drosophila* morphogen dorsal and its inhibitor cactus. *Proc. Natl. Acad. Sci. USA* 91:5350–54

Isoda K, Roth S, Nüsslein-Volhard C. 1992. The functional domains of the *Drosophila* morphogen dorsal: evidence from the analysis of mutants. *Genes Dev.* 6:619–30

Jaffray E, Wood KM, Hay RT. 1995. Domain organization of IκB alpha and sites of interaction with NF-κB p65. *Mol. Cell. Biol.* 15:2166–72

Jiang J, Levine M. 1993. Binding affinities and cooperative interactions with bHLH activators delimit threshold responses to the dorsal gradient morphogen. *Cell* 72:741–52

Jiang J, Rushlow CA, Zhou Q, Small S, Levine M. 1992. Individual dorsal morphogen binding sites mediate activation and repression in the *Drosophila* embryo. *EMBO J.* 11:3147–54

Jones DA, Thomas CM, Hammond KK, Balint KP, Jones JD. 1994. Isolation of the tomato *Cf-9* gene for resistance to *Cladosporium fulvum* by transposon tagging. *Science* 266:789–93

Kappler C, Meister M, Lagueux M, Gateff E, Hoffmann JA, Reichhart JM. 1993. Insect immunity. Two 17 bp repeats nesting a κB-related sequence confer inducibility to the diptericin gene and bind a polypeptide in bacteria-challenged *Drosophila. EMBO J.* 12:1561–68

Keith FJ, Gay NJ. 1990. The *Drosophila* membrane receptor Toll can function to promote cellular adhesion. *EMBO J.* 9:4299–306

Kidd S. 1992. Characterization of the Drosophila *cactus* locus and analysis of interactions between cactus and dorsal proteins. *Cell* 71:623–35

Koj A. 1985. Liver response to inflammation and synthesis of acute-phase plasma proteins. In *The Acute-Phase Response to Injury and Infection. Research Monographs in Cell and Tissue Physiology,* ed. AH Gordon, A Koj, pp. 139–246. Amsterdam: Elsevier

Kopp EB, Ghosh S. 1995. NF-κB and rel proteins in innate immunity. *Adv. Immunol.* 58:1–27

LeClair KP, Blanar MA, Sharp PA. 1992. The p50 subunit of NF-κB associates with the NF-IL6 transcription factor. *Proc. Natl. Acad. Sci. USA* 89:8145–49

Lehming N, McGuire S, Brickman JM, Ptashne M. 1995. Interactions of a Rel protein with its inhibitor. *Proc. Natl. Acad. Sci. USA* 92:10242–46

Lehming N, Thanos D, Brickman JM, Ma J, Maniatis T, Ptashne M. 1994. An HMG-like protein that can switch a transcriptional activator to a repressor. *Nature* 371:175–79

Lemaitre B, Meister M, Govind S, Georgel P, Steward R, et al. 1995. Functional analysis and regulation of nuclear import of dorsal during the immune response in *Drosophila. EMBO J.* 14:536–45

Letsou A, Alexander S, Orth K, Wasserman SA. 1991. Genetic and molecular characterization of *tube,* a *Drosophila* gene maternally required for embryonic dorsoventral polarity. *Proc. Natl. Acad. Sci. USA* 88:810–14

Letsou A, Alexander S, Wasserman SA. 1993. Domain mapping of tube, a protein essential for dorsoventral patterning of the *Drosophila* embryo. *EMBO J.* 12:3449–58

Lin YC, Brown K, Siebenlist U. 1995. Activation of NF-κB requires proteolysis of the inhibitor IκB-alpha: signal-induced phosphorylation of IκB-alpha alone does not release active NF-κB. *Proc. Natl. Acad. Sci. USA* 92:552–56

Lohs-Schardin M, Cremer C, Nüsslein-Volhard C. 1979. A fate map for the larval epidermis of *Drosophila melanogaster:* localized cuticle defects following irradiation of the blastoderm with an ultraviolet laser microbeam. *Dev. Biol.* 73:239–55

Martin GB, Brommonschenkel S, Chunwongse J, Frary A, Gannal MW, et al. 1993. Map-based cloning of a protein kinase gene conferring disease resistance in tomato. *Science* 262:1432–36

Martin GB, Frary A, Wu T, Brommonschenkel S, Chunwongse J, et al. 1994. A member of the *Pto* gene family confers sensitivity to fenthion resulting in rapid cell death. *Plant Cell* 6:1543–52

Martin M, Bol GF, Eriksson A, Resch K, Brigelius FR. 1994. Interleukin-1-induced activation of a protein kinase co-precipitating with the type I interleukin-1 receptor in T cells. *Eur. J. Immunol.* 24:1566–71

Matsusaka T, Fujikawa K, Nishio Y, Mukaida N, Matsushima K, et al. 1993. Transcription factors NF-IL6 and NF-κB synergistically activate transcription of the inflammatory cytokines, interleukin 6 and interleukin 8. *Proc. Natl. Acad. Sci. USA* 90:10193–97

Miyamoto S, Maki M, Schmitt MJ, Hatanaka M, Verma IM. 1994. Tumor necrosis factor alpha-induced phosphorylation of IκB alpha is a signal for its degradation but not dissociation from NF-κB. *Proc. Natl. Acad. Sci. USA* 91:12740–44

Miyamoto S, Verma IM. 1995. Rel/NF-κB/IκB story. *Adv. Cancer Res.* 66:255–92

Morisato D, Anderson KV. 1994. The *spätzle* gene encodes a component of the extracellular signaling pathway establishing the dorsal-ventral pattern of the Drosophila embryo. *Cell* 76:677–88

Morisato D, Anderson KV. 1995. Signaling pathways that establish the dorsal-ventral pattern of the *Drosophila* embryo. *Annu. Rev. Genet.* 29:371–99

Müller CW, Harrison SC. 1995. The structure of the NF-κB p50:DNA-complex: a starting point for analyzing the Rel family. *FEBS Lett.* 369:113–17

Norris JL, Manley JL. 1992. Selective nuclear transport of the *Drosophila* morphogen dorsal can be established by a signaling pathway involving the transmembrane protein Toll and protein kinase A. *Genes Dev.* 6:1654–67

Nose A, Mahajan VB, Goodman CS. 1992. Connectin: a homophilic cell adhesion molecule expressed on a subset of muscles and the motoneurons that innervate them in Drosophila. *Cell* 70:553–67

Pan DJ, Huang JD, Courey AJ. 1991. Functional analysis of the *Drosophila twist* promoter reveals a dorsal-binding ventral activator region. *Genes Dev.* 5:1892–901

Peters JM. 1994. Proteasomes: protein degradation machines of the cell. *Trends Biochem. Sci.* 19:377–82

Petersen UM, Bjorklund G, Ip YT, Engstrom Y. 1995. The dorsal-related immunity factor, Dif, is a sequence-specific trans-activator of *Drosophila* Cecropin gene expression. *EMBO J.* 14:3146–58

Rechsteiner M. 1990. PEST sequences are signals for rapid intracellular proteolysis. *Semin. Cell Biol.* 1:433–40

Reichhart JM, Georgel P, Meister M, Lemaitre B, Kappler C, Hoffmann JA. 1993. Expression and nuclear translocation of the rel/NF-κB-related morphogen dorsal during the immune response of *Drosophila. CR Acad. Sci. III* 316:1218–24

Reichhart JM, Meister M, Dimarcq JL, Zachary D, Hoffmann D, et al. 1992. Insect immunity: developmental and inducible activity of the *Drosophila* diptericin promoter. *EMBO J.* 11:1469–77

Rodriguez MS, Michalopoulos I, Arenzana SF, Hay RT. 1995. Inducible degradation of IκB alpha in vitro and in vivo requires the acidic C-terminal domain of the protein. *Mol. Cell. Biol.* 15:2413–19

Rogers S, Wells R, Rechsteiner M. 1986. Amino acid sequences common to rapidly degraded proteins: the PEST hypothesis. *Science* 234:364–68

Roth S, Hiromi Y, Godt D, Nüsslein-Volhard C. 1991. *Cactus*, a maternal gene required for proper formation of the dorsoventral morphogen gradient in *Drosophila* embryos. *Development* 112:371–88

Roth S, Stein D, Nüsslein-Volhard C. 1989. A gradient of nuclear localization of the dorsal protein determines dorsoventral pattern in the Drosophila embryo. *Cell* 59:1189–202

Rushlow CA, Han K, Manley JL, Levine M. 1989. The graded distribution of the dorsal morphogen is initiated by selective nuclear transport in Drosophila. *Cell* 59:1165–77

Saville KJ, Belote JM. 1993. Identification of an essential gene, *l(3)73Ai*, with a dominant temperature-sensitive lethal allele, encoding a *Drosophila* proteasome subunit. *Proc. Natl. Acad. Sci. USA* 90:8842–46

Scherer DC, Brockman JA, Chen Z, Maniatis T, Ballard DW. 1995. Signal-induced degradation of IκB alpha requires site-specific ubiquitination. *Proc. Natl. Acad. Sci. USA* 92:11259–63

Schneider DS, Hudson KL, Lin TY, Anderson KV. 1991. Dominant and recessive mutations define functional domains of Toll, a transmembrane protein required for dorsal-ventral polarity in the *Drosophila* embryo. *Genes Dev.* 5:797–807

Schneider DS, Jin Y, Morisato D, Anderson KV. 1994. A processed form of the Spätzle protein defines dorsal-ventral polarity in the *Drosophila* embryo. *Development* 120:1243–50

Schwyter DH, Huang JD, Dubnicoff T, Courey AJ. 1995. The *decapentaplegic* core promoter region plays an integral role in the spatial control of transcription. *Mol. Cell. Biol.* 15:3960–68

Sha WC, Liou HC, Tuomanen EI, Baltimore D. 1995. Targeted disruption of the p50 subunit of NF-κB leads to multifocal defects in immune responses. *Cell* 80:321–30

Shelton CA, Wasserman SA. 1993. Pelle encodes a protein kinase required to establish dorsoventral polarity in the Drosophila embryo. *Cell* 72:515–25

Siebenlist U, Franzoso G, Brown K. 1994. Structure, regulation and function of NF-κB. *Annu. Rev. Cell Biol.* 10:405–55

Sipe JD. 1985. Cellular and humoral components of the early inflammatory reaction. In *The Acute-Phase Response to Injury and Infection. Research Monographs in Cell and Tissue Physiology,* ed. AH Gordon, A Koj, pp. 3–21. Amsterdam: Elsevier

Song WY, Wang GL, Chen LL, Kim HS, Pi LY, et al. 1995. A receptor kinase-like protein encoded by the rice disease resistance gene, Xa21. *Science* 270:1804–6

Staskawicz BJ, Ausubel FM, Baker BJ, Ellis JG, Jones JD. 1995. Molecular genetics of plant disease resistance. *Science* 268:661–67

Steward R. 1987. *Dorsal,* an embryonic polarity gene in *Drosophila,* is homologous to the vertebrate proto-oncogene, c-rel. *Science* 238:692–94

Steward R. 1989. Relocalization of the dorsal protein from the cytoplasm to the nucleus correlates with its function. *Cell* 59:1179–88

Sun SC, Åsling B, Faye I. 1991. Organization and expression of the immunoresponsive lysozyme gene in the giant silk moth, *Hyalophora cecropia. J. Biol. Chem.* 266:6644–49

Tatei K, Levine M. 1995. Specificity of Rel-inhibitor interactions in *Drosophila* embryos. *Mol. Cell. Biol.* 15:3627–34

Thisse C, Perrin SF, Stoetzel C, Thisse B. 1991. Sequence-specific transactivation of the Drosophila *twist* gene by the dorsal gene product. *Cell* 65:1191–201

Traenckner EBM, Wilk S, Baeuerle PA. 1994. A proteasome inhibitor prevents activation of NF-κB and stabilizes a newly phosphorylated form of I-κB alpha that is still bound to NF-κB. *EMBO J.* 13:5433–41

Weih F, Carrasco D, Durham SK, Barton DS, Rizzo CA, et al. 1995. Multiorgan inflammation and hematopoietic abnormalities in mice with a targeted disruption of RelB, a member of the NF-κ B/Rel family. *Cell* 80:331–40

Whalen AM, Steward R. 1993. Dissociation of the dorsal-cactus complex and phosphorylation of the dorsal protein correlate with the nuclear localization of dorsal. *J. Cell Biol.* 123:523–34

Whitham S, Dinesh KS, Choi D, Hehl R, Corr C, Baker B. 1994. The product of the tobacco mosaic virus resistance gene *N*: similarity to *Toll* and the interleukin-1 receptor. *Cell* 78:1101–15

Zhou J, Loh YT, Bressan RA, Martin GB. 1995. The tomato gene *Pti1* encodes a serine/threonine kinase that is phosphorylated by Pto and is involved in the hypersensitive response. *Cell* 83:925–35

Annu. Rev. Cell Dev. Biol. 1996. 12:417–39

THE MAMMALIAN MYOSIN HEAVY CHAIN GENE FAMILY

Allison Weiss
Albert Einstein College of Medicine, Department of Microbiology and Immunology, 1300 Morris Park Ave. Bronx, New York 10461

Leslie A. Leinwand
Department of Molecular, Cellular and Developmental Biology, University of Colorado, Boulder, Colorado 80309-0347

KEY WORDS: cloning, gene expression, multigene family, molecular motor, muscle

ABSTRACT

Myosin is a highly conserved, ubiquitous protein found in all eukaryotic cells, where it provides the motor function for diverse movements such as cytokinesis, phagocytosis, and muscle contraction. All myosins contain an amino-terminal motor/head domain and a carboxy-terminal tail domain. Due to the extensive number of different molecules identified to date, myosins have been divided into seven distinct classes based on the properties of the head domain. One such class, class II myosins, consists of the conventional two-headed myosins that form filaments and are composed of two myosin heavy chain (MYH) subunits and four myosin light chain subunits. The MYH subunit contains the ATPase activity providing energy that is the driving force for contractile processes mentioned above, and numerous MYH isoforms exist in vertebrates to carry out this function. The MYHs involved in striated muscle contraction in mammals are the focus of the current review. The genetics, molecular biology, and biochemical properties of mammalian MYHs are discussed below. MYH gene expression patterns in developing and adult striated muscles are described in detail, as are studies of regulation of MYH genes in the heart. The discovery that mutant MYH isoforms have a causal role in the human disease familial hypertrophic cardiomyopathy (FHC) has implemented structure/function investigations of MYHs. The regulation of MYH genes expressed in skeletal muscle and the potential functional implications that distinct MYH isoforms may have on muscle physiology are addressed.

1081-0706/96/1115-0417$08.00

CONTENTS

THE MYH GENE FAMILY

Myosin is a ubiquitous eukaryotic motor protein that interacts with actin to generate the force for cellular movements ranging from cytokinesis to muscle contraction (see Cheney et al 1993; K Ruppel & J Spudich, this volume). The genomes of vertebrates contain many genes encoding distinct classes of myosin including conventional two-headed myosins, as well as unconventional single-headed isoforms that are expressed in a tissue-specific developmentally regulated manner. Conventional myosin exists as a hexameric protein consisting of two myosin heavy chain (MYH) subunits (200 kDa each) and two pairs of non-identical light chain subunits (17–23 kDa). The MYH provides both the motor and filament-forming functions of the intact myosin molecule. The MYH subunit can be divided into two functional domains: a globular, amino-terminal head domain, which contains the motor function, and an elongated alpha-helical coiled-coil carboxyl rod domain, which contains myosin's filament-forming properties. Cloned sequences have been obtained from MYH genes expressed in smooth and striated muscle, as well as in several nonmuscle cell types. The myosins expressed in these cell types comprise three general myosin classes; (*a*) smooth muscle, (*b*) striated muscle, and (*c*) nonmuscle myosins. Although the full extent of the mammalian MYH gene family has undoubtedly not been determined, at least 13 genes have been described. These include 3 nonmuscle MYH genes, 2 smooth muscle MYH genes, and at least 8 striated muscle MYH genes (see Table 1). This review focuses on the mammalian MYH gene family, with an emphasis on those genes expressed in striated muscle.

The striated MYH genes are organized in two chromosomal clusters. Six skeletal MYH genes are clustered on a 300–600-kb segment on human and

Table 1 Cloned mammalian MYHs

MYH Isoform	Clones	
	cDNA	Genomic
Nonmuscle		
NMMHCA	H (1,2)	—
NMMHCB	H (2)	—
Rat brain MYH	R (3)	—
Smooth muscle		
SM1	H (4)	—
SM2	H (4)	—
SMemb	H (4)	—
Striated muscle		
Embryonic	H (5, 6, 30)	H (6)
	M (7)	R (8)
Perinatal	H (9,26)	H (10)
	M (7)	R (11)
Adult 2A	R (12)	—
	M (13)	
Adult 2B	R (12)	—
	M (13)	
Adult 2X/2D	H (14)	H (10)
	R (15, 16)	
Extraocular	H (29)	R (17)
Beta/slow	H (14)	H (20, 21, 23)
	R (18)	R (25)
		M (19)
Alpha cardiac	H (27)	H (20, 21, 24)
	R (22)	R (25)
	M (28)	M (19)

H, human; M, mouse; R, rat. References: 1. Saez et al 1992; 2. Simons et al 1991; 3. Sun & Chantler 1994; 4. Aikawa et al 1993; 5. Eller et al 1989; 6. Karsch-Mizrachi et al 1989; 7. Weydert et al 1985; 8. Strehler et al 1986; 9. Feghali & Leinwand 1989; 10. Yoon et al 1992; 11. Periasamy et al 1984; 12. Nadal-Ginard et al 1982; 13. Weydert et al 1983; 14. Saez & Leinwand 1986; 15. Ogata et al 1993; 16. De Nardi et al 1993; 17. Weiczorek et al 1985; 18. Kraft et al 1989; 19. Gulick et al 1991; 20. Saez et al 1987; 21. Liew et al 1990; 22. McNally et al 1989; 23. Jaenicke et al 1990; 24. Yamauchi-Takihara 1989; 25. Lompre et al 1984; 26. Karsch-Mizrachi et al 1990; 27. Matsuoka et al 1991; 28. Quinn-Laquer et al 1992; 29. A Weiss, unpublished results; 30. Stedman et al 1990.

mouse chromosomes 17 and 11, respectively (Leinwand et al 1983, Weydert et al 1985). A 600-kb yeast artificial chromosome (YAC) contig has been generated for the human skeletal MYH locus. The genes or this contig are arranged in the following order: MYH-embryonic, MYH-2A, MYH-2X, MYH-perinatal, followed by two additional genes (probably corresponding to the MYH-extraocular and MYH-2B genes (Yoon et al 1992; for a review, see Weiss et al 1994). The MYH genes expressed in the heart, MYH-alpha and MYH-beta, are not linked to the skeletal MYH genes and are arranged in tandem with the MYH-beta gene located 4–5 kb upstream from the MYH-alpha gene in human, mouse (chromosome 14; Weydert et al 1985, Saez et al 1987) and rat (Mahdavi et al 1984). The loci for two human nonmuscle genes are found on chromosomes 22 (Saez et al 1990) and 17 (Simons et al 1991), and a smooth muscle MYH gene has been located on chromosome 16 for human and mouse (Deng et al 1993, Matsuoka et al 1993).

The mammalian MYH genes are highly conserved in terms of their primary structure and share a similar intron/exon organization. Evolutionary comparisons among mammalian MYH genes suggest that the gene family arose from multiple duplications of an ancestral MYH gene (Stedman et al 1990, Moore et al 1993). Most of these genes are large (approximately 30 kb) and have either 40 or 41 exons, with a coding region beginning in exon 3 (see Strehler et al 1986 for MYH gene structure). All striated muscle MYHs are highly conserved, ranging from 78–98% amino acid identity. The greatest conservation (> 98% aa identity) is observed between the same isoforms from different species (i.e. human/mouse/rat MYH-alpha comparisons). Figure 1 presents the results of comparison of the amino acid sequences from the head and rod domains of various MYH isoforms from human, rat, and mouse. In all alpha/beta comparisons, it appears that the head domain is more divergent than the rod. The nonmuscle and smooth muscle myosins generally show a closer relationship to each other than either does with the myosin found in striated muscle (Miano et al 1994).

The 5′ and 3′ untranslated sequences of the striated MYH genes tend to be less conserved than coding regions, which makes MYH UTR's useful for generating gene-specific probes for nucleic acid hybridization studies (Lyons et al 1990a,b). Nevertheless, when the same isoforms from different species are compared, MYH 3′UTRs exhibit a significant extent of interspecific sequence conservation (Periasamy et al 1984, Saez & Leinwand 1986). Table 2 shows interspecific comparisons of human 3′UTRs with the 3′UTRs from the corresponding isoforms in rat and mouse. Most MYH 3′UTRs are between 100 and 150 bp in length (with some exceptions), and the 3′UTRs of the same isoforms from different species typically are greater than 64% identical over a region of at least 100 bp. When MYH 3′UTRs are compared within a species (same

Table 2 3′ UTR comparisons of striated muscle MYHs

UTR	Species compared[a]	% Identity	Overlap (bp)
MYH-emb	H/M	83.0	109
	H/R	79.8	109
MYH-peri	H/M	64.4	104
	H/R	64.4	135
MYH-2A	H/M	75.4	130
	H/R	84.9	106
MYH-2X	H/M	84.9	106
	H/R	88.1	101
MYH-beta	H/M	71.0	62
	H/R	74.5	106
MYH-alpha	H/M	68.6	35
	H/R	73.6	38
MYH-EO	H/R	66.0	83

[a]H = human, M = mouse, R = rat

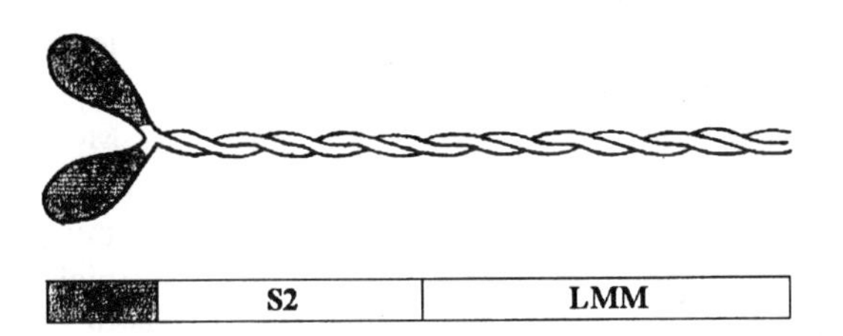

HEAD DOMAIN

halpha	*halpha*					
malpha	96.9	*malpha*				
ralpha	97.3	98.7	*ralpha*			
hbeta	91.0	90.0	92.0	*hbeta*		
rbeta	91.7	91.4	92.1	95.7	*rbeta*	
hemb	79.1	79.0	79.7	78.7	79.2	*hemb*
hperi	80.8	80.3	81.2	80.5	80.6	86.2

ROD DOMAIN

halpha	*halpha*					
malpha	96.6	*malpha*				
ralpha	96.8	98.7	*ralpha*			
hbeta	95.0	93.5	94.1	*hbeta*		
rbeta	94.8	94.1	94.7	98.3	*rbeta*	
hemb	78.1	78.5	78.1	78.6	78.5	*hemb*
hperi	79.5	79.9	79.8	80.3	80.2	83.0

Figure 1 (*upper*) MYH dimer. Shaded region represents proteolytic fragment S1/myosin head domain. Proteolytic fragments S2 and LMM correspond to alpha-helical coiled-coil rod domain. (*lower*) Comparisons of head domains and rod domains among striated MYH isoforms from human (*halpha, hbeta, hemb, hperi*), mouse (*malpha*), and rat (*ralpha, rbeta*). Numbers represent percent of sequence identity within head domain (aa1–840) and rod domain (aa≈ 841–1941).

species, different isoform), long stretches of sequence identity are rare, and any sequence identity occurring over a significant length of DNA (i.e. greater than 50% the length of the UTR) is generally under 64% (data not shown).

MYOSIN ATPase ACTIVITY AND MUSCLE PHYSIOLOGY

The existence of this gene family raises the question of whether MYH isoforms are functionally diverse or whether there is functional redundancy. Several studies agree that the former is most likely the case. Differences in contractile velocities and ATPase activities of skeletal and myocardial fibers have been correlated with variations in their myosin content (Barany 1967). Muscle strips from hearts expressing MYH-alpha cardiac have a more rapid contractile velocity than those expressing MYH-beta. The logical hypothesis arising from these observations is that functional differences among the various MYH isoforms contribute to or result in changes in the physiology of the muscle in which they are expressed. Significant differences in ATPase activity have been demonstrated for a number of myosins. Under conditions of defined cation composition, myosins from different muscle sources hydrolyze ATP at specific rates. The rate of ATP hydrolysis is greatly accelerated in the presence of actin, and thus in vitro measurements of basal and actin-activated ATPase activities can provide a biochemical signature for different myosins. Measurements have been made of alpha and beta cardiac myosin preparations, as well as from myosin purified from rabbit back muscle, which contains a mixture of predominantly fast MYH. Rabbit fast muscle myosin exhibits the highest rate of ATP hydrolysis, and this activity is approximately 10-fold that of MYH-alpha cardiac, and 20–30-fold higher than that of beta myosin (see Schiaffino & Reggiani 1996). Measurement of the ATPase activities of the remaining skeletal MYH isoforms has not yet been possible because skeletal fibers usually contain multiple isoforms, making it difficult to obtain pure isoform preparations. However, embryonic and neonatal muscles contain mixtures of the early developmental isoforms (see section on Developmental Expression of MYH Genes below), and myosins from these muscles have lower ATPase activities than their adult counterparts (Drachman & Johnston 1973, Lowey et al 1993). Sweeney et al (1994) have employed a baculovirus expression system to produce pure populations of mutant and wild-type forms of a portion of the rat MYH-alpha [proteolytic fragment heavy meromyosin (HMM)], enabling actin-activated ATPase and in vitro motility to be measured. These studies demonstrate that the recombinant alpha HMM had biochemical properties similar to the alpha HMM purified directly from cardiac muscle. Similar expression studies of individual skeletal genes would help to assess the extent of functional diversity within the MYH gene family.

MYH SEQUENCE DIFFERENCES RELATE TO FUNCTIONAL DIVERGENCE

The sequences in the MYH responsible for determining its ATPase characteristics have not yet been fully defined. The recent availability of the three-dimensional structure of the myosin head (Rayment et al 1993) has been extremely useful in highlighting sequences of potential functional significance. Although the myosin head is well conserved, clusters of divergent sequences have been identified in certain regions that may contribute to differences in enzymatic activity. Examination of the two closely related cardiac MYH isoforms, MYH-alpha and MYH-beta, has been particularly informative (McNally et al 1989). Despite $\approx 93\%$ aa sequence identity between these isoforms, their ATPase activities differ by greater than twofold (Pope et al 1980). One region containing a notable amount of amino acid divergence between MYH-alpha and MYH-beta lies between amino acids 623–640. This particular region contains a site for trypsin proteolysis that forms the junction between the 50- and 20-kDa fragments and is suggested to form a loop structure that binds actin (Rayment et al 1993). In fact, this region is highly divergent in all myosins examined (Figure 2). Only 3 of 18 amino acids in the loop region are conserved among the four human isoforms compared. In contrast, the flanking amino acids show far less divergence. Experimental evidence testing the contribution of this 18 amino acid loop region to MYH ATPase activity has been obtained by expression studies in *Dictyostelium* (Uyeda et al 1994). Chimeric myosins in which the loop in the *Dictyostelium* MYH was replaced by the corresponding sequences of several vertebrate MYHs were expressed and assayed for ATPase activity and in vitro motility. Substitution of the *Dictyostelium* loop sequence with a sequence from a rabbit fast skeletal isoform increased the ATPase activity of the parent *Dictyostelium* molecule by fivefold and, in fact, the ATPase activity of the chimera more closely resembled that of the rabbit fast myosin. In all the substitutions studied, the ATPase activities of the chimeric molecules reflected those of the donor isoform, indicating that the loop region is an important determinant of MYH ATPase activity.

```
        aa623                                        20/50K Junction                                             aa640
          ↓                                                                                                          ↓
halpha  AGTVDYNILGWLEKNKDPLNETVVGLYQKSSLKLMATLF   SSYATADTGDSGKSKGGK   KKGSSFQTVSALHRENLNKLMTNLRTTHPHFVRCIIPNE
hbeta   ............Q.....................LS...   AN..G..-APIE.G..KA   .........................S.............
hemb    ...............................NR.L.H.Y   ATF....-A....K.VA.   ............F........S.................
hperi   ............D.......D.........AM.TL.S..   .T..S.E-A..SAKK.A.   ............F............S.............
```

Figure 2 Comparisons of the 20/50 K junction sequences for four human sarcomeric MYH genes.

Table 3 Directional responses of cardiac MYH genes

	MYH-alpha	MYH-beta
+T3	↑	↓
−T3	↓	↑
Exercise	↑	↓
Pressure overload	↓	↑
Aging	↓	↑

MYH GENE EXPRESSION

MYH genes are both temporally and spatially regulated, and differential expression of muscle structural proteins, including MYHs, is likely to contribute to the extensive diversity observed in skeletal muscle fiber types (see Schiaffino & Reggiani 1996). The fiber composition of individual muscle groups varies dramatically during development, in response to aging or exercise, and as a function of anatomical location. Skeletal muscle fibers are distinguished by different contractile velocities, metabolic properties, and structural protein compositions, including varying MYH isoform profiles. In contrast, the myocardium is generally more homogeneous and responds more uniformly to a wide variety of stimuli. Developmentally regulated expression of the sarcomeric MYH genes and the responses of MYH genes to certain stimuli are discussed below. Mechanisms that dictate responsiveness to specific factors such as thyroid hormone are fairly well understood, whereas the molecular mechanisms mediating responses to complex stimuli such as exercise are far less understood. For the skeletal genes, the molecular mechanisms for generating fiber-type specification and developmental gene regulation remain important questions. A better understanding of the regulation of cardiac MYH genes exists owing to more extensive transfection and transgenic analyses of these genes (discussed below).

DEVELOPMENTAL EXPRESSION OF MYH GENES

Cardiac MYH Genes

As mentioned above, there are two MYH genes expressed in the myocardium, MYH-alpha and MYH-beta. These genes are differentially regulated during development; have different distribution patterns in the heart (atrial versus ventricular); and also respond differently to various hormonal stimuli, hemodynamic stress, exercise, and cardiac hypertrophy. These directional responses are shown in Table 3.

MYH-alpha is the major isoform expressed in the adult ventricles and atria of small mammals (Lompre et al 1984). Its expression is almost exclusively restricted to the heart, with low levels of expression seen in extraocular and masseter muscles (d'Albis et al 1993, Rushbrook et al 1994), as well as in lung (Subramaniam et al 1991). MYH-beta is expressed in both cardiac and slow (type I) skeletal muscle fibers such as the soleus, and it is the abundant isoform in the ventricles of all mammals during fetal life. The major MYH isoform expressed postnatally in the ventricles of large adult mammals, including humans, is MYH-beta.

In situ hybridization studies on mouse embryos reveal that MYH-alpha and -beta transcripts are found throughout the heart tube between 7.5 and 8 days post coitum [(pc); Lyons et al 1990b] in concordance with the appearance of the first myocardial contractions in the mouse (Sissman 1970, Rugh 1990). As the ventricular chamber forms between days 8 and 9 pc, a process completed by day 10.5 pc, MYH-beta expression begins to be restricted to ventricular myocytes, the outflow tract, and the ventriculoarterial portion. MYH-beta remains the predominant isoform in the ventricles until just before birth. Ventricular MYH-alpha expression begins to decrease gradually at day 10.5 pc. Just before birth, its expression is upregulated until it becomes the predominant adult ventricular isoform (by day 7 after birth, MYH-alpha transcripts have completely replaced those of MYH-beta). In atrial muscle, MYH-alpha is expressed at high levels from day 8 pc through adulthood. However, MYH-beta is expressed only briefly in the atria, and by day 9.5 pc, only a few atrial cells continue to express this gene (see Table 4).

Skeletal MYH Genes

Skeletal muscle is composed of clusters of muscle fibers, each of which is a multinucleated cell formed by successive rounds of myoblast fusion. All skeletal muscles are formed from cells that originate in the somites during embryogenesis. The myotome, one of three layers of the somites, is the first site of MYH gene expression, which occurs at day 9.5 pc, just before myoblasts fuse to form multinucleated myotubes. For limb muscles, the first round of myoblast fusion gives rise to primary myotubes, which are formed before the formation of distinct muscles. A second round of myoblast fusion forms the secondary myotubes juxtaposed to primary myotubes. Primary and secondary myotubes differ in terms of the MYH isoforms they can express and thus can be seen as discrete stages of muscle fiber-type progression. Primary myotubes (except in the soleus) express MYH-emb, MYH-neo, and MYH-beta before becoming restricted to a slow or fast fate, whereas secondary fibers initially express only MYH-emb and MYH-peri simultaneously. A secondary fiber that subsequently expresses MYH-beta (in addition to MYH-emb and MYH-

Table 4 Temporal expression of MYH genes during mouse development

	Genes	Days													
		8.0	8.5	9.5	10.5	11.5	12.5	14.5	15.0	15.5	17.5	18.0	0.0	5.0	7.7
Cardiac muscle[a]	*MYH-alpha*[b]	+++		+++	++	++	++	+		+	++		++		+++
	MYH-alpha[c]	+++		+++	+++	+++	+++	+++		+++	+++		+++		+++
	MYH-beta[b]	+++		+++	+++	+++	+++	+++		+++	+++		++		−
	MYH-beta[c]	+++		+	−	−	−	−		−	−		−		−
Skletal muscle[a]	*MYH-emb*		−	+/−	++	+++	+++	+++		++					
	MYH-peri		−	−	+	++	+++	+++		+++					
	MYH-beta		−	+/−	+	+	+	+		+					
Skeletal muscle[d]	*MYH-emb*								†††			††	†	†/−	
	MYH-peri								†			†††	†	†/−	
	MYH-2B								−			†/−	†/−	†††	
							↑ 1°Myotube formation (d 12 pc)	↑ Innervation of muscle (d 14 pc)		↑ 2°Myotube formation (d 15–18 pc)			↑ Birth		

[a]Data from Lyons et al 1990a,b (in situ hybridization studies)
[b]Ventricular expression
[c]Atrial expression
[d]Data from Weydert et al 1987 (RNase protection study)

peri), is fated to become a slow fiber (for review, see Gunning & Hardeman 1991, Stockdale 1992). The purpose served by expressing multiple isoforms, particularly the early developmental ones, remains unclear.

There are at least seven MYH genes expressed in mammalian skeletal muscle, and the first transcripts for these genes are detected in developing myotomes ≈ 1.5 days after the first MYH-alpha and MYH-beta transcripts are detected in the developing heart. In general, the development of fast myofibers involves early expression of MYH-emb and MYH-peri isoforms, with low levels of MYH-beta expression and the gradual replacement of these isoforms with MYH-2A, MYH-2X, and MYH-2B adult fast isoforms as development proceeds. In formation of adult slow fibers, MYH-beta gene expression persists and ultimately predominates, rather than being replaced by adult fast isoforms. Temporal patterns of expression of these genes are presented in Table 4. In situ hybridization studies have been carried out for MYH-emb, MYH-peri, and MYH-beta genes in developing mouse skeletal muscle and are summarized as follows (Lyons et al 1990a): MYH-emb and MYH-beta are the first genes to be expressed between days 9–10 pc in the myotomes of the most rostral somites (in the cervical region of the embryo as development proceeds rostro-caudally). MYH-peri expression begins at day 10.5 pc, although MYH-emb is still the predominant transcript at this time, and MYH-beta is expressed at low levels as well. At day 12.5 pc, MYH-emb and MYH-peri transcripts are approximately equivalent, but by day 15.5 pc, MYH-peri predominates, although at this time there are regional differences in the distribution of transcripts with MYH-beta in central portions of muscles and MYH-peri located at the muscle periphery. It is notable that MYH-beta can be detected in all muscle masses until becoming downregulated in fibers fated to be fast, or upregulated in fibers that will become slow fibers. These authors suggest that day 15.5 defines the embryonic/fetal boundary because at this time, the MYH-peri transcript becomes restricted to the fibers that will become adult fast fibers, and muscle fibers correspond spatially to the arrangement they will have in the adult. Using an RNase protection assay, Weydert et al (1987) measured the relative mRNA levels of MYH-emb, MYH-peri, and adult MYH-2B in developing rat limb muscles for days 15.0 and 18.0 pc, through postnatal day 5. In accordance with in situ data, these studies show coexpression of MYH-emb and MYH-peri in 1° myotubes (with the ratio of MYH-emb to MYH-peri 2.5:1). Between day 15 and 18 pc, MYH-peri levels increase 20- to 40-fold, resulting in a MYH-peri:MYH-emb ratio of 5:1. At day 18 pc, the adult fast isoform MYH-2B is also detected at low levels. The expression of MYH-2B does not increase further until 3 to 4 days after birth, when the MYH fetal isoforms are decreasing. The replacement of MYH fetal transcripts with MYH-2B transcripts occurs rapidly between birth and postnatal days 3 and 4, marking the first sign of diversification of 1° and 2° generation fibers. Denardi et al (1993) were also able to detect three adult fast mRNAs, MYH-2A, -2X, and -2B, in rat hindlimb muscles a few days after birth. These data, along with immunocytochemical and electrophoretic studies on biochemical preparations of single fibers from rat leg muscles, show that adult fibers may be categorized either as pure, consisting of predominantly a single MYH isoform (beta, 2A, 2X, or 2B), or as a hybrid fiber, consisting of beta/2A, 2A/2X, or 2X/2B mixtures (for review see Schiaffino & Reggiani 1994). It appears that early MYH expression patterns define adult fiber types. However, in adult fibers, changes in isoform composition can occur in response to neuronal (Pette & Vrbova 1985), hormonal (Izumo et al 1986), mechanical (Loughna et al 1990), and electrical stimuli (Gorza et al 1988, Termin et al 1989, Ausoni et al 1990), as well as with aging (Butler-Browne et al 1984, Whalen et al 1984, Larsson et al 1991) and exercise (Andersen et al 1994a,b). Expression patterns observed thus far suggest that there are certain rules that restrict isoform switching, such that switching may only occur in certain obligatory patterns. For example, the pattern beta↔2A↔2X↔2B is also adhered to in electrically stimulated muscles (Ausoni 1990).

MYH GENE REGULATION BY THYROID HORMONE

Both the skeletal and cardiac MYH genes are responsive to circulating levels of thyroid hormone, and the developmental profile of cardiac MYH gene expression discussed above can be accounted for, at least in part, by changes in relative fetal and postnatal levels of thyroid hormone (Lompre et al 1988). Regulation of the MYH gene family by thyroid hormone appears to be quite complex, with the sensitivity and direction of the thyroid hormone response (i.e. an increase versus a decrease in gene expression) dependent on the particular MYH gene in question and on the tissue in which it is expressed (Izumo et al 1986). In the ventricular myocardium, MYH-alpha expression is induced by thyroid hormone, and MYH-beta is repressed. In the atrium, MYH-alpha expression appears to be independent of thyroid hormone levels and is expressed constitutively, demonstrating that the anatomical site of expression can effect a MYH gene's thyroid hormone response, perhaps due to thyroid hormone receptor composition. MYH-alpha and MYH-beta cardiac genes tend to respond antithetically to other stimuli as well (see Table 3).

Directionally variable responsiveness to thyroid hormone occurs with many of the skeletal MYH genes as well. The responses of six rat skeletal MYH genes to thyroid manipulation were shown to depend on the muscle in which each gene was expressed. For example, MYH-beta was always downregulated by thyroid hormone, but the basal level of this gene's expression and the amount of responsiveness varied in different muscles (Izumo et al 1986). The fast MYH-2A gene, however, demonstrates opposing responses to thyroid hormone in different muscles. Differences in metabolism and hormone receptor availability, among other potential secondary effects, probably influence the thyroid hormone responsiveness of MYH genes in different muscles. Studies discussed below have focused on determining the *cis*- and *trans*-acting regulatory sequences that confer thyroid hormone responsiveness.

REGULATORY ELEMENTS OF MYH GENES

Three approaches have been used to define the elements of cardiac MYH genes responsible for tissue-specific, developmental, and hormonal regulation. Initial gene regulation studies were carried out in vitro using transient transfection of reporter genes in fetal and neonatal cardiac myocyte preparations. Such studies delineated the minimal DNA sequences necessary for cardiac-specific expression and thyroid responsiveness of the MYH-alpha gene (Tsika et al 1990, Buttrick et al 1993, Adolph et al 1993, Molkentin & Markham 1993), as well as sequences important for MYH-beta regulation (Thompson et al 1991, Flink et al 1992). Gene transfer into rat hearts by direct DNA injection has provided a means of studying the regulatory elements of genes in vivo (MYH-alpha: Kitsis et al 1991, Buttrick et al 1993), as has the generation of transgenic mice (MYH-alpha: Gulick et al 1991, Subramaniam et al 1991, 1993; MYH-beta: Rindt et al 1993, 1995, Knotts et al 1995).

MYH-alpha

A thyroid hormone-responsive element (TRE) was identified in the promoter of the rat MYH-alpha gene and subsequently shown to confer T3 responsiveness to a reporter gene when cotransfected with the thyroid hormone receptor into both muscle and nonmuscle cell lines (Izumo & Mahdavi 1988). To further define potential upstream regulatory elements of the MYH-alpha gene, a series of human MYH-alpha-CAT reporter constructs containing varying lengths of the sequence located 5′ of MYH-alpha coding region (portions of the beta/alpha intergenic region) were tested for their activity in muscle and nonmuscle cells. A number of positive and negative regulatory elements involved in cardiac-specific gene expression were identified (Tsika et al 1990). T3 responsiveness of transfected promoter constructs was possible only in cardiac myocytes. These studies enabled the identification of the MYH-alpha basal promoter elements, consisting of the canonical TATAA and CAAT sequences, two positive and two negative regulatory elements, and a minimum of two TRE sequences. The human MYH-alpha promoter contains an additional proximal thyroid-responsive element not found in the rat, which may make the human promoter more T3-

sensitive than the rat promoter (Tsika et al 1990). In this study, most of the constructs exhibited constitutive expression in L6E9 cells, a permanent skeletal muscle cell line that can be induced to form myotubes. The pattern of activity of the constructs in these cells was qualitatively comparable to the constitutive profile exhibited in the primary cardiac myocytes. However, the activity in L6E9 cells was always less (between $\approx$ 5–20-fold less) than that observed in primary cardiocytes. Furthermore, addition of T3 did not increase the transcriptional activity of any of the constructs in the L6E9 myoblasts or myotubes, suggesting that although *cis*-acting elements are able to promote expression of the reporter genes in these cells, the additional factors downstream in the biochemical pathway of T3 induction are probably not available. Likewise, none of the reporter constructs in these experiments was transcriptionally active in either of the nonmuscle cell lines tested.

Injections of similar DNA reporter constructs directly into skeletal or cardiac muscle has proven to be an effective means of studying MYH gene regulation in vivo. Wolff et al (1990) made the original observation that naked DNA injected into skeletal muscle could be taken up and expressed by muscle cells, and several groups have subsequently used direct injection of DNA reporter constructs into the rat myocardium in order to study cardiac gene regulation in vivo (for review, see Leinwand & Leiden 1991). When rat MYH-alpha reporter constructs (containing 613-bp upstream of the transcription start site) were injected directly into heart and skeletal muscle, expression of the MYH-alpha-driven reporter gene was detected only in the heart, and its expression was also shown to be modulated by the thyroid state of the animal (Kitsis et al 1991). Human MYH-beta-driven reporter constructs injected into rat ventricles are also expressed and downregulated in hyperthyroid rats (LeBlanc et al 1992).

The results of MYH gene injection studies have differed somewhat from transfection studies. Both studies find basal activity conferred by proximal sequences, which include TATA and CCAAT elements located downstream from TRE sequences, and both studies were able to map a series of upstream thyroid-responsive regions. However, there was differential assignment of the locations of two negative regulatory elements. A more dramatic difference revealed that reporter constructs in the in vitro cardiocyte study were not expressed at all in the absence of T3, whereas all the constructs studied by direct DNA injection in vivo, including shorter constructs without a TRE, were active in hypo- and euthyroid animals. These results indicate that *trans*-acting factors other than thyroid hormone are involved in the regulation of these genes in vivo. Transgenic data have also helped to define portions of the cardiac MYH locus required for proper tissue-specific developmental and hormonal regulation of MYH-alpha in vivo. MYH-alpha-CAT reporter transgenes containing the 3′ end of MYH-beta, and the entire intergenic region between MYH-beta and MYH-alpha, exhibit high levels of tissue-specific expression in adult mice (Gulick et al 1991) and proper ventricular downregulation in the hypothyroid state (Subramaniam et al 1991). This study also showed that a transgene missing 2.5 kb of upstream sequence can also direct proper MYH-alpha expression, but a transgene with only 1.3 kb of the intergenic region that is missing a distal, high-affinity TRE (TRE1) is not expressed at all. Independent mutation of the two TREs in the context of the transgenic reporter carrying the full 5.5 kb of upstream sequence reveals that neither of these sites is necessary for tissue-specific expression. TRE2, however, is required for the developmental upregulation of MYH-alpha expression at birth, as well as for maintenance of expression in the adult. Furthermore, both elements are required for mediating downregulation of MYH-alpha in the hypothyroid state (Subramaniam et al 1993). The same study shows that the alpha 3.0 kb construct discussed above, although sufficient to direct high levels of tissue-specific expression, is not properly downregulated in hypothyroid conditions. This observation is consistent with results of direct injection studies of Kitsis et al (1991) (discussed above) and Ojamaa & Klein (1991) suggesting that upstream sequences are required for complete downregulation in the hypothyroid state. Interestingly, similar constructs are appropriately downregulated in in vitro studies.

MYH-beta

The MYH-beta gene is expressed in both cardiac and skeletal muscle. A molecular understanding of MYH-beta gene regulation should elucidate elements or factors involved in directing gene expression in cardiac versus skeletal muscle. In the embryonic heart, MYH-beta is the predominant MYH gene expressed, but in small rodents it is nearly undetectable in the adult ventricle. It is induced in response to pathological conditions of hemodynamic load and ensuing hypertrophy, and analysis of this genc could provide insight into these regulatory pathways as well. As with MYH-alpha, promoter mapping studies using reporter constructs both in vitro and in transgenic mice have helped identify the locations of several *cis*-elements that regulate gene expression. Despite its expression in the somites and slow fibers of skeletal muscle, MYH-beta appears to be controlled by pathways distinct from those of MyoD (Thompson et al 1991). Transfection of reporter constructs driven by 5′ upstream regions of the human and rat MYH-beta genes into skeletal muscle cells localized two important regions, Be2 and Be3, plus a region containing a C-rich motif (located between Be2 and Be3) that comprise a muscle-specific enhancer (defined as −354 to −182) that can direct expression of MYH-beta (Thompson et al 1991). The Be2 region contains sequences similar to an M-CAT motif, which is involved in mediating muscle-specific expression of the cardiac troponin-T gene (Mar &

Ordahl 1990). This region also has sequences identical to AP5/GT-II sites that are important for regulating the cell type and differentiation stage specificity of the SV-40 viral promoter. Be3 contains a TEF-1 site shown to be important for induction by adrenergic stimuli during the hypertrophic response (Kariya et al 1993, Karns et al 1995). A basal promoter (−186 to +34) was able to restrict the expression of a promiscuous SV-40 promoter to So18 and 1° cardiocytes.

Similarly, a detailed study of the human MYH-beta gene (Flink et al 1992) described a strong positive element located in the 5′ proximal region of this gene that is required for high levels of expression and shown to bind to BF1, a protein found in nuclear extracts of rabbit hearts. This factor is related to the M-CAT factor, which binds the M-CAT motif described above. An M-CAT element was located at −283 to −291, and a C-rich region was also identified in the human gene. Shimizu et al (1992) have identified muscle-specific and ubiquitously expressed *trans*-acting nuclear proteins that both interact with the M-CAT elements of the rabbit MYH-beta gene.

Several transgenic studies have also addressed the importance of the sequence elements implicated in the in vitro transfection studies discussed above in mediating muscle-specific gene expression and transcriptional induction of MYH-beta under hypothyroid conditions (Rindt et al 1993, 1995). A transgene containing 5.6 kb of upstream sequence and another containing 0.6 kb of upstream sequence showed skeletal muscle-specificity, although the 0.6-kb construct directed much lower levels of expression. Constructs containing 5.6 kb of upstream sequence not only are expressed similarly to the endogenous gene, but also exhibit tissue-specific, copy number–dependent and position-independent transgene expression patterns, which suggest the presence of a locus control region (LCR) in the distal 5 kb of these constructs (Knotts et al 1995). An LCR is a regulatory element distinct from classical enhancers in its ability to confer position-independent regulation of gene expression of linked genes (Epner et al 1992). It has been suggested that an LCR or perhaps the factors involved in the thyroid hormone response pathway facilitate coordinate control of the MYH-alpha and -beta genes, as has been shown for the beta globin locus (Grosveld et al 1987, Epner et al 1992) and genes that are regulated by hormone receptors (Knotts et al 1995). More detailed studies of the 5′ distal region may shed light on a mechanism of cardiac-specific induction as well as better understanding of the downregulation of this gene in response to thyroid hormone.

Transgenic mice bearing a mutation in each of the three proximal elements (M-CAT, C-rich, and Be3) were generated within the context of 5.6 kb of mouse MYH-beta upstream sequence (Knots et al 1995). Although the expression patterns varied widely from line to line for all transgenes tested, no single element was shown to independently mediate the hypothyroid response. The 0.6-kb construct, which contains all three proximal elements, is induced only in slow skeletal muscle in the hypothyroid state and not in the heart, where MYH-beta would normally be induced. This is the first example of a distinction between cardiac versus skeletal regulation of a MYH gene. In contrast, the developmental modulation of MYH-beta that typically occurs at birth in the ventricle in response to the surge in circulating thyroid hormone levels does not occur with the 0.6-kb construct.

TRANS-ACTING TRANSCRIPTIONAL REGULATION

Many skeletal muscle genes are transcriptionally activated by members of the myogenic regulator family of proteins (for reviews, see Olson 1992, Rudnicki & Jaenisch 1995), which include MyoD, myogenin, myf5, and MRF4. All members of this family are considered master muscle regulatory proteins because of their ability to induce skeletal myogenesis. These proteins function as heterodimeric transcriptional activators that bind a sequence motif termed an E-box that contains a CANNTG (N, any nucleotide) consensus found in the regulatory regions of many muscle-specific genes. Despite evidence that many other muscle genes are transcriptionally activated by members of this family, no direct regulation of MYH genes by myogenic regulators has been documented. Likewise, none of these skeletal myogenic regulators is expressed in cardiac muscle, which suggests that cardiogenesis may utilize distinct factors or that there are no master regulatory molecules determining cardiogenesis. However, a recent report implicated an E-box-mediated regulation of the MYH-alpha gene (Ojamaa et al 1996). Much progress had been made in identifying *trans*-acting factors that are found specifically in the heart or in both cardiac and skeletal muscle, but a discussion of that field is beyond the scope of this review.

There are also E-box-independent pathways that regulate certain muscle-specific genes, including MYH-alpha (Bouvagnet et al 1987, Mar & Ordahl 1990, Thompson et al 1991). Myocyte-specific enhancer factor-2 (MEF-2), a MADS box transcription factor (Yu et al 1992), binds an A + T-rich sequence found in many muscle-specific genes including muscle creatine kinase (MCK), myosin light chain-2 (MLC-2), cardiac troponin-T, rat atrial neurotrophic factor (ANF), and MYH-alpha. MEF-2 is considered an intermediate myogenic regulatory factor because it is induced by MyoD family members in skeletal muscle (Cserjesi & Olson 1991), but utilizes non-E box sequences to induce transcription. MEF-2 is expressed in cardiac myocytes in the absence of myogenic regulators (Navankasattusas et al 1991, Ianello et al 1991), but the means by which it is induced in the myocardium is unclear. The functional importance of the MEF-2 site in the regulation of MYH-alpha is also unclear. Gene injection and transgenic data suggest that the MEF-2 site may not be important in regulating the MYH-alpha gene in the adult heart (Adolph et al 1993, Buttrick et al 1993). However, a study by Molkentin & Markham (1993) demonstrated that the MEF-2 site was required for MYH-alpha expression in neonatal cardiac myocyte transfection experiments. The reason for this discrepancy could be the in vitro versus in vivo settings or differences in developmental stage.

MYH GENES AND HUMAN DISEASE

Mutations in muscle contractile proteins have been associated with the genetic disease familial hypertrophic cardiomyopathy (FHC). This is a dominant heart disease characterized by left ventricular hypertrophy and myocellular disarray. Other symptoms often include decreased coronary vessel size, arrhythmias, and a high incidence of sudden death, which makes this disease a leading cause of death in young athletes. Thus far, mutations in four different loci, all of which encode contractile proteins, have been associated with FHC. These include alpha-tropomyosin, cardiac troponin-T, myosin-binding protein-C, and MYH-beta gene mutations. Forty different mutant MYH-beta alleles have been described (see Vikstrom & Leinwand 1996) in families with FHC, and all but one bear mutations in the head or head/rod region of the molecule. Of these mutations, 33 of 40 occur within the proteolytic S1 head fragment, which contains myosin's ATPase and actin-binding activities. This fragment is also sufficient to move actin in an in vitro motility assay (Toyoshima et al 1987). Determination of the crystal structure of the MYH head has enabled these mutations to be mapped in a structural context to regions of predicted functional importance. When 29 of these mutations were mapped, many were found to reside in regions involved in actin-myosin interaction and nucleotide binding, as well as in essential light chain binding (Rayment et al 1995). However, many of the mutations are associated with other regions of the motor domain, and it is thought that they result in alteration of some aspect of MYH function without destroying it completely.

It is hypothesized that mutant myosin molecules interact with wild-type myosin proteins and disrupt sarcomere function, acting via a dominant-negative mechanism. Support for this hypothesis comes from assays of mixtures of mutant and wild-type MYHs. Both wild-type rat MYH-alpha and a mutant MYH-alpha allele (arg403gln) were expressed in baculovirus and assayed for ATPase activity and velocity in an in vitro motility assay (Sweeney et al 1994). The mutant allele showed about a fivefold decrease in actin-activated ATPase activity and velocity relative to wild-type. More importantly, when varying proportions of wild-type and mutant myosin mixtures were assayed for motility, the mutant myosin disproportionately reduced the in vitro velocity of the mixture. This biochemical behavior is consistent with the dominant mode of inheritance. Additional studies have been carried out with myosin purified from FHC patients with Arg403⟶Gln and Leu908⟶Val mutations. Myosin from these patients is a mixture of mutant and wild-type, and samples from both mutants exhibit impaired movement in in vitro motility assays (Cuda et al 1993). Also, it has been possible to measure contractile function in soleus muscle from FHC patients with several mutations. The soleus muscle is comprised mostly of slow (type I) fibers expressing the MYH-beta gene. When skinned soleus fibers from patients bearing MYH-beta FHC alleles with mutations in the ATP-binding pocket, the actin-myosin interface, and essential light chain domain (Gly256⟶Glu, Arg403⟶Gln, and Gly741⟶Arg, respectively) were studied, fibers containing MYHs with the latter two mutations had impaired contractile function (Lankford et al 1995).

Although the mechanism by which these alterations result in the FHC clinical phenotype remains unclear, it is nevertheless possible to predict the severity of the clinical manifestation of these mutations based on which residue is mutated (Watkins et al 1992). For example, arg403gln is a particularly malignant allele, and individuals bearing this mutation typically die, on average, at age 33. In contrast, individuals who have a val606met mutation live an almost normal lifespan. It has been suggested that mutations resulting in changes in amino acid charges are associated with a poor prognosis (Watkins et al 1992, Anan et al 1994). Additional biochemical analysis of MYH molecules with these mutations is necessary in order to understand the pathogenesis of FHC. Likewise, an animal model will be very useful in determining the processes leading up to cardiac enlargement and sudden death. To better understand the contribution of MYH mutations in the development of FHC, K Vikstrom et al (manuscript submitted) have generated transgenic mouse lines in which the MYH-alpha gene (the predominant left ventricular MYH isoform in small mammals) has been mutated. Mice bearing the Arg403⟶Gln mutation mentioned above, along with a deletion of amino acids 468–527 in the actin-binding domain, exhibit left ventricular hypertrophy and other histopathology consistent with the human disease. Thus far, studies of these mice have revealed that the presence of a mutant myosin can result in a phenotype resembling that of FHC. Further characterization of these mice should help shed light on the development of this disease at the cellular level.

FUTURE STUDIES

Future areas of investigation should include definition of the full extent of MYH genes in the mammalian gene family, as well as an understanding of the extent of functional diversity that exists. Studies described above reveal biochemical differences between slow and fast MYH isoforms, and in vitro analyses of pure myosin preparations should help to assess the biochemical differences among the many fast isoforms. The generation of knockout mice that lack specific MYH isoforms will be useful for testing the functional redundancy of this large gene family in vivo, as well as for studying the extent of cross-regulation that exists among MYH family members. Preliminary examination of MYH-alpha null mice, which exhibit gestational lethality, reveal that this gene is necessary for normal development (Jones & Robbins 1995). Ongoing experiments in our laboratory suggest that inactivation of the adult fast MYH-2B and MYH-2X genes is not lethal and that only the MYH-2D nulls demonstrate a gross phenotype. It will also be important to determine whether early developmental isoforms are required for normal development.

Literature Cited

Adolph EA, Subramaniam A, Cserjesi P, Olson EN, Robbins J. 1993. Role of myocyte-specific enhancer-binding factor (MEF-2) in the transcriptional regulation of the alpha-cardiac myosin heavy chain gene. *J. Biol. Chem.* 268(8):5349–52

Anan R, Greve G, Thierfelder L, Watkins H, McKenna WJ, et al. 1994. Prognostic implications of novel β-cardiac myosin heavy-chain gene-mutations that cause familial hypertrophic cardiomyopathy. *J. Clin. Invest.* 93:280–85

Andersen JL, Klitgaard H, Bangsbo J, Saltin B. 1994a. Myosin heavy chain isoforms in single fibers from m. vastus lateralis of soccer players: effects of strength training. *Acta Physiol. Scand.* 150:21–26

Andersen JL, Klitgaard H, Saltin B. 1994b. Myosin heavy chain isoforms in single fibers from m. vastus lateralis of sprinters: influence of training. *Acta Physiol. Scand.* 151:135–42

Ausoni S, Gorza L, Schiaffino S, Gunderson K, Lomo T. 1990. Expression of myosin heavy chain isoforms in stimulated fast and slow rat muscles. *J. Neurosci.* 10:153–55

Barany M. 1967. ATPase activity of myosin correlated with speed of muscle shortening. *J. Gen. Physiol.* 50:197–218

Bouvagnet PF, Strehler EE, White GE, Strehler-Page M-A, Nadal-Ginard B. 1987. Multiple positive and negative 5′ regulatory elements control the cell-type-specific expression of the embryonic skeletal myosin heavy chain gene. *Mol. Cell. Biol.* 7:4377–89

Butler-Browne GS, Whalen RG. 1984. Myosin isozyme transitions occurring during the postnatal development of the rat soleus muscle. *Dev. Biol.* 102:324–34

Buttrick PM, Kaplan ML, Kitsis RN, Leinwand LA. 1993. Distinct behavior of cardiac myosin heavy chain gene constructs in vivo: discordance with in vitro results. *Circ. Res.* 72:1211–17

Cheney RE, Riley MA, Mooseker MS. 1993. Phylogenetic analysis of the myosin superfamily. *Cell Motil. Cytoskelet.* 24:215–23

Cserjesi P, Olson EN. 1991. Myogenin induces muscle-specific enhancer binding factor MEF-2 independently of other muscle-specific gene products. *Mol. Cell. Biol.* 11:4854–62

Cuda G, Fananapazir L, Zhu W, Sellers JR, Epstein ND. 1993. Skeletal muscle expression and abnormal function of β-myosin in hypertrophic cardiomyopathy. *J. Clin. Invest.* 91:2861–65

d'Albis A, Anger M, Lompre'A-M. 1993. Rabbit masseter expresses the cardiac alpha-myosin heavy-chain gene. *FEBS Lett.*

324:178–80
DeNardi C, Ausoni S, Moretti P, Gorza L, Velleca M, et al. 1993. Type 2X-myosin heavy chain is coded by a muscle fiber type-specific and developmentally regulated gene. *J. Cell Biol.* 123(4):823–35
Deng Z, Liu P, Marlton P, Claxton DF, Lane S, et al. 1993. Smooth muscle myosin heavy chain locus *(MYHII)* maps to 16p13.13-p13.12 and establishes a new region of conserved synteny between human 16p and mouse 16. *Genomics* 18:156–95
Drachman DB, Johnston DM. 1973. Development of mammalian fast muscle; dynamic and biochemical properties correlated. *J. Physiol.* 234:29–42
Epner E, Kim CG, Groudine M. 1992. What does the locus control region control? *Curr. Biol.* 2(5):262–64
Flink IL, Edwards JG, Bahl JJ, Liew C, Sole M, Morkin E. 1992. Characterization of a strong positive *cis*-acting element of the human β-myosin heavy chain gene in fetal rat heart cells. *J. Biol. Chem.* 267(14):9917–24
Gorza L, Gunderson K, Lomo T, Schiaffino S, Westgaard RH. 1988. Slow-to-fast transformation of denervated soleus muscles by chronic high frequency stimulation in the rat. *J. Physiol.* 402:627–49
Grosveld F, Blomvan Assendelft G, Greaves DR, Kollias G. 1987. Position-independent, high-level expression of the human β-globin gene in transgenic mice. *Cell* 51:975–85
Gulick J, Subramaniam A, Neumann J, Robbins J. 1991. Isolation and characterization of the mouse cardiac myosin heavy chain genes. *J. Biol. Chem.* 266:9180–85
Gunning P, Hardeman E. 1991. Multiple mechanisms regulate fiber diversity. *FASEB J.* 285:3064–70
Ianello RC, Mar JH, Ordahl CP. 1991. Characterization of a promotor element required for transcription in myocardial cells. *J. Biol. Chem.* 266:3309–16
Izumo S, Mahdavi V. 1988. Thyroid hormone receptor isoforms generated by alternative splicing differentially activate myosin HC transcription. *Nature* 334:539–42
Izumo S, Nadal-Ginard B, Mahdavi V. 1986. All members of the MHC multigene family respond to thyroid hormone in a highly tissue-specific manner. *Science* 231:594–600
Jones KW, Robbins J. 1995. Molecular and physiologic analysis of alpha-MyHC gene ablation. *Am. Heart Assoc. 68th Sci. Sess.* 92(8):I-369 (Abstr.)
Kariya K, Farrance IKG, Simpson PC. 1993. Transcriptional enhancer factor-1 in cardiac myocytes interacts with an alpha(1)-adrenergic- and beta-protein kinase C-inducible element in the rat beta-myosin heavy chain promotor. *J. Biol. Chem.* 268:26658–62
Karns LR, Kariya K, Simpson PC. 1995. M-CAT, CArg, and Spl elements are required for alpha(1)-adrenergic induction of the skeletal alpha-actin promoter during cardiac myocyte hypertrophy. Transcriptional enhancer factor-1 and protein kinase C as conserved transducers of the fetal program of cardiac growth. *J. Biol. Chem.* 270:410–17
Kitsis RN, Buttrick PM, McNally EM, Kaplan ML, Leinwand LA. 1991. Hormonal regulation of a gene injected into rat heart in vivo. *Proc. Natl. Acad. Sci. USA* 88:4138–42
Knotts S, Rindt H, Robbins J. 1995. Position independent expression and developmental regulation is directed by the β myosin heavy chain gene's 5′ upstream region in transgenic mice. *Nucleic Acids Res.* 23(16):3301–9
Lankford EB, Epstein ND, Fananapazir L, Sweeney HL. 1995. Abnormal contractile properties of muscle fibers expressing β-myosin heavy chain gene mutations in patients with hypertrophic cardiomyopathy. *J. Clin. Invest.* 95:1409–14
Larsson L, Ansved T, Edstrom L, Gorza L, Schiaffino S. 1991. Effects of age on physiological, immunohistochemical, and biochemical properties of fast-twitch single motor units in the rat. *J. Physiol.* 443:257–75
LeBlanc JM, Kitsis RN, Buttrick PM, Leinwand LA. 1992. Molecular genetic manipulation of myosin. In *Neuromuscular Development and Disease,* ed. AM Kelly, HM Blau, pp. 223–37. New York: Raven
Leinwand LA, Fournier REK, Nadal-Ginard B, Shows TB. 1983. Multigene family for sarcomeric myosin heavy chain in mouse and human DNA: localization on a single chromosome. *Science* 221:766–79
Leinwand LA, Leiden JM. 1991. Gene transfer into myocytes in vivo. *Trends Cardiovasc. Med.* 1(7):271–76
Lompre AM, Nadal-Ginard B, Mahdavi V. 1984. Expression of the cardiac ventricular alpha- and beta-myosin heavy chain genes is developmentally and hormonally regulated. *J. Biol. Chem.* 259:6437–46
Loughna PT, Izumo S, Goldspink G, Nadal-Ginard B. 1990. Disuse and passive stretch cause rapid alteration in expression of developmental and adult contractile protein genes in skeletal muscle. *Development* 109:217–23
Lowey S, Waller GS, Trybus KM. 1993. Function of skeletal muscle myosin heavy and light chain isoforms by an in vitro motility assay. *J. Biol. Chem.* 268:20414–18
Lyons GE, Ontell M, Cox R, Sassoon D, Buckingham M. 1990a. The expression of myosin

genes in developing skeletal muscle in the mouse embryo. *J. Cell Biol.* 111:1465–76

Lyons GE, Schiaffino S, Sassoon D, Barton P, Buckingham M. 1990b. Developmental regulation of myosin gene expression in mouse cardiac muscle. *J. Cell Biol.* 111:2427–36

Mahdavi V, Chambers AP, Nadal-Ginard B. 1984. Cardiac alpha- and β-myosin heavy chain genes are organized in tandem. *Proc. Natl. Acad. Sci. USA* 81:2626–30

Mar JH, Ordahl CP. 1990. M-CAT binding factor, a novel transacting factor governing muscle-specific transcription. *Mol. Cell. Biol.* 10:4271–83

Matsuoka R, Yoshida MC, Furutani Y, Imamura S, Kanda N, et al. 1993. Human smooth muscle myosin heavy chain gene mapped to chromosomal region 16q12. *Am. J. Med. Genet.* 46:61–67

McNally EM, Kraft R, Bravo-Zehnder M, Taylor DA, Leinwand LA. 1989. Full-length rat alpha and beta cardiac myosin heavy-chain sequences. Comparisons suggest a molecular basis for functional differences. *J. Mol. Biol.* 210:665–71

Miano JM, Cserjesi P, Ligon KL, Periasamy M, Olson EN. 1994. Smooth muscle myosin heavy chain exclusively marks the smooth muscle lineage during mouse embryogenesis. *Circ. Res.* 75(5):803–12

Molkentin JD, Markham BE. 1993. Myocyte-specific enhancer-binding factor (MEF-2) regulates α-cardiac myosin heavy chain expression in vitro and in vivo. *J. Biol. Chem.* 268(26):19512–20

Moore LA, Tidyman WE, Arrizubieta MJ, Bandman E. 1993. The evolutionary relationship of avian and mammalian myosin heavy-chain genes. *J. Mol. Evol.* 36:21–30

Navankasattusus S, Zhu H, Garcia AV, Evans SM, Chien KR. 1991. A ubiquitous factor (HF-1) and a distinct muscle factor (HF-lb/Mef2) form an E-box'-independent pathway for cardiac muscle gene expression. *Mol. Cell Biol.* 12:1469–79

Ojamaa K, Klein I. 1991. Thyroid hormone regulation of alpha-myosin heavy chain promotor activity assessed by in vivo DNA transfer in rat heart. *Biophys. Res. Commun.* 179:1269–75

Ojamaa K, Samarel AM, Klein I. 1996. Identification of a contractile-responsive element in the cardiac alpha-myosin heavy chain gene. *J. Biol. Chem.* 270(52):31276–81

Olson EN. 1992. Regulation of muscle transcription by the MyoD family: the heart of the matter. *Circ. Res.* 72(1):1–6

Periasamy M, Wieczorek DF, Nadal-Ginard B. 1984. Characterization of a developmentally regulated perinatal myosin heavy chain gene expressed in skeletal muscle. *J. Biol. Chem.* 259(l):13573–76

Pette D, Vrbova G. 1985. Neural control of phenotypic expression in mammalian muscle fibers. *Muscle Nerve* 8:676–89

Pope B, Hoh JFY, Weeds A. 1980. The ATPase activities of rat cardiac myosin isozymes. *FEBS Lett.* 118:205–11

Rayment I, Holden HM, Sellers JR, Fananapazir L, Epstein ND. 1995. Structural interpretation of the mutations in the β-cardiac myosin that have been implicated in familial hypertrophic cardiomyopathy. *Proc. Natl. Acad. Sci. USA* 92:3864–68

Rayment I, Holden HM, Whittaker M, Yohn CB, Lorenz M, et al. 1993. Structure of the actin-myosin complex and its implications for muscle contraction. *Science* 261:58–65

Rindt H, Gulick J, Knotts S, Neumann J, Robbins J. 1993. In vivo analysis of the murine β-myosin heavy chain gene promoter. *J. Biol. Chem.* 268:5332–38

Rindt H, Knotts S, Robbins J. 1995. Segregation of cardiac and skeletal muscle-specific regulatory elements of the β myosin heavy chain gene. *Proc. Natl. Acad. Sci. USA* 92:1540–44

Rudnicki MA, Jaenisch R. 1995. The MyoD family of transcription factors and skeletal myogenesis. *BioEssays* 17(3):203–9

Rugh R. 1990. *The Mouse.* Oxford, UK: Oxford Univ. Press. 430 pp.

Rushbrook JI, Weiss C, Ko K, Feuerman MH, Carleton S, et al. 1994. Identification of alpha-cardiac myosin heavy chain mRNA and protein in extraocular muscle of the adult rabbit. *J. Muscle Res. Cell Motil.* 15:505–15

Saez CG, Myers JC, Shows TB, Leinwand LA. 1990. Human nonmuscle myosin heavy chain mRNA: generation of diversity through alternative polyadenylation. *Proc. Natl. Acad. Sci. USA* 87:1164–68

Saez L, Leinwand LA. 1986. Characterization of diverse forms of myosin heavy chain expressed in adult human skeletal muscle. *Nucleic Acids Res.* 14:2951–68

Saez LJ, Gianola KM, McNally EM, Feghali R, Eddy R, et al. 1987. Human cardiac myosin heavy chain genes and their linkage in the genome. *Nucleic Acids Res.* 15:5443–59

Schiaffino S, Reggiani C. 1994. Myosin isoforms in mammalian skeletal muscle. *J. Appl. Physiol.* 77(2):493–501

Schiaffino S, Reggiani C. 1996. Molecular diversity of myofibrillar proteins: gene regulation and functional significance. *Physiol. Rev.* In press

Shimizu N, Dizon E, Zak R. 1992. Both muscle-specific and ubiquitous nuclear factors are required for muscle-specific expression of the myosin heavy-chain β gene in cultured cells.

Mol. Cell Biol. 12(2):619–30

Simons M, Wang M, McBride OW, Kawamoto S, Yamakawa K, et al. 1991. Human non-muscle myosin heavy chains are encoded by two genes located on different chromosomes. *Circ. Res.* 69(2):530–39

Sissman N. 1970. Developmental landmarks in cardiac morphogenesis: comparative chronology. *Am. J. Cardiol.* 25:141–48

Stedman HH, Eller M, Julian EH, Fertels SH, Sarkar S, et al. 1990. The human embryonic myosin heavy chain: complete primary structure reveals evolutionary relationships with other developmental isoforms. *J. Biol. Chem.* 265(6):3568–76

Stockdale FE. 1992. Myogenic cell lineages. *Dev. Biol.* 154:284–98

Strehler EE, Strehler-Page M-A, Perriard J-C, Periasamy M, Nadal-Ginard B. 1986. Complete nucleotide and encoded amino acid sequence of a mammalian myosin heavy chain gene: evidence against intron-independent evolution of the rod. *J. Mol. Biol.* 190:291–317

Subramaniam A, Gulick J, Neumann J, Knotts S, Robbins J. 1993. Transgenic analysis of the thyroid-responsive elements in the α-myosin heavy chain promoter. *J. Biol. Chem.* 268(6):4331–36

Subramaniam A, Jones WK, Gulick J, Wert S, Neumann J, Robbins J. 1991. Tissue-specific regulation of the α-myosin heavy chain gene promoter in transgenic mice. *J. Biol. Chem.* 266:24613–20

Sweeney HL, Straceski AJ, Leinwand LA, Faust L. 1994. Heterologous expression of a cardiomyopathic myosin that is defective in its actin interaction. *J. Biol. Chem.* 269: 1603–5

Termin A, Staron RS, Pette D. 1989. Changes in myosin heavy chain isoforms during chronic low-frequency stimulation of rat fast hindlimb muscles. *Eur. J. Biochem.* 186:749–54

Thompson WR, Nadal-Ginard B, Mahdavi V. 1991. A MyoD1-independent muscle-specific enhancer controls the expression of the β-myosin heavy chain gene in skeletal and cardiac muscle cells. *J. Biol. Chem.* 266(33):22678–88

Toyoshima YY, Kron SJ, McNally EM, Niebling KR, Toyoshima C, Spudich JA. 1987. Myosin subfragment-1 is sufficient to move actin-filaments in vitro. *Nature* 328:536–39

Tsika RW, Bahl JJ, Leinwand LA, Morkin E. 1990. Thyroid hormone regulates expression of a transfected human α-myosin heavy-chain fusion gene in fetal rat heart cells. *Proc. Natl. Acad. Sci. USA* 87:379–83

Uyeda TQP, Ruppel KM, Spudich JA. 1994. Enzymatic activities correlate with chimaeric substitutions at the actin-binding face of myosin. *Nature* 368:567–69

Vikstrom KL, Leinwand LA. 1996. Contractile protein mutations and heart disease. *Curr. Opin. Cell Biol.* 8:97–105

Watkins H, Rosenzweig A, Hwang DS, Levi T, McKenna W, et al. 1992. Characteristics and prognostic implications of myosin missense mutations in familial hypertrophic cardiomyopathy. *N. Engl. J. Med.* 326:1108–14

Weiss A, Mayer DCG, Leinwand LA. 1994. Diversity of myosin-based motility: multiple genes and functions. *Soc. Gen. Physiol. Ser.* 49:159–71

Weydert A, Barton P, Harris JA, Pinset C, Buckingham M. 1987. Developmental pattern of mouse skeletal myosin heavy chain gene transcripts in vivo and in vitro. *Cell* 49:121–29

Weydert A, Lazaridis I, Barton P, Gardner I, Leader DP, et al. 1985. Gene for skeletal muscle myosin heavy chains are clustered and are not located on same mouse chromosome as alpha cardiac myosin heavy chain gene. *Proc. Natl. Acad. Sci. USA* 82:7183–87

Whalen RG, Johnstone D, Bryers PS, Butler-Browne MS, Ecob MS, Jaros E. 1984. A developmentally regulated disappearance of slow myosin in fast type muscles of the mouse. *FEBS Lett.* 177:51–56

Wolff JA, Malone RW, Williams P, Chong W, Acsadi G, et al. 1990. Direct gene transfer into mouse muscle in vivo. *Science* 247:1465–68

Yoon SJ, Seller SH, Kucherlapati R, Leinwand LA. 1992. Organization of the human myosin heavy chain gene cluster. *Proc. Natl. Acad. Sci. USA* 89:12078–82

Yu Y-T, Breltbart RE, Smoot LB, Lee Y, Mahdavi V, Nadal-Ginard B. 1992. Human myocyte-specific enhancer factor 2 comprises a group of tissue-restricted MADS box transcription factors. *Genes Dev.* 6:1783–98

Annu. Rev. Cell Dev. Biol. 1996. 12:441–61

TRANSPORT VESICLE DOCKING: SNAREs and Associates

Suzanne R. Pfeffer
Department of Biochemistry, Stanford University School of Medicine, Stanford, California 94305-5307

KEY WORDS: membrane traffic, intracellular transport, protein targeting, secretion

ABSTRACT

Proteins that function in transport vesicle docking are being identified at a rapid rate. So-called v- and t-SNAREs form the core of a vesicle docking complex. Additional accessory proteins are required to protect SNAREs from promiscuous binding and to deprotect SNAREs under conditions in which transport vesicle docking should occur. Because access to SNAREs must be regulated, other proteins must also contain specificity determinants to accomplish delivery of transport vesicles to their distinct and specific membrane targets.

CONTENTS

1081-0706/96/1115-0441$08.00

INTRODUCTION

The SNARE hypothesis currently provides an important framework for a large body of work focused on elucidating the molecular events that underlie vesicular transport and neurotransmitter release. Much of the work utilizes either baker's yeast or vertebrate brain constituents, which therefore link two parallel lines of investigation. With all due respect to the researchers who discovered key neuronal constituents, the nomenclature in this field [synaptotagmin, syntaxin, synapsin, synaptobrevin, synaptophysin, synaptogyrin, synaptoporin (see Südhof 1995)] is, for this author, more difficult to keep track of than Sec1, Sec4, and Sec9. Thus the goal of this review is to cut through "synaptojargon" and describe the current state of the vesicle docking literature. Many excellent, recent reviews also have covered this area from different perspectives (Ferro-Novick & Jahn 1994, Rothman 1994, Bennett 1995, Scheller 1995, Südhof 1995, Rothman & Wieland 1996).

HISTORY—THE SNARE HYPOTHESIS

The SNARE hypothesis, first put forward by Söllner et al (1993a), states that every transport vesicle contains on its surface a protein (or proteins) that specify that vesicle's target, or v-SNARE. Target membranes possess cognate partners (t-SNAREs) to interact with the v-SNAREs; interaction leads to a subsequent fusion event. The concept of targeting molecules is not new; it was clear for many years that different vesicle types would need to contain molecules that specify their targets (Palade 1975). Novel in the SNARE hypothesis was the designation of the SNARE molecules. The first named candidates were participants in synaptic vesicle exocytosis: the synaptic vesicle protein VAMP (or synaptobrevin) was proposed as the v-SNARE, and the presynaptic plasma membrane proteins syntaxin and SNAP-25 were the proposed t-SNAREs.

This story began when Block et al purified NSF [NEM (*N*-ethylmaleimide)-sensitive factor], the first protein that could stimulate an in vitro vesicular transport reaction (Block et al 1988). This NEM-sensitive fusion protein is a trimeric ATPase (Whiteheart et al 1994) that exists bound to membranes and in the cytosol. NSF was shown to be needed for vesicle docking or fusion because vesicles accumulate in its absence in in vitro reactions (Malhotra et al 1988, Orci et al 1989) or in yeast cells bearing a mutation in the corresponding *SEC18* gene (Eakle et al 1988, Wilson et al 1989, Kaiser & Schekman 1990).

NSF requires so-called SNAP (soluble NSF attachment protein) proteins to bind to membranes. Three isoforms have been identified: α, β, and γ (Clary & Rothman 1990). The α-SNAP protein is encoded by the yeast *SEC17* gene (Clary et al 1990, Griff et al 1992); *SEC17* interacts with *SEC18,* which encodes yeast NSF (Kaiser & Schekman 1990). SNAPS are soluble proteins that bind saturably to Golgi membranes (Weidman et al 1989, Whiteheart et al 1992).

A major conceptual breakthrough regarding the molecular events underlying intracellular transport came from a search for the membrane components to which NSF and SNAPs bind. Rothman and coworkers showed that when NSF was mixed together with α-SNAP and a detergent extract of salt-washed Golgi membranes, NSF and α-SNAP became incorporated into a 20S complex (Wilson et al 1992). Importantly, addition of Mg-ATP led to the disassembly of preformed 20S complexes; ATP alone was not sufficient, and ATP hydrolysis was required because Mg-ATPγS failed to trigger complex disassembly (Wilson et al 1992).

These findings set the stage for Söllner et al (1993a) to purify what they initially termed the membrane-derived assembly factor. 20S complexes were preformed on ice by mixing myc-tagged NSF, α-SNAP, and γ-SNAP with a crude, Triton X-100 solubilized, membrane extract from bovine brain. The complexes were stabilized by the addition of ATPγS and EDTA to chelate free magnesium ions and were immobilized using anti-myc antibodies and protein G beads. From their previous work, Rothman and coworkers surmised that only Mg-ATP would be capable of selectively releasing the specific membrane-derived assembly factor. Thus the column could be washed extensively with Mg-ATPγS. Subsequent specific elution could then be accomplished using Mg-ATP. This innovative approach led to the spectacular purification of several polypeptides with surprising identities.

20S complexes were shown to contain several proteins of special neurobiological interest, including syntaxins A and B, the 25 K synaptosome-associated protein (SNAP-25, no relation to α-, β- or γ-SNAP; Oyler et al 1989), VAMP/synaptobrevin 2 (Trimble et al 1988, Baumert et al 1989), and also Hsp70 (Söllner et al 1993a). Syntaxin was first identified by Bennett et al (1992) as a presynaptic plasma membrane protein that interacts with synaptotagmin, a synaptic vesicle protein believed to be important for the calcium dependence of synaptic vesicle release (cf Geppert et al 1994). Given the localization of syntaxin and its strong interaction with a synaptic vesicle protein, Bennett et al (1992) proposed that syntaxin was likely to be important for synaptic vesicle docking.

Complexes of VAMP, syntaxin, and SNAP-25 were also detected in detergent extracts in the absence of α-SNAP and NSF (Söllner et al 1993b). These so-called 7S core complexes could recruit NSF and α-SNAP as long as ATP

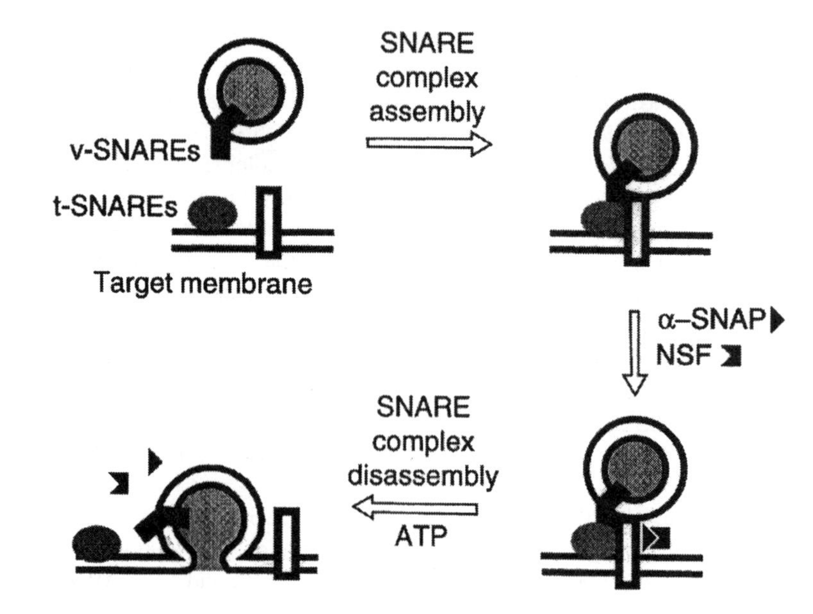

Figure 1 The current model for the role of SNAREs in vesicle targeting and fusion. v-SNAREs target vesicles to t-SNAREs on specific target membranes (Step 1). Prior to fusion, α-SNAP binds the SNARE complex, recruiting NSF (Step 2). ATP hydrolysis by NSF dissociates the SNAREs, permitting subsequent fusion (Step 3) (redrawn from Bennett 1995).

hydrolysis was blocked (Söllner et al 1993b, McMahon & Südhof 1995). This observation suggested that the general transport factors NSF and α-SNAP, first characterized for their roles in intra-Golgi transport and believed to function in many (but not all) transport reactions, also participate in synaptic vesicle exocytosis. Moreover, these observations led to the proposal that NSF and α-SNAP act to disassemble v- and t-SNARE complexes immediately prior to membrane fusion (Figure 1).

SUPPORT FOR THE SNARE HYPOTHESIS

Strong support for the SNARE hypothesis came from the finding that the classic SNAREs, VAMP, syntaxin, and SNAP-25 each had homologues required for many other vesicle transport processes, particularly in yeast. SNARE homologues function in yeast ER-to-Golgi transport, Golgi-to-plasma membrane transport, and also in Golgi-to-vacuole transport (Ferro-Novick & Jahn 1994).

For ER-to-Golgi transport, v-SNARE candidates include Bos1p, Sec22p, Bet1p, and Ykt6p (Lian & Ferro-Novick 1993, Barlowe et al 1994, Søgaard et al 1994), and the t-SNARE Sed5p (Hardwick et al 1992). Vesicles transporting proteins from the Golgi to the plasma membrane contain the v-SNARE homologues Snc1p and Snc1p (Protopopov et al 1993); the cognate t-SNAREs are believed to include Sos1p, Sos2p, and the SNAP-25-relative Sec9p (Aalto et al 1993, Brennwald et al 1994, Couve & Gerst 1994). In each of these transport processes, genetic analyses support the types of v- and t-SNARE interactions observed in neurons. Moreover, the proteins are localized to either vesicle or target, in analogy with VAMP and syntaxin/SNAP-25. In addition, the yeast homologues form stable, isolatable SNARE complexes, particularly when Sec18p (NSF) function is blocked (Søgaard et al 1994, Brennwald et al 1994, Sapperstein et al 1996).

Who Helps Who Find Whom

That SNARE proteins are key players in neuronal exocytosis seems incontrovertible, because they are specific targets for destruction by neurotoxin proteases, and their cleavage leads to immediate blockage of neuronal function (Südhof et al 1993, Südhof 1995). Moreover, certain of the yeast SNARE components are encoded by essential genes. However, do they function in fusion or in docking?

Südhof (1995) has suggested that synaptic SNARE complexes may not function in docking, per se, because syntaxin and SNAP-25 are not restricted to sites of neuronal exocytosis: They are localized along the length of the axon, as well as in nerve terminals (Garcia et al 1995). Moreover, cleavage of SNAREs by botulinum and tetanus toxins does not decrease the number of docked vesicles (see Südhof 1995). These data indicate strongly that SNAREs cannot act alone in vesicle docking; multiple protein-protein interactions are likely required to localize synaptic vesicles to active zones. Moreover, if t-SNAREs are localized to sites other than where they function, they are probably present there in an inactive form. Those not present at active zones may well be bound to a regulatory molecule or exist in an inactive conformation that will not permit v-SNARE interaction (see below). In summary, vesicle docking requires a target-designating protein on the vesicle and cognate proteins at the target that interact with high specificity. v- and t-SNARE proteins are obvious candidates to form the core of such an interaction. Additional, essential factors are also required.

REGULATING SNARE AVAILABILITY

If cognate v- and t-SNAREs could pair at all times, all the organelles in the cytoplasm might become stuck together as part of a giant sandwich. For example, Bos1p, an ER transport vesicle v-SNARE, resides in the ER (Lian &

Ferro-Novick 1993), yet would bind tightly to a Golgi t-SNARE and thereby attach two entire organelles. Similarly, Snc1p and Snc2p, post-Golgi v-SNAREs, would bind the Golgi to a plasma membrane-localized t-SNARE. For this reason, it seems likely that a set of proteins block SNARE accessibility.

It seems reasonable to propose that v-SNAREs become activated during transport vesicle budding. Transport vesicle formation represents a committed event, and there would be no need to regulate subsequent fusion, except in regulated secretion. SNARE activation may involve release of proteins that block SNARE access as well as conformational changes. Interconversion between active and inactive conformations may also be used to modulate the association of putative SNARE-protectors with SNAREs. Finally, the presence of transport vesicle coats may also influence SNARE accessibility (Schekman & Orci 1996).

Sec1p was first identified as a gene product important for late events in the secretory pathway (Novick et al 1980, 1981). Important information regarding possible interactions between Sec1p and a t-SNARE comes from yeast genetic analyses of Aalto et al (1993), who screened for multicopy suppressors of mutations in the *SEC1* gene. Two such suppressors, Sso1p and Sso2p, are homologues of syntaxin that are localized to the yeast plasma membrane. Together, Sso1p and Sso2p perform an essential function in yeast exocytosis, but they cannot suppress a *SEC1* deletion (Aalto et al 1993). These data are consistent with a model in which Sec1p protects the t-SNARE until a v-SNARE on a vesicle (Snc1p, Snc1p) is brought to it for SNARE complex assembly.

The mammalian homologue of yeast Sec1p (also called n-Sec1, munc18, or rbSec1; reviewed in Pfeffer 1994b) has been shown to bind directly to the presynaptic plasma membrane t-SNARE, syntaxin (Hata et al 1993, Pevsner et al 1994b, Garcia et al 1994, Hodel et al 1994). n-Sec1 binds syntaxin 1A with an apparent binding constant in the nanomolar range (Pevsner et al 1994a), and n-Sec1 is not part of SNARE complexes (Söllner et al 1993b, Pevsner et al 1994b, Garcia et al 1995). Thus there appear to be two pools of syntaxin: one that exists as a complex with n-Sec1 and another bound to VAMP and SNAP-25. Indeed, n-Sec1 was shown to inhibit the interaction of VAMP and SNAP-25 with syntaxin 1A (Pevsner et al 1994b).

One prediction of the t-SNARE protection model is that a large fraction of t-SNAREs would exist in a blocked form at steady state on target membranes. Thus Snc1p/Snc1p would occur in direct complex with Sec1p. Similarly, Sed5p would be bound by Sly1p (the Sec1p homologue), rather than free in the Golgi membrane. Indeed, Søgaard et al (1994) detected a large fraction of Sed5p in complex with Sly1p at steady state.

The observation that n-Sec1 is not part of presynaptic SNARE complexes is consistent with the t-SNARE protection model. Sec1p was also missing from SNARE complexes containing Snc1p/Snc1p, Sso1p/Sso2p, and Sec9p (Brennwald et al 1994). However, Sly1p, the Sec1p relative involved in yeast ER-to-Golgi transport, was reported to be part of an ER-Golgi SNARE complex (Søgaard et al 1994). In this case, a mutant *sec18* strain was used to trap the complex. Perhaps Sly1p is not completely released from a potentially dimeric t-SNARE (Banfield et al 1994) under these conditions.

DeCamilli and colleagues looked at the distribution of n-Sec1 in rat brain and endocrine pancreas and detected a significant pool of soluble, cytosolic n-Sec1 (Garcia et al 1995). Moreover, these workers failed to detect a significant pool of t-SNARE-bound n-Sec1. This result was unexpected, given the apparent high-affinity interaction of recombinant n-Sec1 and syntaxin (Pevsner et al 1994b). One possible explanation of these immunoprecipitation experiments is that binding of n-Sec1 blocks key epitopes on syntaxin that are otherwise exposed on the free protein or on syntaxin present in core complexes. Alternatively, the interaction could be detergent sensitive; for example, detergent-specific disruption of VAMP-synaptophysin interactions has been reported (see below; Calakos & Scheller 1994). The use of membrane-permeant, chemical cross-linkers to evaluate quantitatively the status of t-SNARE proteins in intact cells should provide valuable information.

v-SNAREs NEED PROTECTION TOO

Analogous to t-SNARE blockage, v-SNAREs could also be protected, although protection of all t-SNAREs might be sufficient to avert organelle-sandwich formation. Using membrane-permeant chemical cross-linkers in detergent extracts and cultured neurons, Calakos & Scheller (1994) demonstrated that a significant fraction of VAMP occurs in association with another synaptic vesicle protein, synaptophysin. This interaction was resistant to Triton X-100 and CHAPS, but was not stable in octylglucoside. Furthermore, synaptophysin was shown to interfere with VAMP/t-SNARE interactions (Edelmann et al 1995). Thus synaptophysin functions to retain VAMP in a state that cannot interact with a t-SNARE. v-SNARE protection may be a special feature of synaptic vesicles that fuse with their targets in a highly regulated manner. Whether v-SNAREs are similarly blocked in other, less regulated steps of secretion remains to be seen.

An independent indication that v-SNAREs may need protection derives from the fact that they appear to use a posttranslational mechanism for their insertion into the ER; once inserted, they hitchhike along the secretory pathway for transport to their final destinations (Kutay et al 1995). Given this scenario, v-SNAREs are likely to be synthesized in an inactive conformation (and/or be bound to a regulatory molecule) and remain inactive until their incorporation into the correct class of transport vesicle.

If SNAREs are blocked, how do they find each other, and by what mechanisms? Although n-Sec1 binds syntaxin very tightly, VAMP and syntaxin/SNAP-25 interact with an apparent binding constant in the low micromolar range (Calakos et al 1994, Pevsner et al 1994b). This implies that another factor might be needed to displace the high-affinity t-SNARE blocker (n-Sec1) prior to SNARE complex formation. n-Sec1 binding to syntaxin also blocks syntaxin/SNAP-25 association, thus release of Sec1 would permit association of syntaxin with both SNAP-25 and the VAMP v-SNARE. These observations further underscore a need for accessory factors to regulate and catalyze SNARE complex formation.

TRANSPORT STEP-SPECIFIC ACCESSORY FACTORS

Beyond the core SNARE-complex constituents, other proteins are also needed for vesicle docking and fusion. These proteins fall into two classes: transport step-specific factors and general factors. Step-specific components would be expected to interact with other step-specific molecules, especially SNAREs. This group includes the Rab GTPase family, relatives of Sec1p, and an emerging new class of proteins referred to here as Velcro[1] factors. As described below, Velcro factors are likely to act by increasing the efficiency of vesicle docking. In the case of yeast Golgi-to-plasma membrane transport, the putative Velcro factor is a large macromolecular complex located at the docking site (Terbush & Novick 1995) that may have the capacity to bind transport vesicles via Rab proteins (Stenmark et al 1995). The Velcro factors may act as a "molecular fly paper" to trap specific classes of transport vesicles.

General factors might catalyze more generic types of protein rearrangements for a family of related molecules. General factors discovered to date include NSF, α-SNAP, and p97. The functions of the accessory molecules are given below.

RAB PROTEINS IN v-SNARE/t-SNARE COMPLEX FORMATION

Rab proteins are small, membrane-associated GTPases that are rate-limiting for a number of fusion processes in eukaryotic cells (Novick & Brennwald

[1]"Velcro is a trademark used for a fastening tape consisting of a strip of nylon with a surface of minute hooks that fasten to a corresponding strip with a surface of uncut pile, used especially on cloth products such as outerwear and athletic shoes." (from the *American Heritage Dictionary,* 1992. Boston: Houghton Mifflin. 3rd ed.).

1993, Zerial & Stenmark 1993, Nuoffer & Balch 1994, Pfeffer 1994a). Over 30 different Rabs have been identified, each localized to distinct membrane-bound compartments. The Rab GTPases that function in yeast ER-to-Golgi transport and Golgi-to-plasma membrane transport are essential for viability, and temperature-sensitive mutations in these proteins lead to the accumulation of transport vesicles at the nonpermissive temperature. As described below, Rab function appears to be required for SNARE complex assembly, yet the Rabs are not part of SNARE complexes. This family of GTPases may help to recruit additional, step-specific docking factors to facilitate SNARE interaction.

Although the yeast ER-to-Golgi Rab Ypt1p is an essential gene product, Gallwitz and coworkers identified a number of proteins that when overexpressed can compensate for the loss of this protein's function (Dascher et al 1991). These include Sly1p, the Sec1p homologue needed in ER-to-Golgi transport, and two v-SNARE relatives, Sly2p (same as Sec22p) and Sly12p (same as Bet1p). Only a mutant Sly1p could compensate for the loss of Ypt1p function. The wild-type protein failed to suppress for loss of Ypt1p, which suggests that Sly1p acts downstream of Ypt1p. These data are consistent with a model in which Ypt1p acted on Sly1p to release a protected t-SNARE and permit SNARE complex formation. In this case, Ypt1p function would be relieved by a mutant Sec1p homologue that has a decreased affinity for the t-SNARE Sed5p.

Overexpression of the two v-SNAREs (Sec22p and Bet1p) was also sufficient to overcome a loss of the Ypt1p function. Recall that in at least one case, n-Sec1 binds a t-SNARE (syntaxin) with nanomolar affinity, and the v-SNARE/t-SNARE (VAMP-syntaxin-SNAP-25) interaction is in the micromolar range. Overexpression of a v-SNARE could override, at least to some extent, a requirement for a factor that would deprotect a t-SNARE; the presence of multiple, locally concentrated v-SNARE proteins could enable direct binding of the v-SNARE to the t-SNARE by mass action. Overexpression of SNAREs could also bypass a requirement for an efficient vesicle-docking enhancer by a similar mechanism (see below).

Sec4p is the Rab protein that functions in yeast Golgi-to-plasma membrane transport (Goud et al 1988). Brennwald et al (1994) set up a screen for suppressors of a Sec4p effector domain mutation. The most potent suppressor of this *sec4* mutation was *SEC9*, which turns out to encode a SNAP-25 homologue. This indicates that Sec4p acts upstream of this t-SNARE; whether it binds directly to the t-SNARE remains to be seen. It is important to note that overexpression of *SEC9* (SNAP-25) also partially suppressed the temperature sensitivity of *sec1, sec3, sec8,* and *sec15.* A *sec1* mutant might interfere with t-SNARE assembly and the ability of Sec9p to bind to Sso1p/Sso2p t-SNAREs.

Other proteins are likely to represent additional accessory factors needed for this specific process of secretory vesicle docking (see below).

In their analyses of SNARE complex formation for yeast ER-to-Golgi transport, Søgaard et al (1994) found that v-SNARE/t-SNARE interaction was not detected in temperature-sensitive *ypt1* mutant yeast strains at the nonpermissive temperature. These data suggest that the Ypt1p Rab is required for v-SNARE/t-SNARE complex assembly. The SNARE complexes analyzed by these workers contained the v- and t-SNAREs Bos1p and Sed5p, but not the corresponding Rab protein. In summary, Rabs act to facilitate SNARE complex formation but are not core elements of such complexes.

RABS ARE ALSO NEEDED FOR v-SNARE/v-SNARE INTERACTIONS

Ferro-Novick and colleagues have shown that the ER-to-Golgi v-SNAREs Bos1p and Sec22p are associated in detergent-solubilized transport vesicles (Lian et al 1994). This association was not detected in *ypt1* mutant strains, and the authors concluded that Rabs are required to catalyze v-SNARE/v-SNARE assembly and activation. In a sense, assembly of v-SNAREs into an active or unblocked conformation is precisely what Rabs would do by catalyzing the release of a Sec1p homologue from a t-SNARE. If the Rab recruits a protein that mimics the structure of the t-SNARE, that protein would be able to bind both the v-SNARE and Sec1p. Such a protein has not yet been discovered. In any case, Rabs are likely incorporated into transport vesicles during their formation, and this process may occur concomitant with v-SNARE activation.

SNARE-LINKING COMPLEXES

The Velcro Family of Vesicle-Docking Proteins

The yeast plasma membrane t-SNARE genes *SSO1* and *SSO2* suppress mutations in *sec1, sec3, sec5, sec9,* and *sec15* (Aalto et al 1993). Sec9p, the SNAP-25 homologue, would be expected to interact with the syntaxin homologues Sso1p/Sso2p (Brennwald 1994). As for the other gene products, Sec15p and Sec8p and are part of a large, 19.5S-particle complex that includes Sec6p and requires Sec3p, Sec5p, and Sec10p for its assembly (Terbush & Novick 1995).

A spectacular finding is that this large complex is localized to the tips of yeast buds, the site of exocytosis in *Saccharomyces cerevisiae.* This is in contrast to the t-SNAREs Sso1p, Sso2p, and Sec9p, which are distributed uniformly over the plasma membrane (Brennwald et al 1994). It seems likely then that the Sec

6/8/15 complex functions to link transport vesicles with the docking site on the plasma membrane for the ultimate SNARE assembly reaction.

How does a vesicle find the complex? Genetic and biochemical experiments suggest a link between the Sec4p Rab and Sec8p and Sec15p (Salminen & Novick 1989, Bowser et al 1992). The data imply that Rabs on a transport vesicle may bind to the Sec6/Sec8/Sec15 complex in order to position the vesicle for subsequent fusion. In this process, the Sec6/Sec8/Sec15 complex functions as a true docking machine. Studies on Rabaptin 5 described below support such a model.

Rabaptin 5

Rabaptin-5, another candidate Velcro protein, was discovered by Zerial and coworkers, using a two-hybrid screen, to be a protein able to interact with Rab5-GTP in preference to Rab5-GDP (Stenmark et al 1995). (Rab5 is required for endosome-endosome fusion.) Rabaptin-5, a 100-kDa coiled-coil protein, is recruited onto endosomes in the presence of GTP and binds directly to Rab5. Apparently, it also occurs as a large macromolecular complex that may function in vesicle docking by a mechanism analogous to that of the Sec6/Sec8/Sec15 complex (Stenmark et al 1995).

Rabaptin-5's direct link to the active, GTP-bound form of Rab5 represents the first indication of how Velcro factors may interact with transport vesicles. Rab proteins would be incorporated into transport vesicles in their active conformations, thereby recruiting (or being captured by) a docking Velcro factor. This process would increase the efficiency of vesicle targeting. Either the Velcro complex or the Rab could then free the t-SNARE, thus allowing for subsequent fusion.

p115

p115 is a peripheral membrane protein that appears to be concentrated on the Golgi complex (Waters et al 1992). It was purified on the basis of its ability to stimulate intra-Golgi transport (Waters et al 1992) and to associate with transcytotic vesicles (Sztul et al 1993). p115 exists as a parallel homodimer, with two globular heads, an extended coiled-coil tail, followed by an acidic domain at the extreme C terminus (Waters et al 1992, Sapperstein et al 1995). p115 is homologous to a yeast protein Uso1p, which is required for ER-to-Golgi transport (Nakajima et al 1991, Barroso et al 1995, Sapperstein et al 1995). Uso1p is much larger than p115; a larger coiled-coil portion accounts for the size differences between these proteins.

Uso1p and p115 are likely to function in transport vesicle docking and/or fusion (Elazar et al 1994, Sapperstein et al 1995, Barroso et al 1995, Lupashin et al 1996). Accordingly, *USO1* has been shown to interact genetically with genes encoding identified SNAREs. Overexpression of the genes encoding putative ER-to-Golgi v-SNAREs—*BET1, BOS1, SEC22,* and *YKT6*—suppress a temperature-sensitive *uso1* mutation. Overexpression of *BET1* and *SEC22* can rescue Δ*uso1* deletion strains from lethality (Sapperstein et al 1995). Ypt1p (yeast Rab1) overexpression or expression of a mutant of the Sec1p homologue Sly1p rescues the temperature sensitivity of a *uso1* mutation. Moreover, Ypt1p overexpression allows a Δ*uso1* deletion strain to grow, although Uso1p cannot compensate for a loss of Ypt1p (Sapperstein et al 1996). These data suggest strongly that Uso1p acts in SNARE complex formation. This conclusion was substantiated directly by SNARE-complex immunoprecipitations from yeast strains harboring *uso1* mutations (Sapperstein et al 1996).

Perhaps Uso1p/p115 acts as a general, molecular Velcro to help v- and t-SNAREs find each other on different membrane-bound compartments. p115 function may be analogous to the role of the Sec6/Sec8/Sec15 complex in Golgi-to-plasma membrane transport in yeast. It remains to be seen whether Uso1p is also needed for that process.

Another interesting possibility is that p115 could be a Golgi-specific Velcro factor that facilitates arrival of vesicles to the Golgi, departure of vesicles from it, and Golgi-Golgi fusion. Together with NSF and SNAPs, p115 is required for the reassembly of post-mitotic Golgi fragments into Golgi cisternae (Rabouille et al 1995). Velcro factors are expected to be able to facilitate both homotypic and heterotypic fusion events. It is not unreasonable to suggest that the exocytic and endocytic pathways use different sets of Velcro factors.

Another *trans*-Golgi Velcro protein candidate has recently been reported (Erlich et al 1996). *trans*-Golgi p230 is a large, cytoplasmically oriented molecule predicted to contain long stretches of coiled-coil α helices. Its possible interaction with transport vesicles and/or docking factors has yet to be revealed.

OTHER ESSENTIAL BUT GENERAL COMPONENTS

NSF

As discussed above, NSF is a trimeric ATPase required for multiple transport events. NSF dissociates 20S SNARE complexes, in the presence of cellular levels of ATP, rather than catalyzing their assembly (Söllner et al 1993b). Early models suggested that the ability of NSF to disrupt SNARE complexes in an ATP-dependent process might be an important, late pre-fusion event. It was proposed that NSF would relieve the constraint on two tightly bound, coiled-coil v- and t-SNAREs and allow subsequent fusion (Figure 1). Presumably, NSF action would therefore be coupled to a fusion process to permit recycling of SNARE constituents, or to allow SNARE activation. This conclusion is consistent with biochemical and morphological analyses of the function of NSF described earlier.

All the v-SNAREs are oriented with their carboxy-termini inserted into the membrane, and their cytosolic, amino terminal portions are predicted to form coiled-coils. A general factor, such as the NSF ATPase, may have a generic capacity to recognize and rearrange SNARE coiled-coil interactions (Morgan & Burgoyne 1995). Evidence for such an activity was obtained indirectly by Lian et al (1994), who detected a large increase in Sec22p/Bos1p complexes in *sec18* strains. Much more work will be needed to establish the precise role of NSF, as exemplified by the following findings.

Wickner and colleagues have recently investigated the role of NSF, using an in vitro system that reconstitutes vacuole-vacuole fusion (Mayer et al 1996). Similar to other fusion and transport processes, this process requires NSF, α-SNAP, and a Rab homologue (in this case, Ypt7p). As is true for other NSF-dependent reactions, α-SNAP works at the same time during the fusion reaction as does NSF. However, Mayer et al showed that Sec17p (α-SNAP) can be removed from membranes without inhibiting subsequent fusion. Sec18p (NSF) remained membrane-associated for a much longer time, thus α-SNAP cannot be the only membrane-binding site for NSF on the vacuole. Nevertheless, although NSF remained vacuole-bound, it did not appear to be needed except at an early stage in the reaction (Mayer et al 1996). Most importantly, the authors could prepare separate donor and acceptor fractions that had proceeded past an NSF- and α-SNAP-dependent stage in the reaction; when these fractions were mixed later, the vacuoles were still capable of fusion. Thus in this system, NSF and α-SNAP clearly function in a pre-docking and, therefore, pre-fusion process.

In homotypic fusion, there is little reason to block active SNARE proteins as long as specificity in fusion is ensured. ER vesicles and endosomes appear to fuse with one another continuously in living cells. If NSF and α-SNAP function to reactivate (or untangle) v- and t-SNAREs after an earlier, homotypic fusion event, they could act early to regenerate fusion-competent organelles.

Because NSF is used in multiple transport events and is also a member of a larger family of ATPases (Mellman 1995), it seems most likely that this family of proteins catalyzes the rearrangement of a generic class of protein:protein interactions. α-SNAP may serve to recruit the NSF ATPase to an appropriate substrate most efficiently; however, NSF would also be able to recognize such substrates once recruited onto a membrane. Such a model is consistent with NSF membrane-binding sites on membranes lacking α-SNAP.

Although NSF may serve a general function, one must not forget that the yeast NSF gene *SEC18* is essential, and conditional mutations in this gene lead to very rapid and specific blocks of multiple, intracellular transport events (Graham & Emr 1991). Thus NSF is clearly a key player in intracellular transport.

NSF-Relatives

It is now apparent that NSF is a member of a family of related and conserved ATPases. This was first appreciated by Peters et al (1990) upon their discovery of a hexameric, NEM-sensitive ATPase comprised of 97-kDa subunits. p97 is abundant in all eukaryotic cells and is localized to both the nucleus and cytoplasm (Peters et al 1990). Sequence analysis revealed that p97 is identical to "valosin-containing protein" (Koller & Brownstein 1987) and resembles NSF in its ATP-binding domains.

What does p97 do? Homotypic fusion of yeast ER membranes requires functional Cdc48p (Latterich et al 1995) but not NSF (Sec18p) or α-SNAP (Sec17p; Latterich & Schekman 1994). Yeast Cdc48p is 76% identical with p97 and therefore is likely to represent the yeast homologue of p97. It should be noted, however, that p97 alone was not sufficient to reverse a *cdc48* defect in vitro or complement a *CDC48* gene deletion (Latterich et al 1995). This could be because the mammalian proteins were unable to interact with yeast-specific, Cdc48p-interacting proteins. Such proteins will be exciting to discover.

Warren and colleagues have shown that p97 can facilitate the regrowth of mitotic Golgi fragments into longer cisternae (Rabouille et al 1995). When purified Golgi complexes are incubated in mitotic cytosol, the fraction of total membranes detected as cisternae drops from 55 to 20%. Reincubation in buffer or interphase cytosol restored the fraction of membranes occurring as cisternae to 43%. Treatment of membranes with NEM blocked cisternal regrowth, and addition of either p97 or a mixture of NSF, α- and γ-SNAPs and p115 restored the cisternal regrowth process to NEM-treated membranes (Rabouille et al 1995).

Are p97 and NSF redundant ATPases? At least not for ER fusion. Moreover, because the yeast NSF gene is essential, p97 cannot fully replace NSF function. Although p97 and NSF restored Golgi cisternal regrowth to similar extents, Rabouille et al (1995) noted that the reassembled cisternae had different morphologies after incubation with the respective proteins. Nevertheless, addition of both ATPases did not lead to further membrane assembly, possibly because components in addition to the ATPases are required to complete the assembly process.

Acharya et al (1995) studied the reassembly of Golgi stacks in permeabilized cells after their vesiculation with the sponge metabolite illimaquinone. These workers dissected two stages in the reassembly of functional Golgi membranes—vesiculated Golgi membranes first fused in an NSF-dependent reaction to form larger vesicles. In a second p97-dependent reaction, the large vesicles fused to form tubules and short cisternae. These data suggest that two ATPases act in concert to drive Golgi stack assembly.

If the ATPases have partially overlapping roles, p97 should also disassemble v/t-SNARE complexes or lock them together in the presence of ATPγS and EDTA. This prediction is easily tested. p97 has also been found in association with the macromolecular coat protein clathrin (Pleasure et al 1993). Perhaps p97 serves as a chaperone in the assembly and disassembly of the clathrin coat.

The reassembly of a Golgi complex likely involves the independent coalescence of all *cis, medial,* and *trans*-derived vesicles to form larger *cis, medial,* and *trans* cisternae. The finding that p97, rather than NSF, participates in this process, as well as in ER-derived vesicle fusion, led to the proposal that p97 is the homotypic fusion enzyme, whereas NSF is the heterotypic fusion ATPase. This generalization may be premature: For example, homotypic early endosome-early endosome fusion utilizes NSF and Rab proteins (Diaz et al 1989, Zerial & Stenmark 1993). p97 may also further enhance this reaction.

α-SNAP and Complexins

As described earlier, α-SNAP is recruited to assembled SNARE complexes and recruits NSF to those complexes (Söllner et al 1993a, McMahon & Südhof 1995). Recently, McMahon et al (1995) discovered 18–19 kDa, highly charged proteins, termed complexins, that seem to compete with α-SNAP for SNARE-complex binding. Complexins are abundant in brain, although they seem to be present in most tissues. They are largely cytosolic and have been proposed to regulate α-SNAP association with SNARE complexes. Both the generality and nature of complexin function remain to be determined.

REDUNDANCY IN SNARE FUNCTION

SNARE proteins are abundant in nerve tissue, where secretion is tightly regulated. Because of the tight regulation and calcium dependence of neurotransmitter release, a neuron might be expected to use redundant targeting and regulatory factors. Yet SNAREs are also abundant in yeast, where redundancy appears to be a common property of v- and t-SNARE proteins.

SSO1 and *SSO2* plasma membrane t-SNARE genes can each be disrupted without a detectable phenotype (Aalto et al 1993). However, the double deletion strain fails to germinate. The simplest explanation is that the products of these genes act together to perform an essential function. Either t-SNARE alone is sufficient, thus there is no indication of requisite t-SNARE heterodimer formation. Analogously, the cognate *SNC1* and *SNC2* v-SNARE genes can also be disrupted individually without consequence, but the double disruption leads to a conditional lethal phenotype (Protopopov et al 1993). Thus for

Golgi-to-plasma membrane transport, yeast utilizes redundant sets of v- and t-SNAREs.

For ER-to-Golgi transport, the v-SNAREs Bos1p and Sec22p can be cross-linked to each other in detergent-solubilized ER-to-Golgi transport vesicles (Lian et al 1994). Although *BOS1* is an essential gene (Shim et al 1991), *SEC22* is not (Dascher et al 1991). Strains lacking *SEC22* are temperature sensitive for growth, and overproduction of *BOS1* can restore growth to normal levels (Lian et al 1994). However, overproduction of *SEC22* cannot substitute for the loss of *BOS1*. These data suggest that the two v-SNAREs are not functionally equivalent; Sec22p may act by increasing the efficacy of Bos1p as a v-SNARE (Lian et al 1994).

What about the other candidate v-SNAREs for ER-to-Golgi transport? There remains some controversy as to whether transport vesicles contain Bet1p (Lian & Ferro-Novick 1993, Rexach et al 1994), but the ER-to-Golgi t-SNARE Sed5p has been shown to complex with Bet1p, Bos1p, Sec22p, and Ykt6p (Søgaard et al 1994), and each of these gene products could suppress the temperature-sensitive phenotype of a *uso1* (p115) mutant strain (Sapperstein et al 1996), consistent with a role in SNARE complex formation. If all of these proteins represent functional, anterograde v-SNAREs, why is *BOS1* an essential gene? The roles of each of the ER-to-Golgi type-II membrane proteins are not known, and the final word is not yet in as to whether a single, essential *SED5*-encoded t-SNARE is sufficient. *BOS1* is clearly important, and its genetic interaction with *BET1* and *SEC22* suggests that each of the corresponding gene products is important in v-SNARE function. Exactly how *BET1* and *SEC22* enhance *BOS1* function remains to be determined.

In summary, yeast Golgi-to-plasma membrane transport utilizes two v-SNAREs, two syntaxin-like t-SNAREs, and one SNAP-25-like t-SNARE. In contrast, ER-to-Golgi transport has multiple candidate v-SNAREs, but Bos1p seems to be the most important; only a single t-SNARE has been identified to date, and it is encoded by an essential syntaxin-like gene. Why use more than one v- and t-SNARE? Multiple combinations of SNARE interactions could increase the avidity of the interaction; they could also provide additional specificity, if distinct pairwise interactions were utilized. Finally, SNARE-like proteins may regulate v-SNARE accessibility by mimicking t-SNARE structure prior to v-SNARE activation.

SUMMARY AND FUTURE PERSPECTIVES

Data relevant to vesicle docking are spewing forth at a rapid rate. As the dust settles, we should soon know in some detail, how vesicles identify and dock with their targets. At present, the following outline is coming into focus.

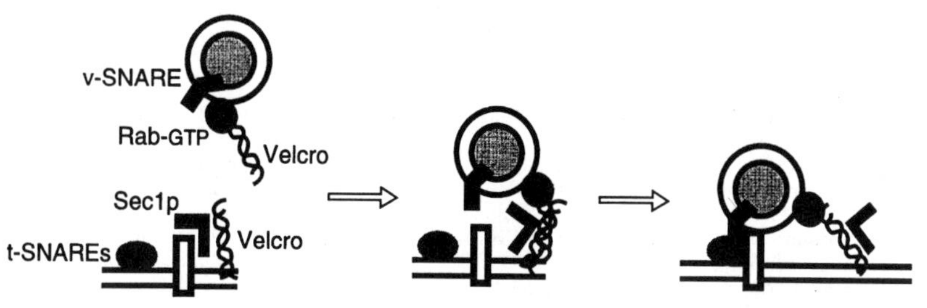

Figure 2 A highly speculative model for vesicle docking. Transport vesicles may find their targets more efficiently via Velcro factors, exemplified by the Sec6/Sec8/Sec15 complex (see text). In this case, the Velcro factor would be located at the fusion target and bind to vesicles by interaction with the vesicle-associated Rab-GTP protein, in analogy with Rabaptin-5. Alternatively, a Velcro factor may be recruited onto Rab-GTP-bearing transport vesicles and interact with a cognate Velcro factor at the target. The now-tethered vesicle can then dock with the t-SNARE if the Velcro can also catalyze release of the t-SNARE blocker, Sec1p. Association with the Velcro may also catalyze GTP hydrolysis by the Rab GTPase.

v- and t-SNAREs form the core of a docked vesicle complex. But SNAREs do not dock alone. SNARE pairing can only be permitted between transport vesicles and their targets; SNAREs present elsewhere must be shielded from each other.

An updated and speculative model for vesicle docking is shown in Figure 2. Reaction-specific accessory factors, including Rab GTPases, and Velcro factors, help direct vesicles to their sites of fusion and increase the efficiency with which vesicles find their targets. Proteins such as Sec1p protect t-SNAREs. At least in the case of regulated synaptic vesicle exocytosis, v-SNAREs will also be protected (not shown). Velcro factors likely represent a complex of multiple, associated polypeptides that may be localized to the fusion site (Terbush & Novick 1995) and/or recruited onto transport vesicles by virtue of an interaction with vesicle-associated Rab proteins in their active, GTP-bound conformations (Stenmark et al 1995). Some element of the Velcro factor complex may act to release Sec1p and its homologues to permit v- and t-SNARE interaction. Finally, general factors that include NSF are likely to catalyze rearrangements of SNARE molecules, and fusion can then ensue (Figure 1). This speculative skeletal outline includes many testable tenets, and a more precise understanding of vesicle docking should soon be at hand.

Literature Cited

Aalto MK, Ronne H, Keränen S. 1993. Yeast syntaxins Sso1p and Sso2p belong to a family of related membrane proteins that function in vesicular transport. *EMBO J.* 12:4095–104

Acharya U, Jacobs R, Peters J-M, Watson N, Farquhar MG, Malhotra V. 1995. The formation of Golgi stacks from vesiculated Golgi membranes requires two distinct fusion events. *Cell* 82:895–904

Banfield DK, Lewis MJ, Rabouille C, Warrren G, Pelham HRB. 1994. Localization of Sed5, a putative vesicle targeting molecule to the *cis*-Golgi network involves both its transmembrane and cytoplasmic domains. *J. Cell Biol.* 127:357–71

Barlowe C, Orci L, Yeung T, Hosobuchi M, Hamamoto S, et al. 1994. COPII: a membrane coat formed by Sec proteins that drive vesicle budding from the endoplasmic reticulum. *Cell* 77:895–907

Barroso M, Nelson DS, Sztul E. 1995. Transcytosis-associated protein (TAP)/p115 is a general fusion factor required for binding of vesicles to acceptor membranes. *Proc. Natl. Acad. Sci. USA* 92:527–31

Baumert M, Maycox PR, Navone F, De Camilli P, Jahn R. 1989. Synaptobrevin: an integral membrane protein of 18,000 daltons present in small synaptic vesicles of rat brain. *EMBO J.* 8:379–84

Bennett MK, Calakos N, Scheller RH. 1992. Syntaxin, a synaptic protein implicated in docking of synaptic vesicles at presynaptic active zones. *Science* 257:255–59

Bennett MK. 1995. SNAREs and the specificity of transport vesicle targeting. *Curr. Opin. Cell Biol.* 7:581–86

Block MR, Glick BS, Wilcox CA, Wieland FT, Rothman JE. 1988. Purification of an *N*-ethylmaleimide-sensitive protein catalyzing vesicular transport. *Proc. Natl. Acad. Sci. USA* 85:7852–56

Bowser R, Muller H, Govindan B, Novick P. 1992. Sec8p and Sec15p are components of a plasma membrane-associated 19.5S particle that may function downstream of Sec4p to control exocytosis. *J. Cell Biol.* 118:1041–56

Brennwald P, Kearns B, Champion K, Keränen S, Bankaitis V, Novick P. 1994. Sec9 is a SNAP-25-like component of a yeast SNARE complex that may be the effector of Sec4 function in exocytosis. *Cell* 79:245–58

Calakos N, Bennett MK, Peterson KE, Scheller RH. 1994. Protein-protein interactions contributing to the specificity of intracellular vesicular trafficking. *Science* 263:1146–49

Calakos N, Scheller RH. 1994. Vesicle associated membrane protein and synaptophysin are associated on the synaptic vesicle. *J. Biol. Chem.* 269:24534–37

Clary DO, Griff IC, Rothman JE. 1990. SNAPs, a family of NSF attachment proteins involved in intracellular membrane fusion in animals and yeast. *Cell* 61:709–21

Clary DO, Rothman JE. 1990. Purification of three related peripheral membrane proteins needed for vesicular transport. *J. Biol. Chem.* 265:10109–17

Couve A, Gerst JE. 1994. Yeast Snc proteins complex with Sec9. *J. Biol. Chem.* 269:23391–94

Dascher C, Ossig R, Gallwitz D, Schmitt HD. 1991. Identification and structure of four yeast genes *(SLY)* that are able to suppress the functional loss of *YPT1*, a member of the *RAS* superfamily. *Mol. Cell. Biol.* 11:872–85

Diaz R, Mayorga LS, Weidman PJ, Rothman JE, Stahl PD. 1989. Vesicle fusion following receptor-mediated endocytosis requires a protein active in Golgi transport. *Nature* 339:398–400

Eakle KA, Bernstein M, Emr SD. 1988. Characterization of a component of the yeast secretion machinery: identification of the *SEC18* gene product. *Mol. Cell. Biol.* 8:4098–109

Edelmann L, Hanson PI, Chapman ER, Jahn R. 1995. Synaptobrevin binding to synaptophysin: a potential mechanism for controlling the exocytotic fusion machine. *EMBO J.* 14:224–31

Elazar Z, Orci L, Ostermann J, Amherdt M, Tanigawa G, Rothman JE. 1994. ADP ribosylation factor and coatomer couple fusion to vesicle budding. *J. Cell Biol.* 124:415–24

Erlich R, Gleeson PA, Campbell P, Dietzsch E, Toh B-H. 1996. Molecular characterization of *trans*-Golgi p230. A human peripheral membrane protein encoded by a gene on choromosome 6p12–22 contains extensive coiled-coil α-helical domains and a granin motif. *J. Biol. Chem.* 271:8328–37

Ferro-Novick S, Jahn R. 1994. Vesicle fusion from yeast to man. *Nature* 370:191–93

Garcia EP, Gatti E, Butler M, Burton J, De Camilli P. 1994. A rat brain Sec1 homologue related to Rop and Unc18 interacts with syntaxin. *Proc. Natl. Acad. Sci. USA* 91:2003–7

Garcia EP, McPherson PS, Chilcote TJ, Takei K, DeCamilli P. 1995. rbSec1A and B colocalize with syntaxin 1 and SNAP-25 throughout the axon, but are not in a stable complex with syntaxin. *J. Cell Biol.* 129:105–20

Geppert M, Goda Y, Hammer RE, Li C, Rosahl

TW, et al. 1994. Synaptotagmin I: a major calcium sensor for transmitter release at a central synapse. *Cell* 79:717–27

Goud B, Salminen A, Walworth NC, Novick P. 1988. A GTP-binding protein required for secretion rapidly associates with secretory vesicles and the plasma membrane in yeast. *Cell* 53:753–68

Graham TR, Emr SD. 1991. Compartmental organization of Golgi-specific protein modification and vacuolar protein sorting events defined in a yeast *sec18* mutant. *J. Cell Biol.* 114:207–18

Griff IC, Schekman R, Rothman JE, Kaiser CA. 1992. The yeast *SEC17* gene product is functionally equivalent to mammalian alpha-SNAP protein. *J. Biol. Chem.* 267:12106–15

Hardwick KG, Pelham HRB. 1992. SED5 encodes a 39 kD integral membrane protein required for vesicular transport between the ER and the Golgi complex. *J. Cell. Biol.* 119:513–21

Hata Y, Slaughter CA, Südhof TC. 1993. Synaptic vesicle fusion complex contains unc18 homologue bound to syntaxin. *Nature* 366:347–51

Hodel A, Schäfer T, Gerosa D, Burger MM. 1994. In chromaffin cells, the mammalian sec1p homolog is a syntaxin-1A-binding protein associated with chromaffin granules. *J. Biol. Chem.* 269:8623–26

Kaiser CA, Schekman R. 1990. Distinct sets of SEC genes govern transport vesicle formation and fusion early in the secretory pathway. *Cell* 61:723–33

Koller KJ, Brownstein MJ. 1987. Use of a cDNA clone to identify a supposed precursor protein containing valosin. *Nature* 325:542–45

Kutay U, Ahnert-Hilger G, Hartmann E, Wiedenmann B, Rapoport TA. 1995. Transport route of synaptobrevin via a novel pathway of insertion into the ER membrane. *EMBO J.* 14:217–23

Latterich M, Fröhlich K-U, Schekman R. 1995. Membrane fusion and the cell cycle: Cdc48p participates in the fusion of ER membranes. *Cell* 82:885–93

Latterich M, Schekman R. 1994. The karyogamy gene KAR2 and novel proteins are required for ER-membrane fusion. *Cell* 78:87–98

Lian JP, Ferro-Novick S. 1993. Bos1p, an integral membrane protein of the endoplasmic reticulum to Golgi transport vesicles, is required for their fusion competence. *Cell* 73:735–45

Lian JP, Stone S, Jiang Y, Lyons P, Ferro-Novick S. 1994. Ypt1p implicated in v-SNARE activation. *Nature* 372:698–701

Lupashin VV, Hamamoto S, Schekman RW. 1996. Biochemical requirements for targeting and fusion of ER-derived transport vesicles with purified yeast Golgi membranes. *J. Cell. Biol.* 132:277–89

Malhotra V, Orci L, Glick BS, Block MR, Rothman JE. 1988. Role of an *N*-ethylmaleimide-sensitive transport component in promoting fusion of transport vesicles with cisternae of the Golgi stack. *Cell* 54:221–27

Mayer A, Wickner W, Haas A. 1996. Sec18p (NSF)-driven release of Sec17p (a-SNAP) can precede docking and fusion of yeast vacuoles. *Cell* 85:83–94

McMahon HT, Missler M, Li C, Südhof TC. 1995. Complexins: cytosolic proteins that regulate SNAP receptor function. *Cell* 83:111–19

McMahon HT, Südhof TC. 1995. Synaptic core complex of synaptobrevin, syntaxin, SNAP-25 forms high affinity alpha-SNAP binding site. *J. Biol. Chem.* 270:2213–17

Mellman I. 1995. Enigma variations: protein mediators of membrane fusion. *Cell* 82:869–72

Morgan A, Burgoyne RD. 1995. Is NSF a fusion protein? *Trends Biochem. Sci.* 5:335–39

Nakajima H, Hirata A, Ogawa Y, Yonehara T, Yoda K, Yamasaki M. 1991. A cytoskeletal-related gene, *USO1,* is required for intracellular protein transport in *Saccharomyces cerevisiae. J. Cell. Biol.* 113:245–60

Novick P, Brennwald P. 1993. Friends and family: the role of Rab GTPases in vesicular traffic. *Cell* 75:597–601

Novick P, Ferro S, Schekman R. 1981. Order of events in the yeast secretory pathway. *Cell* 25:461–69

Novick P, Field C, Schekman R. 1980. Identification of 23 complementation groups required for post-translational events in the yeast secretory pathway. *Cell* 21:205–15

Nuoffer C, Balch WE. 1994. GTPases: multifunctional molecular switches regulating vesicular traffic. *Annu. Rev. Biochem.* 63:949–90

Orci L, Malhotra V, Amherdt M, Serafini T, Rothman JE. 1989. Dissection of a single round of vesicular transport: sequential intermediates for intercisternal movement in the Golgi stack. *Cell* 56:357–68

Oyler GA, Higgins GA, Hart RA, Battenberg E, Billingsley M, et al. 1989. The identification of a novel synaptosomal-associated protein, SNAP-25, differentially expressed by neuronal subpopulations. *J. Cell. Biol.* 109:3039–52

Palade G. 1975. Intracellular aspects of the process of protein secretion. *Science* 189:347–58

Peters JM, Walsh MJ, Franke WW. 1990. An abundant and ubiquitous homo-oligomeric ring-shaped ATPase particle related to putative vesicle fusion proteins Sec18p and NSF. *EMBO J.* 9:1757–67

Pevsner J, Hsu S-C, Braun JEA, Calakos N, Ting AE, et al. 1994a. Specificity and regulation of a synaptic vesicle docking complex. *Neuron* 13:353–61

Pevsner J, Hsu SC, Scheller RH. 1994b. n-Sec1: a neural-specific syntaxin-binding protein. *Proc. Natl. Acad. Sci. USA* 91:1445–49

Pfeffer SR. 1994a. Rab GTPases: master regulators of membrane trafficking. *Curr. Opin. Cell. Biol.* 6:522–26

Pfeffer SR. 1994b. Clues to brain function from baker's yeast. *Proc. Natl. Acad. Sci. USA* 91:1987–88

Pleasure IT, Black MM, Keen JH. 1993. Valosin-containing protein, VCP, is a ubiquitous clathrin-binding protein. *Nature* 365:459–62

Protopopov V, Govindan B, Novick P, Gerst JE. 1993. Homologs of the synaptobrevin/VAMP family of synaptic vesicle proteins function on the late secretory pathway in S. cerevisiae. *Cell* 74:855–61

Rabouille C, Levine TP, Peters J-M, Warren G. 1995. An NSF-like ATPase, p97, and NSF mediate cisternal regrowth from mitotic Golgi fragments. *Cell* 82:905–14

Rexach MF, Latterich M, Schekman R. 1994. Characteristics of endoplasmic reticulum-derived vesicles. *J. Cell. Biol.* 114:219–29

Rothman JE. 1994. Mechanisms of intracellular protein transport. *Nature* 372:55–63

Rothman JE, Wieland FT. 1996. Protein sorting by transport vesicles. *Science* 272:227–34

Salminen A, Novick PJ. 1989. The Sec15 protein responds to the function of the GTP binding protein, Sec4, to control vesicular traffic in yeast. *J. Cell Biol.* 109:1023–36

Sapperstein SK, Lupashin VV, Schmitt HD, Waters MG. 1996. Assembly of the ER to Golgi SNARE complex requires Uso1p. *J. Cell Biol.* 132:755–67

Sapperstein SK, Walter DM, Grosvenor AR, Heuser JE, Waters MG. 1995. p115 is a general vesicular transport factor related to the yeast ER-Golgi transport factor Uso1p. *Proc. Natl. Acad. Sci. USA* 92:522–26

Scheller RH. 1995. Membrane trafficking in the presynaptic nerve terminal. *Neuron* 14:893–97

Schekman R, Orci L. 1996. Coat proteins and vesicle budding. *Science* 271:1526–33

Shim J, Newman AP, Ferro-Novick S. 1991. The *BOS1* gene encodes an essential 27-kD putative membrane protein that is required for vesicular transport from the ER to the Golgi complex in yeast. *J. Cell Biol.* 113:55–64

Søgaard M, Tani K, Ye RR, Geromanos S, Tempst P, et al. 1994. A rab protein is required for the assembly of SNARE complexes in the docking of transport vesicles. *Cell* 78:937–48

Söllner T, Bennett MK, Whiteheart SW, Scheller RH, Rothman JE. 1993a. A protein assembly-disassembly pathway in vitro that may correspond to sequential steps of synaptic vesicle docking, activation and fusion. *Cell* 75:409–18

Söllner T, Whiteheart SW, Brunner M, Erdjument-Bromage H, Geromanos S, et al. 1993b. SNAP receptors implicated in vesicle targeting and fusion. *Nature* 362:318–24

Stenmark H, Vitale G, Ullrich O, Zerial M. 1995. Rabaptin-5 is a direct effector of the small GTPase Rab5 in endocytic membrane fusion. *Cell* 83:423–32

Südhof TC. 1995. The synaptic vesicle cycle: a cascade of protein-protein interactions. *Nature* 375:645–53

Südhof TC, Camilli PD, Niemann H, Jahn R. 1993. Membrane fusion machinery: insights from synaptic proteins. *Cell* 75:1–4

Sztul E, Colombo M, Stahl P, Samanta R. 1993. Control of protein traffic between distinct plasma membrane domains. *J. Biol. Chem.* 268:1876–85

Terbush DR, Novick P. 1995. Sec6, Sec8, and Sec15 are components of a multisubunit complex which localizes to small bud tips in *Saccharomyces cerevisiae. J. Cell Biol.* 130:299–312

Trimble WS, Cowan DM, Scheller RH. 1988. VAMP-1: a synaptic vesicle-associated integral membrane protein. *Proc. Natl. Acad. Sci. USA* 85:4538–42

Waters MG, Clary DO, Rothman JE. 1992. A novel 115-kD peripheral membrane protein is required for intercisternal transport in the Golgi stack. *J. Cell Biol.* 118:1015–26

Weidman PJ, Melancon P, Block MR, Rothman JE. 1989. Binding of an *N*-ethylmaleimide sensitive fusion protein to Golgi membranes requires both a soluble protein(s) and an integral membrane receptor. *J. Cell Biol.* 108:1589–96

Whiteheart SW, Brunner M, Wilson DW, Wiedmann M, Rothman JE. 1992. Soluble *N*-ethylmaleimide-sensitive fusion attachment proteins (SNAPs) bind to a multi-SNAP receptor complex in Golgi membranes. *J. Biol. Chem.* 267:12239–43

Whiteheart SW, Rossnagel K, Buhrow SA, Brunner M, Jaennicke R, Rothman JE. 1994. *N*-ethylmaleimide-sensitive fusion protein: a trimeric ATPase whose hydrolysis of ATP is required for membrane fusion. *J. Cell Biol.* 126:945–54

Wilson DW, Whiteheart SW, Wiedmann M, Brunner M, Rothman JE. 1992. A multisub-

unit particle implicated in membrane fusion. *J. Cell Biol.* 117:531–38
Wilson DW, Wilcox CA, Flynn GC, Chen E, Kuang W-J, et al. 1989. A fusion protein required for vesicle-mediated transport in both mammalian cells and yeast. *Nature* 339:355–59
Zerial M, Stenmark H. 1993. Rab GTPases in vesicular transport. *Curr. Opin. Cell Biol.* 5:613–20

Annu. Rev. Cell Dev. Biol. 1996. 12:463–519

FOCAL ADHESIONS, CONTRACTILITY, AND SIGNALING

Keith Burridge and Magdalena Chrzanowska-Wodnicka
Department of Cell Biology and Anatomy, and Lineberger Comprehensive Cancer Center, University of North Carolina at Chapel Hill, Chapel Hill, North Carolina 27599-7090

KEY WORDS: integrins, ECM, cytoskeleton, Rho, kinases

ABSTRACT

Focal adhesions are sites of tight adhesion to the underlying extracellular matrix developed by cells in culture. They provide a structural link between the actin cytoskeleton and the extracellular matrix and are regions of signal transduction that relate to growth control. The assembly of focal adhesions is regulated by the GTP-binding protein Rho. Rho stimulates contractility which, in cells that are tightly adherent to the substrate, generates isometric tension. In turn, this leads to the bundling of actin filaments and the aggregation of integrins (extracellular matrix receptors) in the plane of the membrane. The aggregation of integrins activates the focal adhesion kinase and leads to the assembly of a multicomponent signaling complex.

CONTENTS

1081-0706/96/1115-0463$08.00

INTRODUCTION

The adhesive interactions between a cell and its surrounding extracellular matrix (ECM) regulate its morphology, migratory properties, growth, and differentiation. One model system for studying adhesion to ECM is the focal adhesion (FA). FAs are specialized sites of adhesion developed by many cells in culture. They consist of aggregated ECM receptors (integrins) that span the plasma membrane, interacting on the outside with ECM components and on the inside with bundles of actin filaments (stress fibers). Many proteins have been identified in FAs, particularly at their cytoplasmic face. Some of these proteins have predominantly a structural role, whereas others are involved in signal transduction.

The field of FA research has grown considerably since one of us participated in writing a review on FAs for this series (Burridge et al 1988). In large part, this growth reflects the increasing interest in integrin-mediated signaling and the widely recognized effect integrin-mediated adhesion has on many aspects of cell behavior. Several excellent recent reviews cover various areas of integrin-mediated signal transduction and FA research (Adams & Watt 1993, Juliano & Haskill 1993, Hemler et al 1994, Pavalko & Otey 1994, Clark & Brugge 1995, Yamada & Miyamoto 1995, Ridley 1995, Richardson & Parsons 1995), including two in Volume 11 of this series (Jockusch et al 1995, Schwartz et al 1995). Our intention here is to avoid repetition of these reviews. Thus we focus on one topic generally neglected: the relationship between contractility/isometric tension and the formation of FAs. We believe that this is an area critical to understanding not only the assembly of these structures but also the signaling that occurs within them. In this context, we discuss the low molecular weight GTP-binding protein Rho, which stimulates the formation of FAs and stress fibers and whose mode of action has generated considerable interest.

STRUCTURAL ORGANIZATION OF FOCAL ADHESIONS

Many of the structural details of FAs have been reviewed recently (Jockusch et al 1995). Structural proteins in FAs are listed in Table 1. Several appear to be

Table 1 Structural components of focal adhesions

Transmembrane	Cytoplasmic		
1. integrins	4. actin	10. zyxin	16. aciculin*
2. syndecan IV	5. α-actinin	11. CRP	17. dystrophin*
3. dystroglycan*	6. filamin	12. radixin	18. syntrophin*
	7. vinculin	13. fimbrin	19. utrophin*
	8. talin	14. profilin	
	9. tensin	15. VASP	

Components marked with an asterisk are found only in the FAs of specialized cells types. 1. reviewed by Hynes 1992; 2. Woods & Couchman 1994; 3. Belkin & Smalheiser 1996; 4. Geiger 1979; 5. Lazarides & Burridge 1975, Wehland et al 1976; 6. Mittal et al 1987, Pavalko et al 1989; 7. Geiger 1979, Burridge & Feramisco 1980; 8. Burridge & Connell 1983; 9. Wilkins et al 1986, Wilkins et al 1987, Bockholt et al 1992; 10. Beckerle 1986; 11. Sadler et al 1992; 12. Sato et al 1992; 13. Bretscher 1981; 14. listed here because of the binding to VASP; Reinhard et al 1995a; 15. Reinhard et al 1992; 16. Belkin et al 1994; 17, 18. Kramarcy & Sealock 1990; 19. Belkin & Burridge 1995.

universal FA constituents, whereas others are cell-type specific (e.g. dystrophin, aciculin, etc). It has not been determined whether some of these components have structural or signaling functions.

Integrins and Other Transmembrane Components

The major transmembrane components in FAs are integrins, a large family of transmembrane heterodimers (Hynes 1992). Integrins are classified into several families according to their β subunits. Most β chains can pair with multiple α subunits, and a few α subunits pair with more than one β subunit (Hynes 1992). The α and β subunits each span the membrane once and have large extracellular and short intracellular domains (with the exception of the β_4 cytoplasmic domain, which is 1000 amino acids long and interacts with the intermediate filament cytoskeleton rather than with microfilaments). Most of the integrins involved in formation of focal adhesions are members of the β_1 or β_3 families. Pairing different α subunits with these β chains generates receptors for many different ECM proteins. A major requirement for FA development is that the ECM provides a ligand for integrins expressed on the cell surface. The type of integrin found within focal adhesions is generally dictated by the ECM to which cells adhere (Singer et al 1988, Dejana et al 1988, Fath et al 1989). However, a situation has been described in which a laminin-binding integrin, $\alpha_6\beta_1$, not known to interact with fibronectin, was detected in the FAs of cells adhering to fibronectin (Cattelino et al 1995). For cells grown in serum, vitronectin is typically the ECM protein adsorbed to the glass or plastic substratum, and cells expressing the vitronectin receptor $\alpha_v\beta_3$ will develop FAs containing this integrin. Probably the most frequently studied FAs are those made by cells plated on surfaces coated with fibronectin. These FAs involve the major fibronectin receptor $\alpha_5\beta_1$.

Both the vitronectin receptor $\alpha_v\beta_3$ and the fibronectin receptor $\alpha_5\beta_1$ recognize the sequence RGD within their respective ligands (Hynes 1992; Ruoslahti, this volume). Massia & Hubbell studied the adhesion of cells plated directly onto RGD peptides covalently attached to substrates. At very low densities, cells attach but do not spread. At higher densities, when the spacing between peptides reaches $\approx$ 440 nm, cells spread but do not form FAs. However, at still higher densities of RGD peptides, with a spacing of $\approx$ 140 nm or less, spreading is accompanied by development of FAs (Massia & Hubbell 1991). This requirement for a high density of integrin ligands suggests that the interacting integrins need to be clustered in order for FAs to form.

The development of FAs has been studied in cells adhering to cell-binding fragments of fibronectin that contain the RGD sequence. Whereas some investigators observed FAs on these domains of fibronectin (Singer et al 1987), others noted their failure to form (Beyth & Culp 1984, Izzard et al 1986, Woods et al 1986) or that the FAs were small and abnormal (Streeter & Rees 1987). Woods and coworkers found that assembly of FAs on the cell-binding fragment of fibronectin was stimulated by addition of a proteoglycan-binding domain of fibronectin (Woods et al 1986). Alternatively, the cells could be stimulated with phorbol esters to activate protein kinase C (PKC) (Woods & Couchman 1992). The reason why FA formation on the cell-binding fragment of fibronectin requires this additional stimulation has been a puzzle, given that simple clustering of integrins with antibodies seems to mimic their formation. A possible explanation has been suggested by Hotchin & Hall (1995) who have found that active Rho is required for FAs to form on intact fibronectin. In some situations, PKC stimulation has been reported to activate Rho (Hall 1994), which raises a question about the state of Rho in the experiments examining FA formation on fibronectin fragments. Was Rho relatively inactive in the cells that failed to form FAs on fibronectin fragments, and did it become activated by the proteoglycan-binding domains of fibronectin or by PKC stimulation? The role of Rho in clustering integrins into FAs is discussed below.

A transmembrane proteoglycan, syndecan 4, has been identified in the FAs of many different cell types adhering to various ECM proteins (Woods & Couchman 1994). In serum-starved cells, syndecan 4 is absent from FAs but can be recruited by stimulating with serum or by activating PKC (Baciu & Goetinck 1995). When cells are plated on fibronectin in the presence of serum, the localization of syndecan 4 to FAs increases with time and is more associated with mature FAs rather than with newly forming ones at the leading edge (Baciu & Goetinck 1995). The late recruitment of syndecan 4 to FAs suggests that it is not involved in the early signaling events associated with FA formation. It will be important to investigate the interactions of the syndecan cytoplasmic domain with structural and signaling components.

Dystroglycan (cranin, laminin-binding protein 120) (Smalheiser & Schwartz 1987, Douville et al 1988, Ibraghimov-Beskrovnaya et al 1992, 1993, Gee et al 1993, Smalheiser & Kim 1995) is a transmembrane receptor for laminin, unrelated to integrins that has been identified in the FAs of several cell types (Belkin & Smalheiser 1996). It has been studied mainly in the context of skeletal muscle, where it binds dystrophin and is in a complex with several dystrophin-associated glycoproteins (Ibraghimov-Beskrovnaya et al 1992, Ervasti & Campbell 1993a,b). However, dystroglycan is quite widely distributed (Smalheiser & Schwartz 1987, Ibraghimov-Beskrovnaya et al 1993). It remains to be determined whether its presence in FAs is due to a direct association with some of the major structural proteins of FAs, such as talin, vinculin and α-actinin, or whether it is associated indirectly via dystrophin or utrophin which, in turn, may be interacting with other components at these sites.

Integrin Cytoplasmic Domains

The cytoplasmic domains of integrins are critical for FA formation. Many of their structural features and interactions have been reviewed by Sastry & Horwitz (1993). The role of the cytoplasmic domains in targeting β_1 and β_3 integrins to FAs has been studied using mutations, deletions, and chimeric constructs. Deletion of the β_1 cytoplasmic domain blocks FA localization of integrin heterodimers (Solowska et al 1989, Marcantonio et al 1990, Hayashi et al 1990), whereas chimeric constructs containing the β_1 cytoplasmic domain fused to unrelated receptors target to FAs (Geiger et al 1992b, LaFlamme et al 1992). Consistent with this, deletion of α cytoplasmic domains drives integrin heterodimers to FAs (Briesewitz et al 1993, Ylanne et al 1993). These results indicate that the β subunit cytoplasmic domain is responsible for targeting an integrin to FAs, whereas the α subunit cytoplasmic sequence inhibits FA localization. Ligand binding relieves this inhibition (LaFlamme et al 1992). The relationship of the α and β subunit cytoplasmic domains has not been fully resolved, but in the nonligand-bound state, the two cytoplasmic domains appear to be in a close proximity (Briesewitz et al 1995). Following ligand binding to the extracellular domain, a conformational change is transmitted to the cytoplasmic domains, allowing the β cytoplasmic region to interact with focal adhesion components. Deletion of the α cytoplasmic domain or expression of the β cytoplasmic domain on its own as a chimeric protein appears to mimic the ligand-bound state.

A key question is whether ligand occupancy increases the affinity of interaction between integrins and cytoskeletal proteins. Miyamoto and colleagues compared the effects of integrin clustering with ligand occupancy (Miyamoto et al 1995a). Integrins were aggregated on the surface of cells using beads coated with non-inhibitory antibodies or with antibodies that block integrin function. Integrin clustering by non-inhibitory antibodies stimulates the colocalization of tensin and focal adhesion kinase (FAK), but not talin, vinculin, α-actinin, or actin. Clustering integrins with inhibitory antibodies (which mimic ligand binding), or with non-inhibitory antibodies in the presence of soluble ligands, stimulates co-clustering of talin, vinculin, α-actinin, and actin. These results argue that, while clustering alone will recruit some proteins, additional proteins are recruited in response to the combination of clustering and ligand occupancy. It may be that ligand occupancy is itself sufficient to induce this association between integrins and these cytoskeletal proteins, but this is difficult to resolve experimentally. Earlier work by this group, in which soluble ligands were seen to drive integrins into FAs formed by other integrins, strongly supports the idea that integrin occupancy is itself sufficient to induce cytoskeletal association with integrins (LaFlamme et al 1992).

Alternatively spliced cytoplasmic domains have been identified for both β_1 and β_3 integrin subunits. These differ in their ability to support FAs or target to these structures. In humans, four alternatively spliced isoforms of β_1 have been identified and are referred to as β_{1A} and β_{1B} (Balzac et al 1993, 1994), β_{1C} (Languino & Ruoslahti 1992), and β_{1D} (van der Flier et al 1995, Zhidkova et al 1995, Belkin et al 1996). Recent genomic analysis of the mouse β_1 integrin gene located only the exons for β_{1A} and β_{1D} (Baudoin et al 1996), which raises the possibility that the β_{1B} and β_{1C} variants emerged late in the evolution of the β_1 integrin family. β_{1A} is the predominant and original isoform identified. In humans, β_{1B} and β_{1C} appear to be relatively minor forms and are not found in FAs.

β_{1B} may be involved in regulation of adhesion and migration (Balzac et al 1993, 1994), whereas β_{1C} exerts a negative effect on growth (Meredith et al 1995, Fornaro et al 1995) (discussed below). In contrast to the low levels of expression of β_{1B} and β_{1C}, the major β_1 integrin isoform in striated muscles is β_{1D} (van der Flier et al 1995, Zhidkova et al 1995, Belkin et al 1996), which is found at all sites of actin filament-membrane attachment (Belkin et al 1996). Unlike the other isoforms, β_{1D} and β_{1A} share significant sequence homology in their alternatively spliced regions. β_{1D} is found in FAs, both in cultured myotubes and when transfected into nonmuscle cells (Belkin et al 1996). It has been speculated that β_{1D} interacts not only with the structural proteins of FAs but also with the specialized cytoskeletal proteins that line the muscle sarcolemma (Belkin et al 1996). β_1 integrins appear to associate with the dystrophin-containing lattice in muscle development (Lakonishok et al 1992). The forces transmitted across the sarcolemma of striated muscles are much greater than those experienced by the plasma membranes of most other cells. These high forces may have generated the need for a unique integrin isoform to provide stronger attachments to the membrane.

Cytoplasmic Proteins

Recently, the number of proteins identified at the FA cytoplasmic face has expanded greatly (Table 1). So too have the number of interactions (Table 2). It has become difficult, if not impossible, to represent this complexity adequately (Figure 1). Several components, as well as interactions, have not been included. The diagram conveys neither the dynamic nature of many of the interactions nor the stoichiometry of the components. The relative abundance of FA constituents varies considerably; FAK, paxillin, tensin, zyxin, and many signaling components are minor compared with vinculin and talin. We are confident that more FA proteins will be discovered and that the complexity of interactions will increase further. Space limitations restrict our discussion to a few topics relevant to the assembly of these structures. More information about FA structural proteins is given elsewhere (Jockusch et al 1995).

One striking observation is that a large number of FA proteins bind actin. This large number hints at redundancy in the linkage of actin filaments to integrins (and possibly to other transmembrane components in FAs). Some of the actin-binding proteins may function at different stages in the life of a FA. It has been suggested, for example, that talin and vinculin may be involved in FA formation, whereas α-actinin may be more important in maintaining or stabilizing microfilament attachment in mature FAs (DePasquale & Izzard 1987, 1991, Pavalko & Burridge 1991, Nuckolls et al 1992). Multiple linkages may also facilitate actin polymerization, permitting addition of actin monomers while

Table 2 Interactions between structural proteins in focal adhesions

	Integrin	Actin	α-actinin	Fimbrin	Profilin	Radixin	Talin	Tensin	VASP	Vinculin	Zyxin
Integrin	—	—	1	—	—	—	2	—	—	—	—
Actin	—	—	3	4	5	6	7	8	9	10	—
α-actinin	1	4	—	—	—	—	—	—	—	11	12
VASP	—	9	—	—	13	—	—	—	—	—	14
Vinculin	—	10	—	—	—	—	15	16	—	—	—

1. Otey et al 1990; 2. Horwitz et al 1986; 3. Burridge & Feramisco 1980; 4. Bretscher 1981; 5. Lassing & Lindberg 1985; 6. Tsukita & Hieda 1989; 7. Muguruma et al 1990, Kaufmann et al 1991; 8. Wilkins et al 1987, Lo et al 1994; 9. Reinhard et al 1992; 10. Menkel et al 1994, Johnson & Craig 1995a; 11. Belkin & Koteliansky 1987, Wachsstock et al 1987; 12. Crawford et al 1992; 13. Reinhard et al 1995a; 14. Reinhard et al 1995b; 15. Burridge & Mangeat 1984; 16. Wilkins et al 1987.

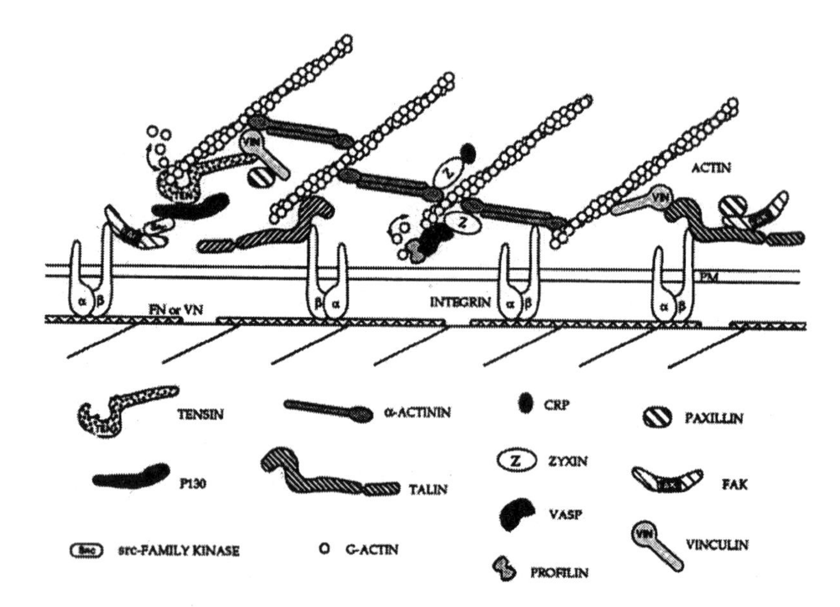

Figure 1 A focal adhesion. The model shows some of the interactions determined in vitro for proteins within focal adhesions formed on a surface coated with fibronectin (FN) or vitronectin (VN). For the sake of clarity, so that linking proteins can be visualized, integrins are shown dispersed rather than clustered. They are linked to actin microfilaments via talin, vinculin and/or α-actinin. Actin polymerization at these sites may involve either tensin or profilin, which interacts with VASP, a zyxin-binding protein. A few of the signaling components (FAK, paxillin, $p130^{cas}$, src family kinases) have also been included; however, several FA proteins and many of the interactions have been omitted. It should be noted that the diagram misrepresents the stoichiometry of components in focal adhesions, i.e. some such as vinculin and talin appear to be many times more abundant than others, such as FAK, paxillin, tensin, and zyxin.

simultaneously maintaining attachment to the membrane. It is noteworthy that some FA proteins, including tensin and talin, have multiple actin-binding sites (Lo et al 1994; L Hemmings et al, manuscript submitted). These may also permit subunit addition while maintaining attachment, as has been suggested for insertin, a fragment of tensin (Weigt et al 1992, Ruhnau et al 1989). Additionally, multiple binding sites may cross-link actin filaments and thus stabilize attachments to the membrane.

The attachment of actin filaments to integrins in FAs remains poorly understood. Both talin and α-actinin were shown to bind to integrins in vitro with relatively low affinity (Horwitz et al 1986, Otey et al 1990). A talin-binding site within the β_{1A} cytoplasmic domain corresponding to residues 780–789 was mapped using synthetic peptides (Tapley et al 1989). Work from this laboratory, however, indicates that sequences along most of the length of the cytoplasmic domain contribute to talin binding (K Simon, unpublished observations). In the case of α-actinin, peptide-binding studies indicate that two regions in the cytoplasmic domain are involved, residues 768–778 and 785–794 (Otey et al 1993). It has been difficult to demonstrate that these interactions also occur in vivo. Working with neutrophils, Pavalko and coworkers demonstrated coimmunoprecipitation of α-actinin and β_2 integrin (Pavalko & LaRoche 1993). In an attempt to get at the in vivo associations of proteins with integrin cytoplasmic domains, Lewis & Schwartz (1995) transfected full-length or truncated chicken β_1 integrins into mouse fibroblasts and examined the colocalization of proteins with these integrins after they had been clustered with antibody-coated beads. This provided a way to map regions of the β_1 cytoplasmic domain responsible for recruiting specific proteins. Deletion of the last 13 residues at the C-terminus inhibits colocalization of talin, actin, and FAK with clustered integrins. Interestingly, α-actinin continues to codistribute with integrins; this is blocked by deletion of an additional 15 residues. These experiments support the idea that talin binding to integrins is required for attachment of actin and that, on its own, α-actinin is insufficient to recruit actin filaments to clustered integrins.

The localization of α-actinin to FAs has been controversial. Although it is usually listed as a FA component, it is often not detected in FAs by immunofluorescence or electron microscopy even though it is prominent along the associated stress fibers (Zigmond et al 1979, Chen & Singer 1982, Samuelsson et al 1993). However, when labeled α-actinin is microinjected, it readily targets to FAs (Feramisco 1979, McKenna et al 1985, Sanger et al 1987). Pavalko and coworkers provided evidence that the inability to detect α-actinin in FAs by immunofluorescence results from inaccessibility of the α-actinin to antibodies (Pavalko et al 1995). FAs are dense structures and this may limit antibody accessibility.

Vinculin is one of the most abundant FA proteins, interacting with talin, α-actinin, tensin, and paxillin. For many years, the reported interaction of vinculin with actin was controversial. The demonstration by Jockusch's group that a vinculin tail domain construct binds actin (Menkel et al 1994) was difficult to reconcile with the inability of several groups to find significant actin binding with intact vinculin (e.g. Evans et al 1984, Otto 1986). A breakthrough came with the discovery by Johnson & Craig of an intramolecular, head-tail interaction within vinculin (Johnson & Craig 1994). This interaction masks binding sites for both talin and actin in vinculin's head and tail domains, respectively (Johnson & Craig 1994, 1995a). It seems likely that exposure of these sites is an important event in the assembly of FAs. It is notable that vinculin binds acidic phospholipids (Niggli et al 1986, Isenberg 1991), in particular PIP_2

(Fukami et al 1994), and that the binding site for PIP_2 is located in the hinge region between the head and tail domains (Johnson & Craig 1995b). These observations prompted examination of the effects of lipids, specifically PIP_2, on the conformation of vinculin. It was shown that the head-tail interaction of vinculin could be dissociated by acidic phospholipids and most effectively by PIP_2 (Weekes et al 1996, Gilmore & Burridge 1996). This exposed the binding sites for actin and talin, and a potential PKC phosphorylation site in the tail of vinculin. Vinculin immunoprecipitated from adherent cells contained PIP_2, but not other phospholipids, consistent with PIP_2 being the physiological effector of the conformational change (Gilmore & Burridge 1996).

Significantly, PIP_2 levels are elevated in response to adhesion in a Rho-dependent manner (Chong et al 1994). These observations are discussed in more detail below in the context of FA assembly.

Zyxin is one of the less abundant proteins present in FAs and adherens junctions (Beckerle 1986). This 82-kDa protein, expressed most prominently in fibroblasts and muscle cells, contains a proline-rich stretch at the C-terminus, involved in binding α-actinin (Crawford et al 1992), and three N-terminal LIM domains (Sadler et al 1992). LIM domains, each composed of two histidine- and cysteine-rich zinc fingers, mediate specific protein-protein interactions (Schmeichel & Beckerle 1994). They have been identified in a number of proteins, some of which are transcription factors, whereas others, such as the cysteine-rich proteins (CRP), are made up almost entirely of LIM domains. Zyxin does not contain an obvious nuclear targeting sequence, but truncation of the proline-rich N-terminus induces nuclear localization of zyxin (Nix & Beckerle 1995). In addition, a 44-amino acid sequence within the proline-rich region localizes zyxin to FAs. This raises the exciting possibility that zyxin shuttles between FAs and the nucleus in response to other factors interacting with this domain. The first LIM domain of zyxin binds to CRP, which is also localized in FAs (Sadler et al 1992, Crawford et al 1994). Expression of CRP proteins is tightly regulated during myogenesis and these proteins have been shown to potentiate muscle differentiation.

Another interesting interaction occurs between zyxin and VASP (M Reinhard et al 1995). VASP (vasodilator-stimulated phosphoprotein) was originally identified in platelets (Halbrugge & Walter 1989) as a major substrate of the cAMP-dependent protein kinase (PKA) and the cGMP-dependent protein kinase (PKG), both of which are stimulated by agents that inhibit platelet activation. Subsequently, VASP was identified in fibroblast FAs and along stress fibers, as well as in membrane ruffles (Reinhard et al 1992, Haffner et al 1995). VASP contains a polyproline-rich sequence that binds profilin (M Reinhard et al 1995), thus indicating a possible function in the control of actin polymerization.

This idea has received strong support from studies on the movement of *Listeria monocytogenes. Listeria* moves inside host cells by rapid polymerization of actin filaments at one pole of the bacterium (Pollard 1995). Both profilin and VASP have been identified at this site (Theriot et al 1994, Chakraborty et al 1995). VASP binds to the *Listeria* protein ActA (Pistor et al 1995), which is responsible for bacterial motility, and suggests that VASP provides a bridge between the bacterium and the microfilament system. Some sequence homology exists between ActA and zyxin, and these proteins even cross react immunologically (R Goldsteyn & E Friederich, personal communication). These observations suggest that zyxin, VASP, and profilin function as a complex, regulating the nucleation of actin polymerization at FAs and equivalent sites in cells. How important actin polymerization is in a stable FA is not clear. It is probably more important in the initial stages of FA development, during the assembly of stress fibers and FAs. This may explain why profilin is generally not detected in mature FAs (Buss et al 1992). An unanswered question concerns the role of VASP phosphorylation by the cyclic nucleotide-dependent protein kinases.

CONTRACTILITY: INFLUENCE ON THE FORMATION OF STRESS FIBERS AND FOCAL ADHESIONS

It had been suggested that stress fibers contract isometrically and that this isometric tension contributes to their assembly (Heath & Dunn 1978, Burridge 1981). Numerous observations contributed to this hypothesis. For example, stress fibers contain many of the contractile proteins found in muscle, including actin and myosin II, and these are arranged in a sarcomeric type of organization (Weber & Groeschel-Stewart 1974, Lazarides & Burridge 1975, Gordon 1978, Kreis & Birchmeier 1980). Although this pattern is suggestive of a contractile function, stress fibers are rarely seen to shorten in living cells. However, they are potentially contractile, as revealed in permeabilized cells exposed to ATP (Isenberg et al 1976, Kreis & Birchmeier 1980, Crowley & Horwitz 1995). It was argued that shortening usually cannot occur because of the strong adhesion at FAs to a rigid substratum, which generates isometric tension. A visible manifestation of tension exerted on the underlying substrate is seen when cells are grown on flexible silicone rubber surfaces. Fibroblasts and other cells that have prominent stress fibers visibly wrinkle these substrates (Harris et al 1980). Detachment of cells from the rubber results in the rapid loss of wrinkles. Evidence that isometric tension can generate bundles of microfilaments comes from studies using cytoplasmic extracts of *Physarum* induced to contract under isotonic or isometric conditions (Fleischer & Wohlfarth-Bottermann 1975).

Isometric contraction generates dense arrays of microfilaments, reminiscent of stress fibers. These structures rapidly disassemble upon the release of tension, just as stress fibers disassemble when cells are detached from a substrate. In contrast, when the cytoplasmic extracts are stimulated to contract under isotonic conditions, shortening is not accompanied by the development of large bundles of microfilaments (Fleischer & Wohlfarth-Bottermann 1975).

Much additional evidence supports the idea that isometric tension contributes to the formation of stress fibers. For example, fibroblasts contract collagen gels in which they are suspended. Free-floating gels may be contracted over several days to as little as 10% of their original size, and the fibroblasts in these gels lack stress fibers. If the gels are anchored, isometric tension is generated and the fibroblasts develop prominent stress fibers (Mochitate et al 1991, Tomasek et al 1992, Tomasek & Haaksma 1991, Grinnell 1994, Halliday & Tomasek 1995). Upon release of tension in anchored gels, there is a rapid contraction of the gel followed by disassembly of the stress fibers (Mochitate et al 1991, Tomasek et al 1992, Grinnell 1994). Application of tension to cells in culture also stimulates the formation of stress fibers (Franke et al 1984). When tension is applied to a localized region of the cell surface, a bundle of actin filaments is induced immediately subjacent to the site of applied tension (Kolega 1986).

Danowski demonstrated that when contractility is stimulated in fibroblasts by microtubule depolymerization, formation of stress fibers is induced (Danowski 1989). Under these conditions, FAs are also stimulated to form (B Geiger & A Bershadsky, personal communication; M Chrzanowska-Wodnicka, unpublished observations). How microtubule depolymerization stimulates contractility has not been established. One set of models has invoked "tensegrity" theories, in which microfilaments are viewed as tension-generating elements and microtubules as compression-resisting struts (Ingber 1993, Buxbaum & Heidemann 1988). The collapse of the compression-resisting elements, the microtubules, should lead to increased contractility. However, an alternative interpretation has been suggested by Kolodney & Elson, who have demonstrated enhanced myosin light chain (MLC) phosphorylation in response to microtubule depolymerization. This suggests that microtubule depolymerization directly activates actomyosin contractility, rather than simply removing an opposing structural element (Kolodney & Elson 1995).

Agents that inhibit contractility promote stress fiber and FA disassembly. For example, stress fibers are disrupted by microinjection of antibodies that bind to the myosin rod domain and inhibit myosin filament assembly (Honer et al 1988). Butanedione monoxime (BDM) inhibits myosin ATPase activity (Higuchi & Takemori 1989, McKillop et al 1994, Cramer & Mitchison 1995), decreases fibroblast contractility, and leads to the reversible disassembly of both stress fibers and FAs (Chrzanowska-Wodnicka & Burridge 1996). It has been long known that high cAMP levels disrupt stress fibers in many cell types. Pursuing the basis for this, Lamb and coworkers related the effects of high cAMP to decreased MLC phosphorylation (Lamb et al 1988). Elevated PKA activity disrupts stress fibers and leads to phosphorylation of the MLC kinase (MLCK), which has been correlated with decreased activity in vitro. MLCK is a key protein regulating nonmuscle actin-myosin interaction and contractility (see below). Further evidence for a role for MLCK in stress fiber stability was obtained by microinjection of antibodies against the MLCK into cells. The antibodies immunoprecipitated MLCK and led to the disassembly of stress fibers (Lamb et al 1988). The FAs were not examined in these experiments, but in a separate study, elevation of cAMP in the same cell type disrupted FAs (Turner et al 1989). Pharmacological agents that inhibit MLCK also inhibit contractility and lead to stress fiber and FA disassembly (Volberg et al 1994, Chrzanowska-Wodnicka & Burridge 1996). MLC phosphorylation can also be modulated by protein phosphatases. Elevating the level of protein phosphatase 1 (PP1) in fibroblasts by microinjection decreases MLC phosphorylation, which leads to a rapid loss of stress fibers (Fernandez et al 1990).

Taken together, this large body of evidence supports the idea that isometric tension induced by contractility drives the formation of stress fibers and FAs. Conversely, these structures disassemble when contractility is inhibited or tension is released. Some agents that stimulate formation or disassembly of FAs may exert their primary effects on contractility. For this reason, and because nonmuscle contractility can be modulated in many different ways, we discuss the regulation of contractility in the next section.

Regulation of Contractility

Much of what is known about the regulation of vertebrate nonmuscle contractility is derived from work on vertebrate smooth muscle. Many, but not all, of the regulatory components are present in both systems. In smooth muscle and nonmuscle cells, the dominant regulatory system involves phosphorylation of the myosin regulatory light chains (20 kDa) by a MLCK (Figure 2). MLCK phosphorylates both Ser-19 and Thr-18 (Sellers 1991, Tan et al 1992). Ser-19 is the major site of phosphorylation induced by agonists that stimulate light chain phosphorylation. Thr-18 is phosphorylated at a slower rate but becomes phosphorylated under conditions of maximal stimulation. The phosphorylation of the regulatory light chain increases the actin-activated ATPase activity of both smooth and nonmuscle myosin II (Adelstein & Conti 1975) and is required for myosin-driven actin filament movement in in vitro motility assays (Sellers 1991, Tan et al 1992). This regulation of myosin motor function may be the primary role of this light chain phosphorylation.

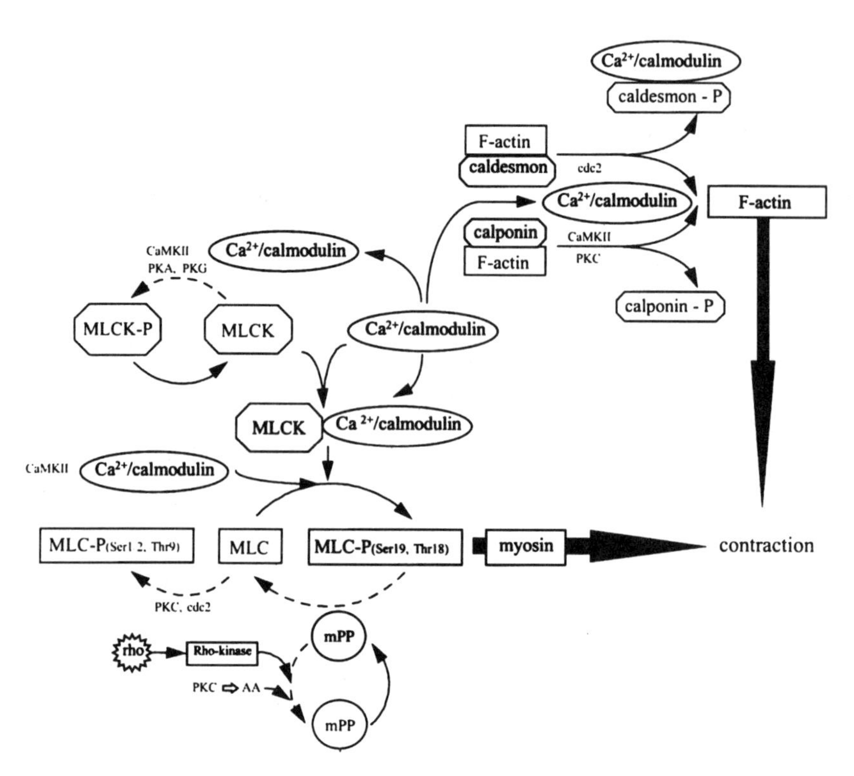

Figure 2 Regulation of contractility in smooth muscle and nonmuscle cells. The interaction of myosin with actin is regulated by phosphorylation of the regulatory myosin light chains (MLC) as well as by the actin-binding proteins, caldesmon and calponin. Active proteins are shaded, inactive are blank. MLC phosphorylation on Ser-19 and Thr-18 is catalyzed by the MLCK, which is activated by binding Ca^{2+}/calmodulin. The same sites on MLCs can also be phosphorylated by CaM kinase II (CaMKII). MLCK is inhibited by phosphorylation by PKA, PKG, or CaM kinase II. MLC dephosphorylation occurs by a myosin phosphatase (mPP), which is inhibited in smooth muscle by activated Rho. Rho stimulates Rho-kinase, which phosphorylates mPP, thereby inhibiting it activity (Kimura et al 1996). The myosin phosphatase is also inhibited by arachidonic acid, which may be generated by activation of PKC. Inhibitory phosphorylation of MLCs occurs primarily on Ser-1 and Ser-2, but also on Thr-9 in vitro, and is catalyzed by PKC and the mitotic kinase $p34^{cdc2}$ (cdc2). The actin-binding proteins, caldesmon and calponin, inhibit the interaction of myosin with actin in the absence of Ca^{2+}. In the presence of Ca^{2+}, the inhibition is relieved either by phosphorylation of caldesmon or calponin, or by binding Ca^{2+}/calmodulin. Phosphorylation of caldesmon by $p34^{cdc2}$ also releases caldesmon from actin.

Regulatory light chain phosphorylation also induces a conformational change in the myosin monomer (Craig et al 1983, Citi & Kendrick-Jones 1986, Ikebe & Reardon 1990) that might be involved in control of thick filament assembly. Nonmuscle myosin II can exist in two conformational states. In the first, the myosin rod is folded back on itself and interacts with the myosin heads. The second conformation is elongated, resembling the classical conformation of isolated skeletal muscle myosin (Citi & Kendrick-Jones 1986). In the folded conformation, myosin cannot assemble into filaments. The equilibrium between the two conformational states is shifted by phosphorylation of Ser-19. This favors the extended conformation and promotes assembly of myosin filaments. Phosphorylation of Thr-18 stabilizes the extended conformation and may also enhance filament formation (Ikebe 1989, Ikebe et al 1988). In lower organisms, such as *Dictyostelium*, phosphorylation of the heavy chains inhibits myosin filament assembly (Tan et al 1992). In vertebrate smooth and nonmuscle cells, heavy chain phosphorylation also occurs. Whether this phosphorylation regulates myosin filament assembly has not been determined, although it seems a likely possibility.

The MLCK from smooth muscle and nonmuscle cells is regulated by Ca^{2+}/calmodulin binding. In the resting state, in the absence of intracellular calcium, an auto-inhibitory domain blocks the kinase domain within MLCK (Tan et al 1992). Elevation of intracellular calcium results in calmodulin binding to MLCK. This causes a conformational shift of the auto-inhibitory domain, which activates the kinase. The interaction of MLCK with calmodulin is additionally regulated by phosphorylation. PKA, PKG, and CaM kinase II can all phosphorylate MLCK at a site that decreases its affinity for Ca^{2+}/calmodulin, thus inhibiting MLCK activity (Nishikawa et al 1984, Tansey et al 1994). A novel isoform of MLCK (embryonic MLCK) has recently been identified in nonmuscle cells (Gallagher et al 1995). This is the predominant or only MLCK expressed by some cells, whereas elsewhere this MLCK is coexpressed with the smooth muscle isoform. This embryonic MLCK is regulated by calcium, but little else is known about its regulation. The MLCK from *Dictyostelium* does not have a Ca^{2+}/calmodulin binding site, and its regulation remains to be elucidated (Tan et al 1992).

Dephosphorylation of MLCs counteracts the effect of their phosphorylation and provides a target for regulating myosin/actin interaction. The predominant myosin phosphatase (mPP), an isoform of PP1, consists of three subunits: a 37-kDa catalytic subunit, a 130-kDa subunit that targets the phosphatase to myosin, and a 20-kDa subunit (Alessi et al 1992, Shimizu et al 1994, Shirazi et al 1994). Dissociation of the catalytic subunit from the other components decreases its activity and is one mechanism by which agonists such as arachidonic acid may stimulate contractility (Gong et al 1992, Somlyo & Somlyo 1994). Another mechanism for regulating mPP activity is by phosphorylation of the 130-kDa myosin-binding subunit (Trinkle-Mulcahy et al 1995, Ichikawa et al 1996). In permeabilized smooth muscle cells, ATP-γ-S promoted contractility and elevated MLC phosphorylation at low calcium concentrations. This was due to decreased mPP activity and correlated with thiophosphorylation of the 130-kDa subunit of mPP (Trinkle-Mulcahy et al 1995). Studying purified mPP, Ichikawa and coworkers (1996) demonstrated the presence of a kinase associated with the phosphatase. This kinase phosphorylated the 130-kDa subunit, decreasing phosphatase activity. Kaibuchi and colleagues have shown that activated Rho stimulates Rho-kinase to phosphorylate the 130-kDa subunit of mPP, thereby inhibiting mPP activity (Kimura et al 1996) (see below).

The regulatory light chains of smooth and nonmuscle myosins are substrates for other kinases in addition to MLCK. CaM kinase II can phosphorylate Ser-19, the same site that is phosphorylated by MLCK (Tan et al 1992). However, because the affinity of CaM kinase II for Ca^{2+}/calmodulin is 100-fold less than that of MLCK, it may not be relevant under most conditions. PKC phosphorylates the regulatory light chain on three sites in vitro, Ser-1, Ser-2, and Thr-9, and the two serines have also been observed to be phosphorylated in vivo (Tan et al 1992). Phosphorylation at these sites decreases the actin-activated ATPase of myosin previously phosphorylated by MLCK and impairs myosin filament stability (reviewed in Tan et al 1992). The action of PKC is essentially antagonistic to MLCK with regard to myosin activity and contractility. $p34^{cdc2}$ also phosphorylates the regulatory light chains on Ser-1 and Ser-2 (Satterwhite et al 1992). The phosphorylation on Ser-1 and Ser-2 is high in mitotically arrested cells and remains high during mitosis until anaphase, when the contractile ring begins to develop. At this point there is decreased phosphorylation on Ser-1 and Ser-2 and increased phosphorylation on Ser-19 (Satterwhite et al 1992, Yamakita et al 1994).

These results have led to the hypothesis that light chain phosphorylation by $p34^{cdc2}$ inhibits contractility during mitosis until the onset of cytokinesis, at which point contractility is important for the contractile ring and the development of the cleavage furrow. The loss of stress fibers and FAs in mitotic cells is striking and correlates well with decreased contractility induced by $p34^{cdc2}$. This also supports the idea that the presence of stress fibers and FAs depends on contractility and that inhibition of contractility leads to disassembly of these structures.

Although light chain phosphorylation is a major regulatory target for smooth muscle and nonmuscle contractility, other regulatory proteins have also been identified (Figure 2). Caldesmon and calponin, two calcium-regulated proteins found on smooth muscle thin filaments, inhibit the actin-activated myosin ATPase in vitro in the absence of calcium. Nonmuscle cells contain a lower molecular weight isoform of caldesmon, which lacks a central domain of the smooth muscle caldesmon. Although the nonmuscle form of caldesmon is significantly smaller, its properties appear to be similar to the smooth muscle form. Caldesmons bind both actin and myosin (Matsumura & Yamashiro 1993). Together with tropomyosin, caldesmon binding to actin filaments increases their stability and resistance to severing by gelsolin in vitro (Ishikawa et al 1989a,b). In reconstituted systems, caldesmon has been shown to inhibit tropomyosin-stimulated, actin-activated myosin ATPase (Yamashiro et al 1995). This effect of caldesmon results from its binding to actin and is lost in the presence of Ca^{2+} and calmodulin, which release caldesmon from the actin-myosin complex and permit actin-myosin interaction. Caldesmon is phosphorylated by several serine/threonine kinases in vitro (Matsumura & Yamashiro 1993), and the phosphorylation of caldesmon has been reported to affect its interactions with the actomyosin system. In particular, phosphorylation of caldesmon by $p34^{cdc2}$ may be one of the events regulating reorganization of the actin cytoskeleton during mitosis (Yamashiro et al 1990, 1991). Phosphorylated caldesmon dissociates from microfilaments, which may promote microfilament severing by gelsolin, thereby contributing to the mitotic disassembly of stress fibers.

Calponin is a 35-kDa protein found primarily in smooth muscle, although it is reported also to be also in some nonmuscle cells (Takeuchi et al 1991). It regulates myosin ATPase activity in a phosphorylation-dependent fashion. When not phosphorylated, it binds to actin and inhibits the Mg ATPase activity of myosin in vitro. Phosphorylation of calponin by PKC or the Ca^{2+}/calmodulin-dependent protein kinase II relieves this inhibition (Winder & Walsh 1990). Calponin phosphorylation increases in parallel with MLC phosphorylation as smooth muscle is induced to contract by activating with carbachol or with okadaic acid, which inhibits the intrinsic phosphatases, consistent with a role of calponin phosphorylation in regulating the contractility of smooth muscle (Winder & Walsh 1990).

RHO

A major regulator of FAs and stress fibers is the small GTP-binding protein Rho (Hall 1994, Takai et al 1995), which is a member of the ras superfamily of proteins. In mammalian cells, it belongs to a subfamily that consists of Rho, rac, cdc42, TC10, and rhoG. Each of these proteins has several closely related isoforms (Hall 1994). In addition to other functions, these Rho-related proteins regulate the organization of the actin cytoskeleton. Whereas Rho regulates FAs and stress fibers (Ridley & Hall 1992), Rac promotes membrane ruffling (lamellipodia) (Ridley et al 1992), and cdc42 promotes the extension of microspikes (filopodia) (Nobes & Hall 1995). Here we concentrate on the effects and mechanism of Rho (Table 3).

The functions of Rho have been studied using two powerful tools: The first is the *Clostridium botulinum* C3 exotransferase (C3), which ADP-ribosylates and inactivates Rho (Narumiya et al 1988, Sekine et al 1989, Aktories & Hall 1989). The second is a recombinant, constitutively activated Rho, which is generated by mutating glycine-14 within rho's effector domain to valine (Val-14rho). This mutation is equivalent to the Val-12 activating mutation in ras, which decreases GTPase activity. The disruption of stress fibers by the C3 exotransferase first indicated a function for Rho in the regulation of actin filaments (Chardin et al 1989). This work was extended by Paterson and colleagues who introduced constitutively activated Rho into cells, stimulating their development of a contracted morphology (Paterson et al 1990). Microinjection of activated Rho into quiescent mouse fibroblasts lacking stress fibers and FAs rapidly induces these structures to form (Ridley & Hall 1992). The quiescent fibroblast system has

Table 3 Effects of Rho

	Cell type	References
Cytoskeletal effects		
Regulation of cell morphology	Vero cells	Chardin et al 1989
	Swiss 3T3 fibroblasts	Paterson et al 1990
Induction of stress fibers	Swiss 3T3 fibroblasts,	Ridley & Hall 1992
	mast cells,	Price et al 1995
	MDCK epithelia	Ridley et al 1995
Induction of actin polymerization	Mast cells	Norman et al 1994
Regulation of cell motility	Neutrophils,	Takaishi et al 1993
	mouse 308R	Takaishi et al 1994
	Swiss 3T3 cells	Stasia et al 1991
Inhibition of SF/HGF-induced motility	MDCK epithelia	Ridley et al 1995
Cell division,	*Xenopus* embryos,	Kishi et al 1993
cleavage furrow formation	sand dollar eggs	Mabuchi et al 1993
Enhancement of contractility	Smooth muscle	Hirata et al 1992
Ca^{+2} sensitization	Smooth muscle	Kokubu et al 1995
Neurite retraction and cell rounding	Neurons	Jalink et al 1994
MLC phosphorylation	Smooth muscle	Noda et al 1995
Regulation of integrins		
Inhibition of integrin-dependent adhesion to fibronectin	Monocytes	Aepfelbacher 1995
Integrin-dependent aggregation	Platelets,	Morii et al 1992
	B lymphocytes	Tominaga et al 1993
Chemoattractant-induced adhesion	Lymphocytes	Laudanna et al 1996

been employed by Rozengurt and colleagues to study the actions of growth factors and neuropeptides. In addition to inducing the reformation of stress fibers and FAs, these agents stimulate the tyrosine phosphorylation of several FA proteins, including FAK, paxillin, and $p130^{cas}$ (Zachary et al 1992, 1993, Kumagai et al 1993, Seckl & Rozengurt 1993, Seufferlein & Rozengurt 1994, 1995). Several of these agents (LPA, bombesin, endothelin, thrombin) mediate their effects via Rho (Ridley & Hall 1992, Jalink et al 1994, Rankin et al 1994). These factors act on receptors that are coupled to heterotrimeric G proteins, including $G_{\alpha 12}$ and $G_{\alpha 13}$ (Buhl et al 1995). The pathway(s) from these to Rho have not been elucidated, although there is evidence for a tyrosine kinase (PTK) upstream of Rho (Nobes et al 1995). Ultimately, these agents must converge to promote the active form of Rho with GTP bound.

Like other G proteins, Rho is active when it has GTP bound. The intrinsic GTPase activity of Rho hydrolyzes the bound GTP, rendering Rho inactive. The cycling between the GTP-bound and GDP-bound forms is directly regulated by several proteins. For the ras family of proteins, the low intrinsic rate of GTP hydrolysis is enhanced by GTPase-activating proteins (GAPs), whereas the exchange of GDP for GTP is enhanced by guanine nucleotide exchange factors (GEFs). Guanine nucleotide dissociation inhibitors (GDIs) slow the rate of GDP dissociation and thereby lock the G protein into the inactive state (Hall 1994). GEFs activate G proteins, whereas GAPs will turn them off, and GDIs maintain them in the inactive state. Several GAPs for Rho have been identified (Ridley 1995). The first, rhoGAP, has a higher activity for cdc42 (Ridley et al 1993). A second GAP for Rho, p190 (Lancaster et al 1994), prevents the formation of stress fibers and FAs when microinjected into cells (Ridley et al 1993). Recently, another form of p190 was discovered (Burbelo et al 1995). An unconventional myosin has been described with rhoGAP activity in the tail domain (J Reinhard et al 1995). One GEF for Rho is the product of the *Dbl* oncogene. The region of Dbl responsible for GEF activity (Dbl homology domain) has been identified in several other proteins including Vav, Ect2, Lbc, Ost, Tim, and Tiam (Ridley 1995), some of which may have GEF activity for Rho or other Rho family members. A rhoGDI has been identified that acts on Rho, Rac, and cdc42. Introduction of rhoGDI into cells inhibits both Rho- and rac-dependent reorganization of actin (Nishiyama et al 1994). The regulation of Rho and related proteins is clearly complex. Adding to this complexity is evidence that various members of the ras superfamily may interact via their GAPs, GEFs, and GDIs. For example, p120 RasGAP interacts with p190 RhoGAP (Settleman et al 1992). Much remains to be learned about the interplay of ras family regulatory proteins.

Downstream Targets of Rho

Several potential targets for Rho have been identified. Cell adhesion regulates the level of PIP_2. Cells put into suspension demonstrate decreased levels of PIP_2. This limits the response of suspended cells to growth factors that exert their growth stimulating effects via PIP_2 hydrolysis (McNamee et al 1993). PIP_2 synthesis from its precursor PIP, catalyzed by PIP 5-kinase, is stimulated by activated Rho (Chong et al 1994), and recently an interaction between Rho and one PIP 5-kinase isoform has been identified (Ren et al 1996, Kimura et al 1996). PIP_2 acts on several cytoskeletal proteins, promoting actin polymerization and enhancing FA formation (see below).

By analogy with ras, which activates the raf/MEK/MAP kinase pathway (Vojtek & Cooper 1995), it was expected that Rho would initiate a kinase cascade. The Rho family members, Rac and cdc42, have been shown to activate kinase pathways that parallel the ras/MAP kinase pathway (Coso et al 1995, Minden et al 1995). In addition, Rho was shown to activate one of the end-points of these Rac and cdc42 pathways (Hill et al 1995). During the past year, two distinct serine/threonine protein kinases that interact with Rho have been identified by several groups. The first was detected initially using GTP-Rho to screen a rat brain expression library (Leung et al 1995). The sequence of this Rho-binding kinase (ROKα) appears to be the truncated form of a bovine brain kinase (Rho-kinase) retained on a GST-Rho affinity column (Matsui et al 1996). A similar kinase was isolated from human platelets using labeled GTP-Rho in ligand overlay assays, cloned and given the name $p160^{ROCK}$ (Rho-associated, coiled-coiled-containing protein kinase) (Ishizaki et al 1996). $P160^{ROCK}$ has significant sequence homology with ROKα/Rho-kinase with the differences possibly reflecting tissue isoforms or species variation. This family of kinases shares homology particularly in its kinase domain with the myotonic dystrophy kinase. Experiments with the purified Rho-kinase indicate that it has little if any kinase activity for many cytoskeletal proteins, with the striking exception of the myosin-binding subunit of the myosin phosphatase (Matsui et al 1996, Kimura et al 1996). As discussed below, this suggests an important link between Rho pathways and contractility, because phosphorylation of this subunit of the myosin phosphatase inhibits the phosphatase activity (Trinkle-Mulcahy et al 1995, Ichikawa et al 1996, Kimura et al 1996). A different serine/threonine protein kinase, PKN, that interacts with Rho was also identified by two of the above groups (Watanabe et al 1996, Amano et al 1996). In addition, the yeast two-hybrid system was used to identify rhophilin, a protein sharing homology with PKN but lacking kinase activity (Watanabe et al 1995). The discovery of these Rho-stimulated serine/threonine kinases is exciting. Each may be the first step in distinct kinase cascades initiated by activated Rho.

Rho may also regulate the affinity of certain integrins for their ligands. Some integrins require activation for binding to their ligands, a phenomenon known as inside-out signaling (for details, see Ginsberg et al 1993, Schwartz et al 1995). The platelet integrin $\alpha_{IIb}\beta_3$ is held in an inactive conformation until platelets are stimulated by thrombin or other agents that trigger blood clot formation. Following platelet stimulation, activation of $\alpha_{IIb}\beta_3$ is a late event leading to platelet aggregation. This is blocked by the C3 exotransferase (Morii et al 1992). The inhibition of platelet aggregation by C3 may reflect rho's action on the cytoskeleton, or on the activation of $\alpha_{IIb}\beta_3$ because both are involved in platelet aggregation. Members of the leukocyte β_2 integrin family similarly require activation. Cell adhesion by $\alpha_L\beta_2$ is blocked by C3 (Tominaga et al 1993). This inhibition differs from the effects of cytochalasin, indicating that the action of Rho cannot be accounted for simply by its effects on the cytoskeleton and suggesting direct effects on integrin activation. The use of C3 has implicated Rho in the activation of β_1 and β_2 integrins in response to chemoattractants (Laudanna et al 1996). The original observations that C3 causes rounding up of cells are also consistent with C3 decreasing adhesion by blocking an effect of Rho on integrins (Chardin et al 1989, Paterson et al 1990). Together, these results suggest that Rho may activate integrins, increasing their interaction with ECM ligands, although the mechanism is far from clear.

Rho and Contractility

Rho is a potent stimulator of stress fibers and focal adhesions. As discussed earlier, contractility and isometric tension also contribute to stress fiber and FA formation. This leads to the question: Does Rho induce formation of these structures by stimulating contractility? Several lines of evidence support this idea. Many of the agents that activate Rho, such as LPA, endothelin, bombesin, and thrombin, are vasoconstrictors, which stimulate smooth muscle contraction. Direct evidence for an effect of Rho on contraction comes from studies with permeabilized smooth muscle. In this system, introduction of activated Rho enhances contractility at given calcium concentrations and lowers the calcium requirement for contractility (Hirata et al 1992). Rho has been implicated in several other contractile events (Table 3). For example, microinjection of C3 inhibits the development of the contractile ring, a transient bundle of microfilaments responsible for generating the cleavage furrow and cytokinesis (Mabuchi et al 1993, Kishi et al 1993). Some Rho-activating agents appear to stimulate contraction of cells other than smooth muscle. For example, thrombin stimulates neurite retraction (Jalink & Moolenaar 1992, Suidan et al 1992) and induces contraction of fibroblasts and endothelial cells (Giuliano & Taylor 1990, Goeckeler & Wysolmerski 1995). LPA stimulates contraction of chicken embryo fibroblasts and human endothelial cells (Kolodney & Elson 1993), as well as neurite retraction (Jalink et al 1994). The contractions induced by LPA and thrombin are Rho-mediated based on their inhibition by C3 (Jalink et al 1994). Induction of stress fibers and FAs by LPA treatment of quiescent Balb/c 3T3 cells is accompanied by an increase in contractility (Chrzanowska-Wodnicka & Burridge 1996). Several agents that inhibit actin-myosin interaction block the LPA-induced formation of stress fibers and FAs, and inhibit the LPA-stimulated tyrosine phosphorylation of FA components. These inhibitors also block FA and stress fiber assembly induced by microinjected Rho, indicating that contractility is downstream of Rho in the LPA pathway (Chrzanowska-Wodnicka & Burridge 1996).

How might Rho stimulate contractility? LPA and thrombin stimulate MLC phosphorylation (Kolodney & Elson 1993, Goeckeler & Wysolmerski 1995, Chrzanowska-Wodnicka & Burridge 1996). This increase in MLC phosphorylation appears to precede the appearance of stress fibers and FAs, which suggests that it is an early event in the assembly of these structures rather than a consequence of their formation (Chrzanowska-Wodnicka & Burridge 1996). MLC phosphorylation could result from stimulation of MLCK or inhibition of the MLC phosphatase. Several studies on smooth muscle point to an inhibition of phosphatase activity by Rho. In smooth muscle, it has been known for some time that excitatory agonists stimulate higher MLC phosphorylation than that induced by KCl depolarization, suggesting that these agonists potentiate phosphorylation not only by elevating intracellular calcium, but also by an additional mechanism (Somlyo & Somlyo 1994). Using permeabilized smooth muscle preparations, MLC phosphorylation and contractility are enhanced at fixed calcium concentrations by GTPγS, which suggests the involvement of a G protein (Kitazawa et al 1991). In this system, GTPγS inhibits MLC dephosphorylation, consistent with inhibition of a phosphatase (Kitazawa et al 1991). C3 blocks the GTPγS-enhanced phosphorylation of MLCs, implicating Rho in this process (Noda et al 1995). In addition, C3 promotes the dephosphorylation of MLCs in permeabilized smooth muscle, further suggesting that Rho inhibits the MLC phosphatase (Noda et al 1995).

Two recent observations suggest a pathway by which Rho may inactivate the MLC phosphatase (mPP). First, there is the discovery that phosphorylation of the 130-kDa-subunit of mPP inhibits its activity (Trinkle-Mulcahy et al 1995, Ichikawa et al 1996, Kimura et al 1996). Second, there is the identification of the Rho-stimulated serine/threonine kinases (Leung et al 1995, Amano et al 1996, Watanabe et al 1996, Ishizaki et al 1996, Matsui et al 1996). One of these, Rho-kinase, phosphorylates the 130-kDa subunit of mPP in vitro (Matsui et al 1996, Kimura et al 1996). It will be important to establish whether Rho-kinase alone or a Rho-activated kinase cascade is involved in the inactivation of mPP in vivo. Potential pathways by which Rho may regulate MLC phosphorylation and other cytoskeletal events such as actin polymerization are indicated in Figure 3.

A MODEL FOR FOCAL ADHESION ASSEMBLY

Based on the observations discussed above, we have proposed a model for how contractility can promote the assembly of FAs (Figure 4) (Chrzanowska-Wodnicka & Burridge 1996). In this model, cells in suspension have integrins dispersed over the cell surface (*A*). Clustering the integrins with antibodies (*B*) coclusters FAK and stimulates tyrosine phosphorylation. In the absence of ligand binding, the integrins are not associated with actin filaments. Cells adhering to ECM via integrins are shown in panels *C* and *D*. In *C*, the cells are quiescent. The integrins mediating adhesion to the underlying ECM are distributed over the ventral surface and are not clustered into FAs. However, because the integrins are bound to their ECM ligands, they are coupled physically to actin microfilaments via proteins such as talin or α-actinin. The actin filaments associated with these integrins are under little or no tension, because the myosin II is in its inactive conformation. In *D*, the cells have been stimulated to contract by agents that activate Rho. Rho activation leads to MLC phosphorylation. This turns on myosin function and also possibly induces a conformational change in the myosin, promoting myosin filament assembly. The resulting force generation will align the actin filaments. Alignment will also be promoted by the bundling action of the myosin filaments, which are very effective at cross-linking F-actin. The tension generated will be transmitted to the integrins in the membrane, leading to their aggregation. This clustering of integrins is a cornerstone in the assembly of FAs. Certainly, clustering of integrins from the outside, in combination with ligand occupancy, induces colocalization of many FA proteins (Miyamoto et al 1995a,b). In earlier models, it was assumed that simply plating cells on ECM proteins leads to the clustering of integrins in FAs. However, this does not occur if Rho is inactive (Hotchin & Hall 1995), which implies that the clustering is driven in some way from the inside rather than from interactions with the ECM on the outside. We have noted that integrins mediating adhesion will aggregate into FAs when contractility is stimulated and, conversely, will disperse from FAs when contractility is inhibited (Chrzanowska-Wodnicka & Burridge 1996). The generally low affinity of integrins for their ECM ligands (Hynes 1992) should result in relatively rapid rates of dissociation. These will permit remodeling of integrin-ECM interactions in response to tension.

In the formation of FAs induced by activated Rho, we envisage a cooperativity between the clustering of integrins driven by contractility and the effects of elevated PIP_2 (Figure 3). Activated Rho stimulates PIP 5-kinase (Chong

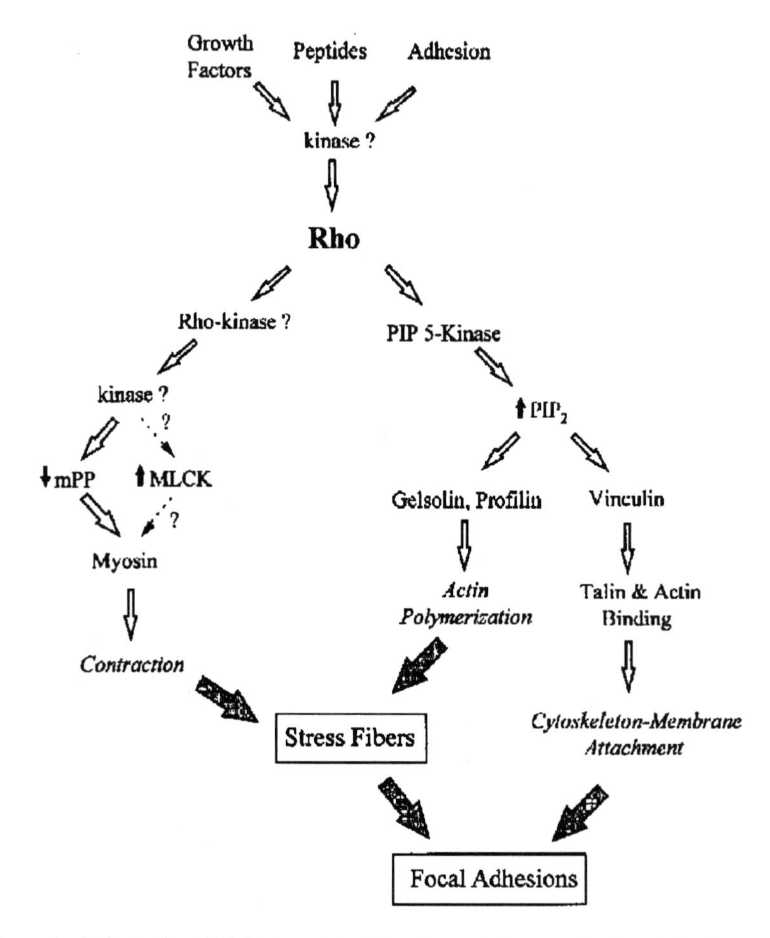

Figure 3 Pathways by which Rho regulates the actin cytoskeleton. Activation of Rho by growth factors, peptides, or adhesion is proposed to regulate the organization of the actin cytoskeleton via two synergistic pathways. In one, MLC phosphorylation is elevated via a kinase cascade that inhibits the myosin phosphatase (mPP). Stimulation of the MLCK (dotted line) may also occur, but there is no evidence for this. Light chain phosphorylation stimulates contractility. In turn, this leads to the bundling of actin filaments into stress fibers and clustering of integrins into focal adhesions. In the second, complementary pathway, Rho stimulates PIP 5-kinase, elevating PIP_2 levels. PIP_2 dissociates gelsolin and profilin from actin, promoting actin polymerization. PIP_2 also binds to vinculin, causing a conformational change that exposes binding sites for talin and actin, enhancing focal adhesion assembly.

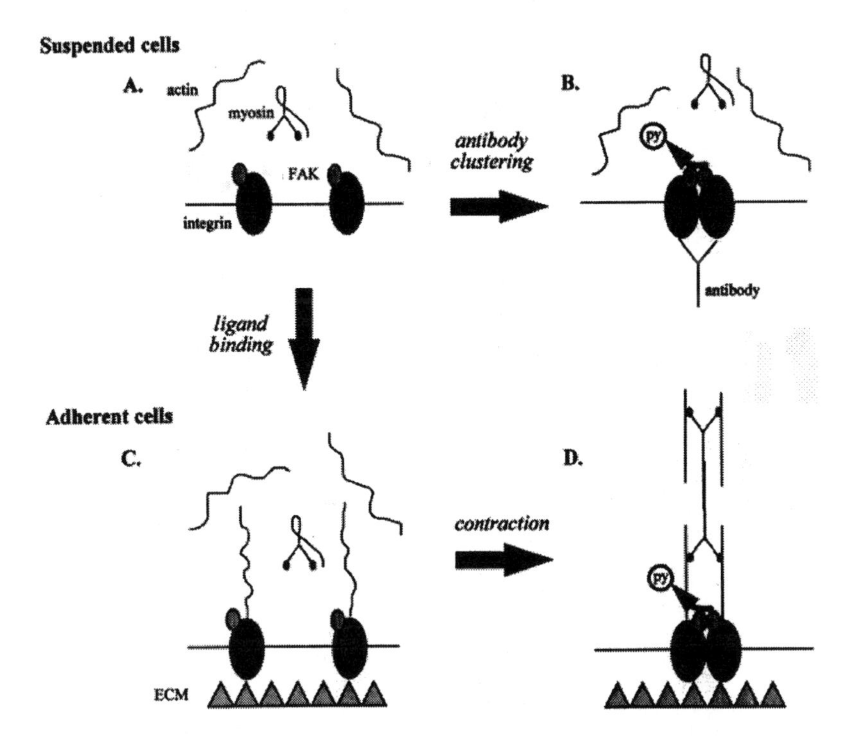

Figure 4 Model for contraction-induced formation of focal adhesions and stress fibers. In suspended cells (*A*), integrins are dispersed over the cell surface and not attached to actin filaments. Integrins can be clustered from the outside with antibodies (*B*), inducing FAK phosphorylation, but not attachment to actin, without ligand binding to the integrins. Adherent cells are shown in (*C*) and (*D*). Their integrins are bound to ECM ligands and because of this ligand occupancy become attached to actin filaments. However, in (*C*) the cells are quiescent, Rho is inactive, and the myosin is in the nonfunctional conformation. In (*D*) Rho activation stimulates contractility by elevating MLC phosphorylation. This leads to myosin filament formation and a force-generating interaction with actin. The tension exerted on the actin filaments bundles them into stress fibers and clusters the integrins to which they are attached. In turn, the integrin clustering stimulates FAK activity, triggering the signaling events associated with tyrosine phosphorylation in FAs.

et al 1994) and elevated PIP_2 stimulates dissociation of both profilin and gelsolin from actin, thereby promoting actin polymerization (Lassing & Lindberg 1985, Janmey & Stossel 1987). In addition, PIP_2 affects the conformation of vinculin, dissociating the intramolecular head/tail interaction, thereby exposing the talin- and actin-binding sites (Gilmore & Burridge 1996). PIP_2 has also been reported to enhance the binding of α-actinin to actin (Fukami et al 1992). Together these actions enhance stress fiber and FA formation. Inhibiting either side of the pathway illustrated in Figure 3 prevents assembly of FAs. Sequestering PIP_2 by microinjection of antibodies against PIP_2 into quiescent cells blocks the formation of FAs and stress fibers induced by Rho (Gilmore & Burridge 1996). Similarly, blocking contractility by a variety of agents with different modes of action inhibits the Rho-mediated assembly of stress fibers and FAs (Chrzanowska-Wodnicka & Burridge 1996).

FA assembly is accompanied by recruitment of signaling components. Evidence has been presented that this recruitment depends on tyrosine phosphorylation (Miyamoto et al 1995b). Signaling in FAs is further discussed below, but some points are relevant here in the context of assembly. One of the most prominent signaling events within FAs is the tyrosine phosphorylation that involves FAK. FAK may associate directly or indirectly with integrins and becomes activated as a result of integrin aggregation (Kornberg et al 1991, Miyamoto et al 1995a). The model predicts that anything promoting contractility and the consequent aggregation of integrins will stimulate FAK activation (if FAK is associated with the clustered integrins), which provides an explanation for the stimulation of FAK by the diverse group of agents that act via Rho. It is generally envisaged that autophosphorylation of FAK recruits src family kinases that, in turn, phosphorylate other components, generating binding sites for signaling proteins with SH2 domains. A multicomponent signaling complex is assembled that depends on the initial activation of FAK (Schaller & Parsons 1994, 1995, Miyamoto et al 1995b, Yamada & Miyamoto 1995, Richardson & Parsons 1995, Schwartz et al 1995). A surprising but consistent observation has been that disruption of the actin cytoskeleton, for example by treatment with cytochalasin D, inhibits the phosphorylation of FAK (Pelletier et al 1992, Haimovich et al 1993, Bockholt & Burridge 1993, Sinnett-Smith et al 1993, Seufferlein & Rozengurt 1994, Shattil et al 1994) as well as various downstream signaling events (Chen et al 1994). In the proposed model, FAK activation depends on the clustering of integrins from the inside by contractility. Disruption of the actin cytoskeleton would consequently block integrin aggregation driven by contractility and would account for the inhibition of FAK activity. Similarly, the model offers a potential explanation for the biphasic effects of PDGF on FAK phosphorylation. At low concentrations, PDGF stimulates FAK phosphorylation, but at higher concentrations it disrupts stress fibers and FAK phosphorylation is inhibited (Rankin & Rozengurt 1994).

This model also offers an explanation for the observation that cells could spread on a low density of integrin ligand but required higher densities in order to form FAs (Massia & Hubbell 1991). At low-ligand density, clustering the integrins would result in their dissociation from ligand. In the absence of bound extracellular ligands, the integrins would likely detach from the cytoskeleton. Therefore, a sufficiently high density of ligand on the substrate would be needed for the aggregated integrins to remain bound.

Does this model for FA formation also account for their assembly as suspended cells are plated on an ECM or during cell migration? We suspect that many aspects of the model are still relevant. Evidence shows that active Rho is required for the formation of FAs when cells are plated on ECM substrates (Hotchin & Hall 1995). FAs appear to form spontaneously when many cells are plated on ECM, but for most cells, Rho remains in an active state for several hours even in the absence of serum. When cells are plated on ECM substrates, inhibitors of contractility block the formation of FAs or result in very small FAs that appear not to have matured into their normal organizational state (M Chrzanowska-Wodnicka & K Burridge, unpublished observations). These results are consistent with a Rho-mediated contractile event contributing to the formation of FAs in cells spreading on ECM proteins.

Tyrosine Phosphorylation and Focal Adhesion Assembly

Several studies have shown that PTK inhibitors block the formation of FAs and stress fibers induced either by adhesion to an ECM-coated surface (Burridge et al 1992, Romer et al 1992, 1994) or by Rho activation (Seckl & Rozengurt 1993, Chrzanowska-Wodnicka & Burridge 1994, Ridley & Hall 1994). Because the major sites of tyrosine phosphorylation in cultured fibroblasts are FAs and because these inhibitors decrease the tyrosine phosphorylation of FA proteins such as FAK and paxillin, the interpretation of these data was that tyrosine phosphorylation of FA components contributes to the assembly of these structures (Burridge et al 1992, Romer et al 1994). However, several lines of evidence indicate that neither the kinase inhibited in these studies nor its substrates reside in FAs. Wilson and coworkers provided evidence that FAK is not required for the formation of FAs in vascular smooth muscle cells (Wilson et al 1995). Similarly, when transformation by temperature-sensitive mutants of v-src was shut off, FAs reassembled in the absence of tyrosine-phosphorylated FAK or paxillin (Fincham et al 1995). Cells from FAK$^-$ mice reveal prominent FAs that contain phosphotyrosine (Ilic et al 1995). Because FAK appears to be the major PTK in FAs in normal cells, this result is unexpected. In the absence of FAK, possibly other kinases are recruited to these structures, or the level or activity of src family kinases in FAs may increase. Alternatively, a homologue of FAK, such as CAKβ/PYK2/RAFTK (Avraham et al 1995, Lev et al 1995, Sasaki et al 1995) may be expressed in these cells, partially compensating for the lack of FAK. If there is compensation by another kinase, it is incomplete because FAK^- mice die as embryos (Ilic et al 1995). The prominent phosphotyrosine in the FAs of the FAK^- cells differs from results obtained using microinjection of a C-terminal FAK construct that contains the FAT sequence but lacks the kinase domain (Gilmore & Romer 1996). This FAK construct displaces endogenous FAK from FAs and abolishes detectable phosphotyrosine at these sites. Significantly, FAs can still be assembled in these cells, but they lack detectable FAK and phosphotyrosine. These experiments suggest that FAK is not required for FA assembly or maintenance, although it appears to be important for other signaling pathways (Gilmore & Romer 1996).

Inhibition of FAK tyrosine phosphorylation has also been observed in cells overexpressing $pp41/43^{FRNK}$ (FRNK), the naturally expressed C-terminal domain of FAK, which lacks the kinase domain (Richardson & Parsons 1996). Overexpression of FRNK also diminished the tyrosine phosphorylation of paxillin and tensin. These cells could still form focal adhesions but at a slower rate. Cell spreading on a fibronectin substrate was significantly retarded (Richardson & Parsons 1996).

These lines of evidence indicate that although tyrosine phosphorylation and FAK activity recruit signaling components to FAs, FAK does not have a role in the assembly of the structural components. If this is the case, where do the PTK inhibitors act to block the formation of FAs and stress fibers? Evidence indicates that they act both upstream and downstream of Rho activation (Ridley & Hall 1994, Chrzanowska-Wodnicka & Burridge 1994, Nobes et al 1995). In the model proposed for FA assembly (Figure 3), Rho initiates a kinase cascade that results in MLC phosphorylation, thereby stimulating contractility. We suspect that one of the kinases regulating contractility may be blocked by PTK inhibitors. Alternatively, a tyrosine kinase may be involved in Rho-stimulated PIP_2 synthesis.

SIGNALING IN FOCAL ADHESIONS

Integrin-mediated cell adhesion triggers tyrosine phosphorylation, ion fluxes, and lipid metabolism (Clark & Brugge 1995, Schwartz et al 1995). Together or individually, these affect many downstream pathways, influencing events such as gene expression, progression through the cell cycle, and apoptosis. Some of the integrin-mediated signal transduction pathways, such as tyrosine phosphorylation, have been shown to be initiated from FAs, whereas others may precede FA formation and be early responses to integrin occupancy or clustering. The

Table 4 Signaling proteins at the cytoplasmic face of focal adhesions

Tyrosine kinases	Heterotrimeric G proteins	Proteases
1. FAK	6. $\gamma 5$	10. calpain II
2. src		
3. csk	Adapters	Others
4. fyn	7. paxillin	11. PI3K
	8. $p130^{cas}$	12. LIP1
	9. Grb2	
Serine kinases		
5. PKC α, δ		

1. Schaller et al 1992; 2. Kaplan et al 1994; 3. Bergman et al 1995; 4. included because of its interaction with FAK; Cobb et al 1994; 5. Jaken et al 1989, Barry & Critchley 1994; 6. Hansen et al 1994; 7. Turner et al 1990; 8. Petch et al 1995; 9. M Kinch, unpublished observation; 10. Beckerle et al 1987; 11. Geiger et al 1992a; 12. Serra-Pages et al 1995.

idea that FAs are major sites of signal transduction is supported by the identification of multiple signaling proteins in FAs (Table 4). Numerous interactions occur between these and other signaling components involved in receptor PTK signaling pathways (Table 5). Using beads coated with antibodies or integrin ligands, Miyamoto and colleagues have identified a striking list of signaling proteins that associate with clustered integrins in a tyrosine phosphorylation-dependent manner (Miyamoto et al 1995b). Many have not been seen in FAs, but the bead technique may allow easier visualization of components that associate transiently with integrins. Alternatively, some of these components may be involved with endocytosis, a process also dependent on tyrosine phosphorylation (Shimo et al 1993, Salamero et al 1995). In the following discussion, we concentrate on signaling involving tyrosine phosphorylation.

Tyrosine Phosphorylation

Early work on tyrosine phosphorylation in FAs has been reviewed (Burridge et al 1988). A link between integrins and tyrosine phosphorylation was first demonstrated in platelets, where thrombin was shown to trigger tyrosine phosphorylation in an integrin-dependent fashion (Ferrell & Martin 1989, Golden & Brugge 1989). These observations were followed by studies with cells in culture, where it was also shown that integrin clustering or integrin-mediated adhesion stimulate tyrosine phosphorylation of a small set of proteins (Guan et al 1991, Kornberg et al 1991, Burridge et al 1992). Coincident with these studies was the discovery of a cytoplasmic PTK that localizes to FAs, the FA kinase ($pp125^{FAK}$ or FAK) (Schaller et al 1992, Hanks et al 1992). The discovery of FAK galvanized the field, and FAK was shown to be the most prominent of the tyrosine-phosphorylated proteins in cells responding to integrin clustering

Table 5 Interactions between signaling proteins and focal adhesion components

	Structural components														
	Integrin	Vinculin	Talin	α-actinin	FAK	src	fyn	csk	Tensin	Paxillin	p130	PI-3 K	C3G	crk	GRB2
FAK	1	—	2	—	—	3	4	5	—	6	7	8	—	—	9
src	—	—	—	—	3	—	—	10	—	11	12	13	—	—	—
fyn	—	—	—	—	4	—	—	14	—	—	15	16	—	—	—
csk	—	—	—	—	5	10	14	—	—	17	—	—	—	—	—
Tensin	—	—	—	—	—	—	—	—	—	—	18	—	—	—	—
Paxillin	19	20	—	—	6	11	—	17	—	—	—	—	—	21	—
p130	—	—	—	—	7	12	15	—	18	—	—	—	22	23	—
PI-3-K	—	—	—	24	8	13	16	—	—	—	—	—	—	—	—
C3G	—	—	—	—	—	—	—	—	—	—	22	—	—	25	—
crk	—	—	—	—	—	—	—	—	—	21	23	—	25	—	—
GRB2	—	—	—	—	9	—	—	—	—	—	—	—	—	—	—
Endonexin	26	—	—	—	—	—	—	—	—	—	—	—	—	—	—
Ilk	27	—	—	—	—	—	—	—	—	—	—	—	—	—	—

1. Schaller et al 1995; 2. Chen et al 1995; 3. Cobb et al 1994; Xing et al 1994; 4. Cobb et al 1994; 5. Birge et al 1993; 6. Turner & Miller 1994, Hildebrand et al 1995; 7. Polte & Hanks 1995; 8. Chen & Guan 1994b, Guinebault et al 1995; 9. Schlaepfer et al 1994; 10. Sabe et al 1994; 11. Weng et al 1993; 12. Reynolds et al 1989; 13. Fukui & Hanafusa 1989, Pleiman et al 1994; 14. reviewed in Sabe et al 1994; 15. L Petch, unpublished data; 16. Prasad et al 1993, Pleiman et al 1994; 17. Sabe et al 1994, Schaller & Parsons 1995; 18. Lo et al 1994; 19. Schaller et al 1995; 20. Turner et al 1990; 21. Birge et al 1993, Schaller & Parsons 1995; 22. Vuori et al 1996; 23. Birge et al 1993; 24. Shibasaki et al 1994; 25. Matsuda et al 1994, Tanaka et al 1994; 26. Shattil et al 1995; 27. Hannigan et al 1996.

or integrin-mediated adhesion (Burridge et al 1992, Hanks et al 1992, Guan & Shalloway 1992, Kornberg et al 1992, Lipfert et al 1992). Several other FA proteins were identified as also becoming tyrosine-phosphorylated in response to integrin-mediated adhesion. These included paxillin (Burridge et al 1992), tensin (Bockholt & Burridge 1993), PI 3-kinase (Chen & Guan 1994a), and $p130^{cas}$ (Petch et al 1995, Vuori & Ruoslahti 1995, Nojima et al 1995). Notably, the more major proteins in FAs—talin, vinculin, integrins—contain little or no phosphotyrosine. This contrasts with the situation in src-transformed cells in which low levels of phosphotyrosine are detected in these proteins.

FAK is a cytoplasmic PTK with three domains. The central catalytic domain separates an N-terminal domain, shown to bind to $\beta 1$ integrin cytoplasmic domain peptides (Schaller et al 1995), and a C-terminal domain, which contains a FA-targeting (FAT) sequence (Hildebrand et al 1993). This latter domain contains binding sites for at least two FA proteins, paxillin and talin (Turner & Miller 1994, Chen et al 1995, Hildebrand et al 1995, Tachibana et al 1995). Lewis & Schwartz (1995) found that association of FAK with clustered integrins in vivo depended on the most C-terminal 13 residues of the β_1 cytoplasmic domain. This was also the sequence required for co-clustering talin and paxillin with integrins, possibly indicating that one or other might contribute to FA targeting. In addition, FAK constructs containing the paxillin-binding sequence target to FAs, but fail to do so if this binding sequence is disrupted (Tachibana et al 1995). In these experiments, the interaction with talin was not examined, but it might also contribute to targeting FAK to FAs. These results support the idea that the interaction of FAK with paxillin or talin targets it to FAs. However, Miyamoto and colleagues found that simple aggregation of integrins was sufficient to co-cluster FAK but not talin or paxillin (Miyamoto et al 1995a). Co-clustering of talin required both aggregation and ligand occupancy, whereas co-clustering of paxillin required tyrosine phosphorylation as well.

The significance of the interaction of the N-terminal domain with integrin peptides is not obvious because this region does not appear to be involved in targeting FAK to FAs. Possibly this interaction functions in the activation of FAK, but most of the evidence indicates that clustering of integrins triggers FAK activation (Kornberg et al 1991, Defilippi et al 1994) and that ligand occupancy is not involved (Miyamoto et al 1995a). However, clustering of one integrin with an activating antibody failed to induce tyrosine phosphorylation (Pelletier et al 1995). Integrin aggregation may dimerize FAK, and thereby stimulate transphosphorylation and activation. This would be consistent with the model for activation of receptor PTKs (Ullrich & Schlessinger 1990).

For many receptor PTKs, ligand binding triggers autophosphorylation. The resulting phosphotyrosines in the receptor cytoplasmic domain bind SH2 domain-containing proteins. A similar situation is found for FAK. Several sites of tyrosine phosphorylation have been identified that recruit specific proteins via their SH2 domains (Schaller et al 1994, Schlaepfer et al 1994). The major site of tyrosine phosphorylation in FAK is Tyr-397, which arises through autophosphorylation (Schaller et al 1994). This generates a binding site for the SH2 domains of $pp60^{src}$ and $pp59^{fyn}$, both of which have been shown to complex with FAK in cells (Cobb et al 1994, Schlaepfer et al 1994, Xing et al 1994). These kinases phosphorylate additional sites in FAK and may stimulate FAK activity (Schlaepfer et al 1994, Calalb et al 1995). One of these sites, Tyr-925, complexes with the adapter protein Grb2 (Schlaepfer et al 1994). In normal cells, the association of Grb2 with FAK is dependent on integrin-mediated adhesion, but in v-src-transformed cells, FAK is constitutively tyrosine-phosphorylated and bound to Grb2. In these src-transformed cells, Grb2 is further associated with Sos, a Ras GTP/GDP exchange factor that activates the MAP kinase pathway (Schlaepfer et al 1994). This pathway may also operate in normal cells because MAP kinase activation has been observed in cells plated onto fibronectin in the absence of growth factors (Chen et al 1994, Schlaepfer et al 1994, Morino et al 1995, Zhu & Assoian 1995), and antibody-clustering of $\alpha_2\beta_1$ integrins in lymphocytes elevates the level of Ras-GTP (Kapron-Bras et al 1993). However, MAP kinase activation may be triggered by other signaling cascades initiated by integrin-mediated adhesion.

The association of $pp60^{src}$ and $pp59^{fyn}$ with FAK raises the question of which PTK is responsible for the tyrosine phosphorylation of specific FA proteins. Examining cells from transgenic mice that had had their *src, fyn,* or *yes* genes individually knocked out revealed that the patterns of tyrosine phosphorylation upon adhesion to fibronectin resembled that in control cells, with the exception of $p130^{cas}$ (Bockholt & Burridge 1995). The tyrosine phosphorylation level of $p130^{cas}$ was reduced in the *fyn*$^-$ and *yes*$^-$ cells, and to a particularly low level in the *src*$^-$ cells, suggesting that its phosphorylation is dependent on src and to a lesser extent on the other src family kinases. Although many of their characteristics resemble normal cells, the src$^-$ cells spread more slowly on fibronectin (Kaplan et al 1995). Besides FAK, one of the major tyrosine-phosphorylated proteins in FAs is paxillin (Turner 1994). Paxillin binds FAK and is a substrate for FAK in vitro and in vivo (Bellis et al 1995, Hildebrand et al 1995, Schaller & Parsons 1995). Autophosphorylation of FAK on Tyr-397 is required for paxillin phosphorylation (Schaller & Parsons 1995), which suggests that src or fyn may be involved, given that these two kinases bind to this site. Paxillin is a substrate in vitro not only for FAK, but also for src and csk (the kinase that regulates src activity by phosphorylation of its C-terminal tyrosine) (Schaller & Parsons 1995). All three kinases phosphorylate sites in vitro that are also phosphorylated in vivo (Schaller & Parsons 1995). These authors suggest a model whereby cell adhesion results in FAK autophosphorylation. This phosphorylation recruits src/fyn or csk by binding to the SH2 domains of these kinases. In turn, these phosphorylate paxillin on multiple sites. The absence of an obvious decrease in paxillin tyrosine phosphorylation in the src$^-$ cells may reflect redundancy in the system with fyn or csk replacing src (Bockholt & Burridge 1995).

Csk is an important regulator of the src family kinases. It inhibits their activity by phosphorylating the C-terminal tyrosine residue. This C-terminal phosphotyrosine then participates in an intramolecular association with its own SH2 domain, leading to a steric inhibition of the kinase domain (Cooper & Howell 1993). Overexpression of csk leads to its detection within FAs (Bergman et al 1995), suggesting that it is normally present but below the level of detection. Deletion of Csk results in src that is not phosphorylated on its C-terminal tyrosine (Tyr-527), and src now becomes detected in focal adhesions (Kaplan et al 1994). Similarly, mutating tyr-527 to phenylalanine, which prevents phosphorylation, results in src becoming prominently associated with FAs. Interestingly, the targeting of src to FAs occurs in the absence of the kinase domain and requires an intact SH3 domain (Kaplan et al 1994). The SH3 domain has been shown previously to bind to paxillin (Weng et al 1993). This may be the interaction that targets src to FAs, whereas the src SH2 interaction with FAK may activate src by displacing the intramolecular association between the SH2 domain and Tyr-527. A src construct lacking the kinase domain, but with intact SH2 and SH3 domains, elevates phosphotyrosine in FAs, particularly in FAK (Kaplan et al 1994). The mechanism for this is uncertain, but it may reflect the src SH2 domain protecting tyrosine-phosphorylated residues from tyrosine phosphatases. In csk$^-$ cells the level of phosphotyrosine in paxillin is increased, consistent with src family kinases being responsible (Nada et al 1994, Thomas et al 1995). The relationship of src and fyn has been investigated in the csk$^-$ background using double knockouts (Thomas et al 1995). These authors observed a decrease in the total level of phosphotyrosine in the src$^-$csk$^-$ cells but not the fyn$^-$csk$^-$ cells, suggesting that src is more critical than fyn to the elevation of phosphotyrosine in the csk$^-$ cells. However, the elevated phosphotyrosine in paxillin was partially decreased by the introduction of either the *src*$^-$ or *fyn*$^-$ mutation, consistent with both these kinases having a role in phosphorylating paxillin.

Another major tyrosine-phosphorylated protein in FAs is p130^{cas} (Petch et al 1995). The data from the src$^-$ cells suggest that much of the tyrosine phosphorylation of this protein is also from the activity of src (Bockholt & Burridge 1995). Like paxillin, this is a protein that interacts with other signaling proteins

(Table 5). The sequence of p130^{cas} reveals that it contains one SH3 domain and numerous tyrosines within the consensus-binding sites for the SH2 domains of src and crk (Sakai et al 1994). Both paxillin and p130^{cas} bind the adapter protein crk, which consists of SH2 and SH3 domains (Birge et al 1992, 1993, Schaller & Parsons 1995). The latter mediates interaction with two guanine nucleotide exchange factors for ras, Sos and C3G (Matsuda et al 1994, Tanaka et al 1994). The multiple interactions of paxillin and p130^{cas} with other signaling proteins has led to the general view that these proteins may be equivalent to the IRS-1 protein in insulin-signaling pathways. IRS-1 becomes tyrosine-phosphorylated on multiple sites in response to insulin binding to the insulin receptor. This provides a scaffold for downstream signaling components to bind via their SH2 domains. Similarly, following their tyrosine phosphorylation, paxillin and p130^{cas} may also act as scaffolds, recruiting other signaling proteins into a complex that activates separate or overlapping signaling pathways.

Very little is known about tyrosine phosphatases (PTPs) in FAs. The transmembrane PTP, LAR has been detected in the FAs of some cells, together with an interacting cytoplasmic protein, LIP (Serra-Pages et al 1995). However, in several cell types LAR has not been detectable in FAs, indicating that other PTPs may be present (G Schneider & K Burridge, unpublished observations).

Does Integrin-Mediated Adhesion Activate Rho?

The formation of stress fibers and FAs requires activated Rho (Ridley & Hall 1992). Cells plated on an ECM in the absence of growth factors also develop FAs, raising the possibility that integrin-mediated adhesion activates Rho, thus providing the stimulus for FA formation. Supporting this idea was the demonstration that PIP_2 levels increase when cells adhere to ECM McNamee et al 1993) and that this depended on Rho (Chong et al 1994). These results indicate that integrin engagement activates Rho. However, this has been challenged by the experiments of Hotchin & Hall (1995), who found that quiescent cells, with low levels of active Rho failed to form FAs or stress fibers when plated on fibronectin substrates for 30–45 min. Under these conditions, FAs formed rapidly upon Rho activation. Similar experiments were performed by Barry and coworkers who also observed a failure to form FAs and stress fibers 30 min after plating quiescent cells on fibronectin (Barry et al 1996). However, in this study FAs and stress fibers were seen 3 h after adhesion to fibronectin, indicating that this interaction does activate Rho, albeit inefficiently or with a delayed time course.

The activation of Rho in response to integrin-mediated adhesion may involve an arachidonic acid/leukotriene pathway. The spreading of cells on ECM substrates stimulates and requires activation of PKC (Chun & Jacobson 1992, 1993,

Vuori & Ruoslahti 1993) and release of arachidonic acid (Chun & Jacobson 1992). Blocking the metabolism of arachidonic acid to leukotrienes prevents HeLa cells spreading on a collagen matrix, and this block can be partially overcome by added leukotrienes (Chun & Jacobson 1992). Notably, leukotrienes have been reported to stimulate Rho (Peppelenbosch et al 1995). Together these results suggest a pathway in which leukotrienes may link integrin-mediated adhesion to Rho activation. Leukotriene synthesis and stimulation of Rho also provide a connection between Rac and Rho in cells responding to EGF (Peppelenbosch et al 1995). The activation of Rac is associated with extensive membrane ruffling (Ridley et al 1992), and this phenotype resembles the behavior of cells freshly plated on ECM. Indeed, the behavior of cells spreading on ECM resembles cells responding sequentially to cdc42, Rac, and finally Rho. Initial adhesion is associated with production of filopodia (the cdc42 phenotype) (Albrecht-Buehler 1976), followed by lamellipodia (the Rac phenotype) and, finally, when cells have become well spread, the development of stress fibers and FAs (the Rho phenotype). A hierarchy has been demonstrated previously, with cdc42 upstream of rac, which is upstream of Rho (Nobes & Hall 1995). It is conceivable, therefore, that this is also the pathway from integrin-mediated adhesion to activation of Rho, going via cdc42, Rac, and the synthesis of leukotrienes.

An alternative pathway by which Rho may become activated during adhesion involves Vav, a possible Rho GEF. In platelets, Vav becomes tyrosine-phosphorylated upon platelet activation by thrombin or collagen (Cichowski et al 1996). The stimulation by collagen occurs via the integrin $\alpha_2\beta_1$, indicating that in some circumstances Vav regulation is integrin-dependent. Vav binds the FA protein zyxin, contains one SH2 domain, two SH3 domains, a pleckstrin homology domain, and a dbl homology domain. The latter is associated with guanine nucleotide exchange. Whether Vav acts on Rho or another ras-related protein is controversial, but the phenotype of cells transformed by Vav suggests that it acts directly or indirectly on Rho (Khosravi-Far et al 1994). Vav is normally confined to hematopoietic cells, but a widely distributed isoform, Vav2, has been identified (Henske et al 1995). Vav itself becomes phosphorylated on serine and tyrosine residues and is probably involved in several signaling cascades (Gulbins et al 1993, Uddin et al 1995, Matsuguchi et al 1995). If Vav2 becomes phosphorylated by integrin engagement, this would suggest one way by which adhesion might activate Rho. Because Rho enhances adhesion and spreading, the activation of Rho by adhesion suggests a positive feedback loop. Positive feedback loops have previously been suggested to occur during adhesion and spreading in the context of PKC activation (Chun & Jacobson 1993).

Focal Adhesions and Growth Control

It has been long known that most normal cells in culture are anchorage dependent, i.e. they require attachment to a substrate in order to go through the cell cycle (Stoker et al 1968). The requirement for anchorage is abolished by many different oncogenes. Suspension of normal cells arrests not only DNA synthesis, but also inhibits RNA and protein synthesis (Otsuka & Moskowitz 1975, Benecke et al 1978, 1980, Farmer et al 1978, Ben-Ze'ev et al 1980). Cell cycle progression is blocked in G_1 by suspension (Otsuka & Moskowitz 1975). Clearly with normal cells, adhesion-dependent signals regulate many activities and permit cells to respond to growth factors. Several integrin-mediated signaling pathways have been implicated. Elevated intracellular pH has been associated with growth of normal cells, and this is regulated by integrins via the Na^+/H^+ antiporter (Schwartz et al 1989, 1990, 1991a,b, Ingber et al 1990). Integrins have also been linked to elevated intracellular Ca^{2+} (Pelletier et al 1992, Leavesley et al 1993, Schwartz et al 1993, Schwartz 1993, Schwartz & Denninghoff 1994). In endothelial cells, the integrin-dependent elevation in intracellular Ca^{2+} involves an integrin-associated protein (IAP) (Schwartz et al 1993). The mechanism by which integrins elevate intracellular Ca^{2+} has not been resolved, but intracellular Ca^{2+} has been associated with many mitogenic signals and pathways. Intracellular Ca^{2+} levels are frequently elevated by release from intracellular stores via a pathway that involves PIP_2 hydrolysis and the generation of inositol trisphosphate (IP_3) (Berridge 1993). As discussed above, PIP_2 levels are elevated by adhesion in a Rho-dependent manner (McNamee et al 1993, Chong et al 1994). One pathway from growth factor receptors involves activation of phospholipase C, which hydrolyzes PIP_2 to generate IP_3 and diacyl glycerol. These act as second messengers elevating intracellular Ca^{2+} and activating PKC. The decrease in PIP_2 that occurs with suspension of cells deprives PLC of its substrate and has been correlated with the inability of cells in suspension to respond to growth factors, even though the growth factors bind to their receptor PTKs (McNamee et al 1993).

Many lines of evidence point to a collaboration between the signals emanating from FAs and growth control. For example, in collagen gels, cells show high proliferation rates if the gel is anchored so that the cells generate isometric tension and develop stress fibers. In contrast, cells in free-floating gels do not develop isometric tension or stress fibers, and the cells become arrested for DNA synthesis and proliferation (Lin & Grinnell 1993, Grinnell 1994). With adherent cells in culture, several mitogens stimulate tyrosine phosphorylation of the same repertoire of FA proteins (FAK, paxillin, $p130^{cas}$, tensin) (Zachary et al 1992, 1993, Kumagai et al 1993, Sinnett-Smith et al 1993, Hordijk et al 1994, Seufferlein & Rozengurt 1994). Most, if not all, of these mitogens stimulate

Rho, contractility, and the formation of FAs in quiescent cells. Introduction of activated Rho into quiescent cells is itself sufficient to stimulate progression through the cell cycle and DNA synthesis (Olson et al 1995).

Adhesion to ECM activates the MAP kinase pathway, which is also triggered by many growth factors. MAP kinase activation may be via Grb2 binding to phosphorylated Tyr-925 in FAK (Schlaepfer et al 1994), but other adapter proteins interact with FA proteins, such as paxillin and $p130^{cas}$ (Table 5) and may provide parallel routes to MAP kinase activation. Adhesion alone is an insufficient stimulus to activate the MAP kinase pathway if Rho is not active (Hotchin & Hall 1995). It is not clear why growth factors and adhesion to ECM should both stimulate MAP kinase activation. The experimental paradigm of plating suspended cells on ECM substrates such as fibronectin is clearly artificial. One can imagine, however, that an equivalent situation is encountered normally when circulating cells adhere to a damaged blood vessel wall or when cells respread after mitosis. Similarly, during migration, the formation of new adhesions may trigger localized activation of MAP kinase pathways.

Is FAK involved in the synergy between integrin and growth factor signaling? Analysis of the role of FAK in growth control has not been possible with the FAK^- cells because they were transfected with activated p53 to promote growth (Ilic et al 1995). However, in an experimental model discussed above, FAK function was analyzed using microinjection of a C-terminal construct to displace endogenous FAK from FAs (Gilmore & Romer 1996). Displacement of FAK from FAs significantly decreased the number of cells entering DNA synthesis in response to stimulation with serum. This finding supports the idea that signaling from FAK contributes to and may be necessary for normal cells to respond to growth factors and to display anchorage-dependent growth. A similar conclusion was reached in experiments where epithelial cells were transfected with constitutively active FAK. This rendered the epithelial cells anchorage independent and even made them tumorigenic (Frisch et al 1996). In the model for FA formation discussed above, FAK activation results from integrin clustering induced by contractility within the microfilament system. This model predicts that disruption of the cytoskeleton should prevent FAK activation, which would resemble loss of adhesion in terms of growth control. This indeed is the case. Disruption of the cytoskeleton with cytochalasin mimics the effect of cell suspension, blocking cells that are anchorage dependent in G_1 (Bohmer et al 1996).

Direct interactions between integrins and growth factor-signaling components have been identified. In response to insulin stimulation, IRS-1 associates with the vitronectin receptor, $\alpha_v\beta_3$. Insulin stimulates growth of cells adhering to vitronectin more effectively than it stimulates cells adhering to other ECM proteins (Vuori & Ruoslahti 1994). The FGF receptor *flg* has been identified in FA complexes isolated on fibronectin-coated beads (Plopper et al 1995). In addition, numerous signaling components associated with receptor PTKs also associate with clustered integrins (Miyamoto et al 1995b).

The interplay between integrins and growth factors is illustrated by the growth and differentiation of myoblasts. Sastry and colleagues have found that ectopic expression of the α_5 integrin subunit inhibits differentiation, promoting proliferation. In contrast, ectopic expression of α_6 promotes differentiation and inhibits proliferation in this system (Sastry et al 1996). Chimeric constructs, in which the cytoplasmic domains were truncated or swapped, demonstrated that these effects depended on the α subunit cytoplasmic domains. The ectopic expression of α_5 changed the response of myoblasts to specific growth factors. For example, normally both TGFβ and bFGF inhibit proliferation and stimulate differentiation, but following transfection with the α_5 subunit, these growth factors inhibited differentiation. However, only bFGF stimulated proliferation, whereas TGFβ now induced apoptosis (Sastry et al 1996). Ectopic expression of α_5 in other situations also affects growth characteristics. For example, expression of α_5 in tumor cells deficient in this integrin subunit has been correlated with restoration of a normal phenotype and anchorage-dependent growth (Giancotti & Ruoslahti 1990). These effects of α_5 expression on tumor cell growth are complex. Ectopic expression of α_5 inhibits anchorage-independent growth, but it promotes growth on an appropriate ECM substrate (Varner et al 1995). These findings imply that a negative growth signal is transmitted by the unengaged $\alpha_5\beta_1$ integrin, but a positive proliferation signal is triggered upon ligand binding to this integrin. Clearly, different α subunits generate distinct signals, although very little is known about their nature. Integrin cytoplasmic domains also generate growth-related signals as evidenced by the inhibition of cell cycle progression and proliferation induced by ectopic expression of the β_{1C} alternatively spliced cytoplasmic isoform (Meredith et al 1995, Fornaro et al 1995). β_{1C} has been identified in platelets, megakaryocytes, and some other blood cells (Languino & Ruoslahti 1992). It is absent from growing endothelial cells but is induced when endothelial cells are growth arrested by exposure to tumor necrosis factor α (Fornaro et al 1995).

The expression of α_5 integrin in transformed cells restores more normal growth characteristics and the formation of FAs. As mentioned above, in normal cells the ability to form FAs correlates with anchorage-dependent growth, and disruption of FAs leads to cell cycle arrest. However, a paradoxical situation is seen in cells in which the levels of vinculin or α-actinin are decreased, for example, by antisense cDNA constructs. Depressing the level of either protein results in decreased FAs, increased motility, and anchorage-independent growth (Rodriguez Fernandez et al 1992, 1993, Gluck et al 1993, Gluck & Ben-Ze'ev

1994). Cells with decreased levels of vinculin or α-actinin form tumors when injected into nude mice. Moreover, several transformed cells or tumors show decreased expression of vinculin or α-actinin. Elevating the levels of these proteins restores a more normal phenotype, including anchorage-dependent growth and suppression of tumorigenicity (Gluck et al 1993, Rodriguez Fernandez et al 1992). A possible explanation for these observations relates to the fact that both vinculin and α-actinin are major PIP_2-binding proteins (Fukami et al 1994). Levels of PIP_2 are reduced when normal cells are suspended, and this has been correlated with growth arrest of cells in suspension (McNamee et al 1993). If vinculin and α-actinin are major sinks for cytoplasmic PIP_2, it is conceivable that reducing their levels increases PIP_2 availability to hydrolysis by PLC and thus permits suspended cells to respond to growth factor stimulation by PIP_2 hydrolysis. It will be important to determine whether vinculin and α-actinin buffer PIP_2 levels. If incorporation of vinculin and α-actinin into FAs releases their bound PIP_2, this would also increase the PIP_2 available for hydrolysis and provide another link between focal adhesions and mitogenic pathways.

Apoptosis

Cooperativity between integrin-mediated adhesion and growth factors is also seen in relation to apoptosis. Loss of adhesion blocks progression through the cell cycle for normal anchorage-dependent cells. But more than that, for some cells it also induces apoptosis (programmed cell death), particularly in the absence of growth factors (Meredith et al 1993, Frisch & Francis 1994, Re et al 1994). Engagement of different integrins prevents apoptosis in different cell types. Thus the α_5 integrin subunit increases survival of CHO cells adhering to fibronectin in the absence of growth factors (Zhang et al 1995). Similarly, in muscle cells expressing ectopic α_5, survival is increased in the presence of bFGF, whereas untransfected cells become apoptotic (Sastry et al 1996). β_1 integrins prevent apoptosis in mammary epithelial cells adhering to basement membranes (Boudreau et al 1995). In contrast, disruption of $\alpha_v\beta_3$ interactions promote apoptosis in colon carcinoma cells, endothelial cells, and melanoma cells (Bates et al 1994, Brooks et al 1994, Montgomery et al 1994). The term *anoikis* (Greek for homelessness) has been coined to describe the apoptosis that results from loss of normal adhesion to ECM (Frisch & Francis 1994).

Is there a role for FAK in integrin-mediated survival versus apoptosis? Meredith and coworkers demonstrated that apoptosis of suspended cells could be prevented by elevating tyrosine phosphorylation, consistent with, but not proving, a role for FAK in this process (Meredith et al 1993). The involvement of FAK in apoptosis/survival has been approached recently in several ways. Reducing the level of FAK in several tumor cell lines with antisense oligonucleotides induced detachment from the substrate and a high incidence of apoptosis (Xu et al 1996). Interestingly, normal cells did not respond by detachment or apoptosis, which led these investigators to propose targeting FAK with antisense oligonucleotides as a potential strategy for treating tumors. However, others have found that perturbations of FAK in normal cells lead to apoptosis. Apoptosis was induced in chicken embryo fibroblasts by microinjection of either an antibody against FAK or integrin cytoplasmic domain peptides that correspond to the region shown to bind FAK in vitro (J Hungerford et al, manuscript submitted). Epithelial cell lines that go into apoptosis upon suspension were rescued by transfection with constitutively activated forms of FAK (Frisch et al 1996). Rescue from apoptosis did not occur if the cells were transfected with a kinase-dead mutant of FAK or with a mutant in which the major autophosphorylation site in FAK had been deleted. Together these results provide strong support for the idea that a major function for FAK is to signal to a cell that it is in contact with ECM. In the absence of this signal, cells will go into apoptosis. With tumor cells, growth becomes anchorage independent owing to activation of oncogenes that short circuit the FAK pathway. It is interesting that many invasive tumors display elevated expression of FAK, which may contribute to their anchorage-independent growth (Owens et al 1995).

TELEOLOGY

Why do cells in culture form FAs? The existence of FAs and stress fibers is closely correlated. They have little to do with cell migration; they are absent from many migratory cells and prominent in the least motile cells (Burridge 1981). In nonmuscle tissues, large bundles of actin filaments, which are contractile, are common in two circumstances: cytokinesis and wound contracture (Gabbiani et al 1973). We suspect that the origin of stress fibers and FAs reflects the response of cells in culture to an apparent wound environment. Many of the agents that activate Rho are released by platelets or other cells in response to wounding (e.g. LPA, thrombin, endothelin). Some of these factors are present in serum; thus cells cultured with serum respond as if in a wound. These factors stimulate Rho-mediated contraction. However, because the underlying substrate is rigid and the adhesions to it are strong, contraction generates isometric tension, leading to the development of stress fibers and FAs. For most tissues, wounding is also a potent mitogenic signal. Cells closing a wound proliferate and normally this ceases once the wound has been repaired. A parallel situation is seen experimentally with cells in collagen gels. If contraction of the gel is physically restricted, the cells develop isometric tension and prominent stress fibers (Tomasek et al 1992, Grinnell 1994). Under these conditions, the cells respond to mitogens by proliferation. Following release of tension, stress fiber disassembly occurs and is accompanied by a return to quiescence and a loss of responsiveness to growth factors (Grinnell 1994). Similarly, when normal cells in culture are detached from a substrate or are prevented from spreading and generating tension, they become refractory to growth factors. Not only does the formation of FAs reflect the response of cells in culture to a wound environment, but the signaling at FAs may also relate to this wound response. The clustering of integrins stimulated by contractility activates FAK and triggers signaling cascades that synergize with growth factors. Together these responses combine in tissues to close a wound both by contraction and cell proliferation. In tissue culture, they contribute to the appearance of stress fibers and FAs and to anchorage-dependent growth.

PERSPECTIVES FOR THE FUTURE

The last few years have witnessed considerable progress in the field of focal adhesion research. Particularly rapid advances have been made identifying downstream pathways from Rho. Because of this, it seems likely that the steps from the activation of Rho to the assembly of focal adhesions will be elucidated in the near future. Many of the structural interactions that occur in FAs may take longer to resolve. We anticipate that FAs will continue to serve as a useful model for studying the signaling that is initiated in response to adhesion to ECM. Many of the presumptive signaling components in FAs, such as paxillin and $p130^{cas}$, have yet to be assigned a function. Experimental strategies aimed at determining their functions are being actively pursued. Perhaps the greater challenge will be to unravel the web of signaling pathways that emerge from FAs. The interactions between integrin-mediated signals and the signals generated in response to growth factors promise to be fertile grounds, relevant to understanding not only anchorage-dependent growth, but also differentiation and apoptosis.

ACKNOWLEDGMENTS

We thank many colleagues for sending us preprints of their work. However, we apologize to the many whose work we have failed to cite because of space limitations. A search generated over 1500 papers in this area in the past five years. We are indebted to Alexey Belkin, Andrew Gilmore, Lew Romer, Patricia Saling, and Mike Schaller for their comments and suggestions on this review. The authors were supported by National Institutes of Health grants GM29860 and HL45100.

Literature Cited

Adams JC, Watt FM. 1993. Regulation of development and differentiation by the extracellular matrix. *Development* 117:1183–98

Adelstein RS, Conti MA. 1975. Phosphorylation of platelet myosin increases actin-activated myosin ATPase activity. *Nature* 256:597–98

Aepfelbacher M. 1995. ADP-ribosylation of Rho enhances adhesion of U937 cells to fibronectin via the $\alpha_5\beta_1$ integrin receptor. *FEBS Lett.* 363:78–80

Aktories K, Hall A. 1989. Botulinum ADP-ribosyltransferase C3: a new tool to study low molecular weight GTP-binding proteins. *Trends Pharmacol. Sci.* 10:415–18

Albrecht-Buehler G. 1976. Filopodia of spreading 3T3 cells: Do they have a substrate-exploring function? *J. Cell Biol.* 69:275–86

Alessi D, MacDougall LK, Sola MM, Ikebe M, Cohen P. 1992. The control of protein phosphatase 1 by targeting subunits. The major myosin phosphatase in avian smooth muscle is a novel form of protein phosphatase 1. *Eur. J. Biochem.* 210:1023–35

Amano M, Mukai H, Ono Y, Chihara K, Matsui T, et al. 1996. Identification of a putative target for rho as the serine-threonine kinase protein kinase N. *Science* 271:648–50

Avraham S, London R, Fu Y, Ota S, Hiregowdara D, et al. 1995. Identification and characterization of a novel related adhesion focal tyrosine kinase (RAFTK) from megakaryocytes and brain. *J. Biol. Chem.* 270:27742–51

Baciu PC, Goetinck PF. 1995. Protein kinase C regulates the recruitment of syndecan-4 into focal contacts. *Mol. Biol. Cell* 6:1503–13

Balzac F, Belkin AM, Koteliansky VE, Balabanov YV, Altruda F, et al. 1993. Expression and functional analysis of a cytoplasmic domain variant of the beta-1 integrin subunit. *J. Cell Biol.* 121:171–78

Balzac F, Retta SF, Albini A, Melchiorri A, Koteliansky VE, et al. 1994. Expression of beta-1B integrin isoform in CHO cells results in a dominant negative effect on cell adhesion and motility. *J. Cell Biol.* 127:557–65

Barry ST, Critchley DR. 1994. The RhoA-dependent assembly of focal adhesions in Swiss 3T3 cells is associated with increased tyrosine phosphorylation and the recruitment of both $pp125^{FAK}$ and protein kinase C-delta to focal adhesions. *J. Cell Sci.* 107:2033–45

Barry ST, Ridley AJ, Flinn HM, Humphries MJ, Critchley DR. 1996. Integrin-mediated assembly of focal adhesions and actin stress fibers is rho-dependent. *Cell Adhesion Commun.* In press

Bates RC, Buret A, van Helden DF, Horton MA, Burns GF. 1994. Apoptosis induced by inhibiton of intercellular contact. *J. Cell Biol.* 125:403–15

Baudoin C, van der Flier A, Borradori L, Sonnenberg A. 1996. Genomic organization of the mouse β1D gene: conservation of the β1D but not of the β1B and β1C integrin splice variants. *Cell Adhesion Commun.* In press

Beckerle MC. 1986. Identification of a new protein localized at sites of cell-substrate adhesion. *J. Cell Biol.* 103:1679–87

Beckerle MC, Burridge K, DeMartino GN, Croall DE. 1987. Colocalization of calcium-dependent protease II and one of its substrates at sites of cell adhesion. *Cell* 51:569–77

Belkin AM, Burridge K. 1995. Localization of utrophin and aciculin at sites of cell-matrix and cell-cell adhesion in cultured cells. *Exp. Cell. Res.* 221:132–40

Belkin AM, Klimanskaya IV, Lukashev ME, Lilley K, Critchley DR, Koteliansky VE. 1994. A novel phosphoglucomutase-related protein is concentrated in adherens junctions of muscle and nonmuscle cells. *J. Cell Sci.* 107:159–73

Belkin AM, Koteliansky VE. 1987. Interaction of iodinated vinculin, metavinculin and α-actinin with cytoskeletal proteins. *FEBS Lett.* 220:291–94

Belkin AM, Smalheiser NR. 1996. Localization of cranin (dystroglycan) at sites of cell-matrix and cell-cell contact: Recruitment to focal adhesions is dependent upon extracellular ligands. *Cell Adhesion Commun.* In press

Belkin AM, Zhidkova NI, Balzac F, Altruda F, Tomatis D, et al. 1996. Beta-1D integrin displaces the beta-1A isoform in striated muscles—localization at junctional structures and signaling potential in nonmuscle cells. *J. Cell Biol.* 132:211–26

Bellis SL, Miller JT, Turner CE. 1995. Characterization of tyrosine phosphorylation of paxillin in vitro by focal adhesion kinase. *J. Biol. Chem.* 270:17437–41

Benecke BJ, Ben-Ze'ev A, Penman S. 1978. The control of mRNA production, translation and turnover in suspended and reattached anchorage-dependent fibroblasts. *Cell* 14:931–39

Benecke BJ, Ben-Ze'ev A, Penman S. 1980. The regulation of RNA metabolism in suspended and reattached anchorage-dependent 3T6 fibroblasts. *J. Cell. Physiol.* 103:247–54

Ben-Ze'ev A, Farmer SR, Penman S. 1980.

Protein synthesis requires cell-surface contact while nuclear events respond to cell shape in anchorage-dependent fibroblasts. *Cell* 21:365–72

Bergman M, Joukov V, Virtanen I, Alitalo K. 1995. Overexpressed Csk tyrosine kinase is localized in focal adhesions, causes reorganization of $\alpha_v\beta_5$ integrin, and interferes with HeLa cell spreading. *Mol. Cell. Biol.* 15:711–22

Berridge MJ. 1993. Inositol trisphosphate and calcium signaling. *Nature* 361:315–25

Beyth RJ, Culp LA. 1984. Complementary adhesive responses of human skin fibroblasts to the cell-binding domain of fibronectin and the heparin sulfate-binding protein, platelet factor-4. *Exp. Cell. Res.* 155:537–48

Birge RB, Fajardo JE, Mayer BJ, Hanafusa H. 1992. Tyrosine-phosphorylated epidermal growth factor receptor and cellular p130 provide high affinity binding substrates to analyze Crk-phosphotyrosine-dependent interactions in vitro. *J. Biol. Chem.* 267:10588–95

Birge RB, Fajardo JE, Reichman C, Shoelson SE, Songyang Z, et al. 1993. Identification and characterization of a high-affinity interaction between v-Crk and tyrosine-phosphorylated paxillin in CT10-transformed fibroblasts. *Mol. Cell. Biol.* 13:4648–56

Bockholt SM, Burridge K. 1993. Cell spreading on extracellular matrix proteins induces tyrosine phosphorylation of tensin. *J. Biol. Chem.* 268:14565–67

Bockholt SM, Burridge K. 1995. An examination of focal adhesion formation and tyrosine phosphorylation in fibroblasts isolated from src(−), fyn(−), and yes(−) mice. *Cell Adhes. Commun.* 3:91–100

Bockholt SM, Otey CA, Glenney JR Jr, Burridge K. 1992. Localization of a 215-kDa tyrosine-phosphorylated protein that cross-reacts with tensin antibodies. *Exp. Cell. Res.* 203:39–46

Bohmer RM, Scharf E, Assoian RK. 1996. Cytoskeletal integrity is required throughout the mitogen stimulation phase of the cell cycle and mediates the anchorage-dependent expression of cyclin D1. *Mol. Biol. Cell* 7:101–11

Boudreau N, Sympson CJ, Werb Z, Bissell MJ. 1995. Suppression of ICE and apoptosis in mammary epithelial cells by extracellular matrix. *Science* 267:891–93

Bretscher A. 1981. Fimbrin is a cytoskeletal protein that crosslinks F-actin in vitro. *Proc. Natl. Acad. Sci. USA* 78:6849–53

Briesewitz R, Kern A, Marcantonio EE. 1993. Ligand-dependent and -independent integrin focal contact localization: the role of the α chain cytoplasmic domain. *Mol. Biol. Cell* 4:593–604

Briesewitz R, Kern A, Marcantonio EE. 1995. Assembly and function of integrin receptors is dependent on opposing α and β cytoplasmic domains. *Mol. Biol. Cell* 6:997–1010

Brooks PC, Montgomery AM, Rosenfeld M, Reisfeld RA, Hu T, et al. 1994. Integrin $\alpha_v\beta_3$ antagonists promote tumor regression by inducing apoptosis of angiogenic blood vessels. *Cell* 79:1157–64

Buhl AM, Johnson NL, Dhanasekaran N, Johnson GL. 1995. G alpha 12 and G alpha 13 stimulate Rho-dependent stress fiber formation and focal adhesion assembly. *J. Biol. Chem.* 270:24631–34

Burbelo PD, Miyamoto S, Utani A, Brill S, Yamada KM, et al. 1995. P190-B, a new member of the rhoGAP family, and rho are induced to cluster after integrin cross-linking. *J. Biol. Chem.* 270:30919–26

Burridge K. 1981. Are stress fibres contractile? *Nature* 294:691–92

Burridge K, Connell L. 1983. A new protein of adhesion plaques and ruffling membranes. *J. Cell Biol.* 97:359–67

Burridge K, Fath K, Kelly T, Nuckolls G, Turner C. 1988. Focal adhesions: transmembrane junctions between the extracellular matrix and the cytoskeleton. *Annu. Rev. Cell Biol.* 4:487–525

Burridge K, Feramisco JR. 1980. Microinjection and localization of a 130K protein in living fibroblasts: a relationship to actin and fibronectin. *Cell* 19:587–95

Burridge K, Mangeat P. 1984. An interaction between vinculin and talin. *Nature* 308:744–46

Burridge K, Turner CE, Romer LH. 1992. Tyrosine phosphorylation of paxillin and $pp125^{FAK}$ accompanies cell adhesion to extracellular matrix: a role in cytoskeletal assembly. *J. Cell Biol.* 119:893–903

Buss F, Temm-Grove C, Henning S, Jockusch BM. 1992. Distribution of profilin in fibroblasts correlates with the presence of highly dynamic actin filaments. *Cell Motil. Cytoskelet.* 22:51–61

Buxbaum RE, Heidemann SR. 1988. A thermodynamic model for force integration and microtubule assembly during axonal elongation. *J. Theor. Biol.* 134:379–90

Calalb MB, Polte TR, Hanks SK. 1995. Tyrosine phosphorylation of focal adhesion kinase at sites in the catalytic domain regulates kinase activity: a role for Src family kinases. *Mol. Cell. Biol.* 15:954–63

Cattelino A, Longhi R, de Curtis I. 1995. Differential distribution of two cytoplasmic variants of the $\alpha_6\beta_1$ integrin laminin receptor in the ventral plasma membrane of embryonic fibroblasts. *J. Cell Sci.* 108:3067–78

Chakraborty T, Ebel F, Domann E, Niebuhr K, Gerstel B, et al. 1995. A focal adhesion factor directly linking intracellularly motile *Listeria monocytogenes* and *Listeria ivanovii* to the actin-based cytoskeleton of mammalian cells. *EMBO J.* 14:1314–21

Chardin P, Boquet P, Madaule P, Popoff MR, Rubin EJ, Gill DM. 1989. The mammalian G protein rhoC is ADP-ribosylated by *Clostridium botulinum* exoenzyme C3 and affects actin microfilaments in Vero cells. *EMBO J.* 8:1087–92

Chen HC, Appeddu PA, Parsons JT, Hildebrand JD, Schaller MD, Guan JL. 1995. Interaction of focal adhesion kinase with cytoskeletal protein talin. *J. Biol. Chem.* 270:16995–99

Chen HC, Guan JL. 1994a. Stimulation of phosphatidylinositol 3′-kinase association with focal adhesion kinase by platelet-derived growth factor. *J. Biol. Chem.* 269:31229–33

Chen HC, Guan JL. 1994b. Association of focal adhesion kinase with its potential substrate phosphatidylinositol 3-kinase. *Proc. Natl. Acad. Sci. USA* 91:10148–52

Chen Q, Kinch MS, Lin TH, Burridge K, Juliano RL. 1994. Integrin-mediated cell adhesion activates mitogen-activated protein kinases. *J. Biol. Chem.* 269:26602–5

Chen WT, Singer SJ. 1982. Immunoelectron microscopic studies of the sites of cell-substratum and cell-cell contacts in cultured fibroblasts. *J. Cell Biol.* 95:205–22

Chong LD, Traynor-Kaplan A, Bokoch GM, Schwartz MA. 1994. The small GTP-binding protein Rho regulates a phosphatidylinositol 4-phosphate 5-kinase in mammalian cells. *Cell* 79:507–13

Chrzanowska-Wodnicka M, Burridge K. 1994. Tyrosine phosphorylation is involved in reorganization of the actin cytoskeleton in response to serum or LPA stimulation. *J. Cell Sci.* 107:3643–54

Chrzanowska-Wodnicka M, Burridge K. 1996. Rho-stimulated contractility drives the formation of stress fibers and focal adhesions. *J. Cell Biol.* 133:1403–15

Cichowski K, Brugge JS, Brass LF. 1996. Thrombin receptor activation and integrin engagement stimulate tyrosine phosphorylation of the proto-oncogene product, $p95^{Vav}$, in platelets. *J. Biol. Chem.* 271:7544–50

Chun JS, Jacobson BS. 1992. Spreading of HeLa cells on a collagen substratum requires a second messenger formed by the lipoxygenase metabolism of arachidonic acid released by collagen receptor clustering. *Mol. Biol. Cell* 3:481–92

Chun JS, Jacobson BS. 1993. Requirement for diacylglycerol and protein kinase C in HeLa cell-substratum adhesion and their feedback amplification of arachidonic acid production for optimum cell spreading. *Mol. Biol. Cell* 4:271–81

Citi S, Kendrick-Jones J. 1986. Regulation in vitro of brush border myosin by light chain phosphorylation. *J. Mol. Biol.* 188:369–82

Clark EA, Brugge JS. 1995. Integrins and signal transduction pathways: the road taken. *Science* 268:233–39

Cobb BS, Schaller MD, Leu TH, Parsons JT. 1994. Stable association of $pp60^{src}$ and $pp59^{fyn}$ with the focal adhesion-associated protein tyrosine kinase, $pp125^{FAK}$. *Mol. Cell. Biol.* 14:147–55

Cooper JA, Howell B. 1993. The when and how of Src regulation. *Cell* 73:1051–54

Coso OA, Chiariello M, Yu JC, Teramoto H, Crespo P, et al. 1995. The small GTP-binding proteins Rac1 and Cdc42 regulate the activity of the JNK/SAPK signaling pathway. *Cell* 81:1137–46

Craig R, Smith R, Kendrick-Jones J. 1983. Light-chain phosphorylation controls the conformation of vertebrate non-muscle and smooth muscle myosin molecules. *Nature* 302:436–39

Cramer LP, Mitchison TJ. 1995. Myosin is involved in postmitotic cell spreading. *J. Cell Biol.* 131:179–89

Crawford AW, Michelsen JW, Beckerle MC. 1992. An interaction between zyxin and α-actinin. *J. Cell Biol.* 116:1381–93

Crawford AW, Pino JD, Beckerle MC. 1994. Biochemical and molecular characterization of the chicken cysteine-rich protein, a developmentally regulated LIM-domain protein that is associated with the actin cytoskeleton. *J. Cell Biol.* 124:117–27

Crowley E, Horwitz AF. 1995. Tyrosine phosphorylation and cytoskeletal tension regulate the release of fibroblast adhesions. *J. Cell Biol.* 131:525–37

Danowski BA. 1989. Fibroblast contractility and actin organization are stimulated by microtubule inhibitors. *J. Cell Sci.* 93:255–66

Defilippi P, Bozzo C, Volpe G, Romano G, Venturino M, et al. 1994. Integrin-mediated signal transduction in human endothelial cells: analysis of tyrosine phosphorylation events. *Cell Adhes. Commun.* 2:75–86

Dejana E, Colella S, Conforti G, Abbadini M, Gaboli M, Marchisio PC. 1988. Fibronectin and vitronectin regulate the organization of their respective Arg-Gly-Asp adhesion receptors in cultured human endothelial cells. *J. Cell Biol.* 107:1215–23

DePasquale JA, Izzard CS. 1987. Evidence for

an actin-containing cytoplasmic precursor of the focal contact and the timing of incorporation of vinculin at the focal contact. *J. Cell Biol.* 105:2803–9

DePasquale JA, Izzard CS. 1991. Accumulation of talin in nodes at the edge of the lamellipodium and separate incorporation into adhesion plaques at focal contacts in fibroblasts. *J. Cell Biol.* 113:1351–59

Douville PJ, Harvey WJ, Carbonetto S. 1988. Isolation and partial characterization of high affinity laminin receptors in neural cells. *J. Biol. Chem.* 263:14964–69

Ervasti JM, Campbell KP. 1993a. Dystrophin and the membrane skeleton. *Curr. Opin. Cell Biol.* 5:82–87

Ervasti JM, Campbell KP. 1993b. A role for the dystrophin-glycoprotein complex as a transmembrane linker between laminin and actin. *J. Cell Biol.* 122:809–23

Evans RR, Robson RM, Stromer MH. 1984. Properties of smooth muscle vinculin. *J. Biol. Chem.* 259:3916–24

Farmer SR, Ben-Ze'ev A, Benecke BJ, Penman S. 1978. Altered translatability of messenger RNA from suspended anchorage-dependent fibroblasts: Reversal upon cell attachment to a surface. *Cell* 15:931–39

Fath KR, Edgell CJ, Burridge K. 1989. The distribution of distinct integrins in focal contacts is determined by the substratum composition. *J. Cell Sci.* 92:67–75

Feramisco JR. 1979. Microinjection of fluorescently labeled α-actinin into living fibroblasts. *Proc. Natl. Acad. Sci. USA* 76:3967–71

Fernandez A, Brautigan DL, Mumby M, Lamb NJ. 1990. Protein phosphatase type-1, not type-2A, modulates actin microfilament integrity and myosin light chain phosphorylation in living nonmuscle cells. *J. Cell Biol.* 111:103–12

Ferrell JE, Martin GS. 1989. Tyrosine-specific phosphorylation is regulated by glycoprotein IIb/IIIa in platelets. *Proc. Natl. Acad. Sci. USA* 86:2234–38

Fincham VJ, Wyke JA, Frame MC. 1995. v-Src-induced degradation of focal adhesion kinase during morphological transformation of chicken embryo fibroblasts. *Oncogene* 10:2247–52

Fleischer M, Wohlfarth-Bottermann KE. 1975. Correlation between tension force generation, fibrillogenesis and ultrastructure of cytoplasmic actomyosin during isometric and isotonic contractions of protoplasmic strands. *Cytobiologie* 10:339–65

Fornaro M, Zheng DQ, Languino LR. 1995. The novel structural motif Gln795-Gln802 in the integrin beta 1C cytoplasmic domain regulates cell proliferation. *J. Biol. Chem.* 270:24666–69

Franke RP, Grafe M, Schnittler H, Seiffge D, Mittermayer C, Drenckhahn D. 1984. Induction of human vascular endothelial stress fibres by fluid shear stress. *Nature* 307:648–49

Frisch SM, Francis H. 1994. Disruption of epithelial cell-matrix interactions induces apoptosis. *J. Cell Biol.* 124:619–26

Frisch SM, Vuori K, Ruoslahti E. 1996. Control of adhesion-dependent cell survival by focal adhesion kinase. *J. Cell Biol.* In press

Fukami K, Endo T, Imamura M, Takenawa T. 1994. α-Actinin and vinculin are PIP_2-binding proteins involved in signaling by tyrosine kinase. *J. Biol. Chem.* 269:1518–22

Fukami K, Furuhashi K, Inagaki M, Endo T, Hatano S, Takenawa T. 1992. Requirement of phosphatidylinositol 4,5-bisphosphate for α-actinin function. *Nature* 359:150–52

Fukui Y, Hanafusa H. 1989. Phosphatidylinositol kinase activity associates with viral $p60^{src}$ protein. *Mol. Cell. Biol.* 9:1651–58

Gabbiani G, Majno G, Ryan GB. 1973. The fibroblast as a contractile cell: the myofibroblast. In *Biology of Fibroblast,* ed. E Kulonen, J Pikkarainen, pp. 139–54. New York: Academic

Gallagher PJ, Garcia JGN, Herring BP. 1995. Expression of a novel myosin light chain kinase in embryonic tissues and cultured cells. *J. Biol. Chem.* 270:29090–95

Gee SH, Blacher RW, Douville PJ, Provost PR, Yurchenco PD, Carbonetto S. 1993. Laminin-binding protein 120 from brain is closely related to the dystrophin-associated glycoprotein, dystroglycan, and binds with high affinity to the major heparin binding domain of laminin. *J. Biol. Chem.* 268:14972–80

Geiger B. 1979. A 130K protein from chicken gizzard: its localization at the termini of microfilament bundles in cultured chicken cells. *Cell* 18:193–205

Geiger B, Ayalon O, Ginsberg D, Volberg T, Rodriguez Fernandez JL, et al. 1992a. Cytoplasmic control of cell adhesion. *Cold Spring Harbor Symp. Quant. Biol.* 57:631–42

Geiger B, Salomon D, Takeichi M, Hynes RO. 1992b. A chimeric N-cadherin/beta 1-integrin receptor which localizes to both cell-cell and cell-matrix adhesions. *J. Cell Sci.* 103:943–51

Giancotti FG, Ruoslahti E. 1990. Elevated levels of the $\alpha_5\beta_1$ fibronectin receptor suppress the transformed phenotype of Chinese hamster ovary cells. *Cell* 60:849–59

Gilmore AP, Burridge K. 1996. Regulation of vinculin binding to talin and actin by phosphatidylinositol-4-5-bisphosphate. *Nature.* 381:531–35

Gilmore AP, Romer LH. 1996. Inhibition of

FAK signalling in focal adhesions decreases cell motility and proliferation. *Mol. Biol. Cell.* 7:1209–24

Ginsberg MH, Xiaoping D, O'Toole TE, Loftus JC, Plow EF. 1993. Platelet integrins. *Thromb. Haemost.* 70:87–93

Giuliano KA, Taylor DL. 1990. Formation, transport, contraction, and disassembly of stress fibers in fibroblasts. *Cell Motil. Cytoskelet.* 16:14–21

Gluck U, Ben-Ze'ev A. 1994. Modulation of α-actinin levels affects cell motility and confers tumorigenicity on 3T3 cells. *J. Cell Sci.* 107:1773–82

Gluck U, Kwiatkowski DJ, Ben-Ze'ev A. 1993. Suppression of tumorigenicity in simian virus 40-transformed 3T3 cells transfected with α-actinin cDNA. *Proc. Natl. Acad. Sci. USA* 90:383–87

Goeckeler ZM, Wysolmerski RB. 1995. Myosin light chain kinase-regulated endothelial cell contraction: the relationship between isometric tension, actin polymerization, and myosin phosphorylation. *J. Cell Biol.* 130:613–27

Golden A, Brugge JS. 1989. Thrombin treatment induces rapid changes in tyrosine phosphorylation in platelets. *Proc. Natl. Acad. Sci. USA* 86:901–5

Gong MC, Fuglsang A, Alessi D, Kobayashi S, Cohen P, et al. 1992. Arachidonic acid inhibits myosin light chain phosphatase and sensitizes smooth muscle to calcium. *J. Biol. Chem.* 267:21492–98

Gordon WE. 1978. Immunofluorescent and ultrastructural studies of "sarcomeric" units in stress fibers of cultured non-muscle cells. *Exp. Cell. Res.* 117:253–60

Grinnell F. 1994. Fibroblasts, myofibroblasts, and wound contraction. *J. Cell Biol.* 124:401–4

Guan JL, Shalloway D. 1992. Regulation of focal adhesion-associated protein tyrosine kinase by both cellular adhesion and oncogenic transformation. *Nature* 358:690–92

Guan JL, Trevithick JE, Hynes RO. 1991. Fibronectin/integrin interaction induces tyrosine phosphorylation of a 120-kDa protein. *Cell Regul.* 2:951–64

Guinebault C, Payrastre B, Racaud-Sultan C, Mazarguil H, Breton M, et al. 1995. Integrin-dependent translocation of phosphoinositide 3-kinase to the cytoskeleton of thrombin-activated platelets involves specific interactions of p85alpha with actin filaments and focal adhesion kinase. *J. Cell Biol.* 129(3):831–42

Gulbins E, Coggeshall KM, Baier G, Katzav S, Burn P, Altman A. 1993. Tyrosine kinase-stimulated guanine nucleotide exchange activity of Vav in T cell activation. *Science* 260:822–25

Haffner C, Jarchau T, Reinhard M, Hoppe J, Lohmann SM, Walter U. 1995. Molecular cloning, structural analysis and functional expression of the proline-rich focal adhesion and microfilament-associated protein VASP. *EMBO J.* 14:19–27

Haimovich B, Lipfert L, Brugge JS, Shattil SJ. 1993. Tyrosine phosphorylation and cytoskeletal reorganization in platelets are triggered by interaction of integrin receptors with their immobilized ligands. *J. Biol. Chem.* 268:15868–77

Halbrugge M, Walter U. 1989. Purification of a vasodilator-regulated phosphoprotein from human platelets. *Eur. J. Biochem.* 185:41–50

Hall A. 1994. Small GTP-binding proteins and the regulation of the actin cytoskeleton. *Annu. Rev. Cell Biol.* 10:31–54

Halliday NL, Tomasek JJ. 1995. Mechanical properties of the extracellular matrix influence fibronectin fibril assembly in vitro. *Exp. Cell. Res.* 217:109–17

Hanks SK, Calalb MB, Harper MC, Patel SK. 1992. Focal adhesion protein-tyrosine kinase phosphorylated in response to cell attachment to fibronectin. *Proc. Natl. Acad. Sci. USA* 89:8487–91

Hannigan GE, Leunghagesteijn C, Fitzgibbon L, Coppolino MG, Radeva G, et al. 1996. Regulation of cell adhesion and anchorage-dependent growth by a new β_1-integrin-linked protein kinase. *Nature* 379:91–96

Hansen CA, Schroering AG, Carey DJ, Robishaw JD. 1994. Localization of a heterotrimeric G protein gamma subunit to focal adhesions and associated stress fibers. *J. Cell Biol.* 126:811–19

Harris AK, Wild P, Stopak D. 1980. Silicone rubber substrata: a new wrinkle in the study of cell locomotion. *Science* 208:177–79

Hayashi Y, Haimovich B, Reszka A, Boettiger D, Horwitz A. 1990. Expression and function of chicken integrin 1 subunit and its cytoplasmic domain mutants in mouse NIH 3T3 cells. *J. Cell Biol.* 110:175–84

Heath JP, Dunn GA. 1978. Cell to substratum contacts of chick fibroblasts and their relation to the microfilament system. A correlated interference-reflection and high-voltage electron-microscope study. *J. Cell Sci.* 29:197–212

Hemler ME, Weitzman JB, Pasqualini R, Kawaguchi S, Kassner PD, Bertichevsky FB. 1994. Structure, biochemical properties and biological functions of integrin cytoplasmic domain. In *Integrins: The Biological Problem,* pp. 242–56. Boca Raton, FL: CRC Press

Henske EP, Short MP, Jozwiak S, Bovey CM,

Ramlakhan S, et al. 1995. Identification of VAV2 on 9q34 and its exclusion as the tuberous sclerosis gene TSC1. *Ann. Hum. Genet.* 59:25–37

Higuchi H, Takemori S. 1989. Butanedione monoxime suppresses contraction and ATPase activity of rabbit skeletal muscle. *J. Biochem.* 105:638–43

Hildebrand JD, Schaller MD, Parsons JT. 1993. Identification of sequences required for the efficient localization of the focal adhesion kinase, $pp125^{FAK}$, to cellular focal adhesions. *J. Cell Biol.* 123:993–1005

Hildebrand JD, Schaller MD, Parsons JT. 1995. Paxillin, a tyrosine phosphorylated focal adhesion-associated protein binds to the carboxyl terminal domain of focal adhesion kinase. *Mol. Biol. Cell* 6:637–47

Hill CS, Wynne J, Treisman R. 1995. The Rho family GTPases RhoA, Rac1, and CDC42Hs regulate transcriptional activation by SRF. *Cell* 81:1159–70

Hirata K, Kikuchi A, Sasaki T, Kuroda S, Kaibuchi K, et al. 1992. Involvement of rho p21 in the GTP-enhanced calcium ion sensitivity of smooth muscle contraction. *J. Biol. Chem.* 267:8719–22

Honer B, Citi S, Kendrick-Jones J, Jockusch BM. 1988. Modulation of cellular morphology and locomotory activity by antibodies against myosin. *J. Cell Biol.* 107:2181–89

Hordijk PL, Verlaan I, van Corven EJ, Moolenaar WH. 1994. Protein tyrosine phosphorylation induced by lysophosphatidic acid in Rat-1 fibroblasts. Evidence that phosphorylation of map kinase is mediated by the Gi-p21ras pathway. *J. Biol. Chem.* 269:645–51

Horwitz A, Duggan K, Buck C, Beckerle MC, Burridge K. 1986. Interaction of plasma membrane fibronectin receptor with talin—a transmembrane linkage. *Nature* 320:531–33

Hotchin NA, Hall A. 1995. The assembly of integrin adhesion complexes requires both extracellular matrix and intracellular rho/rac GTPases. *J. Cell Biol.* 131:1857–65

Hynes RO. 1992. Integrins: versatility, modulation, and signaling in cell adhesion. *Cell* 69:11–25

Ibraghimov-Beskrovnaya O, Ervasti JM, Leveille CJ, Slaughter CA, Sernett SW, Campbell KP. 1992. Primary structure of dystrophin-associated glycoproteins linking dystrophin to the extracellular matrix. *Nature* 355:696–702

Ibraghimov-Beskrovnaya O, Milatovich A, Ozcelik T, Yang B, Koepnick K, et al. 1993. Human dystroglycan: skeletal muscle cDNA, genomic structure, origin of tissue specific isoforms and chromosomal localization. *Hum. Mol. Genet.* 2:1651–57

Ichikawa K, Ito M, Hartshorne DJ. 1996. Phosphorylation of the large subunit of myosin phosphatase and inhibition of phosphatase activity. *J. Biol. Chem.* 271:4733–40

Ikebe M, Koretz J, Hartshorne DJ. 1988. Effects of phosphorylation of light chain residues threonine 18 and serine 19 on the properties and conformation of smooth muscle myosin. *J. Biol. Chem.* 263:6432–37

Ikebe M. 1989. Phosphorylation of a second site for myosin light chain kinase on platelet myosin. *Biochemistry* 28:8750–55

Ikebe M, Reardon S. 1990. Phosphorylation of bovine platelet myosin by protein kinase C. *Biochemistry* 29:2713–20

Ilic D, Furuta Y, Kanazawa S, Takeda N, Sobue K, et al. 1995. Reduced cell motility and enhanced focal contact formation in cells from FAK-deficient mice. *Nature* 377:539–44

Ingber DE. 1993. Cellular tensegrity: defining new rules of biological design that govern the cytoskeleton. *J. Cell Sci.* 104:613–27

Ingber DE, Prusty D, Frangioni JV, Cragoe EJ Jr, Lechene C, Schwartz MA. 1990. Control of intracellular pH and growth by fibronectin in capillary endothelial cells. *J. Cell Biol.* 110:1803–11

Isenberg G. 1991. Actin binding proteins—lipid interactions. *J. Muscle Res. Cell Motil.* 12:136–44

Isenberg G, Rathke PC, Hulsmann N, Franke WW, Wohlfarth-Bottermann KE. 1976. Cytoplasmic actomyosin fibrils in tissue culture cells: direct proof of contractility by visualization of ATP-induced contraction in fibrils isolated by laser micro-beam dissection. *Cell Tissue Res.* 166:427–43

Ishikawa R, Yamashiro S, Matsumura F. 1989a. Differential modulation of actin-severing activity of gelsolin by multiple isoforms of cultured rat cell tropomyosin. Potentiation of protective ability of tropomyosins by 83-kDa nonmuscle caldesmon. *J. Biol. Chem.* 264:7490–97

Ishikawa R, Yamashiro S, Matsumura F. 1989b. Annealing of gelsolin-severed actin fragments by tropomyosin in the presence of Ca^{2+}. Potentiation of the annealing process by caldesmon. *J. Biol. Chem.* 264:16764–70

Ishizaki T, Maekawa M, Fujisawa K, Okawa K, Iwamatsu A, et al. 1996. The small GTP-binding protein Rho binds to and activates a 160 kDa Ser/Thr protein kinase homologous to myotonic dystrophy kinase. *EMBO J.* 15:1885–93

Izzard CS, Radinsky R, Culp LA. 1986. Substratum contacts and cytoskeletal reorganization of BALB/c 3T3 cells on a cell-binding fragment and heparin-binding fragments of plasma fibronectin. *Exp. Cell. Res.* 165:320–

36
Jaken S, Leach K, Klauck T. 1989. Association of type 3 protein kinase C with focal contacts in rat embryo fibroblasts. *J. Cell Biol.* 109:697–704
Jalink K, Moolenaar WH. 1992. Thrombin receptor activation causes rapid neural cell rounding and neurite retraction independent of classic second messengers. *J. Cell Biol.* 118:411–19
Jalink K, van Corven EJ, Hengeveld T, Morii N, Narumiya S, Moolenaar WH. 1994. Inhibition of lysophosphatidate- and thrombin-induced neurite retraction and neuronal cell rounding by ADP ribosylation of the small GTP-binding protein Rho. *J. Cell Biol.* 126: 801–10
Janmey PA, Stossel TP. 1987. Modulation of gelsolin function by phosphatidylinositol 4,5-bisphosphate. *Nature* 325:362–64
Jockusch BM, Bubeck P, Giehl K, Kroemker M, Moschner J, et al. 1995. The molecular architecture of focal adhesions. *Annu. Rev. Cell Dev. Biol.* 11:379–416
Johnson RP, Craig SW. 1994. An intramolecular association between the head and tail domains of vinculin modulates talin binding. *J. Biol. Chem.* 269:12611–19
Johnson RP, Craig SW. 1995a. F-actin binding site masked by the intramolecular association of vinculin head and tail domains. *Nature* 373:261–64
Johnson RP, Craig SW. 1995b. The carboxy-terminal tail domain of vinculin contains a cryptic binding site for acidic phospholipids. *Biochem. Biophys. Res. Commun.* 210:159–64
Juliano RL, Haskill S. 1993. Signal transduction from the extracellular matrix. *J. Cell Biol.* 120:577–85
Kaplan KB, Bibbins KB, Swedlow JR, Arnaud M, Morgan DO, Varmus HE. 1994. Association of the amino-terminal half of c-Src with focal adhesions alters their properties and is regulated by phosphorylation of tyrosine 527. *EMBO J.* 13:4745–56
Kaplan KB, Swedlow JR, Morgan DO, Varmus HE. 1995. c-Src enhances the spreading of src-/-fibroblasts on fibronectin by a kinase-independent mechanism. *Genes Dev.* 9:1505–17
Kapron-Bras C, Fitz-Gibbon L, Jeevaratnam P, Wilkins J, Dedhar S. 1993. Stimulation of tyrosine phosphorylation and accumulation of GTP-bound p21ras upon antibody-mediated $\alpha_2\beta_1$ integrin activation in T-lymphoblastic cells. *J. Biol. Chem.* 268:20701–4
Kaufmann S, Piekenbrock T, Goldmann WH, Barmann M, Isenberg G. 1991. Talin binds to actin and promotes filament nucleation. *FEBS Lett.* 284:187–91
Khosravi-Far R, Chrzanowska-Wodnicka M, Solski PA, Eva A, Burridge K, Der CJ. 1994. Dbl and Vav mediate transformation via mitogen-activated protein kinase pathways that are distinct from those activated by oncogenic Ras. *Mol. Cell. Biol.* 14:6848–57
Kimura K, Ito M, Amano M, Chihara K, Fukata Y, et al. 1996. Regulation of myosin phosphatase by Rho and Rho-associated kinase (Rho-kinase). *Science* 273:245–48
Kishi K, Sasaki T, Kuroda S, Itoh T, Takai Y. 1993. Regulation of cytoplasmic division of *Xenopus* embryo by rho p21 and its inhibitory GDP/GTP exchange protein (rho GDI). *J. Cell Biol.* 120:1187–95
Kitazawa T, Masuo M, Somlyo AP. 1991. G protein-mediated inhibition of myosin light-chain phosphatase in vascular smooth muscle. *Proc. Natl. Acad. Sci. USA* 88:9307–10
Kokubu N, Satoh M, Takayanagi I. 1995. Involvement of botulinum C3-sensitive GTP-binding proteins in α1-adrenoceptor subtypes mediating Ca(2+)-sensitization. *Eur. J. Pharmacol.* 290:19–27
Kolega J. 1986. Effects of mechanical tension on protrusive activity and microfilament and intermediate filament organization in an epidermal epithelium moving in culture. *J. Cell Biol.* 102:1400–11
Kolodney MS, Elson EL. 1993. Correlation of myosin light chain phosphorylation with isometric contraction of fibroblasts. *J. Biol. Chem.* 268:23850–55
Kolodney MS, Elson EL. 1995. Contraction due to microtubule disruption is associated with increased phosphorylation of myosin regulatory light chain. *Proc. Natl. Acad. Sci. USA* 92:10252–56
Kornberg L, Earp HS, Parsons JT, Schaller M, Juliano RL. 1992. Cell adhesion or integrin clustering increases phosphorylation of a focal adhesion-associated tyrosine kinase. *J. Biol. Chem.* 267:23439–42
Kornberg LJ, Earp HS, Turner CE, Prockop C, Juliano RL. 1991. Signal transduction by integrins: increased protein tyrosine phosphorylation caused by clustering of beta 1 integrins. *Proc. Natl. Acad. Sci. USA* 88:8392–96
Kramarcy NR, Sealock R. 1990. Dystrophin as a focal adhesion protein. Colocalization with talin and the M_r 48,000 sarcolemmal protein in cultured *Xenopus* muscle. *FEBS Lett.* 274:171–74
Kreis TE, Birchmeier W. 1980. Stress fiber sarcomeres of fibroblasts are contractile. *Cell* 22:555–61
Kumagai N, Morii N, Fujisawa K, Nemoto Y, Narumiya S. 1993. ADP-ribosylation of rho p21 inhibits lysophosphatidic acid-

induced protein tyrosine phosphorylation and phosphatidylinositol 3-kinase activation in cultured Swiss 3T3 cells. *J. Biol. Chem.* 268:24535–38

Lakonishok M, Muschler J, Horwitz AF. 1992. The $\alpha_5\beta_1$ integrin associates with a dystrophin-containing lattice during muscle development. *Dev. Biol.* 152:209–20

LaFlamme SE, Akiyama SK, Yamada KM. 1992. Regulation of fibronectin receptor distribution. *J. Cell Biol.* 117:437–47. Erratum. 1992. *J. Cell Biol.* 118(2):491

Lamb NJ, Fernandez A, Conti MA, Adelstein R, Glass DB, et al. 1988. Regulation of actin microfilament integrity in living nonmuscle cells by the cAMP-dependent protein kinase and the myosin light chain kinase. *J. Cell Biol.* 106:1955–71

Lancaster CA, Taylor-Harris PM, Self AJ, Brill S, van Erp HE, Hall A. 1994. Characterization of rhoGAP. A GTPase-activating protein for rho-related small GTPases. *J. Biol. Chem.* 269:1137–42

Languino LR, Ruoslahti E. 1992. An alternative form of the integrin beta 1 subunit with a variant cytoplasmic domain. *J. Biol. Chem.* 267:7116–20

Lassing I, Lindberg U. 1985. Specific interaction between phosphatidylinositol 4,5-bisphosphate and profilactin. *Nature* 314:472–74

Laudanna C, Campbell JJ, Butcher EC. 1996. Role of rho in chemoattractant-activated leukocyte adhesion through integrins. *Science* 271:981–83

Lazarides E, Burridge K. 1975. Alpha-actinin: immunofluorescent localization of a muscle structural protein in nonmuscle cells. *Cell* 6:289–98

Leavesley DI, Schwartz MA, Rosenfeld M, Cheresh DA. 1993. Integrin beta 1- and beta 3-mediated endothelial cell migration is triggered through distinct signaling mechanisms. *J. Cell Biol.* 121:163–70

Leung T, Manser E, Tan L, Lim L. 1995. A novel serine/threonine kinase binding the ras-related RhoA GTPase which translocates the kinase to peripheral membranes. *J. Biol. Chem.* 270:29051–54

Lev S, Moreno H, Martinez R, Canoll P, Peles E, et al. 1995. Protein tyrosine kinase Pyk2 involved in Ca^{2+}-induced regulation of ion channel and MAP kinase functions. *Nature* 376:737–45

Lewis JM, Schwartz MA. 1995. Mapping in vivo associations of cytoplasmic proteins with integrin beta 1 cytoplasmic domain mutants. *Mol. Biol. Cell* 6:151–60

Lin YC, Grinnell F. 1993. Decreased level of PDGF-stimulated receptor autophosphorylation by fibroblasts in mechanically relaxed collagen matrices. *J. Cell Biol.* 122:663–72

Lipfert L, Haimovich B, Schaller MD, Cobb BS, Parsons JT, Brugge JS. 1992. Integrin-dependent phosphorylation and activation of the protein tyrosine kinase $pp125^{FAK}$ in platelets. *J. Cell Biol.* 119:905–12

Lo SH, Janmey PA, Hartwig JH, Chen LB. 1994. Interactions of tensin with actin and identification of its three distinct actin-binding domains. *J. Cell Biol.* 125:1067–75

Mabuchi I, Hamaguchi Y, Fujimoto H, Morii N, Mishima M, Narumiya S. 1993. A rho-like protein is involved in the organisation of the contractile ring in dividing sand dollar eggs. *Zygote* 1:325–31

Marcantonio EE, Guan JL, Trevithick JE, Hynes RO. 1990. Mapping of the functional determinants of the integrin beta 1 cytoplasmic domain by site-directed mutagenesis. *Cell Regul.* 1:597–604

Massia SP, Hubbell JA. 1991. An RGD spacing of 440 nm is sufficient for integrin $\alpha_V\beta_3$-mediated fibroblast spreading and 140 nm for focal contact and stress fiber formation. *J. Cell Biol.* 114:1089–100

Matsuda M, Hashimoto Y, Muroya K, Hasegawa H, Kruata T. 1994. Crk protein binds to two guanine nucleotide-releasing proteins for the ras family and modulates nerve growth factor-induced activation of ras in PC12 cells. *Mol. Cell Biol.* 14:5495–500

Matsuguchi T, Inhorn RC, Carlesso N, Xu G, Druker B, Griffin JD. 1995. Tyrosine phosphorylation of p95Vav in myeloid cells is regulated by GM-CSF, IL-3 and steel factor and is constitutively increased by p210BCR/ABL. *EMBO J.* 14:257–65

Matsumura F, Yamashiro S. 1993. Caldesmon. *Curr. Opin. Cell Biol.* 5:70–76

McKenna NM, Meigs JB, Wang YL. 1985. Exchangeability of α-actinin in living cardiac fibroblasts and muscle cells. *J. Cell Biol.* 101:2223–32

McKillop DF, Fortune NS, Ranatunga KW, Geeves MA. 1994. The influence of 2,3-butanedione 2-monoxime (BDM) on the interaction between actin and myosin in solution and in skinned muscle fibres. *J. Muscle Res. Cell Motil.* 15:309–18

McNamee HP, Ingber DE, Schwartz MA. 1993. Adhesion to fibronectin stimulates inositol lipid synthesis and enhances PDGF-induced inositol lipid breakdown. *J. Cell Biol.* 121:673–78

Menkel AR, Kroemker M, Bubeck P, Ronsiek M, Nikolai G, Jockusch BM. 1994. Characterization of an F-actin-binding domain in the cytoskeletal protein vinculin. *J. Cell Biol.* 126:1231–40

Meredith J Jr, Takada Y, Fornaro M, Languino LR, Schwartz MA. 1995. Inhibition of cell cycle progression by the alternatively spliced integrin beta 1C. *Science* 269:1570–72

Meredith JE Jr, Fazeli B, Schwartz MA. 1993. The extracellular matrix as a cell survival factor. *Mol. Biol. Cell* 4:953–61

Minden A, Lin A, Claret FX, Abo A, Karin M. 1995. Selective activation of the JNK signaling cascade and c-Jun transcriptional activity by the small GTPases Rac and Cdc42Hs. *Cell* 81:1147–57

Mittal B, Sanger JM, Sanger JW. 1987. Binding and distribution of fluorescently labeled filamin in permeabilized and living cells. *Cell. Motil. Cytoskelet.* 8:345–59

Miyamoto S, Akiyama SK, Yamada KM. 1995a. Synergistic roles for receptor occupancy and aggregation in integrin transmembrane function. *Science* 267:883–85

Miyamoto S, Teramoto H, Coso OA, Gutkind JS, Burbelo PD, et al. 1995b. Integrin function: molecular hierarchies of cytoskeletal and signaling molecules. *J. Cell Biol.* 131:791–805

Mochitate K, Pawelek P, Grinnell F. 1991. Stress relaxation of contracted collagen gels: disruption of actin filament bundles, release of cell surface fibronectin, and down-regulation of DNA and protein synthesis. *Exp. Cell. Res.* 193:198–207

Montgomery AM, Reisfeld RA, Cheresh DA. 1994. Integrin $\alpha_v\beta_3$ rescues melanoma cells from apoptosis in three-dimensional dermal collagen. *Proc. Natl. Acad. Sci. USA* 91:8856–60

Morii N, Teru-uchi T, Tominaga T, Kumagai N, Kozaki S, et al. 1992. A *rho* gene product in human blood platelets. II. Effects of the ADP-ribosylation by botulinum C3 ADP-ribosyltransferase on platelet aggregation. *J. Biol. Chem.* 267:20921–26

Morino N, Mimura T, Hamasaki K, Tobe K, Ueki K, et al. 1995. Matrix/integrin interaction activates the mitogen-activated protein kinase $p44^{erk-1}$ and $p42^{erk-2}$. *J. Biol. Chem.* 270:269–73

Muguruma M, Matsumura S, Fukazawa T. 1990. Direct interactions between talin and actin. *Biochem. Biophys. Res. Commun.* 171: 1217–23

Nada S, Okada M, Aizawa S, Nakagawa H. 1994. Identification of major tyrosine-phosphorylated proteins in Csk-deficient cells. *Oncogene* 9:3571–78

Narumiya S, Sekine A, Fujiwara M. 1988. Substrate for botulinum ADP-ribosyltransferase, Gb, has an amino acid sequence homologous to a putative *rho* gene product. *J. Biol. Chem.* 263:17255–57

Niggli V, Dimitrov DP, Brunner J, Burger MM. 1986. Interaction of the cytoskeletal component vinculin with bilayer structures analyzed with a photoactivatable phospholipid. *J. Biol. Chem.* 261:6912–18

Nishikawa M, de Lanerolle P, Lincoln TM, Adelstein RS. 1984. Phosphorylation of mammalian myosin light chain kinases by the catalytic subunit of cyclic AMP-dependent protein kinase and by cyclic GMP-dependent protein kinase. *J. Biol. Chem.* 259:8429–36

Nishiyama T, Sasaki T, Takaishi K, Kato M, Yaku H, et al. 1994. rac p21 is involved in insulin-induced membrane ruffling and rho p21 is involved in hepatocyte growth factor- and 12-*O*-tetradecanoylphorbol-13-acetate (TPA)-induced membrane ruffling in KB cells. *Mol. Cell. Biol.* 14:2447–56

Nix DA, Beckerle MC. 1995. Inducible nuclear localization of the focal contact protein, zyxin. *Mol. Biol. Cell* 6:A63

Nobes CD, Hall A. 1995. Rho, rac, and cdc42 GTPases regulate the assembly of multimolecular focal complexes associated with actin stress fibers, lamellipodia, and filopodia. *Cell* 81:53–62

Nobes CD, Hawkins P, Stephens L, Hall A. 1995. Activation of the small GTP-binding proteins rho and rac by growth factor receptors. *J. Cell. Sci.* 108:225–33

Noda M, Yasuda-Fukazawa C, Moriishi K, Kato T, Okuda T, et al. 1995. Involvement of rho in GTPgammaS-induced enhancement of phosphorylation of 20 kDa myosin light chain in vascular smooth muscle cells: inhibition of phosphatase activity. *FEBS Lett.* 367:246–50

Nojima Y, Morino N, Mimura T, Hamasaki K, Furuya H, et al. 1995. Integrin-mediated cell adhesion promotes tyrosine phosphorylation of p130Cas, a Src homology 3-containing molecule having multiple Src homolgy 2-binding motifs. *J. Biol. Chem.* 270:15398–402

Norman JC, Price LS, Ridley AJ, Hall A, Koffer A. 1994. Actin filament organization in activated mast cells is regulated by heterotrimeric and small GTP-binding proteins. *J. Cell Biol.* 126:1005–15

Nuckolls GH, Romer LH, Burridge K. 1992. Microinjection of antibodies against talin inhibits the spreading and migration of fibroblasts. *J. Cell Sci.* 102:753–62

Olson MF, Ashworth A, Hall A. 1995. An essential role for Rho, Rac, and Cdc42 GTPases in cell cycle progression through G1. *Science* 269:1270–72

Otey CA, Pavalko FM, Burridge K. 1990. An interaction between α-actinin and the beta 1 integrin subunit in vitro. *J. Cell Biol.* 111:721–29

Otey CA, Vasquez GB, Burridge K, Erickson BW. 1993. Mapping of the α-actinin binding site within the beta 1 integrin cytoplasmic domain. *J. Biol. Chem.* 268:21193–97
Otsuka H, Moskowitz M. 1975. Arrest of 3T3 cells in G1 phase in suspension culture. *J. Cell. Physiol.* 87:213–20
Otto JJ. 1986. The lack of interaction between vinculin and actin. *Cell Motil. Cytoskelet.* 6:48–55
Owens LV, Xu L, Craven RJ, Dent GA, Weiner TM, et al. 1995. Overexpression of the focal adhesion kinase (p125FAK) in invasive human tumors. *Cancer Res.* 55:2752–55
Paterson HF, Self AJ, Garrett MD, Just I, Aktories K, Hall A. 1990. Microinjection of recombinant p21rho induces rapid changes in cell morphology. *J. Cell Biol.* 111:1001–7
Pavalko FM, Burridge K. 1991. Disruption of the actin cytoskeleton after microinjection of proteolytic fragments of α-actinin. *J. Cell Biol.* 114:481–91
Pavalko FM, LaRoche SM. 1993. Activation of human neutrophils induces an interaction between the integrin beta 2-subunit (CD18) and the actin binding protein α-actinin. *J. Immunol.* 151:3795–807
Pavalko FM, Otey CA. 1994. Role of adhesion molecule cytoplasmic domains in mediating interactions with the cytoskeleton. *Proc. Soc. Exp. Biol. Med.* 205:282–93
Pavalko FM, Otey CA, Burridge K. 1989. Identification of a filamin isoform enriched at the ends of stress fibers in chicken embryo fibroblasts. *J. Cell Sci.* 94:109–18
Pavalko FM, Schneider G, Burridge K, Lim SS. 1995. Immunodetection of α-actinin in focal adhesions is limited by antibody inaccessibility. *Exp. Cell. Res.* 217:534–40
Pelletier AJ, Bodary SC, Levinson AD. 1992. Signal transduction by the platelet integrin $\alpha_{IIb}\beta_3$: induction of calcium oscillations required for protein-tyrosine phosphorylation and ligand-induced spreading of stably transfected cells. *Mol. Biol. Cell* 3:989–98
Pelletier AJ, Kunicki T, Ruggeri ZM, Quaranta V. 1995. The activation state of the integrin $\alpha_{IIb}\beta_3$ affects outside-in signals leading to cell spreading and focal adhesion kinase phosphorylation. *J. Biol. Chem.* 270:18133–40
Peppelenbosch MP, Qiu RG, de Vries-Smits AM, Tertoolen LG, de Laat SW, et al. 1995. Rac mediates growth factor-induced arachidonic acid release. *Cell* 81:849–56
Petch LA, Bockholt SM, Bouton A, Parsons JT, Burridge K. 1995. Adhesion-induced tyrosine phosphorylation of the p130^{src} substrate. *J. Cell Sci.* 108:1371–79
Pistor S, Chakraborty T, Walter U, Wehland J. 1995. The bacterial actin nucleator protein ActA of *Listeria monocytogenes* contains multiple binding sites for host microfilament proteins. *Curr. Biol.* 5:517–25
Pleiman CM, Hertz WM, Cambier JC. 1994. Activation of phosphatidylinositol-3′ kinase by Src-family kinase SH3 binding to the p85 subunit. *Science* 263:1609–12
Plopper GE, McNamee HP, Dike LE, Bojanowski K, Ingber DE. 1995. Convergence of integrin and growth factor receptor signaling pathways within the focal adhesion complex. *Mol. Biol. Cell* 6:1349–65
Pollard TD. 1995. Actin cytoskeleton. Missing link for intracellular bacterial motility? *Curr. Biol.* 5:837–40
Polte TR, Hanks SK. 1995. Interaction between focal adhesion kinase and Crk-associated tyrosine kinase substrate p130^{cas}. *Proc. Natl. Acad. Sci. USA* 92:10678–82
Prasad KV, Janssen O, Kapeller R, Raab M, Cantley LC, Rudd CE. 1993. Src-homology 3 domain of protein kinase p59^{fyn} mediates binding to phosphatidylinositol 3-kinase in T cells. *Proc. Natl. Acad. Sci. USA* 90:7366–70
Price LS, Norman JC, Ridley AJ, Koffer A. 1995. The small GTPases Rac and Rho as regulators of secretion in mast cells. *Curr. Biol.* 5:68–73
Rankin S, Morii N, Narumiya S, Rozengurt E. 1994. Botulinum C3 exoenzyme blocks the tyrosine phosphorylation of p125FAK and paxillin induced by bombesin and endothelin. *FEBS Lett.* 354:315–19
Rankin S, Rozengurt E. 1994. Platelet-derived growth factor modulation of focal adhesion kinase (p125FAK) and paxillin tyrosine phosphorylation in Swiss 3T3 cells. Bell-shaped dose response and cross-talk with bombesin. *J. Biol. Chem.* 269:704–10
Re F, Zanetti A, Sironi M, Polentarutti N, Lanfrancone L, et al. 1994. Inhibition of anchorage-dependent cell spreading triggers apoptosis in cultured human endothelial cells. *J. Cell Biol.* 127:537–46
Reinhard J, Scheel AA, Diekmann D, Hall A, Ruppert C, Bahler M. 1995. A novel type of myosin implicated in signalling by rho family GTPases. *EMBO J.* 14:697–704
Reinhard M, Giehl K, Abel K, Haffner C, Jarchau T, et al. 1995. The proline-rich focal adhesion and microfilament protein VASP is a ligand for profilins. *EMBO J.* 14:1583–89
Reinhard M, Halbrugge M, Scheer U, Wiegand C, Jockusch BM, Walter U. 1992. The 46/50 kDa phosphoprotein VASP purified from human platelets is a novel protein associated with actin filaments and focal contacts. *EMBO J.* 11:2063–70
Reinhard M, Jouvenal K, Tripier D, Walter

U. 1995b. Identification, purification, and characterization of a zyxin-related protein that binds the focal adhesion and microfilament protein VASP (vasodilator-stimulated phosphoprotein). *Proc. Natl. Acad. Sci. USA* 92:7956–60

Ren XD, Bokoch GM, Traynor-Kaplan A, Jenkins GH, Anderson RA, Schwartz MA. 1996. Physical association of the small GTPase rho with a 68-kDa phosphatidylinositol 4-phosphate 5-kinase in Swiss 3T3 cells. *Mol. Biol. Cell* 7:435–42

Reynolds AB, Kanner SB, Wang HC, Parsons JT. 1989. Stable association of activated pp60^{src} with two tyrosine-phosphorylated cellular proteins. *Mol. Cell. Biol.* 9:3951–58

Richardson A, Parsons JT. 1995. Signal transduction through integrins: a central role for focal adhesion kinase? *BioEssays* 17:229–36

Richardson A, Parsons JT. 1996. A mechanism for regulation of the adhesion-associated potein tyrosine kinase pp125^{FAK}. *Nature* 380:538–40

Ridley AJ. 1995. Rho-related proteins: actin cytoskeleton and cell cycle. *Curr. Opin. Genet. Dev.* 5:24–30

Ridley AJ, Comoglio PM, Hall A. 1995. Regulation of scatter factor/hepatocyte growth factor responses by Ras, Rac, and Rho in MDCK cells. *Mol. Cell. Biol.* 15:1110–22

Ridley AJ, Hall A. 1992. The small GTP-binding protein rho regulates the assembly of focal adhesions and actin stress fibers in response to growth factors. *Cell* 70:389–99

Ridley AJ, Hall A. 1994. Signal transduction pathways regulating Rho-mediated stress fibre formation: Requirement for a tyrosine kinase. *EMBO J.* 13:2600–10

Ridley AJ, Paterson HF, Johnston CL, Diekmann D, Hall A. 1992. The small GTP-binding protein rac regulates growth factor-induced membrane ruffling. *Cell* 70:401–10

Ridley AJ, Self AJ, Kasmi F, Paterson HF, Hall A, et al. 1993. rho family GTPase activating proteins p190, bcr and rhoGAP show distinct specificities in vitro and in vivo. *EMBO J.* 12:5151–60

Rodriguez Fernandez JL, Geiger B, Salomon D, Ben-Ze'ev A. 1993. Suppression of vinculin expression by antisense transfection confers changes in cell morphology, motility, and anchorage-dependent growth of 3T3 cells. *J. Cell Biol.* 122:1285–94

Rodriguez Fernandez JL, Geiger B, Salomon D, Sabanay I, Zoller M, Ben-Ze'ev A. 1992. Suppression of tumorigenicity in transformed cells after transfection with vinculin cDNA. *J. Cell Biol.* 119:427–38

Romer LH, Burridge K, Turner CE. 1992. Signaling between the extracellular matrix and the cytoskeleton: tyrosine phosphorylation and focal adhesion assembly. *Cold Spring Harbor Symp. Quant. Biol.* 57:193–202

Romer LH, McLean N, Turner CE, Burridge K. 1994. Tyrosine kinase activity, cytoskeletal organization, and motility in human vascular endothelial cells. *Mol. Biol. Cell* 5:349–61

Ruhnau K, Gaertner A, Wegner A. 1989. Kinetic evidence for insertion of actin monomers between the barbed ends of actin filaments and barbed end-bound insertin, a protein purified from smooth muscle. *J. Mol. Biol.* 210:141–48

Sabe H, Hata A, Okada M, Nakagawa H, Hanafusa H. 1994. Analysis of the binding of the Src homology 2 domain of Csk to tyrosine-phosphorylated proteins in the suppression and mitotic activation of c-Src. *Proc. Natl. Acad. Sci. USA* 91:3984–88

Sadler I, Crawford AW, Michelsen JW, Beckerle MC. 1992. Zyxin and cCRP: two interactive LIM domain proteins associated with the cytoskeleton. *J. Cell Biol.* 119:1573–87

Sakai R, Iwamatsu A, Hirano N, Ogawa S, Tanaka T, et al. 1994. A novel signaling molecule, p130, forms stable complexes in vivo with v-Crk and v-Src in a tyrosine phosphorylation-dependent manner. *EMBO J.* 13:3748–56

Salamero J, Fougereau M, Seckinger P. 1995. Internalization of B cell and pre-B cell receptors is regulated by tyrosine kinase and phosphatase activities. *Eur. J. Immunol.* 25:2757–64

Samuelsson SJ, Luther PW, Pumplin DW, Bloch RJ. 1993. Structures linking microfilament bundles to the membrane at focal contacts. *J. Cell Biol.* 122:485–96

Sanger JM, Mittal B, Pochapin MB, Sanger JW. 1987. Stress fiber and cleavage furrow formation in living cells microinjected with fluorescently labeled α-actinin. *Cell Motil. Cytoskelet.* 7:209–20

Sasaki H, Nagura K, Ishino M, Tobioka H, Kotani K, Sasaki T. 1995. Cloning and characterization of cell adhesion kinase, a novel protein-tyrosine kinase of the focal adhesion kinase subfamily. *J. Biol. Chem.* 270:21206–19

Sastry SK, Horwitz AF. 1993. Integrin cytoplasmic domains: mediators of cytoskeletal linkages and extra- and intracellular initiated transmembrane signaling. *Curr. Opin. Cell Biol.* 5:819–31

Sastry SK, Lakonishok M, Thomas DA, Muschler J, Horwitz AF. 1996. Integrin alpha subunit ratios, cytoplasmic domains, and growth factor synergy regulate muscle proliferation and differentiation. *J. Cell Biol.* 133:169–84

Sato N, Funamaya N, Yonemura S, Tsukita S. 1992. A gene family consisting of ezrin, radixin and moesin. Its specific localization at actin filament/plasma membrane association sites. *J. Cell Sci.* 103:131–43

Satterwhite LL, Lohka MJ, Wilson KL, Scherson TY, Cisek LJ, et al. 1992. Phosphorylation of myosin-II regulatory light chain by cyclin-p34cdc2: a mechanism for the timing of cytokinesis. *J. Cell Biol.* 118:595–605

Schaller MD, Borgman CA, Cobb BS, Vines RR, Reynolds AB, Parsons JT. 1992. pp125^{FAK} a structurally distinctive protein-tyrosine kinase associated with focal adhesions. *Proc. Natl. Acad. Sci. USA* 89:5192–96

Schaller MD, Hildebrand JD, Shannon JD, Fox JW, Vines RR, Parsons JT. 1994. Autophosphorylation of the focal adhesion kinase, pp125^{FAK}, directs SH2-dependent binding of pp60^{src}. *Mol. Cell. Biol.* 14:1680–88

Schaller MD, Otey CA, Hildebrand JD, Parsons JT. 1995. Focal adhesion kinase and paxillin bind to peptides mimicking beta integrin cytoplasmic domains. *J. Cell Biol.* 130:1181–87

Schaller MD, Parsons JT. 1994. Focal adhesion kinase and associated proteins. *Curr. Opin. Cell Biol.* 6:705–10

Schaller MD, Parsons JT. 1995. pp125^{FAK}-dependent tyrosine phosphorylation of paxillin creates a high-affinity binding site for Crk. *Mol. Cell. Biol.* 15:2635–45

Schlaepfer DD, Hanks SK, Hunter T, van der Geer P. 1994. Integrin-mediated signal transduction linked to Ras pathway by GRB2 binding to focal adhesion kinase. *Nature* 372:786–91

Schmeichel KL, Beckerle MC. 1994. The LIM domain is a modular protein-binding interface. *Cell* 79:211–19

Schwartz MA. 1993. Spreading of human endothelial cells on fibronectin or vitronectin triggers elevation of intracellular free calcium. *J. Cell Biol.* 120:1003–10

Schwartz MA, Both G, Lechene C. 1989. Effect of cell spreading on cytoplasmic pH in normal and transformed fibroblasts. *Proc. Natl. Acad. Sci. USA* 86:4525–29

Schwartz MA, Brown EJ, Fazeli B. 1993. A 50-kDa integrin-associated protein is required for integrin-regulated calcium entry in endothelial cells. *J. Biol. Chem.* 268:19931–34

Schwartz MA, Cragoe EJ Jr, Lechene CP. 1990. pH regulation in spread cells and round cells. *J. Biol. Chem.* 265:1327–32

Schwartz MA, Denninghoff K. 1994. Alpha v integrins mediate the rise in intracellular calcium in endothelial cells on fibronectin even though they play a minor role in adhesion. *J. Biol. Chem.* 269:11133–37

Schwartz MA, Ingber DE, Lawrence M, Springer TA, Lechene C. 1991a. Multiple integrins share the ability to induce elevation of intracellular pH. *Exp. Cell. Res.* 195:533–35

Schwartz MA, Lechene C, Ingber DE. 1991b. Insoluble fibronectin activates the Na/H antiporter by clustering and immobilizing integrin $\alpha_5\beta_1$, independent of cell shape. *Proc. Natl. Acad. Sci. USA* 88:7849–53

Schwartz MA, Schaller MD, Ginsberg MH. 1995. Integrins: emerging paradigms of signal transduction. *Annu. Rev. Cell Biol.* 11:549–99

Seckl M, Rozengurt E. 1993. Tyrphostin inhibits bombesin stimulation of tyrosine phosphorylation, c-fos expression, and DNA synthesis in Swiss 3T3 cells. *J. Biol. Chem.* 268:9548–54

Sekine A, Fujiwara M, Narumiya S. 1989. Asparagine residue in the *rho* gene product is the modification site for botulinum ADP-ribosyltransferase. *J. Biol. Chem.* 264:8602–5

Sellers JR. 1991. Regulation of cytoplasmic and smooth muscle myosin. *Curr. Opin. Cell Biol.* 3:98–104

Serra-Pages C, Kedersha NL, Fazikas L, Medley Q, Debant A, Streuli M. 1995. The LAR transmembrane protein tyrosine phosphatase and a coiled-coil LAR-interacting protein co-localize at focal junctions. *EMBO J.* 14:2827–38

Settleman J, Albright CF, Foster LC, Weinberg RA. 1992. Association between GTPase activators for Rho and Ras families. *Nature* 359:153–54

Seufferlein T, Rozengurt E. 1994. Lysophosphatidic acid stimulates tyrosine phosphorylation of focal adhesion kinase, paxillin, and p130. Signaling pathways and cross-talk with platelet-derived growth factor. *J. Biol. Chem.* 269:9345–51

Seufferlein T, Rozengurt E. 1995. Sphingosylphosphorylcholine rapidly induces tyrosine phosphorylation of p125FAK and paxillin, rearrangement of the actin cytoskeleton and focal contact assembly. Requirement of p21rho in the signaling pathway. *J. Biol. Chem.* 270:24343–51

Shattil SJ, Haimovich B, Cunningham M, Lipfert L, Parsons JT, et al. 1994. Tyrosine phosphorylation of pp125^{FAK} in platelets requires coordinated signaling through integrin and agonist receptors. *J. Biol. Chem.* 269:14738–45

Shattil SJ, O'Toole T, Eigenthaler M, Thon V, Williams M, et al. 1995. Beta 3-endonexin, a novel polypeptide that interacts specifically with the cytoplasmic tail of the integrin beta 3 subunit. *J. Cell Biol.* 131:807–16

Shibasaki F, Fukami K, Fukui Y, Takenawa T. 1994. Phosphatidylinositol 3-kinase binds to α-actinin through the p85 subunit. *Biochem. J.* 302:551–57
Shimizu H, Ito M, Miyahara M, Ichikawa K, Okubo S, et al. 1994. Characterization of the myosin-binding subunit of smooth muscle myosin phosphatase. *J. Biol. Chem.* 269:30407–11
Shimo K, Gyotoku Y, Arimitsu Y, Kakiuchi T, Mizuguchi J. 1993. Participation of tyrosine kinase in capping, internalization, and antigen presentation through membrane immunoglobulin in BAL17 B lymphoma cells. *FEBS Lett.* 323:171–74
Shirazi A, Iizuka K, Fadden P, Mosse C, Somlyo AP, et al. 1994. Purification and characterization of the mammalian myosin light chain phosphatase holoenzyme. The differential effects of the holoenzyme and its subunits on smooth muscle. *J. Biol. Chem.* 269:31598–606
Singer II, Kawka DW, Scott S, Mumford RA, Lark MW. 1987. The fibronectin cell attachment sequence Arg-Gly-Asp-Ser promotes focal contact formation during early fibroblast attachment and spreading. *J. Cell Biol.* 104:573–84
Singer II, Scott S, Kawka DW, Kazazis DM, Gailit J, Ruoslahti E. 1988. Cell surface distribution of fibronectin and vitronectin receptors depends on substrate composition and extracellular matrix accumulation. *J. Cell Biol.* 106:2171–82
Sinnett-Smith J, Zachary I, Valverde AM, Rozengurt E. 1993. Bombesin stimulation of p125 focal adhesion kinase tyrosine phosphorylation. Role of protein kinase C, Ca^{2+} mobilization, and the actin cytoskeleton. *J. Biol. Chem.* 268:14261–68
Smalheiser NR, Kim E. 1995. Purification of cranin, a laminin binding membrane protein. Identity with dystroglycan and reassessment of its carbohydrate moieties. *J. Biol. Chem.* 270:15425–33
Smalheiser NR, Schwartz NB. 1987. Cranin: a laminin-binding protein of cell membranes. *Proc. Natl. Acad. Sci. USA* 84:6457–61
Solowska J, Guan JL, Marcantonio EE, Trevithick JE, Buck CA, Hynes RO. 1989. Expression of normal and mutant avian integrin subunits in rodent cells. *J. Cell Biol.* 109:853–61. Erratum. 1989. *J. Cell Biol.* 109(4 Pt. 1):1187
Somlyo AP, Somlyo AV. 1994. Signal transduction and regulation in smooth muscle. *Nature* 372:231–36. Erratum. 1994. *Nature* 372:812
Stasia MJ, Jouan A, Bourmeyster N, Boquet P, Vignais PV. 1991. ADP-ribosylation of a small size GTP-binding protein in bovine neutrophils by the C3 exoenzyme of *Clostridium botulinum* and effect on the cell motility. *Biochem. Biophys. Res. Commun.* 180:615–22
Stoker M, O'Neill C, Berryman S, Waxman V. 1968. Anchorage and growth regulation in normal and virus-transformed cells. *Int. J. Cancer* 3:683–93
Streeter HB, Rees DA. 1987. Fibroblast adhesion to RGDS shows novel features compared with fibronectin. *J. Cell Biol.* 105:507–15
Suidan HS, Stone SR, Hemmings BA, Monard D. 1992. Thrombin causes neurite retraction in neuronal cells through activation of cell surface receptors. *Neuron* 8:363–75
Tachibana K, Sato T, D'Avirro N, Morimoto C. 1995. Direct association of $pp125^{FAK}$ with paxillin, the focal adhesion-targeting mechanism of $pp125^{FAK}$. *J. Exp. Med.* 182:1089–99
Takai Y, Sasaki T, Tanaka K, Nakanishi H. 1995. Rho as a regulator of the cytoskeleton. *Trends Biochem. Sci.* 20:227–31
Takaishi K, Kikuchi A, Kuroda S, Kotani K, Sasaki T, Takai Y. 1993. Involvement of rho p21 and its inhibitory GDP/GTP exchange protein (rho GDI) in cell motility. *Mol. Cell. Biol.* 13:72–79
Takaishi K, Sasaki T, Kato M, Yamochi W, Kuroda S, et al. 1994. Involvement of Rho p21 small GTP-binding protein and its regulator in the HGF-induced cell motility. *Oncogene* 9:273–79
Takeuchi K, Takahashi K, Abe M, Nishida W, Hiwada K, et al. 1991. Co-localization of immunoreactive forms of calponin with actin cytoskeleton in platelets, fibroblasts, and vascular smooth muscle. *J. Biochem.* 109:311–16
Tan JL, Ravid S, Spudich JA. 1992. Control of nonmuscle myosins by phosphorylation. *Annu. Rev. Biochem.* 61:721–59
Tanaka S, Morishita T, Hashimoto Y, Hattori S, Nakamura S, et al. 1994. C3G, a guanine nucleotide-releasing protein expressed ubiquitously, binds to the Src homology domains of CRK and GRB2/ASH proteins. *Proc. Natl. Acad. Sci. USA* 91:3443–47
Tansey MG, Luby-Phelps K, Kamm KE, Stull JT. 1994. Ca(2+)-dependent phosphorylation of myosin light chain kinase decreases the Ca^{2+} sensitivity of light chain phosphorylation within smooth muscle cells. *J. Biol. Chem.* 269:9912–20
Tapley P, Horwitz A, Buck C, Duggan K, Rohrschneider L. 1989. Integrins isolated from Rous sarcoma virus-transformed chicken embryo fibroblasts. *Oncogene* 4:325–33
Theriot JA, Rosenblatt J, Portnoy DA, Goldschmidt-Clermont PJ, Mitchison TJ.

1994. Involvement of profilin in the actin-based motility of L. monocytogenes in cells and in cell-free extracts. *Cell* 76:505–17

Thomas SM, Soriano P, Imamoto A. 1995. Specific and redundant roles of Src and Fyn in organizing the cytoskeleton. *Nature* 376:267–71

Tomasek JJ, Haaksma CJ. 1991. Fibronectin filaments and actin microfilaments are organized into a fibronexus in Dupuytren's diseased tissue. *Anat. Rec.* 230:175–82

Tomasek JJ, Haaksma CJ, Eddy RJ, Vaughan MB. 1992. Fibroblast contraction occurs on release of tension in attached collagen lattices: dependency on an organized actin cytoskeleton and serum. *Anat. Rec.* 232:359–68

Tominaga T, Sugie K, Hirata M, Morii N, Fukata J, et al. 1993. Inhibition of PMA-induced, LFA-1-dependent lymphocyte aggregation by ADP ribosylation of the small molecular weight GTP binding protein, rho. *J. Cell Biol.* 120:1529–37

Trinkle-Mulcahy L, Ichikawa K, Hartshorne DJ, Siegman MJ, Butler TM. 1995. Thiophosphorylation of the 130-kDa subunit is associated with a decreased activity of myosin light chain phosphatase in alpha-toxin-permeabilized smooth muscle. *J. Biol. Chem.* 270:18191–94

Tsukita S, Hieda Y. 1989. A new 82-kD barbed end-capping protein (radixin) localized in the cell-to-cell adherens junction: purification and characterization. *J. Cell Biol.* 108:2369–82

Turner CE. 1994. Paxillin: a cytoskeletal target for tyrosine kinases. *BioEssays* 16:47–52

Turner CE, Glenney JR Jr, Burridge K. 1990. Paxillin: a new vinculin-binding protein present in focal adhesions. *J. Cell Biol.* 111:1059–68

Turner CE, Miller JT. 1994. Primary sequence of paxillin contains putative SH2 and SH3 domain binding motifs and multiple LIM domains: identification of a vinculin and pp125Fak-binding region. *J. Cell Sci.* 107:1583–91

Turner CE, Pavalko FM, Burridge K. 1989. The role of phosphorylation and limited proteolytic cleavage of talin and vinculin in the disruption of focal adhesion integrity. *J. Biol. Chem.* 264:11938–44

Uddin S, Katzav S, White MF, Platanias LC. 1995. Insulin-dependent tyrosine phosphorylation of the vav protooncogene product in cells of hematopoietic origin. *J. Biol. Chem.* 270:7712–16

Ullrich A, Schlessinger J. 1990. Signal transduction by receptors with tyrosine kinase activity. *Cell* 61:203–12

van der Flier A, Kuikman I, Baudoin C, van der Neut R, Sonnenberg A. 1995. A novel beta 1 integrin isoform produced by alternative splicing: unique expression in cardiac and skeletal muscle. *FEBS Lett.* 369:340–44

Varner JA, Emerson DA, Juliano RL. 1995. Integrin $\alpha_5\beta_1$ expression negatively regulates cell growth: reversal by attachment to fibronectin. *Mol. Biol. Cell* 6:725–40

Vojtek AB, Cooper JA. 1995. Rho family members: activators of MAP kinase cascades. *Cell* 82:527–29

Volberg T, Geiger B, Citi S, Bershadsky AD. 1994. Effect of protein kinase inhibitor H-7 on the contractility, integrity, and membrane anchorage of the microfilament system. *Cell Motil. Cytoskelet.* 29:321–38

Vuori K, Hirai H, Aizawa S, Ruoslahti E. 1996. Induction of p130^{cas} signaling complex formation upon integrin-mediated cell adhesion: a role for src family kinases. *Mol. Cell. Biol.* 16:2606–13

Vuori K, Ruoslahti E. 1993. Activation of protein kinase C precedes $\alpha_5\beta_1$ integrin-mediated cell spreading on fibronectin. *J. Biol. Chem.* 268:21459–62

Vuori K, Ruoslahti E. 1994. Association of insulin receptor substrate-1 with integrins. *Science* 266:1576–78

Vuori K, Ruoslahti E. 1995. Tyrosine phosphorylation of p130^{cas} and cortactin accompanies integrin-mediated cell adhesion to extracellular matrix. *J. Biol. Chem.* 270:22259–62

Wachsstock DH, Wilkins JA, Lin S. 1987. Specific interaction of vinculin with α-actinin. *Biochem. Biophys. Res. Commun.* 146:554–60

Watanabe G, Saito Y, Madaule P, Ishizaki T, Fujisawa K, et al. 1996. Protein kinase N (PKN) and PKN-related protein rhophilin as targets of small GTPase rho. *Science* 271:645–48

Weber K, Groeschel-Stewart U. 1974. Antibody to myosin: the specific visualization of myosin-containing filaments in nonmuscle cells. *Proc. Natl. Acad. Sci. USA* 71:4561–65

Weekes J, Barry ST, Critchley DR. 1996. Acidic phospholipids inhibit the intramolecular association between the N- and C-terminal regions of vinculin, exposing actin-binding and protein kinase C phosphorylation sites. *Biochem. J.* 314:827–32

Wehland J, Osborn M, Weber K. 1974. Cell-to-substratum contacts in living cells: a direct correlation between the interference-reflexion and indirect-immunofluorescence microscopy in using antibodies against actin and α-actinin. *J. Cell Sci.* 37:257–73

Weigt C, Gaertner A, Wegner A, Korte H, Meyer

HE. 1992. Occurrence of an actin-inserting domain in tensin. *J. Mol. Biol.* 227:593–95

Weng Z, Taylor JA, Turner CE, Brugge JS, Seidel-Dugan C. 1993. Detection of Src homology 3-binding proteins, including paxillin, in normal and v-Src-transformed Balb/c 3T3 cells. *J. Biol. Chem.* 268:14956–63

Wilkins JA, Risinger MA, Coffey E, Lin S. 1987. Purification of a vinculin binding protein from smooth muscle. *J. Cell Biol.* 104:A130

Wilkins JA, Risinger MA, Lin S. 1986. Studies on proteins that co-purify with smooth muscle vinculin: identification of immunologically related species in focal adhesions of nonmuscle and Z-lines of muscle cells. *J. Cell Biol.* 103:1483–94

Wilson L, Carrier MJ, Kellie S. 1995. pp125^{FAK} tyrosine kinase activity is not required for the assembly of F-actin stress fibres and focal adhesions in cultured mouse aortic smooth muscle cells. *J. Cell Sci.* 108:2381–91

Winder SJ, Walsh MP. 1990. Smooth muscle calponin. Inhibition of actomyosin MgATPase and regulation by phosphorylation. *J. Biol. Chem.* 265:10148–55

Woods A, Couchman JR. 1992. Protein kinase C involvement in focal adhesion formation. *J. Cell Sci.* 101:277–90

Woods A, Couchman JR. 1994. Syndecan 4 heparan sulfate proteoglycan is a selectively enriched and widespread focal adhesion component. *Mol. Biol. Cell* 5:183–92

Woods A, Couchman JR, Johansson S, Hook M. 1986. Adhesion and cytoskeletal organisation of fibroblasts in response to fibronectin fragments. *EMBO J.* 5:665–70

Xing Z, Chen HC, Nowlen JK, Taylor SJ, Shalloway D, Guan JL. 1994. Direct interaction of v-Src with the focal adhesion kinase mediated by the Src SH2 domain. *Mol. Biol. Cell* 5:413–21

Xu L, Owens LV, Sturge GC, Yang X, Liu ET, et al. 1996. Attenuation of the expression of the focal adhesion kinase induces apoptosis in tumor cells. *Cell Growth Diff.* 7:413–18

Yamada KM, Miyamoto S. 1995. Integrin transmembrane signaling and cytoskeletal control. *Curr. Opin. Cell Biol.* 7:681–89

Yamakita Y, Yamashiro S, Matsumura F. 1994. In vivo phosphorylation of regulatory light chain of myosin II during mitosis of cultured cells. *J. Cell Biol.* 124:129–37

Yamashiro S, Yamakita Y, Hosoya H, Matsumura F. 1991. Phosphorylation of non-muscle caldesmon by p34cdc2 kinase during mitosis. *Nature* 349:169–72

Yamashiro S, Yamakita Y, Ishikawa R, Matsumura F. 1990. Mitosis-specific phosphorylation causes 83K non-muscle caldesmon to dissociate from microfilaments. *Nature* 344:675–78

Yamashiro S, Yamakita Y, Yoshida K, Takiguchi K, Matsumura F. 1995. Characterization of the COOH terminus of non-muscle caldesmon mutants lacking mitosis-specific phosphorylation sites. *J. Biol. Chem.* 270:4023–30

Ylanne J, Chen Y, O'Toole TE, Loftus JC, Takada Y, Ginsberg MH. 1993. Distinct functions of integrin α and β subunit cytoplasmic domains in cell spreading and formation of focal adhesions. *J. Cell Biol.* 122:223–33

Zachary I, Sinnett-Smith J, Rozengurt E. 1992. Bombesin, vasopressin, and endothelin stimulation of tyrosine phosphorylation in Swiss 3T3 cells. Identification of a novel tyrosine kinase as a major substrate. *J. Biol. Chem.* 267:19031–34

Zachary I, Sinnett-Smith J, Turner CE, Rozengurt E. 1993. Bombesin, vasopressin, and endothelin rapidly stimulate tyrosine phosphorylation of the focal adhesion-associated protein paxillin in Swiss 3T3 cells. *J. Biol. Chem.* 268:22060–65

Zhang Z, Vuori K, Reed JC, Ruoslahti E. 1995. The alpha-5-beta-1 integrin supports survival of cells on fibronectin and up-regulates Bcl-2 expression. *Proc. Natl. Acad. Sci. USA* 92:6161–65

Zhidkova NI, Belkin AM, Mayne R. 1995. Novel isoform of beta 1 integrin expressed in skeletal and cardiac muscle. *Biochem. Biophys. Res. Commun.* 214:279–85

Zhu X, Assoian RK. 1995. Integrin-dependent activation of MAP kinase: a link to shape-dependent cell proliferation. *Mol. Biol. Cell* 6:273–82

Zigmond SH, Otto JJ, Bryan J. 1979. Organization of myosin in a submembranous sheath in well-spread human fibroblasts. *Exp. Cell. Res.* 119:205–19

Annu. Rev. Cell Dev. Biol. 1996. 12:519–41

SIGNALING BY EXTRACELLULAR NUCLEOTIDES

Anthony J. Brake and David Julius
Department of Cellular and Molecular Pharmacology, Programs in Cell Biology and Neuroscience, University of California, San Francisco, California 94143-0450

KEY WORDS: purinergic receptors, ligand-gated ion channels, synaptic transmission, apoptosis

ABSTRACT

ATP and other nucleotides can be released from cells through regulated pathways, or following the loss of plasma membrane integrity. Once outside the cell, these compounds take on new roles as intercellular signaling molecules that elicit a broad spectrum of physiological responses through the activation of numerous cell surface receptor subtypes. This review summarizes recent advances in the molecular characterization of ATP receptors and discusses roles for cloned receptors in established and novel physiological processes.

CONTENTS

1081-0706/96/1115-0519$08.00

INTRODUCTION

Every biology student knows that adenosine 5′-triphosphate (ATP) is both a ubiquitous carrier of chemical energy and a building block of genetic material in all living organisms. Less well-appreciated is the fact that ATP (and other nucleotides) exerts potent physiological actions on a variety of tissues and cell types through its role as an extracellular signaling molecule. These actions include the modulation of vascular tone, electrolyte transport, mast cell degranulation, cell death, and synaptic transmission (Burnstock 1972, 1976, 1990b, Gordon 1986). Although biological effects of extracellular ATP have been documented for over 60 years (Drury & Szent-Gyorgi 1929), there has been some reluctance to fully accept the idea that nucleotides have the capacity to act as autocrine and paracrine chemical messengers, even though other common cytoplasmic metabolites (e.g. glycine, glutamate, and calcium) are known to function in this manner. Slow progress on this front is largely a reflection of the very limited pharmacopoeia of stable and selective agonists, antagonists, and radioligands with which to identify specific cell surface ATP receptor subtypes and to characterize cellular mechanisms of nucleotide release. ATP also has the unique characteristic of being rapidly hydrolyzed to another extracellular messenger, adenosine, through the action of ectonucleotidases (Pearson & Gordon 1985, Slakey et al 1990). Because adenosine activates its own family of receptors (Bruns 1990), this enzymatic conversion extends the range of signaling possibilities for ATP, but at the same time helps to confound the identification of bona fide ATP receptors in many cellular and physiological systems.

Despite such difficulties, ATP (P_2 purinergic) receptors and adenosine (P_1 purinergic; also known as A_1-A_3) receptors have been distinguished from each other based on the relative potency of various nucleoside and nucleotide agonists and by the differential actions of a limited and somewhat nonselective collection of antagonists (Burnstock 1978, Paton & Taerum 1990). Using similar criteria, ATP receptors have themselves been placed into distinct pharmacological subclasses (Burnstock & Kennedy 1985), and electrophysiological studies have shown that ATP can activate both G protein-coupled receptors that signal through the elaboration of cytoplasmic second messengers (metabotropic receptors) (O'Connor et al 1991) and ligand-gated channels that mediate direct and rapid increases in membrane ion conductance (ionotropic receptors) (Bean 1992). In this regard, ATP resembles classical neurotransmitters such as acetylcholine, GABA, glutamate, and serotonin in its ability to mediate both slow and fast responses via these two primary signaling mechanisms. Indeed, the application of sensitive patch-clamp recording methods to the analysis of membrane currents in brain slice preparations has led to the recent discovery of ATP-evoked synaptic currents in the brain (Edwards et al 1992). These findings have sparked tremendous interest in uncovering roles for ATP as a novel fast excitatory neurotransmitter in synapses of the central nervous system. But the brain is just one of many organs that express ATP receptors, and there is great promise that P_2 purinergic receptors can be targeted for the development of novel drug-based therapies for the management of airway and cardiovascular diseases, inflammation and pain syndromes, and behavioral and cognitive disorders. To realize this potential, it is essential to move beyond the restricted view of purinergic receptors attainable with the limited pharmacological tools available and to acquire a more complete picture of the ATP receptor family using molecular genetic approaches. In the past three years, there has been considerable progress in this area, and it is therefore now appropriate to review these findings in the context of known and proposed physiological roles for extracellular nucleotides.

PHARMACOLOGICAL CLASSIFICATION OF ATP RECEPTORS

P_2 purinoceptors subtypes have been defined according to the relative potency of various nucleotide agonists. By these criteria, five main subtypes with distinct pharmacological profiles have been proposed to exist. Subtype assignments have also relied on differences in signal transduction mechanisms. Electrophysiological studies have shown that P_{2X} receptors mediate rapid excitatory responses through the opening of intrinsic cation-selective channels (Bean & Friel 1990). In contrast, responses attributed to P_{2Y}, P_{2U}, or P_{2T} receptors involve the activation of phospholipase C and the mobilization of intracellular calcium via G protein-dependent mechanisms (Dubyak 1986, Okajima et al 1987). The least well-understood subtype is the P_{2Z} receptor, whose unique response is characterized by the opening of nonselective membrane pores permeable to hydrophilic solutes as large as 900 daltons (Dahlquist & Diamant 1974, Cockcroft & Gomperts 1980, Buisman et al 1988). According to these distinctions, ATP receptors have been categorized as follows:

P_{2Y}: 2-methylthio ATP > ATP $\approx$ ADP > α,β-methylene ATP, β,γ-methylene ATP;

P_{2U}: ATP $\approx$ UTP $\gg$ 2-methylthio ATP;

P_{2T}: 2-methylthio ADP > ADP (ATP is an antagonist);

P_{2X}: α,β-methylene ATP > β,γ-methylene ATP; > ATP > ADP;

P_{2Z}: ATP^{4-}.

Not all physiological responses to ATP can be satisfactorily accounted for by the actions of any one, or combination, of these five principal subtypes. Such observations suggest that the P_2 receptor family extends beyond the putative subtypes that have been detected with the limited collection of agonists, antagonists, and radioligands available. Recent molecular cloning studies have shown that the outline of P_2 receptor subtypes noted above is surprisingly accurate but oversimplified. The isolation of cDNAs encoding P_2 receptors has revealed greater heterogeneity of receptor subtypes within these major subgroups (see below).

PHYSIOLOGICAL ROLES FOR EXTRACELLULAR ATP

The Cardiovascular System

The importance of ATP as an extracellular signaling molecule was first appreciated in the cardiovascular system. ATP affects vascular tone and blood flow by causing either constriction or dilation of peripheral vasculature. These responses are mediated by distinct receptor subtypes located on endothelial and smooth muscle cells (Burnstock 1989, 1990a). Vasoconstriction is mediated through the activation of ionotropic P_{2X} receptors present on vascular smooth muscle cells (Benham 1990). Synaptic release of ATP from sympathetic perivascular neurons activates excitatory P_{2X} receptors to produce the so-called nonadrenergic, noncholinergic component of vasoconstriction that remains after blockade of adrenergic and cholinergic receptors. The purinergic nature of this response is demonstrated by its attenuation following application of α,β-methylene ATP, which desensitizes these receptors. Interestingly, adenosine that is derived from ectonucleotidase-catalyzed degradation of ATP serves to modulate vasoconstriction through inhibitory actions on both perivascular nerves and smooth muscle cells.

Vasodilation is mediated via activation of metabotropic ATP receptors on cells of the endothelial bed (Boeynaems & Pearson 1990). ATP released by platelets or endothelial cells inside blood vessels activates G protein-coupled P_{2Y} and P_{2U} receptors, leading to activation of phospholipase C and the mobilization of intracellular calcium (Motte et al 1995). This, in turn, results in production of endothelium-derived relaxation factor (EDRF; now known to be nitric oxide), which results in relaxation of smooth muscle and thus vasodilation.

Within the cardiovascular system, ATP and ADP also function as paracrine regulators of platelet aggregation. Platelets aggregating at sites of damage in arterial walls release high concentrations of ATP and ADP as constituents of their dense granules. ADP recruits additional platelets through activation of

P_{2T} receptors on their surface. These nucleotides also act on P_{2Y} receptors on endothelial cells to stimulate nitric oxide and prostacyclin release, both of which inhibit platelet aggregation.

Synaptic Transmission

The application of exogenous ATP to a variety of cell types in vitro, including smooth, skeletal, and cardiac muscle cells (Hume & Honig 1986, Nakazawa & Matsuki 1987, Benham & Tsien 1987, Friel 1988, Friel & Bean 1988); parasympathetic cardiac neurons (Fieber & Adams 1991); PC12 pheochromocytoma cells (Nakazawa et al 1990b); and sensory and spinal cord neurons (Jahr & Jessell 1983, Krishtal et al 1983), has demonstrated unequivocally that extracellular nucleotides can elicit direct excitatory actions that lead to rapid membrane depolarization and action potential firing. These effects are mediated primarily through the activation of P_{2X} receptors, which are ATP-gated cation-selective channels. Upon opening, these channels allow sodium and potassium ions to flow across the membrane according to their electrochemical gradients, with little or no selectivity exhibited among small monovalent cations. Differences in agonist selectivity, unitary conductance, permeability to calcium ions, and kinetic parameters of channel desensitization in these cell types suggest that P_{2X} receptors are a heterogeneous collection of ion channel subtypes (Bean & Friel 1990).

These in vitro findings strongly suggested, but did not prove, that neuronally released endogenous ATP can act as a bona fide neurotransmitter at neuron-to-neuron and neuron-to-effector synapses. The quest to validate a role for extracellular ATP in neurotransmission has been played out in regions of the nervous system where synaptic vesicles are known to contain and release ATP in conjunction with other transmitters, particularly norepinephrine and acetylcholine (Sneddon & Burnstock 1984, Richardson & Brown 1987). Within the peripheral nervous system, a role for ATP as an endogenous transmitter has been established through the analysis of ATP-mediated excitatory postsynaptic currents in cultured coeliac ganglion neurons, which normally supply sympathetic input to various visceral organs (Evans et al 1992, Silinsky et al 1992). Within the central nervous system, a role for ATP in synaptic transmission has recently been obtained using an in vitro slice preparation derived from a cholinergic region of the brain called the medial habenula (Edwards et al 1992). In both cases, the identification of synaptic purinergic responses has relied on somewhat indirect pharmacological evidence because selective ATP receptor antagonists are limited. Nevertheless, a convincing case has been made by demonstrating that the stimulation of presynaptic fibers leads to rapid excitatory postsynaptic responses that cannot be blocked by drugs that antagonize the actions of all other known fast excitatory and inhibitory transmitters, including acetylcholine, GABA, glutamate, glycine, and serotonin. Moreover, suramin and α,β-methylene ATP, which are nonselective in their ability to antagonize or desensitize various P_2 receptor subtypes, but which have no known actions on other transmitter systems, were shown to significantly decrease fast excitatory synaptic currents in these preparations. The clear demonstration that ATP acts as an endogenous fast neurotransmitter in mammalian synapses is seen as a significant development because ATP now becomes one of a rather small group of molecules that fulfills this biological role. Thus a substantial component of autonomic control of visceral (urinary bladder, vas deferens, digestive tract) and vascular (peripheral and cardiac) functions may be mediated by ATP acting at synapses between pre- and postganglionic fibers, or at target organs that receive sympathetic or parasympathetic input. Within the brain, glutamate is recognized as the primary fast excitatory transmitter, with acetylcholine and serotonin playing more limited roles. It remains to be seen whether ATP contributes significantly to fast excitatory transmission at central synapses.

Proposed roles for ATP as a transmitter in the CNS can also be extended to include actions within synapses of the spinal cord (Holton & Holton 1954), particularly with regard to the central transduction of pain sensation. Terminals from primary sensory afferent nociceptors (specialized sensory neurons that transduce thermal, mechanical, and noxious chemical stimuli into electrochemical signals) make synaptic connections with interneurons within superficial layers of the spinal cord dorsal horn. These interneurons synapse with both local reflex circuits and ascending pathways that transmit sensory signals to the brain. ATP has been shown to act as a potent fast excitatory agent for a subpopulation of primary sensory neurons and dorsal horn neurons in culture (Jahr & Jessell 1983, Krishtal et al 1983, Fyffe & Perl 1984). These findings suggest that ATP may participate as a chemical messenger at both peripheral and central sites along the pain transduction pathway. ATP released at the site of tissue damage may activate P_{2X} receptors on the peripheral terminals of primary sensory nociceptors to trigger an action potential. ATP may also be released from activated sensory afferents where they make synaptic connections with spinal interneurons of the dorsal horn. Although whole-cell recordings made from neurons within dorsal horn slices indicated that ATP may function primarily to enhance the excitatory action of glutamate within the spinal cord (Li & Perl 1995), a role for ATP as a principal excitatory transmitter at these synapses cannot be ruled out in light of pharmacological limitations and other experimental difficulties inherent in detecting ATP-mediated synaptic responses.

ATP also mediates inhibitory actions in the nervous system through the activation of presynaptic G protein-coupled receptors. Thus ATP can inhibit its own release, together with that of its co-transmitters, via an autoregulatory feedback mechanism (Silinsky et al 1990). In addition, ectonucleotidase conversion of ATP to adenosine can contribute to this effect through the activation of prejunctional inhibitory adenosine receptors. Although these inhibitory mechanisms constitute important physiological roles for purinergic receptors in the nervous system, the primary focus has been on fast excitatory actions mediated by P_{2X} receptors, in large part because this is a more direct and straightforward phenomenon to analyze.

The Immune System

Extracellular ATP is known to act on a variety of cells within the immune system, including mast cells, macrophages, and lymphocytes (Wiley et al 1994, Zambon et al 1994, Di Virgilio 1995). Indeed, the mast cell has provided one of the clearest insights into the molecular mechanisms whereby ATP exerts its physiological effects. During mast cell degranulation, ATP is co-released from secretory granules together with histamine and other mediators of anaphylaxis. This extracellular ATP then functions as a paracrine signaling molecule by activating P_2 receptors on nearby mast cells and potentiating further degranulation through mobilization of intracellular calcium (Osipchuk & Cahalan 1992). By recruiting neighboring mast cells in this manner, amplification of the degranulation response is achieved. Because this response is inhibited by suramin, or by desensitizing concentrations of ATP, it has been concluded that histamine and other signaling molecules are not involved in this process.

One of the more curious actions of ATP is its ability to induce the formation of large, nonselective plasma membrane pores through the activation of P_{2Z} receptors present on mast cells, lymphocytes, macrophages, and related cell lines (Di Virgilio 1995). Exposure of these cells to ATP^{4-} (i.e. Ca^{2+}, Mg^{2+}-free form), which constitutes a low, but appreciable percentage of total ATP in Mg^{2+}- and Ca^{2+}-containing solutions, results in their permeabilization to dyes such as Lucifer Yellow and ethidium bromide and can lead to lytic death following prolonged application. P_{2Z}-mediated pore formation might occur in vivo when macrophages are engulfing their prey, thereby facilitating syncitial formation of macrophage aggregates for the coordinated destruction of large cells by phagocytosis. In this scenario, ATP could be derived from either of the cell types involved.

In addition to the lytic cell death mechanism described above, extracellular ATP has been implicated as an initiator of programmed cell death (apoptosis) in thymocytes, macrophages, and lymphocytes. For example, exposure of mouse macrophages to extracellular ATP leads to nuclear DNA fragmentation and release of processed interleukin-1β, two responses that are associated with apoptotic death in some cell types (Nett-Fiordalisi et al 1995). However provocative these observations, little is known about the specific receptor subtypes or signaling mechanisms that might be involved in ATP-mediated apoptosis, nor is there substantial genetic, biochemical, or pharmacological evidence to support in vivo roles for extracellular ATP in programmed cell death. Further progress in this area will require the molecular identification of ATP receptor subtypes expressed in these model systems. In this regard, a recently cloned ATP-gated ion channel subtype (P^1_{2X}) appears to correspond to a gene (*RP-2*) whose expression is induced in immature thymocytes undergoing programmed cell death (Owens et al 1991, Valera et al 1994). Manipulation of cDNAs encoding this and other cloned ATP receptors in transfected cell systems and transgenic animals will help to clarify potential mechanisms of ATP-mediated apoptosis.

Nonvesicular Mechanisms of ATP Release

An important issue that has yet to be resolved concerns the mechanism whereby ATP is released from intact, nonexcitable cells, particularly those having no regulated exocytotic release pathway. One interesting proposal is that ATP can be pumped out of cells by various members of the ABC transporter family, such as the multidrug resistance (MDR) protein or the cystic fibrosis transmembrane conductance regulator (CFTR) (Abraham et al 1993, Reisin et al 1994). Part of the reason that this model has attracted attention is because it would provide a parsimonious molecular explanation as to why application of exogenous extracellular ATP or UTP can partially reverse epithelial electrolyte transport defects and clinical symptoms associated with cystic fibrosis (al-Awqati 1995). In this scenario, ATP that is transported out of the cell by CFTR might act on nearby P_{2U} receptors to regulate the activity of an outwardly rectifying chloride channel (Schwiebert et al 1995). Accordingly, individuals with cystic fibrosis would exhibit defects in chloride channel function as the result, in part, of a decrease in ATP receptor activation. Although the ability of CFTR to pump ATP from epithelial and endothelial cells has recently been disputed (Reddy et al 1996), the fact remains that P_{2U} receptors are important modulators of chloride and sodium transport in epithelia, and agonists for these receptors are potentially useful as therapeutic agents in the management of cystic fibrosis and other airway diseases (Jiang et al 1993, Knowles et al 1995).

METABOTROPIC ATP RECEPTORS DEFINE A NEW BRANCH OF THE G PROTEIN-COUPLED RECEPTOR SUPERFAMILY

Molecular biology studies have generally confirmed and extended predictions of the structure and diversity of G protein-coupled ATP receptors that emerged from pharmacological, biochemical, and physiological analyses of purinergic signaling systems. Perhaps the most unexpected finding from gene cloning efforts has been the realization that ATP receptors reside in a branch of the G protein-coupled receptor superfamily not closely related to receptors for other purines such as adenosine and cyclic AMP. Additionally, cloning studies have uncovered receptor subtypes displaying novel pharmacological properties that do not conform strictly to the commonly recognized subtypes discussed above. Northern blot and in situ hybridization analyses using receptor cDNAs as probes have begun to delineate patterns of metabotropic P_2 receptor subtype expression, which is turning out to be fairly widespread in most instances.

A functional clone encoding a P_{2U} receptor was isolated from an NG108-15 neuroblastoma-glioma cDNA library by an expression cloning strategy utilizing an electrophysiological assay in *Xenopus* oocytes (Lustig et al 1993). Oocytes injected with RNA transcripts derived from this clone exhibited large inward currents in response to agonist-induced mobilization of intracellular calcium and the consequent opening of endogenous calcium-activated chloride channels. The rank order of agonist potency for these responses (ATP $\simeq$ UTP > ATP-γ-S- $\gg$ 2-*methylthio* ATP, α,β-methylene ATP) is characteristic of the P_{2U} subtype. Using sequence homology methods, the first P_{2Y}-like receptor was isolated from the embryonic chick brain (Webb et al 1993). Oocytes injected with cRNA transcribed from this clone showed inward current responses elicited by nucleotide agonists with the rank order of potency (2-methylthio ATP $\gg$ ATP > ADP) expected for the P_{2Y} subtype. Over a dozen cDNAs encoding metabotropic nucleotide receptors have now been cloned from different tissue and animal sources, as is shown in Table 1 (Barnard et al 1994, Filtz et al 1994, Parr et al 1994, Bowler et al 1995, Chang et al 1995, Communi et al 1995, Henderson et al 1995, Nguyen et al 1995, Rice et al 1995, Tokuyama et al 1995, Ayyanathan et al 1996, Webb et al 1996). Some of these receptors appear to correspond to species variants of the receptors discussed above, whereas others are probably distinct subtypes, as judged by unique pharmacological profiles and tissue distribution. Most of the cloned receptors have been shown to couple to a phospholipase C signal transduction pathway, in agreement with biochemical studies of native metabotropic P_2 receptors (Harden et al 1995).

The deduced protein sequences of cloned P_2 metabotropic receptors share approximately 25% identity with other members of the G protein-coupled receptor family and about 40% identity with one another, indicating that they form a distinct branch of the G protein-coupled receptor superfamily. Interestingly, metabotropic P_2 receptors show greatest similarity to receptors for nonnucleotide or nonnucleoside agonists, including several chemotactic or mitogenic agents such as platelet-activating factor, thrombin, and interleukin-8. In

Table 1 Cloned P2 receptors

P2 receptor subtype	Source	Agonist potency	Other properties	References	Accession number
$P2X_1$	Rat vas deferns	2-MeSATP > ATP > α,β-MeATP > ADP	Rapid desensitization	Valera et al 1994	X80477
	Human bladder	ATP > α,β-MeATP		Evans et al 1994	X83688
$P2X_2$	Rat PC12 Cells	ATP ≈ 2-MeSATP ≈ ATP-γ-S; α,β-MeATP inactive γ	Slow desensitization; Co-assembles with $P2X_3$	Brake et al 1994	U14414
$P2X_3$	Rat SCG	2-MeSATP ≈ ATP > α,β-MeATP	Rapid desensitization; Co-assembles with $P2X_2$	Chen et al 1995	X90651
	Rat DRG			Lewis et al 1995	X91167
$P2X_4$	Rat hippocampus	ATP > 2-MeSATP ≫ α,β-MeATP	Slow desensitization; Insensitive to PPADS, suramin	Bo et al 1995	X91200
	Rat DRG			Buell et al 1996	
	Rat brain			Seguela et al 1996	U32497
$P2X_5$	Rat coeliac ganglia	ATP > 2-MeSATP > ADP	Slow desensitization	Collo et al 1996	X92069
$P2X_6$	Rat SCG	ATP > 2-MeSATP > ADP	Slow desensitization; Insensitive to PPADS, suramin	Collo et al 1996	X92070
$P2Y_1$	Chick brain	2-MeSATP > ATP > ADP		Webb et al 1993	X73268
	turkey brain	UTP, α,β-MeATP inactive		Filtz et al 1994	U09842
	mouse insulinoma cells			Tokuyama et al 1995	U22829
	rat insulinoma cells			Tokuyama et al 1995	U22830
	human placenta			Leon, et al 1995	Z49205
	bovine endothelium			Henderson et al 1995	X87628
	human HEL cells			Ayyanathan et al 1996	U42029
$P2Y_2$	Mouse NG108-15 cells	ATP = UTP ≫ 2-MeSATP		Lustig et al 1993	L14751
	human CF/T43 cells			Parr et al 1994	U07225
	rat lung			Rice et al 1995	U09402
	rat pituitary			Chen et al 1996	L46865
$P2Y_3$	Chick brain	ADP > UTP > ATP = UDP		Barnard et al 1994	
$P2Y_4$		UTP = UDP > ATP = ADP		Communi et al 1996	X91852
				Nguyen et al 1996	U40223
$P2Y_5$	Human T-cells	ATP > ADP > 2-MeSATP ≫ UTP		Webb et al 1996	P32250

order to provide a unique designation for receptor subtypes defined by molecular structure, a revised system of nomenclature for cloned P2 receptors has been proposed in which all metabotropic nucleotide receptors are designated as $P2Y_1$, $P2Y_2$, $P2Y_3$,. . .$P2_n$ (Webb et al 1996). Indeed, as the term nucleotide receptor implies, some of these subtypes appear to show preferential activation by nucleotides other than ATP or ADP. For example, a $P2Y_4$ receptor subtype has been cloned that exhibits a strong preference for UTP as an agonist, with ATP being essentially inactive. Such findings support previous suggestions from physiological studies that nucleotides other than adenine nucleotides act as extracellular messengers.

CLONED P2X RECEPTORS DEFINE A NOVEL CLASS OF LIGAND-GATED ION CHANNELS

Perhaps the most interesting and unexpected finding to emerge from the molecular cloning of ATP receptors has been the discovery of an extended family of P2X receptors that defines a novel structural motif for ligand-gated ion channels. Functional cDNAs encoding the first two members of this family, $P2X_1$ and $P2X_2$ (named according to the revised system of nomenclature for cloned P_2 receptors), were isolated from vas deferens smooth muscle and PC12 pheochromocytoma cells, respectively, using an expression cloning strategy in *Xenopus* oocytes (Brake et al 1994, Valera et al 1994). In each case, expression of a single cDNA clone in oocytes or transfected mammalian cells is sufficient to direct the synthesis of functional, presumably homomeric ATP-gated ion channel complexes on the surface of these cells. $P2X_1$ and $P2X_2$ receptors are clearly related at the level of primary amino acid sequence and predicted secondary structure. Four additional members of this channel family have now been cloned using PCR-based screening strategies (Bo et al 1995, Chen et al 1995, Lewis et al 1995, Buell et al 1996, Collo et al 1996, Seguela et al 1996). All six subtypes share approximately 40% sequence identity distributed fairly evenly over their length, which ranges from 379 to 472 residues (Figure 1).

Subunit Structure

Excitatory and inhibitory ligand-gated ion channels fall into two broad structural classes. Members of one class, including nicotinic acetylcholine, 5-HT3 serotonin, glycine, and $GABA_A$ receptors, share primary sequence similarity and a common molecular organization (Heinemann et al 1990, Unwin 1993a,b). Based largely on studies of the nicotinic acetylcholine receptor, a model has emerged in which each receptor subunit consists of a large amino-terminal extracellular ligand-binding domain followed by four putative transmembrane segments (M1-M4). The channel complex is composed of five such subunits

```
                                               M1
P2X1  -MARRLQDELSAFFFEYDTPRMVLVRNKKVGVIFRLIQLVVLVYVIGWVFVYEKGYQTSS   59
P2X2  -MVRRLARGCWSAFWDYETPKVIVVRNRRLGFVHRMVQLLILLYFVWYVFIVQKSYQDSE   59
P2X3  -------MNCISDFFTYETTKSVVVKSWTIGIINRAVQLLIISYFVGWVFLHEKAYQVRD   53
P2X4  --MAGCCSVLGSFLFEYDTPRIVLIRSRKVGLMNRAVQLLILAYVIGWVFVWEKGYQETD   58
P2X5  -MGQAAWKGFVLSLFDYKTAKFVVAKSKKVGLLYRVLQLIILLYLLIWVFLIKKSYQDID   59
P2X6  MASAVAAALVSWGFLDYKTEKYVMTRNCWVGISQRLLQLGVVVYVIGWALLAKKGYQEWD   60

                                                                *
P2X1  DLISSVSV-KLKGLAVTQLQGLGPQVWDVADYVFPAHGDSSFVVMTNFIVTPQQTQGHCA  118
P2X2  TGPESSIITKVKGITMSEDKV-----WDVEEYVKPPEGGSVVSIITRIEVTPSQTLGTCP  114
P2X3  TAIESSVVTKVKGFG-----RYANRVMDVSDYVTPPQGTSVFVIITKIIVTENQMQGFCP  108
P2X4  SVVSSVTT-KAKGVAVTNTSQLGFRIWDVADYVIPAQEENSLFIMTNMIVTVNQTQSTCP  117
P2X5  TSLQSAVVTKVKGVAYTNTTMLGERLWDVADFVIPSQGENVFFVVTNLIVTPNQRQGICA  119
P2X6  MDPQISVITKLKGVSVTQVKELEKRLWDVADFVRPSQGENVFFLVTNFLVTPAQVQGRCP  120

               *     *                 *          *     *
P2X1  ENPE--GGICQDDSGCTPGKAERKAQGIRTGNCVPFNGTVK-TCEIFGWCPVEVDDKIPS  175
P2X2  ESMRVHSSTCHSDDDCIAGQLDMQGNGIRTGHCVPYYHGDSKTCEVSAWCPVEDG-TSDN  173
P2X3  ENEE--KYRCVSDSQCGPER--FPGGGILTGRCVNYSSVLR-TCEIQGWCPTEVD-TVEM  162
P2X4  EIPD-KTSICNSDADCTPGLRDTHSSGVATGRCVPFNESVK-TCEVAAWCPVENDVGVPT  175
P2X5  EREGIPDGECSEDDDCHAGESVVAGHGLKTGRCLRVGNSTRGTCEIFAWCPVETK-SMPT  178
P2X6  EHPSVPLANCWADEDCPEGEMGTYSHGIKTGQCVAFNGTHR-TCEIWSWCPVESS-AVPR  178

                                                *         *
P2X1  PALLREAENFTLFIKNSISFPRFKVNRRNLVEEVNGTYMKKCLYHKIQHPLCPVFNLGYV  235
P2X2  HFLGKMAPNFTILIKNSIHYPKFKFSKGNIASQKSDYLKH-CTFDQDSDPYCPIFRLGFI  232
P2X3  P-IMMEAENFTIFIKNSIRFPLFNFEKGNLLPNLTDKDIKRCRFHPEKAPFCPILRVGDV  221
P2X4  PAFLKAAENFTLLVKNNIWYPKFNFSKRNILPNITTSYLKSCIYNAQTDPFCPIFRLGTI  235
P2X5  DPLLKDAESFTISIKNFIRFPKFNFSKANVLETDNKHFLKTCHF-SSTNLYCPIFRLGSI  237
P2X6  KPLLAQAKNFTLFIKNTVTFNKFNFSRTNALDTWDNTYFKYCLYDSLSSPYCPVFRIGDL  238

                          *      *
P2X1  VRESGQDFRSLAEKGGVVGITIDWKCDLDWHVRHCKPIYQFHGLYG---EKNLSPGFNFR  292
P2X2  VEKAGENFTELAHKGGVIGVIINWNCDLDLSESECNPKYSFRRLDP--KYDPASSGYNFR  290
P2X3  VKFAGQDFAKLARTGGVLGIKIGWVCDLDKAWDQCIPKYSFTRLDGVSEKSSVSPGYNFR  281
P2X4  VEDAGHSFQEMAVEGGIMGIQIKWDCNLDRAASLCLPRYSFRRLDTRDLEHNVSPGYNFR  295
P2X5  VRWAGADFQDIALKGGVIGIYIEWDCDLDKAASKCNPHYYFNRLDN-KHTHSISSGYNFR  296
P2X6  VAMTGGDFEDLALLGGAVGINIHWDCNLDTKGSDCSPQYSF-QLQE--------RGYNFR  289

                                           (H5)     M2
P2X1  FARHFVQ-NGTNRRHLFKVFGIHFDILVDGKAGKFDIIPTMTTIGSGIGIFGVATVLCDL  351
P2X2  FAKYYKINGTTTTRTLIKAYGIRIDVIVHGQAGKFSLIPTIINLATALTSIGVGSFLCDW  350
P2X3  FAKYYKMENGSEYRTLLKAFGIRFDVLVYGNAGKFNIIPTIISSVAAFTSVGVGTVLCDI  341
P2X4  FAKYYRDLAGKEQRTLTKAYGIRFDIIVFGKAGKFDIIPTMINVGSGLALLGVATVLCDV  355
P2X5  FARYYRDPNGVEFRDLMKAYGIRFDVIVNGKAGKFSIIPTVINIGSGLALMGAGAFFCDL  356
P2X6  TANYWWAASGVESRSLLKLYGIRFDILVTGQAGKFALIPTAITVGTGAAWLGMVTFLCDL  349

P2X1  LLLHILPKRHYYKQKKF-KYAEDMGPGEGEHDPVATSSTLGLQENMRTS-----------  399
P2X2  ILLTFMNKNKLYSHKKFDKVRTPKHPSSRWPVTLALVLGQIPPPPSHYSQDQPPSPPSGE  410
P2X3  ILLNFLKGADHYKARKFEEVTETTLKGTASTNPVFASDQATVEKQSTDSGAYSIGH----  397
P2X4  IVLYCMKKKYYYRDKKY-KYVEDYEQGLSGEMNQ--------------------------  388
P2X5  VLIYLIRKSEFYRDKKFEKVRGQKEDANVEVEANEMEQERPEDEPLERVRQDEQSQELAQ  416
P2X6  LLLYVDREAGFYWRTKYEEARAPKATTNSA------------------------------  379

P2X1  ------------------------------------------------------------  399
P2X2  GPTLGEGAELPLAVQSPRPCSISALTEQVVDTLGQHMGQRPPVPEPSQQDSTSTDPKGLAQL  472
P2X3  ------------------------------------------------------------  397
P2X4  ------------------------------------------------------------  388
P2X5  SGRKQNSNCQVLLEPARFGLRENAIVNVKQSQILHPVKT---------------------  455
P2X6  ------------------------------------------------------------  379
```

arranged in a toroidal fashion to delineate a central ion-conducting pore. Polar side chains projecting from the amphipathic M2 α-helices are believed to project into the pore, where they can interact with ions as they traverse the membrane (Lester 1992). Recent studies of glutamate-gated ion channels suggest that they conform to a different transmembrane topology in which each subunit contains three helical transmembrane domains plus a hydrophobic loop that penetrates, but does not fully traverse the membrane (Gasic & Heinemann 1991, Hollmann et al 1994, Molnar et al 1994, Bennett & Dingledine 1995, Dani & Mayer 1995). This pore loop structure is reminiscent of a similar domain (called H5 or P) in potassium channels that contributes to the ion pore (Jan & Jan 1994, MacKinnon 1995). Because the biophysical properties of ATP-gated channels closely resemble those of other excitatory ligand-gated ion channels, it was expected that P2X receptors would be structurally homologous to one of these two ion channel families. Instead, the cloning of P2X receptors shows that they define yet a third structural motif for ligand-gated ion channels in which each subunit is predicted to have two transmembrane segments (M1 and M2) connected by a large extracellular ligand-binding loop. In this configuration, both amino- and carboxyl-termini are located on the intracellular side of the membrane (Figure 2). Although the deduced amino acid sequences of cloned P2X receptors display no similarity with those of other ion channels, their predicted topology is reminiscent of the subunit structure of a class of ion channels that includes amiloride-sensitive sodium channels, inwardly rectifying potassium channels, and putative mechanosensory channels from *Caenorhabditis elegans* (Kubo et al 1993a,b, Canessa et al 1994, Huang & Chalfie 1994, Suzuki et al 1994). Genetic studies in *C. elegans* suggest that the M2 domain of these channels forms an amphipathic α-helix whose hydrophilic face lines the ion path (Hong & Driscoll 1994). Consistent with this hypothesis is the fact that helical wheel plots of M2 domains from each of the cloned P2X subunits show that they have similar potential to form amphipathic α-helices, despite the limited sequence homology in this region. In addition, some P2X subunits contain a region, just amino-terminal to M2, that resembles the H5 pore loop domain of potassium

Figure 1 The growing family of ATP-gated ion channels. The predicted primary amino acid sequences of cloned $P2X_1$-$P2X_6$ receptor subtypes show that these proteins share approximately 40% sequence identity (*gray shading*) overall. Ten invariant cysteine residues (*) located within the presumptive extracellular loop may be essential for stabilizing a ligand-binding pocket through the formation of specific disulfide bonds. Putative transmembrane α-helices are delimited with black bars labeled M1 and M2. A potential pore loop region akin to that found in potassium channels corresponds to the portion of M2 denoted as (H5).

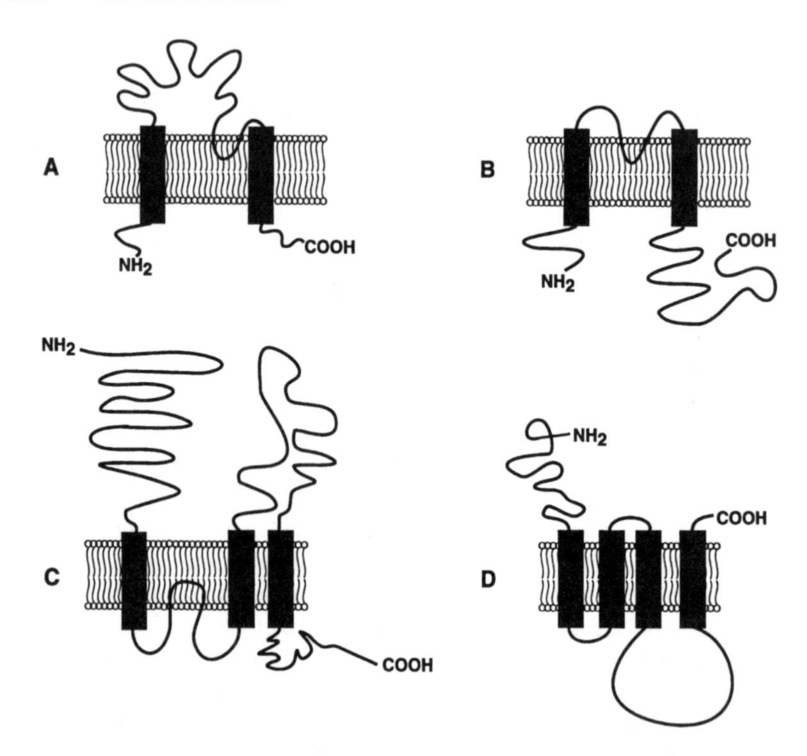

Figure 2 Predicted transmembrane topology of ligand-gated and voltage-insensitive ion channels. The predicted transmembrane topologies of (*A*), P2X receptor; (*B*), inwardly rectifying potassium channel; (*C*), GluR1 glutamate receptor; and (*D*), nicotinic acetylcholine receptor subunits. In each case, the top of the plasma membrane lipid bilayer faces the extracellular milieu and the bottom faces the cytoplasm. Putative transmembrane α-helices are represented as black rectangles, and potential pore loop (H5 or P) domains are depicted as loops that penetrate but do not traverse the bilayer. For ATP, glutamate, and acetylcholine receptors, the presumptive ligand-binding regions correspond to the large extracellular domains.

channels, and it is possible that this segment also contributes to the ion pore of ATP-gated channels. However, the presence of a H5 domain consensus sequence, or its location relative to M2, is not uniform among the six cloned P2X receptor subtypes, making tenuous a functional analogy with potassium channel pore loop structures.

Functional homology to inwardly rectifying potassium channels is also evidenced in the current-voltage relationship of ATP-gated ion channels, which also display a steep inward rectification such that at membrane potentials positive to −40 mV the current passing through the channel declines asymptotically to near zero. In the case of inwardly rectifying potassium channels, this phenomenon is due to blockade of the channel pore by cytoplasmic Mg^{2+} or polyamines (Vandenberg 1987, Lopatin et al 1995, Yang et al 1995a). For P2X channels, the mechanism is not well characterized.

The presumptive ligand-binding extracellular loop of P2X subunits contains many residues that are conserved in all six members of the family. These include ten cysteines, which presumably constrain the protein to assume a ligand-binding pocket through the formation of specific disulfide-bonded pairs. A similar conservation of cysteine residues is seen within the extracellular loop of the α, β, and γ subunits of the amiloride-sensitive sodium channel. P2X receptors also contain 13 invariant glycine residues in this region of the protein, which might be important for allowing formation of essential turns within the polypeptide backbone. Finally, some of the seven invariant lysine residues in the extracellular loop may participate in ligand binding through the formation of ionic interactions with the negatively charged phosphate moieties of nucleotide agonists.

Fully functional amiloride-sensitive sodium channels require the co-assembly of α, β, and γ chains in an as yet undetermined stoichiometry (Canessa et al 1994). Biochemical and electrophysiological studies of inwardly rectifying potassium channels suggest that functional complexes are also multimers composed of four subunits (Yang et al 1995b). By analogy with these structurally similar proteins, ATP-gated channels are likely to exist as oligomeric structures containing four identical or different P2X subunits, but the exact stoichiometry is not yet established.

Functional Heterogeneity of Cloned P2X Receptors

Native P2X receptors in different tissues and cell types exhibit characteristic rates of desensitization and agonist sensitivities. For example, α,β-methylene ATP is a potent agonist for P2X receptors on vas deferens smooth muscle, where it induces a rapid and long-lasting desensitization. By contrast, P2X receptors on PC12 cells are insensitive to α,β-methylene ATP and show little desensitization during prolonged ATP application. In addition, the receptor on vas deferens smooth muscle has an EC_{50} for ATP of 1 μM, whereas the PC12 receptor exhibits lower affinity for agonist with an EC_{50} of 60 μM. These functional properties are recapitulated by the heterologous expression of cloned $P2X_1$ or $P2X_2$ receptor subtypes isolated from these respective sources. The more recently cloned $P2X_3$-$P2X_6$ subtypes can also be divided into these two general functional classes: $P2X_3$ resembles $P2X_1$ in its sensitivity to α,β-methylene ATP and rapid desensitization kinetics, whereas $P2X_4$, $P2X_5$, and $P2X_6$ can be grouped with $P2X_2$ according to their insensitivity to α,β-methylene ATP and slow rates of desensitization. Moreover, $P2X_4$ and $P2X_6$ have the surprising characteristic of being insensitive to blockade by the archetypal P2X receptor antagonists suramin and PPADS (Bo et al 1995, Collo et al 1996, Seguela et al 1996). This antagonist insensitivity is particularly important in light of observations that, among cloned P2X receptors, $P2X_4$ and $P2X_6$ appear to be the subtypes most abundant in the brain (Collo et al 1996). Because the identification of fast, excitatory synaptic responses to ATP has relied largely on the inhibitory actions of these drugs, one must conclude that previously characterized synaptic currents were partially mediated through the actions of novel, as yet uncloned P2X subtypes. In addition, pharmacological criteria for defining ATP-mediated synaptic responses, primarily blockade by suramin, should be reconsidered to accommodate this new molecular information.

Homomeric $P2X_1$ or $P2X_2$ receptors can account for the properties of native ATP receptors studied in vas deferens smooth muscle and some sympathetic neurons, respectively (Nakazawa et al 1990a, Valera et al 1994, Brake et al 1994, Evans et al 1995). However, this is not true in cases such as the brain and sensory neurons, where endogenous, fast excitatory currents elicited by ATP have pharmacological or electrophysiological properties that are not matched by one or a combination of homomeric cloned P2X receptor complexes. These observations suggest that some ATP-mediated responses are elicited through the activation of novel P2X subtypes or by heteromeric channel complexes that exhibit properties distinct from either of the constituent P2X subunits. The formation of heteromeric channels containing $P2X_2$ and $P2X_3$ subunits can, in fact, occur in vitro, and the resulting complexes can be distinguished by a novel functional profile in which α,β-methylene ATP elicits a nondesensitizing response (Lewis et al 1995). ATP-evoked fast excitatory responses of this nature are seen where $P2X_2$ and $P2X_3$ genes are expressed together in vivo, suggesting that $P2X_2$-$P2X_3$ heteromeric receptors represent the native ion channel complex in some sensory ganglia (Khakh et al 1995).

As described above, extracellular ATP has been implicated as an important messenger in the activation of nociceptive neurons that transmit thermal, mechanical, or chemical stimuli from the periphery to the spinal cord. Indeed, all six cloned P2X subtypes are expressed within neurons of the dorsal root and trigeminal and nodose sensory ganglia. With the exception of $P2X_3$, whose expression is restricted to sensory neurons, these receptors are also found in dorsal and ventral regions of the spinal cord. In particular, $P2X_2$, $P2X_4$, and $P2X_6$ are expressed within lamina II of the spinal cord dorsal horn, where many afferent nociceptive fibers terminate. Patterns of P2X receptor expression have thus far been extrapolated from RNA hybridization studies, which identify cell types expressing the genes encoding these proteins, but do not provide information regarding protein abundance or subcellular distribution. To understand how specific P2X receptor subtypes contribute to pain transduction, it will be necessary to determine where these receptors reside at the subcellular level. For example, are $P2X_2$ and $P2X_3$ receptors located on peripheral or central projections of afferent sensory neurons? Are P2X receptors present on dendrites or axons of spinal cord neurons that synapse with incoming sensory afferents? Molecular mechanisms for the specification of subcellular distribution may involve interactions between receptors and other cellular proteins. P2X receptor subtypes show significant divergence in the length and sequence of their carboxyl-terminal tails. This putative cytoplasmic domain might participate in such protein-protein interactions and help determine subtype-selective patterns of subcellular localization. For example, $P2X_2$ receptors have a relatively long, proline-rich carboxyl-terminal tail that could interact with spectrin or PSD-95-like proteins through SH3 domains or other motifs (Rotin et al 1994, Gomperts 1996). In this manner, localization of heteromeric channels could be influenced by protein interaction domains from one or more constituent receptor subunits.

Cloned P2X receptor subtypes are expressed in many tissues and cell types in which endogenous ATP-evoked fast excitatory responses have been previously measured. In some instances, however, receptor expression has been detected in tissues where physiological responses have not been described, suggesting that new roles for extracellular ATP are likely to emerge. One such example is the expression of $P2X_2$ receptor transcripts in the pituitary and adrenal glands, where ATP-gated channels could participate in regulating the release of neurohormones or neurotransmitters, by analogy with its actions on PC12 pheochromocytoma cells. As mentioned previously, $P2X_1$ receptor transcripts are induced in immature thymocytes undergoing apoptosis, thus supporting other tangential observations that implicate extracellular ATP as a cell death signal.

P2X Receptors and Divalent Cations

Many P2X receptors are actually more permeable to Ca^{2+} ions than they are to monovalent cations. Indeed, it has been suggested that the role of P2X receptors, like that of NMDA-type glutamate receptors, in regulating intracellular Ca^{2+} may be more important than its role in depolarizing the cell (Benham 1990, Hille 1992). Measurements carried out in sympathetic neurons have shown that Ca^{2+} carries over 6% of the total current conducted through P2X channels under quasiphysiological conditions (extracellular Na^+ and Ca^{2+} concentrations of 150 and 2.5 mM, respectively). This is compared with a value about 12% for NMDA-type glutamate channels in hippocampal neurons under the same ionic conditions (Rogers & Dani 1995). Whereas the high Ca^{2+} permeability of NMDA receptors is known to be essential in mediating some forms of synaptic plasticity such as long-term potentiation (Mayer & Miller 1990, Malenka 1991, Daw et al 1993), no in vivo phenomena have yet been identified for which Ca^{2+} permeation of ATP-gated channels is required.

Zinc is present in CNS synaptic terminals such as the mossy fiber pathway of the hippocampus and is synaptically released together with glutamate, which suggests that extracellular Zn^{2+} plays a role as a modulator of synaptic transmission. Indeed, in the range of 0.1 to 3 μM, Zn^{2+} has been shown to potentiate NMDA receptor activity by increasing the maximal current response up to twofold, without shifting the agonist dose-response curve (Hollmann et al 1993). Extracellular Zn^{2+} can also potentiate current responses of ATP-gated channels in sympathetic and sensory neurons, but in this case, by increasing agonist potency without altering the maximal response. The same phenomenon has been observed with cloned $P2X_2$ receptor, where Zn^{2+} decreases the EC_{50} for ATP from 60 to 15 μM. A similar potentiation of agonist potency has been reported for the $P2X_4$ subtype. Whether this potentiation reflects an allosteric effect on the receptor-channel complex or an interaction with ATP to increase its agonist potency has not been resolved.

PERSPECTIVES

The cloning and characterization of cDNAs encoding bona fide P2 purinergic receptors has provided irrefutable proof that ATP, and possibly other nucleotides, play important roles as extracellular messengers. This new molecular information not only validates roles for extracellular ATP in the cardiovascular, immune, and nervous systems, but also raises potentially new roles for extracellular nucleotides in processes that include the control of programmed cell death during neuronal or thymocyte development, and the production or secretion of endocrine hormones from the pituitary or adrenal glands. Tests of such hypotheses will come from several directions. First, a more detailed analysis of the temporal and regional patterns of P2 receptor subtype expression may highlight appropriate physiological and developmental systems in which roles for ATP receptors in apoptosis or synaptic transmission can be characterized in greater detail. Second, transfected mammalian cell systems can now be used to study roles for cloned ATP receptors in well-established in vitro models of specific cellular responses, including cell death, hormone secretion, and electrolyte transport. On a more integrative level, receptor gene knockout studies in mice and gene mapping studies in humans will also help to shed light on physiological roles of identified ATP receptor subtypes in the modulation of vascular tone, allergic responses, nociception, inflammation, and pain, to name but a few. Although the availability of genes encoding ATP receptors will greatly facilitate the analysis of extracellular nucleotide function both in vitro and in vivo, the paucity of stable, potent, and subtype-selective P2 purinergic agonists and antagonists is still a limiting factor in these endeavors. A hoped-for contribution from receptor cloning studies will be the development of high throughput screening assays with which to identify new compounds that fulfill these needs. In addition, analysis of receptor function through mutational, biochemical, and structural methods should provide useful information about ligand-binding domains and drug-binding pockets for P2 receptors. A goal of such studies will be, of course, to develop novel drugs that have therapeutic potential for the management of airway and cardiovascular diseases, allergic and inflammatory responses, pain, and cognitive disorders. ATP receptors currently stand as unexploited targets for pharmacological intervention in these and other health-related problems.

On a more basic level of protein structure and function, it is likely that much will be learned from the molecular characterization of P2X receptors. The predicted primary amino acid sequences of these proteins define a third major subclass of ligand-gated ion channels and provide perhaps the clearest evidence to date that ligand-gated channels with similar functional properties can be fashioned from subunits having very different architectural characteristics. It is tempting to speculate that P2X receptors are among the earliest types of ligand-gated channels, having evolved from the family of voltage-insensitive ion channels. Acquisition of a large extracellular ligand-binding domain then extended the functional repertoire of these proteins to include the detection of ATP and possibly other endogenous molecules. Perhaps members of this novel family of ion channels are activated by other well-known transmitters such as acetylcholine, norepinephrine, and dopamine, or newly discovered extracellular messengers. Answers to these questions and hypotheses will require additional phylogenetic and functional studies. Finally, there are interesting questions regarding protein structure and protein-protein interactions. For example, how do P2X receptor subunits interact with one another to form specific homo- or heteromeric complexes? Which segments of these subunits define or contribute to the formation of the ion pore and ligand-binding pocket? Do these signaling complexes interact with other cellular proteins that modulate channel activity or specify the location of ion channel complexes within discreet subcellular locations? Answers to these questions should reveal new concepts in the area of channel function and help us to understand how extracellular ATP affects a wide range of cellular and physiological processes.

NOTE ADDED IN PROOF A recently cloned member of the P2X family, $P2X_7$, displays properties attributed to P_{2Z} receptor. Under conditions of low extracellular divalent cation concentrations, activation of this receptor by ATP or 4-benzoyl benzoyl ATP can lead to the formation of membrane pores permeable to large organic molecules. Under normal divalent cation concentrations, this receptor, similar to other P2X receptors, is permeable only to small cations.

The long carboxyl-terminal tail of this receptor is required for the formation of these large membrane pores (Surprenant et al 1996).

Literature Cited

Abraham EH, Prat AG, Gerweck L, Seneveratne T, Arceci RJ, et al. 1993. The multidrug resistance (mdr1) gene product functions as an ATP channel. *Proc. Natl. Acad. Sci. USA* 90:312–16

al-Awqati Q. 1995. Regulation of ion channels by ABC transporters that secrete ATP. *Science* 269:805–6

Ayyanathan K, Webbs TE, Sandhu AK, Athwal RS, Barnard EA, Kunapuli SP. 1996. Cloning and chromosomal localization of the human $P2Y_1$ purinoceptor. *Biochem. Biophys. Res. Commun.* 218:783–88

Barnard EA, Burnstock G, Webb TE. 1994. G protein-coupled receptors for ATP and other nucleotides: a new receptor family. *Trends Pharmacol. Sci.* 15:67–70

Bean BP. 1992. Pharmacology and electrophysiology of ATP-activated ion channels. *Trends Pharmacol. Sci.* 13:87–90

Bean BP, Friel DD. 1990. ATP-activated channels in excitable cells. *Ion Channels* 2:169–203

Benham CD. 1990. ATP-gated channels in vascular smooth muscle cells. *Ann. NY Acad. Sci.* 603:275–86

Benham CD, Tsien RW. 1987. A novel receptor-operated Ca^{2+}-permeable channel activated by ATP in smooth muscle. *Nature* 328:275–78

Bennett JA, Dingledine R. 1995. Topology profile for a glutamate receptor: three transmembrane domains and a channel-lining reentrant membrane loop. *Neuron* 14:373–84

Bo X, Zhang Y, Nassar M, Burnstock G, Schoepfer R. 1995. A P2X purinoceptor cDNA conferring a novel pharmacological profile. *FEBS Lett.* 375:129–33

Boeynaems J-M, Pearson JD. 1990. P_2 purinoceptors on vascular endothelial cells: physiological significance and transduction mechanisms. *Trends Physiol. Sci.* 11:34–37

Bowler WB, Birch MA, Gallagher JA, Bilbe G. 1995. Identification and cloning of human P2U purinoceptor present in osteoclastoma, bone, and osteoblasts. *J. Bone Miner. Res.* 10:1137–45

Brake AJ, Wagenbach MJ, Julius D. 1994. New structural motif for ligand-gated ion channels defined by an ionotropic ATP receptor. *Nature* 371:519–23

Bruns RF. 1990. Adenosine receptors. Roles and pharmacology. *Ann. NY Acad. Sci.* 603:211–26

Buell G, Lewis C, Collo G, North RA, Surprenant A. 1996. An antagonist-insensitive P2X receptor expressed in epithelia and brain. *EMBO J.* 15:55–62

Buisman HP, Steinberg TH, Fischbarg J, Silverstein SC, Vogelzang SA, et al. 1988. Extracellular ATP induces a large nonselective conductance in macrophage plasma membranes. *Proc. Natl. Acad. Sci. USA* 85:7988–92

Burnstock G. 1972. Purinergic nerves. *Pharmacol. Rev.* 24:509–81

Burnstock G. 1976. Purinergic receptors. *J. Theor. Biol.* 62:491–503

Burnstock G. 1978. A basis for distinguishing two types of purinergic receptors. In *Cell Membrane Receptors for Drugs and Hormones: A Multidisciplinary Approach,* ed. RW Straub, L Bolis, pp. 107–18. New York: Raven

Burnstock G. 1989. Vascular control by purines with emphasis on the coronary system. *Eur. Heart J.* 10(Suppl. F):15–21

Burnstock G. 1990a. Dual control of local blood flow by purines. *Ann. NY Acad. Sci.* 603:31–45

Burnstock G. 1990b. Purinergic mechanisms. *Ann. NY Acad. Sci.* 603:1–18

Burnstock G, Kennedy C. 1985. Is there a basis for distinguishing two types of P2-purinoceptor? *Gen. Pharmacol.* 16:433–40

Canessa CM, Schild L, Buell G, Thorens B, Gautschi I, et al. 1994. Amiloride-sensitive epithelial Na^+ channel is made of three homologous subunits. *Nature* 367:463–67

Chang K, Hanaoka K, Kumada M, Takuwa Y. 1995. Molecular cloning and functional analysis of a novel P2 nucleotide receptor. *J. Biol. Chem.* 270:26152–58

Chen CC, Akopian AN, Sivilotti L, Colquhoun D, Burnstock G, Wood JN. 1995. A P2X purinoceptor expressed by a subset of sensory neurons. *Nature* 377:428–31

Cockcroft S, Gomperts BD. 1980. The ATP4-receptor of rat mast cells. *Biochem. J.* 188:789–98

Collo G, North RA, Kawashima E, Merlo-Pich E, Neidhart S, et al. 1996. Cloning of $P2X_5$ and $P2X_6$ receptors, and the distribution and properties of an extended family of ATP-gated ion channels. *J. Neurosci.* 16:2495–507

Communi D, Pirotton S, Parmentier M, Boeynaems JM. 1995. Cloning and functional expression of a human uridine nucleotide receptor. *J. Biol. Chem.* 270:30849–52

Dahlquist R, Diamant B. 1974. Interaction of ATP and calcium on the rat mast cell: effect on histamine release. *Acta Pharmacol. Toxicol.* 34:368–84

Dani JA, Mayer ML. 1995. Structure and function of glutamate and nicotinic acetylcholine receptors. *Curr. Opin. Neurobiol.* 5:310–17

Daw NW, Stein PS, Fox K. 1993. The role of NMDA receptors in information processing. *Annu. Rev. Neurosci.* 16:207–22

Di Virgilio F. 1995. The P2Z purinoceptor: an intriguing role in immunity, inflammation and cell death. *Immunol. Today* 16:524–28

Drury AN, Szent-Gyorgi A. 1929. The physiological activity of adenine compounds with especial reference to their action upon the mammalian heart. *J. Physiol.* 68:213–37

Dubyak GR. 1986. Extracellular ATP activates polyphosphoinositide breakdown and Ca^{2+} mobilization in Ehrlich ascites tumor cells. *Arch. Biochem. Biophys.* 245:84–95

Edwards FA, Gibb AJ, Colquhoun D. 1992. ATP receptor-mediated synaptic currents in the central nervous system. *Nature* 359:144–47

Evans RJ, Derkach V, Surprenant A. 1992. ATP mediates fast synaptic transmission in mammalian neurons. *Nature* 357:503–5

Evans RJ, Lewis C, Buell G, Valera S, North RA, Surprenant A. 1995. Pharmacological characterization of heterologously expressed ATP-gated cation channels (P2X purinoceptors). *Mol. Pharmacol.* 48:178–83

Fieber LA, Adams DJ. 1991. Adenosine triphosphate-evoked currents in cultured neurones dissociated from rat parasympathetic cardiac ganglia. *J. Physiol.* 434:239–56

Filtz TM, Li Q, Boyer JL, Nicholas RA, Harden TK. 1994. Expression of a cloned P2Y purinergic receptor that couples to phospholipase C. *Mol. Pharmacol.* 46:8–14

Friel DD. 1988. An ATP-sensitive conductance in single smooth muscle cells from the rat vas deferens. *J. Physiol.* 401:361–80

Friel DD, Bean BP. 1988. Two ATP-activated conductances in bullfrog atrial cells. *J. Gen. Physiol.* 91:1–27

Fyffe RE, Perl ER. 1984. Is ATP a central synaptic mediator for certain primary afferent fibers from mammalian skin? *Proc. Natl. Acad. Sci. USA* 81:6890–93

Gasic GP, Heinemann S. 1991. Receptors coupled to ionic channels: the glutamate receptor family. *Curr. Opin. Neurobiol.* 1:20–26

Gomperts SN. 1996. Clustering of membrane proteins: It's all coming together with the PSD-95/SAP90 protein family. *Cell* 84:659–62

Gordon JL. 1986. Extracellular ATP: effects, sources, and fate. *Biochem. J.* 223:309–19

Harden TK, Boyer JL, Nicholas RA. 1995. P2-purinergic receptors: subtype-associated signaling responses and structure. *Annu. Rev. Pharmacol. Toxicol.* 35:541–79

Heinemann S, Boulter J, Deneris E, Conolly J, Duvoisin R, et al. 1990. The brain nicotinic acetylcholine receptor gene family. *Prog. Brain Res.* 86:195–203

Henderson DJ, Elliot DG, Smith GM, Webb TE, Dainty IA. 1995. Cloning and characterisation of a bovine P2Y receptor. *Biochem. Biophys. Res. Commun.* 212:648–56

Hille B. 1992. *Ionic Channels of Excitable Membranes.* Sunderland, Mass: Sinauer. 607 pp. 2nd ed.

Hollmann M, Boulter J, Maron C, Beasley L, Sullivan J, et al. 1993. Zinc potentiates agonist-induced currents at certain splice variants of the NMDA receptor. *Neuron* 10:943–54

Hollmann M, Maron C, Heinemann S. 1994. N-glycosylation site tagging suggests a three transmembrane domain topology for the glutamate receptor GluR1. *Neuron* 13:1331–43

Holton FA, Holton P. 1954. The capillary dilator substances in dry powders of spinal roots: a possible role of adenosine triphosphate in chemical transmission from nerve endings. *J. Physiol.* 126:124–40

Hong K, Driscoll M. 1994. A transmembrane domain of the putative channel subunit MEC-4 influences mechanotransduction and neurodegeneration in *C. elegans*. *Nature* 367:470–73

Huang M, Chalfie M. 1994. Gene interactions affecting mechanosensory transduction in *Caenorhabditis elegans*. *Nature* 367:467–70

Hume RI, Honig MG. 1986. Excitatory action

of ATP on embryonic chick muscle. *J. Neurosci.* 6:681–90

Jahr CE, Jessell TM. 1983. ATP excites a subpopulation of rat dorsal horn neurones. *Nature* 304:730–33

Jan LY, Jan YN. 1994. Potassium channels and their evolving gates. *Nature* 371:119–22

Jiang C, Finkbeiner WE, Widdicombe JH, McCray PB Jr, Miller SS. 1993. Altered fluid transport across airway epithelium in cystic fibrosis. *Science* 262:424–27

Khakh BS, Humphrey PP, Surprenant A. 1995. Electrophysiological properties of P2X-purinoceptors in rat superior cervical, nodose and guinea-pig coeliac neurones. *J. Physiol.* 484:385–95

Knowles MR, Olivier KN, Hohneker KW, Robinson J, Bennett WD, Boucher RC. 1995. Pharmacologic treatment of abnormal ion transport in the airway epithelium in cystic fibrosis. *Chest* 107:S71–S76

Krishtal OA, Marchenko SM, Pidoplichko VI. 1983. Receptor for ATP in the membrane of mammalian sensory neurones. *Neurosci. Lett.* 35:41–45

Kubo Y, Baldwin TJ, Jan YN, Jan LY. 1993a. Primary structure and functional expression of a mouse inward rectifier potassium channel. *Nature* 362:127–33

Kubo Y, Reuveny E, Slesinger PA, Jan YN, Jan LY. 1993b. Primary structure and functional expression of a rat G-protein-coupled muscarinic potassium channel. *Nature* 364:802–6

Lester HA. 1992. The permeation pathway of neurotransmitter-gated ion channels. *Annu. Rev. Biophys. Biomol. Struct.* 21:267–92

Lewis C, Neidhart S, Holy C, North RA, Buell G, Surprenant A. 1995. Coexpression of $P2X_2$ and $P2X_3$ receptor subunits can account for ATP-gated currents in sensory neurons. *Nature* 377:432–35

Li J, Perl ER. 1995. ATP modulation of synaptic transmission in the spinal substantia gelatinosa. *J. Neurosci.* 15:3357–65

Lopatin AN, Makhina EN, Nichols CG. 1995. The mechanism of inward rectification of potassium channels—long-pore plugging by cytoplasmic polyamines. *J. Gen. Physiol.* 106:923–55

Lustig KD, Shiau AK, Brake AJ, Julius D. 1993. Expression cloning of an ATP receptor from mouse neuroblastoma cells. *Proc. Natl. Acad. Sci. USA* 90:5113–17

MacKinnon R. 1995. Pore loops: an emerging theme in ion channel structure. *Neuron* 14:889–92

Malenka RC. 1991. The role of postsynaptic calcium in the induction of long-term potentiation. *Mol. Neurobiol.* 5:289–95

Mayer ML, Miller RJ. 1990. Excitatory amino acid receptors, second messengers and regulation of intracellular Ca^{2+} in mammalian neurons. *Trends Pharmacol. Sci.* 11:254–60

Molnar E, McIlhinney RA, Baude A, Nusser Z, Somogyi P. 1994. Membrane topology of the GluR1 glutamate receptor subunit: epitope mapping by site-directed antipeptide antibodies. *J. Neurochem.* 63:683–93

Motte S, Communi D, Pirotton S, Boeynaems JM. 1995. Involvement of multiple receptors in the actions of extracellular ATP: the example of vascular endothelial cells. *Int. J. Biochem. Cell Biol.* 27:1–7

Nakazawa K, Fujimori K, Takanaka A, Inoue K. 1990a. An ATP-activated conductance in pheochromocytoma cells and its suppression by extracellular calcium. *J. Physiol.* 428:257–72

Nakazawa K, Inoue K, Fujimori K, Takanaka A. 1990b. ATP-activated single-channel currents recorded from cell-free patches of pheochromocytoma PC12 cells. *Neurosci. Lett.* 119:5–8

Nakazawa K, Matsuki N. 1987. Adenosine triphosphate-activated inward current in isolated smooth muscle cells from rat vas deferens. *Pflügers Arch.* 409:644–46

Nett-Fiordalisi M, Tomaselli K, Russell JH, Chaplin DD. 1995. Macrophage apoptosis in the absence of active interleukin-1 beta-converting enzyme. *J. Leukocyte Biol.* 58:717–24

Nguyen T, Erb L, Weisman GA, Marchese A, Heng HH, et al. 1995. Cloning, expression, and chromosomal localization of the human uridine nucleotide receptor gene. *J. Biol. Chem.* 270:30845–48

O'Connor SE, Dainty IA, Leff P. 1991. Further subclassification of ATP receptors based on agonist studies. *Trends Pharmacol. Sci.* 12:137–41

Okajima F, Tokumitsu Y, Kondo Y, Ui M. 1987. P2-purinergic receptors are coupled to two signal transduction systems leading to inhibition of cAMP generation and to production of inositol trisphosphate in rat hepatocytes. *J. Biol. Chem.* 262:13483–90

Osipchuk Y, Cahalan M. 1992. Cell-to-cell spread of calcium signals mediated by ATP receptors in mast cells. *Nature* 359:241–44

Owens GP, Hahn WE, Cohen JJ. 1991. Identification of mRNAs associated with programmed cell death in immature thymocytes. *Mol. Cell. Biol.* 11:4177–88

Parr CE, Sullivan DM, Paradiso AM, Lazarowski ER, Burch LH, et al. 1994. Cloning and expression of a human P(2U) nucleotide receptor, a target for cystic fibrosis pharmacotherapy. *Proc. Natl. Acad.*

Sci. USA 91:3275–79. Erratum. 1994. *Proc. Natl. Acad. Sci. USA* 91(26):13067

Paton DM, Taerum T. 1990. A comparison of P1- and P2-purinoceptors. *Ann. NY Acad. Sci.* 603:165–71

Pearson JD, Gordon JL. 1985. Nucleotide metabolism by endothelium. *Annu. Rev. Physiol.* 47:617–27

Reddy MM, Quinton PM, Haws C, Wine JJ, Grygorczyk R, et al. 1996. Failure of the cystic fibrosis transmembrane conductance regulator to conduct ATP. *Science* 271:1876–79

Reisin IL, Prat AG, Abraham EH, Amara JF, Gregory RJ, et al. 1994. The cystic fibrosis transmembrane conductance regulator is a dual ATP and chloride channel. *J. Biol. Chem.* 269:20584–91

Rice WR, Burton FM, Fiedeldey DT. 1995. Cloning and expression of the alveolar type II cell P2U-purinergic receptor. *Am. J. Respir. Cell Mol. Biol.* 12:27–32

Richardson PJ, Brown SJ. 1987. ATP release from affinity-purified rat cholinergic nerve terminals. *J. Neurochem.* 48:622–30

Rogers M, Dani JA. 1995. Comparison of quantitative calcium flux through NMDA, ATP, and ACh receptor channels. *Biophys. J.* 68:501–6

Rotin D, Bar-Sagi D, O'Brodovich H, Merilainen J, Lehto VP, et al. 1994. An SH3 binding region in the epithelial Na^+ channel (arENaC) mediates its localization at the apical membrane. *EMBO J.* 13:4440–50

Schwiebert EM, Egan ME, Hwang TH, Fulmer SB, Allen SS, et al. 1995. CFTR regulates outwardly rectifying chloride channels through an autocrine mechanism involving ATP. *Cell* 81:1063–73

Seguela P, Haghighi A, Soghomonian JJ, Cooper E. 1996. A novel neuronal P_{2X} ATP receptor ion channel with widespread distribution in the brain. *J. Neurosci.* 16:448–55

Silinsky EM, Gerzanich V, Vanner SM. 1992. ATP mediates excitatory synaptic transmission in mammalian neurones. *Br. J. Pharmacol.* 106:762–63

Silinsky EM, Hunt JM, Solsona CS, Hirsh JK. 1990. Prejunctional adenosine and ATP receptors. *Ann. NY Acad. Sci.* 603:324–34

Slakey LL, Gordon EL, Pearson JD. 1990. A comparison of ectonucleotidase activities on vascular endothelial and smooth muscle cells. *Ann. NY Acad. Sci.* 603:366–79

Sneddon P, Burnstock G. 1984. ATP as a co-transmitter in rat tail artery. *Eur. J. Pharmacol.* 106:149–52

Surprenant A, Rasssendren F, Kawashima E, North RA, Buell G. 1996. Cytolytic P_{2Z} receptor for ATP identified as a P2X receptor ($P2X_7$). *Science* 272:735–38

Suzuki M, Takahashi K, Ikeda M, Hayakawa H, Ogawa A, et al. 1994. Cloning of a pH-sensitive K^+ channel possessing two transmembrane segments. *Nature* 367:642–45

Tokuyama Y, Hara M, Jones EM, Fan Z, Bell GI. 1995. Cloning of rat and mouse P2Y purinoceptors. *Biochem. Biophys. Res. Commun.* 211:211–18

Unwin N. 1993a. Neurotransmitter action: opening of ligand-gated ion channels. *Cell* 72:31–41

Unwin N. 1993b. Neurotransmitter action: opening of ligand-gated ion channels. *Neuron Suppl.* 10:31–41

Valera S, Hussy N, Evans RJ, Adami N, North RA, et al. 1994. A new class of ligand-gated ion channel defined by P2X receptor for extracellular ATP. *Nature* 371:516–19

Vandenberg CA. 1987. Inward rectification of a potassium channel in cardiac ventricular cells depends on internal magnesium ions. *Proc. Natl. Acad. Sci. USA* 84:2560–64

Webb TE, Kaplan MG, Barnard EA. 1996. Identification of 6h1 as a P2Y-purinoceptor-P2Y(5). *Biochem. Biophys. Res. Commun.* 219:105–10

Webb TE, Simon J, Krishek BJ, Bateson AN, Smart TG, et al. 1993. Cloning and functional expression of a brain G-protein-coupled ATP receptor. *FEBS Lett.* 324:219–25

Wiley JS, Chen JR, Snook MB, Jamieson GP. 1994. The P2Z-purinoceptor of human lymphocytes: actions of nucleotide agonists and irreversible inhibition by oxidized ATP. *Br. J. Pharmacol.* 112:946–50

Yang J, Jan YN, Jan LY. 1995a. Control of rectification and permeation by residues in two distinct domains in an inward rectifier K^+ channel. *Neuron* 14:1047–54

Yang J, Jan YN, Jan LY. 1995b. Determination of the subunit stoichiometry of an inwardly rectifying potassium cannel. *Neuron* 15:1441–47

Zambon A, Bronte V, Di Virgilio F, Hanau S, Steinberg TH, et al. 1994. Role of extracellular ATP in cell-mediated cytotoxicity: a study with ATP-sensitive and ATP-resistant macrophages. *Cell Immunol.* 156:458–67

Annu. Rev. Cell Dev. Biol. 1996. 12:543–73

STRUCTURE-FUNCTION ANALYSIS OF THE MOTOR DOMAIN OF MYOSIN

Kathleen M. Ruppel and James A. Spudich
Departments of Biochemistry and of Developmental Biology, Stanford University School of Medicine, Stanford, California 94305

KEY WORDS: mutagenesis, cytoskeletal proteins, molecular motors

ABSTRACT

Motor proteins perform a wide variety of functions in all eukaryotic cells. Recent advances in the structural and mutagenic analysis of the myosin motor has led to insights into how these motors transduce chemical energy into mechanical work. This review focuses on the analysis of the effects of myosin mutations from a variety of organisms on the in vivo and in vitro properties of this ubiquitous motor and illustrates the positions of these mutations on the high-resolution three-dimensional structure of the myosin motor domain.

CONTENTS

1081-0706/96/1115-0543$08.00

INTRODUCTION

Motor proteins are enzymes that transduce the chemical energy of ATP hydrolysis into force and directed movement. One of the most interesting and challenging questions facing cell biology today concerns the cell's use of a growing repertoire of these molecular motors to effect cell shape changes and motility, as well as the intracellular transport of a wide variety of cargoes. These motile processes are accomplished by the actions of three classes of motor proteins operating on cellular filamentous networks—the myosin family of actin-based motors, and the microtubule-based kinesin and dynein families. Over the past several years, numerous family members have been found in virtually all eukaryotic cells in which they have been sought, leading to the estimation that more than 50 different molecular motors may be driving motile processes in any given cell.

Despite the ubiquity and importance of these proteins, the molecular mechanism by which these motors convert chemical energy to mechanical energy has been elusive. The most extensively studied and best understood of the motor proteins is myosin II (or conventional myosin), which drives contraction of the muscle sarcomere, as well as a variety of processes in nonmuscle cells. In addition to a wealth of biochemical and biophysical analyses of this motor, high-resolution structural information has become available for actin (Holmes et al 1990, Kabsch et al 1990), the subfragment 1 motor domain of chicken skeletal muscle myosin (Rayment et al 1993b) and *Dictyostelium* myosin (Fisher et al 1995), and the regulatory domain of scallop myosin (Xie et al 1994). This information, when coupled with in vivo and in vitro studies of genetically altered forms of myosin, provides the basis for the application of the powerful tool of structure-function analysis to the myosin motor.

In this review we focus on the current state of structure-function analysis of myosin. We briefly review myosin structure and its kinetic cycle, describe the different biological systems used for making and studying mutated myosins, and discuss mutations of myosin and their effect on myosin function in vivo and in vitro in the context of the recently solved crystal structures of the myosin motor domain. Because of space limitations, we deal primarily with mutations in the myosin heavy chain portion of the motor domain. The reader is referred to the recent comprehensive review of myosin by Sellers & Goodson (1995) and elsewhere for excellent overviews of myosin light chain mutagenesis (Reinach et al 1986, Goodwin et al 1990, Kendrick-Jones et al 1991, Saraswat et al 1992, Rowe & Kendrick-Jones 1993, Wolff-Long et al 1993, Ikebe et al 1994, Kamisoyama et al 1994, Ostrow et al 1994) and myosin tail mutagenesis (Kubalek et al 1992, Egelhoff et al 1993, Lee et al 1994). In addition, excellent complementary reviews can be found in the Special Topic section Molecular

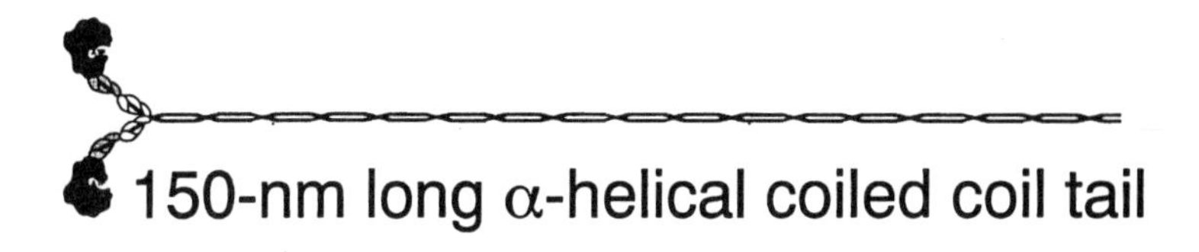

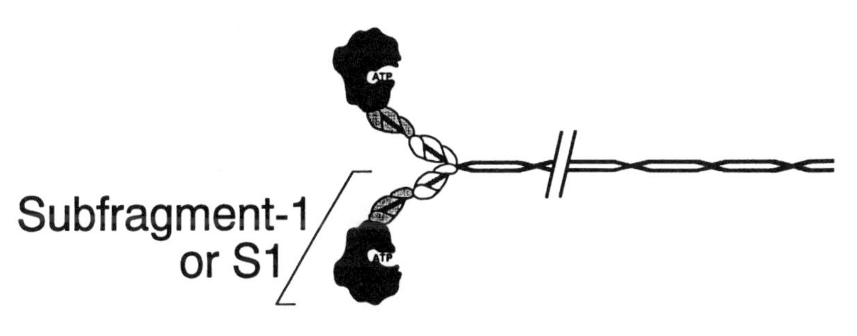

Figure 1 A schematic representation of a hexameric myosin II molecule, showing the light chains (essential light chain, ELC; and regulatory light chain, RLC) that wrap around the long α-helical region of the S1 portion of a heavy chain, and the pair of heavy chains intertwining via their α-helical tail domains to form a coiled-coil rod.

Motor, Volume 58 of the *Annual Review of Physiology*, as well as by A Weiss & L Leinwand, this volume.

MYOSIN STRUCTURE AND FUNCTION—A BRIEF REVIEW

Myosin II (referred to as myosin throughout this review) is a hexamer consisting of two heavy chains and two sets of light chains (Figure 1). The amino-terminal portion of the heavy chain forms a globular head domain to which a set of light chains [essential (ELC) and regulatory (RLC)] bind, and the carboxyl terminus forms a long α-helix that dimerizes to form a coiled-coil rod structure. Proteolytic studies of myosin reveal that the rod domain contains the filament-forming properties of myosin, allowing it to assemble into bipolar, thick filaments, whereas the globular head domain (termed subfragment 1 or S1) possesses the actin- and ATP-binding properties of the molecule (Warrick & Spudich 1987, Vibert & Cohen 1988, Mornet et al 1989). Unconventional myosins generally share this division of labor, i.e. possessing head regions that show considerable homology to the myosin II S1 domain and tail regions that vary greatly in structure (and presumably function) from the myosin II rod and from each other (Korn & Hammer 1990, Hammer 1991, 1994, Pollard et al 1991, Cheney & Mooseker 1992, Mooseker 1993, Titus 1993, Sellers & Goodson 1995).

The swinging cross-bridge model of muscle contraction (Huxley 1969, Huxley & Kress 1985) proposed that the S1 domain of myosin, which gives rise to the cross-bridges connecting the thick (myosin-containing) and thin (actin-containing) filaments in muscle, undergoes a change in conformation while bound to the actin filament that results in force production and contraction. S1 clearly possesses the requisite actin-binding and ATP hydrolytic activities to carry out such a task. With the advent of myosin-driven in vitro sliding-filament assays using purified actin and myosin (Kron & Spudich 1986), it was shown that coverslips coated only with S1 molecules were capable of supporting ATP-dependent sliding of fluorescently labeled actin filaments (Toyoshima et al 1987) and producing force, as measured by deflections of microneedles attached to actin filaments (Kishino & Yanagida 1988). These results showed that S1 alone is a functional motor unit.

Recent research has been heavily focused on S1 as a minimal motor unit, and a broad outline of the mechanism by which the interaction of actin and S1 transduces chemical energy into mechanical energy has taken shape. Based on kinetic studies of the actin-activated myosin ATPase, Lymn & Taylor (1971) proposed a simple model for chemomechanical coupling (Figure 2), which generally has been supported by more recent extensive investigations (Adelstein

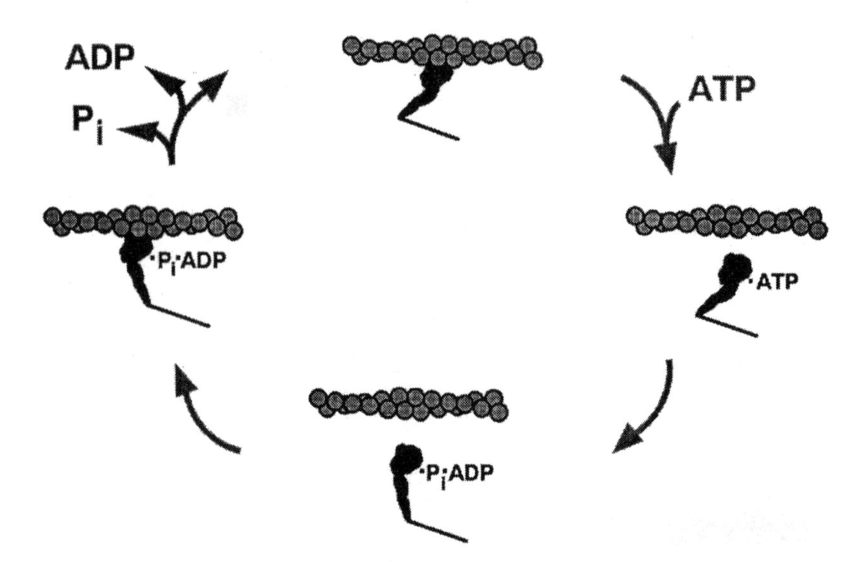

Figure 2 Schematic representation of the actin-S1 chemomechanical cycle (Lymn & Taylor 1971). The top of the cycle represents the rigor complex, and the cycle proceeds in a clockwise fashion as described in the text. For clarity, only one S1 head of myosin is shown attached to a short segment of the α-helical tail.

& Eisenberg 1980, Taylor & Amos 1981, Eisenberg & Hill 1985, Cooke 1986, Hibberd & Trentham 1986, Brenner & Eisenberg 1987, Wray et al 1988, Geeves 1991). In the absence of ATP, S1 and actin form a tight or rigor complex. Binding of ATP to S1 greatly diminishes its affinity for actin, which results in dissociation of the rigor complex. ATP bound to S1 is then rapidly hydrolyzed to ADP and P_i, and the products remain tightly bound. Actin interaction with the myosin-ADP-P_i complex stimulates P_i release, which restores the high affinity of S1 for actin and results in formation of the high-affinity actin-S1-ADP complex (strongly bound state). Subsequent release of ADP returns the cycle to the rigor complex. The power stroke is thought to occur during the high-affinity, strongly bound state. This model provides a useful framework for correlating known kinetic intermediates with putative mechanical states of the actin-S1 interaction.

The high-resolution structures detailing actin (Holmes et al 1990, Kabsch et al 1990), myosin subfragments (Rayment et al 1993b, Xie et al 1994, Fisher et al 1995), and their interaction (Rayment et al 1993a, Schroeder et al 1993) provide the molecular framework necessary to model the physical changes in the actin-S1 interaction that lead to force production. For the purposes of this review, we primarily use the chicken S1 structure of Rayment et al (1993b) to discuss structure-function analysis of mutations, although we refer to both the scallop regulatory domain structure (Xie et al 1994) and the recently solved structures of the *Dictyostelium* catalytic domain with bound nucleotide analogues (Fisher et al 1995) for some mutations.

Chicken skeletal muscle S1 is a highly asymmetric molecule consisting of a globular catalytic domain that contains the ATP- and actin-binding sites and an α-helical neck or regulatory domain that serves to connect the catalytic domain to the tail (see Figure 1; for a stereo view of the high-resolution structure, see Figure 3). The regulatory domain consists of the carboxyl-terminal portion of the heavy chain in the form of a long α-helix around which the calmodulin-like ELC and RLC are clamped. The analogous region of scallop myosin solved by Xie et al (1994) is structurally similar. Docking of the crystal structures of actin (Kabsch et al 1990, Holmes et al 1990) and S1 within the envelope of the density of the actin-S1 complex, as viewed by electron microscopy, led to the identification of the regions on the surface of S1 that interact with actin (Rayment et al 1993a, Schroeder et al 1993). The actin-binding face is located on the side of the catalytic domain opposite from the nucleotide-binding cleft (the larger sphere in Figure 3 indicates the presumed position of the β-phosphate of ATP) (see below) $\approx$ 3.5 nm away. Between these sites is a long narrow cleft with its apex at the base of the nucleotide-binding cleft, which opens out towards the actin-binding face. This narrow cleft splits the central 50-kDa domain (red region of heavy chain in Figure 3) of S1 into upper and lower segments [S1 has traditionally been subdivided into an amino-terminal 25-kDa segment (green region), a central 50-kDa segment, and a carboxyl-terminal 20-kDa segment (dark blue region), as defined by proteolytic digestion of the skeletal muscle myosin flexible loops that connect these segments (Balint et al 1978, Mornet et al 1979)]. Putative actin-binding sites are located on both sides of this cleft.

The recently solved structures of the *Dictyostelium* myosin catalytic domain (the globular portion of S1 minus the light chain binding neck) (Itakura et al 1993), with two different bound nucleotide analogues (Fisher et al 1995), share many of the above features of the chicken S1 structure and also have provided information about possible conformational changes that accompany binding and hydrolysis of nucleotide. This *Dictyostelium* catalytic domain, which has some minimal motor activity in vitro (Itakura et al 1993), was crystallized with either MgADP-beryllium fluoride or MgADP-aluminum fluoride complexed in the active site. Whereas the structure of the *Dictyostelium* protein with bound MgADP-beryllium fluoride was essentially identical to that of the chicken S1 with no bound nucleotide, the MgADP-aluminum fluoride-bound head exhibited several distinct structural changes from the chicken structure. The most striking difference is that the 50-kDa cleft is partially closed in this structure, bringing the upper and lower domains of the 50-kDa segment of the heavy chain into closer proximity. The structure of the carboxyl-terminal portion of the head also appears to have changed. This movement is derived from changes in the active site in the presence of the aluminum fluoride complex. In the beryllium fluoride structure, the bond distance from beryllium to the β-phosphate pseudo-bridging oxygen is 1.57 Å, whereas in the aluminum fluoride structure, the analogous distance is considerably longer (2.0 Å), which suggests that the aluminum fluoride structure may be an analogue of the transition state for ATP hydrolysis. This longer bond distance results in the formation of a new bond between one of the fluoride ions and the amide hydrogen of the highly conserved Gly457 (Gly466 of chicken S1) residue that sits in the loop that drops down from the upper segment of the 50-kDa domain to form the apex and part of the lower border of the 50-kDa cleft (see Figure 3). [Note: All residue numbers referred to in the text denote the position of that residue in the myosin in question—in this case, *Dictyostelium* myosin II. For ease of comparison with the published structures, the analogous residue in the chicken S1 sequence (Maita et al 1991) as defined by sequence alignment (Sellers & Goodson 1995) is given in parentheses. The chicken S1 residue numbers are used in the figures.] Rayment and colleagues hypothesize that changes in the active site cause changes in the nearby so-called reactive thiol region. This region is highly conserved and in skeletal muscle myosin contains two reactive cysteine residues (Cys697 and Cys707 of chicken S1, or SH2 and SH1, respectively) (see Figure 3) that can cross-link to within 3 Å in the presence of nucleotide (Wells & Yount 1979, Wells et al 1980). This cross-linking traps ATP at the active site. None of the structures solved to date accounts for a change in the thiol region that allows this cross-linking of the two thiols, so some of the most interesting crystal structures are yet to come.

Using the information from these crystal structures, combined with previous structural and biochemical studies, a revised swinging cross-bridge model for the actin-myosin cycle has been proposed (Rayment et al 1993a, Fisher et al 1995; for reviews, see Rayment & Holden 1993, 1994, Reedy 1993, Yount 1993). In agreement with many previous models (Cooke 1986, Vibert & Cohen 1988), this model invokes a series of interdomain movements within S1 in which changes in the globular portion of the head with binding of nucleotide and actin are amplified by the extended light chain binding domain acting as a lever arm. According to this model, in the absence of nucleotide, the narrow 50-kDa cleft is closed, with actin tightly bound in the rigor complex. Binding of ATP at the active site results in placement of the γ-phosphate at the apex of the 50-kDa cleft, which causes it to open, disrupting the strong binding interaction between myosin and actin. Thus this cleft provides a mechanism for communication between the spatially distinct actin- and ATP-binding sites. Hydrolysis of the bound ATP to ADP and P_i results in changes in the 50-kDa cleft that are transmitted via the reactive thiol region of the 20-kDa domain to the light chain binding domain, which acts like a lever arm and swings 5 nm or more relative to the actin-binding site into a cocked configuration. These proposed changes are thought to occur during or just after hydrolysis of ATP. Following hydrolysis and priming of the lever arm, the model predicts that rebinding of actin causes further movement of the lower domain of the 50-kDa segment relative to the upper domain, which results in release of P_i from the base of the nucleotide-binding cleft and subsequently the release of ADP. The release of the nucleotide products results in complete closure of the 50-kDa cleft (Fisher et al 1995). The closure of the cleft leads to conformational changes in the 20-kDa reactive thiol domain that are postulated to cause the light chain binding domain to rotate, resulting in a power stroke of 5 nm or more. Thus the rigor actin-S1 complex is reformed.

This model presents testable predictions of the nature of the actin-myosin cross-bridge cycle, some of which have already been analyzed using mutagenic approaches, as described below. In addition, the structures and the derived models suggest functions for the various regions within the molecule that have already been subjected to mutagenesis. The effects of these mutations are discussed in the context of these structural considerations.

MUTAGENESIS SYTEMS

The myosin mutants discussed below have been generated or discovered in a variety of experimental systems ranging from unicellular slime mold to human cardiac muscle. Each system has certain unique advantages for the study of myosin structure-function relationships in vivo and/or in vitro.

Some of the earliest myosin mutations were discovered in classical genetic systems. The nematode *Caenorhabditis elegans* has proven to be an excellent experimental organism for genetic analysis of myosin and sarcomeric function. Nematodes have four different myosin heavy chain (MHC) genes: MHC A and MHC B in the body wall (each assemble as homodimers), and MHC C and MHC D in the pharyngeal muscle. *unc54* is the structural gene for MHC B—the major myosin heavy chain in the body wall musculature (Epstein et al 1974). *unc54* mutations can result in impaired movement or paralysis of the animal but are not lethal because the pharyngeal muscle is unaffected, thus allowing study of the effects of mild and severe myosin mutations on body wall muscle structure, assembly, and function (Brenner 1974, Epstein et al 1974). Genetic analysis of worms with muscle-defective phenotypes has led to the isolation of well over 100 *unc54* mutations (Moerman et al 1982, Waterston et al 1982, Bejsovec & Anderson 1988). These mutations have been divided into several phenotypic classes (MacLeod et al 1977, Moerman et al 1982), including those that lack myosin, those in which the myosin is of abnormal size, and those in which the protein appears to be of normal size. The latter class contains interesting motor domain mutations. A group of strongly dominant mutations has been reported that, as heterozygotes, inhibit normal filament formation of the wild-type MHC B despite the fact that the mutant form accumulates less than 2% of wild-type levels in vivo (Bejsovec & Anderson 1988). These mutations all map to the MHC B motor domain (Bejsovec & Anderson 1990), including several substitutions in and around the ATP-binding site (see below). Another group of mutations are recessive and accumulate normal amounts of MHC B, which assemble into well-ordered sarcomeres; nonetheless these animals move slowly due to mutations in the motor domain that presumably lead to abnormal myosin function (Moerman et al 1982, Dibb et al 1985).

Drosophila melanogaster has a genetic system yielding mutations in two distinct myosins. A single myosin heavy chain II gene in *Drosophila* gives rise to multiple isoforms via alternatively spliced exons (Bernstein et al 1986, Rozek & Davidson 1986). A single isoform is present in the indirect flight muscle (IFM). Mutants defective for this isoform are viable and fertile but have a flightless phenotype that is easy to detect (Fyrberg & Beall 1990). Bernstein and colleagues have identified several mutations in the fly IFM myosin isoform that led to failure to accumulate myosin heavy chain, including one single amino acid substitution in a highly conserved region of the motor domain (Kronert et al 1994). Porter & Montell (1993) have performed extensive site-directed mutagenesis of the *Drosophila ninaC* unconventional myosin. This rhabdomere-specific protein, comprising linked protein kinase and myosin domains (Montell & Rubin 1988), is required both for a normal electroretinogram (Porter et al 1992) and to prevent light and age-dependent retinal degeneration (Matsumoto et al 1987, Porter et al 1992). Mutations in the myosin domain were constructed and expressed in null allele flies by germline transformation. Mutations in several highly conserved regions of the motor domain resulted in distinct phenotypes in vivo (discussed below).

Recently, a single-amino acid substitution in the β-cardiac myosin heavy chain gene was found to be the cause of an inherited form of hypertrophic cardiomyopathy in humans (Geisterfer et al 1990). This autosomal-dominant disease is characterized by variable degrees of ventricular hypertrophy and myofibrillar disarray, and symptoms include angina, arrhythmias, and sudden death. Currently, over 30 different single-amino acid substitutions within the cardiac myosin motor domain and the proximal portion of the rod have been detected in families afflicted with hypertrophic cardiomyopathy (Schwartz et al 1995, Sellers & Goodson 1995). Analysis of the variable severity of the disease between families bearing different mutations suggests that the nature of the amino acid substitution (and by inference, its effect on myosin motor function) is an important predictor of disease penetrance and incidence of sudden death (Epstein et al 1992, Watkins et al 1992, Anan et al 1994, Fananapazir & Epstein 1994, Marian et al 1994, Vosberg 1994). These mutations cluster in interesting regions of the S1 structure (Rayment et al 1995). Unlike the mutations recovered in the *C. elegans* and *Drosophila* systems, the biology of this system demands that these mutations result in myosins that are at least partially functional in vivo because patients with this disease depend on the contractile activity of the β-cardiac myosin for survival. Hence, this system selects for mutations that are likely to have relatively mild biochemical defects.

By contrast, recently discovered mutations that are responsible for a deaf-blindness syndrome in humans (Weil et al 1995) and two forms of deafness in mice (Gibson et al 1995, Avraham et al 1995) most likely have severe effects on the function of the affected unconventional myosins in vivo and in vitro. Patients with the autosomal-recessive disease Usher syndrome type 1 have profound congenital sensorineural hearing loss as well as other defects (Smith et al 1994), whereas mice with mutations in the *shaker-1* gene have hearing impairment and primary abnormalities of the sensory neuroepithelia of the inner ear (Brown & Steel 1994). These defects are caused by mutations in homologous genes in the human and mouse that encode a class VII unconventional myosin. Mice with *Snell's waltzer* recessive deafness exhibit hearing loss and neuroepithelial degeneration secondary to mutation of a class VI unconventional myosin. Previous studies had implicated a myosin motor as an important player in adaptation in the hair cell (Gillespie et al 1993, Hudspeth & Gillespie 1994).

The above mentioned systems are amenable to analysis of the effects of myosin mutations on muscle structure and function in vivo; however, biochemical analysis of the mutated protein is hampered by the presence of multiple isoforms within the organism and/or by the difficulty of obtaining sufficient quantities of the protein for study. One alternative approach is to heterologously express the mutated myosin of interest. Although this approach has worked well for creating mutated light chains in *Escherichia coli*, which subsequently are exchanged onto the conventionally purified myosin (for review, see Sellers & Goodson 1995), attempts at bacterial expression of a functional motor domain have met with little success. Recently, several groups have succeeded in heterologously expressing and purifying wild-type and altered forms of muscle myosins or myosin subfragments in cultured cells using baculoviral vectors (Pato et al 1993, Sweeney et al 1994, Trybus 1994, Onishi et al 1995). This breakthrough should facilitate the analysis of mammalian myosins not previously amenable to mutagenic exploration.

The unicellular slime mold *Dictyostelium discoideum* has emerged as a powerful experimental system for myosin structure-function studies because of the ability to combine in vivo and in vitro analyses of altered myosins (Egelhoff et al 1991, 1993, Kubalek et al 1992, Uyeda & Spudich 1993, Uyeda et al 1994, Ruppel et al 1994). *Dictyostelium* cells can be grown in large quantities in the laboratory and contain a single conventional myosin heavy chain gene that can be removed by homologous recombination (De Lozanne & Spudich 1987, Manstein et al 1989). The resulting myosin null cells are viable but exhibit defects in cytokinesis and development (De Lozanne & Spudich 1987, Knecht & Loomis 1987, Manstein et al 1989) that can be rescued by expression of the wild-type gene on an expression plasmid (Egelhoff et al 1990). The functional state of a mutated myosin can be assessed in vivo by expressing it in this myosin null cell and analyzing the extent to which it complements these characteristic phenotypic defects. The expressed myosins can then be purified in milligram quantities and studied in vitro, and the in vivo and in vitro behaviors of the mutated myosins can be compared. Such an approach has facilitated the characterization of site-directed and random mutations of the myosin motor domain in this organism. Site-directed mutagenesis has been useful for posing specific questions about the roles of different residues and regions of myosin during the actin-myosin cycle. Random mutagenesis, when coupled with a strong in vivo selection for a myosin related phenotype, has the advantage of being model independent and may yield mutations in parts of the protein whose function is unknown. Specific suggestions regarding the roles of particular regions of the myosin motor domain, as studied by mutagenesis of myosin in a variety of systems, are examined below.

MUTAGENESIS

The Light Chain Binding Region Behaves Like a Lever Arm

An unusual structural feature of S1, the motor domain of myosin, is that the carboxyl-terminal portion of the S1 heavy chain forms a long (≈ 8 nm) α-helix around which the calmodulin-like light chains bind (Rayment et al 1993b, Xie et al 1994), and recent models of the myosin chemomechanical cycle couple movement of this helical neck domain or lever to the opening and closing of the nucleotide-binding cleft and the 50-kDa cleft (Rayment et al 1993a, Fisher et al 1995). This swinging neck-lever model is similar to previous models (Vibert & Cohen 1988) contending that the carboxyl-terminal domain of S1 acts as a lever arm that may serve to amplify small conformational changes in the catalytic domain during ATP and actin binding. It has been hypothesized (Rayment et al 1993a, Fisher et al 1995) that such a swing could produce a mechanical stroke similar to that measured by in vitro experiments (Finer et al 1994, Ishijima et al 1994).

Consistent with this model, Lowey et al (1993) chemically removed one or both light chains from smooth muscle myosin and showed a consequent reduction in velocity in vitro of 50–90%. In these studies, the heavy chain ELC- and RLC-binding domains remained but were devoid of light chains. Itakura et al (1993) used molecular genetics and the *Dictyostelium* system to delete both light chain binding sites on the myosin heavy chain, and they reported more than 90% reduction in velocity in vitro of a truncated *Dictyostelium* S1.

To further test this model, the structure was used as a guide to design mutant *Dictyostelium* myosin heavy chain genes of different neck length by deleting or inserting light chain binding site(s) (Uyeda & Spudich 1993, Uyeda et al 1996). Three mutant myosins [ΔBLCBS (lacks both light chain binding sites), ΔRLCBS (lacks the RLC-binding site only), and 2xELCBS (has one extra ELC-binding site)] were created, purified, and tested in vitro (Figure 4). A striking linear relationship was observed between sliding velocities and the number of light chain binding sites. These data support the model because alteration in the length of the lever arm should alter the size of the mechanical stroke per ATP hydrolyzed and, consequently, the velocity of movement. These data are consistent with the movement of a relatively rigid lever arm about a fulcrum

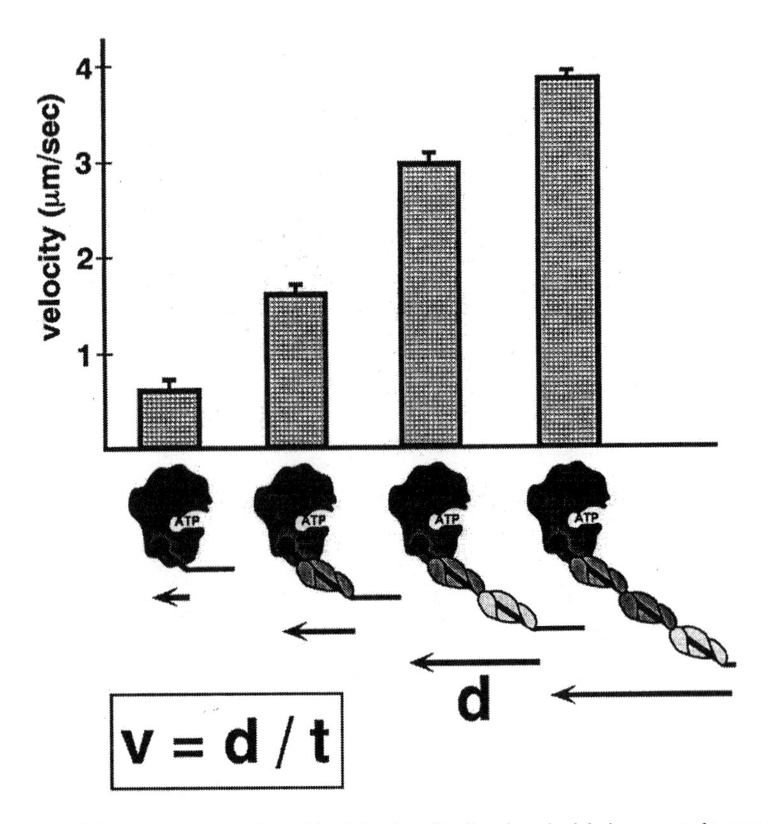

Figure 4 Schematic representation of the light chain binding domain deletions created to test the swinging neck lever model. The wild-type myosin has two light chain binding sites. Deletions of light chain binding sites or addition of a light chain binding site creates mutant forms of myosin that move slower or faster than wild-type myosin, respectively.

point that is very near the reactive thiol region, which is known to undergo structural rearrangement during the ATPase cycle.

Other support for the lever arm model derives from fluorescence polarization experiments with muscle to detect changes in the orientation of the light chain binding region of S1 (Irving et al 1995). Also, three-dimensional reconstruction studies of actin-S1 complexes have shown a considerable change in orientation of the light chain binding domain with respect to the actin filament axis, which depends on whether ADP is bound in the active site (Jontes et al 1995, Whittaker et al 1995).

In addition to the neck-length mutants described above, a large number of point mutations have been localized to the heavy chain portion of the lever arm.

The effects of these mutations on myosin function in vivo suggest important roles for this region in interacting with the ELC, RLC, and the catalytic domain of the heavy chain. The carboxyl-terminal region of the lever arm, an Arg800Pro (Arg821 of chicken S1) *Dictyostelium* cold-sensitive myosin mutation may affect RLC binding (Patterson & Spudich 1996). Other mutations—Asp778Gly, Asp778Asn, and Ser782Asn (Asp780 and Ala784 of chicken S1) in hypertrophic cardiomyopathies (Rayment et al 1995), and Arg779 (Arg800 of chicken S1) in *Dictyostelium* (Patterson & Spudich 1996)—alter residues that are in the region of the heavy chain α-helix that forms the binding site for the ELC. As the putative fulcrum point in the thiol region is approached, multiple mutations cluster to the region of the catalytic domain that lies at the interface of the catalytic domain and the ELC. These mutations include β-cardiac myosin mutations, which lead to hypertrophic cardiomyopathies—Gly716Arg, Arg719Trp, Arg719Gln, Arg723Cys, and Gly741Arg (Ala718, Arg721, Arg725, and Ala743 of chicken S1, respectively) (Rayment et al 1995, Sellers & Goodson 1995), and a *Dictyostelium* myosin mutation that produces a cold-sensitive phenotype in vivo [Gly740 to Asp (Gly761 of chicken S1)] (Patterson & Spudich 1996).

The Reactive Thiol Region, the Fulcrum Point for Movement

The fulcrum point of the lever arm, as suggested from the experiments described above, lies in the lower part of the catalytic domain. This region, which is between the ATP-binding site and the regulatory domain, is highly conserved and in several skeletal muscle myosins contains two reactive cysteine residues (Cys697 and Cys707, or SH2 and SH1, respectively) (Figure 3 and Figure 5*A*,*B*, *upper panels*). An important conformational change must occur in this region given that the crystal structure of S1 in the absence of nucleotide shows the two cysteine residues to be ≈ 18 Å apart, separated by an α-helix, whereas the two thiol groups are cross-linkable to within 3 Å in the presence of nucleotide (Wells & Yount 1979, Wells et al 1980). The location of this region between the active site and the extended regulatory domain makes it a prime candidate for communicating changes in the globular motor domain to the putative lever arm region of the molecule via this conformational change (Fisher et al 1995).

Interestingly, one of two cold-sensitive *Dictyostelium* myosin mutations that map to the reactive thiol area (Patterson & Spudich 1996) appears to interfere with communication between the catalytic domain and the lever arm region. Using a combination of positive and negative selections for the myosin null cell phenotype at two different temperatures, Patterson & Spudich (1995, 1996) screened a population of chemically mutagenized cells and recovered a bank of myosin heavy chains containing single-amino acid substitutions that rendered cells expressing them cold sensitive for myosin function. The temperature shifts used for selection were rapid enough (≈3–5 min) to demand quickly reversible alterations in the mutated proteins upon shift from the permissive to nonpermissive temperatures (and vice versa). This approach yielded myosins with mutations in a variety of important regions, including two mutations in the region that may act as the fulcrum point. Gly691 (Gly710 of chicken) sits at the end of the second helix a few residues away from SH1, and Gly680 (Gly699 of chicken S1) sits in the short stretch of coil that connects the two cysteine-containing helices (Figure 5*A,B upper panels*). Movement about the universally conserved Gly680 may be an important part of the conformational changes that occur in this region during the cycle. Surprisingly, mutagenesis of the analogous glycine in the *Drosophila nina*C motor domain (^{957}GVL^{959} to ^{957}EVR^{959}; ^{699}GVL^{701} of chicken S1), as well as a nearby leucine residue, has no deleterious effects in vivo (Porter & Montell 1993). Mutation of several residues in this region of *nina*C (^{968}GFSSR^{972} to ^{968}EFSSG^{972}; ^{710}GFPSR^{714} of chicken S1) led to temperature-sensitive retinal degeneration in vivo. Finally, two of the hypertrophic cardiomyopathy causing mutations of human β-cardiac myosin map to a region near the reactive thiol helices. The mutations Phe513Cys (Phe515 of chicken S1) (Anan et al 1994) and Gly584Arg (Gly586 of chicken S1) (Watkins et al 1992) alter residues that may interact with the thiol region of the protein and thus affect the ability of this region to undergo conformational changes during the ATPase cycle.

Functions of the ATP-Binding Pocket

The ATP-binding pocket of myosin not only functions to bind and hydrolyze ATP but, importantly, it governs the release of actin from the rigor complex upon nucleotide binding and the reciprocal release of nucleotide products in response to rebinding of actin. Furthermore, the characteristics of the active site must account for the relative stability of the myosin-ADP-P_i complex in the absence of actin.

Early sequence analysis of a variety of myosins revealed the presence of a highly conserved region in the amino-terminal 25-kDa domain that shared sequence similarity with other known nucleotide-binding proteins (Walker et al 1982), including adenylate kinases, ras proteins, elongation factors, ATP-synthase β-subunits, and phosphoglycerate kinases. This general motif is known as a phosphate-binding loop or P-loop (for a review of P-loop-containing proteins, see Saraste et al 1990). In myosin, the P-loop is comprised of the virtually invariant GESGAGKT sequence [universally conserved in all 73 myosins sequenced thus far, with the exception of the *Drosophila nina*C myosin domain (Sellers & Goodson 1995)].

In the chicken S1 structure, the GESGAGKT sequence forms the central loop of a strand-loop-helix motif (Rayment et al 1993b) whose topology is essentially identical to that observed for the nucleotide-binding sites of both adenylate kinase (Muller & Schulz 1992) and ras p-21 (Pai et al 1990). This P-loop and the helix that follows it (Lys185 to Ile199) form the base of the chicken S1 ATP-binding pocket (Figure 5, *middle panels*), whereas the sides are lined with portions of the 25-kDa segment on one side and elements of the 50-kDa segment on the other side that include residues originally identified as active site residues by photoaffinity labeling studies (Yount et al 1992). The pocket sits above the seven-stranded mostly parallel β-sheet that is derived from all three segments of the heavy chain. The ATP-binding pocket in the *Dictyostelium* myosin structure with MgADP-beryllium fluoride bound is virtually superimposable upon the chicken active site pocket and reveals details of the interaction with nucleotide (Fisher et al 1995). In this structure, the purine ring interacts with the 25-kDa segment of the cleft, whereas the α- and β-phosphates and the beryllium fluoride are found between the P-loop and the loop formed by Asn233 to Gly240 (Asn240 to Gly247 of chicken S1) of the 50-kDa domain, with the beryllium fluoride moiety pointing down into the apex of the 50-kDa cleft.

In a variety of systems, the ATP-binding pocket of myosin has been extensively mutated (Figure 5*A,B middle panels*). The *Dictyostelium* system has been used to create several myosins containing single point mutations in the P-loop sequence, including Gly182Glu (Gly182 of chicken S1) and Lys185Arg (Lys185 of chicken S1). Each of these mutated myosins failed to complement the myosin-specific defects of the myosin null cells for cytokinesis (as assayed by growth in suspension) and normal *Dictyostelium* development (K Ruppel & J Spudich, unpublished observations) despite expression levels close to those of the wild-type protein. Biochemical analysis of the mutated proteins revealed that each was deficient in ATP binding and exhibited neither ATPase activity nor movement in sliding filament assays. Presumably, because the actin-binding face of the myosin was not altered by these discrete ATP-binding pocket mutations, the proteins retained the ability to bind to actin filaments in the tightly bound or rigor configuration. However, because binding of ATP to myosin is required to dissociate from actin, these myosins become essentially stuck in the rigor state (so-called rigor-binders).

Similarly, deleterious in vivo effects were observed for mutations of the P-loop of *C. elegans* body wall muscle myosin heavy chain gene *unc54*. Thirty-one dominant *unc54* mutations were recovered following chemical mutagenesis and screening for a paralyzed phenotype—all interfered with thick filament and sarcomere assembly in heterozygotes (Bejsovec & Anderson 1990). The ten most strongly dominant alleles involved point mutations in the three glycines of GESGAGKT: G177E (isolated three times), G180R (isolated three times), G182R (isolated twice), and G182E (isolated twice) (Figure 5*A,B middle panels*). Similar to the substitutions in *Dictyostelium* mutants, these substitutions of glycines by relatively large, charged side chains may incompatible with nucleotide binding. Furthermore, as was the case for the *Dictyostelium* mutants, the in vivo findings correlate well with a nucleotide-binding defect. The thick filament disruption phenotype caused by the nematode *unc54* nucleotide-binding site mutations could be explained by the need for myosins to be able to interact with actin in an ATP-dependent fashion in order for the sarcomere to assemble and function as a contractile unit.

As in the other systems, site-directed mutagenesis of the *Drosophila nina*C myosin P-loop resulted in a phenotype that was indistinguishable from the null allele (Porter & Montell 1993).

Other residues lining the nucleotide-binding pocket have also been subjected to mutagenic analysis in a variety of systems. The highly conserved Lys130 residue of *Dictyostelium* myosin (Lys130 of chicken S1) corresponds to the lysine that is trimethylated in some skeletal muscle myosins (Tong & Elzinga 1983). Mutagenesis of this *Dictyostelium* residue resulted in a myosin with a maximal actin-activated ATPase in solution and motility in vitro that was identical to that of the wild-type protein (Ruppel et al 1994). However, alteration of this lysine did cause a fourfold increase in the apparent K_m of the mutated myosin for ATP, and cells expressing the K130L myosin exhibited temperature-sensitive defects in myosin function in vivo.

Mutations in and around the nucleotide-binding site have also been seen in β-cardiac myosin in cases of human hypertrophic cardiomyopathy, in *C. elegans*, and in *Drosophila nina*C. These mutations, and their relation to the active site pocket, are detailed in Figure 5*A,B* (*middle panels*). The range of in vivo phenotypes observed as a result of these mutations is wide. Little is known about the biochemical effects of these mutations on myosin function in vitro.

Functions of the Actin-Binding Face

Like the nucleotide-binding pocket, the actin-binding face of myosin must account for several properties crucial to the function of this motor. Primarily, it must form a strong interaction with actin during the power stroke that allows putative conformational changes in myosin to result in production of force and translocation of the actin filament. In addition, the interaction of this face of the myosin with actin must signal the release of products from the active site more than 3 nm away in order to achieve this strongly bound state. The actin-binding face also must lower its affinity for actin in response to ATP binding at the active site.

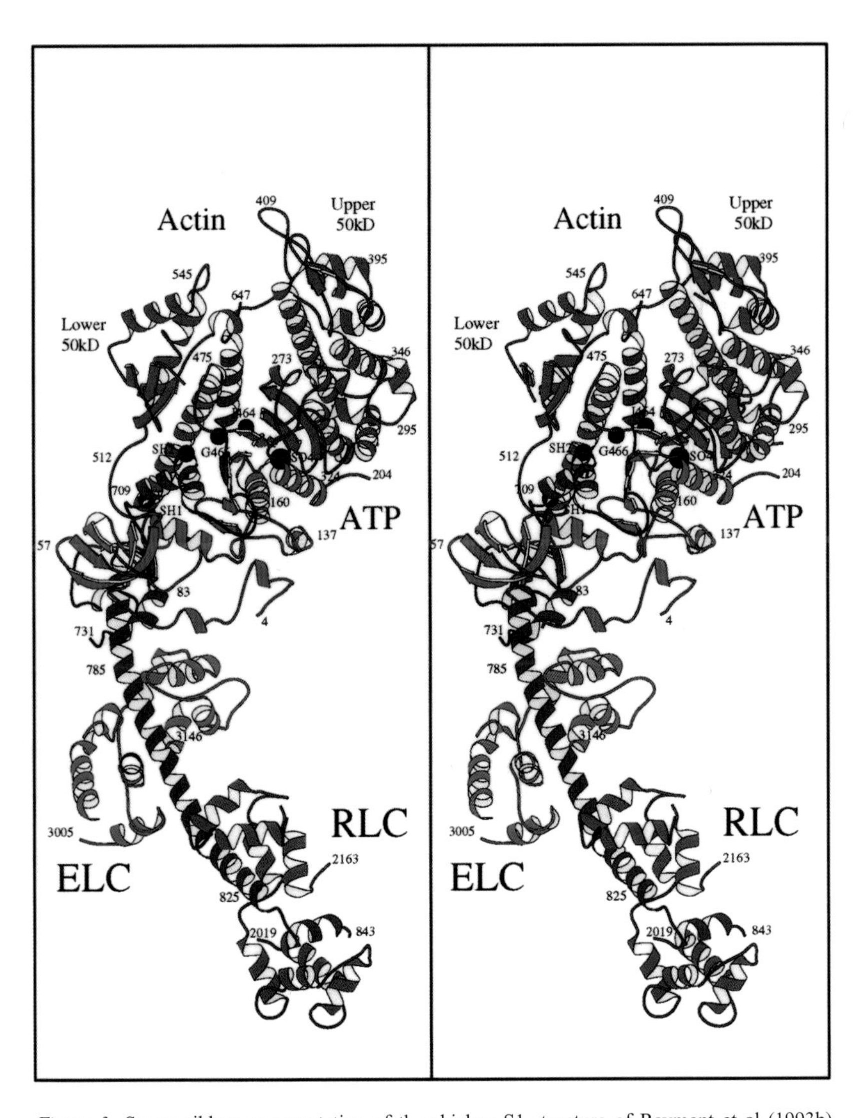

Figure 3 Stereo ribbon representation of the chicken S1 structure of Rayment et al (1993b). The 25-, 50-, and 20-kDa segments of the heavy chain are shown as green, red, and dark blue ribbons, respectively. The upper and lower segments of the 50-kDa domain (based on orientation of the molecule in actin-S1 docking models) are noted. The ELC and RLC are shown in light blue and magenta, respectively. Orientation and labeling shown here are similar to views published by Reedy (1993). The ATP- and actin-binding sites are shown (see text for definitions). Note that they are on opposite sides of the globular portion of the head, whereas the light chains bind to the extended carboxyl-terminal helix, forming the neck or regulatory domain. Within the nucleotide-binding pocket, a large black sphere approximates the location of the sulfate density found in the chicken structure [thought to be the location of β-phosphate binding (Rayment et al 1993b)]. Residue labels are indicated (2000 and 3000 have been added to the residue numbers of the RLC and ELC, respectively) and numbered according to the chicken S1 sequence. Cpk representations of the α-carbons of several key residues are shown, including Gly466 in the 50-kDa cleft and Cys699 (SH2) and Cys707 (SH1) in the reactive thiol region (see text for details). All ribbon diagrams were prepared using the program MOLSCRIPT (Kraulis 1991).

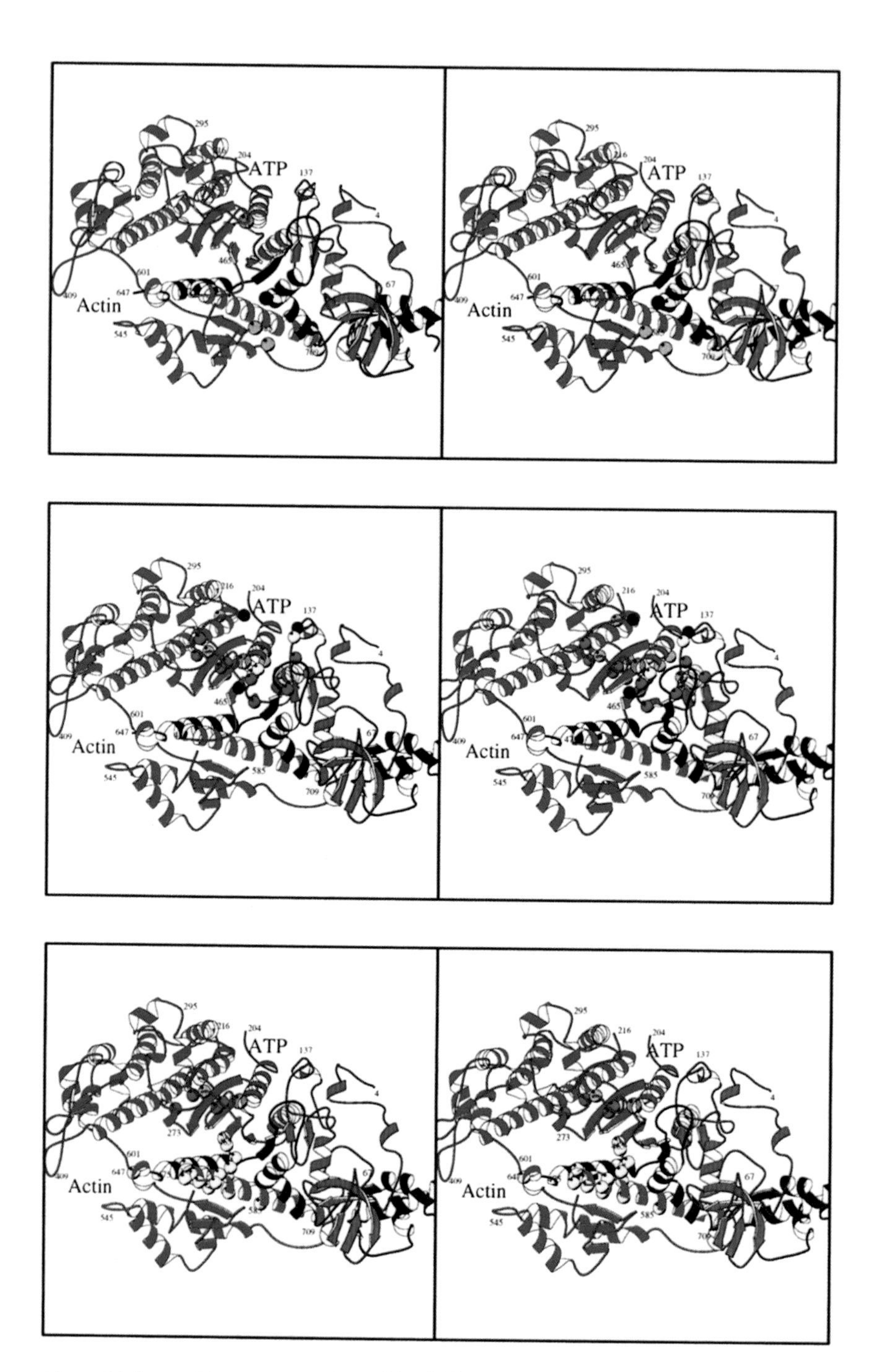

Figure 5A Stereo ribbon representations of the chicken S1 structure showing placement of various groups of mutations by region. Three stereo panels show the globular catalytic domain, oriented ~90° counterclockwise to the view seen in Figure 3. The heavy chain is colored as in Figure 3. Small colored spheres indicate the positions of various mutated residues in and around the reactive thiol region *(upper panel)*, in the nucleotide-binding pocket (*middle panel*), and in the 50-kDa cleft region (*lower panel*).

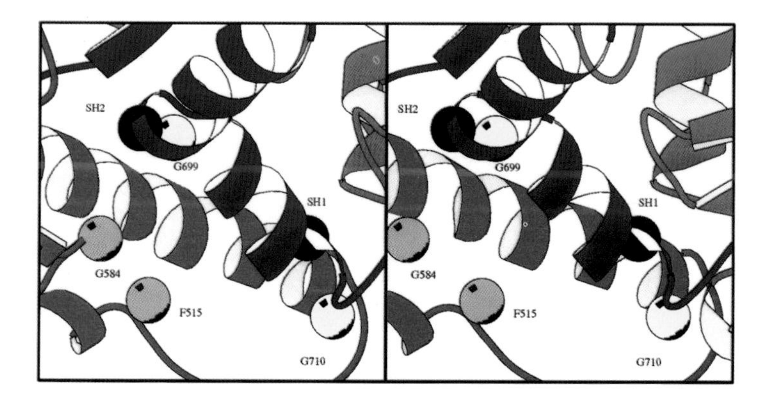
SH2
G699
SH1
G584
F515
G710

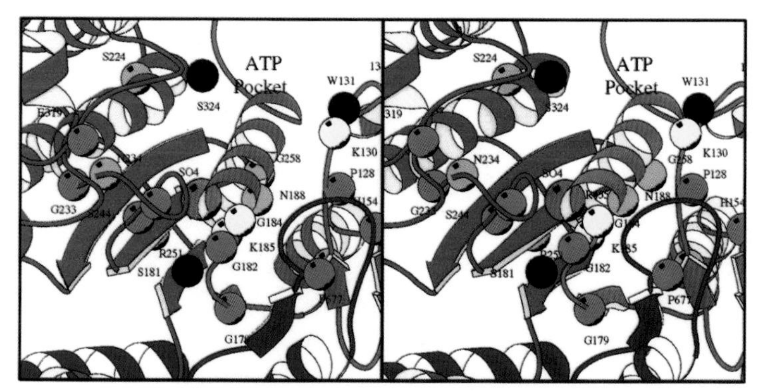
ATP
Pocket
S224
W131
S324
K130
N234
G258
P128
SO4
N188
G233
S244
G184
H154
K185
R251
G182
S181
P677
G179

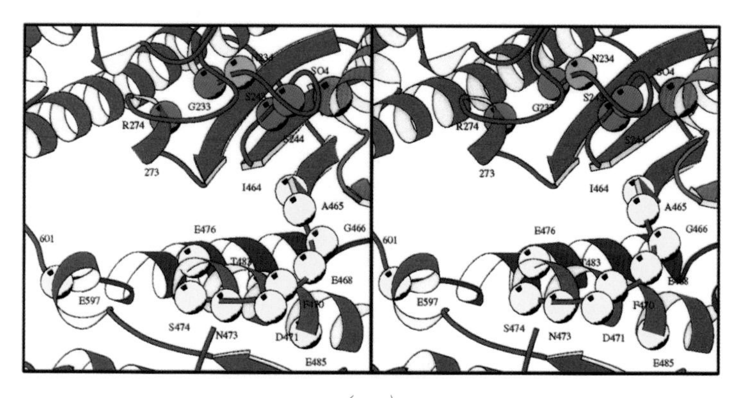
N234
SO4
G233
S243
R274
S244
273
I464
A465
601
E476
G466
T483
E468
E597
F470
S474
N473
D471
E485

(over)

Figure 5B The three stereo panels are in the same orientation as in Figure 5*A*; however, the regions containing the mutations have been magnified. Residues are numbered according to the chicken S1 sequence. *Upper panel*: Enlargement of the reactive thiol region. Amino acid residues (cpk representations of the α-carbons) of the chicken S1 structure that correspond to mutated residues of the human β-cardiac myosin (light purple: F515, G586) and *Dictyostelium* myosin II (yellow: G699, G710) are shown. The reactive thiol cysteines, C697 and C707 (SH2 and SH1, respectively) are shown in black. *Middle panel*: Enlargement of the ATP-binding pocket. Amino acid residues that correspond to mutated residues of *C. elegans unc54* myosin (blue: P128, H154, G179, G182, G233, S234, S243, S244, E319, P677), human β-cardiac myosin (light purple: T125, N188, S224, N234, R251, G258), and *Dictyostelium* myosin (yellow: K130, G184, K185) are shown. Residues that have previously been shown to be active site residues based on photolabeling experiments are shown in black (W131, S181, S324) (Yount et al 1992). The position of the sulfate density in the nucleotide-binding pocket is approximated by a cpk representation of the α-carbon of T186 of the P-loop (grey: SO_4). *Lower panel*: Enlargement of the 50-kDa cleft area. Some of the amino acid residues are shown that correspond to mutated residues of *C. elegans unc54* (blue: G233, S243, S244, R274), human β-cardiac myosin (light purple: N234), and *Dictyostelium* myosin II (yellow: I464, A465, G466, E468, F470, D471, N473, S474, E476, C479, T483, N484, E485, H493, E597).

Prior to the solution of the chicken S1 structure, a variety of approaches had implicated residues in the 50- and 20-kDa domains, as well as the flexible loop connecting these domains, in actin binding (Vibert & Cohen 1988, Tokunaga et al 1991). Modeling the docking of the crystal structures of actin (Holmes et al 1990, Kabsch et al 1990) and chicken S1 (Rayment et al 1993b), using electron density information from three-dimensional reconstructions of the actin-S1 complex as viewed by electron microscopy (Rayment et al 1993a, Schroeder et al 1993), has confirmed and refined the definition of the actin-binding face of S1 (Figure 6).

These studies placed the 50/20-kDa junction loop at the actin-myosin interface, although the loop itself is not seen in the S1 crystal structure presumably

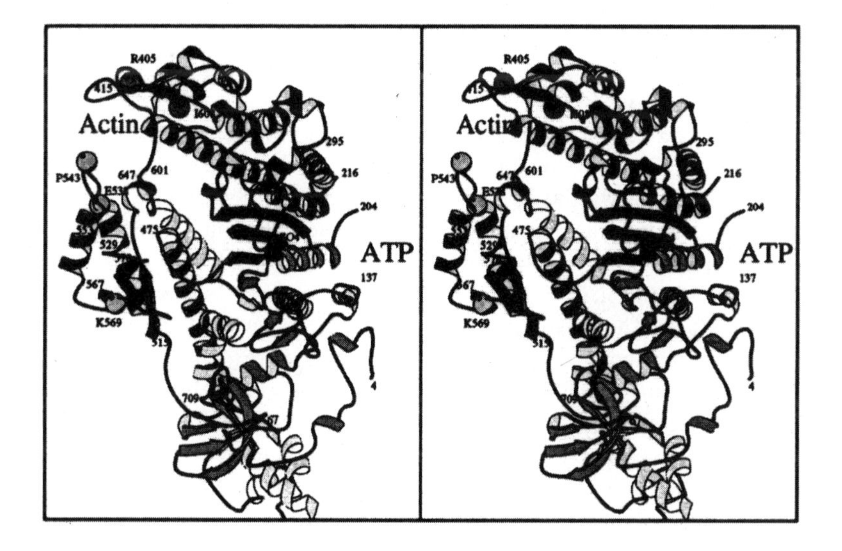

Figure 6 Stereo ribbon representation of the chicken S1 structure showing placement of the actin-binding face mutations. The globular catalytic domain of S1 is oriented as it would be to bind an actin filament whose long axis would be approximately vertical in the plane of the page. Residues are numbered according to the chicken S1 sequence. The R405 residue mutated in several families with hypertrophic cardiomyopathy is shown near the top of the structure on a surface loop (medium grey cpk representation of the α-carbon). The other large surface loop that forms part of the actin-binding face is not seen in the crystal structure (presumably owing to its flexibility) and is not represented here. The three *Dictyostelium* mutations that caused a cold-sensitive phenotype in vivo are shown (light grey cpk representations of the corresponding α-carbons). E538 and P543 form part of the helical domain thought to make stereospecific interactions with actin. K569 is located near a break in density that likely forms an interaction site with the actin monomer, one actin helix turn below the monomer interacting with the other myosin sites mentioned above (see text for details).

because of its flexibility. This placement is consistent with studies showing cross-linking of this region to the highly negatively charged amino-terminus of actin (Sutoh 1983, Yamamoto 1989) and protection of this region from proteolytic cleavage when bound to actin (Mornet et al 1979). Uyeda et al (1994) analyzed the role this loop plays in determining myosin's enzymatic activity by creating a series of chimeras comprised of the *Dictyostelium* myosin backbone with the *Dictyostelium* loop replaced with corresponding loops from mammalian skeletal, α-cardiac, β-cardiac, or chicken smooth muscle myosin. All the chimeras were functional in vivo. Replacing the *Dictyostelium* loop with that from skeletal muscle made the enzyme five- to sixfold more active than the wild-type *Dictyostelium* myosin, whereas swapping in the chicken smooth muscle loop decreased the enzymatic activity approximately twofold. The activities of the cardiac chimeras were intermediate, with the α-isoform having greater activity than the β-isoform. Overall, the actin-activated ATPase activity of the chimeras corresponded well with those of the native homologous myosins from which the loop sequences were derived. These results are consistent with the idea that this surface loop regulates the rate-limiting step of the actin-activated ATPase cycle. Because this loop may be particularly important for the weak binding interaction with actin (DasGupta & Reisler 1989, Sutoh et al 1991), one interpretation of the data is that this loop determines the rate of transition from the weak to strong binding state.

In addition, other *Dictyostelium* mutagenesis studies have identified mutations in this loop. The substitution Gly624Asp (Gly643 in the chicken sequence) conferred a cold-sensitive myosin phenotype on *Dictyostelium* cells expressing this altered myosin (Patterson & Spudich 1996). This glycine is the only glycine residue in the *Dictyostelium* 50/20-kDa junction loop. By contrast, the corresponding loop in chicken skeletal myosin contains nine glycines (based on comparison of 73 myosin heavy chain sequences, the number of glycines ranges from 1 to 9 in this region) (Sellers & Goodson 1995). The relative flexibility of these loops is unknown. Mutating this glycine residue to an aspartate may alter the flexibility of the *Dictyostelium* loop, or the addition of a negative charge in the loop could alter the actin interaction directly.

The crystal-docking studies place another myosin surface loop at the actin-binding face. This loop extends from Arg405 to Lys415 (Figure 6) and forms close contact with residues Pro332 to Glu334 of actin (Rayment et al 1993a). Arg403 (Arg405 in chicken S1) on this loop has been identified as a hot spot for mutations in the human β-cardiac myosin heavy chain gene that result in familial hypertrophic cardiomyopathy. Alteration of this residue to glutamine in three separate families (Geisterfer et al 1990, Epstein et al 1992), or to a leucine (Dausse et al 1993) or tryptophan (Dausse et al 1993) in two others, has been linked to a severe form of this disease. Taking advantage of the fact that the β-cardiac myosin is also expressed in small quantities in slow skeletal muscle fibers, Cuda et al (1993) used a skeletal muscle biopsy sample from a patient carrying the Arg403Gln mutation to isolate the mutated β-cardiac myosin. They found that the mutated protein moved actin filaments at approximately one-fifth the velocity of wild-type myosin controls in an in vitro motility assay. Using a baculovirus expression system, Sweeney and colleagues successfully coexpressed the rat essential and regulatory light chains with a truncated version of the rat α-cardiac myosin heavy chain that had been engineered to include the Arg403Gln mutation (Sweeney et al 1994). The purified mutated protein had normal ATPase activity in the absence of actin, but a decreased actin-activated ATPase (V_{max} decreased > 3.5-fold and K_{app} increased > 3-fold, where K_{app} refers to the apparent K_m of myosin for actin). Similar to the movement of the myosin isolated from a patient with hypertrophic cardiomyopathy, the baculovirus-expressed protein also moved at approximately one-fifth the velocity of the wild-type protein in sliding filament assays. The decreased affinity of this myosin for actin, with the resultant decreases in ATPase activity and motility, correlates well with the location of this residue at the actin-binding face. Sweeney and colleagues (1994) hypothesized that this alteration in the actin-myosin interaction leads to a change in the force-velocity relationship of cardiac muscle: less efficient contractility and hence compensatory hypertrophy, which is characteristic of this disease. Two other hypertrophic cardiomyopathy-causing mutations also map to regions near the actin-binding face [V606M (Watkins et al 1992, Fananapazir & Epstein 1994) and K615N (Nishi et al 1992); I608 and K617 of chicken S1] and may lead to such effects in vivo. Animal modeling of this disease, created by introduction of different cardiac myosin mutations into the germline of mice, has been undertaken (including introduction of the Arg403Gln mutation) and should lead to further understanding of how these mutations result in the pathophysiology characteristic of this disease.

In addition to the aforementioned surface loops, docking studies have also implicated several helical segments on the surface of the myosin molecule in actin binding. The region spanning Pro529 to Lys553 in the lower portion of the 50-kDa domain (Figure 6) contains multiple residues that may participate in stereospecific interactions with residues Ile341-Gln354 and Ala144-Thr148 of actin (Rayment et al 1993a, Schroeder et al 1993). Interaction of exposed hydrophobic residues, as well as complementary pairing of ionic and polar groups, potentially contributes to this interface. Patterson & Spudich (1996) have isolated two cold-sensitive myosins that contain single point mutations in this region. Mutation of Glu531 (Glu538 of chicken S1) to Gln, or of

Pro536 (Pro 543 of chicken S1) to Arg (Figure 6), results in myosins that are conditionally nonfunctional in vivo. The Glu531Gln change may disrupt an ionic interaction, whereas the Pro536Arg mutation inserts a charge at a position implicated in hydrophobic contacts with Ala144 of actin (Rayment et al 1993a).

The three contact regions discussed above involve a single actin monomer and constitute the primary actin-binding site. In the docked structures, another segment of myosin, extending from residues Lys567 to His578 in the chicken structure, forms an exposed loop that may interact with a second actin monomer directly below the first along the filament. There is cross-linking evidence to support the existence of this interaction (Bertrand et al 1988, Andreeva & Borejdo 1992, Andreeva et al 1993). In the chicken S1 structure, the density from residues 571–575 is missing, suggesting that this loop may be somewhat flexible (Rayment et al 1993b). It is hypothesized that this relatively positively charged loop, which makes no direct contact with actin in the models, may extend out to interact ionically with negatively charged residues on actin in the region of Tyr91 to Glu100 (Rayment et al 1993a). Charge-reversion mutagenesis of Glu99 and Glu100 of *Dictyostelium* actin to histidines resulted in actin filaments that were translocated by a truncated version of rabbit skeletal muscle myosin at velocities approximately one-fifth that of wild-type *Dictyostelium* actin in sliding filament assays (Johara et al 1993), suggesting that this interface plays an important role in ATP-driven sliding. Mutation of Arg562 of *Dictyostelium* myosin (equivalent to Lys569 of chicken S1) to either a leucine or a histidine results in a cold-sensitive myosin null phenotype in vivo (Patterson & Spudich 1996). It will be interesting to compare the effects of these myosin mutations on sliding velocity with the effects reported for mutagenesis of the potentially complementary area of actin.

The 50-kDa Narrow Cleft: A Communication Zone Between the ATP- and Actin-Binding Sites

An important feature of the docking studies used to determine the actin-S1 interface is that the best fit of the structures to the EM image reconstructions resulted in an inappropriate collision between the lower portion of the 50-kDa domain and the carboxyl-terminus of actin (Rayment et al 1993a, Schroeder et al 1993). This collision is most likely due to the fact that the chicken structure represents myosin that has neither ATP nor actin bound and thus conceivably may be captured in a structural state that is intermediate between the two, whereas the EM images used for the modeling were of myosin bound tightly to actin in the rigor configuration. It was noted that a better fit to the image reconstructions could be achieved if the long narrow cleft separating the upper and lower portions of the 50-kDa domain were to close (Rayment et al 1993a). Potential actin-S1 interface contact sites are located on both sides of this cleft. That observation, coupled with the fact that this cleft extends from the base of the nucleotide-binding pocket to the actin-binding face, led to the hypothesis that this cleft may provide the necessary pathway for communication between the spatially distinct ATP- and actin-binding sites during the actin-myosin ATPase cycle (Rayment et al 1993a). Subsequent analysis of the structures of *Dictyostelium* myosin with nucleotide analogues bound revealed that this cleft does indeed move in response to changes at the active site of the molecule (Fisher et al 1995). The recently revised version of the swinging cross-bridge model that was described above (Fisher et al 1995) implicates this cleft not only in communicating between the ATP- and actin-binding sites, but also in transmitting changes at these sites to the carboxyl-terminus of the protein to effect the power stroke.

Mutagenic analysis of the residues that comprise this cleft confirm its importance in the actin-myosin chemomechanical cycle. The cleft is lined with a high proportion of evolutionarily conserved residues along both its upper and lower borders. Random mutagenesis of the region of *Dictyostelium* myosin that forms the apex and the lower border of the cleft resulted in distinct phenotypes in vivo and in vitro (Ruppel & Spudich 1996). Twenty-one myosin heavy chain gene alleles containing different point mutations spanning the region from Ile455 to His484 (Ile464 to His493 of chicken S1) (Figure 5, *lower panels*) were introduced into myosin null cells and classified as wild-type, null, or intermediate-like, based on their ability to complement the myosin-specific defects of the null cells (Figure 7). The wild-type-like mutants fully complement the *Dictyostelium* null cell defects in vivo, have near normal to normal solution ATPase and sliding velocities in vitro, and involve either favorable substitutions of highly conserved residues or less favorable substitutions of nonconserved residues. By contrast, mutations that led to a null phenotype (despite wild-type expression levels) are single changes of highly conserved residues, or double- or triple-point mutations. Purification and characterization of several of these myosins revealed that they bind actin in the absence of ATP and dissociate from actin upon addition of nucleotide; however, they are unable to hydrolyze the bound ATP to ADP and P_i. This arrest prior to hydrolysis prevents eventual rebinding to actin in the prestroke state. These myosins are thus unable to interact productively with actin to produce force, which results in no detectable solution ATPase activity, the inability to translocate actin filaments in vitro, and a myosin null phenotype in vivo.

Placement of these mutations onto the S1 structure reveals that they are distributed throughout the region of the lower cleft (Figure 5*A,B lower panels*).

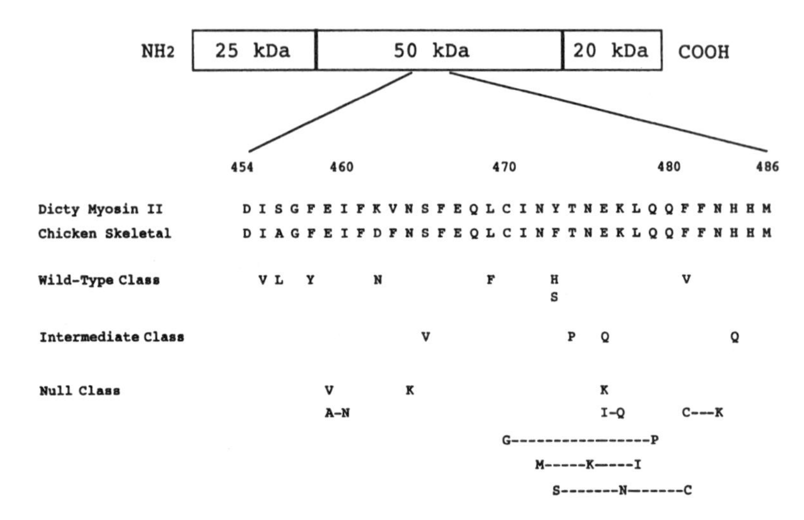

Figure 7 Comparison of the amino acid sequences of the chicken skeletal muscle myosin (Swiss-Prot P13538) and *Dictyostelium* myosin II (PIR A26655) in the highly conserved region of the 50-kDa domain that forms the apex and lower border of the narrow 50-kDa cleft. Below the alignment, a group of mutations in this region of *Dictyostelium* myosin are grouped according to their ability to complement the defects of myosin null cells upon transformation in vivo (see text for details). Each substitution is shown under the residue it replaces; double and triple residue changes in the same mutant are indicated by dashed lines connecting each individual change. Note that the wild-type class contains either conservative substitutions of highly conserved residues or less conservative changes of nonconserved residues, whereas the null class contains less favorable substitutions of highly conserved residues and all of the double and triple mutations. For placement of these mutations on the structure, see Figure *5A,B lower panels*.

These residues are not in a position to play a direct role in ATP hydrolysis; they probably inhibit or alter conformational changes within the cleft that are necessary for hydrolysis to take place. This explanation is supported by the changes seen in the cleft of the *Dictyostelium* myosin head fragment complexed to MgADP-aluminum fluoride (Fisher et al 1995). The main structural change in the molecule upon binding of this transition-state analogue is partial closure of the 50-kDa cleft via rotation around main chain conformational angles of the highly conserved Ile455 (Ile464 of chicken S1) and Gly457 (Gly466 of chicken S1) that results in the formation of new interactions between the upper and lower borders of the cleft. Hence, mutations within this region that either impair this rotation or create unfavorable interactions between the upper and lower borders of the cleft may inhibit the structural changes needed for hydrolysis to take place. Mutations that fall into the null phenotype/ATP nonhydrolyzer class include changes of residues [E459V (E468 of chicken

S1), C470G (C479 of chicken S1), N475I/K (N484 of chicken S1)] that may interact with complementary residues on the other side of the cleft. Interestingly, in vivo random mutagenesis of *Dictyostelium* also has yielded myosin mutants containing substitutions that may affect the interaction of the upper and lower borders of the cleft (Patterson & Spudich 1996). The mutations Glu467Lys (Glu476 of chicken S1) or Glu586Lys (Glu597 of chicken S1), highly conserved residues that form part of the lower border of the 50-kDa cleft (Figure 5*A,B lower panels*), led to a myosin cold-sensitive phenotype in vivo. Several intragenic suppressors of these mutations have already been isolated (B Patterson, unpublished observations), and it will be interesting to see if these suppressors map to potentially complementary residues on the other side of the cleft. Such an approach may lead to an elegant genetic dissection of important interactions within the cleft.

The final class of myosin mutants recovered after mutagenesis of the *Dictyostelium* 50-kDa region were intermediate between the wild-type and null classes, both phenotypically and biochemically. Wild-type expression levels of these mutated myosins only partially complemented the defects of the myosin null cells. Purification of several of these myosins revealed that these mutations resulted in the uncoupling of ATPase activity and motility, as assayed in vitro. For example, the Ser465Val (Ser 474 of chicken S1) myosin only translocates filaments at one-tenth the velocity of wild-type myosin and has an elevated basal MgATPase that is nearly equal to the maximal actin-activated ATPase of wild-type myosin. This alteration of residue Ser465, which sits between the ATP- and actin-binding sites, may affect the communication between them such that the ATP-binding site "sees" a more actin-bound configuration even in the absence of actin. Thus the rate of product release would be activated relative to wild-type, consistent with the elevated basal ATPase rate of this mutant and the relatively low extent of actin activation. The inefficient coupling of product release to actin rebinding in the prestroke state may result in some slippage, which would account for the low sliding velocity of this myosin and its inability to completely rescue the defects of the myosin null cells in vivo.

Mutagenesis of highly conserved residues in the analogous region of myosins from *Drosophila* led to loss of function in vivo. For example, mutation of Glu482 of indirect flight muscle myosin II to lysine, which exactly mimics the Glu476Lys mutation that gave a null phenotype/ATP nonhydrolyzer effect when introduced into *Dictyostelium* myosin, resulted in a null phenotype (Kronert et al 1994), and site-directed mutagenesis of clusters of residues in this region of the *Drosophila nina*C motor domain also resulted in a null phenotype (Porter & Montell 1993).

Relatively less mutagenic analysis has been performed on the residues that form the upper border of the 50-kDa cleft. Like the lower border, the upper border is comprised of evolutionarily highly conserved residues drawn from discontinuous regions of the 50-kDa domain (Figure 5*A,B lower panels*). These regions include the loop from Gly233 to Gly247 that also forms part of the nucleotide-binding pocket, part of the loop that extends out from the seventh strand of the β-sheet (Tyr268 to Pro280), and a portion (Ala428 to Lys431) of the long helix that extends from Val 419 to Leu449 (Rayment et al 1993b, Fisher et al 1995). Mutagenesis of Arg273 (Arg274 of chicken S1) of the upper border of the cleft to a cysteine in *C. elegans* body wall myosin resulted in a more subtle phenotype than the dominant mutations discussed above. Worms homozygous for this mutation were slow and stiff but made normal amounts of myosin heavy chain and had normal muscle ultrastructure (Moerman et al 1982, Dibb et al 1985). The analogous mutation in the *Drosophila nina*C motor had no effect on in vivo function. In contrast, substitution of this arginine by a proline in the unconventional myosin involved in auditory transduction in the mouse that results in the severe *shaker* phenotype (Gibson et al 1995).

CONCLUSIONS AND FUTURE DIRECTIONS

The past few years have witnessed a number of exciting new developments in the field of molecular motors. In particular, the synergistic effects of high-resolution structural data for actin and myosin and the development of key experimental systems for mutagenic study of these proteins have allowed in-depth structure/function analysis of the myosin motor domain that has provided insights into possible mechanisms of chemomechanical transduction. These approaches have refined our understanding of the nucleotide-binding site and the actin-binding face and have begun to dissect the specific residues involved in myosin's interaction with nucleotide and actin. Both structural models and mutagenic studies have confirmed the importance of the central 50-kDa domain for communication between the spatially distinct ATP- and actin-binding sites, a crucial aspect of chemomechanical transduction. The full range of conformational changes of this cleft awaits structural analysis of the motor in other stages of the cycle. The hypothesis that movement of this cleft also results in conformational changes in the carboxyl-terminus of the molecule ultimately leading to the power stroke requires further investigation. Such a conformational change would most likely be transmitted through the reactive thiol region of the 20-kDa domain.

Beyond the basic mechanism of chemomechanical transduction, what elements of the motor affect velocity and enzymatic activity? Chimeric studies of loop 2 (the 50/20-kDa junction) revealed the importance of this region in regulating the turnover rate of the enzyme in solution (Uyeda et al 1994). The suggestion that another strategically placed surface loop may regulate sliding velocity (Spudich 1994) needs to be tested by a similar approach.

Continued advances in several approaches will be critical for gaining an even better understanding of myosin structure and function. The ability to use *Dictyostelium* as a system for random mutagenesis and for selection of cold-sensitive myosin mutants and their suppressors (both intragenic and extragenic) greatly expands our capacity to analyze regions of the motor domain. Successful heterologous expression of myosin and myosin subfragments using baculoviral vector systems opens the door for analysis of a variety of mammalian myosins and others that were previously not amenable to mutagenic exploration, as well as in vitro study of many of the interesting mutations discussed above for which no biochemistry is known. Mutagenic analysis of actin will also lead to a better understanding of the actin-myosin interaction. Recent advances in in vitro motility assays (Svoboda et al 1993, Finer et al 1994, Ishijima et al 1994) that allow the analysis of single motor molecules will facilitate characterization of the biochemical and mechanical defects caused by the introduced mutations. Finally, high-resolution structural studies of myosin in various states of its chemomechanical cycle, produced either by crystallization of myosin with nucleotide analogues bound or by crystallization of mutated myosins that are arrested at various stages, will provide important information about the nature of the physical changes in the molecule during the ATPase cycle that lead to the transduction of chemical energy into mechanical work.

In addition, the stage is now set for structure-function analysis of microtubule-based motors such as kinesin (for recent reviews, see Vale 1992, Goldstein 1993). Structural analysis of kinesin and the kinesin-related *ncd* protein have recently yielded high-resolution structures of both molecules (Sablin et al 1996, Kull et al 1996) and, strikingly, both have a remarkable resemblance to the core of the myosin molecule. The ability to express functional kinesin motors in *E. coli* (Yang et al 1990) will facilitate the production and analysis of altered forms of the protein for biochemical and structural study. Although there are many similarities between the actin and microtubule-based motors, fundamental differences in size and functional requirements in vivo demand certain differences in the working of these motors. A key difference has been revealed by in vitro motility assays showing kinesin to be highly processive—a single kinesin molecule is able to move along a microtubule for many seconds before dissociating (Howard et al 1989, Block et al 1990, Romberg & Vale 1993). In contrast, the conventional myosin motor spends only a small portion of its cycle time in a strongly attached state to actin. This difference reflects the fact that myosin heads tend to work in concert with many other heads on an actin filament, whereas a kinesin molecule may work singly or in small numbers to effect cargo transport along a microtubule. Biochemical analysis of kinesin has revealed thermodynamic and kinetic properties that are distinct from those of myosin (Sadhu & Taylor 1992, Gilbert & Johnson 1994, Hackney 1994), and recent kinetic studies have detailed the processive nature of kinesin's hydrolysis of ATP (Gilbert et al 1995). Despite these differences, similar step sizes have been reported for kinesin (Svoboda et al 1993) and myosin (Finer et al 1994), although there exists an approximately twofold difference in the size of their motor domains. Application of the powerful tool of structure-function analysis to both motor proteins will lead to a greater understanding of the similarities and differences in their chemomechanical transduction mechanisms.

ACKNOWLEDGMENTS

We thank Bruce Patterson, Taro Uyeda, and Jim Sellers for communicating information prior to publication. We thank Ivan Rayment for the secondary structure information of the chicken S1 structure necessary to create the MOLSCRIPT figures, and Dan Chasman for help with figure preparation.

Literature Cited

Adelstein RS, Eisenberg E. 1980. Regulation and kinetics of the actin-myosin-ATP interaction. *Annu. Rev. Biochem.* 49:921–56

Anan R, Greve G, Thierfelder L, Watkins H, McKenna WJ, et al. 1994. Prognostic implications of novel beta-cardiac myosin heavy chain gene mutations that cause familial hypertrophic cardiomyopathy. *J. Clin. Invest.* 93:280–85

Andreeva AL, Andreeva OA, Borejdo J. 1993. Structure of the 265-kilodalton complex formed upon EDC cross-linking of subfragment 1 to F-actin. *Biochemistry* 32:13956–60

Andreeva OA, Borejdo J. 1992. Binding of subfragment-1 to F-actin. *Biochem. Biophys. Res. Commun.* 188:94–101

Avraham KB, Hasson T, Steel KP, Kingsley DM, Russell LB, et al. 1995. The mouse *Snell's waltzer* deafness gene encodes an unconventional myosin required for structural integrity of inner ear hair cells. *Nat. Genet.* 11:369–75

Balint M, Wolf I, Tarcsafavli A, Gergely J, Sreter FA. 1978. Location of SH-1 and SH-2 in the heavy chain segment of heavy meromyosin. *Arch. Biochem. Biophys.* 190:793–99

Bejsovec A, Anderson P. 1988. Myosin heavy-chain mutations that disrupt *Caenorhabditis elegans* thick filament assembly. *Genes Dev.* 2:1307–17

Bejsovec A, Anderson P. 1990. Functions of the myosin ATP and actin binding sites are required for C. elegans thick filament assembly. *Cell* 60:133–40

Bernstein SI, Hansen CJ, Becker KD, Wassenberg DR, Roche ES, et al. 1986. Alternative RNA splicing generates transcripts encoding a thorax-specific isoform of *Drosophila melanogaster* myosin heavy chain. *Mol. Cell. Biol.* 6:2511–19

Bertrand R, Chaussepied P, Kassab R, Boyer M, Roustan C, Benyamin Y. 1988. Cross-linking of the skeletal muscle subfragment 1 heavy chain to the N-terminal actin segment of residues 40–113. *Biochemistry* 27:5728–36

Block SM, Goldstein LSB, Schnapp BJ. 1990. Bead movement by single kinesin molecules studied with optical tweezers. *Nature* 348:348–52

Brenner B, Eisenberg E. 1987. The mechanism of muscle contraction. Biochemical, mechanical, and structural approaches to elucidate cross-bridge action in muscle. *Basic Res. Cardiol.* 82(Suppl. 2):3–16

Brenner S. 1974. The genetics of *Caenorhabditis elegans*. *Genetics* 77:71–94

Brown SDM, Steel KP. 1994. Genetic deafness—progress with mouse models. *Hum. Mol. Genet.* 3(Spec. No. 1):1453–56

Cheney RE, Mooseker MS. 1992. Unconventional myosins. *Curr. Opin. Cell Biol.* 4:27–35

Cooke R. 1986. The mechanism of muscle contraction. *CRC Crit. Rev. Biochem.* 21:53–118

Cuda G, Fananapazir L, Zhu WS, Sellers JR, Epstein ND. 1993. Skeletal muscle expression and abnormal function of beta-myosin in hypertrophic cardiomyopathy. *J. Clin. Invest.* 91:2861–65

DasGupta G, Reisler E. 1989. Antibody against the amino terminus of α-actin inhibits actin-myosin interactions in the presence of ATP. *J Mol. Biol.* 207:833–36

Dausse E, Komajda M, Fetler L, Dubourg O, Dufour C, et al. 1993. Familial hypertrophic cardiomyopathy: microsatellite haplotyping and identification of a hot spot for mutations in the beta-myosin heavy chain gene. *J. Clin. Invest.* 92:2807–13

De Lozanne A, Spudich JA. 1987. Disruption of the *Dictyostelium* myosin heavy chain gene by homologous recombination. *Science* 236:1086–91

Dibb NJ, Brown DM, Karn J, Moerman DG, Bolten SZ, Waterston RH. 1985. Sequence analysis of mutations that affect the synthesis, assembly and enzymatic activity of the unc-54 myosin heavy chain of *Caenorhabditis elegans*. *J. Mol. Biol.* 183:543–51

Egelhoff TT, Brown SS, Spudich JA. 1991. Spatial and temporal control of nonmuscle myosin localization: identification of a domain that is necessary for myosin filament disassembly in vivo. *J. Cell Biol.* 112:677–88

Egelhoff TT, Lee RJ, Spudich JA. 1993. *Dictyostelium* myosin heavy chain phosphorylation sites regulate myosin filament assembly and localization in vivo. *Cell* 75:363–71

Egelhoff TT, Manstein DJ, Spudich JA. 1990. Complementation of myosin null mutants in *Dictyostelium discoideum* by direct functional selection. *Dev. Biol.* 137:359–67

Eisenberg E, Hill TL. 1985. Muscle contraction and free energy transduction in biological systems. *Science* 227:999–1006

Epstein HF, Waterston RH, Brenner S. 1974. A mutant affecting the heavy chain of myosin in *Caenorhabditis elegans*. *J. Mol. Biol.* 90:291–300

Epstein ND, Cohn G, Cyran F, Fananapazir L. 1992. Differences in clinical expression of hypertrophic cardiomyopathy associated with two distinct mutations in the β-myosin heavy chain gene. *Circulation* 86:345–52

Fananapazir L, Epstein ND. 1994. Genotype-phenotype correlations in hypertrophic cardiomyopathy: insights provided by comparisons of kindreds with distinct and identical beta-myosin heavy chain gene mutations. *Circulation* 89:22–32

Finer JT, Simmons RM, Spudich JA. 1994. Single myosin molecule mechanics: piconewton forces and nanometre steps. *Nature* 368:113–19

Fisher AJ, Smith CA, Thoden J, Smith R, Sutoh K, et al. 1995. Structural studies of myosin: nucleotide complexes: a revised model for the molecular basis of contraction. *Biophys. J.* 68:S19–S28

Fyrberg E, Beall C. 1990. Genetic approaches to myofibril form and function in *Drosophila*. *Trends Genet.* 6:126–31

Geeves MA. 1991. The dynamics of actin and myosin association and the crossbridge model of muscle contraction. *Biochem. J.* 274:1–14

Geisterfer LA, Kass S, Tanigawa G, Vosberg HP, McKenna W, et al. 1990. A molecular basis for familial hypertrophic cardiomyopathy: a beta cardiac myosin heavy chain gene missense mutation. *Cell* 62:999–1006

Gibson F, Walsh J, Mburu P, Varela A, Brown KA, et al. 1995. A type VII myosin encoded by the mouse deafness gene *shaker-1*. *Nature* 374:62–64

Gilbert SP, Johnson KA. 1994. Pre-steady-state kinetics of the microtubule-kinesin ATPase. *Biochemistry* 33:1951–60

Gilbert SP, Webb MR, Brune M, Johnson KA. 1995. Pathway of processive ATP hydrolysis by kinesin. *Nature* 373:671–76

Gillespie PG, Wagner MC, Hudspeth AJ. 1993. Identification of a 120-kDa hair-bundle myosin located near stereociliary tips. *Neuron* 11:581–94

Goldstein LS. 1993. With apologies to Scheherazade: tails of 1001 kinesin motors. *Annu. Rev. Genet.* 27:319–51

Goodwin EB, Leinwand LA, Szent-Gyorgyi AG. 1990. Regulation of scallop myosin by mutant regulatory light chains. *J. Mol. Biol.*

216:85–93

Hackney DD. 1994. The rate-limiting step in microtubule-stimulated ATP hydrolysis by dimeric kinesin head domains occurs while bound to the microtubule. *J. Biol. Chem.* 269:16508–11

Hammer JA III. 1991. Novel myosins. *Trends Cell Biol.* 1:50–56

Hammer JA III. 1994. The structure and function of unconventional myosins. *J. Muscle Res. Cell Motil.* 15:1–10

Hibberd MG, Trentham DR. 1986. Relationships between chemical and mechanical events during muscular contraction. *Annu. Rev. Biophys. Biophys. Chem.* 15:119–61

Holmes KC, Popp D, Gebhard W, Kabsch W. 1990. Atomic model of the actin filament. *Nature* 347:44–49

Howard J, Hudspeth AJ, Vale RD. 1989. Movement of mictrotubules by single kinesin motors. *Nature* 342:154–58

Hudspeth AJ, Gillespie PG. 1994. Pulling springs to tune transduction: adaptation by hair cells. *Neuron* 12:1–9

Huxley HE. 1969. The mechanism of muscular contraction. *Science* 164:1356–66

Huxley HE, Kress M. 1985. Crossbridge behaviour during muscle contraction. *J. Muscle Res. Cell Motil.* 6:153–61

Ikebe M, Ikebe R, Kamisoyama H, Reardon S, Schwonek JP, et al. 1994. Function of the N-terminal domain of the regulatory light chain on the regulation of smooth muscle myosin. *J. Biol. Chem.* 269:28173–80

Irving M, St. Claire Allen T, Sabido-David C, Craik JS, Brandmeier B, et al. 1995. Tilting of the light-chain region of myosin during step length changes and active force generation in skeletal muscle. *Nature* 375:688–91

Ishijima A, Harada Y, Kojima H, Funatsu T, Higuchi H, Yanagida T. 1994. Single-molecule analysis of the actin-myosin motor using nano-manipulation. *Biochem. Biophys. Res. Commun.* 199:1057–63

Itakura S, Yamakawa H, Toyoshima YY, Ishijima A, Kojima T, et al. 1993. Force-generating domain of myosin motor. *Biochem. Biophys. Res. Commun.* 196:1504–10

Johara M, Toyoshima YY, Ishijima A, Kojima H, Yanagida T, Sutoh K. 1993. Charge reversion mutagenesis of *Dictyostelium* actin to map the surface recognized by myosin during ATP-driven sliding motion. *Proc. Natl. Acad. Sci. USA* 90:2127–31

Jontes JD, Wilson-Kubalek EM, Milligan RA. 1995. A 32° tail swing in brush border myosin I on ADP release. *Nature* 378:748–51

Kabsch W, Mannherz HG, Suck D, Pai EF, Holmes KC. 1990. Atomic structure of the actin:DNase I complex. *Nature* 347:37–44

Kamisoyama H, Araki Y, Ikebe M. 1994. Mutagenesis of the phosphorylation site (serine 19) of smooth muscle myosin regulatory light chain and its effects on the properties of myosin. *Biochemistry* 33:840–47

Kendrick-Jones J, Rasera da Silva AC, Reinach FC, Messer N, Rowe T, McLaughlin P. 1991. Recombinant DNA approaches to study the role of the regulatory light chains (RLC) using scallop myosin as a test system. *J. Cell Sci.* 155:7–10

Kishino A, Yanagida T. 1988. Force measurements by micromanipulation of a single actin filament by glass needles. *Nature* 334:74–76

Knecht DA, Loomis WF. 1987. Antisense RNA inactivation of myosin heavy chain gene expression in *Dictyostelium discoideum. Science* 236:1081–86

Korn ED, Hammer JA III. 1990. Myosin I. *Curr. Opin. Cell Biol.* 2:57–61

Kraulis PJ. 1991. MOLSCRIPT: A program to produce both detailed and schematic plots of protein structures. *J. Appl. Crystallogr.* 24:946–50

Kron SJ, Spudich JA. 1986. Fluorescent actin filaments move on myosin fixed to a glass surface. *Proc. Natl. Acad. Sci. USA* 83:6272–76

Kronert WA, O'Donnell PT, Bernstein SI. 1994. A charge change in an evolutionarily-conserved region of the myosin globular head prevents myosin and thick filament accumulation in *Drosophila. J. Mol. Biol.* 236:697–702

Kubalek EW, Uyeda TQP, Spudich JA. 1992. A *Dictyostelium* myosin II lacking a proximal 58-kDa portion of the tail is functional in vivo and in vitro. *Mol. Biol. Cell* 3:1455–62

Kull FJ, Sablin EP, Lau R, Fletterick RJ, Vale RD. 1996. Crystal structure of the kinesin motor domain reveals a structural similarity to myosin. *Nature* 380:550–55

Lee RJ, Egelhoff TT, Spudich JA. 1994. Molecular genetic truncation analysis of filament assembly and phosphorylation domains of *Dictyostelium* myosin heavy chain. *J. Cell Sci.* 107:2875–86

Lowey S, Waller GS, Trybus KM. 1993. Skeletal muscle myosin light chains are essential for physiological speeds of shortening. *Nature* 365:454–56

Lymn RW, Taylor EW. 1971. Mechanism of adenosine triphosphate hydrolysis by actomyosin. *Biochemistry* 10:4617–24

MacLeod AR, Waterston RH, Fishpool RM, Brenner S. 1977. Identification of the structural gene for a myosin heavy-chain in *Caenorhabditis elegans. J. Mol. Biol.* 114:133–40

Maita T, Yajima E, Nagata S, Miyanishi T,

Nakayama S, Matsuda G. 1991. The primary structure of the skeletal myosin heavy chain: IV. Sequence of the rod, and the complete 1,938-residue sequence of the heavy chain. *J. Biochem.* 110:75–87

Manstein DJ, Titus MA, DeLozanne A, Spudich JA. 1989. Gene replacement in *Dictyostelium*: generation of myosin null mutants. *EMBO J.* 8:923–32

Marian AJ, Kelly D, Mares AJ, Fitzgibbons J, Caira T, et al. 1994. A missense mutation in the beta myosin heavy chain gene is a predictor of premature sudden death in patients with hypertrophic cardiomyopathy. *J. Sports Med. Phys. Fitness* 34:1–10

Matsumoto H, Isono K, Pye Q, Pak WL. 1987. Gene encoding cytoskeletal proteins in *Drosophila* rhabdomeres. *Proc. Natl. Acad. Sci. USA* 84:985–89

Moerman DG, Plurad S, Waterston RH, Baillie DL. 1982. Mutations in the *unc-54* myosin heavy chain gene of *Caenorhabditis elegans* that alter contractility but not muscle structure. *Cell* 29:773–81

Montell C, Rubin GM. 1988. The *Drosophila ninaC* locus encodes two photoreceptor specific proteins with domains homologous to protein kinases and the myosin heavy chain head. *Cell* 52:757–72

Mooseker M. 1993. A multitude of myosins. *Curr. Biol.* 3:245–48

Mornet D, Bonet A, Audemard E, Bonicel J. 1989. Functional sequences of the myosin head. *J. Muscle Res. Cell. Motil.* 10:10–24

Mornet D, Pantel P, Audemard E, Kassab R. 1979. The limited tryptic cleavage of chymotryptic S-1: an approach to the characterization of the actin site in myosin heads. *Biochem. Biophys. Res. Commun.* 89:925–32

Muller CW, Schulz GE. 1992. Structure of the complex between adenylate kinase from *Escherichia coli* and the inhibitor Ap5A refined at 1.9 Å resolution. *J. Mol. Biol.* 224:159–77

Nishi H, Kimura A, Harada H, Toshima H, Sasazuki T. 1992. Novel missense mutation in cardiac beta myosin heavy chain gene found in a Japanese patient with hypertrophic cardiomyopathy. *Biochem. Biophys. Res. Commun.* 188:379–87

Onishi H, Maeda K, Maeda Y, Inoue A, Fujiwara K. 1995. Functional chicken gizzard heavy mermyosin expression in and purification from baculovirus-infected insect cells. *Proc. Natl. Acad. Sci. USA* 92:704–8

Ostrow BD, Chen P, Chisholm RL. 1994. Expression of a regulatory light chain phosphorylation site mutant complements the cytokinesis and developmental defects of *Dictyostelium* RMLC null cells. *J. Cell. Biol.* 127:1945–55

Pai EF, Krengel U, Petsko GA, Goody RS, Kabsch W, Wittinghofer F. 1990. Refined structure of the triphosphate conformation of H-*ras*-p21 at 1.35 Å resolution: implications for the mechanism of GTP hydrolysis. *EMBO J.* 9:2351–59

Pato MD, Preston YA, Sellers JR, Adelstein RS. 1993. Expression of a truncated form of chicken brain myosin that binds to actin in an ATP-dependent manner using the baculovirus expression system. *Biophys. J.* 64:A144 (Abstr.)

Patterson B, Spudich JA. 1995. A novel positive selection for identifying cold-sensitive myosin II mutants in *Dictyostelium. Genetics* 140:505–15

Patterson B, Spudich JA. 1996. Cold-sensitive mutations of *Dictyostelium* myosin heavy chain highlight functional domains of the myosin motor. *Genetics* 143:801–10

Pollard TP, Doberstein SK, Zot HG. 1991. Myosin-I. *Annu. Rev. Physiol.* 53:653–81

Porter JA, Hicks JL, Williams DS, Montell C. 1992. Differential localizations of and requirements for the two *Drosophila ninaC* kinase/myosins in photoreceptor cells. *J. Cell Biol.* 116:683–93

Porter JA, Montell C. 1993. Distinct roles of the *Drosophila ninaC* kinase and myosin domains revealed by systematic mutagenesis. *J. Cell Biol.* 122:601–12

Rayment I, Holden H. 1993. Myosin subfragment-1: structure and function of a molecular motor. *Curr. Opin. Struct. Biol.* 3:944–52

Rayment I, Holden HM. 1994. The three-dimensional structure of a molecular motor. *Trends Biochem. Sci.* 19:129–34

Rayment I, Holden HM, Sellers JR, Fananapazir L, Epstein ND. 1995. Structural interpretation of the mutations in the beta-cardiac myosin that have been implicated in familial hypertrophic cardiomyopathy. *Proc. Natl. Acad. Sci. USA* 92:3864–68

Rayment I, Holden HM, Whittaker M, Yohn CB, Lorenz M, et al. 1993a. Structure of the actin-myosin complex and its implications for muscle contraction. *Science* 261:58–65

Rayment I, Rypniewski W, Schmidt-Base K, Smith R, Tomchick D, et al. 1993b. The three-dimensional structure of myosin subfragment-1: a molecular motor. *Science* 261:50–58

Reedy MK. 1993. Myosin-actin motors: the partnership goes atomic. *Structure* 1:1–5

Reinach FC, Nagai K, Kendrick Jones J. 1986. Site-directed mutagenesis of the regulatory light chain Ca^{2+}/Mg^{2+} binding site and its role in hybrid myosins. *Nature* 322:80–83

Romberg L, Vale RD. 1993. Chemomechanical cycle of kinesin differs from that of myosin.

Nature 361:168–70

Rowe T, Kendrick-Jones J. 1993. The C-terminal helix in subdomain 4 of the regulatory light chain is essential for myosin regulation. *EMBO J.* 12:4877–84

Rozek CE, Davidson N. 1986. Differential processing of RNA transcribed from the single copy *Drosophila* myosin heavy chain gene produces four mRNAs that encode two polypeptides. *Proc. Natl. Acad. Sci. USA* 83:2128–32

Ruppel KM, Spudich JA. 1996. Structure-function studies of the myosin motor domain: importance of the 50-kDa cleft. *Mol. Biol. Cell.* 7:1123–36

Ruppel KM, Uyeda TQP, Spudich JA. 1994. Role of highly conserved lysine 130 of myosin motor domain: in vivo and in vitro characterization of a site-specifically mutated myosin. *J. Biol. Chem.* 269:18773–80

Sablin EP, Kull FJ, Cooke R, Vale RD, Fletterick RJ. 1996. Crystal structure of the motor domain of the kinesin-related motor ncd. *Nature* 380:555–59

Sadhu A, Taylor EW. 1992. A kinetic study of the kinesin ATPase. *J. Biol. Chem.* 267:11352–59

Saraste M, Sibbald PR, Wiitinghofer A. 1990. The P-loop—a common motif in ATP and GTP-binding proteins. *Trends Biochem. Sci.* 15:430–34

Saraswat LD, Pastra-Landis SC, Lowey S. 1992. Mapping single cysteine mutants of light chain 2 in chicken skeletal myosin. *J. Biol. Chem.* 267:21112–18

Schroeder RR, Manstein DJ, Jahn W, Holden H, Rayment I, et al. 1993. Three-dimensional atomic model of F-actin decorated with *Dictyostelium* myosin S1. *Nature* 364:171–74

Schwartz K, Carrier L, Guicheney P, Komajda M. 1995. Molecular basis of the familial cardiomyopathies. *Circulation* 91:532–40

Sellers JR, Goodson HV. 1995. Motor proteins 2: myosins. *Protein Profiles* 2:1323–81

Smith RJH, Berlin CI, Hejtmancik JF, Keats BJ, Kimberling WJ, et al. 1994. Clinical diagnosis of the Usher syndromes. Usher Syndrome Consortium. *Am. J. Med. Genet.* 50:32–38

Spudich JA. 1994. How molecular motors work. *Nature* 372:515–18

Sutoh K. 1983. Mapping of actin-binding sites on the heavy chain of myosin subfragment-1. *Biochemistry* 22:1579–85

Sutoh Kazuo, Ando M, Sutoh Keiko, Toyoshima YY. 1991. Site-directed mutations of *Dictyostelium* actin: disruption of a negative charge cluster at the N terminus. *Proc. Natl. Acad. Sci. USA* 88:7711–14

Svoboda K, Schmidt CF, Schnapp BJ, Block SM. 1993. Direct observation of kinesin stepping by optical trapping interferometry. *Nature* 365:721–27

Sweeney HL, Straceski AJ, Leinwand LA, Tikunov BA, Faust L. 1994. Heterologous expression of a cardiomyopathic myosin that is defective in its actin interaction. *J. Biol. Chem.* 269:1603–5

Taylor KA, Amos LA. 1981. A new model for the geometry of the binding of myosin crossbridges to muscle thin filaments. *J. Mol. Biol.* 147:297–324

Titus MA. 1993. Myosins. *Curr. Opin. Cell Biol.* 5:77–81

Tokunaga M, Sutoh K, Wakabayashi T. 1991. Structure and structural change of the myosin head. *Adv. Biophys.* 27:157–67

Tong SW, Elzinga M. 1983. The sequence of the NH_2-terminal 204-residue fragment of the heavy chain of rabbit skeletal muscle myosin. *J. Biol. Chem.* 258:13100–10

Toyoshima YY, Kron SJ, McNally EM, Niebling KR, Toyoshima C, Spudich JA. 1987. Myosin subfragment-1 is sufficient to move actin filaments in vitro. *Nature* 328:536–39

Trybus KM. 1994. Regulation of expressed truncated smooth muscle myosins. *J. Biol. Chem.* 269:20819–22

Uyeda TQP, Abramson PD, Spudich JA. 1996. The neck region of the myosin motor domain acts as a lever arm to generate movement. *Proc. Natl. Acad. Sci. USA* 93:4459–64

Uyeda TQP, Ruppel KM, Spudich JA. 1994. Enzymatic activities correlate with chimaeric substitutions at the actin-binding face of myosin. *Nature* 368:567–69

Uyeda TQP, Spudich JA. 1993. A functional recombinant myosin II lacking a regulatory light chain-binding site. *Science* 262:1867–70

Vale RD. 1992. Microtubule motors: many new models off the assembly line. *Trends Biochem. Sci.* 17:300–4

Vibert P, Cohen C. 1988. Domains, motions and regulation in the myosin head. *J. Muscle Res. Cell Motil.* 9:296–305

Vosberg H-P. 1994. Myosin mutations in hypertrophic cardiomyopathy and functional implications. *Herz* 19:75–83

Walker JE, Saraste M, Runswick MJ, Gay NJ. 1982. Distantly related sequences in the α- and β-subunits of ATP synthase, myosin, kinases and other ATP-requiring enzymes and a common nucleotide binding fold. *EMBO J.* 1:945–51

Warrick HM, Spudich JA. 1987. Myosin structure and function in cell motility. *Annu. Rev. Cell Biol.* 3:379–421

Waterston RH, Moerman DG, Baillie D, Lane TR. 1982. Mutations affecting myosin heavy chain accumulation and function in the nematode *Caenorhabditis elegans*. In *Disorders of the Motor Unit*, pp. 747–59. New York: Wiley

Watkins H, Rosenzweig A, Hwang D-S, Levi T, McKenna W, et al. 1992. Characteristics and prognostic implications of myosin missense mutations in familial hypertrophic cardiomyopathy. *N. Engl. J. Med.* 326:1108–14

Weil D, Blanchard S, Kaplan J, Guilford P, Gibson F, et al. 1995. Defective myosin VIIA gene responsible for Usher syndrome type 1B. *Nature* 374:60–62

Wells JA, Knoeber C, Sheldon MC, Werber MM, Yount RG. 1980. Crosslinking of myosin S1. *J. Biol. Chem.* 255:11135–40

Wells JA, Yount RG. 1979. Active site trapping of nucleotides by crosslinking two sulfhydryls in myosin subfragment 1. *Proc. Natl. Acad. Sci. USA* 76:4966–70

Whittaker M, Kubalek EMW, Smith JE, Faust L, Milligan RA, et al. 1995. A 35-angstrom movement of smooth muscle myosin on ADP release. *Nature* 378:748–51

Wolff-Long VL, Saraswat LD, Lowey S. 1993. Cysteine mutants of light chain-2 form disulfide bonds in skeletal muscle myosin. *J. Biol. Chem.* 268:23162–67

Wray JS, Goody RS, Holmes KC. 1988. Towards a molecular mechanism for the crossbridge cycle. *Adv. Exp. Med. Biol.* 226:49–59

Xie X, Harrison DH, Schlichting I, Sweet RM, Kalabokis VN, et al. 1994. Structure of the regulatory domain of scallop myosin at 2.8 Å resolution. *Nature* 368:306–12

Yamamoto K. 1989. Binding manner of actin to the lysine-rich sequence of myosin subfragment 1 in the presence and absence of ATP. *Biochemistry* 28:5573–77

Yang JT, Saxon WM, Stewart RJ, Raff EC, Goldstein LSB. 1990. Evidence that the head of kinesin is sufficient for force generation and motility in vitro. *Science* 249:42–47

Yount RG. 1993. Subfragment-1: The first crystalline motor. *J. Muscle Res. Cell Motil.* 14:547–51

Yount RG, Cremo CR, Grammer JC, Kerwin BA. 1992. Photochemical mapping of the active site of myosin. *Philos. Trans. R. Soc. London Ser. B* 336:55–61

Annu. Rev. Cell Dev. Biol. 1996. 12:575–625

ENDOCYTOSIS AND MOLECULAR SORTING

Ira Mellman
Department of Cell Biology, Yale University School of Medicine, P.O. Box 208002, 333 Cedar Street, New Haven, Connecticut 06520-8002

KEY WORDS: clathrin, endosomes, epithelial cell polarity, Golgi complex, lysosomes, receptors

ABSTRACT

Endocytosis in eukaryotic cells is characterized by the continuous and regulated formation of prolific numbers of membrane vesicles at the plasma membrane. These vesicles come in several different varieties, ranging from the actin-dependent formation of phagosomes involved in particle uptake, to smaller clathrin-coated vesicles responsible for the internalization of extracellular fluid and receptor-bound ligands. In general, each of these vesicle types results in the delivery of their contents to lysosomes for degradation. The membrane components of endocytic vesicles, on the other hand, are subject to a series of highly complex and iterative molecular sorting events resulting in their targeting to specific destinations. In recent years, much has been learned about the function of the endocytic pathway and the mechanisms responsible for the molecular sorting of proteins and lipids. This review attempts to integrate these new concepts with long-established views of endocytosis to present a more coherent picture of how the endocytic pathway is organized and how the intracellular transport of internalized membrane components is controlled. Of particular importance are emerging concepts concerning the protein-based signals responsible for molecular sorting and the cytosolic complexes responsible for the decoding of these signals.

CONTENTS

1081-0706/96/1115-0575$08.00

INTRODUCTION

All eukaryotic cells exhibit one or more forms of endocytosis. Their reasons for doing so are as diverse as their individual functions. At a minimum, endocytosis serves to maintain cellular homeostasis by recovering protein and lipid components inserted into the plasma membrane by ongoing secretory activity. More strikingly, endocytosis is essential for organismal homeostasis, controlling an extraordinary array of activities that every cell must exhibit in order to exist as part of a multicellular community. These activities include the transmission of neuronal, metabolic, and proliferative signals; the uptake of many essential nutrients; the regulated interaction with the external world; and the ability to mount an effective defense against invading microorganisms. Paradoxically, many infectious agents mediate their effects only after contriving to be internalized by their intended hosts.

One hundred years ago, Elie Metchnikoff first recognized that material taken up by endocytosis was degraded after encountering an acidic internal environment. During the past decade, the basic organization of the endocytic pathway was elucidated, particularly in the case of the internalization of protein ligands bound to cell surface receptors (Helenius et al 1983, Steinman et al 1983, Kornfeld & Mellman 1989, Trowbridge et al 1993, Gruenberg & Maxfield 1995). Although discussion of detail and semantics continues, there is fundamental agreement on the most important features of the pathway. These features are summarized in Figure 1 (note that review articles have been cited for most work performed before 1990).

⟶

Figure 1 Organization of the endocytic pathway in animal cells. The major organelles of the endocytic pathway are illustrated and defined in terms of their kinetic relationships, as well as in terms of diagnostic molecular markers. The markers shown include members of the rab family of ras-like GTPases, cation-independent mannose-6-phosphate receptor (MPR), and members of a protein family enriched in lysosomal membranes (lgp/lamp). Endocytosis is typically initiated by the formation of clathrin-coated vesicles at the plasma membrane, which results in the delivery of receptor-ligand complexes to early endosomes in the peripheral cytoplasm. Here, receptor-ligand complexes dissociate, thus resulting in the return of free receptors to the plasma membrane in recycling vesicles, some of which appear to first migrate on microtubule tracks to the perinuclear cytoplasm before reaching the cell surface. Dissociated ligands and other soluble macromolecules are transferred from early to late endosomes and lysosomes where digestable content is degraded. Clathrin-coated vesicles are also thought to be involved in the transport of selected components (e.g. MPR) from the *trans*-Golgi network of the Golgi complex to early endosomes. See text for details.

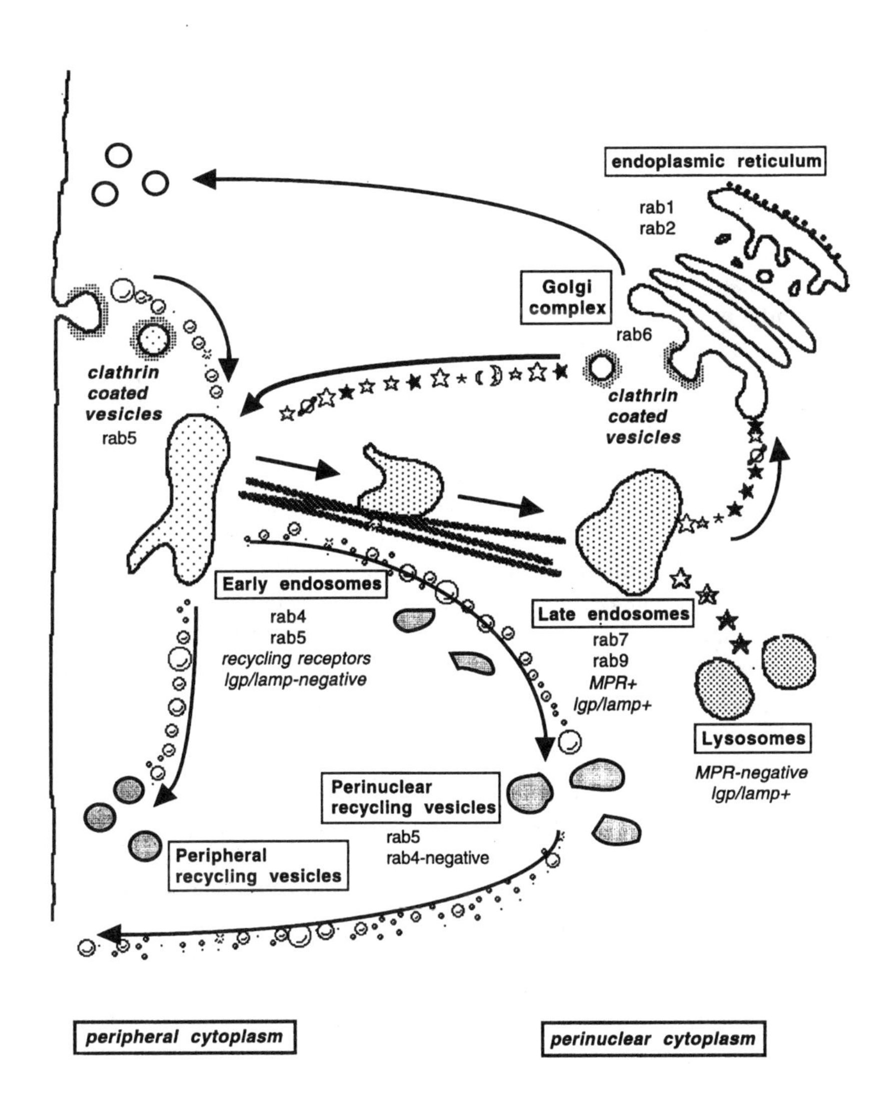

endoplasmic reticulum
rab1
rab2
Golgi complex
rab6
clathrin coated vesicles
clathrin coated vesicles
rab5
Early endosomes
rab4
rab5
recycling receptors
lgp/lamp-negative
Late endosomes
rab7
rab9
MPR+
lgp/lamp+
Lysosomes
MPR-negative
lgp/lamp+
Perinuclear recycling vesicles
rab5
rab4-negative
Peripheral recycling vesicles
peripheral cytoplasm
perinuclear cytoplasm

Most receptor-ligand complexes accumulate at clathrin-coated pits of the plasma membrane, which bud off to yield clathrin-coated vesicles. The vesicles rapidly lose their coats, which facilitates fusion with early endosomes (EEs), a dynamic array of tubules and vesicles distributed throughout the peripheral and perinuclear cytoplasm. Due to a slightly acidic pH (pH $\approx$ 6.0–6.8) maintained by an ATP-driven proton pump (Al-Awqati 1986, Mellman et al 1986, Forgac 1992), EEs host the dissociation of many ligand-receptor complexes. Free receptors selectively accumulate in the early endosome's tubular extensions, which bud off to yield recycling vesicles (RVs) that transport receptors, directly or indirectly, back to the plasma membrane. Dissociated ligands collect in the vesicular portions of the EEs simply because of their high internal volume relative to the endosome's tubular extensions. The vesicular structures pinch off—or are left behind following the budding of RVs—traverse to the perinuclear cytoplasm on microtubule tracks, and fuse with late endosomes (LEs) and lysosomes. Here, ligands are degraded by the even lower pH (pH $\approx$ 5) and the high concentration of lysosomal enzymes. Recycling from lysosomes occurs relatively slowly, which explains why cells are capable of accumulating large amounts of internalized material. In addition, lysosomal enzymes accumulate in LEs and lysosomes by the same pathway: Clathrin-coated vesicles coming from the *trans*-Golgi network (TGN) also fuse with endosomes, delivering newly synthesized enzymes bound to either of two pH-sensitive mannose-6-phosphate receptors (Kornfeld & Mellman 1989). To a first approximation, clathrin-coated vesicles thus comprise a population of transport vesicles specifically addressed for fusion with endosomes, regardless of whether they originated at the plasma membrane or the TGN.

Like the secretory pathway, the endocytic pathway can be thought of as having functionally and physically distinct compartments (Kornfeld & Mellman 1989). Early endosomes are responsible for the dissociation and sorting of receptor and ligands in an environment that minimizes the risk of damaging receptors intended for reutilization. Late endosomes and lysosomes are responsible for accumulating and digesting exogenous and endogenous macromolecules. Unlike the secretory pathway, however, the signposts for these compartments are less clear. The functions of endocytic organelles do not involve the predictable and easily determined enzymatic activities associated with the synthesis, folding, and posttranslational processing of glycoproteins and lipids. Moreover, the physical appearance of endocytic organelles is more pleiomorphic than is true for secretory organelles, which can often be identified on morphological grounds alone. An additional difficulty comes from the fact that truly prolific quantities of membrane are processed by the endocytic pathway (Steinman et al 1983), possibly obscuring compositional differences between compartments.

Nevertheless, biochemical, functional, and genetic probes for monitoring and manipulating the endocytic pathway and its organelles have begun to appear, which have facilitated the dissection of the mechanisms underlying the internalization and recycling of plasma membrane components. In addition, we are beginning to understand the variations responsible for generating specializations of the endocytic pathway found in different cell types. This review summarizes advances in our understanding of endocytosis since the last comprehensive treatments of the topic (Kornfeld & Mellman 1989, Trowbridge et al 1993). Particular attention is paid to the mechanisms underlying the capacity of endocytic organelles for molecular sorting, i.e. the ability to distinguish and selectively transport components to different destinations. Many of these mechanisms are also shared by the secretory pathway. We also consider the growing body of information concerning the regulation of membrane transport during endocytosis. Because much of what has been learned about membrane fusion has emerged from the study of the secretory pathway, this topic will be treated elsewhere (S Pfeffer, this volume).

CELLS EXHIBIT MULTIPLE TYPES OF ENDOCYTOSIS

It has long been clear that there are at least two routes into the cell, generally classified as phagocytosis (cell eating) and pinocytosis (cell drinking) (Silverstein 1977, Steinman et al 1983). Phagocytosis refers to the internalization of large ($>0.5\ \mu$ diameter) particles that must bind to specific plasma membrane receptors capable of triggering their own uptake, usually by causing the formation of F-actin-driven pseudopods that envelop the bound particle. Pinocytosis, more commonly if somewhat less accurately known as endocytosis, typically refers to the constitutive formation of smaller ($<0.2\ \mu$ diameter) vesicles carrying extracellular fluid and macromolecules specifically or nonspecifically bound to the plasma membrane. These vesicles are usually initiated at clathrin-coated pits (see below). Possible alternatives to clathrin-mediated endocytosis include the involvment of caveolae and/or an actin-based mechanism. To greater or lesser degrees, two or more of these mechanisms co-exist in a single cell type.

Phagocytosis

PHAGOCYTIC CELLS The ability to internalize large particles is most often associated with phagocytic protozoa (*Dictyostelium, Acanthamoeba*) or phagocytic leukocytes of the mammalian immune system (macrophages, neutrophils). Uptake is triggered by binding of opsonized particles to cell surface receptors capable of transducing a phagocytic stimulus to the cytoplasm. This stimulus results in the localized polymerization of actin at the site of particle attachment and subsequent pseudopod extension that engulfs the bound particle into a cytoplasmic phagosome (Greenberg et al 1990, 1991). In macrophages, the forming pseudopods are directed by sequential ligand-receptor interactions, which yield a vacuole custom fit to the internalized particle (Silverstein et al 1977).

Under most cirumstances, phagosomes rapidly fuse with endosomes and/or lysosomes exposing their contents to hydrolytic enzymes (Kielian & Cohn 1980, Steinman et al 1983, Rabinowitz et al 1992, Desjardins et al 1994). Receptors that mediate phagocytosis in leukocytes include members of the IgG Fc receptor family as well as some integrins (e.g. complement receptor) and lectins (e.g. mannose receptor) (Kielian & Cohn 1980, Mellman et al 1983, Wright & Silverstein 1983, Ezekowitz et al 1991, Isberg et al 1994). It is thought that lectin-like receptors are responsible for triggering phagocytosis of bacteria in protozoa (Cohen et al 1994), but as yet no receptors have not been identified. In mammals, phagocytosis serves as a first line of defense against microorganisms, as well as providing an important component of the humoral immune response by allowing the processing and presentation of bacterial-derived peptides to antigen-specific T lymphocytes (Harding & Geuze 1992, Pfeifer et al 1993). In protozoa, phagocytosis probably serves a nutritional function; *D. discoideum* mutants conditionally defective in phagocytosis starve to death (Cohen et al 1994).

Phagocytosis can result in the internalization of huge areas of plasma membrane. Classical experiments from Cohn and colleagues demonstrated that more than 50% of a macrophage's surface area could be involved during the uptake of a phagocytic load (Werb & Cohn 1972, Steinman et al 1983). Most membrane proteins are equally susceptible to internalization, although the receptor mediating the uptake can be selectively internalized and degraded (Mellman et al 1983). Phagosomes thus remain at least partly connected to the endocytic pathway, allowing the recycling of many of the bulk-internalized plasma membrane proteins (Muller et al 1983, Joiner et al 1990). During this time, the phagosome membrane is gradually remodeled and comes to resemble the lysosomal membrane in composition (Desjardins et al 1994).

Although the mechanism of phagocytosis remains incompletely characterized, the plasma membrane receptors themselves play a primary role. In macrophages, Fc receptor-mediated phagocytosis of IgG-coated particles by macrophages is associated with localized tyrosine phosphorylation of a variety of cytoplasmic proteins (Greenberg et al 1993, Greenberg 1995). This reflects the recruitment of cytosolic src-family kinases to the Fc receptor cytoplasmic domain via its consensus tyrosine-containing ITAM motif (Greenberg et al 1994, 1996). Although it is widely believed that macrophages and other phagocytes are highly specialized for the uptake of large particles, in fact, transfection of Fc receptor cDNAs into normally non-phagocytic cells also can result in phagocytosis (Joiner et al 1990, Hunter et al 1994, Greenberg et al 1996). Moreover, fibroblasts can be made to ingest IgG-coated particles as efficiently as macrophages by ensuring that the appropriate src-family kinases are co-expressed, either as soluble proteins or as cytoplasmic domain fusions to the expressed Fc receptors (Greenberg et al 1996). Thus given the appropriate receptor coupled to the appropriate signaling molecule, phagocytosis is not a property limited to professional phagocytes but makes use of a mechanism common to many or all cell types.

ENTRY OF INFECTIOUS MICROORGANISMS Although some infectious agents preferentially infect macrophages, many bacterial and protozoan parasites long ago discovered that phagocytosis is not a proprietary feature of professional phagocytes. Some bacteria (*Yersinia, Salmonella, Listeria*) synthesize surface proteins that permit bacterial attachment and stimulation of one or more plasma membrane receptors, which stimulate membrane ruffling and subsequent engulfment of the bound bacteria, even by epithelial cells (Bliska et al 1993, Pace et al 1993, Gálan 1994). This type of phagocytosis differs in some respects from that mediated by Fc receptors in leukocytes in that pseudopod extension is essentially undirected, as opposed to being zippered across the particle surface. Thus bacteria such as *Salmonella* may simply generate an enormous local stimulus that causes sufficient membrane ruffling to allow almost the accidental internalization of bound bacteria. Indeed, *Salmonella* entry is also accompanied by a transient increase in the uptake of extracellular fluid, as would be expected for such a mechanism (Francis et al 1993). The nature of the receptors or signals stimulated by such bacteria are not completely characterized. However, there is a critical role for members of the rho protein family as downstream effectors almost certainly required for the reorganization of the actin cytoskeleton (J Gálan, unpublished results). It is interesting that in leukocytes apparently the same rho family proteins play a critical role in assembling the complex of proteins needed to generate the respiratory burst as well as transcriptional activation of genes involved in mediating the inflammatory responses (Bokoch & Knaus 1994).

Although beyond the scope of this review, it should be mentioned that many bacterial and protozoan pathogens have developed strategies to avoid killing by oxidative mechanisms or lysosomal digestion (Falkow et al 1992). Some (e.g. *Listeria, T. cruzi*) escape from newly formed phagosomes by utilizing the acidic internal pH of the phagosome to activate lytic enzymes that rupture the vacuole membrane (Andrews & Portnoy 1994). *T. cruzi* even appears to actively recruit some lysosomes to the site of entry (Tardieux et al 1992). Finally, protozoan parasites such as *Toxoplasma gondii* heavily modify the phagocytic process, resulting in the formation of vacuoles that are modified such that they are no longer capable of fusing with host cell endosomes and lysosomes (Joiner et al 1990). This mode of entry also avoids triggering a cytotoxic respiratory burst by the phagocyte.

Clathrin-Dependent Endocytosis of Receptors, Ligands, and Extracellular Fluid

The uptake of extracellular fluid and receptor-bound ligands is exhibited to varying extents by all cells. The amounts of membrane and fluid internalized by this ongoing endocytic activity are prolific. Cells such as macrophages and fibroblasts have been estimated to internalize more than 200% of their entire surface area every hour (Steinman et al 1983). The amount of membrane uptake at active presynaptic nerve terminals is probably far greater. Despite the fact that many plasma membrane proteins are subject to endocytosis, most exhibit long half-lives (typically > 24 h) and escape degradation by recycling back to the plasma membrane. Similarly, individual receptors have been estimated to mediate as many as 10 rounds of ligand uptake and recycling each hour yet exhibit long half-lives (Steinman et al 1983). Much has been learned concerning the pathways of endocytosis, and recent work has provided important new insights into the mechanisms and regulation of these events.

THE PRIMARY ROLE OF CLATHRIN IN ENDOCYTOSIS In most animal cell types and under normal conditions, the uptake of receptor-bound ligands and extracellular fluid results mainly from the formation of clathrin-coated vesicles (CCVs). Early evidence in favor of a primary role for clathrin in endocytosis came from quantitative arguments in which the number of CCVs of known diameter formed per minute was measured and used to calculate the volume of fluid they were predicted to contain. Such calculations indicated that a sufficient number of CCVs formed to account for fluid uptake, as determined biochemically (Marsh & Helenius 1980). Indeed, CCV-mediated uptake of virus particles was found to reduce pinocytosis owing to physical displacement of fluid. The primary role of clathrin-coated pits in ligand endocytosis was demonstrated by the correlation between the selective localization of receptor-ligand at coated pits and rapid ligand uptake: Mutant receptors defective in endocytosis were also defective at coated pit localization (Brown & Goldstein 1979).

Recently, additional genetic evidence supports a primary role for clathrin in endocytosis. Deletion of the single gene for the 180-kDa clathrin heavy chain in *Dictyostelium* greatly suppressed the considerable capacity of these cells for fluid pinocytosis, as well as inhibiting the internalization of at least one potential plasma membrane receptor (O'Halloran & Anderson 1992). In *Drosophila, shibire*, a well-known mutation causing temperature-sensitive paralysis, was found to block endocytosis at nerve endings, as well as in nonneuronal cells, apparently by blocking CCV formation from coated pits that accumulated in abundance at the nonpermissive temperature (Koenig & Ikeda 1989, Poodry 1990). The *shibire* locus encodes dynamin, a GTPase initially believed to be a microtubule motor but more likely to be involved in the physical budding of CCVs from clathrin-coated pits (Chen et al 1991, van der Bliek & Meyerowitz 1991). Expression of mutant alleles of dynamin in mammalian cells in culture produced a similar effect, blocking CCV formation from coated pits and inhibiting receptor-mediated and fluid endocytosis (van der Bliek et al 1993, Damke et al 1994, 1995a). Interestingly, the effect on fluid uptake was transient after shifting to the nonpermissive temperature, suggesting that in the absence of a clathrin-mediated pathway, an alternative form of endocytosis could be induced (see below).

STRUCTURE AND FUNCTION OF CLATHRIN-COATED VESICLES The basic composition and structure of CCVs has long been known (Pearse & Robinson 1990). Although a complete consideration of their properties is beyond the scope of this review, it is necessary to discuss their more important features because CCVs serve as an important paradigm for understanding most transport and sorting events on the endocytic pathway.

CCVs were first identified in and isolated from brain, where they were associated with the endocytosis of presynaptic membrane following synaptic vesicle exocytosis. Their coats form lattices of hexagons and pentagons of the protein clathrin (Pearse 1987). The functional unit of these arrays is the triskelion: Three 180-kDa clathrin heavy chains each complexed with a 30–35 kDa light chain. Isolated triskelions can self-assemble into empty cages in vitro in the absence of ATP or other energy sources, suggesting that they exhibit a favorable ΔG for assembly. As such, clathrin itself may provide the energy required for the formation of CCVs from coated pits, although direct evidence for this appealing possibility is still lacking. Indeed, even the formation of invaginated coated pits on plasma membranes in vitro has been found to require energy (Lin et al 1991, Smythe et al 1992), suggesting that however favorable the ΔG for clathrin assembly, it is not sufficient to mediate the entire process.

Coated vesicles have long been thought to form by progressive invagination of planar, hexagonal clathrin lattices, first imaged in the classical electron micrographs of Heuser and colleagues (Heuser & Evans 1980). Although possible, it remains unclear whether such planar arrays directly mature into CCVs, as opposed to clathrin that may be initially recruited to sites already having some degree of curvature. It is not immediately obvious how such a stucture consisting entirely of triskelion-containing hexagons can reassemble into curved lattices consisting of both hexagons and pentagons. For this to happen, significant rearrangements of inter-triskelion interactions must occur (Liu et al 1995). Conceivably, such rearrangements may be catalyzed by molecular chaperones that are not stable components of CCVs or may reflect the partial effect of GTPase activators (e.g. GTPγS) on coated pit invagination in vitro (Carter et al 1993).

Clathrin coats also contain a non-clathrin component, a heterotetrameric adaptor complex (Pearse & Robinson 1990). Originally identified as factors that promoted the assembly of clathrin cages in vitro, adaptors are now known to play a critical role in the attachment of clathrin to membranes (Chang et al 1993, Peeler et al 1993, Robinson 1994, Traub et al 1995). Indeed, it is the adaptor complex that must be first recruited to membranes in order to provide the binding site for clathrin. Conceivably, adaptor binding may induce the membrane curvature that facilitates the clathrin-mediated formation of invaginated coated pits. Thus far, two adaptor complexes have been well characterized. AP-1 and AP-2 consist of four related subunits (adaptins), two of $\approx$ 100-kDa and one of $\approx$ 50 kDa (designated μ for medium chains) and one of $\approx$ 20 kDa (designated σ for small chains). The AP-2 complex consists of either of two closely related (90% identical) 100 kDa α chains (α_A, α_C), a 100-kDa β chain, $\mu 2$ and $\sigma 2$. AP-1 contains 100 kDa γ and β' chains, together with $\mu 1$ and $\sigma 1$ subunits. The β and β' subunits are closely related, but all others are distinct (<50% similarity). All the 100-kDa subunits do, however, have similar domain structures, with a large head domain, a proline-rich hinge or extension, and a COOH-terminal appendage or ear (Heuser & Keen 1988, Ponnambalam et al 1990, Robinson 1993, Page & Robinson 1995). Together with the medium and small chains, the tetramers are assembled into a rectangular box with the two ears extending from the head domain at adjacent corners by their proline-rich hinges (Figure 2).

In addition to mediating clathrin attachment, adaptors are also known to have a second critical function, i.e. recruiting membrane proteins that selectively localize to clathrin-coated regions. This was first suggested by affinity chromatography experiments (Pearse 1988, Glickman et al 1989, Sorkin & Carpenter 1993, Sorkin et al 1995) and, more recently, using the yeast two-hybrid system (Ohno et al 1995). AP-1 and AP-2 adaptors mediate these activities at distinct intracellular sites, however. Under normal conditions, AP-2 complexes are localized to the plasma membrane coated pits; AP-1 complexes, on the other hand, are largely restricted to clathrin-coated buds of the TGN (Robinson & Pearse 1986, Ahle et al 1988). Thus AP-2 complexes must recognize plasma membrane receptors involved in endocytosis, whereas AP-1 complexes interact with proteins in the TGN that exit the Golgi complex via CCVs. Of these, the 205-kDa cation-independent and 47-kDa cation-dependent receptors for mannose-6-phosphate are responsible for targeting newly synthesized lysosomal enzymes from the TGN to endosomes (Geuze et al 1984a; Kornfeld & Mellman 1989).

Adaptor complexes must also selectively include membrane proteins that function in the actual targeting of CCVs to endosomes. Such proteins may include members of the v- or t-SNARE families involved in vesicle docking and/or fusion (Rothman 1994). The only v-SNARE identified in the endocytic pathway (cellubrevin) does not function at the level of CCV fusion with EEs(Link et al 1993) but rather at a later step in recycling (Galli et al 1994). That such proteins are rapidly internalized in CCVs is stongly suggested in neurons where the v-SNAREs synaptobrevin/VAMP-1 and -2 appear to be internalized by CCVs following synaptic vesicle release (Maycox et al 1992).

SIGNALS FOR SELECTIVE INCLUSION IN CLATHRIN-COATED VESICLES The cytoplasmic domains of many receptors, as well as other membrane proteins that selectively accumulate at plasma membrane coated pits, contain specific

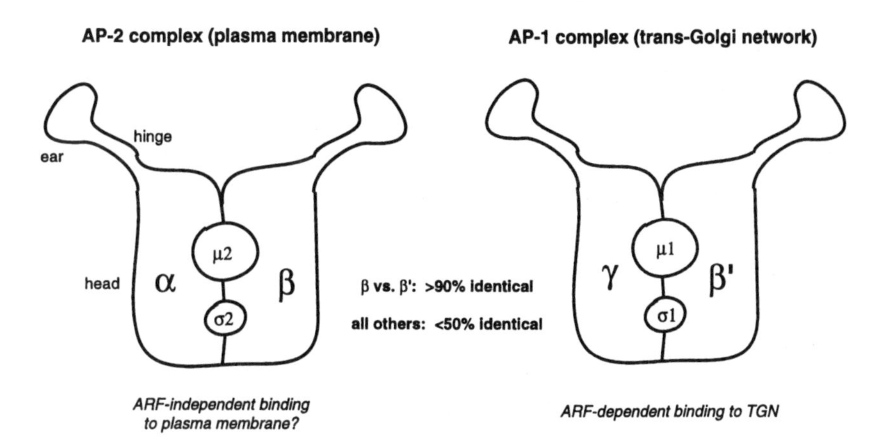

Figure 2 Organization of AP-1 and AP-2 clathrin adaptor complexes. The two known clathrin adaptor complexes (AP-1 and AP-2) are heterotetramers consisting of two ≈ 100-kDa subunits (α, β, β', or γ); two medium chains ($\mu 1$ or AP47; $\mu 2$ or AP50); and two small chains ($\sigma 1$ or AP20; $\sigma 2$ or AP17). The corresponding β and β' subunits are very closely related, whereas all others are related to their counterparts at <50% identity. The complexes consist of a large head domain and a proline-rich hinge region from which extends a COOH-terminal ear. AP-1 binds in an ARF-dependent fashion to the TGN, whereas AP-2 is commonly associated with the plasma membrane, although the ARF (if any) required for its membrane binding is unknown. A third adaptor complex consisting of the proteins β-NAP (≈ 30% identical to α-adaptin) and p47B (≈ 30% identical to the μ subunits) has recently been identified, also possibly associated with the TGN (Simpson et al 1996).

sequence information that facilitates coated pit localization. These signals are degenerate and somewhat variable thus making it difficult to establish precise motifs. Nevertheless, coated pit localization signals usually involve critical aromatic (usually tyrosine) residues placed in a context of one or more amino acids with large hydrophobic side chains (Trowbridge et al 1993) (Figure 3). A second general class of coated pit localization signal, also characterized, involves vicinal leucine residues (or leucine plus another small hydrophobic amino acid) as its most critical feature. Of potential significance is the fact that the di-leucine-based motif seems to occur most commonly among immune receptors expressed in leukocytes (e.g. Fc receptors, CD3, MHC class II-associated invariant chain) (Matter et al 1994, Hunziker & Fumey 1994), whereas the tyrosine-based motifs are far more widely distributed. Although multiple receptors are known to enter the same coated pit, and both tyrosine and di-leucine motifs clearly specify coated pit localization (Miettinen et al 1989, Hunziker et al 1991a, Amigorena et al 1992), whether the two motifs mediate entry into common coated pits, or are decoded by the same adaptor components has not been proven.

The physical structure of coated pit localization signals has been a topic of some interest. Based on an initial screening of crystallographic data of all proteins, the idea first emerged that tyrosine-containing signals adopted a β- or tight-turn conformation exposing the tyrosine on the turn surface (Collawn et al 1990). Evidence favoring this possibility was then provided by two-dimensional NMR analysis of short synthetic peptides, which were found to be capable of adopting such conformations for at least a portion of the time in aqueous solution (Bansal & Gierasch 1992, Eberle et al 1992). Given the limitations of conformational analysis of short peptides in solution, however, much additional information will be required before reaching firm conclusions regarding the actual structure of the coated pit signal. Nor is there any clear indication that the di-leucine motif adopts or participates in a tight-turn conformation. At least one potential coated pit localization domain has been modeled as a nascent helix (Wilde et al 1994).

Additionally, the simple presence of a coated pit localization domain does not ensure that a receptor will localize at clathrin-coated pits. In the case of the major leukocyte Fc receptor FcRII-B, cell type-specific alternative mRNA splicing in B-lymphocytes (FcRII-B1) introduces an in-frame insertion of 47 amino acids at a site membrane-proximal to the di-leucine-type coated pit signal (Miettinen et al 1989). Although the signal itself is not disrupted, its ability to function is inhibited. The effect seems to be one of conformational change from specific intramolecular interactions within the FcRII cytoplasmic tail: The insert does not inhibit coated pit localization of other receptors, nor does its position in the

F	D	N	P	V	Y	LDL receptor
Y	E	N	P	T	Y	β-amyloid precursor protein
F	E	N	T	L	Y	mannose receptor
Y	K	Y	S	K	V	CD-mannose-6-phosphate receptor
		Y	T	R	F	transferrin receptor
		Y	Q	P	L	T-cell receptor (CD3)
	G	Y	Q	T	I	lgp-A/lamp-1
	G	Y	E	Q	F	lgp-B/lamp-2
	G	Y	R	H	V	lysosomal acid phosphatase
		Y	S	K	V	CI-mannose-6-phosphate receptor
		Y	S	A	F	polymeric Ig receptor
		Y	Q	R	L	TGN38
			L	L		Fc receptor
			L	I		MHC class II invariant chain

Figure 3 Consensus sequence motifs for localization at clathrin-coated pits. Although degenerate, the short tetra- or hexapeptide tyrosine-containing sequences demonstrated to be necessary and sufficient for programming localization at clathrin-coated pits have some features in common. Typically, this involves a tyrosine (or phenylalanine) residue followed by one or more hydrophobic or aromatic residues. Many receptors make use of tyrosine-independent signals, usually consisting of a di-leucine-type motif (e.g. Fc receptor, invariant chain). At least three of these motifs (LDL receptor, acid phosphatase, transferrin receptor) have been modeled to adopt a tight-turn conformation; others (TGN38, Ii chain) may not be so configured.

FcRII tail matter (Miettinen et al 1992). Another related example comes from CD4, a membrane protein of T cells that forms a complex with the cytoplasmic src-like kinase lck. In cells expressing lck, CD4 internalization is markedly slowed, presumably owing to masking of the CD4 coated pit localization domain by interaction with the kinase (Pelchen et al 1992). Thus it is difficult to predict the internalization phenotype of a protein from its sequence alone.

Coated pit localization signals are not restricted to plasma membrane receptors; they have also been identified on membrane proteins found in intracellular membranes. Most notable are the major membrane glycoproteins of LEs and lysosomes, lgps or lamps, that have short (usually 10–11 residues) cytoplasmic tails containing a conserved glycine-tyrosine sequence followed by a hydrophobic amino acid two residues towards their COOH-termini (Kornfeld & Mellman 1989). In many cells, lgp/lamps reach their destination by transport directly from the TGN to the endocytic pathway, possibly via CCVs (Kornfeld & Mellman 1989). Some missorting of lgp/lamps to the cell surface does occur, particularly at high expression levels, suggesting that the sorting mechanisms in the TGN are saturable. Nevertheless, cell surface appearance of lgp/lamps is followed by their rapid internalization via plasma membrane-derived CCVs. Thus it might be expected that the coated pit localization signal of lgp/lamps can be recognized by both AP-1 and AP-2 adaptor complexes (see below). Alteration of the conserved glycine residue does not inhibit internalization from the plasma membrane but does reduce sorting of newly synthesized lgp/lamps from the TGN directly to endosomes and lysosomes (Harter & Mellman 1992). Thus the glycine residue may be more important for specifying interactions with AP-1 than with AP-2.

Another example is TGN38, a resident protein of the TGN. TGN38 actually recycles continuously between the TGN and the plasma membrane, presumably via EEs and/or LEs (Reaves et al 1992, 1993, Ladinsky & Howell 1993, Rajasekaran et al 1994, Miesenbock & Rothman 1995). Like lgp/lamps, the TGN38 cytoplasmic domain has a tyrosine-containing motif required for TGN localization as well as for rapid internalization (Wilde et al 1992, Humphrey et al 1993). Although the transport step(s) under the control of the coated pit signal are unclear, given its equilibrium distribution, it appears that TGN38 is likely to interact with AP-1 adaptors (see below).

RECOGNITION OF COATED PIT LOCALIZATION SIGNALS Until recently there was no indication as to which adaptor subunit(s) might be involved. Based on a yeast two-hybrid screen using any of several tyrosine-containing motifs as "bait", Bonifacino and colleagues (Ohno et al 1995) found that the 50-kDa μ subunits are at least partly responsible for interacting with coated pit signals. Indeed, the coated pit signals of lgp/lamps and TGN38 could bind to both the

$\mu 1$ and $\mu 2$ subunits of AP-1 and AP-2 adaptors, respectively. This is consistent with the known or suspected abilities of both proteins to be included in TGN- and plasma membrane-derived CCVs. In contrast, the coated pit signal of the plasma membrane receptor (Tfn receptor) was found to interact with only the AP-2 $\mu 2$ subunit. Thus conventional plasma membrane receptors may avoid exiting the Golgi complex via CCVs because of their relative inability to interact with AP-1 adaptors. It seems likely that this inability is not the result of a fundamental difference is the structure of the coated pit signal, but rather because of relatively modest differences in affinity. Probably, only coated pit signals with the strongest affinities for adaptors can be detected in the two-hybrid assay. A remaining unknown is the behavior of the cation-independent mannose-6-phosphate receptor (MPR), the one receptor definitively shown to exit the TGN via CCVs (Geuze et al 1984a, 1985, Kornfeld & Mellman 1989). As different regions of the cation-independent and -dependent MPR cytoplasmic domains have been associated with endocytosis and sorting of lysosomal enzymes from the TGN (Johnson & Kornfeld, 1992a,b), different regions of the receptor's tail may be involved in AP-2 versus AP-1 binding. There is evidence that coated pit localization or adaptor interaction of the cation-independent MRP is regulated by serine phosphorylation (Le Borgne et al 1993, Meresse & Hoflack 1993). Finally, di-leucine motifs have not yet been shown to interact with either the AP-1 or AP-2 μ chains (Ohno et al 1995), suggesting that another adaptor exists or that there are requirements for μ interaction with di-leucine signals that were not properly reproduced in the two-hybrid system.

These results are important in that they confirm a role for direct and specific protein-protein interactions in receptor localization at clathrin-coated pits and, by extension, in other sorting events. It is remarkable that $\mu 1$ and $\mu 2$, which are <50% identical to each other, bind such similar motifs (e.g. lgp/lamps) while simultaneously distinguishing the closely related targeting signal in the transferrin receptor tail.

RECRUITMENT OF ADAPTOR COMPLEXES TO MEMBRANES How are AP-1 and AP-2 complexes recruited to membranes? One possibility is that the binding site is provided by the ligands with which they interact, i.e. receptor tails bearing coated pit localization motifs. Indeed, overexpression of transferrin receptor was reported to increase the extent of clathrin assembly on the plasma membrane, possibly reflecting an increase in AP-2 binding sites (Miller et al 1991). There is also evidence that AP-1 binding to Golgi membranes may require the cytoplasmic domain of the cation-independent mannose-6-phosphate receptor (Le Borgne et al 1993, 1996). It is also possible that recruitment is receptor independent and may be controlled by specialized adaptor-binding proteins or docking proteins (Mahaffey et al 1990). One possible candidate is the synaptic vesicle component synaptotagmin (Zhang et al 1994). While acting as a calcium sensor during synaptic vesicle exocytosis, synaptotagmin also exhibits a strong affinity in vitro for AP-2 complexes in vitro. Although coated pits do not normally appear at the surface of synaptic vesicles, it is conceivable that synaptotagmin's AP-2 recruitment activity, or the analogous activity of any potential docking protein, is turned off intracellularly and turned on upon insertion into plasma membrane (Wang et al 1993). The phenotype of synaptotagmin mutants in *C. elegans*, which exhibit reduced neuronal endocytosis (Jorgensen et al 1995), supports this possibility. However, knock-outs in mice or *Drosophila* do not have such effects (Neher & Penner 1994). In such cases, this function might be performed by any of the non-neuronal synaptotagmins (Ullrich et al 1994).

At least in the case of AP-1 recruitment to the TGN, it appears that binding is under the control of the ras-like GTPase Arf (ADP-ribosylation factor). In vitro binding is stimulated by the presence of Arf and the non-hydrolyzable GTP (Stamnes & Rothman 1993, Traub et al 1993). Association of AP-1 with TGN membranes is also inhibited by brefeldin A (BFA), which blocks nucleotide exchange on Arf (Robinson & Kreis 1992). In this sense, AP-1 attachment is similar to the binding of at least two other coat proteins, the COP-I and COP-II complexes associated primarily with the ER and Golgi complex (Orci et al 1993a, Barlowe et al 1994). The role of Arf in the assembly of any coat complex remains unknown but may reflect its ability to activate membrane-associated phospholipase D (Brown et al 1993).

Interestingly, AP-2 binding to the plasma membrane is not stimulated by GTPγS or blocked by BFA (Robinson & Kreis 1992). Although this may suggest that AP-2 does not require Arf for membrane attachment, unlike all other coats studied thus far, it is possible that the plasma membrane contains a stably associated Arf protein, or in some other way provides the equivalent of its activity (constitutively active phospholipase D?). Recent results have shown that Arf6 is uniquely plasma membrane-associated and remains tightly bound even in the presence of BFA (Whitney et al 1995, Cavanagh et al 1996). It is not clear, however, that Arf6 is responsible for AP-2 recruitment and is absent from CCVs. Moreover, treatment of permeabilized cells with GTPγS can cause mislocalization of AP-2 adaptors to intracellular endosomal or lysosomal membranes (Seaman et al 1993). Thus potential AP-2 binding sites on endocytic organelles exist. This follows from the likelihood that adaptor-binding sites (receptors or specialized docking proteins) are internalized during CCV formation and delivered to endosomes. It appears that the activity of these binding sites is regulated in some way; recent results indicate that endosomes may host the formation of clathrin coats, although the adaptor complexes involved are not clear (Whitney et al 1995, Stoorvogel et al 1996, Takei et al 1996). As discussed below, such coats may be important for endosome function.

Clathrin-Independent Endocytosis

Clathrin-independent mechanisms for endocytosis are also known to exist (Sandvig & van Deurs 1991, Lamaze & Schmid 1995). In most cell types, such mechanisms are best detected only when CCV formation is inhibited genetically or pharmacologically, or following a hormonal stimulus associated with a transient burst of fluid uptake. However, even under normal conditions, some plant or bacterial toxins are internalized in structures devoid of clathrin coats (Hansen et al 1991). Nevertheless, some of these uncoated vesicles deliver their contents to endosomes and lysosomes (Hansen et al 1993), raising the possibility that they were actually clathrin coated but that the coats were simply not visualized.

Several groups discovered various treatments (depletion of cytosolic K^+, growth in hyperosmotic sucrose, or cytosolic acidification) that effectively blocked the proper assembly of plasma membrane clathrin coats (Heuser & Anderson 1989) and/or arrested the internalization coated pit receptors without inhibiting the uptake of markers such as ricin or IL-2 (Davoust et al 1987, Hansen et al 1993, Subtil et al 1994). Because such treatments also incompletely (or inefficiently) inhibited the endocytosis of extracellular fluid, the idea emerged that cells were capable of considerable endocytic activity in the absence of clathrin. Until recently, it remained unclear as to whether clathrin-independent endocytosis was induced by the experimental manipulations used or coexisted with the clathrin-dependent pathway.

ENDOCYTOSIS IN CLATHRIN-DEFICIENT YEAST The first direct genetic evidence in favor of a clathrin-independent pathway of endocytosis came from early experiments in the yeast *Saccharomyces cerevisiae.* Deletion of the single gene for clathrin heavy chain was found to decrease only partially ($\approx 50\%$) the uptake of receptor-bound α-factor or the marker of fluid phase endocytosis lucifer yellow (Payne & Schekman 1985). This finding was initially controversial because clathrin deletion also greatly slowed cell growth and because endocytosis in yeast was only very poorly characterized at the time (Payne et al 1987, Schekman & Payne 1988). The subsequent isolation and characterization of yeast mutants with conditional defects in α-factor uptake has convincingly demonstrated that yeast exhibit a process reminiscent of endocytosis in animal cells (Chvatchko et al 1986, Zanolari et al 1992, Raths et al 1993, Singer et al 1993, Benedetti et al 1994, Hicke & Riezman 1996). This involves the internalization of receptor-bound α-factor, its transient appearance in a population of non-lysosomal vesicles (endosomes), and final delivery to and degradation in the vacuole (lysosomes). In addition, the intracellular steps of the pathway involve ras-like GTPases (e.g. ypt5, ypt7) that are directly homologous to the rab proteins (rab4, 5, 7, 9) that control the fusion activities of early and LEs in mammalian cells (Schimmoller & Riezman 1993, Singer et al 1995). Thus yeast would appear capable of mediating at least some form of endocytosis in the absence of clathrin or in the presence of conditionally defective clathrin alleles (Seeger & Payne 1992; G Payne, unpublished observation).

What is the mechanism of clathrin-independent endocytosis? Potential insight into this question has been provided by the analysis of genes that control α-factor uptake. Although a coherent picture has yet to emerge, several of these genes have some relationship to the actin cytoskeleton (e.g. actin, fimbrin) (Kubler & Riezman 1993, Benedetti et al 1994). Whether clathrin-dependent uptake or an as yet poorly characterized actin-based system is more important under normal conditions is not clear; nor is it necessarily the case that the two mechanisms work independently of each other. Given other similarities, it is possible that a yeast-like actin-based mechanism plays an underappreciated role in endocytosis in mammalian cells, conceivably accounting for the cell's ability for clathrin-independent uptake. In *Dictyostelium*, where clathrin deletion exerts a more complete effect on endocytosis than in yeast, a role for nonconventional myosins in endocytosis has nevertheless been suggested (Novak et al 1995). Actin-dependence of endocytosis in yeast may also reflect the process of phagocytosis in mammalian cells, which is also strictly actin dependent (see above). On the other hand, it is not at all clear that actin plays a direct role in endocytosis in yeast. Although mutation of type I myosin genes (*Myo3* and *Myo5*) have recently been found to inhibit yeast endocytosis (Gell & Riezman 1996), the effect may be an indirect consequence of a dramatic reorganization of the actin cytoskeleton in these cells (Goodson et al 1996).

ELIMINATION OF CLATHRIN FUNCTION IN MAMMALIAN CELLS DOES NOT ALWAYS BLOCK ENDOCYTOSIS As discussed above, dynamin is a GTPase known from genetic, morphological, and biochemical evidence to provide a critical accessory function in clathrin-dependent endocytosis (De Camilli et al 1995). By complexing around the necks of invaginated coated pits, it may provide the mechanochemical force to accomplish the final step in the budding process (Hinshaw & Schmid 1995, Takei et al 1995) (Figure 4). Expression of mutant dynamin in mammalian cells in culture has been used to inhibit the clathrin-dependent endocytosis of receptor-bound ligands, but also has provided important insight into the clathrin-independent pathway (van der Bliek et al 1993, Damke et al 1994, 1995b). By using a temperature-sensitive mutant allele of human dynamin (predicted from the homologous *Drosophila* sequence), Schmid and colleagues found that shift to the nonpermissive temperature resulted in the rapid reduction of both receptor-mediated and fluid phase endocytosis (Damke et al 1995b). Within 30–60 min, however, fluid uptake was restored to normal, whereas transferrin uptake remained depressed. These results strongly suggest that animal cells have the capacity to induce an apparent clathrin-independent pathway of endocytosis in response to an inactivation of the clathrin (or dynamin)-dependent pathway. They may also explain why animal cells expressing constitutively defective dynamin mutants, cells subjected to K^+ depletion, or clathrin-deficient yeast exhibit fluid phase endocytosis: The cells have already compensated for the loss of the clathrin-dependent pathway.

The nature of the clathrin-independent pathway of endocytosis is unknown, but its existence, at least upon inactivation of the clathrin pathway, is no longer in doubt. At least two options have been suggested to provide endocytic activity in the absence of clathrin: caveolae and macropinocytosis.

CAVEOLAE The existence of plasma membrane pits, independent of clathrin coats, was initially in endothelial cells and adipocytes, where they were eventually termed caveolae (Peters et al 1985, Anderson 1993a). Subsequently observed at varying frequencies in many other cell types, the cytoplasmic faces of caveolae are distinguished by characteristic striations thought to be formed by an accumulation of the integral membrane protein caveolin (Peters et al 1985, Rothberg et al 1992). They are thought also to occur on internal organelles

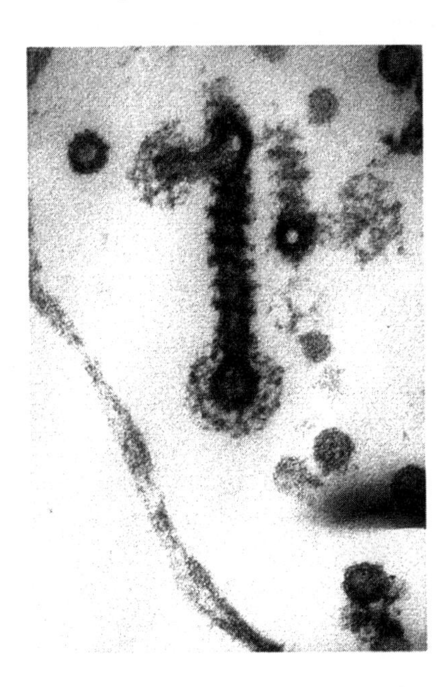

Figure 4 Dynamin is found at the necks of budding clathrin-coated vesicles. Dynamin assembles as regularly spaced rings on the necks of invaginated clathrin-coated pits arrested in the presence of GTPγS at presynaptic nerve endings. Although the precise function of the dynamin array is unknown, it may serve as a mechanochemical GTPase involved in the final budding of coated vesicles from coated pits. Such structures are best observed in synaptosomal preparations, but in somewhat less dramatic fashion are likely to form in nonneuronal cells as well. (Micrograph provided by K Takei & P De Camilli, Yale Univ. Sch. Med.; Takei et al 1995).

(Dupree et al 1993). In endothelium, the formation of caveolae is associated with the endocytic transport of solutes across the cell (transcytosis), as opposed to transport to endosomes and lysosomes, an activity associated with the formation of conventional CCVs (Milici et al 1987). In other cell types, however, it remains unclear the extent to which caveolae mediate fluid uptake or the net internalization of plasma membrane.

Although caveolae have been associated with functions as diverse as endocytosis, polarized sorting in epithelial cells, and signal transduction (Anderson 1993a, Lisanti et al 1994), their actual activities and features have remained somewhat confusing due to difficulties in studying their properties (Parton 1996). They are known to accumulate GPI-anchored membrane proteins such as the folate receptor (Rothberg et al 1992, Ying et al 1992), although even this may occur only after antibody cross-linking (Mayor et al 1994). Uptake of folate via caveolae may not involve endocytosis per se, but rather the budding and immediate reformation of caveolae pits, a process termed potocytosis (Anderson 1993b). Indeed, endocytic uptake of folate via CCV does not appear to result in the productive accumulation of the vitamin (Ritter et al 1995) although folate does reach conventional endosomes after internalization by it physiological receptor (Rijnboutt et al 1996). There is also evidence that caveolin can be transported between the plasma membrane and the secretory pathway and can be induced by certain drug treatments to accumulate in intracellular vesicles (Parton et al 1994, Smart et al 1994, Conrad et al 1995). Nevertheless, it remains highly speculative as to whether caveolae contribute to clathrin-independent fluid phase endocytosis. The existence of viable, caveolin-defective cells (lacking identifiable caveolae) further questions whether they provide an important mechanism for clathrin-independent endocytosis (Fra et al 1994, 1995, Kuliawat et al 1995).

MACROPINOCYTOSIS Cells such as macrophages, dendritic cells, and fibroblasts that can exhibit abundant membrane ruffling activity have a form of fluid phase endocytosis termed macropinocytosis (Steinman & Swanson 1995). This mode of uptake appears to reflect the passive capture of extracellular fluid when the rims of membrane folds fuse back with the plasma membrane. Vesicles or vacuoles formed in this way can be quite large, 1 to 5 μm in diameter. Macromolecules internalized in macropinosomes can be delivered to lysosomes or recycled (Racoosin & Swanson 1992, 1993).

Macropinosome formation is clearly independent of clathrin, although whether it is responsible for endocytosis observed in cells whose clathrin pathway has been inactivated is not clear. However, there are a number of intriguing features of macropinocytosis that link this activity with both phagocytosis and the actin cytoskeleton. For instance, stimulation of fibroblasts with polypeptide hormones such as epidermal growth factor (EGF) causes a transient burst of macropinocytic activity (Haigler et al 1979). A very similar series of events occurs concomitant with the binding and phagocytosis of pathogenic bacteria, particularly *Salmonella* (Pace et al 1993, Alpuche et al 1994). In fact, it seems likely that *Salmonella* entry itself reflects a localized stimulation of membrane ruffling activity in a fashion intimately dependent on actin and the regulation of actin function by GTPases of the rho family (J Gálan, unpublished results). Indeed, a link between signal transduction, membrane ruffling, and rho GTPases has been well established by microinjection experiments (Ridley et al 1992). Further emphasizing the actin dependence of such events is the observation that melanoma cells deficient in actin binding protein (ABP-1) exhibit defects in constitutive membrane ruffling/macropinocytic activity (Cunningham et al 1992).

MOLECULAR SORTING IN ENDOSOMES

Although the selective inclusion of receptors at plasma membrane coated pits is the initial sorting event during endocytosis, no less important are the sorting events that occur in endosomes. Indeed, endosomes are little more than sorting stations whose minimal tasks involve the discharge of internalized ligands from their receptors and the subsequent transfer of ligands to lysosomes and receptors back to the plasma membrane. Recently, it has become clear that endosomes mediate even more sophisticated sorting events, playing an important role in the targeting of membrane proteins to their correct plasma membrane domains in polarized epithelial cells, to a variety of specialized endosome-derived recycling or secretory vesicles, and to the TGN. The mechanisms responsible for these activities remain incompletely understood and are currently the subject of intense interest. A number of important principles have emerged regarding endosome function, however; these are considered below.

Organization and Biogenesis of Endosomes

As illustrated in Figure 1, four classes of endocytic organelles are typically distinguished based largely on their relative kinetics of labeling by endocytic tracers: EEs, LEs, recycling vesicles, and lysosomes (Kornfeld & Mellman 1989). The precise relationship among these structures has yet to be determined, and, in fact, may never be because of the great plasticity and dynamics of the system. It is also almost impossible to recognize these structures on the basis of morphology or position in the cytoplasm alone. Nevertheless, such operational distinctions have proved useful because they reflect important functional specializations in the endocytic pathway and have facilitated much recent analysis. Several biochemical markers of endosomes populations and lysosomes have been identified, allowing these distinctions to be made in more precise terms than is possible using only functional criteria. Among the most important of these are rab proteins: rab4, rab5, and rab11 are associated with EEs; rab7 and rab9 are LE specific (Chavrier et al 1990, van der Sluijs et al 1991, Simons & Zerial 1993, Nuoffer & Balch 1994, Pfeffer 1994). In addition, although recycling plasma membrane receptors are restricted to EEs and recycling vesicles, LEs contain MPR (which recycles predominantly to the TGN) (Kornfeld & Mellman 1989). Both LEs and lysosomes are highly enriched in distinctive, highly glycosylated and conserved lysosomal membrane glycoproteins (lgps, lamps, etc), which are generally depleted from EEs and the plasma membrane (Kornfeld & Mellman 1989).

It is important to emphasize that the organelle populations of the endocytic pathway are intimately interconnected by continuous endocytic and recycling activity. Indeed, much of their existence is a reflection of this activity. Given estimates that the entire cell surface is internalized at least once or twice per hour, the entire endosomal network also must turn over with similar kinetics because the total surface area of endosomes is probably far less than that of the plasma membrane (Steinman et al 1976, Griffiths et al 1989). Indeed, transient cessation of endocytosis in cells expressing a *ts* dynamin mutant is accompanied by a transient disappearance of endosomes (Damke et al 1995a), indicating that much of the bulk membrane area of endosomes is in transient residence, derived by internalization from the plasma membrane (Mellman et al 1980). Finally, the number and appearance of lysosomes can be drastically altered by growth of cells in high concentrations of poorly digestible solutes (Cohn 1966, Cohn et al 1966), which suggests that their existence is regulated by the amount and nature of internalized material. Although it is expected that endosomes and lysosomes have unique membrane components responsible for their special activities, attempting to conceptualize them as pre-existing or stable structures is a difficult and possibly useless task. Nor is it necessary to solve this issue in order to study endosome function or the biochemical mechansims responsible for these functions.

EARLY ENDOSOMES EEs are operationally defined as the first structures labeled by incoming endocytic vesicles (typically CCVs) (Helenius et al 1983). In simple tissue culture cells, EEs represent a dynamic network of tubules and vesicles dispersed throughout the cytoplasm; occasionally, the network has the appearance of a tubular reticulum, particularly in absorptive epithelial cells (Geuze et al 1983, Marsh et al 1986, Hopkins et al 1990, Tooze & Hollinshead

1991, Wilson et al 1991). The primary function of EEs is sorting, and they may represent a common site for all sorting events in the endocytic pathway (see below). For those receptors engaged in rapid recycling, EEs host the initial dissociation of receptor-ligand complexes. Their internal pH is only slightly acidic, usually ranging between pH 6.3–6.8 (Yamashiro et al 1984, Kornfeld & Mellman 1989). It has proved difficult to measure EE pH using fluorescence pH measurements of intact cells because transit through EEs is very rapid (2–3 min), which does not allow for a sufficiently selective accumulation of fluorescent reporter for in vivo measurements. However, the use of enveloped viruses as intracellular pH probes demonstrates that viruses with fusion thresholds of pH 6.3 or higher can penetrate into cells from EEs, whereas those requiring more acidic pH for fusion must await delivery to LEs or lysosomes (Schmid et al 1989). Moreover, this pH range is sufficient to result in the dissociation of ligands such as low density lipoprotein (LDL) (Kornfeld & Mellman 1989).

EEs next engage in the physical separation or sorting of vacant receptors from dissociated ligands, as well as other macromolecules internalized in the fluid phase. Accordingly, EEs have been referred to as sorting endosomes (Mayor et al 1993). The primary mechanism of this event may be geometric: Due to their greater fractional surface area, receptors (and other membrane proteins) accumulate in the EE tubular extensions, whereas due to their greater fractional volume, soluble contents accumulate in vesicular regions. The tubular elements would then pinch off or serve as sites for budding, forming carriers involved in receptor recycling, and the larger vesicular elements would proceed to LEs and lysosomes. Although the recycling carriers would contain some internal fluid that would be returned to the extracellular space, much of it would be retained within the cell.

This general model was well supported by early immunocytochemical evidence (Geuze et al 1983, 1984b). These data also demonstrated that at least some specificity must govern the process because the apparent concentration of receptors in the membrane of EE tubular regions was many fold higher than in the membrane of the vacuolar regions with which they are continuous (Geuze et al 1987). The ability of receptors to concentrate in tubular extensions may reflect a generalized physical property of the membranes involved or the existence of specific signals that direct recycling. There is now considerable evidence in favor of the latter possibility (see below).

RECYCLING VESICLES Recycling vesicles (RVs) are thought to arise from the tubular extensions of EEs, although it is unclear whether they are formed by the pinching off of entire tubules or by the production of buds in a manner analogous to the formation of CCVs (Figure 1). Although some RVs may fuse directly with the plasma membrane, it is clear that at least some translocate to the perinuclear cytoplasm on microtubule tracks where they accumulate as a distinctive population of vesicles and tubules closely apposed to the microtubule organizing center (MTOC) (Hopkins 1983, Hopkins & Trowbridge 1983, Yamashiro et al 1984). These perinuclear recycling vesicles (PNRVs) appear to contain an intracellular pool of recycling receptors, suggesting that there is a rate-limiting step in receptor recycling from PNRVs to the plasma membrane. In the case of transferrin receptor (Tfn-R), up to 75% of the total cell receptor is found intracellularly, much of which in PNRV (van der Sluijs et al 1992). Transit through peripheral EEs requires only 2–3 min, while transit through PNRVs can take 5–10 min (Schmid et al 1988, Daro et al 1996). However, the fraction of receptors that recycle directly from EEs to the plasma membrane versus transit through PNRV is unknown; nor is the function of these structures clear. At least in polarized cells, transport of RVs to the MTOC may reflect the necessity to accumulate vesicles at a site from which physical access to multiple plasma membrane domains can be maximized (Hughson & Hopkins 1990, Hunziker et al 1990, Apodaca et al 1994, Barroso & Sztul 1994, Knight et al 1995).

RVs, particularly those in the MTOC region, have been assumed to represent a distinct compartment, i.e. a population of membrane vesicles functionally and biochemically distinct from all other vesicle populations (Gruenberg & Maxfield 1995). Although evidence for this view is scant, it is based on the observations that RVs are post-sorting and generally do not contain markers destined for lysosomes, that they may have a less acidic pH than EEs, and that they have a characteristic distribution in the cytoplasm (Yamashiro et al 1984). Video microscopy suggests that they also selectively but transiently accumulate ligands such as cross-linked Tfn (Ghosh & Maxfield 1995, Marsh et al 1995); in the absence of biochemical markers for RVs, however, it is impossible to be certain if this is the case.

Very recently, evidence has emerged that some of the RVs in the perinuclear region are not just functionally, but also biochemically distinct from EEs. Although rab4 and rab5 are GTPases characteristic of EEs, rab4 appears to be depleted from PNRVs (Daro et al 1996) (Figure 5). They may also be enriched in a different endosomal rab protein, rab11 (M Zerial, manuscript submitted). It is premature to conclude that these structures comprise a distinct compartment, but these findings do suggest that the membrane of PNRVs is not identical to the EE membrane: i.e. there is specificity in the process that gives rise to RV formation. Nevertheless, much remains to be learned about these structures. They may yet prove to be simple transport vesicles in transit to the plasma membrane

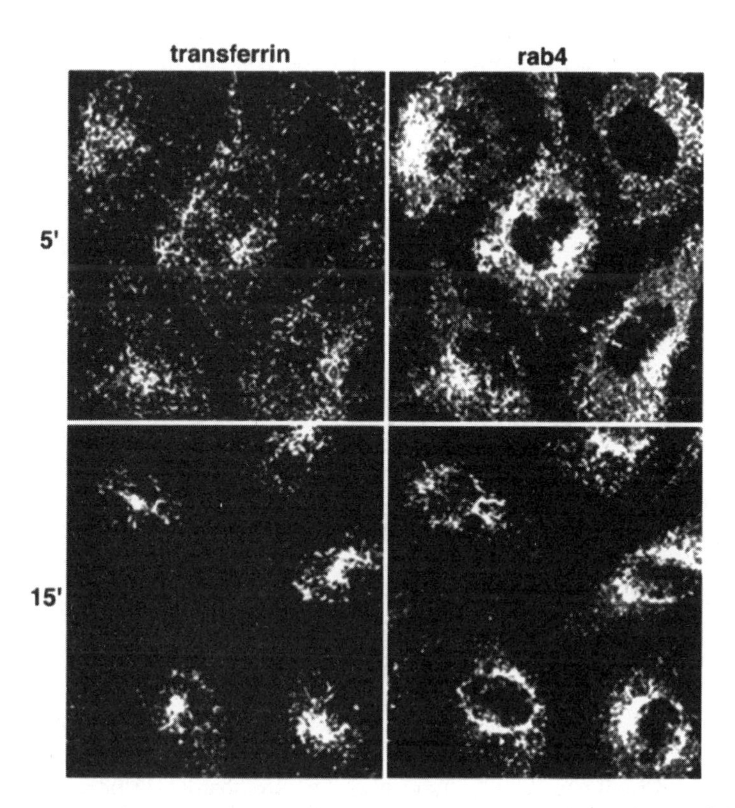

Figure 5 The pathway of receptor recycling consists of at least two distinct vesicle populations: early endosomes and recycling vesicles. CHO cells expressing human transferrin receptor and rab4 were allowed to internalize FITC-labeled transferrin for 5 or 15 min, then processed for immunofluoresence. While initially found in peripheral early endosomes that are strongly positive for rab4, at the 15 min point, a significant fraction of not yet recycled, intracellular transferrin is found in a cluster of vesicles directly over the microtubule organizing center in the perinuclear region of the cell (*lower left panel*). These perinuclear vesicles, commonly referred to as recycling vesicles, are relatively depleted in rab4 and, at least by this criterion, are biochemically distinct from the more peripheral early endosomes.

as opposed to a distinct population or compartment with its own functional attributes. Such evidence in favor of a compartment is the recent observation that some sorting activity occurs in RVs, at least as defined by the retrograde movement of fluorescent Tfn from perinuclear vesicles back to the cell periphery (Ghosh & Maxfield 1995). However, this evidence must be tempered by the likelihood that the endosomal vesicles concentrated in the MTOC region represent a biochemically and functionally heterogeneous array of structures, which makes their study difficult in the absence of distinct markers. Nor is it clear whether PNRVs fuse directly with the plasma membrane, or give rise to yet another transport vesicle population that completes the recycling circuit.

LATE ENDOSOMES AND LYSOSOMES The transfer of material from EEs to LEs involves the dissociation of vacuolar elements from the EE network, and their subsequent migration on microtubules to the perinuclear cytoplasm where they fuse with LEs (Figure 1). Dissociation of the vacuolar elements, sometimes called carrier vesicles (Gruenberg et al 1989, Aniento et al 1993, Clague et al 1994), reflects either an active budding process or, more simply, the removal of tubular elements by the continuous formation of RVs. In either event, formation of the carrier vesicles seems to be accompanied by an increase in the number of internal vesicles they contain. These are derived from invaginations of the limiting endosomal membrane (Cohn et al 1966, McKanna et al 1979). Carrier vesicles and subsequent structures thus take on the general appearance of classical multivesicular bodies (Felder et al 1990, Hopkins et al 1990, Hopkins 1994).

Late endosomes (LEs) are defined as vesicular structures that accumulate and concentrate internalized contents after their passage through EEs. They often have abundant internal membranes and are enriched in lgp/lamps, rab7, rab9, and the cation-independent MPR (Griffiths et al 1988, Geuze et al 1989, Kornfeld & Mellman 1989, Chavrier et al 1991, Lombardi et al 1993, Simons & Zerial 1993, Feng et al 1995). Although still of low density by Percoll gradient centrifugation, LEs also contain hydrolytically active lysosomal hydrolases and likely initiate the degradative process. These lysosome-like characteristics sometimes lead to LEs being referred to as pre-lysosomes or the pre-lysosomal compartment.

The relationship between LEs and lysosomes is dynamic and thus not easily defined. By convention, lysosomes are distinguished from LEs by virtue of the lysosome's higher density on Percoll, the absence of MPR, and the operational feature of being the final site of accumulation of internalized macromolecules (Helenius et al 1983, Kornfeld & Mellman 1989). Because lysosomes almost certainly receive endocytosed material by fusion with LEs (or carrier vesicles), it is useful to view their life history as more of a cycle. As summarized in Figure 6, incoming LEs are active lysosomes that progressively degrade their contents, provide for the recycling of surviving receptors (e.g. MPR), and ultimately increase in density as digestible membrane and content are processed, catabolic products released, and the remaining non-digestible material (lysosomal hydrolases, certain membrane and lipid components) is concentrated. At this stage, they become resting lysosomes, which will again be activated upon fusion with another carrier vesicle or LE. This scheme is consistent with observations that even after extended periods it is impossible to completely chase tracers such as colloidal gold from immunocytochemically defined LEs to lysosomes (Griffiths et al 1988). Such a result suggests the existence of an equilibrium, rather than unidirectional transfer, between LEs and lysosomes.

Specificity of Receptor Sorting in Endosomes

It is well known that most internalized membrane recycles from early stations in the endocytic pathway; the "deeper" into the pathway a marker traverses, the less likely it is to recycle (Besterman et al 1981). The fact that plasma membrane receptors such as the Tfn receptor are found only in EEs and RVs demonstrates that recycling receptors are removed from the pathway at the level of EEs (Stoorvogel et al 1987, Schmid et al 1988, Kornfeld & Mellman 1989, Ghosh et al 1994). How selective is receptor sorting into RVs? The fact that receptors such as those for polypeptide hormones and antibody molecules are often targeted from EEs to LEs and degraded along with their ligands shows that EEs must be able to sort membrane proteins. Early on, it was demonstrated that the stable cross-linking of receptors otherwise capable of rapid recycling was sufficient to prevent their return to the cell surface and instead target them to lysosomes (Mellman & Plutner 1984, Mellman et al 1984, Neutra et al 1985). Phosphorylation-induced changes in oligomeric structure may achieve the same end for tyrosine kinase-containing growth factor receptors subject to down regulation during ligand uptake (Felder et al 1990, Sorkin et al 1992, Cadena et al 1994). Some membrane proteins such as P-selectin and EGF receptor also may contain cytoplasmic domain signals that specify their transport to lysosomes (Green et al 1994, Opresko et al 1995). Accordingly, the view emerged initially that recycling occurred by default, with any membrane protein able to return via RVs to the cell surface unless it was prevented from doing so, for example, by ligand-induced aggregation.

The possibility that receptor recycling occurs in a signal-independent default mode was supported by several considerations. For example, the return of fluorescent sphingolipid analogues from endosomes to the plasma membrane occurs with kinetics indistinguishable from the kinetics of Tfn receptor

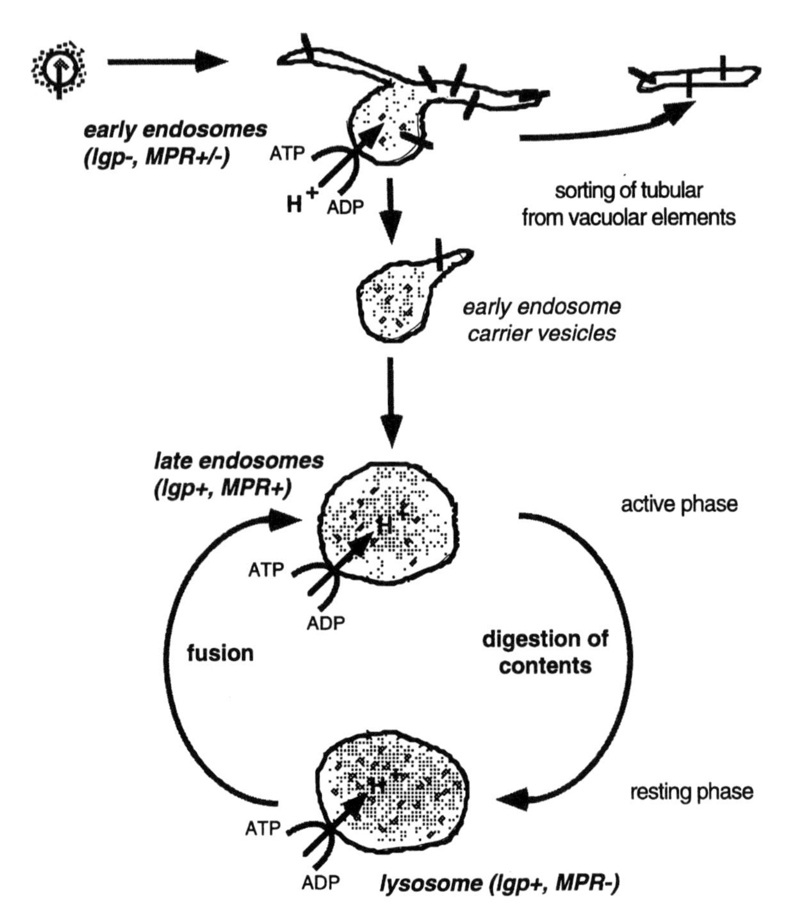

Figure 6 The lysosome cycle. This diagram illustrates a proposed relationship between late endosomes and lysosomes that suggests the two compartments exist in a dynamic equilibrium with each other. After sorting in early endosomes, vacuolar portions of the early endosome network pinch off to yield vesicles (endosome carrier vesicles) containing content destined for lysosomal digestion as well as some receptors (e.g. mannose-6-phosphate receptor, MPR) that may not have been recycled. The carrier vesicles fuse with late endosomes or lysosomes, defined as hydrolase-rich, lgp/lamp-positive structures that, respectively, do or do not contain MPR. The product of this fusion is by definition always a late endosome. As remaining MPR escape by recycling (presumably to the TGN) and as the hydrolases degrade digestible membrane and content, the late endosomes are gradually transformed into lysosomes. By this view, lysosomes consist largely of non-digestible or poorly digestible proteins, lipids, and carbohydrates and thus are hydrolytically inactive due to an absence of substrate. The lysosomes are cyclically converted into late endosomes upon fusion with an incoming early endosome carrier vesicle or with a late endosome. (Model developed by A Helenius & I Mellman, Yale Univ. Sch. Med.)

recycling (Mayor et al 1993). Thus the transport of bulk membrane lipids would appear to occur as rapidly as possible. In addition, relatively rapid recycling was observed for a mutant Tfn receptor lacking a cytoplasmic domain, suggesting that receptors simply followed the pathway defined by membrane lipids (Marsh et al 1995). On the other hand, sphingolipids can engage in lateral hydrogen bonding to form defined membrane domains that themselves are able to sort membrane proteins (Fiedler et al 1993, 1994). Thus kinetic arguments alone cannot be used to establish a case for or against specificity.

TARGETING SIGNALS CONTROL RECEPTOR RECYCLING IN POLARIZED EPITHELIAL CELLS In polarized epithelial cells, receptors internalized at one plasma membrane domain typically recycle back to that domain. This has been clearly demonstrated in Madin-Darby canine kidney (MDCK) cells in which Tfn receptor and LDL receptor, both proteins restricted to the basolateral surface, are continuously internalized and recycled basolaterally with a high degree of fidelity (Mellman et al 1993b). Transport across the cell to the apical domain (transcytosis) is possible as well, although this normally occurs only in the case of specialized transcytotic receptors such as the polymeric Ig receptor (Mostov 1994). It is obvious from the study of epithelial cells that receptor recycling cannot be a default process, but must be governed by one or more targeting signals.

The nature of targeting determinants for polarized transport in epithelial cells have been elucidated recently in some detail. By analyzing the transport of membrane proteins from the TGN to the apical or basolateral domains of MDCK cells, it is now clear that polarity is controlled primarily by discrete determinants found in the cytoplasmic domains of most or all proteins destined for the basolateral surface (Mellman et al 1993a, Matter & Mellman 1994, Mellman 1996). Although degenerate, many sequence motifs responsible for basolateral targeting bear at least a superficial resemblance to coated pit localization signals (Hunziker et al 1991a; Matter et al 1992, 1993, 1994, Prill et al 1993, Rajasekaran et al 1994, Monlauzeur et al 1995). Basolateral signals often depend on critical tyrosine residues for activity, some of which are also shared by a colinear coated pit signal. The tyrosine-dependent signals may also require an adjacent residue as well as a region of three acidic residues, features not required for coated pit localization. A well-characterized example of such signals is provided by the LDL receptor, which contains two distinct tyrosine-dependent determinants (Figure 7). Further emphasizing a possible relationship between basolateral targeting and coated pit determinants is the fact that some receptors (e.g. Fc receptor) are sorted to the basolateral surface by the exact same di-leucine motif that is responsible for rapid endocytosis (Hunziker & Fumey 1994, Matter et al 1994).

The signals are both necessary and sufficient for targeting to the basolateral surface from the TGN. In many cases, they also seem to be hierarchically arranged with a second, cryptic signal for apical targeting. Elimination of the basolateral signal by mutagenesis often results in efficient apical transport, rather than a random distribution on both plasma membrane domains. Thus in the absence of a functional basolateral signal, many proteins may be selectively delivered to the apical surface, which suggests the existence of a second signal for apical transport from the TGN. Although the nature of the apical signal is not known, it is apparently found in the lumenal domain of membrane proteins and may actually be provided by N-linked sugars (Scheiffele et al 1995). Conceivably, a carbohydrate determinant interacts at low affinity with a lectin that facilitates inclusion into nascent apical transport vesicles in the TGN. It is also possible that such interactions can lead to the incorporation of apical proteins into glycolipid rafts serving as differentiated microdomains that concentrate lipid and protein components destined for the apical surface (Fiedler et al 1994).

A striking and important feature of this system is its generality. First, the signals that control basolateral versus apical trasport are exceedingly broadly distributed, present on membrane proteins (e.g. LDL receptor, Tfn receptor) whose expression is not limited to polarized cells. Second, the same signals appear to control polarity in a variety of epithelial cell types and perhaps in non-epithelial polarized cells such as neurons (Dotti & Simons 1990, Dotti et al 1991). Finally, evidence from LDL receptor, Fc receptor, and pIg receptor has demonstrated that the targeting of receptors from endosomes back to the

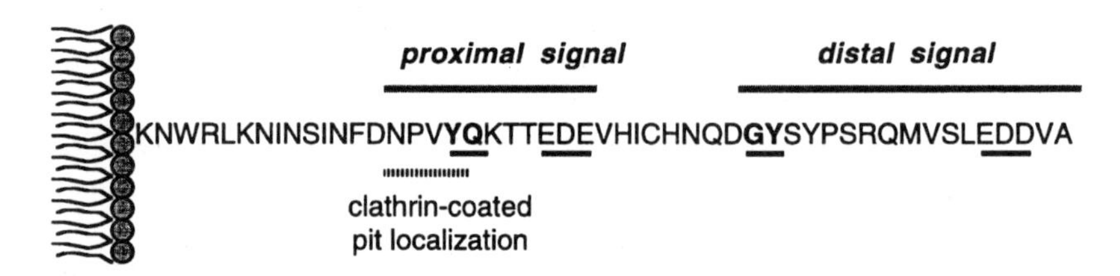

Figure 7 Signals for basolateral sorting in the LDL receptor cytoplasmic domain. The LDL receptor contains two independently acting targeting determinants capable of directing transport from endosomes and the TGN to the basolateral surface of polarized MDCK cells. The membrane proximal determinant shares a dependency on a critical tyrosine residue that is also essential for clathrin-coated pit localization, although no other features of the coated pit signal (minimally, NPXY) are required for optimal basolateral targeting. Removal of the proline residue increases the targeting efficiency of this signal (Matter et al 1993). The membrane distal signal is also tyrosine dependent, but cannot facilitate endocytosis. Both the proximal and distal signals depend on acidic residues downstream from the two tyrosines. Residues identified thus far as being essential for the activity of both signals are shown by solid underlines.

basolateral surface of MDCK cells is ensured by the same signals that are utilized in the TGN (Matter et al 1993, Aroeti & Mostov 1994) (Figure 8). In the case of LDL receptor, for example, partial inactivation of it basolateral targeting signals results in inefficient sorting of receptors internalized at the basolateral surface, causing increasing amounts of basolateral receptor to recycle to the apical surface (transcytosis). This mechanism is probably the basis for the physiological process of transcytosis exhibited by pIg receptor. Here, binding of multimeric ligand (e.g. dimeric IgA) partially inactivates the receptor's constitutively active basolateral targeting signal, which results in its inclusion in RVs targeted to the apical plasma membrane. It seems likely that these pIg receptor-containing transcytotic vesicles (TCVs) form as a consequence of a sorting event in basolateral EEs, although some sorting may also occur in apical EEs (Apodaca et al 1994).

Receptor recycling is thus clearly signal dependent, at least in polarized cells. Remarkably, in hepatocytes, endosomes appear to be the sole sorting site for both internalized and newly synthesized membrane proteins. In these cells, apical and basolateral proteins are not sorted in the TGN but rather are transported together to the basolateral surface from which they are internalized and sorted in endosomes to their sites of final residence (Bartles et al 1987).

SIGNAL-MEDIATED ENDOSOMAL SORTING IN NONPOLARIZED CELLS To what extent might sorting signals play a role in endosome function in non-epithelial cells? To consider this question,the extent to which any cell type is truly nonpolarized should be addressed. Most of the differentiated cells in all complex multicellular organisms are indeed polarized morphologically or functionally, at least at certain stages. Even "spherical" T-lymphocytes are known to polarize during antigen recognition (Poo et al 1988, Stowers et al 1995). Combined with the fact that the sorting signals identified as controlling polarity in epithelial cells are widely distributed on proteins expressed in all cell types, the idea has emerged that perhaps all cells possess cognate apical and basolateral pathways that underlie their ability to adopt polarized phenotypes (Mellman et al 1993a, Matter & Mellman 1994). At least with respect to sorting of apical and basolateral protein in the TGN, there is now some evidence, albeit indirect, to support this possibility (Fiedler & Simons 1995, Müsch et al 1996).

It is thus reasonable to expect that the basolateral targeting signals found in many receptors (e.g. Tfn, LDL receptors) play a role in endosomal sorting and recycling in fibroblasts and other single cells. The existence of such sorting signals need not contribute to the kinetics or efficiency of recycling in simple cells. Even if cognate recycling and transcytotic pathways were to exist, they would mediate transport to a common plasma membrane domain. As discussed above, signals or alterations in conformation may be required for targeting receptors

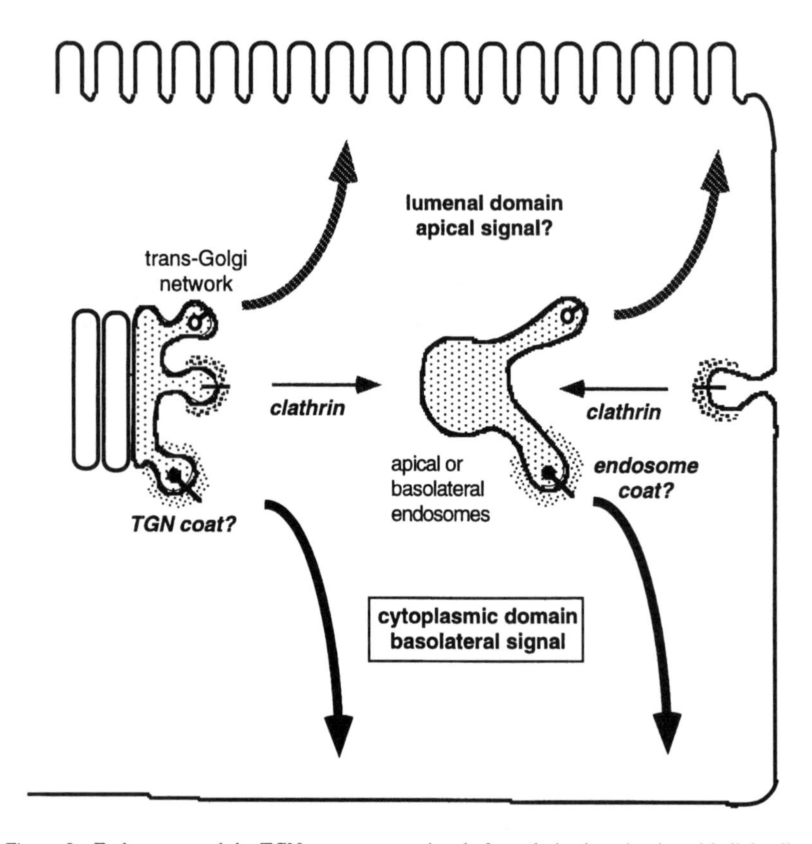

Figure 8 Endosomes and the TGN use common signals for polarized sorting in epithelial cells. Endosomes and the TGN of polarized epithelial cells are capable of decoding a common set of signals that specify transport to the basolateral plasma membrane. This ensures that both newly synthesized proteins and proteins internalized by endocytosis are targeted to their appropriate plasma membrane domains and suggests that biosynthetic and endocytic organelles may use common or at least related mechanisms for polarized sorting. Because the targeting signals are contained entirely within a basolateral protein's cytoplasmic domain, it is presumed that they must interact with one or more adaptor-like complexes recruited from the cytosol.

from early to late endosomes. However, the fact that recycling receptors concentrate in the early endosome's tubular extensions (Geuze et al 1987) implies that they are laterally sorted, which suggests the interaction of a sorting signal with an endosome-associated adaptor complex. Alternatively, it is possible that the recycling tubules represent a differentiated lipid domain (e.g. glycolipid raft) capable of concentrating receptors in an adaptor-independent fashion. In either event, exit from EEs via recycling tubules represents the route taken by the majority of membrane components internalized during endocytosis. Therefore, unless diverted to LEs, internalized membrane proteins would be expected to recycle regardless of whether they possessed a signal or even cytoplasmic domains.

ENDOSOMAL SORTING AND THE FORMATION OF SPECIALIZED RECYCLING VESICLES The ability of endosomes to mediate signal-dependent sorting is further illustrated by their capacity to bud off specialized transport vesicles in addition to the RVs that mediate constitutive receptor recycling. Best characterized is the formation of synaptic vesicles (SVs) from endosomes in neurons and neuroendocrine cells (Kelly 1993, Mundigl & De Camilli 1994). During synaptic transmission, pre-existing SVs undergo regulated exocytosis and fuse with the presynaptic plasma membrane. Vesicle membrane components are then recovered by endocytosis (largely via CCVs) and probably delivered to endosomes, which allows for the budding of newly formed SVs. In neuroendocrine cells, it is clear that the endosomes giving rise to SVs are probably EEs because they contain constitutively recycling receptors such as Tfn receptor (Clift-O'Grady et al 1990, Cameron et al 1991). Such receptors are largely excluded from the nascent SVs, and vice versa. Inclusion of SV components such as the membrane protein synaptobrevin (or VAMP) appears to be controlled by a cytoplasmic domain-sorting determinant, distinct in sequence from synaptobrevin's coated pit localization signal and not obviously related to signals involved in other sorting events (Grote et al 1995). Conceivably, cognate basolateral targeting determinants in the cytoplasmic domains of many constitutively reycling receptors help increase the specificity of SV formation by selectively including such receptors in RVs.

In addition to the formation of SVs in neuronal cells and TCVs in epithelial cells, other examples of specialized post-endosomal recycling vesicles are likely to include insulin-regulated Glut-4 vesicles of adipocytes (Herman et al 1994, Martin et al 1996) and MHC class II-containing vesicles (CIIV) in B-lymphocytes (Amigorena et al 1994, Tulp et al 1994, Wolf & Ploegh 1995). The glucose transporter Glut-4 is well known to be recruited to the plasma membrane from an intracellular pool found in a distinct endosome-like vesicle population in response to insulin stimulation of adipocytes or muscle cells.

Following its endocytosis, sequestration of Glut-4 into this vesicle population is thought to require specific signals found in one or more of the cytoplasmic regions of this polytypic membrane protein (Piper et al 1993, Verhey & Birnbaum 1994, Verhey et al 1993, 1995).

Less is known regarding the entry of MHC class II molecules in to CIIV. Current evidence suggests that newly synthesized MHC class II complexes bound to the class II-associated invariant (Ii) chain are targeted from the TGN to endosomes where Ii chain is proteolytically cleaved. Appearance of class II in CIIV only occurs following Ii cleavage, which suggests that this event requires a determinant to facilitate its transport (Amigorena et al 1995). Although the precise site where these events occur remains to be determined, the fact that CIIV, like Glut-4 vesicles, also contain markers of the receptor reycling pathway suggests that they arise from EEs.

Another potential example of endosomal sorting is provided by TGN38, a membrane protein of unknown function that is enriched in TGN membranes. Rather than being a static TGN marker, TGN38 recycles continuously between the TGN and the plasma membrane, presumably via EEs (Reaves et al 1993, Chapman & Munro 1994). Unlike recycling receptors, however, TGN38 is selectively concentrated in the TGN. Although the TGN38 cytoplasmic domain is known to contain an active coated pit localization domain (with affinity for both AP-1 and AP-2 adaptors) (Wilde et al 1994, Ohno et al 1995), the signal responsible for its transport to and/or retention in the TGN remains unknown. Nor it is definitively established that a putative signal for TGN targeting is decoded in endosomes. Yet, this is the most reasonable possibility, given the fact that TGN38 internalized from the plasma membrane rapidly reaches the TGN, which implies, by definition, an endosomal intermediate (Reaves & Banting 1992). Moreover, treatment of cells with the phosphatidylinositol-3-kinase (PI-3 kinase) inhibitor wortmannin or V-ATPase inhibitors causes missorting of TGN38, which results in its accumulation in endosomes (Chapman & Munro 1994, Reaves & Banting 1994).

Much remains to be learned about the mechanism of endosomal sorting and the extent to which endosomes serve as intermediates in the formation of specialized vesicles involved in transport either to the cell surface or to the Golgi complex. It even remains possible that some or all biosynthetic traffic from the TGN to the plasma membrane passes through endosomes en route. However, what is abundantly clear is that EEs must be capable of serving as sorting platforms responsible for generating a variety of cell type-specific post-endosomal vesicles in addition to vesicles involved in constitutive receptor recycling (Figure 9). As such, endosomes must be capable of decoding the multiple signals likely to specify each of these transport events.

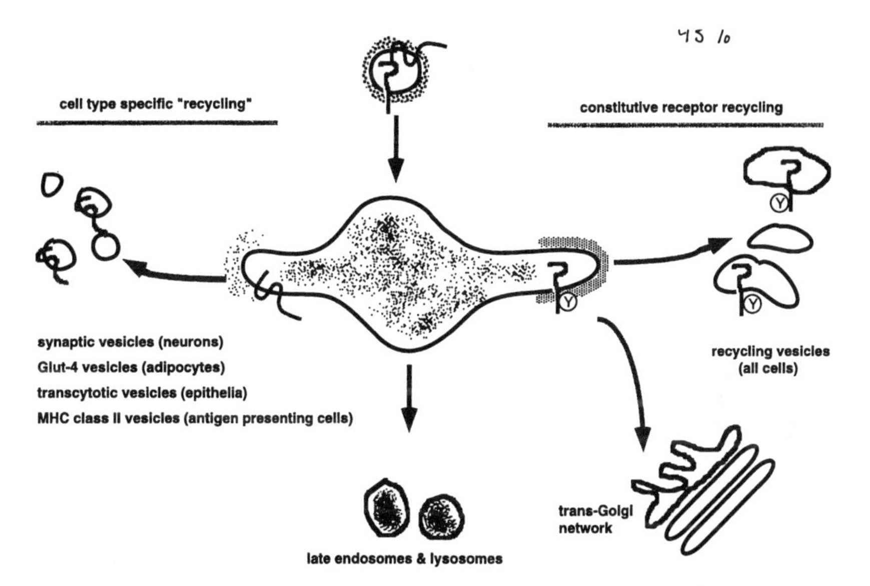

Figure 9 Endosomes serve as a platform that can host multiple sorting events simultaneously. In addition to facilitating the sorting of recycling receptors from dissociated ligands, it is becoming increasingly clear that early endosomes serve as a common sorting site for a variety of transport steps in both specialized and nonspecialized cells. Thus the same early endosomes that give rise to transport vesicles responsible for the return of receptors back to the plasma membrane likely generate a variety of cell type-specific recycling vesicles: e.g. synaptic vesicles in neurons or neuroendocrine cells, Glut-4 vesicles in adipocytes, apically directed transcytotic vesicles in epithelial cells, and possibly class II-containing vesicles in antigen-presenting cells. Early endosomes may also directly give rise to vesicles that mediate transport of Golgi complex proteins (e.g. TGN38) from the cell surface back to the TGN. Because constitutive recycling of receptors such as LDL receptor and transferrin receptor in epithelial cells is restricted to the basolateral surface, it seems likely that inclusion on this pathway involves sorting via specific basolateral targeting determinants (Y). It is possible that such signals help restrict such receptors from entering cell type-specific vesicles even in non-epithelial cells. Thus, cognate apical and basolateral pathways may exist even in cells that are not morphologically polarized in this fashion.

The existence of specific sorting mechanisms not withstanding, it is important to note that the efficiency of sorting in endosomes need not be and probably is not 100%. For example, at least half of the dIgA internalized via pIg receptors in epithelial cells is recycled back to the basolateral plasma membrane rather than being transcytosed to the apical surface (Hunziker et al 1991, Mostov, 1994). Basolateral proteins ectopically inserted into the apical plasma membrane can be recovered following endocytosis and a second round of sorting in endosomes (Matter et al 1993). The system can compensate for such "errors" for the simple reason that it is cyclical: If endosomes fail to correctly sort a receptor in the first cycle, the receptor's re-internalization will afford a second and third chance to achieve the desired targeting result. Accordingly, even slight changes in recycling efficiency might profoundly influence the equilibrium distribution of a given receptor because the system is inherently iterative over time. This strategy ensures both fidelity and flexibility, but also complicates the functional analysis of endosomal sorting signals because their effectiveness may not require total efficiency at each round of endocytosis.

TRANSPORT OF MANNOSE-6-PHOSPHATE RECEPTORS The transport of newly synthesized lysosomal enzymes to lysosomes is largely dependent on the activity of receptors for mannose-6-phosphate (MPR) (Kornfeld & Mellman 1989). Two receptors are known: a large (205-kDa) cation-independent receptor and a small (47-kDa) receptor that bind MPR-containing enzymes in a cation-dependent fashion (Kornfeld 1992). Both receptors bind their ligands in the Golgi complex and exit the TGN via AP-1-containing CCVs. After delivery to endosomes, the enzymes are discharged in a pH-dependent fashion and accumulate in lysosomes. The receptors, on the other hand, return to the Golgi complex with only a small fraction of the receptor reaching the plasma membrane. Based on the fact that the MPRs are at equilibrium concentrated in LEs, the idea emerged that MPRs enter and exit the endocytic pathway at the level of LEs (Griffiths et al 1988, Geuze et al 1989). More recently, however, Ludwig et al demonstrated that a significant fraction of MPR-bound enzyme was delivered to EEs (Ludwig et al 1991). This is an appealing possibility because it suggests that EEs represent a common point of entry, as well as a common sorting site, on both the endocytic and biosynthetic pathways. Because the pH required for lysosomal enzyme from MPRs is somewhat more acidic than for ligand dissociation from plasma membrane receptors (Kornfeld 1992), it is possible that the enzyme-MPR complex must reach LEs to complete enzyme discharge. If this were the case, then LEs would also be capable of sorting and vesicle budding activities in a fashion directly analogous to EEs.

The rationale behind the existence of two distinct receptors remains somewhat unresolved. Clearly, both receptors function in lysosomal enzyme targeting because cells from mice lacking either receptor show defects in enzyme retention. Conceivably, the receptors are specific for different populations of enzymes (differing by posttranslational modification or protein type) (Koster et al 1993, Ludwig et al 1993, 1994, Pohlmann et al 1995).

Mechanisms of Endosome Sorting: Role of Coat Proteins

Endosomes, particularly EEs, are clearly involved in multiple sorting and vesicle transport events in both polarized and nonpolarized cells. Thus far, we can distinguish at least four general classes of events: formation of constitutive recycling vesicles; the formation of specialized post-endosomal vesicles (SVs, TCVs, Glut-4 vesicles, CIIV); the production of vesicles destined for the TGN; and the production of vesicular carriers reponsible for transport to LEs. Given that some or all of these sorting/transport events are governed by targeting determinants on the cytoplasmic domains of the proteins being transported, it seems likely that cytosolic adaptors, either known or novel, play a role in endosomes directly analogous to that played by clathrin adaptors in the TGN or at the plama membranes. The observation that many sorting signals, particularly those involved in polarized transport in epithelial cells, bear some sequence relationship to coated pit signals further strengthens the prediction. Although clathrin was found to be associated with endosomes many years ago (Louvard et al 1983), the functional significance of this finding remains unknown.

ENDOSOMES CONTAIN FUNCTIONAL COP-I AND CLATHRIN COATS The first, albeit indirect, evidence in support of a role for coat proteins in endosome function came from experiments using the macrocyclic antibiotic BFA, an agent previously shown to block transport between the ER and the Golgi complex by interfering with the assembly of coatomer-containing coat complexes (Donaldson et al 1990, Orci et al 1991). Treatment of cells with BFA was found to cause a dramatic morphological alteration of endosomes, inducing extensive tubule formation analogous to the drug's initial effect on Golgi membranes (Hunziker et al 1991b, Lippincott-Schwartz et al 1991, Wood et al 1991). Although BFA did not block endocytosis or recycling per se, it did interfere with at least one endosomal sorting activity, selectively inhibiting the pIg receptor-mediated transcytosis of dIgA from the basolateral to the apical domain of MDCK cells (Hunziker et al 1991b). This effect was mediated at the level of EEs. BFA may also interfere with proper sorting between EEs and the TGN, because it also appears to induce some level of physical continuity between the two compartments (Wood et al 1991, Reaves et al 1993).

Although BFA's mode of action is not entirely clear, it does interfere with the recruitment of Arf proteins to membranes, apparently by blocking GDP-GTP exchange required for the stable attachment of soluble GDP-Arfs (Donaldson et al 1992, Helms & Rothman 1992). As such, the inhibitory effects of BFA might directly or indirectly reflect a defect in the assembly of any Arf-dependent coat complex. By assaying the ability of isolated endosomes to bind cytosolic proteins in a fashion stimulated by GTPγS and blocked by BFA, a number of proteins were found to bind specifically to endosomal membranes. Surprisingly, perhaps, most were members of the COP-I coat complex more commonly associated with transport between the ER and the Golgi, as well as through the Golgi stack (Whitney et al 1995). Coatomer, the cytosolic precursor of

COP-I coats, is thought to consist of seven subunits: α, β, β', δ, γ, ϵ, and ζ (Orci et al 1993b). Interestingly, endosomes appeared to bind only a coatomer subcomplex lacking δ and γ COP. By immuno-EM, coatomer components were found together with EE vacuolar elements, as well as small Tfn-containing vesicles. It appears that this association is of some functional importance. Microinjection of antibodies to β-COP, known to block ER to Golgi transport, also inhibited endosome function by blocking the low pH-dependent penetration of vesicular stomatitis virus (VSV) from endosomes (Whitney et al 1995). Because VSV appears to require pH <6.0 for efficient fusion, this result suggests that β-COP is required either for acidification or, more likely, for interference with the transit of the virus from EEs to LEs. Using an in vitro system designed to reconstitute carrier vesicle formation from EEs, evidence was also obtained that β-COP in fact may be required for this step (Aniento et al 1996).

Although the precise role of COP-I in mediating endosome function remains to be determined, its likely involvement is further suggested by the phenotype of a CHO cell mutant (*ldl-f*) found to bear a temperature-sensitive defect in ϵ-COP, one of the COP-I subunits found to bind to endosomes in vitro. In addition to the expected block in transport between the ER and the Golgi complex, *ldl-f* cells exhibit a temperature-sensitive defect in maintaining LDL receptors on the plasma membrane (Guo et al 1994, Hobbie et al 1994). Biochemical analysis suggests that the receptors are rapidly degraded at the nonpermissive temperature. To be consistent with suggestions from in vitro experiments, the absence of functional COP-I in these cells may interefere with the normal removal of lysosomal enzymes from EEs, thus exposing LDL receptors to a degradative environment.

Also, there is increasing evidence that endosomes are associated with clathrin-containing coats. Recent results have clearly demonstrated the presence of clathrin buds forming from EEs in fibroblasts as well as in neurons (Whitney et al 1995, Stoorvogel et al 1996, Takei et al 1996). These coated buds appear to contain Tfn receptor, suggesting a role in recycling (Stoorvogel et al 1996). Interestingly, however, the buds contain neither γ nor α adaptin, suggesting that they may consist of a novel type of adaptor complex (Stoorvogel et al 1996). In permeabilized cells treated with GTPγS, however, AP-2 adaptors are found to accumulate on endosomal vesicles (Seaman et al 1993). A function for endosomal clathrin or adaptors has yet to be demonstrated.

REGULATION OF COAT PROTEIN BINDING It is possible that endosomal sorting activities are mediated entirely by known coat protein complexes whose recruitment to endosomes may be regulated in some way. Because none of the known Arf proteins exhibits any clear selectivity for endosomes, they are unlikely to be the mediators of such specificity (Cavenaugh et al 1996).

Interestingly, however, in cells overexpressing dominant-negative mutants of the plasma membrane Arf protein Arf6, endosome-like membranes accumulate that are associated with an unidentified cytosolic coat (D'Souza et al 1995, Peters et al 1995).

In the case of COP-I assembly, the fact that endosomes appear to recruit only a subcomplex of coatomer subunits suggests that different subcomplexes exist for attachment to distinct organelles. Although COP-I is known to interact via its α subunit with proteins containing the KKXX-type ER retrieval motif, it is probable that non-KKXX-containing proteins also play a role in coatomer recruitment (Schroder et al 1995). Coatomer subcomplexes do exist, although they have yet to be shown to have specific targets (Cosson & Letourneur 1994, Lowe & Kreis 1995). Phosphorylation of coatomer subunits may also play a role in regulating COP-I binding. Although both β- and δ-COP are known to be phosphorylated, the functional significance of this modification remains unclear (Sheff et al 1996).

Clearly, other factors may regulate the creation or masking of recruitment sites for known coat proteins. Some interesting possibilities include regulation by lipid mediators normally associated with signal transduction mechanisms. Although beyond the scope of this review, it should be pointed out that yeast mutants with defects in sorting of vacuole proteins from the Golgi complex to the vacuole have a well-characterized defect in a PI-3 kinase activity (vps34) (Schu et al 1993). Based on sensitivity to the selective inhibitor of PI-3 kinase wortmannin, transport of lysosomal components, as well as of TGN38, may also be directly or indirectly regulated by alterations in PI-3 kinase activity (Brown et al 1995, Davidson 1995, Shepherd et al 1996).

Various other proteins involved in phosphorylation-dephosphorylation events of both proteins and lipids have been associated with important clathrin-coated pit function. Specifically, it is clear that dynamin physically interacts with a variety of linker proteins and phosphatases (e.g. Grb2, amphiphysin, synaptojanin) closely implicated in signal transduction (McPherson et al 1994, 1996, David et al 1996). Of particular interest is the protein eps15, a substrate for EGF receptor kinase that is associated with the AP-2 clathrin adaptor complex (Benmerah et al 1995). A related protein (End3) also appears to be involved in endocytosis in yeast (Benedetti et al 1994). The precise significance of these interactions remains to be determined. However, these observations appear to presage the existence of an entirely new layer of complexity underlying the regulation of endocytosis and receptor function. These topics have recently been the subject of an excellent review (De Camilli et al 1996).

ALTERNATIVE COAT PROTEINS An alternative possibility is that endosomal sorting, and perhaps even certain sorting events in the TGN (such as basolateral targeting) whose mechanism has yet to be defined, is mediated by novel coat proteins or complexes. Neuronal proteins distantly related (<30% identity) to β-adaptin (β-NAP) and to the medium chains (p47A, p47B) have been cloned and sequenced (Pevsner et al 1994, Newman et al 1995). These proteins have recently been found to form a complex that can be recruited to TGN-like membranes in permeabilized cells (Simpson et al 1996). Because nonneuronal versions of these proteins may also exist, they could be interesting candidates for alternative coat proteins involved in sorting activities. Examination of the developing human cDNA database has revealed an array of similar adaptor homologues exhibiting varying degrees of identity to known adaptor subunits ($\approx$30% identity, 50–60% identity, 75–80% identity). Thus far at least one of these, a γ-adaptin homologue, appears to localize to Golgi membranes (D Lewin et al, unpublished observation).

Although functions for any of the alternative adaptor proteins have yet to be defined, they are attractive candidates for several reasons, not the least of which is the remarkable, perhaps superficial, similarity that exists among the various targeting signals thus far identified. As discussed above, targeting of membrane proteins in polarized epithelial cells often depends on tyrosine-containing or clathrin-coated pit-like signals. Even KKXX-containing signals, normally associated with the binding of COP-I complexes, have been found under some circumstances to mediate rapid endocytosis from the plasma membrane (Kappeler et al 1994, Itin et al 1995). Such similarities may be serendipitous but also may suggest that the adaptors decoding them are fundamentally similar to the authentic adaptors found in clathrin-coated vesicles. At the same time, one must also be cognizant of the possibility that the paradigm defined by the formation of clathrin-coated vesicles is not applicable to all sorting events. Whereas during coated vesicle formation, sorting and budding appear to be effectively and kinetically linked, this need not always be the case. Sorting might occur prior to vesicle budding, a possibility consistent with the pre-existing concentration of recycling receptors in endosomal tubules (Geuze et al 1987) or the existence of morphologically distinct domains in a continuous TGN (Ladinsky et al 1994). In either event, the identification and eventual functional characterization of the protein components involved in sorting, both in endosomes and the TGN, will finally provide the key to understand these organelles in precise biochemical terms.

Literature Cited

Ahle S, Mann A, Eichelsbacher U, Ungewickell E. 1988. Structural relationships between clathrin assembly proteins from the Golgi and the plasma membrane. *EMBO J.* 7:919-29

Al-Awqati Q. 1986. Proton-translocating ATPases. *Annu. Rev. Cell Biol.* 2:179–99

Alpuche AC, Racoosin EL, Swanson JA, Miller SI. 1994. *Salmonella* stimulate macrophage macropinocytosis and persist within spacious phagosomes. *J. Exp. Med.* 179:601–8

Amigorena S, Bonnerot C, Drake JR, Choquet D, Hunziker W, et al. 1992. Cytoplasmic domain heterogeneity and functions of IgG Fc receptors in B-lymphocytes. *Science* 256:1808–12

Amigorena S, Drake J, Webster P, Mellman I. 1994. Transient accumulation of new class II MHC molecules in a novel endocytic compartment in B lymphocytes. *Nature* 369:113–20

Amigorena S, Webster P, Drake J, Newcomb J, Cresswell P, Mellman I. 1995. Invariant chain cleavage and peptide loading in major histocompatibility complex class II vesicles. *J. Exp. Med.* 181:1729–41

Anderson RG. 1993a. Plasmalemmal caveolae and GPI-anchored membrane proteins. *Curr. Opin. Cell Biol.* 5:647–52

Anderson RG. 1993b. Potocytosis of small molecules and ions by caveolae. *Trends Cell Biol.* 3:69–71

Andrews NW, Portnoy DA. 1994. Cytolysins from intracellular pathogens. *Trends Microbiol.* 2:261–63

Aniento F, Emans N, Griffiths G, Gruenberg J. 1993. Cytoplasmic dynein-dependent vesicular transport from early to late endosomes. *J. Cell Biol.* 123:1373–87

Aniento F, Gu F, Parton RG, Gruenberg J. 1996. An endosomal β-COP is involved in the pH-dependent formation of transport vesicles destined for late endosomes. *J. Cell Biol.* 133:29–41

Apodaca G, Katz LA, Mostov KE. 1994. Receptor-mediated transcytosis of IgA in MDCK cells is via apical recycling endosomes. *J. Cell Biol.* 125:67–86

Aroeti B, Mostov KE. 1994. Polarized sorting of the polymeric immunoglobulin receptor in the exocytotic and endocytotic pathways is controlled by the same amino acids. *EMBO J.* 13:2297–304

Bansal A, Gierasch LM. 1992. The NPXY internalization signal of the LDL receptor adopts a reverse-turn conformation. *Cell* 67:1195–201

Barlowe C, Orci L, Yeung T, Hosobuchi M, Hamamoto S, et al. 1994. COPII: a membrane coat formed by sec proteins that drive vesicle budding from the ER. *Cell* 77:895–907

Barroso M, Sztul ES. 1994. Basolateral to apical transcytosis in polarized cells is indirect and involves BFA and trimeric G protein-sensitive passage through the apical endosome. *J. Cell Biol.* 124:83–100

Bartles JR, Feracci HM, Stieger B, Hubbard AL. 1987. Biogenesis of the rat hepatocyte plasma membrane in vivo: comparison of the pathways taken by apical and basolateral proteins using subcellular fractionation. *J. Cell Biol.* 105:1241–51

Benedetti H, Raths S, Crausaz F, Riezman H. 1994. The *END3* gene encodes a protein that is required for the internalization step of endocytosis and for actin cytoskeleton organization in yeast. *Mol. Biol. Cell* 5:1023–37

Benmerah A, Gagnon J, Begue B, Megarbane B, Dautry VA, Cerf BN. 1995. The tyrosine kinase substrates eps15 is constitutively associated with the plasma membrane adaptor AP-2. *J. Cell Biol.* 131:1831–38

Besterman JM, Airhart JA, Woodworth RC, Low RB. 1981. Exocytosis of pinocytosed fluid in cultured cells: kinetic evidence for rapid turnover and compartmentation. *J. Cell Biol.* 91:716–27

Bliska JB, Galan JE, Falkow S. 1993. Signal transduction in the mammalian cell during bacterial attachment and entry. *Cell* 73:903–20

Bokoch GM, Knaus UG. 1994. The role of small GTP binding proteins in leukocyte function. *Curr. Opin. Immunol.* 6:98–105

Brown HA, Gutowski S, Moomaw CR, Slaughter C, Sternweis PC. 1993. ADP-ribosylation factor, a small GTP dependent regulatory protein stimulates phospholipase D activity. *Cell* 75:1137–44

Brown MS, Goldstein JL. 1979. Receptor-mediated endocytosis: insight from the lipoprotein receptor system. *Proc. Natl. Acad. Sci. USA* 76:3330–37

Brown WJ, DeWald DB, Emr SD, Plutner H, Balch WE. 1995. Role for phosphatidylinositol 3-kinase in the sorting and transport of newly synthesized lysosomal enzymes in mammalian cells. *J. Cell Biol.* 130:781–96

Cadena DL, Chan CL, Gill GN. 1994. The intracellular tyrosine kinase domain of the epidermal growth factor receptor undergoes a conformational change upon autophosphorylation. *J. Biol. Chem.* 269:260–65

Cameron PL, Sudhof TC, Jahn R, De CP. 1991. Colocalization of synaptophysin with trans-

ferrin receptors: implications for synaptic vesicle biogenesis. *J. Cell Biol.* 115:151–64
Carter LL, Redelmeier TE, Woollenweber LA, Schmid SL. 1993. Multiple GTP-binding proteins participate in clathrin-coated vesicle-mediated endocytosis. *J. Cell Biol.* 120:37–45
Chang MP, Mallet WG, Mostov KE, Brodsky FM. 1993. Adaptor self-aggregation, adaptor-receptor recognition and binding of alpha-adaptin subunits to the plasma membrane contribute to recruitment of adaptor (AP2) components of clathrin-coated pits. *EMBO J.* 12:2169–80
Chapman RE, Munro S. 1994. Retrieval of TGN proteins from the cell surface requires endosomal acidification. *EMBO J.* 13:2305–12
Chavrier P, Gorvel JP, Stelzer E, Simons K, Gruenberg J, Zerial M. 1991. Hypervariable C-terminal domain of rab proteins acts as a targeting signal. *Nature* 353:769–72
Chavrier P, Parton RG, Hauri HP, Simons K, Zerial M. 1990. Localization of low molecular weight GTP binding proteins to exocytic and endocytic compartments. *Cell* 62:317–29
Chen MS, Obar RA, Schroeder CC, Austin TW, Poodry CA, et al. 1991. Multiple forms of dynamin are encoded by *shibire*, a *Drosophila* gene involved in endocytosis. *Nature* 351:583–6
Chvatchko Y, Howald I, Riezman H. 1986. Two yeast mutants defective in endocytosis are defective in pheromone response. *Cell* 46:355–64
Clague MJ, Urbe S, Aniento F, Gruenberg J. 1994. Vacuolar ATPase activity is required for endosomal carrier vesicle formation. *J. Biol. Chem.* 269:21–24
Clift-O'Grady L, Linstedt AD, Lowe AW, Grote E, Kelly RB. 1990. Biogenesis of synaptic vesicle-like structures in a pheochromocytoma cell line PC-12. *J. Cell Biol.* 110:1693–703
Cohen CJ, Bacon R, Clarke M, Joiner K, Mellman I. 1994. *Dictyostelium discoideum* mutants with conditional defects in phagocytosis. *J. Cell Biol.* 126:955–66
Cohn ZA. 1966. The regulation of pinocytosis in mouse macrophages. *J. Exp. Med.* 124:557–70
Cohn ZA, Hirsch JG, Fedorko ME. 1966. The in vitro differentiation of mononuclear phagocytes. V. The formation of macrophage lysosomes. *J. Exp. Med.* 123:757–66
Collawn JF, Stangel M, Kuhn LA, Esekogwu V, Jing SQ, et al. 1990. Transferrin receptor internalization sequence YXRF implicates a tight turn as the structural recognition motif for endocytosis. *Cell* 63:1061–72
Conrad PA, Smart EJ, Ying YS, Anderson RG, Bloom GS. 1995. Caveolin cycles between plasma membrane caveolae and the Golgi complex by microtubule-dependent and microtubule-independent steps. *J. Cell Biol.* 131:1421–33
Cosson P, Letourneur F. 1994. Coatomer interaction with di-lysine endoplasmic reticulum retention motifs. *Science* 263:1629–31
Cunningham CC, Gorlin JB, Kwiatkowski DJ, Hartwig JH, Janmey PA, et al. 1992. Actin-binding protein requirement for cortical stability and efficient locomotion. *Science* 255:325–27
Damke H, Baba T, van der Bliek Am, Schmid SL. 1995a. Clathrin-independent pinocytosis is induced in cells overexpressing a temperature-sensitive mutant of dynamin. *J. Cell Biol.* 131:69–80
Damke H, Baba T, Warnock DE, Schmid SL. 1994. Induction of mutant dynamin specifically blocks endocytic coated vesicle formation. *J. Cell Biol.* 127:915–34
Damke H, Gossen M, Freundlieb S, Bujard H, Schmid SL. 1995b. Tightly regulated and inducible expression of dominant interfering dynamin mutant in stably transformed HeLa cells. *Meth. Enzymol.* 257:209–20
Daro E, van der Sluijs P, Galli T, Mellman I. 1996. Rab4 and cellubrevin define distinct populations of endosomes on the pathway of transferrin recycling. *Proc. Natl. Acad. Sci. USA*. In press
David C, McPherson PS, Mundigl O, De Camilli P. 1996. A role of amphiphysin in synaptic vesicle endocytosis suggested by its binding to dynamin in nerve terminals. *Proc. Natl. Acad. Sci. USA* 93:331–35
Davidson HW. 1995. Wortmannin causes mistargeting of procathepsin D. Evidence for the involvement of a phosphatidylinositol 3-kinase in vesicular transport to lysosomes. *J. Cell Biol.* 130:797–805
Davoust J, Gruenberg J, Howell KE. 1987. Two threshold values of low pH block endocytosis at different stages. *EMBO J.* 6:3601–9
De Camilli P, Emr SD, McPherson PS, Novick P. 1996. Phosphoinositides as regulators in membrane traffic. *Science.* In press
De Camilli P, Takei K, McPherson PS. 1995. The function of dynamin in endocytosis. *Curr. Opin. Neurobiol.* 5:559–65
Desjardins M, Huber LA, Parton RG, Griffiths G. 1994. Biogenesis of phagolysosomes proceeds through a sequential series of interactions with the endocytic apparatus. *J. Cell Biol.* 124:677–88
Donaldson JG, Finazzi D, Klausner RD. 1992. Brefeldin A inhibits Golgi membrane-catalysed exchange of guanine nucleotide onto ARF protein. *Nature* 360:350–52

Donaldson JG, Lippincott-Schwartz J, Bloom GS, Kreis TE, Klausner RD. 1990. Dissociation of a 110-kD peripheral membrane protein from the Golgi apparatus is an early event in brefeldin A action. *J. Cell Biol.* 111:2295–2306

Dotti CG, Parton RG, Simons K. 1991. Polarized sorting of glypiated proteins in hippocampal neurons. *Nature* 349:158–61

Dotti CG, Simons K. 1990. Polarized sorting of viral glycoproteins to the axon and dendrites of hippocampal neurons in culture. *Cell* 62:63–72

D'Souza SC, Li G, Colombo MI, Stahl PD. 1995. A regulatory role for ARF6 in receptor-mediated endocytosis. *Science* 267:1175–78

Dupree P, Parton RG, Raposo G, Kurzchalia TV, and Simons K 1993. Caveolae and sorting in the trans-Golgi network of epithelial cells. *EMBO J.* 12:1597–605

Eberle W, Sander C, Klaus W, Schmidt B, von Figura K, Peters C. 1992. The essential tyrosine of the internalization signal in lysosomal acid phosphatase is part of a β turn. *Cell* 67:1203–9

Ezekowitz RA, Williams DJ, Koziel H, Armstrong MY, Warner A, et al. 1991. Uptake of *Pneumocystis carinii* mediated by the macrophage mannose receptor. *Nature* 351:155–58

Falkow S, Isberg RR, Portnoy DA. 1992. The interaction of bacteria with mammalian cells. *Annu. Rev. Cell Biol.* 8:333–63

Felder S, Miller K, Moehren G, Ullrich A, Schlessinger J, Hopkins CR. 1990. Kinase activity controls the sorting of the epidermal growth factor receptor within the multivesicular body. *Cell* 61:623–34

Feng Y, Press B, Wandinger NA. 1995. Rab 7: an important regulator of late endocytic membrane traffic. *J. Cell Biol.* 131:1435–52

Fiedler K, Kobayashi T, Kurzchalia TV, Simons K. 1993. Glycosphingolipid-enriched, detergent-insoluble complexes in protein sorting in epithelial cells. *Biochemistry* 32:6365–73

Fiedler K, Parton RG, Kellner R, Etzold T, Simons K. 1994. VIP36, a novel component of glycolipid rafts and exocytic carrier vesicles in epithelial cells. *EMBO J.* 13:1729–40

Fiedler K, Simons K. 1995. The role of N-glycans in the secretory pathway. *Cell* 81:309–12

Forgac M. 1992. Structure and properties of the coated vesicle proton pump. *Ann. NY Acad. Sci.* 671:273–83

Fra AM, Williamson E, Simons K, Parton RG. 1994. Detergent-insoluble glycolipid microdomains in lymphocytes in the absence of caveolae. *J. Biol. Chem.* 269:30745–48

Fra AM, Williamson E, Simons K, Parton RG. 1995. De novo formation of caveolae in lymphocytes by expression of VIP21-caveolin. *Proc. Natl. Acad. Sci. USA* 92:8655–59

Francis CL, Ryan TA, Jones BD, Smith SJ, Falkow S. 1993. Ruffles induced by *Salmonella* and other stimuli direct macropinocytosis of bacteria. *Nature* 364:639–42

Gálan JE. 1994. Interactions of bacteria with non-phagocytic cells. *Curr. Opin. Immunol.* 6:590–95

Galli T, Chilcote T, Mundigl O, Binz T, Niemann H, De CP. 1994. Tetanus toxin-mediated cleavage of cellubrevin impairs exocytosis of transferrin receptor-containing vesicles in CHO cells. *J. Cell Biol.* 125:1015–24

Gell MI, Riezman H. 1996. Role of type I myosins in receptor-mediated endocytosis in yeast. *Science* 272:533–35

Geuze HJ, Slot JW, Schwartz AL. 1987. Membranes of sorting organelles display lateral heterogeneity in receptor distribution. *J. Cell Biol.* 104:1715–23

Geuze HJ, Slot JW, Strous GJAM. 1983. Intracellular site of asialoglycoprotein receptor-ligand uncoupling: double-label immunoelectron microscopy during receptor-mediated endocytosis. *Cell* 32:277–87

Geuze HJ, Slot JW, Strous GJAM, Hasilik A, von Figura K. 1984a. Ultrastructural localization of the mannose 6-phosphate receptor in rat liver. *J. Cell Biol.* 98:2047–54

Geuze HJ, Slot JW, Strous GJAM, Hasilik A, von Figura K. 1985. Possible pathways for lysosomal enzyme delivery. *J. Cell Biol.* 101:2253–62

Geuze HJ, Slot JW, Strous GJ, Peppard K, von Figura K, et al. 1984b. Intracellular receptor sorting during endocytosis: comparative immunoelectron microscopy of multiple receptors in rat liver. *Cell* 37:195–204

Geuze HJ, Stoorvogel W, Strous GJ, Slot J, Zijderhand-Bleekemolen J, Mellman I. 1989. Sorting of mannose 6-phosphate receptors and lysosomal membrane proteins in endocytic vesicles. *J. Cell Biol.* 107:2491–501

Ghosh RN, Gelman DL, Maxfield FR. 1994. Quantification of low density lipoprotein and transferrin endocytic sorting HEp2 cells using confocal microscopy. *J. Cell. Sci.* 107:2177–89

Ghosh RN, Maxfield FR. 1995. Evidence for nonvectorial, retrograde transferrin trafficking in the early endosomes of HEp2 cells. *J. Cell Biol.* 128:549–61

Glickman JA, Conibear E, Pearse BMF. 1989. Specificity of binding of clathrin adaptors to signals on the mannose-6-phosphate/insulin-like growth factor II receptor. *EMBO J.*

8:1041–47
Goodson HV, Anderson BL, Warrick HM, Pon LA, Spudich JA. 1996. Synthetic lethality screen identifies a novel yeast myosin I gene (*myo5*): Myosin I proteins are required for polarization of the actin cytoskeleton. *J. Cell Biol.* In press
Green SA, Setiadi H, McEver RP, Kelly RB. 1994. The cytoplasmic domain of P-selectin contains a sorting determinant that mediates rapid degradation in lysosomes. *J. Cell Biol.* 124:435–48
Greenberg S. 1995. Signal transduction of phagocytosis. *Trends Cell Biol.* 5:93–99
Greenberg S, Burridge K, Silverstein SC. 1990. Co-localization of F-actin and talin during Fc receptor-mediated phagocytosis in mouse macrophages. *J. Exp. Med.* 172:1853–56
Greenberg S, Chang P, Silverstein SC. 1993. Tyrosine phosphorylation is required for Fc receptor-mediated phagocytosis in mouse macrophages. *J. Exp. Med.* 177:529–34
Greenberg S, Chang P, Silverstein SC. 1994. Tyrosine phosphorylation of the gamma subunit of Fc gamma receptors, p72syk, and paxillin during Fc receptor-mediated phagocytosis in macrophages. *J. Biol. Chem.* 269:3897–902
Greenberg S, Chang P, Wang DC, Xavier R, Seed B. 1996. Clustered syk tyrosine kinase domains trigger phagocytosis. *Proc. Natl. Acad. Sci. USA* 93:1103–7
Greenberg S, el Khoury J, di Virgilio F, Kaplan EM, Silverstein SC. 1991. Ca(2+)-dependent F-actin assembly and disassembly during Fc receptor-mediated phagocytosis in mouse macrophages. *J. Cell Biol.* 113:757–67
Griffiths G, Back R, Marsh M. 1989. A quantitative analysis of the endocytic pathway in baby hamster kidney cells. *J. Cell Biol.* 109:2703–20
Griffiths G, Hoflack B, Simons K, Mellman I, Kornfeld S. 1988. The mannose 6-phosphate receptor and the biogenesis of lysosomes. *Cell* 52:329–41
Grote E, Hao JC, Bennett MK, Kelly RB. 1995. A targeting signal in VAMP regulating transport to synaptic vesicles. *Cell* 81:581–89
Gruenberg J, Griffiths G, Howell KE. 1989. Characterization of the early endosome and putative carrier vesicles in vivo and with an assay of vesicle fusion in vitro. *J. Cell Biol.* 108:1301–16
Gruenberg J, Maxfield FR. 1995. Membrane transport in the endocytic pathway. *Curr. Opin. Cell Biol.* 7:552–63
Guo Q, Vasile E, Krieger M. 1994. Disruptions in Golgi structure and membrane traffic in a conditional lethal mammalian cell mutant are corrected by epsilon-COP. *J. Cell Biol.* 125:1213–24
Haigler HT, McKanna JA, Cohen S. 1979. Rapid stimulation of pinocytosis in human carcinoma cells A-431 by epidermal growth factor. *J. Cell Biol.* 83:82–90
Hansen SH, Sandvig K, van Deurs B. 1991. The preendosomal compartment comprises distinct coated and noncoated endocytic vesicle populations. *J. Cell Biol.* 113:731–41
Hansen SH, Sandvig K, van Deurs B. 1993. Molecules internalized by clathrin-independent endocytosis are delivered to endosomes containing transferrin receptors. *J. Cell Biol.* 123:89–97
Harding CV, Geuze HJ. 1992. Class II MHC molecules are present in macrophage lysosomes and phagolysosomes that function in the phagocytic processing of *Listeria monocytogenes* for presentation to T cells. *J. Cell Biol.* 119:531–42
Harter C, Mellman I. 1992. Transport of the lysosomal membrane glycoprotein lgp120 (lgp-A) to lysosomes does not require appearance on the plasma membrane. *J. Cell Biol.* 117:311–25
Helenius A, Mellman I, Wall D, Hubbard A. 1983. Endosomes. *Trends Biochem. Sci.* 8:245–50
Helms JB, Rothman JE. 1992. Inhibition by brefeldin A of a Golgi membrane enzyme that catalyses exchange of guanine nucleotide bound to ARF. *Nature* 360:352–54
Herman GA, Bonzelius F, Cieutat AM, Kelly RB. 1994. A distinct class of intracellular storage vesicles, identified by expression of the glucose transporter GLUT4. *Proc. Natl. Acad. Sci. USA* 91:12750–54
Heuser J, Evans L. 1980. Three-dimensional visualization of coated vesicle formation in fibroblasts. *J. Cell Biol.* 84:560–83
Heuser JE, Anderson RG. 1989. Hypertonic media inhibit receptor-mediated endocytosis by blocking clathrin-coated pit formation. *J. Cell Biol.* 108:389–400
Heuser JE, Keen J. 1988. Deep-etch visualization of proteins involved in clathrin assembly. *J. Cell Biol.* 107:877–86
Hicke L, Riezman H. 1996. Ubiquitination of a yeast plasma membrane receptor signals its ligand-stimulated endocytosis. *Cell* 84:277–87
Hinshaw JE, Schmid SL. 1995. Dynamin self-assembles into rings suggesting a mechanism for coated vesicle budding. *Nature* 374:190–92
Hobbie L, Fisher AS, Lee S, Flint A, Krieger M. 1994. Isolation of three classes of conditional lethal Chinese hamster ovary cell mutants with temperature-dependent defects in low density lipoprotein receptor stability and intracellular membrane transport. *J. Biol.*

Chem. 269:20958–70
Hopkins CR. 1983. Intracellular routing of transferrin and transferrin receptors in epidermoid carcinoma A431 cells. *Cell* 35:321–30
Hopkins CR. 1994. Internalization of polypeptide growth factor receptors and the regulation of transcription. *Biochem. Pharmacol.* 47:151–54
Hopkins CR, Gibson A, Shipman M, Miller K. 1990. Movement of internalized ligand-receptor complexes along a continuous endosomal reticulum. *Nature* 346:335–39
Hopkins CR, Trowbridge IS. 1983. Internalization and procession of transferrin and the transferrin receptor in human carcinoma A431 cells. *J. Cell Biol.* 97:508–21
Hughson EJ, Hopkins CR. 1990. Endocytic pathways in polarized Caco-2 cells: identification of an endosomal compartment accessible from both apical and basolateral surfaces. *J. Cell Biol.* 110:337–48
Humphrey JS, Peters PJ, Yuan LC, Bonifacino JS. 1993. Localization of TGN38 to the *trans*-Golgi network: involvement of a cytoplasmic tyrosine-containing sequence. *J. Cell Biol.* 120:1123–35
Hunter S, Kamoun M, Schreiber AD. 1994. Transfection of an Fc gamma receptor cDNA induces T cells to become phagocytic. *Proc. Natl. Acad. Sci. USA* 91:10232–36
Hunziker W, Fumey C. 1994. A di-leucine motif mediates endocytosis and basolateral sorting of macrophage IgG Fc receptors in MDCK cells. *EMBO J.* 13:2963–67
Hunziker W, Harter C, Matter K, Mellman I. 1991a. Basolateral sorting in MDCK cells involves a distinct cytoplasmic domain determinant. *Cell* 66:907–20
Hunziker W, Măle P, Mellman I. 1990. Differential microtubule requirements for transcytosis in MDCK cells. *EMBO J.* 9:3515–25
Hunziker W, Whitney JA, Mellman I. 1991b. Selective inhibition of transcytosis by brefeldin A in MDCK cells. *Cell* 67:617–27
Isberg RR, Tran Van Nhien G. 1994. Binding and internalization of microorganisms by integrin receptors. *Trends Microbiol.* 2:10–14
Itin C, Kappeler F, Linstedt AD, Hauri HP. 1995. A novel endocytosis signal related to the KKXX ER-retrieval signal. *EMBO J.* 14:2250–56
Johnson KF, Kornfeld S. 1992a. The cytoplasmic tail of the mannose 6-phosphate/insulin-like growth factor-II receptor has two signals for lysosomal enzyme sorting in the Golgi. *J. Cell Biol.* 119:249–57
Johnson KF, Kornfeld S. 1992b. A His-Leu-Leu sequence near the carboxyl terminus of the cytoplasmic domain of the cation-dependent mannose 6-phosphate receptor is necessary for the lysosomal enzyme sorting function. *J. Biol. Chem.* 267:17110–15
Joiner KA, Fuhrman SA, Miettinen H, Kaspar LL, Mellman I. 1990. *Toxoplasma gondii*: fusion competence of parasitophorous vacuoles in Fc receptor-transfected fibroblasts. *Science* 249:641–46
Jorgensen EM, Hartwieg E, Schuske K, Nonet ML, Jin Y, Horvitz HR. 1995. Defective recycling of synaptic vesicles in synaptotagmin mutants of *Caenorhabditis elegans*. *Nature* 378:196–99
Kappeler F, Itin C, Schindler R, Hauri HP. 1994. A dual role for COOH-terminal lysine residues in pre-Golgi retention and endocytosis of ERGIC-53. *J. Biol. Chem.* 269:6279–81
Kelly RB. 1993. Storage and release of neurotransmitters. *Cell* 72:43–53
Kielian MC, ZA Cohn 1980. Phagosome-lysosome fusion: characterization of intracellular membrane fusion in mouse macrophages. *J. Cell. Biol.* 85:754–65
Knight A, Hughson E, Hopkins CR, Cutler DF. 1995. Membrane protein trafficking through the common apical endosome compartment of polarized Caco-2 cells. *Mol. Biol. Cell* 6:597–610
Koenig JH, Ikeda K. 1989. Disappearance and reformation of synaptic vesicle membrane upon transmitter release observed under reversible blockage of membrane retrieval. *J. Neurosci.* 9:3844–60
Kornfeld S. 1992. Structure and function of the mannose 6-phosphate/insulin-like growth factor II receptors. *Annu. Rev. Biochem.* 61:307–30
Kornfeld S, Mellman I. 1989. The biogenesis of lysosomes. *Annu. Rev. Cell Biol.* 5:483–525
Koster A, Saftig P, Matzner U, von Figura K, Peters C, Pohlmann R. 1993. Targeted disruption of the M_r 46,000 mannose 6-phosphate receptor gene in mice results in misrouting of lysosomal proteins. *EMBO J.* 12:5219–23
Kubler E, Riezman H. 1993. Actin and fimbrin are required for the internalization step of endocytosis in yeast. *EMBO J.* 12:2855–62
Kuliawat R, Lisanti MP, Arvan P. 1995. Polarized distribution and delivery of plasma membrane proteins in thyroid follicular epithelial cells. *J. Biol. Chem.* 270:2478–82
Ladinsky MS, Howell KE. 1993. An electron microscopic study of TGN38/41 dynamics. *J. Cell. Sci. Suppl.* 17:41–47
Ladinsky MS, Kremer JR, Furcinitti PS, McIntosh JR, Howell KE 1994. HVEM tomography of the *trans*-Golgi network: structural insights and identification of a lace-like vesicle coat. *J. Cell Biol.* 127:29–38
Lamaze C, Schmid SL. 1995. The emergence of clathrin-independent pinocytic pathways.

Curr. Opin. Cell Biol. 7:573–80

Le Borgne R, Griffiths G, Hoflack B. 1996. Mannose-6-phosphate receptors and ADP-ribosylation factors cooperate for high affinity interaction of the AP-1 Golgi assembly proteins with membranes. *J. Biol. Chem.* 271:2162–70

Le Borgne R, Schmidt A, Mauxion F, Griffiths G, Hoflack B. 1993. Binding of AP-1 Golgi adaptors to membranes requires phosphorylated cytoplasmic domains of the mannose 6-phosphate/insulin-like growth factor II receptor. *J. Biol. Chem.* 268:22552–56

Lin HC, Moore MS, Sanan DA, Anderson RG. 1991. Reconstitution of clathrin-coated pit budding from plasma membranes. *J. Cell Biol.* 114:881–91

Link E, McMahon H, Fischer von Mollard G, Yamasaki S, Niemann H, et al. 1993. Cleavage of cellubrevin by tetanus toxin does not affect fusion of early endosomes. *J. Biol. Chem.* 268:18423–26

Lippincott-Schwartz J, Yuan L, Tipper C, Amherdt M, Orci L, Klausner RD. 1991. Brefeldin A's effects on endosomes, lysosomes, and the TGN suggest a general mechanism for regulating organelle structure and membrane traffic. *Cell* 67:601–16

Lisanti MP, Scherer PE, Vidugiriene J, Tang Z, Hermanowski VA, et al. 1994. Characterization of caveolin-rich membrane domains isolated from an endothelial-rich source: implications for human disease. *J. Cell Biol.* 126:111–26

Liu SH, Wong ML, Craik CS, Brodsky FM. 1995. Regulation of clathrin assembly and trimerization defined using recombinant triskelion hubs. *Cell* 83:257–67

Lombardi D, Soldati T, Riederer MA, Goda Y, Zerial M, Pfeffer SR. 1993. Rab9 functions in transport between late endosomes and the *trans* Golgi network. *EMBO J.* 12:677–82

Louvard D, Morris C, Warren G, Stanley K, Winkler F, Reggio H. 1983. A monoclonal antibody to the heavy chain of clathrin. *EMBO J.* 2:1655–64

Lowe M, Kreis TE. 1995. In vitro assembly and disassembly of coatomer. *J. Biol. Chem.* 270:31364–71

Ludwig T, Griffiths G, Hoflack B. 1991. Distribution of newly synthesized lysosomal enzymes in the endocytic pathway of normal rat kidney cells. *J. Cell Biol.* 115:1561–72

Ludwig T, Munier LH, Bauer U, Hollinshead M, Ovitt C, et al. 1994. Differential sorting of lysosomal enzymes in mannose 6-phosphate receptor-deficient fibroblasts. *EMBO J.* 13:3430–37

Ludwig T, Ovitt CE, Bauer U, Hollinshead M, Remmler J, et al. 1993. Targeted disruption of the mouse cation-dependent mannose 6-phosphate receptor results in partial missorting of multiple lysosomal enzymes. *EMBO J.* 12:5225–35

Mahaffey DT, Peeler JS, Brodsky FM, Anderson RG. 1990. Clathrin-coated pits contain an integral membrane protein that binds the AP-2 subunit with high affinity. *J. Biol. Chem.* 265:16514–20

Marsh EW, Leopold PL, Jones NL, Maxfield FR. 1995. Oligomerized transferrin receptors are selectively retained by a lumenal sorting signal in a long-lived endocytic recycling compartment. *J. Cell Biol.* 129:1509–22

Marsh M, Helenius A. 1980. Adsorptive endocytosis of Semliki forest virus. *J. Mol. Biol.* 142:439–54

Marsh M, Griffiths G, Dean GE, Mellman I, Helenius A. 1986. Three-dimensional structure of endosomes in BHK-21 cells. *Proc. Natl. Acad. Sci. USA* 83:2899–903

Martin S, Tellam J, Livingstone C, Slot JW, Gould GW, James DE. 1996. Glut-4 and Vamp-2 are segregated from recycling endosomes in insulin-sensitive cells. *J. Cell Biol.* In press

Matter K, Hunziker W, Mellman I. 1992. Basolateral sorting of LDL receptor in MDCK cells: The cytoplasmic domain contains two tyrosine-dependent determinants. *Cell* 71:741–53

Matter K, Mellman I. 1994. Mechanisms of cell polarity: sorting and transport in epithelial cells. *Curr. Opin. Cell Biol.* 6:545–54

Matter K, Whitney JA, Yamamoto EM, Mellman I. 1993. Common signals control low density lipoprotein receptor sorting in endosomes and the Golgi complex of MDCK cells. *Cell* 74:1053–64

Matter K, Yamamoto EM, Mellman I. 1994. Structural requirements and sequence motifs for polarized sorting and endocytosis of LDL and Fc receptors in MDCK cells. *J. Cell Biol.* 126:991–1004

Maycox PR, Link E, Reetz A, Morris SA, Jahn R. 1992. Clathrin-coated vesicles in nervous tissue are involved primarily in synaptic vesicle recycling. *J. Cell Biol.* 118:1379–88

Mayor S, Presley JF, Maxfield FR. 1993. Sorting of membrane components from endosomes and subsequent recycling to the cell surface occurs by a bulk flow process. *J. Cell Biol.* 121:1257–69

Mayor S, Rothberg KG, Maxfield FR. 1994. Sequestration of GPI-anchored proteins in caveolae triggered by cross-linking. *Science* 264:1948–51

McKanna JA, Haigler HT, Cohen S. 1979. Hormone receptor topology and dynamics: morphological analysis using ferritin-labeled epidermal growth factor. *Proc. Natl. Acad. Sci. USA* 76:5689–93

McPherson PS, Czernik AJ, Chilcote TJ, Onofri F, Benfenati F, et al. 1994. Interaction of Grb2 via its Src homology 3 domains with synaptic proteins including synapsin I. *Proc. Natl. Acad. Sci. USA* 91:6486–90

McPherson PS, Garcia EP, Slepnev VI, David C, Zhang X, et al. 1996. A presynaptic inositol-5-phosphatase. *Nature* 379:353–57

Mellman I. 1996. Molecular sorting of membrane proteins in polarized and non-polarized cells. *Cold Spring Harbor Symp. Quant. Biol.* LX:745–52

Mellman I, Fuchs R, Helenius A. 1986. Acidification of the endocytic and exocytic pathways. *Annu. Rev. Biochem.* 55:663–700

Mellman I, Plutner H. 1984. Internalization and degradation of macrophage Fc receptors bound to polyvalent immune complexes. *J. Cell Biol.* 98:1170–76

Mellman I, Plutner H, Ukkonen P. 1984. Internalization and rapid recycling of macrophage Fc receptors tagged with monovalent antireceptor antibody: possible role of a prelysosomal compartment. *J. Cell Biol.* 98:1163–69

Mellman I, Yamamoto E, Whitney JA, Kim M, Hunziker W, Matter K. 1993. Molecular sorting in polarized and non-polarized cells: common problems, common solutions. *J. Cell Sci. Suppl.* 17:1–7

Mellman IS, Plutner H, Steinman RM, Unkeless JC, Cohn ZA. 1983. Internalization and degradation of macrophage Fc receptors during receptor-mediated phagocytosis. *J. Cell Biol.* 96:887–95

Mellman IS, Steinman RM, Unkeless JC, Cohn ZA. 1980. Selective iodination and polypeptide composition of pinocytic vesicles. *J. Cell Biol.* 86:712–22

Meresse S, Hoflack B. 1993. Phosphorylation of the cation-independent mannose 6-phosphate receptor is closely associated with its exit from the *trans*-Golgi network. *J. Cell Biol.* 120:67–75

Miesenbock G, Rothman JE. 1995. The capacity to retrieve escaped ER proteins extends to the *trans*-most cisterna of the Golgi stack. *J. Cell Biol.* 129:309–19

Miettinen HM, Matter K, Hunziker W, Rose JK, Mellman I. 1992. Fc receptors contain a cytoplasmic domain determinant that actively regulates coated pit localization. *J. Cell Biol.* 118:875–88

Miettinen HM, Rose JK, Mellman I. 1989. Fc receptor isoforms exhibit distinct abilities for coated pit localization as a result of cytoplasmic domain heterogeneity. *Cell* 58:317–27

Milici AJ, Watrous NE, Stukenbrok H, Palade GE. 1987. Transcytosis of albumin in capillary endothelium. *J. Cell Biol.* 105:2603–12

Miller K, Shipman M, Trowbridge IS, Hopkins CR. 1991. Transferrin receptors promote the formation of clathrin lattices. *Cell* 65:621–32

Monlauzeur L, Rajasekaran A, Chao M, Rodriguez BE, Le BA. 1995. A cytoplasmic tyrosine is essential for the basolateral localization of mutants of the human nerve growth factor receptor in Madin-Darby canine kidney cells. *J. Biol. Chem.* 270:12219–25

Mostov KE. 1994. Transepithelial transport of immunoglobulins. *Annu. Rev. Immunol.* 12:63–84

Muller WA, Steinman RM, Cohn ZA. 1983. Membrane proteins of the vacuolar system. III. Further studies on the composition and recycling of endocytic vacuole membrane in cultured macrophages. *J. Cell Biol.* 96:29–36

Mundigl O, De Camilli P. 1994. Formation of synaptic vesicles. *Curr. Opin. Cell Biol.* 6:561–67

Müsch A, Xu H, Shields D, Rodriguez-Boulan E. 1996. Transport of vesicular stomatitis virus G protein to the cell surface is signal mediated in polarized and non-polarized cells. *J. Cell Biol.* 133:543–58

Neher E, Penner R. 1994. Mice sans synaptotagmin. *Nature* 372:316–17

Neutra MR, Ciechanover A, Owen LS, Lodish HF. 1985. Intracellular transport of transferrin- and asialoorosomucoid-colloidal gold conjugates to lysosomes after receptor-mediated endocytosis. *J. Histochem. Cytochem.* 33:1134–44

Newman LS, McKeever MO, Okano HJ, Darnell RB. 1995. Beta-NAP, a cerebellar degeneration antigen, is a neuron-specific vesicle coat protein. *Cell* 82:773–83

Novak KD, Peterson MD, Reedy MC, Titus MA. 1995. *Dictyostelium* myosin I double mutants exhibit conditional defects in pinocytosis. *J. Cell Biol.* 131:1205–21

Nuoffer C, Balch WE. 1994. GTPases: multifunctional molecular switches regulating vesicular traffic. *Annu. Rev. Biochem.* 63:949–90

O'Halloran TJ, Anderson RG. 1992. Clathrin heavy chain is required for pinocytosis, the presence of large vacuoles, and development in *Dictyostelium. J. Cell Biol.* 118:1371–77

Ohno H, Stewart J, Fournier MC, Bosshart H, Rhee I, et al. 1995. Interaction of tyrosine-based sorting signals with clathrin-associated proteins. *Science* 269:1872–75

Opresko LK, Chang CP, Will BH, Burke PM, Gill GN, Wiley HS. 1995. Endocytosis and lysosomal targeting of epidermal growth factor receptors are mediated by distinct sequences independent of the tyrosine kinase domain. *J. Biol. Chem.* 270:4325–33

Orci L, Palmer DJ, Amherdt M, Rothman JE. 1993a. Coated vesicle assembly in the Golgi requires only coatomer and ARF proteins from the cytosol. *Nature* 364:732–34

Orci L, Palmer DJ, Ravazzola M, Perrelet A, Amherdt M, Rothman JE. 1993b. Budding from Golgi membranes requires the coatomer complex of non-clathrin coat proteins. *Nature* 362:648–52

Orci L, Tagaya M, Amherdt M, Perrelet A, Donaldson JG, Lippincott SJ, et al. 1991. Brefeldin A, a drug that blocks secretion, prevents the assembly of non-clathrin-coated buds on Golgi cisternae. *Cell* 64:1183–95

Pace J, Hayman MJ, Gálan JE. 1993. Signal transduction and invasion of epithelial cells by S. typhimurium. *Cell* 72:505–14

Page LJ, Robinson MS. 1995. Targeting signals and subunit interactions in coated vesicle adaptor complexes. *J. Cell Biol.* 131:619–30

Parton RG. 1996. Caveolae and caveolins. *Curr. Opin. Cell Biol.* In press

Parton RG, Joggerst B, Simons K. 1994. Regulated internalization of caveolae. *J. Cell Biol.* 127:1199–215

Payne GS, Hasson TB, Hasson MS, Schekman R. 1987. Genetic and biochemical characterization of clathrin-deficient *Saccharomyces cerevisiae. Mol. Cell. Biol.* 7:3888–98

Payne GS, Schekman R. 1985. A test of clathrin function in protein secretion and cell growth. *Science* 230:1009–14

Pearse BM. 1987. Clathrin and coated vesicles. *EMBO J.* 6:2507–12

Pearse BMF. 1988. Receptors compete for adaptors found in plasma membrane coated pits. *EMBO J.* 7:3331–36

Pearse BMF, Robinson MS. 1990. Clathrin, adaptors, and sorting. *Annu. Rev. Cell Biol.* 6:151–71

Peeler JS, Donzell WC, Anderson RG. 1993. The appendage domain of the AP-2 subunit is not required for assembly or invagination of clathrin-coated pits. *J. Cell Biol.* 120:47–54

Pelchen MA, Boulet I, Littman DR, Fagard R, Marsh M. 1992. The protein tyrosine kinase p56lck inhibits CD4 endocytosis by preventing entry of CD4 into coated pits. *J. Cell Biol.* 117:279–90

Peters KR, Carley WW, Palade GE. 1985. Endothelial plasmalemmal vesicles have a characteristic striped bipolar surface structure. *J. Cell Biol.* 101:2233–38

Peters PJ, Hsu VW, Ooi CE, Finazzi D, Teal SB, et al. 1995. Overexpression of wild-type and mutant ARF1 and ARF6: distinct perturbations of nonoverlapping membrane compartments. *J. Cell Biol.* 128:1003–17

Pevsner J, Volknandt W, Wong BR, Scheller RH. 1994. Two rat homologs of clathrin-associated adaptor proteins. *Gene* 146:279–83

Pfeffer SR. 1994. Rab GTPases: master regulators of membrane trafficking. *Curr. Opin. Cell Biol.* 6:522–26

Pfeifer JD, Wick MJ, Harding CV, Normark SJ. 1993. Processing of defined T-cell epitopes after phagocytosis of intact bacteria by macrophages. *Infect. Agents Dis.* 2:249–54

Piper RC, Tai C, Kulesza P, Pang S, Warnock D, et al. 1993. GLUT-4 NH2 terminus contains a phenylalanine-based targeting motif that regulates intracellular sequestration. *J. Cell Biol.* 121:1221–32

Pohlmann R, Boeker MW, von Figura K. 1995. The two mannose 6-phosphate receptors transport distinct complements of lysosomal proteins. *J. Biol. Chem.* 270:27311–18

Ponnambalam S, Robinson MS, Jackson AP, Peiperl L, Parham P. 1990. Conservation and diversity in families of coated vesicle adaptins. *J. Biol. Chem.* 265:4814–20

Poo WJ, Conrad L, Janeway CJ. 1988. Receptor-directed focusing of lymphokine release by helper T cells. *Nature* 332:378–80

Poodry CA. 1990. *shibire*, a neurogenic mutant of *Drosophila. Dev. Biol.* 138:464–72

Prill V, Lehmann L, von FK, Peters C. 1993. The cytoplasmic tail of lysosomal acid phosphatase contains overlapping but distinct signals for basolateral sorting and rapid internalization in polarized MDCK cells. *EMBO J.* 12:2181–93

Rabinowitz S, Horstmann H, Gordon S, Griffiths G. 1992. Immunocytochemical characterization of the endocytic and phagolysosomal compartments in peritoneal macrophages. *J. Cell Biol.* 116:95–112

Racoosin EL, Swanson JA. 1992. M-CSF-induced macropinocytosis increases solute endocytosis but not receptor-mediated endocytosis in mouse macrophages. *J. Cell. Sci.* 102:867–80

Racoosin EL, Swanson JA. 1993. Macropinosome maturation and fusion with tubular lysosomes in macrophages. *J. Cell Biol.* 121:1011–20

Rajasekaran AK, Humphrey JS, Wagner M, Miesenbock G, Le BA, et al. 1994. TGN38 recycles basolaterally in polarized Madin-Darby canine kidney cells. *Mol. Biol. Cell* 5:1093–103

Raths S, Rohrer J, Crausaz F, Riezman H. 1993. *end3* and *end4*: two mutants defective in receptor-mediated and fluid-phase endocytosis in *Saccharomyces cerevisiae. J. Cell Biol.* 120:55–65

Reaves B, Banting G. 1992. Perturbation of the morphology of the *trans*-Golgi network following Brefeldin A treatment: redistribution of a TGN-specific integral membrane protein, TGN38. *J. Cell Biol.* 116:85–94

Reaves B, Banting G. 1994. Vacuolar ATPase inactivation blocks recycling to the *trans*-Golgi network from the plasma membrane. *FEBS Lett.* 345:61–66

Reaves B, Horn M, Banting G. 1993. TGN38/41 recycles between the cell surface and the TGN: Brefeldin A affects its rate of return to the TGN. *Mol. Biol. Cell* 4:93–105

Ridley AJ, Paterson HF, Johnston CL, Diekmann D, Hall A. 1992. The small GTP-binding protein rac regulates growth factor-induced membrane ruffling. *Cell* 70:401–10

Rijnboutt S, Jansen G, Posthuma G, Hynes JB, Schornagel JH, Strous GJ. 1996. Endocytosis of GPI-linked membrane folate receptor-alpha. *J. Cell Biol.* 132:35–47

Ritter TE, Fajardo O, Matsue H, Anderson RG, Lacey SW. 1995. Folate receptors targeted to clathrin-coated pits cannot regulate vitamin uptake. *Proc. Natl. Acad. Sci. USA* 92:3824–28

Robinson MK, Pearse BM F. 1986. Immunofluorescent localization of 100K coated vesicle proteins. *J. Cell Biol* 102:48–54

Robinson MS. 1993. Assembly and targeting of adaptin chimeras in transfected cells. *J. Cell Biol.* 123:67–77

Robinson MS. 1994. The role of clathrin, adaptors and dynamin in endocytosis. *Curr. Opin. Cell Biol.* 6:538–44

Robinson MS, Kreis TE. 1992. Recruitment of coat proteins onto Golgi membranes in intact and permeabilized cells: effects of brefeldin A and G protein activators. *Cell* 69:129–38

Rothberg KG, Heuser JE, Donzell WC, Ying YS, Glenney JR, Anderson RG. 1992. Caveolin, a protein component of caveolae membrane coats. *Cell* 68:673–82

Rothman JE. 1994. Mechanisms of intracellular protein transport. *Nature* 372:55–63

Sandvig K, van Deurs B. 1991. Endocytosis without clathrin (a minireview). *Cell Biol. Int. Rep.* 15:3–8

Scheiffele P, Peranen J, Simons K. 1995. N-glycans as apical sorting signals in epithelial cells. *Nature* 378:96–98

Schekman R, Payne G. 1988. Clathrin: a matter of life or death? *Science* 239:919

Schimmoller F, Riezman H. 1993. Involvement of Ypt7p, a small GTPase, in traffic from late endosome to the vacuole in yeast. *J. Cell Sci.* 106:823–30

Schmid S, Fuchs R, Kielian M, Helenius A, Mellman I. 1989. Acidification of endosome subpopulations in wild-type Chinese hamster ovary cells and temperature-sensitive acidification-defective mutants. *J. Cell Biol.* 108:1291–300

Schmid SL, Fuchs R, Måle P, Mellman I. 1988. Two distinct subpopulations of endosomes involved in membrane recycling and transport to lysosomes. *Cell* 52:73–83

Schroder S, Schimmoller F, Singer-Kruger B, Riezman H. 1995. The Golgi-localization of yeast Emp47p depends on its di-lysine motif but is not affected by the ret1–1 mutation in alpha-COP. *J. Cell Biol.* 131:895–912

Schu PV, Takegawa K, Fry MJ, Stack JH, Waterfield MD, Emr SD 1993. Phosphatidylinositol 3-kinase encoded by yeast VPS34 gene essential for protein sorting. *Science* 260:88–91

Seaman MN, Ball CL, Robinson MS. 1993. Targeting and mistargeting of plasma membrane adaptors in vitro. *J. Cell Biol.* 123:1093–105

Seeger M, Payne GS. 1992. Selective and immediate effects of clathrin heavy chain mutations on Golgi membrane protein retention in *Saccharomyces cerevisiae. J. Cell Biol.* 118:531–40

Sheff D, Lowe M, Kreis TE, Mellman I. 1996. Biochemical heterogeneity and phosphorylation of coatomer subunits. *J. Biol. Chem.* 271:7230–36

Shepherd PR, Reaves BJ, Davidson HW. 1996. Phosphinositide 3-kinases and membrane traffic. *Trends Cell Biol.* 6:92–97

Silverstein SC, Steinman RM, Cohn ZA. 1977. Endocytosis. *Annu. Rev. Biochem.* 46:669–722

Simons K, Zerial M. 1993. Rab proteins and the road maps for intracellular transport. *Neuron* 11:789–99

Simpson F, Bright NA, West MA, Newman LS, Darnell RB, Robinson MS. 1996. A novel adaptor-related protein complex. *J. Cell Biol.* In press

Singer KB, Frank R, Crausaz F, Riezman H. 1993. Partial purification and characterization of early and late endosomes from yeast. Identification of four novel proteins. *J. Biol. Chem.* 268:14376–86

Singer KB, Stenmark H, Zerial M. 1995. Yeast Ypt51p and mammalian Rab5: counterparts with similar function in the early endocytic pathway. *J. Cell Sci.* 108:3509–21

Smart EJ, Ying YS, Conrad PA, Anderson RG. 1994. Caveolin moves from caveolae to the Golgi apparatus in response to cholesterol oxidation. *J. Cell Biol.* 127:1185–97

Smythe E, Redelmeier TE, Schmid SL. 1992. Receptor-mediated endocytosis in semiintact cells. *Meth. Enzymol.* 219:223–34

Sorkin A, Carpenter G. 1993. Interaction of activated EGF receptors with coated pit adaptins. *Science* 261:612–15

Sorkin A, Helin K, Waters CM, Carpenter G,

Beguinot L. 1992. Multiple autophosphorylation sites of the epidermal growth factor receptor are essential for receptor kinase activity and internalization. Contrasting significance of tyrosine 992 in the native and truncated receptors. *J. Biol. Chem.* 267:8672–78

Sorkin A, McKinsey T, Shih W, Kirchhausen T, Carpenter G. 1995. Stoichiometric interaction of the epidermal growth factor receptor with the clathrin-associated protein complex AP-2. *J. Biol. Chem.* 270:619–25

Stamnes MA, Rothman JE. 1993. The binding of AP-1 clathrin adaptor particles to Golgi membranes requires ADP-ribosylation factor, a small GTP-binding protein. *Cell* 73:999–1005

Steinman RM, Brodie SE, Cohn ZA. 1976. Membrane flow during pinocytosis. *J. Cell Biol.* 68:665–87

Steinman RM, Mellman IS, Muller WA, Cohn ZA. 1983. Endocytosis and the recycling of plasma membrane. *J. Cell Biol.* 96:1–27

Steinman RM, Swanson J. 1995. The endocytic activity of dendritic cells. *J. Exp. Med.* 182:283–88

Stoorvogel W, Geuze HJ, Strous GJ. 1987. Sorting of endocytosed transferrin and asialoglycoprotein occurs immediately after internalization in HepG2 cells. *J. Cell Biol.* 104:1261–68

Stoorvogel W, Oorschot V, Geuze HJ. 1996. A novel class of clathrin-coated vesicles budding from endosomes. *J. Cell Biol.* 132:21–33

Stowers L, Yelon D, Berg LJ, Chant J. 1995. Regulation of the polarization of T cells toward antigen-presenting cells by Ras-related GTPase CDC42. *Proc. Natl. Acad. Sci. USA* 92:5027–31

Subtil A, Hemar A, Dautry VA. 1994. Rapid endocytosis of interleukin 2 receptors when clathrin-coated pit endocytosis is inhibited. *J. Cell Sci.* 107:3461–68

Takei K, Mundigl O, Daniell L, De Camilli P. 1996. The synaptic vesicle cycle: a single vesicle budding step involving clathrin and dynamin. *J. Cell Biol.* In press

Takel K, McPherson PS, Schmid SL, De CP. 1995. Tubular membrane invaginations coated by dynamin rings are induced by GTP-gamma S in nerve terminals. *Nature* 374:186–90

Tardieux I, Webster P, Ravesloot J, Boron W, Lunn JA, et al. 1992. Lysosome recruitment and fusion are early events required for trypanosome invasion of mammalian cells. *Cell* 71:1117–30

Tooze J, Hollinshead M. 1991. Tubular early endosomal networks in AtT20 and other cells. *J. Cell Biol.* 115:635–53

Traub LM, Kornfeld S, Ungewickell E. 1995. Different domains of the AP-1 adaptor complex are required for Golgi membrane binding and clathrin recruitment. *J. Biol. Chem.* 270:4933–42

Traub LM, Ostrom JA, Kornfeld S. 1993. Biochemical dissection of AP-1 recruitment onto Golgi membranes. *J. Cell Biol.* 123:561–73

Trowbridge IS, Collawn JF, Hopkins CR. 1993. Signal-dependent membrane protein trafficking in the endocytic pathway. *Annu. Rev. Cell Biol.* 9:129–61

Tulp A, Verwoerd D, Dobbestein B, Ploegh HL, Pieters J. 1994. Isolation and characterization of the intracellular MHC class II compartment. *Nature* 369:120–26

Ullrich B, Li C, Zhang JZ, McMahon H, Anderson RG, et al. 1994. Functional properties of multiple synaptotagmins in brain. *Neuron* 13:1281–91

van der Bliek A, Meyerowitz EM. 1991. Dynamin-like protein encoded by the *Drosophila shibire* gene associated with vesicular traffic. *Nature* 351:411–14

van der Bliek A, Redelmeier TE, Damke H, Tisdale EJ, Meyerowitz EM, Schmid SL. 1993. Mutations in human dynamin block an intermediate stage in coated vesicle formation. *J. Cell Biol.* 122:553–63

van der Sluijs P, Hull M, Webster P, Măle P, Goud B, Mellman I. 1992. The small GTP-binding protein rab4 controls an early sorting event on the endocytic pathway. *Cell* 70:729–40

van der Sluijs P, Hull M, Zahraoui A, Tavitian A, Goud B, Mellman I. 1991. The small GTP-binding protein rab4 is associated with early endosomes. *Proc. Natl. Acad. Sci. USA* 88:6313–17

Verhey KJ, Birnbaum MJ. 1994. A leu-leu sequence is esential for a COOH terminal targeting signal of GLUT4 glucose transporter in fibroblasts. *J. Biol. Chem.* 269:2353–56

Verhey KJ, Hausdorff SF, Birnbaum MJ. 1993. Identification of the carboxyterminus as important for isoform specific subcellular targeting of glucose transporter molecules. *J. Cell Biol.* 123:137–48

Verhey KJ, Yeh JI, Birnbaum MJ. 1995. Distinct signals in the GLUT4 glucose transporter for internalization and for targeting to an insulin-responsive compartment. *J. Cell Biol.* 130:1071–79

Wang LH, Rothberg KG, Anderson RG. 1993. Mis-assembly of clathrin lattices on endosomes reveals a regulatory switch for coated pit formation. *J. Cell Biol.* 123:1107–17

Werb Z, Cohn ZA. 1972. Plasma membrane synthesis in the macrophage following phagocytosis of polystyrene latex particles. *J.*

Biol. Chem. 247:2439–46

Whitney JA, Gomez M, Sheff D, Kreis TE, Mellman I. 1995. Cytoplasmic coat proteins involved in endosome function. *Cell* 83:703–13

Wilde A, Dempsey C, Banting G. 1994. The tyrosine-containing internalization motif in the cytoplasmic domain of TGN38/41 lies within a nascent helix. *J. Biol. Chem.* 269:7131–36

Wilde A, Reaves B, Banting G. 1992. Epitope mapping of two isoforms of a *trans* Golgi network specific integral membrane protein TGN38/41. *FEBS Lett.* 313:235–38

Wilson JM, Whitney JA, Neutra MR. 1991. Biogenesis of the apical endosome-lysosome complex during differentiation of absorptive epithelial cells in rat ileum. *J. Cell Sci.* 100:133–43

Wolf PR, Ploegh HL. 1995. How MHC class II molecules acquire peptide cargo: biosynthesis and trafficking through the endocytic pathway. *Annu. Rev. Cell Dev. Biol.* 11:267–306

Wood SA, Park JE, Brown WJ. 1991. Brefeldin A causes a microtubule-mediated fusion of the *trans*-Golgi network and early endosomes. *Cell* 67:591–600

Wright SD, Silverstein SC. 1983. Receptors for C3b and C3bi promote phagocytosis but not the release of toxic oxygen from human phagocytes. *J. Exp. Med.* 158:2016–23

Yamashiro D, Tycko B, Fluss S, Maxfield FR. 1984. Segregation of transferrin to a mildly acidic (pH 6.5) para-Golgi compartment in the recycling pathway. *Cell* 37:789–800

Ying YS, Anderson RG, Rothberg KG. 1992. Each caveola contains multiple glycosyl-phosphatidylinositol-anchored membrane proteins. *Cold Spring Harbor Symp. Quant. Biol.* 57:593–604

Zanolari B, Raths S, Singer KB, Riezman H. 1992. Yeast pheromone receptor endocytosis and hyperphosphorylation are independent of G protein-mediated signal transduction. *Cell* 71:755–63

Zhang JZ, Davletov BA, Sudhof TC, Anderson RG. 1994. Synaptotagmin I is a high affinity receptor for clathrin AP-2: implications for membrane recycling. *Cell* 78:751–60

Annu. Rev. Cell Dev. Biol. 1996. 12:627–61

VIRUS-CELL AND CELL-CELL FUSION[1]

L. D. Hernandez[*#], *L. R. Hoffman*[#], *T. G. Wolfsberg*[*#†], *and J. M. White*[*]
[*]Department of Cell Biology, University of Virginia, Charlottesville, Virginia 22908;
[#]Department of Biochemistry and Biophysics, University of California, San Francisco, California 94143

KEY WORDS: membrane fusion, viral glycoprotein, hemagglutinin, myoblast, ADAM

ABSTRACT

Significant progress has been made in elucidating the mechanisms of viral membrane fusion proteins; both those that function at low, as well as those that function at neutral, pH. For many viral fusion proteins evidence now suggests that a triggered conformational change that exposes a previously cryptic fusion peptide, along with a rearrangement of the fusion protein oligomer, allows the fusion peptide to gain access to the target bilayer and thus initiate the fusion reaction. Although the topologically equivalent process of cell-cell fusion is less well understood, several cell surface proteins, including members of the newly described ADAM gene family, have emerged as candidate adhesion/fusion proteins.

CONTENTS

[†]Current address: National Center for Biotechnology Information, NIH, Bethesda, Maryland 20894.

INTRODUCTION

Membrane fusion events are either exoplasmic or endoplasmic depending on whether the bilayer leaflets that make first contact face the extracellular or cytoplasmic space, respectively (White 1992, Rothman & Warren 1994). This review focuses on exoplasmic fusion events and in particular on virus-cell and cell-cell fusion. Virus-cell fusion events constitute the first step in the infectious process of all enveloped animal viruses, including human pathogens such as influenza and the human immunodeficiency virus (HIV); they are mediated by specific virally encoded membrane fusion glycoproteins. Cell-cell fusion reactions are critical in development; we have proposed that they are mediated by proteins similar to viral fusion proteins. Here we explore the mechanisms of several viral fusion glycoproteins and describe some proteins that have been implicated in cell-cell fusion reactions.

VIRUS-CELL FUSION

Enveloped viruses employ two routes of entry into host cells (Figure 1). One route involves binding to cell surface receptors, endocytosis through clathrin-coated pits and vesicles, delivery to endosomes, and finally fusion with the endosomal membrane induced by the low pH of this compartment (Figure 1*A*). The other route is conceptually simpler: Following binding to cell surface receptors, the virus fuses directly with the plasma membrane at neutral pH (Figure 1*B*). With a few notable exceptions, members of a given virus family use one or the other route of entry into target cells (Table 1).

Table 1 Fusion proteins of enveloped animal viruses

Family	Example	pH of fusion	Fusion protein complex	Fusion peptide
Orthomyxovirus	Influenza	Low	HA1/HA2	Yes*
Togavirus	SFV	Low	E1/E2/E3	Yes*
Flavivirus	TBE	Low	E	Yes*
Rhabdovirus	VSV, rabies	Low	G	Yes*
Bunyavirus	Uukuniemi	Low	G1/G2	Nd
Arenavirus	LCMV	Low	G1/G2	Yes
Retrovirus	HIV	Neutral	SU/TM	Yes*
Paramyxovirus	Sendai	Neutral	F1/F2	Yes*
Herpesvirus	HSV-1	Neutral	gB,gD,gH/L	Nd
Coronavirus	MHV	Neutral	S1/S2	Nd
Hepadnavirus	Hepatitis B	Neutral	S	Yes
Poxvirus	Vaccinia	Neutral	14 kd, 56 kd	Yes
Filovirus	Ebola	Nd	GP	Nd
Iridovirus	ASFV	Nd	Nd	Nd

In column 4, for the cases of multisubunit fusion complexes, the underlined subunit is the one implicated in fusion. In column 5, * indicates that a fusion peptide has been confirmed or strongly predicted (see text for details). ASFV = African Swine Fever virus. Other abbreviations of virus names are given in the text. Nd = not determined.

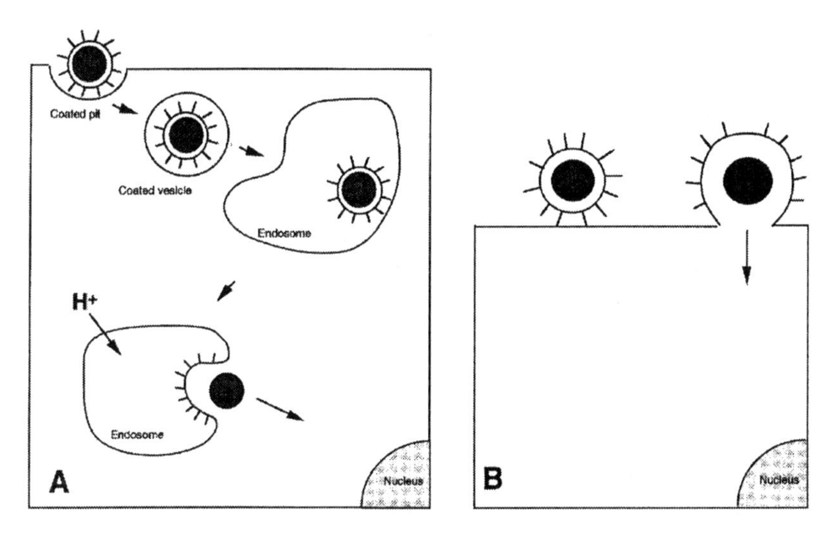

Figure 1 Two routes of enveloped virus entry into cells. (*A*) Endocytosis and fusion in endosomes. (*B*) Fusion at the plasma membrane. The dark sphere denotes the viral nucleocapsid.

VIRAL FUSION GLYCOPROTEINS

Fusion glycoproteins have been identified for most enveloped animal viruses (Table 1; Bentz 1993). A first indication that a glycoprotein is involved in fusion can come from antibody inhibition studies, identification of a sequence with characteristics of a fusion peptide, or mapping of mutations that affect fusion. Confirmation of a fusion protein involves transfecting cDNA encoding the candidate into tissue culture cells and demonstrating fusion activity of the resulting transfectants (or reconstituted liposomes containing the candidate protein). Most viral fusion glycoproteins share a set of ten features (Figures 2, 3). All that have been studied are (*a*) composed of one or two type I integral membrane glycoproteins, (*b*) have the bulk of their mass external to the viral membrane, (*c*) contain N-linked carbohydrates, (*d*) form higher-order oligomers, (*e*) are present on the viral membrane at high surface density and, importantly, (*f*) contain a fusion peptide in a membrane-anchored subunit. Many

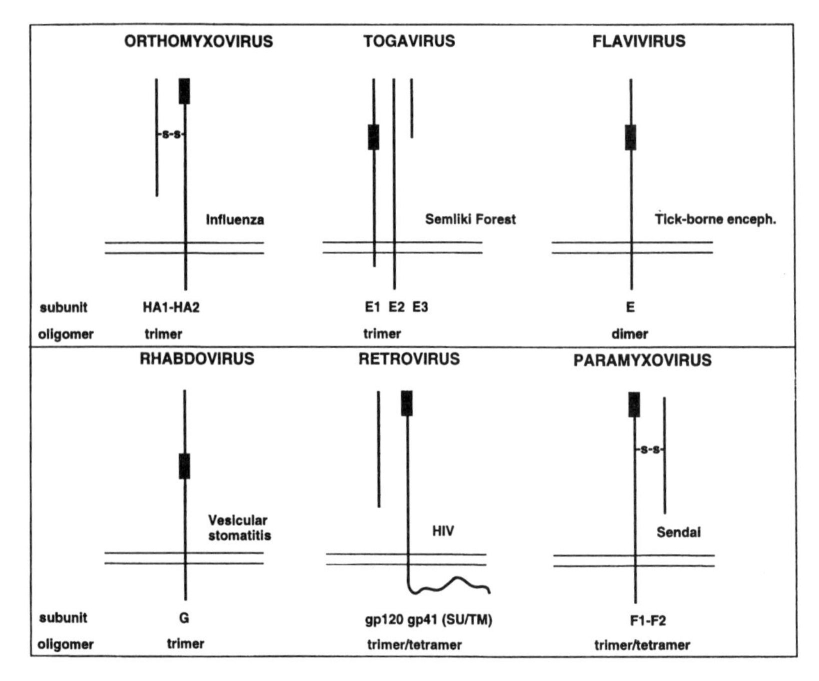

Figure 2 Well-characterized viral fusion proteins. Fusion peptides have been confirmed or strongly predicted for representatives of these six families of enveloped viruses (see text for details). Black rectangle indicates the fusion peptide. Subunits connected by a dash are disulfide bonded. Trimer/tetramer indicates that the oligomer status of the fusion protein is still under consideration.

viral fusion glycoproteins (*g*) are tight complexes of two glycoprotein subunits that (*h*) confer binding as well as fusion activity, and many (*i*) are made as larger precursors that (*j*) require proteolytic processing to potentiate their fusion activity. The most important common denominator among these proteins is the fusion peptide (Figures 2, 3). Fusion peptides are short, relatively hydrophobic sequences conserved within, but not between, virus families. They generally can be modeled as α-helices with one very hydrophobic face, but they may interact with membranes in either an α-helical or a β-sheet structure (Lear & DeGrado 1987, Nieva et al 1994b, Gray et al 1996).

In evaluating the mechanism of a viral fusion protein it is necessary to know what fusion assay has been employed in a given experimental setting. For example, all enveloped viruses fuse with host cell membranes to initiate an infection (Figure 1). Under similar pH conditions, most, but not all, also induce cell-cell fusion, i.e. the formation of polykaryons (also called syncytia). The ability to promote syncytia can depend on the sequence of the fusion peptide (Gething et al 1986), sequences elsewhere in the fusion protein, sequences of internal viral proteins (see below), and the density of the viral fusion glycoprotein in the cell membrane (Gething et al 1986). Hence, the requirements for syncytia formation are more stringent than those for fusion with a small target membrane. In terms of fusion per se, it is important to consider what step of the fusion reaction has been assayed: hydrophobic binding to the target membrane, hemifusion, or fusion pore formation or dilation.

LOW PH-ACTIVATED VIRAL FUSION PROTEINS

Orthomyxo-, toga-, flavi-, rhabdo-, bunya-, and arenaviruses fuse with host cells at low pH. Between and within these virus families, the exact pH profiles for fusion vary, but they generally fall in the pH 4.8 to 6.5 range and resemble titration curves. Low pH is required to elicit a fusion-inducing conformational change. The neutral pH structures of two viral fusion glycoproteins have been determined by X-ray crystallography: the influenza hemagglutinin (HA) and the E glycoprotein of the flavivirus, tick-borne encephalitis (TBE) virus (Figures 3*A*,*B*, respectively) (Wilson et al 1981, Rey et al 1995). In addition, neutral pH structures of the spikes of two togaviruses—Semliki Forest virus (SFV) and Ross river virus—have been gleaned from high-resolution electron microscopic image reconstruction (Cheng et al 1995, Fuller et al 1995, Helenius 1995). High-resolution images are also available for the low pH-induced forms of the influenza HA (Figure 4*B*; Bullough et al 1994) and the SFV spike glycoprotein complex (Fuller et al 1995) after exposure to low pH. From these images, it is clear that different viral fusion glycoproteins possess quite distinct native (pre-fusogenic) structures and that the dynamics of their fusion-inducing conformational changes vary considerably. Nonetheless, we propose that following the fusion-inducing conformational change, subsequent steps in the fusion process are similar.

THE INFLUENZA HEMAGGLUTININ

The influenza HA is the only fusion protein for which atomic resolution structures are available for both its neutral (pre-fusogenic) and its final low-pH form (Figure 4*A*, *B*) (Wilson et al 1981, Bullough et al 1994). Furthermore, much has been learned about individual steps in the HA-mediated fusion reaction (for recent reviews, see Stegmann 1994, Gaudin et al 1995, Hughson 1995, White 1995). We first review what has been learned about conformational changes in HA and its target membrane-binding properties. We then describe our current working model for HA-mediated membrane fusion. Unless stated, the discussion refers to HA from the X:31 strain of influenza, which is the strain whose HA has been subjected to structure analysis. There are important differences in the conformational changes of HAs from different virus strains (Tsurudome et al 1992, Gutman et al 1993, Pak et al 1994).

Fusion-Inducing Conformational Changes in the Influenza HA

The influenza HA is a trimer that projects 130 Å from the target membrane (Figure 3*A*). Each monomer is composed of HA1, a globular receptor-binding subunit, and HA2, a fibrous subunit that houses the fusion peptide. The two subunits

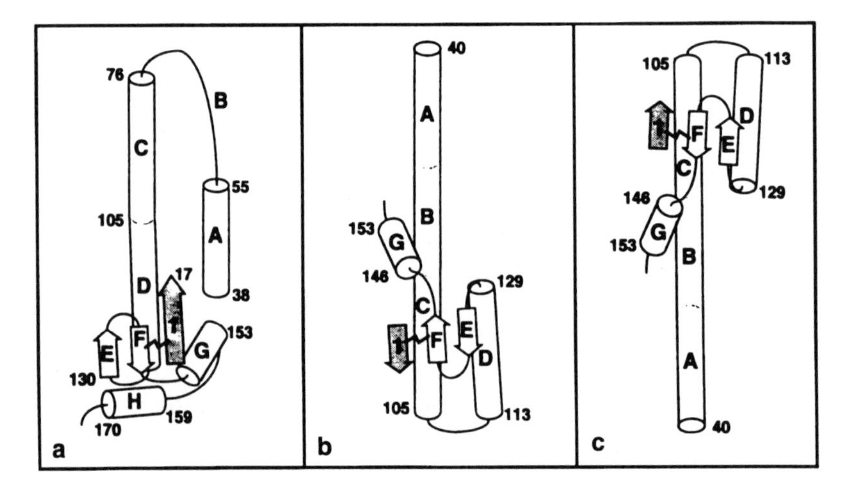

Figure 4 Structure of the HA2 stem of the influenza HA at (*a*) neutral pH and in (*b*) TBHA2, the low-pH structure. In (*c*) the TBHA2 structure has been rotated 180°. In *a* and *c* the viral membrane would be at the bottom of the image. Modified with permission of DC Wiley (Bullough et al 1994).

arise by a proteolytic cleavage event that is essential for fusion activity. In the neutral pH structure, the three fusion peptides, one from each monomer, are located ≈ 100 Å from the target membrane where they are tucked in the trimer interface by a network of hydrogen bonds and lie approximately parallel to the plane of the virus membrane (Wiley & Skehel 1987). In response to low pH, HA changes conformation, exposes its fusion peptides, and binds hydrophobically to target membranes. Many conformational changes, including exposure of the fusion peptides and separation of the globular head domains, were illuminated using biochemical, biophysical, and immunological probes (Doms 1993, Gaudin et al 1995). None, however, foreshadowed the massive rearrangements in HA2 that were revealed by the X-ray structure of a derivative of low-pH HA called TBHA2 (Bullough et al 1994). TBHA2 is made by exposing the bromelain-released HA ectodomain (BHA) to low pH, and then treating it with trypsin to cleave off the globular head domains and, finally, with thermolysin to digest the fusion peptides. The resulting structure (Figure 4*B*) revealed three major changes in the HA2 stem. First, a prominent loop, designated B, has been transformed into an α-helix, thereby connecting the existing helical regions A and C and, presumably, propelling the N-terminal fusion peptide upward toward the target membrane. Second, residues 106–112 convert from a helical form to a loop, thereby causing the rest of the D helix to be flipped 180° and to run antiparallel to the C helix. The third notable feature is that ≈ 20 C-terminal amino acids, which are present in the TBHA2 protein, cannot be seen in the X-ray structure, indicating that they are in a disordered, perhaps extended, conformation. Although the B loop-to-helix transition had been proposed from structure prediction (Ward & Dopheide 1980, Carr & Kim 1993) and synthetic peptide (Carr & Kim 1993) studies, the latter two changes had not been predicted.

A critical question is how the structure reflected by TBHA2, the final low pH and most stable form of HA (Ruigrok et al 1988), participates in fusion. Two studies considered how TBHA2 is oriented vis-à-vis the virus membrane, which in the fused state is united with the target membrane. If X:31 influenza is pretreated at low pH in the absence of a target membrane, it becomes irreversibly inactivated for fusion (White 1995). In other words, it has prematurely undergone its conformational change and is driven into a fusion-incompetent form. In the first study, ^{125}I-labeled photoactivatable phospholipds were incorporated into influenza viral membranes. The virus particles were then exposed to low pH, and the HA isolated and analyzed for radioactive labeling. Label was found only in the fusion peptide. Hence in the fusion-inactive form, the fusion peptide is in the virus membrane, as is its transmembrane domain (Weber et al 1994). If we position TBHA2 so as to account for this finding, it would be oriented as in Figure 4*C* (i.e. 180° rotation of the illustration in Figure 4*B*). Confirmation of this orientation has come from an electron microscopic study. Influenza

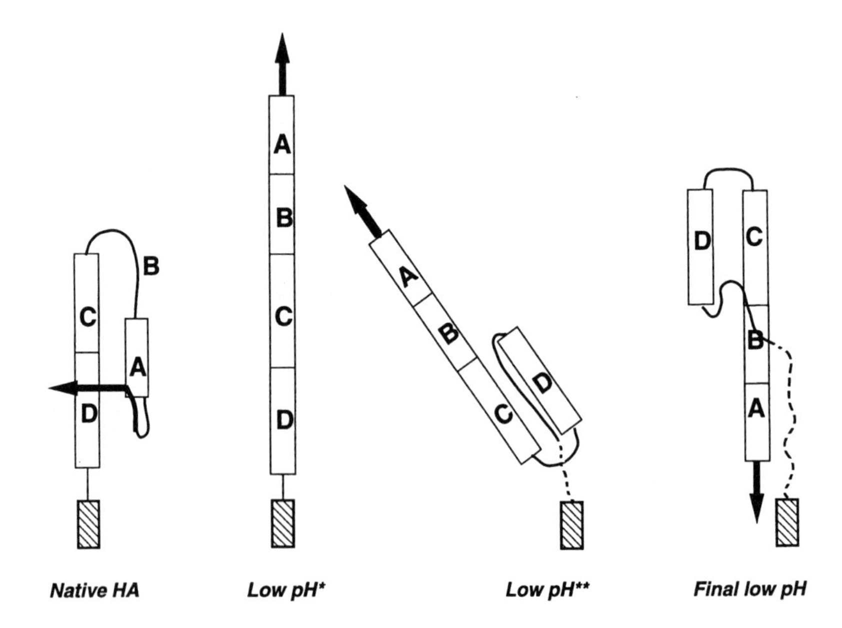

Figure 5 Proposed conformational intermediates in the low-pH-induced conformational change in the HA2 subunit of the influenza HA. Two intermediates (low pH* and low pH**) are proposed to exist between the native neutral pH pre-fusogenic form (*far left*) and its final low-pH form (*far right*). Black arrow denotes the fusion peptide. Hatched box denotes the transmembrane domain. Dotted lines denote the disordered C-terminal residues in TBHA2. A monomer is shown, but the structure remains trimeric.

HA-containing virosomes were pretreated at low pH and then reacted with an antibody that recognizes residues surrounding HA2 107 (i.e. the region that undergoes the helix-to-loop transition). The antibody bound to the membrane distal region of HA, 110 Å from the virosome membrane (Wharton et al 1995).

If at the end of the fusion reaction HA is in a conformation equivalent to the inactive form (Figure 4*C*), we know the structure and orientation of HA (vis-à-vis the bilayers) in its pre- and post-fusogenic forms (Figure 5, far left and far right images, respectively). In considering how HA facilitates fusion, we hypothesize (*a*) that there are, minimally, two conformational intermediates (termed low pH* and low pH**) between native HA and its final low-pH form; (*b*) that low pH* and low pH** resemble structures as diagrammed in Figure 5 (middle two images); and (*c*) that low pH*, low pH**, and the final low-pH form mediate different steps in the fusion reaction.

Target Membrane-Binding Properties of the Influenza HA

Functionally, the most significant low-pH-induced transition in HA is the conversion of its ectodomain from a hydrophilic to a hydrophobic entity. Early studies showed that at low pH, BHA (the HA ectodomain) binds quickly to preformed liposomes of varying lipid compositions (White 1995). Subsequent studies, using target liposomes containing radioactively labeled photoactivatable cross-linking reagents, revealed that the fusion peptide was the only segment of HA labeled (Harter et al 1989). Hence, at low pH, HA binds to target membranes through its fusion peptide.

The exact manner in which the HA fusion peptide associates with target membranes poses some pressing questions. What is the secondary structure of the peptide in the target membrane? Does the fusion peptide assume different structures at different stages of interaction with the target membrane? How deeply does the fusion peptide delve into the target bilayer (at different stages of the fusion reaction)? At what angle does the fusion peptide engage the target membrane (at different stages of the fusion reaction)? Is the interaction of the fusion peptide with the target membrane complete before, during, or at the end of the lag phase that precedes the onset of fusion (see below)? Studies with synthetic fusion peptides have shed some light on these questions. The earliest circular dichroism data indicated that a synthetic HA fusion peptide exists as a random coil in aqueous solution, but assumes a helical conformation at a membrane surface (Lear & DeGrado 1987). Subsequent studies (reviewed in Gray et al 1996) confirmed this general conclusion and noted a rough correlation between the ability of a synthetic fusion peptide to adopt a helical configuration and its ability, in the context of the intact wild-type or mutant HA protein, to promote membrane fusion. A recent study has shown that at the surface of neutral liposomes, the wild-type HA fusion peptide is in $\approx 40\%$ α-helical and $\approx 30\%$ β-sheet structure. Non-fusogenic peptides exhibited lower α-helix to β-sheet ratios (Gray et al 1996).

A Model for HA-Mediated Membrane Fusion

Our working model for HA-mediated fusion is depicted in Figure 6. A major idea embodied in the model is that conformational changes in the HA protein (Figures 4, 5) drive the conformational transitions required to fuse two membranes, i.e. to bring them into close proximity and to induce the bending motions necessary to mediate hemifusion and the opening and subsequent dilation of the fusion pore (Siegel 1993, Helm & Israelachvili 1993, Monck & Fernandez 1994, Chernomordik et al 1995). Before addressing the individual five steps of the model, we consider three basic concepts that have impacted on this and related models for HA-mediated fusion: (*a*) the fusion pore, (*b*) the hemifusion intermediate, and (*c*) the cooperativity of the fusion process.

The concept of the fusion pore was derived from studies of regulated exocytosis and the application of electrophysiological techniques to the process of HA-mediated fusion (Monck & Fernandez 1994, Lindau & Almers 1995). Fibroblasts expressing HA have been patch-clamped and monitored for electrophysiological changes during low-pH-induced fusion with red blood cells

(Spruce et al 1989, 1991). Alternatively, recent experiments have monitored the electrical events associated with fusion of HA-expressing fibroblasts to planar bilayers (Melikyan et al 1993a, 1995a). Similar results have been obtained in both systems: The initial fusion pore is narrow (estimated to be $\approx$ 1–2 nm in diameter), it can flicker (open and close repeatedly), and it can arrest at semi-stable conductances prior to opening (dilating) completely.

The concept of the hemifusion intermediate came from theoretical considerations and from studies in model membrane systems. The first evidence that hemifusion might occur during a protein-mediated fusion event was the observation that when HA was engineered to be anchored in cell membranes via a glycosylphosphatidylinositol (GPI) tail instead of its wild-type transmembrane domain, it induced hemifusion (Figure 6, step 3) but not full fusion (Kemble et al 1994, Melikyan et al 1995b).

The concept of HA cooperativity came from several lines of indirect evidence (Doms & Helenius 1986, Morris et al 1989, Ellens et al 1990) and from the

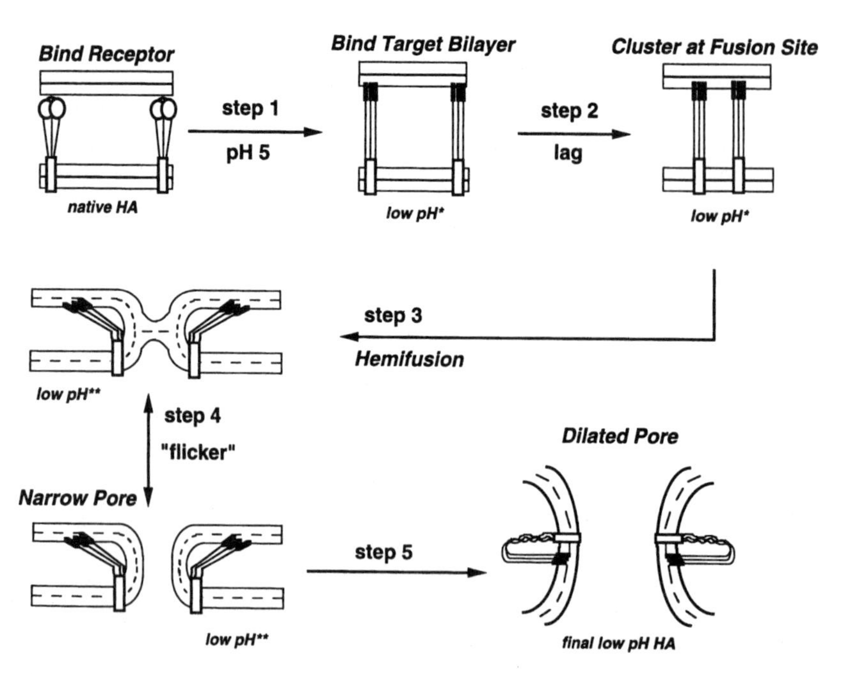

Figure 6 A model for HA-mediated membrane fusion (see text for details). The fusion peptides are denoted as thickened black lines. They are not seen in the first image because they are buried in the native structure. The globular head domains (*circles*) are not seen after the first image because they are proposed to dissociate from the rest of HA prior to target membrane binding.

appeal of models of fusion pores lined by several protein oligomers (Lindau & Almers 1995). The results of a recent study have provided strong support for this notion. Several investigators have observed a lag phase between the application of low pH and the onset of lipid mixing during HA-mediated fusion (Stegmann et al 1990, Clague et al 1991). A recent study (Danieli et al 1996) compared the fusion profiles of cell lines that express HA at different defined surface densities. It was found that the lag time preceding the onset of lipid mixing varied sigmoidally with HA surface density. A Hill plot analysis of the data suggested that a minimum of three or four HA trimers is required to initiate a fusion event. Based on these and other data, a model was proposed in which cooperativity arises at the level of interaction of HA trimers with the target membrane at the nascent fusion site. We now describe individual steps in the fusion model (Figure 6).

STEP 1: CONFORMATIONAL CHANGES AND TARGET MEMBRANE BINDING We propose that step 1 is mediated by a structure resembling low pH* (Figure 5). In this scenario, the spring-loaded conformational change of the B-loop (Carr & Kim 1993, Bullough et al 1994) leads to the formation of an extended coiled-coil and drives the initial interaction of the fusion peptides with the target membrane. Separation of the globular head domains is considered necessary for this step based on the phenotype of a mutant in which these domains were linked by engineered disulfide bonds (Godley et al 1992, Kemble et al 1992). A structure such as low pH* would not, however, be likely to mediate subsequent steps of the fusion reaction because it would still, presumably, keep the two fusing bilayers at a substantial distance from one another.

STEP 2: CLUSTERING OF HA TRIMERS AT THE FUSION SITE Step 2 encompasses the lag phase preceding the onset of lipid mixing, and is proposed to involve the clustering, orientation (vis-à-vis one another), and target membrane binding of (at least) three or four HA trimers. It is further proposed, based largely on the phenotypes of HA fusion peptide mutants (Gething et al 1986, Guy et al 1992, Steinhauer et al 1995, Gray et al 1996; H Qiao & J White, manuscript in preparation), that during this stage there are qualitative and/or quantitative changes in the mode in which the fusion peptide interacts with the target membrane, and that these interactions are cooperative with respect to HA trimers. Establishment of the fusion site may involve further conformational changes in HA (Stegmann et al 1990), as well as changes in its rotational and lateral mobility (Junankar & Cherry 1986, Gutman et al 1993).

STEP 3: HEMIFUSION We propose that in step 3 the HA trimers surrounding the nascent fusion site bend. Bending may be facilitated by a transition to a structure resembling low pH**. If, as we presume, the HA remains firmly tethered to the target membrane via the fusion peptides and firmly tethered to the viral membrane via its transmembrane domains, then if several trimers bend radially outward from the fusion site, they would force the interacting bilayers to bend with them. Such concerted bending action would drive the interacting bilayers into the hemifusion intermediate. Support for the notion that the HA trimers may tilt at the fusion site has come from recent experiments with HA and planar bilayers (Tatulian et al 1995). If such bending of HA trimers brings the membranes into proximity, then there would be no need to invoke an alternative, likely energetically unfavorable (Ruigrok et al 1988), model to achieve membrane-membrane proximity based on the splaying of the top of the A/B/C helix (Yu et al 1994). It is worth noting that at this stage of the reaction, as diagrammed in the hemifusion intermediate, the fusion peptides could insert into the target membrane at an oblique angle (Brasseur et al 1990). Although it is not known how lipids are arranged in the hemifusion junction, recent data favor a lipid stalk rather than an inverted micelle configuration (Siegel 1993, Stegmann 1993).

STEP 4: OPENING OF THE FUSION PORE It is proposed in step 4 that HA imposes tension on the hemifusion diaphragm that eventually causes the hemifusion diaphragm to break. Based on theoretical considerations, it has been proposed that tension is a major driving force for fusion (Nanavati et al 1992). We further propose that rupture of the hemifusion diaphragm requires a transmembrane anchor that spans the viral membrane (Kemble et al 1994, Melikyan et al 1995b). It is not known whether step 4 requires further conformational changes in HA. The observed flickering of the early HA fusion pore (Spruce et al 1989, 1991, Melikyan et al 1993b) suggests that step 4 is reversible.

STEP 5: DILATION OF THE FUSION PORE It is not known what drives the irreversible opening (and subsequent dilation) of the fusion pore. If step 4 is reversible, then we further reason that an additional irreversible conformational change in HA would be needed to drive step 5, the final stage of the fusion reaction. It is tantalizing that this final irreversible step may be driven by the conversion of HA into its final low-pH form, i.e. that which is most stable (Figure 5, *right*) (Ruigrok et al 1988) and is reflected by TBHA2 (Figure 4*B,C*) (Bullough et al 1994). In this scenario, the proposed extension of the C-terminal 20 amino acids of HA2 (which may account for their disorder in the TBHA2 structure) would allow the HA to sweep through the toroid of membrane in the initial fusion pore, thereby relaxing the tension and high bilayer curvature imposed on the membranes in the hemifusion diaphragm and initial fusion pore.

In this manner, upon completion of the fusion reaction, the fusion peptide would reside in the same (the fused) membrane as the transmembrane domain, as it is in its postfusion conformation (Weber et al 1994).

Two additional observations must be accommodated in any model for HA-mediated fusion. The first is that lipid flow is restricted in the initial fusion pore (Tse et al 1993, Zimmerberg et al 1994, Blumenthal et al 1995). Based largely on the properties of GPI-anchored HA, it has been proposed that lipid motion is restricted by the cluster of transmembrane domains in the early fusion pore that presumably acts as a membrane barrier (Kemble et al 1994, Zimmerberg et al 1994, Melikyan et al 1995b). The HAs would then have to move apart from one another, and the membrane curvature would have to be relaxed (i.e. as depicted in Figure 6, step 5) so as to allow unimpeded lipid flow. The second observation is that small aqueous contents move through the fusion pore before large ones (Zimmerberg et al 1994). This could be the result of either progressive widening of the pore or coalescence of several smaller pores.

TOGAVIRUS AND FLAVIVIRUS GLYCOPROTEINS

The togavirus fusion glycoprotein is composed of two type I integral membrane glycoproteins (Figure 2). For some togaviruses (e.g. those of Sindbis and rubella viruses), the spike is composed exclusively of the two integral membrane proteins (E1 and E2). For others (e.g. SFV), the spike contains a third nonmembrane anchored subunit, E3, that arises by proteolytic cleavage of p62, the precursor of E3 and E2. Among the togaviruses, low pH-dependent fusion activity has been most extensively characterized for SFV (Kielian 1995) and thus we focus on the SFV fusion protein E1/E2/E3.

There are several notable features about SFV-mediated membrane fusion. It is fast and efficient and occurs with a pH threshold of $\approx$6.2 to 6.5. As with influenza, SFV is capable of fusing with pure phospholipid vesicles devoid of any proteinaceous receptor structure, and fusion activity is lost if virions are pretreated at low pH. In sharp contrast to influenza, however, there are two critical lipid requirements for SFV fusion. First, SFV fusion requires that the target membranes contain $\approx$30 mol% cholesterol, with the 3 β-hydroxyl group of the sterol being critical for activity (Kielian 1995). Second, SFV fusion occurs optimally with $\approx$1 mol% sphinogomyelin in the target membrane (Nieva et al 1994a); other spingolipids will substitute but with stringent stereochemical restrictions (Moesby et al 1995). Although the molecular basis for the specific lipid requirements is not yet known, it is likely to involve specific interactions between the fusogenic subunit (see below) and the target membrane.

Within the E1/E2/E3 spike glycoprotein complex, E2 likely confers receptor binding activity. E1 has been implicated as the fusogenic subunit because anti-E1 antibodies block fusion; virions largely devoid of E2 support fusion, and E1 houses the likely fusion peptide, a hydrophobic sequence located at an internal position in E1 (Figure 2). The assignment of the candidate fusion peptide was made based largely on its hydrophobicity and on its conservation among the alphavirus genus of togaviruses (Kielian 1995). Recent mutagenesis studies lend credence to this assignment. In particular, a spike glycoprotein with a specific mutation in the candidate fusion peptide, Gly 91 to Asp, assembles properly and undergoes certain low pH-dependent conformational changes, yet is incapable of fusion (Levy-Mintz & Kielian 1991).

In thinking about how the SFV spike mediates fusion, we adhere to a model similar to that depicted in Figure 6: In response to low pH there is a major conformational change that unleashes and repositions the fusion peptide so that it can bind hydrophobically to the target membrane. Subsequently, several conformationally altered trimers associate to establish a fusion site. Further conformational changes lead to hemifusion, the opening of a narrow fusion pore, and finally dilation of the pore. Support for the possibility that such a model may apply to SFV-mediated fusion comes from the following lines of evidence: Both E1 and E2 change conformation at low pH (Wahlberg et al 1992, Wahlberg & Garoff 1992, Bron et al 1993, Justman et al 1993). The receptor-binding subunit E2 dissociates from the fusion subunit E1. The E1 ectodomain binds to target membranes in a reaction that requires cholesterol and sphingomyelin (Klimjack et al 1994). There is a lag phase between treatment with low pH and the onset of lipid mixing (Bron et al 1993). A fusion pore has been detected by electrophysiological techniques (Lanzrein et al 1993) and, lastly, rapid high-resolution studies indicate that there is at least one functional low-pH intermediate in the SFV glycoprotein before formation of its final low-pH form (Fuller et al 1995).

Although a model akin to Figure 6 appears plausible for SFV, it is already clear that there are important mechanistic differences between how HA and the SFV spike cause fusion. The structures of native HA and the native SFV spike are clearly different. Whereas HA has a strongly predicted coiled-coil heptad repeat region (as do retro- and paramyxovirus fusion glycoproteins), the SFV glycoprotein complex does not (Kielian 1995). The dynamics of the low-pH-induced conformational change in the SFV spike are very different from those for HA (see below). The minimum number of SFV spikes needed to initiate fusion may differ, and the SFV fusion pore may differ from the HA fusion pore in terms of its propensity to flicker (no flickering has yet been observed for the SFV pore), its initial conductance, and its residence at semi-stable states. In addition, any model for SFV-induced fusion must account for its specific lipid requirements.

The conformational changes that ensue when the SFV spike glycoprotein is exposed to low pH are equally as dramatic as those that occur in HA, but quite different. In the case of SFV, the spike oligomer is completely reorganized. At neutral pH, the glycoproteins of SFV form a stable dimer of E1 and E2 in sucrose gradients but exist as a trimer of E1/E2 heterodimers on the surface of the virion, with E3 noncovalently attached (Fuller et al 1995). A similar arrangement, a trimer of E1/E2 heterodimers, has been seen for the glycoproteins of Ross river virus (Cheng et al 1995). Biochemical studies have revealed that when SFV is exposed to low pH, E2 dissociates from E1 and an E1 homotrimer forms (Wahlberg et al 1992, Wahlberg & Garoff 1992, Justman et al 1993). High-resolution cryoelectron microscopic studies have revealed that within the first 50 ms following exposure to low pH (Fuller et al 1995) the E1/E2 dimer contacts loosen by swiveling about one another and that E2 begins to dissociate from E1. The cavity at the base of the dimer narrows, and the E1 subunit extends ≈ 20 Å, presumably to position the fusion peptide close to the target membrane. It is inferred that the stable E1 homotrimer forms later. It is thus possible that the 50-ms low-pH structure of the SFV spike is functionally akin to the putative intermediate we depict as low pH* for HA (Figures 5, 6). Similarly, the final stable E1 homotrimer may be likened to the final low-pH form of HA represented by TBHA2. In this manner, the 50-ms structure may drive step 1 of the fusion process—initial interaction with the target membrane; as with HA, separation of the receptor binding subunit (E2) from the fusion subunit (E1) is a necessary prerequisite for step 1. Similarly, formation of the stable E1 homotrimer may drive later stages of the fusion process, perhaps the opening or dilation of the fusion pore.

The combined neutral and low-pH structural data on HA and the SFV spike indicate that (at least these two) viral fusion glycoproteins are sitting on the viral surface in a metastable state. Exposure to low pH drives them into their most thermostable configurations. We propose that intermediates along the transition pathway between the native (metastable) and the final (most thermostable) low-pH forms mediate discrete steps of the fusion process and, in so doing, lower the energy barrier to fusion. Existing data suggest that similar rearrangements likely occur in other togaviruses (Katow & Sugiura 1988, Anthony & Brown 1991, Meyer et al 1992, Paredes et al 1993, Cheng et al 1995).

Flaviviruses are spherical, enveloped viruses similar in size and genome organization to togaviruses. They encode a major transmembrane glycoprotein, E, and a minor glycoprotein, M, which is made as a larger precursor. Fusion of the flavivirus tick-borne encephalitis (TBE) virus occurs at low pH, requires cleavage of the pre M protein, and is mediated by the E glycoprotein, which is evidenced by the ability of monoclonal antibodies against E to inhibit fusion and by the presence of a conserved hydrophobic sequence likely to be a fusion peptide in the E glycoprotein (Figures 2, 3) (Guirakhoo et al 1991). It is not known whether the M protein is required for fusion. The structure of the ectodomain of the E glycoprotein of TBE at neutral pH has recently been solved by X-ray crystallography (Rey et al 1995). From this work, it has been inferred that E lies low and flat along the surface of the virion as an antiparallel homodimer. Biochemical and immunological studies have indicated that the E glycoprotein changes conformation at low pH (Kimura & Ohyama 1988, Roehrig et al 1990, Guirakhoo et al 1991) and, in a manner reminiscent of the SFV spike, becomes reorganized into an E homotrimer (Allison et al 1995). Based on these collective observations and analogies with the elevation of the fusion peptides of HA and SFV, it has been proposed that in response to low pH, the E glycoprotein projects up from the viral membrane and reassociates into a fusogenic homotrimer (Rey et al 1995). Upward lifting of the spike would position the presumed fusion peptide (Figures 2, 3) near the target membrane. This proposed transition is not likely to involve formation of a coiled-coil because sequence analysis of the E glycoprotein does not reveal any regions with coiled-coil propensity (L Hoffman, unpublished observations). Less is known about the next steps in flavivirus fusion. We consider it reasonable, however, that following the fusion-activating conformational change, subsequent steps in fusion (i.e. hemifusion, fusion pore formation, and fusion pore dilation) will occur via a model similar to Figure 6.

OTHER LOW-PH-ACTIVATED VIRAL FUSION PROTEINS

Rhabdoviruses such as vesicular stomatitis virus and rabies virus are bullet-shaped enveloped viruses, and their single glycoprotein, G, is responsible for fusion. Rhabdovirus fusion only occurs at low pH (≈ 6), can occur with pure phospholipid vesicle (i.e. lacking any proteinaceous structures and with no observed specific lipid requirement), and is preceded by a lag phase (Clague et al 1990, Gaudin et al 1993, 1995). Native G is a trimer, but the trimer is further stabilized at low pH (Zagouras & Rose 1993). In response to low pH, G has been observed to undergo conformational changes and to move to the ends of the bullet-shaped particles (Brown et al 1988, Gaudin et al 1993, 1995). At least two sets of conformational changes have been observed. A first set occurs at $\approx$ pH 6.7 and renders the ectodomain hydrophobic. Additional conformational changes presumably account for the lower pH threshold of fusion (Gaudin et al 1995).

High-resolution structural information is not available for G, but advances have been made in defining its fusion peptide. This has been a challenging task because the hydrophobicity plot of G is noisy and because there is not an obvious hydrophobic region that is conserved between VSV and rabies G glycoproteins. (The fact that VSV and rabies are members of different genera of the rhabdovirus family may account for this.) Site-specific mutagenesis work from two laboratories has focused attention on residues 118-139 in VSV G as a likely fusion peptide (Zhang & Ghosh 1994, Fredericksen & Whitt 1995). The conclusion of the mutagenesis studies was recently supported by photoaffinity phospholipid labeling of a CNBr fragment encompassing this region, residues 59-221 (Durrer et al 1995). Although future experiments are necessary to define the labeled region more precisely, it appears that region 118-139 in VSV G is the likely fusion peptide. Our understanding of the fusion mechanism of rhabdovirus G proteins is less developed than is that for the low-pH-activated fusion proteins discussed above. Nevertheless, themes of an initial low-pH-induced conformational change that renders the ectodomain hydrophobic (step 1, Figure 6) and subsequent conformational changes that lead to the formation of a stable trimer reverberate from the other systems. Hence it is plausible that a model akin to Figure 6 will apply to rhabdovirus fusion proteins.

Although less detailed, features of the fusion process of arena- and bunyavirus glycoproteins are beginning to emerge. Arenaviruses synthesize one type I integral membrane glycoprotein, GC, that is processed to G2, a membrane-anchored subunit, and G1, a peripherally associated subunit. Biochemical studies suggest that GP1 and GP2 are further associated as a tetramer and that GP1 forms the head and GP2 the stalk of the spike glycoprotein. Fusion of the arenavirus lymphocytic choriomeningitis virus with pure phospholipid vesicles occurs rapidly at pH < 6.3, and under these conditions, GP1 dissociates from virions but with slower kinetics. As with some strains of influenza, SFV, flaviviruses, and rhabdoviruses, the arenavirus is also inactivated by pretreatment at low pH. These data have been taken in support of a model in which there is an initial low-pH-induced conformational change that activates the virion for fusion and a later change that inactivates it for fusion (Di Simone & Buchmeier 1995). As we have reasoned above, the early and late changes may mediate different stages of fusion. Although the fusion peptide of arenavirus has not been assigned based on mutagenesis and phospholipid analogue labeling studies, a candidate fusion peptide has been suggested (Glushakova et al 1992).

Bunyaviruses contain two type I integral membrane glycoproteins, G1 and G2. They fuse at low pH (≈ 6.3) and conformational changes in the spike glycoproteins have been detected at low pH (Hacker et al 1995, Pekosz et al 1995). Although earlier studies had suggested that G1 and G2 form a heterodimer on the virion surface, a recent study suggests that the Uukuniemi glycoproteins exist as separate homodimers (i.e. a dimer of G1 and a dimer of G2) (Ronka et al 1995). It has not been established unequivocally whether higher-order (hetero-)oligomers form during fusion. Although the G1 glycoprotein has been implicated in virus binding and entry, the precise roles of G1 and G2 remain to be clarified.

NEUTRAL PH VIRAL FUSION PROTEINS

Retro-, paramyxo-, herpes-, corona-, hepadna-, and poxviruses do not require an acidic environment to fuse with host cells (Table 1). These viruses therefore are thought to be are able to fuse directly at the plasma membrane at neutral pH (Figure 1*B*), although this need not preclude fusion in endosomes as well. Compared with what is known about low-pH-dependent viral fusion proteins, less is known about the activation of viral fusion proteins that function at neutral pH. One observation is that viruses that fuse at neutral pH appear to require host cell factors (receptors) in the target membrane. Because of this, we and others have hypothesized that an interaction between the viral fusion protein and its host cell receptor (and/or accessory factors) triggers its fusion-activating conformational change. The transition to a fusogenic state is envisioned to include exposure and repositioning of a previously buried fusion peptide whose interaction with the target membrane would initiate fusion.

RETROVIRAL FUSION PROTEINS

All retroviruses have one envelope glycoprotein (Env) that is responsible for fusion. Env proteins are synthesized as precursors that require proteolytic processing into a surface (SU) and a transmembrane (TM) subunit to manifest fusion activity. The SU subunit has determinants for receptor binding, whereas the TM subunit anchors the protein in the virus membrane and has a hydrophobic domain at or near its N-terminus believed to be the fusion peptide. In some cases (e.g. HIV), cleavage of Env into SU and TM also appears to be important for efficient incorporation into virus particles (Dubay et al 1995).

The Avian Leukosis and Sarcoma Virus Env Glycoprotein

The avian leukosis and sarcoma viruses (ALSV) are grouped based on their host range and interference patterns into five major subgroups designated A through E. The SU and TM subunits remain associated through a disulfide bond and exist as trimers on both the viral and cell surface (Einfeld & Hunter 1988). ALSV is unique among the retroviruses in that its putative fusion peptide is internal to the TM N-terminus. Unlike many retroviruses, ALSV does not induce syncytia except in highly transformed foci (Purchase & Okazaki 1964), but the molecular basis for this inability is not yet clear.

The host cell receptor for ALSV subgroup A, TVA, appears to be necessary and sufficient to elicit fusion (Bates et al 1993). TVA is a small transmembrane protein whose extracellular domain consists of a single low-density lipoprotein receptor repeat motif followed by a short linker region. TVA binds specifically to the subgroup A envelope glycoprotein (Env-A) (Gilbert et al 1994).

Fusion of ALSV with chicken embryo fibroblasts occurs only at temperatures greater than 20°C and at a slow rate (Gilbert et al 1990). It has been shown that both TVA and sTVA (a soluble form of TVA) induce conformational changes in Env-A that correlate with activation of its fusion function. Following interaction of sTVA with Env-A at temperatures greater than 20°C, the SU subunit becomes sensitive to thermolysin (Gilbert et al 1995). Interaction between sTVA and API (a soluble oligomeric form of the Env-A ectodomain) at temperatures greater than 20°C also elicits a receptor-induced conformational change in the envelope glycoprotein. Moreover, it has recently been shown that sTVA binding at fusion permissive temperatures renders the Env-A ectodomain hydrophobic, as manifested by its ability to bind liposomes. Liposome binding is via the TM subunit and appears to involve the putative fusion peptide because mutants in this region have impaired liposome-binding capability (L Hernandez & J White, unpublished data). Thus receptor binding induces conformational changes in Env-A similar to those seen in low-pH activated fusion proteins.

HIV Env Glycoprotein

The HIV envelope glycoprotein is most likely a trimer or a tetramer of SU and TM (Earl et al 1990, Weiss et al 1990). The SU (gp120) and TM (gp41) subunits remain associated only through noncovalent interactions permitting the observed shedding of gp120. gp120 is responsible for binding to CD4, the primary HIV receptor (Weiss 1992). gp41 is responsible for fusion activity with both the amino-terminal fusion peptide and the transmembrane domain playing important roles (Freed et al 1990, Owens et al 1994).

CD4 is a single, membrane-spanning T-cell surface glycoprotein that normally functions in initiating T-cell activation. CD4 induces a number of conformational changes in HIV Env that have been detected with antibody and protease probes (Moore et al 1990, Sattentau & Moore 1991, Clements et al 1991, Hart et al 1991, Sattentau et al 1993); some of these conformational changes are temperature dependent, as is HIV fusion (Frey et al 1995), whereas others are not. Binding of CD4 to Env also induces dissociation of gp120, particularly in laboratory strains. It has been proposed that dissociation of gp120 helps to expose the fusion peptide in gp41.

As in several other viral fusion proteins, the fusion peptide of HIV gp41 is followed by two sequences containing a 4-3 hydrophobic repeat predicted to form a coiled-coil (Delwart et al 1990). Based on studies of a bacterially expressed form of gp41, it has been proposed that gp41 forms a stable α-helical trimer of heterodimers, with the first predicted coiled-coil region forming an interior, parallel homotrimer, and the second packed around the exterior in an antiparallel fashion (i.e. similar to the arrangement of the C and D helices in TBHA2) (Figure 4*B*) (Lu et al 1995). Proline substitutions in the amino-terminal heptad repeat region abolish fusion (Dubay et al 1992a, Cao et al 1993, Chen et al 1993), and synthetic peptides corresponding to the coiled-coil region inhibit fusion (Wild et al 1994a, b). Taken together, it appears that formation of the coiled-coil structure is critical for HIV Env-mediated fusion. However, the stage at which the TBHA2-like coiled-coil structure is required for HIV Env fusion has not been analyzed.

Although sufficient to mediate HIV binding, human CD4 is not sufficient to support HIV entry into nonprimate cells. It has been proposed that an accessory factor(s) must exist to facilitate fusion. The search for an accessory factor has been long and arduous. Very recently, however, compelling evidence has shown that a new member of the family of G-protein coupled transmembrane receptors, termed fusin, fulfills this function for T lymphocytes (Feng et al 1996). How fusin functions as a fusion promoter remains to be clarified. A possibility is that fusin, or another molecule that it regulates, induces a critical conformational change required for fusion.

CD4 and fusin may not be the only receptors that support HIV fusion because HIV is able to infect human cells that have no detectable levels of CD4. It has therefore been postulated that alternate receptors for HIV exist that are responsible for CD4-independent entry. Galactosylceramide (GalCer), a glycosphingolipid, has been shown to bind gp120 (Long et al 1994), and the level of GalCer expression on colonic epithelial cell lines is associated with permissiveness to HIV-1 infection (Fantini et al 1993). Whether GalCer binding is sufficient to trigger conformational changes in HIV Env that would lead to fusion remains to be determined. A number of other alternative receptors for HIV have also been reported (Weiss 1992).

The cytoplasmic tails of the Env glycoproteins of HIV and other lentiviruses in the retrovirus family are long (150-200 amino acids in length) in comparison with those of other viral fusion proteins (Figure 2). Mutations in the tail can affect the conformation of the Env ectodomain (Spies et al 1994), syncytia formation (Mulligan et al 1992), infection kinetics (LaBranche et al 1994), and incorporation of Env into virions (Dubay et al 1992b). Truncations of the SIV tail have resulted in increased cell surface expression and cell-cell fusion (Johnston et al 1993, Zingler & Littman 1993). A variant of SIVmac, with a mutation of a tyrosine to a cysteine at position 723 of ENV, was recently shown to have enhanced surface expression and correspondingly enhanced cell-cell fusion activity caused by the disruption of a tyrosine-based internalization signal (Sauter et al 1996). A similar internalization signal is conserved in HIV and may be important in HIV pathogenesis.

The Murine Leukemia Virus Env Glycoprotein

The murine leukemia viruses (MLV) are divided into five subgroups (ecotropic, amphotropic, polytropic, xenotropic, and 101A) based on their ability to infect rodent and other mammalian cells. The determinants for host range have been mapped to the N-terminal one third of the SU subunit, whereas the N-terminal region of the TM subunit contains the fusion peptide critical in mediating fusion (Jones & Risser 1993). The ecotropic receptor protein (MCAT-1) is a multimembrane-spanning, cationic amino acid transporter. The amphotropic receptor is a phosphate transporter protein related to the gibbon ape leukemia virus receptor (Kavanaugh et al 1994, Wimmer 1994). Expression of MCAT-1 in a variety of cells is sufficient for infection, suggesting that additional host factors are not required for fusion.

Amphotropic MLV and 101A are thought to enter cells by direct fusion with the plasma membrane. Ecotropic MLV appears to be able to produce syncytia in transformed XC and 3T3/DTras cell lines at neutral pH but requires low pH in order to enter these and other cell lines (McClure et al 1990, Wilson et al 1992). As a group, MLVs may use alternate routes of viral entry (Figure 1) into different cell lines.

At the time of virion budding, the cytoplasmic tail of MLV Env is cleaved by a viral protease releasing a 16-amino acid C-terminal peptide called the R peptide. Mutations that block this cleavage decrease virus infectivity, whereas truncation of the peptide by genetic engineering increases fusion activity (Ragheb & Anderson 1994, Rein et al 1994). The R peptide has recently been shown to inhibit fusion when placed on the tail of the SIV envelope glycoprotein (Yang & Compans 1996).

PARAMYXOVIRUS FUSION PROTEINS

All paramyxoviruses contain at least two envelope glycoproteins, an attachment protein (HN, H, or G) and a fusion protein (F). The F protein, which is proteolytically processed to F1 and F2, mediates fusion and has structural similarities to the retrovirus Env and orthomyxovirus fusion proteins (Figure 2). It is most likely a trimer. The fusion peptide is located at the amino-terminus of F1 and is highly conserved among the paramyxoviruses (for review, see Lamb 1993). Some paramyxoviruses (mumps, SV5, RSV) have an additional small hydrophobic envelope protein (SH). SH enhances fusion of RSV but not of SV5 (Heminway et al 1994).

The F protein is absolutely required for fusion, but the role of the HN protein is less clear. For certain paramyxoviruses, e.g. SV5, expression of F alone is sufficient for efficient fusion; for others, coexpression of HN is necessary (Morrison et al 1991, Bousse et al 1994, Bagai & Lamb 1995, Ward et al 1995). In the Sendai virus system, neither influenza HA nor the HN protein from a heterologous paramyxovirus can replace the Sendai HN. These results suggest that there is a specific interaction between the HN and F glycoproteins during fusion. It is postulated that once HN binds to its cellular receptor, it forms an association with F thus recruiting it to contact sites. This association may trigger a conformational change in F, converting it to its fusogenic form. According to this model, the spike glycoproteins of the paramyxoviruses must be rearranged to a new fusogenic oligomer, in a scenario reminiscent of those discussed above for the toga- and flavivirus spike glycoproteins. For viruses such as SV5, which do not require HN for fusion, it has been suggested that their F proteins are hair-triggered, either by contact with the target bilayer or with an unidentified receptor (Lamb 1993).

Similar to retroviral envelope glycoproteins, the F1 glycoproteins of the paramyxoviruses have two heptad repeat regions in their ectodomains (Chambers et al 1990). Heptad repeat A is adjacent to the fusion peptide, whereas heptad repeat B is immediately upstream of the transmembrane domain and has a leucine zipper motif predicted to form a coiled-coil. In Newcastle disease virus (NDV), the introduction of a charged residue in the heptad repeat A region blocked fusion (Sergel-Germano et al 1994). Simultaneous replacement of four of the leucines in the heptad repeat B abolished the fusion activity of measles virus (Buckland et al 1992). Similar mutations in heptad repeat B region of the NDV F protein also abolish fusion without affecting oligomerization (Reitter et al 1995). It is possible that repeats A and B interact in a manner similar to that of analogous heptad repeat regions in HIV Env and the influenza HA.

OTHER NEUTRAL PH VIRAL FUSION PROTEINS

Given the complexity of the envelopes of herpesviruses, it is expected that the fusion mechanism(s) of these viruses also will be complex (Bentz 1993, Spear 1993). Herpes simplex virus (HSV) encodes at least 12 envelope glycoproteins: Of these, gB, gD, gH, and gL are absolutely necessary for infection, and gB, gC, and gD are involved in binding to two sets of viral receptors. gB and gC are believed to be responsible for initial virus attachment (Herold et al 1994), whereas gD and perhaps gB are thought to bind to a secondary virus receptor. gB is absolutely required for infectivity and syncytia formation and is thought to mediate fusion in concert with the gH/L complex (Spear 1993, Roop et al 1993).

HSV-1 rarely causes cell-cell fusion in infected cells. Mutants that do cause extensive syncytia formation (*syn* mutants) have been isolated and mapped to several genetic loci. The best-characterized of these are the *syn1* and *syn3* loci corresponding to the gK and gB genes, respectively. Syncytial mutations in gK are intriguing because this protein does not reach the infected cell surface (Hutchinson et al 1995). *Syn* mutations in the gB gene result in proteins with an altered cytoplasmic tail (Gage et al 1993), which suggests, as described above for retroviral Env proteins, that the tails of fusion glycoproteins are important in regulating fusion and/or syncytia formation. However, it appears that the requirements for virus-cell fusion and cell-cell fusion are different.

Membrane fusion of coronaviruses is mediated by the spike (S) protein, and cell surface expression of S results in extensive syncytia formation (Wimmer 1994). Proteolytic cleavage of S into S1 and S2 is necessary for infectivity and/or cell-cell fusion activity in some strains, whereas it is not required in others. Proteolytic cleavage of S does not position a hydrophobic sequence at the new amino-terminus of S2, as seen in the orthomyxo-, paramyxo- and most retrovirus fusion proteins. S1 is responsible for receptor binding (Cavanagh et al 1986), and S2 appears to be responsible for fusion; several monoclonal antibodies that map to the S2 ectodomain, including a region predicted to form a coiled-coil, block fusion (Daniel et al 1993). Certain variants of mouse hepatitis virus 4 (MHV-4) appear to require low pH to fuse; the requirement for low pH mapped to three amino acid residues in the heptad repeat region (Gallagher et al 1991).

Irreversible conformational changes are induced in MHV by mild alkaline treatment. These changes include release of S1 from virions and viral aggregation possibly by an unmasking of a hydrophobic region on S2 (Sturman et al 1990). Receptor binding might also result in similar structural changes leading to S2-mediated fusion. A number of coronavirus receptors have been identified and this theory can be tested (Wimmer 1994).

Vaccinia virus is the best-studied member of the poxvirus family. Vaccinia has both a single-envelope intracellular form (INV) and a double-envelope extracellular form (EEV). Up to 30 viral envelope-associated proteins have been reported. A 14- and a 56-kDa protein have been implicated in fusion with a putative fusion peptide region located at the N-terminus of the 14-kDa protein (Stern & Dales 1976, Rodriguez et al 1987, Gongetal 1990). Fusion of both the INV and EEV forms do not display increased fusion activity at acid pH (Doms et al 1990).

The best-studied hepadnavirus is hepatitis B (HBV). The HBV envelope contains three envelope glycoproteins L, M, and S. A 180- and a 170-kDa protein present on the surface of duck hepatocytes have been found to bind the duck HBV L protein and may act as receptor or play a role in HBV entry (Kuroki et al 1994, Tong et al 1995). Sequence analysis has revealed a putative fusion peptide at the amino-terminus of S that is highly conserved within the hepadnavirus family and that has a high degree of similarity with the fusion peptides of paramyxoviruses and some retroviruses. Synthetic peptides to this region interact with phospholipids (Rodriguez-Crespo et al 1995). Further analysis of this region must now be performed to test whether it is a true fusion peptide.

CELL-CELL FUSION

Cell-cell fusion reactions are critical in development. Fusion of gametes to form the zygote, myoblasts to form myotubes, and monocytes to form osteoclasts set off new developmental pathways. Fusion of cytotrophoblasts in the placenta generates a protective and nutritive layer for the developing embryo. And fusion of macrophages to form giant cells occurs during certain inflammatory responses. Compared with our knowledge of virus-cell fusion, information about cell-cell fusion is at a rudimentary level. The working hypothesis is that because cell-cell fusion events are exoplasmic, they will be mediated by proteins that resemble viral fusion proteins. More specifically, we propose that proteins involved in cell-cell fusion reactions share some of the ten features common to many of the viral fusion proteins—most notably a fusion peptide. Given that there is virtually no sequence conservation between the fusion proteins of different virus families and that viral fusion proteins assume distinct structures (Figure 3) and undergo highly varied fusion-activating conformational changes, we predict that there may be several different classes of proteins mediating the spectrum of cell-cell fusion events. We submit that once the fusion protein (complex) has been identified and its fusion mechanism deciphered, the fusion reaction will proceed as predicted by the general principles embodied in the working model for virus-cell fusion (Figure 6; steps 2-5). As far as we know, all cell-cell fusion reactions occur at neutral pH. Hence it will be interesting to determine whether proteins involved in cell-cell fusion reactions are more similar to neutral than to low-pH-activated viral fusion proteins. Moreover, it will be interesting to see whether any special proteins or principles are involved in homotypic (e.g. myoblast-myoblast) versus heterotypic (e.g. sperm-egg) cell-cell fusion reactions.

It is premature to describe the molecular mechanism of cell-cell fusion reactions until bona fide cell-cell fusion proteins have been clearly identified. Current efforts are, therefore, focused on identifying cell-cell fusion proteins. This is a much more difficult task than it was for the virus-cell fusion proteins for two major reasons. First, cells have many more surface glyoproteins; they are far more complex than the most complex enveloped viruses. Second, cell-cell fusion events are preceded by cell differentiation and cell-cell adhesion (and in some cases cell migration), making it challenging to determine whether a protein implicated in fusion is involved in differentiation, migration, adhesion, or fusion. Two approaches have been employed to identify candidate cell-cell fusion proteins: (*a*) characterization of antigens recognized by antibodies that block fusion without apparently blocking differentiation or adhesion (Primakoff et al 1987, Saginario et al 1995) and (*b*) characterization of mutants that are defective in cell-cell fusion (Trueheart & Fink 1989, Ferris et al 1996). We now briefly describe several candidate cell-cell fusion proteins. Proof that any of these is truly a fusion protein awaits the types of analyses applied above to confirm the identities of viral fusion proteins.

GAMETE FUSION

The union of two gametes to generate a new organism is a remarkable cell-cell fusion event. Candidate gamete fusion proteins have been identified in mammals (Primakoff et al 1987), marine invertebrates (Glabe 1985, Swanson & Vacquier 1995), *Chlamydomonas* (Ferris et al 1996), and in yeast (Trueheart & Fink 1989, Kurihara et al 1994, Elion et al 1995).

A candidate mammalian sperm fusion protein is fertilin (Figure 7). It is a heterodimer comprised of an α and β subunit that are made as larger precursors and appear to form a higher-order oligomer (Primakoff et al 1987, Blobel et al 1992; S Waters & J White, unpublished data). The β subunit contains a predicted integrin ligand domain and the α subunit contains a candidate fusion peptide. Several lines of evidence have supported a role for the β subunit, in particular its predicted integrin ligand domain, in binding sperm to the egg plasma membrane (Myles et al 1994, Almeida et al 1995, Evans et al 1995). Indirect evidence has supported the notion that the α subunit is involved in fusion: (*a*) All sequenced fertilin α subunits contain a hydrophobic sequence in their ectodomain that models as an amphipathic α-helix or β-sheet (S Waters & J White, unpublished data). (*b*) The candidate fusion peptide from guinea pig fertilin α binds to artificial membranes and induces liposome fusion (Muga et al 1994). (*c*) Fertilin α and β are the prototypes of the newly described ADAM (proteins containing **A** **D**isintegrin **A**nd **M**etalloprotease) gene family (Wolfsberg et al 1995). A new ADAM, meltrin α, has recently been implicated in myoblast fusion (Yagami-Hiromasa et al 1995). However, it remains to be proven that the fertilin α/β complex is truly a fusion protein, either alone or in conjunction with other sperm surface antigens.

Several proteins have been implicated in mating reactions in lower organisms. These include two proteins of sea urchins and abalone sperm that have membrane interactive properties but are not anchored in the sperm membrane via a transmembrane anchor (Glabe 1985, Swanson & Vacquier 1995). A

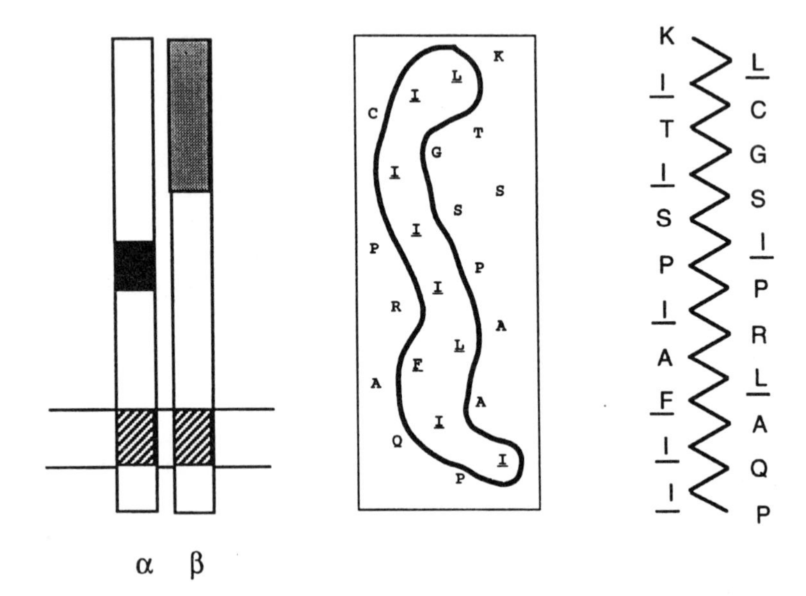

Figure 7 Fertilin, a candidate sperm-egg fusion protein. The left image denotes the subunit organization of the mature fertilin α/β heterodimer. The gray box denotes the predicted integrin ligand domain in the β subunit. The black box denotes the candidate fusion peptide in the α subunit. In the middle and right images, the candidate fusion peptide of guinea pig fertilin α is plotted as an α-helix (*middle*) and as a β-sheet (*right*). Strongly hydrophobic residues are encircled and underlined in the middle image and underlined in the right image.

mutant screen in *Chlamydomonas* revealed a candidate fusion protein called Fus1. Fus1 is a predicted transmembrane-anchored protein that accounts, at least in part, for a fringe of proteins present on the *Chlamydomonas* surface at the site of cell fusion. Fus1 shows no sequence relationship with ADAMs or any other cell surface proteins yet identified. The hydrophobicity plot of Fus1 is noisy, like that of the VSV G glycoprotein, and therefore does not contain an obvious candidate fusion peptide. As described above for VSV G, it will be a challenge to test whether Fus1 is a bona fide fusion protein. In the yeast *Saccharomyces cerevisiae*, several genes, including FUS1, 2, 3, and others (Trueheart & Fink 1989, Elion et al 1995, Kurihara et al 1994), have been implicated in the pathway of cell fusion based on mutant screens. Although the Fus proteins studied are localized at the point of fusion between mating yeast cells and are involved in the pathway of the cell-cell fusion, none is strongly predicted to be a fusion protein that functions like a viral fusion protein. Fus1 is a type I integral membrane glycoprotein with O-linked carbohydrates, but the bulk of the protein is on the cytoplasmic side of the membrane, and the phenotype of the *fus1* mutant is not consistent with a role in membrane merger per se. Fus2 is not predicted to be a membrane protein and is found on the cytoplasmic side of the plasma membrane, perhaps in association with cytoskeletal components. Fus3 is a (cytoplasmic) MAP kinase. Moreover, there do not appear to be any ADAMs in *S. cerevisiae* based on sequence analysis of its genome (T Wolfsberg, unpublished data). Hence the mechanism by which **a** and α yeast cells fuse remains to be clarified.

MYOBLAST FUSION

Meltrin α, a member of the ADAM gene family, is involved in myoblast fusion. A shortened form of meltrin α enhanced the timecourse of myoblast fusion and resulted in the formation of myosacs instead of myotubes, whereas overexpression of full-length meltrin α inhibited myoblast fusion (Yagami-Hiromasa et al 1995). As is found in several other ADAMs, full-length meltrin α is predicted to possess a metalloprotease domain, a disintegrin domain, a candidate fusion peptide, and a signaling domain. Hence, it is not clear whether meltrin α is involved in steps preceding fusion (e.g. cell migration, differentiation, or adhesion) or in fusion per se; the candidate fusion peptide of meltrin α is less hydrophobic than that of fertilin α (TG Wolfsberg, unpublished observation). Nonetheless, these recent findings place members of the ADAMs gene family in the pathways of two important cell-cell fusion events.

OTHER CELL-CELL FUSION REACTIONS

Two other cell surface proteins have been recently implicated in other cell-cell fusion reactions. These are a candidate macrophage fusion protein termed MFP 150 (Saginario et al 1995), as well as the first ADAM found in *Caenorhabditis elegans,* adm-1. The former has rudimentary properties consistent with a role in macrophage fusion, whereas the latter is found in cells that undergo fusion reactions in *C. elegans* (Podbilewicz & White 1994; B Podbilewicz, personal communication). Given their large numbers and widespread distribution, it would not be surprising if ADAMs are involved, either directly or indirectly, in other cell-cell fusion events (Wolfsberg et al 1995).

PERSPECTIVES

It is anticipated that the next five years will witness the elaboration of the mechanisms of several additional viral fusion proteins, ones that function at low as well as at neutral pH. It is also expected that several of the candidate cell-cell fusion proteins will be subjected to functional analyses to test their precise roles in events culminating in cell-cell fusion. Based on what we have learned from the viruses, it is expected that proteins of very different native and fusion-active structures will be able to elicit fusion, but that they will share the basic property of undergoing a triggered conformational change that renders them able to bind to and alter the shape of lipid bilayers. It is further envisioned that later steps in various fusion reactions (e.g. Figure 6; steps 2-5) will have more in common from one system to the next. There are already some hints that this may be the case, even when comparing exoplasmic and endoplasmic fusion events (Chernomordik et al 1993, Monck & Fernandez 1994, Lindau & Almers 1995). We also anticipate that there will be some interesting novelties found in the realms of virus-cell and cell-cell fusion.

Literature Cited

Allison SL, Schalich J, Stiasny K, Mandl CW, Kunz C, Heinz FX. 1995. Oligomeric rearrangement of tick-borne encephalitis virus envelope proteins induced by an acidic pH. *J. Virol.* 69:695–700

Almeida EAC, Huovila A-PJ, Sutherland AE, Stephens LE, Calarco PG, et al. 1995. Mouse egg integrin $\alpha 6 \beta 1$ functions as a sperm receptor. *Cell* 81:1095–104

Anthony RP, Brown DT. 1991. Protein-protein interactions in an alphavirus membrane. *J. Virol.* 65:1187–94

Bagai S, Lamb RA. 1995. Quantitative measurement of paramyxovirus fusion: differences in requirements of glycoproteins between simian virus 5 and human parainfluenza virus 3 or Newcastle disease virus. *J. Virol.* 69:6712–19

Bates P, Young JAT, Varmus HE. 1993. A receptor for subgroup A Rous sarcoma virus is related to the low density lipoprotein receptor. *Cell* 74:1043–51

Bentz J, ed. 1993. *Viral Fusion Mechanisms.* Boca Raton, FL: CRC Press. 529 pp.

Blobel CP, Wolfsberg TG, Turck CW, Myles DG, Primakoff P, White JM. 1992. A potential fusion peptide and an integrin ligand domain in a protein active in sperm-egg fusion. *Nature* 356:248–52

Blumenthal R, Pak CC, Raviv Y, Krumbiegel M, Bergelson LD, et al. 1995. Transient domains induced by influenza haemagglutinin during membrane fusion. *Mol. Membr. Biol.* 12:135–42

Bousse T, Takimoto T, Gorman WL, Takahashi T, Portner A. 1994. Regions on the hemagglutinin-neuraminidase proteins of human parainfluenza virus type-1 and Sendai virus important for membrane fusion. *Virology* 204:506–14

Brasseur R, Vandenbranden M, Cornet B, Burny A, Ruysschaert J-M. 1990. Orientation into the lipid bilayer of an asymmetric amphipathic helical peptide located at the N-terminus of viral fusion proteins. *Biochim. Biophys. Acta* 1029:267–73

Bron R, Wahlberg JM, Garoff H, Wilschut J. 1993. Membrane fusion of Semliki Forest virus in a model system: correlation between fusion kinetics and structural changes in the envelope glycoprotein. *EMBO J.* 12:693–701

Brown JC, Newcomb WW, Lawrenz-Smith S. 1988. pH-Dependent accumulation of the vesicular stomatitis virus glycoprotein at the ends of intact virions. *Virology* 167:625–29

Buckland R, Malvoisin E, Beauverger P, Wild F. 1992. A leucine zipper structure present in the measles virus fusion protein is not required for its tetramerization but is essential for fusion. *J. Gen. Virol.* 73:1703–7

Bullough PA, Hughson FM, Skehel JJ, Wiley DC. 1994. Structure of influenza haemagglutinin at the pH of membrane fusion. *Nature* 371:37–43

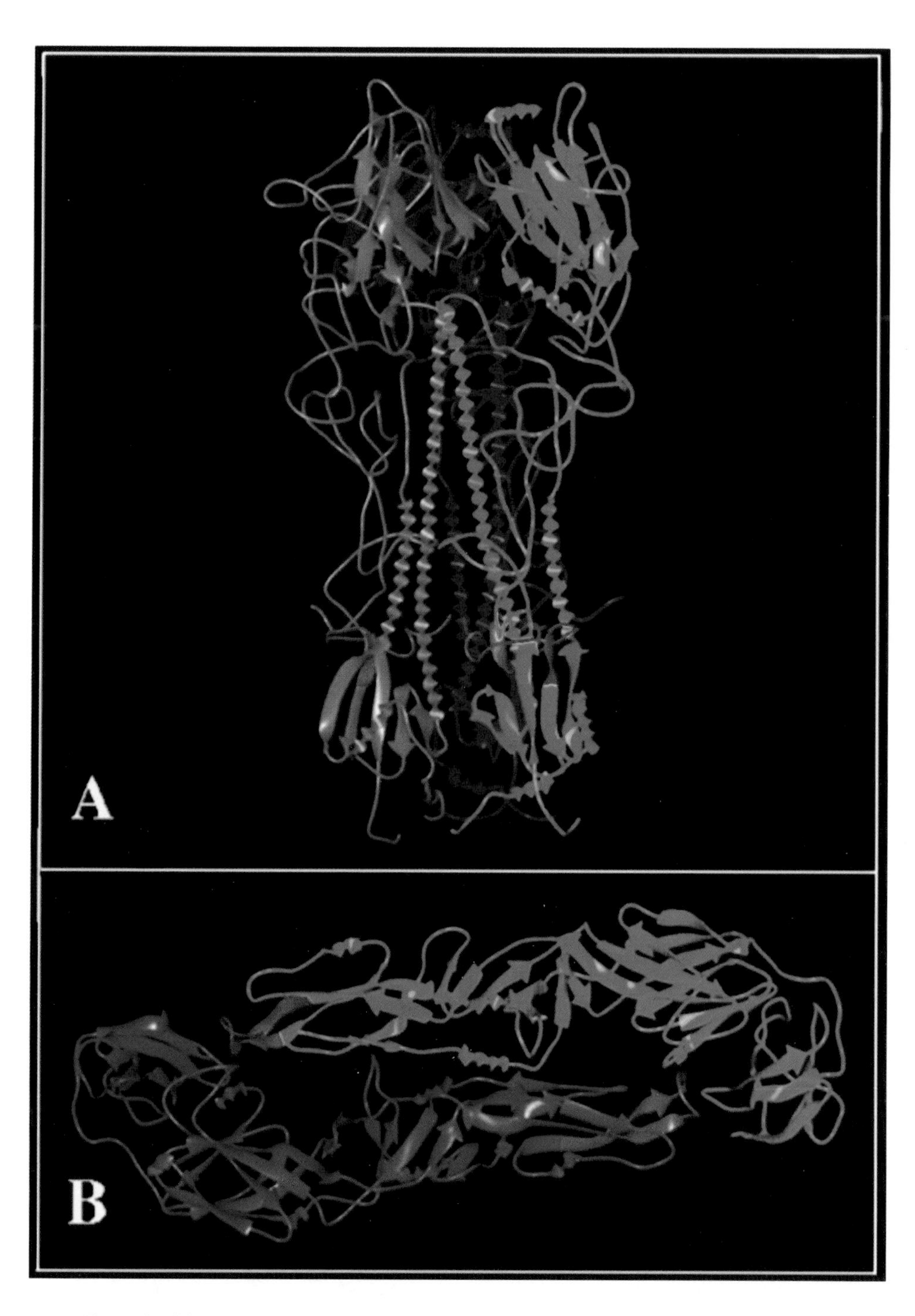

Figure 3 Ribbon diagrams of the neutral pH structures of (*A*) the influenza HA trimer (Wilson et al 1981) and (*B*) the flavivirus E dimer (Rey et al 1995). Red denotes known and predicted fusion peptides of the influenza HA and flavivirus E, respectively.

Cao J, Bergeron L, Helseth E, Thali M, Repke H, Sodroski J. 1993. Effects of amino acid changes in the extracellular domain of the human immunodeficiency virus type 1 gp41 envelope glycoprotein. *J. Virol.* 67:2747–55

Carr CM, Kim PS. 1993. A spring-loaded mechanism for the conformational change of influenza hemagglutinin. *Cell* 73:823–32

Cavanagh D, Davis PJ, Darbyshire JH, Peters RW. 1986. Coronavirus IBV: virus retaining spike glycopolypeptide S2 but not S1 is unable to induce virus-neutralizing or haemagglutination-inhibiting antibody, or induce chicken tracheal protection. *J. Gen. Virol.* 67:1435–42

Chambers P, Pringle CR, Easton AJ. 1990. Heptad repeat sequences are located adjacent to hydrophobic regions in several types of virus fusion glycoproteins. *J. Gen. Virol.* 71:3075–80

Chen SS, Lee CN, Lee WR, McIntosh K, Lee TH. 1993. Mutational analysis of the leucine zipper-like motif of the human immunodeficiency virus type 1 envelope transmembrane glycoprotein. *J. Virol.* 67:3615–19

Cheng RH, Kuhn RJ, Olson NH, Rossman MG, Choi H-K, Baker TS. 1995. Nucleocapsid and glycoprotein organization in an enveloped virus. *Cell* 80:621–30

Chernomordik L, Kozlov MM, Zimmerberg J. 1995. Lipids in biological membrane fusion. *J. Membr. Biol.* 146:1–14

Chernomordik LV, Vogel SS, Sokoloff A, Onaran HO, Leikina EA, Zimmerberg J. 1993. Lysolipids reversibly inhibit Ca^{2+}-, GTP, and pH-dependent fusion of biological membranes. *FEBS Lett.* 318:71–76

Clague MJ, Schoch C, Blumenthal R. 1991. Delay time for influenza virus hemagglutinin-induced membrane fusion depends on hemagglutinin surface density. *J. Virol.* 65:2402–7

Clague MJ, Schoch C, Zech L, Blumenthal R. 1990. Gating kinetics of pH-activated membrane fusion of vesicular stomatitis virus with cells: stopped-flow measuremnts by dequenching of octadecylrhodamine fluorescence. *Biochemistry* 29:1303–8

Clements GJ, Price-Jones MJ, Stephens PE, Sutton C, Schulz TF, et al. 1991. The V3 loops of the HIV-1 and HIV-2 surface glycoproteins contain proteolytic cleavage sites: a possible function in viral fusion? *AIDS Res. Hum. Retroviruses* 7:3–16

Daniel C, Anderson R, Buchmeier MJ, Fleming JO, Spaan WJ, et al. 1993. Identification of an immunodominant linear neutralization domain on the S2 portion of the murine coronavirus spike glycoprotein and evidence that it forms part of a complex tridimensional structure. *J. Virol.* 67:1185–94

Danieli T, Pelletier SL, Henis YI, White JM. 1996. Membrane fusion mediated by the influenza hemagglutinin requires the concerted action of at least three hemagglutinin trimers. *J. Cell Biol.* 133:559–69

Delwart EL, Mosialos G, Gilmore T. 1990. Retroviral envelope glycoproteins contain a "leucine zipper"-like repeat [comment]. *AIDS Res. Hum. Retroviruses* 6:703–6

Di Simone C, Buchmeier MJ. 1995. Kinetics and pH dependence of acid-induced structural changes in the lymphocytic choriomeningitis virus glycoprotein complex. *Virology* 209:3–9

Doms RW. 1993. Protein conformational changes in virus-cell fusion. In *Methods in Enzymology,* ed. N Duzgunes, 221:61–82. San Diego: Academic

Doms RW, Blumenthal R, Moss B. 1990. Fusion of intra- and extracellular forms of vaccinia virus with the cell membrane. *J. Virol.* 64:4884–92

Doms RW, Helenius A. 1986. Quaternary structure of influenza virus hemagglutinin after acid treatment. *J. Virol.* 60:833–39

Dubay JW, Dubay SR, Shin HJ, Hunter E. 1995. Analysis of the cleavage site of the human immunodeficiency virus type 1 glycoprotein: requirement of precursor cleavage for glycoprotein incorporation. *J. Virol.* 69:4675–82

Dubay JW, Roberts SJ, Brody B, Hunter E. 1992a. Mutations in the leucine zipper of the human immunodeficiency virus type 1 transmembrane glycoprotein affect fusion and infectivity. *J. Virol.* 66:4748–56

Dubay JW, Roberts SJ, Hahn BH, Hunter E. 1992b. Truncation of the human immunodeficiency virus type 1 transmembrane glycoprotein cytoplasmic domain blocks virus infectivity. *J. Virol.* 66:6616–25

Durrer P, Gaudin Y, Ruigrok RWH, Graf R, Brunner J. 1995. Photolabeling identifies a putative fusion domain in the envelope glycoprotein of rabies and vesicular stomatitis viruses. *J. Biol. Chem.* 270:17575–81

Earl PL, Doms RW, Moss B. 1990. Oligomeric structure of the human immunodeficiency virus type 1 envelope glycoprotein. *Proc. Natl. Acad. Sci. USA* 87:648–52

Einfeld D, Hunter E. 1988. Oligomeric structure of a prototypic retrovirus glycoprotein. *Proc. Natl. Acad. Sci. USA* 85:8688–92

Elion EA, Trueheart J, Fink GR. 1995. Fus2 localizes near the site of cell fusion and is required for both cell fusion and nuclear alignment during zygote formation. *J. Cell Biol.* 130:1283–96

Ellens H, Bentz J, Mason D, Zhang F, White JM. 1990. Fusion of influenza hemagglutinin-expressing fibroblasts with glycophorin-bearing liposomes: role of hemagglutinin surface density. *Biochemistry* 29:9697–707

Evans JP, Schultz RM, Kopf GS. 1995. Mouse sperm-egg plasma membrane interactions: analysis of roles of egg integrins and the mouse sperm homologue of PH-30 (fertilin) b. *J. Cell Sci.* 108:3267–78

Fantini J, Cook DG, Nathanson N, Spitalnik SL, Gonzalez-Scarano F. 1993. Infection of colonic epithelial cell lines by type 1 human immunodeficiency virus is associated with cell surface expression of galactosylceramide, a potential alternative gp120 receptor. *Proc. Natl. Acad. Sci. USA* 90:2700–4

Feng Y, Broder CC, Kennedy PE, Berger EA. 1996. HIV-1 entry cofactor: functional cDNA cloning of a seven-transmembrane G protein-coupled receptor. *Science* 272:872–77

Ferris PJ, Woessner JP, Goodenough UW. 1996. A sex recognition glycoprotein is encoded by the *plus* mating-type gene *fus1* of *Chlamydomonas reinhardtii. Mol. Biol. Cell.* In press

Fredericksen BL, Whitt MA. 1995. Vesicular stomatitis virus glycoprotein mutations that affect membrane fusion activity and abolish virus infectivity. *J. Virol.* 69:1435–43

Freed EO, Myers DJ, Risser R. 1990. Characterization of the fusion domain of the human immunodeficiency virus type 1 envelope glycoprotein gp41. *Proc. Natl. Acad. Sci. USA* 87:4650–54

Frey S, Marsh M. Gunther S, Pelchen-Matthews A, Stephens P, et al. 1995. Temperature dependence of cell-cell fusion induced by the envelope glycoprotein of human immunodeficiency virus type 1. *J. Virol.* 69:1462–72

Fuller SD, Berriman JA, Butcher SJ, Gowen BE. 1995. Low pH induces swiveling of the glycoprotein heterodimers in the Semliki Forest virus spike complex. *Cell* 81:715–25

Gage PJ, Levine M, Glorioso JC. 1993. Syncytium-inducing mutations localize to two discrete regions within the cytoplasmic domain of herpes simplex virus type 1 glycoprotein B. *J. Virol.* 67:2191–201

Gallagher TM, Escarmis C, Buchmeier MJ. 1991. Alteration of the pH dependence of coronavirus-induced cell fusion: effect of mutations in the spike glycoprotein. *J. Virol.* 65:1916–28

Gaudin Y, Ruigrok RWH, Brunner J. 1995. Low-pH induced conformational changes in viral fusion proteins: implications for the fusion mechanism. *J. Gen. Virol.* 76:1541–56

Gaudin Y, Ruigrok RWH, Knossow M, Flamand A. 1993. Low-pH conformational changes of rabies virus glycoprotein and their role in membrane fusion. *J. Virol.* 67:1365–72

Gething M-J, Doms RW, York D, White JM. 1986. Studies on the mechanism of membrane fusion: site-specific mutagenesis of the hemagglutinin of influenza virus. *J. Cell Biol.* 102:11–23

Gilbert JM, Bates P, Varmus HE, White JM. 1994. The receptor for the subgroup A avian leukosis sarcoma virus binds to subgroup A but not to subgroup C envelope glycoprotein. *J. Virol.* 67:6889–92

Gilbert JM, Hernandez LD, Balliet JW, Bates P, White JM. 1995. Receptor-induced conformational changes in the subgroup A avian leukosis and sarcoma virus envelope glycoprotein. *J. Virol.* 69:7410–15

Gilbert JM, Mason D, White J. 1990. Fusion of Rous sarcoma virus with host cells does not require low pH. *J. Virol.* 64:5106–13

Glabe CG. 1985. Interaction of the sperm adhesive protein, bindin, with phospholipid vesicles. II. Bindin induces the fusion of mixed-phase vesicles that contain phosphatidylcholine and phosphatidylserine in vitro. *J. Cell Biol.* 100:800–6

Glushakova SE, Omelyanenko VG, Lukashevitch IS, Bogdanov AA, Moshnikova AB, et al. 1992. The fusion of artificial lipid membranes induced by the synthetic arenavirus 'fusion peptide'. *Biochim. Biophys. Acta* 1110:202–8

Godley L, Pfeifer J, Steinhauer D, Ely B, Shaw G, et al. 1992. Introduction of intersubunit disulfide bonds in the membrane-distal region of the influenza hemagglutinin abolishes membrane fusion activity. *Cell* 68:635–45

Gong SC, Lai CF, Esteban M. 1990. Vaccinia virus induces cell fusion at acid pH and this activity is mediated by the N-terminus of the 14-kDa virus envelope protein. *Virology* 178:81–91

Gray C, Tatulian SA, Wharton SA, Tamm LK. 1996. Effect of the N-terminal glycine on the secondary structure, orientation and interaction of the influenza hemagglutinin fusion peptide with lipid bilayers. *Biophys. J.* 70:2275–86

Guirakhoo F, Heinz FX, Mandl CW, Holzmann H, Kunz C. 1991. Fusion activity of flaviviruses: comparison of mature and immature (prM-containing) tick-borne encephalitis virions. *J. Gen. Virol.* 72:1323–29

Gutman O, Danieli T, White JM, Henis YI. 1993. Effects of exposure to low pH on the lateral mobility of influenza hemagglutinin expressed at the cell surface. *Biochemistry* 32:101–6

Guy HR, Durell SR, Schoch C, Blumenthal R. 1992. Analyzing the fusion process of influenza hemagglutinin by mutagenesis and molecular modeling. *Biophys. J.* 62:95–97

Hacker JK, Volkman LE, Hardy JL. 1995. Requirement for the G1 protein of California encephalitis virus in infection in vitro and in vivo. *Virology* 206:945–53
Hart TK, Kirsh R, Ellens H, Sweet RW, Lambert DM, et al. 1991. Binding of soluble CD4 proteins to human immunodeficiency virus type 1 and infected cells induces release of envelope glycoprotein gp120. *Proc. Natl. Acad. Sci. USA* 88:2189–93
Harter C, James P, Bächi T, Semenza G, Brunner J. 1989. Hydrophobic binding of the ectodomain of influenza hemagglutinin to membranes occurs through the "fusion peptide." *J. Biol. Chem.* 264:6459–64
Helenius A. 1995. Alphavirus and flavivirus glycoproteins: structure and function. *Cell* 81:651–53
Helm CA, Israelachvili JN. 1993. Forces between phospholipid bilayers and relationship to membrane fusion. *Meth. Enzymol.* 220:130–43
Heminway BR, Yu Y, Tanaka Y, Perrine KG, Gustafson E, et al. 1994. Analysis of respiratory syncytial virus F, G, and SH proteins in cell fusion. *Virology* 200:801–5
Herold BC, Visalli RJ, Susmarski N, Brandt CR, Spear PG. 1994. Glycoprotein C-independent binding of herpes simplex virus to cells requires cell surface heparan sulphate and glycoprotein B. *J. Gen. Virol.* 75:1211–22
Hughson FM. 1995. Structural characterization of viral fusion proteins. *Curr. Biol.* 5:265–74
Hutchinson L, Roop-Beauchamp C, Johnson DC. 1995. Herpes simplex virus glycoprotein K is known to influence fusion of infected cells, yet is not on the cell surface. *J. Virol.* 69:4556–63
Johnston PB, Dubay JW, Hunter E. 1993. Truncations of the simian immunodeficiency virus transmembrane protein confer expanded virus host range by removing a block to virus entry into cells. *J. Virol* 67:3077–86
Jones JS, Risser R. 1993. Cell fusion induced by the murine leukemia virus envelope glycoprotein. *J. Virol.* 67:67–74
Junankar PR, Cherry RJ. 1986. Temperature and pH dependence of the haemolytic activity of influenza virus and of the rotational mobility of the spike glycoproteins. *Biochim. Biophys. Acta* 854:198–206
Justman J, Klimjack MR, Kielian M. 1993. Role of spike protein conformational changes in fusion of Semliki Forest virus. *J. Virol.* 67:7597–607
Katow S, Sugiura A. 1988. Low pH-induced conformational change of Rubella virus envelope proteins. *J. Gen. Virol.* 69:2797–807
Kavanaugh MP, Miller DG, Zhang W, Law W, Kozak SL, et al. 1994. Cell-surface receptors for gibbon ape leukemia virus and amphotropic murine retrovirus are inducible sodium-dependent phosphate symporters. *Proc. Natl. Acad. Sci. USA* 91:7071–75
Kemble GW, Bodian DL, Rosé J, Wilson IA, White JM. 1992. Intermonomer disulfide bonds impair the fusion activity of influenza virus hemagglutinin. *J. Virol.* 66:4940–50
Kemble GW, Danieli T, White JM. 1994. Lipid-anchored influenza hemagglutinin promotes hemifusion, not complete fusion. *Cell* 76:383–91
Kielian M. 1995. Membrane fusion and the alphavirus life cycle. *Adv. Vir. Res.* 45:113–51
Kimura T, Ohyama A. 1988. Association between the pH-dependent conformational change of West Nile flavivirus E protein and virus-mediated membrane fusion. *J. Gen. Virol.* 69:1247–54
Klimjack MR, Jeffrey S, Kielian M. 1994. Membrane and protein interactions of a soluble form of the Semliki Forest virus fusion protein. *J. Virol.* 68:6940–46
Kurihara LJ, Beh CT, Latterich M, Schekman R, Rose M. 1994. Nuclear congression and membrane fusion: two distinct events in the yeast karyogamy pathway. *J. Cell Biol.* 126:911–23
Kuroki K, Cheung R, Marion PL, Ganem D. 1994. A cell surface protein that binds avian hepatitis B virus particles. *J. Virol.* 68:2091–96
LaBranche CC, Sauter MM, Haggarty BS, Vance PJ, Romano J, et al. 1994. Biological, molecular, and structural analysis of a cytopathic variant from a molecularly cloned simian immunodeficiency virus. *J. Virol.* 68:5509–22
Lamb RA. 1993. Paramyxovirus fusion: a hypothesis for change. *Virology* 197:1–11
Lanzrein M, Kasermann N, Weingart R, Kempf C. 1993. Early events of Semliki Forest virus-induced cell-cell fusion. *Virology* 196:541–47
Lear JD, DeGrado WF. 1987. Membrane binding and conformational properties of peptides representing the NH2 terminus of influenza HA-2. *J. Biol. Chem.* 262:6500–5
Levy-Mintz P, Kielian M. 1991. Mutagenesis of the putative fusion domain of the Semliki Forest virus spike protein. *J. Virol.* 65:4292–300
Lindau M, Almers W. 1995. Structure and function of fusion pores in exocytosis and ectoplasmic membrane fusion. *Curr. Opin. Cell Biol.* 7:509–17
Long D, Berson JF, Cook DG, Doms RW.

1994. Characterization of human immunodeficiency virus type 1 gp120 binding to liposomes containing galactosylceramide. *J. Virol.* 68:5890–98

Lu M, Blacklow S, Kim PS. 1995. A trimeric structural domain of the HIV-1 transmembrane glycoprotein. *Nat. Struct. Biol.* 2:1075–82

McClure MO, Sommerfelt MA, Marsh M, Weiss RA. 1990. The pH independence of mammalian retrovirus infection. *J. Gen. Virol.* 71:767–73

Melikyan GB, Niles WD, Cohen FS. 1993a. Influenza virus hemagglutinin-induced cell-planar bilayer fusion: quantitative dissection of fusion pore kinetics into stages. *J. Gen. Physiol.* 102:1131–46

Melikyan GB, Niles WD, Peeples ME, Cohen FS. 1993b. Influenza hemagglutinin-mediated fusion pores connecting cells to planar membranes: flickering to final expansion. *J. Gen. Physiol.* 102:1131–49

Melikyan GB, Niles WD, Ratinov VA, Karhanek M, Zimmerberg J, Cohen FS. 1995a. Comparison of transient and successful fusion pores connecting influenza hemagglutinin expressing cells to planar membranes. *J. Gen. Physiol.* 106:803–19

Melikyan GB, White JM, Cohen FS. 1995b. GPI-anchored influenza hemagglutinin induces hemifusion to both red blood cell and planar bilayer membranes. *J. Cell Biol.* 131:679–91

Meyer WJ, Gidwitz S, Ayers VK, Schoepp RJ, Johnston RE. 1992. Conformational alteration of sindbis virion glycoproteins induced by heat, reducing agents, or low pH. *J. Virol.* 66:3504–13

Moesby L, Corver J, Erukulla RK, Bittman R, Wilschut J. 1995. Sphingolipids activate membrane fusion of Semliki Forest virus in a stereospecific manner. *Biochemistry* 34:10319–24

Monck JR, Fernandez JM. 1994. The exocytic fusion pore and neurotransmitter release. *Neuron* 12:707–16

Moore JP, McKeating JA, Huang YX, Ashkenazi A, Ho DD. 1992. Virions of primary human immunodeficiency virus type 1 isolates resistant to soluble CD4 (sCD4) neutralization differ in sCD4 binding and glycoprotein gp120 retention from sCD4-sensitive isolates. *J. Virol.* 66:235–43

Moore JP, McKeating JA, Weiss RA, Sattentau QJ. 1990. Dissociation of gp120 from HIV-1 virions induced by soluble CD4 [see comments]. *Science* 250:1139–42

Morris S, Sarkar D, White J, Blumenthal R. 1989. Kinetics of pH-dependent fusion between 3T3 fibroblasts expressing influenza hemagglutinin and red blood cells. Measurement by dequenching of fluorescence. *J. Biol. Chem.* 264:3972–78

Morrison T, McQuain C, McGinnes L. 1991. Complementation between avirulent Newcastle disease virus and a fusion protein gene expressed from a retrovirus vector: requirements for membrane fusion. *J. Virol.* 65:813–22

Muga A, Neugebauer W, Hirama T, Surewicz WK. 1994. Membrane interactive and conformational properties of the putative fusion peptide of PH-30, a protein active in sperm-egg fusion. *Biochemistry* 33:4444–48

Mulligan MJ, Yamshchikov GV, Ritter GD Jr, Gao F, Jin MJ, et al. 1992. Cytoplasmic domain truncation enhances fusion activity by the exterior glycoprotein complex of human immunodeficiency virus type 2 in selected cell types. *J. Virol.* 66:3971–75

Myles DG, Kimmel LH, Blobel CP, White JM, Primakoff P. 1994. Identification of a binding site in the disintegrin domain of fertilin required for sperm-egg fusion. *Proc. Natl. Acad. Sci. USA* 91:4195–98

Nanavati C, Markin VS, Oberhauser AF, Fernandez JM. 1992. The exocytic fusion pore modeled as a lipidic pore. *Biophys. J.* 63:1118–32

Nieva JL, Bron R, Corver J, Wilschut J. 1994a. Membrane fusion of Semliki Forest virus requires sphingolipids in the target membrane. *EMBO J.* 13:2797–804

Nieva JL, Nir S, Muga A, Goni FM, Wilschut J. 1994b. Interaction of the HIV-1 fusion peptide with phospholipid vesicles: different structural requirements for fusion and leakage. *Biochemistry* 33:3201–9

Owens RJ, Burke C, Rose JK. 1994. Mutations in the membrane-spanning domain of the human immunodeficiency virus envelope glycoprotein that affect fusion activity. *J. Virol.* 68:570–74

Pak CC, Krumbiegel M, Blumenthal R. 1994. Intermediates in influenza virus PR/8 haemagglutinin-induced membrane fusion. *J. Gen. Virol.* 75:395–99

Paredes AM, Brown DT, Rothnagel R, Chiu W, Schoepp RJ, et al. 1993. Three-dimensional structure of a membrane-containing virus. *Proc. Natl. Acad. Sci. USA* 90:9095–99

Pekosz A, Griot C, Nathanson N, Gonzalez-Scarano F. 1995. Tropism of bunyaviruses: evidence for a G1 glycoprotein-mediated entry pathway common to the California serogroup. *Virology* 214:339–48

Podbilewicz B, White JG. 1994. Cell fusion in the developing epithelial of *C. elegans*. *Dev. Biol.* 161:408–24

Primakoff P, Hyatt H, Tredick-Kline J. 1987.

Identification and purification of a sperm surface protein with a potential role in sperm-egg membrane fusion. *J. Cell Biol.* 104:141–49

Purchase H, Okazaki W. 1964. Morphology of foci produced by standard preparations of Rous sarcoma virus. *J. Natl. Cancer Inst.* 32:579–84

Ragheb JA, Anderson WF. 1994. pH-independent murine leukemia virus ecotropic envelope-mediated cell fusion: implications for the role of the R peptide and p12E TM in viral entry. *J. Virol.* 68:3220–31

Rein A, Mirro J, Haynes JG, Ernst SM, Nagashima K. 1994. Function of the cytoplasmic domain of a retroviral transmembrane protein: p15E-p2E cleavage activates the membrane fusion capability of the murine leukemia virus Env protein. *J. Virol.* 68:1773–81

Reitter JN, Sergel T, Morrison TG. 1995. Mutational analysis of the leucine zipper motif in the Newcastle disease virus fusion protein. *J. Virol.* 69:5995–6004

Rey FA, Heinz FX, Mandl C, Kunz C, Harrison SC. 1995. The envelope glycoprotein from tick-borne encephalitis virus at 2 Å resolution. *Nature* 375:291–98

Rodriguez JF, Paez E, Esteban M. 1987. A 14,000-M_r envelope protein of vaccinia virus is involved in cell fusion and forms covalently linked trimers. *J. Virol.* 61:395–404

Rodriguez-Crespo I, Nunez E, Gomez-Gutierrez J, Yelamos B, Albar JP, et al. 1995. Phospholipid interactions of the putative fusion peptide of hepatitis B virus surface antigen S protein. *J. Gen. Virol.* 76:301–8

Roehrig JT, Johnson AJ, Hunt AR, Bolin RA, Chu MC. 1990. Antibodies to dengue 2 virus E glycoprotein synthetic peptides identify antigenic conformation. *Virology* 177:668–75

Ronka H, Hilden P, von Bonsdorff C-H, Kuismanen E. 1995. Homodimeric association of the spike glycoprotein G1 and G2 of Uukuniemi virus. *Virology* 211:241–50

Roop C, Hutchinson L, Johnson DC. 1993. A mutant herpes simplex virus type 1 unable to express glycoprotein L cannot enter cells, and its particles lack glycoprotein H. *J. Virol.* 67:2285–97

Rothman JE, Warren G. 1994. Implications of the SNARE hypothesis for intracellular membrane topology and dynamics. *Curr. Biol.* 4:220–33

Ruigrok RWH, Aitken A, Calder LJ, Martin SR, Skehel JJ, et al. 1988. Studies on the structure of the influenza virus haemagglutinin at the pH of membrane fusion. *J. Gen. Virol.* 69:2785–95

Saginario C, Qian H-Y, Vignery A. 1995. Identification of an inducible surface molecule specific to fusing macrophges. *Proc. Natl. Acad. Sci. USA* 92:12210–14

Sattentau QJ, Moore JP. 1991. Conformational changes induced in the human immunodeficiency virus envelope glycoprotein by soluble CD4 binding. *J. Exp. Med.* 174:407–15

Sattentau QJ, Moore JP, Vignaux F, Traincard F, Poignard P. 1993. Conformational changes induced in the envelope glycoproteins of the human and simian immunodeficiency viruses by soluble receptor binding. *J. Virol.* 67:7383–93

Sauter MM, Pelchen-Matthews A, Bron R, Marsh M, LaBranche C, et al. 1996. An internalization signal in the simian immunodeficiency virus transmembrane protein cytoplasmic domain modulates expression of envelope glycoproteins on the cell surface. *J. Cell Biol.* 132:795–811

Sergel-Germano T, McQuain C, Morrison T. 1994. Mutations in the fusion peptide and heptad repeat regions of the Newcastle disease virus fusion protein block fusion. *J. Virol.* 68:7654–58

Siegel DP. 1993. Energetics of intermediates in membrane fusion: comparison of stalk and inverted micellar intermediate mechanisms. *Biophys. J.* 65:2124–40

Spear PG. 1993. Entry of alphaherpesviruses into cells (?). *Semin. Virol.* 4:167–80

Spies CP, Ritter GD Jr, Mulligan MJ, Compans RW. 1994. Truncation of the cytoplasmic domain of the simian immunodeficiency virus envelope glycoprotein alters the conformation of the external domain. *J. Virol.* 68:585–91

Spruce AE, Iwata A, Almers W. 1991. The first milliseconds of the pore formed by a fusogenic viral envelope protein during membrane fusion. *Proc. Natl. Acad. Sci. USA* 88:3623–27

Spruce AE, Iwata A, White JM, Almers W. 1989. Patch clamp studies of single cell-fusion events mediated by a viral fusion protein. *Nature* 342:555–58

Stegmann T. 1993. Influenza hemagglutinin-mediated membrane fusion does not involve inverted phase lipid intermediates. *J. Biol. Chem.* 268:1716–22

Stegmann T. 1994. Anchors aweigh. *Curr. Biol.* 4:551–54

Stegmann T, White JM, Helenius A. 1990. Intermediates in influenza-induced membrane fusion. *EMBO J.* 9:4231–41

Steinhauer DA, Wharton SA, Skehel JJ, Wiley DC. 1995. Studies of membrane fusion activities of fusion peptide mutants of influenza virus hemagglutinin. *J. Virol.* 69:6643–51

Stern W, Dales S. 1976. Biogenesis of vaccinia: isolation and characterization of a surface component that elicits antibody suppressing infectivity and cell-cell fusion. *Virology* 75:232–41

Sturman LS, Ricard CS, Holmes KV. 1990. Conformational change of the coronavirus peplomer glycoprotein at pH 8.0 and 37°C correlates with virus aggregation and virus-induced cell fusion. *J. Virol.* 64:3042–50

Swanson WJ, Vacquier VD. 1995. Liposome fusion induced by a M_r 18,000 protein localized to the acrosomal region of acrosome-reacted abalone spermatozoa. *Biochemistry* 34:14202–8

Tatulian SA, Hinterdorfer P, Baber G, Tamm LK. 1995. Influenza hemagglutinin assumes a tilted conformation during membrane fusion as determined by attenuated total reflection FTIR spectroscopy. *EMBO J.* 14:5514–23

Tong S, Li J, Wands JR. 1995. Interaction between duck hepatitis B virus and a 170-kilodalton cellular protein is mediated through a neutralizing epitope of the pre-S region and occurs during viral infection. *J. Virol.* 69:7106–12

Trueheart J, Fink GR. 1989. The yeast cell fusion protein FUS1 is *O*-glycosylated and spans the plasma membrane. *Proc. Natl. Acad. Sci. USA* 86:9916–20

Tse FW, Iwata A, Almers W. 1993. Membrane flux through the pore formed by a fusogenic viral envelope protein during cell fusion. *J. Cell Biol.* 121:543–52

Tsurudome M, Glück R, Graf R, Falchetto R, Schaller U, Brunner J. 1992. Lipid interactions of the hemagglutinin HA2 NH-2-terminal segment during influenza virus-induced membrane fusion. *J. Biol. Chem.* 267:20225–32

Wahlberg JM, Bron R, Wilschut J, Garoff H. 1992. Membrane fusion of Semliki Forest virus involves homotrimers of the fusion protein. *J. Virol.* 66:7309–18

Wahlberg JM, Garoff H. 1992. Membrane fusion process of Semliki Forest virus I: low pH-induced rearrangement in spike protein quaternary structure precedes virus penetration into cells. *J. Cell Biol.* 116:339–48

Ward CD, Paterson RG, Lamb RA. 1995. Mutants of the paramyxovirus SV5 fusion protein: regulated and extensive syncytium formation. *Virology* 209:242–49

Ward CW, Dopheide TA. 1980. Influenza virus haemagglutinin. Structural predictions suggest that the fibrillar appearance is due to the presence of a coiled-coil. *Aust. J. Biol. Sci.* 33:449–55

Weber T, Paesold G, Mischler R, Semenza G, Brunner J. 1994. Evidence for H^+-induced insertion of the influenza hemagglutinin HA2 N-terminal segment into the viral membrane. *J. Biol. Chem.* 269:18353–58

Weiss CD, Levy JA, White JM. 1990. Oligomeric organization of gp120 on infectious human immunodeficiency virus type 1 particles. *J. Virol.* 64:5674–77

Weiss R. 1992. Cellular receptors and viral glycoproteins involved in retrovirus entry. In *The Retroviruses,* ed. J Levy. 2:1–108. New York: Plenum

Wharton SA, Calder LJ, Ruigrok RWH, Skehel JJ, Steinhauer DA, Wiley DC. 1995. Electron microscopy of antibody complexes of influenza virus haemagglutinin in the fusion pH conformation. *EMBO J.* 14:240–46

White JM. 1992. Membrane fusion. *Science* 258:917–24

White JM. 1995. Membrane fusion: the influenza paradigm. *Cold Spring Harbor Symp. Quant. Biol.* 60:581–88

Wild C, Dubay JW, Greenwell T, Baird T Jr, Oas TG, et al. 1994a. Propensity for a leucine zipper-like domain of human immunodeficiency virus type 1 gp41 to form oligomers correlates with a role in virus-induced fusion rather than assembly of the glycoprotein complex. *Proc. Natl. Acad. Sci. USA* 91:12676–80

Wild CT, Shugars DC, Greenwell TK, McDanal CB, Matthews TJ. 1994b. Peptides corresponding to a predictive alpha-helical domain of human immunodeficiency virus type 1 gp41 are potent inhibitors of virus infection. *Proc. Natl. Acad. Sci. USA* 91:9770–74

Wiley DC, Skehel JJ. 1987. The structure and function of the hemagglutinin membrane glycoprotein of influenza virus. *Annu. Rev. Biochem.* 56:365–94

Wilson CA, Marsh JW, Eiden MV. 1992. The requirements for viral entry differ from those for virally induced syncytium formation in NIH 3T3/DTras cells exposed to Moloney murine leukemia virus. *J. Virol.* 66:7262–69

Wilson IA, Skehel JJ, Wiley DC. 1981. Structure of the haemagglutinin membrane glycoportein of influenza virus at 3Å resolution. *Nature* 289:366–72

Wimmer E, ed. 1994. *Cellular Receptors for Animal Viruses.* pp. 403–43. Plainview, NY: Cold Spring Harbor Lab. Press

Wolfsberg TG, Primakoff P, Myles DG, White JM. 1995. ADAM, a novel family of membrane proteins containing a disintegrin and metalloprotease domain: multipotential functions in cell-cell and cell-matrix interactions. *J. Cell Biol.* 131:275–78

Yagami-Hiromasa T, Sato T, Kurisaki T, Kamijo K, Nabeshima Y, Fujisawa-Sehara A. 1995. A

metalloprotease-disintegrin participating in myoblast fusion. *Nature* 377:652–56
Yang C, Compans RW. 1996. Analysis of the cell fusion activities of chimeric simian immunodeficiency virus-murine leukemia virus envelope proteins: inhibitory effects of the R peptide. *J. Virol.* 70:248–54
Yu YG, King DMS, Shin Y-K. 1994. Insertion of a coiled-coil peptide from influenza virus hemagglutinin into membranes. *Science* 266:274–76
Zagouras P, Rose JK. 1993. Dynamic equilibrium between vesicular stomatitis virus glycoprotein monomers and trimers in the Golgi and at the cell surface. *J. Virol.* 67:7533–38
Zhang L, Ghosh HP. 1994. Characterization of the putative fusogenic domain in vesicular stomatitis virus glycoprotein G. *J. Virol.* 68:2186–93
Zimmerberg J, Blumenthal R, Sarkar DP, Curran M, Morris SJ. 1994. Restricted movement of lipid and aqueous dyes through pores formed by influenza hemagglutinin during cell fusion. *J. Cell Biol.* 127:1885–94
Zingler K, Littman DR. 1993. Truncation of the cytoplasmic domain of the simian immunodeficiency virus envelope glycoprotein increases env incorporation into particles and fusogenicity and infectivity. *J. Virol.* 67:2824–31

Annu. Rev. Cell Dev. Biol. 1996. 12:663–95

NUCLEAR FUSION IN THE YEAST *SACCHAROMYCES CEREVISIAE*

Mark D. Rose
Department of Molecular Biology, Princeton University, Princeton, New Jersey 08544-1014

KEY WORDS: karyogamy, microtubules, KAR genes, membrane fusion, spindle pole body, nuclear envelope

ABSTRACT

Karyogamy, or nuclear fusion, is the process during mating by which two haploid yeast nuclei fuse to produce a single diploid nucleus. Karyogamy occurs in two major steps: microtubule-dependent nuclear congression followed by fusion of the nuclear envelope membranes. Many of the proteins required for karyogamy have been discovered to act in related processes during mitotic growth. Accordingly, yeast karyogamy has become an important model system to investigate critical functions of the cytoplasmic microtubules and the microtubule organizing center, the nuclear envelope, and the endoplasmic reticulum.

CONTENTS

1081-0706/96/1115-0663$08.00

INTRODUCTION

In the yeast *Saccharomyces cerevisiae,* as in other fungi, the nuclear envelope remains intact throughout all stages of mitosis, meiosis, and conjugation (Byers 1981a). The permanence of the nuclear envelope makes necessary several adaptations that influence the architecture and function of the nucleus. The first concerns the localization of proteins required to act on intranuclear structures during mitosis. For example, the presence of microtubules in both the nucleus and the cytoplasm implies that yeast tubulin must contain signals for nuclear localization. Clearly, such signals would not be required in organisms that break down the nuclear envelope during mitosis. Furthermore, differential nuclear localization of proteins during the cell cycle provides a powerful regulatory mechanism (Nasmyth et al 1990).

Second, the sole microtubule organizing center, called the spindle pole body (SPB) or spindle plaque, must nucleate both cytoplasmic and nuclear microtubules (Byers et al 1978, Hyams & Borisy 1978, Kilmartin 1994, Snyder 1994). This is accomplished by virtue of the special architecture of the SPB. Consisting of at least three distinct layers, the SPB is a disk embedded within the plane of the nuclear envelope, much in the fashion of a nuclear pore. The outer layer of the SPB nucleates cytoplasmic microtubules whose functions are to mediate nuclear movement and orientation (see Morris et al 1995 for a recent review of the role of cytoplasmic microtubules in fungi). The inner layer of the SPB nucleates the microtubules that mediate spindle assembly, spindle elongation, and chromosome segregation (Page & Snyder 1993, Rose et al 1993, Kilmartin 1994, Snyder 1994). One benefit of the laminar organization of the SPB is that it allows independent regulation of the nuclear and cytoplasmic microtubules.

The third major consequence of the closed mitosis in yeast concerns the sexual processes of conjugation and meiosis. Conjugation between two haploid cells of opposite mating type (**a** and α) is a complex process that is initiated by the mutual exchange of sexual pheromones and culminates in the formation of a diploid zygote (Cross et al 1988, Sprague & Thorner 1994). The diploid nucleus is produced by direct fusion of the two parental haploid nuclei. Nuclear fusion (or karyogamy) occurs in at least two stages; nuclear congression and nuclear envelope fusion (Kurihara et al 1994). The zygote then re-enters the mitotic cycle and produces buds, each of which contains a single diploid nucleus. The buds are normal mitotic cells until nutrient limitation (e.g. nitrogen depletion) signals entry into the meiotic pathway. Meiosis in each diploid cell produces four haploid spores, encased in a sac called an ascus. Upon germination, each spore enters the mitotic cycle as a normal haploid cell.

In recent years, the yeast nuclear fusion pathway has been the subject of particular study because of several distinctive attributes (Rose 1991). Most importantly, the formation of the final diploid nucleus is exquisitely sensitive to the functions of both microtubules and the nuclear envelope, two areas of intensive interest in their own right. However, because mating and nuclear fusion are not essential for viability, mutations were readily isolated that blocked nuclear fusion without leading simultaneously to a serious block in cell division. Some of these mutations cause defects in a nonessential function of otherwise essential cell structures (e.g. SPB, microtubules, and nuclear envelope). Nevertheless, the specificity of the mutant defects suggests that many of the gene products act relatively directly in the nuclear fusion pathway. The promise of this approach was that additional genetic and molecular analysis would expand the set of genes and proteins known to contribute to the essential structures. Similarly, homologues of interesting genes might be found that act in related functions in the mitotic cycle. In many respects, this initial promise has been fulfilled, with progress in both mitosis and nuclear fusion arising from the insights gained in studying yeast nuclear fusion.

YEAST MATING AND NUCLEAR FUSION

A highly detailed description of the yeast mating pathway is beyond the scope of this review and has been well and thoroughly covered (Cross et al 1988, Sprague & Thorner 1994). For current purposes, we touch on some of the salient features of nuclear fusion from earlier studies (Rose 1991). The detailed cytology of conjugation and yeast nuclear fusion have been extensively described (Byers & Goetsch 1975, Byers 1981b, Hasek et al 1987, Rose & Fink 1987, Read et al 1992) and reviewed (Cross et al 1988, Rose 1991, Sprague & Thorner 1994). A schematic view of the pathway is shown in Figure 1. In response to mating pheromone, cells develop a characteristic pointed mating projection. The projection causes the cell to become pear-shaped, and is sometimes called a "shmoo." In wild-type mating, the cell walls of the mating cells adhere and then break down in the region of adhesion. Plasma membrane fusion must occur very rapidly thereafter because unfused cells are not normally found as a distinct intermediate. Prior to cell fusion, the nuclei are positioned close to the tip of the mating projection. As observed by immunofluorescence microscopy, a microtubule bundle emanates from the SPB and terminates close to the tip of the projection. The use of tubulin mutants (Read et al 1992) and microtubule depolymerizing drugs (Delgado & Conde 1984, Hasek et al 1987) indicates that the position of the nucleus at this point is dependent on the microtubules, and

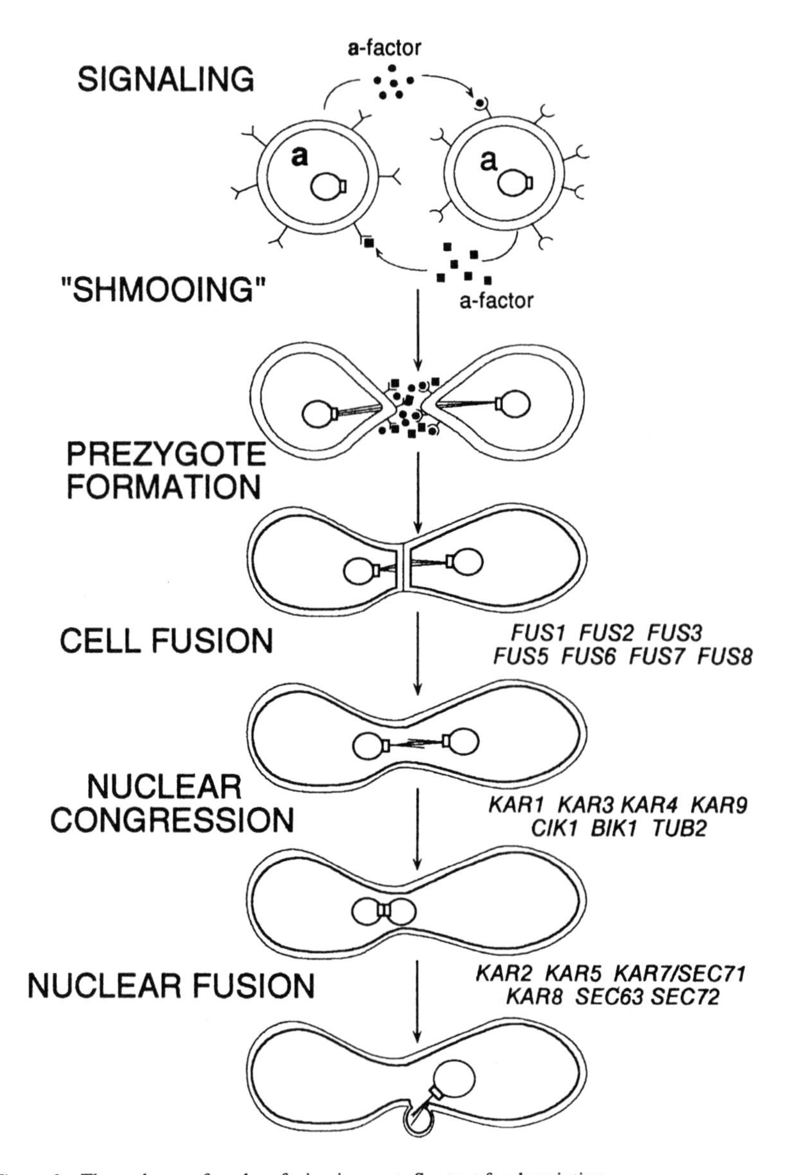

Figure 1 The pathway of nuclear fusion in yeast. See text for description.

it is likely that pre-positioning of the nucleus is important for efficient nuclear fusion. Coincident with plasma membrane fusion, microtubules connect the two nuclei, producing a small spindle-like array. The nuclei move together and nuclear envelope fusion ensues along one edge of the SPB. The resultant single fused SPB has a residual crease along the suture line. As the SPBs fuse, the nuclear membranes fuse such that the lumens of the two nuclear envelopes and the two nucleoplasms become continuous.

Membrane Fusion

The detailed events during nuclear membrane fusion have not been carefully described largely because of the dense staining of the SPB by EM techniques. Therefore, the exact number of membrane fusion events and their topology are not known. Theoretical considerations indicate that either one or three membrane fusion events must occur to give rise to a single nucleus with a single SPB (Figure 2). If it is assumed that the SPB is embedded in the nuclear envelope with a topology similar to a nuclear pore, then the SPB is the site where the inner and outer nuclear envelopes are continuous. As such, a single membrane fusion event along one edge of the pore, followed by membrane flow, is theoretically sufficient to fuse the two nuclei. In this one-step model, the two SPBs would likewise fuse along the same edges. Thus along that edge, the former boundaries between the SPBs and membranes become sites of membrane fusion and SPB fusion. Topological considerations show that this overall fusion event opens a pore between the two nucleoplasms. Subsequent dilation of the pore leads to completion of nuclear fusion.

The alternate model is somewhat easier to visualize. In the three-step model, complete nuclear fusion arises from a series of three fusion events: first between the outer nuclear envelope membranes, then between the inner nuclear envelopes, and finally between the two SPBs. The first fusion event results in continuity between the lumens of the two nuclear envelopes. The second fusion event results in the continuity of the two nucleoplasms. Finally, to produce the single SPB observed in the zygote, the two SPBs must fuse. Given the topology of the embedded SPBs, this last event must also involve membrane fusion. Fusion would be a pinching down between the inner and outer membranes such that the two pores become one. Although there is no strong evidence to support either model, the first model appears to be more consistent with the available data, although there is a strong appeal for the apparently simpler second model.

Regulation of Nuclear Fusion

Several observations have indicated that some components of the nuclear fusion pathway must be regulated by response to mating pheromone. The most important were experiments examining the fate of nuclei during spheroplast fusion

(Curran & Carter 1986, Rose et al 1986). These showed that mitotic nuclei were generally incompetent for nuclear fusion. However, if cells were pretreated with mating pheromone, the nuclei fused with high efficiency, approaching that of normal matings. Other experiments suggest that the two nuclei have only a narrow window of opportunity for fusion. For example, for recessive mutations that block nuclear fusion, subsequent buds are no more likely to be diploid than the initial bud (Lancashire & Mattoon 1979). The simplest interpretation of this observation is that the zygote becomes equivalent to a nonmating **a**/α diploid and therefore refractory to further response to mating pheromones. These data together clearly indicate that the nuclear fusion pathway is likely to depend upon two classes of genes, some of which are induced by mating pheromone and others such as tubulin that are constitutively expressed. These points are discussed further below.

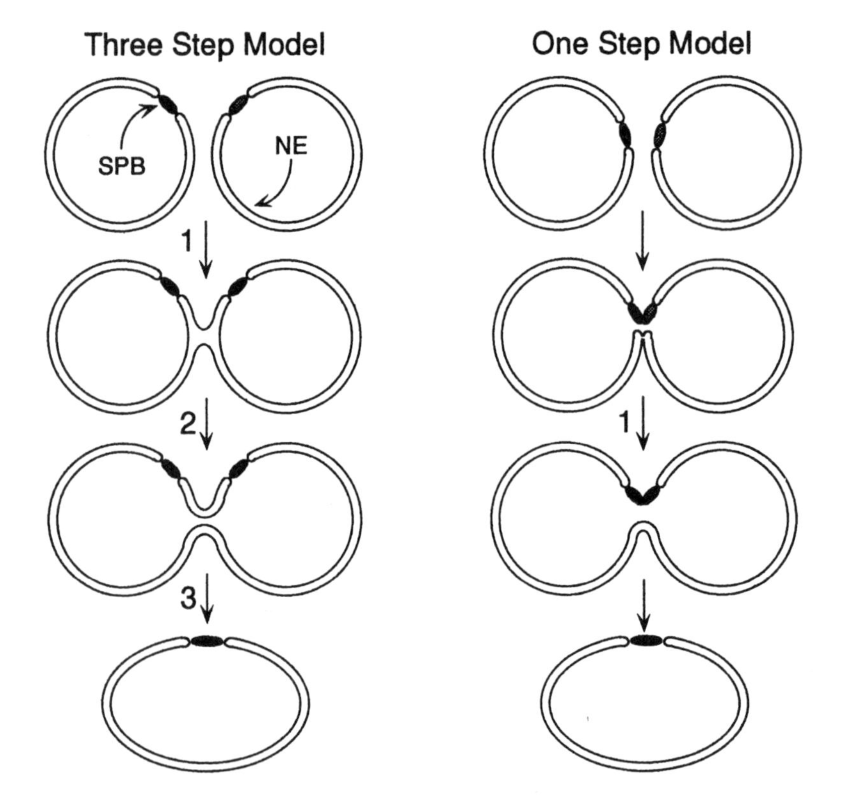

Figure 2 Two models for nuclear envelope fusion. The numbers refer to the specific membrane fusion events. SPB, spindle pole body; NE, nuclear envelope. See text for a detailed description.

APPROACHES TO THE STUDY OF NUCLEAR FUSION

Until recently, nuclear fusion has been studied exclusively by either genetic or cytological methods. The development of cell-free in vitro assays for the fusion of nuclear and ER-derived membranes promises to elucidate the detailed mechanism of the membrane fusion steps (Latterich & Schekman 1994, Latterich et al 1995). Similarly, the development of assays for the functions of individual gene products has helped clarify their roles in nuclear congression.

Karyogamy Mutations

The mutations that block nuclear fusion can be broadly classified into two groups: unilateral mutations that block fusion when mutants are mated to wild-type cells and bilateral mutations that only manifest their defects when two mutant cells are mated together. Of necessity, the first set of karyogamy mutants were either unilateral (e.g. *kar1-1, kar2-1*, and *kar3-1*) (Conde & Fink 1976, Polaina & Conde 1982) or discovered by serendipity (e.g. *bik1-1*) (Berlin et al 1990). The basis of the bilateral mutants can be readily understood as recessive mutations affecting proteins that are diffusible in the zygote. An alternate possibility is that they affect proteins that are not required to act on both nuclei. However, no clear examples of this class exist in nuclear fusion. Nonetheless, there are several mechanisms by which unilateral mutants can arise. One type of unilateral mutation (e.g. *kar3-1*) produces a dominant-negative protein. A second type is comprised of recessive mutations in proteins that are not diffusible between the nuclei or that can not be functionally assembled into the nuclear structure during the brief window of opportunity.

The initial isolation of the unilateral mutants depends on the properties of the unique "cytoductant" cells that arise after a failure in nuclear fusion (Conde & Fink 1976, Lancashire & Mattoon 1979). Cytoductants are defined as the haploid cells that contain the nucleus from one parent and the cytoplasm from the other parent. In practice, the cytoductant is selected to contain the Kar-mutant nucleus by using a pre-existing recessive drug-resistant mutation to select against diploid buds. The cytoplasm of the other parent is selected on the basis of its mitochondrial genome. These methods are reviewed in Berlin et al (1991) and Rose (1991).

Because of the biases in the selection for the unilateral mutants, several screens have been developed to specifically isolate bilateral mutants. In principle, these screens all require that after mutagenesis, the mating types of the candidate mutants can be switched, the candidates are then allowed to mate with themselves, and the mutants are recognized by the consequences of a self-sterile mating defect. One screen used the *HO* gene to switch mating type (Berlin et al 1991). The mating defect was recognized by the presence of residual fertile cells of both mating types in colonies that grew up from mutagenized *HO* cells. However, no new *kar* mutants were reported from that screen.

A novel screen (Kurihara et al 1994) used plasmid segregation to generate cells capable of mating as either **a** or α cells. This system takes advantage of the fact that cells deleted at the mating type locus mate like **a** cells, but the deletion is recessive to the wild-type MATα gene on a plasmid. The initial strain mates as an α; at each cell division, a few percent of the cells missegregate the plasmid, generating **a**-like cells that mate with their sisters. Such matings generate cells of diploid (or higher) ploidy in the population. The number of diploid cells is measured in a semi-quantitative way using a plasmid-loop-out test (Chan & Botstein 1993). In this system, one gene (e.g. *HIS3*) is disrupted by an integrated plasmid bearing the sole wild-type copy of a second gene (e.g. *TRP1*). The plasmid can loop-out by virtue of homologous recombination between repeated sequences from the disrupted gene. The looped-out plasmid lacks an origin of replication and is rapidly lost. Thus in this example, the starting strain is His− and Trp+. The loop-out is His+ and Trp−. Only diploids that carry two copies of the disruption can be both His+ and Trp+ simultaneously. Wild-type strains bearing the construct frequently papillate to His+, Trp+. Mutants with defects anywhere in the mating pathway are detected by their reduced frequency of papillation. Microscopic examination then allows the different types of mutants to be sorted into classes. By this route, mutations in five additional genes—*KAR4, KAR5, KAR7, KAR8,* and *KAR9*—were identified (Kurihara et al 1994).

The new karyogamy mutants fall into two distinct classes, depending upon the final position of the nucleus in unbudded zygotes. The class I mutants contain nuclei whose DNA was separated by approximately 3.3 microns. In contrast, the class II mutants contain nuclei with DNA separated by 0.3 microns or less. The class I genes consist of *KAR1, KAR3, TUB2, BIK1, CIK1, KAR4,* and *KAR9.* This class was determined to be specifically defective in nuclear congression (Figure 1) because of the failure of the nuclei to become closely apposed, the presence of several microtubule-associated proteins in the group, and the detection of aberrant cytoplasmic microtubules in shmoos and zygotes for many of the mutants.

The class II genes are comprised of *KAR2, KAR5, KAR7,* and *KAR8.* Immunofluorescence and EM studies demonstrated that the nuclei become closely apposed in the zygotes but the nuclear envelopes remain separate and unfused over a large region of close contact. Cytoplasmic microtubules appeared normal in both shmoos and zygotes. For these reasons and the results described below, this set of mutants has been recognized as defective specifically in nuclear envelope fusion (Figure 2). As expected, strains containing both class I and class II mutations show a class I defect, indicating that the putative nuclear congression step does occur prior to fusion. Intriguingly, the nuclear envelope in the class II zygotes appears to be connected by thin membranes. Their appearance in mutants blocked in nuclear envelope fusion raises the possibility that the bridges may represent the products of aborted membrane fusion events. The morphology of the bridges varies somewhat between the different mutants, which lends credence to the hypothesis that they arise from different membrane fusion intermediates.

Biochemical Assays of Nuclear/ER Fusion

Recently, Latterich & Schekman (1994) developed a novel cell-free assay for the fusion of membranes derived from the nuclear envelope and/or endoplasmic reticulum. Their assay utilizes a mutation (*gls1*) in the glucosidase responsible for the removal of a terminal glucosyl residue initially present on core oligosaccharides added to proteins in the ER. A precursor protein, prepro-α-factor, is translocated into the ER-lumen of the *gls1* mutant strain. An ER-enriched membrane fraction is then incubated in the presence of a similar membrane fraction from a wild-type (*GLS1+*) strain. When membrane fusion occurs, the lumenal contents of the two sets of membranes mix, leading to trimming of the core oligosaccharides and a 3K reduction in the molecular weight of the prepro-α-factor. Using microsomes, as much as 20% of the precursor protein can be trimmed in the reaction. Isolated nuclei were up to 40% efficient. It is unclear whether these numbers correspond to the true efficiency of membrane fusion, or whether a higher efficiency is masked by unproductive fusion reactions between membranes derived from the same strain. ER membrane fusion requires ATP, intact membranes, and membrane-associated proteins. Unlike other membrane fusion systems, it is independent of added cytosol and Ca^{2+}.

Examination of various secretion-defective mutants revealed the surprising fact that the ER fusion assay was independent of genes required for membrane fusion in the secretory pathway (e.g. *SEC17* and *SEC18*, encoding the yeast SNAP and NSF, respectively). Furthermore, antibodies directed against components required for ER vesicle trafficking, Sec17p, Sar1p, Sec23p, and Sec13p, were ineffective at inhibiting the fusion reaction in vitro. In contrast, mutations in Kar2p, the yeast BiP homologue, produced profound defects in fusion in vitro. Different temperature-sensitive mutations in *kar2* produced allele-specific temperature-sensitive defects in vitro. Depletion of the wild-type protein resulted in a nonconditional defect. Taken together, these results imply that Kar2p plays a direct, albeit unspecified, role in ER membrane fusion.

Several aspects of the in vitro ER fusion assay suggest it as a good model for the events that occur during nuclear fusion. Most importantly, there is a relatively good correlation between the genetic requirements for nuclear fusion and in vitro ER fusion. The remaining class II karyogamy mutants (*kar5, kar7,* and *kar8*) all exhibit an ER fusion defect in vitro (Latterich & Schekman 1994). Furthermore, the in vitro ER fusion is not affected by mutations in several genes not involved in nuclear fusion in vivo (i.e. *sec17* and *sec18*).

However, in other potentially revealing ways, the reaction is not like nuclear fusion. First, ER fusion does not require the activities of the class I genes *KAR1* and *KAR3*. It is reasonable that large nuclei may require microtubule-dependent movement for congression in the dense cytoplasm. In vitro, it is likely that the requirement for active movement can be overcome by the smaller size of the microsomes and the reduced viscosity of the diluted cytosol. Second, when bilateral mutants were examined (*kar5, kar7, kar8*), they were found to exhibit unilateral defects in vitro. That is, mutations in either the donor or the acceptor membrane alone were sufficient to block membrane fusion. This discrepancy can be explained by the observation that all the class II genes encode ER membrane proteins (Ng & Walter 1996) (C Beh, V Brizzio, D Huddler & M Rose, manuscript in preparation), and it is likely that the proteins must be properly localized to function. The conditions of the in vitro fusion assay were not compatible with translocation of proteins into the ER. Therefore, even if wild-type precursor proteins were available for translocation, they could not be transported to their site of function. In vivo, of course, translocation can occur and the relevant nascent protein precursors are available for import into the mutant nuclear envelope. Given that, we must still explain the unilateral nature of recessive *kar2* mutants. The Kar2 protein is present at very high concentrations within the lumen of the ER/nuclear envelope, and the threshold requirement for this protein in ER fusion is high. Therefore, the window of opportunity is likely to be too short for sufficient wild-type Kar2p to accumulate in the mutant nuclear envelope.

Other discrepancies between the two systems are potentially more disquieting. First, the in vitro assay occurs efficiently between membranes from mitotic extracts and is not stimulated by addition of cytosol from pheromone-induced cells (Latterich & Schekman 1994), which raises three distinct possibilities. First, the assay may be measuring a distinct reaction in which some components are shared with nuclear fusion. Second, any putative rate-limiting pheromone-induced component may be a lumenal or integral membrane protein and therefore not present in the cytosol. This possibility is consistent with the observation that thus far all the class II genes encode ER membrane proteins (Ng & Walter 1996) (C Beh, V Brizzio, D Huddler & M Rose, manuscript in preparation). Third, any pheromone-induced components may be restricted to the nuclear congression part of the process. The third possibility is ruled out by the observation that at least one of the class II proteins, Kar5p, is strongly induced by mating pheromone (C Beh & M Rose, manuscript in preparation).

Direct evidence in support of the possibility that the two reactions are distinct came from the discovery that the in vitro reaction is strongly dependent upon a homologue of Sec18p, encoded by the *CDC48* gene (Latterich et al 1995). Mutations in *cdc48* result in a block in mitosis during anaphase, which originally was interpreted as a spindle elongation defect (Frohlich et al 1991). Re-examination of the data suggests that it can also be the result of a defect in nuclear fission. In either case, it is clear that Sec18p is not required in the in vitro reaction because of the existence of an alternate Sec18p-like protein. However, of crucial significance to nuclear fusion, Cdc48p does not seem to be required for nuclear fusion in vivo (C Beh & M Rose, unpublished observations). Thus the two reactions may not be equivalent, but would still share components. This view is consistent with general observations that most, if not all, of the proteins required for nuclear fusion are recruited from other pathways in the cell.

What then is the normal mitotic function of the class II genes? Of particular significance to this question was the subsequent observation that mammalian homologues of Cdc48p are required for the fusion reactions that reform the Golgi after mitosis (Acharya et al 1995, Mellman 1995, Rabouille et al 1995). Possibly, they are also required for fusion reactions that are part of the mechanism of nuclear fission or other types of nuclear envelope/ER remodeling. These observations suggest that there may be another Sec18p homologue that is induced by mating pheromone and is required for nuclear membrane fusion during mating.

NUCLEAR CONGRESSION

As described above, karyogamy can be viewed as having two major steps: nuclear congression and nuclear fusion. Nuclear congression is highly dependent on normal microtubule function and the in vivo assays for nuclear fusion are exquisitely sensitive to perturbations in their functions. As described below, molecular and biochemical characterization of some of the required proteins has allowed the construction of specific models for nuclear congression.

KAR3

Based upon sequence identity, Kar3p was identified as an unusual kinesin-related protein (Meluh & Rose 1990). Similar to kinesin (Yang et al 1989), Kar3p is predicted to have three domains: a motor domain sharing 39% identity with kinesin, a central coiled-coil domain similar to kinesin, and a small globular nonmotor domain unrelated to kinesin. The order of the domains is opposite between the two proteins; the motor domain is N-terminal for kinesin and C-terminal for Kar3p. *KAR3* is most closely related to a handful of other C-terminal kinesins (Endow 1991). Two are notable: ncd from *Drosophila* (McDonald & Goldstein 1990, McDonald et al 1990a, Walker et al 1990) and KlpA from *Aspergillus nidulans* (O'Connell et al 1993). At 46% identity, the ncd motor domain is more closely related to Kar3p than it is to other members of the family. The ncd protein was the first example of a minus-end-oriented kinesin-related protein. Whereas authentic kinesin and certain other members of the kinesin superfamily move preferentially towards the plus-end of the microtubule, ncd moves toward the minus-end. KlpA appears to be most closely related to Kar3p. Although it appears to have no essential function in mating and meiosis in *Aspergillus*, KlpA expressed in *S. cerevisiae* can partially suppress the mating defect of a *kar3* mutation.

Two different types of mutations in Kar3p have been reported to affect nuclear fusion. Several dominant-negative mutations (e.g. *kar3-1*) give rise to a unilateral defect. These all map within the conserved ATP-binding site (Meluh & Rose 1990, Roof et al 1991, Meluh 1992) and, based upon the analogy to kinesin, these should cause the protein to bind tightly to microtubules but be incapable of movement. In contrast, null alleles constructed in vitro are fully recessive and bilateral, showing that wild-type Kar3p is diffusible in the zygote and that the dominant-negative mutants block nuclear fusion. In zygotes made from the null mutants, both nuclei show properly oriented but highly elongated cytoplasmic microtubule bundles that do not interact with each other. In zygotes made from the *kar3-1* mutant, the two elongated microtubule bundles are continuous with each other. These phenotypic observations suggest that Kar3p is required to cross-bridge the two bundles of microtubules and thereby provide the force to move the two nuclei together.

Hybrids of Kar3p fused to β-galactosidase localize to two sites in shmoos, the distal ends of the cytoplasmic microtubules and the SPB (Meluh & Rose 1990a, Vallen et al 1992a, Page et al 1994). The hybrids demonstrated that the N-terminal half of Kar3p contains sequences for microtubule association that are independent of the motor domain. Using antibodies to authentic Kar3p, mutant Kar3-1p was found all along the cytoplasmic microtubules and in zygotes was particularly prominent in the region of overlapping microtubule bundles. The localization data are consistent with the model of Kar3p cross-bridging the cytoplasmic microtubules via its two different domains for microtubule association and thereby providing the force for nuclear congression by sliding them one against another. Localization of Kar3p to the SPB in shmoos is dependent on one specific domain of the SPB protein Kar1p (Vallen et al 1992a) and is not affected by treatment with nocodazole (Meluh & Rose 1990). These results suggest that Kar3p's SPB localization reflects a second function for Kar3p in nuclear fusion and that Kar1p's role in nuclear fusion is mediated by Kar3p.

The directionality of Kar3p movement was investigated using a hybrid protein containing the motor domain fused to glutathione-S-transferase (Endow et al 1994). The hybrid protein, expressed and purified from *Escherichia coli,* exhibited all the properties of an authentic kinesin-like motor protein; it is a microtubule-stimulated ATPase, binds to microtubules in a nucleotide-sensitive fashion, and can move microtubules at 1–2 μm/min in a minus-end directed fashion. Concurrently, authentic Kar3p was identified as a minus-end-oriented motor activity associated with kinetochore protein fractions purified from *S. cerevisiae* (Middleton & Carbon 1994). Thus there can be little doubt that Kar3p is a minus-end-oriented motor similar to ncd. This finding is gratifying in that it is consistent with the simplest model for nuclear congression. It is generally thought that the cytoplasmic microtubules are oriented with their plus-ends extending away from the SPBs. A motor moving the nuclei together by acting on the overlapping microtubules between them must be minus-end-oriented.

A second activity of the GST-Kar3p hybrid was discovered in the in vitro motor assays (Endow et al 1994). It was noticed that the taxol-stabilized microtubules were shrinking at a rate of 0.4 μm/min as they moved during the assay. Microtubules incubated in the absence of GST-Kar3p were stable. Microtubules nucleated from flagellar axonemes were also relatively stable in the presence of GST-Kar3p. Because the axoneme-grown microtubules have free plus-ends, but not free minus-ends, the clear implication is that Kar3p destabilizes microtubules from their minus-ends. This observation is consistent with two, otherwise puzzling, in vivo phenotypes associated with *kar3* and *kar1* mutants. First, the mitotic defects of *kar3* null mutants can be partially suppressed by the presence of low concentrations of benomyl, a microtubule depolymerizing drug (Roof et al 1991, Meluh 1992). Although there are other interpretations for this fact, it suggests that the absence of Kar3p stabilizes microtubules and that Kar3p-mediated depolymerization is required at some stage of the cell cycle. Second, both *kar1* and *kar3* mutants are notable for the excessive length of the cytoplasmic microtubules in shmoos and zygotes (Rose & Fink 1987, Vallen et al 1992a). That the absence of wild-type Kar1p function acts locally on the cytoplasmic microtubules can be seen in matings between a *kar*1 mutant and wild-type (Rose & Fink 1987). In this case, a long microtubule bundle is seen emanating from only one nucleus. The simplest interpretation of these results is that some Kar1p-dependent microtubule length-determining function is present at the SPB. Given that Kar3p-β-galactosidase localization is dependent on Kar1p and that GST-Kar3p destabilizes microtubules from their minus-end, the implication is that a role for Kar3p in congression is to depolymerize the cytoplasmic microtubule bundle.

Depolymerization may seem a superfluous function in congression given the much faster rate of directed movement. However, further considerations suggest that such a function might be essential to nuclear fusion. Without some mechanism for removing the microtubules, nuclei can come no closer than the length of the longest cytoplasmic microtubule in the bundle. Indeed the appearance of the *KAR*$^+$ by *kar1*-zygote suggests that the nuclei are kept separate by the elongated microtubule bundle. Although it is assumed that depolymerization occurs during congression, it is also likely that depolymerization occurs prior to cell fusion. As noted above, *kar1* and *kar3* shmoos already have elongated cytoplasmic microtubules prior to cell fusion (Rose & Fink 1987, Vallen et al 1992a). Therefore, a negative regulator of cytoplasmic microtubule length must be active in shmoos and defective in the mutants.

In wild-type shmoos, the nucleus moves into the shmoo projection with the SPB oriented toward the tip of the projection. In numerous studies, it is clear that the position of the nucleus at this time depends on the integrity of the cytoplasmic microtubules. The *kar1* and *kar3* mutations do not appear to affect nuclear orientation per se, but do affect whether the nucleus migrates into the shmoo projection. Mutations affecting a membrane protein Fus2p affect both nuclear orientation and nuclear fusion (Trueheart et al 1987, Elion et al 1995). As discussed below, Fus2p is required for cell fusion and is localized to the tip of the shmoo projection. In contrast, mutations in cytoplasmic dynein (Eshel et al 1993, Li et al 1993), which is required for mitotic nuclear migration, do not affect nuclear fusion. One model consistent with these data suggests that Fus2p helps anchor and stabilize the plus-ends of the cytoplasmic microtubules at a cortical site in the shmoo tip. Acting at the SPB, Kar3p-dependent depolymerization of the cytoplasmic microtubules would draw the nucleus into the shmoo projection.

CIK1

The *CIK1* gene was identified in a genetic screen designed to find mutations that affect both chromosome stability and nuclear fusion (CIK1: chromosome instability and karyogamy) (Page & Snyder 1992). The *cik1* deletion is viable but temperature-sensitive for growth and exhibits an elevated rate of chromosome loss during mitotic growth. During mating, the deletion strain shows a severe bilateral nuclear fusion defect, similar to that of a *kar3* deletion. Like *kar3* mutants, *cik1* mutants exhibit a congression defect with elongated cytoplasmic microtubules that are oriented properly. The *CIK1* mRNA is strongly induced during mating, indicating that it also plays a specific role during nuclear fusion. Cik1p is 594 amino acids in length and is predicted to possess a central 300-amino acid coiled-coil domain. Authentic Cik1p can be detected at the SPB in both mitotic and mating cells. In shmoos, Cik1-β-galactosidase hybrids show a localization pattern that is identical to that of Kar3p-β-galactosidase hybrids (Page et al 1994). The localization of each hybrid is dependent on the integrity of the other protein (i.e. Cik1-β-galactosidase is dependent on Kar3p and Kar3p-β-galactosidase is dependent on Cik1p). Furthermore, both hybrids are dependent on Kar1p for localization to the SPB but not to the cytoplasmic microtubules. In mitotic cells, the two proteins can be shown to interact in two-hybrid experiments. In mating cells the two proteins coimmunoprecipitate. Taken together, these data strongly suggest that Cik1p is a light chain for Kar3p during mating.

Mitotic Roles for KAR3 and CIK1

Although *CIK1* and *KAR3* are induced by pheromone, mutants both have defects in mitosis. Both exhibit an elevated frequency of chromosome loss, although additional data suggest that Cik1p and Kar3p may have different functions in mitotic cells. First, the two deletion mutants exhibit somewhat different mitotic phenotypes; *kar3* mutants are extremely slow growing at all temperatures (Meluh & Rose 1990); *cik1* mutants are only temperature sensitive (Page & Snyder 1992). Up to 30 to 40% of *kar3* deletion cells are inviable and arrested in the G2/M of the cell cycle (large bud, short spindle, undivided nucleus) (Meluh & Rose 1990). The *cik1* deletion does not result in such a severe defect. Second, mutations in *kar3* suppress mutations in *cin8* and *kip1* (Saunders & Hoyt 1992, Hoyt et al 1993), whereas *cik1* mutations do not (Page et al 1994). Cin8p and Kip1p (Roof et al 1991, Hoyt et al 1992) belong to a subfamily of kinesin-related proteins, some of whose members have been shown to be plus-end-oriented microtubule motors (Endow 1991, Barton & Goldstein 1996). Cin8p and Kip1p (as well as other members of the family) have been implicated in playing essential roles in spindle assembly (Roof et al 1991, Hoyt et al 1992) and spindle elongation (Saunders et al 1995). This observation implies that Cik1p is not required for this aspect of Kar3p function in mitosis. Third, whereas Cik1p-β-galactosidase and Kar3p-β-galactosidase hybrids localize in the nucleus, near the spindle poles in mitotic cells, the localization of Kar3p-β-galactosidase is independent of Cik1p (although the converse does not appear to be true) (Page et al 1994). Finally, although the two proteins are clearly in a complex in mating cells, the two proteins do not readily coimmunoprecipitate in mitotic cells (Page et al 1994). Taken together, these data strongly suggest that Kar3p and Cik1p are not physically associated in mitotic cells and have somewhat different mitotic roles. It seems likely that the two proteins do physically associate as a regulated response to mating pheromone.

During mitosis, Kar3p is in the nucleus where it is engaged in spindle dynamics (Meluh 1992, Page et al 1994). Biochemical data implicate Kar3p as a kinetochore motor (Middleton & Carbon 1994). Its minus-end-oriented motor activity is consistent with a role in anaphase A: the movement of the chromosomes toward the spindle poles. Because yeast chromosomes do not strongly condense in mitosis, clear evidence for anaphase A in *S. cerevisiae* has not been published. However, detailed experiments with fluorescent in situ hybridization to detect the behavior of individual chromosomes promises to provide strong evidence in this regard (D Koshland, personal communication). Given the fact that *KAR3* is not an essential gene, it is somewhat surprising to suggest that Kar3p might be required for anaphase A. However, anaphase B in yeast results in an at least eightfold increase in the length of the mitotic spindle, which is more than sufficient to separate the chromosomes, even without anaphase A. The specific role for Cik1p in mitosis is less clear, but its localization near the SPB clearly reflects a role in spindle function (Page & Snyder 1992, Page et al 1994).

A clear function of the regulated interaction between Kar3p and Cik1p is to effect a change in the specific localization of Kar3p (Page et al 1994). Upon treatment with mating pheromone, both Kar3p-β-galactosidase and Cik1p-β-galactosidase move into the cytoplasm where they localize to the cytoplasmic microtubules and SPB. In *cik1* mutants, Kar3p-β-galactosidase stays in the nucleus, which demonstrates the requirement for Cik1p in relocalization. The specific mechanism by which the two proteins change their interaction and localization is not known, although Cik1p is reported to be modified in a way that suggests phosphorylation. It is not known whether either protein is associated with other specific proteins in mitosis.

KAR1

As discussed above, it is clear that Kar1p is required to localize Kar3p and Cik1p to the SPB. *KAR1* is essential for mitotic growth, and temperature-sensitive mutations result in a block in the initial stages of SPB duplication (Rose & Fink 1987). The Kar1p-β-galactosidase hybrids were localized to the newly replicated SPB (Vallen et al 1992b). This remarkable result indicated that Kar1p was likely to be an SPB component and that the interaction of Kar1p with SPB components might be cell cycle-regulated. Furthermore, the finding that the new SPB is preferentially segregated into the bud implies that the nucleus has an inherent polarity that is aligned with the cell axis. After SPB duplication, the new SPB remains oriented toward the bud and the old SPB moves around to the opposite side of the nucleus during spindle assembly. These results imply that the new SPB must be tethered specifically to a cortical site during the critical movement. Conclusive evidence that Kar1p is a SPB component came from immuno-EM studies showing that the protein is present in the half-bridge structure associated with the SPB (Spang et al 1995).

Kar1p is organized into at least three functional domains (Vallen et al 1992a). A central domain is required for SPB duplication and is both necessary and sufficient to target Kar1p-β-galactosidase hybrid proteins to the SPB. Deletions in this interval do not affect nuclear fusion. An N-terminal domain is required only for nuclear fusion and is necessary for the mating-specific localization of Kar3p-β-galactosidase to the SPB. A third domain of Kar1p, important for both mitosis and karyogamy, includes the hydrophobic carboxyl terminus and is sufficient to target a β-galactosidase-Kar1p hybrid to the nuclear envelope. The essential mitotic regions of KAR1 comprise only a small portion ($\approx$ 20%) of the coding sequence of this 433-residue protein. Overall, these data suggest that Kar1p is composed of two or more protein-binding domains tethered to the nuclear envelope via the hydrophobic tail.

The mitotic function of Kar1p was initially revealed by genetic studies of a unique temperature-sensitive deletion in the essential SPB domain of *KAR1* (Vallen et al 1994). Both dominant and high-copy suppressors were found to be allelic to *CDC31*, which is required for SPB duplication and encodes a calmodulin-like protein most closely related to centrin, a highly conserved protein found in the centrosome of many eukaryotes (Baum et al 1986, Baron & Salisbury 1988, Huang et al 1988). Temperature-sensitive *cdc31* mutations result in a defect in SPB duplication that is identical to the *kar1* mutation (Baum et al 1986), and Cdc31p is also localized to the cytoplasmic side of the SPB half-bridge (Spang et al 1993, Biggins & Rose 1994). One *CDC31* suppressor allele conferred a temperature-sensitive defect in SPB duplication that was counter-suppressed by recessive mutations in *KAR1* (Vallen et al 1994). The strongest *CDC31* alleles suppressed a complete deletion of *KAR1*. These bypass suppressors indicate that Kar1p acts through Cdc31p to initiate SPB duplication and that in the suppressor alleles, Cdc31p must act independently of Kar1p, possibly via a downstream effector.

Proof of a direct interaction between Kar1p and Cdc31p came from in vitro binding studies (Biggins & Rose 1994, Spang et al 1995). Cdc31p, expressed and purified from *E. coli,* binds to Kar1p immobilized on filters. As predicted by the suppressor studies, deletion mapping showed that Cdc31p interacts specifically with a short sequence within Kar1p's essential SPB duplication domain (Biggins & Rose 1994, Spang et al 1995). The Cdc31p-binding sequence contains three acidic amino acids not found in calmodulin-binding peptides, which explains why Kar1p binds Cdc31p but not calmodulin (Spang et al 1995).

Cdc31p was also able to bind to the carboxyl terminus of Nuflp/Spc110p, a major calmodulin-binding component of the SPB (Spang et al 1995). However, the significance of this interaction in vivo is unclear.

In addition to the suppressor studies, immunofluorescence microscopy demonstrated that the interaction between Cdc31p and Kar1p is functionally significant in vivo. In a *kar1* mutant, Cdc31p fails to localize to the SPB (Biggins & Rose 1994, Spang et al 1995). In *kar1* mutants containing the dominant *CDC31* suppressor mutations, Cdc31p again localized to the SPB (Biggins & Rose 1994). Furthermore, the localization of Cdc31p to the SPB was affected by the overexpression of Kar1p and Kar1p-β-galactosidase such that Cdc31p accumulates in aggregates formed by the overexpressed proteins (Biggins & Rose 1994). On the other hand, localization of Kar1p to the SPB is independent of Cdc31p (Spang et al 1995). Taking these data together, we propose that the essential function of Kar1p is to localize Cdc31p to the SPB and that this interaction is normally required for SPB duplication. Given that the suppressor mutant forms of Cdc31p can localize to the SPB in the absence of Kar1p, there must be other Cdc31p interacting proteins in the SPB, and the suppressors may enhance their interaction in some way. Several studies are currently underway to identify these Cdc31p interacting proteins because they are likely critical functional components in the initiation of SPB assembly.

Other Genes Affecting Nuclear Congression

Given the nature of their mutant defects and the functions of the identified gene products, all the nuclear congression mutants probably affect microtubule function. In the following cases, specific functions have not yet been identified for the gene products.

BIK1 Identified serendipitously, *BIK1* is required for the complete array of microtubule-associated functions in yeast (Berlin et al 1990). Although the gene is not essential for viability, deletion causes a severe bilateral defect in nuclear fusion, elevated rates of chromosome loss, and frequent missegregation of the nucleus during mitosis. The *bik1* null mutants contain very short or undetectable cytoplasmic microtubules in both mating and mitosis. Nuclear orientation/migration is defective, and mitosis often occurs within the mother cell, which leads to formation of both binucleate and anucleate cells. Cells overexpressing Bik1p initially have long cytoplasmic microtubules and abnormally short spindle microtubules. After longer periods of overexpression, cells lose all microtubules and arrest at G2/M. In addition, synthetic lethality was observed in double mutant strains containing mutations in *BIK1* and either *TUB1* or *TUB2*, genes that encode α or β-tubulin, respectively. Consistent with these phenotypes, Bik1p colocalizes with both cytoplasmic and nuclear microtubules.

From these results, it has been proposed that Bik1p is required stoichiometrically for the formation or stabilization of microtubules during mitosis and conjugation. That Bik1p may play a direct role in microtubule stability is supported by the finding that it contains a microtubule-binding domain that is conserved in CLIP-170, a mammalian protein implicated in binding endocytic carrier vesicles to microtubules (Pierre et al 1992, 1994), and the *Drosophila* Glued protein, a component of the dynein complex (Waterman-Storer & Holzbaur 1996).

Recent genetic evidence suggests that *BIK1* is not essential because of the presence of a second gene, *ASE1* (Pellman et al 1995). *ASE1* is a nonessential gene encoding a novel protein that localizes specifically to the mitotic spindle midzone. Strains lacking both *ASE1* and *BIK1* are inviable, and a conditional *bik1 ase1* double mutant appears to be defective for anaphase spindle elongation. These results suggest that Bik1p and Ase1p may have partially overlapping functions during spindle elongation. However, *ase1* mutants do not exhibit defects in nuclear migration and congression. Therefore, it is likely that their activities do not overlap for functions associated with the cytoplasmic microtubules.

KEM1 One of the more puzzling genes involved in nuclear fusion is *KEM1* (also known as *XRN1, DST2, SEP1,* and *RAR5*) (Kim et al 1990a, Larimer et al 1992, Bahler et al 1994, Interthal et al 1995, Johnson & Kolodner 1995), originally identified on the basis that *kem1* mutations enhanced the nuclear fusion defect of a *kar1* mutation (Kim et al 1990). The *kem1* mutants exhibit a bilateral nuclear fusion defect on their own and exhibit a number of defects that implicate a role in microtubule-associated functions (Interthal et al 1995). These include benomyl sensitivity, elevated rates of chromosome loss, impaired SPB separation, and defective nuclear migration to the bud neck. Purified Kem1p promotes polymerization of tubulin from porcine brain and from *S. cerevisiae*. Furthermore, Kem1p co-sediments with these microtubules in sucrose gradient centrifugation. Genetic analysis of double mutant strains containing a mutation in *KEM1* and either *TUB1* or *TUB2* further suggests an interaction between Kem1p and microtubules. Taken together, these results are all highly suggestive of a direct role for Kem1p in microtubule function. In spite of these results, several studies have independently identified the Kem1 protein as being a 5′-3′ exoribonuclease (Larimer et al 1992), a DNA strand transferase (Bahler et al 1994, Heyer et al 1995), and a G4 DNA nuclease (Liu & Gilbert 1994, Liu et al 1995). Genetic studies have also identified it as being required for plasmid stability, and *kem1* mutations are synthetically lethal with mutations in the *ski2* and *ski3* genes implicated in the regulation of killer toxin expression (Johnson & Kolodner 1995). *kem1* mutants clearly have a number of pleiotropic defects, but it is unclear which, if any, of the known mutant defects are the direct result of loss of Kem1p function. An intriguing possibility is that the disparate functions of Kem1p belie a functionally relevant coupling of microtubule binding to DNA strand exchange, possibly related to its role in chromosome pairing and exchange.

KAR4 AND KAR9 Two genes recently identified by bilateral nuclear congression mutations are *KAR4* and *KAR9* (Kurihara et al 1994). The phenotype of *kar4* mutants in mating is identical to that of *kar3* and *cik1* mutants, and results in elongated cytoplasmic microtubules that are properly oriented along the zygote axis toward the opposite nucleus. These results suggest that Kar4p may function with or be required for Kar3p/Cik1p function during mating. Recent data indicate that Kar4p is a transcription factor specifically required for the pheromone-dependent induction of *KAR3* and *CIK1* (Kurihara et al 1996).

The *kar9* mutation causes a defect in nuclear orientation in zygotes. Unlike orientation of microtubules in the *bik1* mutation, cytoplasmic microtubules are present although they are not properly oriented. Other mitotic phenotypes indicate that *KAR9* plays a general role in cytoplasmic microtubule function in mitosis (R Miller & M Rose, manuscript in preparation). One attractive model is that Kar9p helps orient the cytoplasmic microtubules in mating and mitosis.

FUS2 Originally identified in a screen for mutants that show a partial defect in cell fusion during mating, *fus2* mutants also show a fairly strong defect in nuclear fusion (Elion et al 1995). The *FUS2* gene is highly induced only in haploid cells treated with mating pheromone. Fus2p localizes to novel vesicular or cytoskeletal structures at the tip of the shmoo projection and at the junction between cells in the course of fusion. The 617-residue protein is predicted to be largely coiled-coil and consistent with a cytoskeletal function, it fractionates in a high-speed pellet, and is not easily extracted by detergents, high salt, or chaotropic agents. The *fus2* mutant defect in cell fusion is partially suppressed by expression (and overexpression) of the unrelated membrane protein Fus1p. The *fus1* mutants exhibit a cell fusion defect similar to that of *fus2*, and the double mutant shows a much stronger cell fusion defect than either mutant alone (McCaffrey et al 1987, Trueheart et al 1987). However, the *fus1* mutant does not show a nuclear fusion defect. Therefore, it is likely that *fus2*'s nuclear fusion defect is not simply a secondary consequence of the cell fusion defect. In *fus2* mutant zygotes where cell fusion but not nuclear fusion has occurred, the defect appears to be in nuclear congression. In these zygotes, the cytoplasmic microtubules are found to be misoriented with respect to the zygote axis. One attractive model is that Fus2p serves as part of the cortical site for directing the attachment and orientation of the cytoplasmic microtubules in the shmoo.

Less clear is the role of Fus2p in cell fusion, because neither Fus1p nor Fus2p is required for the localization of the other protein.

CIN GENES The *CIN* genes were identified in a genetic screen for mutants exhibiting an elevated rate of loss of chromosome loss (Hoyt et al 1990a, Stearns et al 1990). Mutations in *CIN1, CIN2,* and *CIN4* cause cells to be supersensitive to benomyl, cold sensitive, and defective in both nuclear migration and nuclear fusion. At the restrictive temperature, cold-sensitive mutants lose assembled microtubules and arrest in the cell cycle at G2/M. The mutations show a wide variety of genetic interactions with mutations in the tubulin genes. These results suggest that Cin1p, Cin2p, and Cin4p contribute generally to microtubule stability. The finding that *CIN4* encodes a small GTP-binding protein suggests that Cin4p may be acting in a regulatory fashion.

CDC GENES Several cell division cycle genes have been implicated in the nuclear fusion pathway, including *CDC4, CDC28, CDC34,* and *CDC37* (Dutcher & Hartwell 1982, Dutcher & Hartwell 1983a,b). Matings between these mutants and wild-type at the nonpermissive temperature result in significant increases in the frequency of cytoductant formation, which signals failure in nuclear fusion (except for the *cdc4* mutant, most of the defects are much weaker than those of the *kar* mutants). Cdc28p is the *S. cerevisiae* homologue of the MPF cell-cycle division kinase (Beach et al 1982, Ferguson et al 1986, Dunphy et al 1988).

Cdc4p contains limited homology to β-transducin (Peterson et al 1984, Fong et al 1986) and is required for spindle assembly (separation of the duplicated SPBs). The *cdc4* mutants produce multiple rounds of aberrant buds at the nonpermissive temperature (Adams & Pringle 1984). The specific function of Cdc4p in this context is unknown. Recently, a transducin repeat-containing microtubule-associated protein has been identified, raising the possibility that Cdc4p acts through the microtubules (Li & Suprenant 1994). However, the fact that multiple rounds of budding occur suggests that Cdc28 kinase activity remains high at the nonpermissive temperature. If so, the defect may be the result of prolonged activation of the kinase at the nonpermissive temperature.

Mutations in *CDC34* also lead to unseparated SPBs and a multibudded phenotype (Goebl et al 1988). Cdc34p is a ubiquitin conjugating enzyme (Goebl et al 1988, 1994) required for the degradation of the G1 cyclin Cln3 (Tyers et al 1992, Yaglom et al 1995), as well as a potent inhibitor of the B cyclins (Sic40p) (Schwob et al 1994). This mutant is prevented from entering S-phase because the activity of the G1-cyclins remains high while the B-cyclins stay low. Thus this mutant may also be defective for nuclear fusion because Cdc28 remains active for its G1-kinase activity. A specific role for Cdc37p has not yet been

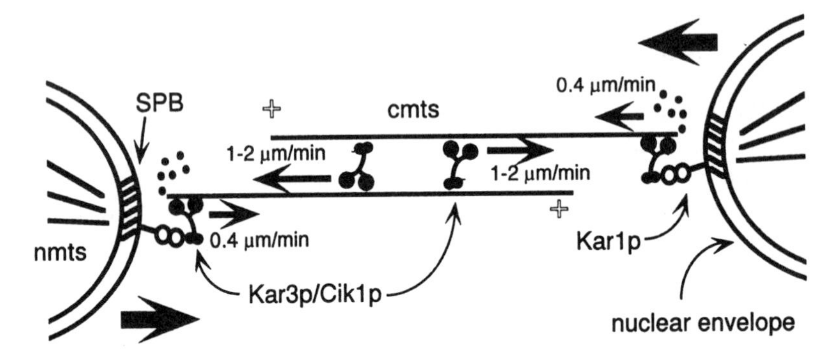

Figure 3 A model for Kar3p/Cik1p in nuclear congression. See text for details. SPB, spindle pole body; nmts, nuclear microtubules; cmts, cytoplasmic microtubules; +, the plus-ends of the microtubules; small dots at the minus-ends of the microtubules depict free tubulin released form the end of the depolymerizing microtubules; small arrows depict the movement of Kar3p/Cik1p relative to the microtubules; large arrows depict the overall movement of the nuclei.

clearly identified; however, a recent report suggests that it is a positive regulator of Cdc28 kinase activity required for the association between the kinase and multiple cyclins (Gerber et al 1995).

These four mutants suggest that there are two types of Cdc28p-associated defects in nuclear fusion: mutations that lead to reduced levels of activity, and mutations that lead to prolonged activation in G1, which lead to nuclear fusion defects. It seems likely that the mechanism of the defects must be distinct. A role for Cdc28p in nuclear fusion is somewhat puzzling, given that the Cdc28p kinase is thought to be inactive during mating (Elion et al 1990a, 1991, Peter et al 1993, Tyers & Futcher 1993). Possibly, the kinase is required prior to cell fusion or in a previous cell cycle to activate a nondiffusible component of the fusion pathway. On the other hand, Cdc28p kinase activity must be inhibited during pheromone arrest by Fus3p and Far1p for normal mating to proceed (Elion et al 1990a, 1991, Peter et al 1993, Tyers & Futcher 1993). Because several components of the nuclear fusion pathway are also utilized for different functions during mitosis, continued activation of the cell cycle may inhibit nuclear fusion.

A Model for Nuclear Congression

A coherent model for the mechanism of nuclear congression and the roles played by specific proteins may now be considered (Figures 1 and 3). Because the cell responds to pheromone, the axis of cell polarity changes such that it becomes oriented toward the mating partner. A shmoo projection forms and grows toward the partner. In the tip of the shmoo projection are cortical proteins that mark the site at which cell fusion and, ultimately, nuclear fusion will occur. One component of this site is likely to be Fus2p. A second component may be Kar9p. Cytoplasmic microtubules are captured by the cortical site and the nucleus moves toward the tip of the shmoo projection. Several proteins are required for the stability of the cytoplasmic microtubules (e.g. Tub2p, Bik1p, Cin1p, Cin2p, and Cin4p), and mutations in these genes affect nuclear fusion at this initial stage of nuclear movement. The force for nuclear movement at this point is not clear, but the mutant phenotypes of these genes suggest that it may come from depolymerization of the cytoplasmic microtubules at the SPB mediated by Kar3p in association with Cik1p and Kar1p. As the cell walls between the two cells and plasma membranes break down and fuse, the cytoplasmic microtubules from the two nuclei interdigitate. The two sets of microtubules are cross-bridged by Kar3p and Cik1p. Kar4p is also required for this step but does not play a direct role. The microtubules slide together by minus-end-oriented motor activity of Kar3p. As a result of this movement and the continued shrinking of the microtubules, the two nuclei become closely apposed. Nuclear envelope fusion then occurs in the vicinity of the SPB.

NUCLEAR MEMBRANE FUSION

In the last few years we have witnessed the exciting coalescence of several lines of research into the mechanism of intracellular membrane fusion. Research in three different systems—transport between Golgi compartments, fusion of synaptic vesicles with the presynaptic membranes, and the identification of genes required for transport from the ER to the Golgi in yeast—demonstrates that one basic mechanism is operative in each type of membrane fusion (Schekman 1992, Bennett & Scheller 1993, 1994, Ferro-Novick & Jahn 1994, Rothman 1994, Rothman & Warren 1994). In each case, an ATPase called NSF (for *N*-ethyl maleimide-sensitive factor; encoded by *SEC18* in yeast), interacts with a protein(s) called SNAP (for soluble NSF attachment protein; encoded by *SEC17*), and the NSF-SNAP complex then interacts with any one of a series of membrane-bound proteins called SNARES (SNAP receptors). The NSF-SNAP complex is a general factor that acts at each stage of the secretory pathway, whereas the SNARES are specific to the stage of secretory or endocytic pathway. Specific v-SNARES are present on the transport vesicles, and the t-SNAREs are present on the target membranes. The formation and function of the complex is then further regulated by interaction with a host of additional proteins, including numerous small GTPases of the Rab family. In yeast, many different SNARES and Rabs have been identified that act at each stage of the secretory pathway (Ferro-Novick & Jahn 1994).

Integral to the SNARE hypothesis model is the idea that the SNAREs and Rabs generate the specificity in the membrane fusion reactions. This seems particularly important given that fusion is occurring between membranes of different compartments (so-called heterotypic fusion reactions) and that a mechanism for ensuring targeting to the appropriate membranes must exist. In the case of nuclear fusion, two membranes of identical origin are involved. In principle, it is possible that the nuclei from the two different mating types could have different SNAREs and thereby conform to the standard model. However, two lines of evidence argue that the nuclei of the same mating type can fuse efficiently during mating. The first evidence came from spheroplast fusion experiments performed with mating type **a** cells treated with alpha factor (Curran & Carter 1986, Rose et al 1986). The second came from the demonstration that the mating type-specific pheromones and pheromone receptors are the major, if not the only, determinants of sexual identity in yeast (Bender & Sprague 1986, Bender & Sprague 1989). The implication is that, aside from the transcription factors that specify mating type-specific genes, internal structures are identical between the two mating types. Thus for nuclear fusion, membranes of identical origin and therefore presumably bearing the identical set of SNAREs must be capable of fusion.

The second difference between the heterotypic membrane fusion events in the secretory pathway and nuclear fusion is the lack of dependence on *SEC18*. As mentioned above, a second NSF-like molecule, Cdc48p, is responsible for the in vitro fusion of nuclear membranes from mitotic cells (Latterich et al 1995). It is possible that the in vitro fusion reaction might also represent a heterotypic reaction, in that it might occur between peripheral ER and nuclear envelope. This would be consistent with the lack of a requirement for Cdc48p in nuclear fusion in vivo. Alternatively, it is possible that an additional NSF-like protein is induced during the mating response. In any event, it will be essential to identify the proteins that are directly responsible for nuclear membrane fusion. Candidates for these proteins are among the set of class II nuclear fusion mutants.

KAR2

The *KAR2* gene encodes the yeast homologue of the ER resident HSP70 protein, also known as BiP (Normington et al 1989, Rose et al 1989). BiP plays a prominent role in the folding of proteins in the secretory pathway (Simons et al 1995). In addition, several in vivo and in vitro experiments demonstrate a critical role for Kar2p in the translocation of proteins into the endoplasmic reticulum (Vogel et al 1990, Sanders et al 1992, Brodsky & Schekman 1993, Brodsky et al 1995). Evidence from in vitro experiments suggests that Kar2p may have two functions during translocation (Sanders et al 1992); one to bind the translocating peptide as it appears in the ER and one to reset components of the translocation apparatus, thereby allowing the initial binding to the nascent chain.

The role for Kar2p in nuclear fusion is less clear. The in vitro ER/nuclear envelope fusion data (Latterich & Schekman 1994) suggest that Kar2p plays a direct role in membrane fusion. It is very unlikely that it is required for karyogamy because of its role in translocation per se. First, there is no correlation between the severity of different *kar2* mutations on karyogamy versus translocation (Vogel et al 1990, Vogel 1993). Second, *kar2* mutations affect the in vitro ER fusion reaction under conditions in which translocation does not occur (Latterich & Schekman 1994). Given the current knowledge of the roles of peripheral and integral membrane proteins in membrane fusion, it is initially somewhat surprising that a lumenal protein would be required (Stegmann et al 1989, White 1992). A possible role for Kar2p comes from the recognition that complete fusion of membranes catalyzed by the influenza hemagglutinin requires its transmembrane domain. A hemagglutin ectodomain construct lacking a transmembrane anchor produces a hemifusion reaction in which lipids but not lumenal contents are mixed (Kemble et al 1994, Melikyan et al 1995). One interpretation of these results is that the ectodomain is required for docking and initiation of membrane fusion, whereas the transmembrane domain plays an essential role in pore dilation. It seems reasonable to suggest that a chaperone activity might be required during pore dilation in some membrane fusion events. That such lumenal proteins have not yet appeared as a requirement for fusion in the secretory pathway suggests either that they are not essential in this pathway or that the existing mutant screens are biased against their isolation.

Other Proteins Interacting with Kar2p

HSP70 proteins generally act in concert with activating proteins sharing homology with the DnaJ protein from *E. coli* (Cyr et al 1994). Kar2p interacts with at least one integral membrane protein containing a lumenal DnaJ domain called Sec63p (Scidmore et al 1993). *SEC63* was identified in two very different genetic screens. The first screen looked for mutations that block the translocation of secreted proteins (Deshaies & Schekman 1987); the second looked for mutations that block the import of proteins into the nucleus (Sadler et al 1989). It seems clear that Sec63p plays a direct role in translocation, possibly by localizing and/or activating Kar2p (Brodsky & Schekman 1993, Scidmore et al 1993). Sec63p has been isolated in complexes with several other proteins required for translocation, including Sec61p, Sec62p, Sec71p, Sec72p, and Kar2p (Deshaies et al 1991, Brodsky & Schekman 1993, Feldheim et al 1993, Kurihara & Silver 1993). Sec63p's function in nuclear import is less clear, but it is interesting that the mutations blocking this process largely map to a cytoplasmic domain of the protein (Feldheim et al 1992, Nelson et al 1993).

Recently a role for Sec63p in nuclear fusion has appeared from the isolation of an unusual allele of Sec63p bearing a C-terminal truncation of 27 amino acids (Ng & Walter 1996). The *sec63-201* allele exhibits a strong bilateral class II nuclear fusion defect. This allele also causes a strong block in translocation, but other alleles of *sec63* that also cause strong blocks in translocation result in either much weaker or no defects in karyogamy. The allele specificity argues that the nuclear fusion defect does not arise from a simple block in translocation.

Given the identification of Kar2p and Sec63p as proteins required for nuclear fusion, it is natural to examine the requirement for other proteins known to interact with Kar2p and Sec63p. Accordingly, deletions of *sec71* and *sec72* were found to result in bilateral nuclear fusion defects (Ng & Walter 1996). Sec71p is an integral ER membrane protein and Sec72p is a peripheral ER membrane protein (Green et al 1992, Feldheim et al 1993, Kurihara & Silver 1993). Deletions in both genes result in viable cells that are temperature sensitive for viability and translocation. At the same time, the *KAR7* gene was cloned and identified as the same gene as *SEC71* (V Brizzio, M Latterich & M Rose, manuscript in preparation). Deletions of *SEC71* and *SEC72* also result in a temperature-sensitive defect in the in vitro ER fusion assay (V Brizzio, M Latterich & M Rose, manuscript in preparation). Thus Sec71p and Sec72p also play roles in nuclear fusion. However, the temperature sensitivity of the deletion strains in the vitro ER fusion assay suggests that these proteins play only an indirect role such as stabilizing the fusion complex. Mutations with no or very mild defects were found in *sec61* and *sec62* (Ng & Walter 1996) (V Brizzio, M Latterich & M Rose, manuscript in preparation). However, it is possible that stronger defects might arise from different mutant alleles.

These results demonstrate that at least four of the six proteins identified as components of the translocation complex are also required for nuclear envelope fusion. The allele specificity of the mutations strongly suggests that their role in nuclear fusion is independent of their role in translocation. Possible functions might include membrane recognition, docking, or catalyzing the membrane fusion reaction. However, it will be difficult to assign specific functions to these proteins in nuclear fusion without detailed biochemical studies of the reaction in vitro.

KAR5 and KAR8

Two other genes identified as being required for nuclear membrane fusion are *KAR7* and *KAR8* (Kurihara et al 1994). Preliminary evidence indicates that Kar5p is a novel lumenal ER membrane protein (C Beh & M Rose, manuscript in preparation). Kar5p is induced during mating, where it presumably plays a critical and specific role. Interestingly, in shmoos, Kar5p is found to cluster near the SPB, the site at which nuclear membrane fusion normally initiates. It is tempting to speculate that Kar5p may play a role in recruiting the proteins required for membrane fusion to the region of the SPB. Although Kar5p may play a mating-specific role in membrane fusion, it is required in the in vitro ER fusion assay made from mitotic extracts. Therefore, it seems likely that Kar5p is also required for a mitotic function involved with ER/nuclear envelope fusion. However, the deletion is viable and the defects are more subtle than those of *cdc48* mutants. The normal mitotic function of this protein remains to be determined.

KAR8 encodes an additional DnaJ-like protein (D Huddler & M Rose, unpublished observations). Sequence analysis suggests that the protein is an integral membrane protein and the DnaJ domain is predicted to be lumenal. From the appearance of the putative nuclear fusion intermediates in *kar8* mutants, it is tempting to speculate that Kar8p is required to chaperone a somewhat downstream step in the membrane fusion reaction.

PERSPECTIVES

In summary, the karyogamy pathway has been separated into at least two major steps, congression and nuclear envelope fusion. Congression is the major function of the cytoplasmic microtubules during conjugation. Several mitotic microtubule-dependent functions are recruited to effect the nuclear movements that constitute congression. Nuclear fusion similarly recruits proteins that have additional and distinct primary roles during mitotic growth. The identification of the roles of *KAR* genes in mitotic growth and the identification of a role in karyogamy for genes known from mitosis has led to a great enrichment of our understanding. However, much work needs to be done before our understanding is complete.

One major question concerns the mechanism of the nuclear membrane fusion pathway. How many membranes actually fuse? If two membranes fuse, are there separate components that mediate the fusion of the internal membranes? Which proteins provide membrane recognition and which catalyze the fusion reaction? Is there an equivalent normal mitotic pathway of nuclear envelope and ER fusion in which *KAR*s participate? How is this pathway related to the pathway of nuclear envelope reassembly that occurs during mitosis in animals and plants?

Although better understood, several issues also remain in the description of the pathway of nuclear congression. One of these concerns the regulation of the pathway. How are Kar3p and Cik1p recruited from nuclear locations to the cytoplasm? Is there additional regulation that prevents their inappropriate induction and relocalization when the cells first respond to pheromone? What resets the functions of these proteins such that they are competent for re-entry into the cell cycle? What is the exact role of Kar1p, and does it act solely through Kar3p and Cik1p? Is there a strong requirement for nuclear migration and orientation prior to cell fusion? What are the mitotic functions of Kar3p and Cik1p? Does Kar3p have different associated light chains at different stages of the cell cycle?

Given the complexity of the process and the fact that the mutant screens have not been saturated, it seems likely that more genes will be discovered that play specific roles in the nuclear fusion pathway. Our understanding of the biochemical steps in the pathway is in its infancy. We look forward to future progress and the continued development of an integrated model of the pathway incorporating both genetic and biochemical breakthroughs.

ACKNOWLEDGMENTS

I wish to thank the many past and current members of my laboratory who have worked on the nuclear fusion pathway and contributed their parts of the puzzle. I also wish to thank our many collaborators, including Martin Latterich, Randy Schekman, Mike Snyder, Sharyn Endow, Mark Winey, and others, whose many contributions and discussions have been critical to continued progress. This work was supported by National Institutes of Health grants 37739 and 52526.

Literature Cited

Acharya U, Jacobs R, Peters J-M, Watson N, Farquhar MG, Malhotra V. 1995. The formation of Golgi stacks from vesiculated Golgi membranes requires two distinct fusion events. *Cell* 82:895–904

Adams AE, Pringle JR. 1984. Relationship of actin and tubulin distribution to bud growth in wild-type and morphogenetic-mutant *Saccharomyces cerevisiae. J. Cell Biol.* 98:934–45

Bahler J, Hagens G, Holzinger G, Scherthan H, Heyer WD. 1994. *Saccharomyces cerevisiae* cells lacking the homologous pairing protein p175SEP1 arrest at pachytene during meiotic prophase. *Chromosoma* 103:129–41

Baron AT, Salisbury JL. 1988. Identification and localization of a novel, cytoskeletal, centrosome-associated protein in PtK2 cells. *J. Cell Biol.* 107:2669–78

Barton NR, Goldstein LSB. 1996. Going mobile: microtubules and chromosome segregation. *Proc. Natl. Acad. Sci. USA* 93:1735–42

Baum P, Furlong C, Byers B. 1986. Yeast gene required for spindle pole body duplication: homology of its product with Ca^{2+}-binding proteins. *Proc. Natl. Acad. Sci. USA* 83:5512–16

Beach D, Durkacz B, Nurse P. 1982. Functionally homologous cell cycle control genes in budding and fission yeast. *Nature* 300:706–9

Bender A, Sprague GF Jr. 1986. Yeast peptide pheromones a-factor and alpha-factor activate a common response mechanism in their target cells. *Cell* 47:929–37

Bender A, Sprague GF Jr. 1989. Pheromones and pheromone receptors are the primary determinants of mating specificity in the yeast *Saccharomyces cerevisiae. Genetics* 121:463–76

Bennett MK, Scheller RH. 1993. The molecular machinery for secretion is conserved from yeast to neurons. *Proc. Natl. Acad. Sci. USA* 90:2559–63

Bennett MK, Scheller RH. 1994. A molecular description of synaptic vesicle membrane trafficking. *Annu. Rev. Biochem.* 63:63–100

Berlin V, Brill JA, Trueheart J, Boeke JD, Fink GR. 1991. Genetic screens and selections for cell and nuclear fusion mutants. *Methods Enzymol.* 194:774–92

Berlin V, Styles CA, Fink GR. 1990. BIK1, a protein required for microtubule function during mating and mitosis in *Saccharomyces cerevisiae,* colocalizes with tubulin. *J. Cell Biol.* 111:2573–86

Biggins S, Rose MD. 1994. Direct interaction between yeast spindle pole body components: Kar1p is required for Cdc31p localization to the spindle pole body. *J. Cell Biol.* 125:843–52

Brodsky JL, Goeckeler J, Schekman R. 1995. BiP and Sec63p are required for both co- and posttranslational protein translocation into the yeast endoplasmic reticulum. *Proc. Natl. Acad. Sci. USA* 92:9643–46

Brodsky JL, Schekman R. 1993. A Sec63p-BiP complex from yeast is required for protein translocation in a reconstituted proteoliposome. *J. Cell Biol.* 123:1355–63

Byers B. 1981a. Cytology of the yeast life cycle. In *The Molecular Biology of the Yeast Saccharomyces: Life Cycle and Inheritance*, ed. JN Strathern, EW Jones, JR Broach, pp. 59–96. Cold Spring Harbor, NY: Cold Spring Harbor Lab. Press

Byers B. 1981b. Multiple roles of the spindle pole bodies in the life cycle of *Saccharomyces cerevisiae.* In *Molecular Genetics of Yeast,* ed. D von Wittstein, J Friis, M Kielland-Brandt, A Stenderup, 16:119–31. Copenhagen: Munksgaard

Byers B, Goetsch L. 1975. Behavior of spindles and spindle plaques in the cell cycle and conjugation in *Saccharomyces cerevisiae. J. Bacteriol.* 124:511–23

Byers B, Shriver K, Goetsch L. 1978. The role of spindle pole bodies and modified microtubule ends in the initiation of microtubule assembly in *Saccharomyces cerevisiae. J. Cell Sci.* 30:331–52

Chan CS, Botstein D. 1993. Isolation and characterization of chromosome-gain and increase-in-ploidy mutants in yeast. *Genetics* 135:677–91

Conde J, Fink GR. 1976. A mutant of *Saccharomyces cerevisiae* defective for nuclear fusion. *Proc. Natl. Acad. Sci. USA* 73:3651–55

Cross F, Hartwell LH, Jackson C, Konopka JB. 1988. Conjugation in *Saccharomyces cerevisiae. Annu. Rev. Cell Biol.* 44:29–57

Curran BP, Carter BL. 1986. Alpha-factor enhancement of hybrid formation by protoplast fusion in *Saccharomyces cerevisiae* II. *Curr. Genet.* 10:943–45

Cyr DM, Langer T, Douglas MG. 1994. DnaJ-like proteins: molecular chaperones and regulators of Hsp70. *Trends Biochem. Sci.* 19:176–81

Delgado MA, Conde J. 1984. Benomyl prevents nuclear fusion in *Saccharomyces cerevisiae. Mol. Gen. Genet.* 193:188–89

Deshaies RJ, Sanders SL, Feldheim DA, Schekman R. 1991. Assembly of yeast Sec proteins involved in translocation into the endoplasmic reticulum into a membrane-bound multisubunit complex. *Nature* 349:806–8

Deshaies RJ, Schekman R. 1987. A yeast mutant defective at an early stage in import of secretory precursors into the endoplasmic reticulum. *J. Cell Biol.* 105:633–45

Dunphy WG, Brizuela L, Beach D, Newport J. 1988. The Xenopus cdc2 protein is a component of MPF, a cytoplasmic regulator of mitosis. *Cell* 54:423–31

Dutcher SK, Hartwell LH. 1982. The role of *S. cerevisiae* cell division cycle genes in nuclear fusion. *Genetics* 100:175–84

Dutcher SK, Hartwell LH. 1983a. Genes that act before conjugation to prepare the Saccharomyces cerevisiae nucleus for caryogamy. *Cell* 33:203–10

Dutcher SK, Hartwell LH. 1983b. Test for temporal or spatial restrictions in gene product function during the cell division cycle. *Mol. Cell. Biol.* 3:1255–65

Elion EA, Brill JA, Fink GR. 1991. FUS3 represses CLN1 and CLN2 and in concert with KSS1 promotes signal transduction. *Proc. Natl. Acad. Sci. USA* 88:9392–96

Elion EA, Grisafi PL, Fink GR. 1990. *FUS3* encodes a cdc2+/CDC28-related kinase required for the transition from mitosis into conjugation. *Cell* 60:649–64

Elion, EA, Trueheart J, Fink GR. 1995. Fus2 localizes near the site of cell fusion and is required for both cell fusion and nuclear alignment during zygote formation. *J. Cell Biol.* 130:1283–96

Endow SA. 1991. The emerging kinesin family of microtubule motor proteins. *Trends Biochem. Sci.* 16:221–25

Endow SA, Kang SJ, Satterwhite LL, Rose MD, Skeen VP, Salmon ED. 1994. Yeast Kar3 is a minus-end microtubule motor protein that destabilizes microtubules preferentially at the minus ends. *EMBO J.* 13:2708–13

Eshel DE, Urrestarazu LA, Vissers S, Jauniaux

JC, van Vliet-Reedijk JC, et al. 1993. Cytoplasmic dynein is required for normal nuclear segregation in yeast. *Proc. Natl. Acad. Sci. USA* 90:11172–76

Feldheim D, Rothblatt J, Schekman R. 1992. Topology and functional domains of Sec63p, an endoplasmic reticulum membrane protein required for secretory protein translocation. *Mol. Cell. Biol.* 12:3288–96

Feldheim D, Yoshimura K, Admon A, Schekman R. 1993. Structural and functional characterization of Sec66p, a new subunit of the polypeptide translocation apparatus in the yeast endoplasmic reticulum. *Mol. Biol. Cell.* 4:931–39

Ferguson J, Ho JY, Peterson TA, Reed SI. 1986. Nucleotide sequence of the yeast cell division cycle start genes *CDC28, CDC36, CDC37,* and *CDC39*, and a structural analysis of the predicted products. *Nucleic Acids Res.* 14:6681–97

Ferro-Novick S, Jahn R. 1994. Vesicle fusion from yeast to man. *Nature* 370:191–93

Fong HKW, Hurley JB, Hopkins RS, Miake R, Johnson MS, et al. 1986. Repetitive segmental structure of the transducin beta subunit homology with the *CDC4* gene and identification of related mRNAs. *Proc. Natl. Acad. Sci. USA* 83:2162–66

Frohlich K-U, Fries HW, Rudiger M, Erdmann R, Botstein D, Mecke D. 1991. Yeast cell cycle protein CDC48p shows full-length homology to the mammalian protein VCP and is a member of a protein family involved in secretion, peroxisome formation, and gene expression. *J. Cell Biol.* 114:443–53

Gerber MR, Farrell A, Deshaies RJ, Herskowitz I, Morgan DO. 1995. Cdc37 is required for association of the protein kinase Cdc28 with G1 and mitotic cyclins. *Proc. Natl. Acad. Sci. USA* 92:4651–55

Goebl MG, Goetsch L, Byers B. 1994. The Ubc3 (Cdc34) ubiquitin-conjugating enzyme is ubiquitinated and phosphorylated in vivo. *Mol. Cell. Biol.* 14:3022–29

Goebl MG, Yochem J, Jentsch S, McGrath JP, Varshavsky A, Byers B. 1988. The yeast cell cycle gene *CDC34* encodes a ubiquitin-conjugating enzyme. *Science* 241:1331–35

Green N, Fang H, Walter P. 1992. Mutants in three novel complementation groups inhibit membrane protein insertion into and soluble protein translocation across the endoplasmic reticulum membrane of *Saccharomyces cerevisiae. J. Cell Biol.* 116:597–604

Hasek J, Rupes I, Svobodova J, Streiblova E. 1987. Tubulin and actin topology during zygote formation of *Saccharomyces cerevisiae. J. Gen. Microbiol.* 133:3355–63

Heyer WD, Johnson AW, Reinhart U, Kolodner RD. 1995. Regulation and intracellular localization of *Saccharomyces cerevisiae* strand exchange protein 1 (Sep1/Xrn1/Kem1), a multifunctional exonuclease. *Mol. Cell. Biol.* 15:2728–36

Hoyt MA, He L, Loo KK, Saunders WS. 1992. Two *Saccharomyces cerevisiae* kinesin-related gene products required for mitotic spindle assembly. *J. Cell Biol.* 118:109–20

Hoyt MA, He L, Totis L, Saunders WS. 1993. Loss of function of *Saccharomyces cerevisiae* kinesin-related *CIN8* and *KIP1* is suppressed by *KAR3* motor domain mutations. *Genetics* 135:35–44

Hoyt MA, Stearns T, Botstein D. 1990. Chromosome instability mutants of *Saccharomyces cerevisiae* that are defective in microtubule-mediated processes. *Mol. Cell. Biol.* 10:223–34

Huang B, Mengersen A, Lee VD. 1988. Molecular cloning of cDNA for caltractin, a basal body-associated Ca^{2+}-binding protein: homology in its protein sequence with calmodulin and the yeast *CDC31* gene product. *J. Cell Biol.* 107:133–40

Hyams JS, Borisy GG. 1978. Nucleation of microtubules in vitro by isolated spindle pole bodies of the yeast *Saccharomyces cerevisiae. J. Cell Biol.* 78:401–14

Interthal H, Bellocq C, Bahler J, Bashkirov VI, Edelstein S, Heyer WD. 1995. A role of Sep1 (= Kem1, Xrn1) as a microtubule-associated protein in *Saccharomyces cerevisiae. EMBO J.* 14:1057–66

Johnson AW, Kolodner RD. 1995. Synthetic lethality of *sep1* (*xrn1*) *ski2* and *sep1* (*xrn1*) *ski3* mutants of *Saccharomyces cerevisiae* is independent of killer virus and suggests a general role for these genes in translation control. *Mol. Cell. Biol.* 15:2719–27

Kemble GW, Danieli T, White JM. 1994. Lipid-anchored influenza hemagglutinin promotes hemifusion, not complete fusion. *Cell* 76:383–91

Kilmartin JV. 1994. Genetic and biochemical approaches to spindle function and chromosome segregation in eukaryotic microorganisms. *Curr. Opin. Cell Biol.* 6:50–54

Kim J, Ljungdahl PO, Fink GR. 1990. *kem* mutations affect nuclear fusion in *Saccharomyces cerevisiae. Genetics* 126:799–812

Kurihara LJ, Beh CT, Latterich M, Schekman R, Rose MD. 1994. Nuclear congression and membrane fusion: two distinct events in the yeast karyogamy pathway. *J. Cell Biol.* 126:911–23

Kurihara LJ, Stewart BG, Gammie AE, Rose MD. 1996. *KAR4,* a karyogamy-specific component of the yeast pheromone response pathway. *Mol. Cell. Biol.* 16:In press

Kurihara T, Silver P. 1993. Suppression of a *sec 63* mutation identifies a novel component of the yeast endoplasmic reticulum translocation apparatus. *Mol. Biol. Cell.* 4:919–30

Lancashire WE, Mattoon JR. 1979. Cytoduction: a tool for mitochondrial genetic studies in yeast. *Mol. Gen. Genet.* 170:333–44

Larimer FW, Hsu CL, Maupin MK, Stevens A. 1992. Characterization of the *XRN1* gene encoding a $5' \longrightarrow 3'$ exoribonuclease: sequence data and analysis of disparate protein and mRNA levels of gene-disrupted yeast cells. *Gene* 120:51–57

Latterich M, Frohlich K-U, Schekman R. 1995. Membrane fusion and the cell cycle: Cdc48 participates in the fusion of ER membranes. *Cell* 82:885–94

Latterich M, Schekman R. 1994. The karyogamy gene *KAR2* and novel proteins are required for ER-membrane fusion. *Cell* 78:87–98

Li Q, Suprenant KA. 1994. Molecular characterization of the 77-kDa echinoderm microtubule-associated protein. Homology to the beta-transducin family. *J. Biol. Chem.* 269:31777–84

Li YY, Yeh E, Hays T, Bloom K. 1993. Disruption of mitotic spindle orientation in a yeast dynein mutant. *Proc. Natl. Acad. Sci. USA* 90:10096–100

Liu Z, Gilbert W. 1994. The yeast *KEM1* gene encodes a nuclease specific for G4 tetraplex DNA: implication of in vivo functions for this novel DNA structure. *Cell* 77:1083–92

Liu Z, Lee A, Gilbert W. 1995. Gene disruption of a G4-DNA-dependent nuclease in yeast leads to cellular senescence and telomere shortening. *Proc. Natl. Acad. Sci. USA* 92:6002–6

McCaffrey G, Clay FJ, Kelsay K, Sprague GF Jr. 1987. Identification and regulation of a gene required for cell fusion during mating of the yeast *Saccharomyces cerevisiae. Mol. Cell. Biol.* 7:2680–90

McDonald HB, Goldstein LSB. 1990. Identification and characterization of a gene encoding a kinesin-like protein in Drosophila. *Cell* 61:991–1000

McDonald HB, Stewart RJ, Goldstein LSB. 1990. The kinesin-like ncd protein of Drosophila is a minus end-directed motor. *Cell* 63:1159–65

Melikyan GB, White JM, Cohen FS. 1995. GPI-anchored influenza hemagglutinin induces hemifusion to both red blood cell and planar bilayer membranes. *J. Cell Biol.* 131:679–91

Mellman I. 1995. Enigma variations: protein mediators of membrane fusion. *Cell* 82:869–72

Meluh PB. 1992. *KAR3, a kinesin-related gene required for nuclear fusion, mitosis and meiosis in Saccharomyces cerevisiae.* PhD thesis. Princeton Univ. 383 pp.

Meluh, PB, Rose, MD. 1990. *KAR3,* a kinesin-related gene required for yeast nuclear fusion. *Cell* 60:1029–41

Middleton K, Carbon J. 1994. *KAR3*-encoded kinesin is a minus-end-directed motor that functions with centromere binding proteins (CBF3) on an in vitro yeast kinetochore. *Proc. Natl. Acad. Sci. USA* 91:7212–16

Morris NR, Xiang X, Beckwith SM. 1995. Nuclear migration advances in fungi. *Trends Cell Biol.* 5:278–82

Nasmyth K, Adolf G, Lydall D, Seddon A. 1990. The identification of a second cell cycle control on the *HO* promoter in yeast: cell cycle regulation of SWI5 nuclear entry. *Cell* 62:631–47

Nelson MK, Kurihara T, Silver PA. 1993. Extragenic suppressors of mutations in the cytoplasmic C terminus of SEC63 define five new genes in *Saccharomyces cerevisiae. Genetics* 134:159–73

Ng DTW, Walter P. 1996. ER membrane protein complex required for nuclear fusion. *J. Cell Biol.* 132:499–510

Normington K, Kohno K, Kozutsumi Y, Gething MJ, Sambrook J. 1989. S. cerevisiae encodes an essential protein homologous in sequence and function to mammalian BiP. *Cell* 57:1223–36

O'Connell MJ, Meluh PB, Rose MD, Morris NR. 1993. Suppression of the *bimC4* mitotic spindle defect by deletion of *klpA*, a gene encoding a *KAR3*-related kinesin-like protein in *Aspergillus nidulans. J. Cell Biol.* 120:153–62

Page BD, Satterwhite LL, Rose MD, Snyder M. 1994. Localization of the Kar3 kinesin heavy chain-related protein requires the Cik1 interacting protein. *J. Cell Biol.* 124:507–19

Page BD, Snyder M. 1992. CIK1: a developmentally regulated spindle pole body-associated protein important for microtubule functions in *Saccharomyces cerevisiae. Genes Dev.* 6:1414–29

Page BD, Snyder M. 1993. Chromosome segregation in yeast. *Annu. Rev. Microbiol.* 47:231–61

Pellman D, Bagget M, Tu H, Fink GR. 1995. Two microtubule-associated proteins required for anaphase spindle movement in *Saccharomyces cerevisiae. J. Cell Biol.* 130:1373–85

Peter M, Gartner A, Horecka J, Ammerer G, Herskowitz I. 1993. *FAR1* links the signal transduction pathway to the cell cycle machinery in yeast. *Cell* 73:747–60

Peterson TA, Yochem J, Byers B, Nunn MF, Duesberg PH, et al. 1984. A relationship between the yeast cell cycle genes *CDC4* and *CDC36* and the ets sequence of oncogenic virus E26. *Nature* 309:556–58

Pierre P, Pepperkok R, Kreis TE. 1994. Molecular characterization of two functional domains of CLIP-170 in vivo. *J. Cell Sci.* 107: 1909–20

Pierre P, Scheel J, Rickard JE, Kreis TE. 1992. CLIP-170 links endocytic vesicles to microtubules. *Cell* 70:887–900

Polaina J, Conde J. 1982. Genes involved in the control of nuclear fusion during the sexual cycle of *Saccharomyces cerevisiae. Mol. Gen. Genet.* 186:253–58

Rabouille C, Levine TP, Peters J-M, Warren G. 1995. An NSF-like ATPase, p97, and NSF mediate cisternal regrowth from mitotic Golgi fragments. *Cell* 82:905–14

Read EB, Okamura HH, Drubin DG. 1992. Actin- and tubulin-dependent functions during *Saccharomyces cerevisiae* mating projection formation. *Mol. Biol. Cell.* 3:429–44

Roof DM, Meluh PB, Rose MD. 1991. Multiple kinesin-related proteins in yeast mitosis. *Cold Spring Harbor Symp. Quant. Biol.* 56:693–703

Rose MD. 1991. Nuclear fusion in yeast. *Annu. Rev. Microbiol.* 45:539–67

Rose MD, Biggins S, Satterwhite LL. 1993. Unravelling the tangled web at the microtubule-organizing center. *Curr. Opin. Cell Biol.* 5:105–15

Rose MD, Fink GR. 1987. *KAR1,* a gene required for function of both intranuclear and extranuclear microtubules in yeast. *Cell* 48:1047–60

Rose MD, Misra LM, Vogel JP. 1989. *KAR2*, a karyogamy gene, is the yeast homolog of the mammalian BiP/GRP78 gene. *Cell* 57:1211–21

Rose MD, Price BR, Fink GR. 1986. *Saccharomyces cerevisiae* nuclear fusion requires prior activation by alpha factor. *Mol. Cell. Biol.* 6:3490–97

Rothman JE. 1994. Mechanisms of intracellular protein transport. *Nature* 372:55–63

Rothman JE, Warren G. 1994. Implications of the SNARE hypothesis for intracellular membrane topology and dynamics. *Curr. Biol.* 4:220–33

Sadler I, Chiang A, Kurihara T, Rothblatt J, Way J, Silver P. 1989. A yeast gene important for protein assembly into the endoplasmic reticulum and the nucleus has homology to DnaJ, an *Escherichia coli* heat shock protein. *J. Cell Biol.* 109:2665–75

Sanders SL, Whitfield KM, Vogel JP, Rose MD, Schekman RW. 1992. Sec61p and BiP directly facilitate polypeptide translocation into the ER. *Cell* 69:353–65

Saunders WS, Hoyt MA. 1992. Kinesin-related proteins required for structural integrity of the mitotic spindle. *Cell* 70:451–58

Saunders WS, Koshland D, Eshel D, Gibbons IR, Hoyt MA. 1995. *Saccharomyces cerevisiae* kinesin- and dynein-related proteins required for anaphase chromosome segregation. *J. Cell Biol.* 128:617–24

Schekman R. 1992. Genetic and biochemical analysis of vesicular traffic in yeast. *Curr. Opin. Cell Biol.* 4:587–92

Schwob E, Bohm T, Mendenhall MD, Nasmyth K. 1994. The B-type cyclin kinase inhibitor p40SIC1 controls the G1 to S transition in S. cerevisiae. *Cell* 79:233–44

Scidmore MA, Okamura HH, Rose MD. 1993. Genetic interactions between *KAR2* and *SEC63,* encoding eukaryotic homologues of DnaK and DnaJ in the endoplasmic reticulum. *Mol. Biol. Cell.* 4:1145–59

Simons JF, Ferro-Novick S, Rose MD, Helenius A. 1995. BiP/Kar2p serves as a molecular chaperone during carboxypeptidase Y folding in yeast. *J. Cell Biol.* 130:41–49

Snyder M. 1994. The spindle pole body of yeast. *Chromosoma* 103:369–80

Spang A, Courtney I, Fackler U, Matzner M, Schiebel E. 1993. The calcium-binding protein cell division cycle 31 of *Saccharomyces cerevisiae* is a component of the half bridge of the spindle pole body. *J. Cell Biol.* 123:405–16

Spang A, Courtney I, Grein K, Matzner M, Schiebel E. 1995. The Cdc31p-binding protein Kar1p is a component of the half bridge of the yeast spindle pole body. *J. Cell Biol.* 128:863–77

Sprague GF Jr, Thorner JW. 1994. Pheromone response and signal transduction during the mating process of *Saccharomyces cerevisiae.* In *The Molecular and Cellular Biology of the Yeast Saccharomyces,* ed. EW Jones, JR Pringle, JR Broach. 2:657–744. Cold Spring Harbor, NY: Cold Spring Harbor Lab. Press

Stearns T, Hoyt MA, Botstein D. 1990. Yeast mutants sensitive to antimicrotubule drugs define three genes that affect microtubule function. *Genetics* 124:251–62

Stegmann T, Doms RW, Helenius A. 1989. Protein-mediated membrane fusion. *Annu. Rev. Biophys. Biophys. Chem.* 18:187–211

Trueheart J, Boeke JD, Fink GR. 1987. Two genes required for cell fusion during yeast conjugation: evidence for a pheromone-induced surface protein. *Mol. Cell. Biol.* 7:2316–28

Tyers M, Futcher B. 1993. Far1 and Fus3 link the mating pheromone signal transduction pathway to three G1-phase Cdc28 kinase complexes. *Mol. Cell. Biol.* 13:5659–69

Tyers M, Tokiwa G, Nash R, Futcher B. 1992. The Cln3-Cdc28 kinase complex of *S. cerevisiae* is regulated by proteolysis and phosphorylation. *EMBO J.* 11:1773–84

Vallen EA, Hiller MA, Scherson TY, Rose MD. 1992a. Separate domains of KAR1 mediate distinct functions in mitosis and nuclear fusion. *J. Cell Biol.* 117:1277–87

Vallen EA, Ho W, Winey M, Rose MD. 1994. Genetic interactions between *CDC31* and *KAR1*, two genes required for duplication of the microtubule organizing center in *Saccharomyces cerevisiae. Genetics* 137:407–22

Vallen EA, Scherson TY, Roberts T, van Zee K, Rose MD. 1992b. Asymmetric mitotic segregation of the yeast spindle pole body. *Cell* 69:505–15

Vogel J. 1993. *KAR2, the yeast homologue of mammalian BiP/GRP78.* PhD thesis. Princeton Univ. 218 pp.

Vogel JP, Misra LM, Rose, MD. 1990. Loss of BiP/GRP78 function blocks translocation of secretory proteins in yeast. *J. Cell Biol.* 110:1885–95

Walker RA, Salmon ED, Endow SA. 1990. The *Drosophila* claret segregation protein is a minus-end directed motor molecule. *Nature* 347:780–82

Waterman-Storer CM, Holzbaur EL. 1996. The product of the *Drosophila* gene, Glued, is the functional homologue of the p150Glued component of the vertebrate dynactin complex. *J. Biol. Chem.* 271:1153–59

White JM. 1992. Membrane fusion. *Science* 258:917–24

Yaglom J, Linskens MH, Sadis S, Rubin DM, Futcher B, Finley D. 1995. p34cdc28-mediated control of Cln3 cyclin degradation. *Mol. Cell. Biol.* 15:731–41

Yang JT, Laymon RA, Goldstein LSB. 1989. A three-domain structure of kinesin heavy chain revealed by DNA sequence and microtubule binding analyses. *Cell* 56:879–89

Annu. Rev. Cell Dev. Biol. 1996. 12:697–715

RGD AND OTHER RECOGNITION SEQUENCES FOR INTEGRINS

Erkki Ruoslahti

La Jolla Cancer Research Center, The Burnham Institute, 10901 North Torrey Pines Road, La Jolla, California 92037

KEY WORDS: peptides, extracellular matrix, cell adhesion, cell migration

ABSTRACT

Proteins that contain the Arg-Gly-Asp (RGD) attachment site, together with the integrins that serve as receptors for them, constitute a major recognition system for cell adhesion. The RGD sequence is the cell attachment site of a large number of adhesive extracellular matrix, blood, and cell surface proteins, and nearly half of the over 20 known integrins recognize this sequence in their adhesion protein ligands. Some other integrins bind to related sequences in their ligands. The integrin-binding activity of adhesion proteins can be reproduced by short synthetic peptides containing the RGD sequence. Such peptides promote cell adhesion when insolubilized onto a surface, and inhibit it when presented to cells in solution. Reagents that bind selectively to only one or a few of the RGD-directed integrins can be designed by cyclizing peptides with selected sequences around the RGD and by synthesizing RGD mimics. As the integrin-mediated cell attachment influences and regulates cell migration, growth, differentiation, and apoptosis, the RGD peptides and mimics can be used to probe integrin functions in various biological systems. Drug design based on the RGD structure may provide new treatments for diseases such as thrombosis, osteoporosis, and cancer.

CONTENTS

1081-0706/96/1115-0697$08.00

Introduction

The arginine-glycine-aspartic acid (RGD) cell adhesion sequence was discovered in fibronectin 12 years ago (Pierschbacher & Ruoslahti 1984a). The surprising finding that only three amino acids would form an essential recognition site for cells in a very large protein was initially received with some skepticism. The observation was, however, soon confirmed with regard to fibronectin and then extended to other proteins; as predicted in the original paper on RGD, it turned out that the RGD sequence is the cell attachment site of many other adhesive proteins (Pierschbacher & Ruoslahti 1984b, Yamada & Kennedy 1984, Gartner & Bennett 1985, Plow et al 1985, Suzuki et al 1985, Gardner & Hynes 1985). These findings and the subsequent discovery of integrins, cell surface receptors that recognize the RGD sequence of various proteins, have given RGD a central role in cell adhesion biology as the prototype adhesion signal. Moreover, numerous pharmaceutical applications are also emerging for RGD. This review describes the current status of the RGD system.

Peptide Mimicry of Adhesion Proteins

Because it is such a small structure, the RGD site can be easily reproduced with peptides, and this is how the site was originally identified. Short peptides containing the RGD sequence can mimic cell adhesion proteins in two ways: When coated onto a surface, they promote cell adhesion, whereas in solution they act as decoys, preventing adhesion. RGD peptides that have not been specifically designed to be selective to bind certain integrins mimic a number of adhesion proteins and bind to more than one receptor. The affinity of these conventional RGD peptides is relatively low; the GRGDSP hexapeptide, which is derived from the cell attachment site of fibronectin, is about 1000 times less effective in cell attachment assays than fibronectin itself (Hautanen et al 1989). However, this peptide is nevertheless quite specific in its activity; changes as small as the replacement of the aspartic acid with a glutamic acid or of the glycine with an alanine reduce the activity of the peptide in cell attachment assays 100-fold or more (Hautanen et al 1989). Conformation of the amino acids is also important; a peptide in which the aspartic acid is in the D-form is inactive. However, RGD peptides in which the arginine residue is a D-arginine, are active (Pierschbacher & Ruoslahti 1987).

The smallest active unit in the peptides is the RGD itself. The fibronectin cell attachment site was originally defined as a tetrapeptide because this was the smallest peptide with measurable activity. However, it was recognized that while the RGD had to remain invariant for cell attachment activity to be preserved, the amino acid following this sequence (a serine in fibronectin) could vary (Pierschbacher & Ruoslahti 1984a,b). It was subsequently found that the

Table 1 Integrin-binding sequence motifs and the integrins recognizing them

KQAGDV	LDV/IDS	RLD/KRLDGS	RGD	L/IET	R...D	YYGDLR/FYFDLR
$(\alpha_5\beta_1)$	$(\alpha_5\beta_1)$		$\alpha_5\beta_1$			
			$\alpha_8\beta_1$			
			$\alpha_v\beta_1$			
		$\alpha_v\beta_3$	$\alpha_v\beta_3$			
			$\alpha_v\beta_5$			
			$\alpha_v\beta_6$			
			$\alpha_v\beta_8$			
$\alpha_{IIb}\beta_3$			$\alpha_{IIb}\beta_3$			
			$(\alpha_2\beta_1)$			$\alpha_2\beta_1$
			$(\alpha_3\beta_1)$			
			$(\alpha_4\beta_1)$			
			$(\alpha_7\beta_1)$			
	$\alpha_4\beta_1$					
		$\alpha_M\beta_2$				
				$\alpha_L\beta_2$		
					$\alpha_1\beta_1$	

Shown are various integrin target sequences and the integrins that bind to them. Binding that is weak or that is only seen under special conditions is indicated by parenthesis. See text for details.

minimal requirement on the C-terminal side of the RGD was a methyl group to block the C-terminal carboxyl group of the aspartic acid residue (Pierschbacher & Ruoslahti 1987).

RGD-Binding Integrins and RGD-Containing Protein

The central role of the RGD sequence in cell recognition was initially inferred from the observation that not only did a fibronectin receptor (later named $(\alpha_5\beta_1)$ bind to its ligand in an RGD-dependent manner (Pytela et al 1985a), but so did another protein that resembled the fibronectin receptor in its polypeptide composition but bound to vitronectin rather than fibronectin (Pytela et al 1985b). At least 8, and possibly as many as 12, of the currently known 20 or so integrins recognize the RGD sequence in their ligands (see Table 1). It is noteworthy that the α subunits of the RGD-binding integrins (α_5, α_8, α_v, and α_{IIb}) show more sequence similarity to one another than to the other α subunits (Pytela et al 1994); these integrins may form an integrin subfamily. A partial list of adhesion proteins with RGD sites includes fibronectin, vitronectin, fibrinogen, von Willebrand factor, thrombospondin, laminin, entactin, tenascin, osteopontin, bone sialoprotein, and, under some conditions, collagens.

In some cases, proteins that are unlikely to serve an integrin binding role in vivo, nevertheless do mediate cell attachment in vitro. Gamma II crystallin and *Escherichia coli* lambda receptor are examples of such proteins (Ruoslahti &

Pierschbacher 1987). It is probable that these proteins are active because they present their RGD sequence in a context that can be recognized by an integrin. The most likely integrin to be involved in such unscheduled recognition is $\alpha_v\beta_3$, which is capable of binding to a large variety of RGD proteins. Although the ability of the *E. coli* lambda receptor to mediate cell attachment may have little biological meaning, many other RGD-containing microbial proteins, such as the penton protein of adenovirus (Huang et al 1995), the foot and mouth virus coat protein (Lea et al 1995), the *Coxsackie* type virus (Roivainen et al 1994), and a surface protein of *Bordetella pertussis* (Leininger et al 1991) serve as ligands through which these microbes bind to integrins on cell surfaces and gain entry into the cell. *Yersinia pseudotuberculosis* binds to the $\alpha_5\beta_1$ integrin through invasin, which is a protein necessary for these bacteria to be able to invade cells. Invasin does not contain an RGD sequence, but fibronectin and RGD peptides inhibit the binding of invasin to the integrin (Van Nhieu & Isberg 1991).

Not all RGD-containing proteins mediate cell attachment; in fact, only a minority of the 2600 (per Swiss Plot database 9/15/95) sequenced proteins that contain the RGD sequence are likely to do so. This is because the RGD sequence may not always be available at the surface of the protein or may be presented in a context that is not compatible with integrin binding. Therefore, even peptides as small as six amino acids can be devoid of cell attachment activity if proline is the residue that follows the RGD sequence (Pierschbacher & Ruoslahti 1987). Thus the surrounding sequences can silence the activity of RGD.

The Tat protein of the human immunodeficiency virus (HIV) is an RGD-containing protein with cell attachment activity. The interaction of Tat with cells is important because Tat can be internalized by cells, thus allowing Tat produced by one cell to enter another cell and turn on the production of latent HIV (Ensoll et al 1990). The role of the RGD sequence in the interactions of Tat with cell surfaces is controversial. Two groups have found the RGD site on Tat to be active (Brake et al 1990, Barillari et al 1993), whereas two others find that the cell attachment activity of the Tat protein is linked to a highly basic sequence in Tat (Vogel et al 1993, Weeks et al 1993). The attachment mediated by the basic sequence was, nevertheless, found to be dependent on an integrin, $\alpha_v\beta_5$, which has both an RGD-directed binding site and a site for a basic sequence. This basic Tat site is characterized by the sequence RQR (Vogel et al 1993, Koivunen et al 1995). The $\alpha_v\beta_5$ integrin, despite its ability to bind Tat, does not appear to play any significant role in the internalization of Tat by cells.

The binding of integrins to an RGD sequence is an evolutionarily conserved system; at least one *Drosophila* integrin binds to its ligand, tiggrin, at an RGD site, and this integrin can also mediate the attachment of *Drosophila* cells to human vitronectin in an RGD-dependent manner (Bunch & Brower 1992, Fogerty et al 1994). Even plant cells may utilize this recognition system. Plant cells have been reported to respond to RGD peptide treatment by detachment from the cell wall (Schindler et al 1989), as well as by loss of gravity-induced polarity (Wayne et al 1992).

RGD Recognition Sites in Integrins

Integrins are heterodimeric proteins in which each of the subunits contributes to the ligand specificity and contains potential binding sites for the ligand. Cross-linking of RGD peptides to the integrins and site-directed mutagenesis have been used to localize a ligand-binding site in the $\alpha_{IIb}\beta_3$ integrin in the first 200-amino acid residues of the β subunit N-terminus (Bajt et al 1992, Bajt & Lotus 1994). The β_3 subunit of the $\alpha_v\beta_3$ integrin becomes cross-linked to an RGD peptide in the same region (Smith & Cheresh 1988). In each case, the ligand-binding site is at or near a binding site for divalent cations. The α subunit also contains one or more ligand-binding sites; as with the β subunits, these sites localize into divalent cation binding sequences (D'Souza et al 1990), and elsewhere in the α_{IIb} subunit (Gartner & Taylor 1990, Calvete et al 1993). In RGD cross-linking of $\alpha_{IIb}\beta_3$, the β subunit provides the primary site for RGD binding, whereas the alternative ligand sequence from the γ chain of fibrinogen, KQAGDV, binds primarily to the α subunit (Santoro & Lawing 1987). The recognition site for collagen binding by the $\alpha_2\beta_1$ integrin and for fibrinogen by the $\alpha_M\beta_2$ integrin is located in the so-called A (or I) domain, which is an insertion into the divalent cation-binding region of the α subunit present in some integrins (Michiashita et al 1993). The crystal structure of an A domain has been determined, and it has been proposed that an equivalent domain is present in the β subunits, corresponding to the ligand-binding site (Lee et al 1995).

The β subunit ligand-binding site has been localized and its function studied with synthetic peptides. A 23-amino acid peptide from the β_3 subunit (amino acids 113–153) is capable of binding both divalent cations and RGD peptides (D'Souza et al 1994). A peptide resembling a portion of the 23-amino acid β_3 peptide has been isolated from a phage-displayed peptide library by selecting for peptides capable of binding to RGD-containing fibronectin fragments (Pasqualini et al 1995). This eight-amino acid cyclic peptide (CWDDGWLC) also binds to short RGD peptides and thus appears to be a mimic of the RGD-binding site of the β_3 subunit. In support of this idea, antibodies prepared against the peptide recognize the β_3 and other β subunits. Similar peptide motifs have been seen in phage that bind to the RGD sequence of adenovirus type 2 penton protein (Hong & Boulanger 1995). The RGD-binding of the

CWDDGWLC peptide is independent of the presence of divalent cations. This observation and the finding by D'Souza et al (1994) that RGD binding causes the extrusion of the divalent cations from the integrin have led us to propose that RGD binds primarily to the site in the integrin that corresponds to the peptide isolated from the phage library and that the primary role of the adjacent binding site for the divalent cation is to keep the site in a conformation receptive for the ligand (Pasqualini et al 1995). The β subunit also contains additional sites that may make contact with the ligand (Charo et al 1991).

Protein Conformation and RGD Recognition by Integrins

Early work with fibronectin fragments and structurally constrained peptides indicated that the presentation of the RGD sequence in the RGD proteins was important in their recognition by integrins. Thus large fibronectin fragments containing the RGD site bound preferentially to the $\alpha_5\beta_1$ integrin, whereas smaller ones, such as a fragment consisting essentially of the 10th type III domain, bound better to a vitronectin receptor ($\alpha_v\beta_3$ integrin) than to a fibronectin receptor ($\alpha_5\beta_1$) (Pytela et al 1985b). As the $\alpha_v\beta_3$ reactivity appeared while the $\alpha_5\beta_1$ reactivity declined, we hypothesized that the truncation of the fibronectin fragments caused an alteration in the conformation of the RGD site, resulting in an altered integrin specificity.

The discovery of auxiliary binding sites in fibronectin for the $\alpha_5\beta_1$ integrin suggested an alternative hypothesis: The poor affinity of the small RGD-containing fibronectin fragments was caused by the loss of such synergistic binding sites on fibronectin (Obara et al 1988). The existence of at least one such site in the III_{10} fibronectin domain has been amply confirmed by mutagenesis studies (Obara & Yoshizato 1995) and, most directly, by demonstrating that peptides whose sequences show similarity with the synergistic site identified in the mutagenesis experiments bind to $\alpha_5\beta_1$ from peptide libraries (Koivunen et al 1994). The synergistic site encompasses the sequence RNS in fibronectin; other possible synergy sites were also tentatively identified in the peptide library searches. These results show that the binding surface on the fibronectin molecule for the $\alpha_5\beta_1$ integrin extends outside the RGD-containing segment. Figure 1 presents a model for this interaction.

The $\alpha_{IIb}\beta_3$ integrin also has at least one synergy site in fibronectin (Bowditch et al 1991). The $\alpha_v\beta_5$ integrin may have an auxiliary binding site on vitronectin that consists of a highly basic sequence and that appears to account for the binding of this integrin to the Tat protein (Vogel et al 1993). A synergistic sequence for $\alpha_v\beta_3$ on vitronectin has also been proposed (Healy et al 1995).

Structural analyses by NMR and crystallography show that the RGD site in fibronectin exists in a loop bound by β strands (Main et al 1992, Dickinson et al 1994) as predicted from the sequence (Pierschbacher & Ruoslahti 1984b). Short

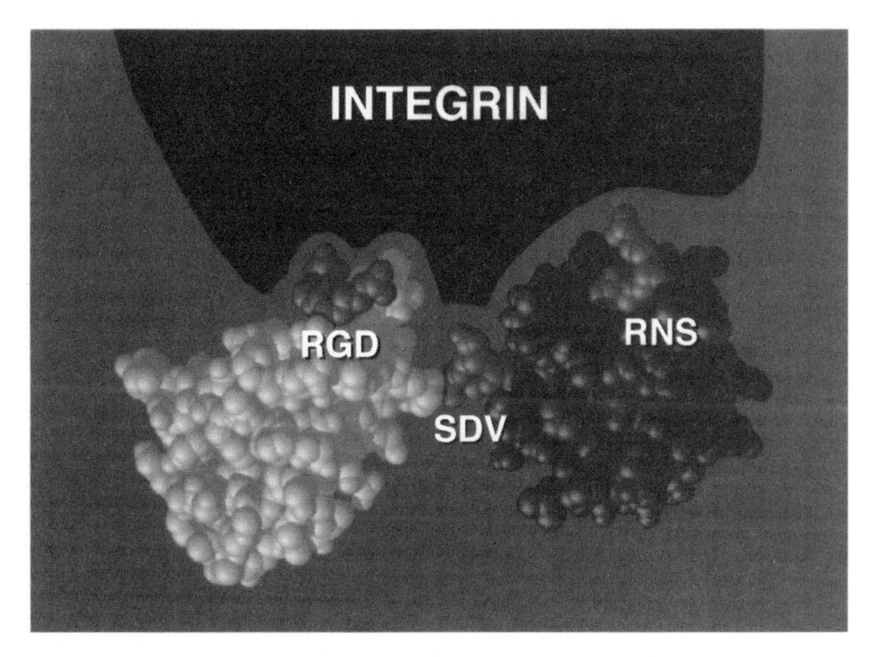

Figure 1 A model for the recognition of the fibronectin RGD site and auxiliary sites by an integrin.

RGD peptides also have the tendency to form a β turn in solution (Pettersson et al 1991). The structures derived for the III_{10} domain of fibronectin have not revealed a precise structure for the RGD sequence: It is too flexible. A similar situation has been encountered with disintegrins, small RGD-containing snake venom proteins that bind to $\alpha_{IIb}\beta_3$ and other integrins and inhibit their function (Senn & Klaus 1993). It may be that the RGD is more rigid in the larger fragments, which favors the binding of $\alpha_5\beta_1$, whereas the flexible form is preferred by $\alpha_v\beta_3$. A similar situation may exist in tenascin-C, the RGD site that is also in a type III fibronectin domain and binds $\alpha_v\beta_3$ in the context of small fragments, but not the intact protein (Leahy et al 1992, Prieto 1993, Joshi et al 1993). The $\alpha_8\beta_1$ integrin may bind to the RGD site of the intact tenascin-C molecule (Bourdon & Ruoslahti 1989, Schnapp et al 1995, Varnum-Finney et al 1995).

Antibodies can mimic integrin ligands; the crystallographic structure of an antibody that binds to the $\alpha_{IIb}\beta_3$ integrin has been reported (Kodandapani et al 1995). The RYD sequence that serves as the integrin-binding site in this antibody is also in a turn, and two possible conformations were established for that turn.

The dependence of the integrin specificity of RGD on the conformation of the RGD site and the nature of the surrounding amino acids has been successfully utilized in the design of integrin-selective peptides. Such synthesis has been particularly successful with the $\alpha_{IIb}\beta_3$ integrin of platelets; peptides that bind to this integrin with affinities 10,000 to 50,000 higher than those of the original RGD peptides from the fibronectin sequence have been designed (Charo et al 1991, Cheng et al 1994). Moreover, these peptides are essentially specific for the platelet integrin among the RGD-directed integrins. Similar, but less impressive, gains in selectivity and specificity have been achieved with some of the α_v integrins and with $\alpha_5\beta_1$ (Gurrath et al 1992, Koivunen et al 1994, 1995).

Two main design principles have made the advances with integrin specificity possible: The peptides are cyclized to provide conformational restraint, and the sequences flanking the RGD are selected to give the best possible affinity and selectivity (Pierschbacher & Ruoslahti 1987, Cheng et al 1994). The two main approaches utilized in this work have been chemical synthesis of cyclic peptides with increasing affinities and selectivities (Samanen et al 1991, Cheng et al 1994), and selection of suitable peptide motifs from peptide libraries (O'Neil et al 1992, Koivunen et al 1993, 1994, 1995, Healy et al 1995). Figure 2

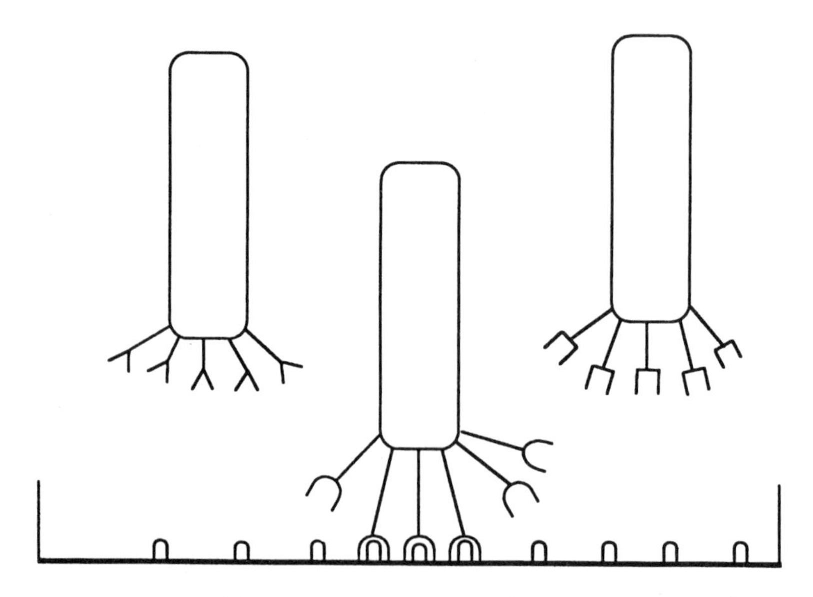

Figure 2 The principle of phage library screening. A library consists of as many as 10^9 different peptides, each carried by an individual phage as an insert in one of its surface proteins (Smith & Scott 1993). The ligand (for example, an integrin) is immobilized onto a surface, and the phage that carry peptides capable of binding to it, are isolated by panning.

illustrates the principle of peptide library selection. The peptide library approach has been particularly successful when the library is designed so that it displays only cyclic peptides. The cyclization can be accomplished in phage-displayed libraries by engineering cysteine residues on both sides of the peptide sequence that is displayed. Such libraries have the general structure CX_nC, where C is a cysteine and X is any amino acid (O'Neil et al 1992, Koivunen et al 1994). Peptide libraries strikingly demonstrate the superior affinities of cyclic peptides for integrins. First, the CX_nC-type libraries yield the highest affinity RGD motifs; second, when libraries displaying peptides with no built-in cysteine residues are screened, the most potent peptides isolated nevertheless contain a disulfide bond (Koivunen et al 1993, Healy et al 1995).

The peptide library experiments have also provided information on the nature of the sequences around the RGD that are most advantageous for the binding of the individual integrins. The amino acid residues on the C-terminal side of the RGD appear to be more significant in integrin binding than the N-terminal ones. This is suggested by the fact that when peptide motifs with higher affinities are selected from the phage libraries by increasing the stringency of the screening, sequence variability on the C-terminal side of the RGD is much reduced. For example, high stringency screening on the $\alpha_5\beta_1$ integrin gives almost exclusively peptides with the amino acids glycine and tryptophan (or phenylalanine) following the RGD (Koivunen et al 1994), whereas the $\alpha_v\beta_3$ and $\alpha_v\beta_5$ integrins show a preference for serine or alanine in these positions, and in a highly potent and specific peptide for $\alpha_{IIb}\beta_3$, they are occupied by a tyrosine and an arginine (Cheng et al 1994). Figure 3 shows the structures of two cyclic peptides selective for different integrins. One of these peptides contains two disulfide bonds within a nine-amino acid peptide and is selective for the α_v integrins. The other is a non-RGD peptide specific for $\alpha_5\beta_1$ (Koivunen et al 1994, 1995).

Cell Adhesion Sequences Other Than RGD

One group of alternative integrin-binding sequences represents slight variations of the RGD sequence. Thus peptides containing the KGD sequence can bind to the $\alpha_{IIb}\beta_3$ integrin (Plow et al 1985) and can be highly potent and specific binders of this integrin (Scarborough et al 1993). The glycine position can be occupied by a number of different individual amino acid residues, or even by two residues, in proteins and conformationally restrained peptides, with retention of integrin-binding activity (Koivunen et al 1993, 1995, Kodandapani et al 1995). One possible example of such a protein is the amyloid peptide. It contains an RHD sequence as its cell attachment site (Ghiso et al 1992, Saporito-Irwin & van Nostrand 1995) and may mediate cell attachment through an integrin. Perhaps an integrin plays a role in amyloid deposition in a similar way that

(a)

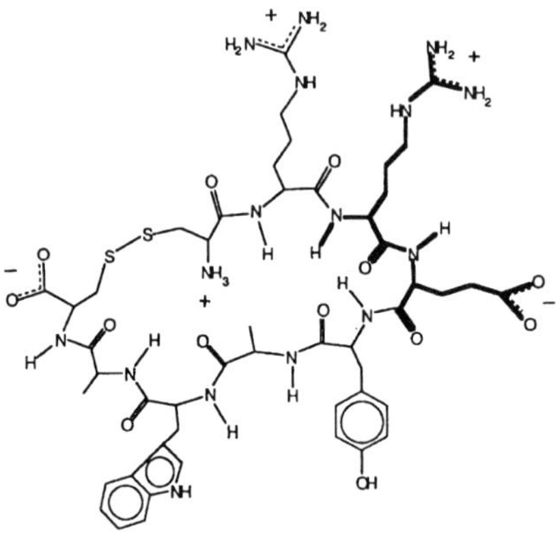

CRRETAWAC

(b)

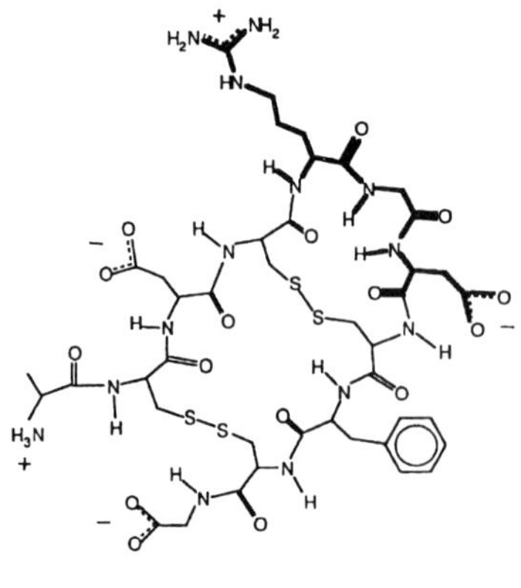

ACDCRGDCFCG

(c)

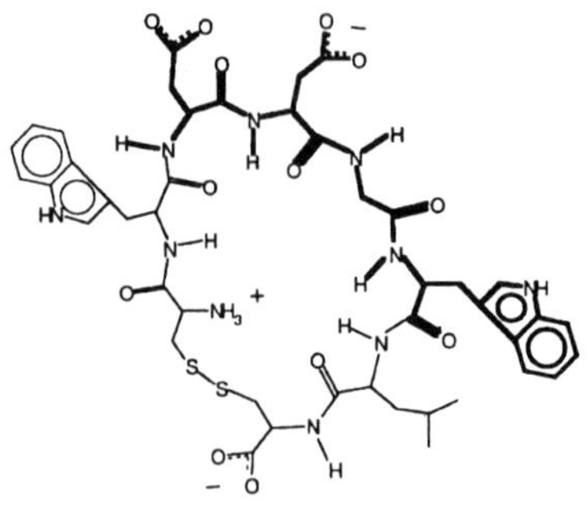

CWDDGWLC

integrins play a role in extracellular matrix assembly. Finally, PECAM, an Ig superfamily adhesion protein in endothelial cells, platelets, and leukocytes, may also contain an alternative to RGD, because PECAM binds to the $\alpha_v\beta_3$ integrin in an RGD-inhibitable manner (Piali et al 1995).

The NGR sequence is another variation of RGD capable of binding to various RGD-directed integrins, albeit with a low affinity. In addition, the peptide SDGR was reported to inhibit cell attachment in a manner similar to RGD peptides (Yamada & Kennedy 1987). However, others have found that DGR-containing peptides do not have any measurable integrin-binding activity if the amino group of the aspartic acid residue is engaged in a peptide bond (Koivunen et al 1994); it may be that the peptide used in the earlier work had lost the amino terminal serine residue.

Another sequence, KQAGDV (Kloczewiak et al 1984), appears to be a mimic of the RGD sequence because its binding to the $\alpha_{IIb}\beta_3$ integrin, for which it is essentially specific, is inhibited by RGD peptides (Lam et al 1987). However, when used in affinity labeling of the integrin, peptides containing these sequences preferentially label different subunits of the integrin (Santoro & Lawing 1987, D'Souza et al 1988, 1990). The $\alpha_4\beta_1$ integrin binds to a sequence centered around an LDV motif (Guan & Hynes 1990, Mould et al 1990), as does $\alpha_4\beta_7$ (Rüegg et al 1992). $\alpha_2\beta_1$ has been reported to recognize DGEA (Staatz et al 1991), but more recent results have implicated a YGYYGDALR sequence in laminin and FYFDLR in type IV collagen as the $\alpha_2\beta_1$ recognition sequence (Underwood et al 1995). Interestingly, both peptides contain an arginine and aspartic acid residue; the DALR/DLR sequence in these peptides is the reverse of the known integrin-binding sequence RLD (Altieri et al 1993, Koivunen et al 1995) and may be the binding site. $\alpha_1\beta_1$ recognizes a configuration of residues formed by arginine and aspartic acid residues on the surface of the type I collagen triple helix (Eble et al 1993). Finally, the $\alpha_M\beta_2$ integrin (CD11b/CD18) recognizes a KRLDGS sequence in fibrinogen (Alfieri et al 1993) and sequences centered around an L/IET sequence serve as β_2 integrin binding sites in I-CAMs (Li et al 1993, Vonderheide et al 1994).

The LDV motif that binds to $\alpha_4\beta_1$ is present in the alternatively spliced II-ICS domain of fibronectin and binds to the $\alpha_4\beta_1$ and $\alpha_4\beta_7$ integrins. The $\alpha_4\beta_1$

Figure 3 Peptide structures. Two integrin-binding peptides selective for different integrins ($\alpha_5\beta_1$ for the peptide in *A*; α_v for the peptide in *B*) (Koivunen et al 1994) are shown. Panel *C* shows a peptide that mimics the binding site of an integrin for RGD (Pasqualini et al 1995). The presumed active sites are highlighted. The disulfide bonding shown for the peptide with two disulfide bonds is hypothetical.

integrin can also recognize two other sequences in that region of fibronectin, IDA(PS) and REDV (Mould et al 1990, Mould & Humphries 1991). The IDA(PS) sequence looks like a variation of LDV, and REDV may be a homologue of RGD; in fact, the position of the REDV sequence of human IIICS corresponds to an RGD in rat IIICS (Norton & Hynes 1987). These similarities suggest a commonality in the ligand recognition specificities of the RGD-directed integrins and $\alpha_4\beta_1$. Indeed, peptides that bind to both $\alpha_5\beta_1$ and $\alpha_4\beta_1$ have been identified (Mould & Humphries 1991, Koivunen et al 1993, Nowlin et al 1993), and highly activated $\alpha_4\beta_1$ can bind to the RGD site of fibronectin. It seems likely that the sites recognized by the various integrins evolved from one primordial recognition site and that this is why the specificities of the present-day integrins are still closely related.

The integrin-binding sequence motifs described above share one amino acid, the aspartic acid (or sometimes the closely related glutamic acid) residue. The aspartic acid may be important because of its potential to contribute to divalent cation binding; one hypothesis regarding the binding of ligands to integrins postulates that the ligand provides a coordination site for divalent cation binding (Edwards et al 1988). The integrin α subunits contain four sequences that resemble the EF-hand divalent cation-binding motif of other proteins. The integrin sequences, however, lack one of the conserved acidic amino acids that serves as a coordination site in canonical EF-hands. The suggestion is that the aspartic acid of RGD and the other adhesion motifs that contain an aspartic acid residue may provide the missing coordination site when an integrin binds to its ligand. On the other hand, the ligand causes extrusion of the divalent cation (D'Souza et al 1994), and a cyclic peptide that mimics the RGD-binding site in the β subunits of integrins binds RGD peptides and proteins in a divalent cation-independent manner (Pasqualini et al 1995). Therefore, if the integrin and ligand share the cation, they may do so transiently.

An interesting cell attachment sequence is the LRE in the $\gamma 2$ subunit of laminin (Hunter et al 1989). The dipeptide RE is quite similar to RGD, except that RE displays the charged side chain groups, which are likely to be critical for integrin binding, one bond distance farther apart than RGD. By comparison, this difference is two bond-distances between RD and RGD. Thus it is possible that the receptor for the LRE sequence will turn out to be an integrin. RE may be the principal active site in the alternative $\alpha_5\beta_1$-binding peptide CRRETAWAC, which was identified from a peptide library (Koivunen et al 1994).

Disintegrins are snake venom proteins capable of binding to integrins and interfering with integrin function. They are particularly effective in preventing blood clotting through high-affinity binding to the $\alpha_{IIb}\beta_3$ integrin in platelets (reviewed in Gould et al 1990). Disintegrins typically have an RGD sequence as their active site, except in one instance in which this sequence is KGD. A family of mammalian proteins termed ADAMs contains a domain that is homologous to the disintegrins (Wolfsberg et al 1995). The ADAMs also contain a metalloprotease domain and a transmembrane domain. Their disintegrin domain does not have the RGD that is typical of the snake venom disintegrins, but at least one ADAM protein, fertilin, serves as a ligand for an integrin, $\alpha_6\beta_1$. The interaction between the fertilin and the peptide modeled after the site in fertilin that corresponds to the RGD in disintegrins inhibits the sperm-egg binding. The key residues in fertilin are TDE (Myles et al 1994, Almeida et al 1995). Thus the ADAMs contain integrin-binding sequences that are analogous to the RGD sequence.

RGD Peptides as Probes of Cell Adhesion

Cell adhesion plays an important role in a variety of basic biological processes, including guiding cells into their appropriate locations in the body, providing cell anchorage, and controlling cell proliferation, differentiation, and apoptosis. Adhesion peptides have found important uses as probes for these phenomena. In addition, there are also practical applications for these peptides.

Adhesive peptides can be used in two different ways: When attached to a surface, they promote cell attachment, whereas when presented in solution, they prevent attachment that would otherwise occur. Both modes of using the peptide have found applications: Surface-coated RGD peptides are being investigated for improvement of tissue compatibility of various implanted devices, and soluble peptides targeted for individual integrins show promise as potential drugs for the treatment of a number of diseases (Pierschbacher et al 1994).

Farthest along among the RGD applications is the use of RGD peptides, or compounds designed to mimic RGD, as anti-thrombotics that act by inhibiting the function of the $\alpha_{IIb}\beta_3$ integrin. Late-stage clinical trials are under way with such compounds. Other applications being explored include the targeting of the $\alpha_v\beta_3$ integrin in osteoporosis. Osteoclasts attach to bone through this integrin (and possibly some other α_v integrins), and inhibiting their attachment with peptides prevents bone degradation in vitro (Reinholt et al 1990) and in vivo (Fisher et al 1993).

An exciting recent development is the finding that $\alpha_v\beta_3$ is selectively expressed on endothelial cells that are engaged in angiogenesis (Brooks et al 1994a,b). RGD peptides and anti-integrin antibodies can prevent tumor growth in vivo by interfering with the angiogenic process that the growing tumor would need to maintain its blood supply (Brooks et al 1994b). RGD peptides can also affect tumor cells in a more direct fashion; it has been known for many years that RGD peptides coinjected with tumor cells into the blood stream can prevent the subsequent growth of tumor nodules in tissues (Humphries et al 1986;

reviewed in Ruoslahti et al 1995). The use of RGD mimetics and long-acting peptides has suggested that spontaneous metastasis can also be curbed with this type of treatment (Saiki et al 1989, Hardan et al 1993).

The mechanism of the antimetastatic action of the peptides in this experimental system is not understood nor is the identity of the integrin targeted by the peptide known. The reason for this is that the peptides that have been used so far have not been specific for individual integrins. Nevertheless, the effect appears to be quite selective because no harmful effects to the study animals have been noted. Immune functions, however, may be affected, as RGD (and LDV) peptides have been shown to interfere with the transmission of hypersensitivity from one animal to another by lymphocyte transfer (Ferguson et al 1991). The $\alpha_5\beta_1$ integrin is a candidate for the target integrin in the experimental metastasis system because interfering with the function of this integrin in cells that are cultured in the absence of growth factors causes tumor cell apoptosis (Zhang et al 1995). Better understanding of the antitumor effects of the RGD peptides could lead to the development of exciting new anticancer compounds.

Finally, insertion of the RGD sequence into a protein can convert the protein or, if the protein is a viral surface protein, the virus to a ligand for integrins (Hart et al 1994, Smith et al 1995). Conversely, adenovirus can be endowed with a new receptor specificity by replacing the penton protein loop that contains the RGD sequence with a different recognition sequence (Wickham et al 1995). Thus protein engineering with RGD can have applications in protein targeting and gene therapy with viruses. Advances in the application of RGD and related sequences to various purposes will depend on detailed understanding of integrin-ligand recognition. Much progress has been made recently in this field.

Future Prospects

Adhesion peptides, RGD in particular, have provided a great deal of information about cell adhesion mechanisms and are serving as a basis for the development of a new group of pharmaceuticals. The next advances in this field are likely to emerge from the design of compounds that are specific for the RGD-directed integrins other than $\alpha_{IIb}\beta_3$, from the discovery of effective peptide ligands for the non-RGD integrins, and from structural studies on integrins and their ligands. A particularly puzzling question relates to the significance of the two (or perhaps three in the integrins that contain an A domain) ligand-binding sites in each integrin. A crystal structure of an entire integrin molecule or its ectodomain, together with a ligand, would go a long way toward solving this puzzle. Such structural studies may also help resolve the mechanisms of integrin signaling, especially inside-out signaling (integrin activation), which may depend on conformation changes. The integrins are a fascinating recognition system that has the potential of becoming one of the best understood protein-protein interaction systems with profound biological and medical significance.

ACKNOWLEDGMENTS

I thank Drs. Kathryn Ely, Eva Engvall, and Kristiina Vuori for their help with the manuscript. The author's work and the preparation of this review were supported by grants CA28896 and Cancer Center Support Grant CA30199 from the NCI, DHHS. This review was partly prepared during the author's tenure as a Nobel Fellow at the Karolinska Institute, Stockholm, Sweden.

Literature Cited

Almeida EAC, Huovila A-PJ, Sutherland AE, Stephens LE, Calarco PG, et al. 1995. Mouse egg integrin $\alpha_6\beta_1$ functions as a sperm receptor. *Cell* 81:1095–104

Altieri DC, Plescia J, Plow EF. 1993. The structural motif glycine 190-valine-202 of the fibrogen-γ chain interacts with CD11β/CD18 integrin ($\alpha_M\beta_2$,Mac-1) and promotes leukocyte adhesion. *J. Biol. Chem.* 258:1847–948

Bajt ML, Ginsberg MH, Frelinger AL III, Berndt MC, Loftus JC. 1992. A spontaneous mutation of integrin $\alpha_{IIb}\beta_3$ (platelet glycoprotein IIb-IIIa) helps define a ligand binding site. *J. Biol. Chem.* 267:3789–94

Bajt ML, Loftus JC. 1994. Mutation of a ligand binding domain of β_3 integrin. *J. Biol. Chem.* 269:20913–19

Barillari G, Gendelman R, Gallo RC, Ensoli B. 1993. The TAT protein of HIV-1, a growth factor for AIDS-Kaposi's sarcoma and cytokine-activated vascular cells, induces adhesion of the same cell-types by using integrin receptors recognizing the RGD amino-acid sequence. *Proc. Natl. Acad. Sci. USA* 90:7941–45

Bourdon M, Ruoslahti E. 1989. Tenascin mediates cell attachment through an RGD-dependent receptor. *J. Cell Biol.* 108:1149–55

Bowditch RD, Halloran CE, Aota S-I, Obara M, Plow EF, et al. 1991. Integrin $\alpha_{IIb}\beta_3$ (platelet GPIIb-IIIa) recognizes multiple sites in fibronectin. *J. Biol. Chem.* 266:23323–28

Brake DA, Debouck C, Biesecker G. 1990. Identification of an Arg-Gly-Asp (RGD) cell adhesion site in human immunodeficiency virus type 1 transactivation protein, tat. *J. Cell Biol.* 111:1275–81

Brooks PC, Clark RAF, Cheresh DA. 1994a. Requirement of vascular integrin $\alpha_v\beta_3$ for angiogenesis. *Science* 264:569–71

Brooks PC, Montgomery AMP, Rosenfeld M, Reisfeld RA, Hu T, et al. 1994b. Integrin $\alpha_v\beta_3$ antagonists promote tumor-regression by inducing apoptosis of angiogenic blood vessels. *Cell* 79:1157–64

Bunch TA, Brower DL. 1992. *Drosophila* PS2 integrin mediates RGD-dependent cell-matrix interactions. *Development* 116:239–47

Calvete JJ, Rivas G, Schafer W, McLane MA, Niewiarowski S. 1993. Glycoprotein IIb peptide-656–667 mimics the fibrinogen gamma-chain-402–411 binding-site on platelet integrin GPIIb/IIIa. *FEBS Lett.* 335:132–35

Charo IF, Nannizzi L, Phillips DR, Hsu MA, Scarborough RM. 1991. Inhibition of fibrinogen binding to GP IIb-IIIa by a GP IIIa peptide. *J. Biol. Chem.* 266:1415–21

Cheng S, Craig WS, Mullen D, Tschopp JF, Dixon D, Pierschbacher MD. 1994. Design and synthesis of novel cyclic RGD-containing peptides as highly potent and selective integrin $\alpha_{IIb}\beta_3$ antagonists. *J. Med. Chem.* 37:1–8

Dickinson CD, Gay DA, Parello J, Ruoslahti E,

Ely KR. 1994. Crystals of the cell-binding module of fibronectin obtained from a series of recombinant fragments differing in length. *J. Mol. Biol.* 237:123–27

D'Souza SE, Ginsberg MH, Burke TA, Plow EF. 1990. The ligand binding site of the platelet integrin receptor GPIIb-IIIa is proximal to the second calcium binding domain of its α-subunit. *J. Biol. Chem.* 265:3440–46

D'Souza SE, Ginsberg MH, Burke TA, Lam SC-T, Plow EF. 1988. Localization of an Arg-Gly-Asp recognition site within an integrin adhesion receptor. *Science* 242:91–93

D'Souza SE, Haas TA, Piotrowicz RS, Byers-Ward V, et al. 1994. Ligand and cation binding are dual functions of a discrete segment of the integrin β_3 subunit: Cation displacement is involved in ligand binding. *Cell* 79:659–67

Eble JA, Golbik R, Mann K, Kahn K. 1993. The $\alpha_1\beta_1$ integrin recognition site of the basement membrance collagen molecule $[\alpha_1(IV)]2\alpha_2(IV)]$. *EMBO J.* 12:4795–802

Edwards J, Hameed H, Campbell G. 1988. Induction of fibroblast spreading by Mn^{2+}: a possible role for unusual binding sites for divalent cations in receptors for proteins containing Arg-Gly-Asp. *J. Cell Sci.* 89:507–13

Ensoli B, Barillari G, Salahuddin SZ, Gallo RC, Wong-Staal F. 1990. Tat protein of HIV-1 stimulates growth of cells derived from Kaposi's sarcoma lesions of AIDs patients. *Nature* 345:84–86

Ferguson TA, Mizutani H, Kupper T. 1991. Two integrin-binding peptides abrogate T cell-mediated immune responses in vitro. *Proc. Natl. Acad. Sci. USA* 88:8072–76

Fisher JE, Caulfield MP, Sato M, Quartuccio HA, Gould RJ, et al. 1993. Inhibition of osteoclastic bone resorption in vivo by echistatin. "An Arginyl-Glycyl-Aspartyl" (RGD)-containing protein. *Endocrinology* 132:1411–13

Fogerty FJ, Fessler LI, Bunch TA, Yaron Y, Perker CG, et al. 1994. Tiggrin, a novel *Drosophila* extracellular matrix protein that functions as a ligand for *Drosophila* $\alpha_{PS2}\beta_{PS}$ integrins. *Development* 120:1747–58

Gardner JM, Hynes RO. 1985. Interaction of fibronectin with its receptor on platelets. *Cell* 42:439–48

Gartner TK, Bennett JS. 1985. The tetrapeptide analog of the cell attachment site of fibronectin inhibits platelet aggregation and fibrinogen binding to activated platelets. *J. Biol. Chem.* 260:11891–94

Gartner TK, Taylor DB. 1990. The amino acid sequence gly-ala-pro-leu appears to be a fibrinogen binding site in the platelet integrin, glycoprotein IIb. *Thromb. Res.* 60:291–309

Ghiso J, Rostagno A, Gardella JE, Liem L, Gorevic PD, Frangione BA. 1992. A 109-amino acid C-terminal fragment of Alzheimer's-disease amyloid percursor protein contains a sequence, RHDS, that promotes cell adhesion. *Biochem. J.* 288:1053–59

Gould RJ, Polokoff MA, Friedman PA, Huang T-F, Holt JC, et al. 1990. Disintegins: a family of integrin inhibitory proteins from viper venoms. *Proc. Soc. Exp. Biol. Med.* 195:168–74

Guan J-L, Hynes RO. 1990. Lymphoid cells recognize an alternativey spliced segment of fibronectin via the integrin receptor $\alpha_4\beta_1$. *Cell* 60:53–61

Gurrath M, Muller G, Kessler H, Aumailley M, Timpl R. 1992. Conformation/activity studies of rationally designed potent anti-adhesive RGD peptides. *Eur. J. Biochem.* 270:911–21

Hardan I, Weiss L, Hershkovitz R, Greenspoon N, Alon R, et al. 1993. Inhibition of metastatic cell colonization in murine lungs and tumor-induced morbidity by non-peptidic Arg-Gly-Asp mimetics. *Int. J. Cancer* 55:1023–28

Hart SL, Harbottle RP, Cooper R, Miller A, Williamson R, Coutelle C. 1995. Gene delivery and expression mediated by an integrin-binding peptide. *Gene Ther.* 2:552–54

Hautanen A, Gailit J, Mann DM, Ruoslahti E. 1989. Effects of modifications of the RGD sequence and its context on recognition by the fibronectin receptor. *J. Biol. Chem.* 264:1437–42

Healy JM, Murayama O, Maeda T, Yoshino K, Sekiguchi K, Kikuchi M. 1995. Peptide ligands for integrin $\alpha_v\beta_3$ selected from random phage display libraries. *Biochemistry* 34:3948–55

Hong SS, Boulanger P. 1995. Protein ligands of the human adenovirus type 2 outer capsid identified by biopanning of a phage-displayed peptide library on separate domains of wild-type and mutant penton capsomers. *EMBO J.* 14:4714–27

Huang S, Endo RI, Nemerow GR. 1995. Upregulation of integrins $\alpha_v\beta_3$ and $\alpha_v\beta_5$ on human monocytes and T lymphocytes facilitates adenovirus-mediated gene delivery. *J. Virol.* 69:2257–63

Humphries MJ, Akiyama SK, Komoriya A, Olden K, Yamada KM. 1986. Identification of an alternatively spliced site in human plasma fibronectin that mediates cell type-specific adhesion. *J. Cell Biol.* 103:2637–47

Hunter DD, Porter BE, Bulock JW, Adams SP, Merlie JP, Sanes JR. 1989. Primary sequence of a motor neuron-selective adhesive site in the synaptic basal lamina protein S-laminin. *Cell* 59:905–13

Joshi P, Chung C-Y, Aukhil I, Erickson HP. 1993. Endothelial cells adhere to the RGD domain and the fibrinogen-like terminal knob of tenascin. *J. Cell Sci.* 106:389–400

Koivunen E, Gay DA, Ruoslahti E. 1993. Selection of peptides binding to the $\alpha_5\beta_1$ integrin from phage display library. *J. Biol. Chem.* 268:20205–10

Koivunen E, Wang B, Ruoslahti E. 1994. Isolation of a highly specific ligand for the $\alpha_5\beta_1$ integrin from a phage display library. *J. Cell Biol.* 124:373–80

Koivunen E, Wang B, Ruoslahti E. 1995. Phage libraries displaying cyclic peptides with different ring size: ligand specificities of the RGD-directed integrins. *Bio/Technology* 13:265–70

Kloczewiak M, Timmons S, Lukas TJ, Hawiger J. 1984. Platelet receptor recognition site on human fibrinogen. Synthesis and structure-function relationship of peptides corresponding to the carboxy-terminal segment of the gamma chain. *Biochemistry* 23:1767–74

Kodandapani R, Veerapandian B, Kunicki TJ, Ely KR. 1995. Crystal structure of the OPG2 Fab. An anti-receptor antibody that mimics an RGD cell adhesion site. *J. Biol. Chem.* 270:2268–73

Lam SC-T, Plow EF, Smith MA, Andrieux A, Ryckwaert JJ, et al. 1987. Evidence that arginyl-glycyl-aspartate peptides and fibrinogen alpha chain peptides share a common binding site on platelets. *J. Biol. Chem.* 262: 947–56

Lea S, Abu-Ghazaleh R, Blakemore W, Curry S, Fry E, et al. 1995. Structural comparison of two strains of foot-and-mouth disease virus subtype 0_1 and a laboratory antigenic variant, G67. *Structure* 3:571–80

Leahy DJ, Hendrickson WA, Aukhil I, Erickson HP. 1992. Structure of a fibronectin type III domain from tenascin phased by MAD analysis of the selenomethionyl protein. *Science* 258:987–91

Lee J-O, Rieu P, Arnaout MA, Liddington R. 1995. Crystal structure of the A domain from the α subunit of integrin CR3 (CD11b/CD18). *Cell* 80:631–38

Leininger E, Roberts M, Kenimer JG, Charle IG, Fairweather N, et al. 1991. Pertactin, an Arg-Gly-Asp-containing *Bordetella pertussis* surface protein that promotes adherence of mammalian cells. *Proc. Natl. Acad. Sci. USA* 88:345–49

Li R, Nortamo P, Kantor C, Kovanen P, Timonen T, Gahmberg CG. 1993. A leukocyte integrin binding peptide from intercellular adhesion molecule-2 stimulates T cell adhesion and natural killer cell activity. *J. Biol. Chem.* 268:21474–77

Main AL, Harvey TS, Baron M, Boyd J, Campbell ID. 1992. The three-dimensional structure of the tenth type III module of fibronectin: an insight into RGD-mediated interactions. *Cell* 71:671–78

Michiashita M, Videm V, Arnaot MA. 1993. A novel divalent cation binding site in the A domain of the β_2 integrin CR3 (CD11b/CD18) is essential for ligand binding. *Cell* 72:857–67

Mould AP, Humphries MJ. 1991. Identification of a novel recognition sequence for the integrin $\alpha_4\beta_1$ in the COOH-terminal heparin-binding domain of fibronectin. *EMBO J.* 10:4089–95

Mould AP, Wheldon LA, Komoriya A, Wayner EA, Yamada KM, Humphries MJ. 1990. Affinity chromatogaphic isolation of the melanoma adhesion receptor for the IIICS region of fibronectin and its identification as the integrin $\alpha_4\beta_1$. *J. Biol. Chem.* 265:4020–24

Myles DG, Kimmel LH, Blobel CP, White JM, Primakoff P. 1994. Identification of a binding site in the distintegrin domain of fertilin required for sperm-egg fusion. *Proc. Natl. Acad. Sci. USA* 91:4195–98

Norton PA, Hynes RO. 1987. Alternative splicing of chicken fibronectin in embryos and in normal and transformed cells. *Mol. Cell. Biol.* 7:4297–307

Nowlin DM, Gorcsan F, Moscinski M, Chiang SL, Lobl TJ, Cardarelli PM. 1993. A novel cyclic pentapeptide inhibits $\alpha_4\beta_1$ and $\alpha_5\beta_1$ integrin-mediated cell adhesion. *J. Biol. Chem.* 268:20352–59

Obara M, Kang M, Yamada KM. 1988. Site-directed mutagenesis of the cell-binding domain of human fibronectin: Separable, synergistic sites mediate adhesive function. *Cell* 53:649–57

Obara M, Yoshizato K. 1995. Possible involvement of the interaction of the α_5 subunit of $\alpha_5\beta_1$ integrin with the synergistic region of the central cell-binding domain of fibronectin in cells to fibronectin binding. *Exp. Cell Res.* 216:273–76

O'Neil K, Hoess RH, Jackson SA, Ramachandran NW, Mousa SA, DeGrado WF. 1992. Identification of novel peptide antagonists for GPIIb/IIIa from a conformationally constrained phage peptide library. *Proteins: Struct. Funct. Genet.* 14:509–15

Pasqualini R, Koivunen E, Ruoslahti E. 1995. A peptide isolated from phage display libraries is a structural and functional mimic of an RGD-binding site on integrins. *J. Cell Biol.* 130:1189–96

Pettersson E, Lüning B, Mickos H, Heinegård D. 1991. Synthesis, NMR and function of an 0-phosphorylated peptide, comprising the RGD-adhesion sequence of osteopontin. *Acta Chem. Scand.* 45:604–8

Piali L, Hammel P, Vherek C, Bachmann F, Gisler RH, et al. 1995. CD31/PECAM-1 is a ligand for $\alpha_v\beta_3$ integrin involved in adhesion of leukocytes to endothelium. *J. Cell Biol.* 130:451–60

Pierschbacher MD, Polarek JW, Craig WS, Tschopp JF, Sipes NJ, Harper JR. 1994. Manipulation of cellular interactions with biomaterials toward a therapeutic outcome: a perspective. *J. Cell. Biochem.* 56:150–54

Pierschbacher MD, Ruoslahti E. 1984a. The cell attachment activity of fibronectin can be duplicated by small fragments of the molecule. *Nature* 309:30–33

Pierschbacher MD, Ruoslahti E. 1984b. Variants of the cell recognition site of fibronectin that retain attachment-promoting activity. *Proc. Natl. Acad. Sci. USA* 81:5985–88

Pierschbacher MD, Ruoslahti E. 1987. Influence of stereochemistry of the sequence Arg-Gly-Asp-Xxx on binding specificity in cell adhesion. *J. Biol. Chem.* 262:17294–98

Plow EF, Pierschbacher MD, Ruoslahti E, Marguerie GA, Ginsberg MH. 1985. The effect of Arg-Gly-Asp containing peptides on fibrinogen and von Willebrand factor binding to platelets. *Proc. Natl. Acad. Sci. USA* 82:8057–61

Prieto AL, Edelman GM, Crossin KL. 1994. Multiple integrins mediate cell attachment to cytotactin/tenascin. *Proc. Natl. Acad. Sci. USA* 90:10154–58

Pytela R, Pierschbacher MD, Ruoslahti E. 1985a. Identification and isolation of a 140 kilodalton cell surface glycoprotein with properties of a fibronectin receptor. *Cell* 40:191–98

Pytela R, Pierschbacher MD, Ruoslahti E. 1985b. A 125/115 kd cell surface receptor specific for vitronectin interacts with the Arg-Gly-Asp adhesion sequence derived from fibronectin. *Proc. Natl. Acad. Sci. USA* 82:5766–70

Pytela R, Suzuki S, Breuss J, Erle DJ, Sheppard D. 1994. Polymerase chain reaction cloning with degenerate primers: homology-based identification of adhesion molecules. *Methods Enzymol.* 245:420–51

Reinholt FP, Haltenly K, Oldberg Å, Heinegård D. 1990. Osteopontin—a possible anchor of osteoclasts to bone. *Proc. Natl. Acad. Sci. USA* 87:4473–75

Roivainen M, Piirainen L, Hovi T, Virtanen I, Riikonen T, et al. 1994. Entry of Coxsackie virus A9 into host-cells: specific interactions with $\alpha_v\beta_3$ integrin, the vitronectin receptor. *Virology* 203:357–65

Rüegg C, Postigo AA, Sikorski EE, Butcher EC, Pytela R, Erle DJ. 1992. Role of integrin $\alpha_4\beta_7/\alpha_4\beta_P$ in lymphocyte adherence to fibronectin and VCAM-1 and in homotypic cell clustering. *J. Cell Biol.* 117:179–89

Ruoslahti E, Pierschbacher MD. 1987. New perspectives in cell adhesion: RGD and integrins. *Science* 238:491–97

Ruoslahti E, Pierschbacher MD, Woods VL Jr. 1995. The $\alpha_5\beta_1$ integrin in tumor suppression. *1st Forum Bull. Inst. Pasteur* 1:242–47

Saiki I, Iida J, Murata J, Ogawa R, Nishi N, et al. 1989. Inhibition of the metastasis of murine malignant melanoma by synthetic polymeric peptides containing core sequences of cell-adhesive molecules. *Cancer Res.* 49:3815–22

Samanen J, Ali F, Romoff T, Calvo R, Sorenson E, et al. 1991. Development of a small RGD peptide fibrinogen receptor antagonist with potent anti-aggregatory activity in vitro. *J. Med. Chem.* 34:3114–25

Santoro SA, Lawing WJ Jr. 1987. Competition for related but nonidentical binding sites on the glycoprotein IIb-IIIa complex by peptides derived from platelet adhesive proteins. *Cell* 48:867–73

Saporito-Irwin SM, Van Nostrand WE. 1995. Coagulation factor XI cleaves the RHDS sequence and abolishes the cell adhesive properties of the amyloid b-protein. *J. Biol. Chem.* 270:26265–69

Scarborough RM, Naughton MA, Teng W, Rose JW, Phillips DR, et al. 1993. Design of potent and specific integrin antagonists with high specificity for glycoprotein IIb-IIIa. *J. Biol. Chem.* 268:1066–73

Schindler M, Meiners S, Cheresh DA. 1989. RGD-dependent linkage between plant cell wall and plasma membrane: consequences for growth. *J. Cell Biol.* 108:1955–65

Schnapp ML, Hatch N, Ramos DM, Klimanskaya IV, Sheppard D, Pytela R. 1995. The human integrin $\alpha_8\beta_1$ functions as a receptor for tenascin, fibronectin, and vitronectin. *J. Biol. Chem.* 270:23196–202

Senn H, Klaus W. 1993. The nuclear magnetic resonance solution structure of Flavoridin, an antagonist of the platelet GPIIb-IIIa receptor. *J. Mol. Biol.* 232:907–25

Smith JW, Cheresh DA. 1988. The arg-gly-asp binding domain of the vitronectin receptor: photoaffinity crosslinking implicates amino acid residues 61–203 of the β subunit. *J. Biol. Chem.* 263:18726–31

Smith JW, Tachias K, Madison EL. 1995. Protein loop grafting to construct a variant tissue-type plasminogen activator that binds platelet integrin $\alpha_{IIb}\beta_3$. *J. Biol. Chem.* 270:30486–90

Staatz WD, Fok KF, Zutter MM, Adams SP, Rodriquez BA, Santoro SA. 1991. Identification of a tretrapeptide recognition sequence for the $\alpha_2\beta_1$ integrin in collagen. *J. Biol. Chem.* 266:7363–67

Suzuki S, Oldberg A, Hayman EG, Pierschbacher MD, Ruoslahti E. 1985. Complete amino acid sequence of human vitronectin deduced from cDNA. Similarity of cell attachment sites in vitronectin and fibronectin. *EMBO J.* 4:2519–24

Underwood PA, Bennett FA, Kirkpatrick A, Bean PA, Moss BA. 1995. Evidence for the location of a binding sequence for the $\alpha_2\beta_1$ integrin of endothelial cells, in the β_1 subunit of laminin. *Biochem. J.* 309:765–71

Van Nhieu GT, Isberg RR. 1991. The *Yersinia pseudotuberculosis* invasion protein and human fibronectin bind to mutually exclusive sites on the $\alpha_5\beta_1$ integrin receptor. *J. Biol. Chem.* 266:24367–75

Varnum-Finney B, Venstrom K, Muller U, Kypta R, Backus C, et al. 1995. The integrin receptor $\alpha_8\beta_1$ mediates interactions of embryonic chick motor and sensory neurons with tenascin-C. *Neuron* 14:1213–22

Vogel BE, Lee S-J, Hildebrand A, Craig W, Pierschbacher M, et al. 1993. A novel integrin specificity exemplified by binding of the $\alpha_v\beta_5$ integrin to the basic domain of the HIV Tat protein and vitronectin. *J. Cell Biol.* 121:461–68

Vonderheide RH, Tedder TF, Springer TA, Staunton DE. 1994. Residues within a conserved amino acid motif of domains 1 and 4 of VCAM-1 are required for binding to VLA-4. *J. Cell Biol.* 125:215–22

Wayne R, Staves MP, Leopold AC. 1992. The contribution of the extracellular matrix gravisensing in characean cells. *J. Cell Sci.* 101:611–23

Weeks BS, Desal K, Lowenstein PM, Klotman ME, Klotman PE, et al. 1993. Identification of a novel cell attachment domain in the HIV-1 Tat protein and its 90-kDa cell surface binding protein. *J. Biol. Chem.* 268:5279–84

Wickham TJ, Carrion ME, Kovesdt I. 1995. Targeting of adenovirus penton base to new receptors through replacement of its RGD motif with other receptor-specific peptide motifs. *Gene Ther.* 2:750–56

Wolfsberg TG, Straight PD, Gerena RL, Huovila A-PJ, Primakoff P, et al. 1995. A widely distribued and developmentally regulated gene family encoding membrane proteins with a disintegrin and metalloprotease domain. *Dev. Biol.* 169:378–83

Yamada KM, Kennedy DW. 1984. Dualistic nature of adhesive protein function: Fibronectin and its biologically active peptide fragments can autoinhibit fibronectin function. *J, Cell Biol.* 99:29–36

Yamada KM, Kennedy DW. 1987. Peptide inhibitors of fibronectin, laminin, and other adhesion molecules: unique and shared features. *J. Cell Physiol.* 130:21–28

Zhang Z, Vuori K, Reed JC, Ruoslahti E. 1995. The $\alpha_5\beta_1$ integrin supports survival of cells on fibronectin and up-regulates Bcl-2 expression. *Proc. Natl. Acad. Sci. USA* 92:6161–65

SUBJECT INDEX

A

B

C

G

H

I

J

K

L

N

R

S

CUMULATIVE INDEXES

CONTRIBUTING AUTHORS, VOLUMES 8–12

CHAPTER TITLES, VOLUMES 8–12